environment
THE SCIENCE BEHIND THE STORIES

7TH EDITION

AP® EDITION

Jay Withgott

Matthew Laposata

Pearson

330 Hudson Street, NY NY 10013

Director, Global Higher Ed Content Management and Strategy, Science & Health Sciences: Jeanne Zalesky
Manager, Global Higher Ed Content Strategy: Life Sciences: Joshua Frost
Content Analyst: Thomas Hoff
Senior Developmental Editors: Susan Teahan, Hilair Chism
Associate Content Analyst: Chelsea Noack
Director, Higher Ed Product Management, Life Sciences: Michael Gillespie
Product Manager: Cady Owens
Managing Producer: Michael Early
Content Producer: Margaret Young
Director, Content Development & Partner Relationships: Ginnie Simione Jutson
Senior Content Developer, Mastering Biology: Sarah Jensen
Project Manager: Sharon Cahill, SPi Global

Content Producers, Mastering Environmental Science: Lucinda Bingham, Mireille Pfeffer
Supervising Media Producer: Tod Regan
Media Producer: Jayne Sportelli
Full-Service Vendor: SPi Global
Design Manager: Maria Guglielmo Walsh
Cover & Interior Designer: Jeff Puda
Illustrators: Imagineering.com, Inc.
Rights & Permissions Project Manager: Eric Schrader, SPi Global
Rights & Permissions Manager: Ben Ferrini
Photo Researcher: Kristen Piljay
Product and Solutions Specialist: Kelly Galli
Senior Product Marketing Manager: Alysun Estes
Manufacturing Buyer: Stacey Weinberger, LSC Communications
Cover Photo Credit: Seng Chye Teo/Moment Unreleased/Getty Images

Library of Congress Cataloging-in-Publication Data
Names: Withgott, Jay author. | Laposata, Matthew, author.
Title: Environment : the science behind the stories / Jay Withgott, Matthew Laposata.
Description: 7th edition. | New York, NY : Pearson, [2019] | Includes bibliographical references and index. | Summary: "Environment: The Science Behind the Stories 7e is written for an introductory environmental science course for non-science majors. The "central case studies" hook students with stories at the beginning of a chapter and are threaded throughout. Related "Science Behind the Stories" boxes are integrated throughout to guide students through scientific discoveries, the ongoing pursuit of questions, and an understanding of the process of science. Unfolding stories about real people and places make environmental science memorable to non-science majors, and engage them in the content"-- Provided by publisher.
Identifiers: LCCN 2019041501 (print) | LCCN 2019041502 (ebook) | ISBN 9780135269145 (Rental Edition) | ISBN 9780135866108 (Loose-Leaf Print Offer Edition) | ISBN 9780136451471 (AP Edition) | 9780136623533: (Instructor's Review Edition)
Subjects: LCSH: Environmental sciences.
Classification: LCC GE105 .B74 2019 (print) | LCC GE105 (ebook) | DDC 363.7--dc23
LC record available at https://lccn.loc.gov/2019041501
LC ebook record available at https://lccn.loc.gov/2019041502
AP® is a trademark registered and/or owned by the College Board, which was not involved in the production of, and does not endorse, this product.

10 2022

Pearson
www.PearsonSchool.com/Advanced

ISBN 10: 0-13-645147-0 (High School Binding)
ISBN 13: 978-0-13-645147-1 (High School Binding)

About the Authors

Jay Withgott has authored *Environment: The Science Behind the Stories* as well as its brief version, *Essential Environment,* since their inception. In dedicating himself to these books, he works to keep abreast of a diverse and rapidly changing field and continually seeks to develop new and better ways to help today's students learn environmental science.

As a researcher, Jay has published scientific papers in ecology, evolution, animal behavior, and conservation biology in journals ranging from *Evolution* to *Proceedings of the National Academy of Sciences.* As an instructor, he has taught university lab courses in ecology and other disciplines. As a science writer, he has authored articles for numerous journals and magazines, including *Science, New Scientist, BioScience, Smithsonian,* and *Natural History.* By combining his scientific training with prior experience as a newspaper reporter and editor, he strives to make science accessible and engaging for general audiences. Jay holds degrees from Yale University, the University of Arkansas, and the University of Arizona.

Jay lives with his wife, biologist Susan Masta, in Portland, Oregon.

Matthew Laposata is a professor of environmental science at Kennesaw State University (KSU). He holds a bachelor's degree in biology education from Indiana University of Pennsylvania, a master's degree in biology from Bowling Green State University, and a doctorate in ecology from The Pennsylvania State University.

Matt is the coordinator of KSU's two-semester general education science sequence titled Science, Society, and the Environment, which enrolls roughly 6000 students per year. He focuses exclusively on introductory environmental science courses and has enjoyed teaching and interacting with thousands of students during his nearly two decades in higher education. He is an active scholar in environmental science education and has received grants from state, federal, and private sources to develop innovative curricular materials. His scholarly work has received numerous awards, including the Georgia Board of Regents' highest award for the Scholarship of Teaching and Learning.

Matt resides in suburban Atlanta with his wife, Lisa, and children, Lauren, Cameron, and Saffron.

ABOUT OUR SUSTAINABILITY INITIATIVES

At Pearson, we know we have a huge opportunity to help create a better future for people and the planet. Our 2020 Sustainability Plan drives our commitment to integrating social and environmental issues into every aspect of our business and advancing the UN Sustainable Development Goals. We are working to improve access to education for those who need it most, support education about sustainability, and implement a range of responsible business practices.

We continue to do our part to address climate change and deforestation, which are the global environmental challenges that are most relevant to our operations and impacts. Within our own operations, we are carbon neutral and use 100% renewable energy. We work with the suppliers who manufacture our books to use responsibly sourced paper, minimize waste, reduce energy consumption, limit the use of harmful chemicals, and uphold labor standards. As we increasingly transition from print to digital products, we are using less paper and at the same time considering the impact of a digital supply chain.

To learn more about our sustainability program, please visit *https://www.pearson.com/sustainability.html.*

Contents

Foundations of Environmental Science — PART 1

1 Science and Sustainability: An Introduction to Environmental Science 2

2 Earth's Physical Systems: Matter, Energy, and Geology 22

3 Evolution and Population Ecology 48

4 Species Interactions and Community Ecology 74

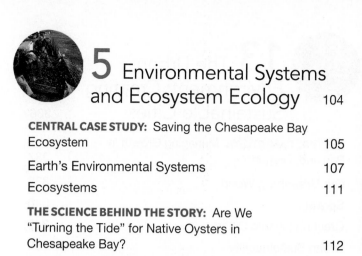

5 Environmental Systems and Ecosystem Ecology 104

6 Ethics, Economics, and Sustainable Development 132

7 Environmental Policy: Making Decisions and Solving Problems 160

Environmental Issues and the Search for Solutions
PART 2

8 Human Population 186

9 The Underpinnings of Agriculture 212

Preface

Dear Student,

You are coming of age at a unique and momentous time in history. Within your lifetime, our global society must chart a promising course for a sustainable future. The stakes could not be higher.

Today we live long lives enriched with astonishing technologies, in societies more free, just, and equal than ever before. We enjoy wealth on a scale our ancestors could hardly have dreamed of. However, we have purchased these wonderful things at a steep price. By exploiting Earth's resources and ecosystem services, we are depleting our planet's ecological bank account. We are altering our planet's land, air, water, nutrient cycles, biodiversity, and climate at dizzying speeds. More than ever before, the future of our society rests with how we treat the world around us.

Your future is being shaped by the phenomena you will learn about in your environmental science course. Environmental science gives us a big-picture understanding of the world and our place within it. Environmental science also offers hope and solutions, revealing ways to address the problems we confront. Environmental science is more than just a subject you study in college. It provides you basic literacy in the foremost issues of the 21st century, and it relates to everything around you throughout your lifetime.

We have written this book because today's students will shape tomorrow's world. At this unique moment in history, the decisions and actions of your generation are key to achieving a sustainable future for our civilization. The many environmental challenges we face can seem overwhelming, but you should feel encouraged and motivated. Remember that each dilemma is also an opportunity. For every problem that human carelessness has created, human ingenuity can devise a solution. Now is the time for innovation, creativity, and the fresh perspectives that a new generation can offer. Your own ideas and energy can, and *will*, make a difference.

—Jay Withgott and Matthew Laposata

Dear Instructor,

You perform one of our society's most vital functions by educating today's students—the citizens and leaders of tomorrow—on the processes that shape the world around them, the nature of scientific inquiry, and the pressing environmental challenges we face. We have written this book to assist you in this endeavor because we feel that the crucial role of environmental science in today's world makes it imperative to engage, educate, and inspire a broad audience of students.

In *Environment: The Science Behind the Stories*, we strive to show students how science informs our efforts to bring about a sustainable society. We also aim to encourage critical thinking and to maintain a balanced approach as we flesh out the vibrant social debate that accompanies environmental issues. As we assess the challenges facing our civilization and our planet, we focus on providing realistic, forward-looking solutions, for we truly feel there are many reasons for optimism.

In crafting the seventh edition of this text, we have incorporated the most current information from this dynamic discipline and have tailored our presentation to best promote student learning. We have examined every line of text and every figure with great care to ensure that all content is accurate, clear, and up-to-date. Moreover, we have introduced a number of changes that are new to this edition.

New to This Edition

This seventh edition includes an array of revisions that enhance our content and presentation while strengthening our commitment to teach science in an engaging and accessible manner.

- **SUCCESS story** This new feature highlights discrete stories (one per chapter) of successful efforts to address environmental problems, ranging from local examples (such as prairie restoration in Chicago) to national and global achievements (such as halting ozone depletion by treaty or removing lead from gasoline). Our book has always focused on positive solutions, but the new emphasis these *Success Stories* bring should help encourage and inspire students by demonstrating how sustainable solutions are within reach. Students can explore data behind these solutions with new *Success Story Coaching Activities* in *Mastering Environmental Science*.

- **DATAGRAPHIC** This new and visually striking feature brings life to key questions in environmental science by presenting data in novel yet intuitive ways. The five *DataGraphics* seek to strengthen student skills in analytical thinking by fostering the ability to draw reasonable conclusions when provided with relevant data. Each *DataGraphic* poses a question, assembles an array of datasets, and leads to a unifying conclusion, guiding students through a synthesis of quantitative information in an inviting and appealing manner.

  - **Chapter 8:** Will Nigeria's population overwhelm its water supply?
  - **Chapter 10:** Can we continue to reduce global hunger?
  - **Chapter 11:** Can we save the world's biodiversity?
  - **Chapter 16:** How can we avoid choking the oceans with plastic?
  - **Chapter 18:** How can we stop global warming?

- **CENTRAL case study** Seven *Central Case Studies* are completely new to this edition, while several others have been thoroughly reshaped to add exciting new angles. All other case studies have been updated as needed to reflect recent developments. These updates provide fresh stories and new ways to frame emerging issues in environmental science. Students will learn how Midwesterners are battling an invasion of Asian carp, how Californian farmers are helping pollinators, how Brazilians are struggling to save the Amazon rainforest, how Texans are balancing water use with oil and gas production, and how researchers are documenting the spread of plastic waste across the world's oceans. Readers will also encounter inspiring new stories of Michigan students running sustainable food programs and of young Americans going to court to challenge their government to tackle climate change.

  - **Chapter 4:** Leaping Fish, Backwards River: Asian Carp Threaten the Great Lakes
  - **Chapter 7:** Young Americans Take On Climate Change in the Courts
  - **Chapter 9:** Bees to the Rescue: By Helping Pollinators, Farmers Help Themselves
  - **Chapter 10:** Sustainable Food and Dining at the University of Michigan
  - **Chapter 12:** Saving the World's Greatest Rainforest
  - **Chapter 15:** Reaching the Tipping Point: Fracking and Fresh Water in West Texas
  - **Chapter 16:** A Sea of Plastic in the Middle of the Ocean

- **connect & continue** Each chapter now concludes with a brief section that provides late-breaking updates to the *Central Case Study* and makes connections outward to related themes, events, or locations, to facilitate further discussion. This new *Connect & Continue* section enhances our long-standing and well-received approach of integrating each *Central Case Study* throughout its chapter.

- **THE SCIENCE behind the story** Fourteen *Science Behind the Story* boxes are new to this edition. These new boxes, along with others that have been updated, provide a current and exciting selection of scientific studies to highlight. Students will follow researchers as they discover a global collapse in insect populations, track chemicals hidden in the food we eat, reveal how global warming creates extreme weather, and much more.

  - **Chapter 2:** Are Yeast the Answer to Cleaning Up Nuclear Waste?
  - **Chapter 4:** How Do Asian Carp Affect Aquatic Communities?
  - **Chapter 7:** What Does the Science on Climate Change Tell Us?
  - **Chapter 8:** Measuring Our "Human Footprint": A Roadmap to Sustainability *and* Prosperity?
  - **Chapter 9:** If We Help Pollinators, Will It Boost Crop Production?
  - **Chapter 11:** Are Insects Vanishing?
  - **Chapter 14:** Are Endocrine Disruptors Lurking in Your Fast Food?
  - **Chapter 14:** What Role Do Pesticides Play in the Collapse of Bee Colonies?
  - **Chapter 15:** What's Killing Smallmouth Bass in the Potomac River Watershed?
  - **Chapter 16:** Inventorying the Great Pacific Garbage Patch
  - **Chapter 18:** Why and How Does Global Warming Lead to Extreme Weather Events?
  - **Chapter 21:** How Well Do Wind and Solar Complement One Another in Texas?
  - **Chapter 22:** Can Campus Research Help Reduce Waste?
  - **Chapter 23:** Mapping Mountaintop Mining's "Footprint" in Appalachia

- **New and revised DATA Q, FAQ, and Weighing the Issues items** Incorporating feedback from instructors across North America, we have examined each example of these three features that boost student engagement and have revised them and added new examples as appropriate.

- **Currency and coverage of topical issues** To live up to our book's hard-won reputation for currency, we've incorporated the most recent data possible throughout, and we've enhanced coverage of emerging issues. As climate change and energy concerns play ever-larger roles in today's world, our coverage has kept pace. This edition highlights the tremendous growth and potential of renewable energy, yet also shows how we continue reaching further for fossil fuels using ever more powerful technologies. The text tackles the complex issue of climate change in depth, while connections to this issue proliferate among topics in every chapter. And in a world newly shaken by dynamic political forces amid concerns relating to globalization, trade, immigration, health care, jobs, racial discrimination, social justice, and wealth inequality, our introduction of ethics, economics, and policy early in the book serves as a framework to help students relate the scientific knowledge they are learning to the complex cultural aspects of the society around them.

- **Enhanced style and design** We have refreshed and improved the look and clarity of our visual presentation throughout the text. A more appealing layout, striking visuals, and an inviting new style all make the book more engaging for students. Over 50% of the photos, graphs, and illustrations in this edition are new or have been revised to reflect current data or for enhanced clarity or pedagogy.

Existing Features

We have also retained the major features that made the first six editions of our book unique and that are proving so successful in classrooms across North America:

- **A focus on science and data analysis** We have maintained and strengthened our commitment to a rigorous presentation of modern scientific research while simultaneously making science clear, accessible, and engaging to students. Explaining and illustrating the *process* of science remains a foundational goal of this endeavor. We also continue to provide an abundance of clearly cited data-rich graphs, with accompanying tools for data analysis. In our text, our figures, and our online features, we aim to challenge students and to assist them with the vital skills of data analysis and interpretation.

- **An emphasis on solutions** For many students, today's deluge of environmental dilemmas can lead them to feel that there is no hope or that they cannot personally make a difference. We have aimed to counter this impression by highlighting innovative solutions being developed around the world—a long-standing approach now enhanced by our new *Success Story* feature. While taking care not to paint too rosy a picture of the challenges that lie ahead, we demonstrate that there is ample reason for optimism, and we encourage action. Our campus sustainability coverage (Chapter 1 and *Central Case Studies* in Chapters 9 and 22) shows students how their peers are applying principles and lessons from environmental science to forge sustainable solutions on their own campuses.

- **Integration of a CENTRAL case study throughout each chapter.** We integrate each chapter's *Central Case Study* into the main text, weaving information and elaboration throughout the chapter. In this way, compelling stories about real people and real places help to teach foundational concepts by giving students a tangible framework with which to incorporate novel ideas. Students can explore the locations featured in each *Central Case Study* with new Case Study Video Tours in *Mastering Environmental Science*.

- **THE SCIENCE behind the story** Because we strive to engage students in the scientific process of testing and discovery, we feature *The Science Behind the Story* boxes in each chapter. By guiding students through key research efforts, this feature shows not merely *what* scientists discovered, but *how* they discovered it.

- **FAQ** The *FAQ* feature highlights questions frequently posed by students, thereby helping to address widely held misconceptions and to fill in common conceptual gaps in knowledge. By also including questions students sometimes hesitate to ask, the *FAQs* show students that they are not alone in having these questions, which helps to foster a spirit of open inquiry in the classroom.

- **WEIGHING the issues** These questions aim to help develop the critical-thinking skills students need to navigate multifaceted issues at the juncture of science, policy, and ethics. They serve as stopping points for students to reflect on what they have read, wrestle with complex dilemmas, and engage in spirited classroom discussion.

- **Diverse end-of-chapter features** *Reviewing Objectives* summarizes each chapter's main points and relates them to the chapter's learning objectives, enabling students to confirm that they have understood the most crucial ideas. *Seeking Solutions* encourages broad creative thinking that supports our emphasis on finding solutions. "Think It Through" questions personalize the quest for creative solutions by placing students in a scenario and empowering them to make decisions. *Calculating Ecological Footprints* enables students to quantify the impacts of their own choices and measure how individual impacts scale up to the societal level.

Mastering™ Environmental Science

With this edition we continue to offer expanded opportunities through *Mastering Environmental Science,* our powerful yet easy-to-use online learning and assessment platform. We have developed new content and activities specifically to support features in the textbook, thus strengthening the connection between these online and print resources. This approach encourages students to practice their science literacy skills in an interactive environment with a diverse set of automatically graded exercises. Students benefit from self-paced activities that feature immediate wrong-answer feedback, while instructors can gauge student performance with informative diagnostics. By enabling assessment of student learning outside the classroom, *Mastering Environmental Science* helps the instructor to maximize the impact of in-classroom time. As a result, both educators and learners benefit from an integrated text-and-online solution.

- **DATA GRAPHIC** The five new DataGraphics from the text come to life in the eText, allowing students to interact with the data. These interactives are assignable in *Mastering Environmental Science.*

- These popular data analysis questions have been moved to *Mastering Environmental Science* in this edition to help students actively engage with graphs and other data-driven figures and allow instructors to assign these questions for practice or homework.

- *GraphIt* activities help students put data analysis and science reasoning skills into practice through a highly interactive and engaging format. Each *GraphIt* prompts students to manipulate a variety of graphs and charts, from bar graphs to line graphs to pie charts, and develop an understanding of how data can be used in decision making about environmental issues. Topics range from agriculture to fresh water to air pollution. These mobile-friendly activities are accompanied by assessment in *Mastering Environmental Science*.

- *Everyday Environmental Science* videos highlight current environmental issues in short (5 minutes or less) video clips and are produced in partnership with BBC News. These videos will pique student interest and can be used in class or assigned as a high-interest out-of-class activity.

- *Dynamic Study Modules* help students study effectively on their own by continuously assessing their activity and performance in real time. Students complete multiple sets of questions for any given topic, to demonstrate concept mastery with confidence. Each *Dynamic Study Module* question set concludes with an explanation of concepts students may not have mastered. They are available as graded assignments prior to class and are accessible on smartphones, tablets, and computers.

- *Process of Science* activities help students navigate the scientific method, guiding them through in-depth explorations of experimental design using *The Science Behind the Story* features from the current and former editions. These activities encourage students to think like a scientist and to practice basic skills in experimental design.

- *Interpreting Graphs and Data: Data Q* activities pair with the in-text *Data Analysis Questions* and coach students to further develop skills related to presenting, interpreting, and thinking critically about environmental science data.

- *"First Impressions" Pre-Quizzes* help instructors determine their students' existing knowledge of environmental issues and core content areas at the outset of the academic term, providing class-specific data that can then be employed for powerful teachable moments throughout the term. Assessment items in the Test Bank connect to each quiz item, so instructors can formally assess student understanding.

- *Video Field Trips* enable students to visit real-life sites that bring environmental issues to life. Students can tour a power plant, a wind farm, a wastewater treatment facility, a site combating invasive species, and more—all without leaving campus.

Environment: The Science Behind the Stories has grown from our collective experiences in teaching, research, and writing. We have been guided in our efforts by input from the hundreds of instructors across North America who have served as reviewers and advisers. The participation of so many learned, thoughtful, and committed experts and educators has improved this volume in countless ways.

We sincerely hope that our efforts are worthy of the immense importance of our subject matter. We invite you to let us know how well we have achieved our goals and where you feel we have fallen short. Please write to us in care of our content analyst, Thomas Hoff (thomas.hoff@pearson.com), at Pearson Education. We value your feedback and are eager to learn how we can serve you better.

—Jay Withgott and Matthew Laposata

Instructor Supplements

A robust set of instructor resources and multimedia accompanies the text and can be accessed through Mastering Environmental Science.

Organized chapter-by-chapter, everything you need to prepare for your course is offered in one convenient set of files. Resources include the following: Video Field Trips, Everyday Environmental Science Videos, PowerPoint Lecture presentations, Instructor's Guide, Active Lecture questions, an image library, TestGen and Test Bank files that include hundreds of multiple-choice questions plus unique graphing and scenario-based questions to test students' critical-thinking abilities.

Support for the AP Curriculum

- A complete correlation chart to the College Board's AP® Environmental Science Curriculum Framework (dated Fall 2019) that corresponds with the chapters and sections within *Environment: The Science Behind the Stories 7e, AP® Edition* can be found online at www.PearsonSchool.com/AdvancedCorrelations.

- **Pearson's Test Prep Series for AP® Environmental Science** This student workbook contains concise content summaries of each chapter that are relevant to the AP® Exam, multiple-choice and free-response practice questions, a practice exam with scoring guidelines, and answers and detailed explanations of all questions and answers. Available for purchase.

Getting Students to See the Data, Connections, and Solutions behind Environmental Issues

Environment: The Science Behind the Stories is known for its student-friendly narrative style, its integration of real stories and case studies, and its presentation of the latest science and research.

ENVIRONMENT
the science behind the stories

7TH EDITION   Jay Withgott | Matthew Laposata

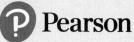

Engage Students with

UPDATED! Central Case Studies

begin and are woven throughout the chapter, drawing students into learning about the real people, real places, and real data behind environmental issues. These provide a contextual framework to make science memorable and engaging.

NEW! Connect & Continue

brings the Central Case Study in each chapter full circle by revisiting it at the end of the chapter and offering specific activities related to the case study. The Case Study Solutions activity encourages students to think about possible ways to address the featured environmental issue. The Local Connections activity allows instructors to relate the case study to their students' local area. And the new Explore the Data links to assignable coaching activities on Mastering Environmental Science.

Interactive Data & Stories

Considering Cost When Saving the Bay

The Chesapeake Bay offers a case study that illustrates the importance of understanding systems, chemistry and the need for taking a systems-level approach to restore ecosystems degraded by human activities. Tools such as landscape ecology, GIS, and ecological modeling aid these efforts by providing a broad view of the Chesapeake Bay ecosystem and how it may react to changes in nutrient inputs and restoration efforts. The Chesapeake Bay Foundation's most recent "State of the Bay" report concluded that the bay's health rating in 2018 was the highest it had been since CBF's founding in 1964, with meaningful improvements in pollution reduction, fisheries recovery, and the restoration of natural habitats in and around the bay (see Figure 5.1, p. 106).

One reason for the recent success is that farmers, residents, resource managers, and local, state, and federal government agencies have embraced a variety of approaches to reduce nutrient inputs into the bay. By educating people about the many inexpensive yet effective steps that can

A forested buffer lining a waterway on agricultural land in Maryland.

be taken in yards, farms, businesses, and local communities to reduce nutrient inputs into the Chesapeake Bay, saving the bay became something for which everyone can do his or her part.

→ **Explore the Data** at Mastering Environmental Science

Family Planning without Coercion: Thailand's Population Program

The government of Thailand, facing many of the same population challenges as its populous neighbor to the north, instituted a comprehensive population program in 1971. At the time, Thailand's population growth rate was 2.3%, and the nation's TFR was 5.4. But unlike the Chinese reproductive program, Thais were given control over their own reproductive choices, provided with family-planning counseling and modern contraceptives, and supported by an engaging public education campaign. Aided by a relatively high level of women's rights in Thai society, Thailand's population program—and the fertility reductions that accompanied the nation's economic development over the past 47 years—reduced its population growth rate to 0.2%, with a TFR of 1.5 children per woman in 2018. The success of this program, and similar initiatives in nations such as Brazil, Cuba, Iran, and Mexico, show that government interventions to reduce birth rates need not be as intrusive as China's to produce similar declines in population growth.

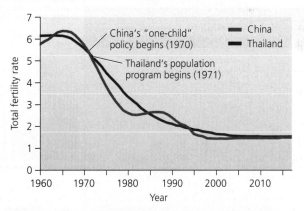

China and Thailand instituted population control programs at roughly the same time and showed similar patterns in fertility declines over the subsequent 45 years, despite utilizing very different approaches. *Data from World Bank, 2017, http://data.worldbank.org.*

→ **Explore the Data** at Mastering Environmental Science

Success Stories, one in every chapter, highlight forward-thinking solutions and successful efforts to address environmental issues. Each Success Story links to new data-analysis activities on Mastering Environmental Science that are available for assignment and grading.

Increase Students' Scientific Literacy

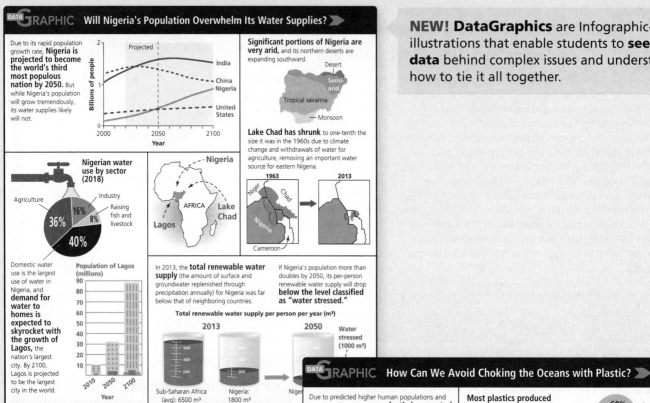

DATA GRAPHIC — Will Nigeria's Population Overwhelm Its Water Supplies?

Due to its rapid population growth rate, **Nigeria is projected to become the world's third most populous nation by 2050.** But while Nigeria's population will grow tremendously, its water supplies likely will not.

Billions of people — Projected — India, China, Nigeria, United States — Year (2000–2100)

Significant portions of Nigeria are very arid, and its northern deserts are expanding southward.

Desert, Semi-arid, Tropical savanna, Monsoon

Lake Chad has shrunk to one-tenth the size it was in the 1960s due to climate change and withdrawals of water for agriculture, removing an important water source for eastern Nigeria.

Niger, Chad, Nigeria, Cameroon (1963 / 2013)

Nigerian water use by sector (2018): Agriculture 36%, Industry 16%, Raising fish and livestock 8%, Domestic 40%

AFRICA — Nigeria — Lagos — Lake Chad

Domestic water use is the largest use of water in Nigeria, and **demand for water to homes is expected to skyrocket with the growth of Lagos,** the nation's largest city. By 2100, Lagos is projected to be the largest city in the world.

Population of Lagos (millions) — Year (2010, 2050, 2100)

In 2013, the **total renewable water supply** (the amount of surface and groundwater replenished through precipitation annually) for Nigeria was far below that of neighboring countries.

If Nigeria's population more than doubles by 2050, its per-person renewable water supply will drop **below the level classified as "water stressed."**

Total renewable water supply per person per year (m³)
2013 — Sub-Saharan Africa (avg): 6500 m³, Nigeria: 1800 m³
2050 — Nigeria: Water stressed (1000 m³)

If Nigeria cannot reduce its projected population growth, water scarcity could slow economic growth and adversely affect the well-being of its people.

P. 191

NEW! DataGraphics are Infographic-style illustrations that enable students to **see the data** behind complex issues and understand how to tie it all together.

DATA GRAPHIC — How Can We Avoid Choking the Oceans with Plastic?

Due to predicted higher human populations and growing affluence, **more plastic is expected to enter oceans in 2050 than today.**

Plastics entering oceans — Each box represents 1 million metric tons — 2019 → 2050 (projected)

Most plastics produced float in seawater, where they can entangle wildlife or where they can fragment and enter the food web.
60% of plastics float
40% of plastics sink

Plastics degrade slowly in seawater. It can take decades, even centuries, before the combined actions of sunlight, waves, and biological activity breaks them down completely.

Styrofoam cup 50
Plastic straw 200
Plastic bottle 450
Fishing line 600
Years to degrade

Keeping plastics out of the ocean is the first step towards recovery:

Some U.S. states **charge a refundable deposit** on plastic drink bottles. This drastically increases recycling rates and keeps these plastics out of the oceans.
Orange H₂O — ME-VT-CT-MA-NY-HI-IA 5¢ OR-MI 10¢ CA CRV (California Refund Value)

We can **eliminate our use of single-use plastic products,** such as straws and cups.
Switch to reusable containers — Fabricate single-use products out of paper, which readily degrades in seawater

We can **enact bans or taxes on plastic grocery bags,** which easily wash or blow into waterways.
Locations with bans or taxes on plastic bags, as of 2020
*Canada has announced plans to ban single-use plastics by the year 2021.

Trends suggest that marine plastic pollution will get worse before it gets better, but by embracing approaches to reuse, recycle, and reduce disposal of plastic items, we can help address plastic pollution in the oceans.

P. 434

DataGraphics include:

- Will Nigeria's population overwhelm its water supply?
- Can we continue to reduce global hunger?
- Can we save the world's biodiversity?
- How can we avoid choking the oceans with plastic?
- How can we stop global warming?

The Science behind the Stories

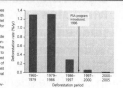

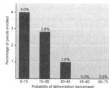

P. 144

New Topics Include:

- Are Endocrine Disruptors Lurking in Your Fast Food?
- What Role Do Pesticides Play in the Collapse of Bee Colonies?
- Inventorying the Great Pacific Garbage Patch
- Why and How Does Global Warming Lead to Extreme Weather Events?
- Can Campus Research Help Reduce Waste?

And more!

P. 362

Bring Environmental Science to Life with Mastering

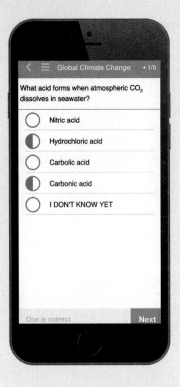

Dynamic Study Modules help students study effectively—and at their own pace. These rely on the latest research in cognitive science, to stimulate learning and improve retention.

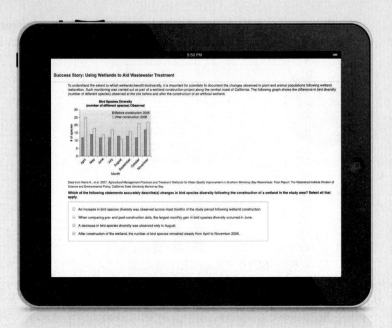

NEW! Data Analysis Activities give students an opportunity to explore data related to the Central Case Studies and Success Stories from the text.

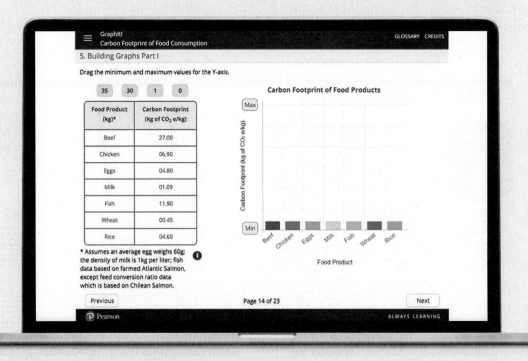

GraphIt! Coaching Activities help students read, interpret, and create graphs that explore real environmental issues using real data.

Engage Students with Active Learning

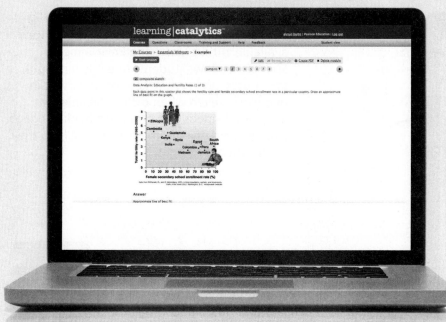

Learning Catalytics enables you to hear from every student in real-time, when it matters most. Choose from a variety of question types that help students recall ideas, apply concepts, and develop critical-thinking skills. Students answer on their own device. Instructors receive analytics in real-time to find out what their students know and don't know and how peer-to-peer learning can be facilitated to help students engage and stay motivated.

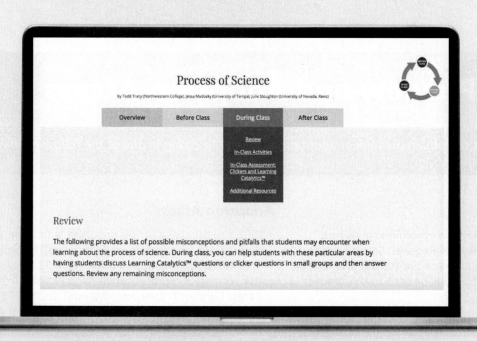

To-Go Teaching Modules help instructors make use of teaching tools before, during, and after class, including new ideas for in-class activities. The modules incorporate the best that the text, Mastering, and Learning Catalytics have to offer. These can be accessed through the Instructor Resources area in Mastering Environmental Science.

A Whole New Reading Experience

Pearson eText is a simple-to-use, mobile-optimized, personalized reading experience available within Mastering Environmental Science. It allows students to easily highlight, take notes, and review key vocabulary all in one place—even when offline. Seamlessly integrated videos and other rich media engage students and give them access to the help they need, when they need it.

The Pearson eText app is available for download in the app store for approved devices.

Mastering Preview and Adoption Access

Upon textbook purchase, students and teachers are granted access to Mastering Environmental Science with Pearson eText. High school teachers can obtain preview or adoption access to Mastering in one of the following ways:

Preview Access

- Teachers can request preview access online by visiting www.PearsonSchool.com/Access_Request. Select Science, choose Initial Access, and complete the form under Option 2. Preview Access information will be sent to the teacher via e-mail.

Pearson reserves the right to change and/or update technology platforms, including possible edition updates to customers during the term of access. This will allow Pearson to continue to deliver the most up-to-date content and technology to customers. Customer will be notified of any change prior to the beginning of the new school year.

Adoption Access

- With the purchase of this program, a Pearson Adoption Access Card with Instructor Manual will be delivered with your textbook purchase. (ISBN: 978-0-13-354087-1)
- Ask your sales representative for a Pearson Adoption Access Card with Instructor Manual. (ISBN: 978-0-13-354087-1)

  OR

- Visit PearsonSchool.com/Access_Request, select Science, choose Initial Access, and complete the form under Option 3—MyLab/Mastering Class Adoption Access. Teacher and Student access information will be sent to the teacher via e-mail.

Students, ask your teacher for access.

Acknowledgments

This textbook results from the collective labor and dedication of innumerable people. The two of us are fortunate to be supported by a tremendous publishing team.

Product manager Cady Owens coordinated our team's efforts for this seventh edition of *Environment: The Science Behind the Stories*. We remain deeply thankful to Cady and appreciate her creative thinking, sound judgment, and committed engagement with our work. Senior content analyst Tom Hoff helped to bring the edition across the finish line. We were excited to welcome back developmental editors Susan Teahan and Sonia Divittorio. Susan had helped pioneer the very first edition of this book years ago, and to Susan's keen eye and resourcefulness Sonia added her own remarkable skills, ensuring this edition a wealth of talented input. Director of content development Ginnie Simione Jutson oversaw our development needs, while content producer Margaret Young effectively managed the innumerable steps in the publishing process.

It was a pleasure to collaborate with senior analyst Hilair Chism, who spearheaded the creation of the new *DataGraphics* that enliven this edition, and with photo researcher Kristin Piljay, who helped us acquire the highest-quality images possible. Mark Mykytiuk of Imagineering oversaw a smooth art production process for hundreds of figures. Kathleen Lafferty and Denne Wesolowski performed meticulous copyediting and proofreading across every page of text. Jeff Puda created this edition's engaging interior and cover design. Associate content analyst Chelsea Noack managed the review process and provided timely assistance as needed. And our thanks go to project manager Sharon Cahill of SPi Global for her excellent work interacting with the compositor to ensure a clean layout and a successful production process.

We welcome and look forward to working with global content manager Josh Frost and director of product management Michael Gillespie as we move collectively toward an exciting future. And we will always be indebted to our former executive editor Alison Rodal, who guided us in the last two editions of *Environment,* and to our former editor-in-chief, Beth Wilbur, for her steadfast support of this textbook across its seven editions.

As always, a select number of top instructors from around North America produced the supplementary materials that support the text. Our thanks go to Sarah Schliemann for updating our Instructor's Guide, to David Serrano for his work with the Test Bank, to James Dauray for revising the PowerPoint lectures, and to Jennifer Biederman for updating the Active Lecture clicker questions.

We give thanks to marketing manager Alysun Burns Estes. And we admire and appreciate the work and commitment of the many sales representatives who help communicate our vision, deliver our product, and collaborate with instructors to ensure their satisfaction.

In the lists of reviewers that follow, we acknowledge the many instructors and outside experts who have helped us to maximize the quality and accuracy of our content and presentation through their chapter reviews, feature reviews, class tests, focus group participation, and other services. The thoughtfulness and thoroughness of these reviewers make clear to us that the teaching of environmental science is in excellent hands.

Finally, we each owe personal debts to the people nearest and dearest to us. Jay thanks his parents and his many teachers and mentors over the years for making his own life and education so enriching. He gives loving thanks to his wife, Susan, who has patiently provided caring support throughout this book's writing and revision over the years. Matt thanks his family, friends, and colleagues, and is grateful for his children, who give him three reasons to care passionately about the future. Most importantly, he thanks his wife, Lisa, for being a beacon of love and support in his life for more than 30 years. The talents, input, and advice of Susan and of Lisa have been vital to this project, and without their support our own contributions would not have been possible.

We dedicate this book to today's students, who will shape tomorrow's world.

—*Jay Withgott and Matthew Laposata*

Reviewers

We wish to express special thanks to the dedicated reviewers who shared their time and expertise to help make this seventh edition the best it could be. Their efforts built on those of the nearly 700 instructors and outside experts who have reviewed material for the previous six editions of this book through chapter reviews, pre-revision reviews, feature consultation, student reviews, class testing, and focus groups. Our sincere gratitude goes out to all of them.

Reviewers for the Seventh Edition

Shamili Ajgaonkar, *College of DuPage*
Scott Benjamin, *Bunker Hill Community College*
Jennifer Biederman, *Winona State University*
Emma Bojinova, *University of Connecticut*
Scott Brames, *Clemson University*
John Brzorad, *Lenoir Rhyne University*
Alyson Center, *Normandale Community College*
James Dauray, *College of Lake County*
Elizabeth Davis-Berg, *Columbia College*
Shannon Davis-Foust, *University of Wisconsin Oshkosh*
Jean DeSaix, *University of North Carolina Chapel Hill*
David Fitzpatrick, *Georgia State University*
Christian George, *High Point University*
David Gillette, *University of North Carolina Asheville*
Heinrich Goetz, *Collin College Preston Ridge*
Richard Grippo, *Arkansas State University Main*
Carl Grobe, *Westfield State University*
Leslie Hendon, *University of Alabama Birmingham*
Joey Holmes, *Rock Valley College*
Jodee Hunt, *Grand Valley State University*
Andrew Lapinski, *Reading Area Community College*
Kim Largen, *George Mason University*
Grace Lasker, *University of Washington Bothell*
Cody Leudtke, *Georgia State University*
Heidi Marcum, *Baylor University*
Terri Matiella, *University of Texas, San Antonio*
Hussein Mohamed, *Dalton State College*
Gregory O'Mullan, *Queen's College*
Thomas Pliske, *Florida International University*
Daniel Ratcliff, *Rose State College*
Eric Sanden, *University of Wisconsin River Falls*
Debra Socci, *Seminole State College*
Julie Stoughton, *University of Nevada Reno*
Phil Yeager, *Fairmont State University*

Reviewers for Previous Editions

Matthew Abbott, *Des Moines Area Community College;* David Aborne, *University of Tennessee–Chattanooga;* Charles Acosta, *Northern Kentucky University;* Shamim Ahsan, *Metropolitan State College of Denver;* Isoken T. Aighewi, *University of Maryland–Eastern Shore;* Jeffrey Albert, *Watson Institute of International Studies;* John V. Aliff, *Georgia Perimeter College;* Mary E. Allen, *Hartwick College;* Karyn Alme, *Kennesaw State University;* Deniz Z. Altin, *Georgia Perimeter College;* Dula Amarasiriwardena, *Hampshire College;* Gary I. Anderson, *Santa Rosa Junior College;* Mark W. Anderson, *The University of Maine;* Corey Andries, *Albuquerque Technical Vocational Institute;* David M. Armstrong, *University of Colorado–Boulder;* David L. Arnold, *Ball State University;* Joseph Arruda, *Pittsburg State University;* Eric Atkinson, *Northwest College;* Thomas W. H. Backman, *Linfield College;* Timothy J. Bailey, *Pittsburg State University;* Stokes Baker, *University of Detroit;* Betsy Bancroft, *Gonzaga University;* Kenneth Banks, *University of North Texas;* Narinder Bansal, *Ohlone College;* Jon Barbour, *University of Colorado–Denver;* Reuben Barret, *Prairie State College;* Morgan Barrows, *Saddleback College;* Henry Bart, *LaSalle University;* James Bartalome, *University of California–Berkeley;* Marilynn Bartels, *Black Hawk College;* Brad Basehore, *Harrisburg Area Community College;* David Bass, *University of Central Oklahoma;* Christy Bazan, *Illinois State University;* Christopher Beals, *Volunteer State Community College;* Laura Beaton, *York College, City University of New York;* Hans T. Beck, *Northern Illinois University;* Richard Beckwitt, *Framingham State College;* Barbara Bekken, *Virginia Polytechnic Institute and State University;* Elizabeth Bell, *Santa Clara University;* Timothy Bell, *Chicago State University;* David Belt, *Johnson County Community College;* Gary Beluzo, *Holyoke Community College;* Bob Bennett, *University of Arkansas;* Terrence Bensel, *Allegheny College;* William B. N. Berry, *University of California, Berkeley;* Jill Bessetti, *Columbia College of Missouri Online;* Kristina Beuning, *University of Wisconsin–Eau Claire;* Jennifer Biederman, *Winona State University;* Peter Biesmeyer, *North Country Community College;* Donna Bivans, *Pitt Community College;* Grady Price Blount, *Texas A&M University–Corpus Christi;* Marsha Bollinger, *Winthrop University;* Lisa K. Bonneau, *Metropolitan Community College–Blue River;* Bruno Borsari, *Winona State University;* Richard D. Bowden, *Allegheny College;* Scott Brame, *Clemson University;* Frederick J. Brenner, *Grove City College;* Nancy Broshot, *Linfield College;* Bonnie L. Brown, *Virginia Commonwealth University;* David Brown, *California State University–Chico;* Evert Brown, *Casper College;* Hugh Brown, *Ball State University;* J. Christopher Brown, *University of Kansas;* Geoffrey L. Buckley, *Ohio University;* Dan Buresh, *Sitting Bull College;* Dale Burnside, *Lenoir-Rhyne University;* Lee Burras, *Iowa State University;* Hauke Busch, *Augusta State University;* Christina Buttington, *University of Wisconsin–Milwaukee;* Charles E. Button, *University of Cincinnati, Clermont College;* John S. Campbell, *Northwest College;* Myra Carmen Hall, *Georgia Perimeter College;* Mike Carney, *Jenks High School;* Kelly S. Cartwright, *College of Lake County;* Jon Cawley, *Roanoke College;* Michelle Cawthorn, *Georgia Southern University;* Linda Chalker-Scott, *University of Washington;* Brad S. Chandler, *Palo Alto College;* Paul Chandler, *Ball State University;* David A. Charlet, *Community College of Southern Nevada;* Sudip Chattopadhyay, *San Francisco State University;* Tait Chirenje, *Richard Stockton College;* Luanne Clark, *Lansing Community College;* Richard Clements, *Chattanooga State Technical Community College;* Philip A. Clifford, *Volunteer State Community College;* Kenneth E. Clifton, *Lewis and Clark College;* Reggie Cobb, *Nash Community College;* John E. Cochran, *Columbia Basin College;* Donna Cohen, *MassBay Community College;* Luke W. Cole, *Center on Race,*

About the Cover

"Supertrees"—decorative vertical gardens that collect rainwater and generate renewable solar energy—tower over the Gardens by the Bay park in Singapore. A tiny and densely populated island city-state in Southeast Asia, Singapore has become a model of urban sustainability, leading the drive to design clean, healthy cities that conserve resources and enrich the lives of their residents. Rigorous central planning has created a city that boasts affordable housing, efficient mass-transit, and a high quality of life. Natural areas are an integral part of this planning: Gardens adorn roofs and balconies, green spaces crisscross the city, and incentives promote the incorporation of plant life into new construction. Singapore is even innovating the growth of food in vertical indoor farms. As our population grows and more people move to urban areas, creating sustainable cities is one of humanity's great challenges, intertwined with pressing issues of health, food security, biodiversity conservation, and energy use in a world of changing climate. Join us in this seventh edition of *Environment: The Science Behind the Stories*, as we discover success stories and creative solutions while exploring the frontiers of environmental science.

Foundations of Environmental Science

PART 1

Science and Sustainability

An Introduction to Environmental Science

Our Island, Earth

Viewed from space, our home planet resembles a small blue marble suspended in a vast inky-black void. Earth may seem enormous to us as we go about our lives on its surface, but the astronaut's view reveals that our planet is finite and limited. With this perspective, it becomes clear that as our population, technological power, and resource consumption all increase, so does our capacity to alter our surroundings and damage the very systems that keep us alive. Learning how to live peacefully, healthfully, and sustainably on our diverse and complex planet is our society's prime challenge today. The field of environmental science is crucial in this endeavor.

Our environment surrounds us

A photograph of Earth from space offers a revealing perspective, but it cannot convey the complexity of our environment. Our **environment** consists of all the living and nonliving things around us. It includes the continents, oceans, clouds, and ice caps you can see in a photo from space, as well as the fields, forests, plants, and animals of the landscapes in which we live. In a more inclusive sense, it also encompasses the towns, cities, farms, buildings, and living spaces that people have created. In its fullest sense, our environment includes the complex webs of social relationships and institutions that shape our daily lives.

People commonly use the term *environment* in the narrowest sense—to mean a nonhuman or "natural" world apart from human society. This is unfortunate, because it masks the vital fact that all of us exist within the environment and are part of nature. As one of many species on Earth, we share dependence on a healthy, functioning planet. The limitations of language lead us to speak of "people and nature" or "humans and the environment" as though they were separate and did not interact. However, the fundamental insight of environmental science is that we are part of the "natural" world and that our interactions with the rest of it matter a great deal.

Environmental science explores our interactions with the world

Understanding our relationship with the world around us is vital because we depend on our environment for air, water, food, shelter, and everything else essential for living. Throughout human history, we have modified our environment. By doing so, we have enriched our lives, improved our health, lengthened our life spans, and secured greater material wealth, mobility, and leisure time. Yet many of the changes we have made to our surroundings have degraded the natural systems that sustain us. We have brought about air and water pollution, soil erosion, species extinction, and other environmental impacts that compromise our well-being and jeopardize our ability to survive and thrive in the long term.

Environmental science is the scientific study of how the natural world works, how our environment affects us, and how we affect our environment. Understanding these interactions helps us devise solutions to society's many pressing challenges. It can be daunting to reflect on the number and magnitude of dilemmas that confront us, but these problems also bring countless opportunities for creative solutions.

Environmental scientists study the issues most centrally important to our future. Right now, global conditions are changing more quickly than ever. Right now, we are gaining scientific knowledge more rapidly than ever. And right now, there is still time to tackle society's biggest challenges. With such bountiful opportunities, this moment in history is an exciting time to be alive—and to be studying environmental science.

◀ **Our island, Earth**

Upon completing this chapter, you will be able to:

+ Describe the field of environmental science

+ Explain the importance of natural resources and ecosystem services to our lives

+ Discuss population growth, resource consumption, and their consequences

+ Explain what is meant by an ecological footprint

+ Describe the scientific method and the process of science

+ Apply critical thinking to judge the reliability of information sources

+ Identify major pressures on the global environment

+ Discuss the concept of sustainability, and describe sustainable solutions being pursued on campuses and across the world

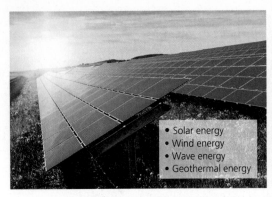
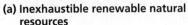
- Solar energy
- Wind energy
- Wave energy
- Geothermal energy

- Fresh water
- Forest products
- Biodiversity
- Soils

- Crude oil
- Natural gas
- Coal
- Minerals

(a) Inexhaustible renewable natural resources

(b) Exhaustible renewable natural resources

(c) Nonrenewable natural resources

FIGURE 1.1 Natural resources may be renewable or nonrenewable. Perpetually renewable, or inexhaustible, resources such as sunlight and wind energy **(a)** will always be there for us. Renewable resources such as timber, soils, and fresh water **(b)** are replenished on intermediate timescales, if we are careful not to deplete them. Nonrenewable resources such as minerals and fossil fuels **(c)** exist in limited amounts that could one day be gone.

We rely on natural resources

Islands are finite and bounded, and their inhabitants must cope with limitations in the materials they need. On our island—planet Earth—there are limits to many of our **natural resources,** the substances and energy sources from our environment that we rely on to survive (**FIGURE 1.1**).

Natural resources that are replenished over short periods are known as **renewable natural resources.** Some renewable natural resources, such as sunlight, wind, and wave energy, are perpetually renewed and essentially inexhaustible. Others, such as timber, water, animal populations, and fertile soil, renew themselves over months, years, or decades. These types of renewable resources may be used at sustainable rates, but they may become depleted if we consume them faster than they are replenished. **Nonrenewable natural resources,** such as minerals and fossil fuels, are in finite supply and are formed far more slowly than we use them. Once we deplete a nonrenewable resource, it is no longer available.

We rely on ecosystem services

If we think of natural resources as "goods" produced by nature, we soon realize that Earth's natural systems also provide "services" on which we depend. Our planet's ecological systems purify air and water, cycle nutrients, regulate climate, pollinate plants, and recycle our waste. Such essential services are commonly called **ecosystem services** (**FIGURE 1.2**). Ecosystem services arise from the normal functioning of natural systems and are not meant for our benefit, yet we could not survive without them. The ways that ecosystem services support our lives and civilization are countless and profound (pp. 120–121, 149, 280).

Just as we sometimes deplete natural resources, we often degrade ecosystem services when, for example, we destroy habitat or generate pollution. The degradation of

ecosystem services can have stark economic consequences. For instance, if a community's drinking water source becomes polluted, it may require a great deal of money to clean up the pollution, restore the water quality, and deal with the health impacts on community members. In recent decades, our depletion of nature's goods and our disruption of nature's services have intensified, driven by rising resource consumption and a human population that grows larger every day.

Cycle nutrients

Purify air

Regulate water flow

Reduce erosion

Prevent flooding

Provide game, wildlife, timber, recreation, and aesthetic beauty

Purify water

FIGURE 1.2 We rely on the ecosystem services that natural systems provide. For example, forested hillsides help people living below by purifying water and air, cycling nutrients, regulating water flow, preventing flooding, and reducing erosion, as well as by providing game, wildlife, timber, recreation, and aesthetic beauty.

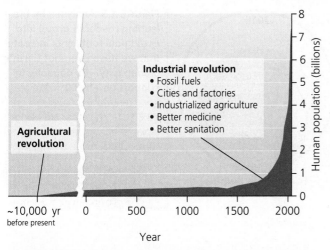

FIGURE 1.3 The global human population increased after the agricultural revolution and then skyrocketed following the industrial revolution. Note that the tear in the graph represents the passage of time and a change in *x*-axis values. *Data compiled from U.S. Census Bureau, U.N. Population Division, and other sources.*

DATA Go to **Interpreting Graphs & Data** on **Mastering Environmental Science**

Population growth amplifies our impact

For nearly all human history, fewer than a million people populated Earth at any one time. Today, our population is approaching 8 *billion* people. For every one person who existed more than 10,000 years ago, several thousand people exist today! **FIGURE 1.3** shows just how recently and suddenly this monumental change has taken place.

Two phenomena triggered our remarkable increase in population size. The first was our transition from a hunter-gatherer lifestyle to an agricultural way of life. This change began about 10,000 years ago and is known as the **agricultural revolution.** As people began to grow crops, domesticate animals, and live sedentary lives on farms and in villages, they produced more food to meet their nutritional needs and began having more children.

The second phenomenon, known as the **industrial revolution,** began in the mid-1700s. It entailed a shift from rural life, animal-powered agriculture, and handcrafted goods toward an urban-centered society provisioned by the mass production of factory-made goods and powered by **fossil fuels** (nonrenewable energy sources including oil, coal, and natural gas; pp. 529–531). Industrialization brought dramatic advances in technology, sanitation, and medicine. It also enhanced food production through the use of fossil-fuel-powered equipment and synthetic pesticides and fertilizers (pp. 215, 246).

The factors driving population growth have brought us better lives in many ways. Yet as our world fills with people, population growth has begun to threaten our well-being. We must ask how well the planet can accommodate the nearly 10 billion people forecast by 2050. Already our sheer numbers are putting unprecedented stress on natural systems and the availability of resources.

Resource consumption exerts social and environmental pressures

Besides stimulating population growth, industrialization increased the amount of resources each of us consumes. By mining energy sources and manufacturing more goods, we have enhanced our material affluence—but have also consumed more and more of the planet's limited resources.

One way to quantify resource consumption is to use the concept of the ecological footprint, developed in the 1990s by environmental scientists Mathis Wackernagel and William Rees. An **ecological footprint** expresses the cumulative area of biologically productive land and water required to provide the resources a person or population consumes and to dispose of or recycle the waste the person or population produces (**FIGURE 1.4**). It measures the total area of Earth's biologically productive surface that a given person or population "uses" once all direct and indirect impacts are summed up.

For humanity as a whole, Wackernagel and his colleagues at the Global Footprint Network calculate that we are now using 69% more of the planet's renewable resources than are available on a sustainable basis. In other words, we are depleting renewable resources by using them 69% faster than they are being replenished. To look at this in yet another way, it would take 1.69 years for the planet to regenerate the renewable resources that people use in just 1 year. The practice of consuming more resources than are being replenished

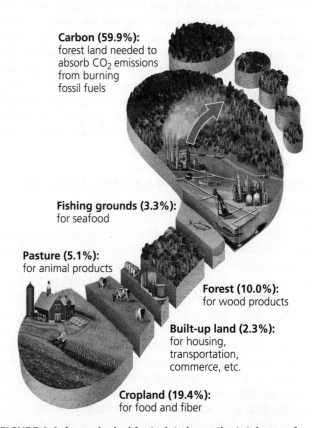

Carbon (59.9%): forest land needed to absorb CO₂ emissions from burning fossil fuels

Fishing grounds (3.3%): for seafood

Pasture (5.1%): for animal products

Forest (10.0%): for wood products

Built-up land (2.3%): for housing, transportation, commerce, etc.

Cropland (19.4%): for food and fiber

FIGURE 1.4 An ecological footprint shows the total area of biologically productive land and water used by a given person or population. Shown is a breakdown of major components of the average person's footprint. *Data from Global Footprint Network, 2019.*

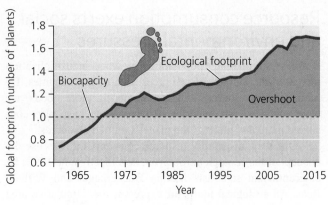

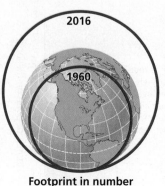

Footprint in number of planets

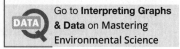

Go to **Interpreting Graphs & Data** on Mastering Environmental Science

is termed **overshoot** because we are overshooting, or surpassing, Earth's capacity to sustainably support us (**FIGURE 1.5**).

Scientists debate how best to calculate footprints and measure overshoot. Yet some things are clear; for instance, people from wealthy nations such as the United States have much larger ecological footprints than do people from poorer nations. Using the Global Footprint Network's calculations, if all the world's people consumed resources at the rate of Americans, humanity would need the equivalent of almost five planet Earths!

We can think of our planet's vast store of resources and ecosystem services—Earth's **natural capital**—as a bank account. To keep a bank account full, we need to leave the principal intact and spend just the interest so that we can continue living off the account far into the future. If we begin depleting the principal, we draw down the bank account. To live off nature's interest—the renewable resources that are replenished year after year—is sustainable. To draw down resources faster than they are replaced is to eat into nature's capital, the bank account for our planet and our civilization. Currently we are drawing down Earth's natural capital—and we cannot get away with it for long.

Environmental science can help us learn from mistakes

Historical evidence suggests that civilizations can crumble when pressures from population and consumption overwhelm resource availability. Historians have inferred that environmental degradation contributed to the fall of the Greek and Roman empires, the Angkor civilization of Southeast Asia, and the Maya, Anasazi, and other civilizations of the Americas. In Syria, Iraq, and elsewhere in the Middle East, areas that in ancient times were lush enough to support thriving ancient societies are today barren desert. Easter Island has long been held up as a society that self-destructed after depleting its resources, although new research paints a more complex picture (see **THE SCIENCE BEHIND THE STORY**, pp. 8–9).

In today's globalized society, the stakes are higher than ever because our environmental impacts are global. If we cannot forge sustainable solutions to our problems, the resulting societal collapse will be global. Fortunately, environmental science holds keys to building a better world. Studying environmental science will help you learn to evaluate the whirlwind of changes taking place around us and to think critically and creatively about ways to respond.

Environmental Science

Environmental scientists examine how Earth's natural systems function, how these systems affect people, and how we influence these systems. Many environmental scientists are motivated by a desire to develop solutions to environmental problems. These solutions (such as new technologies, policies, or resource management strategies) are *applications* of environmental science. The study of such applications and their consequences is, in turn, also part of environmental science.

Environmental science is interdisciplinary

Studying our interactions with our environment is a complex endeavor that requires expertise from many academic disciplines, including ecology, earth science, chemistry, biology, geography, economics, political science, demography, and ethics. Environmental science is **interdisciplinary,** bringing techniques, perspectives, and research results from multiple disciplines together into a broad synthesis (**FIGURE 1.6**).

Traditional established disciplines are valuable because their scholars delve deeply into topics, developing expertise in particular areas and uncovering new knowledge. In contrast, interdisciplinary fields are valuable because their practitioners consolidate and synthesize the specialized knowledge from many disciplines and make sense of it in a broad context to better serve the multifaceted interests of society.

Environmental science is especially broad because it encompasses not only the **natural sciences** (disciplines that examine the natural world) but also the **social sciences** (disciplines that address human interactions and institutions). Most environmental science programs focus more on the natural sciences, whereas programs that emphasize the social sciences often use the term **environmental studies.** Whichever approach one takes, these fields bring together many diverse perspectives and sources of knowledge.

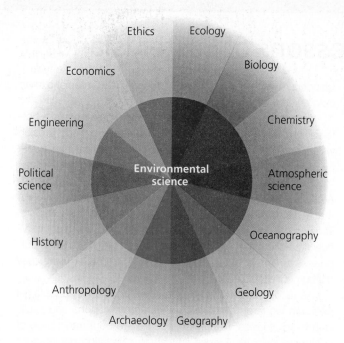

FIGURE 1.6 Environmental science is an interdisciplinary pursuit. It draws from many different established fields of study across the natural sciences and social sciences.

Environmental science is not the same as environmentalism

Although many environmental scientists are interested in solving problems, it would be incorrect to confuse environmental science with environmentalism or environmental activism. They are very different. Environmental science involves the scientific study of the environment and our interactions with it. In contrast, **environmentalism** is a social movement dedicated to protecting the natural world—and, by extension, people—from undesirable changes brought about by human actions.

Of course, like all human beings, scientists are motivated by personal values and interests—and, like any human endeavor, science can never be entirely free of social influence. However, whereas personal values and social concerns may help shape the questions scientists ask, scientists strive to keep their research rigorously objective and free from advocacy. Researchers do their utmost to carry out their work impartially and to interpret their results with wide-open minds, remaining open to whatever conclusions the data demand.

The Nature of Science

Science is a systematic process for learning about the world and testing our understanding of it. The term *science* is also used to refer to the accumulated body of knowledge that arises from this dynamic process of observing, questioning, testing, and discovery.

Knowledge gained from science can be applied to address society's needs—for instance, to develop technology or to inform policy and management decisions (**TABLE 1.1**). From the food we eat to the clothing we wear to the health care we depend on, virtually everything in our lives has been improved by the application of science. Many scientists are motivated by the potential for developing useful applications. Others are inspired simply by a desire to understand how the world works.

Scientists test ideas by critically examining evidence

Science is all about asking and answering questions. Scientists examine how the world works by making observations, taking measurements, and testing whether their ideas are supported by evidence. The effective scientist thinks critically and does not simply accept conventional wisdom from others. The scientist becomes excited by novel ideas but is skeptical and judges ideas by the strength of evidence that supports them.

TABLE 1.1 Examples of Societal Applications of Science

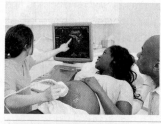 Science **enhances our health and well-being**. Treatments for illness, cures for disease, technologies for better health care (here, **ultrasound imaging** for assessing fetal development)—indeed, all aspects of medicine—rely on scientific research.

 Scientific study leads to **advances in engineering and technology**. Energy-efficient vehicles such as **electric cars** are made possible by research and development in materials science and energy efficiency.

 By revealing our impacts on the atmosphere and climate, scientific research has led to **policies** to fight climate change and to **technology** for producing clean renewable energy, such as these **offshore wind turbines**.

 Science can help us **reduce environmental impacts**. The pursuit of **ecological restoration**—restoring disturbed areas to an earlier, more natural state—is a common land management practice informed by ecological research.

What Are the Lessons of Easter Island?

**Terry Hunt and
Carl Lipo on
Easter Island**

A mere speck of land in the vast Pacific Ocean, Easter Island is one of the most remote spots on the globe. Yet this far-flung island—called Rapa Nui by its inhabitants—has been the focus of an intense debate among scientists seeking to solve its mysteries and decipher the lessons it offers. The debate shows how, in science, new information can challenge existing ideas—and also how interdisciplinary research helps us tackle complex questions.

Ever since European explorers stumbled upon Rapa Nui on Easter Sunday in 1722, outsiders have been struck by the island's barren landscape. Early European accounts suggested that the 2000 to 3000 people living on the island at the time seemed impoverished, subsisting on a few meager crops and possessing only stone tools. Yet the forlorn island also featured hundreds of gigantic statues of carved rock (**FIGURE 1**). How could people without wheels or ropes, on an island without trees, have moved 90-ton statues 10 m (33 ft) high as far as 10 km (6.2 mi) from the quarry where they were chiseled to the sites where they were erected? Apparently, some calamity must have befallen a once-mighty civilization on the island.

Researchers who set out to solve Rapa Nui's mysteries soon discovered that the island had once been lushly forested. Scientist John Flenley and his colleagues drilled cores deep into lake sediments and examined ancient pollen grains preserved there, seeking to reconstruct, layer by layer, the history of vegetation in the region. Finding a great deal of palm pollen, they inferred that when Polynesian people colonized the island (A.D. 300–900, they estimated), it was covered with palm trees similar to the Chilean wine palm—a tree that can live for centuries.

By studying pollen and the remains of wood from charcoal, archaeologist Catherine Orliac found that at least 21 other plant species—now gone—had also been common. Clearly the island had once supported a diverse forest. Forest plants would have provided fuelwood, building material for houses and canoes, fruit to eat, fiber for clothing, and, researchers guessed, logs and fibrous rope to help move statues.

But pollen analysis also showed that trees began declining after human arrival and were replaced by ferns and grasses. Then between 1400 and 1600, pollen levels plummeted. Charcoal in the soil proved that the forest had been burned, likely in slash-and-burn farming. Researchers concluded that the islanders, desperate for forest resources and cropland, had deforested their own island.

With the forest gone, soil eroded away—data from lake bottoms showed a great deal of accumulated sediment. Erosion would have lowered yields of bananas, sugarcane, and sweet potatoes, perhaps leading to starvation and population decline.

Further evidence indicated that wild animals disappeared. Archaeologist David Steadman analyzed 6500 bones and found that at least 31 bird species had provided food for the islanders. Today, only one native bird species is left. Remains from charcoal fires show that early islanders feasted on fish, sharks, porpoises, turtles, octopus, and shellfish—but in later years they consumed little seafood.

As resources declined, researchers concluded, people fell into clan warfare, revealed by unearthed weapons and skulls with head wounds. Rapa Nui appeared to be a tragic case of ecological suicide: A once-flourishing civilization depleted its resources and destroyed itself. In this interpretation—advanced by Flenley and writer Paul Bahn, and popularized by scientist Jared Diamond in his best-selling 2005 book *Collapse*—Rapa Nui seemed to offer a clear lesson: We on our global island, planet Earth, had better learn to use our limited resources sustainably.

When Terry Hunt and Carl Lipo began research on Rapa Nui in 2001, they expected simply to help fill gaps in a well-understood history. But science is a process of discovery, and sometimes evidence leads researchers far from where they anticipated. For Hunt, an anthropologist at the University of Hawai'i at Manoa, and Lipo, an archaeologist at California State University, Long Beach, their work led them to conclude that the traditional "ecocide" interpretation didn't tell the whole story. First, their radiocarbon dating (dating of items using radioisotopes of carbon; p. 26) indicated that people had not colonized the island until about A.D. 1200, suggesting that deforestation occurred rapidly after their arrival. How could so few people have destroyed so much forest so fast?

Hunt and Lipo's answer: rats. When Polynesians settled new islands, they brought crop plants, as well as chickens and other domestic animals. They also brought rats—intentionally as a food source or unintentionally as stowaways. In either case, rats can multiply quickly, and they soon overran Rapa Nui.

Researchers found rat tooth marks on old nut casings, and Hunt and Lipo suggested that rats ate so many palm nuts and shoots that the trees could not regenerate. With no young trees growing, the palm went extinct once mature trees died.

Diamond and others counter that plenty of palm nuts on Easter Island escaped rat damage, that most plants on other islands survived rats introduced by Polynesians, and that more than 20 additional plant species went extinct on Rapa Nui. Moreover, people brought the rats, so even if rats destroyed the forest, human colonization was still to blame.

Despite the forest loss, Hunt and Lipo argue that islanders were able to persist and thrive. Archaeology shows how islanders adapted to Rapa Nui's poor soil and windy weather by developing rock gardens to protect crop plants and nourish the soil. Hunt and Lipo contended that tools viewed by previous researchers as weapons were actually farm implements, that lethal injuries were rare, and that no evidence of battle or defensive fortresses was uncovered.

Hunt, Lipo, and others also unearthed old roads and inferred how the famous statues were transported. It had been thought that a powerful central authority must have forced armies of laborers to roll them over countless palm logs, but Hunt and Lipo concluded that small numbers of people could have moved them by tilting and rocking them upright—much as we might move a refrigerator. Indeed, the distribution of statues on the island suggested the work of family groups. Islanders had adapted to their resource-poor environment by becoming a peaceful and cooperative society, Hunt and Lipo maintained, with the statues providing a harmless outlet for competition over status and prestige.

Altogether, the evidence led Hunt and Lipo to propose that far from destroying their environment, the islanders had acted as responsible stewards. The collapse of this sustainable civilization, they argue, came with the arrival of Europeans, who unwittingly brought contagious diseases to which the islanders had never been exposed. Indeed, historical journals of sequential European voyages depict a society falling into disarray as if reeling from epidemics.

Peruvian ships then began raiding Rapa Nui and taking islanders away into slavery. Foreigners acquired the land, forced the remaining people into labor, and introduced thousands of sheep, which destroyed the few native plants left on the island. Thus, the new hypothesis holds that the collapse of Rapa Nui's civilization resulted from a barrage of disease, violence, and slave raids following foreign contact. Before that, Hunt and Lipo say, Rapa Nui's people boasted 500 years of a peaceful and resilient society.

Hunt and Lipo's interpretation, put forth in a 2011 book, *The Statues That Walked,* would represent a paradigm shift (p. 14) in how we view Easter Island. Debate between the two camps remains heated, however, and interdisciplinary research continues as scientists look for new ways to test the differing hypotheses. This is an example of how science advances, and in the long term, data from additional studies should lead us closer and closer to the truth.

Like the people of Rapa Nui, we are all stranded together on an island with limited resources. What is the lesson of Easter Island for our global island, Earth? Perhaps there are two: Any island population must learn to live within its means, but with care and ingenuity, there is hope that we can.

FIGURE 1 Were the haunting statues of Easter Island (Rapa Nui) erected by a civilization that collapsed after devastating its environment or by a sustainable civilization that fell because of outside influence?

A great deal of scientific work is **descriptive science,** research in which scientists gather basic information about organisms, materials, systems, or processes that are not yet well known. In this approach, researchers explore new frontiers of knowledge by observing, recording, and measuring phenomena to gain a better understanding of them.

Once enough basic information is known about a subject, scientists can begin posing questions that seek deeper explanations about how and why things are the way they are. At this point scientists may pursue **hypothesis-driven science,** research that proceeds in a more targeted and structured manner, using empirical observations or controlled experiments to test hypotheses within a framework traditionally known as the scientific method.

The scientific method is a traditional approach to research

The **scientific method** is a technique for testing ideas with observations. There is nothing mysterious about the scientific method; it is merely a formalized version of the way any of us would naturally use logic to resolve a question. Because science is an active, creative process, innovative researchers may depart from the traditional scientific method when particular situations demand it. Moreover, scientists in different fields approach their work differently because they deal with dissimilar types of information. Nonetheless, scientists of all persuasions broadly agree on fundamental elements of the process of scientific inquiry. As practiced by individual researchers or research teams, the scientific method (**FIGURE 1.7**) typically follows the steps outlined next.

Make observations Advances in science generally begin with the observation of some phenomenon that the scientist wishes to explain. Observations set the scientific method in motion by inspiring questions—and observations also play a role throughout the process.

Ask questions Curiosity is in our human nature. Just observe young children exploring a new environment—they want to touch, taste, watch, and listen to everything, and as soon as they can speak, they begin asking questions. Scientists, in this respect, are kids at heart. Why is the ocean salty? Why are storms becoming more severe? What is causing algae to cover local ponds? When pesticides poison fish or frogs, are people also affected? How can we help restore populations of plants or animals? All these are questions environmental scientists ask.

Develop a hypothesis Scientists pursuing hypothesis-driven science address their questions by devising explanations that they can test. A **hypothesis** is a statement that attempts to explain a phenomenon or answer a scientific question. For example, a scientist wondering why algae are growing excessively in local ponds might observe that chemical fertilizers are being applied on farm fields nearby. The scientist might then propose a hypothesis as follows: "Agricultural fertilizers running into ponds cause the amount of algae in the ponds to increase."

Scientific method

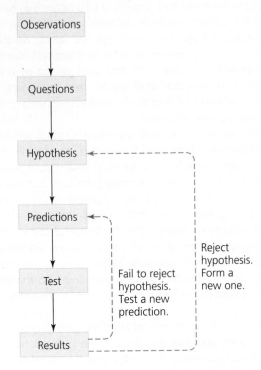

FIGURE 1.7 The scientific method is the traditional experimental approach scientists use to learn how the world works.

Make predictions The scientist next uses the hypothesis to generate **predictions,** specific statements that are logical consequences of the hypothesis and that can be directly and unequivocally tested. In our algae example, a researcher might predict the following: "If agricultural fertilizers are added to a pond, the quantity of algae in the pond will increase."

Test the predictions Scientists test predictions by gathering evidence that could potentially refute the predictions and thus disprove the hypothesis. The strongest form of evidence comes from experiments. An **experiment** is an activity designed to test the validity of a prediction or a hypothesis.

An experiment involves manipulating **variables,** or conditions that can change. For example, a scientist could test the prediction linking algal growth to fertilizer by selecting two identical ponds and adding fertilizer to one of them. In this example, fertilizer input is an **independent variable,** a variable the scientist manipulates, whereas the quantity of algae that results is the **dependent variable,** a variable that depends on the fertilizer input. If the two ponds are identical except for a single independent variable (fertilizer input), any differences that arise between the ponds can be attributed to the independent variable. Such an experiment is known as a **controlled experiment** because the scientist attempts to control for the effects of all variables except for the one that he or she is testing. In our example, the pond left unfertilized serves as a **control,** an unmanipulated point of comparison for the manipulated **treatment** pond.

Whenever possible, it is best to replicate one's experiment—that is, to perform multiple tests of the same comparison.

Our scientist could perform a replicated experiment on, say, 10 pairs of ponds, adding fertilizer to one of each pair.

An experiment in which the researcher actively chooses and manipulates the independent variable is known as a *manipulative experiment*. A manipulative experiment provides strong evidence because it can reveal causal relationships, showing that changes in an independent variable cause changes in a dependent variable. In practice, however, we cannot run manipulative experiments for all questions, especially for processes involving large spatial scales or long timescales. For example, to study global climate change (Chapter 18), we cannot run a manipulative experiment adding carbon dioxide to 10 treatment planets and 10 control planets and then compare the results! Thus, it is common for researchers to run *natural experiments*, which compare how dependent variables are expressed in naturally occurring, but different, contexts. In such experiments, the independent variable varies naturally, and researchers test their hypotheses by searching for **correlation,** or statistical association among variables. For instance, let's suppose our scientist studying algae surveys 50 ponds, 25 of which happen to be fed by fertilizer runoff from nearby farm fields and 25 of which are not. Let's say the scientist finds seven times more algal growth in the fertilized ponds. The scientist may conclude that algal growth is correlated with fertilizer input—that is, that one tends to increase along with the other.

Evidence based on correlation alone is not as strong as the causal evidence that manipulative experiments can provide, but often a natural experiment is the only feasible approach. Because many questions in environmental science are complex and exist on large scales, they must be addressed with correlative data. One benefit of natural experiments is that they preserve the real-world complexity that manipulative experiments often sacrifice. When possible, scientists often try to integrate natural and manipulative experiments to gain the advantages of each.

Analyze and interpret results Scientists record **data,** or information, from their studies (**FIGURE 1.8**). Researchers particularly value quantitative data (information expressed using numbers), because numbers provide precision and are easy to compare. The scientist conducting the fertilization experiment, for instance, might quantify the area of water surface covered by algae in each pond or might measure the dry weight of algae in a certain volume of water taken from each. It is vital, however, to collect data that are representative. Because it is impractical to measure a pond's total algal growth, our researcher might instead sample from multiple areas of each pond. These areas must be selected in a random manner; choosing areas with the most growth or the least growth, or areas most convenient to sample, would not provide a representative sample.

Even with the precision that numerical data provide, experimental results may not be clear-cut. Data from treatments and controls may vary only slightly or replicates may yield different results. Researchers must therefore analyze their data using statistical tests. With these mathematical methods, scientists can determine objectively and precisely the strength and reliability of patterns they find.

FIGURE 1.8 Researchers gather data to test predictions in experiments. Here, a scientist samples algae from a pond.

If experiments disprove a hypothesis, the scientist will revise the hypothesis or may formulate a new hypothesis to replace it. If experiments fail to disprove a hypothesis, such failure lends support to the hypothesis but does not *prove* that the hypothesis is correct. The scientist may choose to generate new predictions to test the hypothesis in different ways and further assess its likelihood of being valid. In this way, the scientific method loops back on itself, giving rise to repeated rounds of hypothesis revision and experimentation (see Figure 1.7).

If repeated tests fail to reject a hypothesis, evidence in favor of it accumulates, and the researcher may eventually conclude that the hypothesis is well supported. Ideally, the scientist would want to test all possible explanations. For instance, our algae researcher might formulate an additional hypothesis, proposing that algae increase in fertilized ponds because chemical fertilizers diminish the numbers of fish or invertebrate animals that eat algae. It is possible, of course, that both hypotheses could be correct and that each may explain some portion of the initial observation that local ponds were experiencing algal blooms.

FAQ

Can science *prove* that an idea is correct?

Technically speaking, science never "proves" an idea to be correct. By following the method and process of experimental scientific research, a scientist may disprove a hypothesis, or the scientist may find support for the hypothesis by testing it and failing to falsify it. But even if a hypothesis attains strong and repeated support from extensive research, it could potentially be shown to be incorrect in the future as a result of newly discovered evidence or of research that is more thorough or better informed. The fact that scientific explanations are always subject to revision, if needed, demonstrates the strength of the scientific process. Unlike opinion or ideology, science is open-ended and tends to be self-corrective and ever-improving, bringing us closer and closer over time to full and accurate understanding.

Scientists use graphs to represent data visually

To summarize and present the data they obtain, scientists often use graphs. Graphs help make patterns and trends in the data visually apparent and easy to understand. **FIGURE 1.9** shows a few examples of how different types of graphs can be used to present data. Each of these types of graphs is illustrated clearly and explained further in **APPENDIX A: HOW TO INTERPRET GRAPHS** at the back of this book. The ability to interpret graphs is a skill you will find useful throughout your life. We encourage you to consult Appendix A closely as you begin your environmental science course.

You also will note that many of the graphs in this book are accompanied by **DATA Q** questions that you can find online at **Mastering Environmental Science**. These questions are designed to help you interpret scientific data and build your graph-reading skills. You can check your answers to these questions in **Mastering Environmental Science**.

The scientific process continues beyond the scientific method

Scientific research takes place within the context of a community of peers. To have impact, a researcher's work must be published and made accessible to this community (**FIGURE 1.10**).

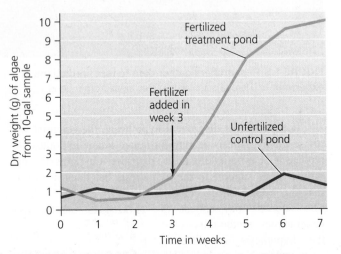

(a) Line graph of algal density through time in a fertilized treatment pond and an unfertilized control pond

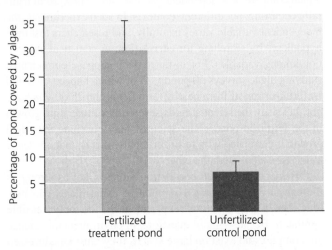

(b) Bar chart of mean algal density in several fertilized treatment ponds and unfertilized control ponds

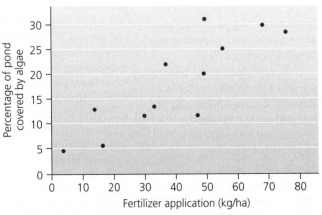

(c) Scatter plot of algal density correlated with fertilizer use on surrounding farmland

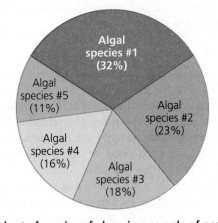

(d) Pie chart of species of algae in a sample of pond water

FIGURE 1.9 Scientists use graphs to present and visualize their data. For example, in **(a)**, a line graph shows how the amount of algae increased when fertilizer was added to a treatment pond in an experiment yet stayed the same in an unfertilized control pond. In **(b)**, a bar chart shows how fertilized ponds, on average, have several times more algae than unfertilized ponds. In **(c)**, a scatter plot shows how ponds with more fertilizer tend to contain more algae. In **(d)**, a pie chart shows the relative abundance of five species of algae in a sample of pond water. See **APPENDIX A** to learn more about how to interpret these types of graphs.

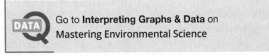

Go to **Interpreting Graphs & Data** on **Mastering Environmental Science**

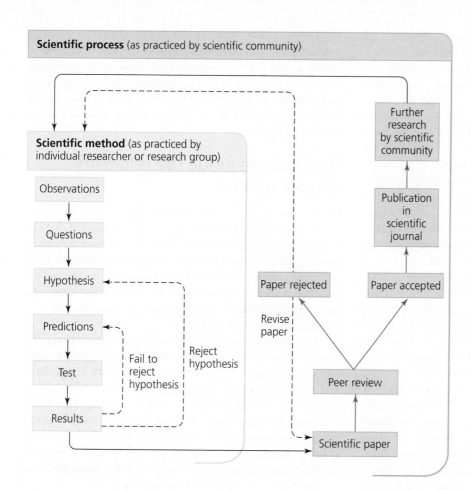

Scientific process (as practiced by scientific community)

Scientific method (as practiced by individual researcher or research group)

Observations

Questions

Hypothesis

Predictions

Test

Results

Fail to reject hypothesis

Reject hypothesis

Scientific paper

Peer review

Revise paper

Paper rejected

Paper accepted

Publication in scientific journal

Further research by scientific community

FIGURE 1.10 The scientific method that many research teams follow is part of a larger framework—the overall process of science carried out by the scientific community. This process includes peer review and publication of research, acquisition of funding, and the elaboration of theory through the cumulative work of many researchers.

Peer review Once a researcher's work is complete, he or she writes up the findings in a research paper and submits the manuscript to a journal (a scholarly publication in which scientists share their work). The journal's editor asks several other scientists who specialize in the subject area to volunteer their time to examine the manuscript, provide comments and criticism (generally anonymously), and judge whether the work merits publication in the journal. This procedure, known as **peer review,** is an essential part of the scientific process.

Peer review is a valuable guard against faulty research contaminating the literature (the body of published studies) on which all scientists rely. It is what makes scientific research papers so much more accurate and reliable than many other forms of publication. Of course, scientists are human, so personal biases and politics can sometimes creep into the review process. Fortunately, just as individual scientists strive to remain objective in conducting their research, the scientific community does its best to ensure fair review of all work.

Conference presentations Scientists frequently present their work at professional conferences, where they interact with colleagues and receive feedback on their research. Such interactions can help improve a researcher's work and foster collaboration among researchers, thus enhancing the overall quality and impact of science.

Grants and funding To fund their research, most scientists need to spend a great deal of time and effort requesting money

from private foundations or from government agencies such as the National Science Foundation. Grant applications undergo peer review just as scientific papers do, and competition for funding is generally intense.

Scientists' reliance on funding sources can occasionally lead to conflicts of interest. A researcher who obtains data showing his or her funding source in an unfavorable light may become reluctant to publish the results for fear of losing funding. This situation can arise, for instance, when an industry funds research to test its products for health or safety. Most scientists resist these pressures, but when you are assessing a scientific study, it is always a good idea to note where the researchers obtained their funding. Because of the potential conflicts of interest that can sometimes arise from private funding, it is important for our society that scientists have access to adequate levels of funding from impartial public sources such as government science agencies.

Repeatability The careful scientist may test a hypothesis repeatedly in various ways. Following publication, other scientists may attempt to reproduce the results in their own experiments. Scientists are inherently cautious about accepting a novel hypothesis, so the more a result can be reproduced by different research teams, the more confidence scientists will gain that it provides a correct explanation.

Theories If a hypothesis survives repeated testing by numerous research teams and continues to predict experimental

Let us say you are a scientist wanting to study the impacts of chemicals released into lakes by pulp-and-paper mills. Obtaining research funding has been difficult. Then a large pulp-and-paper firm contacts you and offers to fund your research examining how the company's chemical effluents affect water bodies. What do you think the benefits and drawbacks of such an offer would be? Would you accept the offer? Why or why not?

outcomes and observations accurately, it may be incorporated into a theory. A **theory** is a widely accepted, well-tested explanation of one or more cause-and-effect relationships that has been extensively validated by a great amount of research. Whereas a hypothesis is a simple explanatory statement that may be disproven by a single experiment, a theory consolidates many related hypotheses that have been supported by a large body of data.

Note that scientific use of the word *theory* differs from popular usage of the word. In everyday language, if we say something is "just a theory," we are suggesting it is a speculative idea without much substance. However, scientists mean exactly the opposite when they use the term. In a scientific context, a theory is a conceptual framework that explains a phenomenon and has undergone extensive and rigorous testing, such that confidence in it is extremely strong. For example, Darwin's theory of evolution by natural selection (pp. 50–51) has been supported and elaborated by many thousands of studies over 160 years of intensive research. Observations and experiments have shown repeatedly and in great detail how plants and animals evolve over generations, attaining characteristics that best promote survival and reproduction. Because of its strong support and explanatory power, evolutionary theory is the central unifying principle of modern biology. Other prominent scientific theories include atomic theory, cell theory, big bang theory, plate tectonics, and general relativity.

Paradigm shifts occur

As the scientific community accumulates data in an area of research, interpretations sometimes may change. Thomas Kuhn's influential 1962 book, *The Structure of Scientific Revolutions,* argued that science goes through periodic upheavals in thought in which one scientific **paradigm,** or dominant view, is abandoned for another. For example, before the 16th century, European scientists believed Earth was at the center of the universe. Their data on the movements of planets fit that concept somewhat well—yet the idea eventually was disproved after Nicolaus Copernicus showed that placing the sun at the center of the solar system explained the data much better.

Another paradigm shift occurred in the 1960s when geologists came to accept plate tectonics (pp. 35–37). By this time, evidence for the movement of continents and the action of tectonic plates had accumulated and become overwhelmingly convincing. Paradigm shifts demonstrate the strength and vitality of science, showing science to be a process that refines and improves itself through time.

Understanding how science works is vital to assessing how scientific interpretations progress through time as information accrues. This is especially relevant in environmental science—a young field that is changing rapidly as we obtain vast amounts of new information, as human impacts on our planet multiply, and as we gather lessons from the consequences of our actions.

How can we judge the reliability of information?

Even if you are not a scientist, every one of us in our everyday lives can benefit from using the type of careful, logical, critical thinking that scientists use. In fact, developing critical thinking skills might well be the single most important thing a student can do to prepare for a lifetime of navigating our society—a society in which far too many uninformed and unscrupulous interests act to confuse, manipulate, and mislead us by distorting information.

In our society today, we are awash in information—and misinformation—from all kinds of sources. To get the day's news, for instance, we can tune into a diversity of television and radio stations, or check countless online websites, or receive social media news feeds, or even open up a good old-fashioned print newspaper. When it comes to news about science, some popularized news coverage is excellent, but some is sorely lacking. And when it comes to coverage of environmental issues, we may receive very different messages from sources with differing political viewpoints.

At a time when some politicians try to discredit any information they do not like as "fake news" and armies of bots and hackers regularly manipulate information online, it can be challenging to know what to trust. However, we know that science—including all the understanding it can bring us—is too valuable to ignore or discard. How, then, can a person find reliable information about scientific research? How can we know which sources to trust and which to be skeptical of?

One key criterion in evaluating the reliability of a source is to determine whether or not the information it offers is solidly based in evidence. It is not always easy to ascertain whether a source is objective, unbiased, and evidence-based, but one can look for a number of clues. **TABLE 1.2** walks you through a checklist of questions to ask yourself whenever you are assessing a source for reliability—for information on science or on anything else.

One question to ask when evaluating sources of information about a scientific matter is how many steps removed from the original research a source is. A person can get the most accurate understanding of a scientific question by reading the scientific literature directly. Your college or university library will have scientific journals full of research papers that are current and peer-reviewed. However, reading original research papers in scientific journals requires a level of specialized education that most of us lack. Thankfully, many of the most prestigious journals, such as *Science* and *Nature,* also feature articles written by their staff that explain the results and relevance of their most groundbreaking and socially relevant papers in language we all can understand.

Another way to learn about important scientific advances is to read articles by science writers published in media outlets that have built solid reputations over years or decades. Magazines like *Scientific American, National Geographic,* and *New Scientist* are examples, as are newspapers such as *The New York Times, Los Angeles Times,* and *Wall Street Journal*—all of which are accessible online as well as in print (and are likely available for free in your campus library). Websites like *Science News* and *Science Daily* and podcasts, videos, television programs, and radio broadcasts from proven media outlets like PBS's *NOVA* and NPR's *Science Friday* are excellent as well. Science journalists inform their articles and broadcasts by reading original research papers, interviewing the scientists who conducted the work, and speaking with other scientists for context and criticism. Most science journalists are trained in science and steeped in their areas of expertise but are also skilled at communicating with a broad audience. They often are able to explain research and bring the wonders of science to life in a way that most scientists cannot.

An original scientific research paper in a journal is an example of a **primary source,** a source that presents novel information and stands on its own. Articles or broadcasts produced by others about the contents of a primary source are examples of a **secondary source.** Secondary sources are often easier to read and understand than primary sources, but as one moves further from the primary source, there is more and more risk of inaccurate information being introduced. Material that has "gone viral" through social media is often the most unreliable because information can easily become altered, just as it is in the classic "telephone game" in which a sentence is whispered from person to person around a circle.

These days the internet is awash in accounts from secondary sources that are distant from the actual research or are not wholly reliable. Writers and broadcasters with little knowledge of science may nonetheless need to produce a news piece on deadline about some new advance. Bloggers who love communicating with their readers and are skillful writers may inadvertently pass along information that is inaccurate.

TABLE 1.2 Questions to Ask When Evaluating a Source for Reliability

WHO is presenting the information?

What are the presenter's profession, credentials, and qualifications?

What are the presenter's organizational affiliations?

What person or institution is financially supporting presentation of the information (e.g., university, private company, government agency, non-profit organization)?

Does the website URL reveal anything about the source (e.g., .com, .org, .edu, .gov)?

WHY is the information being presented?

Does the presenter or sponsor make his or her motivations clear?

Is the presenter trying to teach and inform or to persuade and sell?

Are there advertisements related to the content?

Could the presenter have a financial interest or other vested interest in promoting the argument being made?

HOW reliable is the information?

Is the information based on factual evidence?

Does the source cite *its* sources or explain where *its* information is coming from? (Original research? Personal experience? Other sources?)

Can you find, examine, and verify the information elsewhere?

Do the language and tone seem unbiased and free of emotion?

If an opinion is being expressed, are reasons given to back up the opinion?

Are there spelling, grammar, or typographical errors?

Has the information been reviewed or refereed?

Can you detect political, ideological, cultural, religious, institutional, or personal biases?

Is the information current? When was the information created or last updated?

Have you looked at and compared a variety of sources before choosing this one?

Some table items are based on those in Blakeslee, S., 2010. *Evaluating information—Apply the CRAAP test.* Meriam Library, California State Library, Chico.

FAQ

What about Wikipedia?

Wikipedia can be a tremendous source of information as long as one keeps in mind the way it is created. Wikipedia is a crowd-sourced online encyclopedia whose entries are written by volunteers, and the entries are peer-reviewed by other volunteers. Entries on popular topics that many people with expertise have contributed to are often comprehensive and accurate, but entries on obscure topics that have received less critical attention can sometimes contain unreliable information. Examining the number and quality of sources cited at the end of a Wikipedia entry is one good way to assess the strength of the entry. In fact, a Wikipedia page can often be a helpful way to begin researching a topic because it generally provides citations of some of the most relevant primary sources, which you can then check on your own.

Of most concern is that people with financial conflicts of interest or with strong political ideologies may intentionally sow misinformation to advance their own agendas. Examples in recent years relevant to environmental science have involved the denial of climate change, particularly in U.S. media and politics. Many ideologically driven politicians and many individuals with financial ties to fossil fuel industries have spread doubt about the extent, causes, and consequences of climate change, ignoring or distorting the immense amount of scientific evidence that thousands of researchers worldwide have built up over many decades of careful study.

It is up to each of us to be smart, educated consumers of information and to recognize when sources are misinformed or are trying to sell us their own version of reality. Sorting good information from bad is not always easy, but it is one of the most important life skills any of us can develop.

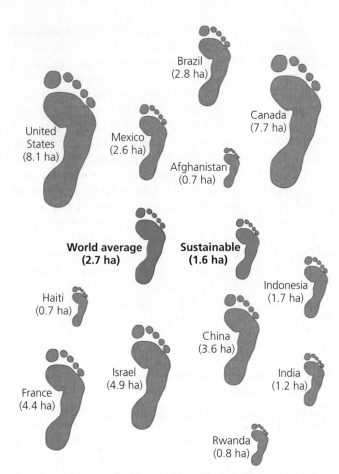

FIGURE 1.11 **People of some nations have much larger ecological footprints than people of others.** Shown are ecological footprints for average citizens of several nations along with the world's average per capita footprint of 2.7 hectares. To achieve sustainable consumption, we would need to have a global per-person average footprint of 1.6 ha. One hectare (ha) = 2.47 acres. *Data from Global Footprint Network, 2019.*

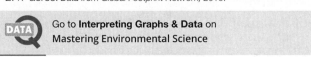

Go to **Interpreting Graphs & Data** on Mastering Environmental Science

Seeking a Sustainable Future

As environmental scientists study the causes and consequences of environmental change, they are helping to address today's primary challenge for society: determining how to live within our planet's means such that Earth and its resources can sustain us—and all life—for the future. This is the challenge of **sustainability,** a guiding principle of modern environmental science and a concept you will encounter throughout this book.

Sustainability is a condition in which our actions do not cause lasting harm to the environment and are socially and economically beneficial as well so that people's needs today are met without impairing future generations' abilities to meet their own needs. Sustainability means leaving our children and grandchildren a world as rich and full as the world we live in now. It means conserving Earth's resources so that our descendants may enjoy them as we have. It means reducing poverty and making a good life possible for all people. It means developing solutions that work in the long term. Sustainability requires maintaining fully functioning ecological systems, because we cannot sustain human civilization without sustaining the natural systems that nourish it.

An examination of the ecological footprints of the world's nations (**FIGURE 1.11**) indicates that we are not (yet) on a sustainable trajectory. Yet as we will see throughout this book, there are many reasons to be optimistic. In each chapter, a **SUCCESS STORY** will feature one specific example of a problem-solving effort that has met with success and produced a sustainable solution. In addition, you will encounter a variety of further solutions to environmental challenges, addressed in many ways, throughout the text.

Population and consumption drive environmental impact

Every day we add more than 200,000 people to the planet. This is like adding a city the size of Augusta, Georgia, on Monday; Akron, Ohio, on Tuesday; Richmond, Virginia, on Wednesday; Rochester, New York, on Thursday; Amarillo, Texas, on Friday; and on and on, day after day. The rate of population growth is now slowing, but our absolute numbers continue to increase. This ongoing rise in human population (Chapter 8) amplifies nearly all our impacts.

Our consumption of resources has risen even faster than our population. The modern rise in affluence has been a positive development for humanity, and our conversion of the planet's natural capital has made life better for most of us so far. However, like rising population, rising per capita (per person) consumption magnifies the demands we make on our environment.

The world's people have not benefited equally from society's overall rise in affluence. Today, the 20 wealthiest nations boast more than 55 times the per capita income of the 20 poorest nations—three times the gap that existed just two generations ago. Within the United States, the richest 10% of people now claim half of the total income and more than 70% of the total wealth. The ecological footprint of the average resident of a developed nation such as the United States is considerably larger than that of the average resident of a developing country (see Figure 1.11).

Our growing population and consumption are intensifying the many environmental impacts we examine in this book, including erosion and other impacts from agriculture (Chapters 9 and 10), deforestation (Chapter 12), toxic substances (Chapter 14), freshwater depletion (Chapter 15), fisheries declines (Chapter 16), air and water pollution (Chapters 15–17), waste generation (Chapter 22), mineral extraction and mining impacts (Chapter 23), and, of course, global climate change (Chapter 18). These impacts degrade our health and quality of life, and they alter the ecosystems and landscapes in which we live. They also are driving the loss of Earth's biodiversity (Chapter 11), which is perhaps our greatest problem because extinction is irreversible. Once a species becomes extinct, it is lost forever.

Energy choices will shape our future

Our reliance on fossil fuels amplifies virtually every impact we exert on our environment. Yet fossil fuels have also helped bring us the material affluence we enjoy. By exploiting the richly concentrated energy in coal, oil, and natural gas, we have been able to power the machinery of the industrial revolution, produce chemicals that boost crop yields, run vehicles and transportation networks, and manufacture and distribute countless consumer products.

However, in extracting coal, oil, and natural gas, we are splurging on a one-time bonanza, because these fuels are nonrenewable and in finite supply. As we pour more money, energy, and resources into extracting fossil fuels that are more difficult to access, we threaten greater environmental and social impacts for relatively less fuel. The energy choices we make today will greatly influence the nature of our lives for the foreseeable future.

WEIGHING the **issues**

Leaving a Large Footprint

What do you think accounts for the variation in per capita (per person) ecological footprints among societies? Do you believe that people with larger footprints have an ethical obligation to reduce their environmental impact so as to leave more resources available for people with smaller footprints? Why or why not?

SUCCESS story Removing Lead from Gasoline

Did you ever wonder why unleaded gas is called "unleaded"? It's because we used to add lead to gasoline to make cars run more smoothly—even though scientific research showed that emissions of this toxic heavy metal from vehicle tailpipes caused severe health problems, including brain damage and premature death. Back in 1970, air pollution was severe in many U.S. cities, and motor vehicle exhaust accounted for 78% of U.S. lead emissions. In response, environmental scientists, medical researchers, auto engineers, and policymakers all merged their knowledge and skills into a process that brought about the removal of lead from gasoline. Because some older vehicles required leaded gas, the ban on lead was phased in gradually but by 1996, all gasoline sold in the United States was unleaded, and the nation's largest source of atmospheric lead pollution was eliminated. As a result, levels of lead in people's blood fell dramatically, producing one of America's greatest public health successes.

→ **Explore the Data** at **Mastering Environmental Science**

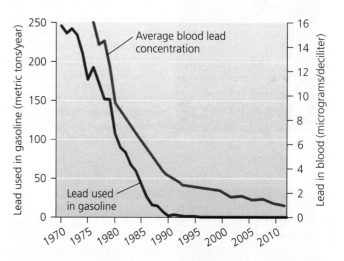

Levels of lead in the blood of U.S. children (ages 1–5) declined as lead use in U.S. gasoline was reduced. *Data from National Health and Nutrition Examination Survey (CDC) and other sources.*

TABLE 1.3 Major Approaches in Campus Sustainability

Waste reduction

Reusing, recycling, and composting offer abundant opportunities for tangible improvements, and people understand and enjoy these activities. Waste reduction events such as trash audits and recycling competitions can be fun and productive. Compost can be applied to plantings to beautify the campus. Some schools aim to become "zero-waste."

Wayne State University

Green buildings

Constructed from sustainable materials, "green buildings" feature designs and technologies to reduce pollution, use renewable energy, and encourage efficiency in water and energy use. These sustainable buildings are certified according to LEED standards (pp. 348–350).

Oberlin College

Water conservation

Indoors, schools are installing water-saving faucets, toilets, urinals, and showers in dorms and classroom buildings to reduce water waste. Outdoors, students are helping to landscape gardens that harvest rainwater and to build facilities to treat and reuse wastewater.

University of South Florida

Energy efficiency

Campuses are installing energy-efficient lighting, motion detectors to shut off lights when rooms are empty, and sensors to record and display a building's energy consumption. Students are mounting campaigns to reduce thermostat settings, distribute efficient bulbs, and publicize energy-saving tips to their peers.

Xavier University of Louisiana

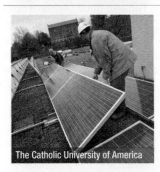

Renewable energy

Many schools are switching from fossil fuels to renewable heating and electricity. Some are installing solar panels, others use biomass in power plants, and a few have built wind turbines on campus. Students have persuaded administrators—and have voted for student fees—to buy "green tags" or carbon offsets to fund renewable energy.

The Catholic University of America

Food and dining

More and more schools grow their own food on campus farms and in gardens, supplying students with local, healthy, organic food. In dining halls, trayless dining cuts down on waste (on average, 25% of food taken is wasted otherwise) because people take only what they really want. Food scraps are composted on many campuses.

Yale University

Transportation

Half the greenhouse gas emissions of the average college or university come from commuting to and from campuses in motor vehicles. To combat pollution, traffic congestion, delays, and parking shortages, schools are investing in bus and shuttle systems, hybrid and alternative-fuel fleet vehicles, and programs to promote carpooling, walking, and bicycling.

University of Virginia

Plants and landscaping

Students are helping to restore native plants and communities, remove invasive species, improve habitat for wildlife, and enhance soil and water quality. Some schools have green roofs, greenhouses, and botanical gardens. Enhancing a campus's natural environment creates healthier, more attractive surroundings, and makes for educational opportunities in ecology and natural resources.

University of Arizona

Carbon-neutrality

To combat climate change, many campuses aim to become carbon-neutral, emitting no net greenhouse gases. These schools seek to power themselves with clean renewable energy, although for most, buying carbon offsets is also necessary. Several hundred college and university presidents have signed the Presidents' Climate Leadership Commitments, with carbon-neutrality as a goal.

Mount Wachusett Community College

Fossil fuel divestment

Endowment money in U.S. higher education is estimated at more than $400 billion, and some of it is invested in stocks of coal, oil, and gas corporations. Since 2012, students have lobbied their administrators, boards of trustees, and fund managers to divest from, or sell off, stocks in fossil fuel companies.

Tufts University

Sustainable solutions abound

Humanity's challenge is to develop solutions that enhance our quality of life while protecting and restoring the environment that supports us. Many workable solutions are at hand:

- Renewable energy sources (Chapters 20 and 21) are beginning to replace fossil fuels.
- Energy-efficiency efforts continue to gain ground (Chapter 19).
- Scientists and farmers are pursuing soil conservation, high-efficiency irrigation, pollinator conservation, and organic agriculture (Chapters 9 and 10).
- Laws and new technologies have reduced air and water pollution in wealthier societies (Chapters 15–17).
- Conservation biologists are protecting habitat and endangered species (Chapter 11).
- Better waste management is helping us conserve resources (Chapter 22).
- Governments, businesses, and individuals are taking steps to reduce emissions of the greenhouse gases that drive climate change (Chapter 18).

These efforts are some of the many sustainable solutions we will examine in the course of this book.

Students are promoting solutions

As a college student, you can help design and implement sustainable solutions on your own campus. Proponents of **campus sustainability** seek ways to help colleges and universities reduce their ecological footprints. Today students, faculty, staff, and administrators on thousands of campuses are working together to make the operations of educational institutions more sustainable (**TABLE 1.3**).

Students are running recycling programs, promoting efficient transportation options, restoring native plants, growing organic gardens, and fostering sustainable dining halls. They are finding ways to improve energy efficiency and water conservation and are pressing for green buildings. To address climate change, students are urging their institutions to reduce greenhouse gas emissions, divest from fossil fuel corporations, and use and invest in renewable energy.

Throughout this book you will encounter examples of campus sustainability efforts (e.g., pp. 241 and 619). Should you wish to pursue such efforts on your own campus, information and links in the **Selected Sources and References for Further Reading** at the back of this book point you toward organizations and resources that can help.

Environmental science prepares you for the future

By taking a course in environmental science, you are preparing yourself for a lifetime in a world increasingly dominated by concerns over sustainability. The course for which you are using this book likely did not exist a generation ago. As society's concerns have evolved, colleges and universities have adapted their curricula. Still, at most schools, fewer than half of students take even a single course on the basic functions of Earth's natural systems, and still fewer take courses on the links between human activity and sustainability. As a result, many educators worry that most students graduate lacking **environmental literacy,** a basic understanding of Earth's physical and living systems and how we interact with them. By taking an environmental science course, you will gain a better understanding of how the world works, you will be better qualified for the green-collar job opportunities of today and tomorrow, and you will be better prepared to navigate the many challenges of creating a sustainable future.

connect & continue

Finding effective ways of living peacefully, healthfully, and sustainably on our diverse and complex planet will require a thorough scientific understanding of both natural and social systems. Environmental science helps us understand our intricate relationship with our environment and informs our attempts to solve and prevent environmental problems. Although many of today's trends may cause concern, a multitude of inspiring success stories give us reason for optimism. Addressing environmental problems can move us toward health, longevity, peace, and prosperity. Science in general, and environmental science in particular, can help us develop balanced, workable, sustainable solutions and create a better world now and for the future.

- **SOLUTIONS** If you could help address three environmental problems in the world today, what would they be? Why did you select each of these three issues? Describe at least one potential solution for each of these challenges. List one obstacle that might stand in the way of each solution. Now describe means by which each of those obstacles might be overcome.

- **LOCAL CONNECTIONS** What environmental issues are people discussing on your campus or in the area where you live? Select one of these issues, and describe two sides of a debate that people are having over it. Where do you stand in this debate and why? Describe how you think the two sides could begin to resolve their differences.

- **EXPLORE THE DATA** Do you know what your ecological footprint is? → Explore Data on **Mastering Environmental Science.**

REVIEWING Objectives

You should now be able to:

+ Describe the field of environmental science

Environmental science is the study of how the natural world works, how our environment affects us, and how we affect our environment. Environmental science uses approaches and insights from many disciplines in the natural sciences and social sciences. (pp. 3, 6–7)

+ Explain the importance of natural resources and ecosystem services to our lives

Renewable resources are unlimited or replenished naturally at a rapid rate. Nonrenewable resources are limited or replenished very slowly. Ecosystem services are benefits we receive from the normal functioning of natural systems. Natural resources and ecosystem services are essential to human life and civilization. (p. 4)

+ Discuss population growth, resource consumption, and their consequences

Today's human population size and per-person resource consumption both are higher than ever. Population growth magnifies our environmental impact by adding to the number of people putting demands on resources. Growing per capita consumption amplifies our environmental impact by increasing the demand each person makes on resources. (pp. 5–6)

+ Explain what is meant by an ecological footprint

An ecological footprint quantifies resource consumption by expressing the total area of biologically productive land and water that is required to provide the resources and waste disposal for a person or population. (pp. 5–6)

+ Describe the scientific method and the process of science

Science is a process of using observations to test ideas. The scientific method consists of making observations, formulating questions, stating a hypothesis, generating predictions, testing predictions, and analyzing results. Scientific research occurs within a larger process of science that includes peer review, journal publication, and interaction with colleagues. (pp. 7, 10–14)

+ Apply critical thinking to judge the reliability of information sources

Critical thinking is a key life skill. By applying strategies to evaluate the reliability of sources and by assessing primary and secondary sources of information, we all can develop ways to judge the trustworthiness of sources in an information-rich media world. (pp. 14–16)

+ Identify major pressures on the global environment

Rising population and intensifying consumption magnify human impacts on the environment, which include resource depletion, air and water pollution, climate change, habitat destruction, and biodiversity loss. (pp. 16–17)

+ Discuss the concept of sustainability, and describe sustainable solutions being pursued on campuses and across the world

Sustainability describes living within our planet's means such that Earth's resources can sustain us— and all life—for the future. Actively developing sustainable solutions (e.g., replacing fossil fuels with renewable energy) can improve our quality of life while protecting and restoring our environment. Many college students are promoting sustainable solutions on campus, such as recycling, energy efficiency, water conservation, and transportation alternatives. (pp. 16–19)

SEEKING Solutions

1. Resources such as soils, timber, fresh water, and biodiversity are renewable if we use them in moderation but can become nonrenewable if we overexploit them (see FIGURE 1.1). For each of these four resources (soils, timber, fresh water, and biodiversity), describe one way we sometimes overexploit the resource and name one thing we could do to conserve the resource. For each, what might constitute sustainable use? (Feel free to look ahead and peruse coverage of these issues throughout this book.)

2. What do you think is the lesson of Easter Island? What more would you like to learn or understand about this island, its history, or its people? What similarities do you perceive between Easter Island and our own modern society? What differences do you see between the predicament of Easter Islanders in 1722 and our situation in the present day?

3. What environmental problem do you feel most acutely yourself? Do you think there are people in the world who do not view your issue as a problem? Who might they be, and why might they take a different view?

4. If the human population were to stabilize tomorrow and never reach 8 billion people, would all our environmental problems be solved? Why or why not? What conditions might get better, and what challenges might remain?

5. Find out what sustainability efforts are being made on your campus. What results have these efforts produced

so far? What further efforts would you like to see pursued on your campus? Do you foresee any obstacles to these efforts? How could these obstacles be overcome? How could you become involved?

6. **THINK IT THROUGH** You have become head of a major funding agency that grants money to scientists pursuing research in environmental science. You must give your staff several priorities to determine what types of scientific research to fund. What environmental problems would you most like to see addressed with research? Describe the research you think would need to be completed to develop workable solutions. What else, beyond scientific research, might be needed to develop sustainable solutions?

CALCULATING Ecological Footprints

Mathis Wackernagel and his colleagues at the Global Footprint Network continue to refine the method of calculating ecological footprints—the amount of biologically productive land and water required to produce the energy and natural resources we consume and to absorb the wastes we generate. According to their most recent data, there are 1.63 hectares (4.0 acres) available for each person in the world, yet we use on average 2.75 ha (6.8 acres) per person, creating a global ecological deficit, or overshoot (p. 6), of 69%.

Compare the ecological footprints of each nation listed in the table. Calculate their proportional relationships to the world population's average ecological footprint and to the area available globally to meet our ecological demands.

NATION	ECOLOGICAL FOOTPRINT (HECTARES PER PERSON)	PROPORTION RELATIVE TO WORLD AVERAGE FOOTPRINT	PROPORTION RELATIVE TO WORLD AREA AVAILABLE
Bangladesh	0.84	0.31 (0.84 ÷ 2.75)	0.48 (0.79 ÷ 1.63)
Tanzania	1.22		
Colombia	2.05		
Thailand	2.49		
Mexico	2.60		
Sweden	6.46		
United States	8.10		
World average	2.75	1.00 (2.75 ÷ 2.75)	1.69 (2.75 ÷ 1.63)
Your personal footprint			

Data from Global Footprint Network, 2019.

1. Why do you think the ecological footprint for people in Bangladesh is so small?

2. Why do you think the ecological footprint for people in the United States is so large?

3. Based on the data in the table, how do you think average per capita income is related to ecological footprints? Name some ways in which you believe a wealthy society can decrease its ecological footprint.

4. Go to an online footprint calculator such as the one at http://www.footprintnetwork.org/en/index.php/GFN/page/personal_footprint and answer the questions to determine your own personal ecological footprint. Enter the value you obtain in the table, and calculate the other values as you did for each nation. How does your footprint compare to that of the average person in the United States? How does it compare to that of people from other nations? Name three actions you could take to reduce your footprint.

Mastering Environmental Science

Students Go to **Mastering Environmental Science** for assignments, an interactive e-text, and the Study Area with practice tests, videos, and activities.

Instructors Go to **Mastering Environmental Science** for automatically graded activities, videos, and reading questions that you can assign to your students, plus Instructor Resources.

CHAPTER 2

Earth's Physical Systems

Matter, Energy, and Geology

case study

What Is the Legacy of the Fukushima Daiichi Nuclear Tragedy?

Fukushima
Daiichi

JAPAN

> " This used to be one of the best places for a business. I'm amazed at how little is left.
>
> Takahiro Chiba, surveying the devastated downtown area of Ishinomaki, Japan, where his family's sushi restaurant was located

> Fukushima should not just contain lessons for Japan, but for all 31 countries with nuclear power.
>
> Tatsujiro Suzuki, Vice-Chairman, Japan Atomic Energy Commission "

At 2:46 p.m. on March 11, 2011, the land along the northeastern coast of the Japanese island of Honshu began to shake violently— and continued to shake for six minutes. These tremors were caused when a large section of the sea-floor along a fault line 125 km (77 mi) offshore suddenly lurched, releasing huge amounts of energy through the earth's crust and generating an earthquake of magnitude 9.0 on the Richter scale (a scale used to measure the strength of earthquakes). The Tohuku earthquake, as it was later named, violently shook the ground in northeastern Japan. In Tokyo, 370 km (230 mi) from the quake's epicenter, commuter trains ground to a stop and skyscrapers swayed to and fro, but major damage to the city was avoided due to earthquake-resistant building codes for structures (see **SUCCESS STORY**, p. 26) and the immediate detection of the offshore earthquake. But even when the earth stopped shaking, the residents of northeastern Japan knew that further danger might still await them—from a tsunami.

A **tsunami** ("harbor wave" in English) is a powerful surge of seawater generated when an offshore earthquake displaces large volumes of rocks and sediment on the ocean bottom, suddenly pushing the overlying ocean water upward. This upward movement of water creates waves that speed outward from the earthquake site in all directions. These waves are hardly noticeable at sea, but can rear up to staggering heights when they enter the shallow waters near shore and can sweep inland with great force. The Japanese had built seawalls to protect against tsunamis, but the Tohoku quake caused the island of Honshu to sink perceptively, thereby lowering the height of the seawalls by up to 2 m (6.5 ft) in some locations. Waves reaching up to 15 m (49 ft) in height then overwhelmed these defenses (**FIGURE 2.1**, p. 24). The raging water swept up to 9.6 km (6 mi) inland; scoured buildings from their foundations; and inundated towns, villages, and productive agricultural land. As the water's energy faded, the water receded, carrying structural debris, vehicles, livestock, and human bodies out to sea.

When the tsunami overtopped the 5.7-m (19-ft) seawall protecting the Fukushima Daiichi nuclear power plant, it flooded the diesel-powered emergency generators responsible for circulating water to cool the plant's nuclear reactors. With the local electrical grid knocked

Upon completing this chapter, you will be able to:

+ Explain the fundamentals of matter and chemistry and apply them to real-world situations

+ Differentiate among forms of energy, and explain the first and second laws of thermodynamics

+ Distinguish photosynthesis, cellular respiration, and chemosynthesis, and summarize their importance to living things

+ Explain how plate tectonics and the rock cycle shape the landscape around us

+ Identify major types of geologic hazards, and describe ways to minimize their impacts

◀ **Destruction in northeastern Japan from tsunami generated by the Tohoku earthquake**

▲ **One of the 300,000 people displaced by the Tohoku earthquake**

FIGURE 2.1 Tsunami waves overtop a seawall following the Tohoku earthquake in 2011. The tsunami caused a greater loss of life and property than the earthquake that generated it and led to a meltdown at the Fukushima Daiichi nuclear power plant.

supplying 30% of Japan's electricity. The once productive fishing industry of the region was shut down over fears of people consuming seafood contaminated with high levels of radioactivity. The world watched and waited to see how the events in Japan would play out over the coming years and took a critical look at the

out by the earthquake and the backup generators off-line, the nuclear fuel in the cores of the three active reactors at the plant began to overheat. The cooling water that normally kept the nuclear fuel submerged within the reactor cores boiled off, exposing the nuclear material to the air and further elevating temperatures inside the cores. As the overheated nuclear fuel melted its containment vessel (an event called a nuclear meltdown), chemical reactions within the reactors generated hydrogen gas. This hydrogen gas then set off explosions in each of the three active reactor buildings, and in an adjacent inactive reactor that was connected to an active reactor through ductwork, releasing radioactive material into the air (**FIGURE 2.2**). To prevent a full-blown catastrophe that could render large portions of their nation uninhabitable, Japanese authorities flooded the reactor cores with seawater pumped in from the ocean.

The 1–2–3 punch of the earthquake–tsunami–nuclear accident left 18,000 people dead and caused hundreds of billions of dollars in material damage. Around 340,000 people were displaced from their homes, and shortly after the accident the evacuation of a 20-km (12-mi) area around the Fukushima Daiichi plant was ordered due to unsafe levels of radioactive fallout in the soil. After the accident, public opposition to nuclear power in Japan ran high, and the government ordered the immediate shutdown and reinspection of its 54 nuclear reactors, which were, at the time,

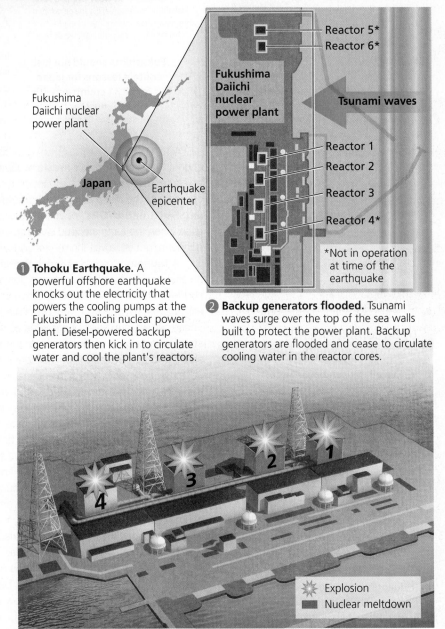

1 Tohoku Earthquake. A powerful offshore earthquake knocks out the electricity that powers the cooling pumps at the Fukushima Daiichi nuclear power plant. Diesel-powered backup generators then kick in to circulate water and cool the plant's reactors.

2 Backup generators flooded. Tsunami waves surge over the top of the sea walls built to protect the power plant. Backup generators are flooded and cease to circulate cooling water in the reactor cores.

3 Nuclear meltdown. No longer bathed in cooling water, the nuclear fuel in reactors 1-3 overheats and melts through the reactor cores. Chemical reactions produce hydrogen gas, which explodes, damaging the reactor buildings and releasing radioactive material into the air.

FIGURE 2.2 Timeline of events in the nuclear accident at the Fukushima Daiichi power plant.

future of nuclear power around the world. Whereas nuclear power does have an inherent danger for large-scale accidents and we have yet to find a safe, sustainable way to store the radioactive wastes it produces, it does have benefits over other methods of generating electricity in that it does not release significant amounts of air pollutants or greenhouse gases into the atmosphere when producing energy.

The destruction at Fukushima following the Tohoku earthquake was the product of natural forces—an earthquake and subsequent tsunami—coupled with an accident involving one of humanity's most advanced technologies: nuclear power. These events highlight why knowledge of matter, energy, and geologic forces is vital to understanding environmental impacts in our complex, modern world.

Matter, Chemistry, and the Environment

The tragic events in northeastern Japan in 2011 were the result of large-scale forces generated by the powerful geologic processes that shape the surface of our planet. Environmental scientists regularly study these types of processes to understand how our planet works. Because all large-scale processes are made up of small-scale components, however, environmental science—the broadest of scientific fields—must also study small-scale phenomena. At the smallest scale, an understanding of matter itself helps us to fully appreciate the processes of our world.

All material in the universe that has mass and occupies space—solid, liquid, and gas alike—is called **matter.** The study of types of matter and their interactions is called **chemistry.** Once you examine any environmental issue, from acid rain to toxic chemicals to climate change, you will likely discover chemistry playing a central role.

Matter is conserved

To appreciate the chemistry involved in environmental science, we must begin with the fundamentals. Matter may be transformed from one type of substance into others, but it cannot be created or destroyed, a principle referred to as the **law of conservation of matter.** In environmental science, this principle helps us understand that the amount of matter stays constant as it is recycled in ecosystems and nutrient cycles (p. 121). It also makes it clear that we cannot simply wish away "undesirable" matter, such as nuclear waste and toxic pollutants. Because harmful substances can't be destroyed, we must take steps to minimize their impacts on the environment.

Elements and atoms are chemical building blocks

The nuclear reactors at Fukushima Daiichi used the element **uranium** to power its reactors. An **element** is a fundamental type of matter, a chemical substance with a given set of properties that cannot be broken down into substances with other properties. Chemists currently recognize 118 elements. Most occur in nature, but more than 20 elements have been created solely in the lab. Elements especially abundant on our planet include **oxygen, hydrogen, silicon, nitrogen,** and **carbon** (**TABLE 2.1**). Elements that organisms need for survival, such as carbon, nitrogen, calcium, and phosphorus, are called **nutrients.** Each element is assigned an abbreviation, or chemical symbol (for instance, "H" for hydrogen and "O" for oxygen). The periodic table of the elements organizes the elements according to their chemical properties and behavior (see **APPENDIX C**).

An **atom** is the smallest unit that maintains the chemical properties of the element. Atoms of each element contain a specific number of **protons,** positively charged particles in the atom's nucleus (its dense center), and the number of

TABLE 2.1 Earth's Most Abundant Chemical Elements, by Mass

EARTH'S CRUST		OCEANS		AIR		ORGANISMS	
Oxygen (O)	49.5%	Oxygen (O)	88.3%	Nitrogen (N)	78.1%	Oxygen (O)	65.0%
Silicon (Si)	25.7%	Hydrogen (H)	11.0%	Oxygen (O)	21.0%	Carbon (C)	18.5%
Aluminum (Al)	7.4%	Chlorine (Cl)	1.9%	Argon (Ar)	0.9%	Hydrogen (H)	9.5%
Iron (Fe)	4.7%	Sodium (Na)	1.1%	Other	<0.1%	Nitrogen (N)	3.3%
Calcium (Ca)	3.6%	Magnesium (Mg)	0.1%			Calcium (Ca)	1.5%
Sodium (Na)	2.8%	Sulfur (S)	0.1%			Phosphorus (P)	1.0%
Potassium (K)	2.6%	Calcium (Ca)	<0.1%			Potassium (K)	0.4%
Magnesium (Mg)	2.1%	Potassium (K)	<0.1%			Sulfur (S)	0.3%
Other	1.6%	Bromine (Br)	<0.1%			Other	0.5%

SUCCESS story

Saving Lives with Building Codes

The Tohoku earthquake was not the first major earthquake to strike Japan. The city of Kobe experienced substantial damage from a quake in 1995 that claimed more than 5500 lives. And in 1923, an earthquake devastated the cities of Tokyo and Yokohama, resulting in more than 142,000 deaths. Losses of life and property from the Tohoku quake were far less extensive than the losses from these earlier events, thanks to new stringent building codes ensuring that the design and construction of structures are such that they resist crumbling and toppling over during earthquakes.

To minimize damage from earthquakes, engineers have developed ways to protect buildings from collapsing while shaking. They do this by strengthening structural components while also designing points at which a structure can move and sway harmlessly with ground motion. Just as a flexible tree trunk bends in a storm while a brittle one breaks, buildings with built-in flexibility are more likely to withstand an earthquake's violent shaking. Such designs continue to figure in the building codes used in California, Japan, and other quake-prone regions.

Such quake-resistant designs are more expensive to build than conventional designs, so many buildings in poorer nations do not have such protections. Consequently, earthquakes in these regions typically result in greater losses of life and property due to the greater numbers of buildings that collapse. For example, Haiti suffered a 7.0 magnitude earthquake in 2010 that devastated huge portions of the capital city of Port-au-Prince and claimed over 220,000 lives. Although the Tohoku earthquake released more than 950 times the energy than the earthquake that struck Haiti, mortality and property damage from the Tohoku quake (not including the damage and loss of life caused by the subsequent tsunami) were minimized because of Japan's earthquake-conscious building codes.

Collapsed buildings in Port-au-Prince, Haiti, following a 7.0 magnitude earthquake in 2010

→ **Explore the Data** at **Mastering Environmental Science**

---

protons is called the element's atomic number. (Elemental carbon, for instance, has six protons in its nucleus; thus, its atomic number is 6.) Most atoms also contain **neutrons,** particles in the nucleus that lack an electrical charge. An element's mass number denotes the combined number of protons and neutrons in the atom. An atom's nucleus is surrounded by negatively charged particles known as **electrons,** which are equal in number to the protons in the nucleus of an atom, balancing the positive charge of the protons (**FIGURE 2.3**).

Isotopes Although all atoms of a given element contain the same number of protons, they do not necessarily also contain the same number of neutrons. Atoms of the same element with differing numbers of neutrons are **isotopes** (**FIGURE 2.4a**). Isotopes are denoted by their elemental symbol preceded by the mass number, the combined number of protons and neutrons in the nucleus of the atom. For example, ^{14}C (carbon-14) is an isotope of carbon with eight neutrons (and six protons) in the nucleus rather than the six neutrons (and six protons) of ^{12}C (carbon-12), the most abundant carbon isotope.

Some isotopes, called **radioisotopes,** are **radioactive** because their chemical identity changes as they shed subatomic particles and emit high-energy radiation. The radiation released by radioisotopes harms organisms because it focuses a great deal of energy in a very small area, which

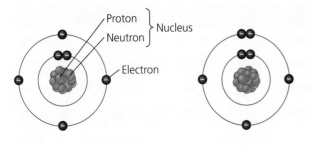

Carbon (C)
Atomic number = 6
Protons = 6
Neutrons = 6
Electrons = 6

Nitrogen (N)
Atomic number = 7
Protons = 7
Neutrons = 7
Electrons = 7

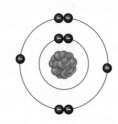

Oxygen (O)
Atomic number = 8
Protons = 8
Neutrons = 8
Electrons = 8

FIGURE 2.3 In an atom, protons and neutrons stay in the nucleus and electrons move about the nucleus. Each chemical element has its own particular number of protons. For example, carbon possesses six protons, nitrogen seven, and oxygen eight. These schematic diagrams are meant to clearly show and compare numbers of electrons for these three elements. In reality, however, electrons do *not* orbit the nucleus in rings as shown; they move through space in more complex ways.

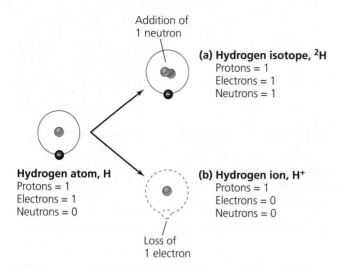

Addition of
1 neutron

(a) Hydrogen isotope, ^{2}H
Protons = 1
Electrons = 1
Neutrons = 1

Hydrogen atom, H
Protons = 1
Electrons = 1
Neutrons = 0

(b) Hydrogen ion, H$^+$
Protons = 1
Electrons = 0
Neutrons = 0

Loss of
1 electron

FIGURE 2.4 Hydrogen has a mass number of 1 because a typical atom of this element contains one proton and no neutrons. Deuterium (hydrogen-2, or ^{2}H), an isotope of hydrogen **(a),** contains a neutron, as well as a proton and thus has greater mass than a typical hydrogen atom; its mass number is 2. The hydrogen ion, H$^+$ **(b),** occurs when an electron is lost; it therefore has a positive charge.

can be damaging to living cells. Intense radiation exposure can kill cells outright or cause changes to the cell's DNA that can increase the probability of the organism developing cancerous tumors.

The greatest danger from radioisotopes occurs when they enter the bodies of organisms through the lungs, skin, or digestive system. It is therefore important after nuclear accidents, like that at Fukushima, to regularly test food and water supplies for radioisotopes and to determine the eventual fate of radioactive particles released into the environment.

A radioisotope decays as it emits radiation, becoming lighter and lighter with the loss of subatomic particles, until it becomes a stable isotope (isotopes that are not radioactive). Each radioisotope decays at a rate determined by its **half-life,** the amount of time it takes for one-half of its atoms to decay. Different radioisotopes have very different half-lives, ranging from fractions of a second to billions of years. The radioisotope uranium-235 (^{235}U), used in commercial nuclear power plants like Fukushima Daiichi, decays into a series of daughter isotopes (atoms formed as radioisotopes that lose protons and neutrons during the process of radioactive decay), eventually forming lead-207 (^{207}Pb). Uranium-235 has a half-life of about 700 million years. Radioisotopes released into the environment from the Fukushima nuclear power plant accident included iodine-131 (half-life of 8 days), cesium-134 (half-life of 2 years), strontium-90 (half-life of 29 years), and cesium-137 (half-life of 30 years). Given these lengthy half-lives, finding ways to safely contain radioisotopes at nuclear waste storage sites and in areas, like Fukushima, that have experienced nuclear accidents is an area of active research (see **THE SCIENCE BEHIND THE STORY,** pp. 30–31).

Ions Atoms may also gain or lose electrons, thereby becoming **ions,** electrically charged atoms or combinations of atoms (**FIGURE 2.4b**). Ions are denoted by their elemental symbol followed by their ionic charge. For instance, a common ion used by mussels and clams to form shells is Ca^{2+}, a calcium atom that has lost two electrons and thus has a charge of positive 2. The damaging radiation emitted by radioisotopes is called **ionizing radiation** (see Figure 2.11, p. 33) because it generates ions when it strikes molecules. These ions affect the stability and functionality of biological important molecules such as enzymes (p. 28) and DNA (p. 28), and this can harm cells, cause them to die, or mutate their genetic code.

Atoms bond to form molecules and compounds

Atoms bond together and form **molecules,** combinations of two or more atoms. Common molecules containing only a single element include those of hydrogen (H$_2$) and oxygen (O$_2$), each of which exists as a gas at room temperature. A molecule composed of atoms of two or more elements is called a **compound.** One compound is **water;** it is composed of two hydrogen atoms bonded to one oxygen atom, and it is denoted by the chemical formula H$_2$O. Another compound is **carbon dioxide,** consisting of one carbon atom bonded to two oxygen atoms; its chemical formula is CO$_2$.

Atoms bond together because of an attraction for one another's electrons. Because the strength of this attraction varies among elements, atoms may be held together in different ways. When electrons are shared between atoms, a **covalent bond** forms. For example, two atoms of hydrogen will share electrons equally as they bind together to form

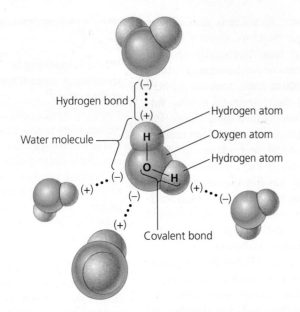

FIGURE 2.5 **By enabling water molecules to adhere loosely to one another, hydrogen bonds give water several unique properties crucial for life.**

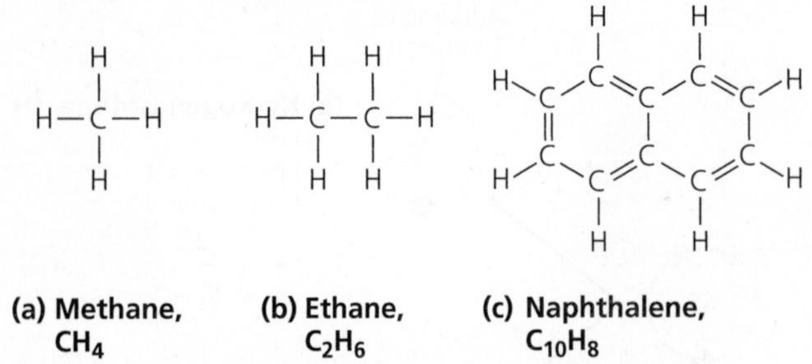

(a) Methane, CH_4 (b) Ethane, C_2H_6 (c) Naphthalene, $C_{10}H_8$

FIGURE 2.6 **Hydrocarbons have a diversity of chemical structures.** The simplest hydrocarbon is methane **(a).** Many hydrocarbons consist of linear chains of carbon atoms with hydrogen atoms attached; the shortest of these is ethane **(b).** The air pollutant naphthalene **(c)** is a ringed hydrocarbon.

hydrogen gas, H_2. However, in a water molecule, oxygen attracts shared electrons more strongly than does hydrogen. The result is that water has a partial negative charge at its oxygen end and partial positive charges at its hydrogen ends. This arrangement allows water molecules to adhere to one another in a type of weakly attractive interaction called a **hydrogen bond** (**FIGURE 2.5**). Hydrogen bonds allow water to dissolve (hold in solution) many other molecules. It is for this reason that water forms the majority of the mass of living cells, as it provides an ideal medium for the many biologically important molecules that exist within cells.

In compounds in which the strength of attraction is sufficiently unequal, an electron may be transferred from one atom to another. This creates oppositely charged ions that form **ionic bonds** due to their differing electrical charges. Such associations are called ionic compounds, or salts. Table salt (NaCl) contains ionic bonds between positively charged sodium ions (Na^+), each of which donates an electron, and negatively charged chloride ions (Cl^-), each of which receives an electron.

Elements, molecules, and compounds can also come together in mixtures without chemically bonding or reacting. Such mixtures are called solutions. Air in the atmosphere is a solution formed of constituents such as nitrogen, oxygen, and water vapor. Other solutions include ocean water, plant sap, petroleum, and metal alloys such as brass.

Matter is composed of organic and inorganic compounds

Beyond their need for water, living things also depend on organic compounds. **Organic compounds** consist of carbon atoms (and generally hydrogen atoms) joined by covalent bonds, and they may also include other elements, such as nitrogen, oxygen, sulfur, and phosphorus. Inorganic compounds, in contrast, lack carbon–carbon bonds.

Carbon's unusual ability to bond together in chains, rings, and other structures to build elaborate molecules has resulted in millions of different organic compounds. One class of such compounds that is important in environmental science is **hydrocarbons,** which consist solely of bonded atoms of carbon and hydrogen (although other elements may enter these compounds as impurities) (**FIGURE 2.6**). Fossil fuels and the many petroleum products we make from them (Chapter 19), such as plastics, consist largely of hydrocarbons.

Macromolecules are building blocks of life

Just as carbon atoms in hydrocarbons may be strung together in chains, organic compounds sometimes combine to form long chains of repeated molecules. These chains are called **polymers.** There are three types of polymers that are essential to life: proteins, nucleic acids, and carbohydrates. Along with lipids (which are not polymers), these types of molecules are referred to as **macromolecules** because of their large sizes.

Proteins consist of long chains of organic molecules called amino acids. The many types of proteins serve various functions. Some help produce tissues and provide structural support. For example, animals use proteins to generate skin, hair, muscles, and tendons. Some proteins help store energy, whereas others transport substances. Some act in the immune system to defend the organism against foreign attackers, such as bacteria and viruses that cause disease. Still others are hormones, molecules that act as chemical messengers within an organism. Proteins can also serve as enzymes, molecules that catalyze, or promote, certain chemical reactions.

Nucleic acids direct the production of proteins. The two nucleic acids—**deoxyribonucleic acid (DNA)** and **ribonucleic acid (RNA)**—carry the hereditary information for

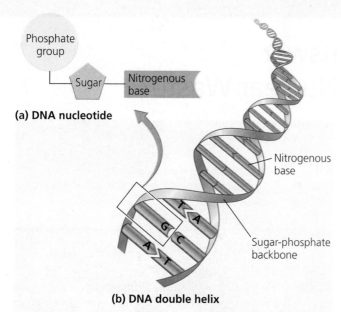

Phosphate group
Sugar
Nitrogenous base

(a) DNA nucleotide

Nitrogenous base

Sugar-phosphate backbone

(b) DNA double helix

FIGURE 2.7 Nucleic acids encode genetic information in the sequence of nucleotides, small molecules that pair together like rungs of a ladder. DNA includes four types of nucleotides **(a),** each with a different nitrogenous base: adenine (A), guanine (G), cytosine (C), and thymine (T). Adenine (A) pairs with thymine (T), and cytosine (C) pairs with guanine (G). In RNA, thymine is replaced by uracil (U). DNA **(b)** twists into the shape of a double helix.

organisms and are responsible for passing traits from parents to offspring. Nucleic acids are composed of a series of nucleotides, each of which contains a sugar molecule, a phosphate group, and a nitrogenous base. DNA contains four types of nucleotides and can be pictured as a ladder twisted into a spiral, giving the molecule a shape called a double helix (**FIGURE 2.7**). Regions of DNA coding for particular proteins that perform particular functions are called **genes.**

Carbohydrates include simple sugars that are three to seven carbon atoms long. Glucose $(C_6H_{12}O_6)$ fuels living cells and serves as a building block for complex carbohydrates, such as starch. Plants use starch to store energy, and animals eat plants to acquire starch. Plants and animals also use complex carbohydrates to build structure. Insects and crustaceans form hard shells from the carbohydrate chitin. Cellulose, the most abundant organic compound on Earth, is a complex carbohydrate found in the cell walls of leaves, bark, stems, and roots.

Lipids include fats and oils (for energy storage), phospholipids (for cell membranes), waxes (for structure), and steroids (for hormone production). Although chemically diverse, the compounds that are categorized as lipids are so grouped because they do not dissolve in water.

Hydrogen ions determine acidity

The functioning of proteins and other biologically important molecules in cells can be influenced by the level of acids and bases in the liquid medium in which they reside. In any aqueous solution, a small number of water molecules split apart, each forming a hydrogen ion (H^+) and a hydroxide ion (OH^-). The product of hydrogen and hydroxide ion concentrations is always the same; as one increases, the other decreases. Pure water contains equal numbers of these ions. Solutions in which the H^+ concentration is greater than the OH^- concentration are **acidic,** whereas solutions in which the OH^- concentration exceeds the H^+ concentration are **basic,** or alkaline.

The **pH** (potential of hydrogen) scale quantifies the acidity or alkalinity of solutions from 0 to 14, according to hydrogen ion concentration (**FIGURE 2.8**). Pure water has a hydrogen ion concentration of 10^{-7} and thus a pH of 7. Solutions with a pH less than 7 are acidic, and those with a pH greater than 7 are basic. The pH scale is logarithmic, so each step on the scale represents a 10-fold difference in hydrogen ion concentration. A substance with a pH of 6, for example, contains 10 times as many hydrogen ions as a substance with a pH of 7 and 100 times as many hydrogen ions as a substance with a pH of 8.

Most biological systems have a pH between 6 and 8, and substances that are strongly acidic (battery acid) or strongly basic (sodium hydroxide) are harmful to living things. Human activities can change the pH of water or soils and make conditions less amenable to life. Examples include the acidification of soils and water from acid rain (pp. 474–477) and from acidic mine drainage (p. 650).

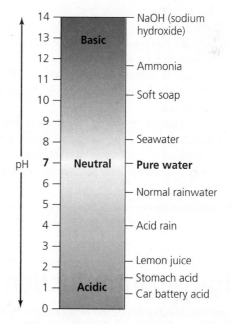

FIGURE 2.8 The pH scale measures how acidic or basic (alkaline) a solution is. The pH of pure water is 7, the midpoint of the scale. Acidic solutions have a pH less than 7, whereas basic solutions have a pH greater than 7.

THE SCIENCE behind the story

Are Yeast the Answer to Cleaning up Nuclear Waste?

Radiation warning signs outside the Hanford Nuclear Reservation

During the Cold War era from 1945 to 1986, the United States generated massive amounts of radioactive waste while producing 46,000 nuclear weapons. A mixture of harmful substances, the radioactive waste includes radioisotopes (such as uranium and plutonium), toxic chemicals, and heavy metals (such as mercury and cadmium) in a highly acidic solution (due to the processing of weapons-grade uranium involving nitric acid). The radioactive waste is stored in underground tanks at 120 locations across the United States, sites administered by the Department of Energy.

As far back as 1959, a leak was detected at the Hanford Nuclear Reservation in Washington, one of the largest nuclear waste storage sites in the country, which houses 60% of the United States' radioactive waste from weapons production since 1943 (**FIGURE 1**). Today, over a third of the 177 tanks on-site are leaking radioactive waste, contaminating an estimated 3 trillion liters (790 billion gal) of surface waters and groundwater and some 70 million cubic meters (2.4 billion cubic feet) of soil. Nearby is the Columbia River, an ecologically important waterway that supplies irrigation water to farms and drinking water to people. Up to 3.7 million liters (977,000 gal) of nuclear waste is estimated to have leaked from tanks at the Hanford site. Needless to say, cleanup of the leakage is long overdue.

Given the dangerous nature of the nuclear waste and the massive quantities of it that are stored at the site, physically removing contaminated soils and waters from the site using human labor is neither safe nor economically feasible.

FIGURE 1 Leaks from some of the 177 underground storage tanks at the Hanford Nuclear Reservation in Washington State are threatening to dangerously contaminate the nearby Columbia River.

A potential alternative that scientists are considering is the use of **bioremediation,** whereby microorganisms clean up toxic waste. Certain microorganisms metabolize toxic chemicals and convert toxic heavy metals into less dangerous forms. Some can also produce biofilms, a weblike matrix of proteins and carbohydrates that surround the cell and to which radioisotope metals and other toxic substances adhere. Biofilm traps these harmful substances, minimizing their movements in the environment.

In 2000, researchers led by Michael Daly of the Uniformed Services University of the Health Sciences sampled the sediments at the Hanford site beneath a tank leaking highly radioactive waste. They were looking for a microorganism to be used in bioremediation of the site and isolated 110 different species of

Energy: An Introduction

Creating and maintaining organized complexity—of a cell, an organism, an ecological system, or a planet—require energy. Energy is needed to organize matter into complex forms, to build and maintain cellular structure, to govern species' interactions, and to drive the geologic forces that shape our planet. Energy is involved in nearly every chemical, biological, and physical phenomenon.

But what is energy? **Energy** is the capacity to change the position, physical composition, or temperature of matter—in other words, a force that can accomplish **work** (when a force acts on an object, causing it to move). As we saw with Japan's 2011 earthquake and tsunami, geologic events involve some of the most dramatic releases of energy in nature. Energy is omnipresent in living things as well. A sparrow in flight expends energy to propel its body through the air (change of position). When the sparrow lays an egg, its body uses energy bacteria. From these bacteria, they selected *Deinococcus radiodurans,* a microbe known for its high tolerance to radioactivity and its ability to neutralize toxic metals, for further research. After six years, the team produced a genetically engineered strain of the bacterium that neutralized toxic metals and metabolized toxic chemicals, all while being exposed to high temperatures and high levels of radioactivity. However, this strain of *D. radiodurans,* nicknamed "Conan the Bacterium" for its heartiness, had a few shortcomings. Although it could tolerate the heat and radioactivity present in contaminated sediments at Hanford, its activity levels dropped off when conditions became moderately acidic. As the nuclear waste stored at Hanford and elsewhere was leaking wastes that were strongly acidic, the bacterium's usefulness was limited, as it was largely ineffective at treating nuclear wastes at locations nearest the tanks, where pH levels were the lowest (the most toxic). *D. radiodurans* also did not produce biofilms under the harshest conditions, which somewhat hindered its effectiveness in trapping contaminants.

To locate other candidates for use in bioremediation at the Hanford sites, Daly and his team looked elsewhere for microorganisms adapted to harsh conditions, sampling 60 different and extreme environments, from arctic ice to underground mines to hot springs. But instead of focusing on the bacteria in their samples, they turned their attention to another kind of microorganism—yeast. Yeast are fungi, not bacteria, and the team found 27 different types of yeast in the samples. Very little research had been conducted on the effectiveness of yeast in bioremediation, but as yeast are generally tolerant of acidic conditions, the team believed these microorganisms had the potential to contain nuclear waste.

As reported in a 2018 paper in the journal *Frontiers in Microbiology,* the researchers screened the 27 types of yeast for their tolerance to ionizing radiation, acidity, and temperature, as well as their resistance to toxic heavy metals. From the screening, a red-pigmented strain of the yeast *Rhodotorula taiwanensis,* which had been collected from an abandoned acid mine drainage (p. 650) facility in Maryland, stood out as an ideal candidate for bioremediation at the Hanford site. Not only did the species tolerate high levels of radioactivity and elevated concentrations of toxic metals, but *R. taiwanesis* did so under highly acidic conditions. It also formed biofilms under the harshest conditions (**FIGURE 2**). Whereas the study of *R. taiwanesis* showed the

promise of fungi in nuclear waste containment, the organism was not the *perfect* fit for cleanup projects like that at Hanford. For example, radioisotopes in nuclear waste generate heat, and *R. taiwanensis* growth slows considerably as temperature increases. In the high-temperature environment beneath a leaking nuclear waste tank, the yeast is therefore not as effective as it is further away from the tank where conditions are cooler.

The pioneering research on bioremediation using *D. radiodurans* and *R. taiwanesis* opens up the possibility of employing a mixture, or "cocktail," of organisms for containment of nuclear waste at the Hanford Nuclear Reservation and similar radioactive sites. By combining a diversity of bacteria and fungi, each with specific tolerances to radioactivity, pH, and temperature—and differing abilities to detoxify harmful chemicals and trap contaminants in biofilms—we may one day be able to effectively contain the nuclear waste leaking at storage sites around the United States and in other nations around the world.

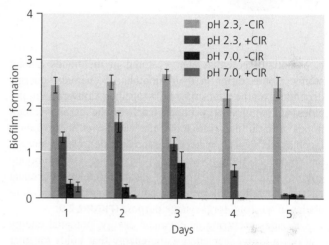

FIGURE 2 **The yeast *Rhodotorula taiwanensis* form biofilms, which aid in the containment of radioactive and toxic wastes, under challenging environmental conditions.** The species form biofilms even when growing on very low pH substrates (pH 2.3) with steady exposure to harmful ionizing radiation (+CIR). *Data from Tkavc, R., et al., 2018. Prospects for fungal bioremediation of acidic radioactive waste sites: Characterization and genome sequence of* Rhodotorula taiwanensis *MD1149.* Front. Microbiol. 8: 2528. doi: 10.3389/fmicb.2017.02528.

to create the calcium-based eggshell and color it with pigment (change in composition). The sparrow sitting on its nest transfers energy from its body to heat the developing chicks inside its eggs (change of temperature).

Energy comes in different forms

Energy manifests itself in different ways and can be converted from one form to another. Two major forms of energy that

scientists commonly distinguish are **potential energy,** energy of position or composition, and **kinetic energy,** energy of motion. Consider river water held behind a dam. By preventing water from moving downstream, the dam causes the water to accumulate potential energy. When the dam gates are opened, the potential energy is converted to kinetic energy as the water rushes downstream.

Energy conversions take place at the atomic level every time a chemical bond is broken or formed. Chemical energy

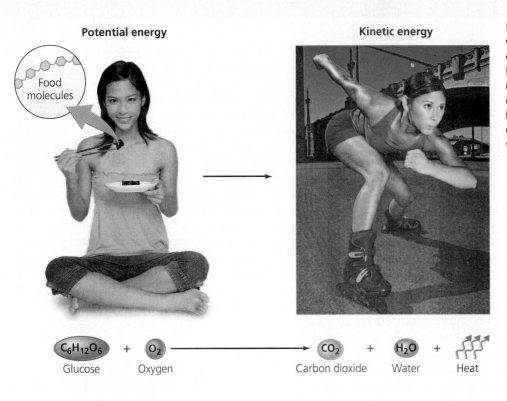

Potential energy

Food molecules

Kinetic energy

FIGURE 2.9 Energy is released when potential energy is converted to kinetic energy. Potential energy stored in sugars (such as glucose) in the food we eat, combined with oxygen, becomes kinetic energy when we exercise, releasing carbon dioxide, water, and heat as by-products.

$C_6H_{12}O_6$ + O_2 ⟶ CO_2 + H_2O + Heat
Glucose   Oxygen   Carbon dioxide   Water   Heat

is essentially potential energy stored in the bonds among atoms. Bonds differ in their amounts of chemical energy, depending on the atoms they hold together. Converting molecules with high-energy bonds (such as the carbon–carbon bonds of fossil fuels) into molecules with lower-energy bonds (such as the bonds in water or carbon dioxide) releases energy and produces motion, action, or heat. Just as automobile engines split the hydrocarbons of gasoline to release chemical energy and generate movement, our bodies split glucose molecules in our food for the same purpose (**FIGURE 2.9**).

Besides occurring as chemical energy, potential energy can occur as nuclear energy, the energy that holds together atomic nuclei. Nuclear power plants use this energy when they break apart the nuclei of large uranium atoms within their reactors. Mechanical energy, such as the energy stored in a compressed spring, is yet another type of potential energy. Kinetic energy can also express itself in different forms, including thermal energy, light energy, sound energy, and electrical energy—all of which involve the movement of atoms, subatomic particles, molecules, or objects.

A nuclear power plant, such as Fukushima Daiichi, illustrates the many ways energy can be used to change the position, physical composition, or temperature of matter and how energy can change form (see Figure 20.5, p. 565, for an illustration of the operation of a commercial nuclear power plant). In the plant's reactor, the potential energy in uranium is released by nuclear fission reactions (the splitting of atom nuclei), producing heat energy. This heat changes water from liquid to steam, and the kinetic energy of steam molecules is harnessed to generate mechanical energy in turbines. This mechanical energy is then sent to an electrical generator, where the spinning of wires within a magnetic field generates electrical energy. Similarly, complex energetic interactions occur in other human uses of energy, as well as in the natural world.

Energy is always conserved, but it changes in quality

The study of the relationships between different forms of energy is termed **thermodynamics,** and the theories that explain energetic interactions are called the *laws of thermodynamics.* The **first law of thermodynamics** states that energy can change from one form to another, but it cannot be created or destroyed. Just as matter is conserved (p. 25), the total energy in the universe remains constant and thus is said to be conserved. The potential energy of the water behind a dam will equal the kinetic energy of its eventual movement downstream. Likewise, we obtain energy from the food we eat and then expend it in exercise, apply it toward maintaining our body and all its functions, or store it in fat. We do not somehow create additional energy or end up with less energy than the food gives us. Any particular system in nature can temporarily increase or decrease in energy, but the total amount in the universe always remains constant.

Although the overall amount of energy is conserved in any conversion of energy, the **second law of thermodynamics** states that the nature of energy will change from a more-ordered state to a less-ordered state as long as no force counteracts this tendency. That is, systems tend to move toward increasing disorder, or entropy. For instance, a log of firewood—the highly organized and structurally complex product of many years of tree growth—transforms in a campfire to a residue of carbon ash, smoke, and gases such as carbon dioxide and water vapor, as well as the light and the heat of the flame (**FIGURE 2.10**). With the help of oxygen, the complex biological polymers that make up the wood are converted into a disorganized assortment of rudimentary molecules and heat and light energy. When energy transforms from a more-ordered state to a less-ordered state, it cannot accomplish tasks as efficiently. For example, the

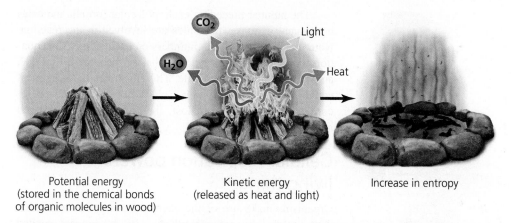

Potential energy
(stored in the chemical bonds
of organic molecules in wood)

Kinetic energy
(released as heat and light)

Increase in entropy

FIGURE 2.10 The burning of firewood demonstrates energy conversion from a more-ordered to a less-ordered state. This increase in entropy reflects the second law of thermodynamics.

potential energy available in ash (a less-ordered state of wood) is far lower than that available in a log of firewood (the more-ordered state of wood).

The second law of thermodynamics specifies that systems tend to move toward disorder. How, then, does any system maintain its order? The order of an object or system can be increased by the input of energy from outside the system. Living organisms, for example, maintain their highly ordered structure by regularly consuming energy. But when organisms die and these inputs of energy cease, they undergo decomposition and revert to a less-ordered state.

Light energy from the sun powers our planet

The energy that powers Earth's biological systems comes primarily from the sun. The sun releases radiation across large portions of the electromagnetic spectrum, although our atmosphere filters out much of this and we see only some of this radiation as visible light (**FIGURE 2.11**). Most of the sun's energy is reflected, or else absorbed and re-emitted, by the atmosphere, land, or water (p. 487). Solar energy drives winds, ocean currents, weather, and climate patterns. A small amount (less than 1% of the total) powers plant growth, and a still smaller amount flows from plants into the organisms that eat them and the organisms that decompose dead organic matter. A minuscule percentage of this energy is eventually

deposited belowground in the chemical bonds in fossil fuels (which are derived from ancient plants; pp. 529–531).

Photosynthesis converts solar energy to chemical energy

Some organisms use the sun's radiation directly to produce their own food. Such organisms, called **autotrophs** or **primary producers,** include green plants, algae, and cyanobacteria. Through the process of **photosynthesis**, autotrophs use sunlight to power a series of chemical reactions that transform molecules with lower-energy bonds—water and carbon dioxide—into sugar molecules with many high-energy bonds. Photosynthesis is an example of a process that moves toward a state of lower entropy, so it requires a substantial input of outside energy, in this case from sunlight.

Photosynthesis occurs within cellular organelles called chloroplasts, where the light-absorbing pigment chlorophyll (the substance that makes plants green) uses solar energy to

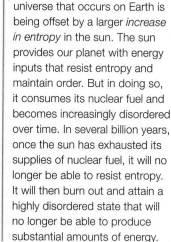

FAQ

What part of the universe is offsetting all the complexity on Earth?

The second law of thermodynamics tells us that systems, such as our universe, tend to move toward entropy. But Earth is a highly ordered planet populated by huge numbers of highly ordered organisms. What part of the universe is offsetting this localized decrease in entropy to satisfy the second law?

It's our local star, the sun. The decrease in entropy in the universe that occurs on Earth is being offset by a larger *increase in entropy* in the sun. The sun provides our planet with energy inputs that resist entropy and maintain order. But in doing so, it consumes its nuclear fuel and becomes increasingly disordered over time. In several billion years, once the sun has exhausted its supplies of nuclear fuel, it will no longer be able to resist entropy. It will then burn out and attain a highly disordered state that will no longer be able to produce substantial amounts of energy.

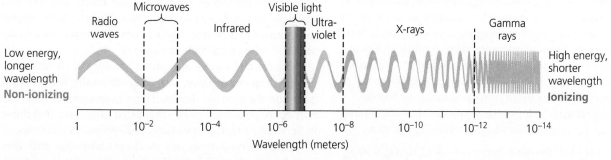

FIGURE 2.11 The sun emits radiation from many portions of the electromagnetic spectrum. Visible light makes up only a small proportion of this energy.

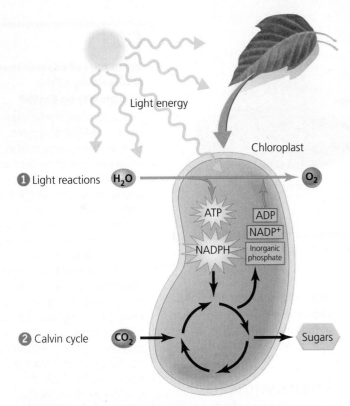

Light energy

Chloroplast

1 Light reactions H_2O O_2

ATP   ADP

NADP$^+$

NADPH   Inorganic phosphate

2 Calvin cycle CO_2 Sugars

FIGURE 2.12 In photosynthesis, autotrophs such as plants, algae, and cyanobacteria use sunlight to convert water and carbon dioxide into oxygen and sugar. In the light reactions, water is converted to oxygen in the presence of sunlight, creating high-energy molecules (ATP and NADPH). These molecules help drive reactions in the Calvin cycle, in which carbon dioxide is used to produce sugars. Molecules of ADP, NADP$^+$, and inorganic phosphate formed in the Calvin cycle, in turn, help power the light reactions, creating an endless loop.

initiate a series of chemical reactions called the light reactions (**FIGURE 2.12**). During these reactions, **1** water molecules split, releasing electrons whose energy is used to produce high-energy molecules. A phosphate is added to the lower-energy molecule adenosine diphosphate (ADP) to form the higher-energy molecule adenosine triphosphate (ATP). A hydrogen ion and a pair of electrons convert the lower-energy molecule nicotinamide dinucleotide phosphate ($NADP^+$) into the higher-energy NADPH. ATP and NADPH are then used to fuel reactions in the Calvin cycle. During **2** the Calvin cycle, carbon atoms—from the carbon dioxide in air that enters the plant through its leaves—are linked together to produce sugars. In photosynthesis, plants draw up water from the ground through their roots, absorb carbon dioxide from the air through their leaves, and harness the power of sunlight with the light-absorbing pigment chlorophyll. With these ingredients, green plants create sugars for their growth and maintenance and, in turn, provide chemical energy to any organism that eats them. Plants also release oxygen as a by-product of photosynthesis, forming the oxygen gas in the air we breathe.

Photosynthesis is a complex process, but the overall reaction can be summarized with the following equation:

$$6CO_2 + 6H_2O + \text{the sun's energy} \longrightarrow C_6H_{12}O_6 + 6O_2$$
(carbon   (water)   (sugar)   (oxygen)
dioxide)

The number preceding each molecular formula indicates how many of those molecules are involved in the reaction. Note that the sums of the numbers on each side of the equation for each element are equal; that is, there are 6 C, 12 H, and 18 O atoms on each side. This illustrates how chemical equations are balanced, with each atom recycled and matter conserved. No atoms are lost; they are simply rearranged among molecules.

Cellular respiration powers living things

Organisms make use of the chemical energy in sugars that they create by photosynthesis in a process called **cellular respiration,** which is vital to life. To release the chemical energy of glucose, cells use oxygen to convert glucose back into its original starting materials, water and carbon dioxide. The energy released during this process is used to power all the biochemical reactions that sustain life. The net equation for cellular respiration is the exact opposite of that for photosynthesis:

$$C_6H_{12}O_6 + 6O_2 \longrightarrow 6CO_2 + 6H_2O + \text{energy}$$
(sugar)   (oxygen)   (carbon   (water)
                             dioxide)

However, the energy released per glucose molecule in respiration is only two-thirds of the energy input per glucose molecule in photosynthesis—a prime example of the second law of thermodynamics. Cellular respiration is a continuous process occurring in all living things and is essential to life. Thus, it occurs in the autotrophs that create glucose and also in **heterotrophs (consumers),** organisms that gain their energy by feeding on other organisms. Heterotrophs include most animals, as well as the fungi and microbes that decompose organic matter.

Geothermal energy also powers Earth's systems

Although the sun is life's primary energy source, it is not the only source of energy for our planet. An additional, although minor, energy source is the gravitational pull of the moon, which in conjunction with the sun's gravitational pull causes ocean tides (p. 429). Another significant energy source is geothermal heating emanating from inside Earth, powered primarily by radioactivity (p. 26). Radiation from radioisotopes deep inside our planet heats the inner Earth, and this heat gradually makes its way to the surface. There, it heats **magma** (rock heated to a molten, liquid state) that erupts from volcanoes and drives plate tectonics (p. 35).

Geothermal energy also powers biological communities. On the deep ocean floor, jets of geothermally heated water gush into the icy-cold depths. These hydrothermal vents can host entire communities of specialized organisms that thrive in the extreme high-temperature, high-pressure conditions.

Hydrothermal vents are so deep underwater that they completely lack sunlight, so the energy flow of these communities cannot be fueled through photosynthesis. Instead, bacteria in deep-sea vents use the chemical-bond energy of hydrogen sulfide (H_2S) to transform inorganic carbon into organic carbon compounds in a process called **chemosynthesis.** Chemosynthesis occurs in various ways, and one way is defined by the following equation:

$$6CO_2 + 6H_2O + 3H_2S \longrightarrow C_6H_{12}O_6 + 3H_2SO_4$$

(carbon dioxide)   (water)   (hydrogen sulfide)   (sugar)   (sulfuric acid)

Energy from chemosynthesis passes through the deep-sea-vent animal community as consumers such as tubeworms, mussels, fish, shrimp, and gigantic clams gain nutrition from chemoautotrophic bacteria and one another.

Geology: The Physical Basis for Environmental Science

A good way to understand how our planet functions is to examine the rocks, soil, and sediments beneath our feet. The physical processes that take place at and below Earth's surface shape the landscape and lay the foundation for most environmental systems and for life.

The physical nature of our planet also benefits our society, because without the study of Earth's rocks and the processes that shape them, we would have no metals for consumer products, no energy from fossil fuels, and no uranium for nuclear power plants. Our planet is dynamic, and this dynamism is what motivates **geology,** the study of Earth's physical features, processes, and history. A human lifetime is just a blink of an eye in the long course of geologic time, and the Earth we experience is merely a snapshot in our changing planet's long history. We can begin to grasp this long-term dynamism as we consider two processes of fundamental importance—plate tectonics and the rock cycle.

Earth consists of layers

Our planet consists of multiple layers (**FIGURE 2.13**). At Earth's center is a dense **core** consisting mostly of iron, solid in the inner core and molten in the outer core. Surrounding the core is a thick layer of less dense, elastic rock called the **mantle.** A portion of the upper mantle called the **asthenosphere** contains especially soft rock, melted in some areas. The harder rock above the asthenosphere is the **lithosphere.** The lithosphere includes both the uppermost mantle and the entirety of Earth's third major layer, the **crust,** the thin, brittle, low-density layer of rock that covers Earth's surface. The intense heat in the inner Earth rises from core to mantle to crust, and it eventually dissipates at the surface.

The heat from the inner layers of Earth also drives convection currents that flow in loops in the mantle, pushing the mantle's soft rock cyclically upward (as it warms) and downward (as it cools), like a gigantic conveyor belt system. As the mantle material moves, it drags large plates of lithosphere along its surface. This movement is known as **plate tectonics,** a process of extraordinary importance to our planet.

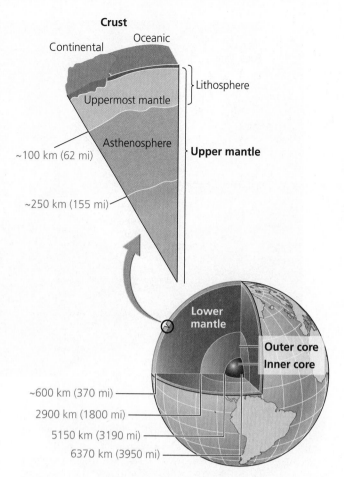

FIGURE 2.13 Earth's three primary layers—core, mantle, and crust—are themselves layered. The inner core of solid iron is surrounded by an outer core of molten iron, and the rocky mantle includes the molten asthenosphere near its upper edge. At Earth's surface, dense and thin oceanic crust abuts lighter, thicker continental crust. The lithosphere consists of the crust and uppermost mantle above the asthenosphere.

Plate tectonics shapes Earth's geography

Our planet's surface consists of about 15 major tectonic plates, which fit together like pieces of a jigsaw puzzle (**FIGURE 2.14**, p. 36). Imagine peeling an orange and then placing the pieces back onto the fruit; the ragged pieces of peel are like the lithospheric plates riding atop Earth's surface. However, the plates are thinner relative to the planet's size, more like the skin of an apple. These plates move at rates of roughly 2–15 cm (1–6 in.) per year. This slow movement has influenced Earth's climate and life's evolution throughout our planet's history as the continents combined, separated, and recombined in various configurations. By studying ancient rock formations throughout the world, geologists have determined that at least twice, all landmasses were joined together in a "supercontinent." Scientists have dubbed the landmass that resulted about 225 million years ago Pangaea (see the inset in Figure 2.14).

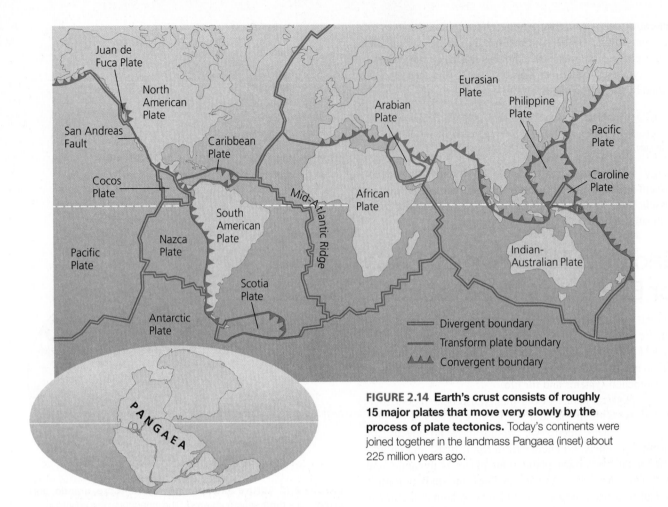

FIGURE 2.14 **Earth's crust consists of roughly 15 major plates that move very slowly by the process of plate tectonics.** Today's continents were joined together in the landmass Pangaea (inset) about 225 million years ago.

There are three types of plate boundaries

The processes that occur at each type of plate boundary all have major consequences.

At **divergent plate boundaries,** tectonic plates push apart from one another as magma rises upward to the surface, creating new lithosphere as it cools (**FIGURE 2.15a**). An example is the Mid-Atlantic Ridge, part of a 74,000-km (46,000-mi) system of divergent plate boundaries slicing across the floors of the world's oceans.

Where two plates meet, they may slip and grind alongside one another, forming a **transform plate boundary** (**FIGURE 2.15b**). This movement creates friction that generates earthquakes (p. 40) along strike-slip faults. The Tohoku earthquake, for example, occurred at such a fault off the eastern coast of Japan. Faults are fractures in Earth's crust, and at strike-slip faults, each landmass moves horizontally in opposite directions. The Pacific Plate and the North American Plate, for example, slide past one another along California's San Andreas Fault. Southern California is slowly inching its way northward along this fault, so the site of Los Angeles will

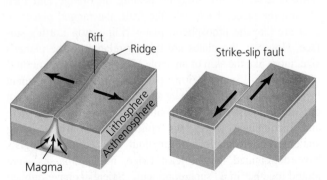

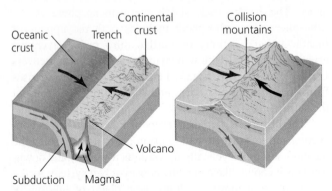

(a) Divergent plate boundary (b) Transform plate boundary (c) Convergent plate boundaries

FIGURE 2.15 **There are three types of boundaries between tectonic plates, generating different geologic processes.**

eventually—in about 15 million years or so—reach that of modern-day San Francisco.

Convergent plate boundaries, where two plates come together, can give rise to different outcomes (**FIGURE 2.15c**). As plates of newly formed lithosphere push outward from divergent plate boundaries, this oceanic lithosphere gradually cools, becoming denser. After millions of years, it becomes denser than the asthenosphere beneath it and dives downward into the asthenosphere in a process called **subduction.** As the lithospheric plate descends, it slides beneath a neighboring plate that is less dense, forming a convergent plate boundary. The subducted plate is heated and pressurized as it sinks, and water vapor escapes, helping to melt rock (by lowering its melting temperature). The molten rock rises, and this magma may erupt through the surface via volcanoes (p. 40).

When one plate of oceanic lithosphere is subducted beneath another plate of oceanic lithosphere, the resulting volcanism may form arcs of islands, such as Japan and the Aleutian Islands of Alaska. Subduction zones may also create deep trenches, such as the Mariana Trench, our planet's deepest abyss, located in the western Pacific Ocean. When oceanic lithosphere slides beneath continental lithosphere, volcanic mountain ranges form that parallel coastlines (Figure 2.15c, left). An example is South America's Andes Mountains, where the Nazca Plate slides beneath the South American Plate.

When two plates of continental lithosphere meet, the continental crust on both sides resists subduction and instead crushes together, bending, buckling, and deforming layers of rock from both plates in a **continental collision** (Figure 2.15c,

right). Portions of the accumulating masses of buckled crust are forced upward as they are pressed together, and mountain ranges result. The Himalayas, the world's highest mountains, resulted from the Indian-Australian Plate's collision with the Eurasian Plate beginning 40–50 million years ago, and these mountains are still rising today as these plates converge.

Tectonics produces Earth's landforms

The processes of plate tectonics build mountains, shape the geography of oceans, islands, and continents, and give rise to earthquakes and volcanoes. The topography created by plate tectonic processes, in turn, shapes climate by altering patterns of rainfall, wind, ocean currents, heating, and cooling—all of which affect rates of weathering and erosion and the ability of plants and animals to inhabit different regions. Thus, the locations of biomes (pp. 94–101) are influenced by plate tectonics. Moreover, plate tectonics has affected the history of life's evolution; the convergence of landmasses into supercontinents such as Pangaea is thought to have contributed to widespread extinctions by reducing the area of species-rich coastal regions and by creating an arid continental interior with extreme temperature swings.

The rock cycle produces a diversity of rock types

Just as plate tectonics shows geology's dynamism on a large scale, the rock cycle shows it on a smaller one. We tend to think of rock as fairly solid stuff. Yet over geologic time, rocks and the minerals that make them up are heated, melted, cooled, broken down, and reassembled in a very slow process called the **rock cycle** (**FIGURE 2.16**).

A **rock** is any solid aggregation of minerals. A **mineral**, in turn, is any naturally occurring solid element or inorganic compound with a crystal structure, a specific chemical composition, and distinct physical properties. Understanding the rock cycle enables us to better appreciate the formation and conservation of soils, mineral resources, fossil fuels, groundwater sources, and other natural resources (all of which will be discussed in later chapters).

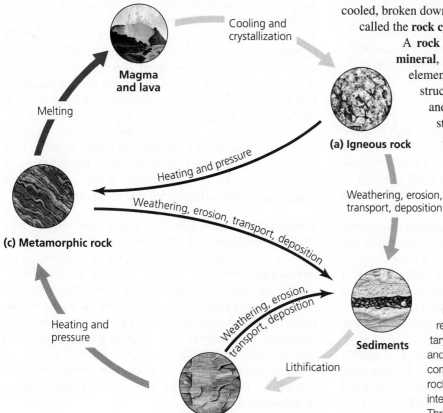

FIGURE 2.16 The rock cycle. Igneous rock **(a)** is formed when rock melts and the resulting magma or lava then cools. Sedimentary rock **(b)** is formed when rock is weathered and eroded, and the resulting sediments are compressed to form new rock. Metamorphic rock **(c)** is formed when rock is subjected to intense heat and pressure underground. Through these processes, each type of rock can be converted into either of the other two types.

(a) Extrusive igneous rock: Basalt in Japan

(b) Sedimentary rock: Sandstone in Arizona

(c) Metamorphic rock: Gneiss in Utah

FIGURE 2.17 **Examples of rock types.** The lava flows in Japan form basalt **(a),** a type of extrusive igneous rock. The layered formation in Paria Canyon, Arizona, is an example of sandstone **(b),** a type of sedimentary rock. Gneiss (pronounced "nice") **(c),** at Antelope Island, Utah, is a type of metamorphic rock.

Igneous rock All rocks can melt. At high enough temperatures, rock will enter the molten, liquid state called magma. If magma is released through the lithosphere (as in a volcanic eruption), it may flow or spatter across Earth's surface as **lava.** Rock that forms when magma or lava cools is called **igneous rock** (from the Latin *ignis*, meaning "fire").

Igneous rock comes in two main classes because magma can solidify in different ways. When magma cools slowly and solidifies while it is below Earth's surface, it forms intrusive igneous rock. Granite is the best-known type of intrusive rock. A slow cooling process allows minerals of different types to aggregate into large crystals, giving granite its multicolored, coarse-grained appearance. In contrast, when molten rock is ejected from a volcano, it cools quickly, so minerals have little time to grow into coarse crystals. This quickly cooled molten rock is classified as extrusive igneous rock, and its most common representative is basalt, the principal rock type of the Japanese islands (**FIGURE 2.17a**).

Sedimentary rock All exposed rock weathers away with time. The relentless forces of wind, water, freezing, and thawing eat away at rocks, stripping off one tiny grain (or large chunk) after another. Through weathering (p. 217) and erosion (pp. 229–230), particles of rock come to rest downhill, downstream, or downwind from their sources, forming **sediments.** Alternatively, some sediments form chemically from the precipitation of substances out of solution. Phosphates dissolved in ocean waters, for example, can precipitate out of the water as solids and settle to the ocean floor.

Over time, deep layers of sediment accumulate, causing the weight and pressure on the layers of sediment below them to increase. **Sedimentary rock** (**FIGURE 2.17b**) is formed as sediments are physically pressed together (compaction) and as dissolved minerals seep through sediments and act as a kind of glue, binding together sediment particles (a process termed lithification). Examples of sedimentary rock include sandstone, made of cemented sand particles; shale, comprising still smaller mud particles; and limestone, formed as dissolved calcite precipitates from water or as calcite from marine organisms settles to the ocean bottom.

These processes also create the fossils of organisms (p. 58), which we use to learn about the history of life on Earth and process for the fossil fuels we use for energy. Because sedimentary layers pile up in chronological order, scientists can assign relative dates to fossils they find in sedimentary rock.

Metamorphic rock Geologic forces may bend, uplift, compress, or stretch rock. When any type of rock is subjected to great heat or pressure, it may alter in form to become **metamorphic rock** (from the Greek word for "changed form"). The forces that metamorphose rock generally occur deep underground, at temperatures lower than the rock's melting point, but high enough to change its appearance and physical properties. Metamorphic rock (**FIGURE 2.17c**) includes rock such as slate, formed when shale is subjected to heat and pressure, and marble, formed when limestone is heated and pressurized. Gneiss, quartzite, and schist are other examples of metamorphic rocks.

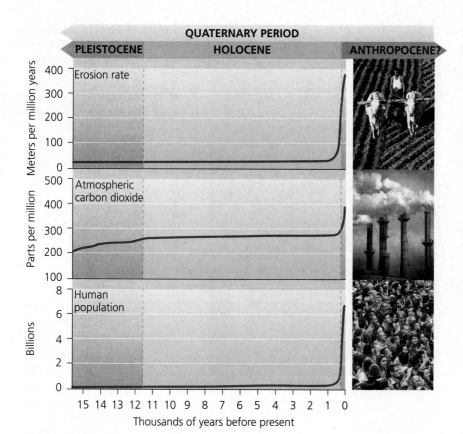

QUATERNARY PERIOD

PLEISTOCENE   HOLOCENE   ANTHROPOCENE?

Meters per million years

400
300
200
100
0

Erosion rate

Parts per million

500
400
300
200
100

Atmospheric carbon dioxide

Billions

8
6
4
2
0

Human population

15 14 13 12 11 10 9 8 7 6 5 4 3 2 1 0
Thousands of years before present

FIGURE 2.18 Global soil erosion rates (top) and atmospheric carbon dioxide concentrations (middle) have increased sharply in just the past few hundred years, along with human population (bottom). These patterns have persuaded some geologists that we should recognize a new epoch in Earth history: the Anthropocene. *Adapted from Zalasiewicz, J., et al., 2008. Are we now living in the Anthropocene? GSA Today 18(2): 4–8, Figure 1.*

Geologic processes occur across "deep time"

Geologic processes occur at timescales that are difficult to conceptualize. But perhaps it is only by appreciating the long time periods within which our planet's geologic forces operate that we can realize how exceedingly slow processes such as plate tectonics or the formation of sedimentary rock reshape our planet. This lengthy timescale is referred to as deep time, or geologic time.

The geologic timescale shows the full span of Earth's history—all 4.5 *billion* years of it—and focuses on the most recent 543 million years (**APPENDIX D**). Geologists have subdivided Earth's history into 3 eras and 11 periods. The Quaternary period, the most recent, occupies a thin slice of time at the top of the scale because this period began "only" 1.8 million years ago.

Geologists divide the geologic timescale using evidence from stratigraphy, the study of strata, or layers, of sedimentary rock. Where scientists find fossil evidence for major and sudden changes in the physical, chemical, or biological conditions present on Earth between one set of layers and the next, they assign a boundary between geologic time periods. For instance, fossil evidence for mass extinctions of species (pp. 60, 284) determines several boundaries, such as that between the Permian and Triassic periods.

We live in the Holocene epoch, the most recent slice of the Quaternary period. The Holocene epoch began about 11,500 years ago with a warming trend that melted glaciers and brought Earth out of its most recent ice age. Since then,

Earth's climate has been remarkably constant, and this constancy provided our species with the long-term stability we needed to develop agriculture and civilization. However, since the industrial revolution, human activity has had major impacts on Earth's basic processes, including a sharp increase in soil erosion from clearing forests and cultivating land; an alteration in the composition of the atmosphere through emitting greenhouse gases, which elevates Earth's average temperature; and a recent explosion in human population, which has intensified all impacts on Earth. All these activities have set into motion a new mass extinction event (p. 285). These realizations led some geologists in 2000 to propose naming a new geologic era, encompassing the past 200 years, after ourselves—the Anthropocene (**FIGURE 2.18**).

Geologic and Natural Hazards

Plate tectonics shapes our planet, but the consequences of tectonic movement can also pose hazards to us. Earthquakes and volcanoes are examples of such geologic hazards. We can see how such hazards relate to tectonic processes by examining a map of the circum-Pacific belt, or "ring of fire" (**FIGURE 2.19**, p. 40). Ninety percent of earthquakes and over half of the world's volcanoes occur along this 40,000-km (25,000-mi) arc of subduction zones and fault systems. Like many locations along the circum-Pacific belt, Japan has experienced earthquakes and volcanism throughout its history.

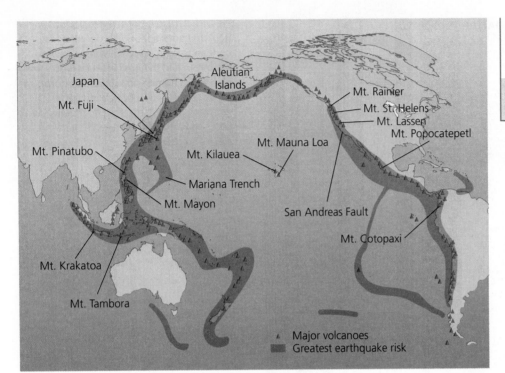

FIGURE 2.19 Most of our planet's volcanoes and earthquakes occur along the circum-Pacific belt, or "ring of fire."

DATA Q Go to **Interpreting Graphs & Data** on Mastering Environmental Science

Major volcanoes
Greatest earthquake risk

Earthquakes result from movement at plate boundaries and faults

Along tectonic plate boundaries and in other places where faults occur, the earth may relieve built-up pressure in fits and starts. Each release of energy causes what we know as an **earthquake.** Most earthquakes are barely perceptible, but as shown by the Tohoku quake of 2011, they are occasionally powerful enough to cause significant losses of human life and property (**TABLE 2.2**). Earthquakes can also occur in the interior of tectonic plates, when faults are formed by continental plates being stretched and pulled apart by geologic forces within the earth. Such earthquakes are not only rare but also poorly understood. The New Madrid seismic zone, which lies beneath the lower Mississippi River basin in the central United States, is one area where such an "intraplate" earthquake may occur (**FIGURE 2.20**). And human activities may also induce earthquakes in areas far from the boundaries of tectonic plates by extracting substances from the earth or injecting liquids into underlying rock layers (see **THE SCIENCE BEHIND THE STORY**, pp. 42–43).

Volcanoes arise from rifts, subduction zones, or hotspots

Where molten rock, hot gas, or ash erupts through Earth's surface, a **volcano** is formed, often creating a mountain over time as cooled lava accumulates. As we have seen, lava can extrude along mid-ocean ridges or over subduction zones as one tectonic plate dives beneath another. Due to its position along subduction zones, Japan has more than 100 active volcanoes, which is 10% of the world total and more than any

TABLE 2.2 Examples of Large Earthquakes

YEAR	LOCATION	FATALITIES	MAGNITUDE[1]
1556	Shaanxi Province, China	830,000	~8
1755	Lisbon, Portugal	70,000[2]	8.7
1906	San Francisco, California	3000	7.8
1923	Kwanto, Japan	143,000	7.9
1964	Anchorage, Alaska	128[2]	9.2
1976	Tangshan, China	255,000+	7.5
1985	Michoacan, Mexico	9500	8.0
1989	Loma Prieta, California	63	6.9
1994	Northridge, California	60	6.7
1995	Kobe, Japan	5502	6.9
2004	Northern Sumatra	228,000[2]	9.1
2005	Kashmir, Pakistan	86,000	7.6
2008	Sichuan Province, China	50,000+	7.9
2010	Port-au-Prince, Haiti	236,000	7.0
2010	Maule, Chile	500	8.8
2011	Northern Japan	18,000[2]	9.0
2015	Kathmandu, Nepal	8900	7.8

[1]Measured by moment magnitude; each full unit is roughly 32 times as powerful as the preceding full unit.
[2]Includes deaths from the resulting tsunami.

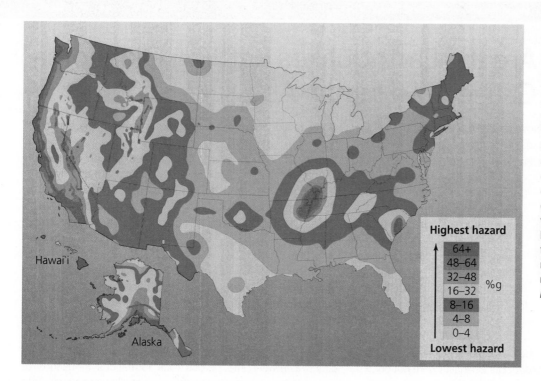

FIGURE 2.20 Many parts of the United States are at elevated risk for earthquakes. The West Coast faces threats from earthquakes due to its position at the boundary of tectonic plates. Portions of the continental interior have elevated risk due to naturally occurring intraplate earthquakes or human-induced earthquakes, typically from wastewater injection or hydraulic fracturing. The units for the figure are %g, a measure of acceleration related to the force of gravity. *Data from* U.S. Geological Survey.

other nation. Mount Fuji, one of Japan's most prominent and recognizable natural features, is one such active volcano.

Lava may also emit at hotspots, localized areas where plugs of molten rock from the mantle erupt through the crust. As a tectonic plate moves across a hotspot, repeated eruptions from this source may create a linear series of volcanoes. The Hawaiian Islands provide an example of this process (**FIGURE 2.21a**).

At some volcanoes, lava flows slowly downhill, such as at Mount Kilauea in Hawai'i (**FIGURE 2.21b**), which has been erupting continuously since 1983! At other times, a volcano may let loose large amounts of ash and cinder in a sudden explosion, such as during the 1980 eruption of Mount Saint Helens (Figure 17.10, p. 457). Very large eruptions can even alter Earth's climate, because the released ash blocks

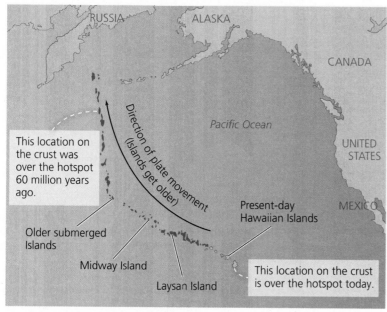

(a) Current and former Hawaiian Islands, formed as crust moves over a volcanic hotspot

(b) Mt. Kilauea erupting

FIGURE 2.21 The Hawaiian Islands are the product of a hotspot on Earth's mantle. The Hawaiian Islands **(a)** have been formed by repeated eruptions from a hotspot of magma in the mantle as the Pacific Plate passes over the hotspot. The Big Island of Hawai'i was most recently formed, and it is still volcanically active. The other islands are older and have already begun eroding. To their northwest stretches a long series of former islands, now submerged. The active volcano Kilauea **(b)**, on the Big Island's southeast coast, is currently located above the edge of the hotspot.

THE SCIENCE behind the story

Are the Earthquakes Rattling Oklahoma Caused by Human Activity?

Geophysicist, Katie Keranen, Cornell University

In November 2011, a series of earthquakes and aftershocks struck the small town of Prague, Oklahoma (population 2300), roughly 100 km (60 mi) east of Oklahoma City. The shaking damaged 14 homes, buckled the pavement of a local highway, and caused several injuries. One of the tremors measured 5.7 on the Richter scale—as of that time, the largest earthquake ever recorded in the state.

Earthquakes are uncommon in Oklahoma, especially ones of such magnitude. But scientists were not completely surprised by the 2011 event, because they had already noted an increase in earthquake activity in the state (**FIGURE 1**). For example, between 1978 and 2008, Oklahoma experienced a yearly average of only about 1.5 earthquakes of 3.0 magnitude or greater. In 2009, that number rose to 20, and by 2016 it had jumped to 641. Scientists have even proposed an explanation for the increase—the injection of wastewater from oil and gas extraction into porous rock layers beneath the surface of the earth.

Spurred by high energy prices, the extraction of crude oil and natural gas from conventional wells (p. 530) and hydraulic fracturing (p. 390) in Oklahoma increased from 2010 to 2013, with gas extraction rising by 17% and oil extraction by 65%. And as Oklahoma's oil and gas output increased, so did its disposal of wastewater by injection, which rose 20% during the same time period.

Oil and gas are the modified remains of ancient marine organisms, and oil and gas deposits often contain briny water that separates from the fuels after they are extracted. The salty wastewater, which can also contain toxic and radioactive compounds, is then typically disposed of by trucking it to a facility far from the oil and gas wells. Disposal involves injection of the salty wastewater into porous rock formations thousands of feet underground, well below the more shallow rock layers that contain groundwater aquifers. This approach is designed to dispose of wastewater in a manner that prevents it from contaminating sources of drinking water, both aboveground and belowground.

Wastewater had been injected into the rocks beneath Oklahoma for decades without any measurable increase in seismic activity, but scientists were aware that continued pumping of wastewater into underground rock formations could lead to

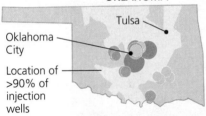

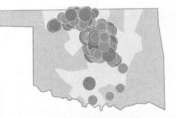

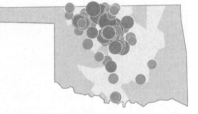

OKLAHOMA

Tulsa

Oklahoma City

Location of >90% of injection wells

2010 43 earthquakes

2014 579 earthquakes of magnitude 3 or higher

2018 196 earthquakes

FIGURE 1 The number of earthquakes in Oklahoma of magnitude 3.0 or greater increased greatly from 2009 to 2015 and then declined once restrictions were placed on wastewater injections into rock layers. Many of these earthquakes are thought to be related to the injection of wastewater from oil and gas extraction into underground rock layers. Each circle indicates an earthquake event. The higher the magnitude of the earthquake, the darker and wider the circle. *Data from Oklahoma Geological Survey, 2019, http://earthquakes.ok.gov/what-we-know/earthquake-map/.*

sunlight, and sulfur emissions lead to a sulfuric acid haze that blocks radiation and cools the atmosphere.

Sometimes a volcano can unleash a pyroclastic flow—a fast-moving cloud of toxic gas, ash, and rock fragments that races down the slopes at speeds up to 725 km/hr (450 mph),

enveloping everything in its path. Such a flow buried the inhabitants of the ancient Roman cities of Pompeii and Herculaneum in A.D. 79, when Mount Vesuvius erupted.

One of the world's largest volcanoes—so large it is called a supervolcano—lies in the United States. The entire basin of earthquakes. As pores within underground rocks become saturated with water, pressure grows in the rocks, causing them to expand. The expanding rock, in turn, expands rock layers, which then push against existing faults in the earth, "lubricating" them and causing them to slip and produce earthquakes.

As seismic activity in Oklahoma increased, scientists grew more and more convinced that the phenomenon was due to wastewater injections. However, persuading legislators, regulators, and the public of the cause proved challenging. Oil and gas extraction is big business in Oklahoma. By some estimates, one in five jobs in the state is connected to the industry. Landowners benefit from royalties earned by fossil fuel extractions on their land. Tax revenue from sales of oil and gas is the state's third-largest revenue source—behind only sales taxes and personal income taxes. With such widespread economic benefits, the oil and gas industry enjoys high levels of support in Oklahoma government. When, following the Prague quake, calls arose to temporarily halt further wastewater injections, the state government urged a cautious approach and echoed the oil and gas industry's position that greater study was needed before decisive action should be taken.

Although the connection between underground fluid injection and earthquakes was well known, studies directly connecting specific events with injection sites were rare. A scientific study published in 2013 changed that, as it directly linked the Prague earthquake to nearby injections of wastewater from oil and gas extraction. The study, headed up by geophysicist Katie Keranen of the University of Oklahoma (now at Cornell University) and published in the journal *Geology*, measured the aftershocks produced by the 2011 earthquake to determine the location of the fault that produced the quake.

When that earthquake hit, researchers reacted quickly, deploying seismic sensors near Prague and gathered detailed readings on two major tremors and the 1183 aftershocks that followed. Analysis of the tremor patterns revealed that the cause was a rupture at the tip of a fault within about 200 m (650 ft) of an active wastewater-injection well at depths consistent with injected rock layers. Keranen's work also showed that it is possible for nearly two decades to pass between the initiation of wastewater injection and a subsequent seismic event, calling into question the safety of many other injection sites. A follow-up study led by Keranen found that earthquakes could be induced as far as 30 km (19 mi) from the fields of injection wells and concluded that up to 20% of the induced seismic activity over an area covering 2000 km^2 (770 mi^2) in the Central United States could be traced to the activity of four high-volume wastewater disposal wells in Oklahoma.

As research continued into the connection between wastewater injection and seismic activity, and the issue gained public attention, the state incrementally increased its regulation of wastewater injection wells. The government mandated regular monitoring of well pressures in injection sites and directed operators to slow injection rates or stop injections altogether if underground conditions were deemed conducive to initiating a seismic event. These actions led to lower levels of wastewater injection starting in 2016 and may have resulted in immediate relief, as the number of earthquakes of magnitude 3.0 or greater in Oklahoma dropped from 903 in 2015 to 623 in 2016, 304 in 2017, and 196 in 2018.

Although recent events seem to show progress in the efforts to reduce earthquakes in Oklahoma, research has shown that seismic activity is likely to occur in the state long after the practice has ceased, even in locations far from the fields of wastewater injection. Therefore, much like the U.S. West Coast with its well-known propensity for earthquakes, the South Central United States will be a hotbed for seismic study in the decades to come (**FIGURE 2**).

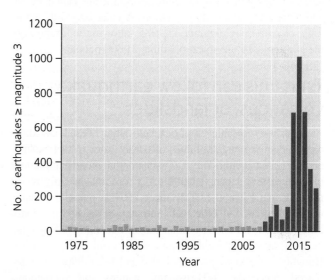

FIGURE 2 Larger earthquakes are becoming more common in the Central and Eastern United States. Much of the increase is centered in Oklahoma, where the local geology, coupled with the use of seismic-inducing activities, such as oil extraction and wastewater injection, has led to more frequent tremors. *Data from Rubinstein, J. L., and A. B. Mahani, 2015. Myths and facts on wastewater injection, hydraulic fracturing, enhanced oil recovery, and induced seismicity. Seismol. Res. Lett.* 86: *1–8 and U.S. Geological Survey, 2016.*

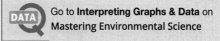

Go to **Interpreting Graphs & Data** on **Mastering Environmental Science**

Yellowstone National Park is an ancient supervolcano that has at times erupted so massively as to cover large parts of the continent deeply in ash. Although another eruption is not expected imminently, the region is still geothermally active, as evidenced by its numerous hot springs and geysers.

Landslides are a form of mass wasting

At a smaller scale than volcanoes and earthquakes, a **landslide** occurs when large amounts of rock or soil collapse and flow downhill. Landslides are a severe and often sudden

WEIGHING
the issues

Assessing Your Risk from Natural Hazards

What types of natural hazards are likely to occur where you live? Name three actions you personally can take to minimize your risk from these hazards. If a natural disaster strikes, should people be allowed to rebuild in the same areas if those areas are prone to experience the hazard again? Should rebuilding costs be supported by taxpayer funds? Should insurance companies be required to insure the rebuilt structures?

manifestation of the phenomenon of **mass wasting,** the downslope movement of soil and rock due to gravity. Mass wasting occurs naturally, but often is brought about by human land use practices that expose or loosen soil, making slopes more prone to collapse.

Most often, mass wasting erodes unstable hillsides, damaging property one structure at a time. Occasionally, mass wasting events can be colossal and deadly; mudslides that followed the torrential rainfall of Hurricane Mitch in Nicaragua and Honduras in 1998 killed more than 11,000 people. Mudslides caused when volcanic eruptions melt snow and send huge volumes of destabilized mud racing downhill are called lahars, and these are particularly dangerous. A lahar buried the entire town of Armero, Colombia, in 1985 following an eruption of the Nevado del Ruiz volcano, killing 21,000 people.

Tsunamis can follow earthquakes, volcanoes, or landslides

Earthquakes, volcanic eruptions, and large coastal landslides can all displace huge volumes of ocean water instantaneously and trigger a tsunami. The 2011 tsunami that inundated portions of northeastern Japan (**FIGURE 2.22**) was not the only recent major tsunami event. In December 2004, a massive tsunami, triggered by an earthquake off Sumatra, devastated the coastlines of countries all around the Indian Ocean, including Indonesia, Thailand, Sri Lanka, India, and several African nations. Roughly 228,000 people were killed and 1–2 million were displaced. Since the 2004 tsunami, nations and international agencies have stepped up efforts to develop systems to give coastal residents advance warning of approaching tsunamis.

Those of us who live in the United States and Canada should not consider tsunamis to be something that occurs only in faraway places. Residents of the Pacific Northwest—such as the cities of Seattle, Washington, and Portland, Oregon—could be at risk if there is a slip in the Cascadia subduction zone that lies 1100 km (700 mi) offshore. The tsunami produced by such a slip would inundate 1.1 million km^2 (440,000 mi^2) of coastal land and cause massive destruction over an area that is currently home to over 7 million people.

We can worsen or lessen the impacts of natural hazards

Aside from geologic hazards, people face other types of natural hazards. Heavy rains can lead to flooding that ravages low-lying areas near rivers and streams (p. 403). Coastal erosion can eat away at beaches. Wildfire can threaten life and property in fire-prone areas. Tornadoes and hurricanes (p. 456) can cause extensive damage and loss of life.

Although we refer to such phenomena as "natural hazards," the magnitude of their impacts on us often depends on choices we make. We sometimes worsen the impacts of so-called natural hazards in various ways. For example, we choose to build homes and businesses in areas that are prone to hazards, such as the floodplains of rivers or in coastal areas susceptible to flooding. People also use and engineer the landscapes around us in ways that can increase the frequency or severity of natural hazards. Damming and diking rivers to control floods can sometimes lead to catastrophic flooding (p. 403), and the clear-cutting of forests on slopes (p. 320) can induce mass wasting and increase water runoff. Human-induced climate change (Chapter 18) can cause sea levels to rise and promote coastal flooding, and can increase the risks of drought, fire, flooding, and mudslides by altering precipitation patterns.

We can often reduce or lessen the impacts of hazards through the thoughtful use of technology, engineering, and policy, informed by a solid understanding of geology and ecology. Examples include building earthquake-resistant structures; designing early warning systems for earthquakes, tsunamis, and volcanoes; and conserving reefs and shoreline vegetation to protect against tsunamis and coastal erosion. In addition, better forestry, agriculture, and mining practices can help prevent mass wasting. Zoning regulations, building codes, and insurance incentives that discourage development in areas prone to landslides, floods, fires, and storm surges can help keep us out of harm's way. Finally, addressing global climate change may help reduce the frequency of natural hazards in many regions.

FIGURE 2.22 A man surveys the destruction caused by the Tohoku earthquake and tsunami in Japan in 2011. Ocean waters pushed miles inland in some locations, scouring everything in their path, and pulling debris out to sea as the waters receded.

TODAY, efforts persist at the Fukushima Daiichi power plant to secure the radioactive material that remains in the crippled reactors. Groundwater beneath the plant continues to be contaminated with radioactive isotopes and leaches into the nearby Pacific Ocean. On land, the government has scoured away the upper soil layers in highly contaminated areas near the power plant in an effort to reduce radioactivity levels. This soil is currently being stored in holding areas in large bags until a comprehensive disposal strategy can be determined. Bioremediation studies are being undertaken—many involving the use of plants or microbes—to find cost-effective ways to decontaminate soils in the region. The remediation efforts at the nuclear power plant and in the surrounding area are anticipated to cost over $112 billion and will take decades to complete. As radiation levels have declined, some of the evacuated areas around the Fukushima nuclear power plant have been reopened to residents. Few people appear interested in doing so, however, as only about 13% of those displaced have returned to their homes.

Opposition to the indefinite use of nuclear power in Japan still remains high, with some 90% of people expressing a desire to phase out the nation's use of nuclear power. But while, shortly after the accident, the government proposed fully eliminating Japan's use of nuclear power by 2040, it has since changed course and views nuclear power as a long-term source of power for the nation. At the time of this writing, 8 of the nation's 54 nuclear power reactors are back online and producing power after completing a recertification process. The Japanese government is also investing heavily in renewable energy and aims to produce, by 2030, nearly half of the nation's power from carbon-free or low-carbon sources of energy, which includes nuclear power.

The nuclear meltdown at Fukushima also forced countries around the world to reassess the role of nuclear power in their societies. Some nations sought to eliminate their use of nuclear power altogether. Germany shut down half of its nuclear power plants and plans to decommission all its reactors by 2022. Switzerland opted to phase out the operation of its five nuclear reactors over the next 20 years. Other nations have chosen to continue their use of nuclear power but with added safeguards. In Japan, a new regulatory agency was formed to inspect and recertify nuclear power plants, as the regulatory apparatus in place before the accident had been criticized for being too closely affiliated with the power companies it was tasked to regulate. In

the United States, a comprehensive review of nuclear power plant safety was ordered immediately following the Fukushima accident, and by 2019, all U.S. nuclear power plants had met new standards designed to make domestic nuclear power plants less susceptible to events like those that caused the meltdown in Fukushima.

As we have seen, the legacy of the Tohoku earthquake and nuclear meltdown at Fukushima Daiichi may be very different from one nation to the next. But one common thread is that our world contains natural geologic hazards, such as earthquakes and tsunamis, as well as hazards posed by human activities, such as nuclear power. And although we cannot ever fully eliminate hazards, we can certainly lessen their potential impacts by thinking carefully about what we build, where we build it, and what we will do should a disaster strike.

- **CASE STUDY SOLUTIONS** Imagine that you live in a coastal community in a region prone to earthquakes and you are a board member of your regional electrical utility. The utility is considering constructing a nuclear power plant 10 miles up the coast from your town. Some of your fellow citizens support the project because it will bring employment to the region and provide a carbon-free source of electricity. However, some residents fear a repeat of the events of 2011 in northeastern Japan if there should be a significant earthquake along one of the onshore or offshore faults. List three specific pieces of information that you would insist on obtaining from geologists and other scientists before casting your vote on the project.

- **LOCAL CONNECTIONS** Nuclear power has both risks and benefits, and people's opinions on its use can differ greatly. Informally interview at least 10 people—friends, family members, or classmates—and solicit their view on the role that nuclear power should play in future U.S. energy production. Should we increase our nation's use of nuclear power, reduce it, or ban it completely? Will people's opinions likely change if a new nuclear power plant is proposed for construction nearby their community?

- **EXPLORE THE DATA** Did the nuclear meltdown in Japan affect energy production from nuclear power in the years since the accident? → **Explore Data** relating to the case study on **Mastering Environmental Science**.

REVIEWING Objectives

You should now be able to:

+ **Explain the fundamentals of matter and chemistry, and apply them to real-world situations.**

Everything in the universe that has mass and occupies space is matter, which can neither be created nor destroyed. Matter comprises atoms and elements, and changes at the atomic level can result in alternate forms of elements, such as ions and isotopes. Atoms bond with one another to form molecules and compounds and are conserved in such reactions. Carbon-based organic compounds are particularly important because they are the building blocks of life and provide energy in the form of fossil fuels (pp. 25–29).

+ **Differentiate among forms of energy, and explain the first and second laws of thermodynamics.**

Energy is the capacity to change the position, physical composition, or temperature of matter. Energy can convert from one form to another—for instance, from potential to kinetic energy, and vice versa—and, like matter, energy is conserved during conversions (the first law of thermodynamics). Systems tend to increase in entropy, or disorder, unless energy is added to build or maintain order and complexity (the second law of thermodynamics) (pp. 30–33).

+ **Distinguish photosynthesis, cellular respiration, and chemosynthesis, and summarize their importance to living things.**

In photosynthesis, autotrophs use carbon dioxide, water, and solar energy to produce oxygen and chemical energy in sugars. In cellular respiration, organisms extract energy from sugars by converting the sugars, in the presence of oxygen, into carbon dioxide and water. Organisms use the energy to combat entropy and sustain life. In chemosynthesis, specialized autotrophs use carbon dioxide, water, and chemical energy from minerals (instead of energy from sunlight) to produce sugars (pp. 33–35).

+ **Explain how plate tectonics and the rock cycle shape the landscape around us.**

The crust of our planet is modified by plate tectonics, which shapes Earth's physical geography and produces earthquakes and volcanoes. On a smaller scale, the rock cycle is the mechanism whereby rocks transform from one type to another (pp. 35–39).

+ **Identify major types of geologic hazards, and describe ways to minimize their impacts.**

Volcanoes, earthquakes, mass wasting, and tsunamis are all natural geologic hazards that affect people and the environment. We can minimize the impact of geologic hazards by making informed decisions about where and how to build (establishing strict zoning regulations and engineering codes within our communities) and by ensuring that our public officials make sound policy decisions when global climate change is a factor (pp. 39–44).

SEEKING Solutions

1. Think of an example of an environmental problem not mentioned in this chapter. How could chemistry help us address the problem?

2. Think about the ways we harness and use energy sources in our society—both renewable sources, such as solar energy, and nonrenewable sources, such as coal, oil, and natural gas. What implications does the first law of thermodynamics have for our energy usage? How is the second law of thermodynamics relevant to our use of energy?

3. Consider rising carbon dioxide levels in the atmosphere and their impact on global climate change (Chapter 18) and then refer to the chemical reactions for photosynthesis and respiration. Provide an argument for why increasing amounts of carbon dioxide in the atmosphere might potentially increase amounts of oxygen in the atmosphere. Now give an argument for why increasing amounts of carbon dioxide might potentially decrease amounts of atmospheric oxygen. What would you need to know to determine which of these two outcomes might occur?

4. Describe how plate tectonics accounts for the formation of (a) mountains, (b) volcanoes, and (c) earthquakes.

5. **THINK IT THROUGH** You serve as city manager and have been asked to develop basic emergency plans for your city in case of natural disaster. For each of the following natural hazards, describe one thing that you would recommend the city do to minimize the impact of such a hazard on lives and property: (a) earthquakes, (b) landslides, (c) flooding.

CALCULATING Ecological Footprints

The second law of thermodynamics has profound implications for human impacts on the environment, as it affects the efficiency with which we produce our food. In ecological systems, a general rule of thumb is that when energy is transferred from plants to plant-eaters or from prey to predator, efficiency is only about 10% (p. 82). Much of this low efficiency is a consequence of the second law of thermodynamics, as some energy is lost to entropy with each transfer. Another way to think of this low efficiency is that eating 1 Calorie of meat from an animal is the ecological equivalent of eating 10 Calories of plant material. So, when we raise animals for meat using grain, it is less energetically efficient than if we ate the grain directly.

Humans are considered omnivores because we can eat both plants and animals. The choices we make about what to eat have significant ecological consequences. With this in mind, calculate the ecological energy requirements for five different diets, each of which provides a total of 2000 dietary Calories per day.

DIET	SOURCE OF CALORIES	NUMBER OF CALORIES CONSUMED	ECOLOGICALLY EQUIVALENT CALORIES	TOTAL ECOLOGICALLY EQUIVALENT CALORIES
100% plant 0% animal	Plant Animal			
90% plant 10% animal	Plant Animal	1800 200	1800 2000	3800
50% plant 50% animal	Plant Animal			
70% plant 30% animal	Plant Animal			
0% plant 100% animal	Plant Animal			

1. How many ecologically equivalent Calories would it take to support you for a year for each of the five diets listed?

2. How does the ecological impact from a diet consisting strictly of animal products (e.g., dairy products, eggs, and meat) compare with that of a strictly vegetarian diet? How many additional ecologically equivalent Calories do you consume each day by including as little as 10% of your Calories from animal sources?

3. What percentages of the Calories in your own diet do you think come from plant versus animal sources? Estimate the ecological impact of your diet, relative to a strictly vegetarian one.

4. List the major factors influencing your current diet (e.g., financial considerations, convenience, access to groceries, taste preferences). Do you envision your diet's distribution of plant and animal Calories changing in the near future? Why or why not?

Mastering Environmental Science

Students Go to **Mastering Environmental Science** for assignments, an interactive e-text, and the Study Area with practice tests, videos, and activities.

Instructors Go to **Mastering Environmental Science** for automatically graded activities, videos, and reading questions that you can assign to your students, plus Instructor Resources.

Evolution and Population Ecology

Saving Hawaii's Native Forest Birds

> *When an island's entire avifauna . . . is devastated almost overnight because of human meddling, it is, quite simply, a tragedy.*
> H. Douglas Pratt, ornithologist and expert on Hawaiian birds

> *To keep every cog and wheel is the first precaution of intelligent tinkering.*
> Conservationist and author Aldo Leopold

HAWAI`I

Pacific Ocean

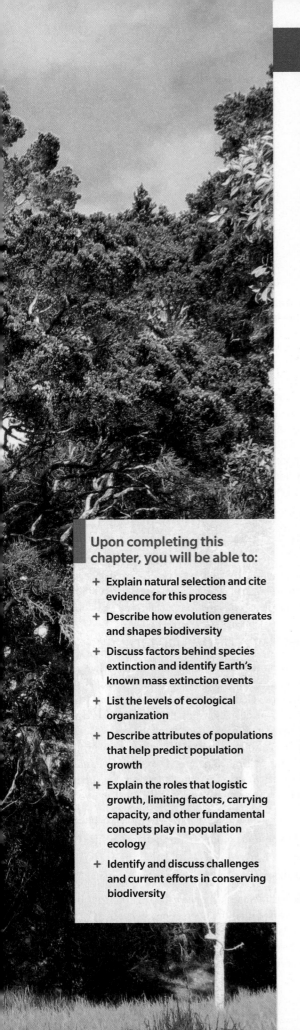

Upon completing this chapter, you will be able to:

+ **Explain natural selection and cite evidence for this process**

+ **Describe how evolution generates and shapes biodiversity**

+ **Discuss factors behind species extinction and identify Earth's known mass extinction events**

+ **List the levels of ecological organization**

+ **Describe attributes of populations that help predict population growth**

+ **Explain the roles that logistic growth, limiting factors, carrying capacity, and other fundamental concepts play in population ecology**

+ **Identify and discuss challenges and current efforts in conserving biodiversity**

Jack Jeffrey stopped in his tracks. "I hear one!" he said. "Over there in those trees!"

Jeffrey led his group of ecotourists through a lush and misty woodland of ferns, shrubs, and vines toward an emphatic chirping sound. They ducked under twisting gnarled limbs covered with moss and lichens, beneath stately ancient 'ōhi'a lehua trees offering bright red flowers loaded with nectar and pollen. At last, in the branches of a koa tree, they spotted the bird—an 'akiapōlā'au, one of fewer than 1500 of its kind left alive in the world.

The 'akiapōlā'au (or "aki" for short) is a sparrow-sized wonder of nature—one of many exquisite birds that evolved on the Hawaiian Islands and exists only here (see inset photo). For millions of years, this chain of islands in the middle of the Pacific Ocean has acted as a cradle of evolution, generating new and unique species. Yet today many of these species are going from the cradle to the grave. More than half of Hawaii's native bird species (95 of 142) have gone extinct in recent times, and most of those that remain—like the aki—teeter on the brink of extinction.

The aki is a type of Hawaiian honeycreeper. The Hawaiian honeycreepers include 18 living species (and at least 38 species recently extinct), all of which originated from a single ancestral species that reached the Hawaiian Islands several million years ago. As new volcanic islands emerged from the ocean and then eroded away, and as forests expanded and contracted, populations were split and, over the millennia, new honeycreeper species evolved.

As honeycreeper species diverged from their common ancestor and from one another, they evolved different colors, sizes, body shapes, feeding behaviors, mating preferences, diets, and bill shapes. Bills in some species became short and straight, allowing birds to glean insects from leaves. In other species, bills became long and curved, enabling birds to probe into flowers to sip nectar. The bills of still other species became thick and strong for cracking seeds. Some birds evolved highly specialized bills: The aki uses the short, straight lower half of its bill to peck into dead branches to find beetle grubs, then uses the long, curved upper half to reach in and extract the insects.

Hawaii's honeycreepers thrived for several million years in the islands' forests, amid unique communities of plants found nowhere else in the world. Sadly, though, these native Hawaiian forests are now under siege. The crisis

◀ **Native Hawaiian forest at Hakalau Forest National Wildlife Refuge**

▲ **The endangered 'akiapōlā'au**

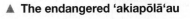

began 750 or more years ago as Polynesian settlers colonized the islands, cutting down trees and introducing non-native animals. Europeans arrived more than 200 years ago and did more of the same. Pigs, goats, and cattle ate their way through the native plants, transforming luxuriant forests into desolate grasslands. Rats, cats, and mongooses destroyed the eggs and young of native birds. Foreign plants from Asia, Europe, and America, whose seeds accompanied the people and animals traveling to the Hawaiian Islands from those regions, spread across the altered landscape.

Foreign diseases also arrived, including strains of pox and malaria that target birds. The native fauna were not adapted to resist pathogens they had never encountered. Avian pox and avian malaria, carried by introduced mosquitoes, killed off native birds everywhere except on high mountain slopes, where it was too cold for mosquitoes. Today, few native forest birds exist anywhere on the Hawaiian Islands below 1500 m (4500 ft) in elevation.

The aki being watched by Jeffrey's group inhabits the Hakalau Forest National Wildlife Refuge (NWR), high atop the slopes of Mauna Kea, a volcano on the Island of Hawai'i, the largest island in the chain (**FIGURE 3.1**). At Hakalau, native birds find a rare remaining patch of disease-free native forest.

Jeffrey was a biologist at Hakalau for 20 years before his retirement and led innovative projects to save native plants and birds from extinction. Staff and volunteers at Hakalau fenced out pigs and other introduced animals, and they planted thousands of native plants in areas deforested by cattle grazing. Young restored native forest is now regrowing on thousands of acres. More and more native birds are using this restored forest every year.

However, global climate change presents new challenges. As temperatures climb, mosquitoes move upslope and malaria

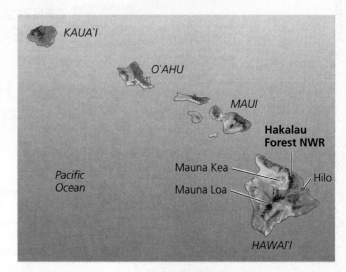

FIGURE 3.1 The Hakalau Forest NWR is located on the slopes of Mauna Kea on the Island of Hawai'i.

and pox spread deeper into the remaining forests so that even protected areas such as Hakalau are not immune. The next generation of managers will need to innovate further to conserve the island's native species.

Island chains like Hawai'i can be viewed as "laboratories of evolution," showing us how populations evolve and how new species arise. But just as islands are crucibles of species formation, they are also hotspots of extinction. As we explore Hawaii's ecological diversity throughout this chapter, we will learn how scientists and conservationists are racing to restore habitats, fight invasive species, and save native plants and animals.

Evolution: The Source of Earth's Biodiversity

The animals and plants native to the Hawaiian Islands help reveal how our world became populated with the remarkable diversity of life we see today—a rich cornucopia of millions of species (**FIGURE 3.2**).

A **species** is a particular type of organism. More precisely, it is a population or group of populations whose members share characteristics and can freely breed with one another and produce fertile offspring. A **population** is a group of individuals of a given species that live in a particular region at a particular time. Over vast spans of time, the process of evolution has shaped populations and species, giving us the vibrant abundance of life that enriches Earth today.

In its broad sense, the term *evolution* means change over time. In its biological sense, **evolution** consists of inherited change in populations of organisms from generation to generation. Changes in genes (p. 29) often lead to modifications in appearance or behavior.

Evolution is one of the best-supported and most illuminating concepts in all of science, and it is the very foundation of modern biology. Perceiving how species adapt to their environments and change over time is crucial for comprehending ecology and for

learning the history of life. Evolutionary processes influence many aspects of environmental science, including agriculture, pesticide resistance, medicine, and environmental health.

Natural selection shapes organisms

Natural selection is a primary mechanism of evolution. In the process of **natural selection,** inherited characteristics that enhance survival and reproduction are passed on more frequently to future generations than characteristics that do not, thereby altering the genetic makeup of populations through time.

Natural selection is a simple concept that offers a powerful explanation for patterns evident in nature. The idea of natural selection follows logically from a few straightforward facts that are readily apparent to anyone who observes the life around us:

- Organisms face a constant struggle to survive and reproduce.
- Organisms tend to produce more offspring than can survive to maturity.
- Individuals of a species vary in their attributes.

Variation is due to differences in genes, the environments in which genes are expressed, and the interactions between genes and environment. As a result of this variation, some individuals of a species will be better suited to their environment than others and therefore will survive longer and be better able to reproduce.

'I'iwi

Nēnē

Haleakala silversword

Happyface spider

FIGURE 3.2 Hawai'i hosts a treasure trove of biodiversity.

Characteristics, or traits, are passed from parent to offspring through the genes, and a parent that produces many offspring will pass on more genes to the next generation than a parent that produces few or no offspring. In the next generation, therefore, the genes of better-adapted individuals will outnumber those of individuals that are less well adapted. From one generation to another through time, traits that lead to better and better reproductive success in a given environment will evolve in the population. This process is termed **adaptation,** and a trait that promotes reproductive success is also called an *adaptation* or an *adaptive trait.*

The concept of natural selection was first proposed in the 1850s by **Charles Darwin** and, independently, by **Alfred Russel Wallace,** two exceptionally keen British naturalists. By this time, scientists and amateur naturalists were widely discussing the idea that populations evolve, yet no one could say how or why. After spending years studying and cataloging an immense variety of natural phenomena—in his English garden and across the world to the Galápagos Islands—Darwin finally concluded that natural selection helped explain the world's great variety of living things. Once he came to this conclusion, however, Darwin put off publishing his findings, fearing that social disruption might ensue if people felt their religious convictions were threatened. Darwin was at last driven to go public when Wallace wrote to him from the Asian tropics, independently describing the idea of natural selection. The two men's shared ideas were presented together at a scientific meeting in 1858, and the next year Darwin published his groundbreaking book *On the Origin of Species.*

With natural selection, humanity at last uncovered a precise and viable mechanism to explain how and why organisms evolve through time. Once geneticists worked out how traits are inherited, this understanding launched evolutionary biology. In the century and a half since Darwin and Wallace, legions of researchers have refined our understanding of evolution, powering dazzling progress in biology that has helped shape our society.

Understanding evolution is vital for modern society

Evolutionary processes play key roles in society and in our everyday lives. As we will see shortly, we depend on a working knowledge of evolution for the food we eat and the clothes we wear, each and every day, as these have been made possible by the selective breeding of crops and livestock. Applying an understanding of evolution to agriculture can also help us prevent antibiotic resistance in feedlots and pesticide resistance in crop-eating insects (p. 252).

Medical advances result from our knowledge of evolution as well. Understanding evolution helps us determine how infectious diseases spread and how they gain or lose potency. It allows scientists to track the constantly evolving strains of influenza (flu), HIV (human immunodeficiency virus, which causes AIDS), and other pathogens. Armed with such information, biomedical experts can predict which flu strains will most likely spread in a given year and then design effective vaccines targeting them. Comprehending evolution also enables us to detect and respond to the evolution of antibiotic resistance by dangerous bacteria.

Additionally, applying our knowledge of evolution informs our technology. From studying how organisms adapt to challenges and evolve new abilities, we develop ideas on how to design technologies and engineering solutions.

FAQ

How strong is the evidence for evolution?

Because Charles Darwin contributed so much to our early understanding of evolution, some people assume the concept itself hinges on his ideas. But scientists and laypeople had been observing nature, puzzling over fossils, and discussing the notion of evolution long before Darwin. Once he and Alfred Russel Wallace independently described the process of natural selection, scientists finally gained a way of explaining how and why organisms change across generations.

Later, geneticists discovered Gregor Mendel's research on inheritance and worked out how traits are passed on—and modern evolutionary biology was born. Twentieth-century scientists such as Ronald Fisher, Sewall Wright, Theodosius Dobzhansky, George Gaylord Simpson, and Ernst Mayr ran experiments, developed sophisticated mathematical models, and documented phenomena with extensive evidence, building evolutionary biology into one of the strongest fields in all of science. Since then, evolutionary research by many thousands of scientists has driven our understanding of biology and has facilitated spectacular advances in agriculture, medicine, and technology.

Selection acts on genetic variation

For an organism to pass a trait along to future generations, genes in its DNA (pp. 28–29) must code for the trait. In an organism's lifetime, its DNA will be copied millions of times by millions of cells. Amid all this copying, sometimes a mistake is made. Accidental changes in DNA, called **mutations,** give rise to genetic variation among individuals. If a mutation occurs in a sperm or egg cell, it may be passed on to the next generation. Most mutations have little effect, but some can be deadly and others can be beneficial. Those that are not lethal provide the genetic variation on which natural selection acts.

Genetic variation also results as organisms mix their genetic material through sexual reproduction. When organisms reproduce sexually, each parent contributes to the genes of the offspring. This process produces novel combinations of genes, generating variation among individuals.

When natural selection acts on genetic variation by favoring certain variants, it can drive a feature in a particular direction (**FIGURE 3.3**). Because such evolutionary change generally requires a great deal of time, a species cannot always adapt to environmental conditions that change quickly. For instance, the current warming of our global climate

(Chapter 18) is occurring too rapidly for most species to adapt, and we may lose many species to extinction as a result.

However, genetic variation can sometimes help protect a population against novel challenges. For example, one of the honeycreeper species of the Hakalau Forest, the 'amakihi, also occurs in certain forests at low elevations where avian malaria has killed off other honeycreepers. Researchers determined that some of the 'amakihis living there when malaria arrived had genes that by chance gave them a natural resistance to the disease. These resistant birds survived malaria's onslaught, and their descendants that carried the malaria-resistant genes reestablished a population that is growing today.

Selective pressures from the environment influence adaptation

Environmental conditions determine what pressures natural selection will exert, and these selective pressures, in turn, affect which members of a population will survive and reproduce. Over many generations, this results in the evolution of traits that enable success within a given environment. Closely related species that live in different environments tend to diverge in their traits as differing selective pressures drive the evolution of different adaptations (**FIGURE 3.4a**). Conversely, sometimes very unrelated species living in similar environments in separate locations may independently acquire similar traits as they adapt to similar selective pressures; this is called **convergent evolution** (**FIGURE 3.4b**).

Of course, environments change over time, and organisms may move to new locations and encounter new conditions. In either case, a trait that promotes success at one time or place may not do so at another. Hawaiian honeycreepers such as the 'apapane and the 'i'iwi fly long distances in search of flowering trees. This behavior had long helped them to find the best resources across a diverse landscape. However, once malaria arrived, the behavior become counterproductive, as birds from malaria-free mountain forests flew downslope into death zones and were bitten by mosquitoes. As environmental conditions vary in time and space, adaptation becomes a moving target.

Evidence of selection is all around us

The results of natural selection are all around us, visible in every adaptation of every organism. Researchers have documented selection through innumerable observations and experiments at field sites from the Galápagos Islands to the rainforests to the oceans to our cities. Scientists also have demonstrated the rapid evolution of traits by selection in countless lab experiments with fast-reproducing organisms such as bacteria, yeast, and fruit flies.

For many of us, the most familiar and obvious evidence for selection is our breeding of domesticated animals. With dogs, cats, and livestock, we have conducted selection under our own direction, that is, **artificial selection.** We have chosen animals possessing traits we like and bred them together, while culling out individuals with traits we do not like. Through such *selective breeding,* we have been able to augment particular traits we prefer.

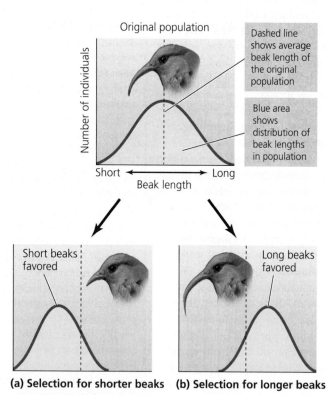

FIGURE 3.3 Natural selection can drive a feature in various directions. Let's consider the 'i'iwi, a Hawaiian honeycreeper, whose population possesses genetic variation for the length of its curved bill. In an environment where flowers grow short nectar tubes **(a),** birds with short bills could feed perfectly well and avoid investing extra energy in growing a long bill, so natural selection would favor a decrease in bill length across the population. In an environment where flowers have long tubes **(b),** birds with long bills could feed more effectively and pass on more genes, causing the population to shift toward longer average bill length.

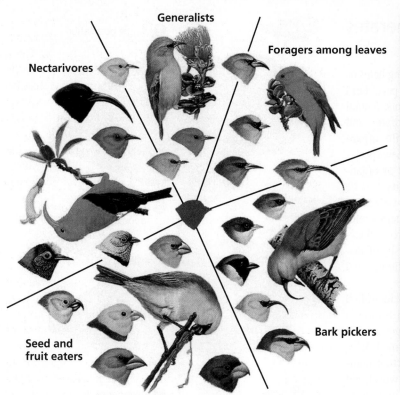

Cactus in Arizona

Euphorb (spurge) in the Canary Islands

(a) Divergent evolution of Hawaiian honeycreepers

(b) Convergent evolution of cactus and spurge

FIGURE 3.4 Natural selection can cause closely related species to diverge or distantly related species to converge. Hawaiian honeycreepers **(a)** diversified as they adapted to different food resources and habitats, as indicated by their diversity of plumage colors and bill shapes. In contrast, cacti of the Americas and euphorbs of Africa **(b)** became similar to one another as they independently adapted to arid environments. These plants each evolved succulent tissues to hold water, thorns to keep thirsty animals away, and photosynthetic stems without leaves to reduce surface area and water loss.
Painting (a) © H. D. Pratt, by permission.

Consider the great diversity of dog breeds (**FIGURE 3.5a**). People generated every type of dog alive by starting with a single ancestral species and selecting for particular desired traits as individuals were bred together. From Great Dane to Chihuahua, all dogs are able to interbreed and produce viable offspring, yet breeders maintain differences among them by allowing only like individuals to breed.

Artificial selection through selective breeding has also given us the many crop plants and livestock we depend on for food and fiber, all of which people domesticated from wild species and carefully bred over years, centuries, or millennia (**FIGURE 3.5b**). Through selective breeding, we have created corn with bigger, sweeter kernels; wheat and rice with larger and more numerous grains; and apples, pears, and oranges with better taste. We have diversified single types into many—for instance, breeding variants of wild cabbage (*Brassica oleracea*) to create broccoli, cauliflower, cabbage, and brussels sprouts. Our entire agricultural system is based on artificial selection.

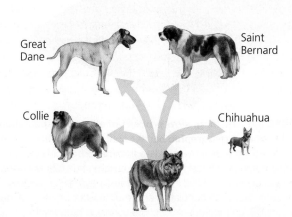

Great Dane

Saint Bernard

Collie

Chihuahua

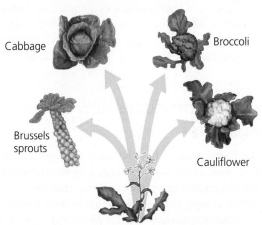

Cabbage

Broccoli

Brussels sprouts

Cauliflower

(a) Ancestral wolf (*Canis lupus*) and derived dog breeds

(b) Ancestral *Brassica oleracea* and derived crops

FIGURE 3.5 Artificial selection through selective breeding has given us many dog breeds and crop varieties.

FAQ

Are humans evolving?

In our modern civilization, it's become easy to feel removed from nature and to believe that human beings are no longer subject to natural selection, only to cultural forces. And our day-to-day environment has indeed changed from the wholly "natural" one of our hunter-gatherer history to a largely constructed one in today's urban-techno-industrial society. But when environments change, new selective pressures come into play. For instance, in the not-too-distant past, many people died prematurely from diseases, injuries, and infection. In today's world, medical advances save and extend our lives. Ironically, as a result, illnesses and disabilities that can be inherited get passed on to more children than in the past, and the frequencies with which these maladies occur increase. At the same time, humanity's old foes continue to have an impact: In developing countries, major killers of young people, like malaria, measles, and tuberculosis, still exert strong selection on the human genome. Natural selection never stops, and all organisms—people included—are always evolving.

Evolution generates biodiversity

Just as selective breeding helps us create new types of pets, farm animals, and crop plants, natural selection can elaborate and diversify traits in wild organisms, helping to form new species and whole new types of organisms. Life's complexity can be expressed as **biological diversity, or biodiversity.** These terms refer to the variety of life across all levels, including the diversity of species, genes, populations, and communities (p. 275).

Scientists have identified and described about 1.8 million species, but many more remain undiscovered or unnamed. Estimates vary for the actual number of species in the world, but they range from 3 million up to 100 million. Hawaii's insect fauna provides one example of how much we have yet to learn. Scientists studying fruit flies in the Hawaiian Islands have described more than 500 species of them, but they have also identified about 500 others that have not yet been formally named and described. Still more fruit fly species probably exist but have not yet been found.

Subtropical islands such as Hawai'i are by no means the only places rich in biodiversity. Step outside just about anywhere and you will find many species within close reach. Plants poke up from cracks in asphalt in every city in the world. A handful of backyard soil may contain an entire miniature world of life, including insects, mites, millipedes, nematode worms, plant seeds, fungi, and millions of bacteria. (We will examine Earth's biodiversity in detail in Chapter 11.)

Speciation produces new types of organisms

How did Earth come to have so many species? The process by which new species are generated is termed **speciation.** Speciation can occur in a number of ways, but the main mode is generally thought to be *allopatric speciation,* whereby species form after populations become physically separated over some geographic distance (**FIGURE 3.6**). Imagine a population of organisms. Individuals within the population possess many similarities that unify them as a species because they are able

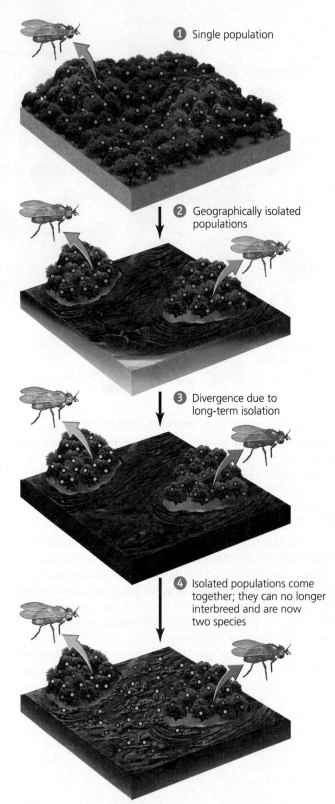

① Single population

② Geographically isolated populations

③ Divergence due to long-term isolation

④ Isolated populations come together; they can no longer interbreed and are now two species

FIGURE 3.6 The long, slow process of allopatric speciation begins when a geographic barrier splits a population—as when forest ① is destroyed by lava flowing from a volcano, but isolated patches of forest ② are left. Hawaiian fruit flies are weak fliers and become isolated in such forested patches, or kipukas. Over centuries, each population accumulates its own set of genetic changes ③, until individuals become unable to breed with individuals from the other population. The two populations now represent separate species and will remain so even if the geographic barrier disappears ④ and the new species intermix.

to breed with one another, sharing genetic information. However, if the population is split into two or more isolated areas, individuals from one area cannot reproduce with individuals from the other areas.

When a mutation arises in the DNA of an organism in one of these newly isolated populations, it cannot spread to the other populations. Over time, each population will independently accumulate its own set of mutations. The populations thereby diverge and may eventually become so different that their members can no longer mate with one another and produce viable offspring. (This can occur because of changes in reproductive organs, hormones, courtship behavior, breeding timing, or other factors.) Populations that no longer exchange genetic information will embark on their own independent evolutionary paths as separate species. The populations will continue diverging in their characteristics as chance mutations accumulate that cause them to differ more and more. And if environmental conditions happen to differ for the two populations, then natural selection may accelerate the divergence.

For speciation to occur, populations must remain isolated for a very long time, generally thousands of generations. Populations can undergo long-term geographic isolation in various ways. Lava flows can destroy forest, leaving small isolated patches intact (as shown in Figure 3.6). Ice sheets may expand across continents during periods of glaciation and split populations in two. Major rivers may change course or mountain ranges may be uplifted, dividing regions and their organisms. Sea level may rise, flooding low-lying regions and isolating areas of higher ground as islands. Drying climate may partially evaporate lakes, subdividing them into smaller bodies of water. Warming or cooling climate may cause plant communities to shift, creating new patterns of plant and animal distribution.

Alternatively, sometimes organisms colonize newly created areas, establishing isolated populations. Hawai'i provides an example. As shown in Figure 2.21 (Chapter 2, p. 41), the Pacific tectonic plate moves over a volcanic "hotspot" that extrudes magma into the ocean, building volcanoes that form islands once they break the water's surface. The plate inches northwest, dragging each island with it, while new islands are formed at the hotspot. The result, over millions of years, is a long string of islands, called an *archipelago*. As each new island is formed, plants and animals that colonize it may undergo allopatric speciation if they are isolated enough from their source population (see **THE SCIENCE BEHIND THE STORY**, pp. 56–57).

We can infer the history of life's diversification

Innumerable speciation events have generated complex patterns of diversity beyond the species level. Scientists represent this history of divergence by using branching diagrams called **phylogenetic trees.** Similar to family genealogies, these diagrams illustrate hypotheses proposing how divergence took place (**FIGURE 3.7**). Phylogenetic trees can show relationships among species, groups of species, populations, or genes. Scientists construct these trees by analyzing patterns of similarity among the

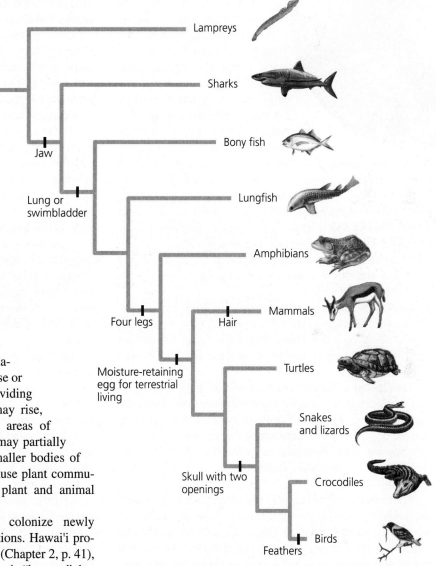

FIGURE 3.7 Phylogenetic trees show the history of life's divergence. The tree here illustrates relationships among groups of vertebrates—just one small portion of the huge and complex "tree of life." Each branch results from a speciation event; as you follow the tree left to right from its trunk to the tips of its branches, you proceed forward in time, tracing life's history. Major traits are "mapped" onto the tree to indicate when they originated.

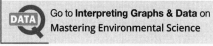 Go to **Interpreting Graphs & Data** on **Mastering Environmental Science**

THE SCIENCE
behind
the story

How Do Species Form in Hawaii's "Natural Laboratory" of Evolution?

**Dr. Heather Lerner,
Earlham College**

For scientists who study how species form, no place on Earth is more informative than an isolated chain of islands. The Hawaiian Islands—the most remote on the planet—are often called a "natural laboratory of evolution."

The key to this laboratory lies in the process that drives Hawaii's geologic history. Turn back to Figure 2.21 (p. 41) and examine it closely. Deep beneath the Pacific Ocean, a volcanic "hotspot" spurts magma as the Pacific Plate slides across it in tectonic motion like a conveyor belt. Mountains of lava accumulate underwater until eventually a volcano rises above the waves, building an island. As the tectonic plate moves northwest, it carries each newly formed island with it, creating a long chain, or *archipelago*. Over several million years, each island gradually subsides, erodes, and disappears beneath the waves. As old islands disappear on the northwest end of the chain, new islands are formed on the southeast end.

Geologists analyzing radioisotopes (p. 26) in the islands' rocks have determined that this process has been ongoing for at least 85 million years. They estimate that Kaua'i was formed about 5.1 million years ago (mya), and the Island of Hawai'i just 0.43 mya.

Despite the remoteness of the Hawaiian archipelago, over time a few plants and animals found their way there, establishing populations. As some individuals hopped to neighboring islands, populations that were adequately isolated evolved into new species. Such speciation by "island-hopping" has driven the radiation of Hawaiian honeycreepers and many other organisms.

For instance, the windswept high volcanic slopes of Hawai'i are graced by some of the most striking flowering plants in the world, the silverswords (see Figure 3.2). These spectacular plants have spiky, silvery leaves and tall stalks that explode into bloom with flowers once in the plant's long life before it dies. Researchers have discovered that Hawaii's 28 species of silverswords all evolved from a modest tarweed plant from California that reached Hawai'i and diversified by island-hopping. University of California, Berkeley, botanist Bruce Baldwin and other researchers analyzed genetic relationships and learned that the silverswords' radiation was rapid (on a geologic timescale), taking place in just 5 million years.

The best-understood radiation has occurred with the Hawaiian fruit flies. Some of these insects speciate within islands in forested patches, called kipukas (see Figure 3.6), but most have done so by island-hopping. By combining genetic analysis and geologic dating, researchers determined that the process began 25 mya on islands that today are beneath the ocean. From a single original fruit fly species, an estimated 1000 species have evolved—fully one-sixth of all the world's fruit fly species.

Other groups have undergone adaptive radiations on the Hawaiian Islands as well, including damselflies, crickets, mirid bugs, spiders, and multiple families of plants. Scientists propose that once a species colonizes an island, it may spread and evolve rapidly because competitors are few and there tend to be unoccupied niches (pp. 62, 77).

The Hawaiian honeycreepers are so diverse that researchers have long puzzled over what type of bird gave rise to their radiation—and whether there was just one colonizing ancestor or many. In 2011, to clarify how the honeycreeper radiation took place, one research team combined genetic sequencing technology with resources from museum collections and our knowledge of Hawaiian geology.

Heather Lerner and five colleagues first took tissue samples from bird specimens in museum collections. Working with Robert Fleischer and Helen James at the Smithsonian Institution, Lerner, now at Earlham College in Indiana, sampled 19 species of honeycreepers plus 28 diverse types of finches from around the Pacific Rim that experts had identified as possible ancestors.

Lerner's team sequenced DNA (pp. 28–29) from each tissue sample, obtaining data from 13 genes and from mitochondrial genomes. (Mitochondria are cell organelles with genetic material inherited from the female parent.) They ran the data through computer programs to analyze how the DNA sequences—and thus the birds—were related to one another, then produced phylogenetic trees (p. 55) showing the relationships (**FIGURE 1**). They published their results in the journal *Current Biology.*

Lerner's team found that the Hawaiian honeycreepers apparently derive from one ancestor and are most related to the Eurasian rosefinches, indicating that honeycreepers evolved after some rosefinch-like bird arrived from Asia. Today's rosefinches are partly nomadic; when food supplies crash, flocks fly long distances to find food. Perhaps a wandering flock of ancestral rosefinches was caught up in a storm long ago and blown to Hawai'i.

Once this common ancestor of today's rosefinches and honeycreepers arrived on an ancient Hawaiian island, its progeny adapted to conditions there by natural selection, resulting in modified bill shape, diet, and coloration. Every once in a great

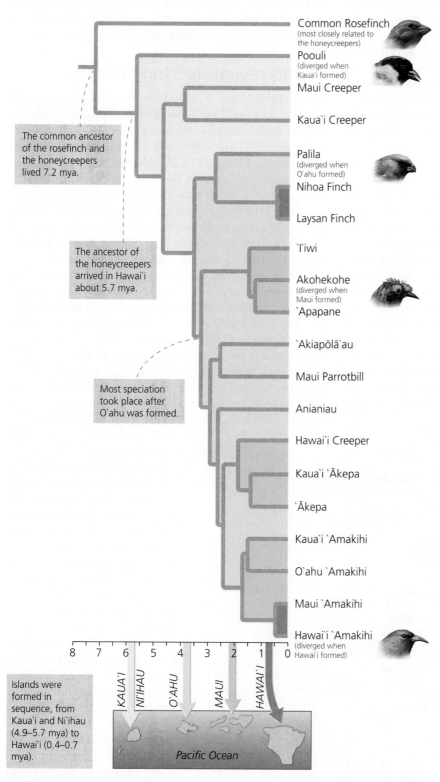

The common ancestor of the rosefinch and the honeycreepers lived 7.2 mya.

The ancestor of the honeycreepers arrived in Hawai'i about 5.7 mya.

Most speciation took place after O'ahu was formed.

Common Rosefinch (most closely related to the honeycreepers)

Poouli (diverged when Kaua'i formed)

Maui Creeper

Kaua'i Creeper

Palila (diverged when O'ahu formed)

Nihoa Finch

Laysan Finch

'I'iwi

Akohekohe (diverged when Maui formed)

'Apapane

'Akiapōlā'au

Maui Parrotbill

Anianiau

Hawai'i Creeper

Kaua'i 'Ākepa

'Ākepa

Kaua'i 'Amakihi

O'ahu 'Amakihi

Maui 'Amakihi

Hawai'i 'Amakihi (diverged when Hawai'i formed)

8 7 6 5 4 3 2 1 0

KAUA'I NI'IHAU O'AHU MAUI HAWAI'I

Islands were formed in sequence, from Kaua'i and Ni'ihau (4.9–5.7 mya) to Hawai'i (0.4–0.7 mya).

Pacific Ocean

FIGURE 1 Using gene sequences, researchers generated this phylogenetic tree showing relationships among the Hawaiian honeycreepers. They then matched the history of the birds' diversification with the known geologic history of the islands' formation. *Adapted from Lerner, H. R. L., et al. 2011. Multilocus resolution of phylogeny and timescale in the extant adaptive radiation of Hawaiian honeycreepers. Curr. Biol. 21: 1838–1844.*

while, wandering birds colonized other islands, founding populations that each adapted to local conditions and might eventually evolve into separate species.

Because the age of each island is known, Lerner's team could calibrate rates of evolutionary change in the birds' DNA sequences, and thus measure the age of each divergence. That is, they could tell how "old" each bird species is. They found that the rosefinch-like ancestor arrived by 5.7 mya, about the time that the oldest of today's main islands (Kaua'i and Ni'ihau) were forming. After O'ahu emerged 4.0–3.7 mya, the speciation process went into overdrive, giving rise to many new species with distinctively different colors, bill shapes, and habits. By the time Maui arose 2.4–1.9 mya, most of the major differences in body form and appearance had evolved (see bottom portion of Figure 1).

Thus, most major innovations arose midway through the island-formation process, when O'ahu and Kaua'i were the main islands in the chain. After this burst of innovation, major changes were fewer, perhaps because most evolutionary possibilities had been explored, or perhaps because the newer islands of Maui and Hawai'i were too close together to isolate populations adequately.

The team's data show that the age of each honeycreeper species does not neatly match the age of the island or islands it inhabits today. Instead, the island-hopping process was complex, with some birds hopping "backward" from newer islands to older ones. Moreover, within each island there is a great deal of variation in climate, topography, and vegetation, because windward slopes catch moisture from trade winds over the ocean and become lush and green, whereas leeward slopes in the rainshadow (p. 96) are arid. The varied habitats and rugged topography create barriers that can lead to speciation within islands.

All these complexities of history, geology, and biology make the Hawaiian Islands especially fascinating. And for all that scientists have learned so far, this "natural laboratory" still has much to teach us about how new species are formed.

genes or external traits of present-day organisms and by inferring which groups share similarities because they are related.

Once we have a phylogenetic tree, we can map traits onto the tree according to which organisms possess them, and we can thereby trace how the traits have evolved. For instance, note in the phylogeny of major vertebrate groups in Figure 3.7 how feathers are shown to have originated in the lineage leading to birds. Because only birds have feathers, it is simplest to conclude that feathers first evolved in this relatively recent lineage. In contrast, almost all vertebrates have jaws, leading us to infer that jaws must have evolved very early in the history of vertebrates, before most divergences took place.

Knowing how organisms are related to one another also helps scientists to classify them and name them. *Taxonomists* use an organism's genetic makeup and physical appearance to determine its species identity. These scientists then group species by their similarity into a hierarchy of categories meant to reflect evolutionary relationships. Related species are grouped together into *genera* (singular, *genus*), related genera are grouped into families, and so on (**FIGURE 3.8**). Each species is given a two-part Latin or Latinized scientific name denoting its genus and species. For instance, the 'akiapōlā'au, *Hemignathus wilsoni,* is similar to other Hawaiian honeycreepers in the genus *Hemignathus*. These species are closely related in evolutionary terms, as indicated by the genus name they share. They are more distantly related to honeycreepers in other genera, but all honeycreepers are classified together in the family Fringillidae. This system of naming and classification was devised by Swedish botanist Carl Linnaeus (1707–1778) long before Darwin's work on evolution. Today, biologists use evolutionary information from phylogenetic trees to help classify organisms under the Linnaean system's rules.

Fossils reveal life's long history

To help decipher the history of life, scientists also study fossils. As organisms die, some become buried by sediment. Under certain conditions, the hard parts of their bodies—such as bones, shells, and teeth—may be preserved, as sediments are compressed into rock (p. 38). Minerals replace the organic material, leaving behind a **fossil,** an imprint in stone of the dead organism (**FIGURE 3.9a**). Over millions of years, geologic processes have buried sediments and later brought sedimentary rock layers to the surface, revealing assemblages of fossilized plants and animals from different time periods. By dating the rock layers that contain fossils, paleontologists (scientists who study the history of Earth's life) can learn when particular organisms lived (**FIGURE 3.9b**). The cumulative body of fossils worldwide is known as the **fossil record.**

The fossil record shows that the number of species existing at any one time has generally increased, but that the species on Earth today are just a small fraction of all species that have ever existed. During life's 3.5 billion years on Earth, complex structures have evolved from simple ones, and large sizes from small ones. At the same time, simplicity and small size have evolved when favored by natural selection; indeed, it is easy to argue that Earth still belongs to the bacteria and other microbes, some of them little changed over eons.

And yet, even enthusiasts of microbes must marvel at the exquisite adaptations of animals, plants, and fungi: the heart that beats so reliably for an animal's entire lifetime; the complex organ system to which the heart belongs; the stunning plumage of a peacock in display; the ability of plants to lift water and nutrients from the soil, gather light from the sun, and produce food; and the human brain and its ability to reason. All these adaptations and more have resulted as evolution has generated new species and whole new branches on the tree of life.

Although speciation generates Earth's biodiversity, it is only part of the equation. The fossil record teaches us that the vast majority of species that once lived are now gone. The disappearance of a species from Earth is called **extinction.** From the fossil record, paleontologists calculate that the average time a species spends on Earth is 1–10 million years.

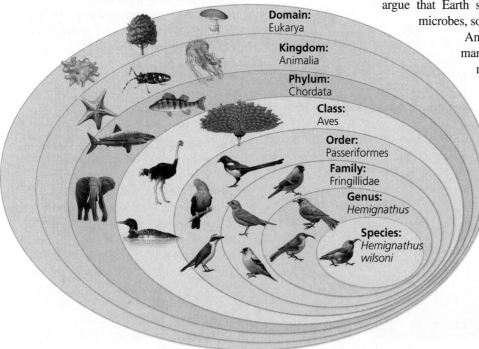

FIGURE 3.8 Taxonomists classify organisms using a nested hierarchical system meant to reflect evolutionary relationships. Species similar in appearance, behavior, and genetics (because they share recent common ancestry) are placed in the same genus. Similar genera are placed in the same family, families are placed within orders, and so on. For example, honeycreepers belong to the class Aves, along with peacocks, loons, and ostriches. However, these birds differ greatly enough (after diverging across millions of years of evolution) that they are placed in different orders, families, and genera.

(a) A fossil of an extinct trilobite

(b) A paleontologist cleaning the teeth of a fossilized mastodon skull

FIGURE 3.9 The fossil record helps reveal the history of life on Earth. Trilobites **(a)** were once abundant, but today we know these extinct animals only from their fossils. Finding, excavating, and preparing fossils is hard work; here, **(b)** a paleontologist cleans the teeth of a mastodon skull.

The number of species in existence at any one time is equal to the number added through speciation minus the number removed by extinction.

Extinction occurs naturally, but human impact can profoundly affect the rate at which it occurs. Today, our planet's biological diversity is being lost at a frightening pace (Chapter 11). This loss affects people directly, because other organisms provide us with life's necessities—food, fiber, medicine, and vital ecosystem services (pp. 4, 120–121, 149). Species extinction brought about by human impact may well be the single biggest problem we face, because the loss of a species is irreversible.

Some species are especially vulnerable to extinction

In general, extinction occurs when environmental conditions change rapidly or drastically enough that a species cannot adapt genetically to the change; the slow process of natural selection simply does not have enough time to work. Small populations are vulnerable to extinction because fluctuations in their size

could, by chance, bring the population size to zero. Small populations also sometimes lack enough genetic variation to buffer them against environmental change. Species narrowly specialized to some particular resource or way of life are also vulnerable, because environmental changes that make that resource or role unavailable can spell doom. Species that are **endemic** to a particular region occur nowhere else on Earth. Endemic species face elevated risks of extinction because when some event influences their region, it may affect all members of the species.

Island-dwelling species are particularly vulnerable to extinction. Because islands are smaller than mainland areas and are isolated by water, fewer species reach and inhabit islands. For this reason, some of the pressures and challenges faced daily by organisms in complex natural systems on a mainland don't exist in the simpler natural systems on islands. For instance, only one land mammal—a bat—ever reached Hawai'i naturally. As a result, Hawaii's birds evolved for millions of years without needing to defend against the threat of predation by mammals. Likewise, Hawaii's plants did not need to invest in defenses (such as thick bark, spines, or chemical toxins) against mammals that might eat them. Because defenses are costly to invest in, through adaptation most island birds and plants lost the defenses their mainland ancestors may have had.

As a result, when people colonized Hawai'i, its native organisms (**FIGURE 3.10a**) were completely unprepared for what people brought with them to the islands (**FIGURE 3.10b**).

(a) Hawaiian Petrel, a native species at risk

(b) Mongoose, an introduced species that preys on natives

FIGURE 3.10 Island-dwelling species that have lost defenses are vulnerable to extinction when enemies are introduced. The Hawaiian petrel **(a),** a seabird that nests in the ground, is endangered as a result of predation by mammals introduced to Hawai'i, such as **(b)** the Indian mongoose.

FIGURE 3.11 Small range sizes can leave species vulnerable to extinction if severe changes occur in their local environment. The Shenandoah salamander (*Plethodon shenandoah*) lives on just three peaks in Shenandoah National Park in Virginia.

Introduced mongooses, cats, and rats preyed voraciously on defenseless ground-nesting seabirds, ducks, geese, and flightless rails. Imported livestock—such as cattle, goats, and pigs—devoured the vegetation, altering entire communities as lush forests gave way to barren hillsides dominated by foreign grasses. Half of Hawaii's native birds and a number of its native plants were driven extinct soon after human arrival.

On a mainland, "islands" of habitat (such as forested mountaintops) can host endemic species that are vulnerable to extinction. In the United States, 40 salamander species are naturally restricted to areas the size of a typical county, and some live only atop single mountains (**FIGURE 3.11**). Many other amphibians are limited to very small ranges, for a variety of reasons both natural and human-caused. The Yosemite toad is restricted to a small region of the Sierra Nevada in California, the Houston toad occupies just a few patches of Texas woodland, and the Florida bog frog lives in a tiny area of Florida wetland.

Earth has seen several episodes of mass extinction

Most extinction occurs gradually, one species at a time, at a rate referred to as the **background extinction rate.** However, the fossil record reveals that Earth has seen at least five events of staggering proportions that killed off massive numbers of species at once. These episodes, called **mass extinction events,** have occurred at widely spaced intervals in our planet's history and have wiped out 50–95% of Earth's species each time (see Figure 11.11, p. 285).

The best-known mass extinction occurred 66 million years ago and brought an abrupt end to dinosaurs and many other types of animals (although birds are descendants of a type of dinosaur that survived). Evidence suggests that the collision of a gigantic asteroid with Earth caused this event.

Still more catastrophic was the mass extinction at the end of the Permian period 250 million years ago (see **APPENDIX D** for Earth's geologic periods). Paleontologists estimate that 75–95% of all species perished during this event. Hypotheses as to what caused the end-Permian extinction include massive volcanism, an asteroid impact, methane releases and global warming, or some combination of factors.

The sixth mass extinction is upon us

Many biologists have concluded that Earth is currently entering its sixth mass extinction event—and that we are the cause. Experts estimate that today's extinction rate is tens to hundreds of times higher than the background rate, and rising (p. 288). Changes to our planet's natural systems set in motion by human population growth, development, and resource depletion have driven many species extinct and are threatening countless more. As we alter and destroy natural habitats; overharvest populations; pollute air, water, and soil; introduce invasive nonnative species; and disrupt climate, we set in motion processes that jeopardize Earth's biodiversity (pp. 288–295). Because we depend directly and completely on the resources and services that nature has to offer, biodiversity loss and extinction ultimately threaten our own survival.

Ecology and the Organism

Extinction, speciation, and other evolutionary processes play key roles in ecology. **Ecology** is the scientific study of interactions among organisms and of relationships between organisms and their environments. Ecology allows us to explain and predict the distribution and abundance of organisms in nature. It is often said that ecology provides the stage on which the play of evolution unfolds. The two are intertwined in many ways.

We study ecology at several levels

Life exists in a hierarchy of levels, from atoms, molecules, and cells (pp. 25–29) up through the **biosphere,** the cumulative total of living things on Earth and the areas they inhabit. **Ecologists** are scientists who study relationships at the higher levels of this hierarchy (**FIGURE 3.12**), namely at the levels of the organism, population, community, ecosystem, landscape, and biosphere.

At the level of the organism, ecology describes relationships between an organism and its physical environment. Organismal ecology helps us understand, for example, what aspects of a Hawaiian honeycreeper's environment are important to it, and why. In contrast, **population ecology** examines the dynamics of population change and the factors that affect the distribution and abundance of members of a population. It helps us understand why populations of some species decline, while populations of others increase.

In ecology, a **community** consists of an assemblage of populations of interacting species that inhabit the same area. A population of 'akiapōlā'au, a population of koa trees, a population of wood-boring grubs, and a population of ferns, together with all the other interacting plant, animal, fungal, and microbial populations in the Hakalau Forest, would be

Biosphere

The sum total of living things on Earth and the areas they inhabit

Landscape

A geographic region including an array of ecosystems

Ecosystem

A functional system consisting of a community, its nonliving environment, and the interactions between them

Community

A set of populations of different species living together in a particular area

Population

A group of individuals of a species that live in a particular area

Organism

An individual living thing

FIGURE 3.12 Green sea turtles are part of a coral reef community that inhabits reef ecosystems along Hawaii's coasts.

considered a community. **Community ecology** (Chapter 4) focuses on patterns of species diversity and on the dynamics of interactions among species.

Ecosystems (p. 111) encompass communities and the abiotic (nonliving) material and processes with which community members interact. Hakalau's cloud-forest ecosystem consists of its community plus the air, water, soil, nutrients, and energy used by the community's organisms. **Ecosystem ecology** (Chapter 5) addresses the flow of energy and nutrients by studying living and nonliving components of systems in conjunction.

Landscape ecology (p. 118) helps us understand how and why ecosystems, communities, and populations are distributed across geographic regions. And as technologies such as satellite imagery help scientists study the complex dynamics of natural systems at a global scale, ecologists are expanding their horizons to the biosphere as a whole.

Each organism has habitat needs

At the level of the organism, each individual relates to its environment in ways that tend to maximize its survival and reproduction. One key relationship involves the specific environment in which an organism lives, its **habitat.** An organism's habitat consists of the living and nonliving elements around it, including rock, soil, leaf litter, humidity, plant life, and more. The 'akiapōlā'au (**FIGURE 3.13**) lives in a habitat of cool, moist, montane forest of native koa and 'ōhi'a trees, where it is high enough in elevation to be safe from avian pox and malaria.

Each organism thrives in certain habitats and not in others, leading to nonrandom patterns of **habitat use.** Mobile organisms actively select habitats in which to live from among the range of options they encounter, a process called **habitat selection.** In the case of plants and of stationary animals (such as sea anemones in the ocean), whose young disperse and settle passively, patterns of habitat use result from success in some habitats and failure in others.

Habitats are scale-dependent. A tiny soil mite may use less than a square meter of soil in its lifetime, whereas an eagle, elephant, or whale may traverse many miles of air, land, or water in just a day. Likewise, the criteria by which organisms assess habitats can vary greatly. A soil mite may focus on the

FIGURE 3.13 The 'akiapōlā'au fills a unique niche by virtue of its odd bill. The bill's straight bottom half and long, curved top half allow the bird to specialize in digging grubs out from native trees in its montane forest habitat.

chemistry, moisture, and texture of the soil. For a whale, water temperature, salinity, and the density of marine microorganisms might be critical characteristics. Organisms also may have different habitat needs in different seasons; many migratory birds use distinct breeding, wintering, and migratory habitats.

Habitat is a vital concept in environmental science. Because habitats provide everything an organism needs, including nutrition, shelter, breeding sites, and mates, the organism's survival depends on the availability of suitable habitats. Often an organism's needs end up in conflict with the desires of people when we choose to alter a habitat for our own purposes.

Organisms play roles in communities

Another way in which an organism relates to its environment is through its **niche,** its functional role in a community. An organism's niche reflects its use of habitat and resources, its consumption of certain foods, its role in the flow of energy and matter, and its interactions with other organisms. The niche is a multidimensional concept, a kind of summary of everything an organism does. The pioneering ecologist Eugene Odum once wrote that "habitat is the organism's address, and the niche is its profession."

Organisms vary in the breadth of their niches. A species with narrow breadth, and thus very specific requirements, is said to be a **specialist.** One with broad tolerances, a "jack-of-all-trades" able to use a wide array of resources, is a **generalist.** A native Hawaiian honeycreeper like the 'akiapōlā'au (see Figure 3.13) is a specialist, because its unique bill is exquisitely adapted for feeding on grubs that tunnel through the wood of certain native trees. In contrast, the common myna (a bird introduced to Hawai'i from Asia) is a generalist; its unremarkable bill allows it to eat many types of foods in many habitats. As a result, the common myna has spread throughout the Hawaiian Islands wherever people have altered the landscape.

Generalists like the myna succeed by being able to live in many different places and withstand variable conditions, yet they rarely thrive in any single situation as well as a specialist adapted for those specific conditions. (A jack-of-all-trades, as the saying goes, is a master of none.) Specialists succeed over evolutionary time by being extremely good at the things they do, yet they are vulnerable when conditions change and threaten the habitat or resource on which they have specialized. An organism's habitat preferences, niche, and degree of specialization each reflect adaptations of the species and are products of natural selection.

Population Ecology

A population, as we have seen, consists of individuals of a species that inhabit a particular area at a particular time. Population ecologists try to understand and predict how populations change over time. The ability to predict a population's growth or decline is useful in monitoring and managing wildlife, fisheries, and threatened and endangered species (see **THE SCIENCE BEHIND THE STORY**, pp. 64–65). It is also crucial for understanding the dynamics of our human population (Chapter 8)—a central element of environmental science and one of the prime challenges for our society today.

Populations show features that help predict their dynamics

All populations—from humans to honeycreepers—exhibit attributes that ecologists use to predict population dynamics. Among these attributes are size, density, distribution, sex ratio, and age structure.

Population size Expressed as the number of individual organisms present in a discrete region or area at a given time, **population size** may increase, decrease, undergo cyclical change, or remain stable over time. Populations generally grow when resources are abundant and natural enemies are few. Populations can decline in response to loss of resources, natural disasters, or impacts from other species.

The passenger pigeon, now extinct, illustrates extremes in population size (**FIGURE 3.14**). Not long ago it was the most abundant bird in North America; flocks of passenger pigeons

(a) Passenger pigeon

(b) 19th-century lithograph of pigeon hunting in Iowa

FIGURE 3.14 In the past, flocks of passenger pigeons literally darkened the skies as billions of birds flew overhead. Hunting and deforestation drove North America's most numerous bird to extinction within decades.

literally darkened the skies. In the early 1800s, ornithologist Alexander Wilson watched a flock of 2 billion birds 390 km (240 mi) long that took 5 hours to fly over and sounded like a tornado. Passenger pigeons nested in gigantic colonies in the forests of the upper Midwest and southern Canada. Once settlers began cutting the forests, the birds became easy targets for market hunters, who gunned down thousands at a time. The birds were shipped to market by the wagonload and sold for food. By 1890, the population had declined to such a low number that the birds could not form the large colonies they evidently needed to breed. In 1914, the last passenger pigeon on Earth died in the Cincinnati Zoo, bringing the continent's most numerous bird species to extinction within just a few decades.

Hawai'i offers a story with a happier ending. Hawaii's state bird is the nēnē (pronounced "nay-nay"), also called the Hawaiian goose (see Figure 3.2). Before people reached the Hawaiian Islands, nēnēs were common throughout the island chain, and the nēnē population is thought to have numbered at least 25,000 birds. After human arrival, the nēnē was nearly driven to extinction by human hunting; livestock and plants that people introduced (which destroyed and displaced the vegetation it fed on); and rats, cats, pigs, and mongooses that preyed on its eggs and young. By the 1950s, these impacts had eliminated nēnēs from all islands except the Island of Hawai'i, where the population size was down to just 30 individuals. Fortunately, dedicated conservation efforts have turned this decline around. Biologists and wildlife managers have labored to breed nēnēs in captivity and have reintroduced them into protected areas. These efforts are succeeding, and today nēnēs live in at least seven regions on four of the Hawaiian Islands, with a population size of more than 2000 birds.

Population density
The flocks and breeding colonies of passenger pigeons showed high population density, another attribute that ecologists assess. **Population density** describes the number of individuals per unit area in a population. High population density makes it easier for organisms to group together and find mates, but it can also lead to competition and conflict if space, food, or mates are in limited supply. Overcrowding can also increase the transmission of infectious disease. In contrast, at low population densities, individuals benefit from more space and resources but may find it harder to locate mates and companions.

Population distribution
Population distribution describes the spatial arrangement of organisms in an area. Ecologists define three distribution types: random, uniform, and clumped (**FIGURE 3.15**). In a *random distribution,* individuals are located haphazardly in no particular pattern. This type of distribution can occur when the resources an organism needs are plentiful throughout an area and other organisms do not strongly influence where members of a population settle.

A *uniform distribution,* in which individuals are evenly spaced, can occur when individuals compete for space. Animals may hold and defend territories. Plants need space for their roots to gather moisture, and they may exude chemicals that poison one another's roots as a means of competing for space. As a result, competing individuals may end up distributed at equal distances from one another.

A *clumped distribution* often results when organisms seek habitats or resources that are unevenly spaced. In arid regions, many plants grow in patches near isolated springs, ponds, or streambeds. Hawaiian honeycreepers tend to cluster near flowering trees that offer nectar. People aggregate in villages, towns, or cities.

Sex ratio
A population's **sex ratio** is its proportion of males to females. In monogamous species (in which each sex takes a single mate), a 1:1 sex ratio maximizes population growth, whereas an unbalanced ratio leaves many individuals without mates. Most species are not monogamous, however, so sex ratios vary from one species to another.

Age structure
Age structure describes the relative numbers of individuals of different ages within a population. By combining this information with data on the reproductive potential of individuals in each age class, a population ecologist can predict how the population may grow or shrink.

Many plants and animals continue growing in size as they age, and in these species, older individuals often reproduce in greater numbers. Older, larger trees in a population produce more seeds than smaller, younger trees of the same

(a) Random: Distribution of organisms displays no pattern.

(b) Uniform: Individuals are spaced evenly.

(c) Clumped: Individuals concentrate in certain areas.

FIGURE 3.15 Individuals in a population can be spatially distributed in three fundamental ways.

THE SCIENCE behind the story

Is Forest Restoration Boosting Bird Populations at Hakalau Forest?

Conservationist Jack Jeffrey in the Hakalau Forest

Are populations of Hawaiian honey-creepers increasing or decreasing? It's a high-stakes question. The answer could tell us whether efforts to save them are on the right track, or whether the birds may be headed for extinction.

At Hakalau Forest National Wildlife Refuge, biologists have been working for years to monitor bird populations and understand their dynamics. The Hakalau Refuge is home to nine species of native forest birds, including four federally endangered ones: the Hawai'i 'ākepa; the Hawai'i creeper; the 'akiapōlā'au; and the 'io, or Hawaiian hawk. A number of non-native birds occur here as well. When the refuge was established in 1985, most of the area within its boundaries was forested but about 15% of it—the area highest up the slopes—was degraded pastureland that had been cleared for cattle ranching years earlier and remained dominated by non-native grasses.

The biologists employed to manage Hakalau after the refuge was established labored to restore the existing forest by planting native plants, removing invasive weeds, and erecting fences to keep out pigs and cattle. Across large regions of the pastureland, they oversaw the planting of 390,000 native koa trees. Gradually, the forest recovered, and stands of young koa trees took root at higher elevations (**FIGURE 1**). But have all these efforts brought higher populations of native forest birds?

In 1987, federal biologists and trained observers began conducting periodic surveys of birds on the refuge. Following standard protocols, they performed "point counts" by walking transects (linear routes), stopping for 8 minutes in predetermined spots, and counting numbers of each species seen or heard. With 343 points along 15 transects across the refuge, this sampling allowed them to estimate population densities for each bird species. Researchers then analyzed changes in these samples over time to draw inferences about changes in population densities and population sizes.

After 26 years of point counts, the biologists—including Richard Camp, Eben Paxton, Jack Jeffrey, and others—summarized their vast trove of data. Raw counts of birds varied from year to year, which is normal for any such survey effort. The team expected that some portion of this variation reflected the real, meaningful biological change that they sought to document. However, the rest of the variation was bound to be due to chance factors or varying conditions, such as weather, differing abilities of observers, and the fact that thicker vegetation makes birds harder to detect. The researchers wanted to find the true biological "signal" and to weed out the "noise," so they employed sophisticated statistical methods to eliminate as much of the non-meaningful variation as possible.

(a) Before restoration, 1992

(b) After restoration, 2008

FIGURE 1 Reforestation at Hakalau Forest NWR. Upper-elevation pastureland in the refuge **(a)** in 1992 and **(b)** in 2008 after reforestation.

FIGURE 2 shows an example of how this was done for data on one common honeycreeper at Hakalau, the 'apapane. Raw count data from the surveys (**FIGURE 2a**) suggested that 'apapanes increased in number from 1987 to 2012. The researchers used these data, taking area and other factors into account, to generate yearly estimates of population size for the entire Hakalau refuge. To interpret trends, they used linear regression, a statistical method that analyzes how values change through time and determines the straight line that, when drawn through data points on a graph, best represents their trend (**FIGURE 2b**). This analysis indicated that the 'apapane population had risen from less than 30,000 in 1987 to more than 40,000 in 2012. The researchers then put the data through another method, called state-space analysis, that aims to remove non-meaningful sources of year-to-year variation. This resulted in estimates that were similar but that the researchers expect are more accurate (**FIGURE 2c**).

All these analyses led researchers to conclude that over the 26-year period at Hakalau, populations of most native bird species

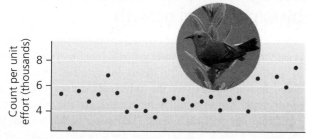

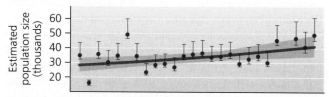

(a) Average raw counts of apapane numbers from transect stops

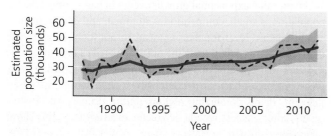

(b) Population estimates of apapanes

(c) Population estimates of apapanes using enhanced data

FIGURE 2 Population data for the 'apapane, a colorful Hawaiian honeycreeper, typify data gathered at Hakalau Forest NWR over 26 years. Raw count data averaged from surveys **(a)** suggest an increase in numbers. Population size estimates for the entire refuge **(b)**, analyzed with linear regression, also indicate an increasing trend. State-space analysis is expected to give the most accurate results, and trends using this method **(c)** confirm that the 'apapane population has risen during the time period. The shaded area indicates 95% confidence intervals. *Data from Camp, Richard J., et al., 2016. Evaluating abundance and trends in a Hawaiian avian community using state-space analysis.* Bird Conservation International *26: 225–242.*

DATA Go to **Interpreting Graphs & Data** on **Mastering Environmental Science**

Only one species showed little change through the years in the reforested areas: The Hawai'i 'ākepa (**FIGURE 3b**) remained very rare throughout the 26-year period. The 'ākepa builds its nests inside cavities in trees. Old gnarled 'ōhi'a trees have cavities, but young koas do not—therefore, although the restored koa forest provides insects for 'ākepas to eat, this species requires mature 'ōhi'a habitat in order to persist.

Findings like these are helping inform managers how best to manage Hakalau's forest and its birds. The data thus far suggest that reforestation and other management actions have been effective, helping to boost populations—or at least to hold them stable—in the face of dire threats.

Of concern, though, is the possibility that challenges from outside the refuge—such as avian malaria and pox moving upslope with climate warming—might eventually overwhelm even the best management efforts. Managers are hoping they can continue planting forest further upslope as the climate warms, trying to keep native birds one step ahead of the upward march of disease. As they do so, researchers will continue to monitor Hakalau's birds, adding to their valuable long-term database of population trends.

were either stable or slowly increasing across most of the refuge. Some species appeared to fare better in the middle-elevation forest being actively managed than in lower-elevation forest.

Meanwhile, in the high-elevation pastures that managers were restoring into forest, population densities of most birds were rising sharply but had not yet reached the densities found in intact forest. When researchers took a close look at patterns of bird colonization of these regenerating koa forests, they found that species responded in different ways, according to their ecological niches and needs:

- Birds that feed mostly on insects, such as the Hawai'i 'elepaio (**FIGURE 3a**), increased in density dramatically, because koa trees host many insects.
- Birds with generalist diets, such as the Hawai'i 'amakihi and the non-native Japanese white-eye, also increased greatly.
- Species that subsist on nectar from flowers, such as the 'apapane and the 'i'iwi, increased moderately; these birds were probably entering the koa forest mostly to feed on flowers of scattered 'ōhi'a trees.
- Fruit-eating birds like the oma'o began to increase only once the fruiting native understory plants that refuge staff had planted became numerous.

The team also found that distance from intact forest seemed to influence the likelihood that forest birds would be found at transect points. For most native forest birds, densities were greater in restored koa forests near the intact native forest than far upslope from it.

(a) Hawai'i 'elepaio   **(b) Hawai'i 'ākepa**

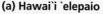

FIGURE 3 The Hawai'i 'elepaio (a), an insect-eating bird, is benefiting from reforestation at Hakalau, but the Hawai'i 'ākepa (b) needs large old 'ōhi'a trees with cavities for its nests.

species. Larger, older fish produce more eggs than smaller, younger fish. Birds use the experience they gain with age to become more successful breeders at older ages.

Human beings are unusual because we often survive past our reproductive years. A human population made up largely of older (post-reproductive) individuals will tend to decline over time, whereas one with many young people (of reproductive or pre-reproductive age) will tend to increase.

Populations may grow, shrink, or remain stable

Let's now take a more quantitative approach to population change by examining some simple mathematical concepts used by population ecologists and by **demographers** (scientists who study human populations). Population change is determined by four factors:

- Natality (births within the population)
- Mortality (deaths within the population)
- Immigration (arrival of individuals from outside the population)
- Emigration (departure of individuals from the population)

We can measure a population's **rate of natural increase** simply by subtracting the death rate from the birth rate:

$$(\text{birth rate}) - (\text{death rate}) = \text{rate of natural increase}$$

To obtain the actual rate of change in a population's size per unit time, called the **population growth rate,** we must also include the effects of immigration and emigration:

$$(\text{birth rate} - \text{death rate}) + (\text{immigration rate} - \text{emigration rate})$$
$$= \text{population growth rate}$$

The rates in these formulas are often expressed in numbers per 1000 individuals per year. For example, a population with a birth rate of 18 per 1000 per year, a death rate of 10 per 1000 per year, an immigration rate of 5 per 1000 per year, and an emigration rate of 7 per 1000 per year would have a population growth rate of 6 per 1000 per year:

$$(18/1000 - 10/1000) + (5/1000 - 7/1000) = 6/1000$$

Given these rates, a population of 1000 in 1 year will grow to 1006 in the next. If the population is 1,000,000, it will reach 1,006,000 the next year. Such population increases are often expressed as percentages, as follows:

$$\text{population growth rate} \times 100\%$$

Thus, a growth rate of 6/1000 would be expressed as

$$6/1000 \times 100\% = 0.6\%$$

By measuring population growth in percentages, we can compare changes in populations of far different sizes. We can also predict amounts of future change. Understanding and predicting such dynamics helps wildlife and fisheries managers regulate hunting and fishing to ensure sustainable harvests, informs conservation biologists trying to protect rare and declining species, and assists policymakers planning for human population growth in cities, regions, and nations.

Unregulated populations increase by exponential growth

When a population increases by a fixed percentage each year, it is said to undergo **exponential growth.** Imagine you put money in a savings account at a fixed interest rate and leave it untouched for years. As the principal accrues interest and grows larger, you earn still more interest, and the sum grows by escalating amounts each year. The reason is that a fixed percentage of a small number makes for a small increase, but the same percentage of a larger number produces a larger increase. Thus, as a savings account (or a population) grows larger, each incremental increase likewise becomes larger in absolute terms. Such acceleration is a characteristic of exponential growth.

The J-shaped curve in **FIGURE 3.16** shows exponential growth. Populations increase exponentially unless they meet constraints. Each organism reproduces, on average, by a certain amount, and as populations grow, there are more individuals reproducing by that amount. If there are adequate resources and no external limits, ecologists expect exponential growth.

Normally, exponential growth occurs in nature only when a population is small, competition is minimal, and environmental conditions are ideal for the organism in question. Most often, these conditions occur when the organism arrives in a new environment that contains abundant resources. Mold growing on a piece of fruit, or bacteria decomposing a dead animal, are cases in point. Plants colonizing regions during primary succession (p. 87) after glaciers recede or volcanoes erupt may also show exponential growth. In Hawai'i, many species that colonized the islands underwent exponential growth for a time after their arrival. One current example of exponential growth in mainland North America is the Eurasian collared dove (see Figure 3.16). Unlike its extinct relative the passenger pigeon, this species arrived here from Europe, thrives in areas disturbed by people, and has spread across the continent in a matter of years.

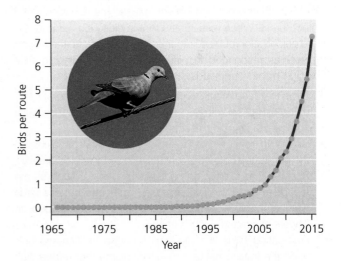

FIGURE 3.16 A population may grow exponentially when colonizing an unoccupied environment or exploiting an unused resource. The Eurasian collared dove is spreading across the United States, propelled by exponential growth. *Data from Pardieck, K. L., et al., 2018. North American Breeding Bird Survey Dataset 1966–2017. v. 2017.0. Laurel, MD: USGS Patuxent Wildlife Research Center.*

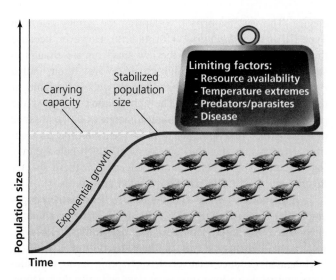

FIGURE 3.17 The logistic growth curve shows how population size may increase rapidly at first, then slow down, and finally stabilize at a carrying capacity.

Limiting factors restrain growth

Exponential growth rarely lasts long. If even a single species were to increase exponentially for very many generations, it would blanket the planet's surface! Instead, every population eventually is constrained by **limiting factors**—physical, chemical, and biological attributes of the environment that restrain population growth. Together, these limiting factors determine the **carrying capacity,** the maximum population size of a species that a given environment can sustain.

Ecologists use the S-shaped curve in **FIGURE 3.17** to show how an initial exponential increase is slowed and eventually brought to a standstill by limiting factors. This phenomenon is called **logistic growth.** A logistic growth curve rises sharply at first but then begins to level off as the effects of limiting factors become stronger. Eventually, the collective influence of these factors stabilizes the population size at its carrying capacity.

We can witness this process by taking a closer look at data for the Eurasian collared dove, as gathered by thousands of volunteer birders and analyzed by government biologists in the Breeding Bird Survey, a long-term citizen science project. The dove first appeared in Florida a few decades ago and then spread west and north. Today, its numbers are growing fastest in western areas it has recently reached but more slowly in southeastern areas where it has been present for longer. In Florida, it has apparently reached carrying capacity (**FIGURE 3.18**). Populations of other European birds that spread across North America in the past,

such as the house sparrow and European starling, have peaked and today are declining.

Many factors influence a population's growth rate and carrying capacity. For animals in terrestrial environments, limiting factors include temperature extremes; prevalence of disease; abundance of predators; and the availability of food, water, mates, shelter, and suitable breeding sites. Plants are often limited by amounts of sunlight and moisture and by soil chemistry, in addition to disease and attack from plant-eating animals. In aquatic systems, limiting factors include salinity, sunlight, temperature, dissolved oxygen, fertilizers, and pollutants. To identify limiting factors, ecologists may conduct experiments in which they increase or decrease a hypothesized limiting factor and observe any effects on population size that result.

The influence of some factors depends on population density

A population's density can enhance or diminish the impact of certain limiting factors. Recall that high population density can help organisms find mates but may also increase competition and the risk of predation and disease. Such limiting factors are said to be **density-dependent** because their influence rises and falls with population density. The logistic growth curve in Figure 3.17 represents the effects of density dependence. The larger the population size, the stronger the influence of the limiting factors.

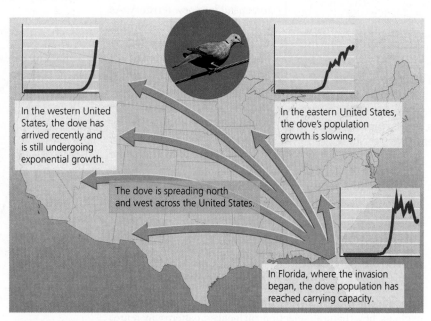

In the western United States, the dove has arrived recently and is still undergoing exponential growth.

In the eastern United States, the dove's population growth is slowing.

The dove is spreading north and west across the United States.

In Florida, where the invasion began, the dove population has reached carrying capacity.

FIGURE 3.18 Exponential growth slows over time and gives way to logistic growth. By breaking down the continent-wide data for the Eurasian collared dove from Figure 3.16, we can track its spread west and north from Florida, where it first arrived. Today, its population growth is fastest in the west, slower in the east (where the species has been present longer), and stable in Florida (where it has apparently reached carrying capacity). *Data from Pardieck, K. L., et al., 2018.* North American Breeding Bird Survey Dataset 1966–2017. *v. 2017.0. Laurel, MD: USGS Patuxent Wildlife Research Center.*

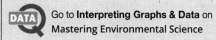 **DATA** Go to **Interpreting Graphs & Data** on Mastering Environmental Science

Density-independent factors are those whose influence is independent of population density. Temperature extremes and catastrophic events such as floods, fires, and landslides are examples of density-independent factors, because they can eliminate large numbers of individuals without regard to their density.

The logistic curve is a simplified model, and real populations in nature can behave differently. Some may cycle above and below the carrying capacity. Others may overshoot the carrying capacity and then crash, destined either for extinction or for recovery.

Carrying capacities can change

Because environments are complex and ever-changing, carrying capacity can vary. If a fire destroys a forest, for example, the carrying capacities for most forest animals will decline, whereas carrying capacities for species that benefit from fire

(such as fire-adapted grasses or trees with specially adapted seeds) will increase. Our own species has proven capable of intentionally altering our environment to raise our carrying capacity. When our ancestors began to build shelters and use fire for heating and cooking, they eased the limiting factors of cold climates and were able to expand into new territory. As human civilization developed, we overcame limiting factors through the development of new technologies and cultural institutions. People have managed so far to increase the planet's carrying capacity for our species, but we have done so by appropriating immense proportions of the planet's natural resources. In the process, we have reduced carrying capacities for countless other organisms that rely on those same resources.

Life history strategies vary among species

Population ecology, organismal ecology, and evolution all come together in **life history theory,** a scientific approach that explains how natural selection influences patterns in reproduction, survival, and life span. In an environment of limited resources and limiting factors, an organism faces trade-offs in how it can apportion its energy. Over time, each species has evolved its own way of allocating its investment among reproduction, parental care, and survival.

For example, many fish, plants, frogs, and insects mature rapidly and devote their energy to producing many offspring in a short time. Most often such species do not provide parental care to these offspring, but simply leave their survival to

chance. The vast majority of offspring die, giving rise to a Type III *survivorship curve* (**FIGURE 3.19**). However, because there are so many offspring, the few survivors are enough to sustain the population. Such species are often called *r-selected,* and they are adapted to do well in variable or unpredictable environments. The abbreviation *r* denotes the per capita rate at which a population increases in the absence of limiting factors. Population sizes of r-selected species fluctuate greatly, such that they are frequently well below carrying capacity. This is why natural selection in these species favors traits that lead to rapid population growth.

Other species are said to be *K-selected* because their populations tend to stabilize near carrying capacity, commonly abbreviated *K.* Large animals such as giraffes, elephants, whales, and humans are typically considered to be K-selected. Such animals produce relatively few offspring during their lifetimes and require a long time to gestate and raise their young. However, the considerable energy and resources they devote to caring for their offspring help give these few offspring a high likelihood of survival, giving rise to a Type I survivorship curve (see Figure 3.19)—especially in stable environments. Because their populations stay close to carrying capacity, these slow-maturing, long-lived organisms must compete to hold their own in a crowded world. Thus, natural selection favors investing in high-quality offspring that can be good competitors.

It is important to note, however, that *r-selected and K-selected species are two extremes on a continuum* and that most species fall somewhere between those extremes. Moreover, many organisms show combinations of traits that do not

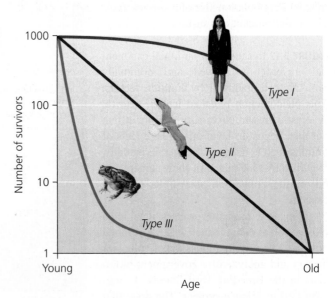

FIGURE 3.19 Survivorship curves show how an individual's likelihood of survival varies with age. In a Type I curve, survival rates are high when organisms are young and decrease sharply when organisms are old. In a Type II curve, survival rates are equivalent regardless of an organism's age. In a Type III curve, most mortality takes place at young ages, and survival rates are higher at older ages.

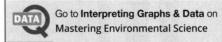

Go to **Interpreting Graphs & Data** on Mastering Environmental Science

correspond to a place on the continuum. A redwood tree, for instance, is large and long-lived, yet it produces many small seeds and offers no parental care.

Conserving Biodiversity

Populations have always been affected by environmental change, but today human development, resource extraction, and population pressure are speeding the rate of change and bringing new types of impacts. Fortunately, committed people are taking action to safeguard biodiversity and to preserve and restore Earth's ecological and evolutionary processes. (We will explore these efforts more fully in our coverage of conservation biology in Chapter 11.) For now, let's see how Hawaiians are confronting the threats to their biodiversity.

Introduced species pose challenges

By introducing species into areas where they do not occur naturally, human beings often set in motion cascades of impacts on native populations, communities, and ecosystems. Some introduced species thrive in their new surroundings, killing or displacing native species (pp. 88–92, 291–294). Island-dwelling organisms are particularly vulnerable to introduced species. Because island inhabitants have evolved in isolation in small areas amid a limited community of other species, they tend to lack defenses against mainland species that are well adapted to deal with a broad array of enemies.

As we have seen, the Hawaiian Islands have been transformed by introduced species. Cattle, goats, sheep, and pigs eat native vegetation, endangering plant populations. Alien grasses, shrubs, and trees spread across the landscapes that livestock have altered. Birds suffering already from habitat loss and predation by non-native mammals now also struggle against diseases like avian pox and malaria transmitted by introduced mosquitoes. Pigs have worsened the malaria problem, because they dig holes in the forest floor, where rainwater forms shallow pools in which mosquitoes breed. As a result, biologists and land managers have found that trying to help a species in trouble often means trying to eradicate or control populations of another that is doing too well. For instance, in many areas in Hawai'i, feral pigs are being hunted and areas are then fenced off once they become free of pigs.

Innovative solutions are working

Amid the challenges of Hawaii's extinction crisis, hard work is resulting in some inspirational success stories, and several species have been saved from imminent extinction. At Hakalau Forest, ranchland is being restored to forest, invasive plants are being removed, native ones are being planted, and existing populations of nēnē are being protected while new populations of the species are being established (see **SUCCESS STORY**).

Early work at Hawai'i Volcanoes National Park inspired the conservation work at Hakalau Forest, as well as efforts by managers and volunteers from the Hawai'i Division of Forestry and Wildlife, The Nature Conservancy of Hawai'i, Kamehameha Schools, and local watershed protection groups. Across Hawai'i, many people are protecting land, removing alien mammals and weeds, and restoring native habitats. Offshore, Hawaiians are striving to protect their

---

SUCCESS story — Restoring Hakalau's Forest

By the time the Hakalau Forest National Wildlife Refuge was established in 1985, much of Hawaii's native forest had been cleared for cattle ranching, while free-roaming pigs and invasive plants had degraded what remained. So Hakalau's managers swung into action. They built fences to keep pigs and cattle out of existing native forest. They labored to remove invasive weeds, while locating and protecting the last remaining individuals of several endangered plant species. They planted half a million native plants, allowing stands of young trees to take root in what had been barren pasture. Gradually, the existing forest recovered, while the reforestation of Hakalau's upper zone is creating new habitat into which birds are moving. Scientists monitoring bird populations have documented increases of up to 10 times greater densities in the restored areas, and today bird populations at Hakalau seem to be faring better than elsewhere on the Island of Hawai'i. This success is a hopeful sign that research and careful management can help undo past ecological damage and conserve populations of endangered island species.

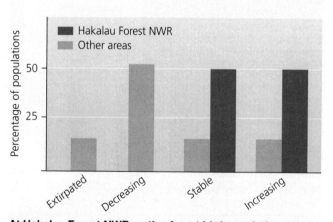

At Hakalau Forest NWR, native forest bird populations are stable or increasing, whereas at other protected areas on the Island of Hawai'i, most populations are decreasing or extirpated (locally extinct). *Data from Camp, Richard J., et al., 2009. Passerine bird trends at Hakalau Forest National Wildlife Refuge, Hawai'i .Hawai'i Cooperative Studies Unit Technical Report HCSU-011.*

→ **Explore the Data** at Mastering Environmental Science

FIGURE 3.20 Hawai'i protects some of its diverse natural areas, helping to stimulate its economy with ecotourism. Here, a scuba diver observes raccoon butterflyfish at a coral reef.

fabulous coral reefs, sea grass beds, and beaches from pollution and overfishing. The northwesternmost Hawaiian Islands are now part of the largest federally declared marine reserve (p. 447) in the world.

Hawaii's citizens are reaping economic benefits from their conservation efforts. The islands' wildlife and natural areas draw visitors from around the world, a phenomenon referred to as **ecotourism** (FIGURE 3.20). A large percentage of Hawaii's tourism is ecotourism, and altogether tourism draws more than 7 million visitors to Hawai'i each year, creates thousands of jobs, and pumps $12 billion annually into the state's economy.

Climate change poses a challenge

Traditionally, people have sought to conserve populations of threatened species by preserving and managing tracts of land (or areas of ocean) designated as protected areas. However, global climate change (Chapter 18) now threatens this strategy. As temperatures climb and rainfall patterns shift, conditions within protected areas may become unsuitable for the species these areas were meant to protect.

Hawaii's natural systems are especially vulnerable. At Hakalau Forest on the slopes of Mauna Kea, mosquitoes and malaria are moving upslope toward the refuge as temperatures rise, exposing more and more birds to disease (**FIGURE 3.21**). Some research suggests that climate change will lower the cloud layer atop Mauna Kea, reducing rainfall at high elevations and pushing the upper limit of the forest downward. If this comes to pass,

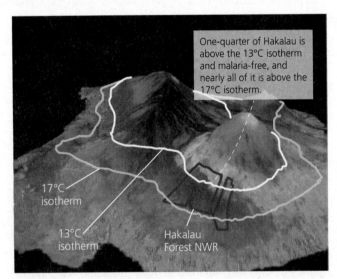

One-quarter of Hakalau is above the 13°C isotherm and malaria-free, and nearly all of it is above the 17°C isotherm.

17°C isotherm

13°C isotherm

Hakalau Forest NWR

(a) Today

Almost no area remains above the 13°C isotherm, so malaria will encompass the whole refuge.

(b) With 2°C of climate warming

FIGURE 3.21 Researchers have modeled how a warming climate will affect the native birds of Hakalau Forest NWR. Avian malaria cannot survive where temperatures dip below 13°C, and it peaks where summer temperatures average 17°C. Today, **(a)** 24% of Hakalau lies above (cooler than) the 13°C isotherm and is free of malaria. If climate warms by 2°C **(b)**, however, then the isotherms move upslope, and only 1% of Hakalau will remain cooler than 13°C and malaria-free. *Data from Benning, T. L., et al., 2002. Interactions of climate change with biological invasions and land use in the Hawaiian Islands: Modeling the fate of endemic birds using a geographic information system. Proc Natl. Acad. Sci. 99: 14246–14249.*

Hakalau's honeycreepers may become trapped within a shrinking band of forest by disease from below and drought from above.

The challenges posed by climate change mean that scientists and managers need to come up with new ways to save declining populations. In Hawai'i, management and ecotourism can help preserve natural systems, but resources to preserve habitat and protect endangered species will likely need to be increased. Restoring altered communities to their former condition—as is being done at Hakalau Forest—will also be necessary. The restoration of ecological communities is one of the topics we will examine in our next chapter, as we shift from the population level to the community level.

CENTRAL CASE STUDY
connect & continue

TODAY, scientists continue to gain insights from the "natural laboratory" of the Hawaiian Islands, while conservationists work diligently to safeguard Hawaii's embattled native flora and fauna. At Hakalau Forest NWR, researchers, staff, and volunteers continue to monitor the birds and plants of the refuge and to restore native forest.

As climate warming worsens, Hakalau's managers plan to continue planting forest further upslope, trying to keep native birds one step ahead of the upward advance of avian pox and malaria. By analyzing data from the bird surveys, researchers hope to determine whether conservation of the birds and forest will be best achieved by expanding forest in a continuous stretch or by planting scattered patches of forest still further upslope and hoping the gaps will fill in naturally as fruit-eating birds spread seeds.

Elsewhere on the Island of Hawai'i, there is good news and bad. A major concern is that a new non-native fungal disease is attacking 'ōhi'a lehua trees, killing thousands of these large, iconic trees that provide the very foundation of native Hawaiian forests. Within just five years, this disease, called rapid 'ōhi'a death, has spread across most of the island and has also been discovered on Kaua'i. There is no known cure, and researchers are racing to learn more about the pathogen and how to stop it. For now, the most Hawaiians can do is to avoid transporting 'ōhi'a wood and to sterilize shoes and boots when walking into and out of forested areas. Because so many native Hawaiian birds rely on flowering 'ōhi'a trees, their widespread loss would be a major blow to conservation efforts.

On the bright side, conservationists are meeting with success as they attempt to save the rarest native bird on the Island of Hawai'i, the endemic and critically endangered 'alalā, or Hawaiian crow. After removing the last few individuals in existence from the wild and breeding them in captivity, biologists have now successfully released 21 'alalā into the wild, where they are being closely monitored.

Challenges are even greater on the Hawaiian island of Kaua'i, where disease-free high-elevation forest is more limited and climate change is shrinking what little remains. Many native birds are disappearing as a result, including the 'akikiki, or Kaua'i Creeper, which numbers fewer than 500 individuals. At the same time, dedicated conservationists are making strides in protecting native seabirds nesting on Kaua'i by erecting fences around breeding areas to keep out predatory mammals; at the newly fenced Kilauea Point NWR, seabird numbers are increasing for the first time in years.

Other islands are seeing success stories. To safeguard the critically endangered Millerbird, which lived only on the tiny island of Nihoa in the northwestern Hawaiian Islands, scientists in 2011 translocated some individuals and established a second population of Millerbirds on Laysan Island 1050 km (650 mi) away. The thriving new Laysan population has grown past 200 birds and will serve as a kind of insurance policy against extinction.

Many challenges remain, but the conservation successes at Hakalau and elsewhere offer hope that we can protect and restore Hawaii's native flora and fauna, preserving the priceless bounty of millions of years of evolution on this extraordinary chain of islands.

- **CASE STUDY SOLUTIONS** Imagine you are the manager of the Hakalau Forest NWR. Despite all the good work your staff and volunteers have done over the years, climate change is threatening the forest's native birds by allowing disease-carrying mosquitoes to move upslope as temperatures grow warmer. Describe several strategies and actions you would consider taking to restore the forest and conserve populations of the native birds.

- **LOCAL CONNECTIONS** The Hawaiian Islands are known for many unique species of plants and animals, and for special natural places (such as volcanoes, waterfalls, and coral reefs). Everywhere in the world, though, has its own natural wonders. Do some research into the region where you live using your library or online sources. Describe several distinctive plants or animals, and several special natural places, that may be found in the region. What plant, animal, or place is most valuable to you, and why? Are any of these organisms or places at risk of disappearing? Are any of them being affected by introduced species or by other human or ecological impacts? What is being done to safeguard the natural wonders of your region? Suggest steps that could be taken in your community to protect populations of species at risk.

- **EXPLORE THE DATA** Find out how we can tell whether a population of endangered birds is recovering or not.
  ➔ **Explore Data** relating to the case study on **Mastering Environmental Science**

REVIEWING Objectives

You should now be able to:

+ Explain natural selection and cite evidence for this process

Natural selection is the process whereby inherited traits that enhance survival and reproduction are passed on more frequently to offspring than traits that do not enhance survival and reproduction. Evidence of natural selection can be found in countless adaptations among wild species. It can also be found in our crop plants, pets, and farm animals, all of which have been bred by artificial selection. (pp. 50–53)

+ Describe how evolution generates and shapes biodiversity

Species can form in various ways; most commonly, geographic isolation over many generations leads to speciation, producing new types of organisms that enhance Earth's biological diversity. Natural selection can be a diversifying force as populations of organisms adapt to their environments. Phylogenetic trees and the fossil record teach us about the history of life by chronicling how organisms have evolved. (pp. 54–59)

+ Discuss factors behind species extinction and identify Earth's known mass extinction events

Although extinction occurs naturally, human impact is profoundly accelerating the rate of extinction. Island species, rare and localized species, and ecologically specialized species are especially vulnerable. When vulnerable species encounter rapid environmental change, this heightens extinction risk. Five episodes of mass extinction are known—caused likely by asteroid impact or geologic factors. Humans are now initiating a sixth mass extinction. (pp. 59–60)

+ List the levels of ecological organization

Ecologists study organisms, populations, communities, ecosystems, landscapes, and the biosphere. Habitat, niche, and specialization are vital ecological concepts. (pp. 60–62)

+ Describe attributes of populations that help predict population growth

Populations are characterized by size, density, distribution, sex ratio, and age structure. Rates of birth, death, immigration, and emigration determine how a population will change. (pp. 62–66)

+ Explain the roles that logistic growth, limiting factors, carrying capacity, and other fundamental concepts play in population ecology

Populations unrestrained by limiting factors undergo exponential growth. Logistic growth results from density dependence; growth slows as population size increases and approaches a carrying capacity. Reproductive strategies differ among species, and carrying capacities can change—all of which affect population ecology. (pp. 66–69)

+ Identify and discuss challenges and current efforts in conserving biodiversity

Habitat loss, the introduction of non-native species, and climate change are among the major challenges to biodiversity. Many people are striving to protect and restore populations, species, and habitats, even as human impacts continue to complicate efforts. (pp. 69–71)

SEEKING Solutions

1. In what ways have artificial selection and selective breeding changed people's quality of life? Give examples. How might artificial selection and selective breeding be used to improve our quality of life further? Can you envision a way in which they could be used to reduce our environmental impact?

2. Compare the passenger pigeon and the nēnē. What factors made each of these species vulnerable to extinction? Why do you think the passenger pigeon succumbed to extinction while the nēnē has survived? What do you think could have been done to save the passenger pigeon?

3. Three of the major reasons for population declines in wild species are habitat loss, introduced species, and climate change. Using examples from this chapter or from your own region, suggest specific ways in which we might reduce each of these impacts on particular populations?

4. What are some advantages of ecotourism for a state like Hawai'i? What might be a potential disadvantage? Describe a source of ecotourism that exists—or that you feel could succeed—in your own region.

5. **THINK IT THROUGH** You are a population ecologist studying animals in a national park, and park managers are asking for advice on how to focus their limited conservation funds. How would you rate the following three species, from most vulnerable (and thus most in need of conservation attention) to least vulnerable? Give the reasons for your choices.

   • A bird that is a generalist in its use of habitats and resources
   • A salamander endemic to the park that lives in high-elevation forest
   • A fish that specializes on a few types of invertebrate prey and has a large population size

CALCULATING Ecological Footprints

Professional demographers delve into the latest statistics from cities, states, and nations to provide us with updated estimates on human populations. The table that follows shows population estimates for two consecutive years that take into account births, deaths, immigration, and emigration. To calculate the population growth rate (p. 66) for each region, divide the 2018 data by the 2017 data, subtract 1.00, and multiply by 100:

$$\text{pop. growth rate} = [(2018 \text{ pop.}/2017 \text{ pop.}) - 1] \times 100$$

Nevada was the fastest-growing U.S. state during this particular 1-year time period, whereas Hawai'i was one of nine states to decline in population (due to a high cost of living, after many years of growth). You can find data for your own state, city, or metropolitan area by exploring the webpages of the U.S. Census Bureau.

REGION	2017 POPULATION	2018 POPULATION	POPULATION GROWTH RATE
Hawai'i	1,424,203	1,420,491	
Nevada	2,972,405	3,034,392	
Your state or city			
United States	325,147,121	327,167,434	0.62%
World	7,536,000,000	7,621,000,000	

Data: U.S. Census Bureau and Population Reference Bureau.

1. Assuming the population growth rate for the United States remained at the calculated rate, what would the population of the United States have been in 2019?

2. The birth rate of the United States in 2017 and 2018 was 11.8 births per 1000 people (or 1.18%), and the death rate was 8.6 deaths per 1000 people (or 0.86%). Using the formula from p. 66:

   (birth rate) − (death rate) = rate of natural increase

   What was the rate of natural increase for the United States between 2017 and 2018, in percentage terms? Now subtract this rate from the population growth rate shown in the table to obtain the net migration rate. Did the United States experience more immigration or more emigration during this period?

3. At the growth rate you calculated for Nevada, the population of that state will double in less than 34 years. What impacts would you expect this to have on (1) food supplies, (2) drinking water supplies, (3) forests and other natural areas, and (4) wildlife populations?

4. How does your own state, city, or metropolitan area compare in its growth rate with other regions in the table? What steps could your region take to lessen the potential impacts of population growth on (1) food supplies, (2) drinking water supplies, (3) forests and other natural areas, and (4) wildlife populations?

Mastering Environmental Science

Students Go to **Mastering Environmental Science** for assignments, an interactive e-text, and the Study Area with practice tests, videos, and activities

Instructors Go to **Mastering Environmental Science** for automatically graded activities, videos, and reading questions that you can assign to your students, plus Instructor Resources.

Species Interactions and Community Ecology

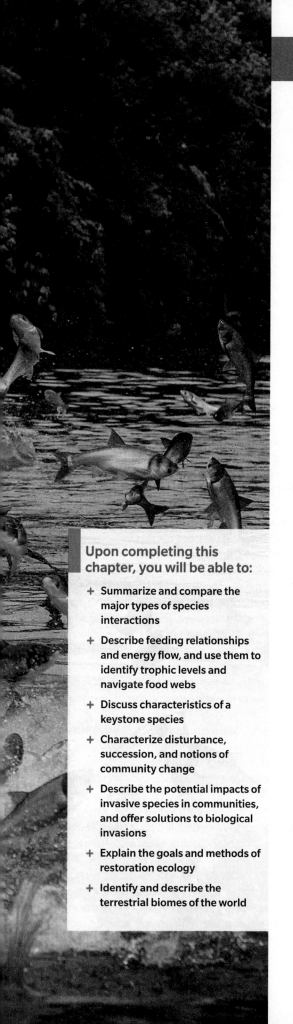

case study

Leaping Fish, Backwards River:

Asian Carp Threaten the Great Lakes

CANADA
Great Lakes

UNITED
STATES

Chicago

> We need to close the Asian carp superhighway, and do it now.
>
> Bill Schuette, Attorney General of Michigan

> This is not about Asian carp. This is about two artificially connected watersheds that many people argue never should have been connected.
>
> Author and Great Lakes expert Peter Annin, Northland College

Call it electroshock therapy for fish. The U.S. Army Corps of Engineers sends electricity into the waters of the Chicago Ship and Sanitary Canal to stun fish swimming toward Lake Michigan, causing them to drift back in the other direction, toward the Mississippi River. These electric barriers are an engineering solution to a biological challenge—to keep invasive Asian carp from reaching the Great Lakes.

But let's rewind the story a bit. It begins far away in China, where for centuries people have raised carp in aquaculture ponds for food. Starting in the 1960s and 1970s, managers of catfish farms and wastewater treatment plants in the southern and central United States began importing several species of these carp to help clean up infestations of algae and parasitic snails. Black carp eat snails and other mollusks. Grass carp scrape leafy aquatic plants off the bottom of ponds. Silver carp and bighead carp consume plankton (microscopic aquatic plants and animals) that they filter from the water.

In time, individuals of these four species, collectively nicknamed "Asian carp," escaped into streams, rivers, ponds, and lakes. Finding plenty of food, these alien fish grew rapidly, becoming too large for predators to capture. Reproducing quickly, their populations grew as they spread to waterways across the southern United States, especially in the lower Mississippi River Valley.

As millions of these non-native fish invaded ecosystems, they competed with native fish for food, preyed on native plants and animals, filtered plankton from the water, and stirred up sediment. These impacts altered water quality, reduced populations of native species, and modified aquatic communities. The silver carp can even injure people: It leaps high out of the water when boats approach—and a hunk of scaly fish 1 m (3.3 ft) long weighing 27 kg (60 lb) can do real damage in a head-on collision!

Before long, Asian carp populations expanded far up the Mississippi River, the Ohio River, the Missouri River, and the tributaries of those rivers, invading

Upon completing this chapter, you will be able to:

+ **Summarize and compare the major types of species interactions**

+ **Describe feeding relationships and energy flow, and use them to identify trophic levels and navigate food webs**

+ **Discuss characteristics of a keystone species**

+ **Characterize disturbance, succession, and notions of community change**

+ **Describe the potential impacts of invasive species in communities, and offer solutions to biological invasions**

+ **Explain the goals and methods of restoration ecology**

+ **Identify and describe the terrestrial biomes of the world**

◀ **Silver carp leap from the waters of the Fox River in Illinois as a boat approaches**

▲ **An invasive Asian carp**

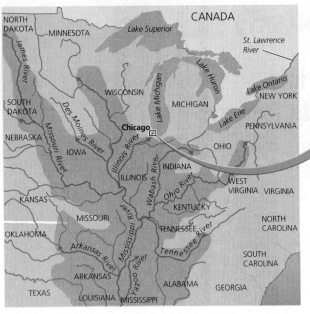

FIGURE 4.1 Asian carp have spread from the southern United States up the Mississippi River and many of its tributaries. Red color indicates stretches of major rivers invaded by carp. The gray-shaded area indicates the region in which carp have invaded smaller rivers, lakes, and ponds. When Chicago **(inset)** built canals and reversed the flow of the Chicago River to flow into the Illinois River, this connected the watersheds of the Great Lakes and the Mississippi River, enabling species to move between them, affecting aquatic communities in each watershed. Today, an electric barrier (see inset) is in place to try to stop Asian carp from reaching the Great Lakes.

waterways across the Midwest (**FIGURE 4.1**). As carp from the Mississippi River advanced up the Illinois River, people around the Great Lakes became alarmed. If the carp moved into Lakes Michigan, Huron, Superior, Ontario, and Erie, they might alter the ecology of the lakes and devastate sport and commercial fisheries worth $7 billion. Millions of people could be affected in Michigan, Wisconsin, Minnesota, Ohio, Indiana, Illinois, Pennsylvania, New York, and Ontario.

The Great Lakes lie in their own watershed, but water (and fish) from the adjacent Mississippi River watershed can occasionally enter at certain locations when major flooding events cause waters from both basins to rise, overtop the divide, and mix. However, the most likely route for Asian carp to invade the Great Lakes is an unintended consequence resulting from an ambitious engineering feat the city of Chicago pulled off a century ago. Chicago had always drawn its drinking water from Lake Michigan, but it also was polluting this drinking water source by dumping its sewage into the Chicago River, which flowed into the lake. So, in 1900 Chicago managed to reverse the flow of the river. It built a canal system that connected the Chicago River to the Illinois River, whose waters flowed naturally to the Mississippi River (see inset in Figure 4.1). As a result, water was now pulled out of Lake Michigan and through the canals, sending Chicago's waste flowing southwest down the Illinois River and into the Mississippi.

In this way, two of North America's biggest natural watersheds—those of the Great Lakes and the Mississippi River—were artificially joined. This was a tremendous boon for ship navigation and commerce, but it also allowed species that had evolved in the Great Lakes to enter the Mississippi Basin, and vice versa. And when species from one place are introduced to another, ecological havoc can result.

Meanwhile, hundreds of miles away at their eastern end, the Great Lakes were connected to the Atlantic Ocean when

engineers completed the Saint Lawrence Seaway in 1959. This 600-km (370-mi) system of locks and canals along the Saint Lawrence River allowed ocean-going ships to travel into the lakes. However, it also allowed access for non-native species, including the sea lamprey (an eel-like creature that attacks the lakes' freshwater fish) and zebra mussels and quagga mussels (which clog pipes and propellers, outcompete native mussels, and transform fisheries). All told, the Great Lakes today host more than 180 non-native species that are blamed for economic losses of hundreds of millions of dollars each year.

Fishermen and scientists concerned that Asian carp might soon add to this toll bent the ear of policymakers, who approved funding for biologists to monitor lakes, rivers, and canals for signs of the alien fish. Policymakers are also funding various efforts to prevent and control the spread of the carp, including catching the carp; poisoning them; deterring their movements; introducing larger fish to eat the carp; and persuading restaurants to put carp on the menu. Altogether, more than $400 million has been spent on trying to keep Asian carp from reaching the Great Lakes.

Against this backdrop, Michigan and other Great Lakes states launched lawsuits against Chicago and the U.S. Army Corps of Engineers in an unsuccessful attempt to force them to re-engineer Chicago's canal system to block this "superhighway" for invasive species. Yet at the same time, some scientists are now questioning whether Asian carp would thrive, or even survive, in the Great Lakes if they made it there: The lake water is nutrient-poor, and the carp need flowing water for their eggs to hatch.

Time will tell whether Asian carp invade the Great Lakes, and if so, what impacts they might have there. We live in a dynamic world, and as species are moved from place to place, we are discovering how the interactions among species can transform the communities around us.

Species Interactions

By interacting with many species in a variety of ways, Asian carp have set in motion an array of changes in the communities they have invaded. Interactions among species are the threads in the fabric of ecological communities. Ecologists organize species interactions into several fundamental categories (**TABLE 4.1**).

Competition can occur when resources are limited

When multiple organisms seek the same limited resource, their relationship is said to be one of **competition.** Competing organisms do not usually fight with one another directly and physically. Instead, competition is often more subtle and indirect, taking place as organisms vie with one another to procure resources. Such resources include food, water, space, shelter, mates, sunlight, and more. Competitive interactions can occur between members of the same species (intraspecific competition) or between members of different species (interspecific competition). As an example of interspecific competition, researchers have found that silver carp and bighead carp compete with native fish for plankton. In the Illinois River, these two species of Asian carp eat so much plankton that native fish have had to shift their diets and have become smaller in size due to reduced nutrition.

When individuals of the same species compete for limited resources, competition intensifies as more individuals occur per unit area (i.e., when populations are denser). This is an example of density dependence (p. 67), and it can limit the growth of a population.

When individuals of different species compete for limited resources, this can affect communities and give rise to different types of outcomes. If one species is a very effective competitor, it may exclude other species from resource use entirely—an outcome called **competitive exclusion.** Alternatively, if no single competitor fully excludes others, the species may continue to live side by side. This result, called **species coexistence,** may produce a stable point of equilibrium, at which the relative population sizes of each remain fairly constant through time.

TABLE 4.1 Species Interactions: Effects on Their Participants

TYPE OF INTERACTION	EFFECT ON SPECIES 1	EFFECT ON SPECIES 2
Mutualism	+	+
Predation, parasitism, herbivory	+	–
Competition	–	–

"+" denotes a positive effect; "–" denotes a negative effect.

Ecologists debate what might happen if silver carp and bighead carp reach the Great Lakes, where invasive zebra and quagga mussels are already filtering most of the plankton resource out of the water. Would the mussels competitively exclude the carp, depriving them of food and preventing them from becoming established? Or, would the carp manage to claim a share of the plankton, carving out a place for themselves to coexist with the mussels? No one knows.

Coexisting species that use the same resources tend to adjust to their competitors to minimize competition with them. Individuals may do this by changing their behavior to use only a portion of the total array of resources they are capable of using. In such cases, individuals do not fulfill their entire niche. A species' niche (p. 62) reflects its functional role in a community, including its resource use, habitat use, food consumption, and other attributes—a niche is a kind of multidimensional summary of everything an organism does.

The full niche of a species is called its **fundamental niche** (**FIGURE 4.2a**). An individual that plays only part of its functional role in a community because of competition or another type of species interaction is said to display a **realized niche** (**FIGURE 4.2b**), the portion of its fundamental niche that is actually "realized," or fulfilled. Asian carp appear to be reducing the realized niches of many native fish as the carp compete with the native fish for food and other resources.

Species experience similar adjustments over evolutionary time in response to competition. Across many generations, the process of natural selection (pp. 50–51) may

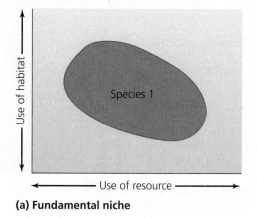

(a) Fundamental niche

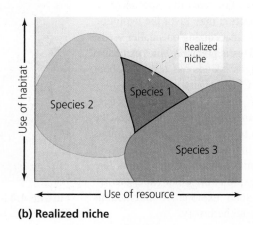

Realized niche

(b) Realized niche

FIGURE 4.2 Competition may force an organism to play a more limited ecological role. With no competitors **(a),** an organism can fulfill its fundamental niche. When competitors restrict what an organism can do—for instance what habitats it can live in or what resources it can use **(b)**—the organism is limited to a realized niche, which covers only part of its fundamental niche.

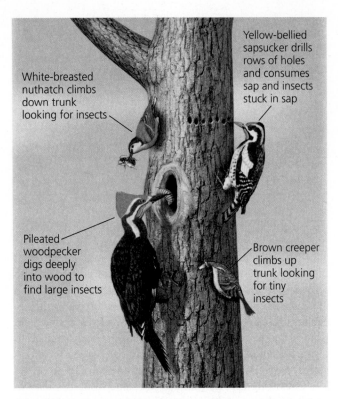

White-breasted nuthatch climbs down trunk looking for insects

Yellow-bellied sapsucker drills rows of holes and consumes sap and insects stuck in sap

Pileated woodpecker digs deeply into wood to find large insects

Brown creeper climbs up trunk looking for tiny insects

FIGURE 4.3 When species compete, they may partition resources. Birds that forage for insects on tree trunks use different portions of the trunk and seek different foods in different ways.

favor individuals that use slightly different resources or that use shared resources in different ways. For example, if two bird species eat the same type of seeds, natural selection might eventually drive one species to specialize on larger seeds and the other to specialize on smaller seeds. Or, one bird species might evolve to become more active in the morning and the other bird species more active in the evening, minimizing interference. This process is called **resource partitioning** because the species partition, or divide, the resources they use in common by specializing in different ways (**FIGURE 4.3**).

Resource partitioning can lead to **character displacement,** in which competing species come to diverge in their physical characteristics because of the evolution of traits best suited to the range of resources they use. For birds that specialize on eating larger seeds, natural selection may favor the evolution of larger bills, whereas for birds specializing on smaller seeds, smaller bills may be favored. This is precisely what extensive recent research on the Galápagos Islands has revealed about the finches first discovered by Charles Darwin (p. 51).

In competitive interactions, each participant exerts a negative effect on other participants by securing resources the others could have used. This is reflected in the two minus signs shown for competition in Table 4.1. In other types of interactions, some participants benefit while others are harmed; that is, one species exploits the other (note the +/– interactions in Table 4.1). Such exploitative interactions include predation, parasitism, and herbivory.

Predators kill and consume prey

Every organism needs to procure food, and for most animals, this means eating other living organisms. **Predation** is the process by which individuals of one species—the *predator*—hunt, capture, kill, and consume individuals of another species, the *prey* (**FIGURE 4.4**).

Black carp were introduced to U.S. fish farms because these carp prey on mollusks, including snails that were harming farmed fish. However, once black carp escaped into the wild, these predators began preying on a wide variety of native snails and mussels.

In response to the spread of Asian carp, fisheries managers are now proposing to reestablish populations of the alligator gar, a fish that would prey on the carp. The alligator gar is native to North America but disappeared from many rivers because in the past, fishermen believed it threatened sport fish, and managers set out to eradicate it. One of the few fish large enough to catch an adult carp, the alligator gar grows up to 3 m (9 ft) long, sometimes weighs over 140 kg (300 lb), and features a long snout with fearsome-looking teeth. People are also preying on Asian carp directly, by fishing for them. Each year several million pounds of Asian carp are removed from the Illinois River by agency personnel and contract fishermen. Carp are tasty but bony, and efforts to establish them as a desirable food fish have not yet succeeded; so far, they have been used mostly for fertilizer and pet food.

Predation can sometimes drive population dynamics, shaping community composition by influencing the numbers of predators and prey. An increase in the population size of prey creates more food for predators, which may survive and reproduce more effectively as a result. As the predator population rises, intensified predation pressure drives down numbers of prey. Diminished numbers of prey support fewer predators, and the predator population declines. This eventually allows the prey population to rise again, starting the cycle anew (**FIGURE 4.5**).

FIGURE 4.4 In predation, a predator kills and consumes prey. This anhinga has speared a fish and will soon swallow it whole.

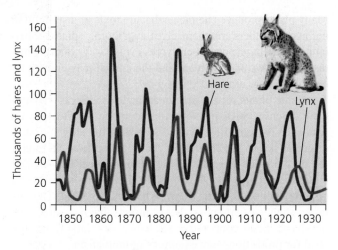

FIGURE 4.5 **Predator–prey systems occasionally show paired cycles.** A classic case is that of snowshoe hares and the lynx that prey on them in Canada. These data come from fur-trapping records of the Hudson Bay Company and represent numbers of each animal trapped. *Data from MacLulich, D. A., 1937. Fluctuation in the numbers of varying hare* (Lepus americanus). Univ. Toronto Stud. Biol. Ser. 43. *Toronto, Canada: University of Toronto Press.*

Predation also has evolutionary consequences. Individual predators that are more adept at capturing prey may live longer lives, reproduce more, and be better providers for their offspring. Natural selection (p. 50) will thereby lead to the evolution of adaptations (p. 51) that enhance hunting skills. Prey, however, face an even stronger selective pressure—the risk of immediate death. As a result, predation pressure has driven the evolution of an elaborate array of defenses against being eaten (**FIGURE 4.6**).

Parasites exploit living hosts

Organisms can exploit other organisms without killing them. **Parasitism** is a relationship in which one organism, the *parasite,* depends on another, the *host,* for nourishment or some other benefit while doing the host harm. Unlike predation, parasitism usually does not result in an organism's immediate death.

Many types of parasites live inside their hosts. For example, tapeworms live in their hosts' digestive tracts, robbing them of nutrition. The larvae of parasitoid wasps burrow into the tissues of caterpillars and consume them from the inside. Other parasites are free-living. Cuckoos of Eurasia and cowbirds of the Americas lay their eggs in other birds' nests and let the host species raise the parasite's young. Still other parasites live on the exterior of their hosts, such as fleas or ticks that suck blood through the skin. The sea lamprey is a tube-shaped vertebrate that grasps the bodies of fish with a suction-cup mouth and a rasping tongue, sucking their blood for days or weeks (**FIGURE 4.7**, p. 80). Sea lampreys invaded the Great Lakes from the Atlantic Ocean after people dug canals to connect the lakes for shipping. The lampreys soon devastated economically important fisheries of chubs, lake herring, whitefish, and lake trout.

Parasites that cause disease in their hosts are called **pathogens.** Common human pathogens include the protists that cause malaria and amoebic dysentery, the bacteria that cause pneumonia and tuberculosis, and the viruses that cause hepatitis and AIDS.

Just as predators and prey evolve in response to one another, so do parasites and hosts, in a reciprocal process of adaptation and counter-adaptation called **coevolution.** Hosts and parasites may become locked in a duel of escalating adaptations, known as an *evolutionary arms race.* Like rival nations racing to stay ahead of one another in military

A gecko's camouflage hides it from predators.

(a) Cryptic coloration (camouflage)

A yellowjacket's coloring signals that it is dangerous.

(b) Warning coloration

This caterpillar's false eyespots startle predators by mimicking a snake's head.

(c) Mimicry

FIGURE 4.6 **Natural selection to avoid predation has resulted in fabulous adaptations.** Some prey species use cryptic coloration **(a)** to blend into their background. Others are brightly colored **(b)** to warn predators they are toxic, distasteful, or dangerous. Others use mimicry **(c)** to fool predators.

FIGURE 4.7 A parasite benefits at the expense of its host. Sea lampreys feed by sucking blood from fish in the Great Lakes.

technology, host and parasite repeatedly evolve new responses to the other's latest advance. In the long run, though, it may not be in a parasite's best interest to do its host too much harm. In many cases, a parasite may leave more offspring by allowing its host to live longer.

Herbivores exploit plants

In **herbivory,** animals feed on the tissues of plants. Insects that feed on plants are the most common type of herbivore; nearly every plant in the world is attacked by insects (**FIGURE 4.8**). Herbivory generally does not kill a plant outright but may affect its growth and reproduction.

Like animal prey, plants have evolved an impressive arsenal of defenses against the animals that feed on them.

FIGURE 4.8 In herbivory, animals feed on plants. This tiny willow sawfly larva feeds on willow leaves.

Many plants produce chemicals that are toxic or distasteful to herbivores. Others arm themselves with thorns, spines, or irritating hairs. In response, herbivores evolve ways to overcome these defenses, and the plant and the animal may embark on an evolutionary arms race.

Some plants recruit certain animals as allies to assist in their defense. Such plants may encourage ants to take up residence by providing swelled stems for the ants to nest in or nectar-bearing structures for them to feed from. In return, the ants protect the plant by attacking insects that land or crawl on it. Other plants respond to herbivory by releasing volatile chemicals when they are bitten or pierced. The airborne chemicals attract predatory insects that may attack the herbivore. Such cooperative strategies—essentially trading food for protection—are examples of mutualism.

Mutualists help one another

Unlike exploitative interactions, **mutualism** is a relationship in which two or more species benefit from interacting with one another. Generally, each partner provides some resource or service that the other needs.

Many mutualistic relationships—like many parasitic ones—occur between organisms that live in close physical contact. Physically close association is called **symbiosis,** and symbiosis can be either mutualistic or parasitic. (Indeed, biologists hypothesize that many mutualistic associations evolved from parasitic ones.) Most terrestrial plant species depend on mutualisms with fungi; plant roots and some fungi together form symbiotic associations that scientists call *mycorrhizae*. In these relationships, the plant provides energy and protection to the fungus, while the fungus helps the plant absorb nutrients from the soil. In the ocean, coral polyps, the tiny animals that build coral reefs (p. 431), provide housing and nutrients for specialized algae known as zooxanthellae in exchange for food the algae produce through photosynthesis (pp. 33–34).

You, too, are engaged in a symbiotic mutualism. Your digestive tract is filled with microbes that help you digest food and carry out other bodily functions. In turn, you provide these microbes with a place to live. Without these mutualistic microbes, none of us would survive for long.

Not all mutualists live in close proximity. **Pollination** (**FIGURE 4.9**) involves free-living organisms that may encounter each other only once. Bees, birds, bats, and other creatures transfer pollen (containing male sex cells) from flower to flower, fertilizing ovaries (containing female sex cells) that grow into fruits with seeds. Most pollinating animals visit flowers for their nectar, a reward the plant uses to entice them. The pollinators receive food, and the plants are pollinated and reproduce. These mutualistic interactions are highly important to agriculture and our food supply (pp. 223–227). Pollination by bees and other insects contributes to the production of three-fourths of our crops and is vital to high yields of melons, squashes, cucumbers, canola, apples, cherries, peaches, pears, mangos, plums, almonds, blueberries, blackberries, raspberries, and even chocolate!

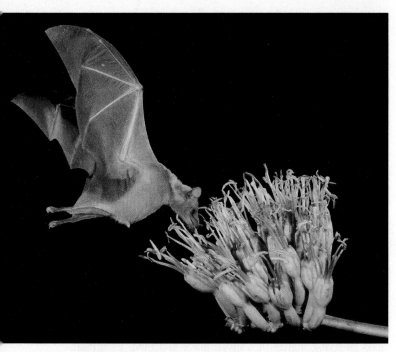

FIGURE 4.9 In mutualism, organisms of different species benefit one another. Lesser long-nosed bats visit flowers of agave plants at night to gather nectar, and in the process, they transfer pollen between flowers, helping the plant to reproduce.

Ecological Communities

A **community** is an assemblage of populations of organisms living in the same area at the same time (as we saw in Figure 3.12, p. 61). Members of a community interact, and these interactions have indirect effects that ripple outward to affect other community members. Species interactions help determine the structure, function, and species composition of communities. Community ecology (p. 61) is the scientific study of species interactions and the dynamics of communities. Community ecologists study which species coexist, how species interact, how communities change through time, and why these community dynamics occur.

Food chains consist of trophic levels

Some of the most important interactions among community members involve who eats whom. As organisms feed on one another, matter and energy move through the community from one **trophic level,** or rank in the feeding hierarchy, to another (**FIGURE 4.10**). As matter and energy are transferred from lower trophic levels to higher ones, they are said to pass up a **food chain,** a linear series of feeding relationships.

Producers Producers, or *autotrophs* ("self-feeders"), make up the first (lowest) trophic level. Terrestrial green plants, cyanobacteria, and algae capture solar energy and use photosynthesis to produce sugars. The chemosynthetic bacteria of hot springs and deep-sea hydrothermal vents use geothermal energy in a similar way to produce food (pp. 34–35).

Consumers Organisms that consume other living organisms are known as **consumers.** Those that consume producers are known as *primary consumers* and make up the second trophic level. Herbivorous grazing animals, such as deer and grasshoppers, are primary consumers. The third trophic level consists of *secondary consumers,* which prey on primary consumers. Wolves that prey on deer act as secondary consumers, as do rodents and birds that prey on grasshoppers. Predators that feed at still higher trophic levels are known as *tertiary consumers.* Examples include hawks and owls that eat rodents that have eaten grasshoppers. Of course, however, nature is not so simple, and many consumers eat organisms on multiple trophic levels.

Detritivores and decomposers Detritivores and decomposers consume nonliving organic matter. **Detritivores,** such as millipedes and soil insects, scavenge the waste products or

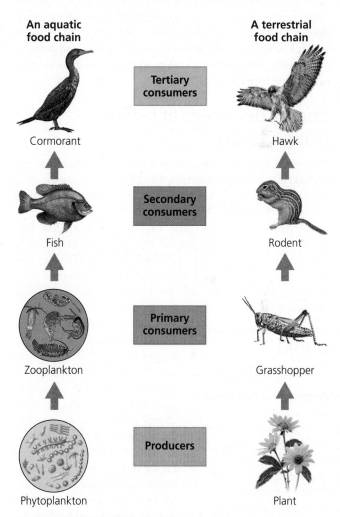

FIGURE 4.10 A food chain organizes species hierarchically by trophic level. The diagram shows aquatic **(left)** and terrestrial **(right)** examples at each level. Arrows indicate the direction of energy flow. Producers synthesize food by photosynthesis, primary consumers (herbivores) feed on producers, secondary consumers eat primary consumers, and tertiary consumers eat secondary consumers. Detritivores and decomposers (not shown) feed on nonliving organic matter and "close the loop" by returning nutrients to the soil or the water column for use by producers.

dead bodies of other community members. **Decomposers,** such as fungi and bacteria, break down leaf litter and other nonliving matter into simpler constituents that can be taken up and used by plants. These organisms enhance the topmost soil layers (pp. 216, 218) and play essential roles as the community's recyclers, making nutrients from organic matter available for reuse by living members of the community.

In Great Lakes communities, the main producers are phytoplankton, consisting of microscopic algae, protists, and cyanobacteria that drift in open water and conduct photosynthesis using sunlight that penetrates the water. The most abundant primary consumers are zooplankton, the tiny aquatic animals that eat phytoplankton. A number of fish eat phytoplankton and thus are primary consumers, whereas fish that eat zooplankton act as secondary consumers. Tertiary consumers include birds and larger fish that feed on zooplankton-eating fish. (The left side of Figure 4.10 shows these relationships in very generalized form.) Asian carp that eat both phytoplankton and zooplankton function on multiple trophic levels. When an organism dies and sinks to the bottom, detritivores scavenge its tissues and decomposers recycle its nutrients.

Energy decreases at higher trophic levels

Within each trophic level in a food chain, organisms use energy in cellular respiration (p. 34). More energy goes toward maintenance than to building new tissues, and most is given off as heat. Only a small portion of the energy is transferred to the next trophic level through predation, herbivory, or parasitism. A general rule of thumb is that each trophic level contains about 10% of the energy of the trophic level below it (although the actual proportion can vary greatly). This pattern can be visualized as a pyramid (**FIGURE 4.11**).

This pyramid pattern illustrates why eating at lower trophic levels—being vegan or vegetarian, for instance—decreases a person's ecological footprint. Each amount of meat or other animal product we eat requires the input of a much greater amount of plant material (see Figure 10.8, p. 248). Thus, when we eat animal products, we use up far more energy per calorie that we gain than when we eat plant products.

In communities, the pyramid-like pattern also tends to hold for numbers of organisms; in general, fewer organisms exist at high trophic levels than at low ones. A grasshopper eats many plants in its lifetime, a rodent eats many grasshoppers, and a hawk eats many rodents. Thus, for every hawk there must be many rodents, still more grasshoppers, and an immense number of plants. Terrestrial systems also tend to

FIGURE 4.11 Lower trophic levels generally contain more energy than higher trophic levels. The 10:1 ratio shown here is typical but varies greatly.

 Go to **Interpreting Graphs & Data** on Mastering Environmental Science

show a pyramid-like relationship for **biomass,** the collective mass of living matter in a given place and time. In aquatic systems, however, low trophic levels may sometimes have less biomass than high trophic levels (because short-lived phytoplankton may reproduce rapidly enough to produce food for a greater biomass of longer-lived zooplankton). Still, energy transfer in such communities is characterized by a pyramid pattern.

Food webs show feeding relationships and energy flow

Thinking in terms of food chains is conceptually useful, but ecological systems are far more complex than simple linear chains. A fuller representation of the feeding relationships in a community is a **food web**—a visual diagram that maps the many paths along which energy and matter flow as organisms consume one another.

FIGURE 4.12 portrays a food web from a temperate deciduous forest of eastern North America. It is greatly simplified and leaves out most species and interactions, yet even within this streamlined diagram we can pick out many food chains involving different sets of species. For instance, grasses are eaten by deer mice that may be consumed by rat snakes—while in another food chain, insects are consumed by spiders that may be eaten by American toads.

Recently, researchers assessed the diets of various fish native to Lake Erie to predict how the food web of the lake might change if Asian carp were to invade it. Their ecological model projected that competition for food from silver and bighead carp would reduce the biomass of plankton-eating fish such as rainbow smelt, gizzard shad, and emerald shiner by 13–37% and that the biomass of predatory adult walleye would, in turn, decrease once they had fewer smelt, shad, and shiners to eat. However, smallmouth bass were predicted to increase by 13–16% because they would benefit from consuming the newly abundant young of the invasive carp.

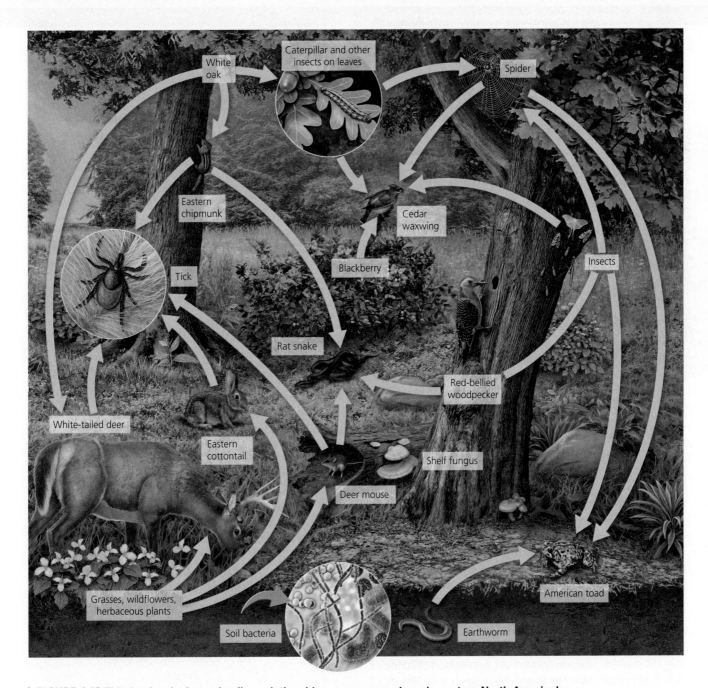

FIGURE 4.12 This food web shows feeding relationships among organisms in eastern North America's temperate deciduous forest. Arrows indicate the direction of energy flow (from prey to predator, from producer to consumer, and from host to parasite). The actual community contains many more species and interactions than can be shown.

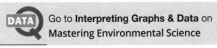 Go to **Interpreting Graphs & Data** on **Mastering Environmental Science**

Overall, the researchers predicted that after 20 years the carp would comprise 30% of the total fish biomass in Lake Erie, after causing populations of most native fish species to decline. Such predictions from models are only as good as the known information that is input to the models, however, and scientists are still actively gathering data on the impacts of competition from Asian carp on native fish (see **THE SCIENCE BEHIND THE STORY**, pp. 84–85).

Some organisms play outsized roles

"Some animals are more equal than others," George Orwell wrote in his classic novel *Animal Farm*. Orwell was making wry sociopolitical commentary, but his remark hints at an ecological truth. In communities, some species exert greater influence than do others. A species that has strong or wide-reaching impact far out of proportion to its abundance is often

THE SCIENCE behind the story

How Do Asian Carp Affect Aquatic Communities?

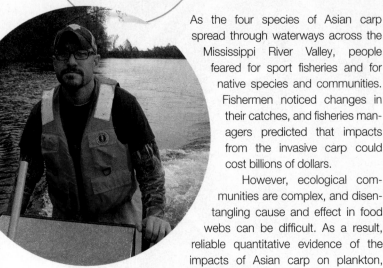

Dr. Quinton Phelps samples for freshwater fish

As the four species of Asian carp spread through waterways across the Mississippi River Valley, people feared for sport fisheries and for native species and communities. Fishermen noticed changes in their catches, and fisheries managers predicted that impacts from the invasive carp could cost billions of dollars.

However, ecological communities are complex, and disentangling cause and effect in food webs can be difficult. As a result, reliable quantitative evidence of the impacts of Asian carp on plankton, plants, and fish has been surprisingly sparse and remarkably slow in coming.

One study, published in 2017 in the journal *PLOS ONE*, showed clear results but also illustrated just how far researchers have to go before gaining a full understanding of the impacts of Asian carp on aquatic communities. In this study, Quinton Phelps of West Virginia University and five colleagues from Missouri, Iowa, and South Dakota set out to determine if silver carp exert population-level effects over large areas on two native plankton-eating fish: gizzard shad and bigmouth buffalo. Previous research in the Illinois River and elsewhere had shown that silver carp compete with certain native fish by eating (and depleting) plankton, but this had not been documented across wide and meaningful scales of space and time.

In the first part of the study conducted by the Phelps team, the researchers assessed data from long-term monitoring of fish in stretches of the Mississippi River Basin from Minnesota to Missouri between 1993 and 2012. Biologists had gathered this data using electrofishing, according to a standardized protocol used by government agencies as part of a federally mandated program. (In electrofishing, an electric current is passed into the water to temporarily shock fish, which float to the surface.) Each fish was identified, measured, and weighed, and counts of each species were made. The team compared data from three northern locations that Asian carp had not reached with data from three southern locations that Asian carp invaded halfway through the study period. In all these sites, the researchers compared data from before 2003 (roughly when the carp became established in the sites they invaded) with

data collected after 2003. Altogether, the project obtained data on more than 9000 fish (**FIGURE 1**).

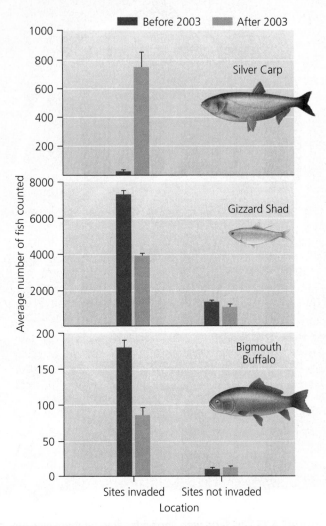

FIGURE 1 Catches of native gizzard shad and bigmouth buffalo decreased as silver carp increased. Bars on the left side of each graph show number of fish counted per unit effort in stretches of the Mississippi River where silver carp invaded around 2003. Bars on the right side of each graph show number of fish counted per unit effort in stretches of river where silver carp did not invade. The red bar of each pairing pertains to the 10 years preceding 2003 (before carp arrival) and the orange bar of each pairing pertains to the 10 years following 2003 (after carp arrival). *Data from Phelps, Q. E., et al., 2017. Incorporating basic and applied approaches to evaluate the effects of invasive Asian carp on native fishes: A necessary first step for integrated pest management. PLOS ONE 12(9): e0184081.*

Results were clear-cut: The team found a significant long-term decrease in catch per unit effort (and thus fish abundance) for both gizzard shad and bigmouth buffalo after the appearance of silver carp in the areas they invaded. The researchers also determined that the physical condition of gizzard shad and bigmouth buffalo declined through time in areas invaded by silver carp. Together, these results indicated that competition for plankton resources was having a harmful effect on the two native species.

In the second portion of the study, researchers sampled four floodplain lakes in Illinois and Missouri by electrofishing after major floods in 2011 sent Mississippi River water into the lakes for the first time since 1937. The scientists sampled the lakes in June and again in November, looking for changes in the fish community across this 5-month period. Silver carp had failed to enter one of the four lakes, and in this lake the researchers found no change in the native fish community. In contrast, a lake that had been invaded by many silver carp showed drastic change: Most of the native fish disappeared. In the remaining two lakes, silver carp were at low to moderate densities, and the team found that native fish densities changed in low to moderate ways. Altogether, the data indicated that more silver carp mean more impact to the native fish community.

The third portion of the research involved controlled laboratory tests in a lab in Missouri. Fish of the three species were captured from the wild and kept in aquaria. For the 14-day experimental trials, fish were placed in tanks in different combinations with limited amounts of plankton, after which growth and survival rates were measured. Silver carp were placed in tanks with gizzard shad and separately with bigmouth buffalo to test for interspecific competition, and all three species were placed in tanks with their own species to obtain comparative data on intraspecific competition.

The data showed clear patterns (**FIGURE 2**). Survival was high in each of the species pairings, but when silver carp and gizzard shad were placed together, only 10% of the gizzard shad survived. Likewise, in all the same-species pairings, the fish grew in biomass, but when silver carp and bigmouth buffalo were placed together, the bigmouth buffalo decreased in biomass. Together, these experimental results showed that silver carp are able to outcompete these two species of native fish when plankton resources are limited.

A strength of this study lay in its three independent approaches. The researchers examined the question of competition between silver carp and two native fish species by analyzing long-term population monitoring data from the field, by taking advantage of a "natural experiment" (p. 11; a major flood) in the field, and by conducting a laboratory experiment under rigorously controlled conditions. The lab experiment demonstrated cause and effect, and the long-term population analysis showed relevance over a large geographic area.

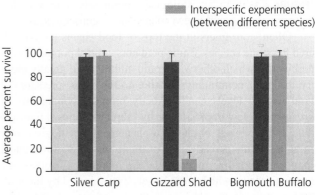

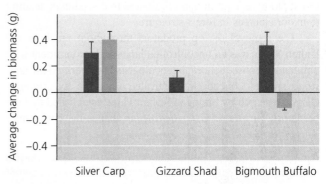

(a) Survival

(b) Growth

FIGURE 2 Controlled lab experiments revealed impacts of competition by silver carp on two native species. In tests of survival **(a),** many fewer gizzard shad survived when silver carp were placed in the same tank with them. In tests of growth **(b),** measured by change in biomass, bigmouth buffalo lost biomass when silver carp were placed in the same tank with them. *Note: Meaningful growth data for gizzard shad could not be obtained because too few survived when in the presence of silver carp.*

Data from Phelps, Q. E., et al., 2017. Incorporating basic and applied approaches to evaluate the effects of invasive Asian carp on native fishes: A necessary first step for integrated pest management. PLOS ONE 12(9): e0184081.

DATA Go to **Interpreting Graphs & Data** on **Mastering Environmental Science**

A great deal more research is needed to answer the many outstanding questions about the impacts of Asian carp in North America's river basins—and the potential impacts of Asian carp in the Great Lakes. The Phelps team's study provides a helpful model for future work.

called a **keystone species.** A keystone is the wedge-shaped piece at the top of a stone arch that holds the structure together. Remove the keystone, and the arch will collapse (**FIGURE 4.13a**). In an ecological community, removal of a keystone species will likewise have major consequences.

Often, consumers at the tops of food chains are considered keystone species. Some predators control populations of herbivores, which otherwise could multiply and greatly modify the plant community (**FIGURE 4.13b**). Thus, predators at high trophic levels can indirectly promote populations of organisms at low trophic levels by keeping species at intermediate trophic levels in check, a phenomenon ecologists refer to as a **trophic cascade.** For example, the U.S. government long paid bounties to promote the hunting of wolves and mountain lions, which were largely exterminated by the mid-20th century. In the absence of these predators, deer populations have grown unnaturally dense and have overgrazed forest-floor vegetation and eliminated tree seedlings, leading to major changes in forest structure.

The removal of top predators throughout much of the United States was an uncontrolled large-scale experiment with unintended consequences, but ecologists have verified the keystone species concept in controlled scientific experiments. Classic research by biologist Robert Paine established that the predatory sea star *Pisaster ochraceus* shapes the community composition of intertidal organisms (p. 429) on North America's Pacific coast. When *Pisaster* is present in this community, species diversity is high, with various types of barnacles, mussels, and algae. When *Pisaster* is removed, the mussels it preys on become numerous and displace other species, suppressing species diversity. A recent outbreak of sea star wasting disease that killed most sea stars along the Pacific coast is replicating Paine's experiment on a massive scale, leading to major changes in coastal communities from Alaska to California.

Animals at high trophic levels—such as wolves, sea stars, sharks, and sea otters (see Figure 4.13)—are often viewed as keystone species that can trigger trophic cascades. However, other types of organisms also exert strong community-wide effects. "Ecosystem engineers" physically modify environments. For example, beavers build dams across streams, creating ponds and swamps by flooding land. Prairie dogs dig burrows that aerate the soil and serve as homes for other animals. Ants disperse seeds, redistribute nutrients, and selectively protect or destroy insects and plants near their colonies. And Asian carp alter the communities they invade in many ways.

Less conspicuous organisms at low trophic levels can exert still greater impact. Remove the fungi that decompose dead matter, or the insects that control plant growth, or the phytoplankton that support the marine food chain, and a community may change rapidly indeed! However, because there

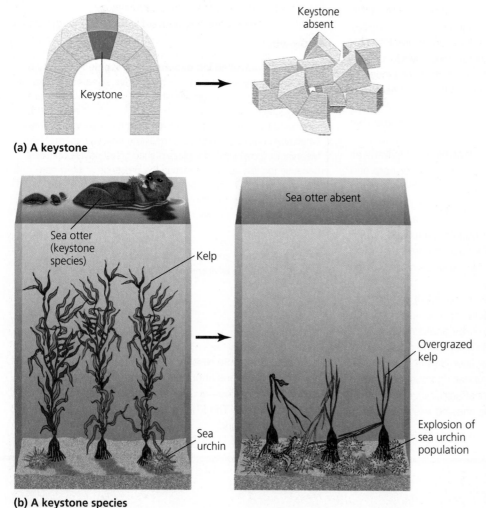

(a) A keystone

(b) A keystone species

FIGURE 4.13 Sea otters are a keystone species. A keystone **(a)** sits atop an arch, holding the structure together. A keystone species **(b)** is a species that exerts great influence on a community's composition and structure. Sea otters consume sea urchins that eat kelp in coastal waters of the Pacific. Otters keep urchin numbers down, allowing lush underwater forests of kelp to grow, providing habitat for many species. When otters are absent, urchins increase and devour the kelp, destroying habitat and depressing species diversity.

are usually more species at lower trophic levels, it is less likely that any single one of them alone has wide influence. Often if one such species is removed, its functions may be performed by other species that remain at the lower trophic level.

Communities respond to disturbance in various ways

The removal of an influential species is just one way an ecological community can be disturbed. In scientific terms, a **disturbance** is an event that has drastic impacts on environmental conditions, resulting in changes to the composition, structure, or function of a community. A disturbance can be localized, such as when a tree falls in a forest, creating a gap in the canopy that lets in sunlight. Or, it can be as large and severe as a hurricane, tornado, or volcanic eruption. Some disturbances are sudden, such as landslides or floods, whereas others are gradual, such as climate change. Some disturbances recur regularly and are considered normal aspects of a system (such as periodic fire, seasonal storms, or cyclical insect outbreaks).

Organisms may adapt to regular and predictable types of disturbance. For instance, many plants that grow in fire-prone regions have evolved ways of surviving fire and have seeds that depend on fire to germinate in the nutrient-rich soil that fire leaves behind. Today, human impacts are major sources of disturbance for ecological communities worldwide—from habitat alteration to pollution to the introduction of non-native species such as Asian carp.

Communities are dynamic systems and may respond to disturbance in several ways. A community that resists change and remains stable despite disturbance is said to show **resistance** to the disturbance. A community that changes in response to disturbance but later returns to its original state is said to show resilience. Alternatively, a community may be modified by disturbance permanently and never return to its original state.

Succession follows severe disturbance

If a disturbance is severe enough to eliminate all or most of the species in a community, the affected site may then undergo a predictable series of changes that ecologists have traditionally called **succession.** In the conventional view of this process, there are two types of succession (**FIGURE 4.14**). **Primary succession** follows a disturbance so severe that no vegetation or soil life remains from the community that had occupied the site. In primary succession, a community is built essentially from scratch. In contrast, **secondary succession** begins when a disturbance dramatically alters an existing community but does not destroy all life and organic matter. In secondary succession, vestiges of the previous community remain, and these building blocks help shape the succession process.

At terrestrial sites, primary succession takes place after a bare expanse of rock, sand, or sediment becomes newly exposed to the atmosphere. This can occur when glaciers retreat, lakes dry up, or volcanic lava or ash covers a landscape. Species that arrive first and colonize the new substrate are referred to as **pioneer species.** Pioneer species are well adapted for colonization; for instance, they generally have spores or seeds that can travel long distances.

The pioneers best suited to colonizing bare rock are the mutualistic aggregates of fungi and algae known as *lichens*. In lichens, the algal component provides food and energy via photosynthesis while the fungal component grips the rock and captures moisture. As lichens grow, they secrete acids that break down the rock surface, beginning the process that forms

PRIMARY SUCCESSION

Bare rock · Lichens

Disturbance: Farming, fire, landslide, etc.

Grasses, small herbs, and forbs

Shrubs and fast-growing trees

Shade-intolerant trees

Forest of shade-tolerant trees

SECONDARY SUCCESSION

FIGURE 4.14 In succession, an area's plant community passes through a series of typical stages. Primary succession begins as organisms colonize a lifeless new surface (two panels at top left). Secondary succession occurs after some disturbance removes most vegetation from an area (panel at bottom left).

soil. Small plants and insects arrive, providing more nutrients and habitat. As time passes, larger plants and animals establish, vegetation increases, and species diversity rises.

Secondary succession begins when a fire, a storm, logging, or farming removes much of the biotic community. Consider a farmed field in eastern North America that has been abandoned. The site first becomes colonized by pioneer species of grasses, herbs, and forbs that disperse well or were already in the vicinity. Soon, shrubs and fast-growing trees such as aspens or poplars begin to grow. As time passes, pine trees rise above the pioneer trees and shrubs, forming a pine forest. This forest gains an understory of hardwood trees, because pine seedlings do not grow well under a canopy but some hardwood seedlings do. Eventually, the hardwoods outgrow the pines, creating a hardwood forest.

Succession occurs in many ecological systems. For instance, a pond may undergo succession as algae, microbes, plants, and zooplankton grow, reproduce, and die, gradually filling the water body with organic matter. The pond acquires further organic matter and sediments from streams and runoff, and eventually it may fill in, becoming a bog (p. 398) or a terrestrial system.

In the traditional view of succession as described, the process leads to a *climax community* that remains in place until some disturbance restarts succession. Early ecologists believed that each region had its own characteristic climax community, determined by climate.

Communities may undergo shifts

Present-day ecologists recognize that community change is far more variable and less predictable than early models of succession suggested. Conditions at one stage may promote progression to another stage, or organisms may, through competition, inhibit a community's progression to another stage. The trajectory of change can vary greatly according to chance factors, such as which particular species happen to gain an early foothold. And climax communities are not determined solely by climate but vary with soil conditions and other factors from one time or place to another. Ecologists came to modify their views about how communities respond to disturbance after observing changes during long-term field studies at locations such as Mount Saint Helens following its eruption (see **THE SCIENCE BEHIND THE STORY**, pp. 90–91).

Once a community is disturbed and changes are set in motion, there is no guarantee that it will return to its original state. Instead, sometimes communities may undergo a **regime shift,** or *phase shift,* in which the character of the community fundamentally changes. This can occur if some crucial climatic threshold is passed, a keystone species is lost, or a non-native species invades. For instance, many coral reef communities have undergone a regime shift and have become dominated by algae after people overharvested fish or turtles that eat algae. Losing sea otters from a nearshore marine community can lead to the loss of kelp forests (p. 86). In some grasslands, livestock grazing and fire suppression have led shrubs and trees to invade, forming shrublands. And across large areas of western North America today, entire forests are undergoing conversions due to catastrophic fires driven by climate change and past forest management. Regime shifts make clear that we cannot always reverse damage caused by human disturbance. Instead, some of the changes we set in motion may become permanent.

Currently, human disturbance, climate change, and the introduction of non-native species are creating wholly new communities that have not previously occurred on Earth. These **novel communities,** or *no-analog communities,* are composed of new combinations of plants and animals; they are communities with no known analog, or precedent, in nature. As we enter more deeply into an age of fast-changing climate, habitat alteration, species extinctions, and species invasions, scientists predict that we will see more and more novel communities.

Introduced species may alter communities

Traditional concepts of communities involve species native to an area. But what if a species arrives from far away? In our age of global mobility and trade, countless organisms are moved from place to place. As a result, most non-native arrivals in communities today are **introduced species,** that is, alien species brought to a new area by people.

In many cases, people have introduced species on purpose. People intentionally imported Asian carp to North America, believing these fish could offer valuable services to aquaculture and wastewater treatment facilities. In contrast, many other introductions of non-native species have occurred by accident. Zebra mussels provide an example. To maintain stability, ships take water into their hulls as they begin their voyage and then discharge it at their destination. Along with this *ballast water* come any plants and animals taken up in the water. In this way, the zebra mussel was brought to America by accident in the 1980s from seas in central Asia. Within a few years, this tiny mussel spread by the trillions through the Great Lakes and into waterways across half the nation. By depleting plankton from the water, depressing fish populations, blocking pipes at power plants, and damaging boat engines, docks, buoys, and fishing gear, the zebra mussel invasion—and that of the closely related quagga mussel that

FAQ

If we disturb a community, will it return to its original state if we just leave the area alone?

Probably not, if the disturbance has been substantial. For example, if soil has become compacted or if water sources have dried up, then the plant species that grew at the site may no longer be able to grow. Different plant species may take their place—and among them, a different suite of animal species may find habitat. Sometimes a whole new community may arise. For instance, as deforestation in the Amazon (p. 315) has reduced rainfall there, unprecedented fires have burned tropical rainforests, converting some of them to scrub-grassland. In the past, people did not realize how permanent such changes could be, because we tended to view natural systems as static and predictable. Today, ecologists recognize that systems are highly dynamic and can sometimes undergo rapid, extreme, long-lasting change.

FIGURE 4.15 **The Burmese python is an invasive species in south Florida.** Individuals of this large Asian snake that escaped from the pet trade established a population in the Florida Everglades that is growing fast. The snakes prey on native birds and mammals.

followed it—have cost Great Lakes economies hundreds of millions of dollars.

Most introduced species fail to establish populations, but some succeed. Over time, successful non-native species may accumulate—and today, many regions support novel communities rich in non-native species. More than 180 non-native species inhabit the Great Lakes. Across much of San Francisco Bay, the majority of species are non-native, while 30% of California's plant species are non-native.

Of those species that successfully persist in their new homes, some may do exceptionally well, spreading widely, pushing aside native species, and coming to dominate communities. Ecologists call such species **invasive species** because they "invade" native communities (**FIGURE 4.15**). The U.S. government defines invasive species as non-native species "whose introduction causes or is likely to cause economic or environmental harm or harm to human health."

Introduced species can become invasive when limiting factors (p. 67) that regulate their population growth are absent. Plants and animals brought to a new area may leave behind the predators, parasites, herbivores, and competitors that had exploited them in their native land. If few organisms in the new environment eat, parasitize, or compete with the introduced species, then it may thrive and spread. As the species proliferates, it may exert diverse influences on other community members (**FIGURE 4.16**). By altering communities, invasive species are one of the central ecological forces in today's world.

Examples abound of invasive species that have had major ecological impacts (see p. 292 for photos of some). The native American chestnut formerly dominated eastern North American forests, but between 1900 and 1930 the chestnut blight, an invasive Asian fungus, killed nearly every mature American chestnut, transforming entire forests. Asian trees had evolved defenses against the fungus over millennia of coevolution, but the American chestnut had not. A different fungus caused Dutch elm disease, destroying most of the American elms that once gracefully lined the streets of many U.S. cities. Grasses introduced in the American West for ranching have overrun entire regions, pushing out native vegetation and encouraging fires that damage native plants like sagebrush and cacti, further promoting dominance of the grasses. Freshwater systems like the Great Lakes have been upended by predatory fishes released from aquaria, bait buckets, stocking, or aquaculture; aquatic plants that become ecosystem engineers; crayfish that disrupt the benthic portions of food webs; and disease pathogens. On the world's islands, hundreds of island-dwelling species have been driven extinct by goats, pigs, rats, and other mammals introduced by people (Chapter 3).

Because of such impacts, most ecologists view invasive species in a negative light, and this often carries over to their perceptions of non-native species in general. Yet many of us enjoy the beauty of introduced ornamental plants in gardens.

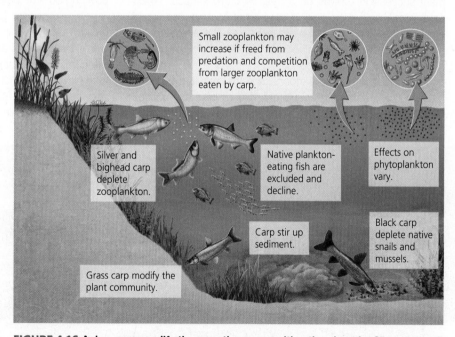

FIGURE 4.16 **Asian carp modify the aquatic communities they invade.** Silver carp and bighead carp outcompete native fish for zooplankton, which, in turn, can affect phytoplankton in various ways. Grass carp and black carp alter water quality by stirring up sediment as they consume native plants and mollusks. *Based on data from Sass, G., et al., 2014. Invasive bighead and silver carp effects on zooplankton communities in the Illinois River, Illinois, USA. J. Great Lakes Research 40: 911–921, and various other studies.*

Small zooplankton may increase if freed from predation and competition from larger zooplankton eaten by carp.

Silver and bighead carp deplete zooplankton.

Native plankton-eating fish are excluded and decline.

Effects on phytoplankton vary.

Carp stir up sediment.

Black carp deplete native snails and mussels.

Grass carp modify the plant community.

THE SCIENCE behind the story

How Do Communities Recover after Catastrophic Disturbance?

Dr. Virginia Dale at Mount Saint Helens

The eruption of Mount Saint Helens offered ecologists a rare opportunity to study how communities recover from catastrophic disturbance. On May 18, 1980, this volcano in the state of Washington erupted in sudden and spectacular violence, with 500 times the force of the atomic blast at Hiroshima.

The massive explosion obliterated an entire landscape of forest as a scalding mix of gas, steam, ash, and rock was hurled outward. A pyroclastic flow (p. 42) sped downslope, along with the largest landslide in recorded history. Rock and ash rained down and mudslides and lahars (p. 44) raced down river valleys, devastating everything in their paths. Altogether, 4.1 km³ (1.0 mi³) of material was ejected from the mountain, affecting 1650 km² (637 mi²), an area twice the size of New York City.

In the aftermath of the blast, ecologists moved in to take advantage of the natural experiment of a lifetime. For them, the eruption provided an extraordinary chance to study how primary succession unfolds on a fresh volcanic surface. Which organisms would arrive first? What kind of communities would emerge? How long would the process take? These researchers set up study plots to examine how populations, communities, and ecosystems would respond.

Today, the barren gray moonscape that resulted from the blast 40 years ago is a vibrant green in many places (**FIGURE 1**), carpeted with shrubs, young trees, and colorful flowers. And what ecologists have learned has modified our view of primary succession and informed the entire study of disturbance ecology.

Given the ferocity and scale of the eruption, most scientists initially presumed that life had been wiped out completely over a large area. Based on traditional views of succession, they expected that pioneer species would colonize the area, spreading slowly from the outside margins inward, and that over many years a community would be built in a systematic and predictable way.

Instead, researchers discovered that some plants and animals had survived the blast. Some were protected by deep snowbanks. Others were sheltered on steep slopes facing away from the blast. Still others were dormant underground when the eruption occurred. These survivors, it turned out, would play key roles in rebuilding the community.

Many of the ecologists drawn to Mount Saint Helens studied plants. Virginia Dale of Oak Ridge National Laboratory in Tennessee and her colleagues examined the debris avalanche, a landslide of rock and ash as deep as a 15-story building. This region appeared barren, yet small numbers of plants of 20 species had survived, growing from bits of root or stem carried down in the avalanche. However, most plant regrowth occurred from seeds blown in from afar. Dale's team used sticky traps to sample these seeds as they chronicled the area's recovery. After 30 years, the number of plant species had grown to 155, covering 50% of the ground surface.

One important pioneer species was red alder. This tree grows quickly, deals well with browsing by animals, and produces many seeds at a young age. As a result, it has become the dominant tree species on the debris avalanche. Because it fixes nitrogen (pp. 125–126), the red alder enhances soil fertility and thereby helps other plants grow. Researchers predict that red alder will remain dominant for years or decades and that

Before eruption

After eruption

Today

FIGURE 1 Mount Saint Helens before its eruption, after the 1980 eruption, and nearly four decades later.

conifers such as Douglas fir will eventually outgrow the alders and establish a conifer-dominated forest.

Patterns of plant growth have varied in different areas. Roger del Moral of the University of Washington and his colleagues compared ecological responses on a variety of surfaces, including barren pumice, mixed ash and rock, mudflows, and the "blowdown zone" where trees were toppled like matchsticks. Numbers of species and percentage of plant cover increased in different ways on each surface, affected by a diversity of factors.

Parts of the blast zone were replanted by people. After the eruption, concerns about erosion led federal officials to disperse seeds of eight plant species (seven of them non-native) by helicopter over large areas in hopes of quickly stabilizing the surface with vegetation. Dale and her colleagues took the opportunity to compare plant growth in areas that were manually seeded versus growth in areas that recovered naturally without manual seeding. They found that manual seeding established plant cover on the ground more quickly, and that this effect was long-lasting: Even after 30 years, manually seeded areas had more plant cover than unseeded areas (**FIGURE 2a**). The same was true, to a lesser extent, with the number of plant species (**FIGURE 2b**). However, the manually seeded areas contained a higher proportion of plants of non-native species than did the unseeded areas, even after 30 years. Researchers predict that the two types of areas will gradually become more similar as conifer forest replaces the pioneer species—and Dale's team is now finding evidence that this is beginning to occur.

Among plants, windblown seeds accounted for most regrowth, but plants that happened to survive in sheltered "refugia" within the impact zone helped to repopulate areas nearby. Indeed, chance played a large role in determining which organisms survived and how vegetation recovered, researchers have found. Had the eruption occurred in late summer instead of spring, there would have been no snow, and many of the plants that survived would have died. Had the eruption occurred at night instead of in the morning, nocturnal animals would have been hit harder.

Animals played major roles in the recovery. In fact, researchers such as Patrick Sugg and John Edwards of the University of Washington showed that insects and spiders arrived in the impact zone in great numbers even before plants did! Insects fly, while spiders disperse by "ballooning" on silken threads, so in summer the atmosphere is filled with an "aerial plankton" of windblown arthropods. Trapping and monitoring at Mount Saint Helens in the months following the eruption showed that insects and spiders landed in the impact zone by the billions. Researchers estimated that more than 1500 species arrived in the first few years, surviving by scavenging or by preying on other arthropods. Most individuals soon died, but the nutrients from their bodies enriched the soil, helping the community to develop.

Once plants took hold, animals began exerting influence through herbivory. Caterpillars fed on plants, occasionally

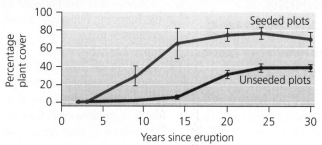

(a) Percentage plant cover

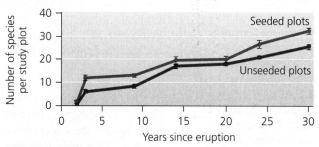

(b) Species richness

FIGURE 2 Plants recovered differently at manually seeded and unseeded sites at Mount Saint Helens. In the 30 years after the eruption, **(a)** percentage of ground covered by plants and **(b)** species richness of plants both increased, with cover and species richness being greater in manually reseeded areas than in unseeded areas recovering naturally. *Data from Dale, V. H. and E. M. Denton, 2018. Plant succession on the Mount St. Helens debris avalanche deposit and the role of non-native species. In C. M. Crisafulli and V. Dale (Eds.),* Ecological responses at Mount St. Helens: Revisited 35 years after the 1980 eruption. *New York: Springer-Verlag.*

 Go to **Interpreting Graphs & Data** on **Mastering Environmental Science**

extinguishing small populations. Researchers led by Charles Crisafulli of the U.S. Forest Service studied mammals and found that their species diversity increased in a linear fashion. Fully 34 of the 45 mammal species present in the larger region colonized the blast area within 35 years—from tiny mice up to elk and mountain goats.

All told, research at Mount Saint Helens has shown that succession is not a simple and predictable process. Instead, communities recover from disturbance in ways that are dynamic, complex, and highly dependent on chance factors. The results also show life's resilience. Even when the vast majority of organisms perish in a natural disaster, a few may survive, and their descendants may eventually build a new community.

Ecological change at Mount Saint Helens will continue for many decades more. All along the way, ecologists will seek to study and learn from this tremendous natural experiment.

WEIGHING
the issues

Are All Non-Native Species Harmful?

If we introduce a non-native species to a community and it greatly modifies the community, do you think that, in itself, is a bad thing? What if it causes a native species to disappear from the community? What if the non-native species arrived on its own, rather than through human intervention? What if it provides economic services, as the European honeybee does? What ethical standard(s) (p. 136) would you apply to assess whether we should reject, tolerate, or welcome a non-native species?

Some introduced species provide economic benefits, such as the European honeybee, which pollinates many of our crops (pp. 223–224). And some organisms are introduced intentionally to control pests through biocontrol (p. 253). While hundreds of millions of dollars are being spent to keep Asian carp out of the Great Lakes, people have largely welcomed the introduction there of Coho and Chinook salmon, two other non-native fish, because they are seen as desirable food fish and do not have as much impact on other species.

In recent years, some scientists have begun pushing back against the conventional wisdom. They point out that the locations and habitats in which non-native species generally establish themselves tend to be those that are already severely disturbed by human impact. In such degraded areas, non-native species can actually be beneficial; they may provide resources to native organisms and, arguably, may help severely degraded ecosystems recover from human impact. Defenders of non-native species also stress that non-natives generally increase, rather than decrease, the species diversity of a community. Some defenders even maintain that in a world shaped by human impact, the species that have proven themselves successful invaders may give biodiversity its best chance of persisting in the face of our disruption.

Whatever view one takes, the changes that introduced species—and particularly invasive species—bring to native populations and communities can be significant. These impacts are growing year by year with our increasing mobility and the globalization of society.

We can respond to invasive species with control, eradication, or prevention

Ever since U.S. policymakers began to recognize the impacts of invasive species in the 1990s, federal funding has become available for their control and eradication. Eradication (total elimination of a population) is extremely difficult (**FIGURE 4.17**). For this reason, managers usually aim merely to control populations—that is, to limit their growth, spread, and impact. Managers have tried to control Asian carp in heavily infested waterways like the Illinois River by netting juvenile fish, hiring contract fishermen to catch adults, bubbling carbon dioxide into the water, broadcasting noise into the water, and testing the effects of hot water and ozone. Some are also aiming to reintroduce the alligator gar as a predator. Meanwhile,

FIGURE 4.17 Fully eradicating an invasive species is difficult. Garlic mustard has displaced native ground cover in many forests of eastern North America. Dedicated volunteers engaged in ecological restoration projects are doing their best to remove it, but it is an uphill battle.

researchers are trying to develop a chemical microparticle that will poison carp while not affecting other fish.

However, most of these attempts at control or eradication are localized or short-term fixes. With one invasive species after another, managers are finding that control and eradication are difficult and expensive. Instead, trying to prevent invasions in the first place represents a better investment. This explains why, with Asian carp, managers are working hard to prevent them from entering the Great Lakes. The Army Corps' electric barriers in the Chicago Ship and Sanitary Canal are one main prevention technique. Engineers have also built various dikes and berms to block floodwaters when heavy rains cause water to overflow from riverbanks and canals.

To meet the challenge of species transport across oceans in ballast water, the U.S. government and the international community each require ships to dump their freshwater ballast at sea (where freshwater organisms are killed by salt water) and exchange it with salt water before entering ports in areas such as the Great Lakes. They also are working with industry researchers to test and approve methods of treating ballast water with chemicals, filters, ultraviolet light, heat, and electricity to kill organisms onboard without exchanging water.

Communities can be restored

Invasive species add to the disturbances that people have forced on natural systems through habitat alteration, pollution, and overhunting. Ecological systems support our civilization and all of life, so when systems become so transformed that they cease to function properly, our very health and well-being are threatened.

This realization gave rise to the science of **restoration ecology.** Restoration ecologists research the historical conditions of ecological communities as they existed before our civilization altered them. These ecologists then try to devise ways to restore altered areas to an earlier condition. In some cases, the intent is to restore the functionality of the system—to reestablish a wetland's ability to filter pollutants and recharge groundwater, for example, or a forest's ability to cleanse the air, build soil, and provide habitat for wildlife. In other cases, the aim is to return a community to its natural "presettlement" condition. Either way, the science of restoration ecology informs the practice of **ecological restoration,** the on-the-ground efforts to carry out these visions and restore communities (see **SUCCESS STORY**).

In general, ecological restoration involves trying to undo impacts of human disturbance and to reestablish species, populations, communities, and natural ecological processes. Specifically, restoration often involves removing invasive species and planting native vegetation that had originally grown on a site. Sometimes it means reintroducing natural processes such as fire or flooding. It may also mean modifying the landscape to reduce erosion or influence patterns of water flow.

The world's largest restoration project is the ongoing effort to restore the Everglades, a vast ecosystem of marshes and seasonally flooded grasslands stretching across southern Florida. This wetland system has been drying out for decades because the water that feeds it has been manipulated for flood control and overdrawn for irrigation and development. Economically important fisheries have suffered greatly as a result, and the region's famed populations of wading birds have dropped by 90–95%. The 30-year, $7.8-billion restoration project intends to restore natural water flow by removing hundreds of miles of canals and levees. Because the Everglades provides drinking water for millions of Florida citizens, as well as considerable tourism revenue, restoring its ecosystem services (pp. 4, 120–121) should prove economically beneficial.

Efforts in Florida have met with some success so far: Canals have been filled in and stretches of the Kissimmee River now flow freely (**FIGURE 4.18**), bringing water to Lake Okeechobee and then to the Everglades. But many challenges remain. The ambitious project has struggled against budget shortfalls and political interference, and invasive species such as the Burmese python (see Figure 4.15) are preying on the native fauna. (We will explore ecological restoration projects further in Chapter 11; p. 302.)

FIGURE 4.18 The largest restoration project today is in Florida. Here, water from a canal has been returned to the Kissimmee River, which now winds freely toward the Everglades.

Restoration projects can sometimes run into complications. One example is Eagle Marsh, a 756-acre wetland near Fort Wayne, Indiana, that happens to play a key role in the Asian carp story. Starting in 2005, volunteers with the Little River Wetland Project began converting farmland back into the marsh that originally existed at the site. They deepened shallow areas, removed pumps and drain tiles, seeded the land with native wildflowers, planted 45,000 native trees, and removed invasive plants in what became the largest inland urban wetland restoration in the nation. Just as they were celebrating their success, however, scientists realized that the site represented a likely entry point for Asian carp to reach the Great Lakes. The marsh sits at the divide between the Wabash River (which flows to the Mississippi) and the Maumee River (which flows to Lake Erie), and in major flooding events, waters from these drainages can comingle, allowing fish from one watershed to slip into the other. Once this was recognized, $4 million was spent to build a 2750-m-long (9000-ft-long) berm to keep the watersheds divided. The berm was finished just in time to thwart three major floods, but this also

SUCCESS story — Restoring Native Prairie Near Chicago

Nearly all the tallgrass prairie in North America was converted to agriculture in the 1800s and early 1900s. Immense regions of land in states like Iowa, Kansas, and Illinois that today grow corn, wheat, and soybeans were once carpeted in a lush and diverse mix of grasses and wildflowers and inhabited by countless birds, bees, and butterflies. Currently, people are trying to recreate areas of these rich native landscapes through ecological restoration. Scientists and volunteers are restoring patches of prairie by planting native vegetation, weeding out invaders and competitors, and introducing prescribed fire (p. 323) to mimic the fires that historically maintained prairie communities. The region

outside Chicago, Illinois, boasts a number of prairie restoration projects, the largest of which is inside the gigantic ring at the Fermilab National Accelerator Laboratory. This project, begun in 1971 by ecologists Robert Betz and Raymond Schulenberg, today involves hundreds of community volunteers and has restored more than 400 ha (1000 acres) of prairie.

Restored prairie at Fermilab, outside Chicago, Illinois.

→ **Explore the Data** at Mastering Environmental Science

WEIGHING
the **issues**

Restoring "Natural" Communities

Practitioners of ecological restoration aim to restore communities to their natural state. But what does "natural" mean? Does it mean the state of a community before modern impacts from industrialization? Before Europeans came to the New World? Before any people laid eyes on the community? Native Americans modified many forests starting thousands of years ago by burning underbrush to improve hunting. Should restorationists try to recreate forests that existed before Native Americans arrived? What values do you think underlie the desire for restoration?

changed the hydrology of the newly restored preserve.

As our population grows and development spreads, ecological restoration is becoming an increasingly vital conservation strategy. However, restoration is difficult, time-consuming, and expensive. It is therefore best, whenever possible, to protect natural systems from degradation in the first place.

Climate change poses new challenges for restoration

Restoration ecologists have long grappled with questions about what baseline conditions they want to aim for, given that most ecological communities change naturally over time and have been influenced by successive waves of people. But today restorationists face still more challenging questions, as a result of global climate change (Chapter 18). As the world warms, as species shift toward the poles and upward in elevation, and as droughts, floods, fires, heat waves, cold snaps, and storms become more severe, it becomes harder to reestablish past communities that can persist in today's modified climatic conditions. Even if we could perfectly re-create a forest community that existed in southern Ohio in 1800, for example, it may not hold together in our new and changing climatic environment. Some species might not survive, while others might begin interacting in new ways. Thus, the traditional approach of trying to restore a well-defined suite of plant and animal species to an area may no longer work. Increasingly, restorationists are instead aiming to restore the functions of communities and ecosystems, regardless of what species may be involved—and to create systems that are resilient and can adapt to climate change in the long term. Some restorationists now are even including non-native species in restoration projects, when these are judged to improve the function and resilience of the community.

Earth's Biomes

Across the world, each location is home to different assemblages of species, leading to endless variety in community composition. However, communities in far-flung places often share strong similarities in their structure and function. This allows us to classify communities into broad types. A **biome** is a major regional complex of similar communities—a large-scale ecological unit recognized primarily by its dominant plant types and vegetation structure. The world contains a number of terrestrial biomes, each covering large geographic areas (**FIGURE 4.19**).

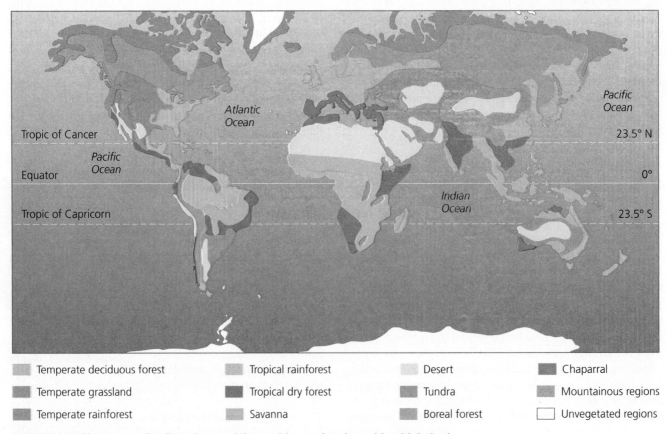

- Temperate deciduous forest
- Temperate grassland
- Temperate rainforest
- Tropical rainforest
- Tropical dry forest
- Savanna
- Desert
- Tundra
- Boreal forest
- Chaparral
- Mountainous regions
- Unvegetated regions

FIGURE 4.19 Biomes are distributed around the world, correlated roughly with latitude.

Climate helps determine biomes

Which biome covers each portion of the planet depends on a variety of physical factors, including temperature, precipitation, soil conditions, wind patterns, and ocean currents. Temperature and precipitation tend to exert the greatest influence (**FIGURE 4.20**). Note also in Figure 4.19 that patches of any given biome tend to occur at similar latitudes. For instance, temperate deciduous forest occurs in Europe, China, and eastern North America. This is due to Earth's north–south gradients in temperature and to atmospheric circulation patterns (p. 455).

Scientists use **climate diagrams,** or *climatographs,* to depict information on temperature and precipitation. As we tour the world's terrestrial biomes, you will see climate diagrams from specific localities. The data in each graph are typical of the climate for the biome the locality lies within.

In mountainous regions, complex mixes of biomes occur because climate varies with elevation. At higher altitudes, temperature, atmospheric pressure, and oxygen all decline, while ultraviolet radiation increases. A hiker scaling one of the great peaks of the Andes in Ecuador, near the equator, might climb from tropical rainforest through cloud forest up to alpine tundra and glaciers. Mountain ranges can also alter

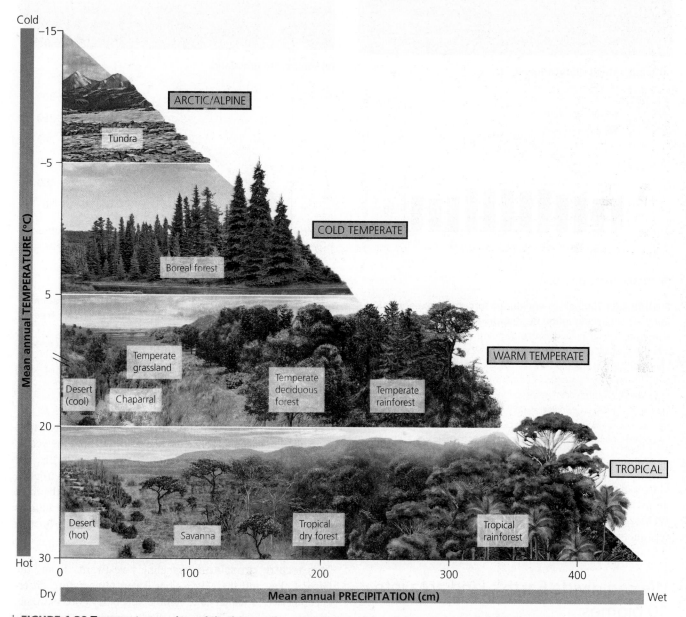

FIGURE 4.20 Temperature and precipitation are the main factors determining where each biome occurs. As precipitation increases, vegetation becomes taller and more luxuriant. As temperature increases, types of plant communities change. For instance, deserts occur in dry regions; tropical rainforests occur in warm, wet regions; and tundra occurs in the coldest regions. *Note:* The break in the *y*-axis indicates a jump in values; the temperate zone has a wider range of temperatures than the other zones.

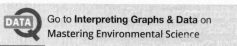

Go to **Interpreting Graphs & Data** on Mastering Environmental Science

(a) Temperate deciduous forest

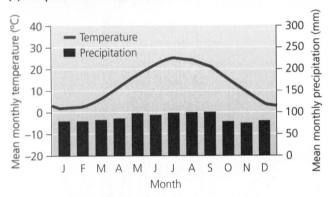

(b) Washington, D.C., USA

FIGURE 4.21 Temperate deciduous forests (a) experience fairly stable precipitation but temperatures that vary with the seasons. Scientists use climate diagrams **(b)** to illustrate average monthly precipitation and temperature. In these diagrams, the blue bars indicate precipitation and the red data lines indicate temperature, from month to month. Summer months are in the center of the x-axis for both Northern Hemisphere and Southern Hemisphere locations. *Climate diagram here and in the following figures adapted from Breckle, S.-W. 2002.* Walter's vegetation of the Earth: The ecological systems of the geo-biosphere, *4th ed. Berlin, Heidelberg: Springer-Verlag.*

climate through the **rainshadow effect.** When moisture-laden air ascends a steep slope, it releases precipitation as it cools. By the time the air flows over the top and down the other side of a mountain, it can be very dry, creating an arid region in the rainshadow.

We can divide Earth's land area into 10 biomes

Temperate deciduous forest The **temperate deciduous forest** (**FIGURE 4.21**) that dominates the landscape around much of the Great Lakes and Mississippi River Valley is characterized by broad-leafed trees that are *deciduous,* meaning that they lose their leaves each fall and remain dormant during winter, when hard freezes would endanger leaves. These midlatitude forests occur in much of Europe and eastern

(a) Temperate grassland

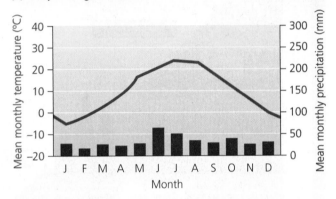

(b) Odessa, Ukraine

FIGURE 4.22 Temperate grasslands experience seasonal temperature variation and too little precipitation for trees to grow. *Climatograph adapted from Breckle, S.-W. 2002.*

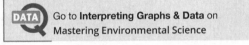
Go to **Interpreting Graphs & Data** on Mastering Environmental Science

China as well as in eastern North America—all areas where precipitation is spread relatively evenly throughout the year.

Soils of the temperate deciduous forest are fertile, but this biome consists of far fewer tree species than do tropical rainforests. Oaks, beeches, and maples are a few of the most common types of trees in these forests. Some typical animals of the temperate deciduous forest of eastern North America are shown in Figure 4.12 (p. 83).

Temperate grassland Traveling westward from the Great Lakes and the Mississippi River, temperature differences between winter and summer become more extreme, rainfall diminishes, and we find **temperate grasslands** (**FIGURE 4.22**). This is because the limited precipitation in the Great Plains region supports grasses more easily than trees. Also known as *steppe* or *prairie,* temperate grasslands were once widespread in much of North and South America and central Asia.

Vertebrate animals of North America's native grasslands include American bison, prairie dogs, pronghorn antelope,

(a) Temperate rainforest

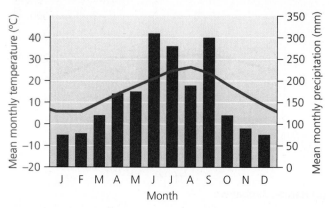

(b) Nagasaki, Japan

FIGURE 4.23 Temperate rainforests receive a great deal of precipitation and have moist, mossy interiors. *Climatograph adapted from Breckle, S.-W. 2002.*

(a) Tropical rainforest

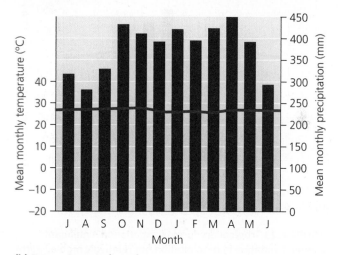

(b) Bogor, Java, Indonesia

FIGURE 4.24 Tropical rainforests, famed for their biodiversity, grow under constant, warm temperatures and a great deal of rain. *Climatograph adapted from Breckle, S.-W. 2002.*

and ground-nesting birds such as meadowlarks and prairie chickens. People have converted most of the world's grasslands for farming and ranching, however, so most of these animals exist today at a small fraction of their historic population sizes.

Temperate rainforest Farther west in North America, the topography becomes varied, and biome types intermix. The coastal Pacific Northwest region, with its heavy rainfall, features **temperate rainforest** (**FIGURE 4.23**). Coniferous trees (cone-producing evergreen trees) such as cedars, spruces, hemlocks, and Douglas fir grow very tall in this biome, and the forest interior is shaded and damp. Moisture-loving animals such as the bright yellow banana slug are common in temperate rainforests.

The soils of temperate rainforests are fertile, but they are susceptible to landslides and erosion if forests are cleared. We have long extracted lumber and other commercially valuable products from temperate rainforests. Timber harvesting has eliminated most old-growth trees in these forests, driving species such as the spotted owl and marbled murrelet, birds that nest in old-growth trees, toward extinction.

Tropical rainforest In tropical regions, we see the same pattern found in temperate regions: Areas of high rainfall grow rainforests, areas of intermediate rainfall support dry or deciduous forests, and areas of low rainfall are dominated by grasses. However, tropical biomes differ from their temperate counterparts in other ways because they are closer to the equator and therefore warmer on average year-round. For one thing, they hold far greater biodiversity.

Tropical rainforest (**FIGURE 4.24**)—found in Central America, South America, Southeast Asia, west Africa, and other tropical regions—is characterized by year-round rain and uniformly warm temperatures. Tropical rainforests have dark and damp interiors, lush vegetation, and highly diverse communities with more species of insects, birds, amphibians, and other animals than any other biome. These forests are not dominated by certain species of trees but instead consist of very high numbers of tree species intermixed, each at a low density. Trees may be draped with vines, enveloped by

(a) Tropical dry forest

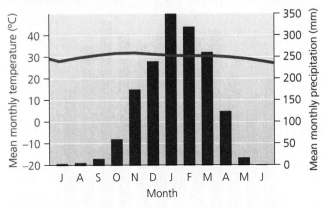

(b) Darwin, Australia

FIGURE 4.25 Tropical dry forests experience significant seasonal variation in precipitation but relatively stable, warm temperatures. *Climatograph adapted from Breckle, S.-W. 2002.*

(a) Savanna

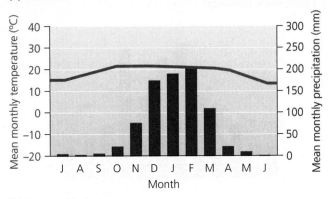

(b) Harare, Zimbabwe

FIGURE 4.26 Savannas are grasslands with clusters of trees. They experience slight seasonal variation in temperature but significant variation in rainfall. *Climatograph adapted from Breckle, S.-W. 2002.*

strangler figs, and loaded with epiphytes (orchids and other plants that grow in trees). Indeed, rainforest trees sometimes collapse under the weight of all the life they support!

Despite this profusion of life, tropical rainforests have poor, acidic soils that are low in organic matter. Nearly all nutrients present in this biome occur in the plants, not in the soil. An unfortunate consequence is that once people clear tropical rainforests, the nutrient-poor soil can support agriculture for only a short time (p. 219). As a result, farmed areas are abandoned quickly, and farmers move on and clear more forest.

Tropical dry forest Tropical areas that are warm year-round but where rainfall is lower overall and highly seasonal give rise to **tropical dry forest,** or *tropical deciduous forest* (**FIGURE 4.25**), a biome widespread in India, Africa, South America, and northern Australia. Wet and dry seasons each span about half a year in tropical dry forest. Organisms that inhabit tropical dry forest have adapted to seasonal fluctuations in precipitation and temperature. For instance, many plants are deciduous and leaf out and grow profusely with the rains, then drop their leaves during dry times of year.

Rains during the wet season can be heavy and, coupled with erosion-prone soils, can lead to severe soil loss where people have cleared forest. Across the globe, we have converted a great deal of tropical dry forest to agriculture. Clearing for farming or ranching is straightforward in regions of tropical dry forest because vegetation is lower and canopies less dense than in tropical rainforest.

Savanna Drier tropical regions give rise to **savanna** (**FIGURE 4.26**), tropical grassland interspersed with clusters of acacias or other trees. The savanna biome is found across stretches of Africa, South America, Australia, India, and other dry tropical regions. Precipitation generally arrives during distinct rainy seasons, whereas in the dry season grazing animals concentrate near widely spaced water holes. Common herbivores on the African savanna include zebras, gazelles, and giraffes. Predators of these grazers include lions, hyenas, and other highly mobile carnivores. Scientific research indicates that the African savanna was the ancestral home of the human species. The open spaces of this biome likely favored the evolution of our upright stance, running ability, and keen vision.

(a) Desert

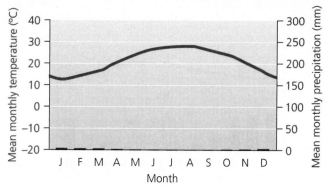

(b) Cairo, Egypt

FIGURE 4.27 Deserts are dry year-round but are not always hot. *Climatograph adapted from Breckle, S.-W. 2002.*

(a) Tundra

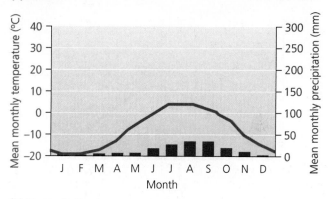

(b) Vaygach, Russia

FIGURE 4.28 Tundra is a cold, dry biome found near the poles. Alpine tundra occurs atop high mountains at lower latitudes. *Climatograph adapted from Breckle, S.-W. 2002.*

Desert Where rainfall is very sparse, **desert** (**FIGURE 4.27**) forms. The driest biome on Earth, most deserts receive well under 25 cm (10 in.) of precipitation per year, much of it during isolated storms months or years apart. Some deserts, such as Africa's Sahara and Namib deserts, are mostly bare sand dunes; others, such as the Sonoran Desert of Arizona and northwest Mexico, receive more rain and are more heavily vegetated.

Deserts are not always hot; the high desert of the western United States is positively cold in winter. Because deserts have low humidity and little vegetation to insulate them from temperature extremes, sunlight readily heats them in the daytime, but heat is quickly lost at night. As a result, temperatures vary greatly from day to night and from season to season. Desert soils can be saline and are sometimes known as lithosols, or stone soils, for their high mineral and low organic-matter content.

Desert animals and plants show many adaptations to deal with a harsh climate. Most reptiles and mammals, such as rattlesnakes and kangaroo mice, are active in the cool of night. Many Australian desert birds are nomadic, wandering long distances to find areas of recent rainfall and plant growth. Desert plants tend to have thick, leathery leaves to reduce water loss, or green trunks so that the plant can conduct photosynthesis without leaves, minimizing the surface

area prone to water loss. The spines of cacti and other desert plants guard them from being eaten by herbivores desperate for the precious water these plants hold. Such traits have evolved by convergent evolution in deserts across the world (see Figure 3.4b, p. 53).

Tundra Nearly as dry as desert, **tundra** (**FIGURE 4.28**) occurs at very high latitudes in northern Russia, Canada, and Scandinavia. Extremely cold winters with little daylight and summers with lengthy days characterize this landscape of lichens and low, scrubby vegetation without trees. The great seasonal variation in temperature and day length results from this biome's high-latitude location, angled toward the sun in summer and away from the sun in winter.

Because of the cold climate, underground soil remains permanently frozen and is called *permafrost*. During winter, surface soil freezes as well. When the weather warms, the soil melts and produces pools of surface water, forming ideal habitat for mosquitoes and other insects. The swarms of insects benefit bird species that migrate long distances to breed during the brief but productive summer. Caribou also migrate to the tundra to breed, then leave for the winter. Only a few animals, such as polar bears and musk oxen, can survive year-round in tundra.

(a) Boreal forest

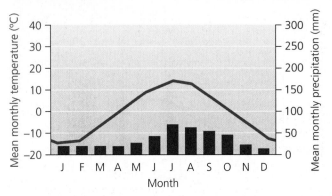

(b) Archangelsk, Russia

FIGURE 4.29 Boreal forest experiences long, cold winters; cool summers; and moderate precipitation. *Climatograph adapted from Breckle, S.-W. 2002.*

(a) Chaparral

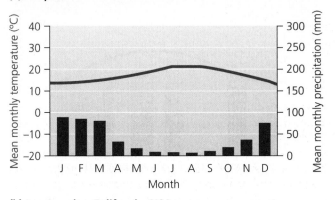

(b) Los Angeles, California, USA

FIGURE 4.30 Chaparral is a seasonally variable biome dominated by shrubs, influenced by marine weather, and dependent on fire. *Climatograph adapted from Breckle, S.-W. 2002.*

Most tundra remains intact and relatively unaltered by human occupation and development. However, global climate change is heating high-latitude regions more intensely than other areas (p. 496), melting sea ice, and altering seasonal cycles to which animals have adapted. Climate warming is also melting permafrost, causing the greenhouse gas methane to seep out of the ground and drive further warming.

Tundra also occurs as alpine tundra at the tops of tall mountains in both temperate and tropical regions. Here, high elevation creates climatic conditions similar to those of high latitude.

Boreal forest The northern coniferous forest, or **boreal forest,** often called *taiga* (**FIGURE 4.29**), extends across much of Canada, Alaska, Russia, and Scandinavia. A few species of evergreen trees, such as black spruce, dominate large stretches of forest, interspersed with bogs and lakes. Boreal forests occur in cooler, drier regions than do temperate forests, and they experience long, cold winters and short, cool summers.

Soils of the boreal forest are typically nutrient-poor and somewhat acidic. As a result of strong seasonal variation in day length, temperature, and precipitation, many organisms compress a year's worth of feeding and breeding into a few warm, wet months. Year-round residents of boreal forest include mammals such as moose, wolves, bears, lynx, and rodents. Some animals hibernate during the harsh winters, and many insect-eating birds migrate here from the tropics to breed during the brief, intensely productive, summers. The boreal forests are vast, but are being lost and modified today as a result of logging, fossil fuel extraction, and fires and pest outbreaks driven by climate change (p. 323–324).

Chaparral In contrast to the boreal forest's broad, continuous distribution, **chaparral** (**FIGURE 4.30**) is limited to small patches widely dispersed around the globe. Chaparral consists mostly of evergreen shrubs and is densely thicketed. This biome is highly seasonal, with mild, wet winters and warm, dry summers—a climate influenced by oceanic waters and often termed "Mediterranean." Besides ringing the Mediterranean Sea, chaparral occurs along the coasts of California, Chile, and southern Australia. Chaparral communities naturally experience frequent fire, and their plants are adapted to resist fire or even to depend on it for germination of their seeds. As a result, people living in regions of chaparral need to pay special attention to managing risks from wildfire.

Aquatic systems resemble biomes

Our discussion of biomes focuses exclusively on terrestrial systems because the biome concept, as traditionally developed and applied, has done so. However, areas equivalent to biomes also exist in the oceans, along coasts, and in freshwater systems. (We will examine freshwater, marine, and coastal systems in the greater detail they deserve in Chapters 15 and 16.) Aquatic systems are shaped not by air temperature and precipitation but by water temperature, salinity, dissolved nutrients, wave action, currents, depth, light levels, and type of substrate (e.g., sandy, muddy, or rocky bottom). Marine communities are also more clearly delineated by their animal life than by their plant life.

One might consider the shallows along the world's coastlines to represent one aquatic system, the continental shelves another, and the open ocean, the deep sea, coral reefs, and kelp forests as still others. Many coastal systems—such as salt marshes, rocky intertidal communities, mangrove forests, and estuaries—share terrestrial and aquatic components. And freshwater systems such as those of the Great Lakes are widely distributed throughout the world.

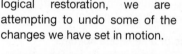

CENTRAL CASE STUDY
connect & continue

TODAY, scientists, policymakers, and managers are trying to limit the spread of Asian carp by controlling their numbers in the Mississippi and Ohio River basins and preventing their establishment in the Great Lakes. The Army Corps is completing a fourth electric barrier in the Chicago Ship and Sanitary Canal, even while admitting that these barriers are an "experimental, temporary fix." Also proposed by the Army Corps is a $275-million upgrade of the Brandon Road Lock and Dam on the canal, near Joliet, Illinois, where a new electric barrier would be built, supplemented with water jets, noise cannons, a flushing system, and more. Illinois state officials and those from Chicago are reluctant to pursue the idea, fearing it will impede shipping traffic and commerce, hurting the economy and forcing goods onto already congested rail and highway networks. Officials from states like Michigan, who fear the arrival of Asian carp into the Great Lakes, are pushing hard for the upgrade and offering to help pay for it. Many of them are urging still more: They would like to see billions of dollars spent to close or reconfigure the canal system to shut off access to non-native species for good.

At the same time, researchers are using eDNA (environmental DNA) monitoring—a technique whereby water can be sampled for traces of genetic material from a given organism—to determine where Asian carp might lurk unseen. And researchers, engineers, and entrepreneurs are continuing to develop innovative new means of deterring Asian carp.

Meanwhile, along the waterways where these invasive fish are established, people bear the costs and adapt as best as they can. Fortunately, all biological invasions eventually slow and populations stop growing. Often, some native species discover the aliens and become their predators, parasites, or competitors. Some long-established invasive species in North America have even begun to decline.

No one knows what the future holds in the case of Asian carp, but these fish and many other species that people have moved from place to place are creating a topsy-turvy world of novel and modified communities. In response, through ecological restoration, we are attempting to undo some of the changes we have set in motion.

- **CASE STUDY SOLUTIONS** A federal agency has put you in charge of devising responses to the invasion of Asian carp. Based on what you know from this chapter, how would you seek to control the spread of these species and reduce their impacts? Describe one or more strategies that you would consider pursuing immediately. Describe one or more strategies for which you would commission further scientific research. What questions would you pose to your scientific advisors? For each of your ideas, name one benefit you would expect it to produce, and identify one obstacle it might face in being implemented.

- **LOCAL CONNECTIONS** Asian carp have invaded waterways across much of the eastern and central United States, and the Great Lakes are home to more than 180 introduced species. But every location in the world by now has its own suite of introduced, non-native species. Research and describe one plant species and one animal species (aside from Asian carp) that have been introduced in the region where you live. What impacts (bad, good, or neutral) have each of these species had on ecological communities in your region? What economic impacts have they had? What specific efforts are being made in your region to deal with invasive species considered harmful? What are some results of these efforts so far?

- **EXPLORE THE DATA** How do scientists determine what impacts Asian carp and other invasive species—such as zebra mussels and quagga mussels—are having on aquatic communities? ➡ **Explore Data** relating to the case study on **Mastering Environmental Science.**

REVIEWING Objectives

You should now be able to:

+ **Summarize and compare the major types of species interactions**

Competition results when organisms vie for limited resources. Competition can occur within or among species and can result in coexistence or exclusion. In predation, an individual of one species kills and consumes an individual of another. In parasitism, an individual of one species derives benefit by harming (but usually not killing) an individual of another species. Herbivory is an exploitative interaction in which an animal feeds on a plant. In mutualism, species benefit from one another (pp. 77–81).

+ **Describe feeding relationships and energy flow, and use them to identify trophic levels and navigate food webs**

Energy is transferred up trophic levels in food chains. Lower trophic levels generally contain more energy. Food webs illustrate feeding relationships and energy flow among species in a community (pp. 81–83).

+ **Discuss characteristics of a keystone species**

Keystone species exert impacts on communities that are far out of proportion to their abundance. Top predators are frequently considered keystone species. Other types of organisms (such as ecosystem engineers) can also exert strong effects on communities (pp. 83, 86–87).

+ **Characterize disturbance, succession, and notions of community change**

Communities experience various types of disturbance, and they may respond in different ways. Succession describes a typical pattern of community change through time. Primary succession begins with an area devoid of life. Secondary succession begins with an area that has been severely disturbed but where remnants of the original community remain. Today's ecologists view succession as being less predictable and deterministic than was thought in the past. If disturbance is severe enough, communities may undergo regime shifts involving irreversible change—or novel communities may form (pp. 87–88, 90–91).

+ **Describe the potential impacts of invasive species in communities, and offer solutions to biological invasions**

People have introduced countless species to new areas. Some non-native species may become invasive if they do not encounter limiting factors on their population growth. Invasive species such as Asian carp have altered the composition, structure, and function of communities. We can respond to invasive species with prevention, control, and eradication measures (pp. 88–89, 92).

+ **Explain the goals and methods of restoration ecology**

Restoration ecology is the science of restoring communities to a previous, more functional or more "natural" condition, variously defined as before human impact or before recent industrial impact. Ecological restoration is the practice of restoring ecological systems. Strategies include removing invasive species, reintroducing native species, and reestablishing ecological processes. Today, climate change poses new challenges to restoration efforts (pp. 92–94).

+ **Identify and describe the terrestrial biomes of the world**

Biomes represent major classes of communities spanning large geographic areas. Their distribution is determined by temperature, precipitation, and other factors. Terrestrial biomes include temperate deciduous forest, temperate grassland, temperate rainforest, tropical rainforest, tropical dry forest, savanna, desert, tundra, boreal forest, and chaparral (pp. 94–101).

SEEKING Solutions

1. Suppose you spot two species of birds feeding side by side, eating seeds from the same plant. You begin to wonder whether competition is at work. Describe how you might design scientific research to address this question. What observations would you try to make at the outset? Would you try to manipulate the ecological system to test your hypothesis that the two birds are competing? If so, how?

2. Using the concepts of trophic levels and energy flow, explain in your own words why the ecological footprint of a vegetarian person is smaller than that of a meat-eater.

3. Spend some time outside on your campus, in your yard, or in the nearest park or natural area. Find at least 10 species of organisms (plants, animals, or others), and observe each one long enough to watch it feed or to make an educated guess about how it derives its nutrition. Now, using Figure 4.12 as a model, draw a simple food web involving all the organisms you observed.

4. Can you think of one organism not mentioned in this chapter as a keystone species that you believe may be a keystone species? For what reasons do you suspect this? How could an ecologist experimentally test whether an organism is a keystone species?

5. **THINK IT THROUGH** You are the executive director of a private land trust in your region, responsible for buying

ecologically valuable tracts of land and managing them so as to preserve their communities of plants and animals. Your land trust has just received a large donation to "preserve and enhance" its main holding, a 1000-hectare tract of prairie wetlands, a rare ecological community in the region. You can choose to spend the money in one of several ways:

(a) to completely eradicate the one most harmful invasive species occurring in this tract.

(b) to completely restore 250 hectares of the tract to the condition it was in before European settlers arrived in the region.

(c) to purchase another 250 hectares of adjacent prairie wetland habitat before it is developed.

(d) to fund trails, boardwalks, recreational access, and education programs to teach people about the value of this ecological community.

Which way would you choose to spend the money, and why? What information do you feel you would need to make your decision? Pose several questions you would ask of scientists, of your board members, and of residents living nearby, in order to make your decision.

CALCULATING Ecological Footprints

Environmental scientists David Pimentel, Rodolfo Zuniga, and Doug Morrison of Cornell University reviewed scientific estimates for the economic and ecological costs inflicted by introduced and invasive species in the United States. They found that approximately 50,000 species had been introduced in the United States and that these accounted for more

than $120 billion in economic costs each year. These costs include direct losses and damage, as well as costs required to control the species. (The researchers did not quantify monetary estimates for losses of biodiversity, ecosystem services, and aesthetics, which they maintained would drive total costs several times higher.) Using data offered in the following table, calculate values missing from the table to determine the number of introduced species of each type of organism and the annual cost that each imposes on our economy.

GROUP OF ORGANISM	PERCENTAGE OF TOTAL INTRODUCED	NUMBER OF SPECIES INTRODUCED	PERCENTAGE OF TOTAL ANNUAL COST	ANNUAL ECONOMIC COST
Plants	50.0	25,000	27.2	
Microbes	40.0		20.2	
Arthropods	9.0		15.7	
Fish	0.28		4.2	
Birds	0.19		1.5	$1.9 billion
Mollusks	0.18		1.7	
Reptiles and amphibians	0.11		0.009	
Mammals	0.04	20	29.4	$37.5 billion
TOTAL	100	50,000	100	$127.4 billion

Source: Pimentel, D., R. Zuniga, and D. Morrison, 2005. Update on the environmental and economic costs associated with alien-invasive species in the United States. *Ecological Economics* 52: 273–288.

1. Of the 50,000 species introduced into the United States, half are plants. Describe two ways in which non-native plants might be brought to a new environment. How might we help prevent non-native plants from establishing in new areas?

2. Organisms that damage crop plants are the most costly of introduced species. Weeds, pathogenic microbes, and arthropods that attack crops together account for

half of the costs documented by Pimentel and his colleagues. What steps can we—farmers, governments, and all of us as a society—take to minimize the impacts of invasive species on crops?

3. How might your own behavior influence the influx and ecological impacts of non-native species like those listed above? Name three actions you could personally take to help reduce the impacts of invasive species.

Mastering Environmental Science

Students Go to **Mastering Environmental Science** for assignments, an interactive e-text, and the Study Area with practice tests, videos, and activities

Instructors Go to **Mastering Environmental Science** for automatically graded activities, videos, and reading questions that you can assign to your students, plus Instructor Resources

Environmental Systems and Ecosystem Ecology

Saving the Chesapeake Bay Ecosystem

> I'm 60. Danny's 58. We're the young ones.
>
> Grant Corbin, Oysterman in Deal Island, Maryland

> The Chesapeake Bay continues to be in serious trouble. And it's really no question why this is occurring. For decades, we simply did not manage the bay as a system the way science told us we must. Now all that is changing, and the bay is responding positively. The challenge is to keep it going!
>
> Will Baker, President, Chesapeake Bay Foundation

A visit to Deal Island, Maryland, on the Chesapeake Bay reveals a situation that is all too common in modern America: The island, which was once bustling with productive industries and growing populations, is falling into decline. Economic opportunities in the community are few, and its populace is shrinking and "graying" as more and more young people leave to find work elsewhere. In 1930, Deal Island had a population of 1237 residents. In 2010, it was a mere 471 people.

Unlike other parts of the country with similar stories of economic decline, the demise of Deal Island and other bayside towns was not caused by the closing of a local factory, steel mill, or corporate headquarters; it was caused by the collapse of the Chesapeake Bay oyster fishery.

The Chesapeake Bay was once a thriving system of interacting microbes, plants, and animals, including economically important blue crabs, scallops, and fish. Nutrients carried to the bay by streams in its roughly 168,000 km² (64,000 mi²) drainage basin, or **watershed**— the land area that funnels water to a given body of water—nourished fields of underwater grasses that provided food and refuge to juvenile fish, shellfish, and crabs. Hundreds of millions of oysters kept the bay clear by filtering nutrients and phytoplankton (microscopic photosynthetic organisms that drift near the surface) from the water column.

Oysters had been eaten locally for thousands of years by Native Americans and later by early European settlers, but the intensive harvest of bay oysters for export began in the 1830s. By the 1880s, the bay boasted the world's largest oyster fishery. People flocked to the Chesapeake to work on oystering ships or in canneries, dockyards, and shipyards. Bayside towns like Deal Island prospered along with the oyster industry and developed a unique maritime culture that defined the region.

But by 2010, the bay's oyster populations had been reduced to a mere 1% of their abundance prior to commercial harvesting, and the oyster industry in the area was all but wiped out. Perpetual overharvesting, habitat destruction, virulent oyster diseases, and water pollution had nearly eradicated this economically and ecologically important organism from bay waters. The monetary losses associated with the oyster fishery collapse have been staggering, costing the economies of Maryland and Virginia an estimated $4 billion over the last 40 years.

Upon completing this chapter, you will be able to:

+ Describe environmental systems

+ Define ecosystems and discuss how living and nonliving entities interact in ecosystem-level ecology

+ Outline the fundamentals of landscape ecology and ecological modeling

+ Explain ecosystem services and describe how they benefit our lives

+ Compare and contrast how water, carbon, nitrogen, and phosphorus cycle through the environment and explain how human activities affect these cycles

◀ Chesapeake Bay oystermen hauling in their catch

▲ Sorting oysters from the Chesapeake Bay

One of the biggest impacts on oyster populations in recent decades is the pollution of the Chesapeake Bay by high levels of nutrient runoff, specifically nitrogen and phosphorus from agricultural fertilizers, animal manure, stormwater, and atmospheric compounds produced by fossil fuel combustion. With so few oysters in the bay, elevated nutrient levels have caused phytoplankton populations to increase.

When excessive amounts of phytoplankton die due to overcrowding and competition, they naturally settle to the bay bottom. Once there, they are decomposed by bacteria, a process that consumes dissolved oxygen, causing chronic oxygen depletion of the water, a condition called **hypoxia.** Vast areas of hypoxia in a body of water are referred to as "dead zones." Grasses, oysters, crabs, fish, and other organisms are perishing in dead zones of the Chesapeake Bay or are forced into other areas of the bay. Hypoxia and other human impacts landed the Chesapeake Bay on the Environmental Protection Agency's list of dangerously polluted waters in the United States.

After 25 years of failed pollution control agreements and nearly $6 billion spent on cleanup efforts, the Chesapeake Bay Foundation (CBF), a non-profit organization dedicated to conserving the bay, sued the Environmental Protection Agency (EPA) in 2009 for failing to use its available powers under the Clean Water Act to clean up the bay. The lawsuit instigated federal action, and in 2010, a comprehensive "pollution budget" and restoration plan was developed and implemented by the EPA with the assistance of the District of Columbia, Delaware, Maryland, New York, Pennsylvania, Virginia, and West Virginia. Unlike previous approaches at restoration that only focused on parts of the Chesapeake Bay, the Chesapeake Bay Watershed Agreement (formalized in 2014) took the holistic approach that a system as complicated as the Chesapeake demanded (**FIGURE 5.1**).

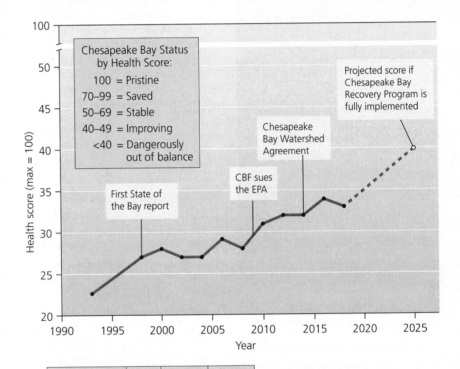

FIGURE 5.1 After a decade of very little improvement, the Chesapeake Bay is finally progressing toward recovery. Numerous parameters that measure the health of the bay—such as nutrient inputs, water quality, and the status of fisheries—are trending upward in recent decades and hope to accelerate in the coming years under a comprehensive management plan. *Data from Chesapeake Bay Foundation, 2016 and 2018 State of the Bay Reports.*

	Indicator	2018 score	Change from 2016	Change from 1998
POLLUTION	Nitrogen	12	−5	−3
	Phosphorus	19	−9	+4
	Dissolved Oxygen	42	+2	+27
	Water Clarity	16	−4	−4
	Toxics	28	0	−2
HABITAT	Forest Buffers	57	0	+4
	Wetlands	42	0	−1
	Underwater Grasses	25	+1	+13
	Resource Lands	33	+1	n/a
FISHERIES	Rockfish	66	0	−4
	Blue Crabs	55	0	+5
	Oysters	10	0	+9
	Shad	10	−1	+8

This systemic approach to restoring the Chesapeake Bay appears to be succeeding as the bay's overall health has improved. The system was challenged in 2018, when record rainfall in its watershed flushed large amounts of plant nutrients into the bay. This elevated nitrogen and phosphorus concentrations in bay waters and reduced water clarity as phytoplankton populations increased and eroded soil clouded the water. But even in the face of these conditions, the bay's fisheries, its fields of underwater grasses, and the concentrations of dissolved oxygen in its waters remained stable or improved from 2016, indicating the ecological functioning of the bay is stronger. The bay is now better able to deal with adverse conditions, like those in 2018, without collapsing. Furthermore, oyster restoration efforts are finally showing promise in the Chesapeake (see **THE SCIENCE BEHIND THE STORY**, pp. 112–113), further enhancing the resiliency of the Chesapeake Bay system. If these initiatives can begin to restore the bay to health, Deal Island and other oyster-fishing communities may again enjoy the prosperity they once did on the scenic shores of the Chesapeake.

Earth's Environmental Systems

Understanding the recovery of the Chesapeake Bay involves comprehending the complex, interlinked systems that make up Earth's environment. A **system** is a network of relationships among parts, elements, or components that interact with and influence one another through the exchange of energy, matter, or information. Earth's natural systems include processes that shape the landscape, affect planetary climate, govern interactions between species and the nonliving entities around them, and cycle chemical elements vital to life. Because we depend on these systems and processes for our very survival, understanding how they function and how human activities affect them is an important aspect of environmental science.

System boundaries are often difficult to delineate

There are many ways to delineate natural systems. For instance, scientists sometimes divide Earth's components into "structural spheres" to help make our planet's dazzling complexity comprehensible. The **lithosphere** (p. 35) is the rock and sediment beneath our feet, the planet's uppermost mantle and crust. The **atmosphere** (p. 452) is composed of the air surrounding our planet. The **hydrosphere** encompasses all water—salt or fresh, liquid, ice, or vapor—in surface bodies, underground, and in the atmosphere. The **biosphere** (p. 60) consists of all the planet's organisms and the abiotic (nonliving) portions of the environment with which they interact.

Beyond these broad spheres, natural systems seldom have well-defined boundaries, so deciding where one system ends and another begins can be difficult. Consider a smartphone. It is certainly a system—a network of circuits and parts that interact and exchange energy and information—but where are its boundaries? Is the system merely the phone itself, or does it include the other phones you call and text, the websites you access on it, and the cellular networks that keep it connected? What about the energy grid that recharges the phone's battery, with its transmission lines and distant power plants?

No matter how we attempt to isolate or define a system, we soon see that it has connections to systems larger and smaller than itself. Systems may exchange energy, matter, and information with other systems, and they may contain or be contained within other systems. Thus, where we draw boundaries may depend on the spatial (space) or temporal (time) scale at which we choose to focus.

Assessing questions holistically by taking a systems approach is helpful in environmental science, in which so many issues are multifaceted and complex. Taking a broad and integrative approach poses challenges, because systems often show behavior that is difficult to predict. However, environmental scientists are rising to the challenge of studying systems holistically, helping us to develop comprehensive solutions to complicated problems such as those faced in the Chesapeake Bay.

Systems exhibit several defining properties

It is difficult to understand systems fully just by focusing on their individual components because systems can have **emergent properties,** characteristics not evident in the components alone. Stating that systems possess emergent properties is a lot like saying, "The whole is more than the sum of its parts." For example, if you were to reduce a tree to its component parts (leaves, branches, trunk, bark, roots, fruit, and so on), you would not be able to predict the entire tree's emergent properties, which include the role the tree plays as habitat for birds, insects, fungi, and other organisms. You could analyze the tree's chloroplasts (photosynthetic cell organelles), diagram its branch structure, and evaluate the nutritional content of its fruit, but you would still be unable to understand the tree as habitat, as part of a forest landscape, or as a reservoir for storing carbon.

Systems involve feedback loops

Earth's environmental systems receive inputs of energy, matter, or information; process these inputs; and produce outputs. As a system, for example, the Chesapeake Bay receives inputs of freshwater, sediments, nutrients, and pollutants from the rivers that empty into it. Oystermen, crabbers, and fishermen harvest some of the bay system's output: matter and energy in the form of seafood. This output subsequently becomes input to the nation's economic system and to the body systems of people who eat the seafood.

Sometimes a system's output can serve as input to that same system, a circular process described as a **feedback loop,** which can be either negative or positive. In a **negative feedback loop** (**FIGURE 5.2**), output that results from a system moving in one direction acts as input that moves the system in the other direction. Input and output essentially neutralize one another's effects, stabilizing the system. As an example, negative feedback regulates body temperature in humans: If we get too hot, our sweat glands pump out water that then evaporates, which cools us down. Or, we just move into the shade. If we get too cold, we shiver, creating heat, or we just move into the sun or put on a sweater. Most systems in nature involve such negative feedback loops. Negative feedback enhances

FAQ

But isn't positive feedback "good" and negative feedback "bad"?

In daily life, positive feedback, such as a compliment, can act as a stabilizing force ("Keep up the good work, and you'll succeed"), whereas negative feedback, such as criticism, can act as a destabilizing force ("You need to change your approach to succeed"). But in environmental systems, it's the opposite! Negative feedback in the environment resists change in systems, enhancing stability and typically keeping conditions within ranges beneficial to life. Positive feedback in the environment, in contrast, exerts destabilizing effects that push conditions to extremes, threatening organisms adapted to the system's normal conditions.

stability, and over time only those systems that are stable will persist.

In a system stabilized by negative feedback, when processes move in opposing directions at equivalent rates so that their effects balance out, they are said to be in **dynamic equilibrium.** Processes in dynamic equilibrium can contribute to **homeostasis,** the tendency of a system to maintain constant or stable internal conditions. A system (such as an organism) in homeostasis keeps its internal conditions within a narrow range that allows it to function. However, the steady state of a homeostatic system may itself change slowly over time. For instance, Earth has experienced gradual changes in atmospheric composition and ocean chemistry over its long history, yet life persists and our planet remains, by most definitions, a homeostatic system.

Rather than stabilizing a system, **positive feedback loops** drive the system further toward an extreme. In positive feedback, increased output from a system results in increased input, leading to further increased output, and so on. Exponential growth in a population (pp. 66–67) is one such example. The more individuals there are, the more offspring that can be produced. Another positive feedback cycle that is of great concern to environmental scientists today involves the melting of glaciers and sea ice in the Arctic as a result of global warming (p. 487). Because ice and snow are white, they reflect sunlight and keep surfaces cool. But if the climate

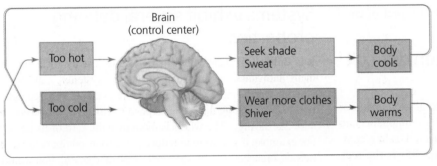

Brain
(control center)

Too hot → Seek shade / Sweat → Body cools

Too cold → Wear more clothes / Shiver → Body warms

(a) Negative feedback

FIGURE 5.2 Feedback loops can stabilize or destabilize systems. (a) The human body's response to heat and cold involves a negative feedback loop that keeps core body temperatures relatively stable. Positive feedback loops **(b)** push systems away from equilibrium. For example, when Arctic glaciers and sea ice melt because of global warming, darker surfaces are exposed, which absorb more sunlight, causing further warming and further melting.

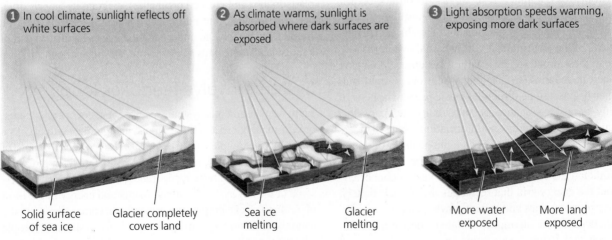

1 In cool climate, sunlight reflects off white surfaces

2 As climate warms, sunlight is absorbed where dark surfaces are exposed

3 Light absorption speeds warming, exposing more dark surfaces

Solid surface of sea ice   Glacier completely covers land   Sea ice melting   Glacier melting   More water exposed   More land exposed

(b) Positive feedback

warms enough to melt the ice and snow, darker surfaces of land and water are exposed, and these darker surfaces absorb sunlight. This absorption warms the surface, causing further melting, which, in turn, exposes more dark surface area, leading to further warming.

Runaway cycles of positive feedback are rare in untouched nature, but they are common in natural systems altered by human activities, and such feedback loops can destabilize those systems. In the Chesapeake Bay, for example, oysters clean the bay's water by filtering phytoplankton, sediments, and nutrients from the water column. When oyster populations were decimated by overharvesting, however, water filtration by oysters declined. With less filtering, concentrations of nutrients and sediments increased in bay waters. Spurred by elevated nutrient levels, phytoplankton populations in the bay exploded, causing hypoxia on the bay bottom and killing many of the remaining oysters. This additional oyster mortality further reduced water filtration in the bay, leading to worsening water quality and even more oyster deaths.

Environmental systems interact

The Chesapeake Bay and the rivers that empty into it provide an example of how systems interact. On a map, the rivers that feed into the bay are a branched and braided network of water channels surrounded by farms, cities, and forests (**FIGURE 5.3**). But where are the boundaries of this system? For a scientist interested in **runoff**—the precipitation that flows over land and enters waterways—and the flow of water, sediment, or pollutants, it may make the most sense to define the bay's watershed as the system. However, for a scientist interested in hypoxia and the bay's dead zones, it may be best to define the watershed together with the bay itself as the system of interest, because their interaction is central to the problem being investigated. Thus, in environmental science, identifying the boundaries of systems depends on the questions being asked.

If the question we are asking about the Chesapeake Bay relates to the dead zones in the bay, then we'll want to define the boundaries of the system to include both the bay's watershed and its **airshed,** the geographic area that produces air pollutants that are likely to end up in a waterway. The hypoxic zones in the bay are due to the extremely high levels of nitrogen and phosphorus delivered to its waters from the 6 states in its watershed and the 15 states in its airshed. In 2017, the bay received an estimated 115 million kg (254 million lb) of nitrogen and 6.8 million kg (15 million lb) of phosphorus. Runoff from agriculture was a major source of these nutrients,

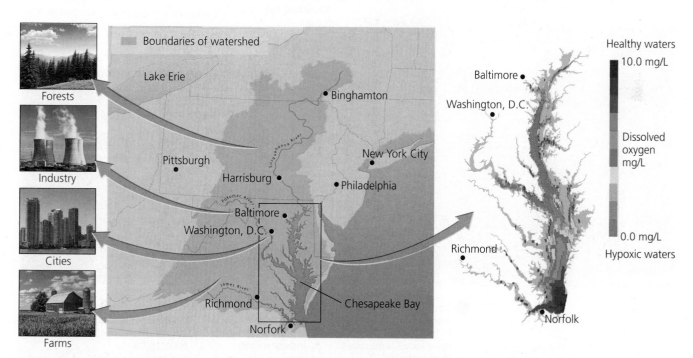

FIGURE 5.3 The Chesapeake Bay watershed encompasses 168,000 km² (65,000 mi²) of land area in 6 states and the District of Columbia. Tens of thousands of streams carry water, sediment, and pollutants from a variety of sources downriver to the Chesapeake, where nutrient pollution has given rise to large areas of hypoxic waters. The zoomed-in map (**at right**) shows dissolved oxygen concentrations in the Chesapeake Bay in 2017. Oysters, crabs, and fish typically require a minimum 3 mg/L of oxygen and are therefore excluded from large portions of the bay where oxygen levels are too low. *Source for figure at right: Marjorie Friedrichs (Virginia Institute of Marine Sciences) and Aaron Bever (Anchor QEA).*

contributing 42% of the nitrogen (**FIGURE 5.4a**) and 56% of the phosphorus (**FIGURE 5.4b**) entering the bay. One-fourth of nitrogen inputs come from atmospheric nitrogenous pollutants released by vehicles and coal-burning power plants.

Elevated nitrogen and phosphorus inputs cause phytoplankton in the bay's waters to flourish. High phytoplankton density leads to elevated mortality as individuals compete for sunlight and nutrients. Dead phytoplankton drift to the bottom of the bay where they are joined by the waste products of zooplankton, tiny creatures that feed on phytoplankton. This increase in organic material on the bay bottom causes an explosion in populations of bacterial decomposers, which deplete the oxygen in bottom waters as they consume the organic matter. Deprived of oxygen, organisms will either flee the area or suffocate. The process of nutrient overenrichment, blooms of phytoplankton, increased production of organic matter, and subsequent ecosystem degradation is known as **eutrophication** (**FIGURE 5.5**).

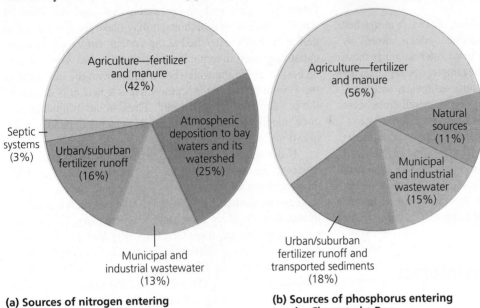

(a) Sources of nitrogen entering the Chesapeake Bay

(b) Sources of phosphorus entering the Chesapeake Bay

FIGURE 5.4 **The Chesapeake Bay receives inputs of (a) nitrogen and (b) phosphorus from many sources in its watershed.** *Data from Chesapeake Bay Program, Watershed Model Phase 5.3.2 (Chesapeake Bay Program Office, 2018). Totals for nitrogen do not equal 100% due to rounding.*

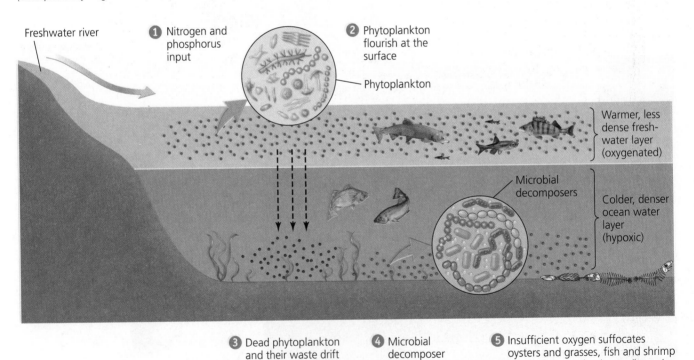

FIGURE 5.5 **Excess nitrogen and phosphorus cause eutrophication in aquatic systems such as the Chesapeake Bay.** Coupled with stratification (layering) of water, eutrophication can severely deplete dissolved oxygen. ❶ Nutrients from river water ❷ boost growth of phytoplankton, ❸ which die and are decomposed at the bottom by bacteria. Stability of the surface layer prevents deeper water from absorbing oxygen to replace ❹ oxygen consumed by decomposers, and ❺ the oxygen depletion suffocates or drives away bottom-dwelling marine life. The process of eutrophication occurs in both freshwater and marine environments and gives rise to large areas of hypoxic waters.

Once oxygen levels at the bottom of the bay are depleted, they are slow to recover. Oxygenated fresh water entering the bay from rivers remains stratified in a layer at the surface and is slow to mix with the denser, saltier bay water, limiting the amount of oxygenated surface water that reaches the bottom-dwelling life that needs it. As a result, sedentary creatures living on the bay bottom, such as oysters, suffocate and die.

Ecosystems

An **ecosystem** consists of all organisms and nonliving entities that occur and interact in a particular area at the same time. Animals, plants, water, soil, nutrients—all these and more compose ecosystems. The ecosystem concept builds on the idea of the biological community (Chapter 4), but ecosystems include abiotic components, as well as biotic ones.

The ecosystem concept originated with scientists who recognized that biological entities are tightly intertwined with the chemical and physical aspects of their environment. For instance, in the Chesapeake Bay **estuary**—a water body where rivers flow into the ocean, mixing fresh water with saltwater—aquatic organisms are affected by the flow of water, sediment, and nutrients from the rivers that feed the bay and from the land that feeds those rivers. In turn, the photosynthesis, respiration, and decomposition that these organisms undergo influence the chemical and physical conditions of the Chesapeake's waters.

Ecologists soon began analyzing ecosystems as an engineer might analyze the operation of a machine. In this view, ecosystems are systems that receive inputs of energy, process and transform that energy while cycling various types of matter, and produce outputs (such as heat, water flow, and waste products) that then enter other ecosystems.

Energy flows and matter cycles through ecosystems

Energy flows in one direction through ecosystems. As autotrophs, such as green plants and phytoplankton, convert solar energy to the energy of chemical bonds in sugar through the process of photosynthesis, they perform **primary production.** The total amount of chemical energy produced by autotrophs is termed **gross primary production.** Autotrophs use most of this production to power their own metabolism by cellular respiration, releasing heat energy to the environment as a by-product. The energy that remains after respiration and that is used

to generate biomass (such as leaves, stems, and roots) is called **net primary production.** Thus, net primary production equals gross primary production minus the energy used in cellular respiration.

Some plant biomass is subsequently eaten by herbivores, which use the energy they gain from plant biomass for their own metabolism or to generate biomass in their bodies (such as skin, muscle, or bone), termed **secondary production.** Herbivores are then eaten by higher-level consumers, which are, in turn, eaten by yet higher-level consumers. Thus, the chemical energy formed by photosynthesis in plants provides energy to higher and higher levels of consumers. The vast majority of this chemical energy is eventually released to the environment as heat when it is metabolized by producers, consumers, or decomposers (**FIGURE 5.6**). Some of the chemical energy from biomass is not released as heat but rather as low energy from chemical bonds, such as when the chemical bonds break between carbon and oxygen atoms in a molecule of carbon dioxide as a by-product of cellular respiration. Then, when producers and consumers die, their biomass is consumed and metabolized by detritivores and decomposers.

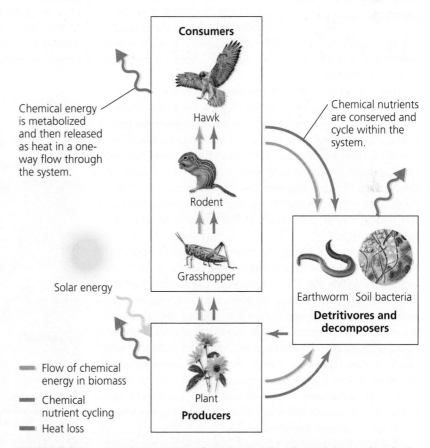

FIGURE 5.6 In ecosystems, energy flows in one direction, whereas chemical nutrients cycle. Light energy from the sun (**yellow arrow**) drives photosynthesis in producers, which begins the transfer of chemical energy in biomass (**orange arrows**) among trophic levels (pp. 81–82) and detritivores and decomposers. Energy exits the system through respiration in the form of heat (**red arrows**). Chemical nutrients (**gray arrows**) cycle within the system. For simplicity, various abiotic components (such as water, air, and inorganic soil content) of ecosystems have been omitted.

THE SCIENCE behind the story

Are We "Turning the Tide" for Native Oysters in Chesapeake Bay?

David Schulte, U.S. Army Corps of Engineers

Back in 2001, the Eastern oyster (*Crassostrea virginica*) was in dire trouble in the Chesapeake Bay. Populations had dropped by 99%, and the Chesapeake's oyster industry, once the largest in the world, had collapsed. Poor water quality, reef destruction, virulent diseases spread by transplanted oysters, and 200 years of overharvesting all contributed to the collapse.

Restoration efforts had largely failed. Moreover, when scientists or resource managers proposed rebuilding oyster populations by significantly restricting oyster harvests or establishing oyster reef "sanctuaries," these initiatives were typically defeated by the politically powerful oyster industry. All this had occurred in a place whose very name (derived from the Algonquin word *Chesepiook*) means "great shellfish bay."

With the collapse of the native oyster fishery and with political obstacles blocking restoration projects for native oysters, support grew among the oyster industry, state resource managers, and some scientists for the introduction of Suminoe oysters (*Crassostrea ariakensis*) from Asia. This species seemed well suited for conditions in the bay and showed resistance to the parasitic diseases that were ravaging native oysters. Proponents argued that introducing Suminoe oysters would reestablish thriving oyster populations in the bay and revitalize the oyster fishery.

Proponents additionally maintained that introducing oysters would also improve the bay's water quality, because as oysters feed, they filter phytoplankton and sediments from the water column. Filter-feeding by oysters is an important ecological service in the bay because it reduces phytoplankton densities, clarifies waters, and supports the growth of underwater grasses that provide food and refuge for waterfowl and young crabs. Because introductions of invasive species can have profound ecological impacts (pp. 88–89), the Army Corps of Engineers was directed to coordinate an environmental impact statement (EIS, p. 172) on oyster restoration approaches in the Chesapeake.

It was in this politically charged, high-stakes environment that Dave Schulte, a scientist with the Corps and doctoral student at the College of William and Mary, set out to determine whether there was a viable approach to restoring native oyster populations. The work he and his team began would help turn the tide in favor of native oysters in the bay's restoration efforts.

One of the biggest impacts on native oysters was the destruction of oyster reefs by a century of intensive oyster harvesting. Oysters settle and grow best on the shells of other oysters, and over long periods this process forms reefs (underwater outcrops of living oysters and oyster shells) that solidify and become as hard as stone. Throughout the bay, massive reefs that at one time had jutted out of the water at low tide had been reduced to rubble on the bottom from a century of repeated scouring by metal dredgers used by oyster-harvesting ships. The key, Schulte realized, was to construct artificial reefs like those that once existed, to get oysters off the bottom—away from smothering sediments and hypoxic waters—and up into the plankton-rich upper waters.

In 2004, armed with the resources available to the Corps, Schulte opted to take a landscape ecology approach to restore patches of reef habitat on 9 complexes of reefs, creating a total of 35.3 hectares (87 acres) of oyster sanctuary near the mouth of the Great Wicomico River in the lower Chesapeake Bay (**FIGURE 1**)—a much larger restoration effort than any previously attempted. Schulte and his team constructed artificial reefs by

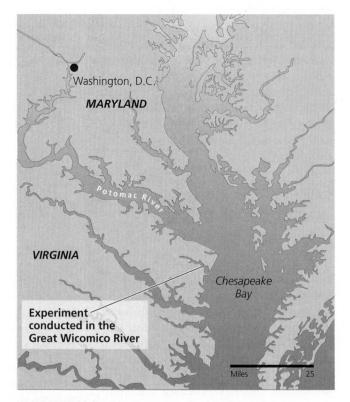

FIGURE 1 Schulte's study was conducted in the Great Wicomico River in Virginia in the lower Chesapeake Bay.

FIGURE 2 A water cannon blows oyster shells off a barge and onto the river bottom to create an artificial oyster reef for the experiment.

spraying oyster shells off barges (**FIGURE 2**). The oyster shells then drifted to the river bottom, forming high-relief reefs in which shells were piled to create a reef that was 25–45 cm (10–18 in.) above the river bottom. They also created low-relief reefs, with shells piled to 8–12 cm (3–5 in.) above the river bottom, that the oysters could colonize, safe from harvesting. Other areas of the river bottom were "unrestored" and left in their natural state.

Oyster populations on the constructed reefs were sampled in 2007, and the results were stunning. The reef complex supported an estimated 185 million oysters, a number nearly as large as the wild population of 200 million oysters estimated to live at that time on the remaining degraded habitat in all of Maryland's waters. Higher constructed reefs supported an average of more than 1000 oysters per square meter—four times more than the lower constructed reefs and 170 times more than unrestored bottom (**FIGURE 3**). Like natural reefs, the constructed reefs began to solidify, providing a firm foundation for the settlement of spat—young, newly settled oysters. In 2009, Schulte's research made a splash when his team published its findings in the journal *Science*, bringing international attention to their study.

After reviewing eight alternative approaches to oyster restoration that involved one or more oyster species, the Corps advocated an approach that avoided the introduction of non-native oysters. Instead, it proposed a combination of native oyster restoration, a temporary moratorium on oyster harvests (accompanied by a compensation program for the oyster industry), and enhanced support for oyster aquaculture in the bay region.

Schulte's restoration project cost roughly $3 million and will require substantial investments if it is to be repeated elsewhere in the bay. This is particularly true in upper portions of the bay, where water conditions are poorer, the oysters are less resistant to disease, and oyster reproduction levels are lower, requiring restored reefs to be "seeded" with oysters. Many scientists

contend that expanded reef restoration efforts are worth the cost because they enhance oyster populations and provide a vital service to the bay through water filtering. Some scientists also see value in promoting oyster farming, in which restoration efforts would be supported by businesses instead of taxpayers.

These efforts are encouraged by the continued success of the project. By the summer of 2016, the majority of high-relief reef acreage was thriving, despite pressures from poachers and several years of hypoxic conditions. Moreover, many of the low-relief reefs that were originally constructed eventually accumulated enough new shell to be as tall as the high-relief reefs in the initial experiment—the reef treatment that showed the highest oyster densities in the original experiment. Furthermore, oyster reproduction rates in 2012 were among the highest Schulte had seen during the project, and a follow-up study in 2013 found that spat from the sanctuary reefs were seeding other parts of the Great Wicomico River and increasing oyster populations outside protected areas.

Protected sites for oyster restoration efforts are now being established elsewhere in the bay. Maryland recently designated 3640 hectares (9000 acres) of new oyster sanctuaries—25% of existing oyster reefs in state waters—and seeded these reefs with more than a billion hatchery-raised spat. This movement toward increased protection for oyster populations, coupled with findings of increased disease resistance in bay oysters, has given new hope that native oysters may once again thrive in the waters of the "great shellfish bay."

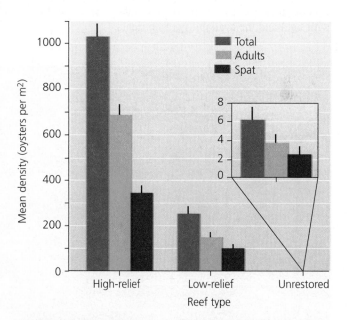

FIGURE 3 Reef height had a profound effect on the density of adult oysters and spat (newly settled oysters). Schulte's work suggested that native oyster populations could rebound in portions of Chesapeake Bay if they were provided with elevated reefs and protected from harvest. *Data from Schulte, D. M., R. P. Burke, and R. N. Lipicus, 2009. Unprecedented restoration of a native oyster metapopulation.* Science 325: 1124–1128.

In contrast to chemical energy, nutrients and other types of matter are generally recycled within ecosystems. Chemical nutrients are recycled because when organisms die and decay, the matter that comprises their body remains in the system.

Ecosystems vary in their productivity

Another way to think of net primary production is that it represents the energy or biomass available for consumption by heterotrophs, organisms that obtain energy by feeding on other organisms. Some plant biomass is eaten by herbivores. Heterotrophs use the energy they gain from plant biomass for their own metabolism, growth, and reproduction. Some of this energy is used by heterotrophs to generate biomass in their bodies (such as skin, muscle, or bone), which is termed secondary production. Plant matter not eaten by herbivores becomes fodder for detritivores and decomposers once the plant dies or drops its leaves.

Ecosystems vary in the rate at which autotrophs convert energy to biomass. The rate at which this conversion occurs is termed **productivity,** and ecosystems whose plants convert solar energy to biomass rapidly are said to have high **net primaryproductivity.** Freshwater wetlands, tropical forests, coral reefs, and algal beds tend to have the highest net primary productivities, whereas deserts, tundra, and open ocean tend to have the lowest (**FIGURE 5.7**). Variation among ecosystems and among biomes (Chapter 4) in net primary productivity results in geographic patterns across the globe (**FIGURE 5.8**). In terrestrial ecosystems, net primary productivity tends to increase with temperature and precipitation. In aquatic ecosystems, net primary productivity tends to rise with light and the availability of nutrients.

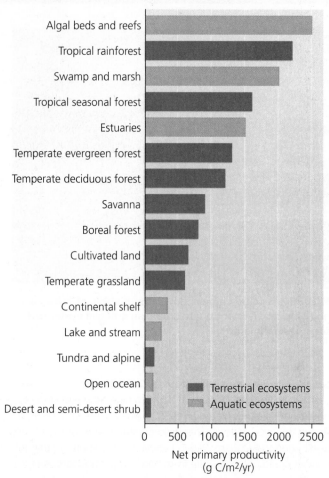

FIGURE 5.7 Net primary productivity varies greatly between ecosystem types. *Data from Whittaker, R. H., 1975.* Communities and ecosystems, *2nd ed. New York, NY: Macmillan.*

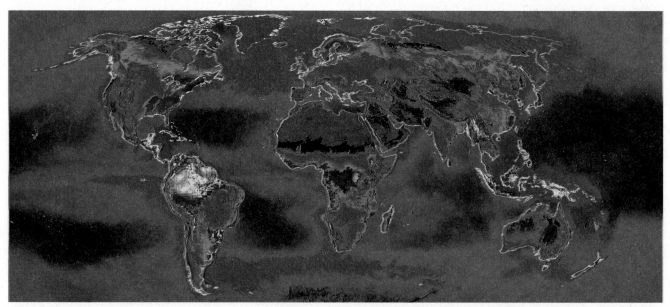

Net primary productivity (kgC/km²/year)

0   1   2   3

FIGURE 5.8 Net primary production on land varies geographically according to temperature, precipitation, and land use (such as for agriculture, urban development, and forestry). In the world's oceans, net primary production is highest around the margins of continents, where nutrients (of both natural and human origin) run off from land. *Data from NASA, https://science.nasa.gov/earth-science/oceanography/living-ocean/remote-sensing.*

Nutrient availability influences productivity

Nutrients (p. 25) are elements and compounds that organisms require for survival. Organisms need several dozen naturally occurring nutrients to survive. Elements and compounds required in relatively large amounts, such as nitrogen, carbon, and phosphorus, are called **macronutrients.** Nutrients needed in small amounts, such as zinc, copper, and iron, are called **micronutrients.**

Nutrients stimulate production by plants, and lack of nutrients can limit production. The availability of nitrogen or phosphorus frequently is a limiting factor (p. 67) for the growth of photosynthetic organisms. When these nutrients are added to a system, producers show the greatest response to whichever nutrient has been in shortest supply.

Canadian ecologist David Schindler and others demonstrated the effects of phosphorus on freshwater systems in the 1970s by experimentally manipulating entire lakes. In one experiment, his team bisected a 16-ha (40-acre) lake in Ontario with a plastic barrier. To one-half the researchers added carbon, nitrate, and phosphate; to the other they added only carbon and nitrate. Soon after the experiment began, they witnessed a dramatic increase in phytoplankton in the half of the lake that received phosphate, whereas the other half (the control for the experiment, p. 10) continued to host phytoplankton levels typical for lakes in the region (**FIGURE 5.9**). This difference held until shortly after they stopped fertilizing seven years later. At that point, phytoplankton decreased to normal levels in the half that had previously received phosphate.

The Chesapeake Bay is not the only water body suffering from eutrophication. Nutrient pollution has led to more than 475 documented hypoxic dead zones (**FIGURE 5.10**), including

After phosphate additions, the water in this portion of the lake turned opaque due to a dramatic and prolonged algal bloom.

FIGURE 5.9 The upper portion of this lake in Ontario was experimentally treated with the addition of phosphate. This treated portion experienced an immediate, dramatic, and prolonged phytoplankton bloom, identifiable by its opaque waters.

one that forms each year near the mouth of the Mississippi River (see **THE SCIENCE BEHIND THE STORY**, pp. 116–117). Some are seasonal (like the Chesapeake Bay's), some occur irregularly, and others are permanent. The increase in the number of dead zones—there were 162 documented in the 1980s and only 49 in the 1960s—reflects how human activities are changing the chemistry of waters around the world.

The good news is that in locations where people have reduced nutrient runoff, hypoxic zones have begun to disappear.

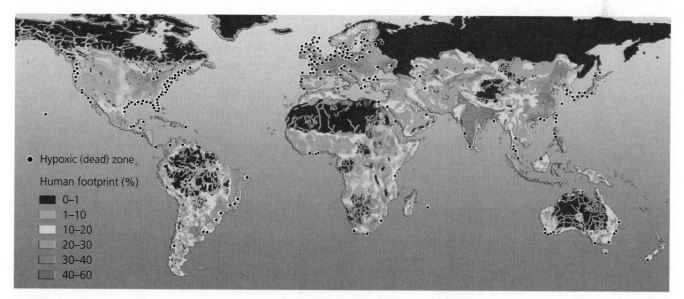

- Hypoxic (dead) zone

Human footprint (%)
- 0–1
- 1–10
- 10–20
- 20–30
- 30–40
- 40–60

FIGURE 5.10 More than 475 marine dead zones have been recorded across the world. Dead zones (shown by dots on the map) occur mostly offshore from areas of land with the greatest human ecological footprints (here, expressed on a scale of 0 to 100, with higher numbers indicating bigger human footprints). *Data from World Resources Institute, 2018, www.wri.org/our-work/project/eutrophication-and-hypoxia, and Diaz, R. and R. Rosenberg, 2008. Spreading dead zones and consequences for marine ecosystems. Science 321: 926–929. Reprinted with permission from AAAS.*

THE SCIENCE behind the story

Are Fertilizers from Midwestern Farms Causing a "Dead Zone" in the Gulf of Mexico?

Dr. Nancy Rabalais, LUMCON

She was prone to seasickness, but Nancy Rabalais cared too much about the Gulf of Mexico to let that stop her. Leaning over the side of an open boat idling miles from shore, she hauled a water sample aboard—and helped launch efforts to breathe life back into the Gulf's "dead zone."

Since that first expedition in 1985, Rabalais, her colleague and husband Eugene Turner, and fellow scientists at the Louisiana Universities Marine Consortium (LUMCON) and Louisiana State University have made great progress in unraveling the mysteries of the region's hypoxia—and in getting it on the political radar screen.

Rabalais and other researchers began by tracking oxygen levels at nine sites in the Gulf every month and continued those measurements for five years. At dozens of other spots near the shore and in deep water, they took less frequent oxygen readings. Sensors, as they are lowered into the water, measure oxygen levels and send continuous readings back to a shipboard computer. Further data come from fixed, submerged oxygen meters that continuously measure dissolved oxygen and store the data.

The team also collected hundreds of water samples, using lab tests to measure levels of nitrogen, salt, bacteria, and phytoplankton. LUMCON scientists logged hundreds of miles in their ships, regularly monitoring more than 70 sites in the Gulf. They also donned scuba gear to view firsthand the condition of shrimp, fish, and other sea life. Such a range of long-term data allowed the researchers to build a "map" of the dead zone, tracking its location and its consequences.

In 1991, Rabalais made that map public, earning immediate headlines. That year, her group mapped the size of the zone at more than 10,000 km² (about 4000 mi²). Bottom-dwelling shrimp were stretching out of their burrows, straining for oxygen. Many fish had fled. The bottom waters, infused with sulfur from bacterial decomposition, smelled like rotten eggs.

The group's years of monitoring also enabled them to explain and predict the dead zone's emergence. As rivers rose each spring (and as fertilizers were applied in the Midwestern farm states), oxygen would start to disappear in the northern Gulf. The hypoxia would last through the summer or fall, until seasonal storms mixed oxygen into hypoxic areas.

The source of the problem, Rabalais said, lay back on land. The Mississippi and Atchafalaya rivers draining into the Gulf were polluted with agricultural runoff, and the nutrient pollution from fertilizers spurred algal blooms whose decomposition by bacteria snuffed out oxygen in wide stretches of ocean water (pp. 109–111). This work had clearly demonstrated the interconnections between freshwater aquatic systems and the Gulf, and how pollutants from farm fields in the upper Midwest could exert effects far away at the mouth of the Mississippi River.

Over time, monitoring linked the dead zone's size to the volume of river flow and its nutrient load. The 1993 flooding of the Mississippi created a zone much larger than the year before, whereas a drought in 2000 brought low river flows, low nutrient loads, and a small dead zone (**FIGURE 1**). Similar relationships between river flow and dead zone size have been seen ever since. In 2005, the dead zone was predicted to be large, but Hurricanes Katrina and Rita stirred oxygenated surface water into the depths, decreasing the dead zone that year.

Many Midwestern farming advocates and some scientists, such as Derek Winstanley, chief of the Illinois State Water Survey, challenged the findings. They argued that the Mississippi naturally carries high loads of nitrogen from runoff and that Rabalais's team had not ruled out upwelling (the wind-driven phenomenon that causes waters from the deep ocean to rise to the surface) in the Gulf as a source of nutrients.

But sediment analyses showed that Mississippi River mud contained many fewer nitrates early in the century, and Rabalais and Turner found that silica residue from phytoplankton blooms increased in Gulf sediments between 1970 and 1989, paralleling rising nitrogen levels. In 2000, a federal integrative assessment team of dozens of scientists laid the blame for the dead zone on nutrients from fertilizers and other sources in the fresh waters emptying into the Gulf.

Then in 2004, while representatives of farmers and fishermen debated political fixes, Environmental Protection Agency water-quality scientist Howard Marshall suggested that to alleviate the dead zone, we'd be best off reducing phosphorus pollution from industry and sewage treatment. His reasoning was this: Phytoplankton need both nitrogen and phosphorus, but there is now so much nitrogen in the Gulf that phosphorus has become the limiting factor on phytoplankton growth.

Since then, research has supported this contention, and scientists now propose that nitrogen and phosphorus should be managed jointly to help reduce the size of the dead zone in the Gulf of Mexico off the Louisiana coast. Moreover, recent research indicates that a federally mandated 30% reduction in

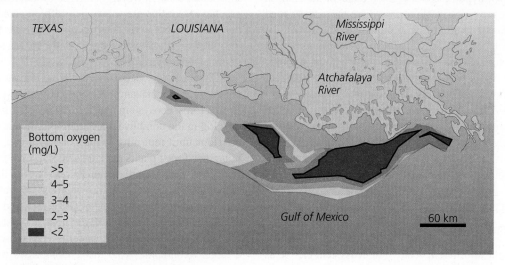

(a) Dissolved oxygen at ocean bottom

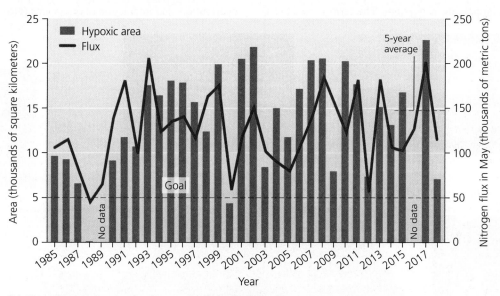

(b) Size of hypoxic zone in the northern Gulf of Mexico

FIGURE 1 The map in (a) shows dissolved oxygen concentrations in bottom waters of the Gulf of Mexico off the Louisiana coast in 2018. The darkest areas indicate the lowest oxygen levels, with regions considered hypoxic (<2 mg/L) outlined in black. The dead zone forms to the west of the mouth of the Mississippi River because prevailing currents carry nutrients in that direction. The graph in **(b)** shows that the size of the hypoxic zone (shown by bars) is correlated with the amount of nitrogen pollution entering from the Mississippi River (shown by line), with nutrient delivery being higher in wetter years and lower in years with Midwest drought. The dead zone in 2018 was 7040 km^2 (2720 mi^2), the fourth-smallest since mapping began in 1985. This was attributed to low levels of precipitation reducing nutrient runoff into the Mississippi River. The average size of the dead zone from 2014 to 2018 was nearly 15,000 km^2 (5800 mi^2), and scientists and policymakers aim to reduce its size to 5000 km^2 (1930 mi^2). *Hypoxic zone data from Nancy Rabalais, LUMICON, and R. Eugene Turner, LSU. Nitrogen flux data from USGS, https://nrtwq.usgs.gov/mississippi_loads/#/.*

nitrogen in the river will not be adequate to eliminate the dead zone. Scientists also maintain that large-scale restoration of wetlands along the river and at the river's delta would best filter pollutants before they reach the Gulf.

All this research is guiding a federal plan to reduce farm runoff, clean up the Mississippi, restore coastal wetlands, and shrink the Gulf's dead zone. It has also led to a better understanding of hypoxic zones around the world.

In New York City, hypoxic zones at the mouths of the Hudson and East rivers were nearly eliminated once the city stopped releasing untreated, nutrient-rich sewage into the river. The Black Sea, which borders Ukraine, Russia, Turkey, and eastern Europe, had long suffered one of the world's worst hypoxic zones. Then in the 1990s, after the Soviet Union collapsed, industrial agriculture in the region declined drastically. With fewer fertilizers draining into it, the Black Sea began to recover, and today fisheries are reviving. However, agricultural collapse is not a strategy anyone would choose to alleviate hypoxia. Rather, scientists are proposing a variety of innovative and economically acceptable ways to reduce nutrient runoff.

Ecosystems interact across landscapes

Ecosystems occur at different scales. An ecosystem can be as small as a puddle of water or as large as a bay, lake, or forest. For some purposes, scientists even view the entire biosphere as a single all-encompassing ecosystem. The term *ecosystem* is most often used, however, to refer to systems of moderate geographic extent that are somewhat self-contained. For example, the tidal marshes in the Chesapeake where river water empties into the bay are an ecosystem, as are the sections of the bay dominated by oyster reefs.

Adjacent ecosystems may share components and interact extensively. For instance, a pond ecosystem is very different from a forest ecosystem that surrounds it, but salamanders that develop in the pond live their adult lives under logs on the forest floor until returning to the pond to breed. Rainwater that nourishes forest plants and picks up nutrients from the forest's leaf litter may eventually make its way to the pond. Areas where ecosystems meet may consist of transitional zones called **ecotones**, in which elements of each ecosystem mix.

Because components of different ecosystems may intermix, ecologists often find it useful to view these ecosystems on a larger geographic scale that encompasses multiple ecosystems. In such a broad-scale approach, called **landscape ecology**, scientists study how landscape structure affects the abundance, distribution, and interaction of organisms (**FIGURE 5.11**). Taking

FIGURE 5.11 Landscape ecology deals with spatial patterns above the ecosystem level. This generalized diagram of a landscape shows a mosaic of patches of five ecosystem types (three terrestrial types, a marsh, and a river). Thick red lines indicate ecotones. A stretch of lowland broadleaf forest running along the river serves as a corridor connecting the large region of forest on the left to the smaller patch of forest alongside the marsh. The inset shows a magnified view of the forest-grassland ecotone and how it consists of patches on a smaller scale.

a view across the landscape is important in studying birds that migrate long distances, mammals that move seasonally between mountains and valleys, and fish such as salmon that swim upriver from the ocean to reproduce.

For a landscape ecologist, a landscape is made up of **patches** (of ecosystems, communities, or habitats) arrayed spatially over a landscape in a **mosaic.** Landscape ecology is of great interest to scientists of **conservation biology** (pp. 295–296), the study of the loss, protection, and restoration of biodiversity. Every organism has specific habitat needs, so when its habitat is distributed in patches across a landscape, individuals may need to expend energy and risk predation traveling from one to another. If the patches are far apart, the organism's population may become divided into subpopulations, each occupying a different patch in the mosaic. Such a network of subpopulations, most of whose members stay within their respective patches but some of whom move among patches or mate with members of other patches, is called a **metapopulation.** When patches are still more isolated from one another, individuals may not be able to travel between them at all. In such a case, smaller subpopulations may be at risk of extinction. Conservation biologists aim to avoid these situations by establishing corridors of habitat (see Figure 5.11) that link patches and allow animals to move among them.

Technology helps us practice landscape ecology

A common tool for research in landscape ecology is the **geographic information system (GIS).** A GIS consists of computer software that takes multiple types of data (for instance, on geology, hydrology, topography, vegetation, plant and animal populations, and human infrastructure from both on-ground studies and satellite imagery) and combines them, layer by layer, on a common set of geographic coordinates (**FIGURE 5.12**). The idea is to create a complete picture of a landscape, analyze how elements of the different data sets are arrayed spatially, and determine how they may be correlated.

GIS has become a valuable tool used by geographers, landscape ecologists, resource managers, and conservation biologists. It is being used to guide restoration efforts in the Chesapeake Bay. The *ChesapeakeStat,* a GIS-enabled website that was launched in 2010, enables scientists, educators, policymakers, and citizens to create customized composite maps that overlay parameters important to the bay's health. This tool is being used to assess the bay's current status, the effects of restoration efforts, and progress toward long-term goals.

Modeling helps ecologists understand systems

Another way in which ecologists seek to make sense of the complex systems they study is by working with models.

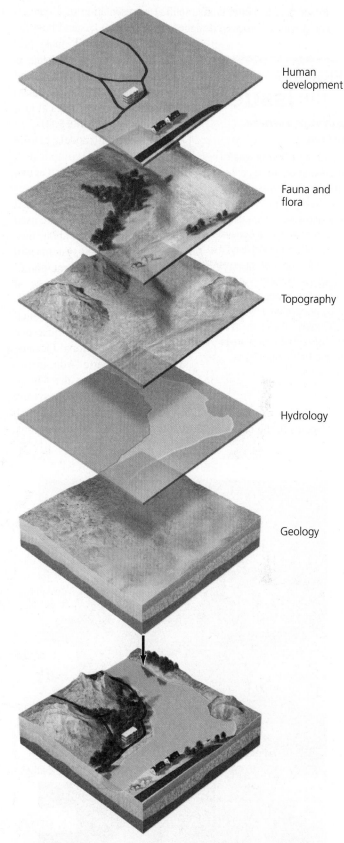

Human development

Fauna and flora

Topography

Hydrology

Geology

FIGURE 5.12 Geographic information systems (GIS) produce maps that layer various types of data on natural landscape features and human land uses. GIS can be used to explore correlations among these data sets and to aid wildlife conservation and regional planning.

In science, a **model** is a simplified representation of a complicated natural process, designed to help us understand how the process occurs and to make predictions. **Ecological modeling** is the practice of constructing and testing models that aim to explain and predict how ecological systems function (**FIGURE 5.13**).

Ecological models can be mathematically complicated, but they are grounded in actual data and based on hypotheses about how components interact in ecosystems. Models are used to make predictions about how large, complicated systems will behave under different conditions. Accordingly, the use of models is a key part of ecological research and environmental regulation today. As just one example, the National Oceanic and Atmospheric Administration (NOAA) uses the Chesapeake Bay Fisheries Ecosystem model to examine predator–prey relationships among fish species in the bay, the effects of hypoxia on fish populations,

WEIGHING
the **issues**

Ecosystems Where You Live

Think about the area where you live and briefly describe its ecosystems. How do these ecosystems interact? Describe the boundaries of watersheds in your region. If one ecosystem were greatly modified (say, if a large apartment complex were built atop a wetland or amid a forest), what impacts on nearby ecosystems might result? (*Note:* If you live in a city, realize that urban areas can also be thought of as ecosystems.)

FIGURE 5.14 Blue crabs (*Callinectes sapidus*) are an ecologically and economically important species in the Chesapeake Bay. Reducing hypoxia in bay waters and reestablishing large expanses of underwater grasses (like those shown) are keys to the species' recovery.

and how the distribution of underwater grasses influences blue crab populations (**FIGURE 5.14**). Data from scientific journal articles and direct measurements are used to establish the model's parameters, which are then used to predict the effects of differing fish harvest levels on species and ecosystems in the Chesapeake Bay.

Ecosystem services sustain our world

Human society depends on healthy, functioning ecosystems. When Earth's ecosystems function normally and undisturbed, they provide goods and services that we could not survive without. As we've seen, we rely not just on natural resources (which can be thought of as goods from nature) but also on the *ecosystem services* (p. 4) that our planet's systems provide (**TABLE 5.1**).

Ecosystem services come from the functioning of healthy ecological systems and the many negative feedback cycles that regulate and stabilize natural systems. Ecosystem services include ecological processes that form the soil that nourishes our crops, purify the water we drink, pollinate the food plants we eat, and break down (some of) the waste and pollution we generate. Ecosystem services also enhance the quality of our lives, ranging from recreational opportunities to pleasing scenery to inspiration and spiritual renewal.

One of the most important ecosystem services is the cycling of chemical nutrients. Through the processes that take place within and among ecosystems, the chemical elements and compounds that sustain life—carbon,

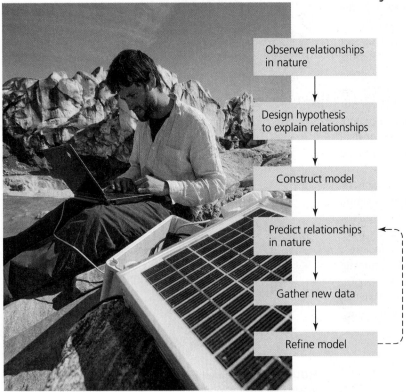

Observe relationships in nature

↓

Design hypothesis to explain relationships

↓

Construct model

↓

Predict relationships in nature

↓

Gather new data

↓

Refine model

FIGURE 5.13 Ecological modelers observe relationships among variables in nature and then construct models to explain those relationships and make predictions. They test and refine the models by gathering new data from nature and seeing how well the models predict those data.

TABLE 5.1 Ecosystem Services

Ecological processes do many things that benefit us:

- Cycle carbon, nitrogen, phosphorus, and other nutrients
- Regulate oxygen, carbon dioxide, stratospheric ozone, and other atmospheric gases
- Regulate temperature and precipitation by phenomena such as ocean currents and cloud formation, and processes such as evaporation and transpiration
- Store and regulate water supplies in watersheds and aquifers
- Form soil by weathering rock, and prevent soil erosion
- Protect against storms, floods, and droughts, mainly by the moderating means of vegetation
- Filter waste, remove toxic substances, recover nutrients, and control pollution
- Pollinate plant crops and control crop pests
- Provide habitat for organisms to breed, feed, rest, migrate, and winter
- Produce fish, game, crops, nuts, and fruits that people eat
- Supply lumber, fuel, metals, fiber crops (such as cotton), feed for livestock, and medicinal compounds
- Provide recreation such as ecotourism, fishing, hiking, birding, hunting, and kayaking
- Provide aesthetic, artistic, educational, spiritual, and scientific amenities

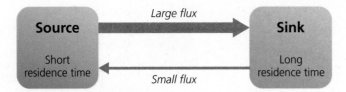

FIGURE 5.15 The main components of a biogeochemical cycle are reservoirs and fluxes. A source releases more materials than it accepts, and a sink accepts more materials than it releases.

nitrogen, phosphorus, water, and many more—cycle through our environment in complex ways.

Biogeochemical Cycles

Just as nitrogen and phosphorus from fertilizer on Pennsylvania corn fields end up in Chesapeake Bay oysters, all nutrients move through the environment in intricate ways. As we have discussed, whereas energy enters an ecosystem from the sun, flows from organism to organism, and dissipates to the atmosphere as heat, the physical matter of an ecosystem is circulated over and over again.

Nutrients circulate through ecosystems in biogeochemical cycles

Nutrients move through ecosystems in **nutrient cycles** (or **biogeochemical cycles**) that circulate chemical elements or molecules through the atmosphere, hydrosphere, lithosphere, and biosphere. A carbon atom in your fingernail today might have been in the muscle of a cow a year ago, may have resided in a blade of grass a month before that, and may have been part of a dinosaur's tooth 100 million years ago. After we die, the nutrients in our bodies will disperse into the environment and could be incorporated into other organisms far into the future.

Nutrients and other materials move from one **reservoir,** or pool, to another, remaining in each reservoir for varying amounts of time (the **residence time**). The dinosaur, the cow, the grass, and your body are each reservoirs for carbon atoms, as are sedimentary rocks and the atmosphere. The rate at which materials move between reservoirs is termed a **flux.** When a reservoir releases more materials than it accepts, it is

called a **source,** and when a reservoir accepts more materials than it releases, it is called a **sink.** **FIGURE 5.15** illustrates these concepts in a simple manner.

As we will see in the following sections, human activities can affect the cycling of nutrients by altering fluxes, residence times, and the relative amounts of nutrients in reservoirs (see **SUCCESS STORY,** p. 124).

The water cycle affects all other cycles

Water is so integral to life and to Earth's fundamental processes that we frequently take it for granted. Water is the essential medium for all manner of biochemical reactions, and it plays key roles in nearly every environmental system, including each of the nutrient cycles we are about to discuss. Water carries nutrients, sediments, and pollutants from the continents to the oceans via surface runoff, streams, and rivers. These materials can then be carried thousands of miles on ocean currents. Water also carries atmospheric pollutants to the surface when they dissolve in falling rain or snow. The **hydrologic cycle,** or *water cycle* (**FIGURE 5.16**, p. 122), summarizes how water—in liquid, gaseous, and solid forms—flows through our environment.

The oceans are the main reservoir in the water cycle, holding more than 97% of all water on Earth. The fresh water we depend on for our survival accounts for the remaining water, and two-thirds of this small amount is tied up in glaciers, snowfields, and ice caps (p. 392). Thus, considerably less than 1% of the planet's water is in forms that we can readily use—groundwater, surface fresh water, and rain from atmospheric water vapor.

Evaporation and transpiration Water moves from oceans, lakes, ponds, rivers, and moist soil into the atmosphere by **evaporation,** the conversion of a liquid to a gaseous form. Warm temperatures and strong winds speed rates of evaporation. Water also enters the atmosphere by **transpiration,** the release of water vapor by plants through their leaves, or by evaporation from the surfaces of organisms, such as sweating in humans. Transpiration and evaporation act as natural processes of distillation, because water escaping into the air as a gas leaves behind its dissolved substances.

Precipitation, runoff, and surface water Water returns from the atmosphere to Earth's surface as **precipitation** when water vapor condenses and falls as rain or snow. This water may be taken up by plants and used by animals, but much of it flows as runoff into streams, rivers, lakes,

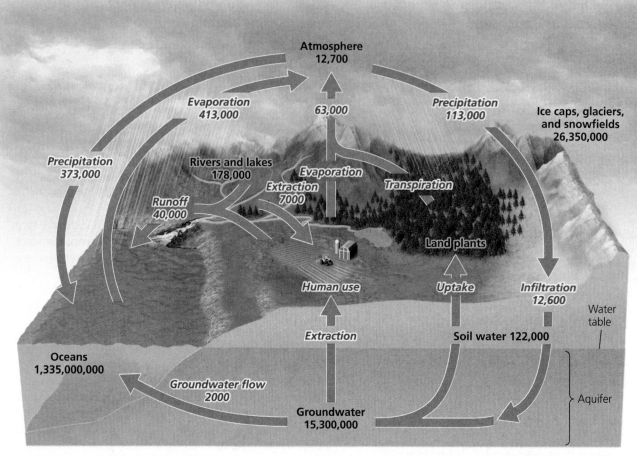

FIGURE 5.16 The water cycle, or hydrologic cycle, summarizes the many routes that water molecules take as they move through the environment. In the figure, reservoir names are printed in black type, and numbers in black type represent reservoir sizes expressed in units of cubic kilometers (km^3). Processes give rise to fluxes (represented by gray arrows), both are printed in italic red type and expressed in km^3 per year. *Data from Schlesinger, W. H., 2013.* Biogeochemistry: An analysis of global change, *3rd ed. London, England: Academic Press.*

ponds, and oceans. Precipitation levels vary greatly from region to region, helping give rise to our planet's variety of biomes (p. 94).

Groundwater Some water soaks down through soil and rock through a process called infiltration, recharging underground reservoirs known as **aquifers** (pp. 392–396). Aquifers are porous regions of rock and soil that hold **groundwater,** water found within the soil. The upper limit of groundwater held in an aquifer is referred to as the **water table.** Aquifers can hold groundwater for long periods of time and can sometimes take hundreds or thousands of years to recharge fully after being depleted. Groundwater becomes surface water when it emerges from springs or flows into streams, rivers, lakes, or the ocean from the soil.

Our impacts on the water cycle are extensive

Human activity affects every aspect of the water cycle. By damming rivers, we slow the movement of water from the land to the sea, and we increase evaporation by holding water in reservoirs. We remove natural vegetation by clear-cutting and developing land, which increases surface runoff, decreases infiltration and transpiration, and promotes evaporation. Our withdrawals of surface water and groundwater for agriculture, industry, and domestic uses deplete rivers, lakes, and streams and lower water tables. This can lead to water shortages and conflict over water supplies (pp. 407–408).

The carbon cycle circulates a vital nutrient

As the definitive component of organic molecules, carbon is an ingredient in carbohydrates, fats, and proteins and occurs in the bones, cartilage, and shells of all living things. The **carbon cycle** describes the routes that carbon atoms take through the environment (**FIGURE 5.17**).

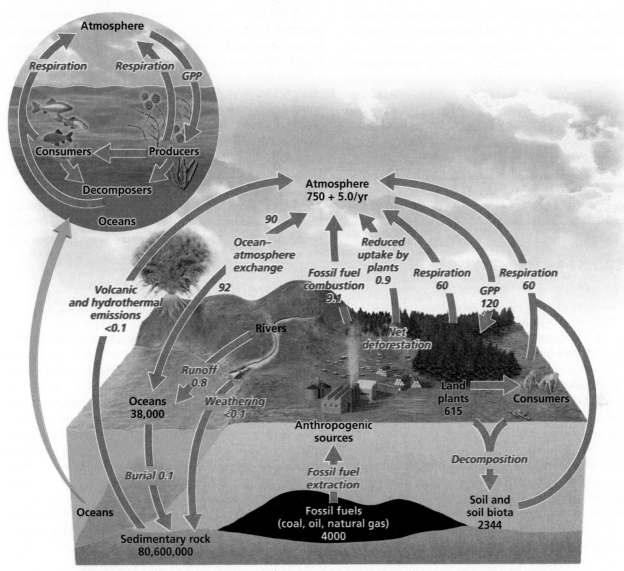

FIGURE 5.17 The carbon cycle summarizes the many routes that carbon atoms take as they move through the environment. In the figure, reservoir names are printed in black type, and numbers in black type represent reservoir sizes expressed in petagrams (units of 10^{15} g) of carbon (note that these values are printed in white type for fossil fuels). Processes give rise to fluxes (represented by gray arrows), both of which are printed in italic red type and expressed in petagrams of carbon per year. In the carbon cycle, plants use carbon dioxide from the atmosphere for photosynthesis (gross primary production, or "GPP" in the figure). *Data from Schlesinger, W. H., 2013.* Biogeochemistry: An analysis of global change, *3rd ed. London, England: Academic Press.*

Photosynthesis, respiration, and food webs Autotrophs pull carbon dioxide out of the atmosphere and out of surface water to use in photosynthesis. They use some of the carbohydrates to fuel cellular respiration, thereby releasing some of the carbon back into the atmosphere and surface water as CO_2. When producers are eaten by primary consumers, which, in turn, are eaten by other animals, more carbohydrates are broken down in cellular respiration, and released as carbon dioxide. The same process occurs as decomposers consume waste and dead organic matter.

Sediment storage of carbon The largest reservoir of carbon, sedimentary rock (p. 38), is formed in ocean basins and

freshwater wetlands. When organisms in these habitats die, their remains may settle in sediments, and as layers of sediment accumulate, the older layers are buried more deeply, and experience high pressure for long periods of time. Over millions of years, these conditions can convert the soft tissues of aquatic organisms into fossil fuels—coal, oil, and natural gas—and transform their shells and skeletons into sedimentary rock, such as limestone.

Carbon trapped in sediments and fossil fuel deposits may eventually be released into the oceans or atmosphere by geologic processes such as uplift, erosion, and volcanic eruptions. It also reenters the atmosphere when we extract and burn fossil fuels.

The oceans Ocean waters are the second-largest reservoir of carbon on Earth. Oceans absorb carbon-containing compounds from the atmosphere, terrestrial runoff, undersea volcanoes, and the detritus of marine organisms. Some carbon atoms absorbed by the oceans—in the form of carbon dioxide, carbonate ions (CO_3^{2-}), and bicarbonate ions (HCO_3^{-})—combine with calcium ions (Ca^{2+}) to form calcium carbonate $(CaCO_3)$, an essential ingredient in the skeletons and shells of microscopic marine organisms. As these organisms die, their $CaCO_3$ shells sink to the ocean floor and begin to form sedimentary rock. The rates at which the oceans absorb and release carbon depend on many factors, including temperature and the numbers of marine organisms converting CO_2 into carbohydrates and carbonates.

We are shifting carbon from the lithosphere to the atmosphere

One major way human activities affect the carbon cycle is through our uses of coal, oil, and natural gas as sources of energy. By mining fossil fuel deposits, we are removing carbon from an underground reservoir, in which it might have otherwise remained for millions of years, and bringing it to surface. By combusting fossil fuels, we release carbon dioxide and greatly increase the flux of carbon from the ground to the air. In addition, cutting down forests removes carbon from vegetation and releases carbon into the air. And if less vegetation is left on the surface, there are fewer plants to draw CO_2 back out of the atmosphere.

As a result, scientists estimate that today's atmospheric CO_2 reservoir is the largest that Earth has experienced in the past 1 million years and likely in the past 20 million years. The ongoing flux of carbon into the atmosphere is the driving force behind today's anthropogenic (human-caused) global climate change (Chapter 18).

Some of the excess CO_2 in the atmosphere is now being absorbed by ocean water. This is causing ocean water to become more acidic, leading to problems that threaten many marine organisms (pp. 437–438).

Our understanding of the carbon cycle is not yet complete. Scientists remain baffled by the so-called missing carbon sink. Of the carbon dioxide we emit by fossil fuel combustion and deforestation, researchers have measured how much goes into the atmosphere and oceans, but there remain roughly 2.3–2.6 billion metric tons unaccounted for. Many scientists think this CO_2 is probably taken up by plants or soils of the temperate and boreal forests (pp. 96–100). They would like to know for sure, though, because if certain forests are acting as a major sink for carbon (and thus restraining global climate change), conserving these ecosystems is particularly vital.

The nitrogen cycle involves specialized bacteria

Nitrogen makes up 78% of our atmosphere by mass and is the sixth most abundant element on Earth. It is an essential ingredient in proteins, DNA, and RNA, and, like phosphorus, is an essential nutrient for plant growth. Thus, the **nitrogen cycle** (FIGURE 5.18) is of vital importance to all organisms. Despite its abundance in the air, nitrogen gas (N_2) is chemically inert and cannot cycle out of the atmosphere and into living organisms without assistance from lightning, highly specialized bacteria, or human intervention. For this reason, the element is relatively scarce in the lithosphere, hydrosphere, and organisms. However, once nitrogen undergoes the right kind of chemical change, it becomes biologically active and available to organisms, and can act as a potent fertilizer.

SUCCESS story **Considering Cost When Saving the Bay**

The Chesapeake Bay offers a case study that illustrates the importance of understanding systems, chemistry and the need for taking a systems-level approach to restore ecosystems degraded by human activities. Tools such as landscape ecology, GIS, and ecological modeling aid these efforts by providing a broad view of the Chesapeake Bay ecosystem and how it may react to changes in nutrient inputs and restoration efforts. The Chesapeake Bay Foundation's most recent "State of the Bay" report concluded that the bay's health rating in 2018 was the highest it had been since CBF's founding in 1964, with meaningful improvements in pollution reduction, fisheries recovery, and the restoration of natural habitats in and around the bay (see Figure 5.1, p. 106).

One reason for the recent success is that farmers, residents, resource managers, and local, state, and federal government agencies have embraced a variety of approaches to reduce nutrient inputs into the bay. By educating people about the many inexpensive yet effective steps that can be taken in yards, farms, businesses, and local communities to reduce nutrient inputs into the Chesapeake Bay, saving the bay became something for which everyone can do his or her part.

A forested buffer lining a waterway on agricultural land in Maryland.

→ **Explore the Data** at **Mastering Environmental Science**

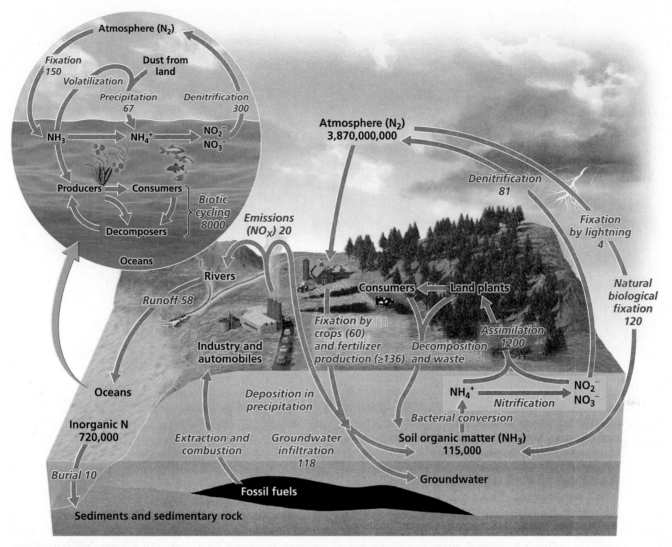

FIGURE 5.18 The nitrogen cycle summarizes the many routes that nitrogen atoms take as they move through the environment. In the figure, reservoir names are printed in black type, and numbers in black type represent reservoir sizes expressed in teragrams (units of 10^{12} g) of nitrogen (note that these values are printed in white type for fossil fuels). Processes give rise to fluxes (represented by gray arrows), both of which are printed in italic red type and expressed in teragrams of nitrogen per year. *Data from Schlesinger, W. H., 2013.* Biogeochemistry: An analysis of global change, *3rd ed. London, England: Academic Press.*

Nitrogen fixation To become biologically available, inert nitrogen gas (N_2) must be "fixed," or combined with hydrogen in nature to form ammonia (NH_3), whose water-soluble ions of ammonium (NH_4^+) can be taken up by plants (**FIGURE 5.19**, p. 125). **Nitrogen fixation** can be accomplished in two ways: by the intense energy of lightning strikes, or when the nitrogen in air comes in contact with particular types of **nitrogen-fixing bacteria** in the soil and water. In the soil, these bacteria live in a mutualistic relationship (pp. 80–81) with many types of plants, including soybeans and other legumes, providing them nutrients by converting nitrogen to a usable form. Some farmers nourish soils by planting crops that host nitrogen-fixing bacteria among their roots.

Nitrification and denitrification Other types of specialized bacteria then perform a process known as **nitrification,** converting NH_4^+ into two nitrite ions, first NO_2^- and then NO_3^-. These water-soluble forms of nitrogen can be taken up by plants and phytoplankton through assimilation and used to fuel their growth. Other sources of these ions are the nitrate-based fertilizers applied to cropland and nitrogenous compounds deposited by air pollution (p. 459).

Animals obtain the nitrogen they need by consuming plants or other animals. Decomposers obtain nitrogen from dead and decaying plant and animal matter, and from the urine and feces of animals. Once decomposers process these nitrogen-rich compounds, they release ammonium ions, a

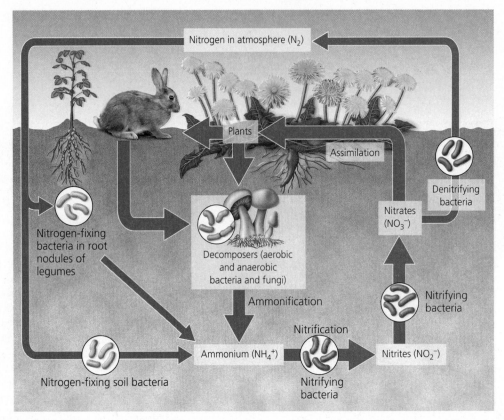

FIGURE 5.19 Specialized bacteria in the soil convert gaseous nitrogen from the air into water-soluble forms that plants can utilize. Other bacteria convert water-soluble forms of nitrogen in the soil back into a gas and return it to the atmosphere.

process called **ammonification,** making the compounds available to nitrifying bacteria to convert again to nitrates and nitrites.

The next step in the nitrogen cycle occurs when **denitrifying bacteria** convert nitrates in soil or water to gaseous nitrogen in a multistep process. Denitrification thereby completes the cycle by releasing nitrogen back into the atmosphere as a gas.

We have greatly influenced the nitrogen cycle

Historically, nitrogen fixation was a **bottleneck,** a step that limited the flux of nitrogen out of the atmosphere and into water-soluble forms. This changed with the research of two German chemists early in the 20th century. Fritz Haber found a way to combine nitrogen and hydrogen gases to synthesize ammonia, a key ingredient in modern explosives and agricultural fertilizers, and Carl Bosch devised methods to produce ammonia on an industrial scale. The **Haber-Bosch process** enabled people to overcome the limits on productivity long imposed by nitrogen scarcity in nature. By enhancing agriculture, the new fertilizers contributed to the past century's enormous increase in human population. Farmers, homeowners, and landscapers alike all took advantage of fertilizers, dramatically altering the nitrogen cycle. Today, using the

Haber-Bosch process, our species is fixing at least as much nitrogen as is being fixed naturally. We have effectively doubled the rate of nitrogen fixation on Earth, overwhelming nature's denitrification abilities.

By fixing atmospheric nitrogen to create fertilizers, we increase nitrogen's flux from the atmosphere to Earth's surface. We also enhance this flux by cultivating legume crops whose roots host nitrogen-fixing bacteria. Moreover, we reduce nitrogen's return to the air when we destroy wetlands whose plants host denitrifying bacteria that convert nitrates to nitrogen gas.

When our farming practices speed runoff and allow soil erosion, nitrogen flows from farms into terrestrial and aquatic ecosystems, leading to nutrient pollution, eutrophication, and hypoxia. These impacts have become painfully evident to oystermen and scientists in the Chesapeake Bay, but hypoxia in waters is by no means the only human impact on the nitrogen cycle. When we burn fossil fuels, we release nitric oxide (NO) into the atmosphere, where it reacts to form nitrogen dioxide (NO_2). This compound is a precursor to nitric acid (HNO_3), a key component of acid precipitation (pp. 474–477). We introduce another nitrogen-containing gas, nitrous oxide (N_2O), when anaerobic bacteria break down the tremendous volume of animal waste produced in agricultural feedlots (p. 249). As these examples show, human activities have affected the nitrogen cycle in diverse and often far-reaching ways.

The phosphorus cycle circulates a limited nutrient

The element phosphorus is a key component of cell membranes and of several molecules vital for life, including DNA, RNA, ATP, and ADP (pp. 28–29, 34). Although phosphorus is indispensable for life, the amount of phosphorus in organisms is dwarfed by the vast amounts in rocks, soil, sediments, and the oceans. Unlike the carbon and nitrogen cycles, the **phosphorus cycle** (**FIGURE 5.20**) has no appreciable atmospheric component besides the transport of tiny amounts in windblown dust and sea spray.

Geology and phosphorus availability The vast majority of Earth's phosphorus is contained within sedimentary rocks and is released only by weathering (p. 217), which releases phosphate ions (PO_4^{3-}) into water. Phosphates dissolved in lakes or in the oceans precipitate into solid form, settle to the bottom, and reenter the lithosphere's phosphorus reservoir in sediments. Because most phosphorus is bound up in rock and only slowly released, environmental concentrations of phosphorus available to organisms tend to be very low. This scarcity explains why phosphorus

is frequently a limiting factor for plant growth and why an influx of phosphorus into an ecosystem can produce immediate and dramatic effects.

Food webs Aquatic producers take up phosphates from surrounding waters, whereas terrestrial producers take up phosphorus from soil water through their roots. Herbivores acquire phosphorus from plant tissues and, when herbivores are consumed, they pass the phosphorus on to their predators. Animals also pass phosphorus to the soil through the excretion of waste. Decomposers break down phosphorus-rich organisms and their wastes and, in so doing, return phosphorus to the soil.

We affect the phosphorus cycle

People increase phosphorus concentrations in surface waters through runoff of the phosphorus-rich fertilizers we apply to lawns and farmlands. A 2008 study determined that an average hectare of land in the Chesapeake Bay region received a net input of 4.52 kg (10 lb) of phosphorus per year, promoting phosphorus accumulation in soils, runoff into waterways, and phytoplankton blooms and hypoxia in the bay. People also add phosphorus to waterways through releases

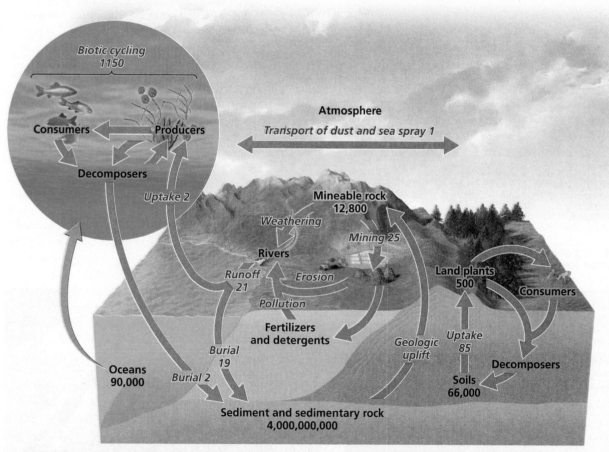

FIGURE 5.20 The phosphorus cycle summarizes the many routes that phosphorus atoms take as they move through the environment. In the figure, reservoir names are printed in black type, and numbers in black type represent reservoir sizes expressed in teragrams (units of 10^{12} g) of phosphorus. Processes give rise to fluxes (represented by gray arrows), both of which are printed in italic red type and expressed in teragrams of phosphorus per year. *Data from Schlesinger, W. H., 2013.* Biogeochemistry: An analysis of global change, *3rd ed. London, England: Academic Press.*

of treated wastewater (pp. 415–416) rich in phosphates from the detergents we use to wash our clothes and dishes.

Tackling nutrient enrichment requires diverse approaches

Given our reliance on synthetic fertilizers for food production and fossil fuels for energy, nutrient enrichment of ecosystems will pose a challenge for many years to come. Fortunately, a number of approaches are available to control nutrient pollution in the Chesapeake Bay and other waterways affected by eutrophication, including:

- Reducing fertilizer use on farms and lawns and timing its application to reduce water runoff
- Using "cover crops," such as clover (pp. 233–234), to keep farmland vegetated after cash crops are harvested
- Planting and maintaining vegetation "buffers" around streams that trap nutrient and sediment runoff
- Using natural and constructed wetlands (p. 417) to filter stormwater and farm runoff
- Improving technologies in sewage treatment plants (pp. 415–416) to enhance nitrogen and phosphorus capture
- Upgrading stormwater systems to capture runoff from roads and parking lots
- Reducing fossil fuel combustion to minimize atmospheric inputs of nitrogen to waterways

Some of these methods cost more than others for similar results. For example, planting vegetation buffers and restoring wetlands can reduce nutrient inputs into waterways at a fraction of the cost of some other approaches, such as retrofitting urban areas to capture stormwater runoff (**FIGURE 5.21**).

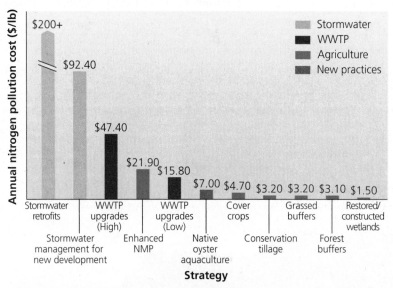

(a) Relative costs of approaches for preventing nutrient inputs to the Chesapeake Bay

FIGURE 5.21 Costs for reducing nitrogen inputs into the Chesapeake Bay vary widely (a). These approaches include cover crops planted on agricultural land **(b).** and retrofitting urban areas with gardens that capture stormwater runoff **(c).**

(b) Cover crops planted on agricultural land in Virginia

(c) Stormwater capture garden in Baltimore, Maryland

Efforts to control nutrient inputs into the Chesapeake Bay are already paying dividends. Levels of nitrogen and phosphorus entering the bay have been declining and are at or near established targets for nutrient additions to bay waters (**FIGURE 5.22**).

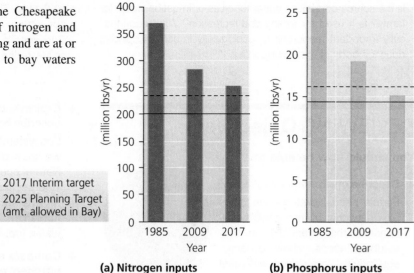

- - - 2017 Interim target
— 2025 Planning Target (amt. allowed in Bay)

(a) Nitrogen inputs   **(b) Phosphorus inputs**

FIGURE 5.22 Inputs of nitrogen (a) and phosphorus (b) to the Chesapeake Bay have been decreasing. Nitrogen inputs in 2017 were slightly higher than the interim target for the year, but phosphorus inputs were lower than the interim target. Further reductions in both nitrogen and phosphorus will be needed to meet the goals for 2025. *Data from Chesapeake Bay Program Watershed Model Phase 5.3.2 (Chesapeake Bay Program Office, 2018).*

CENTRAL CASE STUDY
connect & continue

TODAY, after languishing in a degraded state for decades, the Chesapeake Bay finally has prospects for real recovery, as the federal government and bay states are now managing the bay as a holistic system and working together to save their shared waterway. By studying the environment from a systems perspective and by integrating scientific findings with the policy process, people who care about the Chesapeake Bay and other waterways are working to "bring back" degraded habitats.

While the progress made toward recovery is certainly encouraging, the program's long-term future is uncertain. The federal budget submitted in 2019 by the Trump administration drastically reduced funding for Chesapeake Bay cleanup efforts. Whereas Congress has indicated a willingness to continue the program, if it were to embrace the President's budget, efforts to remedy the bay's will likely collapse. But if the program is continued, the 18 million people living in the bay's watershed have reason to hope that the Chesapeake Bay of tomorrow may be healthier than it is today, thanks to the collaborative efforts of concerned citizens, advocacy organizations, and the federal and bay-state governments.

But as is often the case in environmental science, the challenges faced in places like Deal Island are complex and require equally multifaceted, integrated solutions. Although there are encouraging signs of recovery in bay waters, climate change is causing those same waters to rise (pp. 501–504)—and threaten the island's very existence. The waters of the Chesapeake Bay have risen about 30 cm (1 ft)

in the past century, and if sea levels rise another 90 cm (3 ft) as computer models predict will happen, Deal Island will be completely submerged. Saving Deal Island, and numerous other low-lying areas like it will therefore require that we also take on the challenge of lessening the factors causing these changes in Earth's climatic systems.

- **CASE STUDY SOLUTIONS** You are an oysterman in the Chesapeake Bay, and your income is decreasing because the dead zone is making it harder to harvest oysters. One day your senator comes to town, and you are able to have a one-minute conversation with her. What steps would you urge her and her colleagues in Congress to take to help alleviate the dead zone and bring back the oyster fishery? Now suppose you are a Pennsylvania farmer who has learned that the government is offering incentives to farmers to help reduce fertilizer runoff into the Chesapeake Bay. What types of approaches described in this text might you be willing to try, and why?

- **LOCAL CONNECTIONS** Identify the closest body of water to your area that is suffering from eutrophication. Which of the categories shown in Figure 5.4 would likely be the largest sources of nutrient inputs to this body of water? List three approaches for controlling nutrient inputs—like those used in the Chesapeake Bay—that would be well suited for this body of water. For what human activities is the waterway used?

Is it a source of drinking water for cities or irrigation water for farms? Is it used for fishing and recreation? What economically important resources or ecologically important species does the body of water support?

- **EXPLORE THE DATA** How can we most efficiently focus our efforts on reducing plant nutrient inputs into the Chesapeake Bay? → **Explore Data** relating to the case study on **Mastering Environmental Science**.

REVIEWING Objectives

You should now be able to:

+ **Describe environmental systems**

Earth's natural systems are complex networks of interacting components that generally involve feedback loops, show dynamic equilibrium, and exhibit emergent properties. (pp. 107–111)

+ **Define ecosystems and discuss how living and nonliving entities interact in ecosystem-level ecology**

Ecosystems consist of all organisms and nonliving entities that occur and interact in a particular area at the same time. Matter is recycled in ecosystems, but energy flows in one direction—from producers to higher trophic levels. Input of nutrients can boost productivity—which is the generation of biomass—but an excess of nutrients can alter ecosystems and cause severe ecological and economic consequences. (pp. 111–118)

+ **Outline the fundamentals of landscape ecology and ecological modeling**

Landscape ecology studies how the habitat patches that make up landscape structure influence organisms. Geographic information systems (GIS)—software that overlays multiple types of data on a common set of geographic coordinates—enables landscape ecology to be used increasingly in conservation biology and regional planning. Ecological modeling helps ecologists comprehend the complex systems they study and predict how these systems will react to disturbance. (pp. 118–120)

+ **Explain ecosystem services and describe how they benefit our lives**

Ecosystems provide the "goods" we know of as natural resources. Natural ecological processes provide services that we depend on for everyday living, such as cycling nutrients and purifying drinking water. (pp. 120–121)

+ **Compare and contrast how water, carbon, nitrogen, and phosphorus cycle through the environment and explain how human activities affect these cycles**

Major reservoirs of the hydrologic cycle include the oceans and ice caps, and large fluxes include evaporation, precipitation, and runoff. Most of the carbon on Earth is contained in sedimentary rock, and fluxes include photosynthesis, cellular respiration, and fossil fuel combustion. The major reservoir of nitrogen is the atmosphere, and important fluxes include nitrogen fixation (both by bacteria and by humans) and denitrification. Phosphorus is most abundant in sedimentary rock, and fluxes include the weathering of rocks and human applications of fertilizers. (pp. 121–128)

+ **Explain how human activities affect biogeochemical cycles**

People are affecting Earth's biogeochemical cycles by shifting carbon from fossil fuel reservoirs into the atmosphere, shifting nitrogen from the atmosphere to the planet's surface, and depleting groundwater supplies, among other impacts. Policies that seek to minimize human alterations of cycles, such as the remediation efforts embraced in the Chesapeake Bay, can help us address nutrient pollution. (pp. 122–129)

SEEKING Solutions

1. Once vegetation is cleared from a riverbank, water begins to erode away the bank. This erosion may dislodge more vegetation. Would you expect this to result in a feedback process? If so, which type: negative or positive? Explain your answer. How might we halt or reverse this process?

2. Consider the ecosystem(s) that surround(s) your campus. Describe one way in which energy flows through it and matter is recycled. Now pick one type of nutrient and briefly describe how it moves through this (these) ecosystem(s). Does the landscape contain patches? Can you describe any ecotones?

3. For a conservation biologist interested in sustaining populations of the listed organisms, why would it be helpful to take a landscape ecology perspective? Explain your answer in each case.

   - A forest-breeding warbler that suffers poor nesting success in small, fragmented forest patches
   - A bighorn sheep that must move seasonally between mountains and lowlands
   - A toad that lives in upland areas but travels cross-country to breed in localized pools each spring

4. How do you think we might solve the problem of eutrophication in the Chesapeake Bay? Assess several possible solutions, your reasons for believing they might work, and the likely hurdles we would face. Explain who

should be responsible for implementing these solutions, and why.

5. **THINK IT THROUGH** You are a resource manager assigned to devise a plan to protect a rare species of frog that lives in an area slated for development. You are provided with GIS layers of the area as shown in Figure 5.12. The frogs live in the leaf litter in forests, require ponds for breeding, and cannot travel over large open areas when migrating to breeding ponds. What layers (of those shown in the figure) would be most important for your analysis? Why? How would you use principles of landscape ecology when creating your plan?

CALCULATING **Ecological Footprints**

For many Americans, their desire is to own a suburban home with a weed-free, green lawn. Nationwide, Americans tend about 40.5 million acres of lawn grass. But conventional lawn care involves inputs of fertilizers, pesticides, and irrigation water, not to mention the use of gasoline or electricity for mowing and other care—all of which raise health concerns and affect environmental cycles. Using the figures for a typical lawn in the table, calculate the total amount of fertilizer, water, and gasoline used in lawn care across the nation each year.

	ACREAGE OF LAWN	FERTILIZER USED (LB)	WATER USED (GAL)	GASOLINE USED (GAL)
For the typical quarter-acre lawn	0.25	37	15,700	4.9
For all lawns in your hometown				
For all lawns in the United States	40,500,000			

Data: Chameides, B., 2008. www.huffingtonpost.com/bill-chameides/stat-grok-lawns-by-the-nu_b_115079.html

1. How much fertilizer is applied each year on lawns throughout the United States? Where does the nitrogen for this fertilizer come from? What becomes of the nitrogen and phosphorus that are applied to a suburban lawn but not taken up by grass?
2. Leaving grass clippings on a lawn decreases the need for fertilizer by 50%. What else might a homeowner do to decrease fertilizer use in a yard and the environmental impacts of nutrient pollution?
3. How much gasoline could Americans save each year if they did not take care of their lawns? At today's gas prices, how much money would this save annually?

Mastering Environmental Science

Students Go to **Mastering Environmental Science** for assignments, an interactive e-text, and the Study Area with practice tests, videos, and activities.

Instructors Go to **Mastering Environmental Science** for automatically graded activities, videos, and reading questions that you can assign to your students, plus Instructor Resources.

Ethics, Economics, and Sustainable Development

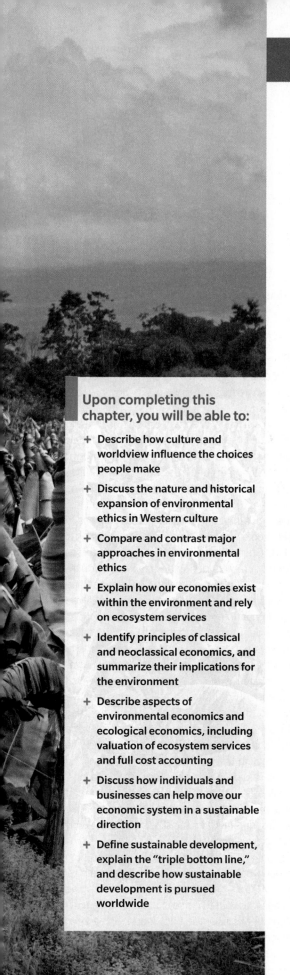

Upon completing this chapter, you will be able to:

+ **Describe how culture and worldview influence the choices people make**

+ **Discuss the nature and historical expansion of environmental ethics in Western culture**

+ **Compare and contrast major approaches in environmental ethics**

+ **Explain how our economies exist within the environment and rely on ecosystem services**

+ **Identify principles of classical and neoclassical economics, and summarize their implications for the environment**

+ **Describe aspects of environmental economics and ecological economics, including valuation of ecosystem services and full cost accounting**

+ **Discuss how individuals and businesses can help move our economic system in a sustainable direction**

+ **Define sustainable development, explain the "triple bottom line," and describe how sustainable development is pursued worldwide**

CENTRAL

case study

Costa Rica Values Its Ecosystem Services

> " Costa Rica's PSA program has been one of the conservation success stories of the last decade.
> Stefano Pagiola, The World Bank

> **In the last 25 years, my home country has tripled its GDP while doubling the size of its forests.**
> Carlos Manuel Rodriguez, former Minister of Energy and the Environment, Costa Rica "

Very few nations have transformed their path of development in just decades—but Costa Rica has. In the 1980s, this small Central American country was losing its forests as fast as any place on Earth. Today, this nation of 5 million people has regained much of its forest cover, boasts a world-class park system, and stands as a global model for sustainable resource management.

Costa Rica took many steps on this impressive road to success. One key step was to begin paying landholders to conserve forest on private land, in a novel government program called *Pago por Servicios Ambientales* (PSA)—Payment for Environmental Services.

Nature provides ecosystem services (pp. 4, 120–121) such as air and water purification, climate regulation, and nutrient cycling. For example, forests in Costa Rica's mountains capture rainfall and provide clean drinking water for farms, towns, and cities below. Ecosystem services are vital for our lives, but historically we have tended to take them for granted, and rarely do we acknowledge their value by paying for them in the marketplace. As a result, these services have diminished as we degrade the natural systems that provide them. For these reasons, many economists believe that it is important to create financial incentives for conserving ecosystem services.

In Costa Rica, which had lost more than three-fourths of its forest, political leaders created financial incentives for conserving ecosystem services through the PSA program—established as part of Forest Law 7575, passed in 1996. Since then, the Costa Rican government has been paying farmers and ranchers to preserve forest on their land, replant cleared areas, allow forest to regenerate naturally, and establish sustainable forestry systems. Payments are designed to approximate potential profits from farming or cattle ranching, and in recent years, these payments have averaged $78/hectare (ha)/yr ($32/acre/yr).

The PSA program recognizes four ecosystem services that forests provide:

- *Watershed protection:* Forests cleanse water by filtering pollutants, and they conserve water and reduce soil erosion by slowing runoff.
- *Biodiversity:* Tropical forests such as Costa Rica's are especially rich in life.

◀ **A Costa Rican banana plantation**

▲ **A keel-billed toucan, one of many species relying on Costa Rica's forests**

133

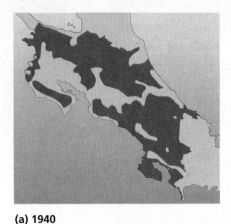

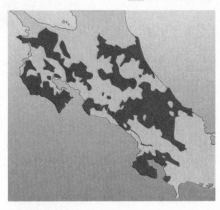

(a) 1940   (b) 1987   (c) 2005

FIGURE 6.1 **Forest cover in Costa Rica decreased between 1940 and 1987, but it increased thereafter.**
Forest cover today is slightly more than that shown for 2005. *Data from FONAFIFO.*

- *Scenic beauty:* Recreation and ecotourism, which bring money to the economy, thrive in areas of scenic beauty.
- *Carbon sequestration:* By pulling carbon dioxide from the atmosphere, forests slow global warming.

To fund the PSA program, Costa Rica's government sought money from people and companies that benefit from these services. With watershed protection, for example, irrigators, bottlers, municipal water suppliers, and utilities that generate hydropower all made voluntary payments into the program, and a tariff on water users was added in 2005. For biodiversity and scenery, the country targeted ecotourism, while international lending agencies provided loans and donations. Because carbon dioxide is emitted when fossil fuels are burned, the nation used a 3.5% tax on fossil fuels to help fund the program. It also sought to sell carbon offsets in international carbon trading markets (p. 514).

Costa Rican landholders rushed to sign up for the PSA program. The agency administering it, *Fondo Nacional de Financiamiento Forestal* (FONAFIFO), signed landowners to contracts and sent agents to advise them on forest conservation and to monitor compliance. Through 2018, FONAFIFO had paid 236 billion colónes ($390 million in today's U.S. dollars) to nearly 18,000 landholders and had registered 1.26 million ha (3.11 million acres)—almost one-fourth of the nation's land area.

Deforestation slowed in Costa Rica in the wake of the program. Forest cover rose by 10% in the decade after 1996, and policymakers, economists, and environmental advocates cheered the PSA program's apparent success. However, some observers argued that forest loss had been slowing for other reasons and that the program itself was having little effect. They contended that payments were being wasted on people who had no plans to cut down their trees. Critics also lamented that large wealthy landowners utilized the program more than low-income small farmers. All these concerns were borne out by researchers (see **THE SCIENCE BEHIND THE STORY**, pp. 144–145).

In response, the government modified its policies, making the program more accessible to small farmers and targeting the payments to locations where forest is most at risk and environmental assets are greatest.

In recent years, forested area in Costa Rica has been rising (**FIGURE 6.1**), from a low of 17% in 1983 to more than 53% today. The nation has thrived economically while protecting its environment. Since the PSA program began, Costa Ricans have enjoyed an increase in real, inflation-adjusted per capita income of 70%—a rise in wealth surpassing the vast majority of nations.

Many factors have contributed to Costa Rica's success in building a wealthier society while protecting its ecological assets. Back in 1948, Costa Rica abolished its armed forces and shifted funds from the military budget to health and education. (The only mainland nation in the world without a standing army, Costa Rica enjoys security from alliances with the United States and other nations and has experienced seven decades of peace.) With a stable democracy and a healthy and educated citizenry, the stage was set for well-managed development, including innovative advances in conservation. The nation created one of the world's finest systems of national parks, which today covers fully one-fourth of its territory. Ecotourism at the parks brings wealth to the country: Each year more than 2 million foreign tourists inject more than $2 billion into Costa Rica's economy. As a result, Costa Ricans understand the economic value of protecting their natural capital.

With the economic value of nature so clear to Costa Ricans, an ethic of conservation has grown and flourished. In today's Costa Rica, an ethical appreciation for nature and an economic appreciation for nature go hand in hand, pointing the way toward truly sustainable development.

As our global society charts a path toward long-term sustainability, we must heed people's economic needs and ethical values. Science can advise us how to conserve resources and maintain ecosystem services, but to achieve a sustainable civilization, we must integrate this scientific knowledge with a keen understanding of ethical and economic concerns.

Environmental Ethics

Environmental science entails a firm understanding of the natural sciences, but to address environmental problems, we also need to understand how people perceive, value, and relate to their environment philosophically and pragmatically. Ethics and economics are two very different disciplines, but each deals with questions of what we value and how those values influence our decisions and actions. **Ethics** is the branch of philosophy that involves the study of good and bad, of right and wrong. The term *ethics* can also refer to the set of moral principles or values held by a person or a society. **Economics** is the study of how people decide to use potentially scarce resources to provide goods and services that are in demand. To find sustainable solutions to environmental problems, we must aim to comprehend not only how natural systems work but also how values shape human behavior.

Culture and worldview influence our decisions

In deciding how to manipulate our environment to meet our needs, we rely on rational assessments of costs and benefits, but our decisions are also influenced by our culture and our worldview (**FIGURE 6.2**). **Culture** can be defined as the ensemble of knowledge, beliefs, values, and learned ways of life shared by a group of people. Together with personal experience and personal circumstance, culture influences each person's perception of the world and his or her place within it—the person's **worldview.** A worldview reflects beliefs about the meaning, operation, and essence of the world.

People with different worldviews can study the same situation yet draw dramatically different conclusions. For example, two Costa Rican ranchers owning identical landholdings may make different decisions about how to manage their property. One might opt to receive payments and conserve forest, whereas the other might opt to clear every hectare for cattle grazing.

Many factors shape people's worldviews and perception of their environment. Among the most influential are spiritual beliefs and political ideology. For instance, one's views on the proper role of government may guide whether one wants government to intervene in a market economy to protect environmental quality. Shared cultural experience is another factor. For example, early European settlers in the Americas, facing the struggles of frontier life, viewed their environment as a hostile force because inclement weather and wild animals frequently destroyed crops and killed livestock. Many people still view nature as a hostile adversary to overcome.

As you progress through this course and through your life, you will encounter scientific data on the environmental impacts of our choices (where to live, how to make a living, what to eat, what to wear, how to spend our leisure time, and so on). You will also see that culture and worldviews play critical roles in such choices. Acquiring scientific understanding is vital in our search for sustainable solutions, but we also need to consider ethics and economics, because these disciplines help us understand how and why we value the things we value.

(a) A Costa Rican guide shows a Hercules beetle to ecotourists

(b) A pineapple farmer makes a living from his crops

FIGURE 6.2 Differences in culture, wealth, and personal circumstance may lead people to interact differently with their environment. In Costa Rica's national parks **(a)**, ecotourism provides visitors exploration, recreation, education, and inspiration while providing quality employment for Costa Ricans, who may see economic advantage in preserving forest as a result. In contrast, a Costa Rican farm worker **(b)** needs agriculture that produces enough to live on and may favor clearing forest if the short-term economic benefits outweigh the costs.

Ethical standards help us judge right from wrong

Ethicists examine how people judge right from wrong by clarifying the criteria that people use in making these judgments. Such criteria are grounded in values—for instance, promoting human welfare, maximizing individual freedom, or minimizing pain and suffering.

People of different cultures or worldviews may differ in their values and thus may disagree about actions they consider to be right or wrong. That is why some ethicists are **relativists** who believe that ethics do and should vary with social context. However, different societies show a remarkable extent of agreement on what moral standards are appropriate. For this reason, many ethicists are **universalists** who maintain that there exist objective notions of right and wrong that hold across cultures and contexts. For both relativists and universalists, ethics is a prescriptive pursuit; rather than simply describing behavior, it prescribes how we *ought* to behave.

The criteria that help differentiate right from wrong are known as **ethical standards.** One classic ethical standard is the *categorical imperative* proposed by German philosopher Immanuel Kant, which advises us to treat others as we would prefer to be treated ourselves. In Christianity this standard is called the "Golden Rule," and most of the world's religions teach this same lesson. Another ethical standard is the *principle of utility*, elaborated by British philosophers Jeremy Bentham and John Stuart Mill. The utilitarian principle holds that something is right when it produces the greatest practical benefits for the most people. We all employ ethical standards as we make countless decisions in our everyday lives.

We value things in two ways

People ascribe value to things in two main ways. One way is to value something for the pragmatic benefits it brings us if we put it to use. This notion is termed **instrumental value** (or *utilitarian value*). The other way is to value something for its intrinsic worth, to believe that something has a right to exist and is valuable for its own sake. This notion is termed **intrinsic value,** or *inherent value*.

A person may ascribe instrumental value to a forest because we can harvest timber from it, hunt game in it, use it for recreation, and drink clean water it has captured and filtered. A person may perceive intrinsic value in a forest because it provides homes for countless organisms that the person feels have an inherent right to live. A forest, an animal, a lake, or a mountain can have both intrinsic and instrumental value. However, different people may emphasize different types of value.

Paying money for ecosystem services is a utilitarian approach that attempts to quantify instrumental values by assigning market prices to them. For people who tend to view nature through the lens of intrinsic values, this approach can make them uneasy. Indeed, scientists who share the goal of conserving natural amenities sometimes disagree on the best means of doing so. For instance, Stanford University biologist Gretchen Daily is a longtime proponent of assigning market values to ecosystem services, with the utilitarian goal of engaging market forces to assist in their conservation. Yet a student in her department, Douglas McCauley, once authored a commentary in the scientific journal *Nature* that argued eloquently against such an approach. McCauley warned that the "commodification of nature" distracted from the environment's intrinsic value, which he maintained was infinite and priceless. "Nature conservation must be framed as a moral issue," McCauley wrote. "We will make more progress in the long run by appealing to people's hearts rather than to their wallets."

Environmental ethics pertains to people and the environment

The application of ethical standards to relationships between people and nonhuman entities is known as **environmental ethics.** Our interactions with our environment can give rise to ethical questions that are difficult to resolve. Consider some examples:

1. Are current generations obligated to conserve resources for future generations? If so, how much should we conserve?

2. Is it ever justified to expose some communities of people to a disproportionate share of pollution? If not, what actions are warranted to prevent this practice?

3. Are humans justified in driving species to extinction? If destroying a forest would drive extinct a little-known insect species but would create jobs for 10,000 people, would that action be ethically admissible? What if it were an owl species? What if only 100 jobs would be created? What if it were a species harmful to people, such as a mosquito or a noxious bacterium or virus?

The first question is central to the notion of sustainability (p. 16) and to the pursuit of sustainable development (p. 155). From an ethical perspective, sustainability means treating future generations as we would prefer to be treated ourselves. The second question goes to the heart of environmental justice (pp. 139–141) and addresses issues of fairness and equity. The third set of questions involves intrinsic values (but also instrumental values) and typifies issues that arise in debates over endangered species management, habitat protection, and conservation biology (Chapter 11).

We have expanded our realm of ethical consideration

Answers to ethical questions depend partly on what ethical standard(s) a person adopts. They also depend on the breadth and inclusiveness of the person's domain of ethical concern. Many traditional non-Western cultures have long granted nonhuman entities intrinsic value and ethical standing. Australian Aborigines view their landscape as sacred and alive. Many indigenous cultures across the Americas feature ethical systems that encompass both people and aspects of their environment.

As the history of Western cultures (European and European-derived societies) has progressed, people have granted intrinsic value and extended ethical consideration to more and more people and things. Today, concern for the welfare of domesticated animals is evident in the great care many people provide for their pets. Animal rights advocates voice concern for animals that are hunted, raised in pens, or used in laboratory testing. Most people now accept that wild animals merit ethical consideration. Increasing numbers of people see intrinsic value in whole natural communities, and some suggest that all of nature—living and nonliving things alike—should be ethically recognized.

What has helped broaden our ethical domain in these ways? Rising economic prosperity has played a role, by making us less anxious about our day-to-day survival. Science has also played a role by demonstrating that people do not stand apart from nature but rather are part of it. For example, ecology makes clear that organisms are interconnected and that what affects wild plants, animals, and ecosystems also

FIGURE 6.3 We can categorize people's ethical perspectives as anthropocentric, biocentric, or ecocentric.

affects people. Evolutionary biology shows that human beings, as one species out of millions, have evolved amid the same pressures as other organisms.

We can simplify our continuum of attitudes toward the natural world by dividing it into three ethical perspectives: anthropocentrism, biocentrism, and ecocentrism (**FIGURE 6.3**).

Anthropocentrism People who have a human-centered view of our relationship with the environment display **anthropocentrism.** An anthropocentrist denies, overlooks, or devalues the notion that nonhuman things can have intrinsic value. An anthropocentrist evaluates the costs and benefits of actions solely according to their impact on people. For example, if cutting down a Costa Rican forest for farming or ranching would provide significant economic benefits while doing little harm to aesthetics or human health, the anthropocentrist would conclude that the action would be worthwhile, even if it would destroy many plants and animals. Conversely, if protecting the forest would provide greater economic, spiritual, or other benefits to people, an anthropocentrist would favor its protection. In the anthropocentric perspective, anything not providing benefit to people is considered to be of negligible value.

Biocentrism The perspective that ascribes intrinsic value to certain living things or to the biotic realm in general is **biocentrism.** In this perspective, human life and nonhuman life both have ethical standing. A biocentrist might oppose clearing a forest if doing so would destroy a great number of plants and animals, even if it would increase food production and generate economic growth for people.

Ecocentrism People who judge actions in terms of their effects on whole ecological systems, which consist of living and nonliving elements and the relationships among them, are

said to show **ecocentrism.** An ecocentrist values the well-being of entire species, communities, or ecosystems over the welfare of a given individual. Implicit in this view is that preserving systems generally protects their components, whereas protecting components may not safeguard the entire system. An ecocentrist would respond to a proposal to clear forest by broadly assessing the potential impacts on water quality, air quality, wildlife populations, soil structure, nutrient cycling, and ecosystem services. Ecocentrism is a more holistic perspective than biocentrism or anthropocentrism. It encompasses a wider variety of entities at a larger scale and seeks to preserve the connections that tie them together into functional systems.

Environmental ethics has ancient roots

Environmental ethics arose as an academic discipline in the 1970s, but people have contemplated our ethical relations with nature for thousands of years. The ancient Greek philosopher Plato argued that humanity had a moral obligation to our environment, writing, "The land is our ancestral home and we must cherish it even more than children cherish their mother."

Some ethicists and theologians have pointed to the religious traditions of Christianity, Judaism, and Islam as sources of anthropocentric hostility toward the environment. They point out biblical passages such as, "Be fruitful and multiply, and fill the earth and subdue it; and have dominion over the fish of the sea and over the birds of the air and over every living thing that moves upon the earth" (Genesis 1:28). Such wording, according to many scholars, has encouraged animosity toward nature.

Others interpret sacred texts of these religions to encourage benevolent human stewardship over nature. Consider the directive, "You shall not defile the land in which you live" (Numbers 35:34). If one views the natural world as God's creation, under this interpretation it would be considered sinful to degrade that creation. Indeed, many people holding the stewardship interpretation see it as their ethical duty to act as responsible stewards of the world in which we live. In recent years, Pope Francis has been an especially powerful voice for the notion that humanity is ethically obliged to care for our environment.

The industrial revolution inspired reaction

As industrialization spread in the 19th century, it amplified human impacts on the environment. In this period of social and economic transformation, agricultural economies became industrial ones, machines enhanced or replaced human and animal labor, and many people moved from farms to cities. Population rose dramatically, consumption of natural resources accelerated, and pollution intensified as we burned coal to fuel railroads, steamships, ironworks, and factories.

Some writers of the time, such as British critic John Ruskin, worried that although people prized the material benefits that nature provided, they no longer appreciated its spiritual and aesthetic benefits. Motivated by such concerns, a number of citizens' groups—forerunners of today's environmental organizations—sprang up in 19th-century England.

In the United States during the 1840s, a philosophical movement called *transcendentalism* flourished, espoused in New England by philosophers Ralph Waldo Emerson and Henry David Thoreau and by poet Walt Whitman. Like Ruskin, the transcendentalists objected to materialism, and they saw natural entities as symbols or messengers of deeper truths. Thoreau viewed nature as divine, but he also observed the natural world closely and came to understand it in the manner of a scientist; indeed, he can be considered one of the first ecologists. His book *Walden,* in which he recorded observations and thoughts while living at Walden Pond away from the bustle of urban Massachusetts, remains a classic of American literature.

Conservation and preservation arose with the 20th century

One admirer of Emerson and Thoreau was **John Muir** (1838–1914), a Scottish immigrant to the United States who made California's Yosemite Valley his wilderness home. Muir lived in his beloved Sierra Nevada for long stretches of time, but he also became politically active and won fame as a tireless advocate for the preservation of wilderness (**FIGURE 6.4**).

Muir was motivated by the rapid deforestation and environmental degradation he witnessed throughout North America and by his belief that the natural world should be treated with the same respect that we give to cathedrals. Today, he is associated with the **preservation ethic,** which holds that we should protect the natural environment in a pristine, unaltered state. Muir argued that nature deserved protection for its own sake (an ecocentrist argument resting on the notion of intrinsic value), but he also maintained that nature promoted human happiness (an anthropocentrist argument based on instrumental value). "Everybody needs beauty as well as bread," he

FIGURE 6.5 Gifford Pinchot was a leading proponent of the conservation ethic. This ethic holds that we should use natural resources in ways that ensure the greatest good for the greatest number of people for the longest time.

wrote, "Places to play in and pray in, where nature may heal and give strength to body and soul alike."

Some of the factors that motivated Muir also inspired **Gifford Pinchot** (1865–1946), the first professionally trained American forester (**FIGURE 6.5**). Pinchot founded what would become the U.S. Forest Service and served as its chief in President Theodore Roosevelt's administration. Like Muir, Pinchot opposed deforestation and unregulated economic development, but Pinchot took a more anthropocentric view of how and why we should value nature. He espoused the **conservation ethic,** which holds that people should put natural resources to use but that we have a responsibility to manage them wisely. The conservation ethic employs a utilitarian standard, holding that we should allocate resources to provide the greatest good to the greatest number of people for the longest time. Whereas preservation aims to preserve nature for its own sake and for our aesthetic and spiritual benefit, conservation promotes the prudent, efficient, and sustainable extraction and use of natural resources for the good of present and future generations.

The contrasting approaches of Pinchot and Muir often pitted them against each other on policy issues of the day. However, both men opposed a prevailing tendency to promote economic development without regard to its negative consequences—and both men left legacies that reverberate today.

These schools of thought have now spread globally, and people worldwide are wrestling with how to balance preservation, conservation, and economic development. In Costa Rica,

FIGURE 6.4 A pioneering advocate of the preservation ethic, John Muir helped establish the Sierra Club, a leading environmental organization. Here Muir **(right)** is shown with President Theodore Roosevelt in Yosemite National Park in 1903. After this wilderness camping trip with Muir, the president expanded protection of areas in the Sierra Nevada.

leaders and citizens felt that development had gone too far and that precious ecosystem services were being lost, threatening the country's future. The nation responded by establishing an extensive system of national parks, eventually protecting 24% of its land area—one of the highest percentages of any nation in the world. Beyond this preservationist policy, many conservationist policies were implemented to encourage the sustainable use of soil, water, and forests. Costa Rica's program to pay for ecological services mixes these approaches, encouraging the preservation of forests on lands actively used for agricultural production to meet overall goals of conservation and sustainable development.

Aldo Leopold's land ethic inspires many people

As a young forester and wildlife manager, **Aldo Leopold** (1887–1949; **FIGURE 6.6**) began his career in the conservationist camp after graduating from Yale Forestry School, which Pinchot had helped found just as Roosevelt and Pinchot were advancing conservation on the national stage. As a forest manager in Arizona and New Mexico, Leopold embraced the government policy of shooting predators, such as wolves, to increase populations of deer and other game animals.

At the same time, Leopold followed the advance of ecological science. He eventually ceased to view certain species as "good" or "bad" and instead came to see that ecological systems depend on protecting all their interacting parts. Drawing an analogy to mechanical maintenance, he wrote, "to keep every cog and wheel is the first precaution of intelligent tinkering."

FIGURE 6.6 Aldo Leopold, a wildlife manager, author, and philosopher, articulated a new relationship between people and the environment. In his essay "The Land Ethic," he called on people to embrace their environment in their ethical outlook.

It was not just science that pulled Leopold from an anthropocentric perspective toward a more holistic one. One day he shot a wolf, and when he reached the animal, Leopold was transfixed by "a fierce green fire dying in her eyes." At that moment, Leopold perceived intrinsic value in the wolf, and the experience remained with him for the rest of his life, helping lead him to an ecocentric ethical outlook. Years later, as a University of Wisconsin professor, Leopold argued that people should view themselves and "the land" as members of the same community and that we are obligated to treat the land in an ethical manner.

Leopold intended this *land ethic* to help guide decision making. "A thing is right," he wrote, "when it tends to preserve the integrity, stability, and beauty of the biotic community. It is wrong when it tends otherwise." Leopold died before seeing his seminal 1949 essay, "The Land Ethic," and his best-known book, *A Sand County Almanac*, in print, but today, many view him as the most eloquent philosopher of environmental ethics.

Environmental justice seeks fair treatment for all people

In recent decades, our society has increasingly recognized inequalities of wealth and privilege, often associated with gender, ethnicity, or race. Addressing these imbalances involves applying standards of fairness and equality and has given rise to the environmental justice movement. **Environmental justice** involves the fair and equitable treatment of all people with respect to environmental policy and practice, regardless of their income, race, or ethnicity.

The struggle for environmental justice has been fueled by the recognition that poor people tend to be exposed to more pollution, hazards, and environmental degradation than richer people (**FIGURE 6.7**, p. 140). Environmental justice advocates also note that racial and ethnic minorities tend to suffer more than their share of exposure to most hazards. Indeed, studies repeatedly document that poor and nonwhite communities each tend to bear heavier burdens of air pollution, lead poisoning, pesticide exposure, toxic waste exposure, and workplace hazards than those in wealthier white communities. This is thought to occur because lower-income and minority communities often have less access to information on environmental health risks, less political power with which to protect their interests, and less money to spend on avoiding or alleviating risks.

A protest in the 1980s by residents of Warren County, North Carolina, against a proposed toxic waste dump in their community helped ignite the environmental justice movement. The state had chosen to site the dump where the highest percentage of African American residents lived. Warren County

(a) Migrant farm workers in Colorado

(b) Homes near a coal-fired power plant

(c) Consumer of lead-tainted drinking water in Flint, Michigan

FIGURE 6.7 Environmental justice efforts are inspired by the fact that the poor are often exposed to more hazards than are the rich. For example, Latino farm workers **(a)** may experience health risks from pesticides, fertilizers, and dust. Low-income white Americans in Appalachia **(b)** may suffer air pollution from nearby coal-fired power plants. African Americans in Flint, Michigan **(c),** were most affected by that city's crisis of lead in its drinking water in 2014–2015. Nationwide monitoring has found for decades that African Americans suffer substantially more lead exposure and other types of water pollution than do white Americans.

protesters lost their battle and the dump was established—but the protest inspired countless efforts elsewhere.

Like African Americans, Native Americans have encountered many environmental justice issues. For instance, uranium mining on lands of the Navajo nation in the Southwest employed many Navajo in the 1950s and 1960s. Although uranium mining had been linked to health problems and premature death, neither the mining industry nor the U.S. government provided the miners information or safeguards against radiation and its risks. Many Navajo families built homes and bread-baking ovens out of waste rock from the mines, not realizing it was radioactive. As cancer began to appear among Navajo miners, a later generation of Americans perceived negligence and discrimination. They pushed Congress to enact the Radiation Exposure Compensation Act of 1990, a federal law compensating Navajo miners who suffered health effects from unprotected work in the mines.

Likewise, low-income white residents of the Appalachian region have long been the focus of environmental justice concerns. Mountaintop coal-mining practices (pp. 538, 652) in this economically neglected region provide jobs to local residents but also pollute water, bury streams, destroy forests, and cause flooding. Low-income residents of affected Appalachian communities continue to have little political power to voice complaints over the impacts of these mining practices.

Today, the world's economies have grown, but the economic and cultural gaps between the wealthy and the working class have widened. And despite historic progress toward racial equality, significant inequities remain while new political demagogues inflame old racial tensions. Environmental laws have proliferated, but many are now under attack, and minorities and the poor continue to suffer substandard environmental conditions. Still, today more people than ever are fighting environmental hazards in their communities and winning.

One ongoing story involves Latino farm workers in California's San Joaquin Valley. These workers harvest much of the U.S. food supply of fruits and vegetables yet suffer some of the nation's worst air pollution. Industrial agriculture generates pesticide emissions, dairy feedlot emissions, and windblown dust from eroding farmland, yet this pollution was not being regulated. Valley residents enlisted the help of organizations, including the Center on Race, Poverty, and the Environment, a San Francisco–based law firm focusing on environmental justice. Together they persuaded California regulators to enforce Clean Air Act provisions and convinced California legislators to pass new legislation regulating agricultural emissions.

Environmental justice is a key component in pursuing the environmental, economic, and social goals of sustainability and sustainable development (pp. 16, 155). As we explore environmental issues from a scientific standpoint throughout this book, we will also encounter the social, political, ethical, and economic aspects of these issues, and the concept of environmental justice will come up again and again.

Questions of environmental justice, like questions of how to value ecosystem services, intertwine ethical issues with economic ones—and friction often develops between people's ethical concerns and their economic desires. Addressing ethical, economic, and environmental concerns together in a mutually productive way is a primary goal of the modern drive for sustainable development (pp. 155–157).

Economics and the Environment

An **economy** is a social system that converts resources into *goods* (material commodities manufactured for and bought by individuals and businesses) and *services* (work done for others as a form of business). The word *economics* and the word *ecology* come from the same Greek root, *oikos*, meaning "household." Economists traditionally have studied the household of human society, whereas ecologists study the broader household of all life.

Economies rely on goods and services from the environment

Economies receive inputs (such as natural resources and ecosystem services) from the environment, process them, and discharge outputs (such as waste) into the environment (**FIGURE 6.8**). These dynamics are readily apparent, yet many mainstream economists still adhere to a worldview that overlooks their importance and largely ignores the environment (instead considering only wages, labor, products, and payments, as depicted in the box in the center of Figure 6.8).

This traditional outlook, which continues to drive many policy decisions, implies that natural resources are free and limitless and that wastes can be endlessly absorbed at no cost. Newer schools of thought, however, recognize that economies exist within the environment and rely on its natural resources and ecosystem services. This is why Costa Rica and other nations have taken steps to better protect their natural assets.

Natural resources (p. 4) are the substances and forces that sustain us—the fresh water we drink, the trees that provide our lumber, the rocks that provide our metals, and energy from the sun, wind, water, and fossil fuels. We can think of natural resources as "goods" produced by nature. Environmental systems also naturally function in a manner that supports our economies. Earth's ecological systems purify air and water, form soil, cycle nutrients, regulate climate, pollinate plants, and recycle waste. Such essential ecosystem services (pp. 4, 120–121) support the very life that makes our economic activity possible. Together, nature's resources and services make up the natural capital (p. 6) on which we depend.

When we deplete natural resources and generate pollution, we degrade the capacity of ecological systems to function. In 2005, scientists from 100 nations concluded in the Millennium Ecosystem Assessment that 15 of 24 ecosystem services they surveyed globally were being degraded or used unsustainably. The degradation of ecosystem services can weaken economies. In Costa Rica, rapid forest loss up through the 1980s was causing soil erosion, water pollution, and biodiversity loss that increasingly threatened the country's economic potential. Low-income small farmers were the first to feel these impacts.

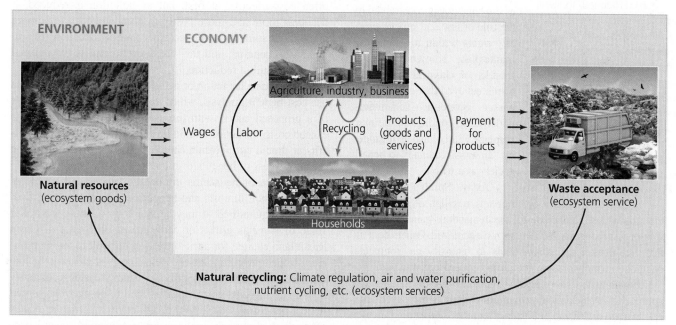

FIGURE 6.8 Economies exist within the natural environment, receiving resources from it, discharging waste into it, and benefiting from ecosystem services. Conventional neoclassical economics has focused on processes of production and consumption between households and businesses **(tan box in middle),** viewing the environment merely as an external factor. In contrast, environmental and ecological economists emphasize that economies exist within the natural environment and depend on all it offers.

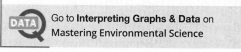
Go to **Interpreting Graphs & Data** on Mastering Environmental Science

We often hear it said that protecting environmental quality costs too much money, interferes with progress, or leads to job loss. However, growing numbers of economists disagree. Instead, they assert that environmental protection tends to *enhance* our economic bottom line and improve our quality of life. The view one takes depends in part on the timeframe. We often make economic judgments on short timescales, and in the short term, many activities that cause environmental damage may be economically profitable. In the longer term, however, environmental degradation often imposes profound economic costs. Moreover, when resource extraction or development degrades environmental conditions, often a few private parties benefit financially while the broader public is harmed. Today, concerns over fossil fuel pollution and climate change have led many people to see immense opportunities in revamping our economies with clean and renewable energy technologies. The jobs, investment, and economic activity that come with building a green energy economy demonstrate how economic progress and environmental protection can go hand in hand.

Indeed, across the world, ecological degradation tends to harm poor and marginalized people before wealthy ones, the Millennium Ecosystem Assessment found. As a result, restoring ecosystem services stands as a prime avenue for alleviating poverty.

Economic theory moved from "invisible hand" to supply and demand

Economics shares a common intellectual heritage with ethics, and practitioners of both study the relationship between individual action and societal well-being. Early philosophers had long believed that individuals acting in their own self-interest tend to harm society. However, Scottish philosopher Adam Smith (1723–1790) argued that self-interested economic behavior often benefits society, as long as the behavior is constrained by the rule of law and private property rights within a competitive marketplace. Known today as a founder of **classical economics,** Smith believed that when people pursue economic self-interest under these conditions, the marketplace will behave as if guided by "an invisible hand" to benefit society as a whole.

Today, Smith's philosophy remains a pillar of free-market thought, which many credit for the tremendous gains in material wealth that industrialized nations have achieved. Others contend that free-market capitalism tends to intensify environmental degradation and worsen inequalities between rich and poor.

Economists subsequently adopted more quantitative approaches. **Neoclassical economics** examines the psychological factors underlying consumer choices, explaining market prices in terms of consumer preferences for units of particular commodities. Standard neoclassical models assume that people behave rationally and have access to full information. In neoclassical economics, buyers desire a low price, whereas sellers desire a high price. This conflict results in a compromise in pricing and in the sale of the "right" quantity of commodities (**FIGURE 6.9**). These dynamics are phrased in terms of *supply,* the amount of a product offered for sale at a given price, and

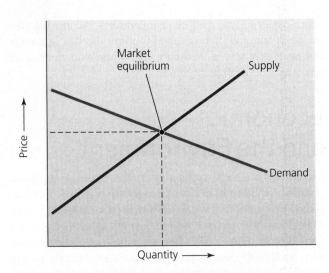

FIGURE 6.9 We can show economic fundamentals in a supply-and-demand graph. The demand curve indicates the quantity of a given good (or service) that consumers desire at each price, and the supply curve indicates the quantity produced at each price. In theory, the market automatically moves toward an equilibrium point at which supply equals demand.

demand, the amount of a product people will buy at a given price if free to do so. Theoretically, the market moves toward an equilibrium point, a price at which supply equals demand.

By applying similar reasoning, economists can determine "optimal" levels of resource use or pollution control. For instance, reducing pollution emitted by a car or factory is often cost-effective at first, but as pollution is reduced, it becomes more and more costly to eliminate each remaining amount. At some point the cost per unit of reduction rises to match the benefits, and this break-even point is the optimal level of pollution reduction.

To evaluate an action or decision, neoclassical economists use **cost-benefit analysis,** which compares the estimated costs of a proposed action with the estimated benefits. If benefits exceed costs, the action should be pursued; if costs exceed benefits, it should not. Given a choice of actions, the one with the greatest excess of benefits over costs should be chosen.

This reasoning seems eminently logical, but problems arise when not all costs and benefits can be easily identified, defined, or quantified. It may be simple to quantify the dollar value of bananas grown or cattle raised on a tract of Costa Rican land cleared for agriculture, yet difficult to assign monetary value to the complex ecological costs of clearing the forest. Because monetary benefits are usually easier to quantify than environmental costs, benefits tend to be overrepresented in cost-benefit analyses. As a result, environmental advocates often think that such analyses are predisposed toward economic development and against environmental protection.

Neoclassical economics has environmental consequences

Today's market systems operate largely in accord with the principles of neoclassical economics. These systems have

TABLE 6.1 Common Examples of External Costs

Health impacts
Physical health problems, stress, anxiety, or other medical conditions caused by air pollution, water pollution, toxic chemicals, or other environmental impacts.

Depletion of resources
Declines in abundance of—or loss of access to—natural resources that provide wealth or sustenance, such as fish, game, and wildlife; timber and forest products; grazing land; or fertile soils.

Aesthetic damage
Degradation of scenery and harm to the enjoyment of one's physical surroundings, as from strip mining; clear-cutting; urban sprawl; erosion; or air, water, noise, and light pollution.

Financial loss
Loss of monetary value caused by pollution or by resource depletion, via declining real estate values, lost tourism revenue, costlier medical expenses, damage from climate change and sea level rise, or other means.

---

generated unprecedented material wealth for market-oriented societies, yet four fundamental assumptions of neoclassical economics often contribute to environmental degradation.

Replacing resources One assumption is that natural resources and human resources (such as workers or technologies) are either infinite or largely substitutable and interchangeable. This implies that once we have depleted a resource, we will always be able to find some replacement for it. As a result, ecosystem goods and services are treated as free gifts of nature, endlessly abundant and resilient, and the market imposes no penalties for depleting them.

It is true that many resources can be replaced. Our societies have transitioned from manual labor to animal labor to steam-driven power to fossil fuel power, and we are beginning a transition to renewable power sources such as wind and solar energy. However, Earth's material resources are ultimately limited. Nonrenewable resources (such as fossil fuels) can be depleted. Many renewable resources (such as soils, fish stocks, and timber) can also be used up, if we exploit them faster than they are replenished. Even seemingly inexhaustible resources can become so polluted that we can no longer use them (such as contaminated water supplies).

External costs A second assumption of neoclassical economics is that all costs and benefits associated with an exchange of goods or services are borne by individuals engaging directly in the transaction. In other words, it is assumed that all costs and benefits are "internal" to the transaction, experienced by the buyer and seller alone.

However, many transactions affect other members of society. When a landholder fells a forest, people nearby suffer from poorer water quality, dirtier air, and a loss of wildlife. When a factory, power plant, or mining operation pollutes the air or water, it harms the health of those who live nearby. In cases like these, people who are not involved in degrading the environment nonetheless end up paying the costs. A cost of a transaction that affects someone other than the buyer or seller is known as an **external cost** (**TABLE 6.1**). Often whole communities suffer external costs while certain individuals enjoy private gain.

If market prices do not take the social, ecological, or economic costs of environmental degradation into account, taxpayers bear the burden of paying them. When economists ignore external costs, a false impression of the consequences of our choices is created, and people are subjected to the impacts of activities in which they did not participate. External costs are one reason governments develop environmental policy (p. 165).

Discounting Third, neoclassical economics grants an event in the future less value than one in the present. In economic terminology, future effects are "discounted." **Discounting** is meant to reflect how people tend to grant more importance to present conditions than to future conditions. Just as you might prefer to have an ice-cream cone today than be promised one next month, market demand is greater for goods and services that are received sooner. Economists quantify this by assigning discount rates when calculating costs and benefits. For example, applying a 5% annual discount rate to forestry decisions means that a stand of trees whose timber is worth $500,000 on the market today would drop in perceived value by 5% each year. From the perspective of a person in today's market, having the timber in 10 years would be worth only $299,368. By this logic, the more quickly the trees are cut, the more they are worth.

Discounting encourages policymakers to play down the long-term consequences of their decisions. Many environmental problems unfold gradually, yet discounting discourages us from addressing resource depletion, pollution buildup, and other cumulative impacts. Instead, discounting shunts the costs of dealing with such problems onto future generations. Discounting has emerged as a flashpoint in the debate over how to respond to global climate change. Economists agree that

THE SCIENCE behind the story

Do Payments Help Preserve Forest?

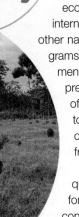

Costa Rican farmers judging whether to clear forest

Costa Rica's program to pay for ecosystem services has garnered international praise and has inspired other nations to implement similar programs, but have Costa Rica's payments actually been effective in preventing forest loss? A number of research teams have sought to answer this surprisingly difficult question by analyzing data from the PSA program.

Some early studies were quick to credit the PSA program for saving forests. A 2006 study conducted for FONAFIFO, the agency administering the program, concluded that PSA payments in the central region of the country had prevented 108,000 ha (267,000 acres) of deforestation—38% of the area under contract. Indeed, deforestation rates fell as the program proceeded; rates of forest clearance during the period 1997–2000 were half what they were in the preceding decade.

However, some researchers hypothesized that PSA payments were not responsible for this decline and that deforestation would have slowed anyway because of other factors. To test this hypothesis, a team led by G. Arturo Sanchez-Azofeifa of the University of Alberta and Alexander Pfaff of Duke University worked with FONAFIFO's payment data, as well as data on land use and forest cover from satellite surveys. They layered these data onto maps using a geographic information system (GIS) (p. 119) and then explored the patterns revealed.

In 2007 in the journal *Conservation Biology,* the researchers reported that only 7.7% of PSA contracts were located within 1 km of regions where forest was at greatest risk of clearance. PSA contracts were only slightly more likely to be near such a region than far from it, which meant, they argued, that PSA contracts were not being targeted to regions where they could have the most impact.

Moreover, because enrollment was voluntary, most landowners applying for payments likely had land unprofitable for agriculture and were not actually planning to clear forest for this purpose (**FIGURE 1**). In a 2008 paper, these researchers compared lands under PSA contracts with similar lands not under contracts. PSA lands experienced no forest loss, whereas the deforestation rate on non-PSA lands was 0.21%/yr. However, their analyses indicated that PSA lands stood only a 0.08%/yr likelihood of being cleared in the first place, suggesting that the program prevented only 0.08%/yr of forest loss, not 0.21%/yr. Other research was bearing out this finding; at least two studies found that many PSA participants, when interviewed, said they would have retained their forest even without the PSA program.

These researchers argued that Costa Rica's success in halting forest loss was likely due to other factors. In particular, Forest Law 7575, which had established the PSA system, had also banned forest clearing nationwide. This top-down government mandate, assuming it was enforceable, in theory made the PSA payments unnecessary. However, the PSA program made the mandate far more palatable to legislators, and Forest Law 7575 might never have passed had it not included the PSA payments.

Despite the PSA program's questionable impact on preserving existing forest, scientific studies show that it has been effective in regenerating new forest. In Costa Rica's Osa Peninsula, Rodrigo Sierra and Eric Russman of the University of

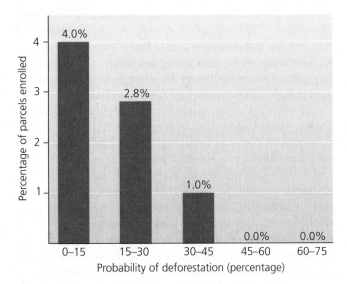

FIGURE 1 In areas at greater risk of deforestation, lower percentages of land parcels were enrolled in the PSA program than in areas of lesser risk. This difference is because land more profitable for agriculture was less often enrolled in the program. *Data from Pfaff, A., et al., 2008.* Payments for environmental services: Empirical analysis for Costa Rica. *Working Papers Series SAN08-05, Terry Sanford Institute of Public Policy, Duke University.*

Texas at Austin found in 2006 that PSA farms had five times more regrowing forest than did non-PSA farms. Interviews with farmers indicated that the program encouraged them to let land grow back into forest if they did not soon need it for production.

In the nation's northern Caribbean plain, a team led by Wayde Morse of the University of Idaho combined satellite data with on-the-ground interviews, finding that PSA payments plus the clearance ban had reduced deforestation rates from 1.43%/yr to 0.10%/yr and that the program encouraged even more forest regrowth. Meanwhile, research by Rodrigo Arriagada of Santiago, Chile, indicated that the regeneration of new forest seemed to be the PSA program's major effect at the national level as well.

Most researchers today hold that Costa Rica's forest recovery results from a long history of conservation policies and economic developments. Indeed, deforestation rates had been dropping steeply before the PSA program was initiated (**FIGURE 2**), for several major reasons:

- Earlier policies (tax rebates and tax credits for timber production) encouraged forest cover.
- The creation of national parks fed a boom in ecotourism, so Costa Ricans saw how conserving natural areas could bring economic benefits.
- Falling market prices for meat discouraged ranching.
- After an economic crisis roiled Latin America in the 1980s, Costa Rica ended subsidies that had encouraged ranchers and farmers to expand into forested areas.

To help the PSA program make better use of its money, many researchers argued that PSA payments should be targeted. Instead of paying equal amounts to anyone who applies, FONAFIFO should prioritize applicants, or pay more money, in regions that are ecologically most valuable or that are at greatest risk of deforestation.

In a 2008 paper in the journal *Ecological Economics,* Tobias Wünscher of Bonn, Germany, and colleagues modeled and tested seven possible ways to target the payments, using data from Costa Rica's Nicoya Peninsula. They found that all seven approaches would improve on the existing program. The best approach paid some landowners more than others while selecting sites based on the perceived value of their ecological services and the apparent risk of deforestation. This approach more than doubled the economic efficiency of the program.

To determine how much money to offer each landowner, Wünscher's team suggested using auctions, in which applicants for PSA funds put in bids stating how much they were requesting. Because applicants have outnumbered available contracts three to one, FONAFIFO could favor the lower bids to keep

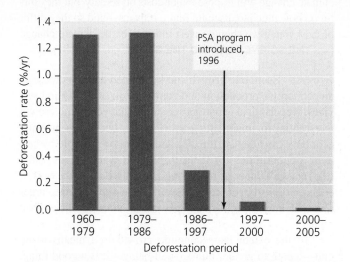

FIGURE 2 Forest recovery was underway in Costa Rica before the PSA program began. Deforestation rates had already dropped steeply, indicating that other factors were responsible.
Data from Sanchez-Azofeifa, G.A., et al., 2007. Costa Rica's payment for environmental services program: Intention, implementation, and impact. Conservation Biology 21: 1165–1173.

costs low while the auction system could make differential payments politically acceptable.

The Costa Rican government has responded to suggestions from researchers by aiming PSA payments toward regions of greater environmental value and by making the program more accessible to small low-income farmers in undeveloped regions. The government has also raised the payment amounts considerably. These actions have had some results; for instance, between 2008 and 2014, the number of small- and medium-sized landholders in the program more than doubled. Following these changes, however, researchers such as David Lansing of the University of Maryland, Baltimore County, noted that even among the new smaller landholders, it was often the older, wealthier individuals with outside sources of income who tended to register for the program, raising the question of how well the program is truly serving the rural poor.

Arriagada surveyed landholders registered in the program and reported in 2015 that they had reduced the number of cattle they grazed and their amount of labor and that they felt they had roughly the same economically and in quality of life as before. These landholders all chose to renew their participation in the program after five years, showing that they were satisfied with it. Many reported feeling more secure about their land tenure than previously and proud to be conserving forests. As Costa Rica's PSA program continues to evolve, researchers—and other nations—will be watching closely to see what lessons they can derive.

climate change will impose major costs on society, but they differ on how much to discount future impacts—and so they differ on how much we should invest today to battle climate change (see **THE SCIENCE BEHIND THE STORY**, pp. 150–151).

Growth **Economic growth** can be defined as an increase in an economy's production and consumption of goods and services. Neoclassical economics assumes that economic growth is essential for maintaining social order, because a growing economy can alleviate the discontent of poorer people by creating opportunities for them to become wealthier. A rising tide raises all boats, as the saying goes; if we make the overall economic pie larger, each person's slice can become larger (even if some people still get much smaller slices than others).

To the extent that economic growth is a means to an end—a path to greater human well-being—it is a good thing. However, when growth becomes an end in itself, it may no longer be the best route toward well-being. Sociologists have coined a word for the way consumption and material affluence often fail to bring people contentment: *affluenza*. Moreover, research shows that people with lower incomes tend to feel worse and worse as the gap between rich and poor widens, even if they gain more absolute wealth themselves.

How sustainable is economic growth?

Our global economy is more than 30 times larger than it was a century ago. All measures of economic activity are greater than ever before. Economic expansion has brought many people much greater material wealth (although not equally, and gaps between rich and poor are wide and growing).

Economic growth can occur in two ways: (1) by an increase in inputs to the economy (such as more labor or natural resources) or (2) by improvements in the efficiency of production due to better methods or technologies (ideas or equipment that enable us to produce more goods with fewer inputs).

As our population and consumption rise, it is becoming clearer that we cannot sustain growth forever using the first approach. Nonrenewable resources are finite, and renewable resources can also be exhausted if we overexploit them (as is happening with many fisheries today). As for the second approach to growth, we have used technological innovation to push back the limits on growth time and again. More efficient technologies for extracting minerals, fossil fuels, and groundwater allow us to mine these resources more fully with less waste. Advanced machinery and robotics in our factories speed up manufacturing. We continue to make computer chips more powerful while also making them smaller (**FIGURE 6.10**). In such ways, we are producing more goods and services with relatively fewer resources.

Can we conclude, then, that human ingenuity and technology will allow us to overcome all environmental constraints and continue growth indefinitely? Answering "yes" are people referred to as **Cornucopians.** (In Greek mythology, a *cornucopia*—literally "horn of plenty"—is a magical goat's

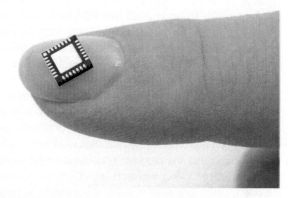

FIGURE 6.10 The miniaturization of computer chips is a striking advance in efficiency. It allows us to store and use far more information with far less input of raw materials than ever before.

horn that overflowed with grain, fruit, and flowers.) Responding "no" are people called **Cassandras,** named after the mythical princess of Troy with the gift of prophecy, whose dire predictions were not believed.

The Cassandra view has been articulated most famously by a group of researchers who published a series of books including *The Limits to Growth* (1972), *Beyond the Limits* (1992), and *Limits to Growth: The Thirty-Year Update* (2004). Starting from the premise that Earth's natural capital is finite and using calculations of resource availability and consumption, they ran simulation models to predict how our economies would fare in the future. Model runs that used data matching society's current consumption patterns consistently predicted economic collapse as resources become scarce (**FIGURE 6.11a**). The team experimented to see what parameters would produce model runs predicting a sustainable civilization (**FIGURE 6.11b**). They used these results to make recommendations for achieving sustainability.

Cornucopians, such as the economist Julian Simon and the statistician Bjorn Lomborg, have countered that Cassandras underestimate the human capacity to innovate and the degree to which technologies can expand our access to resources. They also argue that market forces help avoid resource depletion because as resources become scarce, prices rise, giving individuals and firms incentive to shift to different resources or products and to reuse and recycle. As Simon put it in his 1981 book, *The Ultimate Resource:*

> The natural world allows, and the developed world promotes through the marketplace, responses to human needs and shortages . . . . The main fuel to speed our progress is our stock of knowledge, and the brake is our lack of imagination. The ultimate resource is people—skilled, spirited, and hopeful people who will exert their wills and imaginations for their own benefit, and so, inevitably, for the benefit of us all.

What accounts for the divergence in views between Cornucopians and Cassandras? It may be because we are living in a unique time of transition. Throughout our long history as a species, we have lived in a world with low numbers of people.

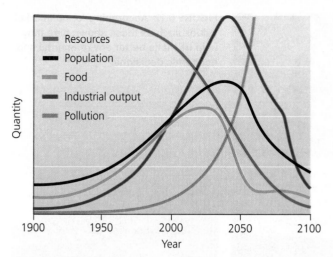

(a) Projection based on status quo policies

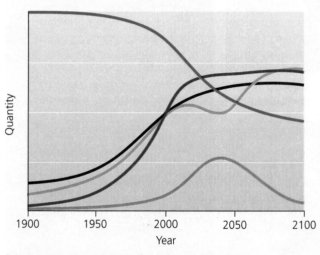

(b) Projection based on policies for sustainability

FIGURE 6.11 Environmental scientists have used data to project future trends in resource availability, human population, food production, industrial output, and pollution. Shown in **(a)** is a projection for a world in which twice as many resources than today have been discovered and extracted but society pursues policies typical of the 20th century. In this projection, population and production rise until declining resources cause them to fall suddenly, while pollution continues to rise. In contrast, **(b)** shows a scenario with policies aimed at sustainability. In this projection, population levels off below 8 billion, production and resource availability stabilize, and pollution declines to low levels. *Data from Meadows, D., et al., 2004.* Limits to growth: The 30-year update. *White River Junction, VT: Chelsea Green Publishing. Used by permission.*

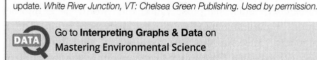

Go to **Interpreting Graphs & Data** on **Mastering Environmental Science**

In such an "empty world," we could always rely on being able to exploit more resources. Today, however, the human population is so large that it is straining Earth's systems and depleting its resources. In this new "full world," we are encountering limits. Because we are still learning what those limits are, people have a wide diversity of viewpoints.

As we will see with many issues, both Cornucopians and Cassandras make valid points. Cassandras have often underestimated our ability to innovate and adapt—yet ultimately, non-renewable resources are finite, and renewable resources can be exploited only at limited rates. If our population and consumption continue to grow and we do not enhance reuse and recycling, we will continue to deplete our natural capital, putting ever-greater demands on our capacity to innovate.

WEIGHING the issues

Cornucopian or Cassandra?

Would you consider yourself more of a Cornucopian or a Cassandra? Which aspect(s) of each point of view do you share? Do you think our society would be better off if everyone were a Cornucopian or if everyone were a Cassandra—or do we need both?

Environmental and ecological economists devise strategies for sustainability

The Cornucopian view has long held sway over mainstream economics, but today many economists are concluding that growth may be unsustainable if we do not reduce our demand for resources. Economists in the field of **environmental economics** believe that we can modify neoclassical economic principles to make resource use more efficient and thereby attain sustainability within our current economic systems. Environmental economists were the first to develop methods to tackle the problems of external costs and discounting.

Economists in the field of **ecological economics** believe that sustainability requires more far-reaching changes. They stress that in nature, every population has a carrying capacity (p. 67) and systems generally operate in self-renewing cycles, not in a linear or progressive manner. Ecological economists maintain that societies, like natural populations, cannot surpass environmental limitations. Many of these economists advocate economies that neither grow nor shrink, but rather are stable. Such a **steady-state economy** is intended to mirror natural ecological systems. Critics of steady-state economies assert that to halt growth would dampen our quality of life. Proponents respond that technological advances would continue, behavioral changes (such as greater use of recycling) would accrue, and wealth and happiness would rise.

Attaining sustainability will certainly require the reforms pioneered by environmental economists and may require the fundamental shifts advocated by ecological economists. One approach they each take is to assign monetary values to ecosystem goods and services, to better integrate them into traditional cost-benefit analyses.

We can assign monetary value to ecosystem goods and services

Ecosystems provide us essential resources and life-support services, including fertile soil, waste treatment, clean water, and clean air. Yet we often abuse the very ecological systems that sustain us. Why? From the economist's perspective, people overexploit natural resources and processes largely because the

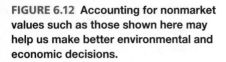
FIGURE 6.12 Accounting for nonmarket values such as those shown here may help us make better environmental and economic decisions.

(a) Use value: The worth of something we use directly

(b) Existence value: The worth of knowing that something exists, even if we never experience it ourselves

(c) Option value: The worth of something we might use later

(d) Aesthetic value: The worth of something's beauty or emotional appeal

(e) Scientific value: The worth of something for research

(f) Educational value: The worth of something for teaching and learning

(g) Cultural value: The worth of something that sustains or helps define a culture

market assigns these entities no quantitative monetary value—or assigns values that underestimate their true worth.

Ecosystem services are said to have **nonmarket values,** values not usually included in the price of a good or service (**FIGURE 6.12**). For example, the aesthetic and recreational pleasure we obtain from natural landscapes is something of real value. Yet because we do not generally pay money for this, its value is hard to quantify and appears in no traditional measures of economic worth. Or consider Earth's water cycle (pp. 121–122): Rain fills our reservoirs with drinking water, rivers give us hydropower and flush away our waste, and water evaporates, purifying itself of contaminants and later falling as rain. This natural cycle is vital to our very existence, yet because we do not pay money for it, markets impose no financial penalties when we disturb it.

In Costa Rica and elsewhere, environmental and ecological economists have sought ways to assign market values to ecosystem services. One technique, **contingent valuation,** uses surveys to determine how much people are willing to pay to protect or restore a resource. Such an approach was used to

assess a proposal to develop a mine near Kakadu National Park in Australia in the 1990s. To determine how much the land was "worth" economically if preserved undeveloped, researchers interviewed Australians and asked how much they would be willing to pay to prevent mine development. On average, respondents said their households would pay $80 to $143 per year to prevent the predicted impacts. Multiplying these figures by the number of households in Australia, the researchers found that preservation was "worth" $435 million to $777 million annually to Australia's population. These numbers exceeded the $102 million in annual economic benefits expected from mine development, so the researchers concluded that it was best to preserve the land undeveloped.

Because contingent valuation relies on survey questions and not actual expenditures, critics point out that people may volunteer idealistic (inflated) values, knowing that they will not actually have to pay the price they name. As a result, many researchers prefer to use methods that measure people's preferences as revealed by data on actual behavior. For example, to gauge how much people value parks, researchers may calculate the amount of money, time, or effort people expend to travel to parks. Economists may compare housing prices for similar homes in different settings to infer the dollar value of landscapes, views, or peace and quiet. They may also calculate how much it costs to restore natural systems that have been damaged, to replace their functions with technology, or to clean up pollution.

In Costa Rica, Taylor Ricketts of Stanford University, working with Gretchen Daily and others, studied pollination (pp. 80, 223) by native bees at a coffee plantation. By carefully measuring how bees pollinated the coffee plants and comparing the resulting coffee production in areas near forest and far from forest, this team calculated that forests were providing the farm with pollination services worth $60,000 per year.

Researchers have even set out to calculate the total economic value of all the services that oceans, forests, wetlands, and other systems provide across the world. Teams headed by ecological economist Robert Costanza have combed the scientific literature and evaluated hundreds of studies that estimated dollar values for 17 major ecosystem services (**FIGURE 6.13**).

The researchers reanalyzed the data using multiple valuation techniques to improve accuracy and then multiplied average estimates for each ecosystem by the global area it occupied. Their analysis in 1997 was groundbreaking, and in 2014 they updated their research. The 2014 study calculated that Earth's biosphere in total provides more than $125 trillion worth of ecosystem services each year, in 2007 dollars. That is equal to $153 trillion in 2019 dollars—an amount that exceeds the global annual monetary value of goods and services created by people!

Costanza also joined Andrew Balmford and 17 other colleagues to compare the benefits and costs of preserving natural systems intact versus converting wild lands for agriculture, logging, or fish farming. After reviewing many studies, they reported in the journal *Science* in 2002 that a global network of nature reserves covering 15% of Earth's land surface and 30% of the ocean would be worth $4.4 trillion to $5.2 trillion—100 times more than the value of those areas were they to be converted and exploited for direct human use. This finding demonstrates, as stated in their research

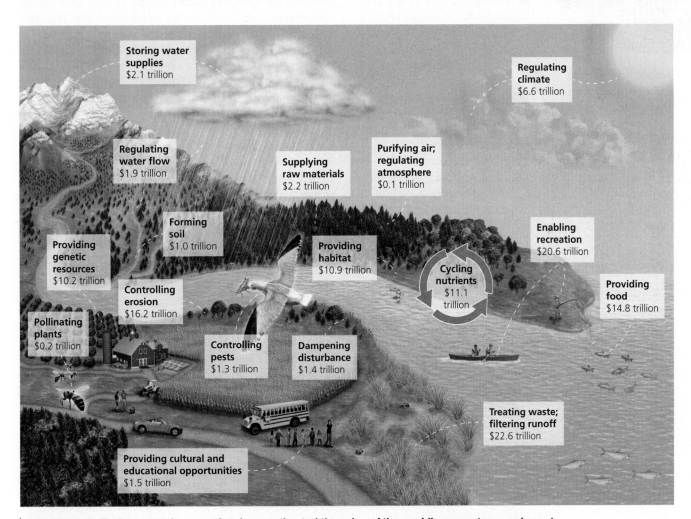

FIGURE 6.13 Environmental economists have estimated the value of the world's ecosystem services at more than $153 trillion (in 2019 dollars). This amount is an underestimate because it does not include ecosystems and services for which adequate data were unavailable. Shown are subtotals for each ecosystem service in 2007 dollars. *Data from Costanza, R., et al., 2014. Changes in the global value of ecosystem services. Global Env. Change 26: 152–158.*

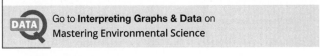

Go to **Interpreting Graphs & Data** on Mastering Environmental Science

THE SCIENCE behind the story

Ethics and Economics: How Much Will Global Climate Change Cost?

Economically costly damage from storms like Hurricane Harvey is forecast to increase with climate change.

Extreme weather. Severe storms. Coastal flooding. Wildfire damage. Health impacts. National security risks. These impacts of global climate change, and many more, are beginning to impose severe costs on our society (see Chapter 18). In the end, how much will the impacts of climate change cost us? Economists poring over scientific data on the consequences of climate change are coming up with quantitative answers to that question. Most of the estimates are high enough to give us cause for strong concern; clearly climate change will cost us many, many billions of dollars.

For example, **FIGURE 1** shows estimates of the biggest sources of economic loss to the United States economy by the end of this century as calculated and summarized from the research literature by scientists and economists for the U.S. government's official *National Climate Assessment* in 2018. It shows well over half a *trillion* dollars in expenses for the United States if we fail to tackle greenhouse gas emissions soon.

Yet even the most careful and comprehensive estimates with the best-researched numbers are subject to a debate that has as much to do with ethics as with economics. At issue is the practice of discounting (p. 143), an economic technique heavy with ethical implications.

To understand, let's reach back to 2006, when the first major effort was made to assess the economic costs of a changing climate. The British government had commissioned economist Nicholas Stern to lead an assessment team that would survey the burgeoning scientific literature on the impacts of rising temperatures, changing rainfall patterns, rising sea level, and increasing storminess (see Chapter 18). Stern's team estimated the economic consequences of each of these climatic changes and put a price tag on the global cost. The *Stern Review on the Economics of Climate Change* concluded that without action to forestall it, climate change would cause annual losses in global Gross Domestic Product (GDP; see p. 152) of 5–20% by the year 2200 (**FIGURE 2**).

Stern's team also calculated that if society invested just 2% of GDP annually to help stabilize atmospheric greenhouse gas concentrations, most future losses could be prevented. The bottom line was that spending a relatively small amount of money right away would save us much larger expenses in the future.

The *Stern Review*'s conclusions, however, depended partly on how it chose to weigh future impacts versus current impacts. Stern used two discount factors. One accounted for the likelihood that people in the future will be richer than we are today

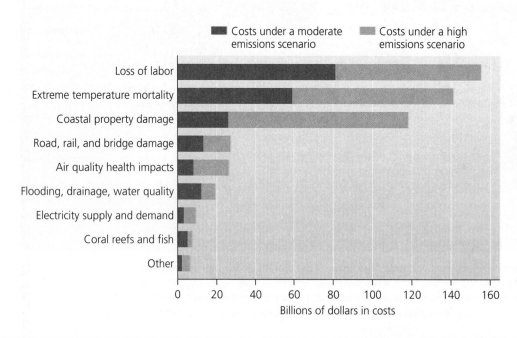

Loss of labor
Extreme temperature mortality
Coastal property damage
Road, rail, and bridge damage
Air quality health impacts
Flooding, drainage, water quality
Electricity supply and demand
Coral reefs and fish
Other

■ Costs under a moderate emissions scenario   ■ Costs under a high emissions scenario

Billions of dollars in costs

FIGURE 1 In the United States alone, climate change will impose costs of hundreds of billions of dollars by the year 2100. The largest types of costs are shown here, for two different emissions scenarios. *Data from U.S. Global Change Research Program, 2018. Impacts, risks, and adaptation in the United States: Fourth national climate assessment, Volume II (Reidmiller, D.R., et al., eds.). USGCRP, Washington, D.C.*

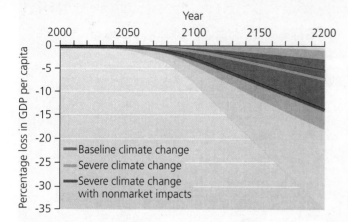

FIGURE 2 Climate change forecast by the IPCC could decrease global per capita GDP by 5.3% annually by the year 2200, the Stern Review calculated. Severe climate change could bring annual losses of 7.3%—and adding nonmarket values (p. 148) raises this figure to 13.8%. Gray-shaded areas show ranges of future values judged statistically to be 95% likely; darkest gray indicates where ranges overlap for all three lines, and lightest gray for just one line. *Data from HM Treasury, 2007.* Stern review on the economics of climate change. *London, U.K.*

and thus better able to handle economic costs. The other considered whether the future should be discounted simply because it is the future. This latter discount factor (called a *pure time discount factor*) is essentially an ethical issue because it assigns explicit values to the welfare of future versus current generations.

The *Stern Review* used a pure time discount rate of 0.1%. This rate means that an impact occurring next year is judged 99.9% as important as one occurring this year. It means that the welfare of a person born 100 years from now is valued at 90% of one born today. This discount rate treats current and future generations *nearly* equally. Future generations are downweighted only because of the (very small) possibility that our species could go extinct (in which case, there would be no future generations to be concerned about).

Many economists viewed this discount rate as too low. One was Yale University economist William Nordhaus, who was the first to model economic costs of climate change back in 1991 and who would go on to win a Nobel Prize for his years of research. Nordhaus proposed a discount rate starting at 3% and falling to 1% in 300 years. He maintained that such numbers are more objective because they reflect how people value things, as revealed by prices we pay in the marketplace. Indeed, when economists assess capital investments or construction projects (say, development of a railroad, dam, or highway), they typically choose discount rates close to those Nordhaus suggested. Nordhaus argued that the *Stern Review*'s near-zero discount rate overweighted the future,

forcing people today to pay too much to address hypothetical future impacts.

Responding to Nordhaus and other critics, Stern's team argued that discount rates of 3% or 1% may be useful for assessing development projects but are too high for long-term environmental problems that directly affect human well-being. A 3% rate means that a person born in 1995 is valued only half as much as a person born in 1970. It means a grandchild is judged to be worth far less than a grandparent simply because of the dates they were born. For various reasons, Stern argued, the market should not be used to guide ethical decisions.

Debates over discounting again flared up over differences between calculations of the social cost of carbon (p. 506) made by the U.S. presidential administrations of Barack Obama and Donald Trump. The social cost of carbon measures the harm to society caused (via climate change) by each ton of carbon dioxide emissions. This number is important because it serves as a basis for deciding when to regulate greenhouse gas emissions, but it is strongly influenced by what discount rate is selected. After a great deal of research and debate, the Obama administration settled on a 3% discount rate, which produced a social cost of carbon of $42 per metric ton.

Once Trump became president, his administration sought to reduce or eliminate regulatory controls over carbon pollution. As such, it chose a much higher discount rate of 7%. It also included only costs to the United States directly and not to the rest of the world. As a result, the federal government's official social cost of carbon fell to just $1 to $7 per metric ton. Applying this much lower figure would mean that regulations on carbon pollution would rarely appear justified.

When federal researchers with the *National Climate Assessment* published the data shown in Figure 1, they were bringing together information from many sources. Most of these sources used a 3% discount rate, although some of them used no discount rate at all. Three years earlier, the U.S. Environmental Protection Agency had estimated that the United States could face $1.3 trillion to 1.5 trillion in costs from climate change by 2100 if it did not reduce greenhouse gas emissions. That EPA report used a 3% discount rate; the costs would have appeared even greater had a lower discount rate been used.

At the end of the day, the choice of a discount rate is an ethical decision on which well-intentioned people may differ. As governments, businesses, and individuals begin to invest in addressing climate change, debates over discount rates used for the social cost of carbon and in research analyses reveal how ethics and economics remain intertwined.

Today, more and more of us are feeling the effects of climate change directly in our personal lives. Hurricane Maria, Hurricane Harvey, flooding in Miami, drought and wildfires in California, and many other climate-related disasters are all driving home the same message: The sooner we address climate change, the better off we'll be.

paper, that "conservation in reserves represents a strikingly good bargain."

Such research has sparked debate. Some ethicists argue that we should not put dollar figures on amenities such as clean air and water because they are priceless and we would perish without them. Others say that arguing for conservation purely on economic grounds risks not being able to justify it whenever it fails to deliver clear economic benefits. Backers of the research counter that valuation does not argue for making decisions on monetary grounds alone, but instead clarifies and quantifies values that we already hold implicitly.

In 2010, researchers wrapped up a large international effort to summarize attempts to quantify the economic value of natural systems. *The Economics of Ecosystems and Biodiversity* study published a number of fascinating reports that you can download online. This effort described the valuation of nature's economic worth as "a tool to help recalibrate [our] faulty economic compass." It concluded that this is useful because "the invisibility of biodiversity values has often encouraged inefficient use or even destruction of the natural capital that is the foundation of our economies."

We can measure progress with full cost accounting

If assigning market values to ecosystem services gives us a fuller and truer picture of costs and benefits, we can take a similar approach to measuring the economic progress we make as a society. For decades, we have assessed each nation's economy by calculating its **Gross Domestic Product** (GDP), the total monetary value of final goods and services a nation produces each year. Governments regularly use GDP to make policy decisions that affect billions of people. However, GDP is a poor measure of economic well-being. It does not account for nonmarket values. It also lumps together all economic activity, desirable and undesirable. Thus, GDP can rise in response to economic activities that harm society.

For example, crime can boost GDP because crime forces people to invest in security measures and to replace stolen items. Oil spills (**FIGURE 6.14**) increase GDP because they require cleanups, which cost money and increase the production of goods and services. Natural disasters such as hurricanes, tornadoes, and earthquakes can boost GDP because of all the costs incurred in emergency measures, cleanup operations, and reconstruction efforts. War can augment GDP because of the economic activity involved in manufacturing weapons and servicing armies. Pollution causes GDP to increase when the polluting substance is manufactured and again when society pays to clean up the pollution.

Environmental economists have developed indicators meant to differentiate desirable from undesirable economic activity and to better reflect our well-being. One such

FIGURE 6.14 Oil spills—like this one along the coast of Santa Barbara County, California, in 2015—actually increase GDP. The many monetary transactions resulting from cleanup efforts mean that pollution can increase GDP—one of many reasons GDP is not a reliable indicator of societal well-being.

alternative to the GDP is the **Genuine Progress Indicator** (GPI). To calculate GPI, we begin with conventional economic activity and add to it positive contributions not paid for with money, such as volunteer work and parenting. We then subtract negative impacts, such as crime and pollution (**FIGURE 6.15a**).

GPI can differ strikingly from GDP: **FIGURE 6.15b** compares these indices internationally across more than half a century. On a per-person basis, GDP rose greatly, but GPI has declined slightly since 1978. Data for the United States show a similar pattern, with per capita GDP more than tripling over the 55 years but GPI remaining flat for the latter 30 years. These discrepancies suggest that people in most nations—including the United States—have been spending more and more money but that their quality of life is not improving.

The GPI is an example of **full cost accounting** (also called *true cost accounting*) because it aims to account fully for all costs and benefits. A variety of full cost accounting indicators have been devised. The Index of

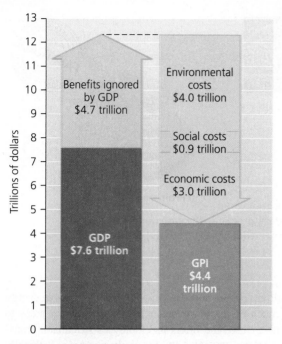

(a) Components of GPI

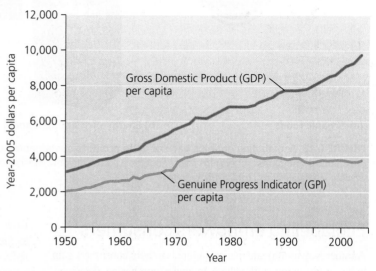

(b) Change in GDP vs. GPI in 17 nations

FIGURE 6.15 Full cost accounting indicators such as the GPI aim to measure progress and well-being more effectively than pure economic indicators such as the GDP. We see in **(a)** how the GPI **(orange bar at right)** adds to the GDP **(reddish bar at left)** benefits such as volunteering and parenting **(upward gold arrow).** The GPI then subtracts external environmental costs such as pollution, social costs such as divorce and crime, and economic costs such as borrowing and the gap between rich and poor **(downward gold arrow).** Shown are values for the United States in 2004, the most recent year available. In **(b),** data from 17 major nations are averaged and show that per capita GDP has increased dramatically since 1950, yet per capita GPI peaked in 1978. *Data **(a)** from Talberth, J., et al., 2007.* The Genuine Progress Indicator 2006: A tool for sustainable development. *Oakland, CA: Redefining Progress; and **(b)** from Kubiszewski, I., 2013. Beyond GDP: Measuring and achieving global genuine progress.* Ecological Economics *93: 57–68. All data are adjusted for inflation using year-2005 dollars.*

DATA Go to **Interpreting Graphs & Data** on **Mastering Environmental Science**

Sustainable Economic Welfare (ISEW) gave rise to the GPI. Net Economic Welfare (NEW) adjusts GDP by adding the value of leisure time and personal transactions while deducting costs of environmental degradation. The Human Development Index assesses standard of living, life expectancy, and education. Several U.S. states have begun using the GPI to measure progress and help guide policy. Critics of full cost accounting argue that the approach is subjective and easily driven by ideology. Proponents respond that making a subjective attempt to measure progress is better than misapplying an indicator such as the GDP to quantify well-being—something it was never meant to do.

Today, attempts are gaining ground to measure happiness (rather than economic output) as the prime goal of national policy. The small Asian nation of Bhutan pioneered this approach with its measure of Gross National Happiness. Another indicator is the Happy Planet Index, which measures how much happiness we gain per amount of resources we consume. By this measure, Costa Rica was recently calculated to be the top-performing nation in the world.

Costa Rica is also one of five nations working with the World Bank (p. 177) in the WAVES program (Wealth Accounting and Valuation of Ecosystem Services) to implement full cost accounting methods. Together they are addressing questions such as how much economic benefit the nation's forests, national parks, and other natural amenities generate through tourism and watershed protection. Data so far show Costa Rica's forests contributing 10 to 20 times more to the economy than had been estimated from timber sales alone.

Markets can fail

When markets do not take into account the positive outside effects on economies (such as ecosystem services) or the negative side effects of economic activity (external costs), the result is what economists call **market failure.** Traditionally, we have tried to counteract market failure with government intervention. Government can help by using laws and regulations to restrain improper individual and corporate behavior. It can tax harmful activities. It can also design financial incentives that use market mechanisms to promote fairness, resource conservation, and economic sustainability. Paying for the conservation of ecosystem services, as Costa Rica does, is one way of deploying financial incentives toward policy goals. (We will examine all these approaches in Chapter 7 in our discussion of environmental policy.)

(a) Organic foods

(b) Energy-efficient appliances

(c) Fair-trade products

FIGURE 6.16 Ecolabeling enables each of us to promote sustainable business practices through our purchasing decisions. Among the many ecolabeled products now widely available are **(a)** organic foods, **(b)** energy-efficient appliances, and **(c)** fair-trade coffee.

Ecolabeling empowers consumers

Another way to mitigate market failure is to help consumers gain better information with which to make purchasing decisions. With **ecolabeling,** sellers who use sustainable practices in growing, harvesting, or manufacturing products advertise this fact on their labels, hoping to win approval from buyers (**FIGURE 6.16**). Examples include labeling recycled paper, organic foods (p. 262), dolphin-safe tuna, fair-trade and shade-grown coffee, and sustainably harvested lumber (p. 324). When labeling information is accurate, each of us as consumers can provide businesses and industries a powerful incentive to shift to more sustainable processes when we favor ecolabeled products.

In a similar vein, individuals who invest money in the stock market can pursue **socially responsible investing,** which entails investing in companies that have met criteria for environmental or social sustainability. A fast-growing trend, by 2018 U.S. investors had sunk $11.6 trillion into socially responsible investing, representing one-fourth of all investment.

Businesses are responding to sustainability concerns

As more consumers and investors express preferences for sustainable products and services, many industries, businesses, and corporations are "greening" their operations. By finding ways to enhance energy efficiency, reduce toxic substances, increase the use of recycled materials, and minimize greenhouse gas emissions, businesses often discover that they reduce costs and increase profit.

Some companies have cultivated an eco-conscious image from the start, such as Ben & Jerry's (ice cream), Patagonia (outdoor apparel), and Seventh Generation (household products). The phone company CREDO donates a portion of its proceeds to environmental and progressive nonprofit groups according to how its customers vote to distribute the funds. At the local level, entrepreneurs are starting thousands of sustainably oriented businesses all across the world.

Today, corporate sustainability has gone mainstream, and many of the world's largest corporations have joined in, including McDonald's, Intel, Ford Motor Company, Toyota, IKEA,

Dow, DuPont, BASF, and IBM. Google uses renewable energy to meet much of its power demand, has taken many steps to reduce energy use, and offers its employees sustainable transportation options (**FIGURE 6.17**). Starbucks purchases "ethically sourced" fair-trade and organic coffee, has reduced its water and electricity consumption, and is seeking to gain LEED green-building certification (pp. 348–350) for all new stores. Hewlett-Packard runs programs to reuse and recycle used toner cartridges, electronics, and plastics; and Dell reuses or recycles 98% of its nonhazardous waste. Sprint aims to recycle 90% of the phones it sells by crediting customers who turn in old phones. Nike, Inc., collects more than 1.5 million used sneakers each year, recycling the materials to create synthetic surfaces for basketball courts, tennis courts, and running tracks. Microsoft became carbon-neutral by charging each of its divisions with monitoring and offsetting its greenhouse gas emissions.

Walmart provides the highest-profile example of corporate greening efforts. Advocates of sustainability had long criticized the world's largest retailer for its environmental and social

FIGURE 6.17 Google offers employees free bicycles to ride to work. Google is just one of thousands of corporations working to make its operations more sustainable.

impacts. The company responded by launching a quest to sell organic products, reduce packaging, use recycled materials, enhance fuel efficiency in its truck fleet, reduce energy use in its stores, power itself with renewable energy, cut carbon dioxide emissions, and preserve 1 acre of natural land for every acre developed. It developed a "sustainability index" to rate products it carries and to help inform eco-conscious consumers. Since then, Walmart has met some of its goals and fallen short on others while saving money through efficiency measures. Walmart's global reach and massive market give it the ability to persuade suppliers to modify their own operations in order to retain its business, which has had far-reaching effects. A recent study of 100 major companies found that the desire to supply products to Walmart was an influential factor driving their own sustainability investments.

Of course, corporations exist to make money for their shareholders, so they cannot be expected to pursue goals that do not turn a profit. Moreover, many corporate greening efforts are more rhetoric than reality, pursued mostly for public relations purposes. Such corporate **greenwashing** can mislead consumers into thinking that a company is acting more sustainably than it actually is. The bottled water industry presents a stark example of greenwashing. Advertising with words such as "pure" and "natural" and images of forests and alpine springs leads us to believe that bottled water is cleaner and healthier for us to drink. In reality, bottled water is often less safe than tap water, the plastic bottles are a major source of waste, considerable amounts of oil are burned to transport the bottles, and the industry depletes aquifers in local communities (p. 403).

In the end, it is up to us all in our roles as consumers to encourage trends in sustainability by rewarding those businesses that truly promote sustainable solutions. In this way, each of us can express the ethical values we cherish through the economic system in which we live.

Sustainable Development

Today's search for sustainable solutions centers on **sustainable development,** economic progress that maintains resources for the future. The United Nations defines sustainable development as development that "meets the needs of the present without sacrificing the ability of future generations to meet their own needs." Sustainable development is an economic pursuit shaped by policy and informed by science. It is also an ethical pursuit because it asks us to manage our resource use so that future generations can enjoy similar access to resources.

Sustainable development involves environmental protection, economic well-being, and social equity

Economists use the term **development** to describe the use of natural resources for economic advancement (as opposed to simple subsistence, or survival). Development involves making purposeful changes intended to improve the quality of human life. It includes the construction of homes, schools, hospitals, power plants, factories, and transportation networks. In the

past, many advocates of development felt that protecting the environment threatened people's economic needs, whereas many advocates of environmental protection felt that development degraded the environment, thereby jeopardizing the improvements in quality of life that were intended. Today, however, people increasingly perceive how successful development depends on a healthy and functional natural environment.

We also now recognize that society's poorer people tend to suffer the most from environmental degradation. As a result, advocates of environmental protection, economic development, and social justice began working together toward common goals. This cooperation gave rise to the modern drive for sustainable development, which seeks ways to promote social justice, economic well-being, and environmental quality at the same time (**FIGURE 6.18**). Governments, businesses, industries, and organizations pursuing sustainable development aim to satisfy a **triple bottom line,** a trio of goals made up of economic advancement, environmental protection, and social equity.

Programs that pay for ecosystem services are one example of a sustainable development approach that seeks to satisfy a triple bottom line. Costa Rica's PSA program aims to enhance its citizens' well-being by conserving the country's natural assets while compensating affected landholders for any economic losses. The intention is to achieve a win-win-win result that pays off in economic, social, and environmental dimensions. Similar programs in other countries are showing mixed results, with some appearing to significantly help alleviate poverty (see **SUCCESS STORY**, p. 156).

Of course, designing sustainable solutions to complex problems is never simple, and people differ in what they mean by "sustainable development." Proponents of a school of thought called *weak sustainability* maintain that we can allow natural capital to decline as long as human-made capital increases to compensate for it. In contrast, proponents of *strong sustainability*

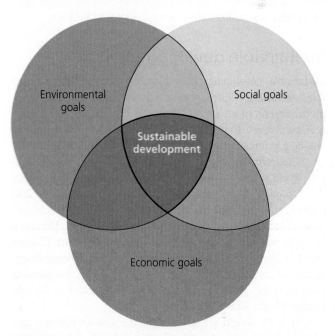

FIGURE 6.18 Sustainable development occurs when social, economic, and environmental goals overlap.

Growing Both Trees and Incomes in Vietnam

One critique of Costa Rica's PSA program has been that its benefits have not flowed to the poorest rural landowners. But halfway across the world in Vietnam, a similar program of payments for ecological services is lifting the poor out of poverty and reducing income inequality while also increasing forest cover. In Vietnam's program, fees are charged to hydropower operators, water supply companies, and tourism firms, and the funds are paid to thousands of farmers and other landowners to encourage them to conserve forests. Vietnam launched its nationwide program in 2011 but began with a pilot project in two provinces in 2009. Researchers studying the program in one of these provinces, Lam Dong province, interviewed farmers, reviewed satellite imagery of tree cover, and reported their findings in 2018. This team found a slight but significant increase in tree cover in the province, from 58% in the period 2000–2008 up to 63% in the period 2009–2014 once the program was established. Tree cover was more constant year to year, and incidents of illegal logging dropped. The program had an even greater effect on people's incomes. The government had targeted the program toward poorer farmers, so the incomes of households participating in the program were lower to begin with than those of households not in the program. However, from 2008 to 2014, the income gap was largely closed, with participants' annual incomes rising by an average of US$1623, or 53% percent. Only $374 of this amount was from the payments themselves; boosts in forestry income and reductions in farming costs helped drive the overall increase in income. If Vietnam's program nationwide can perform as it has in Lam Dong province, payments for ecological services might not only help protect forest resources—they might also lift many thousands of families out of poverty.

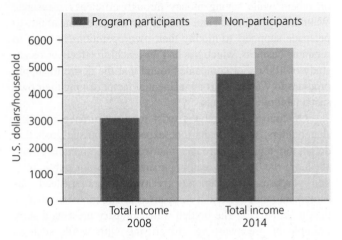

Annual incomes of households participating in Vietnam's payments-for-ecological-services program rose by half in just six years, whereas nonparticipating households showed no change. Data from Phan, T–HD, et al., 2018. Do payments for forest ecosystem services generate double dividends? An integrated impact assessment of Vietnam's PES program. PLOS ONE 13(8): e0200881.

➜ Explore the Data at **Mastering Environmental Science**

---

insist that human-made capital cannot substitute for natural capital and that we must not allow natural capital to diminish.

Sustainable development is global

Sustainable development has blossomed as an international movement. The United Nations, the World Bank, and other global organizations (p. 177) sponsor conferences, fund projects, publish research, and facilitate collaboration across borders among governments, businesses, and nonprofit organizations.

The Earth Summit at Rio de Janeiro, Brazil, in 1992 was the world's first major gathering focused on sustainable development. With representatives from more than 200 nations, this conference gave rise to several notable achievements, including the Convention on Biological Diversity (p. 297) and the Framework Convention on Climate Change (p. 515). Ten years later, nations reconvened in Johannesburg, South Africa, at the 2002 World Summit on Sustainable Development. Then in 2012, the world returned to Rio de Janeiro for the Rio+20 conference.

In 2015, world leaders met at the United Nations and adopted 17 **Sustainable Development Goals** for humanity (**TABLE 6.2**). Each broad goal for sustainable development has

TABLE 6.2 U.N. Sustainable Development Goals

- End poverty in all its forms everywhere
- End hunger, achieve food security, and promote sustainable agriculture
- Promote health and well-being for all
- Ensure quality education for all
- Achieve gender equality and empower women and girls
- Ensure water and sanitation for all
- Ensure access to affordable, reliable, and sustainable energy sources
- Promote jobs and sustainable economic growth
- Build resilient infrastructure, promote sustainable industry, and foster innovation
- Reduce inequality within and among nations
- Make cities safe, resilient, and sustainable
- Ensure sustainable consumption and production
- Take urgent action to combat climate change and its many impacts
- Conserve marine resources
- Protect and restore terrestrial ecosystems and halt biodiversity loss
- Promote peaceful, inclusive, just institutions
- Renew partnerships for sustainable development

a number of specific underlying targets—169 in all—that may be met by implementing concrete strategies. Many of the Sustainable Development Goals were given a 2030 target date, and we are making better progress on some than on others. We still have a long way to go to resolve the many challenges facing humanity. Pursuing solutions that meet a triple bottom line of environmental, economic, and social goals can help pave the way for a truly sustainable world.

The nation of Costa Rica provides one useful model of a pathway toward sustainable development. A series of decisions by its political leaders has enabled the nation to make impressive progress in social, economic, and environmental dimensions, and this progress continues today. By paying farmers and ranchers to preserve and restore forest on private land, for example, the country's citizenry reaps the rewards of a cleaner and healthier environment, which in turn is enhancing economic progress. As Costa Rica's leaders use research-based feedback to refine and improve the PSA program, the program is becoming better able to accomplish its goals. The government, businesses, and people of Costa Rica recognize how economic health depends on environmental protection and seem poised to continue building on their success.

CENTRAL CASE STUDY
connect & continue

TODAY, the concept of paying for ecosystem services has gone global, as Costa Rica's pioneering PSA program has inspired similar approaches throughout the world. In Mexico, Australia, Tanzania, China, Indonesia, Vietnam, and many other nations, these programs are gaining ground as people begin to better appreciate the contributions of ecological systems to human economies. By 2018, according to one scientific assessment, more than 550 active programs were operating in more than 60 nations, involving between $36 billion and $42 billion in transactions annually. These programs are varied but can be broken into three main types: (1) government-financed programs like Costa Rica's and similar ones in China to pay landowners for forest conservation and replanting, (2) user-financed programs whereby users of services pay landholders (e.g., hydroelectric dam operators paying upstream landholders to preserve forest in the watershed), and (3) programs in which regulated entities compensate other parties for conserving ecosystem services for them (e.g., developers paying mitigation fees for wetland restoration or emitters in carbon-trading markets paying carbon offsets for forest conservation).

One example of payments for ecological services in the United States is the federal government's Conservation Reserve Program (p. 236). This program, reauthorized every five years in the Farm Bill, pays farmers to retain natural vegetation on portions of their land to prevent erosion, conserve soil, enhance wildlife habitat, and reduce water pollution. Farmers are thereby compensated for land they do not put into crop production while they (and society as a whole) also benefit from the conservation of ecosystem services. Like Costa Rica's PSA program, the Conservation Reserve Program seeks to defuse a dilemma facing many rural landholders, who often feel short-term economic pressure to clear natural land for agriculture even though they may have an ethical concern for the land's flora and fauna.

In Costa Rica, policymakers have responded to researchers' suggestions and have enhanced the PSA program with the help of funding from new tariffs and fees and from the World Bank. The area conserved each year has grown, and the program has been made more accessible to small landholders, women, and indigenous communities, each of whose representation in the program has increased.

Costa Rica now aims to go further in providing a sustainable model for the world: It plans to become carbon-neutral by 2021. The country already gets 99% of its electricity from renewable sources, and it hopes that carbon dioxide stored by newly conserved forests will help cancel out carbon dioxide emissions from gasoline-burning vehicles. "We are the heirs of a beautiful tradition of innovation and change," Costa Rican President Carlos Alvarado Quesada told Stanford University scientists who are helping his government study and place values on its natural capital. "That's why we're doing this: not because it's fashionable, but because it's an ethical responsibility." In Costa Rica and many other places around the world today for reasons of ethics, economics, and sustainability, public and private parties are engaged in a wide variety of economic transactions that explicitly recognize the importance of natural resources and ecosystem services.

- **CASE STUDY SOLUTIONS** Suppose you are a Costa Rican farmer who needs to decide whether to clear a stand of forest or apply to receive payments to preserve it through the PSA program. Describe all the types of information you would want to consider before making your decision. Now describe what you think each of the following people would recommend to you if you were to go to them for advice: (a) a preservationist, (b) a conservationist, (c) a neoclassical economist, and (d) an ecological economist.

- **LOCAL CONNECTIONS** Costa Rica isn't the only place where forests are threatened, and it isn't the only place where programs have been established to pay people for conserving ecological services. Name and describe several natural resources and ecosystem services that are important in your region. For each of these resources and services, assess whether it is being sustained or whether it is being degraded. How do you think each resource or service could best be conserved? Would you recommend a program of payments to provide incentives for conservation? Why or why not? What other steps might be taken to help conserve each resource or service?

- **EXPLORE THE DATA** How do scientists and economists apply monetary values to natural amenities in a meaningful way? → **Explore Data** relating to the case study on **Mastering Environmental Science.**

REVIEWING Objectives

You should now be able to:

+ Describe how culture and worldview influence the choices people make

Culture and personal experience influence a person's worldview. Factors such as religion, political ideology, and shared experiences shape our perspectives on the environment and the choices we make. (pp. 135)

+ Discuss the nature and historical expansion of environmental ethics in Western culture

Environmental ethics applies ethical standards to relationships between people and aspects of their environments. We value things for utilitarian reasons or for their own sake. Anthropocentrism values people above all else, biocentrism values all life, and ecocentrism values ecological systems. The industrial revolution in Western culture inspired philosophical reactions that fed into environmental ethics. (pp. 135–138)

+ Compare and contrast major approaches in environmental ethics

The preservation ethic values preserving natural systems intact, whereas the conservation ethic promotes responsible long-term use of resources. Environmental justice seeks equal treatment for people of all income levels, races, and ethnicities. (pp. 138–141)

+ Explain how our economies exist within the environment and rely on ecosystem services

Economies depend on the ecological systems around them for natural resources and ecosystem services. The depletion of resources and the degradation of ecosystem services threaten our economic well-being. (pp. 141–142)

+ Identify principles of classical and neoclassical economics and summarize their implications for the environment

Classical economics proposes that individuals acting for their own economic good can benefit society as a whole and provides a philosophical basis for free-market capitalism. Neoclassical economics focuses on supply and demand and quantifies costs and benefits, but four of its assumptions tend to worsen environmental impacts. Conventional economic theory has promoted infinite economic growth, with little regard to resource depletion or environmental impact. (pp. 142–143, 146)

+ Describe aspects of environmental economics and ecological economics, including valuation of ecosystem services and full cost accounting

Environmental economists advocate reforming economic practices to promote sustainability; ecological economists advocate going further by pursuing a steady-state economy. Assigning monetary value to ecosystem goods and services can help reduce external costs and make market prices reflect full costs and benefits. Full cost accounting indicators aim to measure economic progress and human well-being more effectively than GDP. (pp. 147–149, 152–153)

+ Discuss how individuals and businesses can help move our economic system in a sustainable direction

Consumer choice in the marketplace, facilitated by eco-labeling, can encourage businesses to pursue more sustainable practices. (pp. 154–155)

+ Define sustainable development, explain the "triple bottom line," and describe how sustainable development is pursued worldwide

Sustainable development promotes people's economic advancement while using resources to satisfy today's needs without compromising future needs. The United Nations has played a key role in promoting sustainable development globally. Advocates of sustainable development pursue a "triple bottom line" of environmental, economic, and social goals in a coordinated way. (pp. 155–157)

SEEKING Solutions

1. Describe your worldview as it pertains to your relationship with your environment. How do you think your culture has influenced your worldview? How do you think your personal experience has affected it? How do you think your worldview shapes your decisions?

2. Explore the U.S. EPA's online tool for environmental justice data at www.epa.gov/ejscreen. Launch the EJScreen Tool, and go to a map of your own city. Click on

"Add Maps" and then "EJSCREEN Maps." Click on the category "EJ Indexes" and "Add to Map." On the map itself, click on neighborhoods in and around your city. Now go back to EJSCREEN Maps, but this time click on "Demographic Indicators" (then Add to Map) to see which areas of your city have the greatest densities of minority residents and of low-income residents. ("Demographic Index" averages these two variables.) Next click on "Environmental Indicators," and map each of the listed indicators one by one. Describe what patterns you see. Are any environmental indicators most severe in the neighborhoods with high percentages of minority or low-income people? Which indicators show such a pattern? What do you think accounts for the patterns you see? Do you perceive any problems revealed by these data? What could be done to alleviate such problems?

3. Do you think that a steady-state economy is a practical alternative to our current approach that prioritizes economic growth? Why or why not?

4. Do you think we should attempt to quantify and assign market values to ecosystem services? Why or why not? What consequences might arise from such an attempt?

5. **THINK IT THROUGH** You have just returned from serving in the U.S. Peace Corps in Costa Rica where you worked closely with farmers, foresters, ecologists, and policymakers on issues related to Costa Rica's PSA program. You have now been hired as an adviser on natural resource issues to the governor of your state. Think about the condition of the forests, soil, and water in your state. Given what you learned in Costa Rica, would you advise your governor to institute a program to pay citizens to conserve ecosystem services? Why or why not? Describe what policies you would advocate to best conserve your state's resources and ecosystem services while advancing the economic and social condition of its people.

CALCULATING Ecological Footprints

The population of the United States grew from 152,271,000 to 292,892,000 between 1950 and 2004, but the nation's GDP grew even faster. During this period, the per capita Genuine Progress Indicator (GPI) grew as well, but more slowly than GDP.

Given the values provided of the major components of the GPI for the United States in 1950 and 2004 and the U.S population figures for those two years, calculate and enter the per capita rates for these components, as well as the overall GPI values. Calculate the GPI as follows: GDP + Benefits – Environmental costs – Social costs.

COMPONENTS OF GPI	U.S. TOTAL IN 1950 (TRILLIONS OF DOLLARS)	PER CAPITA IN 1950 (THOUSANDS OF DOLLARS)	U.S. TOTAL IN 2004 (TRILLIONS OF DOLLARS)	PER CAPITA IN 2004 (THOUSANDS OF DOLLARS)
GDP	1.153		7.589	
Benefits	1.041		4.746	
Environmental costs	0.407		3.990	
Social and economic costs	0.476		3.926	
GPI				

Data from Talberth, J., C. Cobb, and N. Slattery, 2007. *The Genuine Progress Indicator 2006: A tool for sustainable development*. Oakland, CA: Redefining Progress.

1. How many times greater was the GDP in 2004 than in 1950? By how many times did the GPI increase between 1950 and 2004? What does this comparison tell you?

2. By how many times, respectively, did benefits, environmental costs, and social and economic costs increase between 1950 and 2004? How are trends in each of these components driving the overall trend between the GPI and the GDP? Which component has gotten the worst over the years? How would you account for these trends?

3. There are many ways to define and measure the benefits and costs that go into the GPI. How do you think a person with a biocentric worldview would measure these differently from a person with an anthropocentric worldview? Whose GPI for the year 2005 would likely be higher?

4. Now consider your own life. Very roughly, what would you estimate are the values of the benefits, environmental costs, and social and economic costs you experience? What could you do to help improve these trends in your own personal accounting?

Mastering Environmental Science

Students Go to **Mastering Environmental Science** for assignments, an interactive e-text, and the Study Area with practice tests, videos, and activities.

Instructors Go to **Mastering Environmental Science** for automatically graded activities, videos, and reading questions that you can assign to your students, plus Instructor Resources.

Environmental Policy

Making Decisions and Solving Problems

Young Americans Take on Climate Change in the Courts

• Eugene
OREGON

> " This case is for young people [and] future generations who can't represent themselves in other branches of government.
> Kelsey Juliana, 23, lead plaintiff in the trial

> I want my government to understand that climate change is real, changes are happening right now, and things aren't going to get better on their own. Climate change should be the government's first priority.
> Avery McRae, 13, a plaintiff in the trial "

Kelsey Juliana is a student in environmental studies at the University of Oregon and wants to become a fifth-grade teacher. She's already a celebrity, however: At age 23, she's the eldest of a celebrated group that includes a Native American hip-hop artist, a skier, an animal lover, a punk rocker, a black belt in tae kwon do, and survivors of floods, hurricanes, and oil spills—altogether, 21 individuals from 10 U.S. states ranging in age from 11 to 23 (**TABLE 7.1**, p. 162).

Despite their varied backgrounds, these youth are united by a conviction that global climate change is imperiling their futures. They are frustrated that policymakers have failed to take sufficient action to slow the human-caused increase in worldwide temperatures, extreme weather, intense storms, rising sea levels, ocean acidification, flooding, and drought. So these young "climate warriors" are stepping forward to do something about it—by turning to the judicial system to try to force more effective action out of the U.S. government.

This group of 21 young Americans has launched a landmark lawsuit, *Juliana v. United States*, which asserts that federal policies promoting fossil fuel use have violated their constitutional rights to life, liberty, and property and failed to protect essential natural resources that the government should hold in trust for the public. They are formally demanding that courts order the federal government to stop subsidizing fossil fuels and to swiftly draw down carbon dioxide emissions and take other actions to stabilize the climate system by limiting atmospheric carbon dioxide concentrations to 350 parts per million by the year 2100.

Following a long and tangled history in the court system, the case was scheduled to come to trial in U.S. District Court in Eugene, Oregon, in the fall of 2018. But a last-minute stay by the U.S. Supreme Court stretched the legal battle into 2019.

The lawsuit was first launched in 2015, when Kelsey Juliana connected with the attorney Julia Olson, founder and executive director of the nonprofit group Our Children's Trust. Olson and Juliana assembled a group of like-minded youth from around North America, joined with the nonprofit group Earth Guardians, and put together a team of lawyers who could guide their suit through the courts. Barack Obama was U.S. president at the time, and since then, the stakes have only grown

◄ The 21 youths of the *Juliana v. United States* climate change lawsuit outside the Federal District Courthouse in Eugene, Oregon

▲ Kelsey Juliana, lead plaintiff in the suit

TABLE 7.1 The Youth Plaintiffs in *Juliana v. United States*

Kelsey Juliana
Eugene, Oregon
Lead plaintiff; climate activist since age 10.

Victoria Barrett
White Plains, New York
Experienced Hurricane Sandy; marched in People's Climate March; met with diplomats at Paris climate change conference; spoke at the U.N.

Isaac Vergun
Beaverton, Oregon
Inspired by meeting Bill McKibben, co-founder of 350.org; gathered signatures for fossil fuel divestment for his bar mitzvah project.

Jaime Butler
Flagstaff, Arizona
Navajo Nation member; her family was forced to move due to water scarcity.

Lead attorney Julia Olson speaks outside the federal courthouse in Eugene, Oregon.

Sahara Valentine
Eugene, Oregon
Loves the outdoors, but sees sea level rise and drought; suffers asthma.

Miko Vergun
Beaverton, Oregon
Born in the Marshall Islands, which are threatened by sea level rise.

Levi Draheim
Satellite Beach, Florida
Lives on a barrier island threatened by rising sea levels.

Aji Piper
Seattle, Washington
Works with a tree-planting organization and with Earth Guardians.

Nathan Baring
Fairbanks, Alaska
Is witnessing more storms; wildfires and smoke exacerbated asthma.

Jayden Foytlin
Rayne, Louisiana
Health is threatened by flooding and by air and water pollution from oil infrastructure; experienced Gulf oil spill.

Kiran Oommen
Eugene, Oregon
Student at Seattle University; anarcho-folk-punk musician; community organizer.

Zealand Bell
Eugene, Oregon
Loves wilderness, the outdoors, and skiing; is noting loss of snowpack and drinking water supplies.

Alex Loznak
Oakland, Oregon
Majoring in sustainable development at Columbia University; conducted extensive research for the lawsuit.

Jacob Lebel
Roseburg, Oregon
Born in Quebec; lives on a farm and ran a CSA, so sees effects of climate change on the land; joined the Dakota Access Pipeline protest in 2016.

Xiuhtezcatl Martinez
Boulder, Colorado
Hip-hop musician; director for Earth Guardians; raised in the Aztec tradition.

Avery McRae
Eugene, Oregon
Loves animals and cares for chickens and horses; prizewinner in a science fair.

Tia Hatton
Bend, Oregon
An athlete and skier who is seeing low snowfall and drought.

Journey Zephier
Kapaa, Kaua'i, Hawai'i
Sioux Nation member, born in South Dakota; has seen impacts on beaches and coral reefs in Hawai'i.

Sophie Kivlehan
Allentown, Pennsylvania
Learned about climate change from her grandfather, climate scientist Dr. James Hansen.

Hazel Van Ummersen
Eugene, Oregon
Loves the ocean and coast; concerned about sea-level rise and acidification; has a black belt in tae kwon do.

Nick Venner
Lakewood, Colorado
Active in the Catholic Church, and inspired by Pope Francis's writings; sees impacts on fishing and gardening where he lives.

as his successor Donald Trump pulled the United States out of the Paris climate deal, dismantled Obama's Clean Power Plan, and tried to push the country back toward reliance on coal for energy. Government lawyers and fossil fuel industry representatives tried repeatedly to derail the *Juliana v. United States* lawsuit through a variety of legal maneuvers, but the plaintiffs are hopeful their case will eventually be heard at trial.

If and when a trial is held, experts will present summaries of evidence from decades of scientific research on the known and predicted impacts of climate change (see **THE SCIENCE BEHIND THE STORY**, pp. 166–167). Lawyers will argue that policymakers knew all this information and yet failed to enact policies that could slow the disruption of the climate, instead enacting policies that encouraged the continued use of fossil fuels, whose carbon pollution drives global warming.

The pioneering case raises many novel legal questions. It relies heavily on the public trust doctrine (p. 172), which holds that government is responsible for guarding and maintaining natural resources in trust for the public. Just as private property rights should be protected when land is owned by an individual, the argument goes, the rights of the public should be defended when a commonly held resource (like clean air, clean water, or a stable climate) is crucial to the public's well-being.

The lawsuit makes clear that climate change is not just a scientific issue, policy issue, and economic issue—it is largely a moral issue. Juliana and her fellow plaintiffs maintain that it is ethically unacceptable if the actions (or inactions) of those in power from older generations result in a severely degraded environment and quality of life for younger and future generations.

In response, government lawyers have argued that any harm the plaintiffs may experience from climate change will have resulted from "emissions from every source in the world over decades" and that Americans are not guaranteed under the U.S. Constitution any right to particular climate conditions. They will argue that the courts are not the proper place to decide on policies relevant to climate change.

If the 21 young plaintiffs emerge victorious in the end, the nation will find itself in uncharted waters. A court victory for the youth would create immediate legal pressure on the legislative and executive branches of government to respond with drastically revised energy policies. If the plaintiffs lose their trial, they will have at least had their day in court, made their case to the world, and inspired a broader and deeper discussion about the threats from climate change and the moral dimensions of the crisis.

This legal case will likely go through a variety of appeals and further actions in the federal court system, so we encourage you to follow the twists and turns with your peers and your instructor and see where the story leads. In the meantime, similar lawsuits are springing up elsewhere. In the United States, Our Children's Trust is attempting to sponsor legal actions in all 50 states. Across the world, climate litigation is proceeding in many nations, and one high-profile case has succeeded: In the Netherlands, 886 Dutch citizens and Urgenda, a nonprofit foundation, won a suit in 2015 arguing that the Dutch government had not acted strongly enough to address climate change. The government appealed the verdict but lost in 2018. As a result, the Dutch government may be forced to enact tougher policies to reduce emissions.

It remains to be seen what influence this novel wave of climate litigation will have. What is certain is that a new front has opened in the grassroots struggle against climate change that will bring additional pressure on policymakers to find ways to address the issue.

Environmental Policy: An Overview

The disruption of our planet's climate poses one of the most pressing challenges for our society. When society recognizes a problem, its leaders may try to resolve the problem using policy. **Policy** consists of a formal set of general plans and principles intended to guide decision making. **Public policy** is policy made by people in government, and consists of laws, regulations, orders, incentives, and practices intended to advance societal well-being. **Environmental policy** is policy that pertains to our interactions with our environment. Environmental policy generally aims to regulate resource use or reduce pollution to promote human welfare and/or protect natural systems.

Forging effective policy (**FIGURE 7.1**) requires input from science, ethics, and economics. Science (Chapter 1) provides information and analyses needed to identify and understand problems and devise solutions. Ethics and economics (Chapter 6) offer criteria by which to assess problems and help clarify how society might address them.

The ongoing debates over how to respond to climate change show how science, economics, and ethics can each inform and motivate policymaking. Science has enabled the

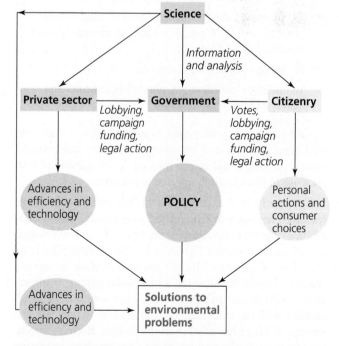

FIGURE 7.1 Policy plays a central role in addressing environmental problems. Government interacts with citizens and the private sector to formulate policy, while science offers information and analysis.

technological advances that allow us to find and extract fossil fuels. Science also helps us understand the impacts exerted by fossil fuel extraction and use, including the emission of carbon dioxide and other greenhouse gases and the global warming that results. In economic terms, fossil fuels have powered our economies for years, but today the impacts of climate change are beginning to impose significant costs, whereas clean renewable energy sources offer new economic opportunities. Ethically, climate change will impose the greatest costs on future generations, whereas past generations reaped most of the benefits of fossil fuels. Policymakers consider all these things as they design environmental policy to address climate change and its impacts on human health, ecological systems, and social well-being.

Environmental policy addresses issues of fairness and resource use

Because market capitalism is driven by incentives for short-term economic gain, it provides businesses and individuals little motivation to minimize environmental impacts, seek long-term social benefits, or equalize costs and benefits among parties. Market prices often do not reflect the value of environmental contributions to economies or the full costs imposed on the public by private parties when their actions degrade the environment (Chapter 6). Such market failure (p. 153) has traditionally been viewed as justification for government involvement. Governments typically intervene in the marketplace for several reasons:

- To provide social services, such as national defense, health care, and education

- To provide "safety nets" (for the elderly, the poor, victims of natural disasters, and so on)

- To eliminate unfair advantages held by single buyers or sellers (i.e., to regulate or break up monopolies that threaten the public good)

- To manage publicly held resources

- To minimize pollution and other risks to health and quality of life

Environmental policy aims to protect people's health and well-being, to safeguard environmental quality and conserve natural resources, and to promote equity or fairness in people's use of resources or exposure to pollution.

The tragedy of the commons When publicly accessible resources are open to unregulated exploitation, they tend to become overused, damaged, or depleted—so argued environmental scientist Garrett Hardin in his 1968 essay, "The Tragedy of the Commons." Basing his argument on an age-old scenario, Hardin explained that in a public pasture (or "common") open to unregulated grazing, each person who grazes animals will be motivated by self-interest to increase the number of his or her animals in the pasture. Because no single person owns the pasture, no one has incentive to expend effort taking care of it. Instead, each person takes what he or she can until the resource is depleted and overgrazing causes

the pasture's food production to collapse, harming everyone. This scenario, known as the **tragedy of the commons,** is widely thought to pertain to many types of resources held and used in common by the public: forests, fisheries, clean air, clean water—even the global climate.

When shared resources are being depleted or degraded, it is in society's interest to develop guidelines for their use. In Hardin's example, guidelines might require pasture users to help restore and manage the resource or they might limit the number of animals each person can graze. These two concepts—management and restriction of use—are central to environmental policy today.

A standard way to alleviate the tragedy of the commons is through management and regulation by government, but we can also address it in other ways. One is to subdivide the resource and sell allotments into private ownership so that each owner gains personal incentive to manage his or her portion. Privatization may be effective if property rights can be clearly assigned (as with land), but it tends not to work with resources such as air or water, which are not confined within property boundaries. Privatization also opens the door to short-term profit taking at the long-term expense of the resource.

Another way to address the tragedy of the commons is for resource users to band together and cooperate voluntarily to prevent overexploitation. Research by the Nobel Prize–winning economist Elinor Ostrom and others has shown that many societies, past and present, have developed ways to manage resources cooperatively and sustainably at the community level. This approach is often effective if the resource is localized, community members are known to one another, and enforcement is straightforward, but these conditions do not always hold.

Our global atmosphere is a common resource for all humanity, and its pollution with greenhouse gas emissions is a tragedy-of-the-commons scenario. Because the atmosphere cannot be privately owned and because voluntary cooperative efforts to restrict emissions have not worked, national policy and international agreements to reduce emissions seem to many people the best means of addressing climate change.

Free riders A second reason we develop policy for publicly held resources is to alleviate the predicament caused by **free riders.** Let's say a community on a river suffers from water pollution that emanates from 10 different factories. In theory, the problem could be solved if every factory voluntarily agrees to reduce its own pollution. However, once all factories begin reducing their pollution, it becomes tempting for any one of them to stop doing so. A factory that avoids the efforts others are making would in essence get a "free ride." If enough factories take a free ride, the whole collective endeavor will collapse. Because of the free-rider problem, private voluntary efforts are often less effective than efforts mandated by public policy. For example, President Trump's announcement that he would pull the United States out of the Paris Agreement, an international treaty to address climate change that the United States had signed in 2016, created a free-rider issue. If the United States were to increase its emissions while the rest of the world tried to decrease theirs, the

United States would be getting a free ride; that is why withdrawal by the United States threatens to unravel the global agreement.

External costs Environmental policy also aims to promote fairness by dealing with external costs (p. 143), harmful impacts suffered by people not involved in the actions that created them. For example, a factory that discharges waste into a river imposes external costs (such as water pollution, health problems, reduced fish populations, and aesthetic impacts) on downstream users of the river.

Government action to force the company that owns the factory to clean up its pollution, pay fees, or reimburse residents for damages helps to "internalize" costs. In these cases, the costs would be paid by the company, which would likely pass them on to consumers by raising the prices of its products. Higher market prices may reduce demand for the products, and consumers may instead come to favor products that impose fewer costs on society as they become relatively less expensive.

In the case of climate change, island-dwelling societies are experiencing disruptive costs from rising sea levels as polar and glacial ice melts from global warming caused by fossil fuel emissions (**FIGURE 7.2**). They are struggling with loss of land from erosion and storms, as well as saltwater intrusion that contaminates drinking water aquifers and inhibits agriculture. Many people have been forced from their homes as "climate refugees." Indeed, entire small island nations that emit very few greenhouse gases (such as the Maldives; p. 502) are finding their long-term existence threatened by emissions, predominantly from the United States, Europe, and China. These small island nations did not cause the problem, but they are suffering the costs. Likewise, as the plaintiffs in the *Juliana v. United States* suit attest, young people will be the ones to suffer the costs of climate change caused by emissions that occurred years or decades before they were born.

These goals of environmental policy—to protect resources against the tragedy of the commons and to promote fairness by eliminating free riders and addressing external costs—are reflected in today's diversity of approaches to environmental policy. As an example, the **polluter-pays principle** specifies that a party responsible for pollution should be held responsible for covering the costs of its impacts. This principle helps protect resources such as clean water and air, promotes just treatment of all parties, and helps minimize external costs by shifting them into the market prices of goods and services.

FIGURE 7.2 Island-dwelling people vulnerable to sea level rise are bearing the brunt of climate change's impacts. They are suffering external costs from carbon pollution from larger more industrialized nations. This boy lives on Ghoramara Island off the coast of India; the island has lost most of its land area and is predicted to disappear in the future.

Various factors can obstruct environmental policy

If environmental policy brings clear benefits, why are environmental laws and regulations often challenged? One reason is the perception that environmental protection requires economic sacrifice (see **FAQ,** p. 142). Businesses and individuals often view regulations as restrictive and costly. Landowners may fear that zoning (p. 343) or protections for endangered species (p. 297) will restrict how they can use their land. Developers complain of time and money lost in obtaining permits, reviews by government agencies, and required environmental controls, monitoring, and mitigation.

Another hurdle for environmental policy stems from the nature of environmental problems, which often develop gradually over long time periods. In contrast, human behavior is geared toward addressing short-term needs, and this approach is reflected in our social institutions. Businesses usually opt for short-term financial gain. The news media focus coverage on new and sudden events. Politicians often act in their short-term interest because they depend on reelection every few years. For all these reasons, environmental policy may be obstructed.

Policy in general can be held up for a variety of reasons, even if a majority of people favor it. Checks and balances in a constitutional democracy seek to ensure that new policy is implemented only after extensive review and debate. However, less desirable factors can also come into play. In democracies such as the United States, each person has a political voice and can make a difference—yet money wields influence. People, organizations, industries, or corporations with enough wealth to buy access to power can exert disproportionate influence over policymakers.

WEIGHING
the **issues**

Internalizing External Costs

Imagine that we were to use policy to internalize all the external costs of gasoline (pollution, health risks, climate change, impacts from oil drilling and transport, etc.) and that, as a result, gas prices rise to $13 per gallon. What effects do you think this would have on the choices we make as consumers (such as driving behavior and types of vehicles purchased)? What influence might it have on the types of vehicles produced and the types of energy sources developed? What effects might it have on our taxes and our health insurance premiums? In the long run, do you think that internalizing external costs in this way would end up costing society more money or saving society money? What factors might be important in determining the outcome?

THE SCIENCE behind the story

What Does the Science on Climate Change Tell Us?

Dr. James Hansen with his granddaughter, Sophie Kivlehan, a litigant in the *Juliana* case

At whatever point *Juliana v. United States* goes to trial, the plaintiffs are expected to call a number of scientists to testify, including well-known figures such as Dr. James Hansen, the former NASA scientist and climate expert who famously testified to Congress on the risks from climate change back in 1988.

The role of these scientists will be to present summaries of the extensive scientific research into global climate change, the impacts that climate disruption is having on people, and the impacts that it is predicted to have in the future. That is no small task, especially considering the voluminous amount of information that thousands of researchers have accumulated over recent decades. How would a scientist even begin?

For many years, large groups of scientists have been pooling their expertise, summarizing research findings on climate change, and presenting them to policymakers and the public. The highest-profile efforts are by the Intergovernmental Panel on Climate Change (IPCC), which issues periodic reports, including comprehensive assessments by hundreds of scientists and governmental representatives that make a global splash every 7 years. (We will learn more about the IPCC and its efforts in our coverage of climate change in Chapter 18).

Other summaries are regularly produced as well. One is the U.S. government's *National Climate Assessment,* which issues periodic reports by hundreds of experts that spell out the known and predicted impacts of climate change in the United States, sector by sector and region by region.

Together, such reports make abundantly clear that temperatures are warming (**FIGURE 1**), that climate is changing, and that these changes result primarily from the carbon dioxide and other greenhouse gases emitted when fossil fuels (coal, oil, and gas) are burned for energy. Research also shows the many ways in

which climate disruption is having ecological, social, and economic impacts. Data from climate modeling make clear that these impacts will grow with time as greenhouse gases continue to warm the atmosphere—until and unless emissions are reduced.

A few of the major trends and impacts are shown in **FIGURE 2**, which uses data from the latest IPCC report. Some of the most significant include:

- Temperatures have been rising rapidly for the past century.
- Precipitation patterns are changing, bringing flooding to some areas and drought to others.

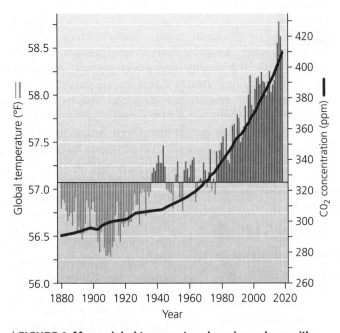

FIGURE 1 Mean global temperature has risen along with the atmospheric concentration of carbon dioxide, the primary greenhouse gas. Bars show mean global temperature each year (blue for below average and red for above average). The dark blue line shows global carbon dioxide concentration. *Data from U.S. National Climate Assessment.*

 Go to **Interpreting Graphs & Data** on **Mastering Environmental Science**

Another reason advances in policy can be stalled is the movement of individuals between government and the private sector, known as the **revolving door.** Some individuals employed in industry gain political influence when they take jobs with the government agencies responsible for regulating their industries. Conversely, businesses often hire former government officials who had regulated their industries. As an example, President Trump selected as his deputy administrator

of the Environmental Protection Agency (EPA) Andrew Wheeler, a lobbyist for the coal industry. With Wheeler in this number-two position, and later in the top position of administrator, the EPA abandoned regulations on the coal industry, despite the prime role of coal emissions in driving climate change. Every U.S. presidential administration, from both parties, has seen dozens or hundreds of appointees pass through the revolving door in one direction or the other.

FIGURE 2 Current trends and future impacts of climate change are extensive. Shown are major trends and impacts (both observed and predicted) as reported by the IPCC. (Mean estimates are shown; the IPCC reports ranges and statistical probabilities as well.) *Data from IPCC, 2013.* Fifth assessment report.

- Storms are becoming stronger, and extreme weather events are becoming more frequent.
- Ice in glaciers and at the poles is melting at unprecedented rates.
- Sea levels are rising as runoff from melting ice enters the ocean and as water volume expands with warmth, flooding islands and coastal communities.
- Ocean water is becoming acidified, threatening to destroy coral reefs.
- Plants, animals, ecosystems, and agriculture are being affected in complex ways.

Such trends are having increasingly severe impacts on people. Floods, droughts, fires, hurricanes, and storm surges are destroying property, harming health, and threatening lives. As quality of life deteriorates for vulnerable people, refugees may stream out of affected areas and into other areas, giving rise to social and political instability.

For the purposes of the *Juliana v. United States* case, the issue is how such climate-related impacts would personally affect the lives and futures of the plaintiffs. Thus, for a typical American aged 11 to 23, the scientific question is: What impacts might you reasonably expect during your lifetime?

Based on the science, the answer would seem to be a significant degree of impact. Yet the judge in this case will also need to weigh other issues, such as whether the judiciary is the proper setting in which to design policy to address climate change. Still, predictions from the science are clear: The average young American can reliably expect to experience many diverse impacts of climate change throughout his or her lifetime.

Defenders of the revolving door assert that corporate executives who take government jobs regulating their own industry bring with them an intimate knowledge of the industry that makes them highly qualified and likely to benefit society with well-informed policy. Critics contend that taking a job regulating your former employer is a clear conflict of interest that undermines the effectiveness of the regulatory process.

Science informs policy but is too often disregarded

Economic interests, ethical values, and political ideology all influence the policy process, but environmental policy that is effective is generally informed by scientific research. For instance, when deciding whether to regulate a substance that may pose a public health risk, regulatory agencies such as the EPA comb the scientific literature for information and may commission new studies to research unresolved questions. When crafting a bill to reduce pollution, a legislator may use data from scientific studies to quantify the cost of the pollution or the predicted benefits of its reduction. In today's world, a nation's strength depends on its commitment to science, which is why governments devote a portion of their tax revenue to fund scientific research.

Unfortunately, sometimes policymakers allow factors other than science to determine policy on scientific matters. Politicians may ignore scientific consensus on well-established matters such as evolution, vaccination, or climate change if it suits their political needs or if they are motivated chiefly by political or religious ideology. Some may reject scientific advice if it helps please campaign contributors or powerful constituencies. For these reasons, taxpayer-funded science often is suppressed or distorted for political ends. As a result, we cannot simply take for granted that science will play a role in policy. Each of us can help ensure that our representatives in government make proper use of our scientific knowledge by reelecting those who do and voting out of office those who do not.

U.S. Environmental Law and Policy

The United States provides a good focus for understanding environmental policy in constitutional democracies worldwide for several reasons. First, the United States has pioneered innovative environmental policy. Second, U.S. policies serve as models—of both success and failure—for other nations and for international bodies. Third, the United States exerts a great deal of influence on the affairs of other nations. Furthermore, understanding U.S. policy at the federal level helps us understand policy at state and local levels, as well as at the international level.

Federal policy arises from the three branches of government

Federal policy in the United States results from actions of the three branches of government—legislative, executive, and judicial—established under the Constitution. Congress creates laws, or **legislation,** by crafting bills that can become law with the signature of the head of the executive branch, the president (**FIGURE 7.3**). Once a law is enacted, its implementation and enforcement are assigned to an administrative agency within the executive branch. Administrative agencies create **regulations,** specific rules intended to achieve the objectives of a law. These agencies also monitor compliance with laws and regulations and may enforce them when they are violated. The president may also issue **executive orders,** specific legal instructions for government agencies. Several dozen administrative agencies influence U.S. environmental policy, ranging from the EPA to the Forest Service to the Food and Drug Administration to the Bureau of Land Management.

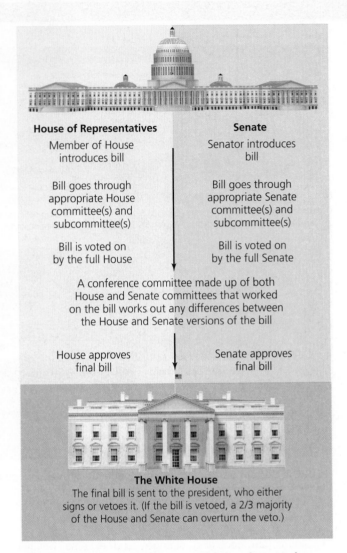

House of Representatives
Member of House introduces bill

Senate
Senator introduces bill

Bill goes through appropriate House committee(s) and subcommittee(s)

Bill goes through appropriate Senate committee(s) and subcommittee(s)

Bill is voted on by the full House

Bill is voted on by the full Senate

A conference committee made up of both House and Senate committees that worked on the bill works out any differences between the House and Senate versions of the bill

House approves final bill

Senate approves final bill

The White House
The final bill is sent to the president, who either signs or vetoes it. (If the bill is vetoed, a 2/3 majority of the House and Senate can overturn the veto.)

FIGURE 7.3 Before a bill becomes U.S. law, it must clear hurdles in both legislative bodies. If the bill passes the House of Representatives and the Senate, a conference committee works out differences between versions before the bill is sent to the president. The president may then sign or veto the bill.

The judiciary, consisting of the Supreme Court and various lower courts, is charged with interpreting law. This is necessary because social norms, societal conditions, and technologies change over time and because Congress writes laws broadly to apply to varied circumstances throughout the nation. Decisions rendered by the courts make up a body of law known as *case law*. Previous rulings serve as *precedents*, or legal guides, for later cases, steering judicial decisions through time. The judiciary is an important arena for environmental policy. Grassroots environmental advocates and organizations use lawsuits to help level the playing field with large corporations and agencies. Conversely, the courts hear complaints from businesses and individuals challenging the constitutional validity of environmental laws that they believe are infringing on their rights. Individuals and organizations also lodge lawsuits against government agencies when they believe that the government is failing to enforce its agencies' own regulations.

Courts interpret the constitutionality of policy

The Constitution lays out several principles that have come to be especially relevant to environmental policy. One is from the Fifth Amendment, which ensures, in part, that private property shall not "be taken for public use without just compensation." Courts have interpreted this clause, known as the *takings clause,* to ban not only the literal taking of private property, but also what is known as regulatory taking. A **regulatory taking** occurs when the government, by means of a law or regulation, deprives a property owner of all or some economic uses of his or her property. People often cite the takings clause in opposing regulations that restrict development on privately owned land.

Another principle enshrined in the Fifth Amendment is that of due process. Due process clauses in both the Fifth and Fourteenth Amendments guarantee citizens that they will receive fair treatment under the law and cannot be arbitrarily deprived of life, liberty, or property. The Fourteenth Amendment also included an equal protection clause that guarantees all citizens "equal protection of the laws." When a U.S. district court in 2016 refused to dismiss *Juliana v. United States,* it based its decision on the public trust doctrine (p. 172), finding that this doctrine was implied by the Constitution's principles of due process and equal protection. Other legal experts, however, questioned whether the public trust doctrine was truly enshrined in the Constitution in these ways. As *Juliana v. United States* and other climate change cases proceed through the courts in coming years, the legal profession will grapple with how to apply constitutional principles to the issues that bear on climate change policy.

State and local governments also make policy

The structure of the federal government is mirrored at the state level with legislatures, governors, judiciaries, and agencies. States, counties, and cities and towns all generate and enforce environmental policy of their own. Indeed, several of the youth represented in *Juliana v. United States* began their civic involvement years earlier by advocating for policies in their hometowns in front of local boards and city councils.

State laws cannot violate principles of the U.S. Constitution, and if state and federal laws conflict, federal laws take precedence. Federal policymakers may influence environmental policy at the level of the states by:

- Supplanting state law to force change. (This is uncommon.)

- Providing financial incentives to encourage change. (This can be effective if federal funding is adequate and if states need the money.)

- Following an approach of *cooperative federalism* whereby a federal agency sets national standards and then works with state agencies to achieve them in each state. (This is most common.)

In recent years, pressure to weaken federal oversight and hand over power to the states has grown, but political scientists say that retaining strong federal control over environmental policy is a good idea for several reasons:

- Citizens of all states should have equitable protection from environmental and health impacts.

- Dealing with environmental problems often requires an "economy of scale" in resources such that one strong national effort is far more efficient than 50 state efforts.

- Many issues involve "transboundary" disputes that cross state lines, and nationwide efforts minimize disputes among states.

States vary in their environmental policy

Ideally, states and localities can act as laboratories for experimenting with novel policy concepts so that ideas that succeed in one place may be adopted elsewhere. For example, many U.S. states in recent years have responded to climate change concerns by adopting mandates that a certain percentage of their energy supply come from renewable sources (such as wind, water, and solar) by certain dates. If some states succeed in achieving these goals while other states fail, we will be able to compare the various results and draw general conclusions about what is required to make such policies effective.

States have always varied in the extent and strength of their environmental policies. In today's politically partisan era, some people are labeling those states with strong policies as "green states" and those with weak policies as "brown states."

California is an example of a state with strong environmental policy, particularly when it comes to climate change and energy issues. California's leadership role in environmental policy goes back to the 1940s when it responded to smog pollution in Los Angeles with pioneering policies to clean up air quality. In 1967, California received a waiver from the federal government that allowed it to set air quality standards that were stricter than the federal standards. As other states began adopting California's standards, automakers were prompted to produce cars that emitted less tailpipe pollution. In addition, California pioneered energy efficiency regulations for appliances and buildings, conserving vast amounts of electricity, and in 2018 it mandated that all new residential construction include rooftop solar panels to generate renewable electricity.

California aimed at climate change directly in 2006 with its Global Warming Solutions Act, signed into law by Republican governor Arnold Schwarzenegger. This law committed the state to bringing its greenhouse

WEIGHING the issues

Green States and Brown States

Do you think your state is a green state or a brown state? Explain your view. What major environmental policies has your state enacted? Can you give an example of an environmental policy that California or another state has in place that your state does not? What environmental policies, if any, would you like to see your state or local political leaders pursue, and why?

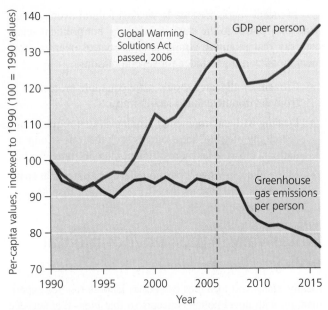

 Go to **Interpreting Graphs & Data** on **Mastering Environmental Science**

FIGURE 7.4 In the decade after passing its Global Warming Solutions Act, California grew economically while reducing its greenhouse gas emissions. In per-person terms, California's inflation-adjusted GDP (gross domestic product) grew 6.8% while its emissions fell 18.0%. (These rates compare to 3.2% GDP growth and 17.0% emissions reductions for the United States as a whole during this period.) In absolute terms, California's inflation-adjusted GDP grew 15.9% while its emissions fell 11.1%. (These rates compare to nationwide rates of 11.6% GDP growth and 10.2% emissions reductions.) Experts say California is experiencing "a virtuous cycle [in which] policy creates market certainty and helps drive investment and technology advancement." *Next10. 2018. 2018 California green innovation index. 10th ed. Next10, San Francisco, CA.*

gas emissions down to 1990 levels by 2020 through a combination of regulations and market-based financial incentives. The act also established one of the world's leading cap-and-trade carbon markets (pp. 181, 514). As it happens, California met its emissions reduction goal four years early, achieving its target in 2016 even while adding 8.5% to its population. To the surprise of skeptics, the state blossomed economically while reducing its emissions (**FIGURE 7.4**). California's leaders are continuing confidently down this path: In 2018, the legislature passed a historic mandate to achieve a zero-carbon electric supply by 2045, while Governor Jerry Brown signed an aspirational executive order proposing carbon-neutrality throughout California's economy by the same year.

California's bold moves have come during a period in which the federal government has mostly been retreating from environmental policy leadership, both domestically and internationally. With Trump as president, extensive rollbacks of federal policy were set in motion. In response, California and like-minded states fought to protect their own policies and to resist changes in federal policy that they felt would harm residents of their states. As we proceed through our discussion of federal policy, keep in mind that a great deal of environmental policy is created and administered at the state and local levels.

Early U.S. environmental policy promoted development

Environmental policy in the United States was created in three periods. Laws enacted during the first period, from the 1780s to the late 1800s, accompanied the westward expansion of the nation and were intended mainly to promote settlement and the extraction and use of the continent's abundant natural resources (**FIGURE 7.5**).

(a) Settlers in Nebraska, circa 1860

(b) Loggers felling an old-growth tree, Washington

FIGURE 7.5 Early U.S. environmental policy promoted settlement and natural resource extraction. The Homestead Act of 1862 allowed settlers **(a)** to claim 65 hectares (160 acres) of public land by paying $16, living there for 5 years, and farming or building a home. The timber industry was allowed to cut the nation's ancient forests **(b)** with little policy to encourage conservation.

Among these early laws were the General Land Ordinances of 1785 and 1787 by which the new federal government gave itself the right to manage the lands it was expropriating from Native Americans. As the young country accrued territory through purchase, treaties with European powers and Native American nations, and military conquest, these laws created a grid system for surveying these lands and readying them for private ownership. Subsequently, the government promoted settlement in the Midwest and West and doled out millions of acres to its citizens and to railroad companies, encouraging settlers, entrepreneurs, and land speculators to move west.

Western settlement was meant to provide U.S. citizens with means to achieve prosperity while relieving crowding in eastern cities. It expanded the geographic reach of the United States at a time when the young nation was still jostling with European powers for control of the continent. It also wholly displaced the millions of Native Americans whose ancestors had inhabited these lands for millennia. U.S. environmental policy of this era reflected a perception that the vast western lands were inexhaustible in natural resources.

The second wave of environmental policy encouraged conservation

In the late 1800s, as the continent became more populated and its resources were increasingly exploited, public policy toward natural resources began to shift. Reflecting the emerging conservation and preservation ethics (p. 138) in American society, laws of this period aimed to alleviate some of the environmental impacts of westward expansion.

In 1872, Congress designated Yellowstone as the world's first national park. In 1891, Congress authorized the president to create forest reserves to prevent overharvesting and protect forested watersheds. In 1903, President Theodore Roosevelt created the first national wildlife refuge. These acts launched the creation of a national park system, national forest system, and national wildlife refuge system that still stand as global models (pp. 318, 326). These developments reflected a new understanding that the continent's resources were exhaustible and required legal protection.

Land management policies continued through the 20th century, targeting soil conservation in the wake of the Dust Bowl (p. 231) and wilderness preservation with the Wilderness Act of 1964 (p. 326).

The third wave responded to pollution

Further social changes in the 20th century gave rise to the third major period of U.S. environmental policy. In a more densely populated nation driven by technology, industry, and intensive resource consumption, Americans found themselves better off economically but living amid dirtier air, dirtier water, and more waste and toxic chemicals than ever before. Events in the 1960s and 1970s triggered greater awareness of environmental problems, bringing about a profound shift in public policy.

FIGURE 7.6 Scientist and writer Rachel Carson revealed the effects of DDT and other pesticides in her 1962 book, *Silent Spring*.

A landmark event was the 1962 publication of *Silent Spring,* a best-selling book by American scientist and writer Rachel Carson (**FIGURE 7.6**). *Silent Spring* awakened the public to the ecological and health impacts of pesticides and industrial chemicals (p. 369). The book's title refers to Carson's warning that pesticides might kill so many birds that few would be left to sing in springtime.

Ohio's Cuyahoga River (**FIGURE 7.7**) also drew attention to pollution hazards. The Cuyahoga was so polluted with oil and industrial waste that the river actually caught fire near Cleveland a number of times in the 1950s and 1960s. Such spectacles, coupled with an oil spill offshore from Santa Barbara, California, in 1969, moved the public to prompt Congress and the president to better safeguard water quality and public health.

FIGURE 7.7 Ohio's Cuyahoga River was so polluted with oil and waste that the river caught fire multiple times in the 1950s and 1960s and would burn for days at a time.

A number of leaders in government and academia were influential in responding to the growing public desire for environmental protection. Stewart Udall, secretary of the Interior Department from 1961 to 1969, shaped key laws and oversaw the creation of more than 100 federal parks and refuges for conservation and public use. Maine Senator Edmund Muskie spearheaded the Clean Air Act and other key environmental laws. Wisconsin Senator Gaylord Nelson helped create Earth Day in 1970, an annual event ever since, which galvanized public support for action to address pollution problems. Legal scholar Joseph Sax in 1970 published a seminal paper developing the **public trust doctrine,** which holds that natural resources such as air, water, soil, and wildlife should be held in trust for the public and that government should protect these resources from exploitation by private parties.

With the help of such leaders, public demand for a cleaner environment during this period inspired a number of major laws that underpin modern U.S. environmental policy (**TABLE 7.2**). You will encounter most of these laws again later in your studies. Each of them has helped shape the quality of your life in fundamental ways.

Why did so many advances in environmental policy occur in the 1960s and 1970s? Scholars of **environmental history,** the study of human interactions with the natural environment through time, suggest it was because (1) environmental problems became readily apparent and were directly affecting people's lives, (2) people could visualize policies to deal with the problems, and (3) citizens were politically active and leaders were willing to act. In addition, photographs from NASA's space program allowed humanity to see, for the first time ever, images of Earth from space (**FIGURE 7.8**). It is hard for us today to comprehend the transformative power those images had at the time, but they revolutionized many people's worldviews by making us aware of the finite nature of our planet.

FIGURE 7.8 This photo of Earth, as seen from the *Apollo 8* space capsule orbiting the moon, made clear that we all live together on a small and beautiful planet with precious and limited resources. The photo, taken on December 24, 1968, by astronaut William Anders, was unprecedented at the time and profoundly transformed many people's worldviews.

Today, largely because of policies enacted since the 1960s, our health is better protected and the nation's air and water are considerably cleaner. Thanks to the many Americans who worked tirelessly in grassroots efforts and to policymakers who listened and chose to make a difference in people's lives, we now enjoy a cleaner environment in which industrial chemicals, waste disposal, and resource extraction are more carefully regulated. Much remains to be done—and we will always need to stand ready to defend our advances—but all of us alive owe a great deal to the dedicated people who inspired policy to tackle pollution during the past several decades.

Passage of NEPA and creation of the EPA were milestones

One of the foremost U.S. environmental laws is the **National Environmental Policy Act** (NEPA), signed into law by Republican president Richard Nixon in 1970. NEPA created an agency called the Council on Environmental Quality and required that an **environmental impact statement** (EIS) be prepared for any major federal action that might significantly affect environmental quality. An EIS summarizes results from studies that assess impacts that could result from a development project undertaken or funded by the federal government.

The EIS process forces government agencies and businesses that contract with them to evaluate impacts using a cost-benefit approach (p. 142) before proceeding with a new dam, highway, or building project. The EIS process rarely halts development projects, but it serves as an incentive to lessen environmental damage. NEPA grants ordinary citizens input into the policy process by requiring that EISs be made publicly available and that policymakers solicit and consider public comment on them.

In 1970, policymakers also created the **Environmental Protection Agency** (EPA). The EPA was charged with conducting and evaluating environmental research, monitoring environmental quality, setting and enforcing standards for pollution levels, assisting the states in meeting the NEPA standards, and educating the public on environmental issues.

Since its inception, the EPA has played a central role in environmental policy. Industries whose practices are regulated by the EPA complain of expense and bureaucracy, but studies that examine how regulations can reduce external costs show that EPA actions bring net benefits worth hundreds of billions of dollars to the American public each and every year (p. 184). EPA regulations are frequently challenged by industry-backed lawsuits, and EPA administrators are continually pressured by policymakers, many of whom rely on industries regulated by the EPA for campaign contributions. As a result, environmental advocates often feel that the EPA falls short in carrying out its mission of protecting the public from environmental hazards.

The social context for policy evolves

In the 1980s, Congress strengthened, broadened, and elaborated on the laws of the 1970s. Major amendments were made

TABLE 7.2 Major U.S. Environmental Protection Laws, 1963–1980

Clean Air Act
1963;
amended 1970 and 1990

Sets standards for air quality, restricts emissions from new sources, enables citizens to sue violators, funds research on pollution control, and established an emissions trading program for sulfur dioxide. As a result, the air we breathe today is far cleaner (pp. 458–463).

Resource Conservation and Recovery Act
1976

Sets standards and permitting procedures for the disposal of solid waste and hazardous waste (p. 630). Requires that the generation, transport, and disposal of hazardous waste be tracked "from cradle to grave."

Endangered Species Act
1973

Seeks to protect species threatened with extinction. Forbids destruction of individuals of listed species or their critical habitat on public and private land, provides funding for recovery efforts, and allows negotiation with private landholders (pp. 296–297).

Clean Water Act
1977

Regulates the discharge of wastes, especially from industry, into rivers and streams (pp. 409, 415). Aims to protect wildlife and human health and has helped to clean up U.S. waterways.

Safe Drinking Water Act
1974

Authorizes the EPA to set quality standards for tap water provided by public water systems and to work with states to protect drinking water sources from contamination.

Soil and Water Conservation Act
1977

Directs the U.S. Department of Agriculture to survey and assess soil and water conditions across the nation and prepare conservation plans. Responded to worsening soil erosion and water pollution on farms and rangeland as production intensified.

Toxic Substances Control Act
1976

Directs the EPA to monitor thousands of industrial chemicals and gives it power to ban those found to pose too much health risk (p. 384). However, the number of chemicals continues to increase far too quickly for adequate testing.

CERCLA ("Superfund")
1980

Funds the Superfund program to clean up hazardous waste at the nation's most polluted sites (p. 638). Costs were initially charged to polluters but most are now borne by taxpayers. The EPA continues to progress through many sites that remain. Full name is the Comprehensive Environmental Response Compensation and Liability Act.

FAQ

Isn't the EPA an advocate for the environment?

Like all administrative agencies, the EPA is part of the executive branch and operates in line with the policies of the presidential administration in power at the time. As such, the EPA under one president may function very differently from the EPA under another. Indeed, sometimes the agency may impede or roll back environmental protections, as occurred under President Trump, whose initial selection to head the agency, Scott Pruitt, had spent years attacking EPA policies and bringing lawsuits against the agency when he was Oklahoma's attorney general. The EPA employs many dedicated scientists who carry out careful research and make scientifically informed policy recommendations. However, they advise administrators appointed by the president, and policy decisions are ultimately made by these politically appointed administrators.

to the Clean Water Act in 1987 and the Clean Air Act in 1990. But the political climate in the United States soon changed. Although public support for the goals of environmental protection remained high, many people began to believe that the regulatory means used to achieve these goals too often imposed economic burdens on businesses or individuals. Attempts were made to roll back environmental policy, particularly by the administrations of President Ronald Reagan and President George W. Bush and during most congressional sessions since 1994. The Obama administration strengthened environmental policy overall, particularly on issues pertaining to climate change, but was repeatedly obstructed by Congress in these efforts. Starting in 2017, the Trump administration brought an aggressive backlash against environmental policy on multiple fronts.

To address climate change, Obama had the EPA develop new regulations to improve fuel efficiency for vehicles and to restrict emissions from coal-fired power plants. These regulations followed a 2007 Supreme Court ruling that the EPA had authority under the Clean Air Act to regulate carbon dioxide and other greenhouse gases as pollutants. Obama's Clean Power Plan (p. 513) aimed to cut coal emissions significantly and was projected to save 3600 lives and bring public benefits worth $54 billion per year. In 2016, however, the Supreme Court halted the program with a 5-to-4 vote, and once Trump came to office, his administration replaced the plan with one that encouraged coal combustion.

Currently in the United States, legal protections for public health and environmental quality remain strong in some areas but have eroded in others. Past policies restricting toxic substances such as lead and DDT have improved public health nationwide, but scientists and regulators cannot keep up with the flood of new chemicals being introduced by industry. As political partisanship and cultural divides intensify between "red" states and "blue" states and urban and rural regions, some states and cities have tried to take up the slack and pursue strong environmental policies, while others have applauded the federal government's retreat. And as climate change concerns move to the fore, people continue to experience impacts from fossil fuel use and extraction while striving to find a path toward clean and renewable energy.

Today's policy debates center on climate change and equity concerns

Currently, a fourth wave of environmental policy is struggling to come to fruition. Today's concerns center on climate change and environmental justice. Climate change is at the forefront because our planet is warming rapidly due to emissions from fossil fuel use and people are now feeling the impacts in their own lives. Beyond being a scientific or economic issue, climate change is increasingly viewed as a moral issue: Young people, it seems, are being asked to make sacrifices in the future to pay for the conveniences of older generations in power now. Climate justice lawsuits like *Juliana v. United States* are giving explicit voice to this moral dimension. So are the growing number of student-led climate strikes across the world, sparked by the young activist Greta Thunberg (p. 510) and the Fridays for the Future movement (**FIGURE 7.9**).

Environmental justice concerns are in the spotlight because our society has increasingly recognized inequalities among groups of people and injustices that often stem from racial, ethnic, or gender identity. Modern currents of social justice—such as debates over immigration, the #MeToo movement, and the Black Lives Matter movement—have

(a) Marching in support of *Juliana v. United States*

(b) Crowds in Hamburg, Germany, one of thousands of cities where people participated in the Global Climate Strike on September 20, 2019

FIGURE 7.9 A surge in youth-led activism is pressuring policymakers to respond to climate change.

shaped our societal discourse and are intersecting with issues of health and the environment. It is increasingly recognized that lower-income people and underserved communities are the ones to suffer first and most from pollution, climate change, and other impacts. The movement for environmental justice (p. 139) has for decades coexisted in parallel with a "mainstream" environmental movement largely supported by white upper-class proponents in an uneasy relationship and often without clear communication. In recent years, however, concerns of these two movements have begun to merge, and social justice and equity have become common goals.

Concerns over climate change and environmental justice came together in the **Green New Deal** proposed in 2019 by U.S. Representative Alexandria Ocasio-Cortez of New York and U.S. Senator Edward Markey of Massachusetts. This non-binding congressional resolution, intended not as legislation but as an aspirational blueprint, put forth ambitious 10-year goals of converting the U.S. power grid to 100% renewable sources, reengineering automotive technology, and reducing greenhouse gas emissions "as much as is technologically feasible." It also set out social justice goals such as reducing poverty with high-paying green energy jobs and addressing the historic oppression of marginalized communities.

Critics were quick to deride the proposal as being unrealistic and expensive. Supporters defended the need for far-reaching goals, even if not all elements would come to pass. Many people debated whether trying to meet a progressive wish list of social justice goals would hinder the achievement of the original idea of a Green New Deal as first proposed by *New York Times* columnist Thomas Friedman in 2007. Friedman's vision was a program to address climate change, while boosting employment and wealth, by converting the economy to clean renewable energy. The idea was modeled on President Franklin Roosevelt's New Deal to create jobs, fight poverty, and build infrastructure during the Great Depression in the 1930s.

Environmental policy advances today on the international stage

Amid the heightened partisanship of current U.S. politics, environmental policy has gotten caught in the political crosshairs. Environmental issues have come to be identified as a predominantly Democratic concern even though some of the greatest conservationists have been Republicans. Indeed, the conservation of environmental amenities should logically be a core value of conservative Republicans given that the words *conservative* and *conservation* share the same root and original meaning. Unfortunately, as a result of current ideological gridlock, significant bipartisan advances now rarely occur, and U.S. leadership in environmental policy internationally has waned.

In the meantime, many other nations are forging ahead with environmental policy. Germany has used policy to make impressive strides with solar energy (p. 589). Sweden maintains a thriving society while promoting progressive environmental policies (p. 561). Costa Rica is bettering its citizens' lives while protecting and restoring its natural capital (p. 133).

Even China, despite becoming the world's biggest polluter, is also taking the world's biggest steps toward renewable energy, reforestation, and pollution control (p. 518).

Worldwide, we have embarked on a fourth wave of environmental policy, defined by two main goals. One is to develop solutions to climate change (Chapter 18), an issue of unprecedented breadth and global reach. The second goal is to achieve sustainable development (p. 155), finding ways to safeguard natural systems while raising living standards for the world's people. International conferences (pp. 156, 516) have been bringing together representatives of the world's nations to grapple with each issue.

International Environmental Policy

Environmental systems pay no heed to political boundaries, and neither do environmental problems. Consider climate change: It is a global issue because carbon pollution from any one nation spreads through the atmosphere and oceans, affecting all nations. Because one nation's laws have no authority in other nations, international policy is vital to solving transboundary problems. However, international law is often more nebulous in its origins and weaker in its authority than national law. As environmental scientist Hilary French has noted, "The world is still composed of nation-states that view themselves as sovereign, meaning that international law has the force of moral suasion, but few if any real teeth."

Globalization makes international institutions vital

We live in an era of rapid and profound change. **Globalization** describes the process by which the world's societies have become more interconnected, linked by trade, diplomacy, and communication technologies in countless ways. Globalization has brought us many benefits by facilitating the spread of ideas and technologies that empower individuals and enhance our lives. Billions of people enjoy a degree of access to news, education, arts, and science that we could barely have imagined in the past, and billions also now live under governments that are more democratic. Our increasingly rich awareness of other cultures has promoted peace and understanding, and warfare has declined.

Yet as globalization proceeds, many people also have come to feel anxious about changing social conditions and threatened by the loss of traditional cultural norms. As people, goods, and ideas flow freely across national borders, debates have flared over trade, jobs, immigration, and identity. Anxieties over change have fueled populist and nationalist reactions against globalization, undercutting social and political stability in North America and Europe.

At the same time, ecological systems are undergoing change at unprecedented rates and scales. People are moving organisms from one continent to another, enabling invasive species to affect ecosystems everywhere. Multinational

TABLE 7.3 Major International Environmental Treaties

CONVENTION OR PROTOCOL	YEAR IT CAME INTO FORCE	NATIONS THAT HAVE RATIFIED IT	STATUS IN UNITED STATES
CITES: Convention on International Trade in Endangered Species of Wild Fauna and Flora (p. 297)	1975	175	Ratified
Ramsar Convention on Wetlands of International Importance	1975	159	Ratified
Montreal Protocol, of the Vienna Convention for the Protection of the Ozone Layer (p. 471)	1989	196	Ratified
Basel Convention on the Control of Transboundary Movements of Hazardous Wastes and Their Disposal (p. 637)	1992	172	Signed but not ratified
Convention on Biological Diversity (p. 297)	1993	168	Signed but not ratified
Stockholm Convention on Persistent Organic Pollutants (p. 384)	2004	152	Signed but not ratified
Paris Agreement, of the UN Framework Convention on Climate Change (p. 516)	2016	181	Signed but not ratified; U.S. plans to withdraw

corporations operate outside the reach of national laws and rarely have incentive to conserve resources or limit pollution while moving from nation to nation. Today our biggest environmental challenges are global in scale (such as climate change, biodiversity loss, ozone depletion, overfishing, and plastic and chemical pollution), yet we lack an adequate global legal framework to address these issues effectively. For all these reasons, in our globalizing world, the institutions that do shape international law and policy play increasingly vital roles.

International law includes customary law and conventional law

International law known as **customary law** arises from long-standing practices, or customs, held in common by most cultures. In contrast, international law known as **conventional law** arises from *conventions,* or *treaties* (written contracts among nations), into which nations enter. One example of a treaty is the United Nations Framework Convention on Climate Change, which in 1994 established a framework for agreements to reduce greenhouse gas emissions that contribute to global climate change. The Kyoto Protocol (a *protocol* is an amendment or addition to a convention) and the Paris Agreement each later specified the agreed-upon details of the emissions limits (pp. 515–516). **TABLE 7.3** shows a selection of major environmental treaties.

Treaties such as the Paris Agreement and the Montreal Protocol (see **SUCCESS STORY**) are truly global in scope and have been ratified (legally approved by a government) by most of the world's nations. However, treaties are also signed among pairs or groups of nations. The United

States, Mexico, and Canada entered into the North American Free Trade Agreement (NAFTA) in 1994 (**FIGURE 7.10**). NAFTA eliminated trade barriers such as tariffs on imports and exports, making goods cheaper to buy. Yet NAFTA also threatened to undermine protections for workers and the environment by steering economic activity to areas where regulations were most lax. Side agreements were negotiated to try to address these concerns, and NAFTA's impacts on jobs and on environmental quality in the three nations have been complex. Many U.S. jobs moved to Mexico, but fears that pollution would soar and regulations would be gutted largely did not come to pass—indeed, some sustainable products and practices spread from nation to nation. In 2018, the Trump administration forced the renegotiation of this treaty, and minor adjustments were agreed to; future

FIGURE 7.10 The North American Free Trade Agreement (NAFTA) eliminated trade barriers to make goods cheaper. However, some U.S. manufacturing jobs moved to Mexico (such as to this garment factory in Tehuacan), where wages are lower and health and environmental regulations are more lax than in the United States.

Using a Treaty to Halt Ozone Depletion

In the 1980s, the world was shocked to learn of a global threat that seemed to come from out of nowhere. Scientific monitoring revealed that ozone in Earth's upper atmosphere was being lost rapidly, stripping our planet of a compound that protects life from dangerous ultraviolet radiation from the sun. Researchers discovered that artificially manufactured chemicals such as chlorofluorocarbons (CFCs)—widely used in aerosol spray cans, as refrigerants, and for other purposes—were destroying ozone once they were released into the atmosphere. Policymakers and industry responded quickly to the science, and in 1987 the world's nations signed the Montreal Protocol, a treaty to cut CFC production in half. Follow-up treaties deepened the cuts and restricted other ozone-depleting substances. Industry successfully switched to alternative chemicals for products. As a result, we stopped atmospheric ozone loss in the 1990s, and Earth's ozone layer is on track to fully recover later this century. Thus, by harnessing international cooperation on policy, humanity succeeded in resolving a major global problem. (Read the full story in Chapter 17, pp. 470–474.) Today, we need the world's nations to act together to tackle other major global problems, such as climate change and biodiversity loss. The Montreal Protocol provides hope that such efforts can succeed.

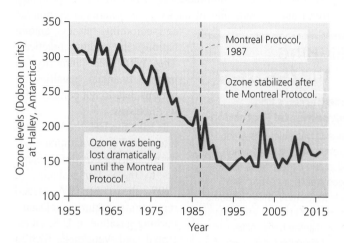

Data from an Antarctic research station revealed the loss of ozone in the stratosphere—and have shown its stabilization following treaties to address the problem.

→ Explore the Data at Mastering Environmental Science

---

consequences for jobs, the economy, and the environment in the three nations will likely be minimal but remain to be seen. With every proposed trade agreement, deliberations recur as negotiators seek ways to gain the benefits of free trade while avoiding environmental damage and economic harm to working people.

Several organizations shape international environmental policy

In our age of globalization, a number of international institutions influence the policy and behavior of nations by providing funding, applying political or economic pressure, and directing media attention.

The United Nations Founded in 1945 and including representatives from virtually all countries of the world, the **United Nations** (UN) seeks to maintain peace, security, and friendly relations among nations; to promote respect for human rights and freedoms; and to help nations cooperate to resolve global economic, social, cultural, and humanitarian challenges. Headquartered in New York City, the United Nations plays an active role in shaping international environmental policy by sponsoring conferences, coordinating treaties, and publishing research. One agency within it, the United Nations Environment Programme (UNEP), promotes sustainability with research and outreach activities that provide information to policymakers and scientists.

The World Bank Established in 1944 and based in Washington, D.C., the **World Bank** is one of the biggest sources of funding for large-scale economic development ventures such as dams, irrigation, and other major infrastructure projects. In fiscal year 2018, the World Bank committed $67 billion in loans and support for projects designed to benefit low-income people in developing countries.

Despite its admirable mission, the World Bank is often criticized for funding unsustainable projects that cause environmental impacts, such as dams that flood valuable forests and farmland to provide electricity. Providing for the needs of growing human populations in poor nations while minimizing damage to the ecological systems on which people depend can be a tough balancing act. Environmental scientists agree that the concept of sustainable development must be the guiding principle for such efforts.

The European Union The **European Union** (EU) seeks to promote Europe's unity and its economic and social progress and to "assert Europe's role in the world." The EU can sign binding treaties on behalf of its member nations and can enact regulations that have the same authority as national laws. The EU's European Environment Agency addresses waste management, noise pollution, water pollution, air pollution, habitat degradation, and natural hazards. The EU also seeks to remove trade barriers among member nations. It has classified some nations' environmental regulations as barriers to trade, arguing that the stricter environmental laws of some northern European nations limit the import and sale of environmentally harmful products from other member nations.

The World Trade Organization Founded in 1995 and based in Geneva, Switzerland, the **World Trade Organization** (WTO) represents multinational corporations. It promotes free trade by reducing obstacles to international commerce and enforcing fairness among nations in trading practices. The WTO has authority to impose financial penalties on nations that do not comply with its directives.

Like the EU, the WTO has interpreted some national environmental laws as unfair barriers to trade. For instance, in 1995, the U.S. EPA issued regulations requiring cleaner-burning gasoline in U.S. cities. Brazil and Venezuela filed a complaint with the WTO, claiming that the new rules discriminated against the petroleum they exported to the United States, which did not burn as cleanly. The WTO agreed, ruling that even though the dirtier-burning South American gasoline posed a threat to human health in the United States, the EPA's rules were an illegal trade barrier. The ruling forced the United States to weaken its regulations.

Nongovernmental organizations A great number of **nongovernmental organizations** (NGOs)—nonprofit and mission-driven organizations not overseen by any government—have become international in scope and exert influence over policy. Those that advocate for aspects of environmental protection are known as environmental NGOs (ENGOs). One example is Urgenda, the NGO that successfully sued the Dutch government to force stronger climate change policies. Some internationally active environmental NGOs, such as the Nature Conservancy, focus on conservation objectives on the ground (such as purchasing and managing land and habitat for rare species) without becoming politically involved. Other such groups, such as Greenpeace, Conservation International, and Population Connection, attempt to shape policy through research, education, lobbying, or protest.

Approaches to Environmental Policy

When most of us think of environmental policy, what comes to mind are major laws, such as the Clean Air Act or the Clean Water Act—or government regulations, such as those limiting what pollutants can be emitted into the air or dumped into a water supply. However, environmental policy is diverse.

WEIGHING the issues

Trade Barriers and Environmental Protection

If Canada has stricter laws for environmental protection than Mexico and if these laws restrict Mexico's ability to export its goods to Canada, then Canada's laws, by WTO policy, could be overruled in the name of free trade. Do you think this policy is fair? Now consider that Canada is wealthier than Mexico and that Mexico could use an economic boost. Does that perspective affect your response?

Policy can follow three approaches

Environmental policy can use a variety of strategies within three major approaches (**FIGURE 7.11**).

Lawsuits in the courts Prior to the legislative push of the late-20th century, most environmental policy questions in the United States were addressed with lawsuits in the courts. Individuals suffering external costs from pollution would sue polluters, one case at a time. The courts sometimes punished polluters by ordering them to stop their operations or pay damages to the affected parties. However, as industrialization proceeded and population grew, pollution became harder to avoid and judges became reluctant to hinder industry.

In 1970, in the case *Boomer v. Atlantic Cement Company,* the U.S. Supreme Court acknowledged that residents of Albany, New York, were suffering pollution from a nearby cement plant, yet the justices allowed the plant to continue operating once it paid financial compensation to the residents. To many people, rulings like these showed that lawsuits alone were no longer a viable avenue for preventing pollution. People began to view legislation and regulation as more effective means of protecting public health and safety.

Today, after decades of attempts by Congress and presidential administrations to weaken and roll back laws and regulations to protect health and environmental quality, the pendulum may be swinging back toward lawsuits. The growing wave of climate litigation, led by *Juliana v. United States,* is trying to achieve policy results through the judiciary that have been hard to achieve through the legislative and executive branches.

Command-and-control policy Most environmental laws and regulations of recent decades use a **command-and-control** approach, in which a regulating agency such as the EPA sets rules, standards, or limits on certain actions and threatens punishment for violations. This simple and direct approach to policymaking has brought residents of the United States and other nations cleaner air, cleaner water, safer workplaces, and many other advances. The relatively safe, healthy, comfortable lives most of us enjoy today owe much to the command-and-control environmental policy of the past several decades.

Even in plain financial terms, putting health and quality of life aside, command-and-control policy has been effective. Each year the White House Office of Management and Budget analyzes U.S. policy to calculate the economic costs and benefits of regulations. These analyses have consistently revealed that benefits far outweigh costs and that environmental regulations have been most beneficial of all. You can explore some of these data in *Calculating Ecological Footprints* (p. 184).

Economic policy tools Despite the successes of command-and-control policy, many people view top-down government mandates to be restrictive. Moreover, whereas command-and-control policy dictates particular solutions to problems, many people believe that private entities competing and innovating in a free market often produce new or better solutions at lower cost. As a result, political scientists, economists, and policymakers have developed policy approaches that aim to channel the innovation and economic efficiency of market capitalism in

PROBLEM
Pollution from factory harms people's health

FIGURE 7.11 Three major policy approaches exist to resolve environmental problems. To address pollution from a factory, we might ➊ seek damages through lawsuits, ➋ limit pollution through legislation and regulation, or ➌ reduce pollution using market-based strategies.

SOLUTIONS
Three policy approaches

➊ **Lawsuits in the courts:** People can sue factories.

➋ **Command-and-control policy:** Governments can regulate emissions.

➌ **Economic policy tools:** Policy can create incentives; a factory that pollutes less (right) will outcompete one that pollutes more (left) through permit trading, avoiding green taxes, or selling ecolabeled products.

directions that benefit the public. Such economic policy tools use financial incentives to encourage desired outcomes, discourage undesired outcomes, and set market dynamics in motion to achieve goals in an economically efficient manner.

All three of these approaches aim to "internalize" external costs suffered by the public by building these costs into market prices. Each approach has strengths and weaknesses, and each is best suited to different conditions. The approaches may also be used together. For instance, government regulation is often needed to frame market-based efforts, and citizens can use the courts to ensure that regulations are enforced. Let's now explore several types of economic policy tools: taxes, subsidies, and emissions trading.

Green taxes discourage undesirable activities

In taxation, money passes from private parties to the government, which uses it to pay for services that benefit the public. Taxing undesirable activities helps internalize external costs by making these costs part of the normal expense of doing business. A tax on an environmentally harmful activity or product is called a **green tax.** When a business pays a green tax, it is essentially reimbursing the public for environmental damage it causes.

Under green taxation, a firm owning a polluting factory might pay taxes on the pollution it discharges—the more pollution, the higher the tax payment. This approach gives firms a financial incentive to reduce pollution while allowing them the freedom to decide how to do so. One polluter might choose to invest in pollution control technology if doing so is more affordable than paying the tax. Another polluter might instead choose to pay the tax—funds the government could use to reduce pollution in some other way.

Green taxes are common in Europe, where many nations have adopted the polluter-pays principle (p. 165). In the United States, similar "sin taxes" on cigarettes and alcohol are long-accepted tools of U.S. social policy. Today, many nations and states are experimenting with carbon taxes—taxes on gasoline, coal-based electricity, and fossil-fuel-intensive products, according to the carbon emissions they produce—to fight climate change (p. 513).

Green taxation provides incentive to reduce pollution not merely to some level specified in a regulation, but to still lower levels. However, businesses generally pass on their tax expenses to consumers, and these increased costs can affect low-income consumers more than high-income ones.

Subsidies promote targeted activities

Another economic policy tool is the **subsidy,** a government giveaway of money or resources that is intended to support or promote an industry or activity. Subsidies take many forms, and one is the *tax break,* which relieves the tax burden on an industry, firm, or individual. Because a tax break deprives the government's treasury of funds it would otherwise collect, it has the same financial effect as a direct giveaway of money.

Subsidies can promote environmentally sustainable activities. One example is the Conservation Reserve program that pays U.S. farmers to conserve areas of natural vegetation around active farmland (p. 236). Others include Costa Rica's payments for ecological services (pp. 133–134) and Germany's feed-in tariffs for solar power (pp. 589–591). However, subsidies often prop up activities that result in environmental impacts. The U.S. Forest Service spends millions of dollars of taxpayer money each year building roads in national forests to allow private timber corporations access to cut trees, which the companies then sell at a profit (**FIGURE 7.12**). Under the General Mining Act of 1872 (p. 657), mining companies extract $500 million to $1 billion in minerals from public lands in the United States each year without paying a penny in royalties to the taxpayers who own these lands. Since this law was enacted, the U.S. government has given away nearly $250 billion of mineral resources, and mining activities have polluted more than 40% of watersheds in the West. The 150-year-old law still allows mining companies to buy public lands for $5 or less per acre.

Fossil fuel industries have been major recipients of subsidies over the years. In the United States from 1950 to 2010, the federal government gave $594 billion of its citizens' money to oil, gas, and coal corporations (most of it in tax breaks), according to one recent compilation (**FIGURE 7.13a**). In comparison, just $171 billion was granted to renewable energy, and most of these subsidies went to hydropower and to corn ethanol, which is not widely viewed as a sustainable

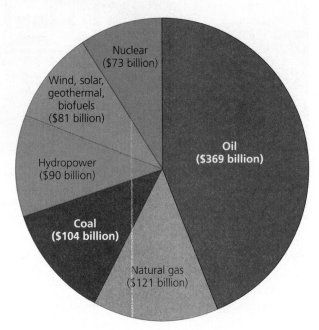

(a) Cumulative U.S. energy subsidies, 1950–2010

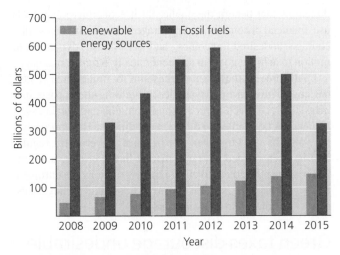

(b) Yearly global energy subsidies, 2008–2015

FIGURE 7.13 The well-established fossil fuel industries receive the majority of energy subsidies. In **(a),** cumulative data for the United States from 1950 to 2010 are shown. In **(b),** global year-by-year data from 2008 to 2015 are shown. *Data from (a) Management Information Services, Inc., 2011.* 60 years of energy incentives: Analysis of federal expenditures for energy development. *Washington, D.C.: Management Information Services; and (b) International Energy Agency, 2016.* World energy outlook 2016. *Paris: IEA.*

 Go to **Interpreting Graphs & Data** on Mastering Environmental Science

FIGURE 7.12 When companies extract timber from U.S. national forests, taxpayers pay the costs of building and maintaining access roads. "By absorbing these costs, the federal government shields the timber industry from the true cost of doing business," the nonprofit group Taxpayers for Common Sense has concluded.

fuel (p. 578). Today, U.S. subsidies are more evenly spread among energy sources, but nonrenewable sources still receive more than renewable sources. In 2016, the most recent year with full data, 54% of $15 billion in subsidies went to nonrenewable sources.

Internationally in recent years, fossil fuel subsidies have outpaced renewable energy subsidies by 2 to 10 times, according to the International Energy Agency (FIGURE 7.13b). In 2009, President Obama and other leaders of the Group of 20 (G-20) nations resolved to gradually phase out their collective $300 billion of annual fossil fuel subsidies. Doing so would hasten a shift to cleaner renewable energy sources and accomplish half the greenhouse gas emissions cuts needed to hold global warming to 2°C. However, since that time, fossil fuel subsidies have not fallen. One reason is successful lobbying by fossil fuel corporations. Another is that consumers have grown used to artificially low prices for gasoline and electricity and might punish policymakers who let these prices rise by lifting the subsidies.

The continuing shower of subsidies lavished on fossil fuel industries is a major motivation for climate litigation like *Juliana v. United States*. These subsidies also strengthen the litigants' arguments that government policy is not simply failing to rein in climate change but, rather, is actively intensifying climate change. Economists today are trying to design politically palatable means of shifting subsidies from fossil fuels to renewable energy sources or toward rebates to taxpayers, and court verdicts could potentially help facilitate such shifts.

Emissions trading uses markets

In a system of **emissions trading,** a government creates a market in permits for the emission of pollutants, and companies, utilities, or industries then buy and sell the permits among themselves. In a **cap-and-trade** emissions trading system, the government first caps the overall amount of pollution it will allow and then grants or auctions off permits to polluters that allow each polluter to emit a certain fraction of that amount. As polluters trade these permits, the government progressively lowers the cap of overall emissions allowed (see Figure 18.33, p. 515).

Suppose you own an industrial plant with permits to release 10 units of pollution, but you find that you can make your plant more efficient and release only 5 units instead. You now have a surplus of permits, which you can sell to some other owner who needs them. Doing so would generate income for you and meet the needs of the other plant, while the total amount of pollution would stay the same. By providing firms an economic incentive to reduce pollution, emissions trading can lower expenses for both industry and the public relative to a conventional regulatory system.

The United States pioneered the cap-and-trade approach with its program to reduce sulfur dioxide emissions, established by the 1990 amendments to the Clean Air Act (p. 458). Through this program, sulfur dioxide emissions from coal-burning power plants and factories declined by 67%, acid rain was reduced, and air quality improved (see Figures 17.25 and 17.26, pp. 476 and 477). The 67% reduction in pollution was greater than the amount actually required by the legislation, offering evidence that cap-and-trade systems can sometimes cut pollution more effectively than command-and-control regulation. Moreover, the cuts were attained at much less cost than was predicted and with no apparent effect on electricity supply or economic growth. Similar cap-and-trade programs have shown success with smog in the Los Angeles basin and with nitrogen oxides emitted by factories and power plants in northeastern states.

Although cap-and-trade programs can reduce pollution, they do allow hotspots of pollution to occur around plants that buy permits to pollute more. Moreover, large firms can hoard permits, deterring smaller new firms from entering the market. Nonetheless, emissions trading shows promise for safeguarding environmental quality while granting industries the flexibility to lessen their impacts in ways that are economically palatable.

To address climate change, European nations are operating a market in greenhouse gas emissions (p. 514). In the United States, carbon-trading markets are running in California and among northeastern states (p. 515). The U.S. House of Representatives passed a bill to establish a nationwide carbon-trading market in 2009, but the Senate did not, so the measure died. Today, a large number of states and nations are experimenting with carbon-trading markets and various forms of carbon taxes. (We will assess some of these efforts in Chapter 18.) If climate litigation such as *Juliana v. United States* succeeds, the United States at the federal level may be forced to consider such steps.

Market incentives are diverse at the local level

You are likely already taking part in transactions involving financial incentives as policy tools. Many municipalities charge residents for waste disposal according to the amount of waste they generate. Some cities place taxes or disposal fees on items whose safe disposal is costly, such as tires and motor oil. Others give rebates to residents who buy water-efficient toilets and appliances because rebates can cost a city less than upgrading its wastewater treatment system. Likewise, power utilities may offer discounts to customers who buy high-efficiency appliances because doing so is less costly than expanding the generating capacity of their plants.

The creative use of economic policy tools is growing at all levels, while command-and-control regulation and legal action in the courts continue to play vital roles in environmental policymaking. As a result, we have a variety of effective policy strategies available to us as we seek sustainable solutions to our society's challenges.

TODAY, command-and-control legislation and regulation remain the most common approaches to policymaking, but market-based policy tools are increasingly being used, and lawsuits in the courts remain influential. Court actions may soon become pivotal in the question of whether the world's nations will develop policies to reduce greenhouse gas emissions and reverse climate change, the preeminent crisis of our age. *Juliana v. United States* is at the leading edge of a wave of climate litigation. "We want to be the catalysts," Kelsey Juliana has said, "in shifting ideals and priorities around the world."

Indeed, as of 2019, there were 26 legal cases worldwide using the public trust doctrine to try to force policymakers to take tougher action on emissions, according to the Sabin Center for Climate Change Law, a think tank based at Columbia University Law School that tracks climate change litigation. Some of these legal cases have been unsuccessful, many others are in progress, and a few are meeting with success. In the United States, Our Children's Trust is assisting or has assisted with lawsuits by young people against state governments in Alaska, Arizona, Colorado, Florida, Iowa, Kansas, Massachusetts, New Mexico, North Carolina, Oregon, Pennsylvania, Texas, and Washington.

Lawsuits in other nations are on similar paths. In Colombia in 2018, a court ruled in favor of 25 youth and ordered their government to address climate change by halting rainforest destruction. In the Netherlands, the successful suit by Urgenda and 886 citizens is forcing the government to revisit its emissions policies. In India and in Pakistan, youth plaintiffs are seeking to force their governments away from reliance on coal, the fossil fuel with the greatest impact on climate.

Meanwhile, an additional 1300 legal cases worldwide (most in the United States) are attempting to tackle climate change in other ways. One way is by making financial claims of corporations in the fossil fuel industry. For instance, the cities of Oakland and San Francisco have taken five oil companies to court, seeking financial damages to help pay for costs the cities will incur as a result of sea level rise. In Peru, a farmer from a town threatened by flooding from a melting glacier is suing a German power company in a court in Germany for financial compensation equal to the percentage that the company's emissions have contributed to global warming. Such cases use as a model the litigation by U.S. attorneys general against tobacco companies in the 1990s that forced those companies to pay more than $200 billion toward anti-smoking initiatives and health care costs related to smoking.

The outcomes of all these climate cases remain to be seen, as do any specific effects they may have on climate change policy in particular nations and across the globe. Simply that they are progressing through the judicial system in so many states and nations, however, is in itself putting pressure on companies and governments to rethink their priorities and try to find ways to reduce their emissions. By opening a new front in the battle against climate change and by highlighting the moral dimensions of climate change, legal cases like those brought by the 21 youth in *Juliana v. United States* are showing promise to help us find political solutions in the near future.

- **CASE STUDY SOLUTIONS** Put yourself in the seat of the judge deciding the *Juliana v. United States* case. What questions would you ask the youth plaintiffs? What questions would you ask the government attorneys? What evidence would you want to see to determine whether the youths' futures were being harmed by climate change? How do you expect you might rule in the case? More broadly, because climate change results not just from the mining and production of fossil fuels and from government policy, but also from everyday people's voluntary use of these fuels, what obligations, if any, do you believe the government should have in addressing emissions and climate change? Should questions about climate change policy be addressed partly in the judicial system or solely by Congress and the president? How best can we as a society ensure that individual people's voices are heard when it comes to climate change?

- **LOCAL CONNECTIONS** What policies or actions have the policymakers in your city, region, or state considered to address greenhouse gas emissions and climate change? Has your state set any emissions targets or enacted any renewable energy mandates? Are there regulations for energy efficiency in buildings or appliances in your state? Give at least one example of a climate change–related policy that has been enacted at your local, regional, or state level and one example of a policy that has failed to be enacted. What results have come about from any enacted policies? What steps would you like to see your state take in terms of policy relating to climate change?

- **EXPLORE THE DATA** What impacts should a typical 20-year-old in the United States expect to experience from climate change during his or her lifetime? ➜ **Explore Data** relating to the case study on **Mastering Environmental Science.**

REVIEWING Objectives

You should now be able to:

+ **Describe environmental policy and discuss its societal context**

Environmental policy aims to protect natural resources and amenities from degradation or depletion and to promote equitable treatment of people. It addresses the tragedy of the commons, free riders, and external costs. In a democracy, everyone has a voice, yet wealth tends to buy influence. (pp. 163–167)

+ **Explain the role of science in policymaking**

Data from scientific research are vital for informing and creating effective policy. However, policymakers sometimes ignore or distort science for ideological reasons or political ends, so we in the public need to remain vigilant. (pp. 167–168)

+ **Discuss the history of U.S. environmental policy, and summarize major U.S. environmental laws**

The legislative, executive, and judicial branches, along with administrative agencies, all play roles in U.S. environmental policy. State and local governments also create a great deal of environmental policy. Federal environmental policy came in three waves. The first promoted frontier expansion and resource extraction. The second aimed to mitigate impacts of the first through conservation. The third targeted pollution and gave us much of today's environmental policy architecture: NEPA,

the EPA, the EIS process, and major laws such as the Clean Air Act, Clean Water Act, and Endangered Species Act. Today, a fourth wave of environmental policy is building around climate change and environmental justice (and internationally around sustainable development). (pp. 168–175)

+ **List institutions that influence international environmental policy, and describe how nations handle transboundary issues**

In our globalizing world, many environmental issues cross political boundaries. International policy includes customary law and conventional law. The United Nations, World Bank, European Union, World Trade Organization, and nongovernmental organizations all help shape international policy. (pp. 175–178)

+ **Compare and contrast the different approaches to environmental policymaking**

Lawsuits have provided a traditional approach to resolving environmental disputes, but legislation from Congress and regulations from administrative agencies make up most U.S. federal policy today. These top-down approaches are referred to as command-and-control laws and regulations. Economic policy tools include green taxes to discourage undesirable activities, subsidies to encourage targeted activities, and market-based approaches such as emissions trading that harness market dynamics to achieve policy goals. (pp. 178–181)

SEEKING Solutions

1. Reflect on causes for the transitions in U.S. history from one type of environmental policy to another. Now peer into the future, and think about how life and society might be different in 25, 50, or 100 years. What would you predict about the environmental policy of the future, and why? What issues might future policy address? Do you predict we will have more or less environmental policy?

2. Compare the roles of the United Nations, the World Bank, the European Union, the World Trade Organization, and nongovernmental organizations. If you could gain the support of just one of these institutions for a policy you favored, which would you choose? Why?

3. Consider the main approaches to environmental policy—litigation, command-and-control laws and regulations, and economic policy tools. Describe an advantage and a disadvantage of each. Do you think

any one approach is most effective? Could we do with just one approach, or does it help to have more?

4. Think of one environmental problem you would like to see solved. From what you've learned in this chapter about policymaking, describe how you think policy could best be created to address this problem.

5. **THINK IT THROUGH** There are many ways to affect political change. You could run for office and help create policy as an elected official. You could serve in a public agency, crafting regulations or managing the implementation of policy. You could become a lawyer or judge and influence policy through the judicial system. You could become a journalist and write about policy and how it affects people's lives. Of course, as a private individual, you can testify at public hearings, attend protests, and write and call decision-makers. What other ways can you think of that you could influence policy? Of all these ways, which appeals to you most, and which do you think you would be best at? Explain why. Describe one pathway you might like to take to make your voice count in the political arena.

CALCULATING Ecological Footprints

Critics of command-and-control policy often argue that regulations are costly to business and industry, yet cost-benefit analyses (p. 142) have repeatedly shown that regulations bring Americans more benefits than costs. Each year, the U.S. Office of Management and Budget assesses costs and benefits of major federal regulations issued by administrative agencies. Results from the most recent report, covering the decade from 2006 to 2016, are shown below. This decade includes periods of both Republican and Democratic control of the presidency and of Congress. Subtract costs from benefits, and enter these values for each agency in the third column. Divide benefits by costs, and enter these values in the fourth column.

Costs and Benefits of Major U.S. Federal Regulations, 2006–2016
(average values from ranges of estimates, in billions of 2015 dollars)

AGENCY	BENEFITS	COSTS	BENEFITS MINUS COSTS	BENEFIT: COST RATIO
Department of Energy	29.8	9.9	19.9	3.0
Department of Health and Human Services	18.8	4.1		
Department of Transportation	31.6	12.1		
Environmental Protection Agency (EPA)	450.8	59.5		
Other departments	85.2	22.6		
Total	**616.2**	**108.2**		

Data from U.S. Office of Management and Budget, 2017. *2017 draft report to Congress on the benefits and costs of federal regulations and agency compliance with the Unfunded Mandates Reform Act.* Washington, D.C.: OMB.

1. For how many of the agencies shown do regulations exert more costs than benefits? For how many do regulations provide more benefits than costs?

2. Which agency's regulations have the greatest excess of benefits over costs? Which agency's regulations have the greatest ratio of benefits to costs?

3. What percentage of total benefits from regulations comes from EPA regulations? Most of the benefits and costs from EPA regulations are from air pollution rules resulting from the Clean Air Act and its amendments. Judging solely by these data, would you say that Clean Air Act legislation has been a success or a failure for U.S. citizens? Explain your answer.

Mastering Environmental Science

Students Go to **Mastering Environmental Science** for assignments, an interactive e-text, and the Study Area with practice tests, videos, and activities.

Instructors Go to **Mastering Environmental Science** for automatically graded activities, videos, and reading questions that you can assign to your students, plus Instructor Resources.

Environmental Issues and the Search for Solutions

PART 2

CHAPTER

8

Human Population

case study

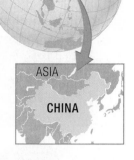

ASIA

CHINA

Will China's New "Two-Child Policy" Defuse Its Population "Time Bomb"?

> We don't need adjustments to the family-planning policy. What we need is a phaseout of the whole system.
>
> Gu Baochang, Chinese demographer at People's University, Beijing, referring to the nation's "one-child" policy in 2013

> As you improve health in a society, population growth goes down. Before I learned about it, I thought it was paradoxical.
>
> Bill Gates, Founder, Microsoft Corporation

The People's Republic of China is the world's most populous nation, home to one-fifth of the more than 7 billion people living on Earth. It is also the site of one of the most controversial social experiments in human history.

When Mao Zedong founded the country's current regime in 1949, roughly 540 million people lived in a mostly rural, war-torn, impoverished nation. Mao's policies encouraged population growth, and by 1970, improvements in food production, food distribution, and public health allowed China's population to swell to 790 million people. At that time, Chinese women gave birth to an average of 5.8 children in their lifetimes and China's population grew by 2.8% annually.

However, the country's burgeoning population and its industrial and agricultural development were eroding the nation's soils, depleting its water, and polluting its air. Realizing that the nation might not be able to continue to feed its people, Chinese leaders decided in 1970 to institute a population control program that prohibited most Chinese couples from having more than one child. The "one-child" program applied primarily to families in urban areas. Many farmers in rural areas were allowed more than one child, because success on the farm often depends on having multiple children. Members of ethnic minority groups were also excluded from the policy, as restrictions on reproduction could threaten the long-term survival of their relatively small populations.

The program encouraged people to marry later and have fewer children and increased accessibility to contraceptives and abortion. Families with only one child were rewarded with government jobs and better housing, medical care, and access to schools. Families with more than one child, meanwhile, were subjected to costly monetary fines, employment discrimination, and social scorn. The experiment was a success in slowing population growth: The nation's growth rate is now down to 0.5%, and Chinese women currently bear, on average, 1.6 children in their lifetimes.

However, the one-child policy also produced a population with a shrinking labor force, increasing numbers of older people, and too few women. These unintended consequences led some demographers to question whether

Upon completing this chapter, you will be able to:

+ Describe the scope of human population growth

+ Discuss divergent views on population growth

+ Explain how human population, affluence, and technology affect the environment

+ Explain the fundamentals of demography

+ Describe the concept of demographic transition

+ Explain how family planning, the status of women, and affluence affect population growth

◀ Crowded street in Shanghai, one of China's largest cities

▲ Will the two-child policy balance China's skewed sex ratio?

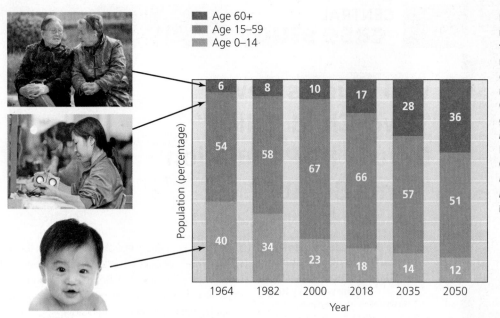

Age 60+
Age 15–59
Age 0–14

FIGURE 8.1 China's restrictive reproductive policies have led to a shrinking workforce and rising numbers of older citizens. Values in the figure represent the percentage of the Chinese population in each age group. *Figure from Population Reference Bureau, 2004.* China's Population: New Trends and Challenges. *Data for 2017–2050 from U.S. Census Bureau International Database,* www.census.gov/population/international/data/idb/.

China's one-child policy simply traded one population problem—overpopulation—for other population problems.

The rapid reduction in fertility that resulted from this policy drastically changed China's age structure (**FIGURE 8.1**). Once consisting predominantly of young people, China's population has shifted, such that the numbers of children and older people are now more equal. This means there will be relatively fewer workers for China's growing economy, which is driving up wages and encouraging companies with factories in China to seek out more inexpensive labor in other nations. The growing number of older Chinese citizens poses problems because the Chinese government lacks the resources to fully support them, putting a heavy economic burden on the millions of only children produced under the one-child policy as they help provide for their retired parents.

Modern China also has too few women. Chinese culture has traditionally valued sons because they carry on the family name, assist with farm labor in rural areas, and care for aging parents. Daughters, in contrast, will most likely marry and leave their parents, as the traditional culture dictates. Thus, when limited to having just one child, many Chinese couples preferred a son to a daughter. Tragically, this led in some instances to selective abortion and the killing of female infants. This has caused a highly unbalanced ratio of young men and women in China, leading to the social instability that arises when large numbers of young men are unable to find brides and remain longtime bachelors.

Until recently, Chinese authorities attempted to address this looming population "time bomb" of an aging population with skewed ratios of men and women by occasionally loosening the one-child policy. For example, the government announced in 2013 that if either member of a married couple is an only child, the couple would be allowed to have a second child—but only 1.5 million of the 11 million citizens eligible for this exemption

applied for it. Faced with the prospect of continued population issues, the Chinese government announced in October 2015 that the former one-child policy would immediately become a two-child policy, and couples would be permitted to have two children without penalty.

It is unclear, however, if Chinese couples, accustomed to the material wealth and urban lifestyle many enjoy, will embrace the opportunity to grow their families—and accept the costs of raising a second child—now that it is allowed. A 2008 survey conducted by China's family-planning commission, for example, reported that only 19% of the people they surveyed wished to have a second child if the one-child policy was relaxed. Birth rates in 2016 and 2017 were higher than in the previous five years, but only modestly so, suggesting that relatively few couples are choosing to have a second child. It therefore remains to be seen if China's relaxing of the one-child policy came too late to defuse the pending time bomb the nation may experience as China's population grays in the midst of its rapid industrialization.

China's reproductive policies have long elicited intense criticism worldwide from people who oppose government intrusion into personal reproductive choices, and such intrusion continues today, albeit with the allowance of a larger family size. The policy, however, has proven highly effective in slowing population growth rates and aiding the economic rise of modern China.

But China is not alone in dealing with population issues. Many of the world's poorer nations experience substantial population growth. Some of these nations are ill-equipped to handle such growth, and this leads to stresses on society, the environment, and people's well-being. As other nations become more crowded and seek to emulate China's economic growth, might their governments also feel forced to turn to drastic policies that restrict individual freedoms?

Our World at Nearly 8 Billion

In our world of more than 7.7 *billion* people, one of our greatest challenges in this century is finding ways to slow human population growth. Ideally, this would be accomplished without restrictive governmental policies, such as those used in China, but rather by establishing societal conditions for all people that lead them to desire having fewer children.

The human population is growing rapidly

Our global population grows by more than 80 million people each year. This means that we add 2.5 people to the planet *every second*. That's more than two additions to the global human population that require food, water, energy, and the host of other natural resources humans need to survive and prosper.

It took until after 1800, virtually all of human history, for our population to reach 1 billion (**FIGURE 8.2**). Yet by 1930 we had reached 2 billion and 3 billion in just 30 more years. Our population added its next billion in only 15 years, and it has taken just 12 years to add each of the next three installments of a billion people. It is expected that the globe will be home to nearly 10 billion people by 2050.

What accounts for our unprecedented population growth? Exponential growth—the increase in a quantity by a fixed percentage per unit time—accelerates the increase in population size over time, just as compound interest accrues in a savings account (p. 66). The reason for this pattern is that a fixed percentage of a small number makes for a small increase but that the same percentage of a large number produces a large increase. Thus, even if the growth *rate* remains steady, population *size* will increase by greater increments with each successive generation.

For much of the 20th century, the growth rate of the human population worldwide rose from year to year. This rate peaked

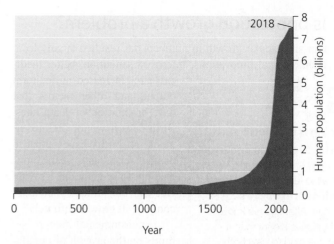

FIGURE 8.2 We have risen from fewer than 1 billion in 1800 to more than 7.5 billion today. *Data from U.S. Census Bureau.*

at 2.1% during the 1960s and has declined to 1.2% since then. Although 1.2% may sound small, a hypothetical population starting with one man and one woman that grows at 1.2% gives rise to a population of 112,695 after just 60 generations.

At a 1.2% annual growth rate, a population doubles in size in just 58 years. We can roughly estimate doubling times with a handy rule of thumb. Just take the number 70 and divide it by the annual percentage growth rate: $70/1.2 = 58.3$. China's current growth rate of 0.5% means it would take roughly 140 years ($70/0.5 = 140$) for its current population to double, but India's current growth rate of 1.5% predicts a doubling in only about 47 years ($70/1.5 = 46.7$). Had China not instituted its one-child policy and its growth rate remained at 2.8%, it would have taken only 25 years ($70/2.8 = 25$) for its population to double in size. Although the global population growth rate is 1.2%, rates vary greatly from region to region (**FIGURE 8.3**) and are typically higher in nations with developing economies.

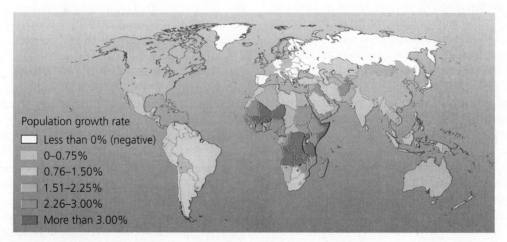

Population growth rate
- Less than 0% (negative)
- 0–0.75%
- 0.76–1.50%
- 1.51–2.25%
- 2.26–3.00%
- More than 3.00%

FIGURE 8.3 Population growth rates vary greatly from place to place. Populations are growing fastest in poorer nations, while populations are beginning to decrease in some highly industrialized nations. *Data from Population Reference Bureau, 2018 World population data sheet.*

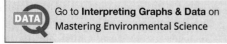

Go to **Interpreting Graphs & Data** on **Mastering Environmental Science**

Is population growth a problem?

Our spectacular growth in numbers has resulted largely from technological innovations, improved sanitation, better medical care, increased agricultural output, and other factors that have brought down death rates. Birth rates have not declined as much, so births have outpaced deaths for many years now, leading to population growth. But can the human population continue to grow indefinitely?

Environmental factors set limits on the growth of populations (p. 67), but environmental scientists who have tried to pin a number to the human carrying capacity (p. 67) have come up with wildly differing estimates. The most rigorous estimates range from 1–2 billion people living prosperously in a healthy environment to 33 billion people living in extreme poverty in a degraded world of intensive cultivation without natural areas.

The difficulty in estimating how many humans our planet can support is that we have repeatedly increased our carrying capacity by developing technology to overcome the natural limits on our population growth.

For example, British economist Thomas Malthus (1766–1834) argued in his influential work *An Essay on the Principle of Population* (1798) that if society did not reduce its birth rate, then rising death rates would reduce the population through war, disease, and starvation. Although his contention was reasonable at the time, agricultural improvements in the 19th century increased food supplies, and his prediction did not come to pass. Similarly, biologist Paul Ehrlich predicted in his 1968 book *The Population Bomb* that human population growth would soon outpace food production and unleash massive famine and conflict in the latter 20th century. However, thanks to the way the "Green Revolution" (pp. 245–246) increased food production in developing regions in the decades after his book, Ehrlich's dire forecasts did not fully materialize.

Does this mean we can disregard the concerns of Malthus and Ehrlich? Some economists say yes. Under the Cornucopian view that many economists hold, population growth poses no problem if new resources can be found or created to replace depleted ones. In contrast, environmental scientists recognize that not all resources can be replaced— such as the ecosystem services provided by a species that was driven to extinction. Thus, the environmental scientists argue, population growth is indeed a problem if it depletes resources,

stresses social systems, and degrades the natural environment, such that our quality of life declines.

To understand the challenges faced by a growing world population, we must consider more than just food. While technological innovations, such as those in the Green Revolution, have increased food production, many of the resources needed by humans exist in finite quantities and cannot be appreciably increased in quantity by technology. Take water, for example. While the human population grows ever larger, the amount of fresh water—the water humans need for their homes, growing crops, and producing energy and manufactured goods—remains essentially the same. So with every passing second, each person's "share" of the world's water decreases. We can convert saline seawater to fresh water through the process of desalination (p. 406), but the process is incredibly energy-intensive, making it impractical for expanded use in a world already dealing with limited supplies of energy and the adverse effects of energy production (notably climate change). To see the potential issues that can arise with rapid population growth and limited natural resources, consider the status of water supplies in Nigeria, a nation whose population is predicted to more than double in size over the next 30 years (see **DATAGRAPHIC**).

Population is one of several factors that affect the environment

One widely used formula gives us a handy way to think about population and other factors that affect environmental quality. Nicknamed the **IPAT model,** it is a variation of a formula proposed in 1974 by Paul Ehrlich and John Holdren. The IPAT model represents how our total impact (I) on the environment results from the interaction among population (P), affluence (A), and technology (T):

$$I = P \times A \times T$$

We can interpret impact in various ways, but generally boil it down to either unsustainable resource consumption or the degradation of ecosystems by pollution. Increased population intensifies impact on the environment as more individuals take up space, use natural resources, and generate waste (see **THE SCIENCE BEHIND THE STORY**, pp. 192–193). Increased affluence magnifies environmental impact. through greater per capita resource consumption, which generally has accompanied enhanced wealth. Technology that enhances our abilities to exploit minerals, fossil fuels, old-growth forests, or fisheries generally increases impact, but technology to reduce smokestack emissions, harness renewable energy, or improve manufacturing efficiency can decrease impact. One reason our population has kept growing, despite limited resources, is that we have developed technology—the T in the IPAT equation—time and again to increase the efficiency of our resource use, alleviate our strain on resources, and allow us to expand our exploitation of resources further.

We might also add a fourth factor, sensitivity (S), to the equation to denote how sensitive a given environment is to human pressures (rewriting the formula as $I = P \times A \times T \times S$). For instance, the arid lands of western China are more

Will Nigeria's Population Overwhelm Its Water Supplies?

Due to its rapid population growth rate, **Nigeria is projected to become the world's third most populous nation by 2050.** But while Nigeria's population will grow tremendously, its water supplies likely will not.

Significant portions of Nigeria are very arid, and its northern deserts are expanding southward.

Lake Chad has shrunk to one-tenth the size it was in the 1960s due to climate change and withdrawals of water for agriculture, removing an important water source for eastern Nigeria.

Nigerian water use by sector (2018)

- Agriculture 36%
- Industry 16%
- Raising fish and livestock 8%
- 40%

Domestic water use is the largest use of water in Nigeria, and **demand for water to homes is expected to skyrocket with the growth of Lagos,** the nation's largest city. By 2100, Lagos is projected to be the largest city in the world.

Population of Lagos (millions)

In 2013, the **total renewable water supply** (the amount of surface and groundwater replenished through precipitation annually) for Nigeria was far below that of neighboring countries.

If Nigeria's population more than doubles by 2050, its per-person renewable water supply will drop **below the level classified as "water stressed."**

Total renewable water supply per person per year (m³)

2013		2050
Sub-Saharan Africa (avg): 6500 m³	Nigeria: 1800 m³	Nigeria (projected): 750 m³

Water stressed (1000 m³)

> If Nigeria cannot reduce its projected population growth, water scarcity could soon slow its economic growth and adversely affect the well-being of its people.

Data from United Nations, Department of Economic and Social Affairs, Population Division, 2017. World population prospects: The 2017 revision. Beck, H.E. et al., 2018. Present and future Köppen-Geiger climate classification maps at 1-km resolution. Scientific Data 5: 180214. The World Bank, 2013. Toward Climate-Resilient Development in Nigeria. Hoornweg, D. and K. Pop, 2014. Socioeconomic Pathways and Regional Distribution of the World's 101 Largest Cities. Global Cities Institute Working Paper No. 04. The Guardian, 2016. The small African region with more refugees than all of Europe. United Nations, Food and Agricultural Organization, 2018. AQUASTAT.

sensitive to human disturbance than are moist regions of southeastern China. Plants grow more slowly in the arid west, making the land more vulnerable to deforestation and soil degradation. Thus, adding an additional person to western China has more environmental impact than adding one to southeastern China. We could refine the IPAT equation further by adding terms for the influence of social factors, such as education, laws, ethical standards, and social stability and cohesion. Such factors all affect how population, affluence, and technology translate into environmental impact.

THE SCIENCE behind the story

Measuring Our "Human Footprint": A Roadmap to Sustainability *and* Prosperity?

Oscar Venter, University of Northern British Columbia

As every person requires natural resources—water, food, and energy, for example—to survive and thrive, over time our growing human population exerts ever-increasing pressure on the natural systems that provide us with these resources. Conceptual frameworks, such as the IPAT formula (p. 190), aim to describe at a theoretical level the connections between human populations, resource consumption, and the impacts people put on the environment. But can we actually quantify and map human stresses on the environment in a measure called the *human footprint*?

One attempt to do just that is the Human Footprint Map. This detailed analysis of human stresses on Earth's terrestrial systems was first published in 2002, using data from the 1990s. It used an approach called *cumulative threat mapping* that combined satellite imaging, which is highly effective at mapping changes in land-use patterns, with ground-based surveys that can detect the impacts of smaller features that go unnoticed by satellites, such as rural roads and the degree of degradation in pasture land. Eight factors that indicate pressure on natural systems from human activity were quantified and mapped for each of 823 terrestrial ecoregional zones, areas with similar ecosystems and environmental resources:

- Built environments (land developed for human activities, such as cities and towns)
- Agricultural land
- Pasture land
- Railways
- Major roads
- Navigable waterways
- Satellite measurements of night-time lights (to estimate human development in rural areas)
- Human population density

The human footprint was reported on a scale from 0 to 50, with a score of 0 representing no impacts from human activities. Note that the human footprint score is a very different measure than an ecological footprint (p. 5), despite their similar-sounding names. An ecological footprint score is a measure of the amount of biologically productive land needed

to sustain a person or a population, whereas the human footprint score is an estimate of the degree of stress placed on a given area by human activities.

The Human Footprint Map proved itself a valuable tool, but as its data grew older year after year, it became clear that a revision of the map was direly needed. That update arrived in 2016, thanks to a team of scientists led by biologist Oscar Venter of the University of Northern British Columbia—and the results were sobering. Using newer data from 2009, they showed that human stresses on the environment were widespread, intense, and had increased overall since 1993. Roughly three-fourths of the planet was significantly altered by human activities, and 97% of the most biodiverse areas on Earth were experiencing stresses from human encroachment (**FIGURE 1a**). In 2009, the global average human footprint was 6.16, representing an increase of 9% from its score of 5.67 in 1993. The degree of change in the human footprint score varied greatly by location (**FIGURE 1b**). Roughly 70% of the 823 ecoregions experienced an increase in average footprint value over the 16-year interval, and many of these areas saw their human footprint score increase by 20% or more. Impacts were especially strong in tropical areas with high levels of biodiversity, such as eastern Brazil, western Africa, and Southeast Asia. Few areas of the planet were untouched by human activity, largely places of remote or inhospitable to human uses. Such areas comprised about 27% of the global land area in 1993 but were nearly 10% smaller by 2009 due to human activities encroaching on once untouched regions.

One of the biggest predictors of high human impacts on a given area in 2009 was the area's suitability for agriculture, as determined by its climate and the quality of its soils. When the human footprint score in 2009 was compared to the agricultural suitability—rated from 0 (land not at all suitable for farming) to 100 (ideal farmland)—a strong positive relationship was observed. That is, as an area's suitability for agriculture increased, so did the degree of human stresses on that area. When the researchers compared the change in human footprint score from 1993 for 2009 versus agricultural suitability, they found an inverted-U shape to the data. This showed that areas that were highly suitable for farming experienced little additional human stress over the 16-year period, as they were likely already being used by humans, and lands highly unsuitable for agriculture also experienced a small change in human impact, as there was little incentive to develop these lands. However, areas that were only moderately suitable for agriculture experienced a large increase in the human footprint score as people expanded agriculture into marginally productive natural areas that were formerly not cultivated.

But among this sea of dispiriting data, there were signs of hope. The first was that while the global population increased by 23% and the global economy grew by 153% from 1993 to 2009, the global human footprint increased by only 9%. This indicated that many nations had found a way to decouple population increases and economic growth from corresponding increases in environmental stresses. Utilizing natural resources more efficiently and more carefully managing land uses are the likely reasons for this outcome. Furthermore, nearly one-quarter of the world's 823 ecoregions experienced a reduction in their human footprint value from 1993 to 2009. These areas were largely in temperate latitudes and included portions of North America, Western Europe, Northern Asia, and southeastern Africa.

Most importantly, careful analysis of the human footprint data provided a "roadmap" toward sustainability by examining the relationship between a society's human footprint scores and its social and economic characteristics. Prosperity has been shown to be an effective mechanism for reducing population growth (p. 206), and a society's level of wealth was found to be related to the change in its human footprint score from 1993 to 2009. The wealthiest nations on Earth were able to both grow their economies and shrink their human footprint over the 16 years between Human Footprint Maps, showing that strong economic growth and environmental sustainability usually go hand in hand. This was not accomplished by fueling their economic growth with imports—and passing environmental damage to another country—as these nations were net exporters of agricultural and forestry products.

When Venter's team compared societal factors with the changes in human footprint scores from 1993 to 2009, they found that nations with a declining human footprint score (that is, reduced stress on the environment) tended to have a high score for human development (a measure of the quality of education and health care in a nation) and urbanization (higher proportions of citizens living in cities versus the countryside) and a low value for governmental corruption. This last factor was highly significant, as it showed that a functioning, objective government enables land uses and natural resources to be regulated and efficiently managed.

Studies such as this one help us understand, across space and time, how humans affect the natural environment. Such information is invaluable to governments, conservation biologists, and other stakeholders as we attempt to limit our stresses on nature and move toward environmental sustainability.

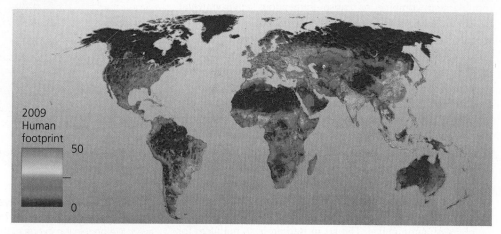

(a) Global human footprint map in 2009

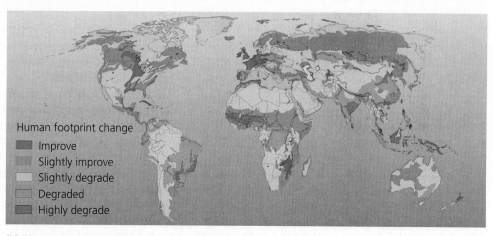

(b) Change in human footprint from 1993 to 2009

FIGURE 1 The global human footprint in 2009 (a) tended to be highest in areas with large human populations and in wealthier societies with high levels of resource use. The change in the human footprint score from 1993 to 2009 **(b)** shows that many ecoregions around the globe are experiencing greater stress than before, leading to environmental degradation. *Data from Venter, O., et al., 2016. Sixteen years of change in the global terrestrial human footprint and implications for biodiversity conservation. Nat. Comm. 7:12558, doi:10.1038/ncomms12558.*

Modern-day China shows how all elements of the IPAT formula can combine to cause tremendous environmental impact in a short time. Although China boasts one of the world's fastest-growing economies, the country is battling unprecedented environmental challenges brought about by this rapid economic development. Intensive agriculture has expanded westward out of the nation's moist rice-growing regions, causing farmland to erode and blow away, much like the Dust Bowl tragedy that befell the U.S. heartland in the 1930s (pp. 231–232). China has overpumped aquifers (groundwater) and drawn so much water for irrigation from the Yellow River that this once-mighty waterway now dries up periodically in many stretches. Although China is reducing its air pollution from industry and charcoal-burning in homes, the country faces new threats to air quality from rapidly rising numbers of automobiles. The air in Beijing is so polluted, for example, that simply breathing it on a daily basis damages the lungs to the same extent as smoking 40 cigarettes (p. 466). As the world's nations with developing economies try to attain the material prosperity that industrialized nations enjoy, China is a window into what much of the rest of the world could soon become.

Demography

People do not exist outside of nature. We exist within our environment as one species out of many. As such, all the principles of population ecology (Chapter 3) that drive population change in the natural world apply to humans as well. The application of principles from population ecology to the study of statistical change in human populations is the focus of **demography.**

Demography is the study of human population

Demographers study human population size, density, distribution, age structure, sex ratio, and rates of birth, death, immigration, and emigration. Each of these factors is useful for predicting human population dynamics and environmental impacts.

Population size Our global human population of more than 7.6 billion is spread among 200 nations with populations ranging up to China's 1.39 billion, India's 1.37 billion, and the 328 million of the United States (**FIGURE 8.4**). The United Nations Population Division estimates that by the year 2050, the global population will surpass 9.8 billion (**FIGURE 8.5**). However, population size alone—the absolute number of individuals—doesn't tell the whole story. Rather, a population's environmental impact depends on that population's density, distribution, and composition (as well as affluence, technology, and other factors outlined earlier).

Population density and distribution People are distributed unevenly over our planet. In ecological terms, our distribution is clumped (p. 63) at all spatial scales. At the global scale, population density is highest in regions with temperate, subtropical, and tropical climates. Population

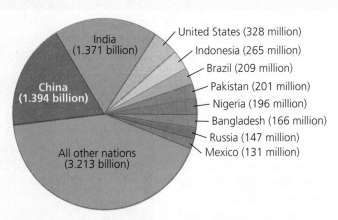

FIGURE 8.4 Almost one in five people in the world lives in China, and more than one of every six live in India. Three of every five people live in one of the 10 most populous nations. *Data from Population Reference Bureau,* 2018 World population data sheet.

density is lowest in regions with extreme-climate biomes, such as desert, rainforest, and tundra. It is highest along seacoasts and rivers. At more local scales, we cluster together in cities and towns.

This uneven distribution means that certain areas bear more environmental impact than others. Just as the Yellow River experiences pressure from Chinese cities and farms, the world's other major rivers all receive more than their share of human impact. At the same time, some areas with low population density are sensitive (a high *S* value in our revised IPAT model) and thus vulnerable to impact. Deserts and arid grasslands, for instance, are easily degraded by development that commandeers too much water.

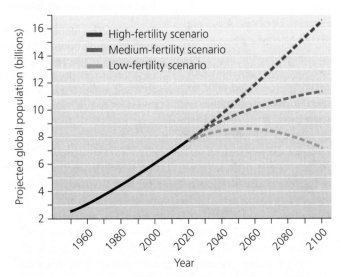

FIGURE 8.5 The United Nations predicts world population growth. In the latest projection, it is estimated that the population will reach 9.8 billion in 2050 and around 11.2 billion in 2100 using a medium-fertility scenario. In the high-fertility scenario, women, on average, have half a child more than in the medium scenario. In the low-fertility scenario, women have half a child fewer than in the medium scenario. *Adapted by permission from Population Division of the Department of Economic and Social Affairs of the United Nations Secretariat, 2017.* World population prospects: The 2017 revision. *http://esa.un.org/unpd/wpp, © United Nations, 2017.*

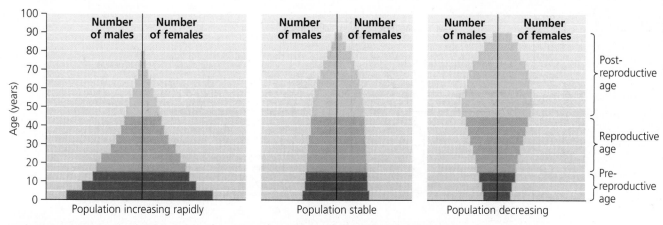

FIGURE 8.6 Age structure diagrams show numbers of males and females of different age classes in a population. A diagram like that on the left is weighted toward young age classes, indicating a population that will grow quickly. A diagram like the one on the right is weighted toward old age classes, indicating a population that will decline. Populations with balanced age structures, like those shown in the middle diagram, will remain relatively stable in size.

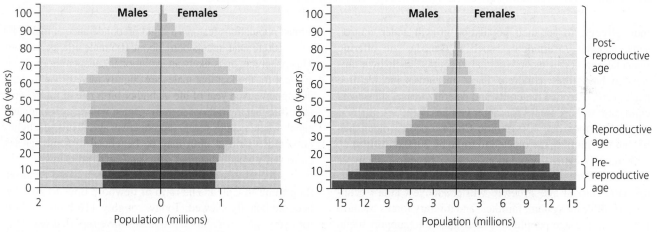

(a) Age structure diagram of Canada in 2019

(b) Age structure diagram of Nigeria in 2019

FIGURE 8.7 Canada (a) shows a fairly balanced age structure, whereas Nigeria (b) shows an age distribution heavily weighted toward young people. Nigeria's population growth rate (2.6%) is over eight times greater than Canada's (0.3%). *Data from U.S. Census Bureau International Database,* http://www.census.gov/population/international/data/idb/.

Age structure Age structure describes the relative numbers of individuals of each age class within a population (pp. 63–66). Data on age structure are especially valuable to demographers trying to predict future dynamics of human populations. A population with many individuals of reproductive age or pre-reproductive age is likely to increase. In contrast, a population made up mostly of individuals past reproductive age will tend to decline over time. A population with an even age distribution will likely remain stable as births keep pace with deaths.

Age structure diagrams, often called population pyramids, are visual tools scientists use to illustrate a population's age structure (**FIGURE 8.6**). The width of each horizontal bar represents the number of people in each age class. A pyramid with a wide base denotes a large proportion of people who have not yet reached reproductive age—and this indicates a population soon capable of rapid growth. As an example, compare age structures for the nations of Canada and Nigeria (**FIGURE 8.7**). Nigeria's large concentration of individuals in

younger age classes predicts a great deal of future reproduction. Not surprisingly, Nigeria has a higher population growth rate than Canada.

Today, populations are aging in many nations, and the global population is "grayer" than in the past (**FIGURE 8.8**, p. 196). The global median age today is 28, but it is predicted that the median will be 38 by the year 2050. By causing dramatic reductions in the number of children born since 1970, China's former one-child policy virtually guaranteed that the nation's population age structure would change. Indeed, in 1970, the median age in China was 20; by 2050, it will likely be 45.

Changing age distributions have caused concerns in some nations that have declining numbers of workers and strong social welfare programs for retirees (which are supported by current workers), such as the Social Security program in the United States. Despite the long-term benefits associated with smaller populations, many policymakers find it difficult to let go of the notion that population growth increases a nation's economic, political, and military strength. So, while China and

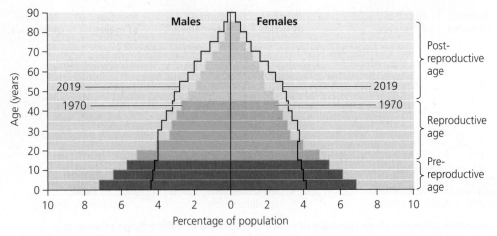

FIGURE 8.8 Comparing the age structure diagrams of the global human population in 1970 and 2019 shows how our population is aging. The more balanced age distribution also predicts slower growth in coming decades than what occurred after 1970. *Adapted by permission from Population Division of the Department of Economic and Social Affairs of the United Nations Secretariat, 2013. World population prospects: The 2012 revision. http://esa.un.org/unpd/wpp/, © United Nations, 2013. Data for 2019 from U.S. Census Bureau International Database,* http://www.census.gov/population/international/data/idb.

India struggle to get their population growth under control, some national governments—such as Canada—are offering financial and social incentives that encourage their own citizens to have more children. These incentives include free health care, extended maternity and paternity leave, subsidized child care, and tax breaks for larger families.

Today, there are 130 million Chinese people older than 65, but that number is expected to triple by 2050 (**FIGURE 8.9**). Although graying populations have benefits—such as increased charitable contributions to society by retirees—the dramatic shift in age structure will challenge China's economy, health care system, families, and military forces because fewer working-age people will be available to support social programs to assist the rising number of older people.

Sex ratios The ratio of males to females also can affect population dynamics. The naturally occurring sex ratio at birth in human populations features a slight preponderance of males; for every 100 female infants born, about 106 male infants are born. This phenomenon is likely an evolutionary adaptation (p. 51) to account for the fact that males are slightly more prone to death during any given year of life. A higher birth rate of males tends to ensure that the ratio of men to women will be approximately equal when people reach reproductive age. Thus, a slightly uneven sex ratio at birth may be beneficial to some species. However, a greatly distorted ratio can lead to problems.

In recent years, demographers have witnessed an unsettling trend in China: The ratio of newborn boys to girls has become heavily skewed. Today, roughly 116 boys are born for every 100 girls. Some provinces have reported sex ratios as high as 138 boys for every 100 girls. As mentioned in the opening Case Study, the leading hypothesis for these unusual sex ratios is that many parents, having learned the sex of their fetuses by ultrasound, are selectively aborting

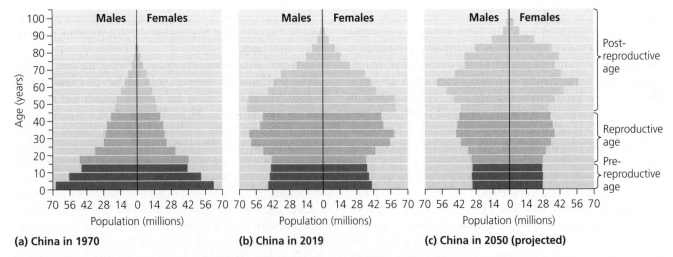

(a) China in 1970   **(b) China in 2019**   **(c) China in 2050 (projected)**

FIGURE 8.9 As China's population ages, older people will outnumber the young. China's one-child policy was highly successful in reducing birth rates but also in significantly changing China's age structure. Population pyramids show the predicted graying of the Chinese population from **(a)** 1970 to **(b)** 2019 to **(c)** what is predicted for 2050. *Data from U.S. Census Bureau International Database,* http://www.census.gov/population/international/data/idb.

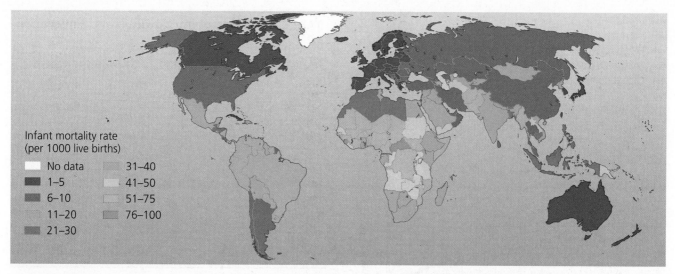

FIGURE 8.10 Infant mortality rates are highest in poorer nations, such as in sub-Saharan Africa, and lowest in wealthier nations. Industrialization brings better nutrition and medical care, which greatly reduce the number of children dying in their first year of life. *Data from Population Reference Bureau,* 2018 World population data sheet.

Infant mortality rate
(per 1000 live births)

- No data
- 1–5
- 6–10
- 11–20
- 21–30
- 31–40
- 41–50
- 51–75
- 76–100

female fetuses. A 2016 report by American researchers has suggested that some of these "missing girls" in rural areas may actually have been born, but they just weren't reported. Local authorities simply "looked the other way" when the girls were born, and then added them to official registries when they became school-aged.

China's skewed sex ratio may further lower population growth rates. However, it has the undesirable social consequence of leaving large numbers of Chinese men single. Without the anchoring effect a wife and family provide, many of these men leave their native towns and find work elsewhere as migrant workers. Living as bachelors far from home, these men often engage in more risky sexual activity than their married counterparts. Researchers speculate that this could lead to a higher incidence of HIV infection in China in the coming decades, as tens of millions of bachelors find work as migrant workers.

Birth and death rates drive population change

The relative rates of birth to deaths determines whether a population grows, shrinks, or remains stable. The formula ecologists use for measuring population growth of organisms is also used by demographers as it pertains to people (p. 66): Birth adds individuals to a population, whereas death removes individuals from a population. Historically, birth and death rates in human populations were similar, resulting in very slow population growth. This is no longer true, as the many technological advances that accompanied the industrial revolution (p. 5) have led to a dramatic decline in human death rates. This acts to widen the gap between birth rates and death rates and results in populations growing larger over time.

These technological improvements have been particularly successful in reducing **infant mortality rates,** the frequency of children dying in infancy. Throughout much of human history, parents needed to have larger families as insurance against the likelihood that one or more of their children would die during infancy. Poor nutrition, disease, exposure to hostile elements, and limited medical care claimed the lives of many infants in their first year of life. As societies have industrialized and become more affluent, infant mortality rates have plummeted as a result of better nutrition, prenatal care, and the presence of medically trained practitioners during birth.

As shown in **FIGURE 8.10**, infant mortality rates vary widely around the world and are closely tied to a nation's level of industrialization. China, for example, saw its infant mortality rate drop from 47 children per 1000 live births in 1980 to 11 children per 1000 live births in 2016 as the nation industrialized and prospered. Many other industrializing nations enjoyed similar success in reducing infant mortality during this time period.

China's growth over its long history illustrates how the relative birth and death rates drive population changes. For most of the past 2000 years, China's population was relatively stable. The first significant increases resulted from enhanced agricultural production and a powerful government during the Qing Dynasty in the 1800s. However, population growth began to outstrip food supplies by the 1850s and quality of life for the average Chinese peasant started to decline (**TABLE 8.1**). Over the next 100 years, China's population grew slowly, at about 0.3% per year, amid food shortages and

TABLE 8.1 Trends in China's Population Growth

MEASURE	1950	1970	1990	2018
Total fertility rate	5.8	5.8	2.2	1.8
Rate of natural increase (% per year)	1.9	2.6	1.4	0.5
Doubling time (years)	37	27	49	140
Population (billions)	0.56	0.83	1.15	1.39

Data from China Population Information and Research Center; and Population Reference Bureau, *2018 World population data sheet.*

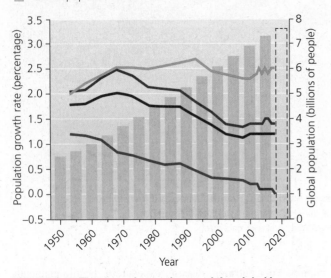

FIGURE 8.11 The annual growth rate of the global human population peaked in the late 1960s and has declined since then. *Data from Population Division of the Department of Economic and Social Affairs of the United Nations Secretariat, 2011. World population prospects: The 2010 revision. http://esa.un.org/unpd/wpp. © United Nations, 2011. Data updates for 2011–2018 from Population Reference Bureau, 2011–2018 World population data sheets.*

political instability. Population growth rates rose again following Mao's establishment of the People's Republic in 1949 but have declined since the advent of the one-child policy.

In recent decades, falling population growth rates in many countries have led to an overall decline in the global population growth rate (**FIGURE 8.11**). This decline has come about, in part, from a steep drop in birth rates. Note, however, that although the rate of growth is slowing, the absolute size of the population continues to increase. This occurs because today's growth rates, although smaller than in previous decades, are still positive (greater than zero), resulting in new people being added to the global population.

Populations also change with immigration and emigration

The term **migration** describes the movements of individuals between populations of organisms. Immigration (people arriving in a country from after leaving another) and emigration (people leaving a country for another) can act to grow or shrink human populations, just as migration affects the population sizes of other organisms (p. 66). People do not leave our planet for another, so while the global human population is unaffected by immigration and emigration, migration can affect the human population dynamics of individual nations.

Immigration and emigration play increasingly important roles in population change in our modern, international world. People regularly relocate from one nation to another to

TABLE 8.2 Rates of Immigration and Emigration by Nation

NATION	NET MIGRATION RATE
Spain	7.8
Canada	5.7
United States	3.9
Germany	1.5
Japan	0
China	−0.4
Mexico	−1.8
Turkey	−4.5
El Salvador	−8.0
Lebanon	−20.3

Net migration rate is the difference in the number of individuals entering a nation to the number leaving a nation in a given year, per 1000 residents. Data from U.S. Central Intelligence Agency, *The world factbook 2017.*

improve their economic opportunities or to flee conflict or environmental degradation in their home nation. Such migration can have significant effects on population growth in nations, such as the United States and Canada, which accept sizable numbers of immigrants. Immigration has become a potent political issue in the United States, Australia, and the European Union, as citizens and governments attempt to humanely deal with the influx of large numbers of individuals seeking refuge and employment. A survey of net migration rates (the difference in the number of immigrants to and emigrants from a given nation) reveals that migration rates vary significantly across nations, with positive rates of net migration for nations with policies that welcome immigrants and negative rates of net migration from nations with civil strife and limited economic opportunities (**TABLE 8.2**).

Migration in the modern world typically occurs when citizens of nations that are experiencing armed conflict, political strife, or difficult economic conditions relocate to other nations. While some of this migration occurs from nations with developing economies to wealthier nations, such as the influx of refugees escaping civil war in Syria and Libya arriving on European shores, a great deal of migration occurs between nations with emerging economies. For example, the collapse of the Venezuelan economy (starting in 2015) generated a humanitarian crisis in Latin America, as nearby nations such as Brazil and Colombia struggled to absorb large numbers of emigrants fleeing Venezuela.

Total fertility rate influences population growth

One key statistic demographers calculate to examine a population's potential for growth is the **total fertility rate (TFR),** the average number of children born per woman during her lifetime. **Replacement fertility** is the TFR that keeps the size of a population stable. For most nations today, replacement fertility roughly equals a TFR of 2.1. (Two children replace the mother and father,

TABLE 8.3 Total Fertility Rates for Major Continental Regions

REGION	TOTAL FERTILITY RATE (TFR)
Africa	4.6
Australia and South Pacific	2.3
Latin America and Caribbean	2.1
Asia	2.1
North America	1.7
Europe	1.6

Data from Population Reference Bureau, *2018 World population data sheet.*

and the extra 0.1 accounts for the risk of a child dying before reaching reproductive age.) If a nation's TFR is above 2.1, it is expected to grow. If its TFR is below 2.1, its population is expected to shrink (in the absence of immigration).

Factors such as industrialization, improved women's rights (p. 203), access to family planning, and quality health care have driven the TFR downward in many nations in recent years. All these factors have come together in Europe, where the region's TFR has dropped from 2.6 to 1.6 in the past half-century. Nearly every European nation now has a fertility rate below the replacement level, and populations are declining in 15 of 45 European nations. In 2016, Europe's overall annual **rate of natural increase** (also called the *natural rate of population change*)—change due to birth and death rates

alone, excluding migration—was between 0.0% and 0.1%. Worldwide by 2015, 84 countries had fallen below the replacement fertility of 2.1. These low-fertility countries make up a sizable portion of the world's population and include China (with a TFR of 1.6). **TABLE 8.3** shows the total fertility rates of major continental regions.

Many nations are experiencing demographic transition

Many nations with lowered birth rates and TFRs are experiencing a common set of interrelated changes. In countries with reliable food supplies, good public sanitation, and effective health care, more people than ever before are living long lives. As a result, over the past 50 years, the life expectancy for the average person globally has increased from 46 to 71 years as the worldwide death rate has dropped from 20 deaths per 1000 people to 8 deaths per 1000 people. Much of the recent increase is due to reductions in the rate of infant mortality. **Life expectancy** is the average number of years that an individual in a particular age group is likely to continue to live, but often people use this term to refer to the average number of years a person can expect to live from birth. Societies going through these changes are generally those that have undergone urbanization and industrialization and have generated personal wealth for their citizens.

To make sense of these trends, demographers developed a concept called the **demographic transition** (**FIGURE 8.12**).

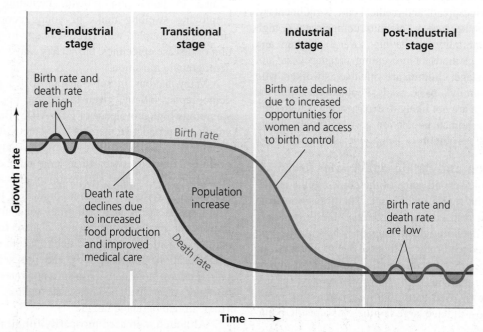

FIGURE 8.12 The demographic transition models a process that has taken some populations from a pre-industrial stage of high birth rates and high death rates to a post-industrial stage of low birth rates and low death rates. In this diagram, the wide green area between the two curves illustrates the gap between birth and death rates that causes rapid population growth during the middle portion of this process. *Adapted from Kent, M., and K. Crews, 1990.* World population: Fundamentals of growth. *By permission of the Population Reference Bureau.*

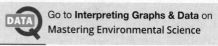
Go to **Interpreting Graphs & Data** on Mastering Environmental Science

This is a model of economic and cultural change proposed in the 1940s and 1950s by demographer Frank Notestein to explain the declining death rates and birth rates that have occurred in Western nations as they industrialized. Notestein argued that nations move from a stable pre-industrial state of high birth and death rates to a stable post-industrial state of low birth and death rates. Industrialization, he proposed, causes these rates to fall by first decreasing death rates and then by lessening the need for large families. Parents thereafter choose to invest in quality of life rather than quantity of children. Because death rates fall before birth rates fall, a period of net population growth results. Thus, under the demographic transition model, population growth is seen as a temporary phenomenon that occurs as societies move from one stage of development to another.

The pre-industrial stage The first stage of the demographic transition model is the **pre-industrial stage,** characterized by conditions that have defined most of human history. In pre-industrial societies, both death rates and birth rates are high. Death rates are high because food supplies are unreliable, disease is widespread, and medical care is rudimentary. Birth rates are high because people must compensate for high rates of infant mortality by having several children and because effective methods of preventing pregnancy are not available. In this stage, children are valuable as workers who can help meet a family's basic needs. Populations within the pre-industrial stage are not likely to experience much growth, which is why the human population grew relatively slowly until the industrial revolution.

Industrialization and falling death rates Industrialization initiates the second stage of the demographic transition, known as the **transitional stage.** This transition from the pre-industrial stage to the industrial stage is generally characterized by declining death rates due to increased food production and reduced rates of disease due to better public sanitation and improved medical care. Birth rates in the transitional stage remain high, however, because people have not yet grown used to the new economic and social conditions and so still have large families. As a result, population growth surges.

The industrial stage and falling birth rates The third stage in the demographic transition is the **industrial stage.** Industrialization increases opportunities for employment outside the home, particularly for women. Children become less valuable, in economic terms, because they are no longer a necessary workforce for the family's survival. If couples are aware of this, and if they have access to birth control, they may choose to have fewer children. Birth rates fall, closing the gap with death rates and reducing population growth.

The post-industrial stage In the final stage of the transition, the **post-industrial stage,** both birth and death rates have fallen to low and stable levels. Population sizes stabilize or decline slightly. Society enjoys the fruits of industrialization without the threat of runaway population growth.

Is demographic transition inevitable?

The demographic transition has occurred in many European countries, the United States, Canada, Japan, and several other developed nations over the past 200–300 years. It is a model that may or may not apply to all developing nations as they industrialize now and in the future. On the one hand, as shown in Figure 8.12 (p. 199), we know that growth rates fell first for industrialized nations, then for less developed nations, and finally for least developed nations, a pattern suggesting that it may merely be a matter of time before all nations experience demographic transition. On the other hand, some developing nations may already be suffering too greatly from the impacts of large populations to replicate the developed world's transition, a phenomenon called **demographic fatigue.** Demographically fatigued governments face overwhelming challenges related to population growth, including educating and employing swelling ranks of young people. When these stresses are coupled with large-scale environmental degradation or disease epidemics, the society may never complete the demographic transition.

Many nations in sub-Saharan Africa are experiencing demographic fatigue, given their large populations and the stunningly high prevalence of HIV/AIDS. Because it removes young and productive members of society, AIDS undermines the ability of poorer nations to develop. Nations lose billions of dollars in productivity when large numbers of its citizens are battling the disease, and treatment puts a huge burden on health care systems. Children orphaned by AIDS further strain social safety nets, requiring societal interventions to prevent the cycle of poverty and disease from claiming yet another generation. Improved public health efforts, including sex education, contraceptives, and intravenous drug abuse policies, have slowed HIV transmission rates in many nations, however, offering hope that these nations may escape the "trap" of demographic fatigue.

Although increased prosperity for all people is a noble goal, it does come with environmental cost. Scientists estimate that for people of all nations to attain the material standard of living that North Americans now enjoy, we would need the natural resources of four and one-half more planet Earths. Hence, whether developing nations (which include the vast majority of the planet's people) pass through the demographic transition is one of the most important questions for the future of our civilization and Earth's environment.

Will China and India complete the demographic transition?

China and India are home to over 2.7 billion people, and this means that one out of every three people on the planet lives in those two nations. China and India have been industrializing their economies over the last several decades, and their population growth patterns have mimicked those of the intermediate stages of the Demographic Transition model. China and India have begun the demographic transition, it seems—but will they finish it?

While China's reproductive policies have gained widespread attention, India was actually one of the first nations to implement national population control policies. India introduced sterilization as a population control mechanism in the 1970s. However, when some policymakers attempted to enforce it, the resulting outcry brought down the government. Since then, India's efforts have been more modest and far less coercive, focusing on family planning and reproductive health care. This has greatly reduced the population growth rate in India, but India will nonetheless likely overtake China and become the world's most populous nation in several decades because of China's more aggressive population initiatives (**FIGURE 8.13**).

A survey of the demographic statistics for the two countries shows that China appears to be further along in the demographic transition than India (**TABLE 8.4**). China's rate of natural increase is roughly one-third that of India, China's total fertility rate has fallen well below replacement level fertility, modern contraceptive use is 75% higher in China than India, and infant mortality rates—a major factor that drives high birth rates—is nearly 2.5 times higher in India than in China. China's population is also significantly more urban than India, and the per-capita income of a typical Chinese citizen is more than twice that of the average Indian.

TABLE 8.4 Demographic Comparison: China and India

MEASURE	CHINA	INDIA
Rate of natural increase (% per year)	0.5	1.4
Total fertility rate	1.8	2.3
Women using modern contraception (%)	84	48
Infant mortality rate (infant death per 1000 live births)	10	34
Urban population (%)	59	34
Gross national income per capita ($)	16,760	7060

Data from Population Reference Bureau, *2018 World population data sheet.*

Most experts predict that China will continue to industrialize, grow its economy, and permanently settle into a stable, post-industrial society with minimal population growth within several decades. What is less clear is whether India, with its large population and continued growth, will be able to complete the transition to an industrialized society or if the suffocating impacts of demographic fatigue will derail the process, causing this populous nation to experience meaningful population growth for decades to come. Given the size of their collective population, the answer to this question has profound impacts for Asia and for all of humanity.

WEIGHING the **issues**

What Are the Consequences of Low Fertility?

In the United States, Canada, and almost every European nation, the total fertility rate is now at or below the replacement fertility rate (although some of these nations are still growing because of immigration). What economic or social consequences do you think might result from below-replacement fertility rates? Would you rather live in a society with a growing population, a shrinking population, or a stable population? Why?

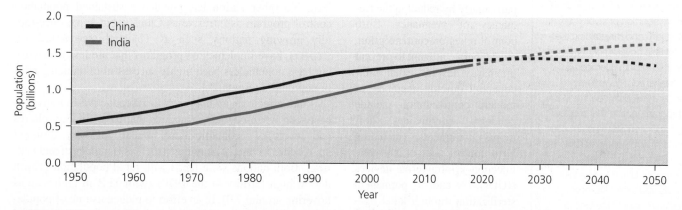

FIGURE 8.13 India will likely soon surpass China as the most populous nation. Both nations have embraced population control initiatives, but China's more stringent reproductive policies will likely lead to its population stabilizing in coming decades, while India's population will continue to grow. China's rate of growth is now lower than India's as a result of China's aggressive population policies. *Data from United Nations, Department of Economic and Social Affairs, Population Division, 2017. World population prospects: The 2017 revision. Custom data acquired via website.*

Population and Society

The demographic transition model links the quantitative study of how populations change with the societal factors that influence (and are influenced by) population dynamics. Many factors affect fertility in a given society. They include public health issues, such as people's access to contraceptives and the rate of infant mortality. They also include cultural factors, such as the level of women's rights, the relative acceptance of contraceptive use, and even cultural influences like television programs (see **THE SCIENCE BEHIND THE STORY**, pp. 204–205). There are also effects from economic factors, such as society's level of affluence, the importance of child labor, and the availability of governmental support for retirees. Let's now examine a few of these societal influences on fertility more closely.

Family planning slows population growth

Perhaps the greatest single factor enabling a society to slow its population growth is the ability of women and couples to engage in **family planning,** the conscious effort to plan the number and spacing of one's children. Family-planning programs and clinics offer information and counseling to potential mothers and fathers on reproductive issues.

An important component of family planning is **birth control,** the effort to control the number of children one bears, particularly by reducing the frequency of pregnancy. Birth control relies on **contraception,** the deliberate attempt to prevent pregnancy despite sexual intercourse. Common methods of modern contraception include condoms, spermicides, hormonal treatments (including birth control pills or hormone injections), intrauterine devices (IUDs), and permanent sterilization through tubal ligation or vasectomy. Many family-planning organizations aid clients by offering free or discounted contraceptives.

Worldwide in 2018, 56% of women aged 15–49 reported using modern contraceptives, with rates of use varying widely among nations. China and the United Kingdom, at 84%, had the highest rate of contraceptive use of any nations. Seven European nations showed rates of contraceptive use of 70% or more, as did Argentina, Brazil, Canada, Chile, Colombia, Costa Rica, Cuba, Ecuador, Nicaragua, Puerto Rico, South Korea, Thailand, and Uruguay. At the other end of the spectrum, 14 African nations had rates at or below 15%.

Low usage rates for contraceptives in some societies are caused by limited availability, especially in rural areas. As the need for contraceptives can be continuous, women in isolated villages can therefore experience "gaps" in birth control. This occurs when couples use up their supply of contraceptives before reproductive counselors once again visit their village. In other societies, low usage may be due to religious doctrine or cultural influences that hinder family planning, denying counseling and contraceptives to people who might otherwise use them. This can result in family sizes that are larger than parents desire and lead to elevated rates of population growth.

In a physiological sense, access to family planning (and the civil rights to demand its use) gives women control over their **reproductive window,** the period of their life—beginning with sexual maturity and ending with menopause—in which they may become pregnant. A healthy woman can potentially bear up to 25 children within this window (**FIGURE 8.14a**), but she may choose to delay the birth of her first child to pursue education and employment. She may also use contraception to delay her first child, space births within the window, and "close" her reproductive window after achieving her desired family size (**FIGURE 8.14b**).

Family-planning programs are working around the world

Data show that funding and policies that encourage family planning can lower population growth rates in all types of nations, even those whose economies are the least industrialized. No other nation has pursued a sustained population control program as intrusive as China's, but some once-rapidly growing nations, such as Thailand (see **SUCCESS STORY**), have implemented programs that are less restrictive but have nonetheless been highly successful in reducing rates of population growth.

The effects of an approach to reproductive initiatives is best seen when comparing nations that have similar cultures and levels of economic development but very different approaches to family planning—such as Bangladesh and Pakistan. Both nations were faced with rapid population growth due to high fertility in the 1970s (with TFR in both nations hovering around 7.0). In an effort to reduce its rate of population growth, Bangladesh instituted a government-supported program to its citizens that improved access to contraception and reproductive counseling. Pakistan took a far less aggressive and coordinated approach, which made access to family

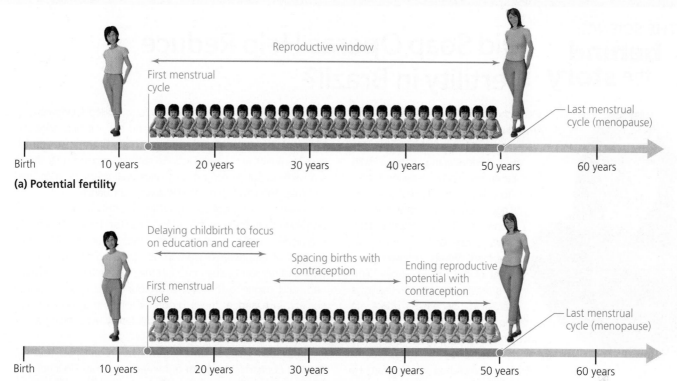

(a) Potential fertility

(b) Fertility reductions by delaying childbirth and contraceptive use

FIGURE 8.14 Women can potentially have very high fertility within their "reproductive window" but may choose to reduce the number of children they bear. They may do this by delaying the birth of their first child, or using contraception to space pregnancies, or ending their reproductive window at a time of their choosing.

SUCCESS story

Family Planning without Coercion: Thailand's Population Program

The government of Thailand, facing many of the same population challenges as its populous neighbor to the north, instituted a comprehensive population program in 1971. At the time, Thailand's population growth rate was 2.3%, and the nation's TFR was 5.4. But unlike the Chinese reproductive program, Thais were given control over their own reproductive choices, provided with family-planning counseling and modern contraceptives, and supported by an engaging public education campaign. Aided by a relatively high level of women's rights in Thai society, Thailand's population program—and the fertility reductions that accompanied the nation's economic development over the past 47 years—reduced its population growth rate to 0.2%, with a TFR of 1.5 children per woman in 2018. The success of this program, and similar initiatives in nations such as Brazil, Cuba, Iran, and Mexico, show that government interventions to reduce birth rates need not be as intrusive as China's to produce similar declines in population growth.

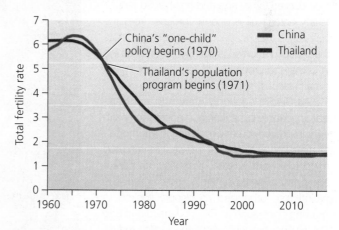

China and Thailand instituted population control programs at roughly the same time and showed similar patterns in fertility declines over the subsequent 45 years, despite utilizing very different approaches. *Data from World Bank, 2017, http://data.worldbank.org.*

➡ **Explore the Data** at Mastering Environmental Science

THE SCIENCE
behind
the story

Did Soap Operas Help Reduce Fertility in Brazil?

Eliana La Ferrara,
Bocconi University

Over the past 50 years, the South American nation of Brazil experienced the second-largest drop in fertility among developing nations with large populations—second only to China. In the 1960s, the average woman in Brazil had six children. Today, Brazil's total fertility rate is 1.8 children per woman, which is lower than that of the United States. Brazil's drastic decrease in fertility is interesting because, unlike in China, it occurred without intrusive governmental policies to control its citizens' reproduction.

Brazil accomplished this, in part, by providing women equal access to education and opportunities to pursue careers outside the home. Women now make up 40% of the workforce in Brazil and graduate from college in greater numbers than men. The Brazilian government also provides family planning and contraception to its citizens free of charge. Eighty percent of married women of childbearing age in Brazil currently use contraception, a rate higher than that in the United States or Canada. Universal access to family planning has given women control over their desired family size and has helped reduce fertility across all economic groups, from the very rich to the very poor. It is interesting to note that induced abortion is not used in Brazil as it is in China; the procedure is illegal except in rare circumstances.

As Brazil's economy grew with industrialization, people's nutrition and access to health care improved, greatly reducing infant mortality rates. Increasing personal wealth promoted materialism and greater emphasis on career and possessions over family and children. The nation also urbanized as people flocked to growing cities such as Río de Janeiro and São Paulo. This brought about the fertility reductions that typically occur when people leave the farm for the city.

It turns out, however, that there may have been a rather unique influence on Brazil's fertility rates over the past several decades—soap operas (**FIGURE 1**). Brazilian soap operas, called *telenovelas,* or *novelas,* are a cultural phenomenon and are watched religiously by people of all ages, races, and incomes. Each telenovela follows the activities of several fictional families, and these TV shows are wildly popular because they have characters, settings, and plot lines with which everyday Brazilians can identify.

Telenovelas do not overtly address fertility issues, but they do promote a vision of the "ideal" Brazilian family. This family is typically middle- or upper-class, materialistic, individualistic, and full of empowered women. By challenging existing cultural and religious values through their characters, telenovelas had, and continue to have, a profound impact on Brazilian society. In essence, these programs provide a model family for Brazilians to emulate—with small family sizes being a key characteristic.

In a 2012 paper in the *American Economic Journal: Applied Economics*, a team of researchers from Bocconi University in Italy, George Washington University, and the Inter-American Development Bank (based in Washington, D.C.) analyzed various parameters to investigate statistical relationships between *telenovelas* and fertility patterns in Brazil from 1965 to 2000. Rede Globo, the Brazilian network that has a virtual monopoly on the most popular telenovelas, increased the number of areas that received its signal over those

FIGURE 1 Telenovelas are a surprising force for promoting lower fertility rates. Here, residents gather outside a café in Río de Janeiro to watch the popular program *Avenida Brasil*.

35 years (**FIGURE 2**); the network now reaches 98% of Brazilian households. By combining data on Rede Globo broadcast range with demographic data, the researchers were able to compare changes in fertility patterns over time in areas of Brazil that received access to telenovelas with areas of Brazil that did not.

The team, led by Dr. Eliana La Ferrara, found that women in areas that received the Globo signal experienced significantly lower fertility than those in areas not served by Rede Globo. La Ferrara and her team also found that fertility declines were age-related, with substantial reductions in fertility occurring in women aged 25–44, but not in younger women (**FIGURE 3**). The authors hypothesized that this effect was likely because women between 25 and 44 were closer in age to the main female characters in telenovelas, who typically are portrayed as having no children or only a single child. The depressive effect on fertility among women in areas served by Globo was therefore attributed to wider spacing of births and earlier ending of reproduction by women over 25, rather than to younger women delaying the birth of their first child.

The researchers determined that access to television alone did not depress fertility. For example, comparisons of fertility rate in areas with access to a different television network, Sistema Brasileiro de Televisão, found no such relationship. The authors of the study concluded that this was likely due to the reliance of Sistema Brasileiro de Televisão on programming imported from other nations, with which everyday Brazilians did not connect as they did with telenovelas from Rede Globo. Television's ability to influence fertility is not limited to Brazil. A 2014 study found that in the United States, tweets and Google searches for terms such as "birth control" increased significantly the day following the airing of new episodes of MTV's *16 and Pregnant*. By correlating geographic patterns in viewership with fertility data, the authors of the U.S. study concluded that MTV's *Teen Mom* series may have been responsible for reducing teenage births in the United States by as many as 20,000 per year.

The factors that affect human fertility can be complex and vary greatly from one society to another. As the Rede Globo study was correlative (p. 11), it does not prove causation between watching telenovelas and reduced fertility. It does show, however, that effects on fertility may derive from intentional factors, such as a government increasing the availability of birth control, and at other times may come from unexpected and unintentional factors—such as popular television shows.

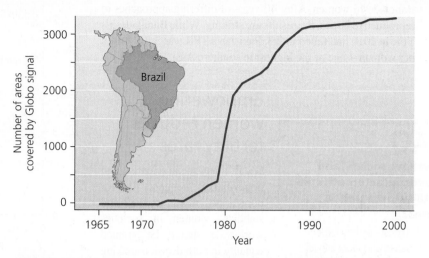

FIGURE 2 The Globo television network expanded over time and now reaches nearly all households in Brazil. Fertility declines were correlated with the availability of Globo, and its telenovelas, over the time periods in the study. *Data from La Ferrara, E., et al., 2012. Soap operas and fertility: Evidence from Brazil. Am. Econ. J. Appl. Econ. 4: 1–31.*

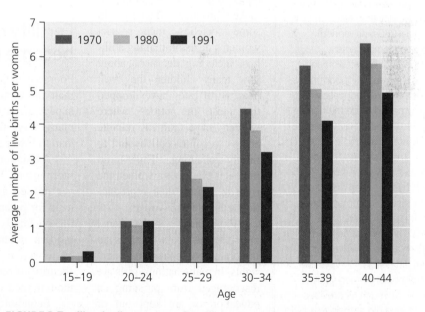

FIGURE 3 Fertility declines among Brazilian women between 1970 and 1991 were most pronounced in later age classes. Some of this decline is attributed to women in those age classes emulating the low fertility of lead female characters in telenovelas. *Data from La Ferrara, E., et al., 2012. Soap operas and fertility: Evidence from Brazil. Am. Econ. J. Appl. Econ. 4: 1–31.*

planning by Pakistani women far less reliable than that for Bangladeshi women. After 40 years of differing approaches to reproductive issues, the results are striking. While Bangladesh's TFR in 2016 had fallen to 2.3, Pakistan's TFR was 3.7 children per woman—one of the highest in southern Asia.

FAQ

How have societal and economic factors affected fertility in the United States?

The United States has experienced peaks and valleys in its TFR over the past century. In 1913, TFR in the United States was around 3.5 children per woman, and then it plunged to around 2.2 during the trying economic times of the Great Depression in the 1930s. Fertility climbed as the nation pulled out of the Great Depression, leveled off during World War II, and then climbed again during the economically prosperous period following the war (the postwar "baby boom"), peaking at 3.7 in 1957. Fertility then fell sharply in the mid-1960s, as modern contraception became widely available and women enjoyed increased opportunities to pursue higher education and employment outside the home.

Fertility rates have been largely stable around the replacement level (2.1) since 1970 but may soon fall. A 2015 report by the Urban Institute found that U.S. women born after the early 1980s (the "millennial" generation) are having children at the lowest rate in American history. Surveys find that millennial women value children as much as previous generations did at their age, but many are delaying marriage and childbirth due to financial insecurity. That is, many young people state that they'd like to marry and have children but simply can't afford to do so when faced with uncertain employment prospects and relatively high levels of debt from college.

Empowering women reduces fertility rates

Today, many social scientists and policymakers recognize that for population growth to slow and stabilize, women in societies worldwide should be granted equality in both decision-making power and access to education and job opportunities. In addition to providing a basic human right, empowerment of women would have many benefits with respect to fertility rates: Studies show that where women are freer to decide whether and when to have children, fertility rates fall, and children are better cared for, healthier, and better educated.

For women, one benefit of equal rights is the ability to make reproductive decisions. In some societies, men restrict women's decision-making abilities, including decisions about how many children they will bear. Birth rates have dropped the most in nations where women have gained reliable access to contraceptives and to family planning. This trend indicates that giving women the right to control their reproduction reduces fertility rates.

Expanding educational opportunities for women is an important component of equal rights. In many nations, girls are discouraged from pursuing an education or are kept out of school altogether. Worldwide, more than two-thirds of people who cannot read are women. And data clearly show that as women receive educational opportunities, fertility rates decline (**FIGURE 8.15**). Education encourages women to delay childbirth as they pursue careers

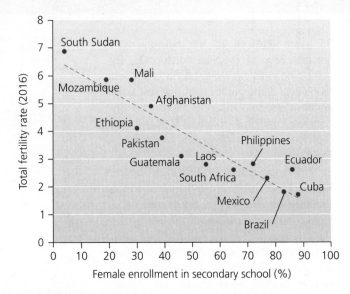

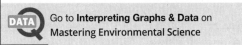

FIGURE 8.15 Increasing female literacy is strongly associated with reduced fertility in many nations. *Data for 2016 from UNESCO Institute of Statistics, http://uis.unesco.org, and Population Reference Bureau, 2016 World population data sheet.*

DATA Go to **Interpreting Graphs & Data** on **Mastering Environmental Science**

and gives them more knowledge of reproductive options and greater say in reproductive decisions.

Fertility decreases as people become wealthier

Poorer societies tend to show higher population growth rates than do wealthier societies (**FIGURE 8.16**), as one would expect given the demographic transition model. There are many ways that growing affluence and reducing poverty lead to lower rates of population growth.

As nations industrialize, they become wealthier and more urban. This depresses fertility, as children are no longer needed as farmhands. Better health care reduces the need for parents to account for infant mortality when deciding on family size. Women move into the workforce, and modern contraception becomes available and affordable. Moreover, if a government provides some form of social security to retirees, parents need fewer children to support them in their old age.

Economic factors are tied closely to population growth. Poverty exacerbates population growth, and rapid population growth worsens poverty. This connection is important because a vast majority of the next billion people to be added to the global population will be born into nations in Africa, Asia, and Latin America that have emerging, industrializing economies (**FIGURE 8.17**). This is unfortunate from a social standpoint, because in some cases these countries, many of which already are strained by rapid population growth, will be unable to provide for their people. It is also unfortunate from

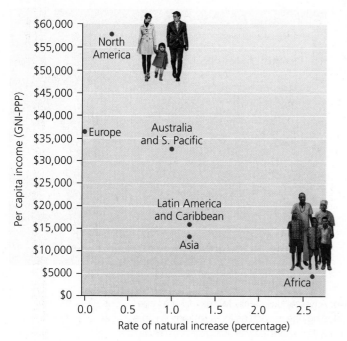

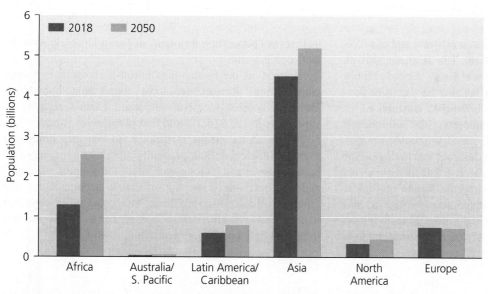

FIGURE 8.16 Poverty and population growth show a fairly strong correlation, despite the influence of many other factors. Regions with the lowest per capita incomes tend to have the most rapid population growth. Per capita income is here measured in GNI PPP, or "gross national income in purchasing power parity." GNI PPP is a measure that standardizes income among nations by converting it to "international" dollars, which indicate the amount of goods and services one could buy in the United States with a given amount of money. *Data from Population Reference Bureau,* 2018 World population data sheet.

FIGURE 8.17 The vast majority of future population growth will occur in developing regions. Africa will experience the greatest population growth of any region in coming decades. It is predicted that the highly industrialized regions of Europe and North America will experience only minor population change. *Data from Population Reference Bureau,* 2018 World population data sheet.

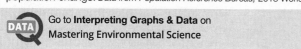

DATA Go to **Interpreting Graphs & Data** on **Mastering Environmental Science**

an environmental standpoint, because poverty often results in environmental degradation. People who depend on agriculture and live in areas of poor-quality farmland, for instance, may need to farm even if doing so degrades the soil and is unsustainable. Poverty also drives people to cut forests and to deplete biodiversity as they seek to support their families. For example, impoverished settlers and miners hunt large mammals for "bush meat" in Africa's forests, including the great apes that are now heading toward extinction.

Expanding wealth can escalate a society's environmental impacts

Poverty can lead people into environmentally destructive behavior, but wealth can produce even more severe and far-reaching environmental impacts. The affluence of a society such as the United States, Japan, or Germany is built on levels of resource consumption unprecedented in human history. Consider that the richest one-fifth of the world's people possess

Average per U.S. resident:
• Economic activity: $60,200
• Footprint: 8.4 hectares
• CO_2: 16.2 metric tons

Average per India resident:
• Economic activity: $7,060
• Footprint: 1.1 hectares
• CO_2: 1.8 metric tons

(a) A family living in the United States　　　　　　**(b) A family living in India**

FIGURE 8.18 Material wealth varies widely from nation to nation. A typical U.S. family **(a)** may own a large house with a wealth of material possessions. A typical family in a developing nation such as India **(b)** may live in a much smaller home with far fewer material possessions. One hectare = 2.471 acres. *Data for ecological footprints are from National Footprint Accounts: 2018 edition, Global Footprint Network, http://www.footprintnetwork.org; data for carbon dioxide emissions are from Global Carbon Project, 2017, http://www.globalcarbonproject.org.*

over 80 times the income of the poorest one-fifth and use 85% of the world's resources (**FIGURE 8.18**). This is meaningful, as the environmental impact of human activities depends on not only the number of people involved but also the way those people live. (Recall the *A* for *affluence* in the IPAT equation.)

An *ecological footprint* represents the cumulative amount of Earth's surface area required to provide the raw materials a person or population consumes and to dispose of or recycle the waste produced (p. 5). Individuals from affluent societies leave considerably larger per capita ecological footprints (p. 16). In this sense, the addition of 1 American to the world has as much environmental impact as the addition of 3.4 Chinese, 8 Indians, or 14 Afghans. This fact reminds

us that the "population problem" does not lie solely with the developing world.

Indeed, as the human population is rising so is human consumption. Researchers have found that humanity's global ecological footprint surpassed Earth's capacity to support us in 1971 (p. 6), and that our species is now living 50% beyond its means (**FIGURE 8.19**). We are running a global ecological deficit, gradually draining our planet of its natural capital and its long-term ability to support our civilization. The rising consumption that is accompanying the rapid industrialization of China, India, and other populous nations makes it all the more urgent for us to reverse this trend and find a path to global sustainability.

―――― Ecological footprint
■ ■ ■ Projected ecological footprint
- - - - Biocapacity

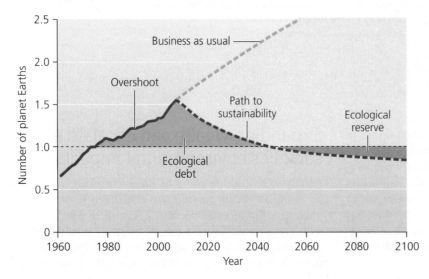

FIGURE 8.19 The global ecological footprint of the human population is estimated to be 50% greater than what Earth can bear. If population and consumption continue to rise **(orange dashed line)**, we will increase our ecological deficit, or degree of overshoot, until natural systems degrade severely and populations crash. If, instead, we pursue a path to sustainability **(red dashed line)**, we can eventually repay our ecological debt and sustain our civilization. *Adapted from WWF, 2008. Living planet report 2008. Gland, Switzerland: WWF International.*

CENTRAL CASE STUDY
connect & continue

TODAY, the number of births each year in China continues to decline. There were 17.9 million births in China in 2016—the first year after the relaxation of the "one-child" policy—but by 2017 the number of births had fallen to 17.2 million, despite more people having government permission to expand their families. The trend continued, and accelerated, for in 2018 the number of births in China fell by a stunning 12% to only 15.2 million. It appears that while Chinese citizens are now permitted to have larger families, this change in reproductive policy may not be sufficient to temper the major demographic changes in China's population that are predicted, as a result of the nearly 40-year implementation of the "one-child" policy. The Chinese government is now considering a complete scrapping of all government-instituted reproductive policies.

Globally, the human population is larger than at any time in the past. Our growing population and our growing consumption affect the environment and our ability to meet the needs of all the world's people. However, there are at least two reasons to be encouraged. First, although the global population is still rising, the rate of growth has decreased nearly everywhere, and some countries are even seeing population declines. Many nations, including China, are progressing or have through the demographic transition, showing that it is possible to create prosperous societies with little or no population growth. Second, progress has been made in expanding rights for women worldwide. Although there is still a long way to go, women are obtaining better education, more economic independence, and greater ability to control their reproductive decisions. Aside from the clear ethical progress these developments entail, they are also acting to slow population growth.

The human population cannot continue to grow forever. The question is how it will stop rising: Will it occur through the gentle and benign process of demographic transition as is happening in India, through restrictive governmental intervention such as China's reproductive policies, or through the miserable Malthusian-described checks of disease and social conflict that are caused by overcrowding and competition for scarce resources? How we answer this question today will determine not only the quality of the world in which we live but also that which we leave to our children and grandchildren. Although nations learn from one another on how to deal with rising (or shrinking) populations, it has become clear that there is no "one size fits all" solution to population issues. Population programs must be as unique as the nations they serve.

- **CASE STUDY SOLUTIONS** The reduction in birth rates that accompanied China's population control initiatives is leading to significant change in the nation's age structure. Review Figure 8.9, which shows that the population is growing older, leading to the top-heavy population pyramid for the year 2050. If you were tasked with maximizing the contributions of the growing number of retirees to Chinese society, what sorts of programs would you devise?

- **LOCAL CONNECTIONS** The U.S. Census Bureau provides demographic statistics, based on census data, on its website for every U.S. state and county. Look up the county and state where you currently live and compare its current population with its population 10 years ago. Did your county and state increase or decrease in population? Based on your experiences or research, propose two or three reasons why your county showed the population change patterns it did. Do the same for your state. What sort of population change pattern would you prefer for where you live, more people or fewer? With your classmates and friends, share your explanations on why population changed in your state and your opinions on local population change. Do your classmates and friends share your conclusions and opinions on local population changes, or do their thoughts on local population changes differ from yours?

- **EXPLORE THE DATA** Practice using current age-distribution diagrams to predict the future population growth of select nations. How good are your skills? ➜ **Explore Data** relating to the case study on **Mastering Environmental Science.**

REVIEWING Objectives

You should now be able to:

+ Describe the scope of human population growth

The human population currently stands at more than 7.6 billion and increases by 1.2% annually. It is predicted that this number will rise to around 9.8 billion people by 2050. (p. 189)

+ Discuss divergent views on population growth

Because growing populations can deplete resources, strain food supplies, and stress social systems, Thomas Malthus and Paul Ehrlich warned that overpopulation could greatly harm humanity. The contrasting Cornucopian view holds that innovation will replace consumed resources and thus population growth can continue. (p. 190)

+ Explain how human population, affluence, and technology affect the environment

The IPAT model summarizes how environmental impact (I) results from interactions among population size (P), affluence (A), and technology (T): $I = P \times A \times T$. Rising population and rising affluence may each increase consumption and environmental impact. Technology has frequently worsened environmental degradation, but it can also help lessen the impacts on us. (pp. 190–194)

+ **Explain the fundamentals of demography**

Demography applies principles of population ecology to the statistical study of human populations. Demographers study size, density, distribution, age structure, and sex ratios of populations, as well as rates of birth, death, immigration, and emigration. (pp. 194–199)

+ **Describe the concept of demographic transition**

The demographic transition model explains why population growth slows as nations experience the process of industrialization. Economic and societal factors in agrarian societies that favored large family sizes are replaced with those that favor smaller families with industrialization and urbanization. The demographic transition may or may not proceed to completion in all of today's developing nations. (pp. 199–201)

+ **Explain how family planning, the status of women, and affluence affect population growth**

Family-planning programs, reproductive education, and access to modern contraceptives have reduced population growth in many nations. Fertility rates also tend to fall in societies that grant women equal rights to men, as women delay childbirth to pursue education and employment. Poorer societies tend to show faster population growth than do wealthier societies. (pp. 202–208)

SEEKING Solutions

1. The World Bank estimates that half the world's people survive on less than $2 per day. How do you think this situation affects the political stability of the world? Explain your answer.

2. Apply the IPAT model to the example of China provided in the chapter. How do population, affluence, technology, and ecological sensitivity affect China's environment? Now consider your own country or your own state. How do population, affluence, technology, and ecological sensitivity affect your environment? How can we minimize the environmental impacts of growth in the human population?

3. Do you think that all of today's developing nations will complete the demographic transition and come to enjoy a permanent state of low birth and death rates? Why or why not? What steps might we as a global society take to help ensure that they do? Now think about developed nations such as the United States and Canada. Do you think these nations will continue to lower and stabilize their birth and death rates in a state of prosperity? What factors might affect whether they do so?

4. **THINK IT THROUGH** India's prime minister puts you in charge of that nation's population policy. India has a population growth rate of 1.4% per year, a TFR of 2.3, a 48% rate of contraceptive use, and a population that is 66% rural. What policy steps would you recommend, and why?

5. **THINK IT THROUGH** Now suppose that you have been tapped to design population policy for Germany. Germany is losing population at an annual rate of 0.2%, has a TFR of 1.6, a 69% rate of contraceptive use, and a population that is 77% urban. What policy steps would you recommend, and why?

CALCULATING Ecological Footprints

A nation's population size and the affluence of its citizens each influence its resource consumption and environmental impact. As of 2018, the world's population passed 7.6 billion.

In 2017, the average per capita income was $16,927 per year, and the latest estimate for the world's average ecological footprint was 2.8 hectares (ha) per person. The sampling of data in the table below will allow you to explore patterns in how population, affluence, and environmental impact are related.

NATION	POPULATION (MILLIONS OF PEOPLE)	AFFLUENCE (PER CAPITA INCOME)[1]	PERSONAL IMPACT (PER CAPITA FOOTPRINT, IN HA/PERSON)	TOTAL IMPACT (NATIONAL FOOTPRINT, IN MILLIONS OF HA)
Brazil	202.8	$15,900	3.1	628.7
China	1364.1	$13,130	3.7	
Ethiopia	95.9	$1,500	1.1	
France	64.1	$39,720	4.7	
India	1296.2	$5760	1.1	
Japan	127.1	$37,920	4.7	
Mexico	119.7	$16,710	2.5	
Russia	143.7	$24,710	5.6	
United States	317.7	$55,860	8.4	2668.7

[1]Measured in GNI PPP (gross national income in purchasing power parity), a measure that standardizes income among nations by converting it to "international" dollars, the amount of goods and services one could buy in the United States with a given amount of money.
Sources: Population and affluence data are from Population Reference Bureau, *2014 and 2015 World population data sheets*. Footprint data are from Global Footprint Network, http://www.footprintnetwork.org/ecological_footprint_nations/. All data are for 2014.

1. Calculate the total impact (national ecological footprint) for each country.

2. Draw a graph illustrating each country's per capita impact (on the *y* axis) versus its affluence (on the *x* axis). What do the results show? Explain why the data look the way they do.

3. Draw a graph illustrating the total impact of each country (on the *y* axis) in relation to its population (on the *x* axis). What do the results suggest to you?

4. Draw a graph illustrating the total impact of each country (on the *y* axis) in relation to its affluence (on the *x* axis). What do the results suggest to you?

5. You have just used three of the four variables in the IPAT equation. Now give one example of how the *T* (technology) variable could potentially increase the total impact of the United States and one example of how it could potentially decrease the U.S. impact.

Mastering Environmental Science

Students Go to **Mastering Environmental Science** for assignments, an interactive e-text, and the Study Area with practice tests, videos, and activities.

Instructors Go to **Mastering Environmental Science** for automatically graded activities, videos, and reading questions that you can assign to your students, plus Instructor Resources.

The Underpinnings of Agriculture

Bees to the Rescue:
By Helping Pollinators,
Farmers Help Themselves

Sran Family
Orchards   Fresno

CALIFORNIA

> "Because we understand the importance of bees and their diversity within the food system, we [do] our part in giving the bees the most beneficial diet.
> Jason Hickman, of Sran Family Orchards

> There are two spiritual dangers in not owning a farm. One is the danger of supposing that breakfast comes from the grocery, and the other that heat comes from the furnace.
> Conservationist and philosopher
> Aldo Leopold

M anaging a 3000-acre network of 10 almond orchards is no easy task. Luckily for almond grower Lakhy Sran and his family, he's got help—in the form of millions of bees.

By protecting and nurturing the honeybees and native bees that pollinate his almond trees in California's San Joaquin Valley, Sran has enabled his family business to thrive. In fact, Sran Family Orchards is the world's largest grower and processor of organic almonds.

When Sran's great aunt first came to California in the 1950s and started the family's farming business, the landscape outside of Fresno still resembled a lush and verdant paradise. Since then, the San Joaquin Valley has become one of the most intensively farmed landscapes on Earth. Decades of industrial farming have stripped the region of much of its fertile topsoil; polluted its soil, water, and air with chemical pesticides and fertilizers; and rendered its fields increasingly parched and drought-prone. Today the region's farmers need to input more water and nutrients than ever before, because these natural resources have been lost or degraded over time.

Among the most vital natural resources on which agriculture depends are the bees, flies, butterflies, and other insects that pollinate our crops, helping flowering plants produce the fruits, vegetables, nuts, and seeds we harvest. More than 150 food crops in North America—ranging from apples to tomatoes to blueberries to strawberries to squash to almonds—rely to a significant degree on pollinators. Introduced European honeybees pollinate more than $15 billion of crops annually in the United States, and the many native species of bees account for well over $3 billion of America's agriculture income. We can thank pollinators for one of every three bites of food we eat.

Yet many pollinators are in steep decline today. They have lost the habitat and diverse natural flowering resources they need as industrialized production has extended vast single-crop fields end-to-end across the landscape. Valuable pollinating insects are commonly killed accidentally by applications of chemical insecticides meant to target crop pests. Weakened and depleted by these impacts, many pollinators suffer from the spread of new pathogens and diseases.

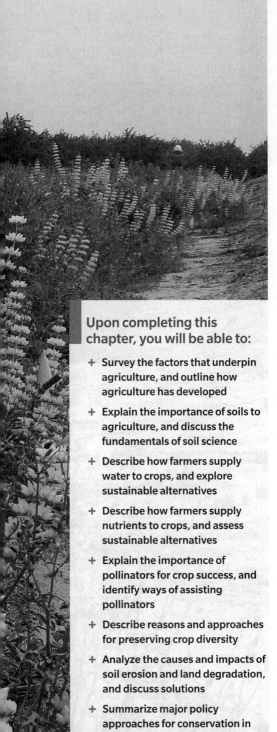

◀ **Wildflowers planted for bees, bordering an almond orchard in California**

▲ **A blue orchard bee pollinating an almond flower**

FIGURE 9.1 A cover crop of red clover planted between rows of almond trees in a San Joaquin Valley orchard helps nourish the soil, conserve water, and provide resources for pollinators.

That's where farmers like Lakhy Sran come in. They know that any help they lend to bees will be amply paid back. That's why Sran Family Orchards planted 24 acres of hedgerows filled with 13 species of native, drought-tolerant, flowering shrubs and wildflowers that bloom at different times of year. Because pollinators need pollen and nectar sources throughout the season but almonds bloom for only 3 weeks of the year, this diversity of plant species sustains the bees. Between their rows of almond trees, the Sran family has planted a carpet of flowering understory cover crops—116 acres so far—that provide resources for bees while enhancing soil fertility and water availability (**FIGURE 9.1**). An innovative drip irrigation system keeps the plants healthy and gives bees a year-round water source. Since initiating these practices, the Srans have reduced water use by 18% and fertilizer application by 26%.

Sran Family Orchards keeps its bees healthy by using no chemical pesticides at all in its certified organic orchards and by minimizing pesticide use while pursuing integrated pest management (p. 254) in its conventional orchards. This approach helps sustain the domesticated honeybees (*Apis mellifera*) that almond growers pay beekeepers to bring to their orchards each year, and it also supports a wide variety of wild native bees that pollinate almonds. In particular, the Srans are encouraging the blue orchard bee (*Osmia lignaria*) by providing hollow bamboo stalks in which this type of mason bee will nest. This native California bee is efficient and can work in cooler conditions than honeybees. By diversifying the spectrum of pollinators, the Srans hope to maximize their almond production and ensure the reliability of their yields.

All these efforts led Sran Family Orchards to become the first grower to win designation as a bee-friendly farm from Bee Better Certified, a new certification program. Bee Better Certified is spearheaded by the Xerces Society (pronounced *zerkseas*), the leading nonprofit organization dedicated to the conservation of invertebrate animals, including pollinating insects. The Xerces Society works with farmers and ranchers across North America, sending scientifically trained experts to help growers conserve the bees and butterflies that enhance their production.

In California, Xerces staff are working with several major growers in addition to Sran Family Orchards. Olam (one of the world's largest almond growers) and Muir Glen Organic Tomatoes are both running large programs to create pollinator habitat and reduce pesticide use. So far, more than 60 miles of pollinator hedgerows have been established in California's Central Valley. Across the continent, Xerces has helped hundreds of farmers and ranchers restore nearly 700,000 acres of land to pollinator-friendly conditions. In projects from Montana to Massachusetts, these growers are seeing better pollination and fruit production, while using fewer insecticides, after planting hedgerows and wildflower strips amid their orchards and fields.

Just as farmers and ranchers are recognizing the importance of pollination, people in cities and towns are pitching in to help pollinators in parks and backyard gardens. A program called Bee City USA encourages communities to create sustainable habitats for pollinators. This program has given rise to Bee Campus USA, a similar initiative for college and university campuses.

Today, close to 60 campuses are participating in Bee Campus USA. Students, faculty, and staff at these schools are developing campus habitat plans and planting native flowering plants. Schools are sponsoring service-learning projects, integrating pollination into the academic curriculum, and hosting events, workshops, and presentations to educate their campuses and the broader community about pollinators. Many, like Mineral Area College in Missouri and Morehead State University in Kentucky, are working with groundskeepers to reduce pesticide use on campus. Some, like Georgia Institute of Technology and East Georgia State College, are using apiaries and honeybee hives to educate and inspire people about bees. You can read about the many efforts and achievements of Bee Campus USA schools on the program's website. You might just want to get your own campus involved, too.

Resources and Services for Agriculture

Most of us now live in cities and towns rather than on farms or ranches, so it is easy not to think about agriculture from day to day. Yet agriculture provides our most fundamental daily needs. Most of the food we eat originates with the efforts of farmers or ranchers working the land to produce crops or raise animals. Many of the clothes we wear derive from cotton crops tended by farmers. Our lives depend directly on agriculture—and agriculture is responsible for some of our biggest impacts on the environment, as well. For these reasons, it is important to recognize where our food and fiber come from and to understand how they are produced. And for these same reasons, it will benefit us to devise ways to reduce the environmental impacts that agriculture exerts on the naturally occurring resources and processes that support it. By conserving the resources and processes that underpin agriculture, we can begin to make our production of food and fiber truly sustainable.

Natural resources and ecosystem services underpin agriculture

We can define **agriculture** as the practice of raising crops and livestock for human use and consumption. We obtain most of our food and fiber from **cropland,** land used to raise plants for human use, and from **rangeland,** or pastureland, land used for grazing livestock.

Growing crops and raising animals require inputs of natural resources. Crops require soil, sunlight, water, nutrients, and mechanisms for pollination. Livestock require food, water, space, and shelter. Both crops and livestock benefit from genetic diversity. Each of these resources is made available by the naturally occurring processes we call ecosystem services (pp. 4, 120–121, 149, 280). Soil formation, water cycling, nutrient cycling, pollination, and the maintenance of genetic diversity are among the most essential ecosystem services that nature provides and that we rely on. All these natural resources and ecosystem services are the factors that underpin agriculture—and making agriculture sustainable means safeguarding their availability and quality.

As our human population has grown, so have the amounts of natural resources and ecosystem services we exploit for agriculture. Today we commandeer fully 38% of Earth's land surface to produce food and fiber—far more land area than any other human activity. Rangeland covers 26% of Earth's land surface, and cropland covers 12%. The percentages are even greater in the United States, where nearly half the land is devoted to agriculture. Here, rangeland covers 27% and cropland covers 19% of the land area.

Agriculture has industrialized

For thousands of years, the work of cultivating, harvesting, storing, and distributing crops was performed by human and animal muscle power, along with hand tools and simple machines—an approach known as **traditional agriculture.**

The industrial revolution (p. 5) introduced large-scale mechanization and fossil fuel combustion to agriculture, just as it did to industry. Farmers replaced horses and oxen with machinery that provided faster and more powerful means of cultivating, harvesting, transporting, and processing crops. Farmers boosted yields by intensifying irrigation and introducing synthetic fertilizers, while the advent of chemical pesticides reduced herbivory by crop pests and competition from weeds. The use of machinery necessitated highly organized approaches to farming, leading farmers to plant vast areas with single crops in straight orderly rows. Such **monocultures** ("one type") are distinct from the **polycultures** ("many types") typical of traditional agriculture, such as Native American farming systems that mixed maize, beans, squash, and peppers.

All these changes were facilitated by the rise of large agricultural corporations that sold farmers and ranchers goods and equipment and influenced the methods that these growers used. All together this array of new approaches to producing food and fiber came to be known as **industrial agriculture.**

Industrial agriculture spread worldwide with the advent of the Green Revolution (p. 245), which introduced new technology, crop varieties, and farming practices to the developing world. These advances dramatically increased yields and helped millions avoid starvation. Yet despite its successes, industrial agriculture is exacting a price. The intensive cultivation of monocultures using pesticides, irrigation, and chemical fertilizers has many consequences, among them the degradation of soil, water, and pollinators that we rely on for our terrestrial food supply.

Sustainable agriculture safeguards resources and services

Sustainable agriculture describes agriculture that maintains the healthy soil, clean water, pollinators, and other resources and services essential to long-term crop and livestock production. It is agriculture that can be practiced in the same way in the same place far into the future while maintaining high yields. Making agriculture sustainable involves limiting pollution from fossil fuel inputs, and it means taking a circular, recycling-oriented approach that seeks to treat agricultural systems as ecosystems, much as the Sran family employs native plants and wildflowers to encourage beneficial insects that then enhance their almond production. Today, like the Srans, many farmers, ranchers, and scientists are creating agricultural systems that better mimic the way natural ecosystems function. (We will explore a variety of approaches for making agriculture sustainable in Chapter 10.)

In the sections that follow, we will examine several of the main factors that underpin agriculture—soil, water, nutrients, pollinators, and genetic diversity. We will study the threats to these essential resources and services, and we will see how farmers, ranchers, scientists, and policymakers are working to conserve them.

Soil: A Foundation of Agriculture

Soil supports all of Earth's terrestrial ecosystems, and healthy soil is vital for agriculture. Because every one of us relies directly on agriculture for the meals we eat and the clothing we wear, the quality of our lives is closely tied to the quality of our soil.

Soil is a living system

Soil is a multifaceted system consisting of disintegrated rock, organic matter, water, gases, nutrients, and microorganisms. Although it is derived from rock, soil is shaped by living organisms (**FIGURE 9.2**). By volume, soil consists roughly of 50% mineral matter and up to 5% organic matter, with these amounts varying by soil type. The rest consists of the space between soil particles (pore space) taken up by air or water. The organic matter in soil includes living and dead microorganisms as well as decaying material from plants and animals. The composition of a region's soil strongly influences the character of its ecosystems. In fact, because soil is composed of living and nonliving components that interact in complex ways, soil itself meets the definition of an ecosystem (pp. 61, 111).

Soil supports agriculture

Agriculture relies on healthy soil in several ways (**FIGURE 9.3**). Crop plants depend on soil that contains organic matter to provide the nutrients they need for growth. Plants also need soil with a structure and texture that holds water and allows roots to penetrate deeply, which assists with uptake of water and dissolved nutrients, allowing for proper growth. And plants depend on living organisms in the soil. In particular, many fungi form symbiotic mutualistic associations with plant roots, called *mycorrhizae* (p. 80). The dense network of fungal tissue in the soil helps draw up water and minerals, some of which are passed to the plant, and the plant provides to the fungus the carbohydrates produced by photosynthesis. Wheat and many other crops rely on mycorrhizae for proper growth. Animal agriculture also depends on soil with all these characteristics; because animals subsist on plants, the livestock we raise rely indirectly on healthy soil.

Healthy soil has sustained agriculture for thousands of years. When people first began

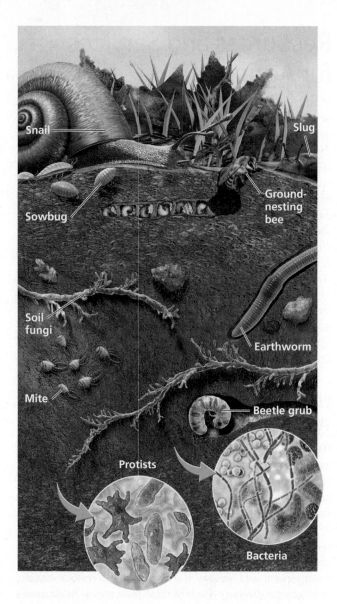

FIGURE 9.2 Soil, a complex mixture of organic and inorganic components, is full of living organisms. Most soil organisms decompose organic matter. Some, such as earthworms, help aerate the soil. Ground-nesting bees dig tunnels, lay eggs in compartments they seal off with mud, and provision their larvae with pollen until the larvae mature and emerge at the surface.

FAQ

Isn't soil just lifeless dirt?

Don't let appearances fool you—healthy soil is a complex ecological system full of life! A single teaspoon of soil can contain millions of bacteria and thousands of fungi, algae, and protists. Soil provides habitat for earthworms, insects, mites, millipedes, centipedes, nematodes, sow bugs, and other invertebrates, as well as for burrowing mammals, amphibians, and reptiles. These organisms improve the nutrient content of soil, as well as its texture for retaining water and helping plants' roots to grow. As a rule, where we find soil rich in life, we find thriving ecological communities aboveground and excellent conditions for agriculture.

farming, they were able to take advantage of deep, nutrient-rich soils that had built up over vast spans of time. Today we face the challenge of producing immense amounts of food from soil that has been farmed many times while also conserving soil fertility for the future.

Whether soil is considered a renewable natural resource depends on one's timeframe. Soil depleted of its fertility may renew itself over time, but this renewal occurs very slowly and generally not within a human lifetime. Thus, for most practical purposes it is best to consider soil a nonrenewable resource. Most farming and grazing practiced so far has depleted soils faster than they form, so it is imperative for our civilization's future that we develop sustainable methods of working with soil. Habitat restoration projects like those that

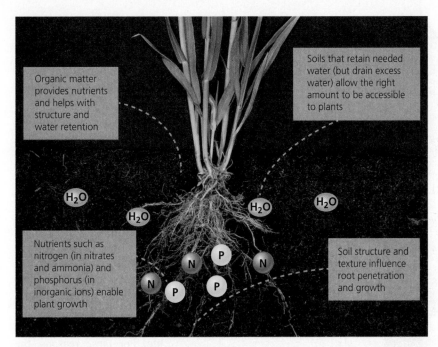

FIGURE 9.3 Crop plants such as wheat depend on healthy soil for nutrients, organic matter, water retention, and proper root growth.

Within the image:
- Organic matter provides nutrients and helps with structure and water retention
- Soils that retain needed water (but drain excess water) allow the right amount to be accessible to plants
- Nutrients such as nitrogen (in nitrates and ammonia) and phosphorus (in inorganic ions) enable plant growth
- Soil structure and texture influence root penetration and growth

complex organic molecules are broken down into simpler ones, which plants can take up through their roots. Partial decomposition of organic matter creates **humus,** a dark, spongy, crumbly mass of material made up of complex organic compounds. Soils with high humus content hold moisture well and are productive for plant life.

Weathering and the accumulation and transformation of organic matter are influenced by five main factors:

- *Climate:* Soil forms faster in warm, wet climates, because heat and moisture speed most physical, chemical, and biological processes.
- *Organisms:* Organisms contribute to weathering, and plants and decomposers add organic matter to soil.
- *Topography:* Hills and valleys affect exposure to sun, wind, and water, and they influence how soil moves.
- *Parent material:* Its attributes influence properties of the soil.
- *Time:* Soil formation can take decades, centuries, or millennia.

Soil generally forms so slowly that it cannot readily be regained once it has been lost. Because forming just 1 inch of soil can require hundreds or thousands of years, we would be wise to conserve the soil we have.

Sran Family Orchards and other growers working with the Xerces Society have carried out enhance soil quality; growing cover crops and native plants adds organic matter and nutrients to the soil and improves soil texture, enabling it to hold more water and provide homes for beneficial insects like ground-nesting bees.

Soil forms slowly

The formation of soil plays a key role in terrestrial primary succession (p. 87), which begins when the lithosphere's parent material is exposed to the effects of the atmosphere, hydrosphere, and biosphere. **Parent material** is the base geologic material in a particular location. It can be hardened lava or volcanic ash; rock or sediment deposited by glaciers, wind, or water; or **bedrock,** the mass of solid rock that makes up Earth's crust. Parent material is broken down by **weathering,** the physical, chemical, and biological processes that convert large rock particles into smaller particles. Physical weathering results from wind, rain, freezing, and thawing. Chemical weathering occurs as water or gases chemically alter rock. Biological weathering involves living things; for example, lichens (p. 87) produce acid that eats away at rock, and trees' roots rub against rock.

Once weathering has produced fine particles, biological activity contributes to soil formation through the deposition, decomposition, and accumulation of organic matter. As plants, animals, and microbes die or deposit waste, this material is incorporated amid the weathered rock particles, mixing with minerals. For example, the deciduous trees of temperate forests drop their leaves each fall, making leaf litter available to detritivores and decomposers (p. 81–82) that break it down and incorporate its nutrients into the soil. In decomposition,

A soil profile consists of horizons

As wind, water, and organisms move and sort the fine particles that weathering creates, distinct layers of soil eventually develop. Each layer is known as a **soil horizon,** and the cross-section as a whole, from the surface to the bedrock, is known as a **soil profile.**

The simplest way to categorize soil horizons is to recognize A, B, and C horizons—or topsoil, subsoil, and parent material, respectively. However, soil scientists often recognize at least three additional horizons (**FIGURE 9.4,** p. 218). Soils vary by location, and few soil profiles contain all six horizons, but any given soil contains at least some of them.

Generally, the degree of weathering and the concentration of organic matter decrease as one moves downward in a soil profile. Minerals are transported downward by **leaching,** the process whereby solid particles suspended or dissolved in liquid are transported to another location. Soil that undergoes leaching is a bit like coffee grounds in a drip filter. When it rains, water infiltrates the soil, dissolves some of its components, and carries them downward. Minerals commonly leached from the E horizon include iron, aluminum, and silicate clay. In some soils, minerals may be leached so rapidly that plants are deprived of nutrients. Leached minerals may enter groundwater, and some can pose human health risks when the water is extracted for drinking.

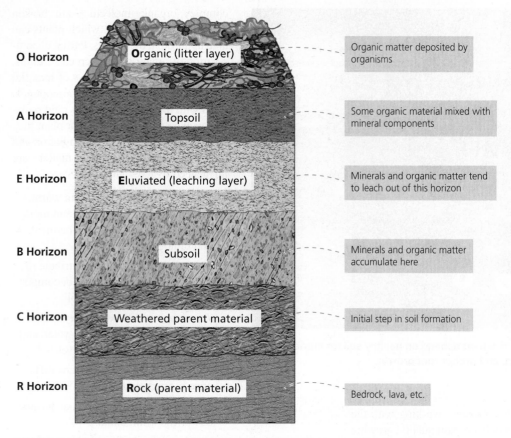

O Horizon	**O**rganic (litter layer)	Organic matter deposited by organisms
A Horizon	Topsoil	Some organic material mixed with mineral components
E Horizon	**E**luviated (leaching layer)	Minerals and organic matter tend to leach out of this horizon
B Horizon	Subsoil	Minerals and organic matter accumulate here
C Horizon	Weathered parent material	Initial step in soil formation
R Horizon	**R**ock (parent material)	Bedrock, lava, etc.

FIGURE 9.4 Mature soil consists of layers, or horizons, that have different attributes.

A crucial horizon for agriculture and ecosystems is the A horizon, or **topsoil.** Topsoil consists mostly of inorganic mineral components, with organic matter and humus from above mixed in. Topsoil is the portion of the soil that is most nutritive for plants, and it takes its loose structure, dark coloration, and strong water-holding capacity from its humus content. The O and A horizons are home to most of the organisms that give life to soil. Topsoil is vital for agriculture, but agriculture practiced unsustainably over time will deplete organic matter, reducing the soil's fertility and ability to hold water.

Soils differ in quality

Scientists classify soils—and farmers judge soil quality for farming—based on properties such as color, texture, structure, and pH.

Color A soil's color can indicate its composition and its fertility. Black or dark brown soils are usually rich in organic matter, whereas a pale color often indicates a history of leaching or low organic content.

Texture Soil texture is determined by the size of particles (**FIGURE 9.5**). **Clay** particles are less than 0.002 mm in diameter; **silt,** 0.002–0.05 mm; and **sand,** 0.05–2 mm. Sand grains, as any beachgoer knows, are large enough to see individually and do not adhere to one another. Clay particles, in contrast, readily adhere to one another and give clay a sticky feeling

when moist. Silt is intermediate, feeling powdery when dry and smooth when wet. Soil with an even mixture of the three particle sizes is known as **loam.**

Soils with large particles are porous and allow water to pass through quickly, so crops planted in sandy soils require frequent irrigation. Conversely, soils with fine particles have small pore spaces because particles pack closely together, making it difficult for water and air to pass through—in clay soils, water infiltrates slowly but is held tightly, while less oxygen is available to soil life. For these reasons, the best soils for plant growth and agriculture tend to be silty soils with medium-sized pores, or loamy soils with a mix of pore sizes.

Structure Soil structure is a measure of the "clumpiness" of soil. An intermediate degree of clumpiness is generally best for plant growth. Soil can be compacted by excessive foot traffic or by repeated plowing; compaction reduces the soil's ability to absorb water and inhibits the penetration of plants' roots.

pH Plants can die in soils that are too acidic or too alkaline, so soils of intermediate pH values (p. 29) are best for most plants. Soil pH influences the availability of nutrients for plants' roots. For instance, acids from organic matter may leach some nutrients from the sites of exchange between roots and soil particles.

Regional soil differences affect agriculture

Soil characteristics vary from place to place, and they are affected by climate and other variables. For example, it may surprise you to learn that soils of the Amazon rainforest are much less productive than soils in Iowa or Kansas. That is because the enormous amount of rain that falls in tropical regions such as the Amazon basin readily leaches minerals and nutrients out of the topsoil and E horizon, below the reach of plants' roots. At the same time, warm temperatures in the Amazon speed the decomposition of leaf litter and the uptake of nutrients by plants. Consequently, most nutrients become tied up in the forest's lush vegetation, and only small amounts of humus remain in the thin topsoil layer.

As a result, when tropical rainforest (p. 97) is cleared for farming, cultivation quickly depletes the soil's fertility. That is

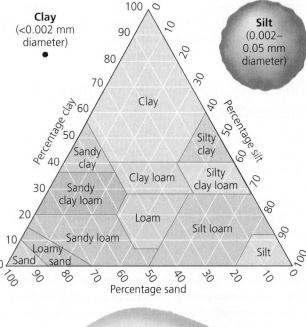

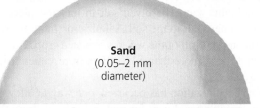

Clay
(<0.002 mm
diameter)

Silt
(0.002–
0.05 mm
diameter)

Percentage clay

Percentage silt

Clay

Silty clay

Sandy clay

Clay loam

Silty clay loam

Sandy clay loam

Loam

Sandy loam

Silt loam

Loamy sand

Sand

Silt

Percentage sand

Sand
(0.05–2 mm
diameter)

FIGURE 9.5 The texture of soil depends on its mix of particle sizes. Using this diagram, scientists classify soil texture according to the proportions of sand, silt, and clay. After measuring the percentage of each particle size in a soil sample, a scientist can trace the appropriate white lines inward from each side of the triangle to determine texture. Loam is generally best for plant growth, although some plants grow better in other textures.

DATA Go to **Interpreting Graphs & Data** on Mastering Environmental Science

why the traditional form of agriculture in tropical forested areas is *swidden* agriculture, in which the farmer cultivates a plot for one to a few years and then moves on to clear another plot, leaving the first to grow back to forest. Plots are often burned before planting, in which case the practice is called **slash-and-burn** agriculture (**FIGURE 9.6a**). Farmers burn the fallen rainforest trees on site, then enrich the soil by tilling into it the nutrient-rich ash, providing the soil sufficient fertility to grow crops. Unfortunately, these nutrients are depleted quickly, causing farmers to move deeper into the forest and slash and burn another swath of land. At low population densities this process can be sustainable, but with today's dense human populations, soils are often not given adequate time to regenerate—and increasingly, farmed plots are converted to pasture for ranching. As a result, agriculture has degraded the soils of many tropical areas.

In contrast, on the grasslands of North America, which have been almost entirely converted to agriculture (**FIGURE 9.6b**), there is less rainfall and therefore less leaching, so nutrients remain within reach of plants' roots. Plants return nutrients to the topsoil as they die, maintaining its fertility. This process creates the thick, rich topsoil of temperate grasslands, which can be farmed repeatedly with minimal loss of fertility as long as farmers guard against loss of soil.

Water for Agriculture

Just as soil is a crucial resource for farming and ranching, so is water. Plants require water for growth, and we have long provided our crops and our livestock with supplemental water when needed to boost production.

Irrigation boosts productivity

The artificial provision of water beyond that which plants receive naturally from rainfall is known as **irrigation**. Irrigation allows people to grow crops in regions that might

(a) Slash-and-burn agriculture on nutrient-poor soil in the tropics

(b) Industrial agriculture on rich topsoil in Iowa

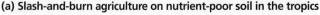

FIGURE 9.6 Regional soil differences affect how people farm. In tropical forested areas such as Laos **(a),** farmers pursue swidden agriculture by the slash-and-burn method because tropical rainforest soils **(inset)** are nutrient-poor and easily depleted. On Iowa's farmland **(b),** less rainfall means fewer nutrients are leached from the topsoil, and more organic matter accumulates, forming a thick, dark topsoil layer **(inset).**

(a) Flood-and-furrow irrigation of cotton in the southern California desert near the Colorado River

(b) Center-pivot irrigation in the southern California desert

FIGURE 9.7 Irrigating water-thirsty crops in arid regions causes us to lose a great deal of water to evaporation. In arid southern California, cotton crops **(a)** are bathed generously in irrigation water by the flood-and-furrow method. With center-pivot irrigation **(b),** crops are watered from overhead by huge pipes that rotate around a central point, creating irrigated green circles on the landscape.

otherwise be too dry (**FIGURE 9.7**). It also can help farmers maintain yields in times of drought. Some crops, such as rice and cotton, use large amounts of water and generally require irrigation, whereas others, such as beans and wheat, need relatively little water. The amount of water a crop requires also is influenced by the rate of evaporation and by the soil's ability to hold water and make it available to plant roots.

Fully 70% of all fresh water that people withdraw from rivers, lakes, and underground aquifers is used for irrigation. Irrigated acreage has increased dramatically worldwide along with the adoption of industrial farming methods, reaching 324 million ha (800 million acres), roughly the area of India. In many cases, withdrawing water for irrigation has dried up rivers and lakes and has depleted groundwater in important aquifers like the Ogallala Aquifer that underlies the Great Plains (p. 396). In California's San Joaquin Valley where the Sran family grows almonds, aquifers are being depleted as growers withdraw far more water in most years than is replaced by rainfall. During California's historic 2007–2016 drought, more than 5000 new wells were drilled in the valley,

some of them thousands of feet deep, to reach falling water tables. Since 2003, the San Joaquin Valley has lost an average of 780 billion gallons of groundwater each year. (We will examine irrigation further in Chapter 15.)

Salinization and waterlogging pose challenges

Although plants need water, it is possible to apply too much water. **Waterlogging** occurs when overirrigation saturates the soil and causes the water table to rise to the point that water drowns plant roots, depriving them of access to gases and essentially suffocating them.

A more frequent problem is **salinization,** the buildup of salts in surface soil layers. In dryland areas where precipitation and humidity are low, the evaporation of water from the soil's A horizon may pull water with dissolved salts up from lower horizons. As the water evaporates at the surface, the salts remain, often turning the soil surface white. Irrigation in arid areas generally hastens salinization, and irrigation water often contains some dissolved salt in the first place, which introduces new salt to the soil. Salinization inhibits crop production on one-fifth of the world's irrigated cropland, causing farmers to lose more than $11 billion in revenues from crop sales each year.

Once salinization has occurred, one way to alleviate it is to stop irrigating and wait for rain to flush salt from the soil. However, in the dryland areas where salinization is most often a problem, the precipitation that occurs is rarely adequate to remove salt. In theory, one could flush the soil with large quantities of less-saline irrigation water—if such water is available and if one is careful to avoid waterlogging. A better solution, however, may be to plant salt-tolerant plants, such as barley, that can be used as food or pasture.

Because remedying salinization once it has occurred is expensive and difficult, it is better to prevent it in the first place. The best way to prevent salinization is to avoid planting crops that require a great deal of water in dryland areas. A second way is to irrigate with water that is low in salt content. A third way is to irrigate efficiently, supplying no more water than a crop requires.

Sustainable approaches to irrigation maximize efficiency

A good way to reduce water use in agriculture is to better match crops and climate. Many arid regions have been converted into productive farmland through extensive irrigation, often with the support of government subsidies. Some farmers in these areas cultivate crops that require large amounts of water, such as rice and cotton. This leads to extensive water loss from evaporation in the arid climate. Choosing other crops that require far less water could enable these areas to remain agriculturally productive while greatly reducing water use.

Another approach is to embrace technologies that improve efficiency in water use. Currently, irrigation efficiency worldwide is low, as plants end up using only about 40% of the

(a) Conventional irrigation

(b) Drip irrigation

FIGURE 9.8 Irrigation methods vary in their water use.
Conventional methods **(a)** are inefficient, because most water is lost to evaporation and runoff. In drip irrigation systems **(b)**, hoses periodically drip water directly into soil near plants' roots, so much less is wasted.

water that we apply. The rest evaporates or soaks into the soil away from plant roots (**FIGURE 9.8a**). Drip irrigation systems that target water directly toward plant roots through hoses or tubes can increase efficiencies to more than 90% (**FIGURE 9.8b**).

Other ways to conserve water while irrigating include lining canals to prevent leakage, covering them to reduce evaporation, and adopting farming methods (such as conservation tillage; p. 233) that reduce water loss. On farms and at residential homes alike, rainwater can be gathered in barrels or tubs in an approach called *rainwater harvesting*. As systems for drip irrigation and rainwater harvesting become more affordable, more farmers, gardeners, and homeowners are turning to them.

In California's San Joaquin Valley, state policies enacted in the wake of the 2007–2016 drought will require many growers to implement conservation measures to reduce water consumption. Almond growers are among the heaviest water users, relying on generous irrigation throughout spring and summer—so they will need to be part of the solution. Almond growers like the Srans may be asked to take part in water-trading markets and to flood their orchards in winter with excess water to help recharge aquifers. Growers of less-profitable field crops may be required to leave more of their land fallow (unplanted) in some years.

Nutrients for Plants

Along with water and soil, nutrients (p. 25) are vital for plant growth and thus for agriculture. Plants require nitrogen, phosphorus, and potassium to grow, as well as smaller amounts of more than a dozen other nutrients. Plants remove these nutrients from soil as they grow, and nutrients may also be removed by leaching. If farmland soil is depleted of nutrients, crop yields decline. For this reason, farmers enhance nutrient-limited soils by adding **fertilizer,** substances that contain nutrients essential for plant growth.

Fertilizers boost crop yields but can be overapplied

There are two main types of fertilizers (**FIGURE 9.9**). **Inorganic fertilizers** are mined or synthetically manufactured mineral supplements. **Organic fertilizers** consist of the remains or

(a) Inorganic fertilizer

(b) Organic fertilizer

FIGURE 9.9 Two main types of fertilizer exist. Inorganic fertilizer **(a)** consists of synthetically manufactured granules. Organic fertilizer **(b)** includes substances such as compost crawling with natural decomposers like earthworms.

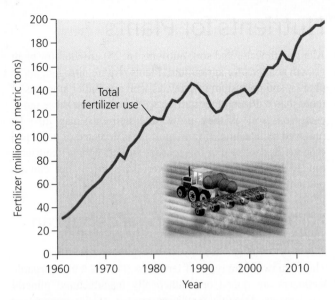

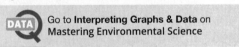

FIGURE 9.10 Use of synthetic, inorganic fertilizers has risen sharply over the past half-century. Today, usage stands at nearly 200 million metric tons annually. *Data from UN Food and Agriculture Organization (FAO).*

DATA Q Go to **Interpreting Graphs & Data** on Mastering Environmental Science

wastes of organisms and include animal manure; crop residues; charcoal; fresh vegetation; and **compost,** a mixture produced when decomposers break down organic matter such as food and crop waste in a controlled environment.

Historically, people relied on organic fertilizers to replenish soil nutrients, but during the latter half of the 20th century, many farmers embraced inorganic fertilizers (**FIGURE 9.10**). The wide use of inorganic fertilizers has boosted our food production, but their overapplication has triggered increasingly severe pollution problems. Because inorganic fertilizers tend to be more susceptible to leaching and runoff than are

organic fertilizers, they more readily contaminate surface water bodies and groundwater supplies.

Nutrients from inorganic fertilizers can have impacts far beyond the fields where they are applied (**FIGURE 9.11**). For instance, nitrogen and phosphorus runoff from farms and other sources spurs phytoplankton blooms in the Chesapeake Bay (p. 106), the Gulf of Mexico (p. 116), and other marine and coastal regions (p. 115), creating oxygen-depleted "dead zones" that kill fish and shellfish. Such eutrophication (pp. 110, 410) occurs at countless river mouths, lakes, and ponds throughout the world. Moreover, nitrates are readily leached through soil and contaminate groundwater. Components of some nitrogen fertilizers can even volatilize (evaporate) into the air, contributing to photochemical smog (p. 468) and acid deposition (p. 474). In these ways, unnatural amounts of nitrates and phosphates spread through ecosystems and can pose human health risks, including cancer and methemoglobinemia, a disorder that can asphyxiate and kill infants. Indeed, human inputs of nitrogen have modified the nitrogen cycle and now account for one-half the total nitrogen flux on Earth (p. 126).

Sustainable fertilizer use involves targeting and monitoring nutrients

Sustainable approaches to fertilizing crops with inorganic fertilizers target the delivery of nutrients to plant roots and avoid the overapplication of fertilizer. Farmers using drip irrigation systems can add fertilizer to irrigation water, thereby releasing it only near plant roots. Growers practicing no-till farming or conservation tillage (p. 233) often inject fertilizer along with seeds. Farmers can also avoid overapplication of fertilizer by regularly monitoring soil nutrient content and applying fertilizer only when nutrient levels are too low. These types of approaches are examples of **precision agriculture,** which involves using technology to precisely monitor crop conditions, crop needs, and resource use, to maximize production

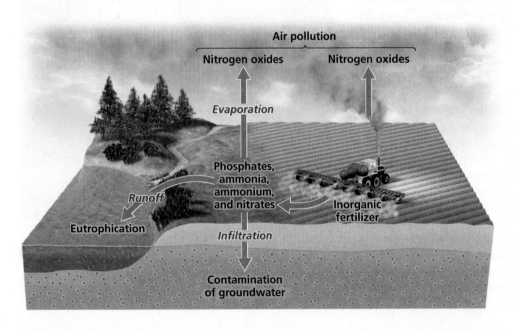

FIGURE 9.11 Overapplication of fertilizers has impacts beyond the farm field, because nutrients not taken up by plants end up elsewhere. Nitrates can leach into aquifers and contaminate drinking water. Runoff of phosphates and nitrogen compounds can alter the ecology of waterways by eutrophication. Compounds such as nitrogen oxides may pollute the air.

while minimizing waste of resources. In addition, by planting buffer strips of vegetation along field edges and watercourses, growers can help capture nutrient runoff before it enters streams and rivers.

Sustainable agriculture embraces the use of organic fertilizers, because they provide not only nutrients but also organic matter that improves soil structure and helps soils retain water and nutrients. One example is *biochar,* a term for charcoal (wood heated in the absence of oxygen) used as a soil amendment. Biochar was mixed into soil by Amazonian people for centuries to increase soil fertility and crop productivity, and today it is also being considered as a means of carbon sequestration (p. 548).

There are many ways to use organic fertilizer, and college campuses are pioneering some of them. For example, Kennesaw State University in Georgia runs a closed-loop system for recycling wastes. Uneaten food from the dining halls and scraps from food preparation are placed in a large aerobic digester tank. Over time, the food items break down inside the digester, generating roughly 1900 L (500 gal) of nutrient-rich "compost tea" each day. This liquid is trucked to Kennesaw's nearby campus farms, where it is used as an organic fertilizer.

Using organic fertilizers is not problem-free, though. Generally, organic fertilizers must be applied more often than inorganic fertilizers, and finding adequate supplies can sometimes be difficult. When manure is applied in amounts needed to supply sufficient nitrogen for a crop, it may introduce excess phosphorus, which can run off into waterways. Accordingly, most sustainable approaches do not rely solely on organic fertilizers but instead integrate them with the targeted use of inorganic fertilizer.

Pollination for Farming

Soil, water, and nutrients provide mostly physical and chemical resources for agriculture, but many crops need the biological resources and ecosystem services provided by pollinators, those insects and other animals that pollinate plants, allowing them to set seed and produce fruit.

Pollination (p. 80) is the process by which male sex cells of a flowering plant (pollen) fertilize female sex cells of a flowering plant (ova, or egg cells). Pollination can occur in different ways. Plants such as grasses and conifer trees are pollinated thanks to the wind. Millions of minuscule pollen grains are blown long distances, and by chance a small number land on the female parts of other plants of their species. In contrast, the many kinds of plants that sport flowers are typically pollinated by animals, such as hummingbirds, bats, and insects (see Figure 4.9, p. 81). Flowers are evolutionary adaptations that function to attract pollinators; the sweet smells and bright colors of flowers are signals that advertise the sugary nectar and protein-rich pollen within. Animals seeking nectar and pollen are drawn to flowers, and they end up transferring pollen from flower to flower as they visit

TABLE 9.1 A Few of the Many of Types of Pollinators

European honeybee
The honeybee raised by beekeepers and used commercially for crop pollination was introduced to America from Europe centuries ago. It sometimes competes with native bees.

Bumblebees
Bumblebees are large and conspicuous native bees that nest in small groups in the ground. Some of the many species pollinate tomatoes and other crops.

Small native bees
Most of the thousands of species are small, are solitary, nest in the ground, and pollinate a diversity of plants, including fruit and vegetable crops. Some are colorful, with yellows and iridescent greens.

Bee . . . or fly?
Many insects mimic bees and wasps for protection against predators. This hoverfly looks like a bee, and along with many other flies it pollinates a diversity of flowers and crops.

them. This process enables flowering plants to reproduce, set seed, and create fruits.

Many crops rely on pollinators

Our staple grain crops, such as wheat and corn, are derived from grasses and are wind-pollinated, but many fruit, vegetable, and nut crops depend on insects for pollination (**TABLE 9.1**). The bees, butterflies, hoverflies, and other insects that pollinate our crops provide ecosystem services estimated to be worth as much as $577 billion worldwide each year. Without them, our diets would be much less diverse, and we would have trouble producing enough food to feed our population. Pollinators are the unsung heroes of agriculture.

A thorough survey on pollination undertaken by tropical bee biologist Dave Roubik documented 800 types of cultivated plants that rely on bees and other insects for pollination. An estimated 73% of these types are pollinated by bees, 19% by flies, 5% by wasps, 5% by beetles, and 4% by moths and butterflies. Bats pollinate 6.5% and birds 4%. Overall, the 4000 native species of bees in the United States are estimated

FIGURE 9.12 Beekeepers bring hives of honeybees to crops and orchards when it is time for flowers to be pollinated.

to provide more than $3 billion of pollination services each year to U.S. crops, ranging from cherries to cranberries to sunflower to avocados to peppers to watermelon.

The European honeybee (*Apis mellifera*)—introduced to North America from the Old World many decades ago—underpins modern agriculture like no other pollinator. Farmers and orchard growers regularly hire beekeepers to bring hives of domesticated honeybees to their fields and orchards when it is time to pollinate crops (**FIGURE 9.12**). Across the United States, honeybees pollinate more than 100 crops that make up fully one-third of the American diet, providing an estimated $17 billion in annual services.

Protecting pollinators helps agriculture

When we harm pollinating insects, our crop yields are lowered—but by encouraging pollinators, we can enhance our yields (see **THE SCIENCE BEHIND THE STORY**, pp. 226–227). As one example of many, the U.S. Great Basin states are a world center for the production of alfalfa seed, and alfalfa flowers are pollinated mostly by native alkali bees that live in the soil as larvae. In the mid-20th century, many farmers began plowing the soil and increasing chemical pesticide use in an effort to boost yields. These measures killed vast numbers of the soil-dwelling bees, and alfalfa seed production plummeted in areas where alkali bees were lost. Fortunately, farmers who realized the importance of the bees began transplanting divots of soil containing bee larvae to establish and manage new populations near their crops. By encouraging these native bees, these farmers were able to raise yields of alfalfa seed from 300–600 lb/acre to 1000–2400 lb/acre.

Another key relationship is the one between bumblebees and tomatoes. Bumblebees have a trick up their sleeve, called *buzz pollination:* When visiting a tomato flower, a bumblebee vibrates its wings at about 400 Hz. This shakes pollen loose, some of which the bee collects and some of which pollinates the flower. Because tomatoes (and peppers and some other crops) benefit from this vibration and because honeybees do not provide it, tomato yields are much greater when bumblebees visit the plants. For years, companies growing tomatoes in greenhouses were hiring people to walk around with a vibrating device to manually buzz-pollinate the plants. Now, certain bumblebee species are reared and sold commercially for use in greenhouses, resulting in better yields and less expense.

Bee declines make pollinator conservation urgent

Today, many pollinating insects are in trouble. Scientific data indicate that populations of many wild native bees are declining steeply across North America and that species are disappearing, region by region. At the same time, domesticated European honeybees are suffering dramatic declines. Scientists studying the pressures on bees, butterflies, and other pollinators are concluding that they are suffering a "perfect storm" of stresses, many of which result from industrial agriculture. A direct source of mortality is the vast arsenal of chemical insecticides we apply to crops, lawns, and gardens to kill pest insects (pp. 251–252). All insects are vulnerable to these poisons, so when we try to control pests, we also end up killing beneficial insects such as bees. Because sprayed chemicals can be carried by the wind, *pesticide drift* sometimes kills beneficial insects long distances away from where the poisons were meant to be applied.

An indirect source of mortality is habitat loss, which for pollinators like bees may mean losing flowers they forage on, undisturbed areas of vegetation with plants they rest on, and logs, stumps, stems, and unpolluted soil in which they nest and lay eggs. Pollinators have suffered the loss of habitat and flower resources for decades, but it has grown worse recently as chemical herbicides (weed-killers) have eliminated non-crop plants from farm monocultures, depriving pollinators of a diversity of nectar and pollen sources.

Bees are also being attacked by novel parasites and pathogens that, like many invasive species (pp. 89, 292), have been moved around the world by human travel and trade. In particular, two accidentally introduced parasitic mites have swept through honeybee populations in recent years, decimating hives and pushing beekeepers toward financial ruin.

Researchers are finding that these multiple sources of stress seem to interact and cause more damage than the sum of their parts. For example, pesticide exposure and difficulty finding food might weaken a bee's immune system, making it more vulnerable to parasites. Any or all of these factors may possibly play a role in **colony collapse disorder,** a mysterious malady that for the past decade or two has destroyed up to one-third of all honeybees in the United States each year.

Today pollinators face a new threat in the form of neonicotinoid pesticides (p. 252), a novel type of chemical pesticide designed to be systemic in a plant, spread throughout its tissues. These pesticides are intended to kill crop pests devouring the leaves and stems of a plant, but bees and butterflies collecting pollen or nectar from flowers of a "neonic"-treated plant ingest the pesticide as well. Even many nursery plants today are treated with neonics, meaning that unsuspecting homeowners are poisoning beneficial insects in their own yards without even knowing it. The full effects of neonicotinoids on bees and other beneficial insects are not yet understood, and scientific research has shown a diversity of results thus far, so further research is urgently needed. The European Union has concluded that some neonicotinoids pose severe risks to pollinators and has issued a partial ban on their sale and use.

We can help pollinators in many ways

Fortunately, we have solutions at hand to help restore populations of bees and other pollinators. By retaining or establishing wildflowers and flowering shrubs in or near farm fields, orchards, or pastures, growers can provide bees with refuge and diverse food resources (FIGURE 9.13). In Minnesota, with help from the Xerces Society, organic farmer Sarah Woutat planted wildflower strips along her field edges, along with flowering cover crops and a quarter-acre of native wildflowers. At North Carolina A&T State University, students and staff designed and planted a 160-ft hedgerow of native flowering plants at the school's research farm (FIGURE 9.14). Along the nation's highways, the U.S. Federal Highway Administration is working with the Xerces Society to encourage flowers that provide resources to pollinators while beautifying roadsides.

Farmers, ranchers, and land managers can also help by eliminating or decreasing the use of chemical insecticides. Most pest problems can be addressed effectively by using biocontrol or integrated pest management (pp. 253–254) instead, thus reducing chemical pollution and the resulting health impacts on people and wildlife.

Among large commercial growers, Sran Family Orchards is one of many taking steps to help the insects that pollinate their crops. Some growers are working with the Xerces Society, some are working with public agencies and receiving support from conservation programs (pp. 235–237), and some are experimenting on their own. Just an hour's drive from the Sran family's orchards, almond grower Olam worked with

FIGURE 9.14 **This hedgerow that students established at North Carolina A&T State University supports pollinators, helps insects that are natural enemies of pest insects, shelters crops from wind, and serves to educate visitors.**

Xerces to install 9 km (5.6 mi) of hedgerow and wildflower habitat among its 3000 ha (7400 acres) of orchards. It planted wildflower strips near its reservoirs and 13 ha (32 acres) of cover crops within its orchards. Olam is also modifying its pest control methods to reduce chemicals, adjust the timing of application, and encourage natural enemies of pests instead. Meanwhile, in the northern half of California's Central Valley, Muir Glen Tomatoes worked with Xerces to transform a barren roadside into a lush corridor of manzanita, elderberry, coyote brush, California lilac, milkweed, lupine, poppies, and other native plants. The new habitat is helping boost Muir Glen's production of organic tomatoes, and scientists monitoring the habitat strip found a doubling of bees in just the first year after installation.

Indeed, each and every one of us who tends a yard while owning or renting a home or apartment can help pollinators by reducing or eliminating our use of pesticides and by planting native flowering plants. Providing nesting sites for bees is easy, too: For mason bees, tubes can be purchased, holes can be created by drilling into wood, or hollow plant stems can be left untouched over the winter and into the spring. For bumblebees and ground-nesting bees, often the best thing to do is to do nothing at all—simply leave fallen leaves on the ground, the soil undisturbed, and brush piles, logs, and tree stumps in place. Every residential yard planted with flowers and native plants instead of a lawn will enjoy visits by butterflies, bees, and other beneficial insects.

Preserving Genetic Diversity

Pollinators represent one biological resource vital for agriculture, but another biological resource lies within the crop plants and livestock themselves: the genetic diversity of the animals and plants we raise.

FIGURE 9.13 **Farmers and orchardists can help conserve pollinating insects by planting buffer strips of flowering native plants along the edges of their cultivated fields and orchards.**

THE SCIENCE behind the story

If We Help Pollinators, Will It Boost Crop Production?

A researcher examines flowers in an experiment to measure pollination rates

Making a living from farming is not easy, so nobody should expect a farmer to go out of the way to help pollinators if it doesn't make good economic sense. A key question, then, is whether farmers who invest in assisting bees and other pollinating insects will see that investment paid back by increased crop yields.

Scientists addressing this question are building on a small mountain of research since the 1990s that has quantified the contributions of insects in pollinating various crop plants. (Many crops can produce *some* fruit by wind or self-fertilization but produce *more* if pollinated by insects.) Researchers have often begun by monitoring visits to flowers by insects and measuring "visitation rates" by each species. However, not all visits result in the successful transfer of pollen, so visitation rate alone doesn't reveal how often pollination actually occurs. Researchers determine pollination frequency experimentally by excluding pollinators from some flowers (by wrapping flowers, plants, or plots in mesh, for instance) while leaving other flowers accessible, and then measuring any differences in fruit or seed production.

Such research involves meticulous fieldwork, but it has paid off with a wealth of detailed knowledge for dozens of crops. Researchers have learned that wild native bees outperform honeybees in pollinating most crops and that sometimes the diversity (species richness; p. 275) of pollinators matters more than the visitation rate.

If having a diversity of wild insects often helps crop production, this suggests that farms near areas of natural habitat (forests, shrublands, or flowery fields) that nourish pollinators might perform better than farms within vast regions of industrial monoculture far from natural habitat. Researchers have tested this hypothesis now for many crops in many parts of the world. One of the best early tests was a 2002 study on watermelon in California by a team led by Claire Kremen of the University of California, Berkeley.

Kremen's team compared visitation and pollination by native bees and by honeybees on organic and conventional fields, near and far from natural habitat. The results were clear: Bees were most numerous, diverse, and effective on farms nearest to natural habitat (**FIGURE 1**). Indeed, the data revealed that organic farms near natural habitat were the only ones gaining enough pollination from wild bees to produce marketable fruit without having to pay for managed honeybees to be introduced.

Similar results soon followed from other researchers. For coffee in the tropics and for cherries, field beans, winter wheat,

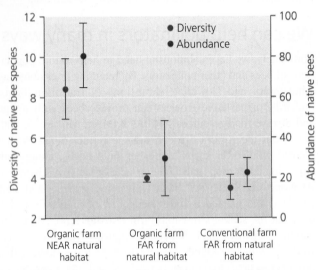

(a) Bee diversity and abundance at three types of farms

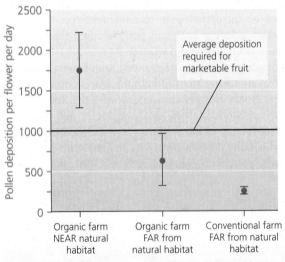

(b) Pollination deposition rates at three types of farms

FIGURE 1 Organic farms near natural habitat received the most bees and pollination in an experiment on watermelon in California. Diversity and abundance of native bees **(a)** were highest on organic farms near natural habitat. Pollen deposition **(b)** was also greatest on organic farms near natural habitat. Horizontal line in (b) indicates amount needed for marketable fruit production. *Data from Kremen, C., et al., 2002. Crop pollination from native bees at risk from agricultural intensification.* Proc. Nat. Sci. USA *99: 16812–16816.*

pumpkins, canola, strawberries, and buckwheat in Europe, proximity to wild flowering resources in natural habitat seemed to enhance pollination and crop yields as a result of visits by native insects. In the studies that did *not* find this pattern, it appeared to be because habitat and pollination were not truly limited; such studies occurred in landscapes with plenty of natural habitat, with crop systems that were not industrialized, or in areas with plenty of managed honeybees to compensate for shortfalls in native bees.

A 2008 review by Taylor Ricketts of the University of Vermont and other researchers summarized trends from 23 studies on 16 crops from 5 continents. Overall, both visitation and pollinator species richness increased with proximity to natural or semi-natural habitat. Fruit and seed production also rose with proximity to habitat but varied widely, possibly because of self-pollination or because honeybees sometimes compensate for losses of native bees.

Most recently, researchers have set out to test with manipulative experiments whether creating new habitat for pollinators will enhance crop yields. In these studies, small plots or buffer strips of wildflowers, native shrubs, or grasses are established in or near agricultural fields, and pollination and fruit production in these treated fields are compared with those of similar control fields that did not receive flowering habitat plots.

One result so far comes from England, where researchers led by Richard Pywell of the Centre for Ecology and Hydrology at the Natural Environment Research Council studied wheat, canola, and field beans. They tested the effects of replacing 3% and 8% of the cropped area (around field edges) with wildflower beds. In 2015, Pywell's team reported that yields for the crops with wildflower beds were significantly higher than for the control crops—and that overall production was higher for the experimental fields as a whole, even after considering the loss of 3% and 8% of area to the flowering beds. Moreover, the enhancement effect of wildflower beds on crop yields grew with time over the five years of the study (**FIGURE 2**).

A different result came from Switzerland, where Dominik Ganser of the University of Bern and his team worked with strawberries.

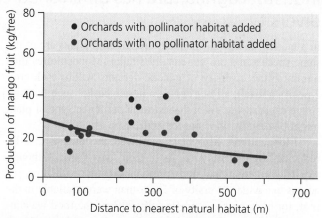

FIGURE 3 Mango production was higher for plantations with pollinator habitat patches added. The graph shows data from one site of several, where patterns were strong. The red data and trend line show how yields were higher in plantations that were closer to natural habitat in the landscape. The blue data show how plantations that received habitat patches to encourage pollinators averaged even greater yields. *Data from Carvalheiro, L.G., et al., 2012. Creating patches of native flowers facilitates crop pollination in large agricultural fields: Mango as a case study. J. Appl. Ecol. 49: 1373–1383.*

They found that fields with wildflower strips added at the edges received more pollination near the edges but less pollination in the centers so that overall there was no significant difference in yield between wildflower-strip fields and control fields. The team concluded in 2018 that establishing wildflower strips within the fields, not just at the edges, would be needed to boost yield throughout.

Creating wildflower habitat patches is not free, so even if yields are increased, is the effect great enough to make money for the farmer? In a four-year study (2009–2012) on Michigan blueberry farms, professors of entomology Brett Blaauw (now of the University of Georgia) and Rufus Isaacs (of Michigan State University) documented that wildflower plantings increased native bees and hoverflies. Blaauw and Isaacs observed that amid more of these native pollinators, the blueberry yield increased in years 3 and 4. Revenue and profits increased, too, making the plantings financially worthwhile: By year 4, income from the extra yield had covered the cost of the plantings, and future yields were expected to remain high, bringing additional profits.

Another study involved mangoes, the most popular fruit in the world. In plantations of mango trees in South Africa, researchers led by Luisa Carvalheiro of the Federal University of Goias in Brazil found that mangoes in plantations with newly created habitat patches received a larger number and diversity of pollinating insects than those in control plantations (**FIGURE 3**). On average, the plantations with habitat patches produced 1.5 kg/tree more in marketable fruit, leading to extra profit of US$342–$373/ha. Once expenses from establishing the patches were deducted, the net profit attributable to the habitat patches was $285–$317/ha.

Collectively, these studies point to benefits for crop production from maintaining natural noncrop areas within agricultural landscapes—and from creating novel patches of habitat and resources near farms and orchards to nourish the insects that pollinate our crops.

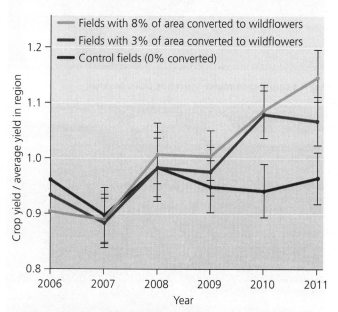

FIGURE 2 Crop yields were higher for fields with wildflower beds added, and increased through time. Lines show trends for combined yields of three crops in England, for fields with 0%, 3%, and 8% of area in flower beds. *Data from Pywell, R.F., et al., 2015. Wildlife-friendly farming increases crop yield: Evidence for ecological intensification. Proc. R. Soc. B 282: 20151740.*

Industrial agriculture has diminished diversity

In a modern industrial monoculture, all crop plants are of the same species and are genetically similar. Monocultures thus minimize the diversity of genetic variants within each crop type. As a result, all our eggs are in one basket such that any single catastrophic event that harms a particular type of plant might potentially wipe out the entire crop.

Monocultures also have narrowed the human diet by reducing the diversity of food crops we grow. Although expanded international trade provides most individual people access to a wider diversity of foods than were available in the past, for humanity as a whole fully 90% of the food we consume now comes from just 15 crop species and 8 livestock species—a drastic reduction in diversity from earlier times. One can find examples with every food type in every region of the world. Only 30% of the maize varieties that grew in Mexico in the 1930s exist today. The number of wheat varieties in China dropped from 10,000 in 1949 to 1000 by the 1970s. In the United States, apples and other fruit and vegetable crops have decreased in diversity by 90% in less than a century. Mass-market forces contribute to this trend toward lesser diversity because commercial food processors prefer items to be uniform in size and shape for convenience and because consumers often opt for familiar-looking foods.

A prime example of reduced diversity is the familiar banana we see in every grocery store. Nearly all the bananas sold and eaten in developed nations are of one variety, the Cavendish banana. All the world's Cavendish banana plants are genetically identical clones, so there is no genetic variation to protect against disease. As a result, growers douse banana plants with enormous amounts of chemical fungicides to ward off the fungal diseases to which they are vulnerable—resulting in expense, pollution, health risks to workers, and the likelihood of the fungi evolving resistance to the chemicals. This situation is a rerun of history, for the Cavendish rose to prominence only after a previous (and superior-tasting) dominant variety, the Gros Michel, was wiped out in the 1960s by a fungal disease. Fortunately, hundreds of other types of bananas still exist in the world's tropical nations; if we lose the Cavendish, we may need to make use of them soon.

Crop diversity provides insurance

Preserving the integrity of diverse native variants gives us a bulwark against the potential failure of our homogenized commercial crops and livestock. Each type of plant and animal we cultivate today was domesticated from a wild ancestral species, and most crops have had a complex evolutionary history, with people creating many *landraces* (variants adapted to local conditions) over the centuries. The wild relatives of crop plants and their local landraces contain a diversity of genes that we may someday need to introduce into our commercial crops (through crossbreeding or genetic engineering) to confer resistance to disease or pests or to meet other unforeseen challenges.

For instance, all the world's maize (corn) arose from one or more wild grass species from the highlands of Mexico. For generations, Mexican farmers had selectively bred varieties of maize, creating hundreds of local landraces (**FIGURE 9.15a**). When researchers in 2001 presented evidence that genetically engineered corn might be interbreeding with these landraces, the world's agricultural scientists refocused attention on the importance of preserving genetic diversity of the Mexican landraces, as a form of insurance for the world's corn.

Likewise, the potato blight that devastated Ireland in the 1840s and sparked the emigration of more than a million Irish farmers to the United States occurred because every potato in Ireland had derived from just one or a few strains. These strains originated in the Andes Mountains in South America, where potatoes were first domesticated. Today, thousands of amazingly diverse potato varieties persist in the Andes (**FIGURE 9.15b**), cultivated in backyard gardens and serving as a reservoir of genetic diversity that can protect the world's potato crops. Safeguarding the regions and cultures that maintain a wealth of crop diversity is the best way to conserve the genetic resources so vital to our long-term success with agriculture.

(a) A sampling of maize varieties from Mexico

(b) A sampling of potato varieties from the Andes

FIGURE 9.15 Local landraces preserve genetic diversity for crop plants. Mexico hosts numerous varieties of maize **(a)** bred by farmers over centuries, whereas Peru and other Andean nations host thousands of types of potatoes **(b)** showing striking variety.

FIGURE 9.16 The "doomsday seed vault" in Arctic Norway stores seed samples as insurance against global agricultural catastrophe.

population, we will need to modify our diets or increase agricultural production (or both)—and do so sustainably, without depleting our resources or degrading their ecosystem services. We cannot simply keep expanding farming and grazing into new areas, because land suitable and available for agriculture is running out. When we try to farm or graze on unsuitable lands, we allow fertile soil to be blown and washed away, turn grasslands into deserts, destroy forests, diminish biodiversity, encourage invasive species, and pollute soil, air, and water with toxic chemicals. Instead, we must find ways to improve the efficiency of food production in areas already under cultivation while pursuing agricultural methods that exert less impact on natural systems.

Seed banks are living museums

Another way to preserve genetic assets for agriculture is to collect and store seeds from diverse crop varieties. This is the work of **seed banks,** institutions that preserve seed types as a kind of living museum of genetic diversity. These facilities keep seed samples in cold, dry conditions to keep them viable, and the seeds are planted and harvested periodically to renew the stocks.

Examples of seed banks include the Millennium Seed Bank in the United Kingdom; the U.S. National Seed Storage Laboratory at Colorado State University; Seed Savers Exchange in Iowa; Native Seeds/SEARCH in Tucson, Arizona; and the Wheat and Maize Improvement Center in Mexico. In total, 1400 such facilities house 1–2 million distinct types of seeds worldwide.

The most renowned seed bank is the so-called doomsday seed vault established in 2008 on the island of Spitsbergen in Arctic Norway. The internationally funded Svalbard Global Seed Vault (**FIGURE 9.16**) stores millions of seeds from around the world (spare sets from other seed banks) as a safeguard against global agricultural calamity—"an insurance policy for the world's food supply." This secured, refrigerated facility is built deep into a mountain in an area of permanently frozen ground. The site has no tectonic activity, has little natural radiation or humidity, and is high enough above sea level to stay dry even if climate change melts all the planet's ice. The doomsday seed vault is an admirable effort, but we would be well advised not to rely on it to save us. It would be far better to manage our agriculture wisely and sustainably so that we never need to break into the vault.

Conserving Soil and Land Resources

We face challenges with each of the key resources needed for agriculture: soil, water, nutrients, pollinators, and genetic diversity. If we are to feed the world's rising human

Damage to land and soil makes conservation vital

Each year, our planet gains more than 80 million people yet loses 5–7 million ha (12–17 million acres, about the size of West Virginia) of productive cropland to land degradation. **Land degradation** refers to a general deterioration of land that diminishes its productivity and biodiversity, impairs the functioning of its ecosystems, and reduces the services these ecosystems provide to us. Land degradation is caused by the cumulative impacts of unsustainable agriculture, deforestation, and urban development. It is manifested in processes such as soil erosion, nutrient depletion, water scarcity, salinization, waterlogging, chemical pollution, changes in soil structure and pH, and loss of organic matter from the soil.

Within the broad problem of land degradation, agricultural concerns often focus on **soil degradation,** the process by which soils deteriorate in quality and decline in productivity. Soil degradation results primarily from forest removal, intensive cropland agriculture, and overgrazing of livestock (**FIGURE 9.17**, p. 230). In many places throughout the world, especially in drier regions, it has become more difficult to raise crops and graze livestock as soil continues to degrade. By midcentury, there will likely be 2 billion more mouths to feed (p. 194), so it is imperative that we begin practicing agriculture in sustainable ways that conserve soil and other agricultural resources.

Erosion threatens ecosystems and agriculture

Erosion is the removal of material from one place and its transport to another by the action of wind or water. When eroded material is deposited at a new location, we call this **deposition.** Erosion and deposition are natural processes, and in the long run, deposition helps create new soil. For example,

(a) Farmer in Ghana, West Africa, with degraded soil

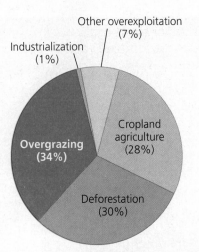

FIGURE 9.17 Soil degradation is posing challenges for agriculture in many areas of the world. Most soil degradation results from overgrazing by livestock, deforestation, and cropland agriculture. *Data from Wali, M.K., et al., 1999. Assessing terrestrial ecosystem sustainability: Usefulness of regional carbon and nitrogen models.* Nature and Resources 35: 21–33.

(b) Causes of soil degradation

flowing water can deposit freshly eroded nutrient-rich sediment across river valleys and deltas, helping to form fertile soils. That is why floodplains are excellent for farming.

However, erosion can be a problem for ecosystems and for agriculture because it tends to occur much more quickly than soil is formed. Erosion also tends to remove topsoil, the most valuable soil layer for living things (**FIGURE 9.18**). Windy regions experience the most wind erosion, whereas areas with steep slopes and intense bouts of precipitation suffer the most water erosion. In all types of landscapes, vegetation helps protect against erosion, whereas a lack of plant cover makes soil vulnerable to erosion.

People are the primary cause of soil erosion today, and we have accelerated it to unnaturally high rates. In a 2004 study, geologist Bruce Wilkinson analyzed prehistoric erosion rates from the geologic record and compared them with modern rates. He concluded that human activities move more

than 10 times more soil than all natural processes combined. We commonly worsen erosion in three ways:

- Overcultivating fields through poor planning or excessive plowing

- Overgrazing rangeland with more livestock than the land can support

- Clearing forests on steep slopes or with clear-cuts (p. 320)

Intensified soil erosion has consequences for agriculture. One study estimated that U.S. croplands have been losing about 2.5 cm (1 in.) of topsoil every 15–30 years, reducing corn yields by 4.7–8.7% and wheat yields by 2.2–9.5%. America's farmlands lose roughly 5 tons of soil for every ton of grain harvested. Worldwide, more than 19 billion ha (47 billion acres) of croplands suffer from erosion and other forms of soil degradation resulting from human activity.

Thankfully, solutions exist. A 2007 study by soil scientist David Montgomery found that land farmed under conservation approaches (which we will soon discuss) erodes far less than land under conventional industrial farming (**FIGURE 9.19**). To minimize erosion in discrete locations, we can erect physical barriers that capture soil. In the long term and across large areas, the growth of vegetation is what prevents soil loss. Vegetation slows wind and water flow, while plant roots hold soil in place and take up water.

Desertification reduces productivity of arid lands

Much of the world's population lives and farms in *drylands*, arid and semi-arid environments that cover about 40% of Earth's land surface. These dry areas are prone to **desertification,** a form of land degradation in which more than 10% of productivity is lost as a result of erosion, soil compaction, forest removal, overgrazing, drought, salinization, climate change, water depletion, and other factors. Severe desertification can expand existing desert areas and create new ones. This process has occurred most dramatically in areas of the Middle East that have been inhabited, farmed, and grazed for long periods—including the Fertile Crescent region, where agriculture first originated (p. 244).

FIGURE 9.18 Water erosion can readily remove soil from areas where soil is exposed, such as farmland.

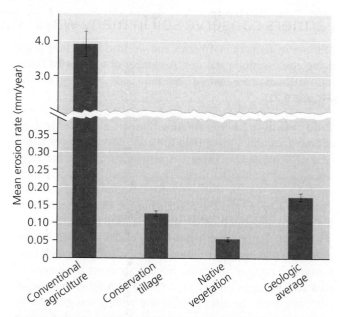

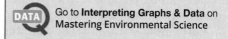

FIGURE 9.19 Erosion rates from conventional (industrial) agriculture are high. They greatly exceed rates in fields farmed under conservation tillage (p. 233), rates in areas covered by native vegetation, and rates averaged over the geologic record. *Data from Montgomery, D.R., 2007. Soil erosion and agricultural sustainability. Proc. Natl Acad. Sci. USA 104: 13268–13272.*

DATA Q Go to **Interpreting Graphs & Data** on **Mastering Environmental Science**

By some estimates, desertification endangers the food supply or well-being of more than 1 billion people in more than 100 countries and costs tens of billions of dollars in income each year. Along the border of Iran and Afghanistan, a formerly moist oasis that supported a million livestock

turned barren in just five years, and windblown sand buried more than 100 villages. Everywhere, soil degradation forces ranchers to crowd onto poorer land and farmers to reduce the fallow periods during which land lies unplanted and can regain nutrients. In positive feedback cycles (p. 108), both of these actions worsen soil degradation further. Desertification is expected to increase as climate change alters rainfall patterns, and the United Nations estimates that tens of millions of people could be displaced.

As a result of desertification, gigantic dust storms from denuded land in China now sometimes blow across the Pacific Ocean to North America. Such massive dust storms occurred in the United States during the early 20th century, when desertification shook American agriculture and society to their very roots.

The Dust Bowl prompted soil conservation in the United States

Prior to the large-scale cultivation of North America's Great Plains, native prairie grasses of this temperate grassland region held soils in place. In the late 19th and early 20th centuries, many homesteading settlers arrived in Oklahoma, Texas, Kansas, New Mexico, and Colorado, hoping to make a living there as farmers. Between 1879 and 1929, cultivated area in the region soared from 5 million ha (12 million acres) to 40 million ha (99 million acres). Farmers grew abundant wheat and ranchers grazed many thousands of cattle, sometimes expanding onto unsuitable land and causing erosion by removing native grasses and altering soil structure.

In the early 1930s, a drought worsened the ongoing human impacts, and the region's strong winds began to erode millions of tons of topsoil (**FIGURE 9.20**). Massive dust storms traveled up to 2000 km (1250 mi), blackening

(a) Kansas dust storm, 1930s

FIGURE 9.20 Drought and poor agricultural practices devastated millions of U.S. farmers in the 1930s in the Dust Bowl. The photo (a) shows towering clouds of dust approaching homes in Rolla, Kansas. The map (b) shows the Dust Bowl region. Eroded soil from this region blew eastward all the way to the Atlantic Ocean.

Severity of Erosion

Severe — Most severe

(b) Dust Bowl region

rain and snow as far away as New York and Washington, D.C. Some areas lost 10 cm (4 in.) of topsoil in just a few years. The most affected region in the southern Great Plains became known as the **Dust Bowl,** a term now also used for the historical event itself. The "black blizzards" of the Dust Bowl forced many thousands of farmers off their land, driving them into poverty and westward as refugees and adding to the economic hardship of America's Great Depression.

In response, the U.S. government, along with state and local governments, funded research into soil conservation. Congress passed the Soil Conservation Act of 1935, which established the Soil Conservation Service (SCS). This agency worked closely with farmers to develop conservation plans for individual farms, using science to assess the land's resources and condition and collaborating with landowners to ensure that the plans harmonized with landowners' objectives. SCS teams included soil scientists, forestry experts, engineers, economists, and biologists and were among the earliest examples of interdisciplinary environmental problem solving. Today, the SCS's county-based conservation districts (organized by the states but operating with federal direction, authorization, and funding) continue to implement soil conservation programs and empower local residents to plan and set priorities.

In 1994, the SCS was renamed the Natural Resources Conservation Service (NRCS), and its responsibilities were expanded to include water quality protection and pollution control. The NRCS and most state universities employ agricultural extension agents (**FIGURE 9.21**), trained experts who assist farmers by providing information on new research and techniques and by helping them apply this knowledge to implement conservation measures on their land.

FIGURE 9.21 Agricultural extension agents assist farmers by providing information on new research and techniques. In this photo, an extension agent from the National Resources Conservation Service inspects potato plants grown by a Montana farmer.

Farmers conserve soil in many ways

Millions of farmers in America and worldwide are now practicing conservation measures. A number of well-tested farming methods can reduce erosion and soil degradation (**FIGURE 9.22**).

Crop rotation In **crop rotation,** farmers alternate the type of crop grown in a given field from one season or year to the next (**FIGURE 9.22a**). Many American farmers who grow wheat or corn alternate these crops with soybeans. Soybeans are legumes, plants with specialized bacteria on their roots that fix nitrogen (p. 125), revitalizing the soil with nutrients. In this way, crop rotation returns nutrients to the soil.

Rotating crops also helps break cycles of disease from pathogens and damage from insect pests associated with continuous planting of the same crop; if an insect is adapted to feed and lay eggs on one crop, planting a different type of crop will leave the insect's offspring with nothing to eat. In addition, crop rotation helps minimize the erosion that occurs when farmers let fields lie fallow.

Contour farming Water running down a hillside can carry soil away, so farmers have developed methods for cultivating slopes. **Contour farming** (**FIGURE 9.22b**) consists of plowing furrows sideways across a hillside, perpendicular to its slope and following the natural contours of the land. In contour farming, the side of each furrow acts as a small dam that slows runoff and captures eroding soil. Contour farming is most effective on gradually sloping land.

Terracing On steep terrain, the most effective method for conserving both soil and water is terracing (**FIGURE 9.22c**). Terraces are level platforms, sometimes with raised edges, that are cut into hillsides to retain water from irrigation or precipitation. **Terracing** transforms slopes into a series of steps like a staircase, enabling farmers to cultivate hilly land without losing large amounts of soil to water erosion. Terracing is labor-intensive to establish, but it is likely the only sustainable way to farm in mountainous terrain, where farmers have used the technique for centuries.

Intercropping Farmers practice **intercropping** by planting different crops in alternating bands or other spatially mixed arrangements (**FIGURE 9.22d**). Intercropping helps slow erosion by providing more ground cover than does a single crop. Like crop rotation, intercropping reduces vulnerability to insects and disease and, when a nitrogen-fixing legume is included, replenishes the soil with nutrients. Intercropping with flowering plants that provide resources to pollinators can enhance pollination rates of the main crop.

Shelterbelts To reduce erosion from wind, farmers establish **shelterbelts,** or *windbreaks* (**FIGURE 9.22e**). These are rows of trees or tall shrubs planted along the edges of fields to slow the wind. On America's Great Plains, fast-growing species such as poplars are often used. Shelterbelts can be combined with intercropping; mixed crops are planted in rows surrounded by, or interspersed with, rows of trees that

(a) Crop rotation

(b) Contour farming

(c) Terracing

(d) Intercropping

(e) Shelterbelts

(f) No-till farming

FIGURE 9.22 Farmers have adopted various strategies to conserve soil. Rotating crops **(a)** such as soybeans and corn helps restore soil nutrients and reduce impacts of pests. Contour farming **(b)** reduces erosion on hillsides. Terracing **(c)** minimizes erosion in mountainous areas. Intercropping **(d)** reduces soil loss and maintains soil fertility. Shelterbelts **(e)** protect against wind erosion. In **(f),** soybeans grow through stubble remaining from a wheat crop, in no-till agriculture.

protect from wind while also providing fruit, wood, and wildlife habitat.

Conservation tillage Conservation tillage encompasses approaches that reduce the amount of tilling (plowing) relative to conventional farming. Turning the earth by tilling aerates the soil and works weeds and old crop residue into the soil to nourish it, but tilling also leaves the surface bare, allowing wind and water to erode away precious topsoil.

No-till farming (**FIGURE 9.22f**)—the ultimate form of conservation tillage—eliminates tilling altogether. Rather than plowing after each harvest, farmers leave crop residues atop their fields or plant *cover crops*, keeping the soil covered with plant material at all times to protect against erosion. To plant the next crop, they use a "no-till drill" mounted on a tractor to cut a thin, shallow groove into the soil surface, drop in seeds, and cover them. By planting seeds of the new crop through the residue of the old, less soil erodes away, organic material accumulates, and the soil soaks up more water and holds it longer—all of which encourage better plant growth.

By adding organic matter to the soil, no-till farming and other forms of conservation tillage capture and store carbon that otherwise would make its way to the atmosphere and contribute to climate change (Chapter 18). Moreover, because conservation tillage reduces tractor use, farmers burn less gasoline. Currently across the United States, nearly one-fourth of farmland is under no-till cultivation, and more than 40% is farmed using conservation tillage. Forty percent of soybeans, 21% of corn, and 18% of cotton receive no-till treatment.

All six of these conservation approaches are having positive effects: According to U.S. government figures, soil erosion on cropland in the United States declined from 3.05 billion tons in 1982 to 1.72 billion tons in 2010. Each of these approaches pays multiple benefits, and landowners can use many of them at once. In Montana, Doug and Anna Crabtree of Vilicus Farms grow sunflower, safflower, flax, and a dozen other crops across 600 ha (1500 acres) of semi-arid land. The buffer strips of pollinator habitat they created with the help of the Xerces Society not only sustain beneficial insects, but also serve as shelterbelts to reduce wind erosion and, like contours or terraces, catch runoff and reduce water erosion. In the buffers, they use seeds of drought-tolerant species that grow quickly for effective control of erosion and weeds. The Crabtrees rotate their crops, have experimented with different types of tillage, and plant cover crops that enhance soil quality while providing extra resources to pollinators.

FIGURE 9.23 Overgrazing occurs when livestock deplete plants faster than they can regrow, reducing the vegetation that covers the soil. The area to the right of this fence is being grazed by cattle; the area to the left is ungrazed.

WEIGHING
the issues

How Would You Farm?

You are a farmer who owns land on both sides of a steep ridge. You want to plant a sun-loving crop on the sunny, but very windy, south slope of the ridge and a crop that needs a great deal of irrigation on the north slope. What farming techniques might best conserve your soil? What factors might you want to know about before you decide to commit to one or more methods?

Grazing practices affect soil and land

We have focused in this chapter largely on the cultivation of crops, but raising livestock (discussed more fully in Chapter 10) is a major component of agriculture. People across the globe tend more than 3.4 billion cattle, sheep, and goats, most of which graze on grasses on the open range. Grazing allows us to obtain food and fiber from grasslands, but it can take a toll on soils and ecosystems. As long as livestock populations do not exceed the rangeland's carrying capacity (p. 67) and do not consume grass faster than it can regrow, grazing can be sustainable. Moreover, human use of rangeland does not necessarily exclude its use by wildlife or its continued functioning as a grassland ecosystem. However, grazing too many livestock that destroy too much plant cover impedes plant regrowth. Without adequate regeneration of plant biomass, the result is **overgrazing** (**FIGURE 9.23**).

When livestock remove too much plant cover or churn up the soil with their hooves, soil is exposed and made vulnerable to erosion. In a positive feedback cycle, soil erosion makes it difficult for vegetation to regrow, a problem that perpetuates the lack of cover and gives rise to more erosion (**FIGURE 9.24**). Too many livestock trampling the ground can also compact soil and alter its structure. Soil compaction makes it more difficult for water to infiltrate, for soil to be aerated, and for plants' roots to expand. All these effects further decrease plant growth and survival. In the new, modified environment that overgrazing creates, non-native weedy plants often invade and outcompete native vegetation. Many of these weeds are unpalatable or poisonous to livestock, leaving the rancher with a novel plant community the animals cannot utilize.

Worldwide, overgrazing causes as much soil degradation as cropland agriculture does, and it causes more desertification. Degraded rangeland costs an estimated $23 billion per year in lost productivity. Grazing exceeds the sustainable supply of grass in India by 30% and in parts of China by up to 50%. To relieve pressure on rangelands, both nations are now beginning to feed crop residues to livestock.

Subsidies and conservation measures each influence ranching

In the United States, range managers assess the carrying capacity of publicly owned rangelands and advise livestock owners of these limits so that herds are rotated from site to site to conserve grass cover and soil integrity. Most U.S. rangeland is federally owned and managed by the Bureau of Land Management (BLM). The BLM is the nation's largest

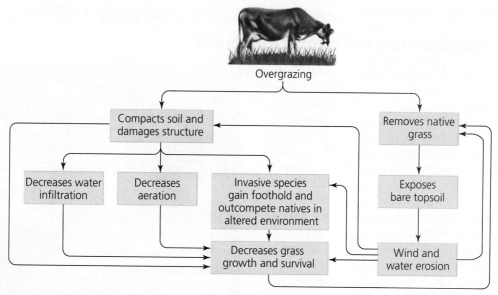

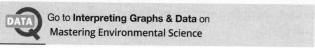

FIGURE 9.24 Overgrazing has ecological consequences. When grazing by livestock exceeds the carrying capacity of rangelands and their soil, this can set in motion a series of consequences and positive feedback loops (p. 108) that degrade soils and grassland ecosystems.

DATA Q Go to **Interpreting Graphs & Data** on **Mastering Environmental Science**

landowner, overseeing more than 100 million ha (248 million acres), mostly in 12 western states (see Figure 12.14, p. 319). Ranchers are granted rights to graze livestock on BLM lands for inexpensive fees; a grazing permit in 2019 was just $1.35 per month per "animal unit" (one horse, one cow plus calf, five sheep, or five goats).

As a result of these low fees, ranchers in the United States have traditionally had little incentive to conserve rangelands. Because most grazing has taken place on public lands leased from the government and because taxpayers have subsidized grazing, a "tragedy of the commons" situation (p. 164) has developed, and overgrazing has affected large regions of the American West.

In recent years, however, some ranchers and environmental advocates have teamed up to preserve ranchland against something each of them views as a common threat— the encroaching commercial and residential development of suburban sprawl (p. 340). Although a rancher may be able to make a great deal of money by selling land to a developer, many ranchers do not wish to see the loss of the wide-open spaces and the ranching lifestyle they cherish. Many ranchers are now collaborating with public agencies, scientists, and even environmental advocates to find ways to save ranchland, raise livestock sustainably, and safeguard the health of the land (see **SUCCESS STORY**, p. 236).

Policy can promote conservation measures in agriculture

Governments have long sought ways to encourage agricultural production, but some long-standing policies have worsened land degradation. Many nations spend billions of dollars in subsidies to fund practices that are not economically and environmentally sustainable, such as growing water-thirsty crops in desert regions. In the United States, roughly one-fifth of the income of the average farmer comes from subsidies. Proponents of such subsidies stress that the uncertainties of weather make profits and losses from farming unpredictable from year to year. To persist, these proponents say, farmers need some way of being compensated in bad years. This may be the case, but subsidies can encourage people to cultivate land that would otherwise not be farmed; to produce more food than is needed, driving down prices for other producers; and to practice methods that damage or deplete soil, water, nutrients, and pollinators. Opponents of environmentally destructive subsidies suggest that a better model is for farmers to buy insurance to protect themselves against production shortfalls.

In recent years, more and more public policy has sought to lessen the environmental impacts and external costs (pp. 143, 165) of agriculture. In theory, the marketplace should discourage people from farming and grazing using intensive methods that degrade the land they own if such practices are not profitable in the long run. But land degradation often unfolds gradually, whereas farmers and ranchers generally cannot afford to go without profits in the short term, even if they know conservation is in their long-term interests. For this reason, we have increasingly developed public policy to encourage conservation measures. In current U.S. policy, financial incentives are used to influence agricultural land use.

Collaborating to Restore Grasslands

The high desert grasslands of southern Arizona and New Mexico are gorgeous country and were long ideal for ranching, but by the 1990s, cattle ranchers in the region found themselves struggling as soil eroded and as trees and brush began to crowd out the grass. The ranchers knew that overgrazing was part of the problem, but they also suspected that decades of efforts to suppress wildfire were to blame. So in 1993, innovative ranchers launched the Malpai Borderlands Group, designating 325,000 ha (800,000 acres) of their land for protection and study. They enlisted scientists from government agencies who ran landscape-level experiments to test the impacts of grazing and burning on vegetation cover and plant and animal communities. By comparing areas where fire was suppressed to areas where fire burned, these researchers documented how the suppression of fire leads brush and trees to encroach on grass—confirming that past firefighting efforts had altered the ecosystem. With these results in hand, the Malpai Borderlands Group began to return periodic fire to the landscape by conducting carefully controlled prescribed burns (p. 323) and by allowing natural fires to run their course. Well over 100,000 ha (250,000 acres) have been burned since then. Scientists and ranchers have also collaborated to reseed damaged areas with native grasses, build structures of rock and brush to slow water erosion, rotate cattle among shared plots of land to help overgrazed areas recover, and restore native animals like jackrabbits, prairie dogs, bison, and bighorn sheep.

Ranchers of the Malpai Borderlands Group worked with ecologists and brought back fire as a landscape process, conducting controlled burns to help restore grassland.

All these efforts are restoring the grasslands and have made the Malpai Borderlands Group a model of collaboration between landowners and scientists working together toward conservation goals.

➡ **Explore the Data** at **Mastering Environmental Science**

WEIGHING
the **issues**

Soil, Subsidies, and Sustainability

Do you think financial incentive programs such as the Conservation Reserve Program and the Wetlands Reserve Program are wise policies? Are financial incentives more effective than government regulation for promoting certain land use goals? Do you think they can help lead us toward agriculture that is truly sustainable?

About every five years, Congress has passed comprehensive legislation that guides agricultural policy. The 2018 Farm Bill funded 15 programs that encourage the conservation of soil, grasslands, wetlands, wildlife habitat, and other natural resources on agricultural lands. Some of these programs take the place of earlier programs that had promoted plowing land at any cost. For example, the Wetlands Reserve Program offers payments to landowners who protect, restore, or enhance wetland areas on their property. Many of the provisions promoting soil conservation require farmers to adopt soil conservation plans and practices before they can receive government subsidies.

The **Conservation Reserve Program,** established in the 1985 Farm Bill, pays farmers to stop cultivating highly erodible cropland and instead to place it in conservation reserves planted with grasses and trees (**FIGURE 9.25**). This program was a response to previous subsidy programs that had encouraged farmers to put such marginal lands into production regardless of the environmental consequences. Farmers apply for the Conservation Reserve Program, and the U.S. Farm Service Agency contracts with them after assessing the predicted benefits to wildlife, water quality, air quality, and the farm through reduced erosion. Each year the federal government pays farmers about $2 billion for the conservation of these lands, which currently cover an area the size of Virginia. The U.S. Department of Agriculture estimates that each dollar invested in the program saves nearly 1 ton of topsoil. Besides reducing erosion, the Conservation Reserve Program generates income for farmers, improves water quality, and provides habitat for wildlife. It is a prime example of a program of "payment for ecological services" (Chapter 6)—public funds going to private individuals who take action to conserve resources and ecosystem services that benefit society as a whole.

Conservation measures for pollinators were included in the 2018 Farm Bill through the Conservation Reserve Program, the Environmental Quality Incentives Program, and three other programs. These programs provide financial incentives for farmers and ranchers to conserve the insects beneficial to agriculture, including planting hedgerows and buffer strips of wildflowers, using cover crops and intercropping, and adopting pollinator-friendly pest control measures. In this way, public policy is pointing the way toward a more sustainable model for farming and ranching that safeguards the natural resources and ecosystem services that underpin our agriculture.

FIGURE 9.25 Buffer strips of natural vegetation between farmed fields reduce erosion and improve water quality. The Conservation Reserve Program pays farmers to establish these areas.

CENTRAL CASE STUDY
connect & continue

TODAY, it has become more vital than ever to conserve the resources and services on which crops and livestock rely, including soil, water, nutrients, genetic diversity, and pollinator populations. Most bees and other pollinating insects are declining in the face of daunting challenges, but there is plenty we can do to help them—and thereby help agriculture. Thousands of farmers are now working with Xerces Society experts, NRCS extension agents, and others to make their lands friendlier for pollinators. Some of these efforts are being funded through 2021 by $4 million pledged by General Mills and the U.S. Department of Agriculture.

New challenges are mounting from climate change, however (Chapter 18). Researchers are finding evidence that droughts and hot spells—like those that have hurt California almond growers like Sran Family Orchards in recent years—are harming some bee populations. Extreme weather events are posing challenges for restoration efforts, too. In California, Xerces staff are having to adjust their wildflower seed mixes to make them more drought-tolerant. In Oregon, a large meadow Xerces was restoring was unexpectedly hit with two "hundred-year" floods, so the seed mix had to be modified toward more flood-tolerant species.

In some cases, species declines are prompting stronger restoration efforts. After scientists petitioned to list the monarch butterfly for protection under the Endangered Species Act in 2015, government agencies responded by helping restore more than 1 million acres of habitat. Similarly, the listing of the rusty-patched bumblebee under the Endangered Species Act led to the protection of 570,000 acres of habitat, as well as potentially improved management on 2.5 million acres more.

The best news is that every person can help conserve pollinators, even in cities and backyards. Creating habitat for beneficial insects is easier than for large animals because far less area is required. Reducing chemical pesticides is easily done and also brings health benefits to people. Some communities are banning neonicotinoid pesticides—23 communities in eight states have enacted bans since 2014. Indeed, recent research indicates that pollinators are faring better in cities than on industrial farmland. At colleges and universities, efforts are growing fast, and there are many ways to help. Becoming involved with Bee Campus USA can provide you with ideas and support.

With the help of science, technology, policy, and hard-won experience on the land, our society is finding ways to boost agricultural yields while conserving vital resources and reducing environmental impacts. In light of our growth in population and consumption, we will likely need still wider adoption of conservation measures to achieve sustainable agriculture and feed the nearly 10 billion people expected to crowd our planet by midcentury.

- **CASE STUDY SOLUTIONS** Pollinating insects are facing a "perfect storm" of threats in industrial agricultural landscapes, including loss of flowering resources, loss of nesting habitat, mortality from traditional chemical insecticides and newer neonicotinoids, and the spread of parasites and pathogens. List each step that Sran Family Orchards is taking to help pollinators in its almond orchards. Now add other such steps farmers are taking that are mentioned elsewhere in the chapter. If you were to run a farm or an orchard, what approaches do you think you would take to ensure that you have an adequate and healthy supply of pollinators?

- **LOCAL CONNECTIONS** What are the major agricultural products grown or processed in your state or region? Which of them rely on pollinators? What type(s) of pollinators do these agricultural products depend on? Describe three steps that are being taken, or that could be taken, to help conserve these pollinators in your state or region.

- **EXPLORE THE DATA** What are researchers learning about bee declines? ➜ **Explore Data** relating to the case study on **Mastering Environmental Science.**

REVIEWING Objectives

You should now be able to:

+ Survey the factors that underpin agriculture, and outline how agriculture has developed

Over the past century or so, industrial agriculture has been replacing traditional agriculture. As a result, today we produce more food than ever before but also face widespread environmental impacts. Conserving the natural resources and ecosystem services that support farming and ranching is crucial to attaining sustainable agriculture. (p. 215)

+ Explain the importance of soils to agriculture, and discuss the fundamentals of soil science

Productive agriculture requires healthy soil, and soil forms extremely slowly. Soil is a complex system full of organisms that decompose organic matter. Soil formation begins with weathering and is influenced by climate, organisms, topography, parent material, and time. Soil profiles consist of layers (such as the A horizon, or topsoil) with characteristic properties. We can characterize soil by color, texture, structure, and pH. Soil properties differ among regions, affecting plant growth and the choice of farming approaches. (pp. 216–219)

+ Describe how farmers supply water to crops, and explore sustainable alternatives

Conventional irrigation boosts crop yields but uses water inefficiently. Overirrigation can cause salinization and waterlogging, which lower crop yields. Farmers can conserve water by using efficient techniques and choosing crops to match soils and climates. (pp. 219–221)

+ Describe how farmers supply nutrients to crops, and assess sustainable alternatives

Growers can use inorganic or organic fertilizers to supplement nutrients for crops. Overapplying fertilizers can lead to pollution when leaching or runoff transport nutrients that affect water supplies, ecosystems, and human health. Fertilizer use can be made more sustainable by targeting fertilizers directly to plants and monitoring when they are needed. (pp. 221–223)

+ Explain the importance of pollinators for crop success, and identify ways of assisting pollinators

Many crops require pollination by insects or other animals to set seed and form fruit. Today many pollinators are dwindling due to habitat loss, chemical pesticides, parasites, and disease. Conserving pollinators is vital to our food security. By providing habitat and resources and by reducing pesticide use, farmers, ranchers, land managers, and homeowners can help conserve pollinators. (pp. 223–227)

+ Describe reasons and approaches for preserving crop diversity

Protecting diversity of native crop varieties provides insurance against failure of major crops. Seed banks preserve rare and local varieties of seed, acting as storehouses for genetic diversity. (pp. 225, 228–229)

+ Analyze the causes and impacts of soil erosion and land degradation, and discuss solutions

Some agricultural practices cause erosion, which reduces crop yields. As industrial agriculture intensifies, soils become degraded and we lose millions of acres of productive cropland each year. Soil degradation is the major component of land degradation globally, and desertification affects large areas of the world's arid regions. Farming techniques such as crop rotation, contour farming, intercropping, terracing, shelterbelts, and conservation tillage help reduce soil erosion and boost crop yields. Overgrazing can degrade soil and affect native communities, but herd rotation, grazing limits, and collaboration with scientists on conservation measures can all contribute to more sustainable grazing. (pp. 229–235)

+ Summarize major policy approaches for conservation in agriculture

Some existing policies worsen land degradation—for example, government subsidies to farmers that encourage overcultivation. Policies that promote conservation include the Conservation Reserve Program and other programs funded in U.S. farm bills. (pp. 235–237)

SEEKING Solutions

1. Select two specific techniques or approaches described in this chapter that you think are especially effective in sustaining agricultural resources—for instance, in conserving soil, conserving water, or minimizing nutrient pollution—and describe how these techniques or approaches accomplish the goal of sustaining resources and ecosystem services.

2. How do you think a farmer can best help conserve soil? How do you think a scientist can best help conserve soil? How do you think a national government can best help conserve soil? Give an example of each.

3. What pollinating insects exist in your area? Do a bit of research online to find several species of bees or other insects that pollinate crops in your region. Describe each of them and explain the benefits they bring. Now research three steps that could be taken (by yourself, homeowners, farmers, or others) to help conserve or increase populations of these beneficial insects.

4. **THINK IT THROUGH** You are a land manager with the U.S. Bureau of Land Management (BLM, p. 326) and have just been put in charge of 500,000 acres of public grasslands that have been degraded by decades of overgrazing. Soil is eroding, creating large gullies. Invasive weeds are replacing native grasses. Shrubs are encroaching on grassland areas because fire was suppressed. Environmental advocates want an end to ranching on the land. Ranchers want grazing to continue. What steps would you take to assess the land's condition? What steps would you take to begin restoring its soil and vegetation? Would you allow grazing? If so, would you set limits on it? How might you decide what those limits should be?

5. **THINK IT THROUGH** You are the head of an international granting agency that assists farmers with soil conservation and sustainable agriculture. You have $10 million to disburse. Your agency's staff has decided that the funding should go to (1) farmers in an arid area of Africa that is prone to salinization, (2) farmers in a fast-growing area of Indonesia where swidden agriculture is practiced, (3) farmers in Argentina practicing no-till agriculture, and (4) farmers in a dryland area of Mongolia undergoing desertification. What types of projects would you recommend funding in each of these areas? How would you apportion your funding among them, and why?

CALCULATING Ecological Footprints

In the United States, approximately 5 tons of topsoil are lost for every ton of grain harvested. Erosion rates vary greatly with soil type, topography, tillage method, and crop type. For simplicity, let us assume that the 5:1 ratio applies to all plant crops and that a typical diet includes 1 pound of plant material or its derived products (sugar, for example) per day. In the first two columns of the table, calculate the annual topsoil losses associated with growing this food for you and for other groups, assuming the same diet.

	PLANT PRODUCTS CONSUMED (LB)	SOIL LOSS (LB) AT 5:1 RATIO	SOIL LOSS (LB) AT 3:1 RATIO	SOIL LOSS PREVENTED (LB) AT 3:1 RELATIVE TO 5:1 RATIO
You	365	1825	1095	730
Your class				
Your state				
United States				

1. Improved soil conservation measures reduced erosion in the United States by 41% from 1982 to 2010. If additional measures were again able to reduce the current rate of soil loss by this percentage, the ratio of soil lost to grain harvested would fall from 5:1 to about 3:1. Calculate the soil losses associated with food production at a 3:1 ratio, and record your answers in the third column of the table. How many times your own weight is the amount of soil lost at a 5:1 ratio? How many times your own weight is the amount of soil lost at a 3:1 ratio?

2. Now calculate the amount of topsoil hypothetically saved by the additional conservation measures in Question 1, and record your answers in the fourth column of the table. How many times your own weight is the amount of soil erosion prevented by these conservation measures?

3. Define a "sustainable" rate of soil loss. Describe how you might determine whether a given farm was practicing sustainable use of soil.

Mastering Environmental Science

Making Agriculture Sustainable

Sustainable Food and Dining at the University of Michigan

The students are involved from seed to plate.
Frank Turchan, Executive Chef for Michigan Dining

What we eat has changed more in the last 40 years than in the previous 40,000.
Eric Schlosser, *Fast Food Nation* (2005)

t's not unusual to see phrases such as "farm-to-table" and "think globally, eat locally" when you're dining at a trendy restaurant, but would you expect to see them at your campus dining hall? Increasingly, college and university dining services are leading the way in culinary sustainability, a pursuit that embraces using fresh, healthy, locally produced foods to provide delicious, nutritious meals.

One leader in sustainable dining is the University of Michigan. The 46,000 students at the University of Michigan's Ann Arbor campus enjoy food grown and tended by fellow students on their own campus farm—just one aspect of a thriving sustainable food program across campus.

Michigan's journey toward dining sustainability began in the 1990s when professors established a course in sustainable agriculture. Students soon formed a gardening club and planted a demonstration garden on campus. In 2011, students proposed creating a campus farm next to the university's Matthaei Botanical Garden and won a $42,000 grant from the university's sustainability program, Planet Blue, to get it up and running.

As the farm got underway in 2012, the university was hiring faculty and creating classes, an undergraduate minor, and a graduate certificate program in Sustainable Food Systems. Michigan Dining (MDining) was incorporating the latest trends in sustainability into its campus dining halls. And the University of Michigan Sustainable Food Program was established, acting as an umbrella organization for 11 food-related student organizations, 7 campus gardens, and the Campus Farm.

The farm grew, and in 2016 the university hired a farm manager, Jeremy Moghtader, who had managed Michigan State University's organic farm for 12 years. Along with 3 full-time student managers, Moghtader coordinates 20–50 student volunteers per week, as well as 10–15 faculty and 20–30 staff from the botanical garden and from the university's Nichols Arboretum. Together this team constructed hoop houses (simple tunnel-like greenhouses), allowing them to grow vegetables year-round.

In 2017, the Campus Farm was certified by the U.S. Department of Agriculture to sell its produce, and MDining began buying it. In that first year, the farm supplied more than $35,000 worth of tomatoes,

Upon completing this chapter, you will be able to:

+ Explain the challenge of feeding a growing human population

+ Identify the goals, methods, and consequences of the Green Revolution

+ Discuss how we raise animals for food, and assess the impacts of meat consumption

+ Explore strategies for pest and weed management

+ Describe the science behind genetic engineering

+ Compare the benefits and costs of genetically modified foods, and assess the public debate over them

+ Analyze the nature, growth, and potential of organic agriculture

+ Summarize potential pathways toward sustainable agriculture

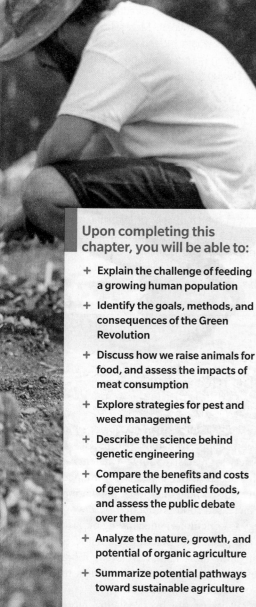

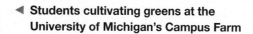

◀ **Students cultivating greens at the University of Michigan's Campus Farm**

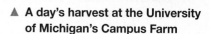

▲ **A day's harvest at the University of Michigan's Campus Farm**

peppers, cucumbers, kale, and other vegetables to the dining halls. "It creates a personal interaction with the food," sophomore Blake McWatters told the campus newspaper. "Instead of food coming off a truck, it means the food is coming from students."

Once food reaches the dining halls, MDining takes the baton, aiming to maximize nutrition and taste while minimizing impacts and waste. Dining hall staff prepare foods to order (such as sandwiches, omelets, and stir-fries), giving students exactly what they want without wasting food. To reduce food waste, MDining also uses trayless dining and smaller portion sizes, letting students return for seconds as needed. Each dining hall offers vegetarian and vegan selections for every meal and encourages certified sustainable seafood over red meat, helping cut down on the carbon emissions and other environmental impacts of red meat. Even the dining halls themselves have green building features. Appliances and lighting are energy-efficient, low-emissions paints and carpeting are used, and chairs, tables, paneling, and tiles are largely made from recycled or reclaimed materials.

To further reduce food waste, scraps from food preparation (such as carrot tops and potato peelings) are sent to a compost center to be turned into organic mulch and topsoil. Leftover food from the dining halls is composted as well. A pulping machine extracts water from the waste to compact it, helping with transport to the compost center. Used cooking oil is recycled into biodiesel (p. 578) and animal feed.

On Mondays of each week, the dining halls serve no red meat and highlight local produce. The campus also puts on special events, including farmers' markets, a fall harvest festival, food drives for low-income students, culinary conferences for chefs from other schools, a sustainable dinner for incoming students that emphasizes local Michigan foods, and zero-waste meal events with only recyclable and compostable items. MDining staff also engage in community outreach, visiting local elementary schools to talk about food sustainability.

The university's goal is for 20% of its food to be sustainably sourced (local and third-party-certified) by 2025—which is quite a challenge given Michigan's short growing season. The university also set a goal to reduce its waste sent to landfills by 40% by 2025—and because one-third of campus waste comes from food, MDining is playing a big part in this effort.

Across North America, more and more campus food service operations are pursuing culinary sustainability. Michigan State University engages students in organic farming through teaching, research, and production. The University of Massachusetts Amherst runs the largest collegiate dining program sourced from local farmers. At California State University, Chico, more than 35 students are employed at the school's 800-acre farm, supplying fruits, vegetables, meat, and dairy products to the dining halls, farmers' markets, and local restaurants. Altogether, about 70% of the largest colleges and universities in the United States now have a campus farm or garden from which their dining halls can source food directly, according to a recent survey of major institutions. Meanwhile, hundreds of schools have set food waste reduction initiatives in motion, and many campus dining halls are gaining Leadership in Energy and Environment Design (LEED) certification (p. 349) as green buildings.

In all these ways, campuses are helping in the global quest for sustainable agriculture—they are microcosms of society at large and can serve as laboratories for sustainable practices. If our planet is to support nearly 10 billion people by midcentury, we will need to adopt agricultural approaches that ensure robust yields, reduce pollution and waste, and conserve the resources and ecosystem services that support our food production. Collectively, efforts at the University of Michigan and hundreds of other schools are reducing the ecological footprints of colleges and universities and pointing the way to a more sustainable food future—all while supplying students with healthy and delicious meals.

The Race to Feed the World

A student enjoying a meal at one of the University of Michigan's dining halls, surrounded by a diverse bounty of food options, might find it hard to imagine that many people in the world struggle each day to get enough to eat. Sadly, though, this is the case—and as our human population continues to grow, the challenge of feeding the world's people will only intensify. Our numbers are expected to swell to nearly 10 billion by the middle of this century; for every four people living today, there will be five people in 2050. Feeding more than 2 billion additional mouths while protecting the natural resources that underpin agriculture will require a large-scale embrace of farming and grazing practices that are more sustainable than those we currently use.

We are producing more food per person

Providing **food security**—the guarantee of an adequate, safe, nutritious, and reliable food supply available to all people at all times—is one of society's greatest ongoing challenges. The good news is that for decades, our capacity to produce food has been growing even faster than our global population (**FIGURE 10.1**). We have increased food production by devoting more fossil fuel energy to agriculture; intensifying our use of irrigation, fertilizers, and pesticides; planting and harvesting more frequently; cultivating more land; developing more productive crop and livestock varieties; and finding ways to make more efficient use of natural resources.

Improving people's quality of life by producing more food per person is a monumental achievement of which humanity can be proud. However, ensuring that our food

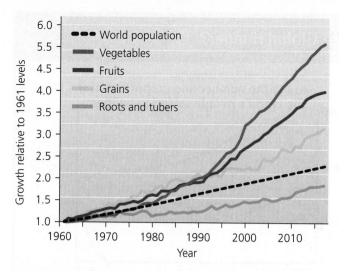

FIGURE 10.1 **Global production of most foods has risen more quickly than world population.** This means that we have produced more food per person each year. The data lines show cumulative increases relative to 1961 levels (e.g., a value of 2.0 means twice the 1961 amount). Food is measured by weight. *Data from UN Food and Agriculture Organization (FAO).*

production can be sustained depends on conserving soil, water, and biodiversity by adopting responsible agricultural practices (Chapter 9). Today, many soils are in decline, and most of the planet's arable land has already been claimed. Hence, even though food production has outpaced population growth so far, we have no guarantee that this will continue.

We face nutritional challenges

Despite our rising food production, more than 800 million people—one of every nine people in the world—do not have enough to eat. These people are said to be undernourished or to suffer from **undernutrition,** commonly defined as receiving too few nutrients (calories, proteins, vitamins, minerals, or carbohydrates) such that health problems result. At the same time, hundreds of millions of other people consume *more* than is healthy and suffer from **overnutrition,** receiving too many calories, proteins, vitamins, minerals, or carbohydrates each day. Undernutrition and overnutrition are each forms of **malnutrition,** which describes a general condition of impaired health resulting from a diet with an imbalance of nutrients.

Undernutrition can lead to disease (**FIGURE 10.2**). For example, people who eat a diet high in starch but deficient in protein or essential amino acids (p. 28) can develop *kwashiorkor.* Children who have recently stopped breast-feeding and are no longer getting protein from breast milk are most at risk for developing kwashiorkor, which causes bloating of the abdomen, deterioration and discoloration of hair, mental disability, immune suppression, developmental delays, and reduced growth. Protein deficiency together with a lack of calories can lead to *marasmus,* which causes wasting or shriveling among millions of children in the developing world. A deficiency of iron in the diet can lead to anemia, which causes fatigue and developmental disabilities. Iodine deficiency can

cause swelling of the thyroid gland and brain damage. Vitamin A deficiency can lead to blindness. Severe undernutrition can be fatal; every 5 seconds, somewhere in the world, a child dies because he or she lacks access to a nutritious diet.

In most cases, undernutrition results from poverty that limits the amount of food people can buy. One in every seven of the world's people lives on less than $1.25 per day, and one in three lives on less than $2 per day, according to World Bank estimates. Political obstacles, regional conflict and wars, and inefficiencies in distribution contribute significantly to hunger as well. Often people are caught in a tragic cycle of poverty and hunger: Undernutrition inhibits one's ability to find or perform work, which leads to a lack of income, which in turn leads to further undernutrition.

Most people who are undernourished live in the developing world. However, hunger is also a problem in the United States, where the U.S. Department of Agriculture (USDA) has classified 49 million Americans as "food insecure," lacking the income required to reliably procure sufficient food.

Globally, the number of people suffering from undernutrition has been falling since the 1960s, and the percentage of people who are undernourished has fallen even more steeply. This reduction of hunger represents one of humanity's greatest triumphs. However, in the past few years these trends appear to have reversed: The number and percentage of undernourished people is now rising again, likely due to impacts of resource degradation, climate change, and instability in war-torn regions. Despite immense progress, we still have a long way to go to eliminate hunger (see **DATAGRAPHIC,** p. 244).

Overnutrition is a problem particularly in industrialized nations such as the United States, where food is abundant, junk food is cheap, and people often lead sedentary lives with little exercise. As a result, more than one in three American adults are obese, according to the Centers for Disease Control and Prevention. Across the world as a whole, the World

FIGURE 10.2 **Millions of children suffer from forms of malnutrition, such as kwashiorkor and marasmus.**

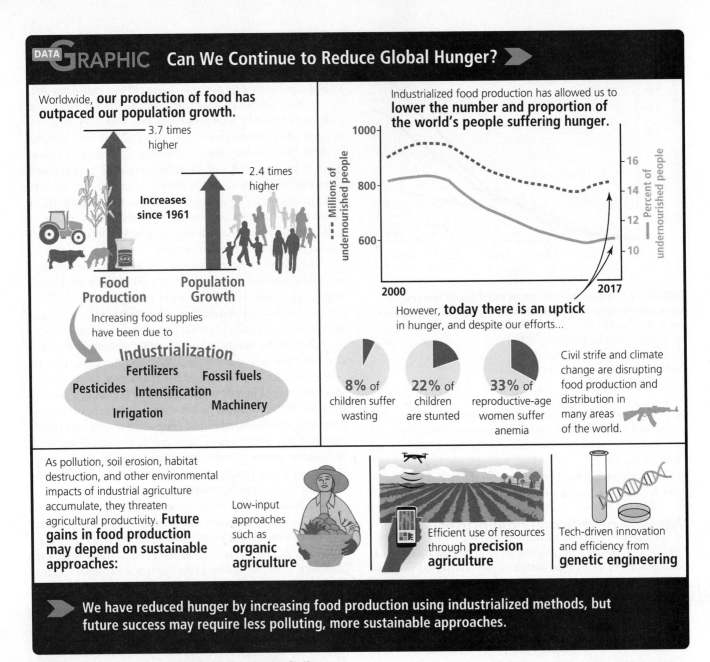

Worldwide, **our production of food has outpaced our population growth.**

3.7 times higher

2.4 times higher

Increases since 1961

Food Production

Population Growth

Increasing food supplies have been due to

Industrialization

Fertilizers   Fossil fuels
Pesticides   Intensification
Irrigation   Machinery

Industrialized food production has allowed us to **lower the number and proportion of the world's people suffering hunger.**

1000

800

600

- - - Millions of undernourished people

Percent of undernourished people

16

14

12

10

2000   2017

However, **today there is an uptick** in hunger, and despite our efforts...

8% of children suffer wasting

22% of children are stunted

33% of reproductive-age women suffer anemia

Civil strife and climate change are disrupting food production and distribution in many areas of the world.

As pollution, soil erosion, habitat destruction, and other environmental impacts of industrial agriculture accumulate, they threaten agricultural productivity. **Future gains in food production may depend on sustainable approaches:**

Low-input approaches such as **organic agriculture**

Efficient use of resources through **precision agriculture**

Tech-driven innovation and efficiency from **genetic engineering**

We have reduced hunger by increasing food production using industrialized methods, but future success may require less polluting, more sustainable approaches.

Data from Food and Agriculture Organization of the United Nations (FAO).

Health Organization estimates that 39% of adults are overweight and that, of these, one-third are obese. Excessive weight can lead to heart disease, diabetes, stroke, some types of cancer, and other health problems. The growing availability of highly processed foods (which are often calorie-rich, nutrient-poor, and inexpensive) suggests that overnutrition will remain a challenge.

Agriculture has changed over time

To provide people proper nutrition and enhance global food security, we will need to find ways to make our agricultural methods more sustainable. This will be challenging, but we can take heart that across human history our means of producing food has already undergone many transformations.

During most of the human species' existence, we were hunter-gatherers, depending on wild plants and animals for food and fiber. Then about 10,000 years ago, as the climate warmed and glaciers retreated, people in some cultures began to raise plants from seed and to domesticate animals. By sowing the seeds of wild plants whose fruits, grains, and nuts were especially large and delicious and by breeding the most productive plants with one another, our ancestors managed to create generation after generation of plants with increasingly large and flavorful produce. This practice of selective breeding (p. 52) resulted in the hundreds of crops we enjoy today, all of which are artificially selected versions of wild plants. People followed the same process of selective breeding with animals, creating livestock from wild species. Evidence from archaeology suggests that agriculture was invented independently by different cultures in at least five areas of the world, and possibly in 10 or more (**FIGURE 10.3**).

Once our ancestors learned to cultivate crops and raise animals, they began to settle in more permanent camps and

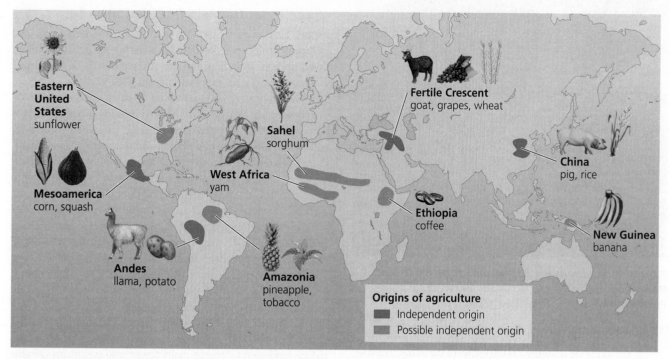

FIGURE 10.3 Agriculture originated independently in multiple regions of the world as different cultures domesticated plants and animals from wild species. *Data from Diamond, J., 1997. Guns, germs, and steel. New York, NY: W.W. Norton; and Goudie, A., 2000. The human impact, 5th ed. Cambridge, MA: MIT Press.*

villages. The need to harvest crops kept people sedentary, and once they were sedentary, it made sense to plant more crops. As food supplies grew and became more reliable, carrying capacities (p. 67) increased and populations rose. More people meant more food was needed, which led to the intensification of agriculture. The capacity to grow excess produce also enabled some people to live off the food that others produced, which eventually led to the development of professional specialties, commerce, technology, densely populated urban centers, social stratification, and politically powerful elites. For better or worse, the advent of agriculture may be largely responsible for the civilization we know today.

During these past few thousand years, most people practiced traditional agriculture (p. 215), tending crops and livestock using human and animal muscle power, hand tools, and simple machines. Most traditional farmers produced only enough food for their own subsistence, but today many have integrated into market economies and produce excess food to sell, using teams of animals for labor and significant quantities of irrigation water and fertilizer.

In the past century or two, the industrialization of agriculture has boosted our production of food and fiber to extraordinary levels. Industrial agriculture (p. 215) has increased yields by introducing advanced technology and mechanized approaches, by using fossil fuels to run machines and produce synthetic fertilizers and chemical pesticides, and by expanding irrigation across huge areas of land. By intensifying our inputs of energy and resources and by managing production efficiently at vast scales, we have enhanced our agricultural output beyond what any farmer or rancher of the past could have imagined.

The Green Revolution boosted yields worldwide

As the human population rose and as more and more arable land was put to use, scientists realized that we could not go on increasing food production forever simply by converting additional land to farming and ranching. In response, they devised methods and technologies to increase crop yields (production per unit area) on existing cultivated land. As a result, yields rose dramatically in industrialized nations. For instance, the average acre of a U.S. corn field raised its corn output fivefold during the 20th century. Many people came to view such growth in production and efficiency as key to ending starvation in developing regions. The desire to address food insecurity for the world's growing population led in the mid- and late-20th century to the **Green Revolution,** a movement to introduce new technology, crop varieties, and farming practices to developing nations.

The transfer of technology and practices to the developing world that marked the Green Revolution began in the 1940s, when American agricultural scientist Norman Borlaug introduced Mexico's farmers to a specially bred type of wheat (**FIGURE 10.4**, p. 246). This strain of wheat produced large seed heads, was resistant to diseases, had short stature to resist wind, and produced high yields. Within two decades of planting this new crop, Mexico tripled its wheat production and began exporting wheat. The stunning success of this program inspired others. Borlaug—who won the Nobel Peace Prize for his work—took his wheat to India and Pakistan and helped transform agriculture there as well.

FIGURE 10.4 Norman Borlaug helped launch the Green Revolution. The high-yielding, disease-resistant wheat that he bred helped boost yields in developing countries.

Industrialized agriculture has brought mixed consequences

Along with the new grains, developing nations imported the methods of industrial agriculture. They began applying large amounts of synthetic fertilizers and chemical pesticides, irrigating crops generously with water, and using more machinery powered by fossil fuels. From 1900 to 2000, people increased energy inputs into agriculture by 80 times while expanding the world's cultivated area by just one-third. This high-input agriculture succeeded in producing more corn, wheat, rice, and soybeans from each hectare of land. Intensified agriculture saved millions in India from starvation in the 1970s and turned that nation into a net exporter of grain (**FIGURE 10.5**).

The environmental and social impacts of these developments have been mixed. On the positive side, intensifying the use of already-cultivated land reduced pressures to convert additional lands for cultivation. Between 1961 and 2013, global food production more than tripled and per-person food production rose 48%, while area converted for agriculture increased only 11%. In this way, the Green Revolution helped preserve biodiversity and ecosystems by preventing a great deal of deforestation and habitat destruction. On the negative side, the intensified use of fossil fuels, water, fertilizers, and pesticides has worsened topsoil erosion and the pollution of soil, air, and water (Chapter 9).

In today's conventional industrial agriculture, crops are planted in monocultures, large expanses of single crop types (p. 215; **FIGURE 10.6**). This method makes planting and harvesting efficient and thereby increases output. However, monocultures reduce biodiversity because many fewer wild organisms are able to live in monocultures than in native habitats or in traditional small-scale polycultures. Moreover, when all plants in a field are of the same species (and thus genetically similar), they are susceptible to viral diseases, fungal pathogens, and insect pests that can multiply and spread quickly from plant to plant. For this reason, monocultures bring risks of catastrophic failure.

Today, yields are declining in some regions targeted by the Green Revolution, likely due to soil degradation from the heavy use of fertilizers, pesticides, and irrigation. Because wealthier farmers with larger holdings of land were best positioned to invest in Green Revolution technologies, many low-income farmers who could not afford the technologies were driven out of business. Many of them moved to cities, adding to the extensive migration of poor rural people to urban areas of the developing world (p. 338).

Soon many developing countries were doubling, tripling, or quadrupling their yields using selectively bred strains of wheat, rice, corn, and other crops from industrialized nations. When Borlaug died in 2009 at age 95, he was widely celebrated as having saved more lives than anyone in history—estimated to be as many as a billion.

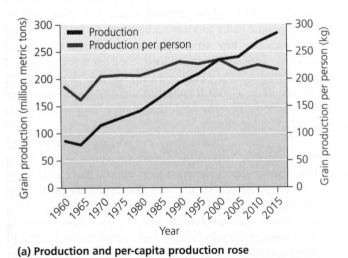

(a) Production and per-capita production rose

(b) Imports turned to exports

FIGURE 10.5 Green Revolution technology enabled India to boost its grain production. India's production grew faster than its population **(a)**, such that grain per person increased from the 1960s to the 1990s. As a result **(b)**, India was able to stop importing grain and begin exporting it to other nations. *Data from UN Food and Agriculture Organization (FAO).*

FIGURE 10.6 Monocultures improve the efficiency of planting and harvesting but are susceptible to outbreaks of pests. Armyworms (**inset**) attack corn fields like this one.

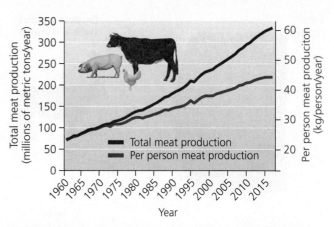

FIGURE 10.7 Total production of meat from farmed animals and amounts produced per person have each risen steadily worldwide for several decades. *Data from UN Food and Agriculture Organization (FAO).*

How can we achieve sustainable agriculture?

Most experts believe that to sustain our growing population we will need to begin raising crops and livestock in ways that are less polluting and less resource-intensive. **Sustainable agriculture** (p. 215) consists of farming and grazing that maintain the healthy soil, clean water, pollinators, genetic diversity, and other resources and ecosystem services needed for the production of crops and livestock over the long term. The world's farmers and ranchers have already adopted many strategies and conservation methods to this end (explored in Chapter 9). Reducing fossil fuel inputs and the pollution these inputs cause is a key aim of sustainable agriculture, and to many, this means moving away from the industrial model and toward more traditional organic and low-input models. Yet plenty of experts today believe that technology offers our best hope of making agriculture sustainable and that the genetic engineering of crops and livestock is a vital component of this effort. Competing visions exist for agriculture in our world today, and there are compelling arguments on all sides.

Raising Animals for Food

Food from cropland agriculture makes up a large portion of the human diet, but most of us also eat animal products. Just as farming methods have changed over time, so have the ways we raise animals for food.

Consumption of animal products is growing

As population, wealth, and global commerce have increased, so have humanity's production and consumption of meat, milk, eggs, and other animal products (**FIGURE 10.7**). Worldwide,

more than 30 billion domesticated animals are raised and slaughtered for food each year. Global meat production has increased more than fivefold since 1950, and per capita meat consumption has doubled. The Food and Agriculture Organization of the United Nations (FAO) predicts that as developing nations become wealthier, per capita meat consumption could nearly double again by the year 2050.

The rising consumption of meat and other animal products has come to have significant environmental, social, and economic impacts. For instance, the FAO has estimated that livestock in the United States account for 55% of the nation's soil erosion, 37% of pesticide applications, 50% of antibiotics consumed, and one-third of the nitrogen and phosphorus pollution in U.S. waterways. How we respond to demand for animal products affects our society's ecological footprint and our quest for sustainable agriculture.

Our food choices have consequences

What we choose to eat has ramifications for how we use energy, land, and water. Every time one organism consumes another, only about 10% of the energy moves from one trophic level up to the next (p. 82). As a result, if we feed grain to a cow and then eat beef from the cow, we lose most of the grain's energy to the cow's metabolism. Energy was used up as the cow converted the grain to tissue as it grew and as the cow conducted cellular respiration (p. 34) on a constant basis to maintain itself. For this reason, eating meat is far less energy-efficient than relying on a vegetarian diet, and a meat-rich diet leaves a far greater ecological footprint than a plant-rich diet.

In contrast, if we eat lower on the food chain (a plant-rich diet with fewer animal products), we put a greater proportion of the sun's energy to use as food. The lower on the food chain we eat, the smaller is our ecological footprint and the more of us Earth can support. That is one reason so many people have turned to vegetarian and vegan diets or have simply decreased their meat consumption. Doing so is increasingly easy at colleges and universities like the University of Michigan, where delicious vegetarian and vegan options are always available in the dining halls.

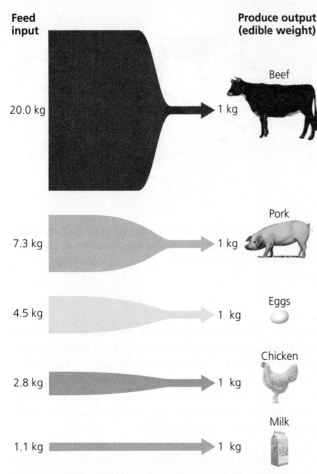

Feed input

20.0 kg

7.3 kg

4.5 kg

2.8 kg

1.1 kg

Produce output (edible weight)

Beef
1 kg

Pork
1 kg

Eggs
1 kg

Chicken
1 kg

Milk
1 kg

FIGURE 10.8 Producing different animal food products requires different amounts of animal feed. Twenty kilograms of feed must be provided to cattle to produce 1 kg of beef. *Data from Smil, V., 2001. Feeding the world: A challenge for the twenty-first century. Cambridge, MA: MIT Press.*

Another thing to consider is that some animals convert grain feed into milk, eggs, or meat more efficiently than others (**FIGURE 10.8**). Scientists have calculated relative energy-conversion efficiencies for different types of animals and have used these to infer the area of land (**FIGURE 10.9a**) and weight of water (**FIGURE 10.9b**) required to produce 1 kg (2.2 lb) of edible protein from milk, eggs, chicken, pork, and beef. The research illustrates that producing eggs and chicken meat requires the least land and water, whereas producing beef requires the most. These differences in efficiency demonstrate that when we choose what to eat, we are also choosing indirectly how to use resources such as land and water. University dining services like MDining are becoming attentive to this fact and are encouraging meal options that minimize or eliminate red meat.

Our diets also have consequences for climate change, because livestock are a major source of greenhouse gases that lead to global warming (p. 488). **FIGURE 10.9c** shows the amount of emissions (in carbon dioxide equivalents) released in the production of 1 kg (2.2 lb) of edible protein from milk, eggs, chicken, pork, and beef. Again, beef exerts the largest footprint. Livestock release methane and nitrous oxide in their metabolism and waste. Nitrous oxide is also released from certain crops used to feed animals and from fertilizers applied

to those crops. Carbon dioxide is released to the atmosphere when forests are cleared for ranching or for growing feed and when fossil fuels are burned to grow feed or transport animals. The FAO reports that across the world, livestock agriculture contributes 5% of our carbon dioxide emissions, 44% of our methane emissions, and 53% of our nitrous oxide emissions, accounting altogether for 14.5% of the emissions driving climate change—a larger share than automobile transportation! On a brighter note, the FAO also concludes that we can reduce greenhouse gas emissions from livestock by 30% by more widely adopting clean and efficient technologies and practices already being used by the most resourceful producers.

Feedlots have benefits and costs

In traditional agriculture, livestock are kept by farming families near their homes or are grazed on open grasslands by nomadic herders or sedentary ranchers. These traditions survive, but industrial agriculture and the rising demand for meat, milk, and eggs have brought an additional method of raising animals for food. **Feedlots,** also known as *factory farms* or *concentrated animal feeding operations,* are huge pens designed to deliver energy-rich food to animals living at extremely high densities (**FIGURE 10.10**). Today, nearly half the world's pork and most of its poultry come from feedlots.

Feedlots allow for economic efficiency and greater food production, which make animal products affordable for more people. Concentrating cattle and other livestock in feedlots also removes them from rangeland, thereby reducing the grazing impacts (pp. 234–235) these animals would exert across large areas of the landscape.

Intensified animal production through the industrial feedlot model has some negative consequences, though. One-third of the world's cropland is now devoted to growing feed for animals, and 45% of global grain production goes to feed livestock and poultry. As a result, supplies of staple grains for people can become constricted, elevating prices and endangering food security for the poor.

Livestock also produce prodigious amounts of manure and urine, and the highly concentrated waste from feedlots can pollute surface water and groundwater. Rich in nitrogen and phosphorus, livestock waste is a common cause of eutrophication (pp. 110, 410). This waste may also release bacterial and viral pathogens that can sicken people, including *Salmonella, Escherichia coli, Giardia, Microsporidia, Pfiesteria,* and pathogens that cause diarrhea, botulism, and parasitic infections.

The crowded conditions under which animals are kept in feedlots necessitate heavy use of antibiotics to control disease.

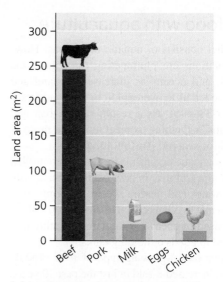

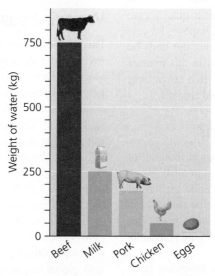

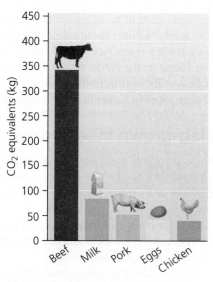

(a) Land required to produce 1 kg of protein

(b) Water required to produce 1 kg of protein

(c) Greenhouse gas emissions released in producing 1 kg of protein

FIGURE 10.9 Producing different types of animal products requires different amounts of (a) land and (b) water—and releases different amounts of (c) greenhouse gas emissions. Raising cattle for beef exerts the greatest impacts in all three ways. *Data (a, b) from Smil, V., 2001.* Feeding the world: A challenge for the twenty-first century. *Cambridge, MA: MIT Press; and (c) from FAO, 2015.* Global Livestock Environmental Assessment Model (GLEAM).

DATA Q Go to **Interpreting Graphs & Data** on **Mastering Environmental Science**

Overuse of antibiotics can cause microbes to evolve resistance to the antibiotics (just as pests evolve resistance to pesticides; p. 252), which makes the drugs less effective and leads feedlot managers to increase dosages and switch to ever-stronger antibiotics. Hormones are administered to livestock as well, and metals are added to feed to spur growth. Livestock excrete most of these chemicals, which end up in wastewater and may be transferred up the food chain in downstream ecosystems. Some chemicals that remain in livestock are transferred to people when we eat their meat.

Federal and state agencies regulate U.S. feedlots to try to minimize health and environmental impacts, but these efforts are limited. Legal actions brought by industry groups coupled with recent executive-branch decisions have restricted the Environmental Protection Agency's authority to control feedlot waste. Manure from feedlots can be applied on farms as fertilizer, reducing the need for chemical fertilizers. However, overapplication can pollute soil and water supplies and can cause eutrophication in freshwater bodies. Improper operation or spills from waste-holding lagoons can result in similar pollution problems.

Not all livestock on modern farms are raised in intensive facilities. Meat, eggs, and dairy products are also produced from *free-range* animals that are not always confined in pens.

(a) Cattle feedlot in Colorado

(b) Chicken factory farm in Arkansas

FIGURE 10.10 Most meat eaten in the United States comes from animals raised in feedlots or factory farms. These facilities house thousands of animals such as **(a)** cattle or **(b)** chickens at high densities.

The certification of an operation as free-range is quite broad. For example, the only requirement the USDA imposes for certifying poultry operations as "free-range" is that the birds be given access to the outdoors. Raising animals in this manner can be costlier than in feedlots, so prices for free-range meat and eggs are often higher than for feedlot-raised animals.

Is lab-grown meat a solution?

One potential way we could reduce the ecological footprint of high-impact foods like beef is to grow the food in a lab. Using techniques of *in vitro* tissue culture like those used in areas of research and medicine, animal muscle cells can be cultivated in the lab to form **lab-grown meat**, also called *cultured meat*. This meat is genetically identical to animal meat and could potentially be healthier to eat than animal meat.

In 2013, Dutch scientists created the first lab-grown hamburger patty (**FIGURE 10.11**). Today, the technology for producing lab-grown meat of many types at a large scale seems to be ready, with costs coming down quickly. A number of start-up companies are angling to begin production, and government regulators are struggling to determine whether and how to regulate such foods. It remains uncertain whether many consumers would choose to eat lab-grown meat, however, and some major corporations in the meat industry view it as competition and oppose it, so lab-grown meat has not yet been made commercially available.

FIGURE 10.11 The world's first lab-grown hamburger patty was unveiled in 2013: "meat" in a petri dish and **(inset)** on a bun.

We raise seafood with aquaculture

Part of the human diet consists of aquatic organisms. However, wild fish populations are plummeting throughout the world's rivers, lakes, and oceans as increased demand and new technologies have led us to overexploit marine and freshwater fisheries (pp. 438–446). As a result, raising fish and shellfish on "fish farms" is helping us meet more and more of our growing demand for seafood (**FIGURE 10.12**).

The cultivation of aquatic organisms for food in controlled environments, called **aquaculture,** is now being pursued with more than 220 freshwater and marine species, ranging from fish to shrimp to clams to seaweeds (**FIGURE 10.13**). Many aquatic species are raised in open water in large, floating net-pens. Others are raised in ponds or holding tanks.

Aquaculture is our fastest-growing type of food production; global output has increased 8-fold in just the past 30 years. Most widespread in Asia, aquaculture today produces $250 billion worth of food and provides half of the fish and two-thirds of the shellfish that people eat.

Aquaculture brings benefits and costs

Aquaculture can improve food security by enhancing the supply of reliable protein sources. Aquaculture also helps reduce fishing pressure on overharvested and declining wild fish stocks. Aquaculture consumes fewer fossil fuels and provides a safer work environment than does commercial fishing. Fish farming can also be remarkably energy-efficient, producing up to 10 times more fish per unit area than is harvested from waters of the continental shelf and up to 1000 times more than from the open ocean.

When practiced on a small scale by families or villages, as in much of the developing world, community-based aquaculture can be sustainable. Moreover, it can be compatible with other activities; for instance, uneaten fish scraps make excellent fertilizers for crops, and food waste can be fed to fish.

At large scales, industrialized aquaculture produces ample amounts of food but also exerts environmental impacts.

FIGURE 10.12 People practice many types of aquaculture. Here, fish-farmers tend their animals at a fish farm in China.

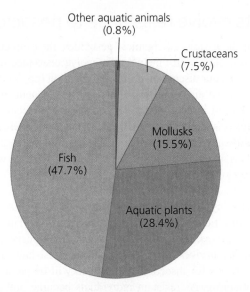

FIGURE 10.13 Aquaculture involves many types of fish but also a wide diversity of other marine and freshwater organisms. *Data from UN Food and Agriculture Organization (FAO).*

Pie chart labels: Other aquatic animals (0.8%), Crustaceans (7.5%), Mollusks (15.5%), Fish (47.7%), Aquatic plants (28.4%)

Industrial-scale shrimp farming along tropical coastlines has polluted coastal waters and destroyed large areas of ecologically valuable mangrove forest. Aquaculture can produce remarkable amounts of waste, both from the farmed organisms and from feed that goes uneaten and decomposes in the water. As with feedlot animals, commercially farmed fish often are fed grain, which is energy-inefficient and can reduce food supplies for people. Farmed fish are also sometimes fed fish meal made from wild ocean fish such as herring and anchovies, whose harvest places additional stress on wild populations. For all these reasons, industrialized aquaculture can leave a large ecological footprint.

Dense concentrations of farmed animals can incubate disease, which may lead to antibiotic treatment. If farmed aquatic organisms escape into ecosystems where they are not native, they may spread disease to native organisms or may outcompete them for food or habitat. These problems are occurring with several species of Asian carp farmed in the United States that have escaped and whose growing populations are altering ecosystems and harming fisheries (pp. 75–76). Such concerns delayed government approval of genetically engineered Atlantic salmon, which in 2015 became the first genetically modified animal approved for sale as food.

Controlling Pests and Weeds

Anyone who has ever farmed crops or raised animals has struggled to limit damage from unwelcome organisms that can weaken or destroy crops and livestock. Some insects feed on crop plants, some microbial or fungal pathogens attack livestock or crops, and some plants compete with crop plants for sun, water, and nutrients. Any pursuit of sustainable agriculture will need to find safe and effective ways of limiting losses from these natural adversaries.

"Pests" and "weeds" hinder agriculture

What people term a **pest** is any organism that damages crops or livestock. What we term a **weed** is any plant that competes with our crops. These are subjective terms that we define entirely by our own economic interests. There is nothing inherently malevolent about a pest or a weed. These organisms are simply growing and behaving naturally, adapted to survive and reproduce, like any other organism. They just happen to affect our agricultural productivity in doing so.

Throughout the history of agriculture, the insects, mites, rodents, fungi, and viruses that attack our crops have taken advantage of the ways we cluster food plants in agricultural fields. Pests pose an especially great threat to monocultures, because a pest adapted to specialize on a crop can move easily from plant to plant (see Figure 10.6). From the perspective of an insect that feeds on corn, grapes, or apples, encountering a grain field, vineyard, or orchard is like discovering an endless buffet. In a natural ecosystem, each organism's population is kept in check by its predators, competitors, parasites, and pathogens—but in an industrial monoculture, the abundance of one type of food and the lack of other habitats can allow a pest population to flourish, unhindered by its natural enemies.

We have developed thousands of chemical pesticides

In the past half-century, most farmers have turned to commercially produced chemicals to suppress pests and weeds. In that time, companies have developed thousands of synthetic chemicals to kill insects (*insecticides*), plants (*herbicides*), and fungi (*fungicides*). Such poisons are collectively termed **pesticides.** All told, more than 535 million kg (1.18 billion lb) of active ingredients from conventional pesticides are applied in the United States each year—almost 4 pounds per person. Ninety percent of this total is applied on agricultural land (**FIGURE 10.14**). Since 1990, pesticide use has risen nearly 80% worldwide. Usage in industrialized nations has leveled off, but it continues to rise in the developing world. Today, more than $56 billion is expended annually on pesticides across the world.

FIGURE 10.14 Synthetic chemical pesticides are widely applied to crops in conventional industrial agriculture.

Pesticides boost food production but also have negative impacts

Chemical pesticides have served to greatly increase our food production. They also help us control pests in our homes, and they protect millions of people in developing nations from insect-borne diseases such as malaria. However, exposure to synthetic pesticides can have health consequences for people (Chapter 14). Especially vulnerable are farm workers, who experience high levels of pesticide exposure, and children, whose brains and bodies are growing and developing. We all encounter and ingest chemical pesticide residues when we eat non-organic produce that we buy at the grocery store, and pesticides leave residues and breakdown products in the soil, water, and air, affecting organisms and ecosystems in many ways. Although not every pesticide has effects on people, in the United States and most other nations, new chemical products are not comprehensively tested for health effects before being brought to market (pp. 383–384), so a pesticide may cause health impacts yet still be widely used.

An additional problem is that pesticides commonly kill nontarget organisms, including the predators and parasites of the pests they are meant to target. Once these valuable natural enemies are eliminated, pest populations become even harder to control. Moreover, chemical pesticides affect bees and other beneficial insects that pollinate our crops (p. 224). For example, use of the herbicide glyphosate in North America has been so widespread that milkweed plants have gone from being abundant in farm country to being scarce. Monarch butterflies rely on milkweed for food, so as glyphosate application (and intensified monocultural farming that limits borders between fields) have eliminated milkweed from the landscape, monarch populations have plummeted (p. 294).

Today, a new class of chemical insecticide has become common: **neonicotinoids.** Seed companies treat crop seed with neonicotinoids, and these poisons become systemic in the plant, dispersing throughout its tissues as it grows and ending up in leaves, stems, flowers, fruit, and pollen. "Neonics" make a plant toxic to insects; they kill insects that feed on the plant (as intended) but also harm bees pollinating the plant (p. 224–225) and predators that eat insects that feed on the plant (unintended consequences). Neonicotinoids are also applied to plants chemically, and they may enter soil, water, and even plants that grow from treated soil, continuing to kill a diversity of organisms that pose no threat to crops.

Pests evolve resistance to pesticides

Despite the toxicity of chemical pesticides, their effectiveness tends to decline with time as pests evolve resistance to them. Recall from our discussion of natural selection (pp. 50–53) that organisms within populations vary in their genetic makeup. Because most insects, weeds, and microbes can occur in huge numbers, it is likely that a small fraction of individuals may by chance already have genes that enable them to metabolize and detoxify a given pesticide (p. 371). These individuals will survive exposure to the pesticide, whereas individuals without these genes will not (**FIGURE 10.15**).

Let's say an insecticide application kills 99.99% of the insects in a field. That is a lot, but it means that 1 in 10,000 insects survives. If an insect that is genetically resistant to an insecticide survives and mates with other resistant individuals, the genes for insecticide resistance will be passed on to their offspring. As resistant individuals become prevalent in the pest population, insecticide applications will cease to be effective, the pest population will grow, and crop losses will escalate.

1 Pests attack crops.

2 Pesticide is applied.

4 Survivors breed and produce a pesticide-resistant population. When they next attack crops, a stronger, more toxic pesticide must be applied.

3 Most pests are killed. A few with innate resistance survive.

FIGURE 10.15 Through natural selection, crop pests often evolve resistance to the poisons we apply to kill them. This process spurs the development and use of ever-more-toxic pesticides, creating an endless cycle called the "pesticide treadmill."

In many cases, industrial chemists are caught up in an evolutionary arms race (p. 79) with the pests they battle, racing to intensify or retarget the toxicity of their chemicals while the armies of pests evolve ever-stronger resistance to their efforts. Because we frequently become stuck in this cyclical process, it has been nicknamed the "pesticide treadmill." Among arthropods (insects and their relatives) alone, there are currently more than 10,000 known cases of resistance by nearly 600 species to more than 330 insecticides. Hundreds more weed species and plant diseases have evolved resistance to herbicides and other pesticides. Many species, including insects such as the green peach aphid, Colorado potato beetle, and diamondback moth, have evolved resistance to multiple chemicals.

Biological control pits one organism against another

Because of pesticide resistance, toxicity to nontarget organisms, and human health risks from some synthetic chemicals, agricultural scientists have sought alternatives to chemical pesticides. The prime alternative has been to battle pests and weeds with organisms that eat or infect them. This strategy, called **biological control**—often referred to as *biocontrol*—operates on the principle that "the enemy of one's enemy is one's friend." Biological control essentially aims to reestablish the restraining influence that predators and parasites exert over populations in nature. For example, parasitoid wasps (p. 79) are natural enemies of many caterpillars. These wasps lay eggs on a caterpillar, and the larvae that hatch from the eggs feed on the caterpillar, eventually killing it. Parasitoid wasps are frequently used as biocontrol agents and have often succeeded in controlling pests and reducing chemical pesticide use.

One classic case of successful biological control is the introduction of the cactus moth, *Cactoblastis cactorum,* from Argentina to Australia in the 1920s to control invasive prickly pear cactus that was overrunning rangeland. Larvae of the moth devour cactus tissue, and within just a few years, the species managed to free millions of hectares of Australian rangeland from the cactus (**FIGURE 10.16**).

A widespread biocontrol approach is the use of *Bacillus thuringiensis* (Bt), a naturally occurring soil bacterium that produces a protein that is nontoxic to people but kills many caterpillars and some fly and beetle larvae. Farmers spray Bt spores on their crops to protect against insect attack. For instance, the Campus Farm at the University of Michigan uses Bt on some of its crops as a means of controlling insect pests without resorting to chemical pesticides. Scientists have also managed to isolate the gene responsible for this bacterium's poison and engineer it into crop plants so that the plants produce the poison (p. 257).

Biocontrol agents themselves can become pests

When a pest is not native to the region where it is damaging crops, scientists may consider introducing a natural enemy (a predator, parasite, or pathogen) of the pest from its native range, expecting that the enemy will attack it. Alternatively,

(a) Before cactus moth introduction

(b) After cactus moth introduction

FIGURE 10.16 Photos from the 1920s show an Australian ranch before (a) and after (b) introduction of the cactus moth to clear invasive non-native prickly pear cactus from rangeland.

scientists may consider importing a biocontrol agent from abroad that the pest has never encountered, reasoning that the pest has not evolved ways to avoid the biocontrol agent. In either case, this approach involves introducing an animal or microbe from a foreign ecosystem into a new ecological context. Carrying out such a foreign introduction is risky, because no one can know for certain what effects the biocontrol agent might have. In some cases, biocontrol agents have turned invasive and become pests themselves. When that happens, biocontrol organisms are more difficult to manage than chemical controls, because they cannot be "turned off" once they are set loose. Following the cactus moth's success in Australia, for example, the moth was introduced in other countries to control non-native prickly pear. However, moths introduced to Caribbean islands spread to Florida on their own and are now eating their way through rare native cacti in the southeastern United States.

In Hawai'i, wasps and flies have been introduced to control pests at least 122 times over the past century, and biologists Laurie Henneman and Jane Memmott suspected that some of these might be harming native Hawaiian caterpillars that were not pests. They sampled parasitoid wasp larvae from 2000 caterpillars of various species in a remote mountain swamp far from farmland. In this wilderness preserve, they found that fully 83% of the parasitoids were biocontrol agents that had been intended to combat lowland agricultural pests.

Because of concerns about unintended impacts, researchers study biocontrol proposals carefully before putting them into action, and government regulators must approve these efforts. If biological control works as planned, it can be a permanent, effective, and environmentally benign solution. Yet there will never be a sure-fire way of knowing in advance whether a given biocontrol program will work as planned.

Integrated pest management combines varied approaches

Because both biocontrol and chemical control methods pose risks, agricultural scientists and farmers have developed more sophisticated strategies, trying to combine the best attributes of each approach. **Integrated pest management** (IPM) incorporates numerous techniques, including biocontrol, use of chemicals when needed, close monitoring of populations, habitat alteration, crop rotation (p. 232), alternative tillage methods (p. 233), and mechanical pest removal. This nuanced approach aims to maximize control over pests while minimizing chemical pollution, through careful observation and thoughtful strategy.

IPM works especially well in landscapes at small to moderate scales, such as small farms and university campuses, where a grower can gain a good comprehension of needs and challenges and where successes and failures can be observed and understood. Many college and university landscaping policies and campus farm strategies today revolve around IPM. However, increasing numbers of large industrial producers have adopted IPM as well (such as many of the pollinator-conscious growers profiled in Chapter 9). At a nationwide scale, successes in countries such as Indonesia (see **SUCCESS STORY**) have led to the wide adoption of IPM in many parts of the developing world that are embracing sustainable agriculture.

Fighting Pests with IPM in Indonesia

When reading the Success Stories in this book, keep in mind that successes can be reversed. Although frustrating, such reversals can sometimes reinforce our understanding of what leads to successful outcomes.

For a time, Indonesia stood as an exemplary case of success with integrated pest management (IPM). This rice-dependent Asian nation had subsidized chemical pesticide use heavily for years but came to understand that pesticides were actually making pest problems worse. They were killing the natural enemies of the brown planthopper, which began to devastate rice fields as its populations exploded. Concluding that pesticide subsidies were costing money, causing pollution, and decreasing yields, the Indonesian government in 1986 banned the import of 57 pesticides, slashed pesticide subsidies, and promoted IPM. Within just four years, pesticide production fell by half, pesticide imports fell by two-thirds, and subsidies were phased out (saving taxpayers $179 million annually). Rice yields rose 13%. A national IPM program was put in place, and "Farmer Field Schools" empowered farmers to learn, teach, and make decisions based on their own observation and knowledge. Farmers using IPM gained slightly higher yields, higher overall returns, and more economic stability than farmers still using pesticides. The brown planthopper ceased to be a problem, and another pest, the white stemborer, was defeated by IPM farmers as well.

After the Asian financial crisis and the fall of the autocratic President Suharto in 1998, however, everything changed. Amid the (mostly good) democratic reforms and political decentralization, the national IPM program lost funding and ended in 1999. Pesticide corporations rushed in and marketed

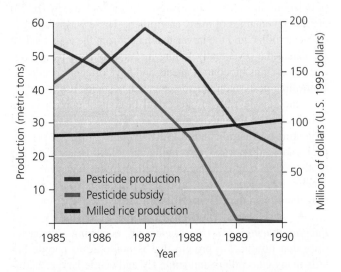

Once Indonesia threw its weight behind IPM in 1986, pesticide production, imports, and subsidies fell, and yields of rice increased. *Data from the World Bank.*

their products in the newly deregulated environment, and pesticide use began to soar. In 2009, the brown planthopper returned with a vengeance, leading since then to pest outbreaks as severe as the ones that had inspired the IPM program in the first place.

It remains to be seen if Indonesia will return to IPM, but its success served as a model, and today IPM and Farmer Field Schools have spread to many other Asian nations.

➔ **Explore the Data** at Mastering Environmental Science

Genetically Modified Food

Industrial agriculture has enabled us to feed a greater number and proportion of the world's people, but relentless population growth and intensified environmental impacts are demanding further innovation. Modern advances in genetics have enabled scientists to directly alter the genes of organisms, including crop plants and livestock. The genetic modification of organisms that provide us food and fiber holds considerable promise to enhance nutrition and the efficiency of agriculture while reducing impacts on ecosystems.

Any process whereby scientists directly manipulate an organism's genetic material in the laboratory by adding, deleting, or changing stretches of DNA (p. 28) is referred to as **genetic engineering.** Genetic engineering is one form of **biotechnology,** the material application of biological science to create products derived from organisms. Biotechnology has brought our society a dazzling array of advances, helping us develop medicines, clean up pollution, understand the causes of cancer and other diseases, dissolve blood clots after heart attacks, and even make better beer and cheese. Yet many people worry that genetically engineered organisms may pose risks that are not yet fully understood. Moreover, the expanding role of biotechnology in agriculture has strengthened the influence of multinational corporations over farmers and our food supply. For these reasons, agricultural biotechnology, particularly genetic engineering, has faced resistance from many consumer advocates, small farmers, and critics of big business.

We have many ways of modifying foods

Since the dawn of agriculture, people have been working to improve the foods we grow and eat. As technology has advanced, scientists have developed a number of ways to alter crops and livestock, generally with the aim of increasing food security or reducing the environmental impacts of agriculture.

Traditional selective breeding For millennia, people have improved crops and livestock through selective breeding (p. 52), the practice of choosing individuals with the best characteristics and breeding them together, hoping that offspring will inherit the desired characteristics. In this approach, we simply observe differences among individuals that have resulted from naturally occurring genetic mutations and then breed individuals together to produce more of those with the traits we desire. Over many generations, this allowed our ancestors to produce plants and animals with better and better properties, eventually bringing us the many types of crops and livestock we rely on today.

Mutagenesis About 100 years ago, scientists realized they could use radiation (p. 27) or certain chemicals to encourage the emergence of heritable mutations, thus speeding up the generation of a wider array of plant and animal variants from which to choose for selective breeding. Various efforts at such *mutagenesis,* or *mutation breeding*, over the years have resulted in many current varieties we eat today—by some estimates, 1000 or more crops and livestock.

Genetic engineering to create GMOs Organisms that are genetically engineered are commonly referred to as **genetically modified organisms** (GMOs). Foods derived from GMOs are referred to as **genetically modified (GM) foods.** A genetically modified organism may be derived from the transfer of genetic material from various sources, including organisms of the same species, other species, or artificial synthesis in the lab. An organism genetically engineered to contain DNA from another species is called a **transgenic** organism, and a gene transferred between species is called a **transgene.**

One major genetic engineering approach uses *recombinant DNA*—DNA patched together from multiple sources to create novel genetic combinations. **FIGURE 10.17a** (p. 256) shows the steps typically followed using recombinant DNA technology. The goal is to place genes that code for certain desirable traits (such as rapid growth, disease resistance, or high nutritional content) into the genomes of organisms lacking those traits.

For instance, to create the world's most abundant genetically modified (GM) crop (Roundup Ready soybeans, which survive spraying with the herbicide Roundup), scientists worked with a strain of the soil bacterium *Agrobacterium tumefaciens.* They transferred to the soybean a version of a bacterial gene that codes for an enzyme that is insensitive to glyphosate, the active ingredient in Roundup. Expression of the gene is regulated by three additional stretches of DNA that were introduced—one from *Agrobacterium,* one from a type of petunia, and one derived from a common plant virus, the cauliflower mosaic virus.

Gene editing In just the last several years, researchers have brought genetic engineering to a new level: They have gained the ability to make targeted changes in the genomes of plants and animals with great precision. Using an array of new techniques for **gene editing,** or *genome editing,* scientists can

FAQ

Is genetic engineering radically new and different, or not?

In principle, the genetic alteration of plants and animals by people is nothing new. Through artificial selection and selective breeding (p. 52), our ancestors were influencing the genetic makeup of livestock and crop plants for thousands of years by preferentially mating individuals with favored traits (such as fast growth, higher yields, or disease resistance) so that offspring would inherit these traits.

The techniques geneticists use to create engineered organisms, however, differ from traditional selective breeding in several ways. For one, selective breeding mixes genes from individuals of the same or similar species, whereas scientists working with recombinant DNA routinely mix genes of organisms as different as viruses and plants. For another, selective breeding involves whole organisms living in the field, whereas genetic engineering works with genetic material in the lab. For yet another, traditional breeding selects from combinations of genes that come together on their own, whereas genetic engineering creates novel combinations directly in a more controlled way. Thus, in traditional breeding, people use a process of selection (pp. 50–53) acting on random mutations, whereas in genetic engineering, scientists use a process of mutation (p. 52) that is nonrandom and precisely directed.

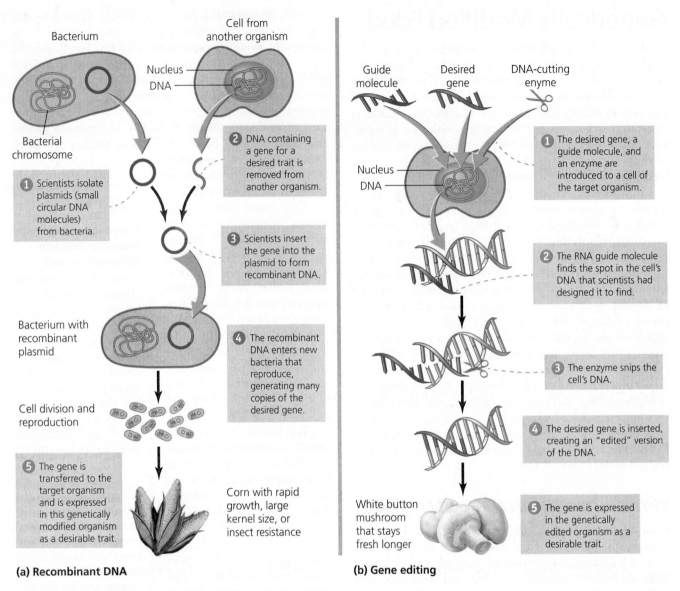

(a) Recombinant DNA

(b) Gene editing

FIGURE 10.17 Genetic engineering involves a diversity of methods. The creation of recombinant DNA **(a)** is a well-established process that scientists have been following for decades. Newer techniques of gene editing **(b)** (here, a simplified representation of CRISPR) are diverse and undergoing rapid improvement.

now insert, delete, or change short stretches of DNA—even single nucleotides (p. 29)—to turn particular genes on or off, or to modify their effects (**FIGURE 10.17b**). One popular such approach, called CRISPR (**C**lustered **R**egularly **I**nterspaced **S**hort **P**alindromic **R**epeats), uses genetic material from bacteria that the bacteria had used to fight infection from viruses. Many other techniques and variations are rapidly being developed. Because researchers have lately been learning the specific functions of more and more genes, gene editing is promising rapid advances in our ability to make precise and controlled changes to crops and livestock.

In 2018, the world's first gene-edited food was approved for sale in the United States. Pennsylvania State University researchers had used CRISPR to edit a gene in the white button mushroom commonly sold at grocery stores, disabling production of an enzyme that turns the flesh brown with exposure to the air. The gene-edited version of the mushroom will stay white longer, thus extending its shelf life and reducing the number of mushrooms that come to look unappealing and go to waste, unsold.

Biotechnology is transforming the products around us

In less than four decades, GM foods have gone from science fiction to big business. When recombinant DNA technology was first developed in the 1970s, scientists collectively regulated and monitored their own research. Once the scientific community declared confidence in the 1980s that the technique was safe, industry leaped at the chance to develop hundreds of applications, from improved medicines (such as hepatitis B vaccine and insulin for diabetes) to designer plants and animals (including glow-in-the-dark pet goldfish). A selection of notable developments in GM foods is shown in

TABLE 10.1 Several Notable Examples of Genetically Modified Food Technologies

CROP	DESCRIPTION AND STATUS	CROP	DESCRIPTION AND STATUS
Golden rice 	Engineered to produce beta-carotene to fight vitamin A deficiency in Asia and the developing world. May offer only moderate nutritional enhancement despite years of work. Currently undergoing field trials in Asia.	**Bt cotton**	Engineered with genes from bacterium *Bacillus thuringiensis* (Bt) that kills insects. Has increased yield, decreased insecticide use, and boosted income for 14 million small farmers in India, China, and other nations.
Virus-resistant papaya 	Resistant to ringspot virus and grown in Hawai'i. In 2011 became the first biotech crop approved for consumption in Japan.	**Roundup-Ready alfalfa** 	Engineered to tolerate Monsanto's Roundup herbicide (glyphosate), so the chemical can be applied in great quantities to kill weeds. Many weeds are evolving resistance as a result. Planted in the United States, 2005–2007, then banned once a lawsuit forced better assessment of environmental impact. Reapproved in 2011.
GM salmon 	Engineered for fast growth, large size. Is the first GM animal approved for sale as food. To prevent fish from breeding with wild salmon and spreading disease, Aqua Bounty promised to make their fish sterile and raise them in inland pens.	**Roundup-Ready sugar beet** 	Tolerant of Monsanto's Roundup herbicide (glyphosate). Swept to dominance (95% of U.S. crop) in just 2 years. A lawsuit forced more environmental review, after it had already become widespread. Reapproved in 2012.
Biotech potato 	Resistant to late blight, the pathogen that caused the 1845 Irish Potato Famine and that still destroys $7.5 billion of potatoes each year. Being developed by European scientists, but struggling with EU regulations on research.	**Biotech soybean**	The most common GM crop in the world, covering nearly half the cropland devoted to biotech crops. Engineered for herbicide tolerance, insecticidal properties, or both.
Bt corn 	Engineered with genes from bacterium *Bacillus thuringiensis* (Bt) that kills insects. One of many Bt crops developed.	**Sunflowers and superweeds**	Research on Bt sunflowers suggests that transgenes might spread to their wild relatives and turn them into vigorous "superweeds" that compete with the crop or invade ecosystems. This is most likely with crops like squash, canola, and sunflowers that can breed with wild relatives.

TABLE 10.1. The stories behind GM foods illustrate both the promises and pitfalls of food biotechnology.

GM crops have been adopted across the world with remarkable speed. Worldwide, more than three of every four soybean and cotton plants are now transgenic, as are nearly one of every three corn and canola plants. Globally in 2018, nearly 16 million farmers grew GM crops on 192 million ha (474 million acres) of farmland—about 12% of all cropland in the world. More than half the world's GM crops are now grown in developing nations (**FIGURE 10.18**).

In the United States today, nearly 95% of corn, soybeans, and cotton consist of genetically modified strains, while 100% of canola and sugar beets are GM varieties. Lower percentages of alfalfa, squash, potato, papaya, and apples are genetically modified. It is conservatively estimated that more than 70% of processed foods in U.S. stores contain GM ingredients (such as corn syrup from GM corn). Thus, it is extremely likely that you consume GM foods on a daily basis.

Most GM crops today are engineered to tolerate herbicides so that farmers can apply herbicides to kill weeds

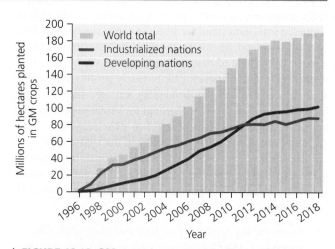

FIGURE 10.18 GM crops have spread with remarkable speed since their commercial introduction in 1996. *Data from the International Service for the Acquisition of Agri-biotech Applications.*

 Go to **Interpreting Graphs & Data** on **Mastering Environmental Science**

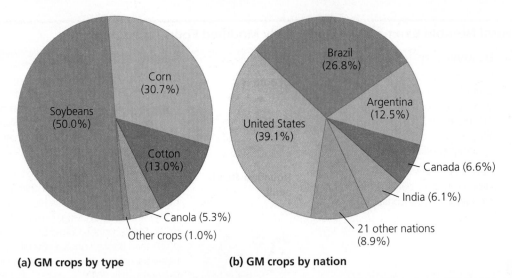

(a) GM crops by type

Soybeans (50.0%)
Corn (30.7%)
Cotton (13.0%)
Canola (5.3%)
Other crops (1.0%)

(b) GM crops by nation

Brazil (26.8%)
Argentina (12.5%)
United States (39.1%)
Canada (6.6%)
India (6.1%)
21 other nations (8.9%)

FIGURE 10.19 So far, genetic engineering has mainly involved common crops grown in industrialized nations. Of the world's GM crops **(a),** soybeans are the most common. Of global acreage planted in GM crops **(b),** the United States devotes the most area. *Data from the International Service for the Acquisition of Agri-biotech Applications (ISAAA).*

without having to worry about killing their crops. Other crops are engineered to resist insect attack. An increasing number are "stacked," or modified for multiple traits. Tolerance of herbicides and resistance to pests enables large-scale commercial farmers to grow crops more efficiently, and this is largely why sales of GM seeds to farmers have risen so quickly.

Soybeans account for half the world's GM crops (**FIGURE 10.19a**). Of the 26 nations growing GM crops in 2018, five (the United States, Brazil, Argentina, Canada, and India) accounted for more than 90% of production, with the United States alone growing 39% of the global total (**FIGURE 10.19b**).

What are the benefits of GM crops?

Genetically engineered foods have long been promoted as a way to assist poor people in developing nations. If foods can be made more nutritious, then malnutrition can be fought more effectively. If crops can be made tolerant of drought or degraded soil, then small farmers can more easily grow food on marginal land. However, most of these noble intentions have not yet come to pass, largely because the corporations that develop GM varieties cannot easily profit from selling seed to small farmers in developing nations. Instead, most biotech crops have been engineered for insect resistance and herbicide tolerance, which improve efficiency for large-scale industrial farmers who can afford the technology.

Regardless, proponents of genetic engineering maintain that GM foods bring substantial environmental and social benefits and assist the pursuit of sustainable agriculture in several major ways:

- By increasing production of food and fiber while lowering costs, GM crops enhance food security and reduce poverty and hunger for millions of people.

- By raising yields on existing farmland, GM crops alleviate pressure to clear forests for agriculture, thus helping conserve biodiversity, habitat, and ecosystem services.

- Crops engineered for drought tolerance help conserve water by reducing the need for irrigation.

- Crops engineered for better nutrition—such as golden rice (p. 257)—could help fight malnutrition and enhance children's health in developing nations.

- Foods made more visually appealing to consumers (like non-browning apples and potatoes or the new gene-edited white button mushroom) can reduce food waste.

- Crops engineered for insect resistance allow farmers to reduce or eliminate use of chemical insecticides.

- GM crops that reduce fossil fuel use during cultivation help address climate change. For example, herbicide-tolerant crops free farmers from having to plow to control weeds, thus cutting down on gasoline use while reducing erosion and increasing carbon storage in the soil. In 2016, GM crops were estimated to have reduced carbon dioxide emissions by the equivalent of taking 17 million cars off the road for a year.

For reasons like these, millions of people growing GM crops feel that they are contributing toward sustainable agriculture. Scientific research has largely supported most of these assertions. In 2014, researchers conducted a **meta-analysis,** an effort that gathers results from all scientific studies on a particular research question and statistically analyzes their data for significant patterns or trends. This meta-analysis calculated that, on average, researchers had found that the adoption of GM crops increased yields by 22% and boosted farmers' profits by 68% while reducing chemical pesticide use by 37% (see **THE SCIENCE BEHIND THE STORY,** pp. 260–261).

To understand the influence on pesticide use, we need to consider insecticides and herbicides separately. Farmers who adopt insect-resistant Bt crops (pp. 253, 257) tend to use fewer insecticides because their crops do not need them; the insect resistance is already built into the plant. In contrast, herbicide-tolerant GM crops tend to result in *more* use of chemical herbicides because farmers often apply more

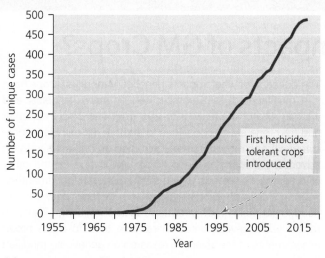

(a) Known cases of herbicide resistance

FIGURE 10.20 Weeds are evolving resistance to herbicides. Documented cases of herbicide resistance are approaching **(a)** 500 biotypes involving more than 250 species of plants. In little over a decade, weed resistance to glyphosate **(b)** spread across North America. *Data from Heap, I., 2019. International Survey of Herbicide-Resistant Weeds, www.weedscience.com.*

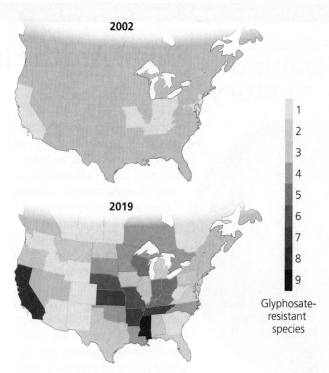

(b) Spread of glyphosate resistance

herbicide (to make sure weeds are killed) if their crops can withstand heavier applications.

Herbicide use is rising in part because weeds are evolving resistance to herbicides, causing farmers to apply even more of them. Just as insects can evolve resistance to insecticides (see Figure 10.15), weeds can evolve resistance to the chemicals that target them. Worldwide so far, nearly 500 varieties of more than 250 weed species are known to have evolved resistance to herbicides (**FIGURE 10.20a**).

One of the most widely used herbicides is the Monsanto company's Roundup, which contains glyphosate as its active ingredient. Besides producing Roundup, Monsanto produces GM seeds for crops that survive spraying with Roundup. For years farmers have loved these Roundup Ready crops because they reduce labor and boost yields: Simply spray fields with Roundup, and the weeds die while the crops thrive. However, widespread use of Roundup and other glyphosate-based herbicides is resulting in the evolution of resistance (**FIGURE 10.20b**). Scientists have confirmed glyphosate resistance in 41 weed species so far and resistance to the most widespread herbicide, atrazine, in 66 species. As resistant weeds fill their fields, farmers douse their plants with more chemicals, including additional herbicides that are more powerful and more polluting.

What are the risks of GM crops?

Many people worry about health impacts from eating GM foods, but these concerns have not been corroborated by research. A 2013 review of 1783 scientific studies found that "the scientific research conducted so far has not detected any significant hazards directly connected with the use of [genetically engineered] crops." Committees from many organizations representing the world's scientists, including the U.S. National Academy of Sciences, the American Association for the Advancement of Science, the American Medical Association, the European Commission, the World Health Organization, and others, have concluded that GM foods present no clear evidence of danger.

Most scientists who harbor concerns over GMOs focus on potential ecological impacts. Many conventional crops can interbreed with their wild relatives (e.g., domesticated rice can breed with wild rice), so there seems little reason to believe that transgenic crops might not do the same. In one confirmed case of transgene escape, GM oilseed rape (a relative of canola) was found hybridizing with wild mustard. In another case, creeping bentgrass engineered for use on golf courses—a GM plant not yet approved by the USDA—pollinated wild grass up to 21 km (13 mi) away from its test growing site. One concern is that if genes for herbicide resistance are transferred to other plants, the result would be the creation of "superweeds." Another concern is that transgenes will "contaminate" the genetic purity of local landraces of crops. In the region of Mexico where corn was first domesticated, years of scientific debate

FAQ

Is it safe to eat genetically modified foods?

In principle, there is nothing about the process of genetic engineering that should make genetically modified food any less safe to eat than conventional food. The fact that technology (rather than traditional breeding) is used to move or edit a gene does not in itself make the gene unsafe. To determine whether a GM food poses any health risks, long-term studies would need to be performed comparing that food with conventional versions, one by one—just as researchers would study any other substance for health risks (Chapter 14). Researchers will never be able to test all foods for all potential effects, so we can never guarantee that GM foods pose no risks—but thus far, no study has shown clear evidence of human health impacts from any GM food.

THE SCIENCE behind the story

Dr. Wilhelm Klümper (left) and Dr. Matin Qaim (right)

What Are the Impacts of GM Crops?

Most of us are confronted with such a deluge of information—and misinformation—about genetic engineering and GM crops that it can be hard to know what to believe. Even for a trained scientist, it's not easy to wade through hundreds of research papers in a complex and fast-changing field and come up with a simple answer to a question such as, "What impacts do genetically modified crops have?"

In cases like this, scientists often turn to meta-analyses (p. 258). The prefix *meta* means beyond, or encompassing and transcending, or at a higher level—and a *meta-analysis* is essentially an analysis of analyses. For instance, if your doctor wants to prescribe a treatment for your medical condition, he or she probably won't rely on just one research paper, because different studies may come up with different results (through error, by chance, or because of varying conditions). Instead, your doctor would prefer to assess the entirety of the research literature to come up with the best information—and consulting a published meta-analysis can help him or her do so.

The question of what impacts genetically modified crops have is a multifaceted issue complicated by many variables, and we can expect the answer to vary from one crop to another and from one set of conditions to another. Nonetheless, policymakers and regulators are confronted on the one hand with advocates demanding bans on all GMOs and on the other with agronomic experts urging them to speed and strengthen support for crop biotechnology—so general answers are needed to help with informed decision-making. Thus, in recent years several researchers have tried to make broad assessments of the scientific research on genetically modified foods.

The most widely cited work so far is that of Wilhelm Klümper and Matin Qaim of Georg-August-University in Göttingen, Germany, published in the journal *PLOS ONE* in 2014. They found 24,000 potential studies and screened them for suitability until arriving at 147 that compared GM crops to non-GM crops and that met all requirements of the meta-analysis. Klümper and Qaim then used variants on the statistical technique known as regression (p. 64) to test for patterns in types of impacts the studies had documented.

Klümper and Qaim's major results can be seen in **FIGURE 1**. According to their analysis, research so far finds that GM crops produce yields that average 22% higher than non-GM crops. The analysis also found that GM crops involve 37% less chemical pesticide use than non-GM crops. Thus, the analysis indicated that GM crops help farmers produce more food while applying fewer pesticides.

---

and controversy followed the apparent contamination of local landraces of maize by transgenic corn imported from the United States. Most scientists suspect that many transgenes will inevitably make their way from GM crops into wild plants, but the consequences of this are open to debate.

Because biotechnology is rapidly changing, and because the large-scale introduction of GMOs into our environment is recent, we cannot yet know everything about the consequences. Some experts believe that we should proceed with caution, adopting the **precautionary principle,** the idea that one should not undertake a new action until its ramifications are well understood. Others believe that plenty of research shows benefits outweighing costs and that it is time to double down on our bet on biotechnology.

Public debate over GMOs continues

Although U.S. consumers have largely accepted GM foods (or remain unaware that most food they eat now contains GM products), consumers in Europe, Japan, and other nations have expressed widespread unease about genetic engineering. Opposition to GM foods in Europe blocked the import of hundreds of millions of dollars in U.S. agricultural

products from 1998 to 2003. In 2013, Japan and South Korea suspended purchases of U.S. wheat after unapproved GM wheat plants from field experiments were found on one American farm.

More than 60 nations require that foods with genetically engineered ingredients be labeled so that consumers know what they are buying. The United States does not label GM foods, but Vermont, Connecticut, and Maine have passed GMO labeling laws. Proponents of labeling argue that consumers have a right to know what's in the food they buy. Opponents argue that labeling implies that labeled foods are dangerous, whereas research has not shown that to be the case. They say that labeling will entail expense and that consumers wishing to avoid GM foods can do so by buying certified organic foods or those that voluntarily apply "non-GMO" labels. In nations where labeling has been adopted, stores have ended up eliminating some GM foods from their shelves because of consumer avoidance.

WEIGHING the **issues**

Do You Want Your Food Labeled?

How would you vote if your state held a ballot initiative to require that genetically modified foods be labeled? Would you personally choose among foods based on such labeling? Why or why not?

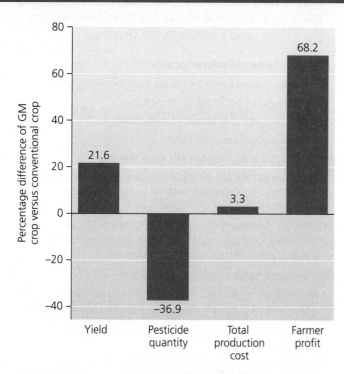

FIGURE 1 **GM crops showed higher yields, lower pesticide use, and higher farmer profits than conventional non-GM crops.** *Data from Klümper, W., and M. Qaim, 2014. A meta-analysis of the impacts of genetically modified crops. PLOS ONE 9(11) e111629: 1–7.*

The analysis also showed that GM crops increase farmers' production costs by 3%. GM seeds are more expensive, but this is balanced by the fact that farmers save money on pesticides. The 3% higher overall production costs, in turn, are far surpassed by the 22% yield gains. Because farmers have so much more crop to sell, their profits rise by 68% with GM crops, according to the analysis.

In exploring the data further, the researchers found that the pesticide savings were entirely due to reductions in insecticide use due to insect-resistant crops such as Bt corn and cotton (p. 257). Insect-resistant crops showed a 42% reduction in chemical application, whereas herbicide-tolerant crops such as Roundup Ready canola showed no significant difference in chemical application compared with non-GM crops. Yields were also substantially greater with insect-resistant crops than with herbicide-tolerant ones.

Critics took issue with the study's methods. Some complained that Klümper and Qaim had included results from conference presentations and studies by advocacy organizations and said they should have included only peer-reviewed studies published in respected journals. However, the Klümper and Qaim analysis found that GM crop yield gains were actually greater in peer-reviewed papers; had they included only the data in those papers, their results would have been even more favorable to GM crops. Some critics faulted Klümper and Qaim for including mostly surveys of farmers rather than experimental field trials. Qaim responded that both types of studies are useful and that surveys have the advantage of reflecting results in real-world conditions on working farms rather than idealized settings on smaller-scale research plots.

Further scientific research on the impacts of biotechnology in agriculture will continue to be published at a rapid pace. Only time will tell if future studies modify the big picture painted by Klümper and Qaim: that GM crops can bring greater food production and economic benefits to farmers, with less chemical pollution.

As GM crops have spread, so has the influence of a handful of large corporations that develop GM technologies; among them are Monsanto, Syngenta, Bayer CropScience, Dow, DuPont, and BASF. Critics say these multinational corporations threaten the independence and well-being of the family farmer and that government regulators generally side with big business rather than small farmers. Critics of biotechnology (**FIGURE 10.21**) also voice concern that much of the research into the impacts of GMOs is funded, conducted, or influenced by the corporations that stand to benefit if their products are approved. Whereas the Green Revolution was largely a public venture, the "gene revolution" brought by genetic engineering is largely driven by the financial interests of corporations selling proprietary products.

Indeed, corporations patent the transgenes they develop and go to great lengths to protect their investments. The Monsanto Company has threatened legal action against thousands of farmers, accusing them of growing patented plants without a contract, or for saving seeds from one harvest to another—something farmers have done from time immemorial yet is now illegal with seeds that contain patented genes. The risk of accidental contamination with GM material is especially acute for organic farmers. Many of them are concerned that if they experience an unwanted

FIGURE 10.21 **Grassroots opposition to GM foods is widespread.** At this demonstration in the Philippines, protestors contended that Monsanto's GM corn increases costs to farmers and harms the environment while enriching the corporation.

influx of pollen or seed from GM plants from a neighbor's property, they could lose their organic certification.

Organic Agriculture

Concerns over genetic engineering, chemical pesticides, fossil fuel pollution, and other aspects of high-input industrial agriculture have led many people to prefer *low-input agriculture:* farming and ranching that use lesser amounts of pesticides, fertilizers, growth hormones, antibiotics, water, and fossil fuel energy than in industrial agriculture. The low-input approach seeks to reduce both the environmental impacts and the financial costs of food production by allowing nature to provide ecosystem services (such as pollination, natural pest control, and organic fertilizer) that farmers using industrial methods must pay for.

Campus farms at colleges and universities generally take a low-input approach, and the University of Michigan's farm is a good example. Here staff and students use organic fertilizer such as chicken manure, biocontrol such as Bt, and drip irrigation to conserve water. They manage honeybee hives and maintain surrounding natural vegetation to encourage pollinating insects. And they aim to avoid the overuse of synthetic chemicals and fossil fuels whenever possible.

Food-growing practices that use absolutely no synthetic fertilizers, pesticides, hormones, or antibiotics but instead rely solely on biological approaches such as composting (p. 624) and biocontrol are known as **organic agriculture.**

Organic approaches reduce inputs and pollution

The bounty of organic agriculture is increasingly available to us (**FIGURE 10.22**). But what exactly is meant by the term *organic?* In 1990, Congress passed the Organic Food Production Act to establish national standards for organic products and facilitate their sale. Under this law, the USDA in 2000

FIGURE 10.22 The USDA organic logo shows that a product is certified organic under the National Organic Program.

TABLE 10.2 USDA Criteria for Certifying Crops and Livestock as Organic

For crops to be considered organic . . .

- The land must be free of prohibited substances for 3 years.
- Crops must not be genetically engineered.
- Crops must not be irradiated to kill bacteria.
- Sewage sludge cannot be used.
- Organic seeds and planting stock are preferred.
- Farmers must not use synthetic fertilizers. Only crop rotation, cover crops, animal or crop wastes, or approved synthetic materials are allowed.
- Most conventional pesticides are prohibited. Pests, weeds, and diseases should be managed with biocontrol, mechanical practices, or approved synthetic substances.

For livestock to be considered organic . . .

- Mammals must be raised organically from the last third of gestation (time in the womb); poultry, from the second day of life.
- Feed must be 100% organic, although vitamin and mineral supplements are allowed.
- Dairy cows must receive 80% organic feed for 9 months, followed by 3 months of 100% organic feed.
- Hormones and antibiotics are prohibited; vaccines are permitted.
- Animals must have access to the outdoors.

Adapted from the National Organic Program, 2002. *Organic production and handling standards.* Washington, DC: U.S. Department of Agriculture.

issued criteria by which crops and livestock can be officially certified as organic (**TABLE 10.2**). These standards went into effect in 2002 as part of the National Organic Program. California, Washington, and Texas established stricter state guidelines for labeling foods organic, and today many U.S. states and more than 80 nations have laws spelling out organic standards.

For farmers, organic farming can bring many benefits: lower input costs, enhanced income from higher-value produce, and reduced chemical pollution and soil degradation (see **THE SCIENCE BEHIND THE STORY**, pp. 264–265). Surveys reveal that farmers adopt organic techniques primarily because they want to practice stewardship toward the land and they want to safeguard their family's health. Farmers face obstacles to adopting organic methods, however. Foremost among these are the risks and costs of shifting to new methods, particularly during the transition period. In the United States, farmers need to meet standards for three years before their products can be certified.

Many consumers favor organic food out of concern about health risks posed by the pesticides, hormones, and antibiotics used in conventional industrial agriculture. Consumers also buy organic products out of a desire to improve environmental quality. The main obstacle for consumers is price. Organic products tend to be 10–30% more expensive than conventional ones, and some (such as milk) can cost twice as much. However, enough consumers have proved willing to pay a premium for organic products that grocers and other businesses are making them widely available.

Organic agriculture is on the rise

Just a generation ago, few farmers grew organic food, few consumers demanded it, and the only place to buy it was in specialty stores. That has changed; more than 80% of Americans buy organic food at least occasionally, most retail groceries offer it, and almost 14% of fruits and vegetables sold in the United States are organic. Americans now spend nearly $50 billion annually on organic products, amounting to almost 6% of all food sales (**FIGURE 10.23**). Worldwide, sales of certified organic food remain just 1% of all food sales, but they grew five times between 2000 and 2016, when sales reached $90 billion.

Production of organic foods is increasing along with demand. Although organic agriculture takes up just 1% of agricultural land worldwide (58 million ha, or 143 million acres, as of 2016), this area is rapidly expanding. Two-thirds of this area is in developed nations, and nearly two-thirds is grazing land. In the United States, the number of certified operations has more than tripled since the 1990s, while acreage devoted to organic crops and livestock has quadrupled: today, 2.0 million ha (5.0 million acres) are under organic management, representing 0.6% of U.S. agricultural land. Europe boasts far more: 13.5 million ha (33.3 million acres) as of 2016, most in nations of the European Union, where 6.7% of the agricultural area is in organic production.

The European Union supports farmers financially during conversion—an example of a subsidy (p. 180) aimed at reducing the external costs (pp. 143, 165) of industrial agriculture. The United States offers no such subsidy, which may explain why U.S. organic production lags behind that of Europe. However, recent farm bills have enhanced U.S. funding for research on organic agriculture and to assist growers with the costs of certification. Government support is helpful, because conversion often means a temporary loss in income for farmers. Once conversion is complete, though, studies suggest that reduced inputs and higher market prices generally make organic farming more profitable for the farmer than conventional methods. Indeed, a comprehensive summary in 2015 of the scientific and economic literature found that on average across the world, organic farming is 22–35% more profitable than conventional farming. Data from the USDA also support this finding; recent figures for corn, soy, wheat, and milk indicate that although organic yields are somewhat lower than conventional yields, price premiums are great enough that organic approaches are more profitable for the farmer.

Can organic agriculture feed the world?

Although the benefits of organic approaches in reducing pollution are clear, people have long wondered whether organic agriculture can produce enough food for the world's growing population. In recent years, a number of researchers have reviewed the growing literature of published scientific studies comparing yields of organic agriculture and conventional industrial agriculture. These meta-analyses—analyzed summaries of many studies (p. 258)—have found that organic agriculture, on average, tends to produce yields of roughly 80% of those of conventional agriculture (**FIGURE 10.24**). The gap between organic and conventional yields is highly variable, however, and depends on many factors.

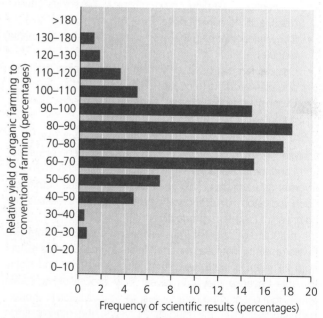

FIGURE 10.24 On average, crop yields on organic farms are about 80% of yields on conventional industrial farms. This graph shows the results of a meta-analysis (p. 258); each bar represents the frequency with which scientific results (for organic yields relative to conventional yields) fell into each interval shown on the *y* axis. *Data from de Ponti, T., 2012. The crop yield gap between organic and conventional agriculture. Agricultural Systems 108: 1–9.*

Go to **Interpreting Graphs & Data** on Mastering Environmental Science

FIGURE 10.23 Organic agriculture is growing. Sales of organic food in the United States have increased rapidly, both in total dollar amounts **(bars)** and as a percentage of the overall food market **(line).** *Data from Organic Trade Association's 2019 Organic Industry Survey, and previous years.*

Go to **Interpreting Graphs & Data** on Mastering Environmental Science

THE SCIENCE behind the story

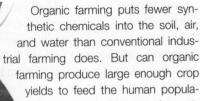

How Productive Is Organic Farming?

Organic farming puts fewer synthetic chemicals into the soil, air, and water than conventional industrial farming does. But can organic farming produce large enough crop yields to feed the human population? The world's two longest-running field experiments on the topic—in Switzerland and in Pennsylvania—suggest that the answer is yes.

Let's first visit Switzerland, where 1 in every 8 hectares of agricultural land is managed organically (the sixth-highest rate among nations) and where people consume more organic food per capita than anywhere else in the world. Back in 1977, Swiss researchers established experimental farms at Therwil, near the city

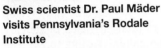

Swiss scientist Dr. Paul Mäder visits Pennsylvania's Rodale Institute

of Basel. At the research site, wheat, potatoes, and other crops are grown in plots cultivated in different treatments:

- Conventional farming using chemical pesticides, herbicides, and inorganic fertilizers
- Conventional farming that also uses organic fertilizer (cattle manure)
- Organic farming using only manure, mechanical weeding, and plant extracts to control pests
- Organic farming that also adds natural boosts, such as herbal extracts in compost

Researchers record crop yields at harvest each year. They analyze the soil regularly, measuring nutrient content, pH, structure, and other variables. They also measure the biological diversity and activity of microbes and invertebrates in the soil. Such indicators of soil quality help researchers assess the potential for long-term productivity.

In 2002, soil scientist Paul Mäder and colleagues from two Swiss research institutes reported in the journal *Science* results from 21 years of data. Long-term studies are rare and highly valuable because they can reveal slow processes or subtle effects that get stamped out by year-to-year variation in shorter-term studies. Over the study's 21 years, the organic fields yielded 80% of what the conventional fields produced. Organic potato crops averaged just 58–66% of conventional yields because of nutrient deficiency and disease, but organic crops of winter wheat produced 90% of conventional yields.

Although the organic plots produced 20% less on average, they received 35–50% less fertilizer than the conventional fields

and 97% fewer pesticides. Thus, Mäder's team concluded, the organic plots were highly efficient and represent "a realistic alternative to conventional farming."

How can organic fields produce decent yields without relying on synthetic chemicals? The answer lies in the soil. Mäder's team found that soil in the organic plots had better structure, better supplies of some nutrients, and much more microbial activity and invertebrate biodiversity than soil in conventional fields. Even when compared with conventional fields that were supplemented with manure, organic fields still outperformed them in these respects (**FIGURE 1**).

Studies are continuing at the Swiss plots today, producing new research results. As one example, environmental chemist Magdalena Necpalova and colleagues modeled data from Therwil and three other Swiss study sites in 2018 and found that conventional farming was a source of greenhouse gas emissions, but that organic farming plus reduced tillage (p. 233) sequestered carbon and reduced emissions by up to 128%, thereby helping address climate change.

Organic farming has proved even more successful in Pennsylvania, where agricultural scientists at the Rodale Institute have compared organic and conventional fields of corn and soybeans in a large-scale experiment running since 1981 on its 330-acre farm. In 2011, the institute released results from 30 years' worth of data (**FIGURE 2**).

Averaged across the 30 years, yields of organically grown crops equaled yields of conventionally grown crops. Moreover, the organic crops required 30% less energy input, and raising them released 35% fewer greenhouse gas emissions. Because of lower energy inputs and higher crop prices for organic produce, farming organically resulted in three times the profit for the farmer.

As with the Swiss experiment, the secret lies in the soil. At the Rodale Institute's farm over the years, the soil of the organic

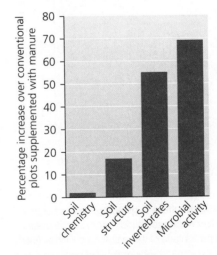

FIGURE 1 Organic fields developed better soil quality than conventional fields supplemented with manure. Values for soil chemistry (6 variables), structure (3 variables), invertebrates (5 variables), and microbial activity (6 variables) were compared. *Data from Mäder, P., et al., 2002. Soil fertility and biodiversity in organic farming. Science 296: 1694–1697.*

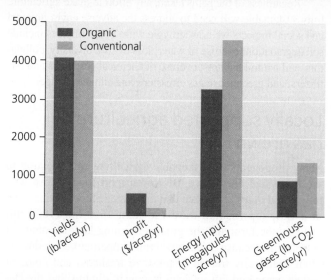

FIGURE 2 **Organic crops equaled conventional crops in yields while producing more profit, requiring less energy input, and releasing fewer greenhouse gas emissions.** *Data from Rodale Institute, 2011. The farming systems trial: Celebrating 30 years. Kutztown, PA: Rodale Institute.*

 Go to **Interpreting Graphs & Data** on **Mastering Environmental Science**

FIGURE 3 **Organically grown corn did better than conventionally grown corn during periods of drought in the Rodale Institute experiment.** Better-quality soil (shown in left half of the central inset) was part of the reason.

fields became visibly darker and better textured than the conventional fields' soil (**FIGURE 3**). By 2016, the organic fields' soil contained 18% more organic matter than the conventional fields' soil. This has helped corn crops in the organic fields outperform those in the conventional fields during times of drought.

Given such differences in soil quality, researchers expect that organic fields should perform better and better relative to conventional fields as time goes by—in other words, organic fields will prove to be more sustainable. Moreover, the striking yield results of the Rodale experiment suggest that perhaps organic agriculture can feed the world's people every bit as reliably as today's conventional industrial agriculture.

Several other long-term experiments comparing organic and conventional farming systems are currently running in the United States (**TABLE 1**). Their results so far parallel those found in Pennsylvania and Switzerland: Organic yields approach or match conventional yields, while reducing pollution, building better soil quality, and producing higher profits for farmers.

TABLE 1 Five Additional Long-Term Studies of Organic vs. Conventional Farming

STUDY AND LOCATION	YEAR STARTED	KEY RESULTS SO FAR
University of California, Davis: Davis, California	1988	Organic beans and safflower show yields comparable to those of conventional plots. Crops that require more nitrogen show lower yields in organic plots.
University of Minnesota: Lamberton, Minnesota	1989	Organic and conventional corn yields are equivalent. Soil quality has increased in organic plots. Organic plots are more resilient to stress.
University of Wisconsin–Madison: Arlington, Wisconsin	1989	Organic corn and soybeans yield ~80% of conventional plots in rainy years when managing weeds is difficult. Forage crops and milk production are equivalent between plot types. Better technology improves organic yields.
USDA Agricultural Research Service: Beltsville, Maryland	1996	Organic plots accumulate more nitrogen and soil organic carbon and can improve soil quality even more than no-till farming.
Iowa State University: Greenfield, Iowa	1998	Organic yields are equivalent to or greater than conventional yields for corn and soybeans. Organic farming improves soil quality and produces higher profits for farmers.

Summarized from Delate, K., et al. 2015. A review of long-term organic comparison trials in the U.S. *Sustainable Agriculture Research* 4(3): 5–14.

So, if yields from organic agriculture average 20% less than those of conventional industrial agriculture, can we feed the world with organic agriculture? The mainstream view has long been "no," and critics point out that lower yields mean that more land would need to be converted to agriculture to produce the same amount of food. However, increasing numbers of scientists, farmers, and other experts are more optimistic. They point out that organic yields have been rising as farmers gain more experience, and yields could rise still further if organic agriculture were to receive the research funding that goes toward conventional agriculture. (The USDA sends only 2% of its research funding to organic agriculture.) In addition, they say, by protecting soil quality, organic farming keeps more acreage productive for farming over the long term than does conventional farming.

Experts also point out that overall food production is not what limits our ability to feed the world. Only 43% of the global grain crop actually feeds people directly; the rest goes to livestock feed, biofuels (p. 576), and processed products like high-fructose corn syrup. Politics and the logistics of transport limit food distribution, and one-third of all food intended for our consumption ends up going to waste. Thus, by reducing waste, eating less meat, dealing with distribution problems, and prioritizing food for people, the world could easily be fed by organic agriculture even with lower yields, supporters maintain.

Sustainable Food Production

Growing adequate food and fiber for our burgeoning population while maintaining the environmental systems that support our agriculture is a tremendous challenge. Sustainable agriculture, like sustainability itself, involves a triple bottom line of social, economic, and environmental dimensions (p. 155). Sustainable agriculture must produce enough to provide food security to society, be profitable enough to provide farmers and ranchers a viable living, and conserve resources adequate to support future agriculture.

We can consider multiple paths toward sustainable agriculture

A variety of approaches exist as pathways to sustainable agriculture. On one end of the spectrum is conventional industrial agriculture. It produces high yields thanks to intensive inputs of fossil fuels, pesticides, and fertilizers to monocultures. At the other end of the spectrum is organic agriculture, which rejects chemical pesticides and inorganic fertilizers and takes a low-input approach, accepting lower yields but protecting natural resources that support agriculture in the long term.

Into the debate between the high-input industrial approach and the low-input organic approach has stepped biotechnology. Using genetic engineering to increase crop yields while reducing environmental impacts is one strategy that can theoretically lead toward sustainable food production. Genetic engineering has tremendous potential for the future if we can harness and direct it toward safely serving people's needs, especially those of small farmers and consumers.

Regardless of the paths taken, any effort to make agriculture truly sustainable will need to address the adverse environmental and social impacts we now struggle with. These impacts include soil degradation; overuse of water; loss of crop diversity, pollinators, and natural habitats; overuse of chemical pesticides and fertilizers; and greenhouse gas emissions and climate change.

Locally supported agriculture has grown

Many proponents of sustainable agriculture call attention to the fossil fuel energy we use to transport food. Those who tally "food miles" calculate that the average food product sold in a U.S. supermarket travels at least 1600 km (1000 mi) between the farm and the grocery, burning large amounts of petroleum. Because of the travel time, supermarket produce is often chemically treated to preserve freshness and color or picked green and left to ripen in transit, diminishing the flavor and nutritional value.

In response, increasing numbers of farmers and consumers are supporting local small-scale agriculture and adopting the motto "think globally, eat locally." Countless families tend backyard gardens to meet a portion of their own food needs. In cities across the world, urban agriculture is booming: Neighborhood-scale community gardens (p. 348) offer city dwellers small plots of land on which to grow their own food. In more rural areas, many families buy produce at roadside stands or pay to pick their own berries, fruits, and vegetables at "U-Pick" farms. And in towns and cities throughout North America, farmers' markets **(FIGURE 10.25)** have sprung up as people rediscover the joys of fresh, locally grown produce. At **farmers' markets,** consumers buy meats, dairy products, and fresh fruits and vegetables in season from local producers who gather there to sell their goods. These markets generally offer a wide choice of organic items and unique local varieties not found in supermarkets. All these efforts and trends are helping serve people living in so-called food deserts—areas that traditionally have lacked access to fresh, affordable, nutritious food, whether they be dense urban neighborhoods or remote rural communities.

FIGURE 10.25 Farmers' markets are flourishing as consumers rediscover the benefits of buying fresh, locally grown produce.

College and university dining services are increasingly aiming to source their food locally, as the University of Michigan is doing. To add to the bounty from the university's Campus Farm, MDining purchases food from more than 30 Michigan farmers and growers. Despite the region's short growing season, the university has been able to obtain products grown or processed in Michigan for nearly 40% of its food. In the dining halls and at its farmers' markets, MDining provides information on where its food comes from, offering biographies of its vendors.

Some consumers are partnering with local farmers in **community-supported agriculture** (CSA). In a CSA program, consumers pay farmers in advance for a share of their yield, usually a weekly delivery of produce. Consumers receive fresh seasonal produce, and farmers get a guaranteed income stream up front to invest in their crops—a welcome alternative to taking out loans and being at the mercy of the weather. Colleges and universities with large campus farms are now starting to run CSA programs. Clemson University, Dickinson College, Case Western Reserve University, and the University of Massachusetts Amherst are among dozens of schools that provide their communities with food through CSA programs.

Farmers' markets and community-supported agriculture help strengthen local economies while giving consumers access to fresher foods. They also often cut down on food miles. However, the number of miles a given item has traveled is not in itself a reliable measure of fossil fuel consumption. In many cases, shipping a food item far away as part of a large and efficient shipment by barge or rail may result in *less* fossil fuel use per item than transporting the item a short distance in small quantities by a gas-guzzling truck.

To determine which alternative involves less energy use overall, one needs to conduct a **life-cycle analysis,** a quantitative analysis of the inputs and outputs across all stages of an item's production, transport, sale, and use. Performing a full life-cycle analysis (p. 633) is a complex endeavor, and we can expect results to vary from case to case. In general, research so far has shown that fruits and vegetables consume the least fossil fuel energy, grains use more, eggs and chicken use still more, and beef and dairy products use the most.

In the most comprehensive life-cycle analysis of U.S. food production and delivery so far, researchers found that food miles from producer to retailer contributed just 4–5% of total greenhouse gas emissions of the entire process. Fully 83% of emissions resulted from production at the farm or feedlot. As a result, the average consumer can reduce his or her ecological footprint more effectively through dietary choices (such as eating more fruits and vegetables and less meat and dairy) than by eating locally sourced food, these researchers maintain. Thus, although eating locally is helpful in lowering carbon emissions, eating fewer animal products makes a far bigger difference. In *Calculating Ecological Footprints* (p. 271), you will work with some of these data yourself.

Hydroponics conserves key resources

Today, a number of innovative growing methods are beginning to produce food in creative ways that conserve resources. Many fall into the category of **hydroponics,** the practice of growing plants without using soil. In hydroponic systems, nutrients are supplied to plant roots directly via water-based solutions. Hydroponic systems come in many designs, but plants are generally grown inside sealed greenhouses and supplied with light, water, and nutrients from an array of computer-controlled equipment. In the hydroponic approach called *aeroponics,* plants are suspended in air and the roots are misted with nutrient solution. Because no soil is used, hydroponic systems enable vertical farming, with plants suspended above other plants or arrayed on shelving (**FIGURE 10.26**).

Hydroponics boasts several advantages over conventional soil-based farming:

- Less land is used (up to 3–5 times less) because plants can be packed close together.

- Less water is used (up to 90% less) because water is circulated and reused within the greenhouse.

- Few or no pesticides are needed because there is no soil for weeds and because most pest insects are sealed out. If pests occur, they can be addressed with biocontrol by introducing natural enemies.

- Less gasoline is needed for transporting produce because greenhouses can be built in regions of market demand—even within cities in empty factory, warehouse, or office buildings.

Hydroponic systems can provide fresh food for urban dwellers year-round, can operate in regions of poor soil or harsh climate, and are insulated from extreme weather events.

FIGURE 10.26 Using hydroponic techniques, we can grow vegetables in greenhouses without soil, without pesticides, and while conserving water.

Hydroponics has disadvantages, too. It requires substantial inputs of electricity (for lighting, timers, water pumps, temperature and humidity control, and so on), and this energy consumption makes for expense and a large carbon footprint. This drawback is being mitigated, however, as technology advances and as renewable electricity sources come online. In addition, not all foods can be grown hydroponically. However, those that can—like leafy greens and tomatoes—tend to be perishable, high-value crops that benefit most from the reduced transport distances.

The University of Michigan is now seeking to pursue hydroponics, aeroponics, and rooftop gardens as alternative ways to grow fresh local produce. Recently added to the dining halls are vertical tower gardens—4-foot-tall lighted towers growing leafy greens using small amounts of water and nutrients. They originated with student Carly Rosenberg and her project in her environmental science course. Rosenberg had interned for the summer at an urban farm in Los Angeles and saw how the University of Southern California and the University of California at Los Angeles were using vertical tower gardens on campus to supply food to their dining halls. She sold MDining officials on the idea, and today, towers in two University of Michigan dining

halls serve as educational displays that inspire discussion about fresh produce and local food.

Sustainable agriculture mimics natural ecosystems

Hydroponics may seem like futuristic farming, but the best way to pursue sustainable agriculture may be to get back to basics: to mimic the way a natural ecosystem functions. Ecosystems naturally operate in cycles and are internally stabilized with negative feedback loops (p. 108). In this way, they provide a useful model for agriculture.

One example comes from Japan, where rice farmers are reviving ancient traditions and finding them superior to modern industrial methods. Takao Furuno is one such farmer. Starting in the late 1980s, he and his wife added a crucial element to their rice paddies: the crossbred aigamo duck. Each spring after they plant rice, the Furunos release hundreds of aigamo ducklings into their paddies (**FIGURE 10.27**). The ducklings eat weeds that compete with the rice, and they eat insects and snails that attack the rice. The ducklings also fertilize the rice plants with their waste and oxygenate the water

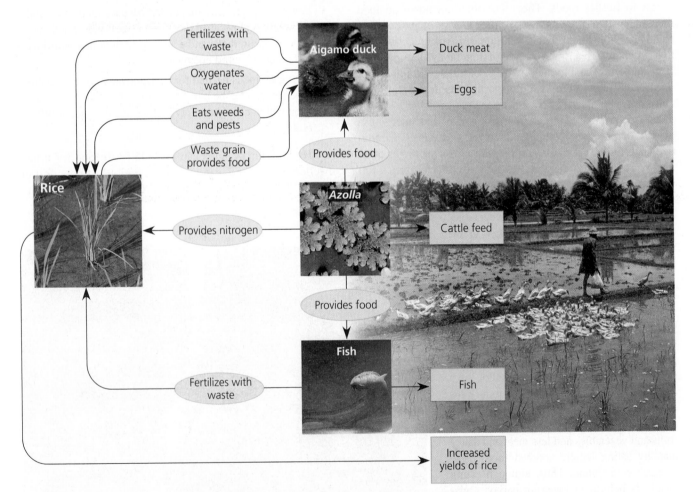

FIGURE 10.27 Aigamo rice-farming in Japan is an example of a sustainable agricultural system. Ducks, fish, and a water fern help create an ecosystem with the rice that makes each of them more productive. In this diagram, benefits that one organism provides to another are shown in ovals, with arrows pointing in the direction the benefits flow. In the rectangles are types of food that result as outputs from the system.

by paddling. Noting his success, scientists have worked with Furuno to determine what he does that works so well. They have found that in paddies that have ducks, rice plants grow larger and yield far more rice than in paddies without ducks. Once the rice grains form, the ducks are removed from the paddies (because they would eat the rice grains) and kept in sheds, where they are fed waste grain. The ducks mature, lay eggs, and can be sold at market.

Besides ducks, the Furunos raise fish in the paddies, which provide food and fertilizer as well. They also let the aquatic fern *Azolla* cover the water surface. This plant fixes nitrogen, feeds the ducks, and provides habitat for insects, plankton, and aquatic invertebrates, which provide additional food for the fish and ducks. Because fast-growing *Azolla* can double its biomass in 3 days, surplus plant matter is harvested and used as cattle feed. The end result is a productive ecosystem in which pests and weeds are transformed into resources and which yields multiple types of organic food. From only 2 ha of paddies and 1 ha of organic vegetables, the Furunos annually produce 7 tons of rice, 300 ducks, 4000 ducklings, and enough vegetables to feed 100 people. At this rate—twice the productivity of the region's conventional farmers—just 2% of Japan's people could supply the nation's food needs.

Takao Furuno wrote a book to popularize the "aigamo method," and an estimated 75,000 Asian farmers are now using it, increasing their yields 20–50% and regaining huge amounts of time they used to spend manually weeding. Whether farming rice with ducks in Asia or cultivating healthy soil, pollinators, and natural enemies of crop pests to help grow fruits and vegetables in places like Michigan, treating agricultural systems as ecosystems is a key aspect of making agriculture sustainable.

CENTRAL CASE STUDY
connect & continue

TODAY, at the University of Michigan, more students than ever before are engaged in learning about—and participating in—sustainable agriculture. The Sustainable Food Program is thriving, and the dining halls continue to incorporate more local foods as MDining pursues new partnerships with its region's farmers. As a result, the university is fast closing in on its goal of 20% sustainably sourced local food by 2025.

The university also recently surpassed its goal of reducing waste sent to landfills by 40%. Efforts to reduce food waste played a major role in this achievement. For instance, the school put on "VegWeek" with a series of lectures, films, panel discussions, and a "waste dinner" made from food items that usually get discarded (such as stew made from vegetable scraps and desserts made from spent grain from beer-brewing). This dinner aimed to raise awareness of food waste, and so many people attended that the event ran out of food!

Meanwhile, the university's Campus Farm is planning to expand and provide still more food to the dining halls. Staff and students running the farm plan to introduce new vegetables such as cabbage and squash, build additional hoop houses for growing produce in cold weather, and construct a root cellar to keep root vegetables fresher for longer. With additional funding, they hope to perform more education and community outreach and may consider running a food truck across campus to make fresh produce more accessible for students. Through their many efforts, students on campuses like the University of Michigan are experiencing the centrality of agriculture in our lives. They are becoming food ambassadors to society at large and taking personal steps to help make agriculture more sustainable.

• **CASE STUDY SOLUTIONS** Each college or university has many ways it could help enhance the sustainability of agriculture—on its campus or across society at large. It could operate an organic campus farm. It could focus on making its dining services a model of sustainability. It could tackle food waste issues. It could pioneer non-chemical means of pest control on its grounds. It could use such efforts to educate students and the community. It could also focus on supporting its faculty to do research into genetic engineering with the goal of creating crops and livestock that reduce impacts on land and resources. Likewise, it could support faculty research into how to increase production using organic methods. Of all these approaches, which do you think would be most useful and effective? Explain why. Now consider that you as a student could become personally involved as a volunteer on a campus farm or in a dining hall or as a research assistant in an agriculture professor's lab. What would you most like to do in this regard? Why?

• **LOCAL CONNECTIONS** What initiatives is your college or university pursuing along the lines of sustainable food and dining? Does your school have a farm on campus? If so, research what is grown there and what students do on the farm and then report on what you find. Does your dining hall source food from your campus farm? From local farmers? Does it take any steps to encourage healthy eating? Is it easy to find vegetarian or vegan meal options? What does your dining hall do to reduce food waste? Does it use trayless dining or compost its waste? Describe what steps your school's dining services are taking toward sustainable dining. Is your school doing anything beyond what the University of Michigan is doing? If you wished to, how could you get involved in food or dining sustainability issues on campus?

• **EXPLORE THE DATA** How are the University of Michigan and other institutions doing in pursuing their sustainability goals? → Explore Data relating to the case study on **Mastering Environmental Science.**

REVIEWING Objectives

You should now be able to:

+ **Explain the challenge of feeding a growing human population**

Our food production has outpaced our population growth thus far, yet more than 800 million people still go hungry each year. Undernutrition and overnutrition each challenge the goal of food security, as the human population and resource consumption continue to grow. (pp. 242–245)

+ **Identify the goals, methods, and consequences of the Green Revolution**

The Green Revolution aimed to enhance agricultural productivity in developing nations. Scientists used selective breeding to develop crop strains that grew quickly, were more nutritious, or were resistant to disease or drought. The increased efficiency and higher yields fed more people while reducing the amount of natural land converted for farming. However, the expanded use of fossil fuels, chemical fertilizers, and synthetic pesticides increased pollution and soil degradation. (pp. 245–247)

+ **Discuss how we raise animals for food, and assess the impacts of meat consumption**

As consumption of animal products has increased, we have turned to intensive production through feedlots and aquaculture. Feedlots create waste and pollution but also relieve pressure on lands that could otherwise be overgrazed. Aquaculture provides food security and relieves pressures on wild fish stocks but, at industrial scales, can give rise to pollution and habitat loss. Eating animal products leaves a considerably greater ecological footprint than eating plant products. The prospect of lab-grown food may possibly offer a means of producing meat with fewer environmental impacts (pp. 247–251)

+ **Explore strategies for pest and weed management**

We kill "pests" and "weeds" with synthetic chemicals that can pollute and pose health hazards. Pests and weeds can evolve resistance to chemical pesticides or herbicides, leading us to design more toxic poisons. Biological control uses natural enemies of pests against them, whereas integrated pest management combines various techniques and minimizes the use of synthetic chemicals. (pp. 251–254)

+ **Describe the science behind genetic engineering**

Most genetic engineering uses recombinant DNA to move genes for desirable traits from one type of organism to another, creating genetically modified organisms. Recently gene editing has shown promise to generate many new types of foods. These processes are both like and unlike traditional selective breeding. (pp. 255–258)

+ **Compare the benefits and costs of genetically modified foods, and assess the public debate over them**

Biotech crops offer a number of major benefits for agriculture and society. These crops have not yet reached their potential but could help make our agriculture sustainable. There is no clear evidence for human health effects, but genetically modified crops may have ecological impacts, including the spread of transgenes and an increase in herbicide pollution. Many people have ethical qualms about altering food through genetic engineering. Development of biotech foods has been controlled by multinational corporations, which critics view as a threat to small farmers. (pp. 258–262)

+ **Analyze the nature, growth, and potential of organic agriculture**

Because it reduces chemical and fossil fuel inputs, organic agriculture exerts fewer environmental impacts than industrial agriculture. Organic produce makes up a small part of the market but is growing rapidly. Scientific studies show that organic agriculture produces good crop yields while conserving natural resources and is a realistic global alternative to industrial agriculture. (pp. 262–266)

+ **Summarize potential pathways toward sustainable agriculture**

We can work toward sustainable agriculture in a variety of ways. Conventional industrial agriculture produces high yields but at the cost of pollution. Organic agriculture protects resources but generally has somewhat lower yields. Biotechnology and genetic engineering aim to offer novel ways to enhance production while minimizing environmental impact. Locally supported agriculture minimizes food miles and fossil fuel use. Hydroponics promises resource conservation and fresh produce for urban consumers. Mimicking natural ecosystems is a key approach to making agriculture sustainable. (pp. 266–269)

SEEKING Solutions

1. Assess several ways in which high-input industrial agriculture can be beneficial for people and for environmental quality, as well as several ways in which it can be detrimental. Now suggest several ways in which we might modify industrial agriculture to mitigate its impacts on people and the environment.

2. What factors make for an effective biological control strategy of pest management? What risks are involved in biocontrol? If you had to decide whether to use

biocontrol against a particular pest, what questions would you want to have answered before you decide?

3. People who view genetically modified foods as solutions to world hunger and insecticide overuse often want to speed their development and approval. Others adhere to the precautionary principle and want extensive testing for health and environmental safety. How much caution do you think is warranted before a new genetically modified crop is introduced? Describe what you feel the ideal process should be.

4. **THINK IT THROUGH** You are a farmer in your local area, and you have 500 acres to farm as you wish. What crops and animals would you raise, what farming approaches would you pursue, and why? Would you farm organically or by conventional industrial methods? Would you choose to grow genetically engineered crops? How would you manage for pests and weeds? Give reasons for each of your answers.

5. **THINK IT THROUGH** You are a USDA official and must help decide whether to allow the commercial planting of a new genetically modified strain of cabbage that is tolerant to a best-selling herbicide and has twice the vitamin content of regular cabbage. What questions would you ask of scientists before deciding whether to approve the new crop? What scientific data would you want to see? Would you also consult nonscientists? Would you consider ethical, economic, and social factors?

CALCULATING Ecological Footprints

Many people who want to reduce their ecological footprint have focused on how much energy is expended (and how many climate-warming greenhouse gases are emitted) in transporting food from its place of production to its place of sale. The typical grocery store item is shipped by truck, air, or sea for many hundreds of miles before reaching the shelves, and this transport consumes oil. This concern over "food-miles" has helped drive the "locavore" movement to buy and eat locally sourced food. However, food's transport from producer to retailer, as measured by food-miles, is just one source of carbon emissions in the overall process of producing and delivering food.

In 2008, environmental scientists Christopher Weber and H. Scott Mathews conducted a thorough life-cycle analysis (p. 267) of U.S. food production and delivery. By filling in the table below, you will get a better idea of how our dietary choices contribute to climate change.

FOOD TYPE	TOTAL EMISSIONS[1] ACROSS LIFE CYCLE	EMISSIONS[1] FROM DELIVERY[2]	PERCENTAGE EMISSIONS FROM DELIVERY[2,3]
Fruits and vegetables	0.85	0.10	11.8
Cereals and carbohydrates	0.90	0.07	
Dairy products	1.45	0.03	
Chicken/fish/eggs	0.75	0.03	
Red meat	2.45	0.03	
Beverages	0.50	0.04	

[1] Emissions are measured in metric tons of carbon dioxide equivalents, per household per year.
[2] "Delivery" means transport from producer to retailer.
[3] "Percentage emissions from delivery" is calculated by dividing "Emissions from delivery" by "Total emissions across life cycle" and then multiplying by 100 (to convert proportion to percentage).
Source: Weber, C.L., and H.S. Mathews, 2008. Food-miles and the relative climate impacts of food choices in the United States. *Environmental Science and Technology* 42: 3508–3513.

1. Which type of food is responsible for the most greenhouse gas emissions across its whole life cycle? Which type is responsible for the least emissions?

2. What is the range of values for percentage of emissions from transport of food to retailers (delivery)? Weber and Mathews found that 83% of food's total emissions came from its production process on the farm or feedlot. What do these numbers tell you about how you might best reduce your own footprint with regard to food?

3. After measuring mass, energy content, and dollar value for each food type, the researchers calculated emissions per kilogram, calorie, and dollar. In every case, red meat produced the most emissions, followed by dairy products and chicken, fish, and eggs. They then calculated that shifting one's diet from meat and dairy to fruits, vegetables, and grains for just one day per week would reduce emissions as much as eating 100% locally (cutting food-miles to zero) all the time. Knowing all this, how would you choose to reduce your own food footprint? By how much do you think you could reduce it?

Mastering Environmental Science

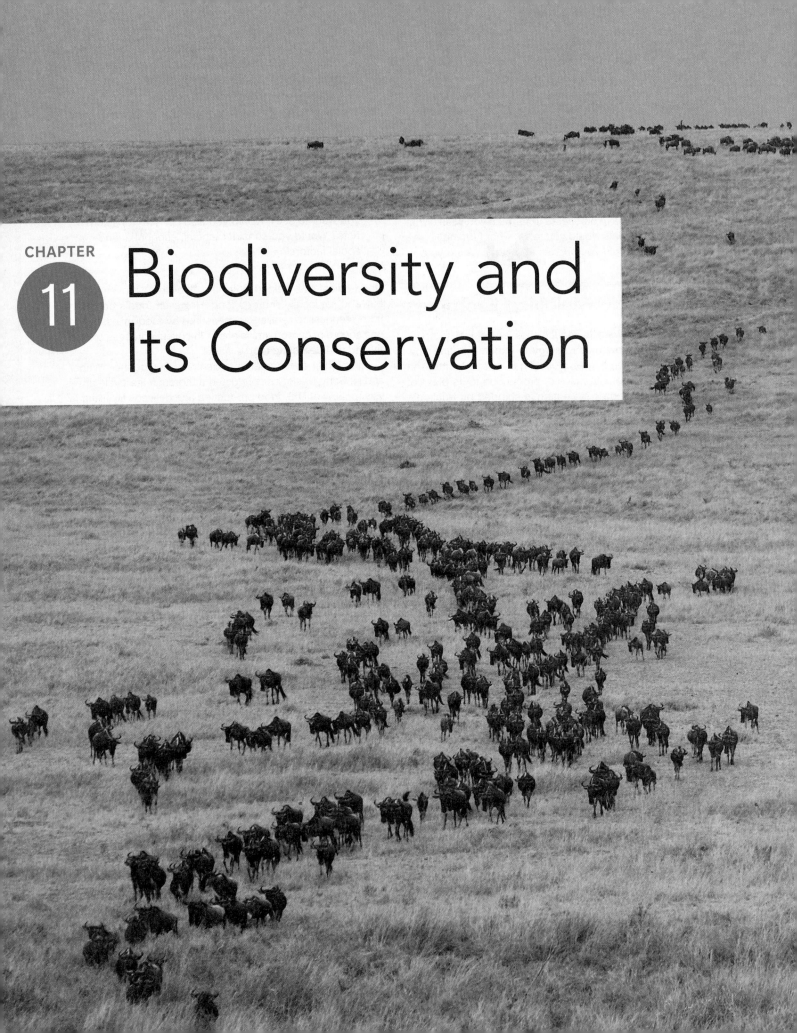

Biodiversity and Its Conservation

Will We Slice through the Serengeti?

Lake Victoria KENYA

Serengeti National Park TANZANIA

" **Construction of the road will be a huge relief for us. We will sell ... maize and horticultural products to our colleagues in Arusha and they will bring us cows and goats.**
Bizare Mzazi, a small farmer outside Serengeti National Park

If we construct this road, all our rhinos will disappear. ... We should strive to conserve our heritage for future generations.
Sirili Akko, executive officer of the Tanzania Association of Tour Operators "

t's been called the greatest wildlife spectacle on Earth. Each year more than 1.2 million wildebeest migrate across the vast plains of the Serengeti in East Africa, along with more than 700,000 zebras and hundreds of thousands of antelope. The herds can stretch as far as the eye can see. Lions, hyenas, leopards, and cheetahs track the procession and pick off the weak and the unwary, while hungry crocodiles wait in ambush at river crossings. After bearing their calves during the wet season, the wildebeest journey north to find fresh grass. The great herds spend the dry season at the northern end of the Serengeti ecosystem and then head back south to complete their annual cyclical journey.

This epic migration, with its dramatic interplay of predators and prey, has cycled on for millennia. Yet today, the entire phenomenon may be threatened by a proposal to build a commercial highway across the Serengeti, slicing straight across the animals' migratory route.

Before examining the highway proposal, let's step back for a broad view of the Serengeti. The people native to this region, the Maasai, are semi-nomadic herders who have long raised cattle on the grasslands and savannas. Because the Maasai subsist on their cattle and have lived at low population densities, wildlife thrived here long after it had declined in other parts of Africa.

When East Africa was under colonial rule, the British created game reserves to conserve wildlife for their own hunting. After Tanzania, Kenya, and other African nations gained independence in the mid-20th century, the British reserves became the basis for today's national protected areas. Serengeti National Park was established in Tanzania in 1951, and the Maasai Mara National Reserve was later created just across the border in Kenya. These two protected areas, together with several adjacent ones, encompass the Serengeti ecosystem. This 30,000-km² (11,500-mi²) region is one of the last places on the planet where an ecosystem remains nearly intact and functional over a vast area.

Today, 2 million people from around the world visit Tanzania and Kenya each year, most of them ecotourists who visit the parks and protected areas. Serengeti National Park alone receives 800,000 annual visitors. Tourism injects close to $3 billion into these nations' economies and

Upon completing this chapter, you will be able to:

+ **Characterize the scope of biodiversity on Earth**

+ **Specify the benefits that biodiversity brings us**

+ **Discuss today's extinction crisis in geologic context**

+ **Evaluate the primary causes of biodiversity loss**

+ **Assess the science and practice of conservation biology**

+ **Analyze efforts to conserve threatened and endangered species**

+ **Compare and contrast conservation efforts above the species level**

◀ **Wildebeest crossing the vast plains of the Serengeti**

▲ **A Maasai man herding his cattle**

creates jobs for tens of thousands of local people. Because the region's people see that functional ecosystems full of wildlife bring foreign dollars into their communities, many support the parks.

However, many inhabitants of northern Tanzania remain desperately poor and live without electricity or medical care. Farmers, villagers, and townspeople along the shores of Lake Victoria feel isolated from the rest of Tanzania by a poor road system. Walled off by Serengeti National Park to their east (which does not allow commercial truck traffic on its few dirt roads), these people have little access to outside markets to buy and sell goods. In response, Tanzania's former president Jakaya Kikwete promised to build a paved commercial highway across the Serengeti, connecting Lake Victoria communities with cities to the east and ports on the Indian Ocean. The World Bank and the German government offered to finance the $480-million project, and Chinese contractors stood ready to build it.

Around the world, conservationists reacted with alarm. The proposed highway would slice right through the middle of the wildebeest migration path (**FIGURE 11.1**). Scientists predicted that the road would block migration routes and that vehicles would kill countless animals in collisions. A highway would also provide access for poaching (the illegal killing of wildlife for meat or body parts) and would allow an entry corridor for exotic plant species that could invade the ecosystem. A highway would encourage human settlement right up to the park boundary, making the park an island of habitat hemmed in by agriculture, housing, and commerce. And by boosting development, a highway could spur the towns on Lake Victoria to grow into large cities, creating demand for even larger transportation corridors in the future. For all these reasons, experts predicted that the highway would diminish animal populations and possibly destroy the migration spectacle.

Such an outcome could devastate tourism, so the region's tourism operators opposed the highway. So did most Kenyans, who feared that the highway would prevent migratory animals from reaching Kenya's Maasai Mara Reserve. In 2010, a Kenyan nongovernmental organization, the African Network for Animal Welfare, sought to stop the highway with a lawsuit in the East Africa Court of Justice, a body that adjudicates international matters in the region. After various hearings and appeals, this court in 2014 issued a final ruling prohibiting the project from going forward.

Meanwhile, international pressure rose on Tanzania to abandon its plans. Highway opponents proposed an alternative route that would wrap around the Serengeti's southern end (see Figure 11.1), passing through more towns and serving five times as many people along the way. The World Bank and the German government offered to help fund this alternative route instead.

In 2016, John Magufuli took office as Tanzania's new president. He resurrected the original highway proposal and authorized construction of the highway through the park. Again, the diverse coalition of highway opponents raised an international outcry, and within weeks, Magufuli backed down.

Soon a new plan kicked off debate. The nation of Uganda wanted to export oil from its recently developed oilfields eastward to ports on the Indian Ocean. In 2016, three routes for an oil pipeline were proposed, with the most direct option cutting straight across Serengeti National Park. Such a pipeline (accompanied by a road) would create a barrier to migratory animals just as a highway would. It would bring all the impacts of a highway, along with risks of oil spills. So, when news of the pipeline project became public, scientists and conservationists again rushed to object to a development corridor through the park. Faced with the outcry, Magufuli's government announced that any pipeline route would avoid passing through game reserves and national parks. Officials began studying routes around the Serengeti.

In Tanzania, poaching is on the rise and animal populations are falling. The Serengeti is one of our planet's last intact large ecosystems, so impacts here have global ramifications for biological diversity on Earth. We would all be impoverished if the Serengeti's biodiversity were lost, so we must hope that Africans can find ways to build wealth and improve their standard of living while conserving their wildlife and natural systems. East Africa has helped pioneer globally acclaimed, win-win solutions in conservation thus far, so there is hope that it will show the way yet again.

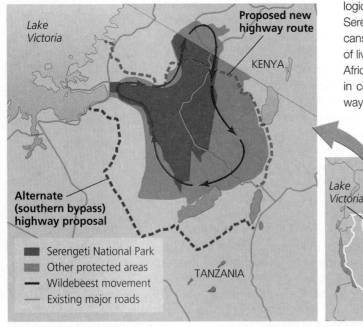

Proposed new highway route

Lake Victoria

KENYA

Alternate (southern bypass) highway proposal

■ Serengeti National Park
■ Other protected areas
— Wildebeest movement
— Existing major roads

TANZANIA

Lake Victoria

KENYA

TANZANIA

Indian Ocean

FIGURE 11.1 A proposed highway would slice through Serengeti National Park. It would increase commerce and connect Tanzanian people on each side, but it would also cut across the migration route for wildebeest and other animals. Highway opponents suggest an alternate route around the park's southern edge.

Life's Diversity on Earth

Rising human population and resource consumption are putting ever-greater pressure on the flora and fauna of our planet. We are diminishing the ultimate source of our civilization's wealth and happiness: Earth's diversity of life, the very quality that makes our planet unique in the known universe. Thankfully, many people around the world are working tirelessly to save threatened animals, plants, and ecosystems in efforts to stem the loss of our planet's priceless biological diversity.

Biodiversity encompasses multiple levels

Biological diversity, or **biodiversity,** is the variety of life across all levels of biological organization. It includes diversity in genes, populations, species, communities, and ecosystems (**FIGURE 11.2**). The level of biodiversity that people find easiest to visualize and that we refer to most commonly is species diversity.

Species diversity A **species** is a distinct type of organism, a set of individuals that uniquely share certain characteristics and can breed with one another and produce fertile offspring (p. 50). Species form by the process of speciation (p. 54) and may disappear by extinction (p. 58). **Species diversity** describes the number or variety of species found in a particular region. One component of species diversity is *species richness,* the number of species inhabiting an area. Another is *evenness* or *relative abundance,* the extent to which species in a given area differ in numbers of individuals. (Greater evenness means they differ less.)

Biodiversity exists below the species level in the form of *subspecies,* populations of a species that occur in different geographic areas and differ from one another in slight ways. Subspecies arise by the same processes that drive speciation but result when divergence stops short of forming separate species. As an example, the black rhinoceros diversified into about eight subspecies, each inhabiting a different part of Africa. The eastern black rhino, which is native to Kenya and Tanzania, differs slightly in its attributes from each of the other subspecies.

Genetic diversity Scientists designate subspecies when they recognize substantial genetically based differences among

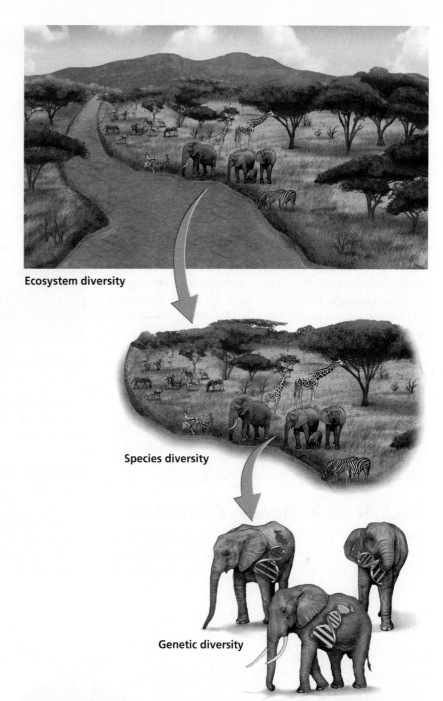

Ecosystem diversity

Species diversity

Genetic diversity

FIGURE 11.2 The concept of biodiversity encompasses multiple levels in the hierarchy of life.

individuals from different populations of a species. However, all species consist of individuals that vary genetically from one another to some degree, and this variation is another important component of biodiversity. **Genetic diversity** encompasses the differences in DNA composition (p. 28) among individuals, and these differences provide the raw material for adaptation to local conditions. In the long term, populations with more genetic diversity may be more likely to persist, because their variation better enables them to cope with environmental change.

Populations with little genetic diversity are vulnerable to environmental change if they lack genetic variants to help them adapt to changing conditions. Populations with low genetic diversity may also show less vigor, be more vulnerable to disease, and suffer *inbreeding depression,* which occurs when genetically similar individuals mate and produce weak or defective offspring. Scientists have sounded warnings over low genetic diversity in species that have dropped to low population sizes, including American bison, elephant seals, and the cheetahs of the East African plains. Diminished genetic diversity in our crop plants is a prime concern to humanity (p. 225).

Ecosystem diversity The number and variety of ecosystems (pp. 61, 111) is referred to as **ecosystem diversity,** but biologists may also refer to the diversity of communities (pp. 60, 81) or habitats (p. 61). Scientists may also consider the geographic arrangement of habitats, communities, or ecosystems across a landscape, including the sizes and shapes of patches and the connections among them (p. 118). Under any of these concepts, a seashore of beaches, forested cliffs, offshore coral reefs, and ocean waters would hold far more biodiversity than the same acreage of a monocultural cornfield. A mountain slope where vegetation changes with elevation from desert to forest to alpine meadow would hold more biodiversity than a flat area the same size consisting of only desert, forest, or meadow. The Serengeti region holds a diversity of habitats, including savanna (p. 98), grassland, hilly woodlands, seasonal wetlands, and rock outcroppings. This habitat diversity contributes to the rich diversity of species in the region.

Biodiversity is unevenly distributed

Some groups of organisms include more species than others. For example, in numbers of species, insects show a staggering predominance over all other forms of life (**FIGURE 11.3**;

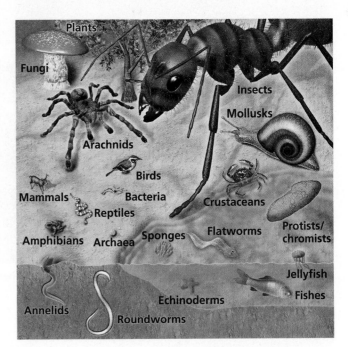

FIGURE 11.3 Some groups contain more species than others. This illustration shows organisms scaled in size to the number of species known from each group, giving a visual sense of their species richness. *Data from Roskov, Y., et al. (eds.), 2019. Species 2000 & ITIS catalogue of life, 25 Mar. 2019. Digital resource at www.catalogueoflife.org/col. Leiden, the Netherlands: Species 2000: Naturalis.*

FIGURE 11.4). Among insects, about 40% are beetles, and beetle species alone outnumber all non-insect animal species and all plant species. No wonder the British biologist J.B.S. Haldane famously quipped that God must have had "an inordinate fondness for beetles."

In some groups, large numbers of species formed rapidly as populations spread into a variety of environments and

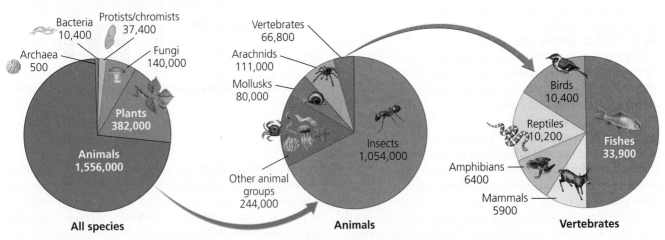

FIGURE 11.4 Most species are animals. Nearly two-thirds of animals are insects. In comparison, vertebrates make up only 4%. *Data from Roskov, Y., et al. (eds.), 2019. Species 2000 & ITIS catalogue of life, 25 Mar. 2019. Digital resource at www.catalogueoflife.org/col. Leiden, the Netherlands: Species 2000: Naturalis.*

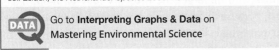

DATA Go to **Interpreting Graphs & Data** on **Mastering Environmental Science**

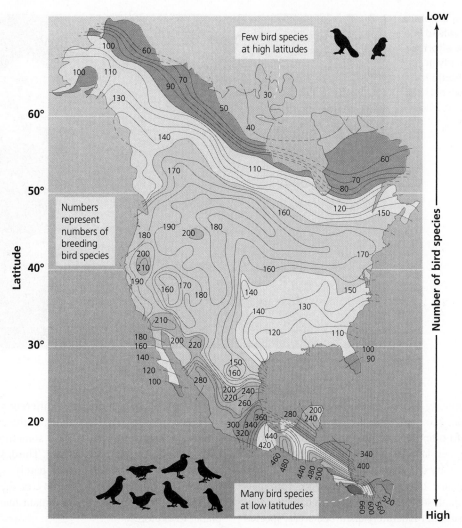

FIGURE 11.5 Species richness tends to increase toward the equator. Birds show a clear latitudinal gradient in species richness. Regions of arctic Canada and Alaska are home to just 30–100 breeding species of birds, whereas areas of Costa Rica and Panama host more than 600. *Adapted from Cook, R.E., et al., 1969. Variation in species density in North American birds. Systematic Biology 18: 63–84. (Originally published as Systematic Zoology.) By permission of Oxford University Press.*

adapted to local conditions. Other groups diversified because of a tendency to become subdivided by barriers that promote speciation (pp. 54–55). Still other groups accumulated species through time because of low rates of extinction.

Biodiversity is also greater in some places than in others. For instance, species richness generally increases as one approaches the equator (**FIGURE 11.5**). This pattern of variation with latitude, called the *latitudinal gradient in species richness,* is one of the most obvious patterns in ecology yet one of the most difficult for scientists to explain. Hypotheses abound for the cause of the latitudinal gradient. Some focus on geographic area, arguing that the tropics provide more room for speciation. Some focus on solar energy, arguing that more plant growth makes areas nearer the equator more productive. Others focus on the relative stability of tropical climates, arguing that this discourages small numbers of generalist species from dominating ecosystems and instead allows numerous specialist species to coexist. Still others focus on history, arguing that polar and temperate regions are

relatively species-poor because glaciation events repeatedly forced organisms out of these regions and toward tropical latitudes.

Regardless of the explanation, the latitudinal gradient influences the species diversity of Earth's biomes (pp. 94–100). Tropical dry forests and tropical rainforests support far more species than do tundra and boreal forests, for instance. Tropical biomes typically show more evenness as well, whereas in high-latitude biomes with low species richness, some species greatly outnumber others. For example, Canada's boreal forest is dominated by vast expanses of black spruce, whereas Panama's tropical forest contains hundreds of tree species, none of which greatly outnumber others.

Within a given habitat, biodiversity varies according to the attributes of the habitat. Structurally diverse habitats tend to allow for more ecological niches (pp. 62, 77) and thereby support greater species richness and evenness. For instance, forests generally support more diversity than grasslands, and coral reefs support more diversity than the open ocean.

FIGURE 11.6 Joining a bioblitz is one way to get involved in a local citizen science effort to learn about biodiversity in your area and contribute to knowledge. Here students help experts survey insects in Glacier National Park, Montana.

Each habitat supports a somewhat different mix of organisms. For this reason, ecotones (transition zones where habitats intermix; p. 118) often tend to support high biodiversity. Likewise, for any given large region, species diversity tends to increase with diversity of habitats.

Human disturbance often creates patchwork combinations of habitats at small scales, which increases habitat diversity locally. As a result, in moderately disturbed areas, species diversity tends to rise. However, at larger scales, human disturbance generally replaces a variety of regionally unique habitats with one or a few homogenized disturbed habitats. This reduction in habitat diversity results in a decrease in species diversity because it causes specialist species reliant on the regionally unique habitats to disappear, being replaced by a smaller variety of widespread generalist species. Moreover, species that rely on large expanses of habitat disappear when those habitats are fragmented by human disturbance.

Many species await discovery

Despite centuries of exploration and study by professional biologists and amateur naturalists, Earth is home to so much life that humanity remains profoundly ignorant of the number of species that exist. So far, scientists have identified and described more than 1.8 million species of plants, animals, fungi, and microorganisms. However, estimates for the total number that actually exist range from 3 million to 100 million, with the most widely accepted estimates in the neighborhood of 10 million.

Our knowledge of species numbers is incomplete for several reasons. First, many species are tiny and easily overlooked. They include bacteria, fungi, protists, nematodes (roundworms), and soil-dwelling arthropods (insects, spiders, and other invertebrate animals). Second, many organisms are difficult to identify; sometimes, organisms thought to be of the same species turn out to be different species once examined more closely. That is frequently the case with microbes, fungi, and small insects, but also sometimes with organisms as large as birds, trees, and whales. Third, some areas of Earth remain little studied. We have barely sampled the ocean depths, hydrothermal vents (p. 34), or the tree canopies and soils of tropical forests. There remain many frontiers on our planet to explore!

What's most exciting is that a person doesn't need to be a professional biologist to contribute to humanity's knowledge of biodiversity. Indeed, much of our knowledge about the world's species has come from amateur naturalists—laypeople with the passion to learn about the natural world around them, study their favorite organisms in the field, develop deep expertise, and share their knowledge with others.

Such expert naturalists today are contributing vast amounts of useful data to science through the pursuit of **citizen science** (also called *community science*), the study and collection of data for scientific purposes by nonprofessional volunteers. Programs like Monarch Watch and the Christmas Bird Count enlist thousands of volunteers each year to help with observation, monitoring, and conservation. Hundreds of parks, nature centers, agencies, and organizations host *bioblitzes,* events in which volunteers, with the help of experts, search for organisms and document biodiversity on a site-by-site basis (**FIGURE 11.6**).

Moreover, today's digital technology and capacities for online communication and databases have opened up a whole new world of information sharing. Millions of people now use, enjoy, and collectively build socially interactive online data repositories like eBird and iNaturalist to document life everywhere, from backyard gardens to international parks. There is a world of natural wonder out there, endless amounts to learn, and many ways to get involved for anyone who takes the time to get outdoors and look around.

Benefits of Biodiversity

These days most of us live in cities and suburbs, spend nearly all our time indoors, and pass hours each day staring at electronic screens. It's no wonder that we often fail to appreciate how biodiversity enriches our lives! Yet we benefit from biodiversity and it supports our society in fundamental ways. Indeed, our cities, homes, and technology would simply not exist without the resources and services that Earth's living species provide us.

Biodiversity enhances food security

Biodiversity provides the food we eat. Throughout our history, human beings have used at least 7000 plant species and several thousand animal species for food. Today, industrial agriculture has narrowed our diet. Globally, we now get 90% of our food from just 15 crop species and 8 livestock species, and this lack of diversity leaves us vulnerable to crop failures. In a world in which 800 million people go hungry, we can improve food security (the guarantee of an adequate, safe, nutritious, and reliable food supply; p. 242) by finding sustainable ways to harvest or farm wild species and rare crop varieties.

TABLE 11.1 shows a selection of promising wild food resources from just one region of the world—Central and South America. Plenty of additional new or underused food sources exist there and elsewhere worldwide. As examples, the babassu palm of the Amazon produces more vegetable oil than any other plant. The serendipity berry generates a sweetener 3000 times sweeter than table sugar. Some salt-tolerant grasses and trees are so hardy that farmers can irrigate them with salt water to produce animal feed and other products.

Moreover, the wild relatives of our crop plants hold reservoirs of genetic diversity that can help protect the crops we grow in monocultures by providing helpful genes for cross-breeding or genetic engineering (p. 255). We have already received tens of billions of dollars' worth of disease resistance from the wild relatives of potatoes, wheat, corn, barley, and other crops.

Organisms provide drugs and medicines

People have made medicines from plants and animals for centuries, and about half of today's pharmaceuticals are derived from chemical compounds from wild plants (**TABLE 11.2**, p. 280). A well-known example is aspirin, which was derived from chemicals found in willows and in meadowsweet, but there are countless others. The rosy periwinkle of Madagascar produces compounds that treat Hodgkin's disease and a deadly form of leukemia. The Pacific yew of North America's Pacific Northwest produces a compound that forms the basis for the anti-cancer drug paclitaxel (Taxol). In Australia, a rare species of cork, *Duboisia leichhardtii,* provides hyoscine, a compound that physicians use to treat cancer, stomach disorders, and motion sickness. Each year, pharmaceutical products owing

---

TABLE 11.1 Potential New Food Sources*

Amaranths
(three species of Amaranthus)

Grain and leafy vegetable; livestock feed; rapid growth, drought resistant

Capybara
(Hydrochoeris hydrochaeris)

World's largest rodent; meat esteemed; easily ranched in open habitats near water

Buriti palm
(Mauritia exuosa)

"Tree of life" to Amerindians; vitamin-rich fruit; pith as source for bread; palm heart from shoots

Vicuna
(Lama vicugna)

Related to llama; source of meat, fur, and hides; can be profitably ranched

Maca
(Lepidium meyenii)

Cold-resistant root vegetable resembling radish, with distinctive flavor; near extinction

Chachalacas
(Ortalis, many species)

Tropical birds; adaptable to human habitations; fast-growing

*The wild species shown here are just some of the many plants and animals that could supplement our food supply.
Source: Adapted from Wilson, E.O., 1992. *The diversity of life.* Cambridge, MA: Belknap Press.

TABLE 11.2 Natural Plant Sources of Pharmaceuticals*

Pineapple
(Ananas comosus)

Drug: Bromelain
Application: Controls tissue inflammation

Pacific yew
(Taxus brevifolia)

Drug: Taxol
Application: Anti-cancer agent (especially ovarian cancer)

Autumn crocus
(Colchicum autumnale)

Drug: Colchicine
Application: Anti-cancer agent

Velvet bean
(Mucuna deeringiana)

Drug: L-dopa
Application: Parkinson's disease suppressant

Yellow cinchona
(several species of *Cinchona*)

Drug: Quinine
Application: Anti-malarial agent

Common foxglove
(Digitalis purpurea)

Drug: Digitoxin
Application: Cardiac stimulant

*Shown are just a few of the many plants that provide chemical compounds of medical benefit.
Source: Adapted from Wilson, E.O., 1992. *The diversity of life.* Cambridge, MA: Belknap Press.

their origin to wild species generate up to $150 billion in sales and save thousands of human lives.

The world's biodiversity holds a still-greater treasure chest of medicines yet to be discovered. For this reason, pharmaceutical companies hire scientists to engage in **bioprospecting,** searching for organisms that might provide new drugs, medicines, foods, or other valuable products. There is urgency to such searches. For one thing, a great deal of traditional knowledge of the attributes of local plants and animals in species-rich tropical regions is being lost as communities of indigenous people succumb to outside pressures from the industrial world. For another, many animals that show particular promise are at risk of being lost to extinction before we can profit from what they have to offer **(TABLE 11.3)**. We already lost such an opportunity in two species of gastric brooding frogs discovered in the rainforests of Queensland, Australia (see top photo in Table 11.3). Females of these bizarre frogs raised their young inside their stomachs, where in any other animal, stomach acids would soon destroy them! Apparently, the young frogs exuded substances that neutralized their mother's acid production. Any such substance could be of immense use for treating human stomach ulcers, which affect 25 million Americans alone. Sadly, both frog species went extinct in the 1980s, taking their medical secrets with them forever. With every species that goes extinct, we lose one more potential opportunity to find cures and treatments.

Biodiversity provides ecosystem services

Contrary to popular opinion, some things in life can indeed be free—as long as we protect the ecological systems that provide them. Forests cleanse air and water while buffering us against floods. Native crop varieties provide insurance against disease and drought. Wildlife can attract tourism that generates income for people. Intact ecosystems provide these and other valuable processes, known as ecosystem services (pp. 4, 120–121), for us all, free of charge. According to scientists, biodiversity helps:

- Provide food, fuel, fiber, and shelter.
- Purify air and water.
- Detoxify and decompose wastes.
- Stabilize Earth's climate.
- Moderate floods, droughts, and temperatures.
- Cycle nutrients and renew soil fertility.
- Pollinate plants, including many crops.
- Control pests and diseases.
- Maintain genetic resources for crop varieties, livestock breeds, and medicines.
- Provide cultural and aesthetic benefits.

In these ways, organisms and ecosystems support vital processes that people cannot replicate or would need to pay for if nature did not provide them. The economic value of just 17 of the world's ecosystem services has been estimated at more than $153 trillion per year (p. 149)—more than the gross domestic product of all national economies combined.

Biodiversity helps maintain functioning ecosystems

Ecological research demonstrates that biodiversity tends to enhance the stability of communities and ecosystems. Research has also found that biodiversity tends to increase the resilience (p. 87) of ecological systems—their ability to withstand disturbance, recover from stress, or adapt to change. Thus, when biodiversity is lost, this typically diminishes a natural system's ability to function and to provide services to our society.

Will the loss of a few species really make much difference in an ecosystem's ability to function? Consider a metaphor first offered by Paul and Anne Ehrlich (p. 190): The loss of one rivet from an airplane's wing—or two, or three—may not cause the plane to crash. But as rivets are removed, the structure will be compromised, and eventually, the loss of just one more rivet will cause it to fail. This metaphor suggests that we would be wise to preserve as many components of our ecosystems as possible to make sure these systems continue to function.

Research shows that removing a keystone species (p. 86) such as a top predator can significantly alter an ecological system. Think of lions, leopards, and cheetahs on the Serengeti—or wolves, mountain lions, and grizzly bears at Yellowstone National Park (a place sometimes called "America's Serengeti"). These predators prey on herbivores that consume many plants. The removal of a top predator can have consequences that multiply as they cascade down the food chain.

Likewise, losing an "ecosystem engineer" (such as ants or earthworms; p. 86) can have major effects. For example, elephants eat and trample many young plants, helping maintain the open structure of Africa's savannas. Scientists have found that when elephants are removed (as by illegal hunting), the landscape fills in with scrubby vegetation, converting the savanna into a dense scrub forest and affecting countless other species.

Ecosystems are complex, however, and it is difficult to predict which species may be most influential. Thus, many people prefer to apply the precautionary principle (p. 260) in the spirit of Aldo Leopold (p. 139), who advised, "To keep every cog and wheel is the first precaution of intelligent tinkering."

Biodiversity boosts economies through tourism and recreation

When people travel to observe wildlife and explore natural areas, they create economic opportunities for area residents. Visitors spend money at local businesses, hire local people as guides, and support parks that employ

TABLE 11.3 Major Types of Animals at Risk That Offer Potential Medical Uses

SPECIES AT RISK	POTENTIAL MEDICAL USES
Amphibians 40% of all species are threatened with extinction. 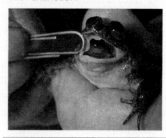	• Known compounds provide antibiotics; chemicals for painkillers, heart disease, and high blood pressure; and adhesives for treating tissue damage. • Ability to regenerate organs and tissues could suggest how we might, too. • "Antifreeze" compounds that allow frogs to survive freezing might help us preserve organs for transplants.
Sharks Overfishing has reduced populations of most species. Some risk extinction. 	• Squalamine from sharks' livers could lead to novel antibiotics, appetite-suppressants, drugs to shrink tumors, and drugs to fight vision loss. • Study of salt glands is helping address kidney diseases.
Horseshoe crabs Overfishing is sharply diminishing populations. 	• A number of antibiotics are being developed. • The compound T140 may treat AIDS, arthritis, and several cancers. • Cells from blood can help detect cerebral meningitis in people.
Bears Nine species are at risk of extinction. 	• An acid from bears' gallbladders treats gallstones and liver disease and prevents bile buildup during pregnancy. • While hibernating, bears build bone mass. If we learn how, it might help us treat osteoporosis and hip fractures. • Hibernating bears excrete no waste for months. Learning how could help treat renal disease.
Cone snails Most live in coral reefs, which are threatened ecosystems. 	• One compound may prevent death of brain cells from head injuries or strokes. • Another is a painkiller 1000 times more potent than morphine. So far, just a few hundred of the 70,000–140,000 compounds these snails produce have been studied.

Source: Adapted from Chivian, E., and A. Bernstein, 2008. *Sustaining life: How human health depends on biodiversity.* New York, NY: Oxford University Press.

local residents. Ecotourism (p. 70) can thereby bring jobs and income to areas that might otherwise suffer poverty.

The parks and wildlife of Kenya and Tanzania provide prime examples of ecotourism lifting economies. Ecotourism brings in fully a quarter of all foreign money entering Tanzania's economy each year. Leaders in both nations who recognize biodiversity's economic benefits have managed their parks and reserves diligently. Ecotourism is a vital source of income for many nations, including Costa Rica, with its tropical forests; Australia, with its Great Barrier Reef; and Belize, with its caves, coral reefs, and forested Mayan ruins. The United States, too, benefits from ecotourism; its national parks draw millions of visitors from around the world.

Popular sites for ecotourism can sometimes become victims of their own success. Excessive development of infrastructure for tourism can damage an area's natural assets, and too many visitors to natural areas can degrade the outdoor experience and disturb wildlife. Anyone who has been to Yosemite, the Grand Canyon, or the Great Smoky Mountains on a crowded summer weekend can attest to this. Still, ecotourism commonly provides a powerful financial incentive for nations, states, and local communities to preserve natural areas and reduce impacts on the landscape and on native species.

People value connections with nature

Not all of biodiversity's benefits to people can be expressed in the hard numbers of economics or the practicalities of food and medicine. Some scientists and philosophers argue that people find a deeper value in biodiversity. Harvard University biologist Edward O. Wilson popularized the notion of **biophilia**, asserting that human beings share an instinctive love for nature and feel an emotional bond with other living things (**FIGURE 11.7**). As evidence of biophilia, Wilson and others cite our affinity for parks and wildlife, our love

FIGURE 11.7 A girl meets a chameleon in Madagascar. Biophilia holds that human beings have an instinctive love and fascination for nature and a deep-seated desire to affiliate with other living things.

for pets, the high value of real estate with a view of natural landscapes, and our interest in hiking, bird-watching, fishing, hunting, backpacking, and similar outdoor pursuits.

Indeed, a love for nature and biodiversity appears to be good for us: We thrive mentally and physically when we have access to nature, and we suffer when we do not. As children in recent years have been increasingly deprived of outdoor experiences and contact with wild organisms, writer Richard Louv argues that they suffer from what he calls "nature-deficit disorder." In his 2005 book, *Last Child in the Woods,* Louv maintained that an alienation from nature and biodiversity damages childhood development and may lie behind much of the angst and anxiety felt by young people today.

In the wake of Louv's insight, many researchers have begun to study this question scientifically. In a 2017 book, *The Nature Fix,* writer Florence Williams surveyed the growing scientific evidence that exposure to nature revitalizes our bodies and our brains. Through interviewing scientists, assessing research, and presenting stories from around the world, Williams shows how access to wildlife and green spaces can relieve stress, bring us happiness, make us mentally sharper, and improve our physical health.

Do we have ethical obligations toward other species?

Aside from biodiversity's pragmatic benefits, many people believe that living organisms simply have an inherent right to exist. Like any other animal, human beings need to use resources and consume other organisms to survive. However, we also have conscious reasoning ability and can make deliberate decisions. Our ethical sense has developed from this intelligence and ability to choose. As our society's sphere of ethical consideration has widened over time, and as more of us take up biocentric or ecocentric worldviews (p. 137), more people have come to feel that other organisms have intrinsic value and an inherent right to exist. In this view, the conservation of biodiversity is justified on ethical grounds alone.

Biodiversity Loss and Extinction

Despite our society's expanding ethical breadth and despite the many clear benefits that biological diversity brings us, the future of many species remains far from secure. In today's fast-changing world, every corner of our planet has been touched in some manner by human impact, and biological diversity is rapidly being lost.

Human disturbance creates winners and losers

We affect ecosystems and landscapes in many ways, creating both "winners" and "losers" among the world's plants and

TABLE 11.4 Characteristics of Winning and Losing Species

WINNERS TEND TO BE	LOSERS TEND TO BE
• Generalists, using many resources or habitats • Geographically widespread • Users of open, early successional habitats • Able to cope with fast-changing conditions • Small and fast-reproducing (r-selected) • Low on the food chain • Not in need of large areas of habitat • Mainland species	• Specialists on certain resources or habitats • Limited to a small range • Users of mature, dense habitats • Needing stable conditions • Large and slow-reproducing (K-selected) • High on the food chain • Needing large areas of habitat • Island species

animals. In general, when we alter natural systems, we tend to make each area more similar to other areas by spreading into diverse natural environments and shaping them to our own species' particular tastes and needs. We also tend to make landscapes more open in structure by clearing vegetation away to make room for farms, pastures, towns, and cities. And we frequently create pollution. Because the general nature of our impacts is similar across regions and cultures, certain types of organisms tend to do well in our wake, whereas other types tend not to do well. As a result, the species that benefit from the changes we make—and the species that are harmed—each tend to show predictable sets of attributes (**TABLE 11.4**).

"Winning" species tend to be generalists able to fill many niches, tolerate disturbance, and use open habitats or edges. The house mouse (*Mus musculus*) is one example. This small, fast-reproducing mammal thrives by living and feeding in and near our buildings. In contrast, "losing" species tend to be those that specialize on certain resources, have trouble coping with change, and rely on mature and well-vegetated habitats. The tiger (*Panthera tigris*) is such a species. Large, slow-reproducing, and high on the food chain, it needs huge areas of mature habitat full of prey and free of people. Geographically widespread species stand a much better chance to succeed in a changing world undergoing human impact than species limited to small areas, and mainland species tend to do better than island species.

Many populations are declining

People often focus on the extinction of species when they think of biodiversity loss, but in today's world, the greatest impacts are occurring through gradual declines in the sizes of populations among many—and probably *most*—species on Earth. This widespread steady loss has been given haunting names like "defaunation" and "the great thinning." The phenomenon is indeed troubling—many of these declines are slow enough that we easily fail to notice them yet are rapid enough to devastate ecological systems.

As a population shrinks, it encounters two problems: It loses genetic diversity, and the geographic area over which it occurs tends to become smaller as the species disappears from parts of its range. Both problems make a population

vulnerable to further declines. Many species today are less numerous than they once were and occupy less area than they once did. In the Serengeti and elsewhere in East Africa, scientific studies have documented significant population declines among large mammals (**FIGURE 11.8**). Such declines mean that species diversity, genetic diversity, and ecosystem diversity are all being lost.

To quantify such change in populations at the global scale, scientists at the World Wildlife Fund and the United Nations Environment Programme (UNEP) developed the *Living Planet Index*. This index expresses how large the average population size of a species of vertebrate animal is now, relative to its size in the baseline year of 1970. The most recent compilation summarized trends from 16,704 populations of 4005 vertebrate species that are sufficiently monitored to provide reliable data. Between 1970 and 2014, the Living Planet Index fell by 60%—meaning that, on average, population sizes of fishes, birds, mammals, reptiles, and

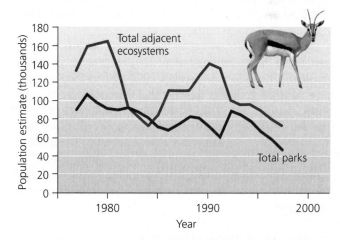

FIGURE 11.8 Populations of large mammals have declined across Kenya. Data from 30 years of aerial surveys (1977 to 2007) show downward trends in the populations of most species, both inside and outside parks. Species such as wildebeest, zebras, and some antelope migrate into and out of parks, so factors that affect their populations outside the parks will affect numbers counted within park boundaries as well. *Data from Western, D., et al., 2009. The status of wildlife in protected areas compared to non-protected areas of Kenya.* PLOS ONE 4(7): e6140.

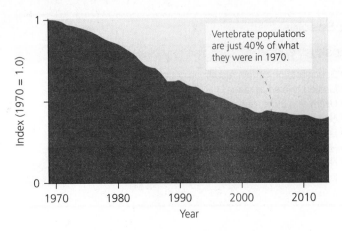

FIGURE 11.9 Populations of vertebrate animals average less than half the size they were just half a century ago. Between 1970 and 2014, the Living Planet Index, an indicator of the status of global biodiversity, fell by 60%. *Data from WWF, 2018. Living planet report 2018. Gland, Switzerland: WWF International.*

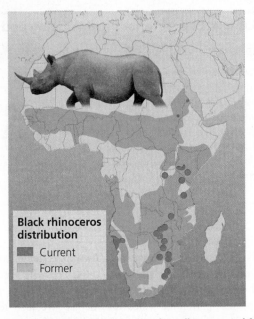

FIGURE 11.10 The black rhinoceros has disappeared from most of its range across Africa. *Based on data from Deon Furstenburg, wildliferanching.com, and other sources.*

amphibians became 60% smaller than they were just over four decades earlier (**FIGURE 11.9**). In the mere lifetimes of the authors of this textbook, Earth may have lost the majority of its vertebrate animals.

This shocking decline in biodiversity has been driven by losses in freshwater species more than by losses in terrestrial or marine species, according to the data. And it has been driven primarily by losses in tropical regions, where species richness is high and deforestation since 1970 has been severe. In temperate regions, losses have been lower because forests are now regrowing, pollution is better controlled, and ecological restoration is taking hold in industrialized countries (and because many populations had already declined greatly before 1970). Still, declines everywhere have been steep. A high-profile 2019 study calculated that since 1970, North America has lost 29% of its birds and now holds 3 billion fewer individual birds.

In recent years, attention has focused not just on large vertebrates like tigers and elephants, but on the countless small and inconspicuous creatures that support food webs. A new wave of studies is showing that insects appear to be in steep decline (see **THE SCIENCE BEHIND THE STORY**, pp. 286–287). Some documented declines in insects are considerably more severe than those for vertebrates. The declines are leading to headlines about an "Insect Armageddon" or "Insect Apocalypse"—and it would indeed be catastrophic for our world if we lost so many insects that ecosystems were to collapse.

Extinction is irreversible

When a population declines to a very low level, extinction becomes a possibility. **Extinction** (p. 58) occurs when the last member of a species dies and the entire species ceases to exist. The disappearance of a particular population from a given area, but not the entire species globally, is referred to as "local extinction" or **extirpation**. Extirpation is an erosive process that can, over time, lead to extinction. The black

rhinoceros has been extirpated from most of its historic range across Africa (**FIGURE 11.10**). As a species it is not yet extinct, but at least three of its subspecies are extinct.

Human impact is responsible for most extirpation and extinction today, but these processes also occur naturally at a much slower rate. If species did not naturally go extinct, our world would be filled with dinosaurs, trilobites, ammonites, and the millions of other creatures that vanished from Earth during the immense span of time before humans appeared. Paleontologists estimate that roughly 99% of all species that ever lived are now extinct.

Most extinctions preceding the appearance of human beings occurred singularly for independent reasons, at a pace referred to as the **background extinction rate.** By studying traces of organisms preserved in the fossil record (p. 58), scientists infer that for mammals and marine animals, on average, 1 species out of every 1–10 million has vanished each year.

Earth has experienced five mass extinction events

Extinction rates rose far above the background extinction rate at several discrete points in Earth's history. In the past 440 million years, our planet has experienced five **mass extinction events** (p. 60). Each event eliminated more than one-fifth of life's families (p. 58) and at least half of its species (**FIGURE 11.11**). The most severe episode occurred at the end of the Permian period (see **APPENDIX D**). At this time, about 250 million years ago, close to 90% of all species went extinct. The best-known episode occurred 66 million years ago at the end of the Cretaceous period, when evidence points to an asteroid impact (and possibly widespread volcanism) that brought an end to the dinosaurs and many other groups.

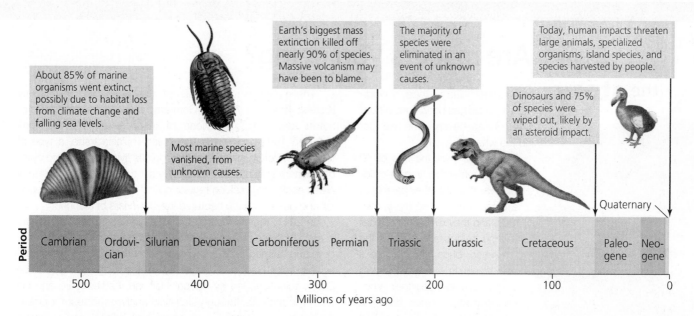

About 85% of marine organisms went extinct, possibly due to habitat loss from climate change and falling sea levels.

Most marine species vanished, from unknown causes.

Earth's biggest mass extinction killed off nearly 90% of species. Massive volcanism may have been to blame.

The majority of species were eliminated in an event of unknown causes.

Dinosaurs and 75% of species were wiped out, likely by an asteroid impact.

Today, human impacts threaten large animals, specialized organisms, island species, and species harvested by people.

Quaternary

| Period | Cambrian | Ordovi-cian | Silurian | Devonian | Carboniferous | Permian | Triassic | Jurassic | Cretaceous | Paleo-gene | Neo-gene |

500   400   300   200   100   0
Millions of years ago

FIGURE 11.11 Scientists have documented five mass extinction events in the past 500 million years. Ongoing study of the fossil record (p. 58) is revealing clues about the causes and consequences of each event.

If current trends continue, our modern era, known as the Quaternary period, may see the extinction of more than half of all species. Although similar in scale to previous mass extinctions, today's ongoing mass extinction is different in two primary respects. First, humans are causing it. Second, we will suffer as a result.

We are setting the sixth mass extinction in motion

In the past few centuries alone, we have recorded hundreds of instances of species extinction caused by people. Sailors documented the extinction of the dodo on the Indian Ocean island of Mauritius in the 17th century, for example, and today only a few body parts of this unique bird remain in museums. Among North American birds (**TABLE 11.5**) in the past two centuries, we have driven into extinction the Carolina parakeet, great auk, Labrador duck, and passenger pigeon (p. 62); almost certainly the Bachman's warbler and Eskimo curlew; and likely the ivory-billed woodpecker. Several more species, including the whooping crane, Kirtland's warbler, and California condor (p. 298), teeter on the brink of extinction.

People have been hunting species to extinction for thousands of years. Archaeological evidence shows that in case

TABLE 11.5 Selected Extinct and Endangered North American Birds

Carolina parakeet
(*Conuropsis carolinensis*)

Extinct. World's northernmost parrot, common across much of the United States until mid-1800s. Ate crops in noisy flocks. Deforestation, human hunting, and possibly disease played roles in its extinction.

Ivory-billed woodpecker
(*Campephilus principalis*)

Extinct? An iconic symbol of the South. Forest clearing eliminated the old-growth trees it needed. Recent fleeting, controversial observations raised hope that it persists, but proof has been elusive.

Whooping crane
(*Grus americana*)

Endangered. North America's tallest bird. Breeds in Canada, winters in Texas; dwindled from habitat loss and hunting. Captive breeding and reintroduction programs have raised numbers from just 21 to more than 800.

Kirtland's warbler
(*Setophaga kirtlandii*)

Endangered. Depends on young jack pine habitat in Michigan. Staging an impressive recovery thanks to management of vegetation and invasive cowbirds, which parasitize their nests.

THE SCIENCE
behind
the story

Are Insects Vanishing?

One major clue to today's biodiversity crisis has been right in front of our faces—every time we get behind the wheel of a car.

Ask anyone over the age of 40 what they remember from a childhood vacation with their parents or a cross-country road trip in their 20s, and they will probably recall having to stop and clean all the smashed bugs off their car's windshield. In years past, drivers might need to wash their windshields multiple times during a summertime trip of a few hundred miles because so many smeared insects were obscuring their vision.

But when was the last time, if ever, that you have had to do this? Today, a person can drive from New York to Chicago or from San Francisco to Los Angeles and

A windshield smeared with insects hit while driving

still have a clear windshield. Entomologists—the scientists who study insects—have nicknamed this the "windshield effect," and they see it as an indicator of something tragic and terrifying: a global collapse in the numbers of insects.

Edward O. Wilson (p. 282) calls insects "the little things that run the world." Indeed, insects decompose waste, pollinate plants, and play key roles in food webs and ecosystems, supporting the existence of most vertebrate animals. Without insects, the world as we know it would cease to exist.

We've long known that we poison insects with pesticides, rob insects of habitat, and pollute the soil and water insects live in, but historically it's proven difficult to measure what impacts we actually have on insect populations. Insects are numerous and diverse, and their numbers naturally fluctuate from year to year, making them challenging to monitor. Yet for years, entomologists have sensed that the numbers of insects everywhere were falling. Targeted studies were finding that particular insect species were getting rarer, and experts have noted declines in birds, fish, and mammals that rely on insects for food. So, when the first scientific study came out showing clear evidence of long-term decline across many types of insects, readers were shocked, but not surprised.

The data in this initial report did not emanate from well-funded professionals at a world-class university, but rather from members of a century-old entomology club in Germany, a group of amateur naturalists who through their own dedication and hard-won expertise had been voluntarily monitoring insects in their free time at local nature reserves for decades.

These German naturalists of the Entomological Society Krefeld, based in Krefeld, Germany, and led by Martin Sorg, made systematic collections of insects from 63 locations between 1989 and 2016. They used malaise traps, a type of tent-like structure in which netting guides flying insects into cylinders where they are captured. They set these traps up in natural areas each year, making regular collections from spring through fall and preserving the captured insects in jars of alcohol.

In 2017, they measured the total biomass (p. 82) of insects in each of these preserved samples using a standardized procedure and teamed up with professional entomologists and statisticians from Radboud University in the Netherlands. These Dutch scientists, led by Caspar Hallman, helped Sorg and his team run the data through statistical analyses to test for possible influences of factors such as weather, habitat, and changing land use on trends through time.

Once all factors were corrected for, the analyses showed that the average biomass of flying insects in these German samples had decreased by 76% across the 27 years of data collection (**FIGURE 1**). The decline was even steeper—82%—for midsummer, when insects are most abundant. The scientists published their results in the journal *PLOS ONE,* and the scientific community took notice.

Because malaise traps capture a wide variety of flying insects and because the team had measured overall biomass (without regard to species), these results were viewed as likely reflecting trends in insects as a whole. As to what was causing the declines, hypotheses abound. However, the nature reserves

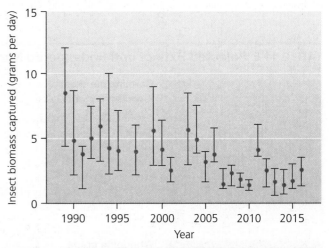

FIGURE 1 Biomass of flying insects decreased by 76% from 1989 to 2016 in nature reserves in Germany. These valuable long-term data exist thanks to the efforts of volunteer amateur entomologists. *Data from Hallman, C.A., 2017. More than 75 percent decline over 27 years in total flying insect biomass in protected areas. PLOS ONE 12: e0185809.*

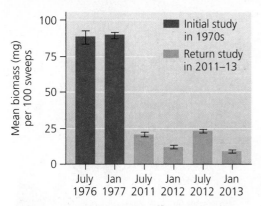

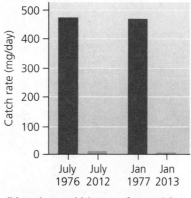

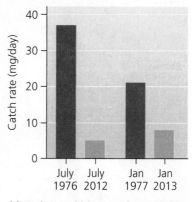

(a) Arthropod biomass from sweep-netting

(b) Arthropod biomass from sticky traps on ground

(c) Arthropod biomass from sticky traps in canopy

FIGURE 2 **Arthropod biomass fell steeply between 1976 and 2013 in the Luquillo rainforest in Puerto Rico.** Declines were found in insects and spiders **(a)** collected by sweep-netting, **(b)** collected with sticky traps on the ground, and **(c)** collected with sticky traps in the canopy. *Data from Lister, B.C., and Garcia, A., 2018. Climate-driven declines in arthropod abundance restructure a rainforest food web.* Proc. Nat. Acad. Sci. USA *115: E10397–E10406.*

were small areas within a fragmented landscape consisting largely of agricultural fields, typical of western Europe. Thus, impacts from intensive agriculture, such as habitat loss and chemical pesticides, likely helped drive the declines.

One year later, another study made similar waves. Bradford Lister, a biologist at Rensselaer Polytechnic University in New York, decided to revisit Puerto Rico's Luquillo rainforest, where he had collected data on the abundance of insects and insect-eating vertebrates back in the 1970s. In 2011–2013, he returned with Andres Garcia of the National Autonomous University of Mexico and resampled arthropods (insects and spiders) in exactly the same ways at exactly the same sites as he had done in 1976 and 1977.

The results of the Luquillo study showed dramatic declines: For arthropods collected by sweeping nets through vegetation, the biomass had declined by 4 to 8 times (**FIGURE 2a**). For arthropods caught in sticky traps on the ground and in the canopy, the biomass had fallen 30 to 60 times (**FIGURES 2b, 2c**). And tellingly, Lister and Garcia also documented steep declines at Luquillo in lizards, frogs, and birds that eat arthropods. Taken together, the data, published in *Proceedings of the National Academy of Sciences* in 2018, indicate a food web in the midst of collapse.

Lister and Garcia proposed that climate change may be behind the Puerto Rican declines. At Luquillo, average temperatures between 1976 and 2012 rose by 2°C (3.6°F). In their paper, the researchers explained how changes in the warming, drying forest have likely exerted major impacts on the arthropod fauna within it.

Soon after the Luquillo study was published, Australian entomologists Francisco Sánchez-Bayo and Kris Wyckhuys published a review in 2019 of all 73 studies in the literature that have reported insect declines from various locations around the

globe. They summarized major causes behind the declines (**FIGURE 3**), and they projected that 40% of the world's insect species could go extinct over the next few decades if agricultural practices and chemical pollution problems were not addressed.

As a result of the findings in Germany, Puerto Rico, and elsewhere, scientists are searching for neglected data sets from past decades that might be compared to new surveys. They are also seeking to start new studies to establish baselines against which to measure future changes. As the data come in, the rest of us can only hope that by taking steps to tackle habitat loss, pesticide use, climate change, and other threats to the little things that run the world, we can avoid the horror of an insect Armageddon.

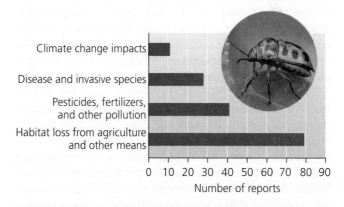

FIGURE 3 **Half of all documented insect declines are thought to be due to habitat loss.** Habitat loss results largely from agriculture but also from urbanization, deforestation, wetland loss, and other changes. Declines also result from pesticide pollution, biological interactions, and climate change. *Data from Sanchez-Bayo, F., and K.A.G. Wyckhuys. 2019. Worldwide decline of the entomofauna: A review of its drivers.* Biological Conservation *232: 8–27.*

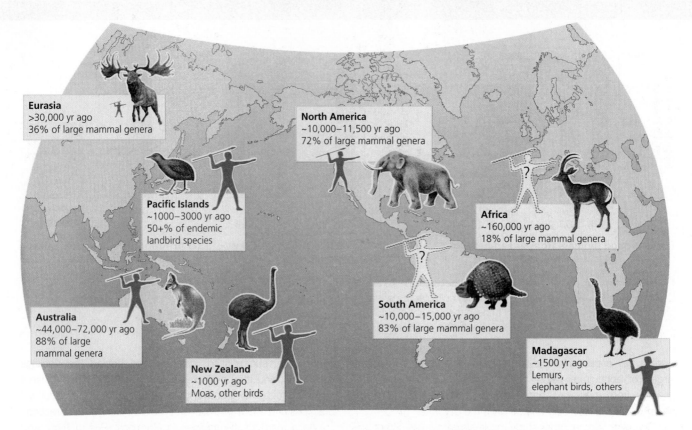

FIGURE 11.12 This map shows when humans arrived in each region and the extent of extinctions that followed. One extinct animal from each region is illustrated. Larger human hunter icons indicate more evidence and certainty that hunting (as opposed to climate change or other factors) was a primary cause of extinctions. Data on hunting for South America and Africa are so far too sparse to be conclusive. *Adapted from Barnosky, A.D., et al., 2004. Assessing the causes of late Pleistocene extinctions on the continents.* Science 306: 70–75; *and Wilson, E.O., 1992.* The diversity of life. *Cambridge, MA: Belknap Press.*

after case, a wave of extinction followed close on the heels of human arrival on islands and continents (**FIGURE 11.12**). After Polynesians reached Hawai'i, half its birds went extinct. Birds, mammals, and reptiles vanished following human arrival on many other oceanic islands, including large island masses such as New Zealand and Madagascar. Dozens of species of large vertebrates died off in Australia after people arrived roughly 50,000 years ago. North America lost 33 genera of large mammals (such as camels, lions, horses, mammoths, mastodons, saber-toothed cats, and giant ground sloths) after people arrived more than 13,000 years ago.

Today, species loss is accelerating as our population growth and resource consumption put increasing strain on habitats and wildlife. In 2019, scientists with the Intergovernmental Science-Policy Platform on Biodiversity and Ecosystem Services, a U.N.-sponsored international expert review panel, calculated that the current global extinction rate is tens to hundreds of times greater than the background extinction rate—and increasing.

To monitor threatened and endangered species, the International Union for Conservation of Nature (IUCN) maintains the **Red List,** a regularly updated list of species facing high risks of extinction. As of 2019, the Red List reported that at least 28% of the 98,512 species evaluated by scientists thus far were judged to be at risk of extinction.

Estimates for the best-studied groups indicated that 25% of mammal species, 14% of bird species, and 40% of amphibian species are threatened with extinction. In the United States alone during the past 500 years, 237 animal species and 38 plant species are known to have gone extinct.

Several major causes of biodiversity loss stand out

Scientists have identified five primary causes of population decline and species extinction: habitat loss, pollution, overharvesting, invasive species, and climate change. Each of these factors is intensified by human population growth and by our increasing per capita consumption of resources. By understanding how these pressures are diminishing biodiversity, we might begin to address these root causes and reverse biodiversity loss (see **DATAGRAPHIC**, p. 293).

Habitat loss A species' habitat is the specific environment in which it lives (p. 61), and habitat loss is the single greatest threat to biodiversity. Because organisms have adapted to their habitats over thousands or millions of years of evolution, any sudden, major change in their habitat will likely render it less suitable for them. Habitat is lost when it is destroyed outright but also when it becomes fragmented or degraded.

If a mass extinction is happening, why don't I notice species disappearing all around me?

There are two reasons most of us don't personally sense the scale of biodiversity loss. First, if you live in a town or city, the plants and animals you see from day to day are generalist species that thrive in disturbed areas. You don't see the species most in trouble: those that rely on less-disturbed habitats far from developed areas.

Second, a human lifetime is very short! The loss of populations and species may seem a slow process to us, but on Earth's timescale it is sudden. Because each of us is born into a world that has already lost species, we don't recognize what's already vanished. Likewise, our grandchildren won't appreciate what we lose in our lifetimes. Scientists call this the shifting baseline syndrome: Each human generation experiences just a portion of the overall phenomenon, so we have difficulty sensing the big picture. Nonetheless, researchers and naturalists who spend their time outdoors observing nature see biodiversity loss around them all the time—and that's precisely why they feel so passionate about preventing it.

Many human activities alter, degrade, or destroy habitat. Urban and suburban development and sprawl (p. 340) supplant natural ecosystems, driving many species from their homes (**FIGURE 11.13**). Farming replaces diverse natural communities with simplified ones of only a few plant species. Grazing modifies grasslands, and can lead to desertification (p. 230). Clearing forests removes the food and shelter that forest-dwelling organisms need to survive. Damming rivers creates reservoirs upstream while affecting water conditions and floodplain communities downstream.

Habitat loss occurs most commonly through gradual, piecemeal degradation, such as **habitat fragmentation** (**FIGURE 11.14**). When farming, logging, road building, or development intrude into an unbroken expanse of forest or grassland, the once-continuous area of habitat is broken into fragments, or patches. As habitat fragmentation proceeds across a landscape, animals and plants requiring the habitat disappear from one fragment after another. Fragmentation can also block animals from moving from place to place, which is the concern of opponents of the proposed highway through the Serengeti. In response to habitat fragmentation, conservationists try to link fragments together with corridors of habitat along which animals can travel. (In Chapter 12, we will learn more about the fragmentation of forests, the effects on wildlife, and potential solutions.)

Habitat loss is the primary cause of population decline for terrestrial vertebrates. Most migratory songbirds of North America have declined in recent decades, largely as a result of forest loss and fragmentation both on their North American breeding grounds and their Latin American wintering grounds. And North America's Great Plains are today almost entirely converted to agriculture; less than 1% of original prairie habitat

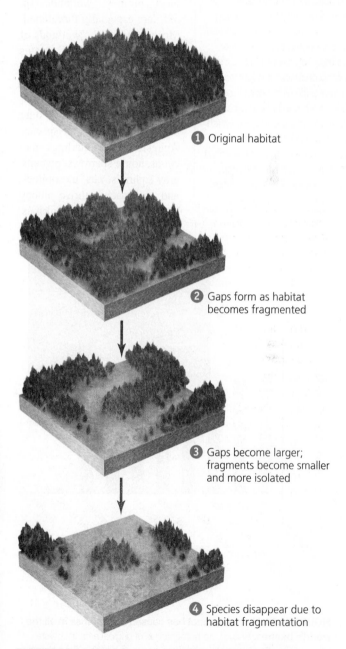

① Original habitat

② Gaps form as habitat becomes fragmented

③ Gaps become larger; fragments become smaller and more isolated

④ Species disappear due to habitat fragmentation

FIGURE 11.14 Habitat fragmentation occurs as human impact creates gaps that expand and eventually come to dominate the landscape, stranding islands of habitat. As habitat becomes fragmented, fewer populations can persist, and numbers of species in the fragments decline.

FIGURE 11.13 Development of land for housing is one way in which habitat is altered or destroyed.

WEIGHING the issues

Habitat Fragmentation in Your Region

Examine satellite imagery of the region where you live using an online application such as Google Earth or Google Maps. How much of the area around you is developed with roads, buildings, cropland, and pastureland? How much appears to be forested or otherwise covered by natural vegetation? Are the naturally vegetated areas connected to one another, or are they fragmented? Discuss how easy or difficult it might be for an animal such as a deer or a bear to move from one habitat patch to another. Do you see ways in which natural areas could be better connected using corridors? Now zoom in or zoom out a bit on the map, and address these same questions at a different landscape scale.

remains. As a result, grassland bird populations have fallen by an estimated 82–99%.

Habitat loss affects all the world's biomes (**FIGURE 11.15**). More than half of Earth's temperate forests, grasslands, and shrublands had been converted by 1950 (mostly for agriculture). Today, habitat is being lost most rapidly in tropical rainforests, tropical dry forests, and savannas. Within most biomes, wetlands (p. 398) are especially threatened. More than half the wetlands of the 48 contiguous U.S. states and Canada have been drained for agriculture (see photo, Figure 11.15).

Of course, when we alter habitat, we benefit some species. Animals such as starlings, raccoons, house sparrows, pigeons, gray squirrels, rats, mosquitoes, and cockroaches thrive among us in farmland, towns, and cities.

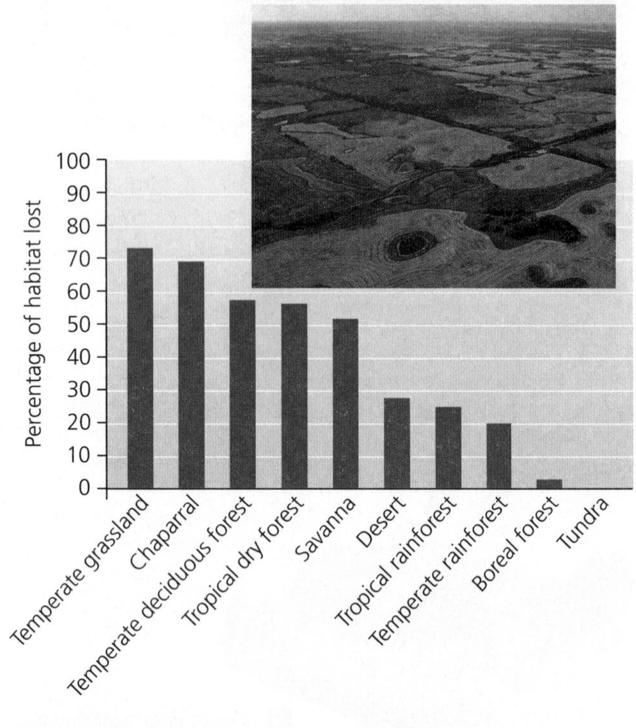

FIGURE 11.15 Human impact has caused habitat loss in all the world's biomes. Shown are percentages of original area that were fully and directly converted for human use through 1990. The photo shows "prairie pothole" wetlands (lakes and ponds once abundant in the northern Great Plains) that have been partly drained for farming.
Data adapted from Millennium Ecosystem Assessment, 2005. Ecosystems and human well-being: Biodiversity synthesis. *Washington, D.C.: World Resources Institute.*

However, these "winning" species that do well in our midst tend to be weedy generalists that are in little danger of disappearing any time soon. The concern is that far too many other species are "losing" as a result of the way we alter habitats.

Pollution Pollution can harm organisms in many ways. Air pollution degrades forest ecosystems and affects the atmosphere and climate. Noise pollution and light pollution impinge on the behavior of animals. Water pollution impairs fish and amphibians. Agricultural runoff containing fertilizers, pesticides, and sediments harms many species. Heavy metals, polychlorinated biphenyls (PCBs), endocrine-disrupting compounds, and other toxic chemicals poison people and wildlife. Plastic garbage in the ocean can strangle, drown, or choke marine creatures. The effects of oil spills on wildlife are dramatic and well known. We examine all these impacts in other chapters of this book.

Pollution can cause extensive damage to wildlife and ecosystems, yet it tends to be less significant as a cause of population-wide decline for plants and for vertebrate animals than public perception holds it to be, and it is far less influential than habitat loss. For insects and other arthropods, pollution from chemical pesticides is likely a major cause of population-wide declines, although habitat loss is a significant factor as well.

Overharvesting We have always hunted and harvested animals and plants from nature, but our growth in population and consumption is now leading us to remove many species more quickly than they are able to reproduce. As forests around the world are lost (p. 312), tree species prized for their wood, such as teak and mahogany, are disappearing. In Africa, gorillas and other primates killed for "bush meat" could soon face extinction. In the oceans, many fish stocks are now overharvested (p. 438). Whaling drove the Atlantic gray whale extinct and has left several other types of whales threatened or endangered. Thousands of sharks are killed each year merely for their fins, which are used in soup. Altogether, our oceans today contain only 10% of the large animals they once did (p. 441), which has far-reaching impacts on marine food webs.

For most species, being hunted or harvested will not in itself pose a threat of extinction, but for some species it can. Animals that are large in size, are long-lived, and raise few young in their lifetimes (K-selected species; p. 68) are most vulnerable to hunting. For example, people have long killed elephants to extract their tusks for ivory (**FIGURE 11.16**). By 1989, 7% of African elephants were being slaughtered each year, and the world's nations enacted a global ban on the commercial trade of ivory. Elephant numbers recovered following the ban, but after 2005, **poaching** (the unlawful killing of wildlife for meat or body parts) reached an all-time high in 2011, driven by high black-market prices for ivory paid by wealthy overseas buyers. From 2011 through 2018, more than 240,000 African elephants were killed—enough to send populations downward and threaten the species' future. In response, the United States and China each enacted national bans on the ivory trade.

With the illegal global trade in wildlife products surpassing $20 billion per year, poaching has led to steep population declines for many animals. Rhinoceros populations have crashed as poachers slaughter rhinos for their horns, which

FIGURE 11.16 Poachers slaughter elephants to sell their tusks for ivory. Here, Kenyan officials at Maasai Mara National Reserve prepare to set fire to tusks confiscated from poachers in an effort to discourage the trade.

are ground into powder and sold to ultra-wealthy Asian consumers as cancer cures, party drugs, or hangover treatments, even though the horns have no such properties and consist of the same material as our fingernails. Across Asia, tigers are threatened by poaching as well as habitat loss; body parts from one tiger can fetch a poacher $15,000 on the black market, where they are sold as aphrodisiacs in China and other Asian countries. Today half the world's tiger subspecies are extinct, and most of the remaining animals are crowded onto just 1% of the land they occupied historically.

In much of Africa today, protecting wildlife is a dangerous job. Poaching is conducted with brutal efficiency by organized crime syndicates using helicopters, night-vision goggles, and automatic weapons. Park rangers are heavily armed, yet are routinely outgunned in firefights with poachers, and many courageous rangers have lost their lives. In this way, the demand for luxury goods by wealthy foreign consumers in Asia, Europe, and America has grave consequences for Africans living in regions like the Serengeti.

Scientists are assisting efforts to curb poaching and save wildlife. Researchers in the field are tracking elephants and ivory by putting radio collars on animals, satellite-tracking tusks with microchips, and flying drones (unmanned surveillance aircraft) over parks to capture real-time video of poachers. Meanwhile, researchers in the lab are conducting genetic analyses to expose illegal hunting and wildlife trade. For instance, forensic DNA testing can reveal the geographic origins of elephant ivory, helping authorities focus on hotspots of illegal activity (see **THE SCIENCE BEHIND THE STORY**, pp. 300–301).

Invasive species Another major cause of biodiversity loss is invasive species **(TABLE 11.6,** p. 292). When non-native species are introduced to new environments, most perish, but the few that survive may do very well, especially if they find themselves freed from the predators, parasites, and competitors that

had kept their populations in check back home. Once released from such limiting factors (p. 67), an introduced species may proliferate and become invasive (p. 89), displacing native species—and occasionally pushing native species toward extinction.

Some introductions are accidental. Examples include animals that escape from the pet trade, weeds whose seeds cling to our socks as we travel from place to place, and aquatic organisms, such as zebra mussels (p. 88), transported in the ballast water of ships. If a highway is built through the Serengeti, ecologists fear that passing vehicles would introduce weed seeds. Several American plants, such as datura, parthenium, and prickly poppy, have already spread through other African grasslands and are toxic to native herbivores.

Other introductions are intentional. In Lake Victoria, west of the Serengeti, the Nile perch was introduced as a food fish to supply people much-needed protein (see Table 11.6). It soon spread throughout the vast lake, however, preying on and driving extinct dozens of native species of cichlid fish from one of the world's most spectacular evolutionary radiations of animals. The Nile perch is providing people food, but at a significant ecological cost. Throughout history, people everywhere have brought food crops and animals with them as they colonized new places, and today we continue international trade in exotic pets and ornamental plants, often heedless of the ecological consequences.

Species native to islands are especially vulnerable to introduced species. Island species have existed in isolation for millennia with relatively few parasites, predators, and competitors. As a result, native island species have not evolved the defenses necessary to resist invaders. For instance, Hawaii's native plants and animals have been under siege from invasive organisms such as rats, pigs, and cats, which has led to a number of extinctions (Chapter 3).

TABLE 11.6 Invasive Species

European gypsy moth
(*Lymantria dispar*)

Introduced to Massachusetts in the hope that it could produce silk. The moth failed to do so and instead spread across the eastern United States, where its outbreaks defoliate trees over large regions every few years.

Asian long-horned beetle
(*Anoplophora glabripennis*)

Since the 1990s, has repeatedly arrived in North America in imported lumber. These insects burrow into wood and can kill the majority of trees in an area. Chicago, Seattle, Toronto, New York City, and other cities have cleared thousands of trees to eradicate these invaders.

European starling
(*Sturnus vulgaris*)

Introduced to New York City in the 1800s by Shakespeare devotees intent on bringing every bird mentioned in Shakespeare's plays to America. Outcompeting native birds for nest holes, within 75 years starlings became one of North America's most abundant birds.

Emerald ash borer
(*Agrilus planipennis*)

Discovered in Michigan in 2002, this wood-boring insect reached 12 U.S. states and Canada by 2010, killing millions of ash trees in the upper Midwest. Billions of dollars will be spent trying to control its spread.

Cheatgrass
(*Bromus tectorum*)

After introduction to Washington state in the 1890s, cheatgrass spread across the western United States. It crowds out other plants, uses up the soil's nitrogen, and burns readily. Fire kills many native plants, but not cheatgrass, which grows back stronger without competition.

Sudden oak death
(*Phytophthora ramorum*)

This disease has killed more than 1 million oak trees in California since the 1990s. The pathogen (a water mold) was likely introduced via infected nursery plants. Scientists are concerned about damage to eastern U.S. forests if it spreads to oaks there.

Brown tree snake
(*Boiga irregularis*)

Nearly every native forest bird on the South Pacific island of Guam has disappeared, eaten by these snakes, which arrived from Asia as stowaways on ships and planes after World War II. Guam's birds had not evolved with snakes and had no defenses against them.

Nile perch
(*Lates niloticus*)

A large fish from the Nile River introduced to Lake Victoria in the 1950s. It proceeded to eat its way through hundreds of species of native cichlid fish, driving a number of them extinct. People value the perch as food, but it has radically altered the lake's ecology.

Kudzu
(*Pueraria montana*)

A Japanese vine that can grow 30 m (100 ft) in a single season, the U.S. Soil Conservation Service introduced kudzu in the 1930s to help control erosion. Kudzu took over forests, fields, and roadsides throughout the southeastern United States.

Polynesian rat
(*Rattus exulans*)

One of several rat species that have followed human migrations across the world. Polynesians transported this rat to islands across the Pacific, including Easter Island (pp. 8–9). On each island it caused ecological havoc and has driven birds, plants, and mammals to extinction.

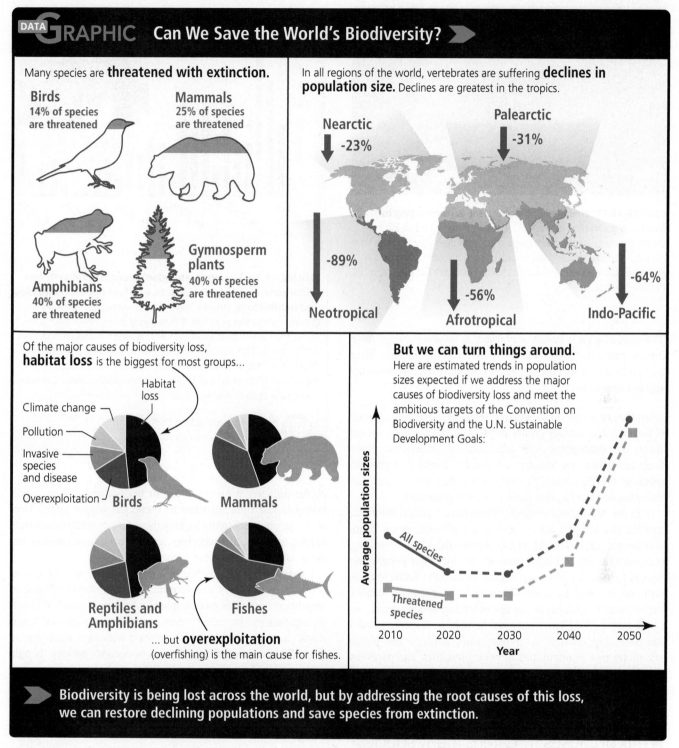

Many species are **threatened with extinction.**

Birds
14% of species are threatened

Mammals
25% of species are threatened

Amphibians
40% of species are threatened

Gymnosperm plants
40% of species are threatened

In all regions of the world, vertebrates are suffering **declines in population size.** Declines are greatest in the tropics.

Nearctic -23%

Palearctic -31%

-89% **Neotropical**

-56% **Afrotropical**

-64% **Indo-Pacific**

Of the major causes of biodiversity loss, **habitat loss** is the biggest for most groups...

Habitat loss
Climate change
Pollution
Invasive species and disease
Overexploitation

Birds

Mammals

Reptiles and Amphibians

Fishes

... but **overexploitation** (overfishing) is the main cause for fishes.

But we can turn things around.

Here are estimated trends in population sizes expected if we address the major causes of biodiversity loss and meet the ambitious targets of the Convention on Biodiversity and the U.N. Sustainable Development Goals:

Average population sizes

All species

Threatened species

2010 2020 2030 2040 2050
Year

Biodiversity is being lost across the world, but by addressing the root causes of this loss, we can restore declining populations and save species from extinction.

Data from IUCN 2019, The IUCN Red List of Threatened Species, Version 2019-1; and WWF, 2018. Living planet report 2018. Gland, Switzerland: WWF International.

Some of the most devastating invasive species are microscopic pathogens that cause disease. In Hawai'i, malaria and avian pox transmitted by introduced mosquitoes are killing off the islands' native birds, which lack immunity to these foreign diseases. Indeed, some scientists classify disease separately as an additional major cause of biodiversity loss.

Experts debate the role of introduced species in today's world. For decades, most biologists focused on the economic damage and the negative impacts on native ecosystems exerted by introduced species that turned invasive. However, many introduced species, such as the European honeybee (p. 92), provide economic benefits. And on our planet today there is no truly pristine ecosystem—all have been touched in some way by human impact, and many contain novel communities (p. 88), newly formed mixtures of native and non-native species. In some cases, these communities host greater species diversity than the communities they replaced and may function just as well in providing ecosystem services.

FIGURE 11.17 This starving polar bear is rummaging for food in a trash dump. As Arctic warming melts the sea ice from which they hunt seals, polar bears must endure longer summertime periods without easy access to food. The polar bear became the first species listed under the Endangered Species Act as a result of climate change.

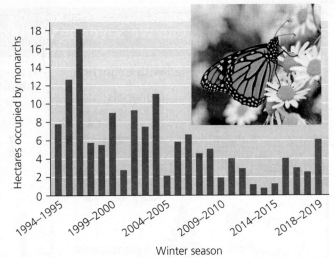

FIGURE 11.18 The once-abundant monarch butterfly has undergone alarming declines due to herbicides, pesticides, and habitat loss. Annual surveys on its Mexican wintering grounds show the population occupying many fewer hectares of forest than in the past (although with an encouraging rebound in 2018–2019). The situation is worse for the smaller population of West Coast monarchs: 2018–2019 data showed a decline of more than 99% since the 1980s. *Data from MonarchWatch, collected by the Monarch Butterfly Biosphere Reserve and World Wildlife Fund Mexico.*

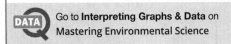

Go to **Interpreting Graphs & Data** on **Mastering Environmental Science**

Thus, although it is undeniably true that invasive species have driven extensive losses of native biodiversity in case after case, introduced species sometimes increase overall biodiversity at local scales.

Climate change Our manipulation of Earth's climate (Chapter 18) is having global impacts on biodiversity. As we warm the atmosphere with emissions of greenhouse gases from fossil fuel combustion, we modify climate patterns and increase the frequency of extreme weather events, such as droughts and storms, that put stress on populations.

In the Arctic, where temperatures have warmed the most, melting sea ice and other impacts are affecting polar bears and people alike (**FIGURE 11.17**). Across the world, warming temperatures are forcing organisms to shift their geographic ranges toward the poles and higher in altitude. Some species will not be able to adapt. Mountaintop organisms cannot move farther upslope to escape warming temperatures, so many may perish. Trees may not be able to disperse toward the poles fast enough. As ranges shift, animals and plants encounter new communities of prey, predators, and parasites to which they are not adapted. In a variety of ways, scientists predict that climate disruption will put many thousands of the world's plants and animals at increased risk of extinction.

A mix of causes threatens many species

For many species, multiple factors are combining to cause declines. The monarch butterfly, once familiar to every American schoolchild, is today in precipitous decline (**FIGURE 11.18**). On its breeding grounds in the United States and Canada, industrial agriculture and chemical herbicides have eliminated most of the milkweed plants on which monarchs depend. Our highly efficient monocultures leave no natural habitat remaining, while insecticides intended for crop pests also kill monarchs and other beneficial insects. Each fall, the entire population of monarchs from eastern and central North America migrates south and funnels into a single valley in Mexico, where the butterflies cluster

by the millions in groves of tall fir trees where climate conditions are just right for them to spend the winter safely. Here some people are illegally logging these forests while others fight to save the trees, the butterflies, and the ecotourism dollars that these natural marvels bring to the community.

Reasons for the decline of a population or species can be complex and difficult to determine. The worldwide collapse of amphibians provides an example of a "perfect storm" of bewildering factors. In recent years, entire populations of frogs, toads, and salamanders have vanished without a trace, and at least 170 species studied just years or decades ago are thought to be extinct. As these creatures disappear before our eyes, scientists are racing to discover why, and studies implicate a wide array of causes (**FIGURE 11.19**). These causes include habitat destruction, chemical pollution, invasive species, climate change, and a disease called chytridiomycosis caused by an invasive fungal pathogen, *Batrachochytrium dendrobatidis,* that has swept the globe. Biologists suspect that multiple factors are interacting and multiplying the effects.

As researchers learn more, they are designing responses to amphibian declines. An IUCN conservation action plan recommends that we protect and restore habitat, crack down on illegal harvesting, enhance disease monitoring, and establish captive breeding programs.

Many people around the world are striving to save vanishing species. The search for solutions to our biodiversity crisis is dynamic and inspiring, and scientists are developing innovative approaches to sustain Earth's diversity of life.

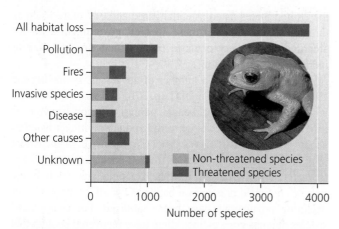

FIGURE 11.19 The world's amphibians are vanishing. The golden toad is one of at least 170 species of amphibians that have suddenly gone extinct in recent years. This brilliant orange toad of Costa Rican cloud forests disappeared due to drought, climate change, and/or disease. Habitat loss is the main reason for amphibian declines, but many declines remain unexplained. *Data from IUCN, 2008.* Global amphibian assessment.

DATA Q Go to **Interpreting Graphs & Data** on **Mastering Environmental Science**

Conservation Biology: The Search for Solutions

The urge to act as responsible stewards of natural systems, and to use science as a tool in this endeavor, sparked the rise of **conservation biology.** This scientific discipline is devoted to understanding the factors, forces, and processes that influence the loss, protection, and restoration of biological diversity. Conservation biologists choose questions and pursue research with the aim of developing solutions to such problems as habitat degradation and species loss (**FIGURE 11.20**). Conservation biology is thus an applied and goal-oriented science, with implicit values and ethical standards.

Conservation biology responds to biodiversity loss

Conservation biologists integrate an understanding of ecology and evolution as they use field data, lab data, theory, and experiments to study our impacts on other organisms. They also frequently incorporate philosophy, law, economics, sociology, public health, and other disciplines to develop a full understanding of our relationship with the natural world. In designing, testing, and implementing responses to human impacts on nature, these researchers address the challenges facing biological diversity at all levels, from genes to species to ecosystems.

At the genetic level, conservation geneticists study genetic attributes of organisms to infer the status of their populations. If two populations of a species are genetically distinct, they may have different ecological needs and may require different types of management. Moreover, as a population dwindles, genetic variation is lost.

Conservation geneticists investigate how small a population can become and how much genetic variation it can lose before running into problems such as inbreeding depression (p. 276), whereby genetic similarity causes parents to produce weak or defective offspring. By determining a population's *minimum viable population size,* conservation geneticists help wildlife managers decide how vital it may be to increase the population.

(a) Sampling insects in Madagascar

(b) Drawing blood from a Seychelles Magpie Robin

(c) Checking camera traps in Africa

FIGURE 11.20 Conservation biologists use many approaches to study the loss, protection, and restoration of biodiversity, seeking to develop scientifically sound solutions.

Studies of genes, populations, and species inform conservation efforts with habitats, communities, ecosystems, and landscapes. As landscape ecologists know, organisms may be distributed across a landscape in a metapopulation, or network of subpopulations (p. 119). Because small and isolated subpopulations are most vulnerable to extirpation, conservation biologists pay special attention to them. By examining how organisms disperse from one habitat patch to another, and how their genes flow among subpopulations, conservation biologists try to learn how likely a population is to persist or succumb in the face of environmental change.

Endangered species are a focus of conservation efforts

The primary legislation for protecting biodiversity in the United States is the **Endangered Species Act** (ESA). Enacted in 1973, the ESA offers protection to species (or subspecies or populations below the species level) that are judged to be **endangered** (in danger of becoming extinct in the near future) or **threatened** (vulnerable to becoming endangered soon). The ESA forbids the government and private citizens from taking actions that kill individuals of these species or destroy the habitats critical to their survival. The ESA also forbids trade in products made from threatened and endangered species. The aim is to prevent extinctions and enable declining populations to recover. As of 2019, there were 1274 species in the United States listed as endangered and 387 more listed as threatened. For most of these species, government agencies are running recovery plans to protect them and stabilize or increase their populations.

The ESA has had a number of successes. Following the 1973 ban on the pesticide DDT (p. 370) and years of effort by wildlife managers, the bald eagle, peregrine falcon, brown pelican, and other birds have recovered and are no longer listed as endangered (see **SUCCESS STORY**). Intensive management programs with species such as the red-cockaded woodpecker (see Figure 12.19, p. 322) have held populations steady in the face of continued pressure on habitat. Overall, roughly 40% of declining populations have been stabilized. For every listed species that has gone extinct, three have recovered enough that they have been removed from the endangered species list.

These successes have come even though the U.S. Fish and Wildlife Service and the National Marine Fisheries Service, the agencies that administer the ESA, are perennially underfunded. Federal authorization for spending under the ESA expired in 1992, so Congress appropriates funds for its administration year by year. A number of species today are judged by scientists to need protection but have not been added to the endangered species list because funding is inadequate to help recover them. Such species are said to be "warranted but precluded"—meaning that their listing is warranted by scientific research but precluded by lack of resources. This has led some environmental advocacy groups to sue the government for failing to enforce the law.

SUCCESS story Bringing Back Endangered Birds

In the early 1970s, things looked dire for a number of iconic North American bird species whose populations had collapsed, including the majestic peregrine falcon (the world's fastest bird), the stately brown pelican, and several kinds of hawks and owls. Even the United States' national bird, the bald eagle, was threatened with extinction. The falcon, pelican, and eagle had almost completely disappeared from the 48 contiguous U.S. states before scientists discovered what was threatening them. One hint was that they were all predators atop their food chains and thus were receiving heavy doses of toxic chemicals accumulated from the many smaller animals they ate over time (p. 373). Research eventually revealed that the chemical insecticide DDT (p. 385)—which had become widely used in the mid-20th century—was causing these birds' eggshells to become thin and break too early, killing the young. U.S. leaders banned DDT in 1973, the same year they enacted the Endangered Species Act. Together these two actions led to spectacular recoveries of the peregrine falcon, brown pelican, and bald eagle. Biologists began running programs to assist recovery, and the populations of these birds roared back. Today, all three species are thriving across large portions of North America.

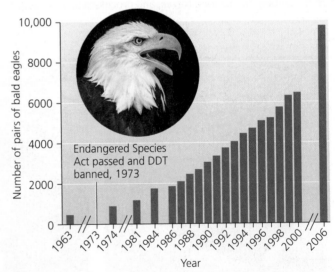

Numbers of bald eagles rebounded following protection of the species under the Endangered Species Act. *Data from U.S. Fish and Wildlife Service, based on annual volunteer surveys. Data after 2000 are scarce because surveys were discontinued once it became clear that the eagle was recovering.*

→ **Explore the Data** at **Mastering Environmental Science**

Polls repeatedly show that most Americans support protecting endangered species. Some opponents, however, believe that the ESA can imperil people's livelihoods. This perception has been common in the Pacific Northwest, where protection for the northern spotted owl and the marbled murrelet—birds that rely on old-growth forest—have slowed timber harvesting. In addition, many landowners worry that federal officials will restrict the use of private land on which threatened or endangered species are found. This has led to a practice described as "shoot, shovel, and shut up" among landowners who want to conceal the presence of such species on their land.

In fact, however, the ESA has stopped very few development projects, and a number of its provisions and amendments promote cooperation with landowners. A landowner and the government can agree to a *habitat conservation plan,* an arrangement that grants the landowner an "incidental take permit" to harm some individuals of a species if he or she voluntarily improves habitat for the species. Likewise, in a *safe harbor agreement,* the government agrees not to mandate additional or different management requirements if the landowner acts to assist a species' recovery.

Recent efforts to conserve the greater sage grouse (**FIGURE 11.21**) exemplify the cooperative public-private approach. This species had been judged "warranted but precluded," but cattle ranchers and the oil and gas industry opposed listing the species as endangered because restrictions on land use across vast sagebrush regions of the West could complicate ranching and drilling activities. As a result, federal agencies embarked on a campaign with ranchers and the energy industry across 13 western states to design voluntary agreements to lessen impacts on sage grouse populations. In 2015, the bird was denied listing, with federal officials saying that the collaborative agreements were adequate to conserve the species. This announcement produced celebration and relief in many quarters but also criticism both from development advocates, who believed that the agreements were too restrictive, and from environmental advocates, who judged that the strategy would fail to save the species.

More recently, the Trump Administration has undermined the sage grouse agreements while issuing new rules weakening the enforcement of the ESA. It remains to be seen whether these rule changes will survive legal challenges in court.

International treaties promote conservation

On the global stage, the United Nations (p. 177) has facilitated international treaties to protect biodiversity. The 1973 **Convention on International Trade in Endangered Species of Wild Fauna and Flora** (CITES) protects rare species by banning the international transport of their body parts. The 1990 global ban on the ivory trade may be this treaty's biggest accomplishment so far. When nations enforce its provisions, CITES can protect rhinos, elephants, tigers, and other species whose body parts are traded internationally.

In 1992, leaders of most nations agreed to the **Convention on Biological Diversity,** a treaty that aims to help nations conserve biodiversity, use it in a sustainable manner, and ensure the fair distribution of its benefits. By providing funding and incentives for conservation and by promoting public education and scientific cooperation, this treaty has produced many success stories. It has helped African nations gain economic benefits from ecotourism at wildlife preserves such as Serengeti National Park, has prompted nations worldwide to protect more area in reserves, has enhanced global markets for sustainable crops such as shade-grown coffee, and has encouraged movement away from pesticide-intensive farming practices. Yet the treaty's overall goal—"to achieve, by 2010, a significant reduction of the current rate of biodiversity loss at the global, regional and national level"—was not met.

Currently, the convention's signatory nations aim to achieve 20 new biodiversity targets by 2020. Goals include:

- Cutting the loss of natural habitats in half—and, where feasible, bringing this loss close to zero.
- Conserving 17% of land areas and 10% of marine and coastal areas.
- Restoring at least 15% of degraded areas.
- Alleviating pressures on coral reefs.

Captive breeding, reintroduction, and cloning are being pursued

To help save species at risk, some zoos and botanical gardens have become centers for **captive breeding,** in which individuals are bred and raised in controlled conditions with the intent of reintroducing their progeny into the wild. The IUCN counts 69 plant and animal species that now exist *only* in captivity or cultivation.

FIGURE 11.21 **This male greater sage grouse is doing a courtship display in its sagebrush habitat.**

FIGURE 11.22 We can reestablish populations by reintroducing them to areas where they were extirpated. Black rhinos have been helicoptered in to Serengeti National Park from other areas where populations are healthy.

Reintroducing species into areas they used to inhabit is expensive and resource-intensive, but it can pay big dividends. In 2010, the first of 32 black rhinos was translocated from South Africa to Serengeti National Park to help restore a former population (**FIGURE 11.22**). This translocation followed similar reintroduction projects elsewhere in Africa. Such efforts do not always work—the tragic deaths of 11 rhinos in 2018 after drinking salty water in a new location that they were unfamiliar with is a case in point—but in general, carefully thought-out reintroductions have met with success.

A prime example of captive breeding and reintroduction is the program to save the California condor, North America's largest bird (**FIGURE 11.23**). Condors are harmless scavengers

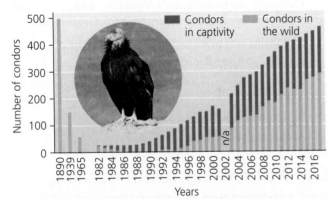

FIGURE 11.23 California condors are being bred in captivity and released to the wild, gradually rebuilding their population. *Data from California Condor Recovery Program, U.S. Fish and Wildlife Service.*

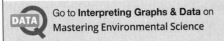

Go to **Interpreting Graphs & Data** on **Mastering Environmental Science**

of dead animals, yet people a century ago used to shoot them for sport. Condors also collided with electrical wires and succumbed to lead poisoning after scavenging carcasses of animals killed with lead shot. By 1982, only 22 condors remained, and biologists made the wrenching decision to take all the birds into captivity.

A collaborative program between the Fish and Wildlife Service and several zoos has boosted condor numbers. As of 2018, there were 176 birds in captivity and 312 birds living in the wild. Condors have been released at sites in California, Arizona, and Baja California (Mexico), where they thrill people lucky enough to spot the huge birds soaring through the skies. Unfortunately, many of these long-lived birds still die of lead poisoning, and wild populations will likely not become sustainable until hunters of animals on which condors feed convert from lead shot to nontoxic shot made of copper or steel. That change may give condors a fighting chance by helping stop the accumulation of a highly toxic substance in the environment (while also making it safer for hunters to feed their families the game animals they hunt). California has now banned lead shot for all hunting.

One idea for saving species from extinction is to create individuals by cloning them. In this technique, DNA from an endangered species is inserted into a cultured egg without a nucleus, and the egg is implanted into a female of a closely related species that acts as a surrogate mother. Several mammals have been cloned in this way, with mixed results. Some scientists even talk of resurrecting extinct species from DNA recovered from preserved body parts, and researchers may soon attempt this method with the woolly mammoth, the passenger pigeon, and other extinct species. In 2009, a subspecies of Pyrenean ibex (a type of mountain goat) was cloned from cells taken from the last surviving individual, which had died in 2000. However, the cloned baby ibex died shortly after birth, possibly from effects of cloning not yet well understood. Even if cloning can succeed from a technical standpoint, it is not an adequate response to biodiversity loss. Cloning does nothing to protect genetic diversity, and without ample habitat and protection in the wild, having cloned animals in a zoo does little good.

Forensics can help protect species

To counter poaching and other illegal harvesting, scientists have an advanced tool at their disposal. **Forensic science,** or *forensics,* involves the scientific analysis of evidence to make an identification or answer a question relating to a crime or an accident. Now enterprising conservation biologists are using forensics to protect species at risk (see **The Science behind the Story,** pp. 300–301). By analyzing DNA from organisms or their tissues sold at market, researchers can often determine the species or subspecies—and sometimes the geographic origin. This information can help catch perpetrators of illegal hunting.

For example, to determine whether whales that are legally protected by international agreement are being illicitly killed, researchers have genetically analyzed samples of whale meat sold in Asian markets. By comparing their genetic data with data from known whale populations, these conservation geneticists have identified the species and geographic origin of hundreds of market samples and have documented numerous instances of illegal hunting. They even found that some meat labeled as whale meat came instead from dolphins, orcas, porpoises, sheep, and horses! After calculating that Japanese and South Korean ships were killing so many minke whales that they would eventually wipe out this region's population, the scientists used their findings to urge the international community to monitor catches more closely.

Several strategies address habitats, communities, and ecosystems

Scientists know that protecting species does little good if the larger systems these species rely on are not also sustained, yet no law or treaty exists to protect communities or ecosystems. For these reasons, conservation advocates pursue several strategies for conserving ecological systems at broader scales.

Umbrellas and flagships As with endangered species under the ESA, particular species can often serve as tools to conserve habitats, communities, and ecosystems. Such species are nicknamed *umbrella species* because they act as a kind of umbrella to protect many others. Umbrella species often are large animals that roam great distances, like the Serengeti's lions and wildebeest. Because such animals require large areas, meeting their habitat needs helps meet

those of thousands of less charismatic animals, plants, and fungi that might never elicit as much public interest.

Advocacy organizations often use large charismatic vertebrates as *flagship species* to promote biodiversity conservation. For instance, the symbol of the World Wide Fund for Nature (in North America, the World Wildlife Fund) is the panda, an endangered animal requiring large stands of undisturbed bamboo forest. The panda's lovable appearance helps solicit public support for conservation efforts that protect far more than just the panda.

Biodiversity hotspots To prioritize regions for conservation efforts, scientists have mapped biodiversity hotspots (**FIGURE 11.24a**). A **biodiversity hotspot** is a region that supports an especially great number of species that are endemic (p. 59), found nowhere else in the world (**FIGURE 11.24b**). To qualify as a hotspot, a region must harbor at least 1500 endemic plant species (0.5% of the world's plant species). In addition, a hotspot must have already lost 70% of its habitat to human impact and be at risk of losing more.

The ecosystems of the world's biodiversity hotspots together once covered 15.7% of the planet's land surface. Today, because of habitat loss, they cover only 2.3%. This small amount of land is the exclusive home for half the world's plant species and 42% of terrestrial vertebrate species. The hotspot concept motivates us to focus on these valuable regions, where the greatest number of unique species can be protected.

Parks and protected areas A prime way to conserve habitats, communities, ecosystems, and landscapes is to set aside areas of land to be protected from development. Some such preserves are privately owned; for example, The Nature

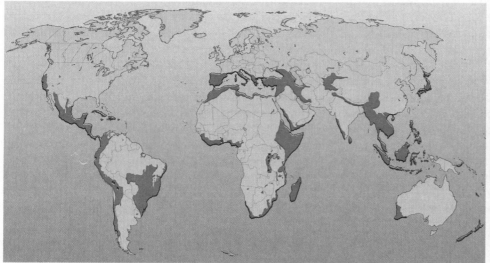

(a) The world's biodiversity hotspots

(b) Ring-tailed lemur

FIGURE 11.24 Biodiversity hotspots are priority areas for habitat preservation. Highlighted in color **(a)** are the 34 hotspots mapped by Conservation International, a nongovernmental organization. (Only 15% of the highlighted area is actually habitat; most is developed.) These regions are home to species such as the ring-tailed lemur **(b)**, a primate endemic to Madagascar that has lost more than 90% of its forest habitat as a result of human population growth and resource extraction. *Data from Conservation International.*

THE SCIENCE
behind
the story

Can Forensic DNA Analysis Help Save Elephants?

As any television buff knows, forensic science is a crucial tool in solving mysteries and fighting crime. In recent years, conservation biologists have been using forensics to unearth secrets and catch bad guys in the multi-billion-dollar illegal global wildlife trade. One such detective story centers on the poaching of Africa's elephants for ivory.

Each year, tens of thousands of elephants are slaughtered illegally by poachers simply for the ivory derived from elephants' tusks (**FIGURE 1**). Customs agents and law enforcement authorities manage to discover and confiscate tons of tusks being shipped internationally in the ivory trade, yet only a small percentage of tusks are found and confiscated. Poachers are rarely apprehended, so the organized international crime syndicates that run these lucrative illicit operations have remained largely unhindered.

Confiscated tusks being destroyed in Kenya to discourage poaching

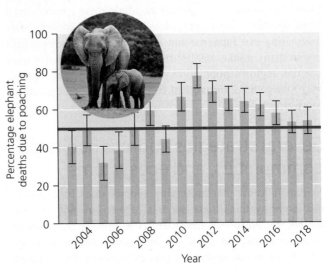

FIGURE 1 Since 2010, more than half of African elephant deaths have been due to poaching, a level scientists conclude is unsustainable. Values above the horizontal line in this graph are thought to cause declines in the population. *Data from MIKE (Monitoring the Illegal Killing of Elephants).*

Enter conservation biologist Samuel Wasser of the University of Washington in Seattle. By bringing the tools of genetic analysis, he and his students and colleagues have shed light on where elephants are being killed and where tusks are being shipped, thereby helping law enforcement efforts. In 2015, Wasser's team published a summary of nearly 20 years of work in the journal *Science,* revealing two major "poaching hotspots" in Africa.

The researchers began by accumulating 1350 reference samples of DNA from elephants at 71 locations across 29 African nations. Two subspecies of African elephant exist: savanna elephants, which live in open savannas, and forest elephants, which live in the forests of West and Central Africa. Of the 1350 genetic samples (from tissues or dung), 1001 came from savanna elephants and 349 came from forest elephants.

Wasser's teams sequenced DNA from the reference samples and compiled data on 16 highly variable stretches of DNA. For each of the 71 geographic locations, they measured frequencies of alleles (different versions of genes) in these variable

DNA stretches. By this process, they created a map of allele frequencies for 71 geographic areas for both types of elephants. This map would act as a kind of reference library they could use to assess any samples from tusks confiscated by law enforcement from the ivory trade.

Working with law enforcement officials, Wasser was able to access 20% of all ivory seizures made internationally between 1996 and 2005, 28% made between 2006 and 2011, and 61% made between 2012 and 2014 (**FIGURE 2**). Taking samples from these tusks and sequencing the DNA, Wasser's teams were then able to compare the results to their library of 71 locations and look for matches. In this way, with the help of sophisticated statistical techniques, they were able to determine the geographic origin of each of the tusks they tested, within an estimated distance of about 300–400 km (185–250 mi).

For instance, ivory seized in the Philippines from 1996 to 2005 all appeared to come from forest elephants in an area of eastern Democratic Republic of Congo—just one small portion of their

Conservancy purchases lands featuring important habitats and seeks to connect them with other tracts of protected land so that ecological processes can function across broad regions. Most protected areas, however, are publicly owned and managed.

Currently, people have set aside nearly 15% of the world's land area in national parks, state parks, provincial parks, wilderness areas, biosphere reserves, and other publicly held protected areas (pp. 325–328). Many of these lands are managed for recreation, water quality protection, or other

FIGURE 2 Dr. Samuel Wasser worked with law enforcement officials to obtain samples of confiscated ivory.

large geographic range. This region was difficult to patrol, however, due to its remoteness and to warfare occurring at the time, so little could be done with the information.

In contrast, when customs agents seized 6.5 tons of tusks in Singapore in 2002, Wasser's team determined that their DNA matched known samples from savanna elephants in Zambia, indicating that many more elephants were being killed there than Zambia's government had realized. The Zambian government responded by replacing its wildlife director and imposing harsher sentences on poachers and ivory smugglers.

For shipments seized between 2006 and 2014, the genetic research indicated a surprisingly clear pattern of two main poaching hotspots. Most forest elephant tusks seized originated from a small region of West-Central Africa where two protected areas overlap the boundaries of four nations (**FIGURE 3a**). As for savanna elephant tusks, most came from animals in southern Tanzania and northern Mozambique during the early portion of the 2006–2014 period. Later in the period, tusks originated from throughout Tanzania (**FIGURE 3b**), pointing to a shift northward and an increase in poaching in the parks of central and northern Tanzania.

In most cases, shipments seized at ports in places such as Hong Kong, Malaysia, Taiwan, or Sri Lanka were labeled with their shipping origin (often a coastal port in Kenya, Tanzania, Togo, or other African countries). With the additional help of Wasser's research, authorities could learn the entire route of the ivory shipments, from where the elephants were killed to where the tusks were exported to where they were imported.

Other researchers are now using similar forensic approaches to combat poaching of additional species. In a 2018 paper in the journal *Current Biology,* a team led by conservation biologist Cindy Harper of the University of Pretoria, South Africa, assessed genetic material from 120 rhinoceros killings, matching the data to a reference database compiled from 4000 wild rhinos. Data from this study led to nine convictions of poachers and traffickers, with sentences totaling 123 years of prison time. Combined with international survey efforts, DNA forensic studies are painting a clearer picture of crime networks than ever before and are giving law enforcement authorities increasingly strong information with which to help protect wildlife.

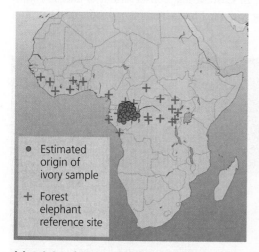

(a) Origin of tusks confiscated in Hong Kong

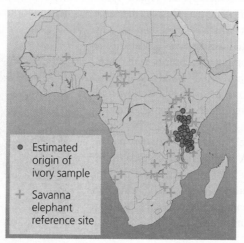

(b) Origin of tusks confiscated in Uganda

FIGURE 3 Genetic analysis of confiscated ivory reveals where elephants were killed. For example, analysis of a shipment confiscated in Hong Kong in 2013 **(a)** shows that it came from forest elephants killed in a small area of West-Central Africa. Likewise, tusks from a shipment confiscated in Uganda in 2013 **(b)** were found to have come from various areas within Tanzania. *Adapted from Wasser, S., et al., 2015. Genetic assignment of large seizures of elephant ivory reveals Africa's major poaching hotspots. Science 349: 84–87.*

purposes (rather than for biodiversity), and many suffer from illegal logging, poaching, and resource extraction. These areas offer animals and plants a degree of protection from human persecution, however, and some are large enough to preserve entire natural systems that otherwise would be fragmented, degraded, or destroyed. Increasingly, areas of ocean are also being protected in marine reserves and marine protected areas (p. 447).

Serengeti National Park and the adjacent Maasai Mara National Reserve are two of the world's largest and most

FIGURE 11.25 Ecological restoration attempts to restore communities to their original state prior to human disturbance. Here people plant trees in an effort to restore the Mau Forest Complex in Kenya.

celebrated parks, but Tanzania and Kenya have each set aside a number of other protected areas. Some of the best known include (in Kenya) Amboseli, Tsavo, Mount Kenya, and Lake Nakuru National Parks, and (in Tanzania) Ngorongoro Conservation Area, Selous Game Reserve, Kilimanjaro National Park, and Gombe Stream National Park. Altogether roughly 25% of Tanzania's land area and 12% of Kenya's land area is protected.

Alas, setting aside land is not always enough to ensure effective conservation. In Kenya and Tanzania, pressures from outside the reserves (settlements, hunting, competition with livestock, and habitat loss to farming) are reducing populations of migratory wildlife within the reserves (see Figure 11.8). Similar situations occur in North America: Despite the large size of Yellowstone National Park (which is larger than the state of Delaware), elk, bears, bison, and wolves roam seasonally in and out of the park, sometimes coming into conflict with ranchers. As a result, conservationists have tried to find ways to protect animals and habitats across the Greater Yellowstone Ecosystem, the broader region over which the animals roam.

Moreover, as global warming (Chapter 18) drives species toward the poles and upward in elevation (p. 505), they can be forced out of protected areas. A major challenge today is to link protected areas across the landscape with corridors of habitat so that species like wildebeest or grizzly bears can move in response to climate change. (We will explore parks and protected areas more fully in Chapter 12.)

Ecological restoration Protecting natural areas before they become degraded is the best way to safeguard biodiversity and ecological systems. In some cases, however, we can restore degraded natural systems to a semblance of their former condition through the practice of ecological restoration (p. 93). Ecological restoration aims not simply to bring back populations of animals and plants but to reestablish the processes—the cycling of matter and the flow of energy—that make an ecosystem function. By restoring complex natural systems such as the Illinois prairie (p. 93), the Florida Everglades (p. 93), or the southeastern longleaf pine forest (p. 322), restoration ecologists aim

to recreate functioning systems that filter pollutants, cleanse water and air, build soil, and recharge groundwater, providing habitat for native wildlife and services for people.

In Kenya, efforts are being made to restore the Mau Forest Complex, Kenya's largest remaining forested area and a watershed that provides water for the Maasai Mara Reserve and for the people of the region (**FIGURE 11.25**). Over the years, so much forest in this densely populated area has been destroyed by agriculture, settlement, and timber extraction that the water supply for the Serengeti's wildlife and for millions of Kenyan people is now threatened. Kenya's government is working with international agencies and funding to replant and protect areas of the forest.

Community-based conservation is growing

Helping people, wildlife, and ecosystems all at the same time is the focus of many current conservation efforts. In the past, conservationists from cities and from industrialized nations, in their zeal to preserve ecosystems in rural areas or developing nations, often neglected the needs of people in the areas they wanted to protect. Today, in contrast, many conservationists actively engage local people in efforts to protect land and wildlife—a cooperative approach called **community-based conservation** (**FIGURE 11.26**). One-fourth of the world's protected areas are now being managed using community-based conservation.

In several African nations, the African Wildlife Foundation funds community-based conservation programs to help local people conserve elephants, lions, rhinos, gorillas, and other animals. In East Africa, conservationists and scientists began working with the Maasai and other people of the region

FIGURE 11.26 Scientists and conservation advocates work cooperatively with local people to conserve wildlife. Biologist Alayne Cotterill of the conservation research group Living with Lions works with Maasai warriors to monitor lion populations near the Serengeti.

years ago, understanding that to conserve animals and ecosystems, local people need to be stewards of the land and feel invested in conservation. This work has proven challenging because the parks and reserves were created on land historically used by local people. Residents had been forcibly relocated; by some estimates, 50,000 Maasai were evicted to create Serengeti National Park.

In the view of many local people, the parks were a government land grab, and laws against poaching deprive them of a traditional right to kill wildlife. As human population grew in the region, conflicts between people and wildlife increased. Ranchers worried that wildebeest and buffaloes might spread disease to their cattle. Farmers lost produce when elephants ate their crops. And the economic benefits of ecotourism were not being shared with all people in the region.

In response, proponents of conservation have tried to reallocate tourist dollars to local villages and to transfer some authority over wildlife management to local people. In the regions around the Maasai Mara Reserve, the Kenya Wildlife Service and international non-profit organizations have been helping farmers and ranchers build strong electric fences to keep wildlife away from their crops and livestock. These efforts are reducing conflicts between people and wildlife and are fostering more favorable attitudes toward conservation.

Working cooperatively to make conservation beneficial for local people requires patience, investment, and trust on all sides. Setting aside land for preservation may deprive local people of access to exploitable resources, but it also helps ensure that those resources can be sustainably managed and will not be used up or sold to foreign corporations. If tourism revenues are adequately distributed, people can gain direct economic benefits from conserving wildlife. Community-based conservation has not always been successful, but in a world of rising human population and consumption, sustaining biodiversity will require locally based management that sustainably meets people's needs.

CENTRAL CASE STUDY
connect & continue

TODAY, animals in the Serengeti continue to stage their epic migrations while tourists visit, park rangers guard against poaching, and the region's people weigh the value of wildlife and ecotourism dollars versus prospects for development. The original proposal for a highway cutting through the park remains alive, and plans are underway for construction to begin at the eastern end, with Chinese work crews scheduled to complete the eastern stretch in 2020. Under this plan, the road would be paved outside of Serengeti National Park and unpaved within it; however, progress remains uncertain. Likewise, routes for Uganda's oil pipeline are still being debated, but so far it appears that the pipeline will most likely not cross the park. Amid these debates, scientists continue monitoring wildlife populations and are documenting declines in areas where human pressures are greatest. Wildebeest numbers in the Serengeti have held stable so far, but four other migratory wildebeest populations in the region have crashed.

In the meantime, a new issue has arisen. On the Kenyan side of the border, there are proposals to build a series of seven dams on the Mara River, the river that flows from the Mau Forest through the Serengeti, providing crucial water to the migrating animals. Farmers in Kenya want the water for agriculture, and more and more people have moved into the Mau Forest, where deforestation has already lowered water levels in the river. The non-profit group Serengeti Watch calls the Mara River "the lifeblood of the Serengeti" and warns that if the dams are built and water withdrawn, even greater harm would come to the Serengeti's animals and ecosystem than would come through the highway proposal.

None of these issues are easy to resolve. It can be challenging to achieve economic development while also protecting natural resources. The debates over highways, pipelines, and dams in the Serengeti bring these issues into stark relief, but there is hope that if local people can obtain economic benefits from ecotourism, they will be inspired to help protect the Serengeti ecosystem and achieve a win-win solution for economic development and biodiversity conservation.

- **CASE STUDY SOLUTIONS** You are an adviser to the president of Tanzania, who is seeking to develop a formal policy on potential highways and pipelines through Serengeti National Park. Given what you know from our Central Case Study, what would be your preliminary advice to the president, and why? To better inform your advice, what further facts and perspectives would you seek to learn from the region's residents? From scientists? From conservation advocates? From tourism operators? From international sources whose funding might help determine the types of projects that could get built?

- **LOCAL CONNECTIONS** Few places on Earth feature wildlife spectacles like those of the Serengeti, but nearly every place on Earth can boast interesting plants and animals and valuable ecosystems. What are some of the plants and animals living in your region that interest you? Are there any that are limited to your region or considered iconic of your region? Describe at least one major type of ecosystem in your area. How do people interact with it? What pressures do its plants and animals experience from human impacts? Can you think of ways in which people could treat natural areas in your region with more care while continuing to thrive socially and economically?

- **EXPLORE THE DATA** What are researchers discovering about populations of African animals? → **Explore Data** relating to the case study on **Mastering Environmental Science.**

REVIEWING Objectives

You should now be able to:

+ Characterize the scope of biodiversity on Earth

Biodiversity can be considered at three levels: genetic diversity, species diversity, and ecosystem diversity. Some groups (such as insects) are more diverse than others, and diversity is unevenly distributed across habitats, biomes, and regions of the world. More than 1.8 million species have been described so far, but scientists agree that the world holds millions more. (pp. 275–278)

+ Specify the benefits that biodiversity brings us

Wild species are sources of food, medicine, and economic development, and biodiversity supports functioning ecosystems and the services they provide us. In addition, many people believe that we have a psychological need to connect with the natural world, as well as an ethical duty to preserve nature. (pp. 278–282)

+ Discuss today's extinction crisis in geologic context

Most populations of wild species are declining, and today's extinction rate is far higher than the natural background rate. Earth has experienced five mass extinction events in the past 440 million years, and human impact is now initiating a sixth mass extinction. (pp. 282–288)

+ Evaluate the primary causes of biodiversity loss

Habitat loss (through destruction, alteration, or fragmentation) is the main cause of current biodiversity loss.

Pollution, overharvesting, invasive species, and climate change are also important causes. Many organisms (such as amphibians) face impacts from a mix of factors. (pp. 288–295)

+ Assess the science and practice of conservation biology

Conservation biologists study biodiversity loss and seek ways to protect and restore biodiversity. These scientists integrate research at the genetic, population, species, ecosystem, and landscape levels. (pp. 295–296)

+ Analyze efforts to conserve threatened and endangered species

The U.S. Endangered Species Act has been largely effective, despite debate over its merits and limitations in funding. Internationally, CITES and the Convention on Biological Diversity aim to safeguard biodiversity. Modern recovery strategies include captive breeding and reintroduction programs, and forensics can help us trace products from illegally poached animals. (pp. 296–301)

+ Compare and contrast conservation efforts above the species level

Species cannot truly be conserved without conserving habitats, communities, and ecosystems. Charismatic species are often used in popular appeals. Biodiversity hotspots help prioritize regions globally for conservation. Parks and protected areas conserve biodiversity at the landscape level. Ecological restoration efforts are restoring degraded ecosystems. Community-based conservation empowers people to invest in conserving their local species and ecosystems. (pp. 299–303)

SEEKING Solutions

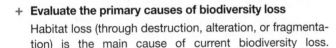

1. Many arguments have been advanced for the importance of preserving biodiversity. Which argument do you find most compelling, and why? Which argument do you find least compelling, and why?

2. Some people declare that we shouldn't worry about endangered species because extinction has always occurred. How would you respond to this view?

3. Compare the approach of setting aside protected areas with the approach of community-based conservation. What are the advantages and disadvantages of each? Can—and should—we follow both approaches?

4. **THINK IT THROUGH** You are a legislator in a nation with no endangered species act. You want to introduce a law to protect your nation's biodiversity. Consider the U.S.

Endangered Species Act, as well as international efforts such as CITES and the Convention on Biological Diversity. What strategies would you write into your legislation? How would your proposed law be similar to and different from each of these efforts? Explain your answers.

5. **THINK IT THROUGH** As a resident of your community and a parent of two young children, you attend a town meeting called to discuss the proposed development of a shopping mall and condominium complex. The development would eliminate a 100-acre stand of forest, the last sizeable forest stand in your town. The developers say the forest loss will not matter because plenty of 1-acre stands still exist, scattered throughout the region. Consider the development's possible impacts on the community's biodiversity, children, and quality of life. What will you choose to tell your fellow citizens and the town's decision-makers at this meeting, and why?

CALCULATING Ecological Footprints

Research shows that much of humanity's footprint on biodiversity comes from our use of grasslands for livestock grazing and forests for timber and for other resources. Grasslands and forests contribute different amounts to each nation's biocapacity (a region's natural capacity to provide resources and absorb our wastes; p. 6), depending on how much of these habitats each nation has. Likewise, the per capita biocapacity and per capita ecological footprints of each nation vary further according to their populations. When footprints are equal to or below biocapacity, resources are being used sustainably. When footprints surpass biocapacity, resources are being used unsustainably.

In the table, fill in the proportion of each nation's per capita footprint accounted for by use of grazing land and forest land. Then fill in the proportion of each nation's per capita biocapacity provided by grazing land and forest land.

FOOTPRINTS AND BIOCAPACITIES IN HECTARES PER PERSON	KENYA	TANZANIA	UNITED STATES	CANADA
Footprint from grazing land	0.21	0.29	0.32	0.27
Footprint from forest land	0.24	0.21	0.86	1.27
Total ecological footprint	1.02	1.22	8.10	7.74
Percentage footprint from grazing and forest	44			
Biocapacity of grazing land	0.21	0.29	0.28	0.27
Biocapacity of forest land	0.02	0.14	1.53	8.63
Total biocapacity	0.49	1.02	3.65	15.12
Percentage biocapacity from grazing and forest				59

Data from Global Footprint Network, 2019.

1. In which nations is grazing land being used sustainably? In which nations is forest land being used sustainably?

2. Do forest use and grazing make up a larger part of the footprint for temperate-zone industrialized nations such as Canada and the United States or for tropical-zone developing nations such as Kenya and Tanzania? What do you think accounts for this difference between these two types of nations? What else besides land use of forests and grasslands contributes to an ecological footprint? (You may wish to consult p. 5.)

3. The Living Planet Index declined 60% between 1970 and 2014 (pp. 283–284), but its temperate and tropical components differed. During this period, indices for temperate regions decreased by 23–31%, whereas indices for tropical regions declined by 56–89%. Based on this information, do you predict that biodiversity loss has been steepest in Kenya and Tanzania or in Canada and the United States? Explain your answer.

Mastering Environmental Science

Students Go to **Mastering Environmental Science** for assignments, an interactive e-text, and the Study Area with practice tests, videos, and activities.

Instructors Go to **Mastering Environmental Science** for automatically graded activities, videos, and reading questions that you can assign to your students, plus Instructor Resources.

Forests, Forest Management, and Protected Areas

Amazon
rainforest

BRAZIL

Saving the World's Greatest Rainforest

> **Imagine a man without lungs. Imagine Earth without Amazon rainforest.**
> Vinita Kinra, Indian Canadian author
>
> **Destroying rainforest for economic gain is like burning a Renaissance painting to cook a meal.**
> Edward O. Wilson, biologist and Pulitzer Prize–winning author

B y any measure, the Amazon rainforest is enormous—and enormously important. It encompasses most of the rainforest on Earth. It captures water, regulates climate, and absorbs much of our planet's carbon dioxide while releasing oxygen. It hosts many thousands of plant and animal species—as well as tribes of indigenous people, some not yet contacted by outside civilization.

The Amazon rainforest plays all these vital roles in the health of our planet, yet we are losing it as people clear its trees for agriculture and settlement. Fully one-fifth of the enormous forest that was standing just half a century ago is now gone.

Nine South American nations claim portions of the Amazon rainforest, but nearly two-thirds of it lies within Brazil. Just as the United States in the 1800s pushed its frontier west, Brazil has pushed into the Amazon, seeking to relieve crowding and poverty in its cities by encouraging urban-dwellers to settle there. While foreign investors helped fund development, Brazil's government provided incentives for frontier settlement. In the 1970s, Brazil built the Trans-Amazonian highway across the forest and promised each settler 100 hectares (250 acres) of land, along with loans and 6 months' salary. People flooded into the region and cleared vast areas for small farms.

Most settlers were unable to make a living at farming, and many plots were abandoned. However, once land is cleared, it can be sold for other purposes. Land speculators bought and sold parcels, and vast tracts ended up going to wealthy landowners overseeing large-scale farming and ranching ventures. Since that time, cattle ranching has grown and now accounts for three-quarters of the forest loss in the Brazilian Amazon (**FIGURE 12.1**, p. 308).

Starting in the 1990s, rising global demand for soy, sugar, rice, corn, and palm oil fueled the destruction of Amazonian forest for industrial-scale farming. In particular, vast monocultures of soybeans spread across the land. Most soy was exported for biodiesel and animal feed.

Meanwhile, each new highway, road, and path cut into the forest for farming or ranching also allowed access for miners, loggers, and poachers—many of whom were exploiting resources illegally. More roads and openings ensued, dividing the forest into smaller and smaller fragments. Soon, dam construction and oil and gas drilling caused more forest loss. As a result of all these activities, each year from the 1970s through 2008 an average of nearly 17,000 km² (6600 mi²) were cleared

Upon completing this chapter, you will be able to:

+ Summarize the ecological and economic contributions of forests

+ Outline the history and current scale of deforestation

+ Assess dynamics of timber harvest management

+ Discuss forest management in relation to fire, pests, and climate change, and evaluate sustainable forestry certification

+ Identify federal land management agencies and the lands they manage

+ Recognize types of parks and protected areas

+ Evaluate challenges to the effectiveness of parks and protected areas

◀ **Aerial view of a stretch of Amazon rainforest**

▲ **Member of the Xingu tribe from the Amazon**

(a) An area of rainforest cleared for agriculture

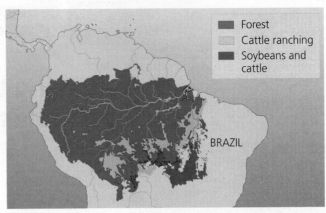

Forest
Cattle ranching
Soybeans and
cattle

BRAZIL

(b) Amazon rainforest and regions of cleared forest

(c) Farmers setting fires to clear land

(d) A jaguar, one of many species dependent on rainforest

FIGURE 12.1 Large areas of Amazon rainforest have been cleared. Most clearance **(a)** has been for large-scale industrial cattle ranching and soybean farming, especially **(b)** in the southern and eastern regions of the Amazon. Small farmers burn forest **(c)** to clear it, and these fires sometimes grow out of control in the dry season. Wild creatures **(d)** are displaced and may die when their forest habitat is destroyed. *Map adapted from Nepstad, D., et al., 2014. Slowing Amazon deforestation through public policy and interventions in beef and soy supply chains. Science 344: 1118–1123.*

in Brazil (**FIGURE 12.2**)—an area larger than Connecticut. The Brazilian Amazon was losing more forest than anywhere else in the world.

In response, an international outcry arose from scientists and the public. People worldwide lamented the loss of the Amazon's biodiversity and the fate of its indigenous tribespeople, who were dying from disease and conflict introduced by loggers, miners, and poachers trespassing on their land. As the threat of global climate change grew worse, people also dreaded the loss of one of Earth's biggest carbon sinks. For all these reasons, the Amazon rainforest came to be perceived as an international treasure that was vital to protect.

This new international scrutiny put pressure on Brazil to curb deforestation. Brazil's leaders responded in 2003–2004 by strengthening enforcement of the nation's Forest Code, a rarely enforced 1965 law mandating that landowners in the Amazon conserve 80% of their land as forest. The government also established large new protected areas to shelter indigenous tribes, conserve water, and prevent soil erosion. Advances in satellite imagery enabled better monitoring of forest loss by remote sensing.

Meanwhile, international pressure convinced banks to begin requiring landowners to document compliance with environmental regulations to get loans. And it persuaded purchasers of beef,

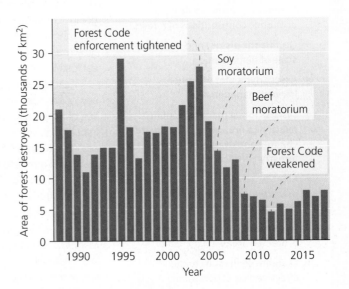

FIGURE 12.2 Annual forest loss in the Brazilian Amazon slowed dramatically after 2004. However, forest loss has been rising since policymakers weakened the Forest Code in 2012. Bars show area of primary forest cleared directly by people each year. *Data from PRODES, Instituto Nacional de Pequisas Espaciais (INPE).*

leather, and soy to buy only those certified as sustainable by third-party certifiers. In 2006, soy traders responded to international pressure by agreeing not to buy Brazilian soybeans planted in recently deforested areas. This "soy moratorium" gave growers in Brazil incentive to plant on previously cleared land rather than cutting down forest. A similar moratorium on beef from recently deforested areas went into effect in 2009. During this period, global commodity prices were affecting Brazilian agriculture, and a fall in beef prices made cattle ranching less profitable.

As a result of all these factors, deforestation rates in the Brazilian Amazon fell sharply after 2004. Since 2009, they have averaged 6500 km^2 (2500 mi^2) per year (see Figure 12.2). Scientists and environmental advocates around the world celebrated this achievement and hailed Brazil as a global model for slowing forest loss.

Today, however, deforestation is rising again. Policymakers weakened the Forest Code in 2012, loosening restrictions on forest clearance and granting amnesty to people who had illegally cleared forest. Then, in 2018, as Brazil struggled through economic troubles and corruption scandals engulfing its political leaders, weary and frustrated voters handed the presidency to Jair Bolsonaro, an autocratic right-wing populist who vowed to open up the Amazon to a new wave of development. Under Bolsonaro, deforestation began rising as farming, ranching, logging, and mining interests all felt emboldened to push deeper into the Amazon.

The Amazon rainforest now faces broader challenges. Global climate change is promoting drought, drying out the eastern and southern portions of the Amazon. As a result, fires are burning increasingly large areas (see Figure 12.11, p. 315), and regions of rainforest are undergoing an ecological regime shift (p. 88), morphing into more open communities. With less vegetation transpiring moisture, the air dries further, putting the remaining rainforest at risk. Scientists now warn that the entire Amazon rainforest may be nearing a "tipping point" beyond which it will no longer remain a rainforest.

As Brazil's current government acts to encourage development at the expense of its forests, observers worry that the nation will jettison the promises it made at the Paris Climate Conference in 2015 to reduce forest loss. Brazil's decisions in coming years will shape the fate of the Amazon, which in turn bears significance for forests, biodiversity, climate, and society across the globe.

Forest Ecosystems and Forest Resources

A **forest** is any ecosystem with a high density of trees. Forests provide habitat for countless organisms; help maintain the quality of soil, air, and water; and play key roles in our planet's biogeochemical cycles (p. 121). Forests also provide us with wood for fuel, construction, paper production, and more.

There are many types of forests

Most of the world's forests occur as boreal forest (p. 100), a biome that stretches across much of Canada, Scandinavia, and Russia; or as tropical rainforest (p. 97), a biome in Latin America, Indonesia, Southeast Asia, and equatorial Africa. Temperate deciduous forests, temperate rainforests, and tropical dry forests (pp. 96, 97, 98) also cover large regions.

Within each forest biome, the plant community varies from region to region because of differences in soil and climate. As a result, ecologists classify forests into **forest types,** categories defined by their predominant tree species (**FIGURE 12.3**). The eastern United States contains 10 forest types, ranging from spruce-fir to oak-hickory to longleaf–slash pine. The western United States has 11 forest types, ranging from Douglas fir and hemlock–sitka spruce forests of the moist Pacific Northwest to ponderosa pine and pinyon-juniper woodlands of the drier interior. In the Amazon rainforest, two of the main forest types are *várzea* (forest that gets flooded by swollen waters of the Amazon River or its tributaries for weeks or months each year) and *terra firme* (forest on land high enough to avoid seasonal flooding).

(a) Maple-beech-birch forest, Michigan

(b) Ponderosa pine forest, Arizona

(c) Redwood forest, California

FIGURE 12.3 Shown are 3 of the 21 forest types found in the contiguous United States.

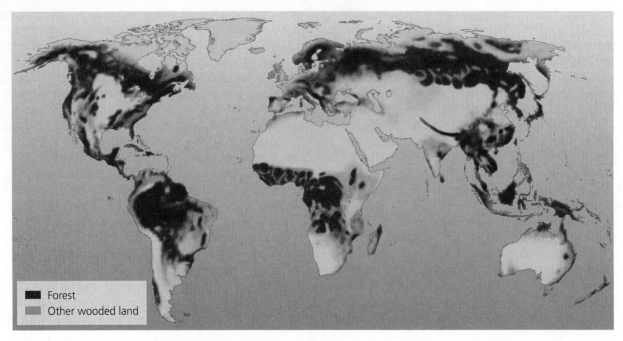

FIGURE 12.4 Forests cover 31% of Earth's land surface. Most widespread are boreal forests in the north and tropical forests near the equator. In addition, "other wooded land" (pale green in map) supports trees at sparser densities. *Data from Food and Agriculture Organization of the United Nations, 2010.* Global forest resources assessment 2010.

Altogether, forests currently cover 31% of Earth's land surface (**FIGURE 12.4**). Forests occur on all continents except Antarctica.

Forests are ecologically complex

Because of their structural complexity and their capacity to provide many niches for organisms, forests comprise some of the richest ecosystems for biodiversity (**FIGURE 12.5**). The leaves, fruits, and seeds of trees, shrubs, and forest-floor plants furnish food and shelter for an immense diversity of insects, birds, mammals, and other animals. Plants are also colonized by an extensive array of fungi and microbes, in both parasitic and mutualistic relationships (pp. 79–80).

In a forest's **canopy**—the upper level of leaves and branches in the treetops—beetles, caterpillars, and other

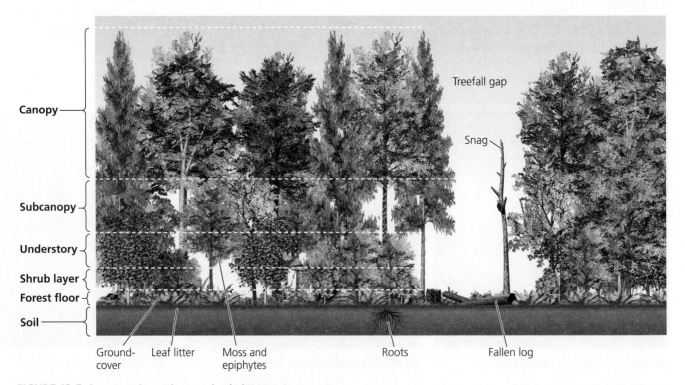

FIGURE 12.5 A mature forest is complex in its structure.

leaf-eating insects abound, providing food for birds such as warblers and tanagers, while arboreal mammals from squirrels to sloths to monkeys consume fruit and leaves. Animals also live and feed in the **subcanopy** (the middle portion of a forest beneath the tree crowns of the canopy), in shrubs and small trees of the **understory** (the shaded lower level of a forest), and among groundcover plants on the forest floor. Other animals use the bark, branches, and trunks of trees or the vines, lichens, and mosses that grow on them.

Cavities in trunks and limbs provide nest and shelter sites for a wide variety of animals. Dead and dying trees (called **snags**) are especially valuable; insects that break down the wood provide food for woodpeckers, which create cavities that other animals may later use. Fallen trees create openings called *treefall gaps,* letting sunlight through the canopy and down to the forest floor, thus encouraging the growth of early successional plants (see secondary succession, p. 87) and creating habitat diversity within the forest. Much of a forest's biodiversity resides on the forest floor, where decaying logs and the fallen leaves and branches of the leaf litter nourish the soil. A multitude of soil organisms helps decompose plant material and cycle nutrients (pp. 81–82).

In tropical rainforests, particularly tall individual trees called emergent trees protrude here and there above the canopy. Epiphytes, plants specialized to grow atop other plants, add to the biomass and species diversity at all levels of a rainforest. Epiphytes include many ferns, mosses, lichens, orchids, and bromeliads.

Forests with a greater diversity of plants tend to host a greater diversity of organisms overall. As forests change by the process of succession, their species composition changes along with their structure. In general, old-growth forests contain more structural diversity, microhabitats, and resources for more species than young forests. Old-growth forests also are home to more species that today are threatened, endangered, or declining, because today old forests have become rare relative to young forests.

Forests provide ecosystem services

Besides hosting biodiversity, forests supply us with many vital ecosystem services (pp. 4, 120; **FIGURE 12.6**). As plants grow, their roots stabilize the soil and help prevent erosion. Trees' roots draw minerals up from deep soil layers and deliver them to surface soil layers where other plants can use them. Plants also return organic material to the topsoil when they die or drop their leaves. When rain falls, leaves and leaf litter slow runoff by intercepting water. This process helps water soak into the ground to nourish roots and recharge aquifers, thereby preventing flooding, reducing soil erosion, and helping keep streams and rivers clean. Forest plants also filter pollutants and purify water as they take it up from the soil and release it to the atmosphere by the process of transpiration (p. 121). Plants release the oxygen that we breathe, regulate moisture and precipitation, and moderate climate.

Because forests perform all these ecological functions, they are indispensable for our health and survival. Forests also enhance our quality of life by providing us with cultural, aesthetic, and recreational "nonmarket" values (p. 148). People seek out forests for adventure and for spiritual solace alike—to admire beautiful trees, to observe wildlife, to enjoy clean air, and for many other reasons.

Stores carbon

Supports biodiversity

CO_2

O_2

Produces oxygen

Provides fuel wood, lumber, paper, medicines, dyes, foods, fibers

H_2O

Purifies water, filters pollution

Promotes and provides health, beauty, recreation

Returns organic matter to soil

Slows runoff, prevents flooding

Transports minerals to soil surface

Stabilizes soil, prevents erosion

FIGURE 12.6 Forests provide us a diversity of ecosystem services, as well as resources that we can harvest.

FAQ

How do forests affect climate?

It seems obvious that climate would affect forests—rainy climates lead to rainforest rather than deserts, for instance. But forests also affect climate in several ways. At the small scale of "microclimate," the complex vegetation structure of a forest serves as a windbreak and moderates temperature, keeping forests cooler than open areas on hot days and warmer than open areas on cold days. At a large scale, forests can regulate humidity and temperature across an entire region. For example, leaves release water vapor through transpiration (p. 121), making the air more humid. That is how the rainforest helps keep the Amazon region humid with plenty of rainfall—and it is why deforestation in the Amazon is drying out the remaining forest. At the planetwide scale, forests help prevent global warming and climate change. Leaves absorb carbon dioxide in photosynthesis (p. 33) to build plant tissues, and carbon dioxide is our atmosphere's main greenhouse gas. So, when we lose forests, carbon dioxide is released to the atmosphere, worsening climate change.

Carbon storage limits climate change

Of all the services that forests provide, their storage of carbon is eliciting the greatest interest as nations debate how to control global climate change (Chapter 18). Because plants absorb carbon dioxide from the air during photosynthesis (p. 33) and then store carbon in their tissues, forests serve as a major reservoir for carbon. Scientists estimate that the world's forests store about 296 billion metric tons of carbon in living tissue, which is more than the atmosphere contains. Each year, forests absorb about 2.4 billion metric tons of carbon from the air, with the Amazon rainforest taking care of one-fourth of that total.

Conversely, when plant matter is burned or when plants die and decompose, carbon dioxide is released—and thereafter, less vegetation remains to soak it up. Carbon dioxide is the primary greenhouse gas driving global climate change (p. 488). Therefore, when we destroy forests, we worsen climate change. Forest destruction and degradation account for 17% of the world's greenhouse gas emissions—more than all the world's motor vehicles emit. The more forests we preserve or restore, the more carbon we keep out of the atmosphere and the more effectively we can address climate change.

Forests provide us valuable resources

Carbon storage and other ecosystem services alone make forests priceless to our society, but forests also provide many economically valuable resources. Among these are plants for medicines, dyes, and fibers; and fruits, nuts, mushrooms, and game animals for food. Amazonian forests abound with resources ranging from Brazil nuts, chocolate, and açaí berries to latex for making rubber (from the sap of rubber trees). And of course, forest trees provide us with wood. For millennia, wood has fueled the fires to cook our food and keep us warm. It has built the homes that keep us sheltered. It built the ships that carried people and cultures between continents. And it gave us paper, the medium of the first information revolution. Forests and their resources have played a key role in helping our society achieve the standard of living we enjoy today.

In recent decades, industrial harvesting has allowed us to extract more timber than ever before, supplying all these needs of a rapidly growing human population and its expanding economy. Most commercial timber extraction today takes place in Canada, Russia, and other nations with large expanses of boreal forest and in tropical nations with large areas of rainforest, such as Brazil and Indonesia. In the United States, most logging takes place in pine plantations of the South and conifer forests of the West.

Forest Loss

When trees are removed more quickly than they can regrow, the result is **deforestation,** the clearing and loss of forests. Deforestation has altered landscapes across much of our planet. In the time it takes you to read this sentence, 2 ha (5 acres) of tropical forest will have been cleared. As we modify, fragment, and eliminate forests, we lose biodiversity, degrade soil, accelerate climate change, and disrupt the ecosystem services that support our societies.

Agriculture and demand for wood put pressure on forests

For millennia, people have cleared forests to make way for agriculture and to extract wood products. A recent study based on satellite imagery examined loss of tree cover globally in 2017 and found it resulted from four causes in nearly equal proportions: (1) farming, ranching, and mining; (2) logging and timber management; (3) slash-and-burn farming; and (4) wildfire. However, the main causes varied greatly from one continent to another (**FIGURE 12.7**). The 27% due to farming, ranching, and mining represented permanent deforestation, whereas the other three main causes represented potentially temporary forest loss.

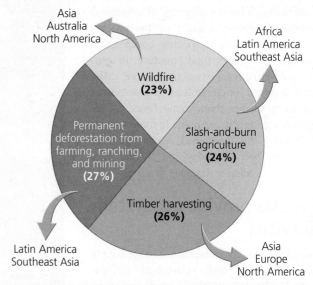

FIGURE 12.7 A study of satellite imagery showed four main causes of the loss of tree cover, but found that they varied by region. Listed are the regions where each cause predominated. Urban development (not shown) accounted for less than 1% of forest loss. *Data are for 2017, from Curtis, P.G., et al., 2018. Classifying drivers of global forest loss.* Science 361: 1108–1111.

In 2015, the Food and Agriculture Organization (FAO) of the United Nations released its latest *Global Forest Resources Assessment.* In this report, researchers combined remote sensing data from satellites, analysis from forest experts, questionnaire responses, and statistical modeling to form a comprehensive picture of the world's forests. The assessment concluded that we are eliminating 7.6 million ha (18.8 million acres) of forest each year. Subtracting annual regrowth from this amount makes for an annual net loss of 3.3 million ha (8.2 million acres)—an area about the size of Maryland. This rate (for the period 2010–2015) is lower than deforestation rates for earlier years. In the 1990s, the world had been losing 1.8% of its forest each year; in 2010–2015, it lost just 0.8% each year.

Another major research effort has calculated higher rates of forest loss. In 2013, this research team analyzed satellite data, published their work in the journal *Science*, and worked with staff from Google to create interactive maps of forest loss and gain across the world. They found much more loss of tree cover than the FAO had found: From 2000 to 2012, this team calculated annual losses of 19.2 million ha and annual gains of 6.7 million ha, for a net loss per year of 12.5 million ha (30.9 million acres—an area larger than New York State). This effort, continued today under the name Global Forest Watch (a non-profit organization sponsored by the World Resources Institute), relies on the latest high-resolution satellite imagery and uses different methodology and definitions than does the FAO. Both approaches are useful and give different perspectives on the complex issue of forest loss across the globe. You can visit the interactive website of Global Forest Watch and explore its data on forest loss and regrowth for any region of the world.

We deforested much of North America

Deforestation for farmland and timber propelled the expansion of the United States and Canada westward across the North American continent. The vast deciduous forests of the East were cleared by the mid-1800s, making way for countless small farms. Timber from these forests built the cities of the Atlantic seaboard. Cities of the upper Midwest, such as Chicago, Detroit, and Milwaukee, were constructed from timber felled in the vast pine and hardwood forests of Wisconsin and Michigan.

As a farming economy shifted to an industrial one, wood was used to stoke the furnaces of industry. Logging operations moved south to the Ozarks of Missouri and Arkansas and then to the pine woodlands and bottomland hardwood forests of the South. Once mature trees were removed from these areas, timber companies moved west, cutting the continent's biggest trees in the Rocky Mountains, the Sierra Nevada, the Cascade Mountains, and the Pacific Coast ranges. Exploiting forest resources fed the U.S. economy, but the forests were not harvested sustainably. Instead, our store of renewable resources for the future was being depleted.

By the 20th century, very little **primary forest**—natural forest uncut by people—remained in the 48 contiguous U.S. states, and today even less is left (**FIGURE 12.8**). Nearly all the large oaks and maples found in eastern North America today, and even most redwoods of the California coast, are *second-growth* trees: trees that sprouted after old-growth trees were cut. Second-growth trees characterize **secondary forest,** which contains smaller, younger, trees than does primary forest. The species composition, structure, and nutrient balance of a secondary forest often differ markedly from the primary forest that it replaced.

Forests are being cleared most rapidly in developing nations

Uncut primary forests still remain in many less-industrialized nations in tropical regions. These nations are in the position the United States and Canada enjoyed a century or two ago: having a resource-rich frontier that they can develop. Today's powerful industrial technologies allow these nations to exploit their resources and push back their frontiers even faster than

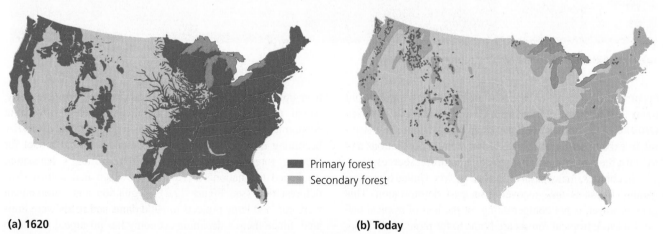

Primary forest
Secondary forest

(a) 1620   **(b) Today**

FIGURE 12.8 Areas of primary (uncut) forest have been dramatically reduced. When Europeans first colonized North America **(a)**, much of what is now the United States was covered in primary forest **(dark green).** Today, nearly all this primary forest is gone **(b),** cut for timber and to make way for agriculture. Much of the landscape has become reforested with secondary forest **(pale green).** *Adapted from (a) U.S. Forest Service; and (b) Hansen, M.C., et al., 2013. High-resolution global maps of 21st-century forest cover change. Science 342: 850–853, and George Draffan, Endgame Research (www.endgame.org).*

FIGURE 12.9 Tropical forests are being lost. Many are burned to farm soybeans or to create grazing land for cattle, as seen here **(a)** in Brazil. Most of the soybeans and beef are exported to consumers in wealthier nations. Africa, Latin America, and Indonesia are losing the most forest **(b)**, whereas Europe and the United States are slowly gaining secondary forest. In Asia, tree plantations are increasing, but natural primary forests are still being lost. *Data from FAO, 2015.* Global Forest Resources Assessment. *FAO, Rome. By permission.*

(a) Cattle on burned and cleared land

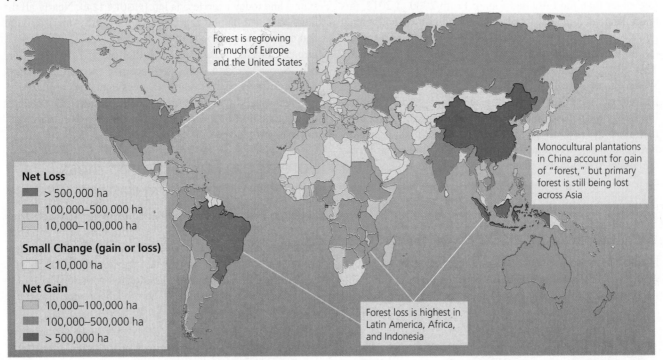

Forest is regrowing in much of Europe and the United States

Monocultural plantations in China account for gain of "forest," but primary forest is still being lost across Asia

Net Loss
- > 500,000 ha
- 100,000–500,000 ha
- 10,000–100,000 ha

Small Change (gain or loss)
- < 10,000 ha

Net Gain
- 10,000–100,000 ha
- 100,000–500,000 ha
- > 500,000 ha

Forest loss is highest in Latin America, Africa, and Indonesia

(b) Annual change in forest area by region (1990-2015)

occurred in North America. As a result, deforestation is rapid today in Indonesia, Africa, and Central and South America (**FIGURE 12.9**). In these regions, developing nations are striving to expand settlement for their burgeoning populations and to boost their economies by extracting natural resources.

In contrast, many parts of Europe and the United States are gaining forest as they recover from past deforestation. This gain, however, is not compensating for the loss of tropical forests, because tropical forests are home to far more biodiversity than the temperate forests of North America and Europe.

As we have seen in the Central Case Study, Brazil was losing forests faster than any other country in the 1980s and 1990s (**FIGURE 12.10**). Its government was promoting settlement on the forested frontier and was subsidizing an expansion of large-scale cattle ranching and soybean farming to meet demand

from consumers in the United States and Europe. More recently, Brazil reduced deforestation significantly while advancing economically and politically as a stable democracy, thereby becoming an inspiring success story and a global model for fighting forest loss. However, in 2012, Brazil's legislature weakened the nation's Forest Code, which had helped slow deforestation (see Figure 12.2). Regulation and enforcement were cut, and large projects to build dams and roads were initiated. Since then, a declining economy has prompted people to clear forests illegally for short-term financial gain, and deforestation in the Amazon and elsewhere in Brazil is rising. Moreover, most of Brazil's neighbors that share portions of the Amazon rainforest—Bolivia, Peru, Ecuador, Colombia, Venezuela, Guyana, Surinam, and French Guiana—have not yet managed to slow their deforestation rates.

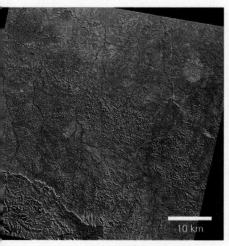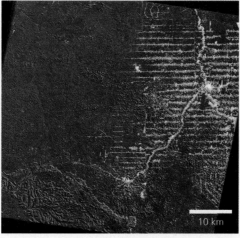

(a) Rondonia, Brazil, 1975   **(b) Rondonia, Brazil, 1992**   **(c) Rondonia, Brazil, 2018**

FIGURE 12.10 Deforestation of Amazonian rainforest has been rapid in recent decades. Satellite images of part of the state of Rondonia in Brazil show extensive clearing resulting from settlement in the region. *Landsat data courtesy of U.S. Geological Survey, Department of the Interior/USGS. Landsat product IDs: (a) LM02-L1TP-249067-19750619-20180425-01-T2, (b) LT04-L1TP-232067-19920622-20170122-01-T1, and (c) LC08-L1TP-232067-20180724-20180731-01-T1.*

Climate change (Chapter 18) is also posing new challenges for the Amazon. Rising temperatures and drought are causing fires (set by farmers each year in the dry season to clear forest by slash-and-burn methods (p. 219)) to spread out of control, burning large areas. In 2016 and 2017, fires burned such wide regions that the resulting loss in tree cover spiked far higher than that caused by direct forest clearance by people.

We can assess this by comparing data sets. In **FIGURE 12.11**, the red bars, reprised from Figure 12.2, show the official data on deforestation provided by the Brazilian government, indicating the area of primary forest cleared directly by people each year. The dark blue line shows data from University of Maryland researchers working with Global Forest Watch. The latter data show overall loss of tree cover (from secondary forest as well as primary forest) from all causes (including fires) at all scales (including clear-cuts smaller than the government data measures). Both data sets use satellite imagery, but they measure different aspects of forest loss. The government dataset is valuable because it is long term and focuses on loss of primary forest, the most ecologically valuable type of forest. The Global Forest Watch data set is valuable because it includes more kinds of forest degradation and captures changes in overall canopy cover, which may be a better indicator of carbon storage capacity.

FAQ

Is the Amazon all primary forest?

People long assumed that the entire Amazon rainforest was pristine, untouched primary forest, but recent scientific research is revealing that portions of the Amazon had been cleared for farms, orchards, and villages centuries ago and that the region evidently supported thousands, or perhaps millions, of people. Most of them died in epidemics that swept the Americas in the wake of early visits by Europeans, who unknowingly brought diseases against which Native American people had evolved no defenses. Once these Amazonian civilizations disappeared, the regions they had cleared regrew into the forest we see today.

Worldwide, the Global Forest Watch data indicate that tree cover loss in tropical forests has been increasing year by year. One reason is that developing nations frequently are desperate enough for economic development and foreign capital that they impose few restrictions on logging, mining, or other extractive activities. Often they allow their timber to be

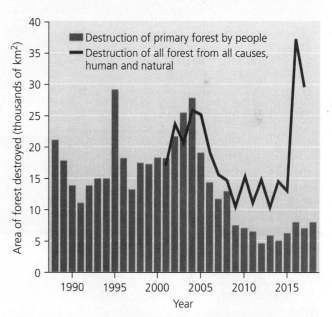

FIGURE 12.11 Tree cover losses in the Amazon have exceeded official estimates of primary forest clearance in recent years. The blue data line shows overall loss of tree cover—secondary as well as primary—from all causes, including loss from naturally occurring fires and loss of areas smaller than those monitored by the government. The red data bars show only loss of primary forest due to direct clearance by humans. Fires were extensive in 2016 and 2017. However, some burned areas will recover, and due to a change in methodology, fire data from before 2011 is underestimated. *Data (red bars) from PRODES, Instituto Nacional de Pequisas Espaciais (INPE), and (blue line) from Global Forest Watch.*

WEIGHING
the issues

Logging Here or There

Suppose you are an activist protesting the logging of old-growth trees near your hometown. Now let's say you know that if the protest is successful, the company will move to a less-industrialized tropical country and cut that country's primary forest instead. Would you still protest the logging in your hometown? Would you pursue any other approaches?

extracted by foreign multinational corporations, which pay fees for a **concession,** or right to extract the resource. Once a concession is granted, the corporation has little or no incentive to manage forest resources sustainably. Local people may receive temporary employment from the corporation, but once the timber is gone, they no longer have the forest and the ecosystem services it provided. As a result, most economic benefits are short term and are reaped by the foreign corporation. Much of the wood extracted in developing nations is exported to Europe and North America. In this way, our consumption of high-end furniture and other wood products can drive forest destruction in poorer nations.

Throughout Southeast Asia and Indonesia today, vast areas of tropical rainforest are being cut to establish plantations of oil palms (**FIGURE 12.12**). Oil palm fruit produces palm oil, which has come to be a common ingredient in snack foods, soaps, and cosmetics and is beginning to be used as a biofuel. In Indonesia, the world's largest palm oil producer, oil palm plantations have displaced more than 9 million ha (22 million acres) of rainforest. Few animals live in oil palm monocultures, so the destruction of primary forest means a staggering loss of biodiversity. Moreover, much of the forest being cleared is *peat swamp forest,* a type of moist forest that forms deposits of carbon-rich peat (p. 530). Peat forests store huge amounts of carbon, and when they are cleared, drained, and burned, carbon dioxide released to the air worsens climate change. On top of this, clearing for plantations encourages further development and eases access for people to enter the forest and conduct logging illegally, leading to further degradation and biodiversity loss.

The palm oil boom represents a conundrum for environmental advocates. Many people eager to fight climate change had urged the development of biofuels (p. 576) to replace fossil fuels. Yet grown at the large scale that our society is demanding, monocultural plantations of biofuel crops such as oil palms are causing severe environmental impacts by displacing natural forests.

Solutions are emerging

Many avenues are being pursued to address deforestation in less-industrialized nations. Some conservation proponents are running community-based conservation projects (p. 302) that empower local people to act as stewards of their forest resources (see **SUCCESS STORY**). In other cases, conservation organizations are buying concessions and using them to preserve forest rather than to cut it down. In such a **conservation concession,** the nation receives money *and* keeps its natural resources intact. The South American nation of Surinam entered into such an agreement with the nongovernmental organization Conservation International and virtually halted logging while gaining $15 million.

Another approach is the **debt-for-nature swap.** In this approach, a conservation organization or other entity offers to pay off a portion of a developing nation's international debt in exchange for a promise by the nation to set aside forest reserves, fund environmental education, and better manage protected areas. The U.S. government committed itself to a program of debt-for-nature swaps through its 1998 Tropical Forest Conservation Act. Under this law, deals were struck with 14 developing nations, enabling more than $339 million of national funds intended for debt payments to go to conservation efforts instead. In the largest such deal, the U.S. government forgave Indonesia $30 million of debt and two conservation groups paid Indonesia $2 million. In return, Indonesia promised to preserve forested areas that are home to the Sumatran tiger and other species.

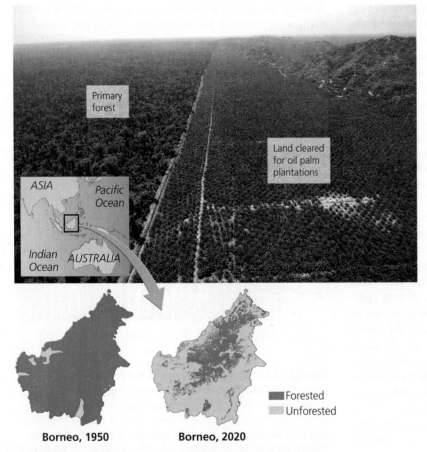

FIGURE 12.12 Oil palm plantations are replacing primary forest across Southeast Asia and Indonesia. Since 1950, the immense island of Borneo (maps at bottom) has lost most of its forest. *Data from Radday, M., WWF-Germany, 2007. Designed by Hugo Ahlenius, UNEP/GRID-Arendal. Extent of deforestation in Borneo 1950–2005, and projection towards 2020; www.grida.no/resources/8324.*

SUCCESS story

Planting Trees, Empowering People: Wangari Maathai and the Green Belt Movement

When an area is deforested, all is not lost. People can reforest the land by planting trees. One success story has unfolded in the African nation of Kenya, which has long experienced some of the world's fastest population growth. By the 1970s, forest had disappeared across vast areas of Kenya as rural people grew crops and harvested wood for fuel and building material. Communities began running out of water and sinking deeper into poverty as a result. Stepping forward to tackle this problem was Wangari Maathai, East Africa's first female professor and later a member of Kenya's parliament. In 1977, she founded a grass-roots organization, the Green Belt Movement, to teach people how forests help conserve water and to organize people to plant trees near where they live. Since then, farmers and villagers have planted 51 million trees at 6500 sites across southern Kenya. Rural women are paid for each tree

they plant—a valuable source of income in an impoverished region. Besides restoring forests and improving water supplies, the work has helped empower individuals, particularly women and girls, while enhancing livelihoods and making communities more sustainable. For her efforts, Maathai was recognized internationally with the Nobel Peace Prize in 2004. Maathai died in 2011, but her legacy lives on. Today, her institute, her foundation, her four books, and the ongoing work of the Green Belt Movement inspire countless individuals to carry on her work with trees, people, and communities.

Dr. Wangari Maathai

→ **Explore the Data** at **Mastering Environmental Science**

Forest proponents are also trying to focus new agricultural development on lands that are already cleared. Brazil's 2006 soy moratorium, negotiated by soy producers, international conservationists, and Brazilian policymakers, successfully reduced clearing for soy 30-fold. People elsewhere in the world are now trying similar approaches.

In Indonesia, a moratorium on drainage of peat forests began in 2016 and may be paying dividends: In 2017, primary forest loss there fell by 60%. This moratorium came atop earlier efforts, such as one by the World Resources Institute (WRI). WRI works with palm oil companies that own concessions to clear primary rainforest and steers them instead to plant their plantations on land that is already logged. WRI then protects the forests that were slated for conversion or allows the forests to be used for certified sustainable forestry (p. 324). Because primary forests store more carbon than oil palm plantations, these land-use swaps reduce greenhouse gas emissions and can gain Indonesia credit via carbon offsets (p. 518).

Carbon offsets and other financial incentives are central to international plans to curb deforestation and climate change. Across years of international climate conferences (p. 516), negotiators developed a program called **Reducing Emissions from Deforestation and Forest Degradation** (REDD; changed to REDD+ as the program expanded in scope) whereby wealthy industrialized nations pay poorer developing nations to conserve forest. The aim was to make forests more financially valuable when saved than when cut down. Under this plan, poor nations would gain income while rich nations received carbon credits to offset their emissions in an international cap-and-trade system (pp. 181, 514).

The REDD+ plan has not yet been formally agreed to, and the $100 billion per year that leaders of rich nations had proposed to transfer to poor nations by 2020 now seems unlikely to materialize. The approach gained momentum at the Paris Climate Conference in 2015, where Brazil gained

praise by promising to reduce its emissions, compared with its 2005 levels, by 37% by 2025 and 43% by 2030 through aggressive steps to curb deforestation. However, its policy shifts since then may put these goals out of reach.

Forest Management

As our demand for forest resources and amenities intensifies, we will benefit by managing forests with care. Foresters are professionals who manage forests through the practice of **forestry.** Foresters must balance our society's demand for forest products against the central importance of forests as ecosystems.

Forest management is one type of resource management

Debates over how to manage forest resources reflect broader questions about how to manage natural resources in general. Resources such as fossil fuels and many minerals are nonrenewable, whereas resources such as the sun's energy are perpetually renewable (p. 4). Between these extremes lie resources that are renewable if they are not exploited too rapidly. They include timber, as well as soils, fresh water, rangeland, wildlife, and fisheries.

Resource management describes the use of strategies to manage and regulate the harvest of renewable resources. Sustainable resource management involves harvesting these resources in ways that do not deplete them. Resource managers help conserve soil resources with improved farming practices, safeguard the supply and quality of surface water and groundwater, encourage low-impact grazing on rangeland, and protect fish and wildlife from overharvesting. Resource managers are guided by research in the natural sciences and by social, political, and economic factors.

Resource managers follow several strategies

A key question in managing resources is whether to focus strictly on the resource of interest or to look more broadly at the environmental system of which it is a part. Taking a broader view often helps avoid degrading the system and thereby helps sustain the resource in the long term.

Maximum sustainable yield A guiding principle in resource management has traditionally been **maximum sustainable yield.** Its aim is to achieve the maximum amount of resource extraction without depleting the resource from one harvest to the next. Recall the logistic growth curve (see Figure 3.17, p. 67), which reflects how limiting factors slow exponential population growth and cap it at a carrying capacity. The logistic curve indicates that a population grows most quickly when it is at an intermediate size—specifically, at one-half of carrying capacity. For this reason, a fisheries manager aiming for maximum sustainable yield will likely prefer to keep fish populations at intermediate levels so that they rebound quickly after each harvest. Theoretically, doing so should result in the greatest amount of fish harvested over time while sustaining the population (**FIGURE 12.13**).

This management approach, however, keeps the fish population at only half its carrying capacity—well below the size it would attain in the absence of fishing. Suppressing population size in this way will likely affect other species and alter the food web dynamics of the community. From an ecological point of view, management for maximum sustainable yield may set in motion significant ecological changes.

In forestry, maximum sustainable yield argues for cutting trees shortly after they go through their fastest stage of growth. Because trees often increase in biomass most quickly at an intermediate age, they are generally cut long before they grow as large as they would in the absence of harvesting. This practice maximizes timber production over time, but it also alters forest ecology and eliminates habitat for species that depend on mature trees.

Ecosystem-based management Because of these dilemmas, more and more managers espouse **ecosystem-based management,** which aims to minimize impact on the ecosystems and ecological processes that provide the resource. Under this approach, foresters may protect certain forested areas, restore ecologically important habitats, and consider patterns at the landscape level (p. 118), allowing timber harvesting while preserving the functional integrity of the forest ecosystem. This approach entails fostering the area's ecological processes, including succession (p. 87), in which the forest community naturally changes over time. Ecosystems are complex, however, so it can be challenging to determine how best to implement this type of management. As a result, ecosystem-based management has come to mean different things to different people.

Adaptive management Some management actions will succeed, and some will fail. A wise manager will try new approaches if old ones are ineffective. **Adaptive management** involves systematically testing different approaches and aiming to improve methods through time. For managers, it entails monitoring the results of one's practices and adjusting them as needed, based on what is learned. Adaptive management is intended to be a fusion of science and management, because it explicitly tests hypotheses about how best to manage resources.

Adaptive management was featured in the 1994 Northwest Forest Plan, a set of federal policies and guidelines crafted by the administration of President Bill Clinton to resolve disputes between loggers and preservationists over the last remaining old-growth temperate rainforests in the contiguous United States. This plan sought to allow limited logging to continue in the Pacific Northwest, with adequate protections for old-growth-dependent species such as the northern spotted owl, and to let science guide management.

Fear of a "timber famine" inspired national forests

The United States began formally managing forest resources more than a century ago in response to rampant deforestation. The depletion of forests throughout the eastern United States had prompted widespread fear of a "timber famine." This led the federal government to form a system of forest reserves: public lands set aside to grow trees, produce timber, and protect water quality. Today's system of **national forests** consists of 77 million ha (191 million acres) managed by the U.S. Forest Service and covering more than 8% of the nation's land area (**FIGURE 12.14**).

The U.S. Forest Service was established in 1905 under the leadership of Gifford Pinchot (p. 138) during the Progressive Era, a time of social reform when people began using science to inform public policy. In line with Pinchot's conservation ethic (p. 138), the Forest Service aimed to manage the forests for "the greatest good of the greatest number in the long run." Pinchot believed that the nation should extract and use resources from its public lands, but that wise and careful management of timber resources was imperative.

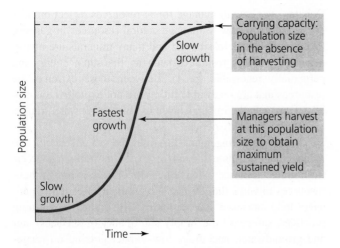

FIGURE 12.13 Maximum sustainable yield maximizes the amount of resource that can be harvested while sustaining the harvest in perpetuity.

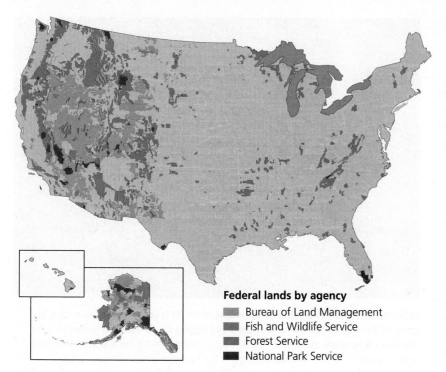

FIGURE 12.14 U.S. residents enjoy more than 250 million ha (600 million acres) of public lands open to every American. Of this total, 77 million ha (191 million acres) are administered as national forests. Other public lands include national parks and monuments, national wildlife refuges, and land managed by the Bureau of Land Management. *Data from U.S. Geological Survey.*

Federal lands by agency
- Bureau of Land Management
- Fish and Wildlife Service
- Forest Service
- National Park Service

We extract timber from private and public lands

Almost 90% of the timber harvesting in the United States today takes place on private land owned by the timber industry or by small landowners. To obtain maximum profits year after year, timber companies generally pursue maximum sustainable yield on their land.

Private timber companies also extract timber from the U.S. national forests and from publicly held state and local forests. In the national forests, U.S. Forest Service staff conduct timber sales and build roads to provide access for logging companies. The Forest Service sells timber below the costs it incurs for marketing and administering the harvest and for building access roads, while the companies go on to sell the timber they harvest for profit. In this way, taxpayers subsidize private timber harvesting on public land (p. 180). These subsidies also tend to inflate harvest levels beyond what would occur in a free market.

Across all types of forests in the United States, net tree growth (growth minus removal and mortality) has been positive for more than 60 years (**FIGURE 12.15**), leading to larger amounts of tree biomass year by year. Extraction of timber began to increase in the 1950s as the nation underwent a postwar economic boom, paper consumption rose, and the growing population moved into newly built suburban homes. Harvests then began to decrease after the 1990s as economic trends shifted, public concern over logging grew, and management philosophy evolved. Today tree growth continues to outpace tree removal, but tree mortality (death from natural causes) is rising (see Figure 12.15). As trees age, as the climate changes, and as pests invade, more and more trees in American forests are succumbing to drought, wildfire, and insect infestation.

Even when the regrowth of trees outpaces their removal, and even when mortality is low, the character of a forest may change. Once primary forest is replaced by younger secondary forest or by single-species plantations, the resulting community may be very different, and generally it is less ecologically valuable.

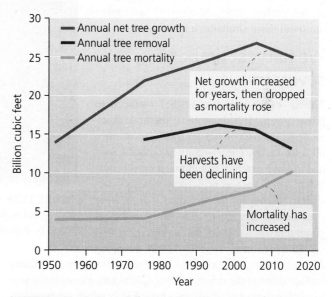

FIGURE 12.15 Annual net tree growth rose from 1952 to 2006, but has fallen since then. The decrease has occurred because tree mortality from causes such as wildfire, drought, and pest outbreaks (p. 323) has risen sharply, even while amounts of timber harvested have fallen. *Data from Oswalt, S.N., et al., coords., 2019. Forest resources of the United States, 2017: A technical document supporting the Forest Service 2020 RPA Assessment. Gen. Tech. Rep. WO-97. Washington, D.C.: USDA Forest Service, Washington Office.*

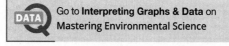

Go to **Interpreting Graphs & Data** on **Mastering Environmental Science**

Plantation forestry has grown

Today's timber industry focuses on production from plantations of fast-growing tree species planted in single-species monocultures (p. 215). All trees in a stand are planted at the same time, so the stands are **even-aged**; that is, all trees are the same age (**FIGURE 12.16**). Stands are cut after a certain number of years (called the *rotation time*), and the land is replanted with seedlings. Plantation forestry is growing, and 7% of the world's forests are now plantations. One-fourth of these plantations feature non-native tree species.

Ecologists and foresters alike view plantations as akin to crop agriculture. Because there are few tree species and little variation in tree age, plantations do not offer habitat to many forest organisms. Even-aged single-species plantations lack the structural complexity that characterizes a mature natural forest as seen in Figure 12.5. Plantations are also vulnerable to outbreaks of pest species such as bark beetles, as we shall soon see. For all these reasons, some harvesting methods aim to maintain **uneven-aged** stands, in which a mix of ages (and often a mix of tree species) creates greater structural diversity and makes the stand more similar to a natural forest.

We harvest timber in several ways

Timber companies use several methods to harvest trees. In the simplest method, **clear-cutting,** all trees in an area are cut at once (**FIGURE 12.17**). Clear-cutting is cost-efficient, and to some extent it can mimic natural disturbance events such as fires, tornadoes, or windstorms. However, the ecological impacts of clear-cutting are severe. An entire ecological community is removed, soil erodes away, and sunlight penetrates to ground level, changing microclimatic conditions. As a result, new types of plants replace those of the original forest. Clear-cutting essentially sets in motion a process of succession (p. 87) in which the resulting climax community may be quite different from the original climax community. For example, when we clear-cut an eastern U.S. oak-hickory forest, the forest that will grow to replace it may be dominated by maples, beeches, and tulip trees.

Concerns about clear-cutting led foresters and the timber industry to develop alternative harvesting methods. Clear-cutting (**FIGURE 12.18a**) remains the most widely practiced method, but alternative approaches involve cutting some trees while leaving others standing. In the **seed-tree** approach (**FIGURE 12.18b**), small numbers of mature and vigorous seed-producing trees are left

FIGURE 12.16 Even-aged management, with all trees of equal age, can be seen in the foreground in a plantation of young trees that are growing after mature trees were clear-cut. Uneven-aged management maintains a mix of tree ages, as seen in the more mature forest in the background.

FIGURE 12.17 Clear-cutting is cost-efficient for timber companies but has ecological consequences. These consequences include soil erosion, water pollution, and altered community composition.

standing so that they can reseed the logged area. In the **shelterwood** approach (also represented by Figure 12.18b), small numbers of mature trees are left in place to provide shelter for seedlings as they grow. These three methods all lead to largely even-aged stands of trees in the next generation.

In contrast, **selection systems** (**FIGURE 12.18c**) allow uneven-aged stand management, because only some trees are cut at any one time. In *single-tree selection,* widely spaced trees are cut one at a time, whereas in *group selection,* small patches of trees are cut. The stand's overall rotation time may be the same as in an even-aged approach, because multiple harvests are made, but the stand remains mostly intact between harvests. Selection systems can maintain much of a

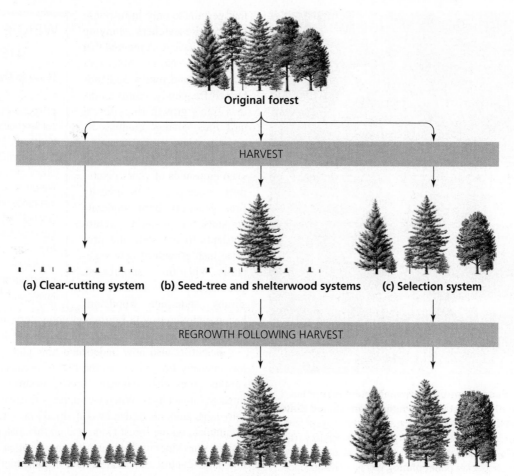

Original forest

HARVEST

(a) Clear-cutting system   **(b) Seed-tree and shelterwood systems**   **(c) Selection system**

REGROWTH FOLLOWING HARVEST

FIGURE 12.18 Foresters have devised several methods to harvest timber. In clear-cutting **(a)**, all trees are cut, extracting a great deal of timber inexpensively but leaving a vastly altered landscape. In seed-tree systems and shelterwood systems **(b)**, a few large trees are left in clear-cuts to reseed the area or provide shelter for seedlings. In selection systems **(c)**, a minority of trees is removed at any one time, while most are left standing.

forest's structural diversity, but they still involve ecological impacts, because moving trucks and machinery over a network of roads and trails to access individual trees compacts the soil and disturbs the forest floor. Timber companies often resist selection methods because they are expensive, and selection methods are unpopular with loggers because they pose more safety risks than clear-cutting.

All timber-harvesting methods disturb soil, alter habitat, affect plants and animals, and modify forest structure and composition. Most methods speed runoff, raise flooding risk, and worsen soil erosion, thereby degrading water quality. When steep hillsides are clear-cut, landslides can result. Finding ways to minimize these impacts is important, because timber harvesting is necessary to obtain the wood products that we all use.

Forest management has evolved over time

For more than half a century, the U.S. Forest Service has nominally been guided by the policy of **multiple use,** meaning that the national forests were to be managed for recreation, wildlife habitat, mineral extraction, and various other uses. In practice, however, timber production was very often the primary use. In recent decades, as people became more aware of the impacts of logging and as development spread across the landscape, many Americans began to plead that public forests be managed for recreation, wildlife, and ecosystem integrity, as well as for timber.

In 1976, Congress passed the **National Forest Management Act.** This law mandated that every national forest draw up plans for renewable resource management based on the concepts of multiple use and maximum sustainable yield and subject to public input under the National Environmental Policy Act (p. 172). Guidelines specified that these plans:

- Consider environmental factors as well as economic ones so that profit alone does not determine harvesting decisions.

- Provide for diverse ecological communities and preserve regional diversity of tree species.

- Ensure research and monitoring of management practices.

- Allow increased harvests only if sustainable.

- Ensure that timber is extracted only where impacts on tree regeneration and soil, watershed, fish, wildlife, recreation, and aesthetic resources can be assessed and minimized.

Bands mark woodpecker nest and roost trees

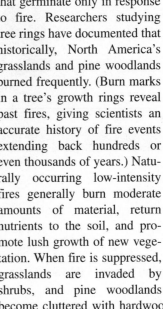

FIGURE 12.19 Ecosystem-based management is practiced in this longleaf pine forest in the southeastern United States. Foresters and biologists restore and nurture mature longleaf pine trees while burning and removing brush from the understory. This habitat is home to unique species such as the endangered red-cockaded woodpecker **(inset photo)**, whose trees used for nesting and roosting are identified by bands painted on the trunks.

Following passage of the National Forest Management Act, Forest Service scientists and managers developed new programs to manage non-game wildlife and launched ecological restoration projects to recover plant and animal communities that had been lost or degraded (**FIGURE 12.19**). Timber-harvesting methods were brought more in line with ecosystem-based management goals. A set of approaches called **new forestry** called for timber cuts that mimic natural disturbances. For example, "sloppy clear-cuts" that leave a variety of trees standing were intended to mirror the changes a forest might experience if hit by a severe windstorm.

In recent years, forest management priorities have varied. Some presidential administrations have freed forest managers from requirements of the National Forest Management Act, granting them more flexibility in managing forests but loosening environmental protections and restricting public oversight, whereas other presidential administrations have restored or enhanced protections for forests and the public.

Fire can hurt or help forests

A major element of forest management involves how to handle wildfire. For more than a century, the Forest Service and other agencies suppressed fire whenever and wherever it broke out. Yet fire is a natural process on which many species and ecological communities depend. Some plants have seeds that germinate only in response to fire. Researchers studying tree rings have documented that historically, North America's grasslands and pine woodlands burned frequently. (Burn marks in a tree's growth rings reveal past fires, giving scientists an accurate history of fire events extending back hundreds or even thousands of years.) Naturally occurring low-intensity fires generally burn moderate amounts of material, return nutrients to the soil, and promote lush growth of new vegetation. When fire is suppressed, grasslands are invaded by shrubs, and pine woodlands become cluttered with hardwood understory. Invasive plants move in, and animal diversity and abundance decline.

Scientists also now understand how putting out frequent low-intensity fires increases the risk of occasional large catastrophic fires that damage forests, destroy property, and threaten human lives. When we suppress fire, we allow unnaturally large amounts of dead wood, dried grass, and leaf litter to accumulate on the forest floor, and all this material eventually can become kindling for a catastrophic fire that grows too big and hot to control. Severe fires have become more numerous in recent years (**FIGURE 12.20**), and fire control expenses now eat up the majority of the U.S. Forest Service's budget.

At the same time, increased residential development alongside forested land—in the **wildland-urban interface**—is placing more homes in fire-prone situations. Some fires in the wildland-urban interface have brought death and tragedy.

WEIGHING the issues

How to Handle Fire?

Can you suggest solutions to help protect people's homes in the wildland-urban interface while improving the ecological condition of forests? Should people who choose to live in homes in fire-prone areas pay a premium for insurance against fire damage? Should homeowners in fire-prone areas be fined if they don't adhere to fire-prevention maintenance guidelines? Should new home construction be allowed in areas known to be fire-prone?

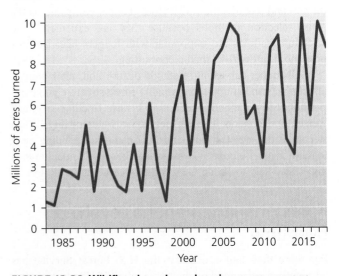

FIGURE 12.20 Wildfires have been burning more acreage across the United States. Fuel buildup from decades of fire suppression has contributed to this trend. *Data from National Interagency Fire Center.*

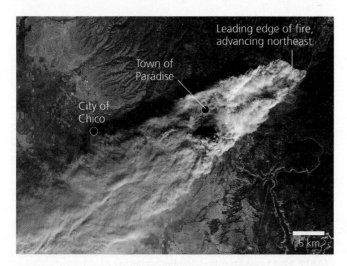

FIGURE 12.21 The Camp Fire destroyed Paradise, California, a town in the wildland-urban interface, killing at least 85 people. Satellite imagery from November 8, 2018 (natural colors with infrared highlights), shows the fire advancing northeast after ravaging Paradise, with smoke blowing southwest. *Landsat data courtesy of U.S. Geological Survey, Department of the Interior/USGS. Landsat product ID: LC08-L1TP-044032-20181108-20181116-01-T1.*

In 2018, the fast-moving Camp Fire destroyed the town of Paradise, California, and damaged several other communities, killing at least 85 people and destroying 19,000 buildings worth more than $16 billion (**FIGURE 12.21**).

To reduce fuel loads, protect property, and improve the condition of forests, land management agencies now burn areas of forest deliberately with low-intensity fires under carefully controlled conditions (**FIGURE 12.22**). Such **prescribed fire** clears away fuel loads, nourishes the soil with ash, and encourages the vigorous growth of new vegetation. Because prescribed burning (also called controlled burning) is time-intensive and sometimes misunderstood by politicians and the public, prescribed fires are conducted on only a small proportion of land (just more than 2 million acres per year). As a result, vast areas of forests in the United States remain vulnerable to catastrophic fires.

In the wake of major fires in California in 2003, the U.S. Congress passed the Healthy Forests Restoration Act. Although this law encouraged some prescribed fire, it primarily promoted the physical removal of small trees, underbrush, and dead trees by timber companies. The removal of dead trees, or snags, following a natural disturbance (such as fire, windstorm, insect damage, or disease) is called **salvage logging.** From a short-term economic standpoint, salvage logging may seem to make good sense. However, snags have immense ecological value; for example, the insects that decay them provide food for wildlife, and many animals rely on holes in snags for nesting and roosting. Removing timber from recently burned land can also cause soil erosion, impede forest regeneration, and promote additional wildfire.

Climate change and pest outbreaks are altering forests

Global climate change (Chapter 18) is now worsening wildfire risk in North America, just as it is in the Amazon, by bringing warmer and drier weather (**FIGURE 12.23a**, p. 324). California, for example, has suffered a series of uncontrollable and destructive forest fires in recent years amid long-term drought in a changing climate. Parts of Europe, Australia, and Asia are also seeing more frequent and severe fires. In the Amazon in recent years, unprecedented fires have destroyed large areas of rainforest—a wet forest biome that is *not* adapted to fire. Tropical rainforests in Indonesia and Central America have also suffered severe fire damage during hot, dry spells resulting from strong El Niño conditions (p. 427) and climate change. Likewise, Canada's boreal forests have suffered vast and unprecedented fires in recent years.

Deforestation can worsen the impacts of climate change, because when forests are removed across a large region, humidity and transpiration are reduced and less precipitation falls, making the region drier. Many areas of the Amazon rainforest, particularly its edges, are much drier than they used to be, which intensifies the risk of rainforest destruction by wildfire. In this way, deforestation and climate change create a dangerous positive feedback loop.

Climate change is also promoting outbreaks of pest insects—and pests such as bark beetles, which feed within the bark of conifer trees, are adding to wildfire risk. Bark beetles attract one another to weakened trees and attack in large numbers, eating tissue, laying eggs, and bringing with them a small army of fungi, bacteria, and other pathogens (**FIGURE 12.23b**, p. 324). Bark beetle infestations can wipe out vast numbers of trees with amazing speed. Since the 1990s, beetle outbreaks have devastated tens of millions of acres of forest in western North America, killing tens of *billions* of conifer trees and leaving them as fodder for fires.

Today's unprecedented outbreaks result from milder winters that allow beetles to overwinter farther north and from warmer summers that speed up their feeding and reproduction. In Alaska, beetles have switched from a two-year life cycle to a one-year cycle. In parts of the Rocky Mountains, they now produce two broods per year instead of one. Moreover, drought has stressed and weakened trees in many regions, making them vulnerable to attack.

FIGURE 12.22 Prescribed fire helps promote forest health and prevent larger damaging fires. Forest Service staff are shown here conducting a carefully controlled low-intensity burn.

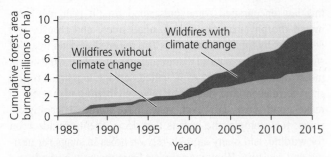

(a) Climate change drives wildfires

(b) Bark beetles (inset) kill more trees in a warmer climate

FIGURE 12.23 Forests face intensifying threats as temperatures rise. Nearly one-half of wildfires in the western United States **(a)** are now thought to be caused by climate change. Warm summers and mild winters favor bark beetles **(b)**, which are killing vast areas of trees throughout the West, especially in even-aged plantations. *Source: (a) Abatzoglou, J.T., and A.P. Williams, 2016. Impact of anthropogenic climate change on wildfire across western US forests. Proc Nat Acad Sci 113: 11770–11775, as portrayed by U.S. Global Change Research Program, 2017. Climate science special report: Fourth national climate assessment, Volume I (Wuebbles, D.J., et al., eds.). USGCRP, Washington, D.C .*

On top of the effects of climate change, past forest management has resulted in large areas of even-aged plantation forests dominated by single species that the beetles prefer, and many trees in these forests are now at a prime age for beetle infestation.

As climate change interacts with pests, diseases, fire, deforestation, and management strategies, our forests are being altered in profound ways. Already many dense, moist forests—from British Columbia to the Amazon—have undergone ecological regime shifts (p. 88) and have been replaced by drier woodlands, shrublands, or grasslands. Further changes to our forests could create novel types of ecosystems not seen today.

Sustainable forestry is gaining ground

Each of us can help address the many challenges to our forests by choosing certified sustainable items when we shop for wood or paper products. Certification organizations examine timber-harvesting practices and rate them against criteria for sustainability. These organizations then grant **sustainable forest certification** to forests, companies, and products produced using practices they judge to be sustainable. Such practices reduce impacts on land, water, and wildlife; conserve resources rather than depleting them; and enable people to make a living from forest products over the long term. As consumers, we can also choose to buy reclaimed or salvaged wood, which lowers the market demand for cutting trees.

Among certification organizations, the Forest Stewardship Council (FSC) is considered to have the strictest standards. To gain FSC certification, timber-harvesting operations are required to protect rare species and sensitive habitats, safeguard water sources, control erosion, minimize pesticide use, and maintain the diversity of the forest and its ability to regenerate after harvesting. FSC certification rests on 10 general principles (**TABLE 12.1**) and 56 more-detailed criteria.

As you turn the pages of this textbook, you are handling paper made from trees that were grown, managed, harvested, and processed using FSC-certified practices from sustainably managed forests. Moreover, the paper for this textbook is "chain-of-custody certified," meaning that all steps in the lifecycle of the paper's production—from timber harvest to transport to pulping to production—have met strict standards.

The number of FSC-certified forests, companies, and products is growing, tied in part to the increasing construction of green buildings (pp. 348–350). As of 2019, slightly more than 6% of the world's forests managed for timber production—200 million ha (494 million acres) in 84 nations—were FSC-certified. The range of certification programs is also growing. In Brazil, for example, a new program by the Sustainable Agriculture Network certifies cattle ranches that meet sustainability guidelines that help protect forest.

Research indicates that certification can be effective. One recent study showed that FSC certification in Indonesia reduced deforestation there by 5% between 2000 and 2008 while also improving air quality, respiratory health, firewood availability, and nutrition for people living near timber-harvesting areas.

TABLE 12.1 Ten Principles of Forest Stewardship Council (FSC) Certification

To receive FSC certification, forest product companies must:
1. Comply with all laws and treaties.
2. Show uncontested, clearly defined, long-term land rights.
3. Recognize and respect indigenous peoples' rights.
4. Maintain or enhance long-term social and economic well-being of forest workers and local communities, and respect workers' rights.
5. Use and share benefits derived from the forest equitably.
6. Reduce environmental impacts of logging and maintain the forest's ecological functions.
7. Continuously update an appropriate management plan.
8. Monitor and assess forest condition, management activities, and social and environmental impacts.
9. Maintain forests of high conservation value.
10. Promote restoration and conservation of natural forests.

Source: Forest Stewardship Council.

Pursuing sustainable forestry practices is often more costly for producers, but they recoup these costs when consumers pay more for certified products. And in the long term, sustainable practices conserve the resource base, thus holding down costs for everyone. When we ask businesses if they carry certified wood or paper, we make them aware of our preferences and help drive demand for these products. For example, consumer demand led Home Depot and other major retailers to carry sustainable wood, and these retailers' purchasing decisions influence timber-harvesting practices around the world. If certification standards are kept strong, then we as consumers can help promote sustainable forestry practices by exercising choice in the marketplace.

Parks and Protected Areas

As our world fills with more people consuming more resources, the sustainable management of forests and other ecosystems becomes ever more important. So does our need to preserve functional ecosystems by setting aside tracts of undisturbed land to remain forever undeveloped. For ethical reasons as well as pragmatic ecological and economic ones, people worldwide have chosen to set aside areas of land in perpetuity to be protected from development. Today, nearly 15% of the world's land area has been granted some degree of protection in various types of parks and reserves.

Why create parks and reserves?

People establish parks and protected areas because these areas offer us:

- Inspiration from scenic beauty.
- Hiking, fishing, hunting, kayaking, bird-watching, and other recreation.
- Revenue from ecotourism (pp. 70, 281–282).
- Health, peace of mind, exploration, wonder, and spiritual solace.
- Utilitarian benefits and ecosystem services (such as how forested watersheds provide cities and towns clean drinking water and a buffer against floods).
- Refuges for biodiversity, helping maintain populations, habitats, communities, and ecosystems.

Parks and reserves were pioneered in the United States

Preservation has been part of the American psyche ever since John Muir rallied support for saving scenic lands in the Sierras (p. 138). Today, Americans enjoy several types of lands managed by federal agencies for the good of the public.

National parks The majestic scenery of the American West persuaded U.S. leaders to create the world's first **national parks,** public lands protected from resource extraction and development but open to nature appreciation and recreation (**FIGURE 12.24**). Yellowstone National Park

(a) Crater Lake National Park, Oregon

(b) Arches National Park, Utah

(c) Great Smoky Mountains National Park, North Carolina and Tennessee

FIGURE 12.24 The awe-inspiring beauty of America's diverse national parks draws millions of people for recreation and wildlife-watching.

was established in 1872, followed by Sequoia, General Grant (now Kings Canyon), Yosemite, Mount Rainier, and Crater Lake National Parks. The Antiquities Act of 1906 gave the U.S. president authority to declare selected public lands as *national monuments,* which may later become national parks.

The National Park Service was created in 1916 to administer the growing system of parks and national monuments, which today numbers 418 sites totaling more than 34 million ha (84 million acres) and includes national historic sites, national recreation areas, national seashores, and other areas. The parks receive more than 300 million reported recreation visits each year—almost one per U.S. resident. Because U.S. national parks are open to everyone and showcase the nation's natural beauty in a democratic way, writer Wallace Stegner famously called them "the best idea we ever had."

National wildlife refuges Another type of protected area in the United States is the **national wildlife refuge** (**FIGURE 12.25**). The system of national wildlife refuges, begun in 1903 by President Theodore Roosevelt, now totals more than 560 sites comprising 39 million ha (96 million acres), plus tens of millions of acres of ocean and islands owned or co-managed in marine refuges and national monuments. There are wildlife refuges within an easy drive of nearly every major U.S. city.

The U.S. Fish and Wildlife Service administers the national wildlife refuges, which serve as havens for wildlife but also in many cases encourage hunting, fishing, wildlife observation, photography, environmental education, and other public uses. Some wildlife advocates find it ironic that hunting is allowed at many refuges, but hunters have long been in the forefront of the conservation movement and have traditionally supplied the bulk of funding for land acquisition and habitat management. Many refuges are managed for waterfowl and other game animals, but managers increasingly consider non-game species and work to maintain and restore habitats and ecosystems.

Wilderness areas In response to the public's desire for undeveloped areas of land, Congress in 1964 passed the Wilderness Act, which allowed some areas of existing federal lands to be designated as **wilderness areas.** These areas are off-limits to development but are open to hiking, nature study, and other low-impact public recreation (**FIGURE 12.26**). Congress declared that wilderness areas were needed

"to assure that an increasing population, accompanied by expanding settlement and growing mechanization, does not occupy and modify all areas . . . leaving no lands designated for preservation and protection in their natural condition." Despite these words, some preexisting extractive land uses, such as grazing and mining, were allowed to continue within some wilderness areas as a political compromise so the act could be passed.

Wilderness areas may be established within national forests, national parks, national wildlife refuges, and land managed by the Bureau of Land Management (BLM). They are overseen by the agencies that administer those areas. Overall, the nation has 765 wilderness areas totaling 44 million ha (109 million acres), covering 4.5% of U.S. land area (2.7% if Alaska is excluded).

FIGURE 12.25 National wildlife refuges offer safe havens for wildlife. For example, thousands of sandhill cranes spend the winter at Bosque del Apache National Wildlife Refuge in New Mexico.

FIGURE 12.26 Wilderness areas are preserved and protected from development. Here visitors enjoy kayaking on Lake Superior at Pictured Rocks National Lakeshore in Michigan's Upper Peninsula.

Not everyone supports setting land aside

The restriction of activities in parks, refuges, and wilderness areas has generated some opposition to U.S. land protection policies, especially among residents and policymakers in western states. When these former territories were admitted to the union as states, the federal government retained jurisdiction over much of the acreage within their borders. Idaho, Oregon, and Utah today control less than half the land within their borders, and in Nevada, 80% of the land is federally managed. Some western state leaders have sought to obtain public federal land and to facilitate private resource extraction and development on it. They have been encouraged by industries that extract timber, minerals, and fossil fuels, as well as by farmers, ranchers, trappers, and mineral prospectors at the grassroots level. Both wealthy advocates of privatization and people who make their living off the land have supported efforts to secure local private control of public lands, limit government regulation, promote the extraction of resources, and expand motorized recreation on public lands.

Occasionally, debates over these issues have led to violence. In 2016, a ragtag group of armed militants staged a takeover of the Malheur National Wildlife Refuge in Oregon and held the headquarters for 41 days. The occupation was meant to protest government control over land, but the events traumatized the local community, and most people both locally and nationally opposed the occupation and voiced support for public lands.

In 2017, the Trump administration moved to greatly reduce the size of Bears Ears and Grand Staircase–Escalante National Monuments, two national monuments in Utah that had been created by Presidents Bill Clinton and Barack Obama. Utahns were evenly split over the action, and most Americans opposed the loss of national monument acreage.

Similar debates take place wherever parks are established. In East Africa, for example, many Maasai people living near Serengeti National Park resent the park because it displaced them from their traditional land (pp. 302–303). Many other people living near the park want a highway to be built through it to provide them better mobility and access to trade (pp. 273–274).

Parks and protected areas regularly bring substantial economic benefits to people who live nearby through ecotourism. However, individuals who do not have jobs related to parks and tourism may feel that a park restricts their economic opportunities. For this reason, proponents of protected areas increasingly try to ensure that the economic benefits of preserving natural land are spread equitably among people in nearby communities.

Internationally, protected areas often serve the interests of indigenous people. In Brazil and other Latin American nations, some rainforest parks and reserves encompass regions occupied by indigenous tribes. The tribes are thereby granted some protection from conflict with miners, farmers, and settlers, and this protection helps them continue their traditional ways of life. In the most remote reaches of the Amazon, government agencies have gone to great lengths to insulate and protect some of the world's last remaining uncontacted indigenous tribespeople.

Many agencies and groups protect land

Efforts to set aside and manage public land at the national level are paralleled at state, regional, and local levels. Each U.S. state has agencies that manage resources and provide for recreation on public lands, as do many counties and municipalities. When Mackinac Island, the nation's second national park, was transferred to the state of Michigan, it became the first officially designated state park in the nation. Another example is Adirondack State Park in New York. In the 19th century, the state of New York established this park to protect land in a mountainous area where streams converge to form the Hudson River, which thereafter flows south past Albany to New York City. State leaders saw the need for river water to power industries, keep canals filled, and provide drinking water, and their far-sighted decision to create the park has paid dividends through the years. Today, there are nearly 7000 state parks across the United States, along with regional parks, county parks, and others.

Private non-profit groups also preserve land. **Land trusts** are local or regional organizations that purchase ecologically important tracts of land to preserve them for future generations. The Nature Conservancy, the world's largest land trust, uses science to select areas and ecosystems in greatest need of protection. Smaller land trusts are diverse in their missions and methods. Nearly 1400 local and state land trusts in the United States together own 3.3 million ha (8.1 million acres) and have helped preserve an additional 19.6 million ha (48.3 million acres), including scenic areas such as Big Sur on the California coast, Jackson Hole in Wyoming, and Maine's Mount Desert Island. Moreover, thousands of local volunteer groups (often named "Friends of" an area) have organized from the grassroots to help care for protected lands.

Protected areas are increasing internationally

Many nations have established protected areas and are benefiting from ecotourism as a result—from Kenya and Tanzania (Chapter 11) to Costa Rica (Chapter 6) to Belize to Ecuador to India to Australia. Worldwide area in protected parks and reserves has increased more than tenfold since 1970, and today the world's 239,000 protected areas cover 14.9% of the planet's land area. However, parks in developing nations do not always receive funding adequate to manage resources, provide for recreation, and protect wildlife from poaching and trees from logging. As a result, many of the world's protected areas are merely "paper parks"—protected on paper but not in reality.

Brazil has designated more land area for protection than any other nation. Most of it lies deep in the Amazon rainforest, in large tracts set aside for indigenous tribes of people who live in the forest. Altogether about 30% of the Brazilian Amazon is afforded—on paper, at least—some degree of protection. A recent scientific study of the Brazilian Amazon found that protected areas suffered 10 times less deforestation than did non-protected areas. However, policymakers tend to locate protected areas in regions that are remote and that face fewer development pressures. Thus, to obtain a fair comparison, the researchers controlled for location variables and found that

FIGURE 12.27 Biosphere reserves couple preservation with sustainable development. Each biosphere reserve comprises three zones: core, buffer, and transition. The photo illustrating the transition zone shows one example of sustainable use of forest products: Women process and sell Maya nuts harvested from rainforest trees in the Maya Biosphere Reserve in Guatemala. Here, management of forest concessions by local communities has helped reduce illegal logging.

Biodiversity protection

Core area

Buffer zone

Transition zone

Sustainable agriculture, small settlements

Limited development, research, education, ecotourism

only about half the difference in deforestation rates was due to protected status, with the other half due to location.

Many protected areas today incorporate sustainable use to benefit local people. The United Nations oversees **biosphere reserves (FIGURE 12.27)**, areas with exceptional biodiversity that couple preservation with sustainable development (p. 155). Each biosphere reserve consists of (1) a core area that preserves biodiversity; (2) a buffer zone that allows local activities and limited development; and (3) an outer transition zone where agriculture, human settlement, and other land uses are pursued sustainably.

The Maya Biosphere Reserve in Guatemala is an example. Rainforest here is protected in core areas, while in the transition zone timber harvesting takes place in concessions, some of which are FSC-certified. A study by the non-profit Rainforest Alliance found that FSC certification in the Maya Reserve gave people economic incentives to conserve the forest. In fact, the study found that rates of deforestation and wildfire were lower in the FSC-certified areas where sustainable logging occurred than in the core area that was supposed to be fully protected.

World heritage sites are another type of international protected area. Nearly 1100 sites in more than 160 countries are listed for their natural or cultural value. One such site is a mountain gorilla reserve shared by three African countries. This reserve, which integrates national parklands of Rwanda, Uganda, and the Democratic Republic of Congo, is also an example of a transboundary park, a protected area overlapping national borders. A North American example of a transboundary park is Waterton–Glacier National Park on the Canadian–U.S. border. Some transboundary reserves function as "peace parks," helping ease tensions by acting as buffers between nations with historical boundary disputes. That is the case with Peru and Ecuador, as well as Costa Rica and Panama. Some people hope that peace parks might also help resolve conflicts between Israel and its neighbors.

Beyond all these efforts on land, the importance of conserving the oceans' natural resources is leading us to establish protected areas and reserves in the oceans and along coasts (pp. 446–447). Today, 7.3% of the world's marine and coastal waters fall within designated protected areas.

How much protected area is needed?

There is no single right or wrong answer to the very complicated question of how much area should be protected. The current official target for 2020 among nations under the Convention on Biological Diversity (p. 297) is for 17% of land area and 10% of marine and coastal area to be set aside for protection. (The convention also calls for restoring at least 15% of degraded areas, reducing the loss of natural habitats, and taking other measures.) However, as biodiversity loss proceeds, climate change intensifies, and human population and consumption continue to grow, many scientists and biodiversity advocates are calling for more ambitious targets.

Most notably, the esteemed scientist E. O. Wilson (p. 282) is helping lead a campaign for one-half of Earth's area to be protected. Citing extensive research in conservation biology, Wilson and other scientists with the Half-Earth Project argue that "the only way to reverse the extinction crisis is through a conservation moonshot . . . to keep half the land and half the sea of the planet as wild and protected from human intervention or activity as possible." Such an effort, they recognize, will require collaboration with and support from local people.

Habitat fragmentation makes preserves more vital

Protecting large areas of land has taken on new urgency now that scientists understand the threats posed to species and ecosystems by habitat fragmentation (p. 289; see **THE SCIENCE BEHIND THE STORY**, pp. 330–331). Expanding agriculture, spreading cities, highways, logging, and other impacts routinely divide large contiguous expanses of habitat into small, disconnected ones (see Figure 11.14; p. 289). Forests are being fragmented everywhere today as a result of logging (**FIGURE 12.28a**) and other factors, including agriculture and residential development (**FIGURE 12.28b**). Even in areas where forest cover is increasing (such as the northeastern United States), forests are becoming fragmented into ever-smaller parcels.

When forests are fragmented, many species suffer. Bears, mountain lions, and other animals that need large areas of habitat may disappear. For other species, the problem may lie with **edge effects,** impacts that result because the conditions along a fragment's edge differ from conditions in the interior. Birds that thrive in the interior of forests may fail to reproduce near the edge of a fragment (**FIGURE 12.28c**). Their nests often are attacked by predators and parasites that favor open habitats or travel along habitat edges. Because of edge effects, avian ecologists judge forest fragmentation to be a main reason populations of many North American songbirds are declining.

A 2015 research study in the journal *Science* concluded that across the world, 70% of forested area now lies within 1 km (0.62 mi) of an edge. As a result, the vast majority of the world's forested area is now being directly influenced by human activities, modified microclimates, and the incursion of non-forest species.

Insights from islands warn us of habitat fragmentation

In assessing the impacts of habitat fragmentation on populations, ecologists and conservation biologists have leaned on concepts from **island biogeography theory.** Introduced by ecologist Robert MacArthur and biologist E. O. Wilson in 1963, this theory explains how species come to be distributed among oceanic islands. Since MacArthur and Wilson's work, researchers have applied it widely on land to "habitat islands"—patches of one habitat type isolated within "seas" of others.

Island biogeography theory explains how the number of species on an island results from a balance between the number added by immigration and the number lost through local extinction (extirpation; p. 284). It predicts an island's

FIGURE 12.28 Forest fragmentation has ecological consequences. Fragmentation results from **(a)** clear-cutting and **(b)** agriculture and residential development. Fragmentation affects forest-dwelling species such as the **(c)** wood thrush, whose nests are parasitized by cowbirds that thrive in surrounding open country. *Source: (b) Curtis, J.T., 1956. The modification of mid-latitude grasslands and forests by man. In Thomas, W.L., Jr. (Ed.), Man's role in changing the face of the earth. ©1956. Used by permission of the publisher, University of Chicago Press.*

(a) Fragmentation from clear-cuts in British Columbia

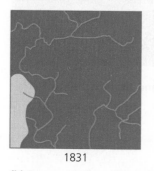

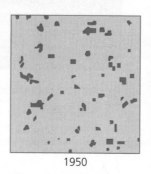

| 1831 | 1882 | 1950 |

(b) Fragmentation of wooded area (green) in Cadiz Township, Wisconsin

(c) Wood thrush

THE SCIENCE behind the story

What Happens When a Forest Is Fragmented?

Dr. Thomas Lovejoy

What happens to animals, plants, and ecosystems when we fragment a forest? A massive experiment in the middle of the Amazon rainforest is helping scientists learn the answers.

Stretching across 1000 km² (386 mi²) of tropical rainforest north of the Brazilian city of Manaus, the Biological Dynamics of Forest Fragments Project (BDFFP) is the world's largest and longest-running experiment on forest fragmentation. For years, hundreds of researchers have muddied their boots here, publishing more than 700 scientific research papers, graduate theses, and books.

The story begins back in the 1970s as conservation biologists debated the SLOSS question (p. 332) and how to apply island biogeography theory to forested landscapes being fragmented by development. Biologist Thomas Lovejoy decided that some good, hard data were needed. He conceived a huge experiment to test ideas about forest fragmentation and established it in the heart of the biggest primary rainforest on the planet—South America's Amazon rainforest.

Farmers, ranchers, loggers, and miners were streaming into the Amazon then, and deforestation was rife. If scientists could learn how large fragments had to be to retain their species, they could work more productively with policymakers to preserve forests in the face of development pressures.

Lovejoy's team of Brazilians and Americans worked out a deal with Brazil's government: Ranchers could clear some forest within the study area if they left square plots of forest standing as fragments within those clearings. By this process, 11 fragments of three sizes (1 ha [2.5 acres], 10 ha [25 acres], and 100 ha [250 acres]) were left standing, isolated as "islands" of forest, surrounded by "seas" of cattle pasture (**FIGURE 1**). Each fragment was fenced to keep cattle out. Then, 12 study plots (of 1, 10, 100, and 1000 ha) were established

within the large expanses of continuous forest still surrounding the pastures to serve as control plots against which the fragments could be compared.

Besides comparing treatments (fragments) and controls (continuous forest), the project also surveyed populations in fragments before and after they were isolated. These data on trees, birds, mammals, amphibians, and invertebrates showed declines in the diversity of most groups (**FIGURE 2**).

As researchers studied the plots over the years, they found that small fragments lost more species, and lost them faster, than large fragments—just as island biogeography theory predicts. To slow down species loss by 10 times, researchers found that a fragment needs to be 1000 times bigger. Even 100-ha fragments were not large enough for some animals and lost half their species in less than 15 years. Monkeys died out because they need large ranges. So did colonies of army ants and the birds that follow them to eat insects scared up by the ants as they swarm across the forest floor.

Fragments distant from continuous forest lost more species, but data revealed that even very small openings can stop organisms from dispersing to recolonize fragments. Many understory birds adapted to deep interior forest would not traverse cleared

FIGURE 1 Experimental forest fragments of 1, 10, and 100 ha were created in the BDFFP.

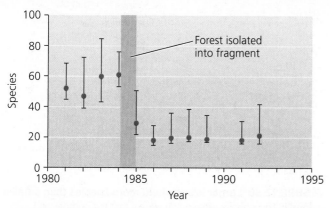

FIGURE 2 Species richness of understory birds declined in this 1-ha forest plot after it was isolated as a fragment in 1984. Error bars show statistical uncertainty around mean estimates. *Data from Ferraz, G., et al., 2003. Rates of species loss from Amazonian forest fragments.* Proc. Natl. Acad. Sci. USA *100: 14069–14073. ©2003 National Academy of Sciences. By permission.*

Go to **Interpreting Graphs & Data** on **Mastering Environmental Science**

areas of only 30–80 m (100–260 ft). Distances of just 15–100 m (50–330 ft) were insurmountable for some bees, beetles, and tree-dwelling mammals.

Soon, a complication ensued: Ranchers abandoned many of the pastures because the soil was unproductive, and these areas began filling in with secondary forest, making the fragments less like islands. However, this change led to new insights. Researchers learned that secondary forest can act as a corridor for some species, allowing them to disperse from mature forest and recolonize fragments where they had disappeared. By documenting which species did this, scientists learned which may be more resilient to fragmentation.

The secondary forest habitat also introduced new species—generalists adapted to disturbed areas. Frogs, leaf-cutter ants, and small mammals and birds that thrive in second-growth areas soon became common in adjacent fragments. Open-country butterfly species moved in, displacing interior-forest butterflies.

The invasion by open-country species illustrated one type of edge effect (p. 329), but there were more. Edges receive more sunlight, heat, and wind than interior forest, which can kill trees adapted to the dark, moist interior. Tree death releases carbon dioxide, worsening climate change. Sunlight also promotes growth

of vines and shrubs that create a thick, tangled understory along edges. As BDFFP researchers documented these impacts, they found that many edge effects extended deep into the forest (**FIGURE 3**). Small fragments essentially became "all edge."

The results on edge effects are relevant for all Amazonia, because forest clearance and road construction create an immense amount of edge. A satellite image study in the 1990s estimated that for every 2 acres of land deforested, 3 acres were brought within 1 km of a road or pasture edge.

BDFFP scientists emphasize that impacts across the Amazon will be more severe than those revealed by their experiments. That is because most real-life fragments (1) are not protected from hunting, logging, mining, and fires; (2) do not have secondary forest to provide connectivity; (3) are not near large tracts of continuous forest that provide recolonizing species and maintain humidity and rainfall; and (4) are not square in shape and thus feature more edge. Scientists say their years of data argue for preserving numerous tracts of Amazonian forest that are as large as possible.

The BDFFP has inspired other large-scale, long-term experiments on forest fragmentation elsewhere in the world; such studies are now running in Kansas, South Carolina, Australia, Borneo, Canada, France, and the United Kingdom. These projects are providing data showing how smaller area, greater isolation, and more edge are affecting species, communities, and ecosystems. Thus far, they indicate that fragmentation decreases biodiversity by 13–75% while altering nutrient cycles and other ecological processes—and that the effects magnify with time. Insights from these projects are helping scientists and policymakers strategize how best to conserve biodiversity amid continued pressures on forests.

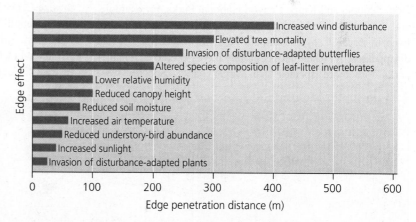

FIGURE 3 Edge effects can extend far into the interior of forest fragments. *Data from studies summarized in Laurance, W.F., et al., 2002. Ecosystem decay of Amazonian forest fragments: A 22-year investigation.* Conservation Biology *16: 605–618, Fig. 3. Adapted by permission of John Wiley & Sons, Inc.*

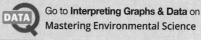 Go to **Interpreting Graphs & Data** on **Mastering Environmental Science**

species richness based on its size and its distance from the mainland:

- *Distance effect:* The farther an island lies from a continent, the fewer species tend to find and colonize it. Thus, remote islands host few species because of low immigration rates (**FIGURE 12.29a**).

- *Target size:* Larger islands have higher immigration rates because they present fatter targets for dispersing organisms to encounter (**FIGURE 12.29b**).

- *Differential extinction:* Larger islands have lower extinction rates because more space allows for larger populations, which are less likely to drop to zero by chance (**FIGURE 12.29c**).

Together, the latter two trends give large islands more species than small islands. Large islands also contain more species because they tend to possess more habitats than smaller islands. Very roughly, the number of species on an island is expected to double as island size increases 10-fold. This phenomenon is illustrated by a **species-area curve** (**FIGURE 12.30**).

The same patterns also hold true for terrestrial habitat islands, such as forests fragmented by logging and road building. Small "islands" of forest lose diversity fastest, starting with large animal species that were few in number to begin with. One of the first researchers to show this pattern experimentally was William Newmark, who as a graduate student at University of Michigan in 1983 examined historical records of mammal sightings in North American national parks. The parks, increasingly surrounded by development, are islands of natural habitat isolated by farms, ranches, roads, and cities. Newmark found that most parks were missing species they had held previously. The red fox and river otter had vanished from Sequoia and Kings Canyon National Parks, for example, and the white-tailed jackrabbit and spotted skunk no longer lived in Bryce Canyon National Park. In all, 42 species had disappeared. As island biogeography theory predicted, smaller parks lost more species than larger parks. Species were

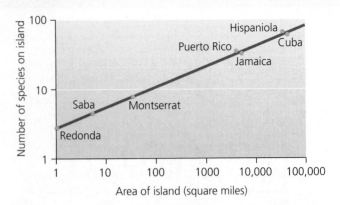

FIGURE 12.30 Larger islands hold more species than smaller islands. This species-area curve shows that the number of amphibian and reptile species on Caribbean islands increases with island area. The increase is not linear, but logarithmic; note the scales of both the axes. *Data from MacArthur, R.H., and E.O. Wilson. The theory of island biogeography. ©1967 Princeton University Press, 1995 renewed PUP. Reprinted by permission of Princeton University Press.*

disappearing because the parks were too small to sustain their populations, Newmark concluded, and because the parks had become too isolated to be recolonized by new arrivals.

Reserve design has consequences for biodiversity

Because habitat fragmentation is a central issue in biodiversity conservation and because there are limits on how much land can feasibly be set aside, conservation biologists have argued heatedly about whether it is better to make reserves large in size and few in number or many in number but small in size. This so-called **SLOSS** debate (for "**s**ingle **l**arge **o**r **s**everal **s**mall") is complex, but it seems clear that large species that roam great distances, such as wildebeest and zebras (Chapter 11), benefit more from the "single large" approach to reserve design than from the "several small" approach.

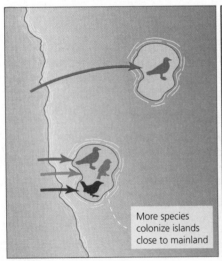

More species colonize islands close to mainland

(a) Distance effect

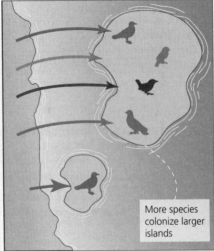

More species colonize larger islands

(b) Target size

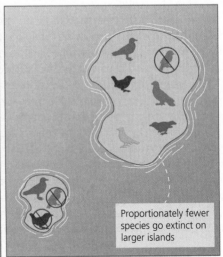

Proportionately fewer species go extinct on larger islands

(c) Differential extinction

FIGURE 12.29 Islands that are larger or closer to a mainland tend to support more species than islands that are smaller or farther away.

In contrast, creatures such as insects that live as larvae in small areas may do just fine in a number of small isolated reserves—if they can disperse as adults by flying from one reserve to another. The SLOSS debate was one motivation for establishing the Amazonian forest fragmentation project described in this chapter's **Science behind the Story.**

A related issue is how effectively **corridors** of protected land allow animals to travel between islands of habitat (**FIGURE 12.31**, and also see Figure 5.11, p. 118). In theory, connections between fragments provide animals access to more habitat and encourage gene flow to maintain populations over the long term. For these reasons, many land managers now try to connect new reserves to existing reserves.

Climate change threatens protected areas

Today, global climate change (Chapter 18) threatens our investments in protected areas. By increasing wildfire risk and intensifying droughts, floods, and storms, climate change is affecting organisms and ecosystems both inside and outside of reserve boundaries. Moreover, as temperatures become warmer, species ranges are shifting toward cooler climes: toward the poles and upward in elevation (p. 505). In a landscape of fragmented habitat, some organisms may be unable to travel or disperse from one fragment to another. Species we had hoped to protect in parks may, in a warming world, become trapped in them. High-elevation species face the most risk

FIGURE 12.31 We can connect two areas of forest with a corridor. Shown is a wildlife overpass in Canada's Banff National Park that allows animals to cross over the Trans-Canada Highway.

from climate change, because they have nowhere to go once a mountaintop becomes too warm or dry (p. 505). Viewed in this light, corridors to allow movement from place to place become still more important. In response to these challenges, conservation biologists are now looking beyond parks and protected areas as they explore strategies for saving biodiversity.

CENTRAL CASE STUDY
connect & continue

TODAY, we continue to lose forests in the Amazon and around the world. The slowing of deforestation in the Brazilian Amazon during the decade following 2004 was encouraging, but renewed commitment will be needed to reduce forest loss further. Atop the traditional impacts of human clearance for agriculture, drought and fires intensified by climate change are now threatening to push the Amazon toward a tipping point beyond which rainforest may be converted into a non-forest ecosystem.

Currently, political events are suggesting a grim near-term future for Amazonian forest in Brazil. Jair Bolsonaro, elected Brazil's president in November 2018, is acting to reverse virtually all Brazil's recent achievements in environmental protection. Bolsonaro is promoting the cutting of forest to support industrial soy farming and cattle ranching interests, and he has signaled that he will not enforce environmental protection laws. His administration is weakening its own environmental protection agencies or folding them into the Agriculture Ministry so that agricultural production will take precedence over environmental concerns. And Bolsonaro is working to eliminate protections for indigenous peoples living in the rainforests and open their reserves to mining. Meanwhile, his allies in the congress are pushing to break up protected areas and further weaken the Forest Code. As a result, as of 2019, deforestation was increasing.

Moreover, following Bolsonaro's election, the nation cancelled its commitment to host an international climate policy summit, and Bolsonaro may yet withdraw Brazil from the historic 2015 Paris Climate Accord (p. 516). If Brazil abandons its commitment to help address climate change by tackling deforestation of the Amazon, the ramifications will be felt throughout the world.

Regardless of what may eventually come to pass during Bolsonaro's presidency, the condition of the Amazon rainforest today raises questions relevant for forests and people everywhere: How can we best safeguard areas of high biodiversity? How can economies transition from resource exploitation to conservation and sustainable use? How can protected areas function as intended? Can climate change policies and financial incentives protect and restore forests? Can sustainable-certification programs help preserve forests? And as development spreads across the landscape and forests are fragmented, how can we conserve species, communities, and ecosystems while benefiting local people?

• **CASE STUDY SOLUTIONS** You run a non-profit environmental advocacy organization and are trying to save a tract of rainforest in the Brazilian Amazon. You have worked in this region for years and care for the local people, who want to save the forest and its animals but also need to make a living and use the forest's resources. Brazil's government plans to sell a concession to a multinational timber corporation to log the entire tract of land unless your group can work out some other plan. Describe what solution(s) you would try to arrange for the land in question. Consider the range of options discussed in this chapter, including protected areas, land trusts, biosphere reserves, forest management techniques, carbon offsets, and certified sustainable forestry. Explain reasons for your choice(s).

- **LOCAL CONNECTIONS** Identify an area of forest in your region, and learn what you can about its history. Is it primary forest or secondary forest? Was it always the size it is now, or has some area been lost or gained? Is forest (or other natural habitat) in your region becoming more fragmented over time? If so, what could be done to prevent further fragmentation while improving quality of life for people living in the area? Now consider a park or protected area in your region. What benefits does it bring to people living nearby? Would you like to see more or fewer, larger or smaller, parks and protected areas in your region? Why?

- **EXPLORE THE DATA** What is driving trends of forest loss in the Amazon? → **Explore Data** relating to the case study on **Mastering Environmental Science**.

REVIEWING Objectives

You should now be able to:

+ **Summarize the ecological and economic contributions of forests**

    Forests are ecologically complex, support a wealth of biodiversity, store carbon, and contribute ecosystem services. Forests also provide us timber and other economically important products and resources. (pp. 309–312)

+ **Outline the history and current scale of deforestation**

    We have lost forests as a result of timber harvesting and clearance for agriculture. The United States deforested much of its land as settlement, farming, and industrialization proceeded. Today, deforestation is occurring most rapidly in developing nations. (pp. 312–317)

+ **Assess dynamics of timber harvest management**

    Resource managers have long managed for maximum sustainable yield and have begun to implement ecosystem-based management and adaptive management. The U.S. national forests were established to conserve timber and allow its sustainable extraction, although most U.S. timber today comes from private lands. Plantation forestry, featuring single-species, even-aged stands, is widespread and growing. Harvesting methods include clear-cutting and other even-aged techniques, as well as selection strategies that maintain uneven-aged stands that more closely resemble natural forest. (pp. 317–322)

+ **Discuss forest management in relation to fire, pests, and climate change, and evaluate sustainable forestry certification**

    Foresters are now managing for recreation, wildlife habitat, and ecosystem integrity. However, increasingly severe fires are posing challenges. Fire suppression encourages eventual catastrophic fires. One solution is to reduce fuel loads by conducting prescribed burns. Climate change and outbreaks of bark beetles are also affecting forests. Certification of sustainable forest products allows consumer choice in the marketplace to influence forestry practices. (pp. 322–325)

+ **Identify federal land management agencies and the lands they manage**

    The U.S. Forest Service, National Park Service, Fish and Wildlife Service, and Bureau of Land Management manage national forests, national parks, national wildlife refuges, and BLM land, respectively. (pp. 318–319, 325–326)

+ **Recognize types of parks and protected areas**

    Public demand for preservation and recreation led to the creation of national parks, reserves, and wilderness areas. States, municipalities, and private land trusts all manage protected areas at the state, regional, and local levels. At the international level, biosphere reserves are one of several types of protected lands. (pp. 325–328)

+ **Evaluate challenges to the effectiveness of parks and protected areas**

    Some people oppose land protection policies for economic or ideological reasons. Because habitat fragmentation affects wildlife, conservation biologists are using island biogeography theory to learn how best to design systems of parks and reserves and implement solutions such as corridors. However, climate change is now posing new threats to protected areas. (pp. 327, 329–333)

SEEKING Solutions

1. People in industrialized nations are fond of warning people in industrializing nations to stop destroying rainforest. People of industrializing nations often respond that this approach is hypocritical, because the industrialized nations became wealthy by deforesting their land and exploiting their resources in the past. What would you say to the president of an industrializing nation, such as Indonesia or Brazil, in which a great deal of forest is being cleared? What strategies might you

suggest to help that nation achieve wealth while also conserving forests?

2. What might you tell an opponent of parks and preserves to help him or her understand why a wilderness hiker wants scenic land in Utah to be federally protected? What might you tell a wilderness hiker to help him or her understand why the park opponent disapproves of protection? How might you help them find common ground?

3. Consider the impacts that climate change may have on species' ranges. If you were trying to save an endangered mammal species whose range had been reduced to a small area and you had generous funding to acquire land to help restore its population, how would you design a protected area? Would you use corridors? Would you include a diversity of elevations? Would you design few large reserves or many small ones? Explain your answers.

4. **THINK IT THROUGH** You are the publisher of a magazine with a circulation of 1 million readers. Recently, you have begun receiving queries from readers about whether you use certified sustainable paper. As you decide whether to switch to FSC-certified paper, what

information would you want to learn (e.g., in terms of your readership, finances, and effects on forests)? Name three beneficial impacts of FSC certification that you would feature in your magazine to impress on readers why such a switch makes a positive difference. Why do you think each of these three things might resonate with your readers?

5. **THINK IT THROUGH** You have just become the supervisor of a national forest. Timber companies are asking to cut as many trees as you will let them, while environmental advocates want no logging at all. Ten percent of your forest is old-growth primary forest, and 90% is secondary forest. Your forest managers are split among preferring maximum sustainable yield, ecosystem-based management, and adaptive management. What management approach(es) will you take, and why? Will you allow logging of all old-growth trees, some, or none? Will you allow logging of secondary forest? If so, what harvesting strategies will you encourage? What would you ask scientists before deciding on policies on fire management and salvage logging?

CALCULATING Ecological Footprints

We all rely on forest resources. The average North American consumes 211 kg (466 lb) of paper and paperboard each year. Using the estimates of paper and paperboard consumption for each region of the world, calculate the per capita consumption for each region using the population data in the table. Note: 1 metric ton = 2205 pounds.

	POPULATION (MILLIONS)	TOTAL PAPER CONSUMED (MILLIONS OF METRIC TONS)	PER CAPITA PAPER CONSUMED (POUNDS)
Africa	1203	8	15
Asia	4437	199	
Europe	740	93	
Latin America	637	28	
North America	360	76	
Oceania	40	4	
World	7418	407	121

Data are for 2016, from Population Reference Bureau and UN Food and Agriculture Organization (FAO).

1. How much paper would be consumed if everyone in the world used as much paper as the average North American?

2. How much paper would North Americans save each year if they consumed paper at the rate of Europeans? At the rate of Africans?

3. North Americans have been reducing their per-person consumption of paper and paperboard by more than

1% annually in recent years by recycling, shifting to online activity, and reducing packaging of some products. Name three specific things you personally could do to reduce your own consumption of paper products.

4. Describe three ways in which consuming FSC-certified paper rather than conventional paper can reduce the environmental impacts of paper consumption.

Mastering Environmental Science

The Urban Environment

Creating Sustainable Cities

Upon completing this chapter, you will be able to:

+ **Describe the scale of urbanization**

+ **Define sprawl and discuss its causes and consequences**

+ **Outline city and regional planning and land use strategies**

+ **Evaluate transportation options, urban parks, and green buildings**

+ **Analyze environmental impacts and advantages of urban centers**

+ **Assess urban ecology and the pursuit of sustainable cities**

case study

Managing Growth in Portland, Oregon

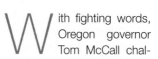

> Sagebrush subdivisions, coastal condomania, and the ravenous rampage of suburbia . . . all threaten to mock Oregon's status as the environmental model for the nation.
> Oregon Governor Tom McCall, 1973

> We have planning boards. We have zoning regulations. We have urban growth boundaries and 'smart growth' and sprawl conferences. And we still have sprawl.
> Environmental scientist Donella Meadows, 1999

With fighting words, Oregon governor Tom McCall challenged his state's legislature in 1973 to take action against runaway sprawling development, which many Oregonians feared would ruin the communities and landscapes they loved. McCall was echoing the growing concerns of state residents that farms, forests, and open space were being gobbled up and paved over.

Foreseeing a future of subdivisions, strip malls, and traffic jams engulfing the pastoral Willamette Valley, Oregon acted. The state legislature passed Senate Bill 100, creating a sweeping land use law that would become the focus of acclaim, criticism, and careful study for years afterward by other states and communities trying to manage their own urban and suburban growth.

Oregon's land use law required every city and county to draw up a comprehensive land use plan in line with statewide guidelines that had gained popular support from the state's electorate. As part of each land use plan, each metropolitan area had to establish an **urban growth boundary** (UGB), a line on a map separating areas desired to be urban from areas desired to remain rural. Development for housing, commerce, and industry would be encouraged within these urban growth boundaries but restricted beyond them. The intent was to revitalize city centers; prevent suburban sprawl; and protect farmland, forests, and open landscapes outside urban areas.

Residents of the area around Portland, the state's largest city, established a new regional planning entity to apportion land in their region. The Metropolitan Service District, or Metro, represents 24 municipalities and three counties. Metro adopted the Portland-area urban growth boundary in 1979, enclosing 92,000 ha (227,000 acres) of land, and has tried to focus growth in existing urban centers and to build communities where people can walk, bike, or take mass transit between home, work, and shopping. These policies have largely worked as intended. Portland's downtown and older neighborhoods have thrived, regional urban centers are becoming denser and more community oriented, mass transit has expanded, and farms and forests have been preserved on land beyond the UGB. Portland began attracting international attention for its "livability."

◀ **Flower festival at Pioneer Courthouse Square in downtown Portland, Oregon**

▲ **Mount Hood overlooking downtown Portland**

337

To many Portlanders today, the UGB remains the key to maintaining quality of life in city and countryside alike. In the view of its critics, however, the "Great Wall of Portland" is an elitist and intrusive government regulatory tool. In 2004, Oregon voters approved a ballot measure that threatened to eviscerate their state's land use rules. Ballot Measure 37 required the state to compensate certain landowners if government regulation had decreased the value of their land. For example, regulations prevented landowners outside UGBs from subdividing their lots and selling them for housing development. Under Measure 37, the state had to pay these landowners to make up for theoretically lost income—or else allow them to ignore the regulations. Because state and local governments did not have enough money to pay such claims, the measure was on track to gut Oregon's zoning, planning, and land use rules.

Landowners filed more than 7500 claims for payments or waivers affecting 295,000 ha (730,000 acres). Although the measure had been promoted to voters as a way to protect the rights of small family landowners, most claims were filed by large developers. Neighbors suddenly found themselves confronting the prospect of massive housing subdivisions, gravel mines, strip malls, or industrial facilities being developed next to their homes, and many who had voted for Measure 37 came to have misgivings.

The state legislature, under pressure from opponents and supporters alike, settled on a compromise: to introduce a new ballot measure. Oregon's voters passed Ballot Measure 49 in 2007. It restricts development outside the UGB that is on a large scale or that degrades sensitive natural areas, but it protects the rights of small landowners to gain income from their property by developing small numbers of homes.

In 2010, Metro finalized a historic agreement with its region's three counties to determine where urban growth will be allowed over the next 50 years. Metro and the counties considered 121,000 ha (300,000 acres) of undeveloped land and apportioned some into "urban reserves" open for development and most into "rural reserves" where farmland and forests would be preserved. Boundaries were precisely mapped to give clarity and direction for landowners and governments.

People are confronting similar issues in communities everywhere, and debates like those in Oregon will determine how our cities and landscapes will change in the future.

Our Urbanizing World

In 2009, we passed a turning point in human history. For the first time ever, more people were living in urban areas (cities and suburbs) than in rural areas. As we undergo this historic shift from the countryside into towns and cities—a process called **urbanization**—two pursuits become ever more important. One is to make our urban areas more livable by meeting residents' needs for a safe, clean, healthy urban environment and a high quality of life. The other is to make urban areas sustainable by creating cities that can prosper in the long term while minimizing ecological impacts.

Industrialization drove urbanization

Since 1950, the world's urban population has multiplied by more than five times, whereas the rural population has not even doubled. Urban populations are growing because the human population overall is growing (Chapter 8) and because more people are moving from rural areas to urban areas than are moving from urban areas to rural areas.

This shift from country to city began long ago. Agricultural harvests that produced surplus food freed a proportion of citizens from farm life and allowed the rise of specialized manufacturing professions, class structure, political hierarchies, and urban centers (p. 245). The industrial revolution (p. 5) spawned technological innovations that created jobs and opportunities in urban centers for people who were no longer needed on farms and ranches. Industrialization, urbanization, and technology increased production efficiencies, and economic opportunities grew faster in cities. This process of positive feedback continues today.

United Nations demographers project that in 2050, more than two of every three people will live in urban areas.

Between now and then, they estimate, the urban population will increase by 53%, whereas the rural population will decline by 10%. Trends differ between more developed and less developed regions, however (**FIGURE 13.1**). In more developed

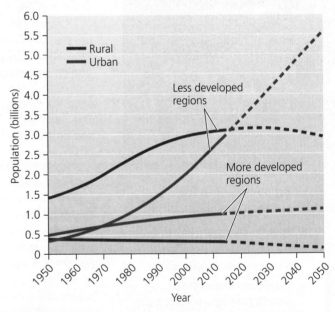

FIGURE 13.1 Population trends differ between poor and wealthy nations. In less developed regions, urban populations are growing quickly, and rural populations will soon begin declining. More developed regions are already largely urbanized, so their urban populations are growing slowly, whereas rural populations are falling. Solid lines indicate past data, and dashed lines indicate future projections. *Data from UN Population Division, 2018. World urbanization prospects: The 2018 revision. By permission.*

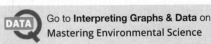
Go to **Interpreting Graphs & Data** on **Mastering Environmental Science**

nations such as the United States, urbanization has slowed because four of every five people already live in cities, towns, and **suburbs,** the smaller communities that ring cities. Back in 1850, the U.S. Census Bureau classified only 15% of U.S. citizens as urban-dwellers. That percentage now stands at 82%. Most U.S. urban-dwellers reside in suburbs; fully half the U.S. population today is suburban.

In contrast, today's less developed nations, where most people still reside on farms and in rural villages, are urbanizing rapidly. As industrialization diminishes the need for farm labor while increasing urban commerce and jobs, rural people are streaming to cities. Sadly, across the globe, wars, conflict, and ecological degradation are also driving millions of people from the countryside into urban centers. For all these reasons, most fast-growing cities today are in the less developed world. In cities such as Delhi, India; Lagos, Nigeria; and Karachi, Pakistan, population growth often exceeds economic growth, and the result is overcrowding, pollution, and poverty. United Nations demographers estimate that urban areas of less developed nations will absorb nearly all of the world's population growth from now on.

Environmental factors influence the location of urban areas

Real estate agents use the saying, "Location, location, location" to stress how a home's setting determines its value. Location is vital for urban centers as well. Think of any major city, and chances are it's situated along a major river, seacoast, railroad, or highway—some corridor for trade that has driven economic growth (**FIGURE 13.2**).

Well-located cities often serve as linchpins in trading networks, funneling in resources from agricultural regions, processing them, manufacturing products, and shipping those products to other markets. Portland, Oregon, got its start in the mid-19th century as pioneers arriving by the Oregon Trail settled where the Willamette River joins the Columbia River, just upriver from where the Columbia flows into the Pacific Ocean. With this strategic location for trade, Portland grew as it received, processed, and shipped overseas the agricultural products from farms of the river valleys, and as it imported goods shipped in from other ports.

Another example of this geographic pattern is Chicago, which grew with extraordinary speed in the 19th and early 20th centuries. At that time, railroads funneled through it the resources from the vast lands of the Midwest and West on their way to consumers and businesses in the populous cities of the East. Chicago became a center for grain processing, livestock slaughtering, meatpacking, and much else.

Today, powerful technologies and cheap transportation enabled by fossil fuels have allowed cities to thrive even in resource-poor regions. The Dallas–Fort Worth area prospers from oil-fueled transportation by interstate highways and a major international airport. Southwestern cities such as Los Angeles, Las Vegas, and Phoenix flourish in desert regions by appropriating water from distant sources. Whether such cities can sustain themselves as oil and water become increasingly scarce in the future is an important question.

(a) St. Louis, Missouri

(b) Fort Worth, Texas

FIGURE 13.2 Cities tend to develop along trade corridors. St. Louis **(a)** is situated on the Mississippi River near its confluence with the Missouri River, where river trade drove its growth in the 19th and early 20th centuries. Fort Worth, Texas **(b),** grew in the late 20th century as a result of the interstate highway system and a major international airport.

In recent years, many cities in the southern and western United States have grown as people have moved there in search of warmer weather, more space, new economic opportunities, or places to retire. Between 1990 and 2018, the population of the Denver metropolitan area grew by 81%, that of the Dallas–Fort Worth metropolitan area by 87%, that of the Houston area by 88%, that of the Atlanta region by 101%, that of the Phoenix area by 117%, and that of the Las Vegas metropolitan area by a whopping 162%.

WEIGHING the **issues**

What Made Your City?

Consider the town or city in which you live or the major urban center located nearest you. Why do you think it developed where it did? What physical, social, or environmental factors may have aided its growth? Do you think it will prosper in the future? Why or why not?

People moved to suburbs

U.S. cities grew rapidly in the 19th and early 20th centuries as a result of immigration from abroad and increased trade as the nation expanded west. The bustling economic activity of downtown districts held people in cities despite growing crowding, poverty, and crime. However, by the mid-20th century, many affluent city-dwellers were choosing to move outward to cleaner, less crowded suburban communities. These people were pursuing more space, better economic opportunities, cheaper real estate, less crime, and better schools for their children.

The exodus to the suburbs in 20th-century America was aided by the rise of the automobile, an expanding road network, and inexpensive and abundant oil. Millions of people could now commute by car to downtown workplaces from new homes in suburban "bedroom communities." By facilitating transport, highway networks also made it easier for businesses to import and export resources, goods, and waste. The federal government's development of the interstate highway system was pivotal in promoting these trends.

As affluent people moved out into the expanding suburbs, jobs followed. This hastened the economic decline of downtown districts, and American cities stagnated. Chicago's population declined to 80% of its peak as residents moved to its suburbs. Philadelphia's population fell to 76% of its peak, Washington, D.C.'s to 71%, and Detroit's to just 55%.

Portland followed this trajectory: Its population growth stalled in the 1950s through the 1970s as crowding and deteriorating economic conditions drove city-dwellers to the suburbs. But the city bounced back. Policies to revitalize the city center helped restart Portland's growth (**FIGURE 13.3**).

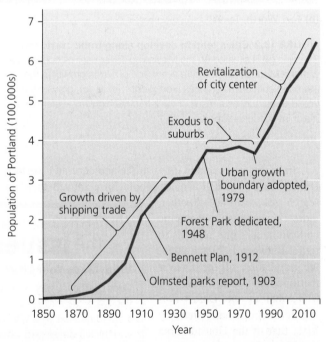

FIGURE 13.3 Portland's population grew, stalled, and then grew again. The shipping trade helped boost Portland's economy and population in the 1890s–1920s. City residents left for the suburbs in the 1950s–1970s, but policies to enhance the city center revitalized Portland's growth. *Data from U.S. Census Bureau.*

Today, the picture is complex. On the one hand, many cities have revitalized themselves as thriving cultural centers, as Portland has, and have reclaimed their allure. As a result, young people in particular are eagerly streaming back into city centers. At the same time, many affluent individuals and families seeking more space and privacy are moving still farther out into **exurbs,** communities beyond the suburbs. In our age of the Internet, cell phones, and videoconferencing, people can easily communicate from far-flung locations, so living and working in a city is no longer as vital to business or career success.

In most ways, suburbs and exurbs have delivered the qualities people have sought in them. The wide spacing of homes, with each on its own plot of land, gives families room and privacy. However, by allotting more space to each person, suburban and exurban growth spreads human impacts across the landscape. Natural areas disappear as housing developments are constructed. Road networks ease travel, but people find themselves needing to climb into a car to get anywhere. People commute longer distances to work and spend more time stuck in traffic. As expanding rings of suburbs and exurbs grow larger than the cities they surround, towns merge into one another. These aspects of growth inspired a new term: *sprawl.*

Sprawl

The term *sprawl* has become laden with meanings and suggests different things to different people, but we can begin our discussion by giving **sprawl** a simple, nonjudgmental definition: the spread of low-density urban, suburban, or exurban development outward from an urban center.

Urban areas spread outward

The spatial growth of urban and suburban areas is clear from maps and satellite images of rapidly spreading cities such as Las Vegas (**FIGURE 13.4**). Another example is Chicago, whose metropolitan area spreads over a region 40 times the size of the city. All in all, new houses and roads supplant more than 2700 ha (6700 acres) of U.S. land every day.

Sprawl results from development approaches that place homes on spacious lots in residential tracts that spread over large areas but are far from urban centers and commercial amenities (**FIGURE 13.5**). Such approaches allot each person more space than in cities. For example, the average resident of Chicago's suburbs takes up 11 times more space than a resident of the city. As a result, the outward spatial growth of suburbs and exurbs across the landscape generally outpaces growth in numbers of people.

In fact, many researchers define *sprawl* as the physical spread of development at a rate that exceeds the rate of population growth. For instance, the population of Phoenix grew 12 times larger between 1950 and 2000, yet its land area grew 27 times larger. Between 1950 and 1990, the population of 58 major U.S. metropolitan areas rose by 80%, but the land area they covered rose by 305%. Even in 11 metro areas where population declined between 1970 and 1990 (as with Rust

(a) 1972   (b) 1997   (c) 2018

FIGURE 13.4 Satellite images of Las Vegas, Nevada, show the rapid urban and suburban expansion referred to as sprawl. Las Vegas is one of the fastest-growing cities in North America. From **(a)** 1972 to **(b)** 1997 to **(c)** 2018, its population and its developed area each grew immensely. *Landsat data courtesy of U.S. Geological Survey, Department of the Interior/USGS. Landsat product IDs: (a) LM01-L1TP-042035-19720913-20180429-01-T2, (b) LT05-L1TP-039035-19970417-20160923-01-T1, and (c) LC08-L1TP-039035-20180411-20180417-01-T1.*

Belt cities such as Detroit, Cleveland, and Pittsburgh), the amount of land covered increased.

Sprawl has several causes

There are two main components of sprawl. One is human population growth—there are simply more of us alive each year (Chapter 8). The other is per capita land consumption—each person is taking up more land than in the past. The amount of sprawl is a function of the number of people added to a region times the amount of land each person occupies.

A study of U.S. metropolitan areas between 1970 and 1990 found that these two factors contribute about equally to sprawl but that cities vary in which is more influential. The Los Angeles metro area increased in population density by 9% between 1970 and 1990, becoming the nation's most densely populated metro area. Increasing density should be a good recipe for preventing sprawl, yet L.A. grew in size by a whopping 1021 km² (394 mi²) because of an overwhelming influx of new people. In contrast, the Detroit metro area lost 7% of its population between 1970 and 1990, yet it expanded in area by 28%. In this case, sprawl was caused solely by increased per capita land consumption.

Each person is taking up more space these days in part because of factors mentioned earlier: Better highways, inexpensive gasoline, telecommunications, and the Internet freed businesses from reliance on the centralized infrastructure a major city provides and gave workers greater flexibility to live where they desire.

Economists and politicians long encouraged the unbridled spatial expansion of cities and suburbs. The conventional assumption has been that growth is always good and that attracting business, industry, and residents will enhance a community's economic well-being, political power, and cultural influence. Today, this assumption is being challenged as growing numbers of people feel negative effects of sprawl on their lifestyles.

What's wrong with sprawl?

To some people, the word *sprawl* evokes strip malls, traffic jams, homogeneous commercial development, and tracts of cookie-cutter houses encroaching on farmland, ranchland, or forests. For other people, sprawl is simply the collective result of choices made by millions of well-meaning people trying to make a better life for their families. What can scientific research tell us about the impacts of sprawl?

FIGURE 13.5 Sprawl is characterized by the spread of development across large areas of land. This kind of development requires people to drive cars to reach commercial amenities or community centers.

Transportation Most studies show that sprawl constrains transportation options, essentially forcing people to own a vehicle, drive it most places, drive greater distances, and spend more time in it. Sprawling communities suffer more traffic accidents and have few or no mass transit options. Across the United States since 1980, the average length of the commute to work has risen by 24%, and total vehicle miles driven has risen far faster than population growth. A car-oriented culture encourages congestion and increases dependence on oil.

Pollution By promoting automobile use, sprawl increases pollution. Carbon dioxide emissions from vehicles contribute to climate change (Chapter 18), and air pollutants containing nitrogen and sulfur lead to tropospheric ozone, urban smog, and acid precipitation (Chapter 17). Runoff of water from roads and parking lots may be polluted by motor oil that has leaked from vehicles and by road salt applied to combat ice. Paved areas produce 16 times more runoff than do naturally vegetated areas, and polluted runoff that reaches waterways can pose risks to ecosystems and human health.

Health Beyond the health impacts of pollution, some research suggests that sprawl promotes physical inactivity because driving cars largely takes the place of walking during daily errands. Physical inactivity increases obesity and high blood pressure, which can lead to other ailments. A 2003 study found that people from the most-sprawling U.S. counties show higher blood pressure and weigh 2.7 kg (6 lb) more for their height than people from the least-sprawling U.S. counties.

Land use As more land is developed, less is left as forests, fields, farmland, or ranchland. Of the estimated 1 million ha (2.5 million acres) of U.S. land converted each year, roughly 60% is agricultural land and 40% is forest. These lands provide vital resources, recreation, aesthetic beauty, wildlife habitat, and air and water purification. Today an alarming number of children grow up without the ability to roam through woods, fields, and open space, which used to be a normal part of childhood. Being deprived of regular access to nature as a child, many experts maintain, can inflict psychological and emotional harm on an individual, with consequences for society (p. 282).

Economics Sprawl drains tax dollars from communities and funnels money into infrastructure for new development on the fringes of those communities. Funds that could be spent maintaining and improving downtown centers are instead spent on extending the road system, water and sewer system, electricity grid, telephone lines, police and fire service, schools, and libraries. The costs of extending public infrastructure are generally not charged to developers but are paid by taxpayers of the community. In theory, fees on developers or property taxes on new homes and businesses can pay back the public investment, but studies find that in most cases existing taxpayers end up subsidizing new development.

Creating Livable Cities

To respond to the challenges presented by sprawl, architects, planners, developers, and policymakers are trying to revitalize city centers and to plan and manage how urbanizing areas develop. They aim to make cities safer, cleaner, healthier, and more pleasant for their residents.

Planning helps us create livable urban areas

How can we design cities to maximize their efficiency, functionality, and beauty? This question is central to **city planning** (also known as **urban planning**). City planners advise policymakers on development options, transportation needs, public parks, and other matters.

Washington, D.C., is the earliest example of city planning in the United States. President George Washington hired French architect Pierre Charles L'Enfant in 1791 to design a capital city for the new nation on undeveloped land along the Potomac River. L'Enfant laid out a baroque-style plan of diagonal avenues cutting across a grid of streets, with space allotted for majestic public monuments, and the city was built largely according to his plan (**FIGURE 13.6**). A century later, as the city became crowded and dirty, a special commission in 1901 undertook new planning efforts to beautify the city while staying true to the intentions of L'Enfant's original plan. These planners imposed a height restriction on new buildings to keep the magnificent government edifices and monuments from being dwarfed by modern skyscrapers, thus preserving the spacious, stately feel of the city.

City planning in North America came into its own at the turn of the 20th century as urban leaders sought to beautify and impose order on fast-growing, unruly cities. Landscape architect Daniel Burnham's 1909 *Plan of Chicago* was perhaps the grandest effort of this time. As implemented over decades, Burnham's plan expanded Chicago's parks and playgrounds, streamlined traffic, improved neighborhood living conditions, and cleared industry and railroads from the shore of Lake Michigan to provide public access to the water.

Portland gained its own comprehensive plan just three years later, in 1912. Edward Bennett's *Greater Portland Plan* proposed to rebuild the harbor; dredge the river channel; construct new docks, bridges, tunnels, and a waterfront railroad; superimpose wide radial boulevards on the old city street grid; establish civic centers downtown; and greatly expand the number of parks. Voters approved the plan by a two-to-one margin, but they defeated a bond issue that would have paid for park development. As the century progressed, other major planning efforts were conducted, and some ideas, such as establishing public squares downtown, came to fruition.

In today's world of sprawling metropolitan areas, **regional planning** has become at least as important as city planning. Regional planners deal with the same issues as

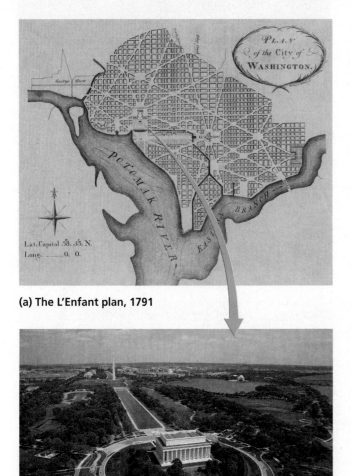

(a) The L'Enfant plan, 1791

(b) Washington, D.C., today

FIGURE 13.6 Washington, D.C., is a prime example of early city planning. The 1791 plan **(a)** for the new U.S. capital laid out splendid diagonal avenues cutting across gridded streets, allowing space for the magnificent public monuments **(b)** that grace the city today.

city planners, but they work on broader geographic scales and coordinate their work with multiple municipal governments. In some places, regional planning has been institutionalized in formal government bodies; the Portland area's Metro is such a regional planning entity. A historic accomplishment in regional planning was sealed when Metro and its region's three counties in 2010 announced their collaborative plan apportioning undeveloped land into "urban reserves" and "rural reserves." The agreement enables homeowners, farmers, developers, and policymakers to feel informed and secure knowing what kinds of land uses lie in store on and near their land over the next half-century.

Zoning is a key tool for planning

One tool that planners use is **zoning,** the practice of classifying areas for different types of development and land use

(FIGURE 13.7). For instance, to preserve the cleanliness and tranquility of residential neighborhoods, industrial facilities may be kept out of districts zoned for residential use. By specifying zones for different types of development, planners can guide what gets built where. Zoning also gives home buyers and business owners security because they know in advance what types of development can and cannot be located nearby.

Zoning involves government restriction on the use of private land and represents a constraint on personal property rights. For this reason, some people consider zoning a regulatory taking (p. 169) that violates individual freedoms. Most people defend zoning, however, saying that government has a proper and useful role in setting limits on property rights for the good of the community.

When Oregon voters passed Ballot Measure 37 in 2004 (see Central Case Study, p. 338), it shackled government's ability to enforce zoning regulations with landowners who bought their land before the regulations were enacted. However, many Oregonians soon began witnessing new development they did not condone, so in 2007 they passed Ballot Measure 49 to restore public oversight over development. The passage of Oregon's Measure 37 spawned similar ballot measures in other U.S. states, but voters defeated most of them. In general,

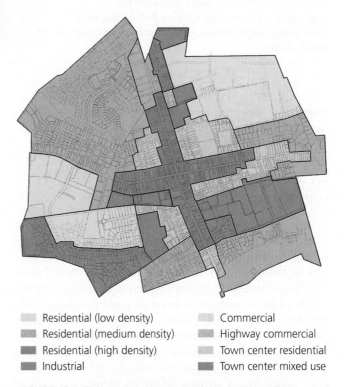

Residential (low density)
Residential (medium density)
Residential (high density)
Industrial

Commercial
Highway commercial
Town center residential
Town center mixed use

FIGURE 13.7 This zoning map for Littletown, Pennsylvania, shows several typical zoning patterns. Public and institutional uses are clustered together in "mixed-use" areas in the center of town and along major roads. Industrial zones tend to be located away from most residential areas. Residential zones vary in the density of homes allowed.

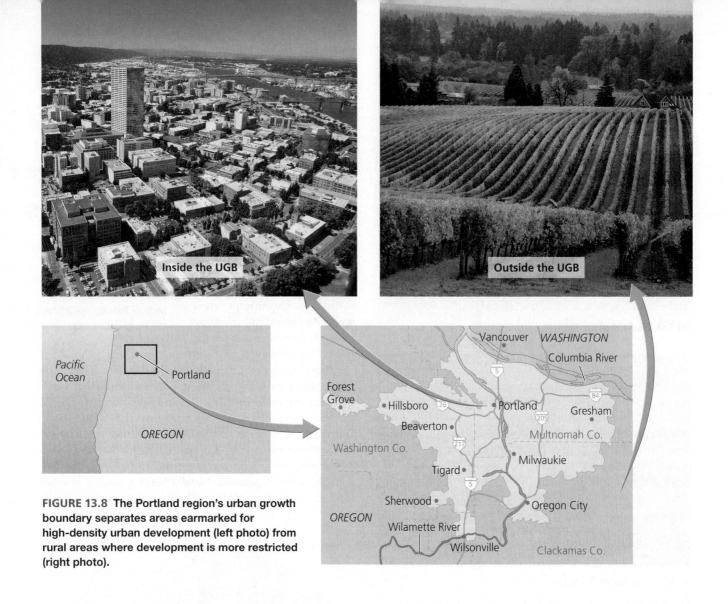

FIGURE 13.8 The Portland region's urban growth boundary separates areas earmarked for high-density urban development (left photo) from rural areas where development is more restricted (right photo).

people have supported zoning over the years because the public good it produces for communities is widely felt to outweigh the restrictions on private use.

Urban growth boundaries are now widely used

Planners in Oregon sought to curb sprawl by containing growth largely within existing urbanized areas. They did so by establishing urban growth boundaries (UGBs), legally binding lines on a map that separate areas zoned to be high density and urban from areas intended to remain low density and rural. Oregon's UGBs aimed to revitalize downtowns; protect working farms, orchards, ranches, and forests; and ensure urban-dwellers access to open space (**FIGURE 13.8**). UGBs also save taxpayers money by reducing the amounts that municipalities need to pay for infrastructure. Since Oregon instituted its policies, many other regions have adopted UGBs, including the states of Washington, Tennessee, and California, as well as a number of cities, including Boulder, Colorado; Lancaster, Pennsylvania; Honolulu, Hawai'i; Miami, Florida; Minneapolis–St. Paul, Minnesota; and Virginia Beach, Virginia.

In most ways, the Portland region's UGB has worked as intended. It has lowered prices for land outside the UGB while raising prices within it. It has preserved farms and forests by restricting development outside the UGB. It has increased the density of new housing inside the UGB by more than 50% as homes are built on smaller lots and as multistory apartments replace low-rise structures. Downtown employment has grown as businesses and residents invest anew in the central city. Through it all, Portland has been able to absorb considerable immigration while avoiding rampant sprawl.

Nonetheless, the Portland region's urbanized area grew by 101 km^2 (39 mi^2) in the decade after its UGB was established, because 146,000 people were added to the population. Intensifying population pressure has led Metro to enlarge the UGB three dozen times since its establishment, adding 32,000 acres to its original 227,000. In addition, UGBs tend to increase housing prices within their boundaries. In Portland, housing has become far less affordable than it was, leading the mayor and city council in 2015 to declare a "housing emergency," which has not yet been lifted. Today in the city, demand for housing exceeds supply, rents are soaring, and low- and middle-income people are being forced out of neighborhoods they have lived in for years as these neighborhoods experience **gentrification,** a transformation to conditions that cater to wealthier people. These trends suggest that relentless population growth may thwart even the best anti-sprawl efforts and that livable cities can fall victim to their own success if they are in high demand as places to live.

Smart growth and new urbanism aim to counter sprawl

As more people feel impacts of sprawl on their everyday lives, efforts to manage growth are springing up throughout North America. Oregon's Senate Bill 100 (p. 337) was one of the first, and since then, dozens of states, regions, and cities have adopted similar land use policies. Urban growth boundaries and other approaches from these policies have coalesced under the concept of **smart growth** (**TABLE 13.1**).

Proponents of smart growth aim to rejuvenate the older existing communities that so often are drained and impoverished by sprawl. Smart growth means "building up, not out," focusing development and economic investment in existing urban centers and favoring denser, mixed-use architecture—multistory housing blended with commercial space.

A related approach among architects, planners, and developers is **new urbanism,** which seeks to design walkable neighborhoods with homes, businesses, schools, and other amenities all nearby for convenience. The aim is to create functional neighborhoods in which families can meet most of their needs close to home without using a car. Trees, green spaces, a mix of architectural styles, and creative street layouts add to the visual interest of new-urbanist developments. These developments mimic the traditional urban neighborhoods that existed before the advent of suburbs.

New-urbanist neighborhoods are often served by public transit systems. In **transit-oriented development,** compact communities in the new-urbanist style are arrayed around stops on a rail line, enabling people to travel most places they need to go by train and foot alone. Several lines of the Washington, D.C., Metro system have been developed in this manner.

Among the 600 communities in the new-urbanist style across North America are Seaside, Florida; Kentlands in Gaithersburg, Maryland; Addison Circle in Addison, Texas; Mashpee Commons in Mashpee, Massachusetts; Harbor Town in Memphis, Tennessee; Celebration in Orlando, Florida; and Orenco Station, west of Portland.

Transit options help cities

Traffic jams on roadways cause air pollution, stress, and countless hours of lost time. They cost Americans an estimated $305 billion each year—almost $1000 per person—in fuel and lost productivity. To encourage more efficient transportation, policymakers can raise fuel taxes, charge trucks for road damage, tax inefficient modes of transport, and reward carpoolers with carpool lanes. But a key component of improving the quality of urban life is to give residents alternative transportation options.

Bicycle transportation is one key option (**FIGURE 13.9**). Portland has embraced bicycles like few other American cities,

TABLE 13.1 Ten Principles of "Smart Growth"

1. Mix land uses.
2. Take advantage of compact building design.
3. Create a range of housing opportunities and choices.
4. Create walkable neighborhoods.
5. Foster distinctive, attractive communities with a strong sense of place.
6. Preserve open space, farmland, natural beauty, and critical environmental areas.
7. Strengthen existing communities, and direct development toward them.
8. Provide a variety of transportation choices.
9. Make development decisions predictable, fair, and cost-effective.
10. Encourage community and stakeholder collaboration in development decisions.

Source: U.S. Environmental Protection Agency.

and today more than 5% of its commuters ride to work by bike (the national average is 0.5%). The city has developed nearly 400 miles of bike lanes and paths, 6500 public bike racks, and special markings at intersections to enhance safety for bicyclists and pedestrians. Amazingly, all this infrastructure was created for the typical cost of just 1 mile of urban freeway. Portland also has a bike-sharing program similar to programs in cities such as Montreal, Toronto, Denver, Minneapolis, Miami, San Antonio, Boston, New York City, and Washington, D.C.

Like many other major cities, Portland is now experimenting with electric scooters. Proponents of e-scooters say that they help reduce car traffic by enabling people needing to travel only short distances to take scooters instead. Opponents complain that the scooters clutter sidewalks and cause safety concerns. In 2018, Portland ran a four-month pilot program, allowing three companies to rent scooters downtown. About 34% of Portland

FIGURE 13.9 Bicycles provide a healthy alternative to transportation by car. This Portland bicycle lot accommodates riders who commute downtown by bike and is conveniently located at a streetcar stop.

residents and 48% of visitors used a scooter instead of driving at some point during this period, according to the Portland Bureau of Transportation—but people also lodged 6000 complaints to the agency. The city established a further 1-year trial period before it decides on whether to allow them permanently.

Other transportation options include **mass transit** systems: public systems of buses, trains, subways, or *light rail* (smaller rail systems powered by electricity). Mass transit systems move large numbers of passengers while easing traffic congestion, taking up less space than road networks, and emitting less pollution than cars. A 2005 study (**FIGURE 13.10**) calculated that each year, rail systems in U.S. metropolitan areas save

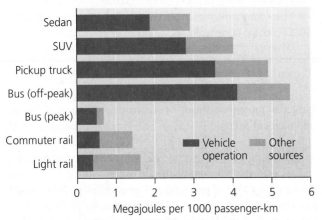

(a) Energy consumption for different modes of transit

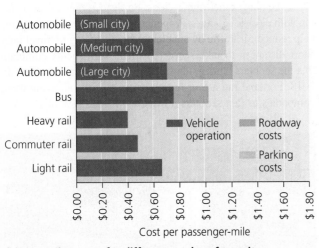

(b) Operating costs for different modes of transit

FIGURE 13.10 Public transit tends to (a) consume less energy and (b) cost less than automobile transit. Bus transit is highly efficient in places and at times of high use ("peak" in figure), but much less so when and where use is low ("off-peak" in figure). Data presented for light rail are averages of systems in Boston and San Francisco. Data for greenhouse gas emissions (not shown) are very similar to those for energy consumption.
Data from (a, b) Chester, M., and Horvath, A., 2009. Environmental assessment of passenger transportation should include infrastructure and supply chains. Environmental Research Letters 4: 024008 (8 pp.); and (c) Litman, T., 2005. Rail transit in America: A comprehensive evaluation of benefits. © 2005 T. Litman.

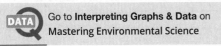 Go to **Interpreting Graphs & Data** on **Mastering Environmental Science**

taxpayers $67.7 billion in costs related to congestion, parking, road maintenance, and accidents—far more than the $12.5 billion that governments spend to subsidize rail systems each year. As long as an urban center is large enough to support the infrastructure necessary, rail systems are cheaper, more energy-efficient, and less polluting than roadways choked with cars—as are bus systems when they are heavily used.

In Portland, buses and light rail (together with *streetcars,* trolleys that serve short trips downtown and share rights-of-way with cars) carry 100 million riders per year (**FIGURE 13.11**). The most-used train systems in the United States are the "heavy rail" systems of its largest cities, such as New York City's subways, Washington, D.C.'s Metro, the T in Boston, and the San Francisco Bay area's BART. Each of these rail systems carries more than one-fourth of its city's daily commuters.

In general, however, the United States lags behind most nations in mass transit. Many countries, rich and poor alike, have extensive and accessible bus systems that ferry citizens within and between towns and cities cheaply and effectively (see **SUCCESS STORY**). And whereas Japan, China, and many European nations have developed entire systems of modern high-speed "bullet" trains (**FIGURE 13.12**), the United States has only one such train, Amtrak's *Acela Express*. This train connects Boston and Washington, D.C., via New York, Philadelphia, and Baltimore, and it travels more slowly than most bullet trains.

The United States chose instead to invest in road networks for cars and trucks largely because (relative to most other nations) its population density was low and gasoline was cheap. As energy costs and population rise, however, mass transit becomes increasingly appealing, and residents begin to clamor for train and bus systems in their communities. Americans may eventually see more high-speed rail; the 2009 stimulus bill passed by Congress set aside $8 billion for developing high-speed rail, and the Obama administration identified 10 potential corridors for its development. Projects are proceeding in several of these corridors, including in California, where the state is debating how fully to fund construction of a system connecting Los Angeles, San Francisco, San Diego, and Sacramento.

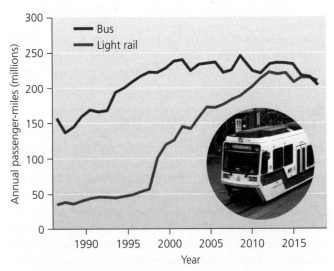

FIGURE 13.11 Ridership has grown on Portland's buses and on its MAX light rail system. *Data from Trimet.*

FIGURE 13.12 "Bullet trains" of high-speed rail systems in Europe and Asia can travel at 150–220 mph. This Chinese train is speeding through the city of Qingdao.

FIGURE 13.13 Central Park in New York City was one of America's first city parks, and it remains one of the largest and finest.

Urban residents need park lands

City-dwellers often desire some escape from the noise, commotion, and stress of urban life. Natural lands, public parks, and open space provide greenery, scenic beauty, freedom of movement, and places for recreation. In addition, these green spaces keep ecological processes functioning by helping regulate climate, purify air and water, and provide wildlife habitat. The animals and plants of urban and suburban parks also serve to satisfy our natural affinity for contact with other organisms (sometimes called biophilia; p. 282). In the wake of urbanization and sprawl, protecting natural lands and establishing public parks have become vital as most of us come to feel increasingly disconnected from nature.

City parks in North America arose in the late 19th century as urban leaders yearning to make crowded and dirty cities more livable established public spaces using aesthetic ideals borrowed from European parks, gardens, and royal hunting grounds.

The lawns, shaded groves, curved pathways, and pastoral vistas of many American city parks grew from these European ideals, as interpreted by America's leading landscape architect, Frederick Law Olmsted. Olmsted designed Central Park in New York City (**FIGURE 13.13**) and many other urban parks.

East Coast cities such as New York, Boston, and Philadelphia developed parks early on, but western cities were not far behind. In San Francisco, William Hammond Hall transformed 2500 ha (1000 acres) of the peninsula's dunes into Golden Gate Park, a verdant playground of lawns, trees, gardens, and sports fields. Portland's quest for parks began in 1900, when city leaders created a parks commission and hired Olmsted's son, John Olmsted, to design a park system. His 1904 plan proposed acquiring land to ring the city generously with parks, but no action was taken. A full 44 years later, residents pressured city leaders to create Forest Park along a forested ridge on the northwest side of the city. At 11 km (7 mi) long, it is one of the largest city parks in North America.

SUCCESS story **Creating a Global Model for Bus Transit**

Establishing a mass transit system often requires strong and visionary political leadership. Such was the case in Curitiba, a metropolis of 2.5 million people in southern Brazil. Faced with an influx of immigrants from outlying farms in the 1970s, city leaders led by Mayor Jaime Lerner undertook an aggressive planning process so that they could direct their city's growth rather than being overwhelmed by it. They established a fleet of hundreds of public buses, took steps to encourage bicycles and pedestrians, and reconfigured Curitiba's road system to maximize its efficiency. The buses were given dedicated lanes, while pre-boarding fare purchase in futuristic tube stations sped things along, giving this innovative "bus rapid transit" system the speed and efficiency of a rail system. Three-fourths of Curitiba's population began using the buses daily, car usage dropped, and surveys showed that residents were happier than people in other Brazilian cities. The city's system began to attract international acclaim, and eventually more than 160 other cities worldwide followed Curitiba's lead and adopted bus rapid transit systems. Today, this much-celebrated city is—like Portland, Oregon—becoming a victim of its own success as rapid population growth begins

Commuters boarding a bus in Curitiba, Brazil

to overwhelm its infrastructure and car congestion grows. Curitiba's new generation of leaders will need to innovate further, but already Curitiba has shown the rest of the world how investing thoughtfully in well-planned transportation infrastructure can pay big dividends.

→ **Explore the Data** at **Mastering Environmental Science**

Park lands come in various types

Large city parks are vital to a healthy urban environment, but even small spaces make a big difference. Playgrounds give children places to be active outdoors and interact with their peers. Community gardens allow people to grow vegetables and flowers in a neighborhood setting.

Greenways are strips of land that connect parks or neighborhoods. They often run along rivers, streams, or canals and provide access to walking trails. Greenways can protect water quality, boost property values, and serve as corridors for the movement of wildlife. Across North America, the Rails-to-Trails Conservancy has helped convert more than 38,000 km (23,500 mi) of abandoned railroad rights-of-way into greenways for walking, jogging, and biking.

One newly developed linear park along an old rail line is the High Line Park in Manhattan in New York City (**FIGURE 13.14**). An elevated freight line running above the city's streets was going to be demolished, but a group of citizens saw its potential for a park and pushed the idea until city leaders came to share their vision. Today, more than 13,000 people per day use the 23-block-long High Line for recreation or on their commute to work.

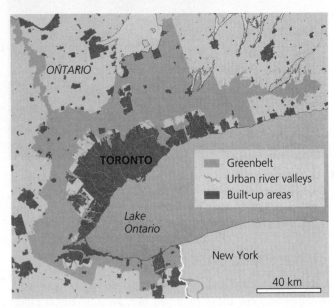

FIGURE 13.15 Toronto, Ontario, boasts the world's largest greenbelt. This immense swath of natural land circling the city acts as an urban growth boundary and provides recreation and ecosystem services for Toronto-area residents.

The concept of the corridor is sometimes implemented on a large scale. **Greenbelts** are long and wide corridors of park lands, often encircling an entire urban area. The world's largest greenbelt surrounds the city of Toronto, Ontario, and its major suburbs (**FIGURE 13.15**). At nearly 800,000 ha (2 million acres) in size, this massive system of parkland, forested land, and urban rivers is calculated to provide $3.2 billion in ecosystem services each year to the people of the region. Toronto and certain other Canadian cities such as Ottawa and Vancouver employ greenbelts as urban growth boundaries, containing sprawl while preserving open space for urban residents.

In the United States, a major greenbelt is the Chicago area's forest preserve system, a 40,000-ha (100,000-acre) network of oak woodlands, prairies, and wetlands that stretches through Chicago's suburbs like a necklace. These natural lands—the largest system of urban green space in the United States—accommodate 62 million recreation visits each year.

Green buildings bring benefits

Although we need park lands, we spend most of our time indoors, so the buildings in which we live and work affect our health and well-being. Buildings also consume 40% of our energy and 70% of our electricity, contributing to the greenhouse gas emissions that drive climate change. As a result, there is a thriving movement in architecture and construction to design and build **green buildings,** structures meant to minimize the ecological footprint of their construction and operation.

Green buildings are built from sustainable materials, limit their use of energy and water, minimize adverse health impacts, control pollution, and recycle waste (**FIGURE 13.16**). Constructing or renovating buildings using new efficient

FIGURE 13.14 The High Line Park was created thanks to a visionary group of Manhattan citizens. They pushed to make a park out of an abandoned elevated rail line.

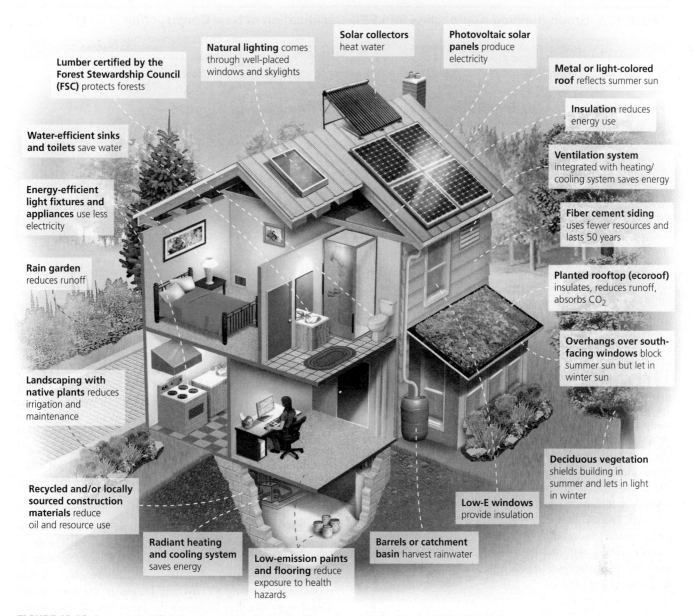

Lumber certified by the Forest Stewardship Council (FSC) protects forests

Natural lighting comes through well-placed windows and skylights

Solar collectors heat water

Photovoltaic solar panels produce electricity

Metal or light-colored roof reflects summer sun

Insulation reduces energy use

Water-efficient sinks and toilets save water

Ventilation system integrated with heating/cooling system saves energy

Energy-efficient light fixtures and appliances use less electricity

Fiber cement siding uses fewer resources and lasts 50 years

Rain garden reduces runoff

Planted rooftop (ecoroof) insulates, reduces runoff, absorbs CO_2

Landscaping with native plants reduces irrigation and maintenance

Overhangs over south-facing windows block summer sun but let in winter sun

Recycled and/or locally sourced construction materials reduce oil and resource use

Deciduous vegetation shields building in summer and lets in light in winter

Low-E windows provide insulation

Radiant heating and cooling system saves energy

Low-emission paints and flooring reduce exposure to health hazards

Barrels or catchment basin harvest rainwater

FIGURE 13.16 A green building incorporates design features to minimize its ecological footprint.

technologies is probably the most effective way cities can reduce energy consumption and greenhouse gas emissions.

The U.S. Green Building Council promotes sustainable building efforts through the **Leadership in Energy and Environmental Design** (LEED) certification program. Builders apply for certification (for new buildings or renovation projects) and, depending on their performance, may be granted silver, gold, or platinum status (**TABLE 13.2**, p. 350).

Green building techniques add expense to construction, but the added cost generally ranges only from 0 to 10%. Moreover, this cost is recouped over time as money is saved on utility bills. Studies indicate that over a building's lifetime, green buildings save up to 10 times more money than their extra construction cost. In addition, studies suggest that workers tend to be more productive in green buildings.

Today, LEED certification is booming. Portland features several dozen LEED-certified buildings, including the nation's

first LEED Gold–certified sports arena (the Moda Center, where the Trailblazers basketball team plays). Savings on energy, water, and waste at the Moda Center paid for the cost of its LEED upgrade after just one year.

Schools, colleges, and universities are leaders in sustainable building. In Portland, the Rosa Parks Elementary School was built with locally sourced and nontoxic materials, uses 24% less energy and water than comparable buildings, and diverted nearly all its construction waste from the landfill. Schoolchildren learn about renewable energy by watching a display of the electricity produced by their building's photovoltaic solar system. Portland State University, the University of Portland, Reed College, and Lewis and Clark College are just a few of the many colleges and universities nationwide constructing green buildings as part of their campus sustainability efforts (pp. 18–19).

TABLE 13.2 Green Building Approaches for LEED Certification of New Construction

APPROACHES THAT ARE REWARDED (POINTS MAY TOTAL UP TO 100):	MAXIMUM POINTS*
ENERGY: Monitor energy use; use efficient design, construction, appliances, systems, and lighting; use clean, renewable energy sources	37
THE SITE: Build on previously developed land; minimize erosion, runoff, and water pollution; use regionally appropriate landscaping; integrate with transportation options	21
INDOORS: Improve indoor air quality; provide natural daylight and views; improve acoustics	17
MATERIALS: Use local or sustainably grown, harvested, and produced products; reduce, reuse, and recycle waste	14
WATER USE: Use efficient appliances inside; landscape for water conservation outside	11
THEN, UP TO 10 BONUS POINTS MAY BE AWARDED FOR:	
INNOVATION: New and innovative technologies and strategies to go beyond LEED requirements	6
THE REGION: Addressing environmental concerns most important for one's region	4

*Out of 110 possible points, 40 are required for LEED certification, 50 for silver, 60 for gold, and 80 for platinum levels.

Urban Sustainability

Most of our efforts to make cities safer, cleaner, healthier, and more beautiful are also helping make them more sustainable. A sustainable city is one that can function and prosper over the long term, providing generations of residents a good quality of life far into the future. In part, this entails minimizing the city's impacts on the natural systems and resources that nourish it. It also entails viewing the city as an ecological system (see **THE SCIENCE BEHIND THE STORY**, pp. 352–353). Urban centers exert both positive and negative environmental impacts. The extent and nature of these impacts depend on how we use resources, produce goods, transport materials, and deal with waste.

Urban resource consumption brings a mix of environmental effects

You might guess that urban living has a greater environmental impact than rural living. However, the picture is not that simple; instead, urbanization brings a complex mix of consequences.

Resource sinks Cities and towns are sinks (p. 121) for resources, needing to import from source areas beyond their borders nearly everything they need to feed, clothe, and house their inhabitants and power their commerce. Urban and suburban areas rely on large expanses of land elsewhere to supply food, fiber, water, timber, metal ores, and mined fuels. Urban centers also need areas of natural land to provide ecosystem services, including purification of water and air, nutrient cycling, and waste treatment. Indeed, for their day-to-day survival, major cities such as New York, Boston, San Francisco, and Los Angeles depend on water they pump in from faraway watersheds (**FIGURE 13.17**).

The long-distance transportation of resources and goods from countryside to city requires fossil fuel use and thereby has considerable environmental impact. However, imagine that the world's 4.3 billion urban residents were instead spread evenly across the landscape. What would the transportation requirements be, then, to move all those resources and goods around to all those people? A world without cities would likely require *more* transportation to provide people the same degree of access to resources and goods.

Efficiency Once resources arrive at an urban center, the concentration of people allows goods and services to be delivered efficiently. For instance, providing electricity for densely packed urban homes and apartments is more efficient than providing electricity to far-flung homes in the countryside. The density of cities facilitates the provision of medical services, education, water and sewer systems, waste disposal, public transportation, and more.

More consumption Because cities draw resources from afar, their ecological footprints are much greater than their actual land areas. For instance, urban scholar Herbert Girardet calculated that the ecological footprint of London, England, extends 125 times larger than the city's actual area. By another estimate, cities take up only 2% of the world's land surface but consume more than 75% of its resources.

However, the ecological footprint concept is most meaningful when used on a per-person basis. So, in asking whether urbanization intensifies resource consumption, we must ask whether the average urban- or suburban-dweller has a larger footprint than the average rural-dweller. The answer is yes, but urban and suburban residents also tend to be wealthier than rural residents, and wealth correlates with resource consumption. Thus, although urban and suburban citizens tend to consume more than their rural counterparts, the reason could simply be that they tend to be wealthier.

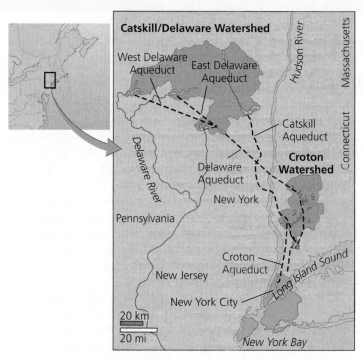

Catskill/Delaware Watershed

West Delaware Aqueduct
East Delaware Aqueduct
Hudson River
Massachusetts
Catskill Aqueduct
Croton Watershed
Delaware Aqueduct
Connecticut
Delaware River
New York
Pennsylvania
Croton Aqueduct
New York City
Long Island Sound
New Jersey
20 km
20 mi
New York Bay

FIGURE 13.17 New York City pipes in its drinking water from reservoirs in two upstate watersheds. The city acquires, protects, and manages upstate watershed land to minimize pollution of these water sources.

Urbanization preserves land

Because people pack together densely in cities, more land outside cities is left undeveloped. Indeed, that is the very idea behind urban growth boundaries. If cities did not exist and all 7 billion of us were evenly spread across the planet's land area, no large blocks of land would be left uninhabited, and we would have far less room for agriculture, wilderness, biodiversity, or privacy. The fact that half the human population is concentrated in discrete locations helps allow space for natural ecosystems to continue functioning and provide the ecosystem services on which all of us, urban and rural, depend.

Cities suffer and export pollution

Just as cities import resources, they export wastes, either passively through pollution or actively through trade. In so doing, urban centers transfer the costs of their activities to other regions—and mask the costs from their own residents. Citizens of Indianapolis, Columbus, or Buffalo may not recognize that pollution from the coal-fired power plants that supply them electricity worsens acid precipitation hundreds of miles to the east. New York City residents may not realize how much garbage their city produces if it is shipped elsewhere for disposal.

However, not all waste and pollution leave the city. Urban-dwellers are exposed to heavy metals, industrial compounds, and chemicals from manufactured products that accumulate in soil and water. Airborne pollutants cause smog and acid precipitation (Chapter 17). Fossil fuel combustion releases greenhouse gases as well as pollutants that pose health risks.

City residents suffer thermal pollution as well because cities tend to have ambient temperatures that are several degrees higher than those of surrounding areas. This **urban heat island effect** (**FIGURE 13.18**) results from the concentration of heat-generating buildings, vehicles, factories, and

FAQ

Aren't cities bad for the environment?

Stand in the middle of a big city and look around. You see concrete, cars, and pollution. Environmentally bad, right? Not necessarily. The widespread impression that urban living is less sustainable than rural living is largely a misconception. Consider that in a city you can walk to the grocery store instead of driving. You can take the bus or the train. Police, fire, and medical services are close at hand. Water and electricity are easily supplied to your entire neighborhood, and waste is easily collected. In contrast, if you live in the country, resources must be used to transport all these services for long distances, or you need to burn gasoline traveling to reach them. By clustering people together, cities distribute resources efficiently while also preserving natural lands outside the city.

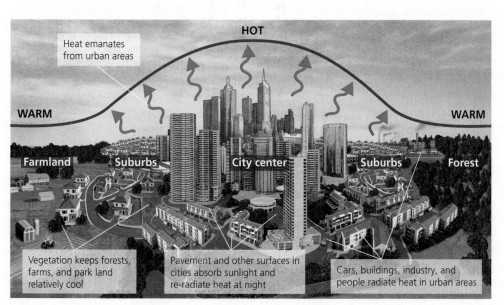

FIGURE 13.18 Cities produce urban heat islands, creating temperatures warmer than surrounding areas.

HOT

Heat emanates from urban areas

WARM
WARM

Farmland
Suburbs
City center
Suburbs
Forest

Vegetation keeps forests, farms, and park land relatively cool

Pavement and other surfaces in cities absorb sunlight and re-radiate heat at night

Cars, buildings, industry, and people radiate heat in urban areas

THE SCIENCE behind the story

How Do Baltimore and Phoenix Function as Ecosystems?

Sampling water beneath an overpass in Baltimore

Researchers in urban ecology examine how ecosystems function in cities and suburbs, how natural systems respond to urbanization, and how people interact with the urban environment. Today, Baltimore and Phoenix are centers for urban ecology.

These two cities are very different: Baltimore is an Atlantic port city on the Chesapeake Bay with a long history, whereas Phoenix is a young and fast-growing southwestern metropolis that sprawls across the desert. Each was selected by the U.S. National Science Foundation as a research site in its Long Term Ecological Research program, which funds multidecade ecological research. Since 1997, hundreds of researchers have studied Baltimore and Phoenix explicitly as ecosystems, examining nutrient cycling, biodiversity, air and water quality, environmental health risks, and more.

Research teams in both cities are combining old maps, aerial photos, and new remote sensing satellite data to reconstruct the history of landscape change. In Phoenix, one group showed how urban development spread across the desert in a "wave of advance," affecting soils, vegetation, and microclimate as it went. In Baltimore, mapping showed that development fragmented the forest into smaller patches over the past 100 years, even while the overall amount of forest remained the same.

For each city, study regions encompass both heavily urbanized central city areas and outlying rural and natural areas. To measure the impacts of urbanization, many research projects compare conditions in these two types of areas.

Baltimore scientists can see ecological effects of urbanization by comparing the urban lower end of their site's watershed with its less developed upper end. In the lower end, pavement,

rooftops, and compacted soil prevent rain from infiltrating the soil, so water runs off quickly. The rapid flow cuts deep streambeds into the earth while leaving surrounding soil drier. As a result, wetland-adapted trees and shrubs are vanishing, replaced by dry-adapted upland trees and shrubs.

The fast flow of water also worsens pollution. In natural areas, streams and wetlands filter pollution by breaking down nitrogen compounds. But in urban areas, where wetlands dry up and runoff from pavement creates flash floods, the filtering capacities of streams can be overwhelmed by high volumes of nitrates. In Baltimore, the resulting pollution ends up in the Chesapeake Bay, which suffers eutrophication and a large hypoxic dead zone (pp. 110, 410). Baltimore scientists studying nutrient cycling (p. 121) found that urban and suburban watersheds suffer far more nitrate pollution than natural forests, yet also found that the filtering ability of urban park lands helps keep pollution much lower than in agricultural landscapes (**FIGURE 1**).

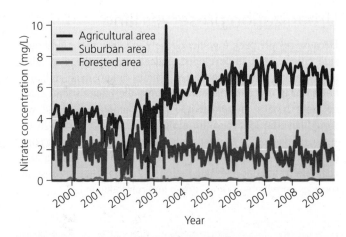

FIGURE 1 **Streams in Baltimore's suburbs contain more nitrates than streams in nearby forests, but fewer than those in agricultural areas, where fertilizers are applied liberally.**
Data from Baltimore Ecosystem Study, www.lternet.edu/research/keyfindings/urban-watersheds.

people. It also results from the way buildings and dark paved surfaces absorb sunlight throughout the day and then release the energy slowly at night as heat. In addition, buildings block cooling wind currents, and impervious surfaces cause rainwater to run off into sewers rather than evaporating and cooling the air. Much of the energy a city gathers in the daytime is released as heat at night, which warms the nighttime air and interferes with patterns of convective circulation that would otherwise cool the city. To minimize the urban heat island

effect, we can plant more vegetation and can paint rooftops pale colors to reflect sunlight.

Urban residents also suffer noise pollution and light pollution. **Noise pollution** consists of undesired ambient sound. Excess noise can cause stress, and at intense levels (such as with prolonged exposure to the sounds of leaf blowers, lawn mowers, and jackhammers) can harm hearing. The glow of **light pollution** from city lights may impair sleep and obscures the night sky, impeding the visibility of stars.

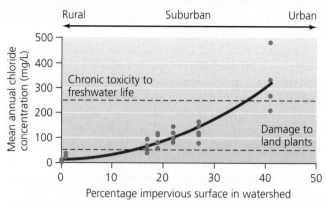

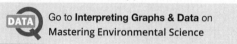

FIGURE 2 Salt concentrations in Baltimore-area streams are high enough to damage plants in the suburbs and to kill aquatic animals in urban areas. Data samples (red dots) were obtained over a five-year period. *Adapted from Kaushal, S.S., et al., 2005. Increased salinization of fresh water in the northeastern United States.* Proc. Natl. Acad. Sci. USA *102: 13517–13520, Fig 2. ©2005 National Academy of Sciences, U.S.A. By permission.*

DATA Q Go to **Interpreting Graphs & Data** on **Mastering Environmental Science**

Baltimore research also reveals impacts of applying salt to icy roads in winter. Road salt makes its way into streams, which become up to 100 times saltier. Such high salinity kills organisms (**FIGURE 2**), degrades habitat and water quality, and impairs streams' ability to remove nitrate.

To study contamination of groundwater and drinking water, researchers are using isotopes (p. 26) to trace where salts in the most polluted streams are coming from. Baltimore is now improving water quality substantially with a $900-million upgrade of its sewer system.

Urbanization also affects species and ecological communities. Cities and suburbs facilitate the spread of non-native species, because people introduce exotic ornamental plants and because urbanization's impacts on the soil, climate, and landscape favor weedy generalist species over more specialized native ones. In Baltimore, non-native plant species are most abundant in urban areas. In Phoenix's dry climate, pollen from some non-native plants causes allergy problems for city residents.

Community ecologists studying the wild animals and plants that persist within Phoenix are finding that urbanization alters relationships among them. Compared with natural landscapes, cities offer steadier and more reliable food resources—think of people's bird feeders or food scraps from dumpsters.

Growing seasons are extended and seasonal variation is buffered in cities as well. The urban heat island effect (p. 351) raises nighttime temperatures and makes temperatures more similar year-round. Buildings and ornamental vegetation shelter animals from extreme conditions, and irrigation in yards and gardens provides water. In a desert city like Phoenix, watering boosts primary productivity and lowers daytime temperatures. Together, all these changes lead to higher population densities of animals but lower species diversity as generalists thrive and displace specialists.

Urban ecologists in Phoenix and Baltimore are also studying social and demographic aspects of the urban environment. Some research measures how natural amenities affect property values. One study found that proximity to a park increases a home's property values—unless crime is pervasive. If the robbery rate surpasses 6.5 times the national average (as it does in Baltimore), proximity to a park begins to depress property values.

Other studies focus on environmental justice concerns (pp. 139–141). These studies have repeatedly found that sources of industrial pollution tend to be located in neighborhoods that are less affluent and that are home primarily to people of racial and ethnic minorities. Phoenix researchers mapped patterns of air pollution and toxic chemical releases and found that minorities and the poor are exposed to a greater share of these hazards. As a result, they suffer from higher rates of childhood asthma.

In Baltimore, researchers found a more complex pattern. Toxic release sites were more likely to be in working-class white neighborhoods than in African American neighborhoods. This, the researchers concluded, was a result of historical inertia. In the past, living close to one's workplace—the factories that release toxic chemicals—was something people preferred, and white workers claimed the privilege of living near their workplaces.

Whether addressing the people, natural communities, or changing ecosystems of the urban environment, studies on urban ecology like those in Phoenix and Baltimore will be vitally informative in our ever more urban world.

These various forms of pollution and the health risks they pose are not evenly shared among urban residents. Those who receive the brunt of the pollution are often those who are too poor to live in cleaner areas. Environmental justice concerns (pp. 139–141) center on the fact that a disproportionate number of people living near, downstream from, or downwind from factories, power plants, and other polluting facilities are people who are poor and, often, people of racial or ethnic minorities.

Urban centers foster innovation

Cities promote a flourishing cultural life and, by mixing together diverse people and influences, spark innovation and creativity. The urban environment can promote education and scientific research, and cities have long been viewed as engines of technological and artistic inventiveness. This inventiveness can lead to solutions to societal problems, including ways to reduce environmental impacts. For instance, research into renewable energy is helping us develop ways to replace fossil fuels. Technological advances have helped us reduce pollution. Wealthy and educated urban populations also provide markets for low-impact goods, such as organic produce. Recycling programs help reduce the solid waste stream. Environmental education is helping people choose their own ways to live cleaner, healthier, lower-impact lives. All these phenomena are facilitated by the education, innovation, science, and technology that are part of urban culture.

Urban ecology helps cities move toward sustainability

Cities that import all their resources and export all their wastes have a linear, one-way metabolism. Linear models of production and consumption tend to destabilize environmental systems and are ultimately not sustainable. Proponents of sustainability for cities stress the need to develop circular systems, akin to systems found in nature, which recycle materials and use renewable sources of energy.

Researchers in the field of **urban ecology** hold that cities can be viewed explicitly as ecosystems and that the fundamentals of ecosystem ecology and systems science (Chapter 5) apply to the urban environment. Major urban ecology projects are ongoing in Baltimore and Phoenix, where researchers are studying these cities as ecological systems (see **The Science behind the Story,** pp. 352–353).

Planners and visionary leaders have come up with designs for entire "eco-cities" built from scratch that follow cyclical, sustainable patterns of resource use and waste recycling. So far, none of these efforts have come to fruition, but existing cities across the world are adopting ecologically sustainable strategies by maximizing the efficient use of resources, recycling waste and wastewater, and developing green technologies. Urban agriculture that recycles organic waste and produces locally consumed food is thriving in many places, from Portland to Cuba to Japan. Curitiba, Brazil (p. 347), shows the kind of success that can result when a city invests in well-planned infrastructure.

In 2007, New York City unveiled an ambitious plan that then-mayor Michael Bloomberg hoped would make it "the first environmentally sustainable 21st-century city." PlaNYC was a 132-item program to reduce greenhouse gas emissions, improve mass transit, plant trees, clean up polluted land and rivers, and enhance access to park land (**FIGURE 13.19**). In 2015, Mayor Bill de Blasio continued the program under a new name, OneNYC, while adding new dimensions to promote environmental justice and economic equity.

Under the plan thus far, according to city documents, New York City has enhanced energy efficiency in hundreds of buildings, installed thousands of solar panels, planted 1.5 million trees, and opened or renovated several hundred parks, playgrounds, and community gardens. It expanded curbside recycling and composting while reducing landfill disposal by 10%. To fight water pollution, it installed green infrastructure, upgraded wastewater treatment, and cleaned up 750 polluted **brownfield** sites (p. 638) while acquiring 36,000 acres to protect upstate drinking water supplies. To promote cleaner transportation, city leaders installed bike lanes and racks and launched a bike-sharing program, retrofitted ferries to reduce pollution, converted hundreds of taxis to hybrid vehicles, expanded biodiesel use, and introduced more than 1200 electric vehicles and 500 charging stations. The city also divested its pension plan from fossil fuel stocks and is providing green job training to several thousand workers. Altogether, these actions have helped reduce the city's greenhouse gas emissions since 2005 by 15%. Today, New York City's air is the cleanest in 50 years and its water is the cleanest in more than a century.

New York City's Sustainability Goals
- Reduce greenhouse gas emissions 80% by 2050
- Send zero waste to landfills by 2030
- Enhance parks
- Convert contaminated land to safe, beneficial use
- Manage water resources and alleviate flooding
- Achieve the best air quality of all large U.S. cities

FIGURE 13.19 New York City is making impressive strides in urban sustainability under its OneNYC program. The newly completed One World Trade Center tower that now dominates the skyline has many energy-saving and water-saving green building features.

Steps toward livability enhance sustainability

Most steps being taken to make cities more livable also help make them more sustainable. Planning and zoning are pursuits that specifically entail a long-term vision. By projecting further into the future than political leaders or businesses generally do, planning and zoning are powerful forces for sustaining urban communities. The principles and practices of smart growth and new urbanism cut down on energy consumption, helping us address the looming challenge of climate change (Chapter 18). Encouraging mass transit reduces gasoline consumption and carbon emissions. Parks offer ecosystem services while promoting residents' health. And green buildings bring a diversity of health and environmental benefits.

Successes from Portland to Curitiba to New York City show how we can make cities more sustainable. Indeed, because they affect the environment in many positive ways and can promote efficient resource use, urban centers are a key element in achieving progress toward global sustainability.

CENTRAL CASE STUDY
connect & continue

TODAY, as the human population shifts from rural to urban lifestyles, our impacts become less direct but more far-reaching. Making urban and suburban areas both more livable and sustainable will be vital for our future. Portland, Oregon, is one city that has enhanced the quality of life for its residents while making strides toward environmental sustainability. In a great many ways, Portland's reputation as a green city and international model of livability is well justified.

However, as people stream to Portland in droves, the city risks becoming a victim of its own success. Growth forecasts estimate that the number of households in Portland will jump by 44–57% as early as 2035 and that households throughout the region will rise in number by 56–74%. As density increases inside the urban growth boundary, new challenges—such as rising rents, highway traffic jams, parking congestion on residential streets, and debates over gentrification—are beginning to strain the smart-growth vision that has worked so well thus far.

In a search for solutions amid an ongoing declared "housing emergency," Portland's leaders have been engaging citizens in planning processes to keep their city "prosperous, healthy, equitable, and resilient." In 2016, they formally adopted the 2035 Comprehensive Plan to help guide decision making through 2035. Currently debate is swirling over the Residential Infill Project, a proposed program of changes to boost population density by allowing more multifamily dwellings in single-family residential neighborhoods. Similar changes are being enacted several hours away in Bend, Oregon's fastest-growing city.

To deal with traffic congestion, city, regional, and state leaders are planning to construct new light rail lines between Portland and its suburbs. They are also exploring adding more lanes to the freeways and charging tolls to discourage rush-hour traffic and encourage carpooling.

Portland is just one of many urban centers seeking to expand economic opportunity and enhance quality of life while protecting environmental quality. Planning and zoning, smart growth and new urbanism, mass transit, parks, and green buildings are all ingredients in sustainable cities, and we should be encouraged about our progress in these endeavors. Ongoing experimentation throughout our urbanizing world will help us determine how to continue creating better and more sustainable communities in which to live.

- **CASE STUDY SOLUTIONS** After you earn your college degree, you decide to settle in the Portland, Oregon, region, where you are being offered three equally desirable jobs in three very different locations. If you accept the first, you will live in downtown Portland, amid commercial and cultural amenities but where population density is high and growing. If you take the second, you will live in one of Portland's suburbs where you have more space but where commute times are long and sprawl may soon surround you for miles. If you select the third, you will live in a rural area outside the urban growth boundary with plenty of space and scenic beauty but few cultural amenities. You are a person who aims to live in an ecologically sustainable way. Where would you choose to live? Why? What considerations will you factor into your decision?

- **LOCAL CONNECTIONS** Consider the region where you live or attend school, including your nearest large city. For each of the following pursuits, describe what this area possesses, and suggest at least one way you think the area could enhance its quality of life and its sustainability: (a) planning and zoning policies; (b) walkable neighborhoods as promoted by smart growth and new urbanism; (c) transportation options (mass transit, bike lanes, ways to avoid car congestion); (d) parks, green spaces, and natural lands; and (e) green buildings.

- **EXPLORE THE DATA** How do experts measure the pros and cons of mass transit in urban areas? → **Explore Data** relating to the case study on **Mastering Environmental Science.**

REVIEWING Objectives

You should now be able to:

+ Describe the scale of urbanization

The world's population has become predominantly urban. Our ongoing shift from rural to urban living is driven largely by industrialization and is proceeding fastest in the developing world. The location and growth of cities have always been influenced by environmental factors, but the geography of urban areas changes as cities decentralize and suburbs grow and expand. (pp. 338–340)

+ Define sprawl and discuss its causes and consequences

Sprawl covers large areas of land with low-density development. Both population growth and increased per capita land use contribute to sprawl. Sprawl results from the home-buying choices of individuals who prefer suburbs to cities, but has been facilitated by government policy and technological developments. Sprawl may lead to negative impacts involving transportation, pollution, health, land use, natural habitat, and economics. (pp. 340–342)

+ Outline city and regional planning and land use strategies

City planning and regional planning, along with zoning, are key tools for improving the quality of urban life. Urban growth boundaries, smart growth, and new urbanism attempt to re-create compact and vibrant urban spaces. (pp. 342–345)

+ Evaluate transportation options, urban parks, and green buildings

Mass transit systems can enhance the efficiency and sustainability of urban areas. Urban park lands provide recreation, soothe the stress of urban life, and give people access to natural areas. Green buildings minimize their ecological footprints by using sustainable materials, limiting the use of energy and water, minimizing health impacts on their occupants, controlling pollution, and recycling waste. (pp. 345–350)

+ Analyze environmental impacts and advantages of urban centers

Cities are resource sinks with high per capita resource consumption, and they create substantial waste and pollution. However, cities also maximize efficiency, help preserve natural lands, and foster innovation. (pp. 350–354)

+ Assess urban ecology and the pursuit of sustainable cities

Linear modes of consumption and production are unsustainable, and more circular modes will be needed to create truly sustainable cities. Many cities worldwide are taking steps to decrease their ecological footprints. Most steps taken for urban livability also function to enhance sustainability. (pp. 352–355)

SEEKING Solutions

1. Describe the causes of the spread of suburbs and outline the environmental, social, and economic impacts of sprawl. Overall, do you think the spread of urban and suburban development that is commonly labeled *sprawl* is predominantly a good thing or a bad thing? Do you think it is inevitable? Give reasons for your answers.

2. Would you personally want to live in a neighborhood developed in the new-urbanist style? Why or why not? Would you like to live in a city or region with an urban growth boundary? Why or why not?

3. All things considered, do you think cities are a positive thing or a negative thing for environmental quality? How much do you think we may be able to improve the sustainability of our urban areas?

4. **THINK IT THROUGH** You are the facilities manager on your campus, and your school's administration has committed funds to retrofit one existing building with sustainable green construction techniques so that it earns LEED certification. Consider the various buildings on your campus, and select one that you believe is unhealthy or that wastes resources in some way and that you would like to see retrofitted. Describe for an architect three specific ways in which green building techniques might be used to improve this particular building.

5. **THINK IT THROUGH** You are the president of your college or university, and students are clamoring for you to help create the world's first fully sustainable campus. Considering how people enhance livability and sustainability in cities, what lessons might you try to apply to your college or university? You are scheduled to give a speech to the campus community about your plans and will need to name five specific actions you plan to take to pursue a sustainable campus. What will they be, and what will you say about each choice to describe its importance?

CALCULATING Ecological Footprints

One way to reduce your ecological footprint is with alternative transportation. Each gallon of gasoline is converted during combustion to approximately 20 pounds of carbon dioxide (CO_2), which is released into the atmosphere. The table lists typical amounts of CO_2 released per person per mile through various modes of transportation, assuming typical fuel efficiencies.

For an average North American person who travels 12,000 miles per year, calculate and record in the table the CO_2 emitted yearly for each transportation option. Then calculate and record the reduction in CO_2 emissions, relative to single-occupant automobile driving, that one could achieve by relying solely on each option.

MODE OF TRANSPORT	CO_2 PER PERSON PER MILE	CO_2 PER PERSON PER YEAR	CO_2 EMISSION REDUCTION	YOUR ESTIMATED MILEAGE PER YEAR	YOUR CO_2 EMISSIONS PER YEAR
Automobile (driver only)	0.825 lb	9900 lb	0		
Automobile (2 persons)	0.413 lb				
Bus	0.261 lb				
Walking	0.082 lb				
Bicycle	0.049 lb				
				Total = 12,000	

1. Which transportation option provides the most miles traveled per unit of carbon dioxide emitted?

2. Clearly, it is unlikely that any of us will walk or bicycle 12,000 miles per year. In the last two columns, estimate what proportion of the 12,000 annual miles you think you actually travel by each method, and then calculate the CO_2 emissions that you are responsible for generating over the course of a year. Which transportation option accounts for the most emissions for you?

3. How could you reduce your CO_2 emissions? How many pounds of emissions do you think you could realistically eliminate over the course of the next year by making changes in your transportation decisions?

Mastering Environmental Science

Environmental Health and Toxicology

Bisphenol A: Worldwide

Are We Being Poisoned by Our Food Packaging?

> This chemical is harming snails, insects, lobsters, fish, frogs, reptiles, birds, and rats, and the chemical industry is telling people that because you're human, unless there's human data, you can feel completely safe.
>
> Dr. Frederick vom Saal, BPA researcher

> There is no basis for human health concerns from exposure to BPA.
>
> The American Chemistry Council

Thanks to a lifetime of nutritional education, we are aware that a diet high in sugars, fats, and processed foods can lead to harmful health impacts such as obesity, diabetes, and high blood pressure. Hence, we are taught to choose our foods wisely to avoid the dangers of an unhealthy diet. But, it turns out, the beverage bottles, food cans, and wrappers that surround our food may also pose a risk to our health by leaching harmful chemicals into the foods we eat and the liquids we drink.

Plastics are the predominant source of these hazardous chemicals. Plastics are polymers, or chains of molecules, that contain a number of synthetic chemicals. When plastics break down, the synthetic chemicals release into the surrounding environment. Some conditions that promote the breakdown of plastics are extreme temperatures, exposure to ultraviolet light from the sun, and prolonged contact with highly acidic liquids, such as soda and fruit juices.

One chemical of concern in plastics is **bisphenol A (BPA** for short). BPA is present in polycarbonate plastic, a hard, clear type of plastic used in water bottles, food containers, CDs and DVDs, electronics, baby bottles, and children's toys. Epoxy resins, plastic materials resistant to corrosion that are strongly adhesive, also contain BPA. Epoxy resins are used to line the insides of metal food and drink cans and the insides of pipes for our water supply, as well as in enamels, varnishes, adhesives, and even dental sealants for our teeth.

Many plastic products also contain another class of harmful chemicals, called **phthalates.** Used to soften plastics, they are found in plastics that come in contact with food, such as plastic gloves worn during food preparation, as well as in perfumes and children's toys (**FIGURE 14.1**, p. 360; see **THE SCIENCE BEHIND THE STORY**, pp. 362–363).

Unfortunately, BPA and phthalates can leach out of many products into our food, water, air, and bodies. Over 90% of Americans carry detectable concentrations of BPA and phthalates in their urine, according to data collected by National Health and Nutrition Examination Survey (NHANES) at the Centers for Disease Control and Prevention (CDC). Because these chemicals only reside in the human body for a few hours, these data suggest that we are receiving almost continuous exposure.

What, if anything, are BPA and phthalates doing to us? To address such questions, scientists run experiments on laboratory animals, administering

Upon completing this chapter, you will be able to:

+ Explain the goals of environmental health and identify major environmental health hazards

+ Describe the types of toxic substances in the environment, the factors that affect their toxicity, and the defenses that organisms have against them

+ Explain the movements of toxic substances and how they affect organisms and ecosystems

+ Discuss the approaches used to study the effects of toxic chemicals on organisms

+ Summarize risk assessment and risk management

+ Compare philosophical approaches to risk and how they relate to regulatory policy

◀ A hiker drinks from a "BPA-free" water bottle

▲ Takeout food in a plastic carry-out container

FIGURE 14.1 Some plastics, such as gloves worn during food preparation, can introduce phthalates into people's food.

known doses of the substance and measuring the health impacts that result. Hundreds of studies with rats, mice, and other animals have shown many apparent health impacts from BPA and phthalates, including a wide range of reproductive abnormalities. Studies have also connected BPA and phthalates to impacts on human health. For example, a 2013 review of the scientific literature found 91 studies that examined the relationship between the level of BPA in subjects' urine or blood and a variety of health problems (**FIGURE 14.2**). Phthalates have also been associated with a range of adverse health impacts, such as birth defects, breast cancer, reduced

sperm counts, diabetes, and cognitive impairment in children exposed to phthalates in the womb.

Many of these adverse health effects occur at extremely low doses of these chemicals—much lower than the exposure levels set so far by regulatory agencies for human safety. Scientists say this is because BPA and phthalates mimic certain hormones, such as male and female sex hormones; that is, they are structurally similar to sex hormones and can induce some of their effects in animals (see Figure 14.10, p. 371). Sex hormones function at minute concentrations, so when a synthetic chemical that is similar to the hormone reaches the body in a similarly low concentration, it can fool the body into responding. Other hormonal systems, such as the thyroid system that regulates growth and development, can also be affected by hormone-mimicking chemicals. Collectively, these chemicals alter the functioning of the endocrine system in humans, the system that regulates growth, reproduction, embryonic development, and a host of other physiological processes (p. 371). BPA, phthalates, and other hormone-mimicking chemicals are therefore called **endocrine-disrupting chemicals.**

In reaction to research linking BPA and phthalates to adverse health effects on humans, a growing number of researchers, doctors, and consumer advocates have called on governments to regulate BPA, phthalates, and other endocrine-disrupting chemicals and for manufacturers to stop using them. The chemical industry has long insisted that BPA and phthalates are safe, pointing to industry-sponsored research that finds no direct health impacts on people exposed to these chemicals. But some governments, including the United States, have not adopted the industry's view and have taken steps to regulate the use of BPA and phthalates in consumer products.

Canada has banned BPA completely and France outlawed its use in food packaging. In many other nations, including the United States, its use in products for babies and small children has been restricted. Accordingly, concerned parents can now

Cardiovascular disease

Hypertension

Sperm quality

Male sexual performance

Male genital abnormalities

Behavioral issues

Type 2 diabetes

Obesity

Asthma

Blood hormone levels

Kidney functioning

Thyroid function

Immune function

Egg development and maturation

FIGURE 14.2 Studies have linked elevated blood/urine BPA concentrations to numerous health impacts in humans. Although these correlative studies do not conclusively prove that BPA causes each observed ailment, they indicate topics for further research.

more easily find BPA-free items for their infants and children, but the rest of us remain exposed through most food cans, many drink containers, and thousands of other products. The European Union (EU) and nine other nations have banned some types of phthalates, and in 2008, the United States banned six types of phthalates in toys. Still, across North America, many routes of exposure remain.

In the face of mounting public concern about the safety of these chemicals, many U.S. companies voluntarily removed them from their products, even in the absence of stringent regulation by the U.S. government. Walmart, for example, decided to stop carrying children's products that contained BPA several years before the U.S. Food and Drug Administration (FDA) banned BPA use in baby bottles in 2012. Canned goods and containers that do not contain BPA are now available from large food companies, such as ConAgra, Campbell's, and Tyson Foods, and are typically identified as BPA-free on their label. There is precedent for such efforts: BPA was voluntarily phased out of can liners by Japanese companies starting in the late 1990s.

In a modern world replete with anthropogenic human impacts on ecosystems and human health, the adverse effects on human and ecosystem health from exposure to BPA and phthalates aren't likely to be among our greatest environmental health threats. They do, however, provide a timely example of how we as a society identify and assess health risks and decide how to manage them.

Environmental Health

Examining the impacts of human-made chemicals such as BPA is just one aspect of the broad field of **environmental health,** which assesses environmental factors that influence our health and quality of life. These factors include both natural and anthropogenic (human-caused) factors. Practitioners of environmental health seek to prevent adverse effects on human health and on the ecological systems that are essential to our well-being.

We face four types of environmental hazards

We can categorize them into four main types: physical, chemical, biological, and cultural. Although some amount of risk is unavoidable, much of environmental health consists of taking steps to minimize the risks of encountering hazards and to lessen the impacts of the hazards we do encounter.

Physical hazards Arising from processes that occur naturally in our environment, **physical hazards** pose risks to human life or health. Some are ongoing natural phenomena, such as ultraviolet (UV) radiation from sunlight (**FIGURE 14.3a**). Excessive exposure to UV radiation damages DNA in cells and has been tied to skin cancer, cataracts, and immune suppression. We can reduce these risks by shielding our skin from intense sunlight with clothing and sunscreen.

Other physical hazards are discrete events such as earthquakes, volcanic eruptions, fires, floods, landslides, hurricanes, and droughts. We cannot prevent many of these physical hazards, but we can minimize our risks of physical injury from

(a) Physical hazard

(b) Chemical hazard

(c) Biological hazard

(d) Cultural hazard

FIGURE 14.3 Environmental health hazards come in four types. The sun's ultraviolet radiation is an example of a physical hazard **(a).** Chemical hazards **(b)** include both synthetic and natural chemicals. Biological hazards **(c)** include diseases and the organisms that transmit them. Cultural, or lifestyle, hazards **(d)** include the behavioral decisions we make, such as smoking, as well as the socioeconomic constraints that may be forced on us.

THE SCIENCE behind the story

Are Endocrine Disruptors Lurking in Your Fast Food?

Dr. Ami Zota, George Washington University

Fast food may expose people to higher levels of endocrine disruptors than other types of food. A 2016 study, headed by Dr. Ami Zota of George Washington University and published in the journal *Environmental Health Perspectives,* embraced an epidemiological approach to answer a simple question: Did people who had recently eaten fast food have higher levels of BPA and phthalates in their bodies than people who had not recently eaten fast food?

To find out, the team dove into a treasure trove of data, the National Health and Nutrition Examination Survey (NHANES). This survey, conducted every two years by the Centers for Disease Control and Prevention (CDC), gathers detailed information from people across the United States by asking participants to undergo a physical examination by a medical professional, providing blood and urine samples and completing a detailed questionnaire about their lifestyle, including a description of the foods they have recently consumed.

The team scoured the surveys from 2003 to 2010 and found that approximately one-third of participants reported having eaten fast food in the 24 hours preceding their examination. (For the sake of the study, fast food was defined as food from restaurants that lack wait service, carryout and delivery food options, and pizza.) The team then used a correlational approach to look for relationships in the subjects' reported consumption of fast food and their urinary concentrations of BPA and two types of phthalates used in food packaging and processing: di(2-ethylhexyl) phthalate (DEHP) and diisononyl phthalate (DiNP).

After analyzing more than 8800 individuals, Dr. Zota's team found a positive correlation between the consumption of fast food and urinary concentrations of both types of phthalates, demonstrating that people who had recently eaten fast food had measurably higher levels of phthalates than people who had not eaten fast food (**FIGURE 1**). The quantity of fast food eaten was also related to concentrations of the two phthalates in subjects. People who consumed less than 35% of their calories from fast food ("low consumers") had DEHP levels 15% higher than people who had not eaten fast food ("nonconsumers"), while those who consumed more than 35% of their calories from fast food ("high consumers") had DEHP levels 23% higher than people who had not eaten fast food. Similar results were observed for DiNP, where low consumers and high consumers had urinary concentrations 24% and 39% higher, respectively, than nonconsumers. The researchers hypothesized that plastics in gloves from people preparing fast-food meals and from food-processing equipment were releasing phthalates into foods, particularly hot foods and foods that contain high levels of fat to which phthalates can bind. Unlike phthalates, BPA did not show a statistically significant correlation with fast-food consumption. The researchers hypothesize that BPA exposure from fast-food consumption may be minor relative to exposure from other

them by preparing ourselves with emergency plans and avoiding practices that make us vulnerable to certain physical hazards. For example, scientists can map geologic faults to determine areas at risk of earthquakes, engineers can design buildings to resist damage, and governments and individuals can create emergency plans to prepare for a quake's aftermath.

Chemical hazards Some of the synthetic chemicals that humanity manufactures, such as pharmaceuticals, disinfectants, and pesticides, are categorized as **chemical hazards** (**FIGURE 14.3b**). Some substances produced naturally by organisms, such as venoms, also can be chemically hazardous, as can many substances that we find in nature and then process for our use, such as hydrocarbons, lead, and asbestos. Following our overview of environmental health, much of this chapter will focus on chemical health hazards and the ways we study and regulate them.

Biological hazards **Biological hazards** result from ecological interactions among organisms that cause harm to people

(**FIGURE 14.3c**). When we become sick from a virus, bacterial infection, or other pathogen, we are suffering parasitism (p. 79). This is what we call **infectious disease.** Some infectious diseases are spread when pathogenic microbes attack us directly. With others, infection occurs through a **vector,** an organism, such as a mosquito, that transfers the pathogen to the host. Infectious diseases such as malaria, cholera, HIV, and influenza (flu) are major biological hazards, especially in developing nations with widespread poverty and limited health care. As with physical and chemical hazards, it is impossible for us to avoid risk from biological hazards completely, but through monitoring, sanitation, and medical treatment we can reduce the likelihood and impacts of infection.

Cultural hazards Risks posed to our life and health from our place of residence, the circumstances of our socioeconomic status, our occupation, or our behavioral choices can be thought of as **cultural hazards,** or lifestyle hazards. We can minimize or prevent some of these cultural hazards, whereas others may be beyond our control. For instance, people can

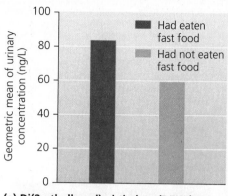
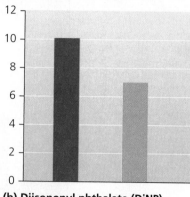
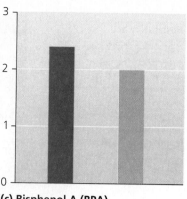

(a) Di(2-ethylhexyl) phthalate (DEHP) (b) Diisononyl phthalate (DiNP) (c) Bisphenol A (BPA)

FIGURE 1 **People who had recently eaten fast food showed significantly higher levels of (a) di(2-ethylhexyl) phthalate (DEHP) and (b) diisononyl phthalate (DiNP) in their urine (measured as nanograms of chemical per liter of urine) than people who had not eaten fast food.** Urinary concentrations of **(c)** BPA were not significantly different in the two groups. Note that the scale on the *y* axis differs in each of the three figure parts.

Data from Zota, A. R., et al., 2016. Recent fast food consumption and bisphenol A and phthalates exposures among the U.S. population in NHANES, 2003–2010. Environmental Health Perspectives *124: 1521–1528.*

DATA Go to **Interpreting Graphs & Data** on **Mastering Environmental Science**

sources of BPA, such as eating canned foods and drinking beverages from cans and bottles, resulting in any added BPA from fast-food items simply being "washed out" by larger exposures from other sources.

Industry groups such as the National Restaurant Association and the American Chemistry Council point out that the concentrations of phthalates detected in subjects were below the levels deemed dangerous for humans by the EPA and argue that phthalate exposures from fast food pose no risk to human health. Critics counter that the guidelines established by the EPA have not been revised since 1988, despite recent research

showing impacts on reproductive and developmental systems at levels similar to those seen in the study.

The regulation of chemicals that operate at extremely low concentrations, such as phthalates, will continue to pose regulatory challenges. Studies such as this one conducted by Dr. Zota and her team, however, can identify major sources of exposure to phthalates people may experience and prompt further study and action by scientists, government regulators, and fast-food companies to reduce such exposures—and safeguard human health from the far-reaching health impacts of endocrine-disrupting chemicals.

choose whether or not to smoke cigarettes (**FIGURE 14.3d**), but exposure to secondhand smoke in the home or workplace may be beyond one's control. Much the same might be said for diet and nutrition, workplace conditions, and drug use. Environmental justice advocates (pp. 139–141) argue that "forced" risks from cultural hazards, such as living near a hazardous waste site, are often elevated for people with fewer economic resources or less political clout.

Toxicology is the study of chemical hazards

Although most indicators of the state of overall human health are improving as the world's wealth increases, modern society is exposing us to more and more synthetic chemicals. Some of these substances pose threats to human health, but figuring out which of them do—and how, and to what degree—is a complicated scientific endeavor. **Toxicology** is the science that examines the effects of poisonous substances on humans and other organisms. Toxicologists assess and

compare substances to determine their **toxicity,** the degree of harm a chemical substance can inflict. A toxic substance, or poison, is called a **toxicant,** but any chemical substance may exert negative impacts if we ingest or expose ourselves to enough of it. Conversely, if the quantity is small enough, a toxicant may pose no health risk at all. These facts are often summarized in the catchphrase, "The dose makes the poison." In other words, a substance's toxicity depends not only on its chemical properties but also on its quantity.

In recent decades, our ability to produce new chemicals has expanded, concentrations of chemical contaminants in the environment have increased, and public concern for health and the environment has grown. These trends have driven the rise of **environmental toxicology,** which deals specifically with toxic substances that come from or are discharged into the environment. Toxicologists generally focus on human health, using other organisms as models and test subjects. Environmental toxicologists study animals and plants to determine the ecological impacts of toxic substances and to see whether other organisms can serve as indicators of health threats that could soon affect people.

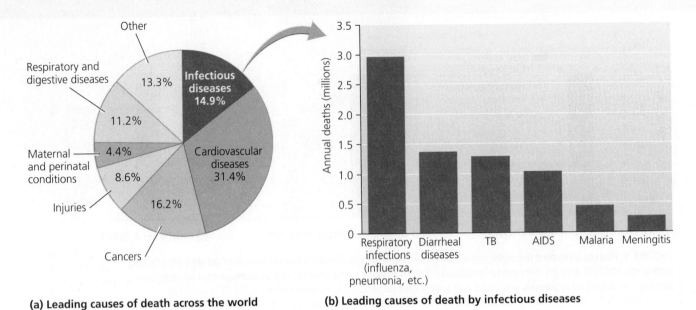

(a) Leading causes of death across the world

(b) Leading causes of death by infectious diseases

FIGURE 14.4 Infectious diseases are the second-leading cause of death worldwide. Six types of diseases account for 80% of all deaths from infectious disease. *Data are for 2016 from World Health Organization. Geneva, Switzerland: WHO, http://www.who.int.*

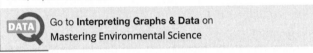 Go to **Interpreting Graphs & Data** on Mastering Environmental Science

The biological hazard of infectious disease continues to plague humanity

Despite all our technological advances, we still find ourselves battling disease, which causes the vast majority of human deaths worldwide (**FIGURE 14.4**). Infectious diseases have ravaged human populations throughout history, sometimes claiming huge numbers of lives in massive, widespread epidemics called *pandemics*. Examples of such diseases include cholera, bubonic plague, tuberculosis, malaria, smallpox, and various strains of flu—just to name a few. Although infectious disease accounts for fewer deaths than noninfectious disease, infectious disease robs society of more years of human life in its people, because it strikes all ages, including the very young.

Infectious diseases spread when a pathogenic organism enters a host through the skin, via the respiratory system, or by the consumption of contaminated food or water. Once established in the host, the pathogen uses energy from the host's tissues and the favorable internal conditions inside the body (a warm, wet environment that is ideal for bacteria and viruses) to produce huge numbers of offspring. Because these offspring will need new hosts to survive and reproduce, the pathogen helps to expel its offspring by inducing vomiting or diarrhea (for pathogens that inhabit the digestive tract, such as *Salmonella*) or coughing and sneezing (for those that inhabit the respiratory system, such as influenza viruses) in the host. These pathogen-containing body fluids or aerosols are then taken in by other people, or contaminate local food and water sources, completing the cycle of infection (**FIGURE 14.5**).

In our modern world of global mobility and dense human populations, novel infectious diseases (or new strains of old

diseases) that emerge in one location spread quickly to other locations. Recent examples include the H5N1 avian flu ("bird flu"), which first appeared in humans in 1997; severe acute respiratory syndrome (SARS), where the first outbreak in humans was reported in 2003; the H1N1 swine flu that spread across the globe in 2009–2010; and the outbreak of Ebola in West Africa in 2014. Diseases like influenza are caused by pathogens that mutate readily, giving rise to a variety of strains of the disease with slightly different genetics. As a pathogen's genes determine its virulence, a measure of how fast a disease spreads and the harm it does to infected individuals. Frequent genetic mutations make a pathogen more likely to become a highly virulent strain, which may arise to infect swaths of a population and threaten a global pandemic at any time.

We fight infectious disease with diverse approaches

Although infectious disease has troubled human civilizations since the dawn of time, it wasn't until the late 1800s that scientists firmly established the connection between diseases and pathogenic viruses, bacteria, and protists. This discovery gave rise to antibiotics, chemicals that combat disease by killing pathogens. It also produced societal approaches that help to minimize the spread of disease. These include sterilizing drinking water, providing sanitary facilities (such as indoor plumbing or communal latrines) that prevent fecal contamination of drinking water sources, ensuring nutritious diets for all people to strengthen the immune system, and providing medical care that identifies infected individuals and breaks the cycle of

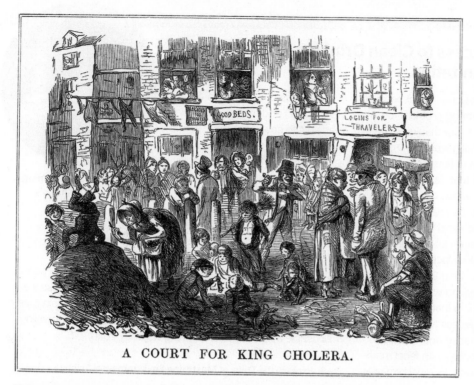

FIGURE 14.5 The infectious disease cholera ravaged London in the 19th century. Feces, both animal and human, were ever-present in city streets; the disease was spread through direct contact with feces and contamination of local drinking water sources.

infection through treatment, quarantine, and early intervention (see **SUCCESS STORY**, p. 366). Collectively, these approaches illustrate the ways we can combat infectious disease in the modern world. Besides providing food security (p. 242), this means ensuring access to safe drinking water and improving sanitation for all people, especially the poor.

Another important tool in fighting infectious disease is expanding access to health care. In developing nations, this includes opening clinics, immunizing children against diseases, providing prenatal and postnatal care for mothers and babies, and making generic and inexpensive pharmaceuticals available.

Education campaigns play a vital role in rich and poor nations alike. Public service announcements and government programs that educate the public about proper hygiene and safe food-handling procedures can minimize the incidence of disease (**FIGURE 14.6**). And as some infectious diseases are spread by sexual contact, such as HIV/AIDS, education on sex and reproductive health is helping men and women reduce risk from disease and avoid unwanted pregnancies.

Environmental health experts continue to work, along with governments and nongovernmental organizations such as the World Health Organization (WHO), to implement programs around the world that act to reduce the incidence of infectious disease and prevent widespread outbreaks. But given that the pathogens that infect us will not stop evolving, we cannot afford to let up on our vigilance in acting proactively to head off pandemics before they begin.

Noninfectious disease is affected by genes, environment, and lifestyle

We've seen that infectious diseases are caused by a pathogenic organism infecting a host. **Noninfectious diseases,** such as cancer and heart disease, develop without the action of a foreign organism. You don't "catch" noninfectious diseases—people develop them through a combination of their genetics, coupled with environmental and lifestyle factors. For instance, whether a person develops lung cancer depends not only on his or her genes but also on environmental conditions, such as the individual's exposure to airborne cancer-inducing chemicals, and to lifestyle choices, such as whether or not he or she chooses to smoke.

The incidence of noninfectious diseases can be altered by cultural trends and lessened by wider adoption of healthy lifestyles. In the United States over the last 30 years, for example, the percentage of Americans who smoke

FIGURE 14.6 Public education programs, like this one in India aimed at combatting the spread of swine flu, can reduce the incidence of infectious disease.

Improving Access to Clean Drinking Water and Sanitation

Potentially fatal water-borne diseases, such as cholera and dysentery, have been a vexing problem for many of the world's developing countries. But thanks to the efforts of the United Nations (UN), national governments, and aid organizations the threat from these infectious diseases—spread by the ingestion of food or water that has been contaminated with human or animal feces—is abating. Public health education campaigns have taught people how to sterilize their drinking water, as well as how to avoid contaminating the source of it. Infrastructure projects, such as the drilling of water wells and the construction of modern facilities to treat drinking water and wastewater, have provided 91% of people across the globe with access to safe drinking water—an increase of 2.6 billion people since 1990. In that same time period, some 2.1 billion people gained access to sanitary facilities as people in cities and villages embraced approaches to minimize the contamination of water supplies with fecal material, such as by installing composting toilets (as piped water is sometimes unavailable to the very poor) and through improved access to sewer systems, septic systems, and public latrines.

Initiatives to provide people with clean drinking water have been very successful, enabling the UN to meet its Millennium Development Goal (enacted in 2000) of providing 88% of the world's population with clean water by 2015. While these efforts represent a significant improvement, many people in rural areas in poorer nations continue to lack access to safe drinking water and advanced sanitation. Efforts are currently underway to address these deficiencies, and the United Nations in 2017 announced the goal of providing all people access to clean drinking water by 2030.

A young woman sampling water from her village's new well

→ **Explore the Data** at **Mastering Environmental Science**

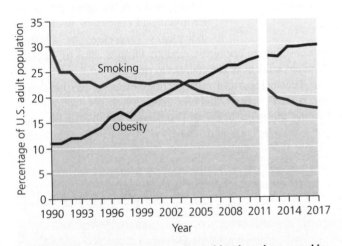

FIGURE 14.7 The prevalence of smoking has decreased in the United States in recent decades, but obesity is on the rise. Data from 1990 to 2011 are not directly comparable to data from 2012 to 2015 for both smoking and obesity due to methodology changes implemented after collecting the 2011 data. *Data from United Health Foundation, 2017.* America's health rankings, *2017 edition. Minnetonka, MN: United Health Foundation.*

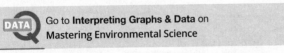

Go to **Interpreting Graphs & Data** on **Mastering Environmental Science**

cigarettes has decreased while obesity rates have increased (**FIGURE 14.7**). This has reduced the prevalence of factors that induce lung cancer but increased those that induce heart disease and type 2 diabetes. Obesity is typically due to low levels of physical activity coupled with a diet high in calories and fat, but other factors can be involved. Studies have found that exposing pregnant mice to high levels of BPA increases the production of fat cells in her developing embryos. This causes mice that were prenatally exposed to BPA to store more fat as adults than mice that were not exposed to BPA in the womb, even if both groups of mice have identical diets.

Many cultural hazards exist indoors

Americans spend roughly 90% of their lives indoors. But the spaces inside homes and workplaces, like the outdoors, can be rife with environmental hazards (pp. 477–480).

Cigarette smoke and radon are leading indoor hazards and are the top two causes of lung cancer in developed nations. Cigarette smoke contains substances that can harm the respiratory system and induce cancer. **Radon** is a highly toxic radioactive gas that is colorless and undetectable without specialized kits. It seeps up from the ground in areas with certain types of bedrock and can accumulate in basements and homes with poor air circulation.

Walls in homes and offices can harbor toxic compounds. Exposure to mold, which can flourish in wall spaces when moisture levels are high, can lead to serious respiratory infections. **Asbestos,** used in the past as insulation in walls and other products, is dangerous when its fibers are inhaled. Long-term exposure to asbestos scars the lung tissue, impairs lung function, and leads to a respiratory disorder called **asbestosis.**

Another indoor health hazard is **lead,** a heavy metal. When ingested, lead can cause **lead poisoning,** damage to the brain, liver, kidney, and stomach; learning problems and behavioral abnormalities in children; anemia; hearing loss; and even death. Lead poisoning among U.S. children has greatly declined in recent years, as a result of education

LeeAnne Walters holding samples of tap water from her home

Walters with her son Gavin at a 2015 rally in Flint

Walters testifying before Congress with professor Marc Edwards of Virginia Tech

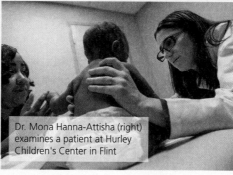

Dr. Mona Hanna-Attisha (right) examines a patient at Hurley Children's Center in Flint

Worker replacing a lead water-supply pipe outside a home in Flint

FIGURE 14.8 In the face of inaction by state regulators, the persistence of citizens, scientists, and physicians during the water crisis in Flint, Michigan, eventually led to meaningful change.

campaigns and the U.S. federal government mandating the phaseout of lead-based paints and leaded gasoline (p. 17) in the 1970s. Today, lead poisoning can result from drinking water that has passed through the lead pipes common in older homes or from ingesting or inhaling lead-containing dust produced by the slow wearing-away of leaded paint.

Lead poisoning captured headlines recently due to tragic events in Flint, Michigan (**FIGURE 14.8**). In April 2014, as a cost-cutting measure, Flint ceased purchasing its drinking water from the city of Detroit and switched its water supply to the Flint River. Almost immediately after the switch people began noticing that their water looked, smelled, and tasted odd.

Reports of children developing rashes and strange illnesses started to emerge across Flint, but state regulators concluded that the city's water remained safe to drink, even after tests revealed that lead levels in some homes were 1000 times higher than the allowable limit as set by the U.S. Environmental Protection Agency (EPA). Over the next 18 months, the state continued to claim that the problems with lead were not widespread, despite researchers estimating that up to 40% of Flint's homes were experiencing elevated lead levels in their tap water and local pediatricians reporting a doubling of the number of children with high levels of lead in their blood.

Eventually, the Michigan Department of Environmental Quality concluded that lead contamination was indeed widespread in Flint and was likely the result of the city's water not being properly treated with chemicals that prevent the corrosion of the lead pipes found in many older homes in Flint. In October

2015, after a year and half of exposure to lead-tainted water, Flint city officials agreed to reconnect Flint's water main to the water supply of the city of Detroit—and the EPA and the state of Michigan awarded the city nearly $200 million to upgrade its drinking water infrastructure and replace lead pipes in older homes.

One recently recognized indoor chemical hazard is **polybrominated diphenyl ethers (PBDEs).** These toxic compounds are used as fire retardants in computers, televisions, plastics, and furniture, and they may evaporate at very slow rates throughout the lifetime of the product. Like bisphenol A, PBDEs appear to act as hormone mimics. Animal testing suggests that PBDEs may also cause cancer, affect thyroid hormones, and influence the development of the brain and nervous system. The U.S. government's National Health and Nutrition Examination Survey in 2009 tested the blood and urine of a cross-section of Americans and found one type of PBDE was detected in nearly every person in the study—showing how widespread exposure to PBDEs is in modern America. The EU banned PBDEs in 2003, but in the United States, there has so far been little movement on government regulation of these chemicals.

The dangers posed by fire retardants such as PBDEs have caused some to question the stringent flammability standards that often make the use of such chemicals necessary. As stated by Linda Birnbaum of the National Institute of Environmental Health Sciences in an interview on PBDEs, "I don't question the need for flame retardants in airplanes, but do we need them in nursing pillows and babies' strollers?"

Risks must be balanced against rewards

It is important to keep in mind that artificially produced chemicals have played a crucial role in giving us the high standard of living we enjoy today. These chemicals have helped create the industrial agriculture that produces our food, the medical advances that protect our health and prolong our lives, and many of the modern materials and conveniences that we use every day. With most hazards, there is some tradeoff between risk and reward. Plastic bottles, can liners, and wrappers serve a useful purpose in safely containing and preserving our food, and their usefulness means that despite their health risks, we may as a society choose to continue including them in products when we judge their benefits to outweigh their impacts on human health. Thus, for BPA, phthalates, and numerous other toxic chemicals, it is appropriate to remember their benefits as we examine some of the unfortunate side effects that these substances may elicit.

Toxic Substances and Their Effects on Organisms

Our environment contains countless natural substances that may pose health risks. These include petroleum, oozing naturally from the ground; radon gas, seeping up from bedrock; and **toxins,** toxic chemicals manufactured in the tissues of living organisms. For example, toxins can be chemicals that plants use to ward off herbivores or that insects use to defend themselves from predators. In addition, we are exposed to many synthetic (human-made) chemicals, some of which have toxic properties.

Synthetic chemicals are all around us—and in us

Tens of thousands of synthetic chemicals have been manufactured (**TABLE 14.1**), and synthetic chemicals surround us in our daily lives. Each year in the United States, we manufacture or import over 100 kg (220 lb) of chemical substances for every man, woman, and child. Many of these substances find their way into soil, air, and water—and into humans and other organisms (**FIGURE 14.9**).

The **pesticides** we use to kill insects and weeds on farms, lawns, and golf courses are some of the most widespread synthetic chemicals in the environment. For example, a U.S. Geological Survey (USGS) National Water Quality Assessment (NAWQA) study published in 2018 examined 100 streams in 11 Midwestern states and found, on average, that each stream contained traces of 52 pesticides or pesticide by-products. A total of 94 different pesticides and 89 pesticide by-products were detected in waterways, and in 50% of the streams, at least one pesticide exceeded doses expected to

TABLE 14.1 Estimated Numbers of Chemicals in Commercial Substances

TYPE OF CHEMICAL	ESTIMATED NUMBER
Chemicals in commerce	100,000
Industrial chemicals	72,000
New chemicals introduced per year	2000
Pesticides (21,000 products)	600
Food additives	8700
Cosmetic ingredients (40,000 products)	7500
Human pharmaceuticals	3300

Data are for the 1990s, from Harrison, P., and F. Pearce, 2000. *AAAS atlas of population and environment.* Berkeley, CA: University of California Press.

cause harm to aquatic life. One-fourth of the streams had 2–4 pesticides that exceeded safe standards. Pesticide concentrations were at levels expected to be toxic for fish in five of the 100 streams. The greatest contributors to toxicity were from widely used agricultural **herbicides,** pesticides such as atrazine, acetochlor, and metachlor applied in crop fields to target and kill undesirable plants and weeds—and organophosphate **insecticides,** pesticides that target insects specifically. While streams in agricultural areas had the highest potential impacts on aquatic plants due to herbicide runoff from farm fields, streams in urban areas had elevated toxicity for stream invertebrates (such as the larvae of various species of flies) due to higher concentrations of insecticides. This study found some of the most complex mixtures of pesticides ever detected in U.S. waters, and similar analyses of pesticide concentrations in surface waters are planned in coming years for other U.S. regions.

As a result of all pesticide exposure, every one of us carries traces of hundreds of synthetic chemicals in our bodies. The U.S. government's latest NHANES (the health survey that found 93% of Americans showing traces of BPA in their urine; p. 359) gathered data on 148 foreign compounds in Americans' bodies. Among these were several toxic, persistent, organic pollutants restricted by international treaty (pp. 384–385). Depending on the pollutant, these were detected in 41% to 100% of the people tested. Smaller-scale surveys have found similar results.

Our exposure to synthetic chemicals began in the womb, as substances our mothers ingested while pregnant were transferred to us. A 2009 study by the non-profit Environmental Working Group found 232 chemicals in the umbilical cords of 10 newborn babies it tested. Nine of the 10 cords contained BPA, leading researchers to note that we are born "pre-polluted."

Synthetic chemical exposure should not necessarily be cause for alarm. Not all synthetic chemicals pose health risks, and relatively few are known with certainty to be toxic. However, of the roughly 100,000 synthetic chemicals on the market today, very few have been thoroughly tested. For the vast

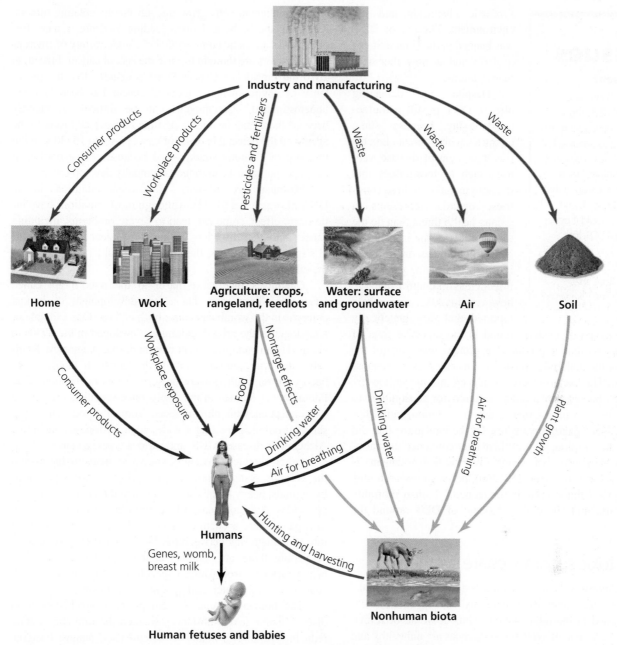

FIGURE 14.9 Synthetic chemicals take many routes in traveling through the environment. People take in only a tiny proportion of these compounds, and many compounds are harmless. However, people receive small amounts of toxicants from many sources, and developing fetuses and babies are particularly sensitive.

majority, we simply do not know what effects, if any, they may have on our health.

Silent Spring began the public debate over synthetic chemicals

It was not until the 1960s that people began to seriously consider the risks of exposure to pesticides. One event behind this growing awareness was the publication of Rachel Carson's 1962 book *Silent Spring*. At the time, pesticides were indiscriminately sprayed over residential neighborhoods and public areas, on an assumption that the chemicals would do no harm to people.

Carson brought together a diverse collection of scientific studies, medical case histories, and other data to contend that the insecticide DDT in particular, and synthetic pesticides in general, were hazardous to people, wildlife, and ecosystems. At the time, most consumers had no idea that the store-bought chemicals they used in their houses, gardens, and crops might be toxic.

The chemical industry challenged Carson's book vigorously, attempting to discredit the author's science and personal reputation. Carson suffered from cancer as she finished *Silent Spring* (p. 171), and she lived only briefly after its publication. However, the book was a best-seller and helped generate significant social change in views and actions toward

Although many nations have banned the use of DDT, the compound is still manufactured in India and exported to nations that lack such bans. How do you feel about this? Is it unethical for a company to sell a substance that has been deemed toxic by so many nations? Or, would it be unethical *not* to sell DDT to tropical nations if they desire it for improving public health, such as controlling mosquitoes that transmit malaria or the Zika virus?

synthetic chemicals and the environment. The use of DDT was banned in the United States in 1973 and is now illegal in many nations.

Despite its damaging effects, DDT is still manufactured today because some countries with tropical climates use it to control disease vectors, such as mosquitoes that transmit malaria. In these countries, malaria represents a greater health threat than do the toxic effects of the insecticide. New technologies are promising to reduce the need for insecticides to control malaria, however. In 2012, researchers reported that they genetically modified a type of bacteria found in the digestive tract of mosquitoes so that it produced a protein that impairs the hatching of the malarial parasite from its egg sacs in the mosquito's gut. The bacteria were introduced into mosquitoes by mixing the bacteria in a sugar solution for mosquitoes to drink. Of the mosquitos that drank the solution, only 20% later contained malarial parasites. Of the mosquitos that did not drink the solution, 90% of them later contained the harmful parasites. The use of genetically modified organisms is not itself without risk (pp. 255–260), but such research may provide nonchemical options for reducing human mortality from malaria and speed the phaseout of DDT around the world.

Not all toxic substances are synthetic

Although many toxicologists focus on synthetic chemicals, toxic substances also exist naturally in the environment around us and in the foods we eat. Thus, it would be a mistake to assume that all synthetic substances are unhealthy and that all natural substances are healthy. In fact, the plants and animals we eat contain many chemicals that can cause us harm. Recall that many plants produce toxins to ward off animals that eat them. In domesticating crop plants, we have selected for strains with reduced toxin content, but traces of naturally occurring plant toxins can still be found in many of the plants we eat. Furthermore, when we consume animal meat, we ingest toxins the animals obtained from plants or animals they ate. Scientists are actively debating just how much risk natural toxicants pose, and it is clear that more research is required to answer the question.

Toxic substances come in different types

Toxic substances can be classified based on their particular effects on health. The best-known toxicants are **carcinogens,** which are substances or types of radiation that cause cancer. In cancer, malignant cells grow uncontrollably, creating tumors, damaging the body, and often leading to death. Cancer frequently has a genetic component, but a wide variety of environmental factors are thought to raise the risk of cancer. Indeed, in 2010 the President's Cancer Panel concluded that the prevalence of environmentally induced cancer has been "grossly underestimated." Carcinogens can be difficult to identify because there may be a long lag time between exposure to the agent and the detectable onset of cancer—up to 15–30 years in the case of cigarette smoke—and because only a portion of people exposed to a carcinogen eventually develop cancer.

Mutagens are substances that cause mutations in the DNA of organisms (p. 28). Although most mutations have little or no effect, some can lead to severe problems, including cancer and other disorders. If mutations occur in an individual's sperm or egg cells, then the individual's offspring suffer the effects.

Chemicals that cause harm to the unborn are called **teratogens.** Teratogens that affect development of human embryos in the womb can cause birth defects. One example of a teratogen is the drug thalidomide, developed in the 1950s to aid in sleeping and to prevent nausea during pregnancy. Tragically, the drug caused severe birth defects in thousands of babies whose mothers were prescribed this medication. Thalidomide was banned in the 1960s once scientists recognized its connection with birth defects. Ironically, today the drug shows promise in treating a wide range of diseases, including Alzheimer's disease, AIDS, and various types of cancer.

Other chemical toxicants known as **neurotoxins** assault the nervous system. Neurotoxins include venoms produced by animals, heavy metals such as lead and mercury, and some pesticides. A famous case of neurotoxin poisoning occurred in Japan, where a chemical factory dumped mercury waste into Minamata Bay between the 1930s and 1960s. Thousands of people there ate mercury-contaminated fish and soon began suffering from slurred speech, loss of muscle control, sudden fits of laughter, and in some cases death.

The human immune system protects our bodies from disease. Some toxic substances weaken the immune system, reducing the body's ability to defend itself against bacteria, viruses, and other threats. Allergy-causing agents, called **allergens,** overactivate the immune system, causing an immune response when one is not necessary. One hypothesis for the increase in the number of people diagnosed with asthma in recent decades is that allergenic synthetic chemicals are more prevalent today in our environment than in the past. Allergens are not universally considered toxicants, however, because they affect some people but not others and because one's response does not necessarily correlate with the degree of exposure.

Pathway inhibitors are toxicants that interrupt vital biochemical processes in organisms by blocking one or more steps in a biochemical pathway. Rat poisons, for example, cause internal hemorrhaging in rodents by interfering with the biochemical pathways that create blood clotting proteins. Some herbicides, such as atrazine, kill plants by chemically blocking steps in photosynthesis. Cyanide kills by interrupting chemical pathways that produce energy in mitochondria, thereby depriving cells of life-sustaining energy.

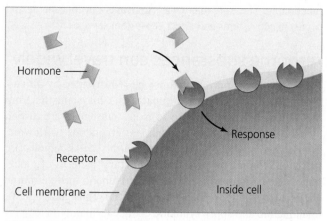

(a) Normal hormone binding

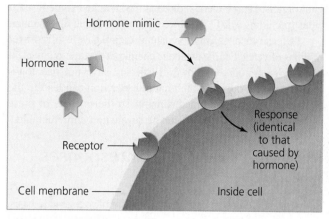

(b) Hormone mimicry

FIGURE 14.10 Many endocrine-disrupting substances mimic the chemical structure of hormone molecules. Like a key similar enough to fit into another key's lock, the hormone mimic binds to a cellular receptor for the actual hormone, causing the cell to react as though it had encountered the actual hormone.

Most recently, scientists have recognized endocrine-disrupting chemicals, toxic substances that interfere with the endocrine system. The endocrine system consists of chemical messengers, known as hormones, that travel through the bloodstream at extremely low concentrations and have many vital functions. They stimulate growth, development, and sexual maturity, and they regulate brain function, appetite, sex drive, and many other aspects of our physiology and behavior. Some hormone-disrupting toxicants affect an animal's endocrine system by blocking the action of hormones or accelerating their breakdown. Others are so similar to certain hormones in their molecular structure and chemistry that they "mimic" the hormone by interacting with receptor molecules just as the actual hormone would (**FIGURE 14.10**).

Among other effects, both BPA and phthalates appear to act as endocrine disruptors on the reproductive system. BPA is one of many endocrine-disrupting chemicals that seem to mimic the female sex hormone estrogen and bind to estrogen receptors. Indeed, emerging research is indicating that BPA might not be the only estrogen-mimicking compound in plastics, calling into question the safety of all the plastics that are ubiquitous in our lives. With their diverse impacts on human health, BPA and phthalates show how a substance can be a carcinogen, a mutagen, and an endocrine disruptor all at the same time. To make matters worse, other classes of chemicals in plastics may be mimicking hormones. A 2017 study, for example, found that one-third of fast-food packaging (mostly items with coatings designed for grease resistance) contained fluorinated chemicals that may also act as endocrine disruptors.

Organisms have natural defenses against toxic substances

Although synthetic toxicants are new, organisms have long been exposed to natural toxicants. Mercury, arsenic, cadmium, and other harmful substances are found naturally in the environment. Some organisms produce biological toxins to avoid predators or capture prey. Examples include venom in poisonous snakes and spiders, toxins in sea urchins, and the natural insecticide pyrethrin found in chrysanthemums. Over time, organisms able to tolerate these harmful substances have gained an evolutionary advantage.

Skin, scales, and feathers are the first line of defense against toxic substances because they resist uptake from the surrounding environment. However, toxicants can circumvent these barriers and enter the body through eating, drinking, and breathing. Once inside the organism, the toxic substances are distributed widely by the circulatory and lymph systems in animals, and by the vascular system in plants.

Organisms possess biochemical pathways that use enzymes to detoxify harmful chemicals that enter the body. Some pathways break down, or metabolize, toxic substances to render them chemically inert. Other pathways make toxic substances water soluble so they are easier to excrete through the urinary system. In humans, many of these pathways are found in the liver. As a result, this organ is disproportionately affected by intake of harmful substances, such as excessive alcohol.

Some toxic substances cannot be effectively detoxified or made water soluble by detoxification enzymes. Instead, the body sequesters these chemicals in fatty tissues and cell membranes to keep them away from

FAQ

Why do some insects survive exposure to a pesticide, while others are killed by it?

When a population of organisms is exposed to a toxicant, such as a pesticide, a few individuals often survive while the vast majority of the population is killed. The individuals that survive do so because they possess genes (which others in the population do not) that code for enzymes that counteract the toxic properties of the toxicant. Because the effects of these genes are expressed only when the pesticide is applied, many people think the toxicant "creates" detoxification genes by mutating the DNA of a small number of individuals. This is not the case. The genes for detoxifying enzymes were present in the DNA of resistant individuals from birth, but the effects of those genes were seen only when pesticide exposure caused selective pressure for resistance to the pesticide.

vital organs. Heavy metals, dioxins, PBDEs, and some insecticides (including DDT) are stored in body tissue in this manner.

Defense mechanisms for natural toxins have evolved over millions of years. For the synthetic chemicals that are so prevalent in today's environment, however, species have not had long-term exposure, during which natural selection can lead to the evolution of resistance mechanisms, so the impacts of these toxic substances on species can be severe and unpredictable.

Individuals vary in their responses to hazards

Some of the defenses described previously have a genetic basis. As a result, individuals may respond quite differently to identical exposures to hazards because they happen to have different combinations of genes. Poorer health also makes an individual more sensitive to biological and chemical hazards. Sensitivity also can vary with sex, age, and weight. Because of their smaller size and rapidly developing organ systems, younger organisms (for example, fetuses, infants, and young children) tend to be much more sensitive to toxicants than are adults. Regulatory agencies such as the EPA typically set human chemical exposure standards for adults and extrapolate downward for infants and children. However, many scientists contend that these linear extrapolations often do not offer adequate protection to fetuses, infants, and children due to their greater sensitivity to toxicants.

The type of exposure can affect the response

The risk posed by a hazard often varies according to whether a person experiences high exposure for short periods of time, known as **acute exposure,** or low exposure over long periods of time, known as **chronic exposure.** Incidences of acute exposure are easier to recognize, because they often stem from discrete events, such as accidental ingestion, an oil spill, a chemical spill, or a nuclear accident. Toxicity tests in laboratories generally reflect acute toxicity effects. However, chronic exposure is more common—and more difficult to detect and diagnose. Chronic exposure often affects organs gradually, as when smoking causes lung cancer or when alcohol abuse leads to liver damage. Because of the long time periods involved, relationships between cause and effect may not be readily apparent.

Toxic Substances and Their Effects on Ecosystems

When toxicants concentrate in environments and harm the health of many individuals, populations of the affected species become smaller. This decline in population size can then affect other species. For instance, species that are prey of the organism affected by toxicants could experience population growth because predation levels are lower. Predators of the poisoned species, however, would decline as their food source became less abundant. Cascading impacts can cause changes in the composition of the biological community and threaten ecosystem

functioning. There are many ways toxicants can concentrate and persist in ecosystems and affect ecosystem services.

Airborne substances can travel widely

Toxic substances may sometimes be redistributed by air currents (Chapter 17), exerting impacts on ecosystems far from their site of release. Because so many substances are carried by the wind, synthetic chemicals are ubiquitous worldwide, even in seemingly pristine areas. Earth's polar regions are particularly contaminated, because natural patterns of global atmospheric circulation (pp. 455–456) tend to move airborne chemicals toward the poles (**FIGURE 14.11**). Thus, although we manufacture and apply synthetic chemicals mainly in temperate and tropical regions, contaminants are strikingly concentrated in the tissues of Arctic polar bears, Antarctic penguins, and people living in Greenland. Polychlorinated biphenyls (PCBs), for example, were a class of toxic industrial chemicals used in electrical equipment, paints, and plastics from 1929 until they were banned in the United States in 1979. Yet, wildlife in the Arctic to this day are found to possess dangerously high levels of these chemicals in their tissues.

Effects can also occur over relatively shorter distances. Pesticides, for example, can be carried by air currents to sites far from agricultural fields in a process called *pesticide drift*. The

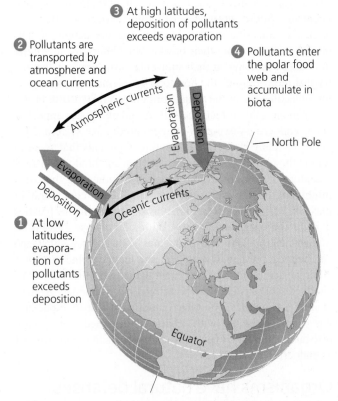

FIGURE 14.11 Air and water currents direct pollutants to the poles. In the process of global distillation, pollutants that evaporate and rise high into the atmosphere at tropical and temperate latitudes are carried toward the poles by atmospheric currents, whereas ocean currents carry pollutants deposited in the ocean toward the poles. This process exposes polar organisms to unusually concentrated levels of toxic substances.

Central Valley of California is the world's most productive agricultural region, and the region's frequent winds often blow airborne pesticide spray—and dust particles containing pesticide residue—for long distances. In the nearby mountains of the Sierra Nevada, research has associated pesticide drift from the Central Valley with population declines in four species of frogs.

Toxic substances may concentrate in water

Water running off from land often collects toxicants from large areas and concentrates them in small volumes of surface water. The NAWQA findings on water quality reflect this concentrating effect. Wastewater treatment plants also add pharmaceuticals and toxicants from consumer products to waterways. If chemicals are able to persist in the soil, they can leach into groundwater and contaminate drinking water supplies.

Many chemicals are soluble in water and enter organisms' tissues through drinking or absorption. For this reason, aquatic animals such as fish, frogs, and stream invertebrates are effective indicators of pollution. If scientists find that low concentrations of pesticides are harming frogs, fish, and invertebrates, it is viewed as a warning that people could be next. The contaminants that wash into streams and rivers also flow and seep into the water we drink. Once concentrated in waters, toxic substances can move long distances through aquatic systems and affect a diversity of organisms and ecosystems (p. 392). For example, in 2017 scientists reported that small crustaceans collected from some of the deepest trenches in the Pacific—depths up to 9.5 km (6 mi) below the surface—contained levels of PCBs and PBDEs that were up to 50 times higher than concentrations found in crustaceans from some of the most polluted rivers on Earth.

Some toxicants persist in the environment

A toxic substance that is released into the environment may degrade quickly and become harmless, or it may remain unaltered and persist for many months, years, or decades. The rate at which a given substance degrades depends on its chemistry and on factors such as temperature, moisture, and sun exposure. The *Bt* toxin (p. 253) used in biocontrol and genetically modified crops has a very short persistence time, whereas chemicals such as DDT and PCBs can persist for decades. Atrazine, one of our most widely used herbicides, is highly variable in its persistence, depending on environmental conditions.

Persistent synthetic chemicals exist in our environment today because we have designed them to persist. The synthetic chemicals used in plastics, for instance, are used precisely because they resist breakdown. Sooner or later, however, most toxicants degrade into simpler compounds called **breakdown products.** Often these are less harmful than the original substance, but sometimes they are just as toxic as the original chemical, or more so. For instance, DDT breaks down into DDE, a highly persistent and toxic compound in its own right. Atrazine produces a large number of breakdown products whose effects have not been fully studied.

Toxic substances may accumulate and move up the food chain

Within an organism's body, some toxic substances are quickly excreted, and some are degraded into harmless breakdown products. Others persist intact in the body. Substances that are fat soluble or oil soluble (including organic compounds such as DDT and DDE) are absorbed and stored in fatty tissues. Substances such as methylmercury may be stored in muscle tissue. Such persistent toxicants accumulate in an animal's body in a process termed **bioaccumulation,** such that the animal's tissues have a greater concentration of the substance than exists in the surrounding environment.

Toxic substances that bioaccumulate in an organism's tissues may be transferred to other organisms as predators consume prey, resulting in a process called **biomagnification** (**FIGURE 14.12**). When one organism consumes another, the predator takes in any stored toxicants and stores them in its

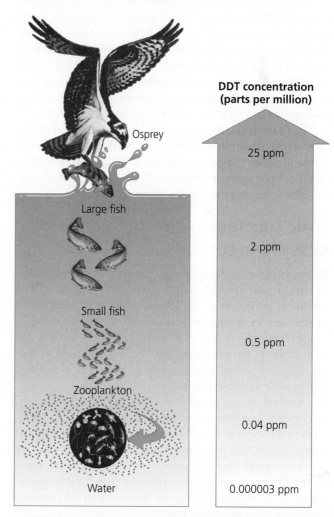

FIGURE 14.12 In a classic case of biomagnification, DDT becomes highly concentrated in fish-eating birds such as ospreys. Organisms at the lowest trophic level take in fat-soluble compounds such as DDT from water. As animals at higher trophic levels eat organisms lower on the food chain, each organism passes its load of toxicants up to its consumer, such that organisms on all trophic levels bioaccumulate the substance in their tissues.

own body. Thus, bioaccumulation takes place on all trophic levels. Moreover, each individual predator consumes many individuals from the trophic level beneath it, so with each step up the food chain, concentrations of toxicants become magnified.

The process of biomagnification occurred throughout North America with DDT. Top predators, such as birds of prey, ended up with high concentrations of the pesticide because concentrations became magnified as DDT moved from water to algae to plankton to small fish to larger fish and finally to fish-eating birds.

Biomagnification of DDT caused populations of many North American birds of prey to decline precipitously from the 1950s to the 1970s. The peregrine falcon was nearly wiped out in the eastern United States, and the bald eagle, the U.S. national bird, was virtually eliminated from the lower 48 states. Eventually, scientists determined that DDT was causing these birds' eggshells to grow thinner, so that eggs were breaking in the nest and killing the embryos within. In a remarkable environmental success story, populations of all these birds have rebounded (pp. 296–297) since the United States banned DDT.

Impacts from biomagnification still exist, though. Unfortunately, DDT continues to impair wildlife in parts of the world where it is still used. Mercury bioaccumulates in some commercially important fish species, such as tuna. Polar bears of Svalbard Island in Arctic Norway show extremely high levels of PCB contamination from biomagnification and polar bear cubs suffer immune suppression, hormone disruption, and high mortality when they receive PCBs in their mothers' milk.

In all the aforementioned cases, biomagnification affects ecosystem composition and the ecosystem services that nature provides. When populations of top predators such as eagles or polar bears are reduced, species interactions change, and effects cascade through food webs (p. 82).

Toxic substances can threaten ecosystem services

Toxicants can alter the biological composition of ecosystems and the manner in which organisms interact with one another and their environment. In so doing, harmful compounds can threaten the ecosystem services (pp. 4, 120) provided by nature. For example, pesticide exposure has been implicated as a factor in the recent declines in honeybee populations (pp. 224–225), affecting the ecosystem service of pollination they provide to wild plants and agricultural crops.

Nutrient cycling is one of the many services that healthy, functioning ecosystems provide. Decomposers and detritivores in the soil (pp. 216–217) break down organic matter and replenish soils with nutrients for plants to use. When soils are exposed to pesticides or fungicides (antifungal agents), nutrient cycling rates are altered. This can make nutrients less available to producers, affecting their growth and causing impacts that cascade throughout the ecosystem.

Studying Effects of Hazards

Determining health effects of particular environmental hazards is a challenging job, especially because any given person or organism has a complex history of exposure to many hazards throughout life. Scientists rely on several different methods with people and with wildlife to study the effects of environmental hazards, ranging from correlative surveys to manipulative experiments (p. 11).

Wildlife studies integrate work in the field and lab

Scientists study the impacts of environmental hazards on wild animals to help conserve animal populations and to understand potential risks to people. Just as placing canaries in coal mines helped miners determine whether the air within the mine was safe for them to breathe, studying how wild animals respond to pollution and other hazards can help us detect environmental health threats before they do us too much harm.

Often wildlife toxicologists work in the field with animals to take measurements, document patterns, and generate hypotheses, before heading to the laboratory to run controlled manipulative experiments to test their hypotheses. The work of two of the pioneers in the study of endocrine disruptors illustrates the approaches embraced in wildlife studies.

Biologist Louis Guillette studied alligators in Florida (FIGURE 14.13a) and discovered that many showed bizarre reproductive problems. Females had trouble producing viable eggs, young alligators had abnormal gonads, and male hatchlings had too little of the male sex hormone testosterone while female hatchlings had too much of the female sex hormone estrogen. Because certain lakes received agricultural runoff that included insecticides, such as DDT and dicofol, and herbicides, such as atrazine, Guillette hypothesized that chemical contaminants were disrupting the endocrine systems of alligators during their development in the egg. Indeed, when Guillette and his team compared alligators in polluted lakes with those in cleaner lakes, they found the ones in polluted lakes to be suffering far more developmental problems. Moving into the lab, the researchers found that several contaminants detected in alligator eggs and young could bind to receptors for estrogen and reverse the sex of male embryos. Their experiments showed that atrazine appeared to disrupt hormones by inducing production of aromatase, an enzyme that converts testosterone to estrogen.

Following Guillette's work, researcher Tyrone Hayes (FIGURE 14.13b) found similar reproductive problems in frogs and attributed them to atrazine. In lab experiments, male frogs raised in water containing very low doses of the herbicide became feminized and hermaphroditic, developing both testes and ovaries. Hayes then moved to the field to look for correlations between herbicide use and reproductive abnormalities in the wild. His field surveys showed that leopard frogs across North America experienced hormonal problems in areas of heavy atrazine usage. His work indicated that atrazine, which kills plants by blocking biochemical pathways in photosynthesis, can also act as an endocrine disruptor.

Human studies rely on case histories, epidemiology, and animal testing

In studies of human health, we gain much knowledge by directly studying sickened individuals. This process of observation and

(a) Louis Guillette with an American aligator (*Alligator mississippiensis*)

(b) Tyrone Hayes with researchers Laura Meehan and Young Kim-Parker at the University of California, Berkeley

FIGURE 14.13 Wildlife studies examine the effects of toxic substances in the environment. Researchers Louis Guillette **(a)** and Tyrone Hayes **(b)** found that alligators and frogs, respectively, show reproductive abnormalities that they attribute to endocrine disruption by pesticides.

analysis of individual patients is known as a **case history** approach. Case histories have advanced our understanding of human illness, but they do not always help us infer the effects of rare hazards, new hazards, or chemicals that exist at low environmental concentrations and exert minor, long-term effects. Case histories also tell us little about probability and risk, such as how many extra deaths we might expect in a population due to a particular cause.

For such questions, which are common in environmental toxicology, we need **epidemiological studies,** large-scale comparisons among groups of people, usually contrasting a group that was exposed to some hazard against a group that has not. Epidemiologists track the fate of all people in the study for a long period of time (often years or decades) and measure the rate at which deaths, cancers, or other health problems occur in each group. The epidemiologists then analyze the data, looking for observable differences between the groups, and statistically tests hypotheses accounting for differences.

When a group exposed to a hazard shows a significantly greater degree of harm, it suggests that the hazard may be responsible. The epidemiological process is akin to a natural experiment (p. 11), in which an event creates groups of subjects that researchers can study (for example, people exposed to carcinogenic compounds in their drinking water versus those not similarly exposed). Consider asbestos that is mined from the ground, much like other minerals. Epidemiologists have tracked asbestos miners for evidence of asbestosis, lung cancer, and mesothelioma (cancer of the cells that line the body's internal organs). Similarly, survivors of the Chernobyl and Fukushima nuclear disasters have been closely monitored by health professionals for thyroid cancer and other illnesses (pp. 570–571), and Canadian epidemiologists are now

tracking people to determine the impacts of long-term exposure to BPA.

Epidemiological studies measure a statistical association between a health hazard and an effect, but they do not confirm that the hazard causes the effect. To establish causation, manipulative experiments are needed. However, subjecting people to massive doses of toxic substances in a lab experiment would clearly be unethical. This is why researchers have traditionally used animals—such as laboratory strains of rats, mice, and other mammals—as test subjects (**FIGURE 14.14**).

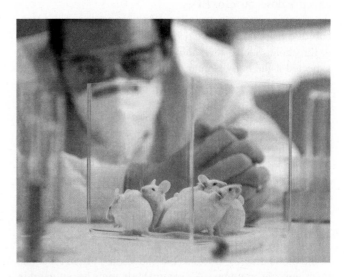

FIGURE 14.14 Animal testing is used to study toxic substances in the laboratory. Tests with specially bred strains of mice, rats, and other animals allow researchers to study the toxicity of substances, develop safety guidelines, and make medical advances in ways they could not achieve without these animals.

Because of shared evolutionary history, substances that harm mice and rats are reasonably likely to harm us. Some people feel the use of animals for testing is unethical, but animal testing enables scientific and medical advances that would be impossible or far more difficult otherwise.

Dose-response analysis is a mainstay of toxicology

The standard method of testing with lab animals in toxicology is **dose-response analysis,** wherein scientists quantify the toxicity of a substance by measuring the strength of its effects or the number of animals affected at different doses. The **dose** is the amount of substance the test animal receives, and the **response** is the type or magnitude of negative effects the animal exhibits as a result. The response is generally quantified by measuring the proportion of animals exhibiting negative impacts. The data are plotted on a graph, with dose on the x axis and response on the y axis (**FIGURE 14.15a**). The resulting curve is called a **dose-response curve.**

Once they have plotted a dose-response curve, toxicologists can calculate a convenient shorthand gauge of a substance's toxicity: the amount of the substance it takes to kill half the population of study animals used. This lethal dose for 50% of individuals is termed the lethal-dose–50%, or **LD_{50}**. A high LD_{50} means a great deal of the toxic substance is required to kill 50% of the test organisms and this indicates low toxicity for a substance. A low LD_{50} indicates very little of the toxic substance is needed to kill 50% of the test subjects, and this indicates the substance has high toxicity.

If the experimenter is interested in adverse health impacts that are nonlethal, he or she may want to document the level of toxicant at which 50% of a population of test animals is negatively affected in some way other than death (for instance, the level of toxicant that causes 50% of lab mice to develop reproductive abnormalities). Such a level is called the effective dose–50%, or **ED_{50}**.

Some substances can elicit effects at any concentration, but for others, responses may occur only above a certain dose, or threshold. Such a **threshold dose** (**FIGURE 14.15b**) might be expected if the body's organs can fully metabolize or excrete a toxicant at low doses but become overwhelmed at high concentrations.

Sometimes a response may *decrease* as a dose increases. Toxicologists are finding that some dose-response curves are U-shaped, J-shaped, or shaped like an inverted U (**FIGURE 14.15c**). Such counterintuitive curves contradict toxicology's traditional assumption that "the dose makes the poison." These unconventional dose-response curves often occur with endocrine disruptors, such as BPA and phthalates, likely because the hormone system is geared to respond to minute concentrations of substances (normally, hormones in the bloodstream). Because the endocrine system responds to minuscule amounts of chemicals, it may be vulnerable to disruption by contaminants that are dispersed through the environment and that reach our bodies in very low concentrations.

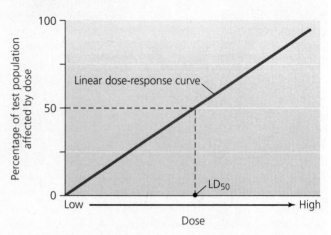

(a) Linear dose-response curve

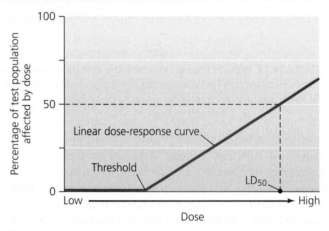

(b) Dose-response curve with threshold

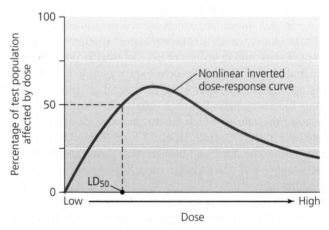

(c) Unconventional dose-response curve

FIGURE 14.15 Dose-response curves show that organisms' responses to toxicants may sometimes be complex. In a classic linear dose-response curve **(a)**, the percentage of animals killed or otherwise affected by a substance rises with the dose. The point at which 50% of the animals are killed is labeled the lethal dose–50, or LD_{50}. For some toxic substances, a threshold dose **(b)** exists, below which doses have no measurable effect. Some substances—in particular, endocrine disruptors—show unconventional, nonlinear dose-response curves **(c)** that are U-shaped, J-shaped, or shaped like an inverted U.

Researchers generally give lab animals much higher doses relative to body mass than people would receive in the environment. This is so that the response is great enough to be measured and differences between the effects of small and large doses are evident. Data from a range of doses give shape to a dose-response curve. Once the data from animal tests are plotted, researchers can extrapolate downward to estimate responses to still-lower doses from a hypothetically large population of animals. This way, researchers can come up with an estimate of, say, what dose of a toxicant causes cancer in 1 in 1 million mice. A second extrapolation is required to estimate the effect on humans, with our greater body mass. Because these two extrapolations stretch beyond the actual data obtained, they introduce uncertainty into the interpretation of what doses are safe for people.

Chemical mixes may be more than the sum of their parts

It is difficult enough to determine the impact of a single environmental hazard, but the task becomes astronomically more difficult when multiple environmental hazards interact. Chemical substances, when mixed, may act together in ways that cannot be predicted from the effects of each in isolation. Mixed toxicants may sum each other's effects, cancel out each other's effects, or multiply each other's effects. Whole new types of impacts may arise when toxicants are mixed together. Interactive impacts that are greater than the simple sum of their constituent effects are called **synergistic effects** (see THE SCIENCE BEHIND THE STORY, pp. 378–379).

With Florida's alligators, lab experiments have indicated that the DDT breakdown product DDE can either promote or inhibit sex reversal, depending on the presence of other chemicals. Mice exposed to a mixture of nitrate, atrazine, and aldicarb have been found to show immune, hormone, and nervous system effects that were not evident from exposure to each of these chemicals alone.

Traditionally, environmental health has tackled effects of single hazards one at a time. In toxicology, the complex experimental designs required to test interactions, and the sheer number of chemical combinations, have meant that single-substance tests have received priority. This approach is changing, but the interactive effects of most chemicals are unknown.

Endocrine disruption poses challenges for toxicology

As today's emerging understanding of endocrine disruption leads toxicologists to question their assumptions, unconventional dose-response curves are presenting challenges for scientists studying toxic substances and for policymakers trying to set safety standards for them. Knowing the shape of a dose-response curve is crucial if one is using it to predict responses at doses below those that have been tested. Because so many novel synthetic chemicals exist in very low concentrations over wide geographic areas, many scientists suspect that we may have underestimated the dangers of toxicants that exert impacts at low concentrations.

Scientists first noted endocrine-disrupting effects in some chemicals decades ago, but the idea that synthetic chemicals might be altering the hormones of animals was not widely appreciated until the 1996 publication of the book *Our Stolen Future*, by Theo Colburn, Dianne Dumanoski, and J. P. Myers. Like *Silent Spring*, this book integrated scientific work from various fields and presented a unified view of the hazards posed by endocrine-disrupting chemicals. And, like *Silent Spring*, the book brought into the public eye a chemical hazard that was previously unknown to most people.

Today, thousands of studies have linked hundreds of substances to effects on reproduction, development, immune function, brain and nervous system function, and other hormone-driven processes. Evidence is strongest so far in nonhuman animals, but many studies suggest impacts on humans (see Figure 14.2, p. 360). Some researchers argue that the sharp rise in breast cancer rates—one in eight U.S. women today develops breast cancer—may be due to hormone disruption, because an excess of estrogen appears to feed tumor development in older women. Other scientists attribute male reproductive problems to elevated BPA exposure. For example, studies found that workers in Chinese factories that manufactured BPA had elevated rates of erectile dysfunction and reduced sperm counts when compared to workers in factories manufacturing other products. Endocrine-disrupting chemicals have been implicated as a possible contributing factor to recent declines in the sperm quality in men from industrialized nations. A study, published in 2017, compiled data from 43,000 men in 50 industrialized nations and found that total sperm counts declined by 59.3% from 1973 to 2011. The declines showed no sign of slowing, a potentially worrying sign for future fertility should such declines continue.

Much of the research into hormone disruption has brought about strident debate. This is partly because scientific uncertainty is inherent in any developing field. Another reason is that negative findings about chemicals pose an economic threat to the manufacturers of those chemicals, who stand to lose many millions of dollars in revenue if their products were to be banned or restricted in the United States.

Are synthetic chemicals underappreciated as agents of global change?

Given the profound effects synthetic, toxic chemicals can have on individuals, species, and ecosystems, why is it we still know so little about their impacts—especially when many have been part of the environment for more than half a century?

In recent decades, the growth in the diversity and application of synthetic chemicals, especially pesticides, has rivaled or exceeded the growth of other factors that cause global-level environmental changes—such as emissions of

THE SCIENCE behind the story

What Role Do Pesticides Play in the Collapse of Bee Colonies?

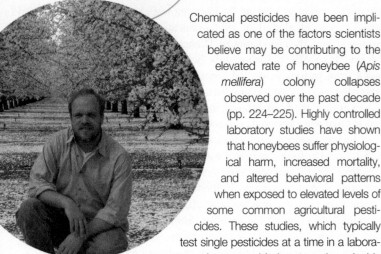

Dr. Dennis vanEngelsdorp, University of Maryland

Chemical pesticides have been implicated as one of the factors scientists believe may be contributing to the elevated rate of honeybee (*Apis mellifera*) colony collapses observed over the past decade (pp. 224–225). Highly controlled laboratory studies have shown that honeybees suffer physiological harm, increased mortality, and altered behavioral patterns when exposed to elevated levels of some common agricultural pesticides. These studies, which typically test single pesticides at a time in a laboratory, have provided extremely valuable insight into the toxicity levels of some pesticides for bees but do not capture the stresses faced by bees in natural settings, where they can be exposed to hundreds of potentially harmful chemicals.

A recent study—led by Dennis vanEngelsdorp of the University of Maryland and published in 2016 in *Nature Scientific Reports*—took on the challenge of determining how the multitude of pesticides in the environment may be affecting honeybee colonies. The researchers embraced an approach by which they treated the highly organized, communal beehive as a single "super-organism" and measured the cumulative pesticide exposures experienced by colonies during a typical agricultural season. The researchers then compared the pesticide exposures of individual colonies with measures of colony health—such as whether the colony survived the season or if the colony lost its queen—to identify factors related to pesticide exposure that may predict bee colony collapses.

Treating a bee colony as a single organism makes sense, as a thriving honeybee colony depends on the collective efforts of hundreds or thousands of individual bees, so stressors that impair even one of the critical functions that sustain the hive can cause the entire colony to collapse. Hives typically have a single queen, a fertile female whose main function is to deposit eggs into honeycomb cells within the hive; the queen is the mother of most, if not all, bees in the hive. *Worker bees*, the most numerous class of bees in the hive, are sterile females that have many responsibilities. *Nurse bees* are workers that care for the queen and the larval bees that hatch from her eggs. *Foragers* leave the hive to forage for food, such as nectar and pollen, which they bring back to the hive. Other workers maintain, clean, and defend the hive. Chemicals secreted by the queen, called pheromones, prevent the creation of a new queen while the reigning queen is healthy and active. When a queen becomes old and dies, her pheromone releases decline and eventually cease. The lack of pheromones drives worker bees to begin feeding several larval worker bees a high-nutrient food called *royal jelly*. The royal jelly–fed larva develop into queen bees, who embark on a mating flight, during which they are fertilized by male bees, called *drones*. Each new queen is then able to begin her life anew, as the queen of her own hive.

As part of the study, vanEngelsdorp's team surveyed a total of 91 bee colonies from March 2007 to January 2008. The colonies were part of three migratory commercial beekeeping operations that, starting in Florida, moved northward up the East Coast of the United States, pollinating farmers' crops (a service for which the farmers paid the beekeepers). Between pollination jobs, the bees foraged in natural areas for pollen to create honey, or rested in holding areas where the hives were maintained and prepared for their next job.

Pesticides can become concentrated in beehives because bees forage over a wide area in their search for pollen and nectar, often covering 100 km^2 (39 mi^2) around the hive. Forager bees bring back pollen, some of it containing agricultural pesticides, to the hive and store it in honeycomb cells near developing bee larvae. Nurse bees consume this pollen-rich substance, called "beebread," and use specialized glands to convert it to a protein-rich and calorically dense secretion that is then fed to the larvae (**FIGURE 1**). Nectar, which can also be contaminated with pesticides, is similarly deposited in honeycomb cells and used to create honey, which sustains the worker bees and the queen throughout their adult lives. Thus, contaminated pollen and nectar from up to 100 km^2 (39 mi^2) can become concentrated inside a single beehive, potentially affecting all life stages of the colony. This is akin to bioaccumulation (p. 373), where toxic substances attain higher concentrations in the tissues of organisms than are found in the surrounding environment.

FIGURE 1 Nurse bees tending to larvae in the hive. Nearby honeycomb cells are used to store beebread, while others store honey.

To assess the colony's cumulative pesticide exposure, the researchers collected samples of beebread throughout the season, and took samples of honeycomb wax (in which larvae were brooded and honey and beebread were stored) at the beginning and end of the season. The samples were then sent to a laboratory and screened for the presence of 171 common agricultural pesticides. At the end of the season, the researchers noted which hives survived the entire season, as well as those that experienced a "queen event"—the death of the queen bee. The death of the queen can occur naturally but is also seen in stressed hives where the worker bees, perceiving something is wrong with the queen, kill her and attempt to replace her with a new queen. These events are predictors of hive survival, as studies have shown that as the number of queen events increases in a hive, so does the likelihood of the hive dying.

For each of the beebread and honeycomb wax samples collected, the researchers determined the total number of pesticides detected, the number of those pesticides that exceeded accepted toxicity levels, and the colony's hazard quotient (HQ), a measure of the cumulative toxicity the hive experienced from all the detected pesticides. These three measures were then subjected to correlation analysis, wherein the researchers measured hive health by observing how the bee colonies responded to the combination of chemical pesticides in their surrounding agricultural environment.

In all, 93 different pesticides were detected in the colonies over the course of the season. More than half of the bee colonies perished during the study period, with hive mortality rates increasing as the season progressed. Correlation analysis revealed that the likelihood of a colony dying was positively correlated with the *total number* of pesticides in beebread and honeycomb wax but not with the *concentration* of the pesticides, as measured by HQ. This was an unexpected result; it indicated that as the number of detected pesticides increased, so did the chances of the hive dying, regardless of the concentration of

each pesticide in the hive. In the words of vanEngelsdorp, "Our results fly in the face of one of the basic tenets of toxicology—that the dose makes the poison" (pp. 376–377). The researchers hypothesized that as the sheer number of toxic compounds increased in the hive, the bees experienced extreme physiological stress as they tried to detoxify such a diverse combination of contaminants within their bodies. In some cases, this stress was simply too much for the bees to overcome, and more and more bees perished, eventually collapsing the hive.

When examining surviving and collapsed hives, vanEngelsdorp and his team determined that hives experiencing a queen event contained more total pesticides in their beebread and wax and had a higher HQ in wax than hives that maintained their queen throughout the season (**FIGURE 2**). The concentration of pesticides in the hive, as measured by HQ, therefore did not correlate with the survival of the hive. However, HQ did correlate with queen events, suggesting that hives with higher cumulative concentrations of pesticides were more likely to experience the loss of their queen.

Additionally, the team found that fungicides, which are sprayed on crops to reduce damage from fungal infections, correlated with increased risk of hive death. This result was unexpected. Although laboratory studies had shown that bee larvae experience higher mortality when exposed to elevated concentrations of some fungicides, the relatively lower concentrations of fungicides bees experience in agricultural settings was believed to be safe.

It is important to note that correlative studies (p. 11), such as this one conducted by vanEngelsdorp and his team, do not establish causation. Rather, correlative studies highlight connections between observations, providing a roadmap of promising avenues for future research. This study, and others like it to come, may play a valuable role in aiding scientists to one day unravel the mystery of what is killing one of the most important organisms in modern agriculture—and help us find solutions to save our bees.

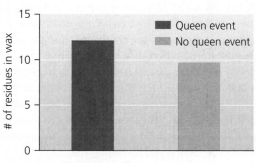
(a) **Total number of pesticides in wax**

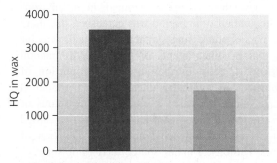
(b) **Hazard quotient (HQ) in wax**

FIGURE 2 **Hives that experienced a "queen event" had significantly more pesticides in their wax honeycomb and higher hazard quotient (HQ) values in their wax than did hives whose queen survived the season.** The HQ is calculated by dividing the total concentration of pesticides in the wax by the lethal dose of each detected pesticide required to kill 50% of the bees in a sample (the pesticide's LD_{50}; p. 376). *Figure from Traynor, K., et al., 2016. In-hive pesticide exposome: Assessing risks to migratory honey bees from in-hive pesticide contamination in the eastern United States.* Nature Scientific Reports *6: 33207, http://www.nature.com/articles/srep33207.*

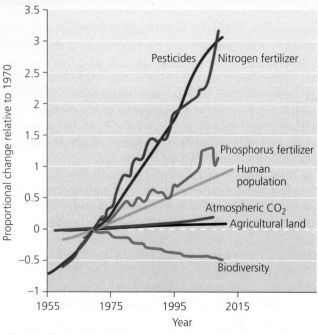

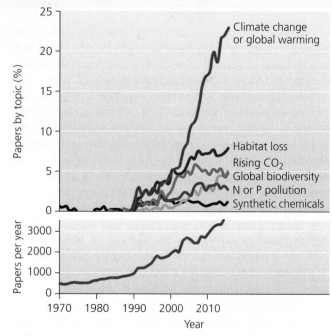

(a) Changes in the magnitude of major drivers of global environmental change since 1955

(b) Papers examining major drivers of global environmental change, published in the 20 top ecology journals

FIGURE 14.16 The growth in the global application of pesticides (a) since 1970 has rivaled that of other factors contributing to global environmental change, but the number of scientific papers published on synthetic chemicals (b) has lagged far behind those of other topics. *Data from Bernhardt, E. S., et al., 2017. Synthetic chemicals as agents of global change. Frontiers in Ecology and the Environment 15: 84–90.*

carbon dioxide, losses of biodiversity, and our introductions of nitrogen and phosphorus into ecosystems from fertilizers (**FIGURE 14.16a**). And yet the attention given in scientific circles to the study of synthetic chemicals in ecosystems lags far behind these other stressors.

This has likely occurred because research on the effects of synthetic chemicals has been far less extensive than research into other major environmental stresses, as noted in a 2017 study by American and German scientists. The researchers analyzed all of the scientific papers published in the top 20 ecological journals from 1970 to 2015 and found that while the growth of synthetic chemicals in the environment over the studied time period was extensive, the share of scientific papers addressing synthetic chemicals was far lower than that of other agents of global environmental change (**FIGURE 14.16b**). Furthermore, the proportion of papers published on synthetic chemicals had remained roughly steady since the early 1990s, while the shares of other topics increased significantly. A review of awarded grants by the National Science Foundation's Division of Environmental Biology, a major source of funding for ecological research, found that in 2015 less than 3% of funded projects mentioned synthetic chemicals in their title or abstract and only one study (accounting for 0.006% of awarded funds) explicitly studied the effects of synthetic chemicals on ecosystems. In comparison, studies mentioning terms such as *climate change* or *nitrogen* accounted for 23.4% of total funding.

The researchers note that in order to determine the effects synthetic chemicals are having on the environment, we must

greatly expand our study of these chemicals using ecological methods and models. Only then will we be able to truly predict how humanity's extensive introductions of pesticides, pharmaceuticals, and industrial chemicals are affecting ecological systems—and how best to manage those introductions to preserve biodiversity and human health.

Risk Assessment and Risk Management

Policy decisions on whether to ban chemicals or restrict their use generally follow years of rigorous testing for toxicity. Likewise, strategies for combating disease and other health threats are based on extensive scientific research. However, policy and management decisions also incorporate economics and ethics, and all too often, the decision-making process is heavily influenced by pressure from powerful corporate and political interests. The steps between the collection and interpretation of scientific data and the formulation of policy involve assessing and managing risk.

We express risk in terms of probability

Exposure to an environmental health threat does not invariably produce a given consequence. Rather, it causes some probability of harm, some statistical chance that damage will result. To understand a health threat, a scientist must know more than just its identity and strength. He or she must also know the chance that one will encounter it, the frequency with which one may encounter it, the amount of substance or degree of threat to which one is exposed, and one's sensitivity to the threat. Such factors help determine the overall risk posed.

Risk can be measured in terms of **probability,** a quantitative description of the likelihood of a certain outcome. The probability that some harmful outcome (for instance, injury, death, environmental damage, or economic loss) will result from a given action, event, or substance expresses the **risk** posed by that phenomenon.

Our perception of risk may not match reality

Every action we take and every decision we make involves some element of risk, some (generally small) probability that things will go wrong. We typically try to behave in ways that minimize risk, but our perceptions of risk do not always match statistical reality. People often worry unduly about small risks yet happily engage in activities that pose high risks. For instance, most people perceive flying in an airplane as a riskier activity than driving a car, but plane travel is actually statistically safer (**FIGURE 14.17**).

Psychologists argue that this disconnect occurs because we feel more at risk when we are not controlling a situation and safer when we are "at the wheel"—regardless of the actual risk involved.

This psychology may help account for people's anxiety over exposure to BPA, nuclear power, toxic waste, and pesticide residues on foods—environmental hazards that are invisible or little understood and whose presence in our lives is largely outside our personal control. In contrast, people are more ready to accept and ignore the risks of smoking cigarettes, overeating, and not exercising—voluntary actions or inactions statistically shown to pose far greater risks to health.

Risk assessment analyzes risk quantitatively

The quantitative measurement of risk and the comparison of risks involved in different activities or substances together are termed **risk assessment.** Risk assessment is a way to identify and outline problems. In environmental health, it helps ascertain which substances and activities pose health threats to people or wildlife and which are largely safe.

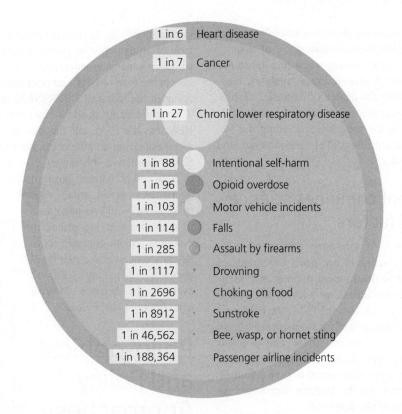

1 in 6	Heart disease
1 in 7	Cancer
1 in 27	Chronic lower respiratory disease
1 in 88	Intentional self-harm
1 in 96	Opioid overdose
1 in 103	Motor vehicle incidents
1 in 114	Falls
1 in 285	Assault by firearms
1 in 1117	Drowning
1 in 2696	Choking on food
1 in 8912	Sunstroke
1 in 46,562	Bee, wasp, or hornet sting
1 in 188,364	Passenger airline incidents

FIGURE 14.17 Our perceptions of risk do not always match the reality of risk. Listed here are several leading causes of death in the United States, along with a measure of the risk each poses. The larger the area of the circle in the figure, the greater the risk of dying from that cause. *Data are for 2017, from* Injury Facts, *2018. Itasca, IL: National Safety Council.*

Go to **Interpreting Graphs & Data** on **Mastering Environmental Science**

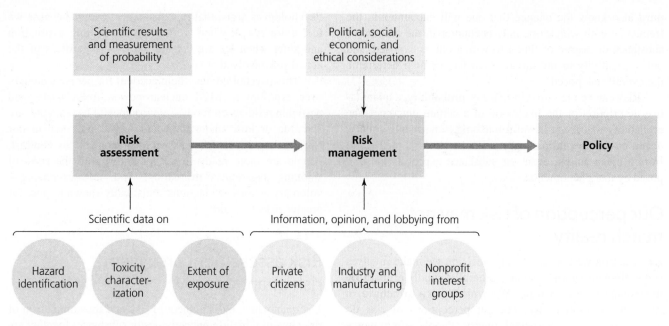

FIGURE 14.18 The first step in addressing risks from an environmental hazard is risk assessment. Once science identifies and measures risks, then risk management can proceed. In risk management, economic, political, social, and ethical issues are considered in light of the scientific data from risk assessment.

Assessing risk for a chemical substance involves several steps. The first steps involve the scientific study of toxicity we examined previously—determining whether a substance has toxic effects and, through dose-response analysis, measuring how effects vary with the degree of exposure. Subsequent steps involve assessing the individual's or population's likely extent of exposure to the substance, including the frequency of contact, the concentrations likely encountered, and the length of every encounter.

Risk management combines science and other social factors

Accurate risk assessment is a vital step toward effective **risk management,** which consists of decisions and strategies to minimize risk (**FIGURE 14.18**). In most nations, risk management is handled largely by federal agencies. In the United States, these include agencies such as the FDA, the EPA, and the CDC. In risk management, scientific assessments of risk are considered in light of economic, social, and political needs and values. Risk managers assess costs and benefits of addressing risk in various ways with regard to both scientific and nonscientific concerns before making decisions on whether and how to reduce or eliminate risk.

In environmental health and toxicology, comparing costs and benefits (p. 142) can be difficult because the benefits are often economic, whereas the costs often pertain to health. Moreover, economic benefits are generally known, easily quantified, and of a discrete and stable amount, whereas health risks are hard-to-measure probabilities, often involving a small percentage of people likely to suffer greatly and a large majority likely to experience little effect. Because of the lack of equivalence in the way costs and benefits are measured, risk management frequently tends to stir up debate.

In the case of BPA and phthalates, eliminating food packaging in the name of safety could do more harm than good. The plastic lining inside metal cans, for example, can release BPA into the food, but the plastic lining also helps prevent metal corrosion and the contamination of food by pathogens. Some alternative substances exist to those that expose users to BPA and phthalates, but replacing BPA and phthalate with alternatives would entail economic costs to industry, and those costs would be passed on to consumers in the prices of products. Such complex considerations can make risk management decisions difficult even if the science of risk assessment is fairly clear.

Philosophical and Policy Approaches

Because we cannot know a substance's toxicity until we measure and test it, and because so many untested chemicals and combinations exist, science will never eliminate the many

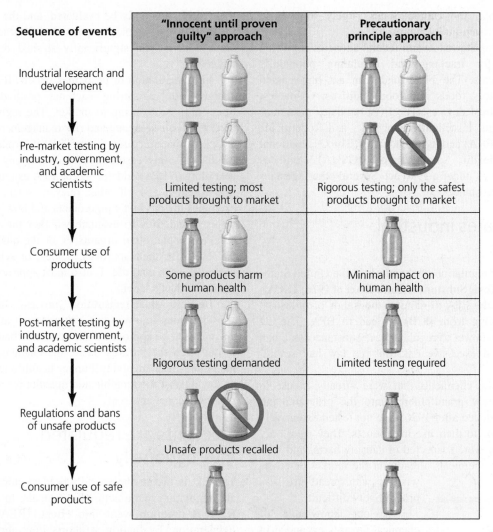

Sequence of events	"Innocent until proven guilty" approach	Precautionary principle approach
Industrial research and development		
Pre-market testing by industry, government, and academic scientists	Limited testing; most products brought to market	Rigorous testing; only the safest products brought to market
Consumer use of products	Some products harm human health	Minimal impact on human health
Post-market testing by industry, government, and academic scientists	Rigorous testing demanded	Limited testing required
Regulations and bans of unsafe products	Unsafe products recalled	
Consumer use of safe products		

FIGURE 14.19 Two main approaches can be taken to introduce new substances to the market. In one approach, substances are "innocent until proven guilty"; they are brought to market relatively quickly after limited testing. Products reach consumers more quickly, but some fraction of them **(white jug in the diagram)** may cause harm to some fraction of people. The other approach is to adopt the precautionary principle, bringing substances to market cautiously, only after extensive testing. Products that reach the market should be safe, but many perfectly safe products **(gray bottle in the diagram)** will be delayed in reaching consumers.

uncertainties that accompany risk assessment. In such a world of uncertainty, there are two basic philosophical approaches to categorizing substances as safe or dangerous (**FIGURE 14.19**).

One approach is to assume that substances are harmless until shown to be harmful. This is nicknamed the "innocent-until-proven-guilty" approach. Because thoroughly testing every existing substance (and combination of substances) for its effects is a hopelessly long, complicated, and expensive pursuit, the innocent-until-proven-guilty approach has the virtue of facilitating technological innovation and economic activity. However, it has the disadvantage of putting into wide use some substances that may later turn out to be dangerous.

The other approach is to assume that substances are harmful until shown to be harmless. This approach follows the precautionary principle (p. 260). This more cautious approach should enable us to identify troublesome toxicants

before they are released into the environment, but it may also impede the pace of technological and economic advance.

These two approaches are actually two ends of a continuum of possible approaches. The two endpoints differ mainly in where they lay the burden of proof—specifically, whether product manufacturers are required to prove a product is safe or whether government, scientists, or citizens are required to prove a product is dangerous.

Philosophical approaches are reflected in policy

This choice of philosophical approach has direct implications for policy, and nations vary in how they blend the two approaches. European nations have recently embarked on a policy course that largely incorporates the precautionary principle, whereas the United States largely follows an innocent-until-proven-guilty approach.

In the United States, several federal agencies apportion responsibility for tracking and regulating potentially harmful chemicals. The FDA, under an act first passed in 1938, monitors foods and food additives, cosmetics, drugs, and medical devices. The EPA regulates pesticides under the Federal Insecticide, Fungicide, and Rodenticide Act of 1947 (FIFRA) and its amendments. The Occupational Safety and Health Administration (OSHA) regulates workplace hazards under a 1970 act. Several other agencies regulate other substances.

EPA regulates industrial chemicals

The widespread regulation of chemicals in the United States began with the **Toxic Substances Control Act of 1976 (TSCA),** which directed the EPA to monitor thousands of industrial chemicals, ranging from PCBs to lead to BPA. The act gave the agency power to regulate these substances and ban those that posed excessive risk, but the law had several weaknesses.

For example, chemicals that were already in use in 1976 were simply grandfathered into the program, and chemicals introduced after 1976 were not tested extensively for toxicity *prior* to their use in products. They were only tested *after* they were suspected of causing harm, and banning dangerous chemicals already on the market or otherwise in widespread use was prohibitively difficult.

As our knowledge of chemical hazards expanded in recent decades, it became clear that the TSCA was in dire need of update. Legislators, scientists, and public health advocates lobbied for change, and these efforts came to fruition in 2016 with the passage of the **Frank R. Lautenberg Chemical Safety for the 21st Century Act.** The act improves upon TSCA by directing the EPA to review the potential toxicity of *all* industrial chemicals currently in use and by subjecting new chemicals to rigorous safety testing *before* they are used in products. While successful implementation of this act would enhance human health by limiting exposure to toxic chemicals, critics argue that many dangerous chemicals will remain untested. There are, for example, some 80,000 chemicals that could potentially be evaluated, and the funds provided to the EPA—the agency responsible for these toxicity evaluations—have been significantly slashed in recent federal budgets.

In its regulation of pesticides under FIFRA, the EPA is charged with "registering" each new pesticide that manufacturers propose to bring to market. The registration process involves risk assessment and risk management. The EPA first asks the manufacturer to provide information, including results of safety assessments the company has performed according to EPA guidelines. The EPA examines the company's research and all other relevant scientific research. It examines the product's ingredients and how the product will be used and tries to evaluate whether the chemical poses risks to people, other organisms, or the quality of water or air. The EPA then approves, denies, or sets limits on the chemical's sale and use. It also must approve language used on the product's label.

Because the registration process takes economic considerations into account, critics say it allows hazardous chemicals to be approved if the economic benefits are judged to outweigh the hazards. Here, the challenges of weighing intangible risks involving human health and environmental quality against the tangible and quantitative numbers of economics become apparent.

Toxicants are regulated internationally

The EU is taking the world's boldest step toward testing and regulating manufactured chemicals. In 2007, the EU's **REACH** program went into effect. (REACH stands for *R*egistration, *E*valuation, *A*uthorisation, and Restriction of *Ch*emicals.) REACH largely shifts the burden of proof for testing chemical safety from national governments to industry and requires that chemical substances produced or imported in amounts of more than 1 metric ton per year be registered with the European Chemicals Agency. This agency evaluates industry research and decides whether the chemical seems safe and should be approved, whether it is unsafe and should be restricted, or whether more testing is needed.

The world's nations have also sought to address chemical pollution with international treaties. The *Stockholm Convention on Persistent Organic Pollutants (POPs)* came into force in 2004 and has been ratified by over 150 nations. POPs are toxic chemicals that persist in the environment, bioaccumulate and biomagnify up the food chain, and can travel long distances. The PCBs and other contaminants found in polar bears are a prime example. Because contaminants often cross international boundaries, an international treaty seemed the best way to deal fairly with such transboundary pollution. The Stockholm Convention aims first to end the use and release of 12 POPs shown to be most dangerous, a group nicknamed the "dirty dozen" (**TABLE 14.2**). It sets guidelines for phasing out these chemicals and encourages transition to safer alternatives.

TABLE 14.2 The "Dirty Dozen" Persistent Organic Pollutants (POPs)

TOXICANT	DESCRIPTION	TOXICANT	DESCRIPTION
Aldrin	Insecticide to kill termites and crop pests	Furans	By-product of processes that release dioxins; also present in commercial mixtures of PCBs
Chlordane	Insecticide to kill termites and crop pests	Heptachlor	Broad-spectrum insecticide
DDT	Insecticide to protect against insect-spread disease; still applied in some countries to control malaria	Hexachlorobenzene	Fungicide for crops; released by chemical manufacture and processes that release dioxins and furans
Dieldrin	Insecticide to kill termites, textile pests, crop pests, and disease vectors	Mirex	Household insecticide; fire retardant in plastics, rubber, and electronics
Dioxins	By-product of incomplete combustion and chemical manufacturing; released in metal recycling, pulp and paper bleaching, auto exhaust, tobacco smoke, and wood and coal smoke	PCBs	Industrial chemical used in heat-exchange fluids, electrical transformers and capacitors, paints, sealants, and plastics
Endrin	Pesticide to kill rodents and crop insects	Toxaphene	Insecticide to kill crop insects and livestock parasites

Data from United Nations Environment Programme (UNEP), 2001.

CENTRAL CASE STUDY
connect & continue

TODAY, continued research on endocrine-disrupting chemicals is helping to determine the many ways we are exposed to these chemicals and how they may negatively affect human health. Nations have banned some endocrine-disrupting chemicals, such as BPA, but regulation in many nations is minimal or nonexistent. But growing consumer concern over the presence of BPA, brought about by media attention, has spurred a number of companies to remove BPA from their products, even in the absence of governmental regulation. Additionally, a 2017 study by the federal National Institute for Occupational Safety and Health (NIOSH) opened a new avenue of regulation for BPA as a workplace hazard, when 78 workers from six companies that use BPA in their products were found to have urine BPA levels nearly 70% higher than the average American.

An emerging field of research is examining how endocrine-disrupting chemicals are affecting fish and other aquatic organisms in waterways. Endocrine disruptors wash into rural waterways from pesticide runoff from farm fields and are introduced into urban waterways from wastewater treatment plants and industrial facilities—so no waterway is safe from the long reach of these chemicals. Researchers have recently correlated exposure to endocrine-disrupting chemicals with fish kills in the rivers in rural Virginia and West Virginia that feed the Potomac River (pp. 412–413). The hormone-mimicking chemicals appear to suppress the immune system of some fish, causing normally manageable infections by pathogens to become lethal.

It is important to remember that synthetic chemicals such as phthalates and BPA, while exposing people to some risk, have brought us innumerable modern conveniences, a larger food supply, and medical advances that extend human lives. The plastic lining of cans that contain BPA, for example, prevent corrosion and the contamination of canned goods with harmful microorganisms. A safer future, one that safeguards the well-being of both people and the environment, therefore depends on knowing the risks that some hazards pose, assessing those risks, and having means in place to phase out harmful substances and replace them with safer ones.

- **CASE STUDY SOLUTIONS** You work for a public health organization and have been asked to educate the public about BPA and to suggest ways to minimize exposure to the chemical. You begin by examining your lifestyle and finding ways to use alternatives to BPA-containing products. Create a list of five ways you are exposed daily to BPA, and then list changes you could make in your daily life that would allow you to avoid or minimize these exposures. Do these steps require more time and/or money? What are some costs associated with your embracing these changes? What would you tell an interested person about BPA?

- **LOCAL CONNECTIONS** As shown in a **THE SCIENCE BEHIND THE STORY** feature in the chapter, the consumption of fast food can lead to elevated exposure to phthalates. It follows that states that have the most fast-food locations per-capita may have higher per-capita exposure to phthalates. How does your state stack up versus others when it comes fast food? How do the various regions of the United States compare in the number of fast-food restaurants per capita?

- **EXPLORE THE DATA** What exposure to BPA elicits genetic damage in mammals, and how does that damage change with increasing dosage? → **Explore Data** relating to the case study on **Mastering Environmental Science**.

REVIEWING Objectives

You should now be able to:

+ **Explain the goals of environmental health and identify major environmental health hazards**

The study of environmental health assesses environmental factors that affect human health and quality of life. Environmental health threats include physical, chemical, biological, and cultural hazards that occur both indoors and outside. Disease, both infectious and noninfectious, remains a major threat to human health and is being addressed with a diversity of approaches. (pp. 359–366)

+ **Describe the types of toxic substances in the environment, the factors that affect their toxicity, and the defenses that organisms have against them**

Toxicant types include carcinogens, mutagens, teratogens, allergens, pathway inhibitors, neurotoxins, and endocrine disruptors. The toxicity of a substance may be influenced by the nature of exposure (acute or chronic) and individual variation in the strength of any given organism's defenses, such as detoxifying enzymes, against the toxin. (pp. 366–370)

+ **Explain the movements of toxic substances and how they affect organisms and ecosystems**

Toxic substances may travel long distances through the atmosphere, waterways, or groundwater. Some toxic substances bioaccumulate and move up the food chain, poisoning consumers at high trophic levels through the process of biomagnification and impairing ecosystem services. (pp. 370–372)

+ **Discuss the approaches used to study the effects of toxic chemicals on organisms**

Scientists use wildlife toxicology, case histories, epidemiology, animal testing, and dose-response analysis to assess the toxicity of chemicals. Toxicity is affected by dosage, but some chemicals have unconventional dose-response curves and synergistic interactions with other chemicals. (pp. 372–380)

+ **Summarize risk assessment and risk management**

Risk assessment involves quantifying and comparing risks involved in different activities or substances. Risk management integrates science with political, social, and economic concerns to design strategies to minimize risk. (pp. 380–382)

+ **Compare philosophical approaches to risk and how they relate to regulatory policy**

An innocent-until-proven-guilty approach assumes that a chemical substance is safe unless shown to be harmful after release to the public, whereas a precautionary approach assumes that a substance may be harmful unless proven safe by its manufacturer prior to sale. (pp. 382–385)

SEEKING Solutions

1. Describe some environmental health hazards you may be living with indoors in your home, workplace, or elsewhere. Describe outdoor hazards you are aware of being exposed to. How may you have been affected by indoor or outdoor hazards in the past? How could you best deal with these hazards in the future?

2. Do you feel that laboratory animals should be used in experiments in toxicology? Why or why not?

3. Why has research on endocrine disruption spurred so much debate? What steps do you think could be taken to help establish greater consensus on how such research is conducted among scientists, industry, regulators, policymakers, and the public?

4. **THINK IT THROUGH** In your public speaking class, you have been asked to create a presentation that supports either the policy approach of the United States or that of the European Union concerning the study and management of the risks of synthetic chemicals. Which position would you advocate? Explain the major points you would raise in your presentation to support your position.

5. **THINK IT THROUGH** You are the parent of two young children, and you want to minimize the environmental health risks your kids are exposed to. Name five steps that you could take in your household and in your daily life that would minimize your children's exposure to environmental health hazards.

CALCULATING Ecological Footprints

In 2012, the last year the EPA reported data on pesticide use (pp. 251–253), Americans used 1.18 billion pounds of pesticide active ingredients, and global use totaled 5.82 billion pounds. That same year, the U.S. population was 314 million, and the world's population was 7.06 billion. In the table below, calculate your share of pesticide use as a U.S. citizen in 2012 and the amount used by (or on behalf of) the average citizen of the world.

Annual Pesticide Use

CONSUMER	PESTICIDE ACTIVE INGREDIENTS (LB)
United States (total)	1.18 billion
World (total)	5.82 billion
Your share of U.S. total	
Your share of world total	

1. Calculate your share of U.S. pesticide use (factoring in annual U.S. pesticide use and the U.S. population) and your share of global pesticide use (factoring in annual global pesticide use and the global human population). Does the per-capita pesticide use for a U.S. citizen seem reasonable for you personally? Why or why not? Do you think the area in which you live would have more than, less than, or about the U.S. average for pesticide use? Explain your answer.

2. Approximately how many times larger is your share of U.S. pesticide use compared to your share of global pesticide use? Pesticides applied to farm fields, lawns, and golf courses can wash into local waterways and expose people that utilize such water to potentially higher levels of harmful chemicals than people in areas where pesticide use is lower. Does it concern you that Americans may be experiencing many times the exposure to toxicants from pesticides than people in other nations, or are the risks outweighed by other benefits? Explain your answer.

Mastering Environmental Science

Students Go to **Mastering Environmental Science** for assignments, an interactive e-text, and the Study Area with practice tests, videos, and activities.

Instructors Go to **Mastering Environmental Science** for automatically graded activities, videos, and reading questions that you can assign to your students, plus Instructor Resources.

Freshwater Systems and Resources

Upon completing this chapter, you will be able to:

+ **Describe the distribution of fresh water on Earth and the major types of freshwater systems**

+ **Discuss how humans use water and alter freshwater systems**

+ **Assess problems of water supply and propose solutions to address depletion of fresh water**

+ **Describe the major classes of water pollution and propose solutions to address water pollution**

+ **Explain how we treat drinking water and wastewater**

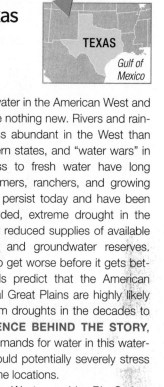

TEXAS

Gulf of Mexico

Reaching the Tipping Point:

Fracking and Fresh Water in West Texas

" **Oil is all right, but the oil isn't going to do any good if we don't have water to drink.**
Charles Phillips, resident of Howard County, Texas

The water use for horizontal oil and gas drilling has doubled in the past five years per well. In wet regions like the Marcellus (in Pennsylvania), I don't think water use is that big a deal. *In dry areas it can be a big deal.* **The strongest effects come through increased pumping of groundwater.**
Rob Jackson, Professor of Earth System Science, Stanford University, speaking in 2015 "

Conflicts over water in the American West and Southwest are nothing new. Rivers and rainfall are far less abundant in the West than the much wetter eastern states, and "water wars" in the West over access to fresh water have long occurred between farmers, ranchers, and growing cities. Those conflicts persist today and have been exacerbated by extended, extreme drought in the region that has greatly reduced supplies of available fresh water in rivers and groundwater reserves. The situation is likely to get worse before it gets better, as climate models predict that the American Southwest and Central Great Plains are highly likely to experience long-term droughts in the decades to come (see **THE SCIENCE BEHIND THE STORY**, pp. 394–395). New demands for water in this water-scarce environment could potentially severely stress water resources in some locations.

To illustrate the water challenges faced in the American West, consider Big Spring, Texas, a community of some 28,000 people in the oil-producing region of west Texas. The arid region in which Big Spring lies receives only about 50 cm (20 in.) of rainfall a year, and surface waters are not plentiful. Local farmers and ranchers largely rely on groundwater supplies to irrigate their crops and provide water to cattle. Groundwater also supplies drinking water to rural homes and the nearby Midland-Odessa metropolitan area—the fastest-growing metropolitan region in the United States from 2010 to 2015. Big Spring is located at the southernmost end of the Ogallala Aquifer (p. 396), a huge expanse of groundwater that stretches northward across the Great Plains. The geology of the aquifer is such that it refills very slowly, and a century of extensive extraction has caused water levels in the aquifer to drop significantly. In nearby areas in north Texas, for example, water levels in the aquifer have dropped 30 m (100 ft) in only the last 30 years.

This community, like many others in the region, is intimately tied to the oil industry. Petroleum (crude oil) extracted from the nearby, highly productive oil fields has long

◄ **Tanker trucks delivering water to a hydraulic fracturing operation in Texas**

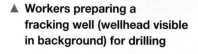

▲ **Workers preparing a fracking well (wellhead visible in background) for drilling**

389

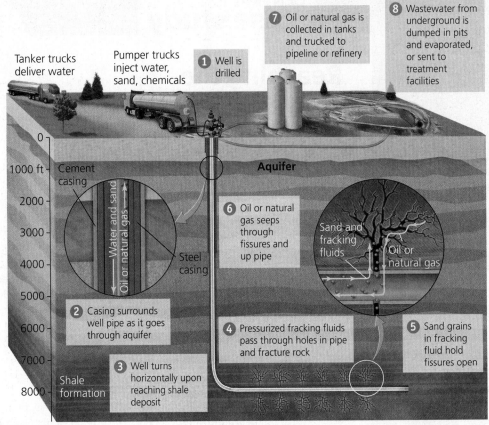

FIGURE 15.1 Hydraulic fracturing is used to extract oil or natural gas trapped in shale deposits deep underground. **①** A well is drilled with **②** protective casing and **③** is turned horizontally on reaching a shale deposit. **④** Pressurized fluids fracture the rock, and **⑤** sand lodges in the cracks, holding them open for **⑥** oil or natural gas to seep into them and rise through the pipe to **⑦** tanks at the surface, where it is collected in tanks. **⑧** Wastewater also rises to the surface and is piped to treatment facilities or to wastewater pits, where it is left to evaporate.

powered the local economy. In Howard County, in which Big Spring is located, the oil industry today employs about 2800 people and generates an estimated $4.5 billion in annual economic activity. This history of economic development makes the area very friendly to the oil and gas industry, but a new oil boom is forcing some residents to reexamine their views.

Big Spring sits atop the Permian Basin, an ancient formation of shale rock that holds an estimated $1 trillion of crude oil and $300 billion of natural gas in "unconventional" fossil fuel reserves (such fossil-fuel-rich shale formations are called *shale plays*). Conventional oil and gas reserves are located within porous rock deep underground and brought to the surface by pumping. In unconventional reserves, petroleum (**shale oil**) and natural gas (**shale gas**) are tightly "locked up" within non-porous shale but can be "freed" from shale formations through the process of **hydraulic fracturing** (also called *hydrofracking*, or simply *fracking*).

Hydraulic fracturing involves drilling deep into the earth and then angling the drill horizontally once a shale formation is reached. An electric charge sets off targeted explosions that perforate the drilling pipe and create fractures in the shale. Drillers then pump a slurry of water, sand, and chemicals down the pipe under great pressure. The sand lodges in the fractures and holds them open, while some of the liquids return to the surface as wastewater. Oil and natural gas trapped in the

shale migrate into the fractures and are carried through the drilling pipe to the surface, where it is collected in tanks (**FIGURE 15.1**).

But fracking requires a great deal of water, and many of these shale rock formations are located in areas already experiencing water stress—such as west Texas. Some 115 billion L (30 billion gal) of water were used for fracking in the Permian Basin in 2017, and it is predicted that water demand for fracking will only continue to grow. This added demand for water arises as new fracking wells are drilled, and existing wells increase their lateral drilling length. As wells age, they also require increasing amounts of water to extract oil and gas from within the shale rock. These new demands for water are further stressing water supplies in the region and putting some residents at odds with the fracking industry.

Water use in fracking does not seem extensive when viewed at the state or national levels—fracking consumes less than 1% of industrial water use in the United States and only 1% of total water use in Texas, for example. But when viewed at the local level, the water demands of fracking become more apparent. In Howard County, home to Big Spring, water used for fracking represents 20% of annual water usage in this community. And Big Spring is not alone. From 2011 to 2014, three-fourths of the 40,000 fracking wells drilled in the United States were located in water-scarce areas, and more than half were located in regions suffering from drought (**FIGURE 15.2a, b**).

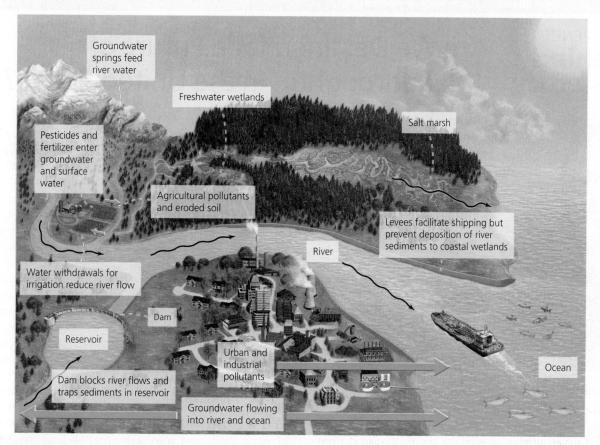

FIGURE 15.4 Water flows through freshwater systems and marine and coastal aquatic systems that interact extensively with one another. People affect the components of the system by constructing dams and levees, withdrawing water for human use, and introducing pollutants. Because the systems are closely connected, these impacts can cascade through the system and cause effects far from where they originated. In the figure, orange arrows indicate inputs into water bodies and black lines the direction of water flow.

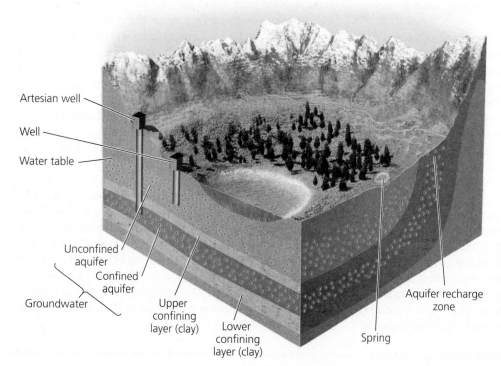

FIGURE 15.5 Groundwater occurs in unconfined aquifers above, or in confined aquifers sandwiched between, impermeable layers. Water may rise to the surface at springs, in wetlands, and through wells. Artesian wells tap into confined aquifers to mine water under pressure.

THE SCIENCE behind the story

Are We Destined for a Future of "Megadroughts" in the United States?

Benjamin I. Cook, NASA

From 2012 to 2016, record-low levels of precipitation coupled with record-high heat kept California and other western states in severe drought conditions, prompting the state to act aggressively to implement a series of far-reaching measures to promote water conservation in agriculture, industry, and homes. While near-record levels of precipitation bathed the state in 2016 and 2017—refilling reservoirs and expanding snow-pack in the mountains—state officials maintained many of the water conservation measures enacted during the drought, as climatic studies suggest that California may very well experience severe drought in the future.

This was the sobering conclusion reached by a team of researchers led by Benjamin Cook, research scientist at NASA Goddard Institute for Space Studies, and advanced in a paper published in the journal *Science Advances* in 2015. In an effort to frame the current drought in the western United States in the proper context, the group compared the climate of the Central Plains and Southwest over the past 1000 years with predictions of their climates over the next 100 years based on computer simulations. "We are the first to do this kind of quantitative comparison between the projections and the distant past," said co-author Jason Smerdon, "and the story is a bit bleak."

Researchers used the North American Drought Atlas (NADA)—a reconstruction of climates based on data from tens of thousands of samples of tree rings from the United States, Canada, and Mexico—to estimate the past climates of the Central Plains and Southwest regions. Trees are a natural archive of climate data, because they grow at varying rates depending on moisture, temperature, and other factors. Using data from the recent past, scientists determine the relationship between tree growth patterns and climatic factors, and they then use this relationship to reconstruct past climates for which they have data on tree growth from tree rings.

To predict the future climate of the southwestern United States and the Central Plains, the researchers used 17 climate models and ran simulations based on a "business as usual" scenario, in which the growth in greenhouse gas emissions followed current trends, as well as on a "moderate reduction" scenario, in which growth in greenhouse gas emissions was more modest.

The study also used three indicators of drought, measuring the level of soil moisture available to plants from the surface to

30-cm (12-in.) depth; from the surface to 2-m (6.6-ft.) depth; and by using the Palmer Drought Severity Index (PDSI), a measure of the difference between soil moisture supply (from precipitation) and soil moisture demand (from evaporation and uptake by plants). In the PDSI, negative values mean drier conditions, or drought, and positive values mean wetter conditions.

When the researchers combined data from the past, present, and future, the results were stunning—and troubling. The climate models predicted unprecedented levels of drought in the Central Plains and Southwest regions through 2100, with high levels of agreement between the 17 climate models and the three measures of soil moisture (**FIGURE 1**). The study concluded that the drying of soils was not being driven by drastic reductions in precipitation but rather by increased evaporative pressure; that is, warmer temperatures leading to higher rates of evaporation and elevated rates of soil water uptake by plants.

The models also concluded that human-induced climate change, and not natural variability in climate, would be the driving force for drier soils in the Central Plains and Southwest. A comparison with the past indicated that these future conditions would be far worse than those seen during the "Medieval megadrought period" from A.D. 1100 to 1300. This extended drought is thought to have led to the decline of the ancient Pueblos (the Anasazi), who lived along the Colorado Plateau. Smerdon described their results as follows: "Even when selecting for the worst megadrought-dominated period, the 21st Century projections make the mega-droughts seem like quaint walks through the Garden of Eden."

Comparing the probability of extended droughts in the latter half of the 20th century (1950–2000) with the probability of such droughts in the latter half of the 21st Century (2050–2099) provided further cause for concern. According to the simulations, the probability that the two regions would experience a decade-long drought essentially doubled, and the chance that they would experience a multidecade drought increased from around 10% to more than 80% (**FIGURE 2**). If greenhouse gases are reduced, the chance of a multidecade drought drops to 60–70% for the Central Plains but remains above 80% for the Southwest, providing yet another incentive for combating global climate change.

Studies like these help us to prepare for the future by providing an idea of what to expect. They show that it would be wise to maintain indefinitely the water conservation strategies embraced by California during its historic drought—even in times of unusually high levels of precipitation. These studies also suggest that, given the future predictions of dire drought, implementing water conservation in other states in the Central Plains and Southwest would be a prudent course of action. In the words of study co-author Toby Ault, "The time to act is now. The time to start planning for adaptation is now. We need to assess what the rest of this century will look like for our children and grandchildren."

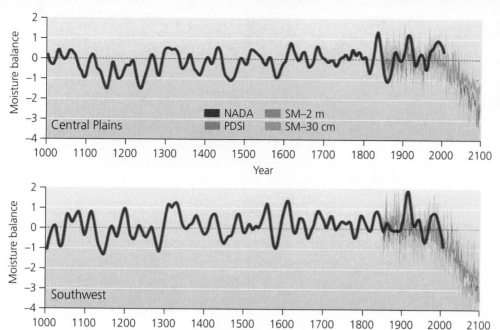

FIGURE 1 Soil moisture levels for the Central Plains and Southwest regions, as predicted by the North American Drought Atlas (NADA) and computer climate models. Negative values for moisture balance indicate drier soils (drought) and positive values indicate wetter soils. The predicted drought conditions in the late 21st century are unprecedented in the last 1000 years. (PDSI = Palmer Drought Severity Index, SM-30 cm = Soil moisture to 30-cm depth, and SM-2 m = Soil moisture to 2-m depth. The pale blue–shaded areas represent the variability in model PDSI values across computer climate models.) *Data from Cook, B. I., et al., 2015. Unprecedented 21st century drought risk in the American Southwest and Central Plains, Science Advances 1(1): e1400082.*

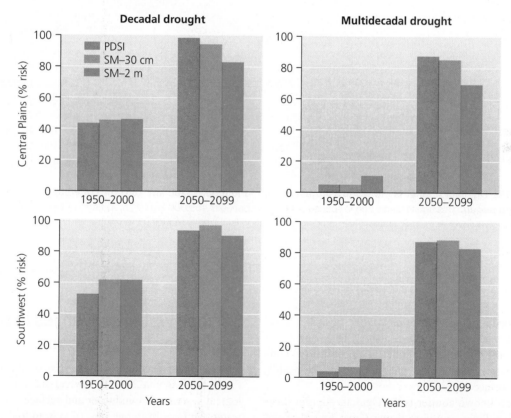

FIGURE 2 Risk of decadal (11-year) and multi-decadal (35-year) drought in the Southwest and Central Plains in the late 20th century and late 21st century for three measures of soil moisture. Due to human-induced climate change, both regions are far more likely to experience long-term drought in the future than in the past. *Data from Cook, B. I., et al., 2015. Unprecedented 21st century drought risk in the American Southwest and Central Plains, Science Advances 1(1): e1400082.*

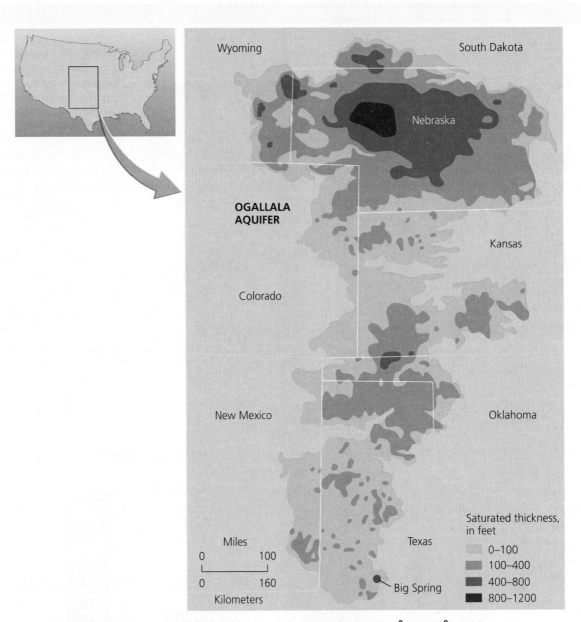

FIGURE 15.6 The Ogallala Aquifer is the world's largest aquifer, and it held 3700 km³ (881 mi³) of water before pumping began. This aquifer (indicated in shades of blue) lies beneath 453,000 km² (175,000 mi²) of the Great Plains, stretching over eight U.S. states. Overpumping for irrigation is currently reducing the volume and extent of this aquifer.

called a **confined aquifer,** or *artesian aquifer.* In such a situation, the water is under great pressure. In contrast, an **unconfined aquifer** lacks an impermeable upper layer to confine it, so its water is under less pressure and can be readily recharged by surface water.

The largest known aquifer is the Ogallala Aquifer (also called the High Plains aquifer) in the Great Plains of the United States (**FIGURE 15.6**). Water from this massive aquifer has permitted American farmers to create the most productive grain-producing region in the world. However, unsustainable water withdrawals are threatening long-term use of the aquifer for agriculture.

Surface water converges in river and stream ecosystems

Surface water accounts for just 1% of the fresh water on Earth, but it is vital for our survival and for the planet's ecological systems. Groundwater and surface water interact, and water can flow from one type of system to the other. Surface water becomes groundwater by infiltration. Groundwater becomes surface water through springs (and human-drilled wells), often keeping streams flowing or wetlands moist when surface conditions are otherwise dry. Each day in the United States, 1.9 trillion L (492 billion gal) of groundwater are released into surface waters—nearly as much as the daily flow of the Mississippi River.

Water that falls from the sky as rain, emerges from springs, or melts from snow or a glacier and then flows over the land surface, is called **runoff.** As it flows downhill, runoff converges where the land dips lowest, forming streams, creeks, or brooks. These small watercourses may merge into rivers, whose water eventually reaches a lake or ocean. A smaller river flowing into a larger one is called a *tributary*. The area of land drained by a *river system*—a river and all its tributaries—is that river's **watershed** (p. 105), also called a **drainage basin.**

Landscapes determine where rivers flow, but rivers shape the landscapes through which they run as well. A river that runs through a steeply sloped region and carries a great deal of sediment may flow as an interconnected series of watercourses called a *braided river* (**FIGURE 15.7a**). In flatter regions, most rivers are *meandering rivers* (**FIGURE 15.7b**). In a meandering river, the force of water rounding a bend gradually eats away at the outer shore, eroding soil from the bank. Meanwhile, sediment is deposited along the inside of the bend, where water currents are weaker. Over time, river bends become exaggerated in shape, forming oxbows (**FIGURE 15.7c**). If water erodes a shortcut from one end of the loop to the other, pursuing a direct course, the oxbow is cut off and remains as an isolated, U-shaped water body called an *oxbow lake*.

Over thousands or millions of years, a meandering river may shift from one course to another, back and forth over a large area, carving out a flat valley and picking up sediment that is later deposited in coastal wetlands. Areas nearest a river's course that are subject to periodic flooding are said to be within the river's **floodplain.** Frequent deposition of silt (eroded soil) from flooding makes floodplain soils especially fertile. As a result of deposition, agriculture thrives in floodplains, and *riparian* (riverside) forests are productive and species-rich. A river's meandering course is often driven by large-scale flooding events that scour new channels. However, extensive damming on many of Earth's major rivers has reduced the rate of river meandering significantly from historic rates. Instead of coursing downriver, floodwaters are often trapped in reservoirs behind dams or contained in river channels by levees.

Rivers and streams host diverse ecological communities. Algae and detritus support many types of invertebrates, from water beetles to crayfish. Insects as diverse as dragonflies, mayflies, and mosquitoes develop as larvae in streams and rivers before maturing into adults that take to the air. Fish and amphibians consume aquatic invertebrates and plants, and birds such as kingfishers, herons, and ospreys dine on fish and amphibians.

Lakes and ponds are ecologically diverse systems

Lakes and ponds are bodies of standing surface water. The largest lakes, such as North America's Great Lakes, are essentially inland seas, because of the large volumes of water they hold. Lake Baikal in Asia is the world's deepest lake, at 1637 m (just over 1 mile) deep. The Caspian Sea is the

(a) Braided river in Nebraska

(b) Meandering river in Colorado

(c) Oxbow along a river in Arizona

FIGURE 15.7 Rivers are classified according to their flow across the landscape. The Platte River in Nebraska **(a)** is an example of a braided river, whereas the East River in Colorado **(b)** meanders across the landscape. The Horseshoe Bend portion of the Colorado River in Arizona **(c)** demonstrates an oxbow.

world's largest body of fresh water, covering nearly as much land area as Montana or California. Although lakes and ponds can vary greatly in size, scientists have described several zones common to these waters (**FIGURE 15.8**).

Around the nutrient-rich edges of a water body, the water is shallow enough that aquatic plants grow from the mud and reach above the water's surface. This region, named the *littoral zone,* abounds in invertebrates—such as insect larvae, snails, and crayfish—that fish, birds, turtles, and amphibians feed on. The *benthic zone* extends along the bottom of the lake or pond, from the shore to the deepest point. Many invertebrates live in the mud, feeding on detritus or on one another. In the open portion of a lake or pond, far from shore, sunlight penetrates shallow waters of the *limnetic zone.* Because light enables photosynthesis, the limnetic zone supports phytoplankton (algae, protists, and cyanobacteria), which in turn support zooplankton, both of which are eaten by fish. Below the limnetic zone lies the *profundal zone,* the volume of open water that sunlight does not reach. This zone lacks photosynthetic life and is lower in dissolved oxygen than are the upper waters.

Ponds and lakes change over time as streams and runoff bring them sediment and nutrients. **Oligotrophic** lakes and ponds, which are low in nutrients and high in oxygen, may slowly transition to the high-nutrient, low-oxygen conditions of **eutrophic** water bodies (p. 410). Eventually, water bodies may fill in completely by the process of aquatic succession (p. 87). These changes occur naturally, but eutrophication can also result from human-caused nutrient pollution.

Freshwater wetlands include marshes, swamps, bogs, and vernal pools

Wetlands are systems in which the soil is saturated with water, and generally feature shallow standing water with ample vegetation. There are many types of wetlands, and most are enormously rich and productive. In *freshwater marshes,* shallow water allows plants such as cattails and bulrushes to grow above the water surface. *Swamps* also consist of shallow water rich in vegetation, but they occur in forested areas. *Bogs* are ponds covered with thick floating mats of vegetation and can

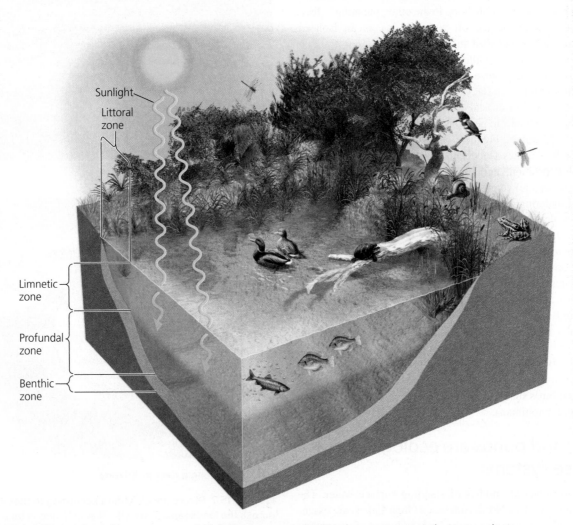

FIGURE 15.8 Lakes and ponds are composed of distinct zones. In the littoral zone, emergent plants grow along the shoreline. The limnetic zone is the layer of open, sunlit water, where photosynthesis takes place. Sunlight does not reach the deeper profundal zone. The benthic zone, at the bottom of the water body, often is muddy, rich in detritus and nutrients, and low in oxygen.

represent a stage in aquatic succession. *Vernal pools* are seasonal wetlands that form in early spring from rain and snowmelt, and then dry up once weather becomes warmer and drier. Numerous animals, such as frogs and salamanders, have evolved to take advantage of the ephemeral time windows in which these seasonal pools exist each year.

Wetlands are extremely valuable habitat for wildlife and provide important ecosystem services (pp. 120–121) by slowing runoff, reducing flooding, recharging aquifers, and filtering pollutants. Despite the vital roles played by wetlands, people have drained and filled them extensively for agriculture. Many wetlands are lost when people divert and withdraw water, channelize rivers, and build dams. The United States and southern Canada, for example, have lost more than half their wetlands since European colonization.

Human Activities Affect Waterways

So far, our tour of aquatic systems has shown the ecological and economic value of freshwater ecosystems. We will now see how these systems are affected when we withdraw water for human use, build dams and levees, and introduce pollutants that alter water's chemical, biological, and physical properties.

Although fresh water is a limited resource, it is also a renewable resource as long as we manage our use sustainably. However, people are withdrawing water at unsustainable levels and depleting many sources of surface water and groundwater. Already, one-third of the world's people are affected by chronic shortages of fresh water.

In addition to our overextraction of fresh water, people have intensively engineered freshwater waterways with dams, levees, and diversion canals to satisfy demands for water supplies, transportation, and flood control. An estimated 60% of the world's 227 largest rivers (77% of those located in North America and Europe), for example, have been strongly or moderately affected by human engineering.

Fresh water and human populations are unevenly distributed across Earth

World regions differ in the size of their human populations and, as a result of climate and other factors, possess varying amounts of groundwater, surface water, and precipitation. Hence, not every human around the world has equal access to fresh water (**FIGURE 15.9**). Because of the mismatched distribution of water and population, human societies have always struggled to transport fresh water from its source to where people need it. The western United States is one example. Growing demand for water in recent decades from population growth and

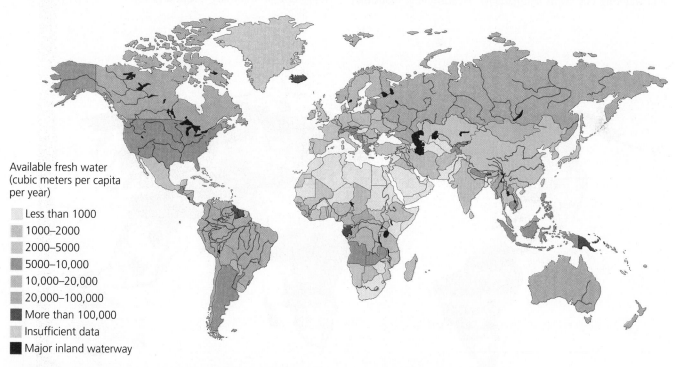

Available fresh water (cubic meters per capita per year)

- Less than 1000
- 1000–2000
- 2000–5000
- 5000–10,000
- 10,000–20,000
- 20,000–100,000
- More than 100,000
- Insufficient data
- Major inland waterway

FIGURE 15.9 Nations vary tremendously in the amount of fresh water per capita available to their citizens. For example, with over 100,000 cubic meters per capita per year, Iceland, Papua New Guinea, Gabon, and Guyana each have more than 100 times more water per person than do many Middle Eastern and North African countries. *Data from Harrison, P., and F. Pearce, 2000. AAAS atlas of population and the environment, edited by the American Association for the Advancement of Science, © 2000 by the American Association for the Advancement of Science.*

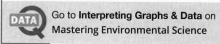

DATA **Go to Interpreting Graphs & Data on Mastering Environmental Science**

extensive agriculture—and now fracking for fossil fuels—has made water an ever-more precious resource. This has led to cities, farmers, and industrial operations meeting their demands for water by extensively diverting water from the regions' rivers and overdrawing groundwater supplies in many parts of the West, where water availability is relatively low.

Fresh water is distributed unevenly in time as well as space. For example, India's monsoon storms can dump half of a region's annual rain in just a few hours. Rivers have seasonal differences in flow because of the timing of rains and snowmelt. For this reason, people build dams to store water from wetter months for use in drier times of the year, when river flow diminishes.

As if the existing mismatches between water availability and human need were not enough, global climate change (Chapter 18) has and will continue to worsen conditions in many regions by altering precipitation patterns, melting glaciers, causing early-season runoff, and intensifying droughts and flooding. A 2009 study, for example, found that one-third of the world's 925 major rivers experienced reduced flow from 1948 to 2004, with the majority of the reduction attributed to effects of climate change.

Water supplies households, industry, and agriculture

We all use water at home for drinking, cooking, cleaning, and watering our lawns and gardens. Most mining, industrial,

and manufacturing processes require water. Farmers and ranchers use water to irrigate crops and water livestock. Globally, we allot about 70% of our annual use of fresh water to agriculture. Industry accounts for roughly 20%, and residential and municipal uses account for only 10%.

The removal of water from an aquifer or from a body of surface water without returning it to its source is called **consumptive use.** Our primary consumptive use of water is for agricultural irrigation (pp. 219–220). **Nonconsumptive use** of water, in contrast, does not remove, or only temporarily removes, water from an aquifer or surface water body. Using water to generate electricity at hydroelectric dams is an example of nonconsumptive use; water from a river is passed through dam machinery to turn turbines and released downstream. As noted in the opening case study, wastewaters from fracking are often injected into porous rocks deep underground, essentially making its use consumptive as the water is not returned to the aquifer or stream from which it was drawn.

The large amount of water used for agriculture is due to our rapid population growth, which requires us to feed and clothe more and more people each year. Overall, we withdraw 70% more water for irrigation today than we did 50 years ago, and have doubled the amount of land under irrigation. This expansion of irrigated land has helped food and fiber production to keep up with population growth, but many irrigated areas are using water unsustainably, threatening their long-term productivity (**FIGURE 15.10**).

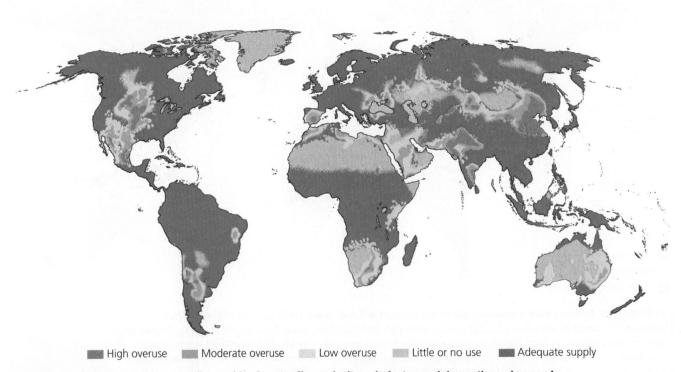

■ High overuse ■ Moderate overuse ■ Low overuse ■ Little or no use ■ Adequate supply

FIGURE 15.10 Regions where overall use of fresh water (for agriculture, industry, and domestic use) exceeds the annual available supply, thus requiring groundwater depletion or diversion of water from other regions. Agriculture accounts for 70% of total usage. The map actually understates the problem, because it does not reflect seasonal shortages. *Data from UNESCO, 2006. Water: A shared responsibility. World Water Development Report 2. UNESCO and Berghahn Books.*

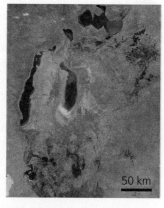

(a) Satellite image of Aral Sea, 1987

(b) Satellite image of Aral Sea, 2018

(c) A ship stranded by the Aral Sea's fast-receding waters

FIGURE 15.11 **The Aral Sea in central Asia was once the world's fourth largest lake.** However, it has been shrinking **(a, b)** because so much water was withdrawn to irrigate cotton crops. Ships were stranded **(c)** along the former shoreline of the Aral Sea because the waters receded so far and so quickly. Today, restoration efforts are beginning to reverse the decline in the northern portion of the sea; the waters there are slowly rising.

Excessive water withdrawals can drain rivers and lakes

In many places, we are withdrawing surface water at unsustainable rates and drastically reducing river flows. Because of excessive water withdrawals, many of the world's major rivers—such as the Colorado and Rio Grande rivers in western North America, the Yellow River in China, and the Nile River in Egypt—regularly run dry before reaching the sea. This reduction in flow not only threatens the future of the cities and farms that depend on the river, but also drastically alters the ecology of the rivers and their deltas, changing plant communities, wiping out populations of fish and invertebrates, and devastating commercially important fisheries.

Worldwide, roughly 15–35% of water withdrawals for irrigation are thought to be unsustainable. In areas where agriculture is demanding more fresh water than can be sustainably supplied, *water mining*—withdrawing water faster than it can be replenished—is taking place. In these areas, aquifers are being depleted or surface water is being piped in from other regions to meet the demands for irrigation.

Nowhere are the effects of surface water depletion so evident as in the Aral Sea. Once the fourth-largest lake on Earth, just larger than Lake Huron in the Great Lakes, it lost more than four-fifths of its volume in only 45 years (**FIGURE 15.11a**). This dying inland sea, on the border of present-day Uzbekistan and Kazakhstan, is the victim of poor irrigation practices. Many decades ago, the former Soviet Union instituted industrial cotton farming in this dry region by flooding cropland with water from the two rivers that supplied the Aral Sea with its water. For a few decades, this practice boosted Soviet cotton production, but it shrank the Aral Sea, and the irrigated soil eventually became salty and waterlogged. Today, 60,000 fishing jobs are gone, winds blow pesticide-laden dust up from the dry lake bed (**FIGURE 15.11b**), and little cotton grows on the blighted soil (**FIGURE 15.11c**).

Groundwater can also be depleted

Groundwater is more easily depleted than surface water because most aquifers recharge very slowly. Today, we are mining groundwater, extracting 160 km^3 (5.65 trillion ft^3) more water each year than returns to the ground. If we compare an aquifer to a bank account, we are making more withdrawals than deposits, and the balance is shrinking. Groundwater depletion is a major problem because one-third of Earth's human population—including 99% of the rural population of the United States—relies on groundwater for its needs.

As aquifers are mined, water tables drop deeper underground. This deprives freshwater wetlands of groundwater inputs, causing them to dry up. Groundwater also becomes more difficult and expensive to extract. In parts of Mexico,

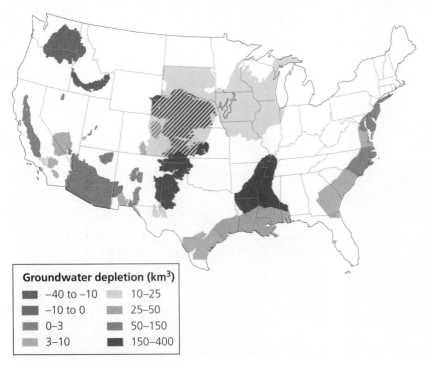

Groundwater depletion (km³)

■ −40 to −10	▨ 10–25
■ −10 to 0	▨ 25–50
▨ 0–3	■ 50–150
▨ 3–10	■ 150–400

FIGURE 15.12 A map of cumulative groundwater depletion from 1900 to 2008 shows that many U.S. aquifers have been overexploited. Aquifers in the Midwest and Southwest have been particularly affected by excessive withdrawals. *Data from Konikow, L. F., 2013. Groundwater depletion in the United States (1900–2008). U.S. Geological Survey Scientific Investigations Report 2013–5079, http://pubs.usgs.gov/sir/2013/5079.*

well water undrinkable. This occurs when overpumping of groundwater near the coast greatly reduces the volume of fresh groundwater flowing from porous rocks under the land into connected rock layers that lie beneath the ocean. Saline groundwater, the seawater that exists in the pores of rocks beneath the ocean, can then penetrate into inland aquifers. Typically, saline groundwater does not penetrate far inland because it is constantly being "pushed back" by fresh groundwater flowing into the sea. But when fresh groundwater flows are reduced by overextraction, saline groundwater experiences less resistance to inland movements and can penetrate deep into coastal freshwater aquifers. This causes coastal wells to draw up saline groundwater instead of the desired fresh groundwater. This problem has occurred in California, Florida, India, the Middle East, and many other locations where groundwater is heavily extracted.

As aquifers lose water, they become less able to support overlying strata, and the land surface above may sink, compact, or collapse, creating **sinkholes.** Venice, Bangkok, Beijing, and Mexico City are slowly sinking, causing streets to buckle and underground pipes to rupture. Once the ground sinks, soil and rock become compacted, losing the porosity that enabled them to hold water. Recharging a depleted aquifer thereafter becomes more difficult. Compaction of rock layers due to groundwater overextraction caused some agricultural regions in California to sink by 9 m (30 ft) from 1925 to 1977, and many areas continue to sink today as overextraction continues.

India, China, and multiple Asian and Middle Eastern nations, water tables are falling 1–3 m (3–10 ft) per year. In the United States, overpumping has lowered water tables in many aquifers, particularly those in the Midwest and West (**FIGURE 15.12**).

In west Texas, falling water tables have tempered the enthusiasm of many local communities for the growing fracking industry during the recent "shale oil" boom. Local residents rely almost exclusively on slowly regenerating groundwater for use in their homes and on their farms. Extracting large quantities of water from the aquifer for fracking endangers the sustainability of the water system for humans and agriculture in the entire region. And to make matters more tenuous, in Texas—unlike in many other states—landowners have the right to extract as much groundwater on their property as they choose. The limited supply of water and high demand for it among fracking companies have therefore led some landowners to drill additional groundwater wells on their property so they can sell any extracted water to fracking companies for use in drilling (**FIGURE 15.13**). This increased demand for groundwater, coupled with other demands for water and long-term drought, has drastically reduced groundwater levels in the region. As groundwater levels drop, wells run dry and communities struggle to find new sources of water. Nearby Barnhart, Texas, gained national attention in 2013 when the well that supplied the town with its drinking water ran dry, a situation some local residents blamed on fracking's added demands on the local aquifer.

When groundwater is overextracted in coastal areas, saltwater from the ocean can intrude into inland aquifers, making

FIGURE 15.13 A landowner in Texas advertises groundwater for sale to fracking companies.

Groundwater supplies our bottled water

These days, our groundwater is being withdrawn for a new purpose—to be packaged in plastic bottles and sold on supermarket shelves. Bottled water is a booming business. The average American drinks over 136 L (36 gal) of bottled water a year, and annual sales top $14 billion in the United States and $160 billion worldwide.

Bottled water exerts substantial ecological impact. The energy costs of bottled water are estimated to be 1000–2000 times greater than the energy costs of tap water, mostly as a result of transportation costs. After their contents are consumed, at least three out of four bottles in the United States are not recycled, and end up interred in landfills. That's 30–40 billion containers per year (5 containers for every human being on Earth) and close to 1.5 million tons of plastic waste. Many people who buy bottled water do so because they believe it is superior to tap water. However, in blind taste tests people think tap water tastes just as good, and chemical analyses show that bottled water is no safer or healthier than tap water.

As a result of the environmental impacts of bottled water, dozens of colleges and universities in the United States and Canada have banned the sale of bottled water or restricted the use of plastic water bottles on campus. Similarly, major U.S. cities, including New York, San Francisco, and Seattle, prohibit using government funds to purchase bottled water, in part due to the cost savings of drinking tap water instead of bottled water.

People build levees along rivers to control floods

Among the reasons we control the movement of fresh water, flood prevention ranks high. People have always been attracted to riverbanks for their water supply and for the flat topography and fertile soil of floodplains. But if one lives in a floodplain, one must be prepared to face flooding. **Flooding** is a normal, natural process that occurs when snowmelt or heavy rain swells the volume of water in a river so that water spills over the river's banks. In the long term, floods are immensely beneficial to both natural systems and human agriculture, because floodwaters build and enrich soil by spreading nutrient-rich sediments over large areas.

In the short term, however, floods can do tremendous damage to the farms, homes, and property of people who choose to live in floodplains. To protect against floods, communities and governments have built **levees** (also called *dikes*) along banks of rivers to hold water in main channels. These structures prevent flooding at most times and places, but can sometimes worsen flooding because they force water to stay in channels and accumulate, building up enormous energy and leading to occasional catastrophic overflow events.

We divert surface water to suit our needs

People have long diverted water from rivers and lakes to farms, homes, and cities with artificial rivers called **aqueducts** or

FIGURE 15.14 Aqueducts, like this example in California's agricultural Central Valley, function as man-made rivers to redistribute water.

canals (**FIGURE 15.14**). Water from the Colorado River in the western United States is heavily diverted and utilized as the river flows toward the Pacific Ocean. Early in its course, some Colorado River water is piped through a mountain tunnel and down the Rockies' eastern slope to supply the city of Denver. More is removed for Las Vegas and other cities, and for farmland, as the water proceeds downriver. When the river reaches Parker Dam on the California–Arizona state line, large amounts of water are diverted into the Colorado River Aqueduct, which brings water to the Los Angeles and San Diego areas. Arizona also draws water from Parker Dam, transporting it in the canals of the Central Arizona Project. Farther south, water is diverted into the Coachella and All-American canals, destined for agriculture, mostly in California's Imperial Valley.

The world's largest water diversion project is currently under way in China. Three sets of massive aqueducts, totaling 2500 km (1550 mi) in length, are being built to move trillions of gallons of water from the Yangtze River in southern China, where water is plentiful, northward to China's capital city of Beijing and the Yellow River. This will aid water levels in the Yellow River, which today routinely dries up at its mouth due to northern China's relatively dry climate and because of excessive withdrawals of the river's water for farms, factories, and homes. Many scientists say the estimated $62 billion project won't transfer enough water to make a difference in the arid north, and will

WEIGHING the issues

Reaching for Water

The diversion of water from Mono Lake to Los Angeles is not the only controversial diversion of fresh water in the United States. The rapidly growing Las Vegas metropolitan area is exceeding its allotment of water from the Colorado River and has proposed a $15 billion project that would divert groundwater from 450 km (280 mi) away to meet the growing demand in Nevada's largest city. Do you think such diversions are ethically justified? If the project ends up destroying rural communities and wetland ecosystems at the diversion site in eastern Nevada, will this be an acceptable cost given the economic activity generated in Las Vegas? How else might cities like Las Vegas meet their future water needs?

cause extensive environmental impacts, all while displacing hundreds of thousands of people.

In the past, large-scale diversion projects have enabled politically strong yet water-poor regions to forcibly appropriate water from communities too weak to keep it for themselves. For example, the city of Los Angeles grew by commandeering water from the rural Owens Valley 350 km (220 mi) away. In so doing, it turned the environment of this region of eastern California to desert, creating dustbowls and destroying its economy. Then in 1941, city leaders in Los Angeles decided to divert streams feeding into Mono Lake, over 565 km (350 mi) away in northern California. As the lake level fell 14 m (45 ft) over 40 years, the lake decreased in size by up to 5000 ha (12,500 acres) and salt concentrations in lake waters nearly doubled.

Most of Earth's major rivers are dammed

A **dam** is any obstruction placed in a river or stream to block its flow. Dams create **reservoirs,** artificial lakes that store water for human use. We build dams to prevent floods, provide drinking water, facilitate irrigation, and generate electricity (pp. 582–585).

Worldwide, we have erected more than 45,000 large dams (greater than 15 m, or 49 ft, high) across rivers in more than 140 nations and have built tens of thousands of smaller dams. Only a few major rivers in the world remain undammed and free-flowing; these run through the tundra and taiga of Canada, Alaska, and Russia, and in remote regions of Latin America and Africa. **FIGURE 15.15** summarizes the mix of benefits and costs of damming for people and the environment.

As an example of mixed consequences, we can consider the world's largest dam project. The Three Gorges Dam on China's Yangtze River, 186 m (610 ft) high and 2.3 km (1.4 mi) wide, was completed in 2008. Its reservoir stretches for 616 km (385 mi; as long as Lake Superior in the Great Lakes). This project provides flood control, enables boats and barges to travel farther upstream on the Yangtze, and generates enough hydroelectric power to replace dozens of large coal or nuclear power plants.

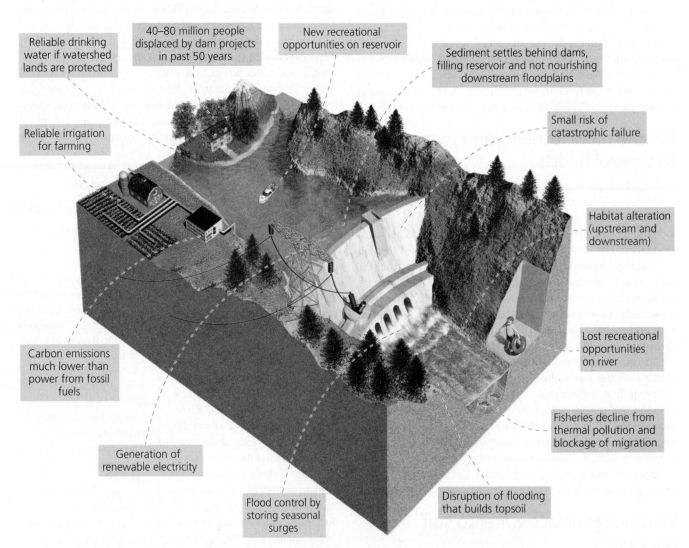

FIGURE 15.15 Damming rivers has diverse consequences for people and the environment. The generation of clean and renewable electricity is one of several major benefits **(green boxes)** of hydroelectric dams. Habitat alteration is one of several negative impacts **(orange boxes).**

However, the Three Gorges Dam cost $37 billion to build, and its reservoir flooded 22 cities, forcing the relocation of more than 1.2 million people. The reservoir slows the river's flow so that suspended sediment settles behind the dam. Because the river downstream is deprived of sediment, the tidal marshes where the Yangtze empties into the sea are eroding away. This has left the city of Shanghai with a degraded coastal environment and less coastal land to develop. Many scientists worry that the Yangtze's many pollutants will also become trapped in the Three Gorges Dam reservoir, eventually making the water undrinkable. To avoid such a buildup of pollutants, the Chinese government plans to spend $5 billion building hundreds of sewage treatment and waste disposal facilities. On top of all these worries, earthquakes in southern China in 2008 and again in 2012 raised fears that a future quake could damage the dam, perhaps even leading to its collapse. Such a dam failure would be catastrophic, as it could generate a wall of water that breaches downstream dams and creates a domino effect of dam failures. This occurred in 1975, when record flooding along the Yangtze caused breaches at 62 separate dams, and the subsequent flooding claimed 26,000 lives.

Some dams are being removed

People who believe that the costs of some dams outweigh their benefits are pushing for such dams to be dismantled. By removing dams and letting rivers flow free, they say, we can restore riparian ecosystems, reestablish economically valuable fisheries, and revive river recreation such as fly-fishing and rafting. Increasingly, private dam owners and the Federal Energy Regulatory Commission (FERC), the U.S. government agency charged with renewing licenses for dams, have agreed. Many aging dams are in need of costly repairs or have outlived their economic usefulness, making them suitable candidates for removal. Some 400 dams have been dismantled in the United States over the past decade, and more will come down within the next 10 years, when the licenses of over 500 dams come up for renewal.

The drive to remove dams first gathered steam in 1999 with the dismantling of the Edwards Dam on Maine's Kennebec River. FERC had determined that the environmental benefits of removing the dam outweighed the economic benefits of relicensing it. Within a year after the 7.3-m (24-ft) high, 279-m (917-ft) long dam was removed, large numbers of 10 species of migratory fish, including salmon, sturgeon, shad, herring, alewife, and bass, ventured upstream and began using the 27-km (17-mi) stretch of river above the dam site. Some property owners along the former reservoir who had opposed the dam's removal had a change of heart once they saw the healthy and vibrant river that now ran past their property.

In 2014, the world's largest dam removal project was completed when the last section of the 64-m (210-ft) high Glines Canyon Dam on the Elwha River in Washington State was demolished. Built in 1914 to supply power for local wood mills, the dam decimated local fisheries by preventing salmon from migrating upriver to spawn, imperiling the livelihood of Native

Americans who had long harvested the river's salmon and shellfish. But since the dam's removal, the Elwha River system has rebounded. Sediments that were once held behind the dam have flooded downstream, rebuilding riverbanks, beaches, and estuaries. Habitats for shellfish and small fish are being restored in and around the river's mouth. And as salmon migrate upriver to spawn in the Elwha's tributaries, it is hoped that an entire functional ecosystem will emerge along with their return.

In California, a movement is under way to remove the O'Shaughnessy Dam on the Tuolumne River in Yosemite National Park. Such a removal would drain the Hetch Hetchy Valley, which is currently submerged beneath the dam's reservoir (**FIGURE 15.16**), and restore the native plants and animals that occupied the area prior to the dam's completion in 1923. This action is strongly opposed by the city of San Francisco, however, as the reservoir provides drinking water for the city.

Tuolumne River

(a) Hetch Hetchy Valley before dam (1907)

(b) Hetch Hetchy Valley after dam

FIGURE 15.16 Calls for the removal of the O'Shaughnessy Dam on the Tuolumne River in the Hetch Hetchy Valley region of Yosemite National Park have increased as more and more dams have been retired in the United States. Proponents of this dam's removal seek to restore the valley to the natural ecosystem **(a)** that existed before the dam's reservoir **(b)** submerged it.

Solutions to Depletion of Fresh Water

In the past 50 years, population growth, expansion of irrigated agriculture, and industrial development doubled our annual use of fresh water and stressed supplies of fresh water. To address shortages of fresh water, we can aim either to increase supply or to reduce demand.

Increasing water supplies for humans carries ecological costs

Constructing large dams was a common solution to water shortages in the past. However, large dams have already been constructed at sites most suitable for them, and most of the remaining locations are in regions that make such construction projects prohibitive. Building more dams, therefore, does not appear to be a viable solution to meet people's increasing demands for fresh water.

We can increase supply temporarily through more intensive extraction of surface and groundwater, but overextracting rivers, lakes, and aquifers is not sustainable in the long term. Diversion projects may solve supply problems in one area while causing water shortages in others.

A supply strategy with some potential for sustainability is to generate fresh water by **desalination,** or **desalinization,** the removal of salt from seawater or other water of marginal quality. One method of desalination mimics the hydrologic cycle by heating and evaporating ocean water and then condensing the vapor—essentially distilling fresh water. Another method forces water through membranes with tiny pores to filter out salts; the most common such process is reverse osmosis. The process converts saline water with up to 35,000 parts per million (ppm) of dissolved salts to fresh water with less than 1000 ppm dissolved salts.

More than 20,000 desalination facilities operate worldwide. However, desalination is expensive, requires large inputs of fossil fuel energy, kills aquatic life at water intakes, and generates concentrated salty waste. As a result, large-scale desalination is pursued mostly in wealthy oil-rich nations where water is extremely scarce and few options exist other than desalination to meet demand for water. In Saudi Arabia, for example, desalination produces half the nation's drinking water. The wastewater produced from fracking typically contains very high salt concentrations, so efforts to find ways to reuse fracking wastewaters for agricultural irrigation or other uses incorporate desalination as part of the treatment process.

We can decrease our demand for water by using it more efficiently

Because supply-based strategies do not hold great promise for increasing water supplies, people are embracing demand-based solutions. Strategies for reducing the demand for fresh water include conservation and efficiency measures. Such strategies require changes in individual behaviors and can therefore be politically difficult, but they offer better economic returns and cause less ecological and social damage. Our existing shift from supply-based to demand-based solutions is already paying dividends. The United States, for example, decreased its water consumption by 16% from 1980 to 2010 thanks to conservation measures, even while its population grew 36%. Let's examine approaches that can conserve water in agriculture, households, industry, and municipalities.

Agriculture Because most water is used for agriculture, it makes sense to look first to agriculture for ways to decrease demand. Farmers can improve efficiency by adopting more efficient irrigation methods. "Flood and furrow" irrigation, in which furrows running through crops are flooded with water, accounts for 90% of irrigation worldwide. However, crop plants end up using only 40% of the water applied, as much of the water evaporates or seeps into the ground, away from crops. Other methods of irrigation are far more efficient. Low-pressure spray irrigation squirts water downward toward plants, and drip irrigation systems target individual plants, introducing water directly onto the soil (p. 221). Experts estimate that drip irrigation—in which as little as 10% of water is wasted—could cut water use in half while raising crop yields, and could produce as much as $3 billion in extra annual income for farmers of the developing world. Researchers are currently experimenting with various materials and approaches to develop reliable, inexpensive drip irrigation systems that could convey these benefits to poorer farmers.

Another way to reduce agricultural water use would be to eliminate government subsidies for the irrigation of crops that require a great deal of water—such as cotton and rice—in arid areas. Biotechnology may play a role by producing crop varieties that require less water through selective breeding (pp. 52–53) and genetic modification (pp. 255–258).

Households In our households, we can reduce water use by installing low-flow faucets, showerheads, washing machines, and toilets. Automatic dishwashers, studies show, use less water than does washing dishes by hand. Catching

FAQ

When fresh water is scarce, why don't we irrigate crops with seawater?

Freshwater resources are often strained by agricultural irrigation. Some agriculture occurs in coastal areas, where huge volumes of seawater are available a short distance away, yet farmers do not use it to water their crops, even in times of extreme drought. Why?

The crops we grow for food, such as wheat and broccoli and strawberries, are terrestrial plants and are evolutionarily adapted to use fresh water to grow. If the cells of a terrestrial plant, such as corn, are placed in a solution of seawater, the water inside the cells will be "pulled out" into the surrounding saltwater. This occurs because there are fewer dissolved substances (solutes such as sugars, proteins, and salts) within the cell than there are dissolved substances (solutes such as salts) in the seawater. Water will move to the region of higher solute—by osmosis—in order to dilute it. With continued exposure to saltwater, the plant will dehydrate and die. Marine organisms have evolved physiological mechanisms to cope with water loss from their cells, so they are able to survive in seawater, unlike the terrestrial plants that supply us with so much of our food.

FIGURE 15.17 Xeriscaping lets homeowners and businesses reduce water consumption by landscaping with attractive, drought-tolerant plants.

rain runoff from your roof in a barrel, or *rainwater harvesting,* will reduce the amount of water you need to use from a hose. Replacing exotic vegetation with native plants adapted to the region's natural precipitation patterns can also reduce water demand. For example, more and more people in the United States are practicing **xeriscaping,** a type of landscaping that uses plants adapted to arid conditions (**FIGURE 15.17**). Xeriscaping is particularly popular in arid regions, such as the Southwest, and in other parts of the country where water supplies are frequently stressed, often by frequent drought.

Industry and municipalities Industry and municipalities can take water-saving steps as well. Manufacturers are shifting to processes that use less water and, in doing so, are reducing their costs. Fracking companies, faced with the reality that many shale gas and oil reserves are in areas that have very little water, are attempting to develop technologies that lessen the water use in fracking, including approaches that "recycle" fracking wastewater for use in active fracking operations.

Las Vegas is one of many cities that are recycling treated municipal wastewater for irrigation and industrial uses. Finding and patching leaks in pipes have saved some cities and companies large amounts of water—and money. Boston and its suburbs reduced water demand by 43% over 30 years by patching leaks, retrofitting homes with efficient plumbing, auditing industry, and promoting conservation. This program enabled Massachusetts to avoid an unpopular $500 million river diversion scheme.

Market-based approaches to water conservation are being debated

Economists who want to use market-based strategies to achieve sustainable water use have suggested ending government subsidies of inefficient practices, thereby letting water become a commodity whose price reflects the true costs of its extraction. Others worry that turning water into a fully priced commodity would make it less available to the world's poor and increase income disparity. Because the industrial use of water can be 70 times more profitable than agricultural use, market forces alone might favor uses that would benefit wealthy and industrialized people, companies, and nations at the expense of the rural poor.

Similar concerns surround another potential solution, the privatization of water supplies. During the 1990s, many public water systems were partially or wholly privatized, as governments transferred construction, maintenance, management, or ownership to private companies. This was done to enhance efficiency, but firms have little incentive to allow equitable access to water for rich and poor alike. Already in some developing countries, rural residents without access to public water supplies find themselves forced to buy water from private vendors and pay up to 12 times more than those connected to public supplies. Privatization issues are present in west Texas fracking country, where individual landowners have the right to sell as much groundwater as they can pump on their land, even if potential overextraction of local aquifers threatens the water supply of their neighbors.

Other experiences indicate that decentralization of control over water, from the national level to the local level, can help conserve water. In Mexico, the effectiveness of irrigation systems improved dramatically once they were transferred from national ownership to the control of 386 local water user associations.

Regardless of how demand is addressed, the shift from supply-side to demand-side solutions is paying dividends. In Europe, a new focus on demand (through government mandates and public education) has decreased public water consumption, and industries are becoming more water-efficient.

Nations often cooperate to resolve water disputes

We've seen that nations have unequal access to freshwater supplies, and there are fears that scarcity of this vital resource could lead to conflict. For example, a 2012 report by the intelligence agencies of the U.S. government concluded that in the subsequent decade, many nations vital to American interests will experience political and economic instability due to water shortages, making freshwater supplies one of the greatest threats to U.S. national security interests. In the face of such instability, one recourse would be to commandeer the water resources of other nations by force, sparking international military conflict.

A total of 261 major rivers—whose watersheds cover 45% of the world's land area—cross national borders, and transboundary disagreements are common. Water is already a key element in the hostilities among Israel, the Palestinian people, and neighboring nations in the arid Middle East. A potential flashpoint for hostilities is northern Africa. Ethiopia is constructing a large dam on an upstream portion of the Nile River in its nation, and the downstream nation of Egypt—the country most associated with the river—is promising military action if its water supplies are lowered once the dam begins creating a massive reservoir in 2020. Egypt is entirely dependent on the Nile for agricultural irrigation and to supply

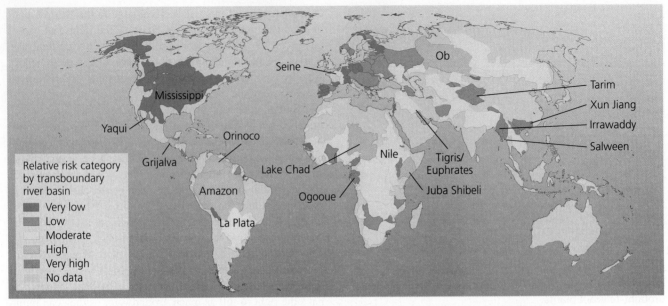

FIGURE 15.18 Some transboundary water basins are governed by water use treaties among the nations through which the river flows; other basins have weak agreements or none at all. Basins without comprehensive water-sharing agreements are those most at risk to experience conflict over shared water supplies. *Data from UNEP-DHI and UNEP, 2016. Transboundary River Basins: Status and Trends. United Nations Environment Programme (UNEP), Nairobi.*

water to its growing population, and the United Nations predicts that Egypt could begin suffering water shortages as early as 2025, making military conflict a real possibility if negotiations over shared water use fail.

The United States has its share of conflicts over water. Water allocations from the Colorado River have long been a source of conflict between farms and growing cities in the West. In the East, the states of Georgia, Alabama, and Florida have been embroiled in disputes over water withdrawals from shared rivers.

On the positive side, many nations have cooperated to resolve water disputes. India has struck agreements to co-manage transboundary rivers with Pakistan, Bangladesh, Bhutan, and Nepal. In Europe, nations along the Rhine and Danube rivers have signed water-sharing treaties. **FIGURE 15.18** categorizes transboundary water basins according to the strength of the agreements that govern water sharing among the nations in its watershed. Nearly 40% of the transboundary basins lack water use agreements entirely, or are governed by treaties to which one or more nations in its watershed have not ratified. Basins with risk scores of moderate and above are those for which future conflict over water supplies may arise, particularly in instances—such as in the Nile Basin—where large dams or water diversions projects by one nation will affect the water supplies of downstream countries.

Freshwater Pollution and Its Control

We have seen that people affect aquatic systems by withdrawing too much water and by erecting dams, diversions, and levees that alter natural processes in aquatic systems. Introducing toxic substances and disease-causing organisms into surface waters and groundwater is yet another way that people adversely affect aquatic ecosystems—and threaten human health. As noted in the opening case study, the overuse of water supplies is not the only impact of fracking for oil and natural gas, as the process can introduce wastewater that contains compounds that are harmful to organisms into aquatic ecosystems.

The World Commission on Water, an organization that focuses on water issues, concluded in 1999 that more than half the world's major rivers were overdrawn and seriously polluted, threatening the people and ecosystems that rely on those waterways. Levels of impairment are similar in U.S. waterways. In 2013, the Environmental Protection Agency (EPA) reported that 55% of the 2000 U.S. streams and rivers sampled in 2008–2009 were in poor condition to support aquatic life. The largely invisible pollution of groundwater, meanwhile, has been termed a "covert crisis." Preventing pollution is easier and more effective than treating it later. Many of our current solutions to pollution therefore embrace preventive strategies rather than "end-of-pipe" strategies such as treatment and cleanup.

Water pollution comes from point sources and non-point sources

Water pollution—changes in the chemical, physical, or biological properties of waters caused by human activities—comes in many forms and can have diverse impacts on aquatic ecosystems and human health. Most forms of water pollution are not conspicuous, so scientists and technicians measure water's chemical properties, such as pH, plant nutrient concentrations, and dissolved oxygen levels; physical characteristics, such as temperature and turbidity—the density of suspended particles in a water sample; and biological properties, such as the presence of harmful microorganisms or the species diversity in aquatic ecosystems.

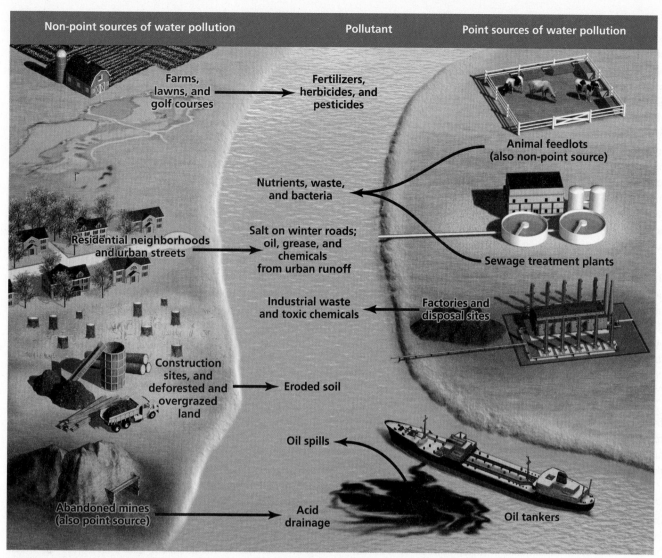

Farms, lawns, and golf courses → Fertilizers, herbicides, and pesticides

Animal feedlots (also non-point source)

Nutrients, waste, and bacteria ← Sewage treatment plants

Residential neighborhoods and urban streets → Salt on winter roads; oil, grease, and chemicals from urban runoff

Industrial waste and toxic chemicals ← Factories and disposal sites

Construction sites, and deforested and overgrazed land → Eroded soil

Oil spills ←

Abandoned mines (also point source) → Acid drainage   Oil tankers

FIGURE 15.19 Point-source pollution (on the right) comes from discrete facilities or locations, usually from single outflow pipes. Non-point-source pollution **(on the left)**, such as runoff from streets, residential neighborhoods, lawns, and farms, originates from numerous sources spread over large areas.

Some water pollution is emitted from **point sources**—discrete locations, such as a factory or sewer pipe, or a fracking well. Fracking, for example, generates substantial quantities of briny (salty) wastewater that contains proprietary chemicals used as part of the fracking fluids, and naturally occurring elements that may be radioactive (such as radium) or toxic (such as arsenic). This wastewater is difficult to treat and safely dispose of in surface waters, so it is often injected into porous rocks deep underground. This disposal method has, however, been linked to the increased risk of induced earthquakes (pp. 42–43), so it is not a sustainable solution. As of 2019, fracking for oil and gas has produced around 1 trillion L (265 billion gal) of wastewater requiring disposal in the United States.

In contrast to point sources, pollution from **non-point sources** is cumulative, arising from multiple inputs over larger areas, such as farms, city streets, and residential neighborhoods (**FIGURE 15.19**). The U.S. Clean Water Act (pp. 173–174) has addressed point-source pollution since 1972 with some success

by targeting industrial discharges and other point sources of pollution. As a result, water quality in the United States today suffers most from non-point-source pollution resulting from countless common activities such as applying fertilizers and pesticides to farms and lawns, adding salt to roads in winter, and leaking oil from vehicles. To minimize the non-point-source pollution of drinking water, governments are not powerless, as they can limit development on watershed land surrounding reservoirs that supply water to people's homes.

Water pollution takes many forms

Water pollution comes in many forms that impair waterways, and threaten people and organisms that drink the water or live in or near affected waters. Let's survey the major classes of water pollutants affecting waters in the world today.

Toxic and endocrine-disrupting chemicals Our waterways have become polluted with many toxic organic substances of our own making, including pesticides, industrial chemicals, petroleum products, fracking fluids, and other synthetic chemicals (pp. 368–370). Many of these can poison animals and plants, alter aquatic ecosystems, and cause an array of human health problems. In addition, toxic and radioactive metals such as arsenic, lead, mercury, and radium that enter waterways can also damage human health and the environment. Wastewater from conventional natural gas wells, for example, has been found to increase radioactivity levels in stream sediments at release sites to 650 times that of normal sediment radioactivity levels. This occurs when radioactive radium in the wastewater binds strongly to sediments at the release site and accumulates over time. Acidic compounds from acid rain (p. 474) and from acid drainage from mining sites (p. 650) can lower the pH of waters and prove toxic to various forms of aquatic life.

Endocrine disruptors—chemicals that mimic the hormones that regulate physiology and stimulate growth, development, and sexual maturity through the endocrine system—are of particular concern in aquatic systems. Some pesticides, herbicides, industrial chemicals, pharmaceuticals, and constituents of consumer products can enter waterways from agricultural and urban runoff and wastewater treatment plants and act as endocrine disruptors, exerting negative effects on aquatic organisms (see **THE SCIENCE BEHIND THE STORY**, pp. 412–413).

Strategies for reducing concentrations of toxic chemicals in freshwater systems include issuing and enforcing more stringent regulations on industry and modifying our industrial processes and our purchasing decisions to rely less on these substances so their opportunities to enter waterways are limited.

Pathogens and waterborne diseases

Disease-causing organisms (pathogenic viruses, protists, and bacteria) can enter drinking water supplies that become contaminated with human waste from inadequately treated sewage or with animal waste from feedlots, chicken farms, or hog farms (p. 248). Specialists monitoring water quality can identify fecal contamination when their tests detect fecal coliform bacteria, which live in the intestinal tracts of people and other vertebrates. These bacteria are usually not pathogenic themselves, but they serve as indicators of fecal contamination, alerting us that the water may hold other pathogens that can cause ailments such as giardiasis, typhoid, or hepatitis A.

Biological pollution by pathogens causes more human health problems than any other type of water pollution. In the United States, an estimated 20 million people fall ill each year from drinking water contaminated with pathogens. Worldwide, the United Nations estimates that 3800 children die every day of diseases associated with unsafe drinking water, such as cholera, dysentery, and typhoid fever. Irrigation water contaminated with pathogens can also make some produce unsafe, opening another route by which waterborne pathogens can impair human health. Pathogenic fecal coliform bacteria in 2018 contaminated romaine lettuce grown in Arizona and California, for example, leading to a nationwide recall of lettuce from these regions.

We reduce the risks posed by waterborne pathogens by using chemical or other means to disinfect drinking water (p. 415) and by treating wastewater (pp. 415–417) to prevent fecal contamination of waterways. Other measures to lessen health risks include public education to encourage good personal hygiene to reduce people's contact with pathogens and government enforcement of regulations to ensure the cleanliness of food production, processing, and distribution.

Nutrient pollution

The dead zones in the Chesapeake Bay (p. 109) and coastal Louisiana (pp. 116–117) show how nutrient pollution from fertilizers and other sources can lead to eutrophication and hypoxia (waters dangerously low in dissolved oxygen aquatic organisms need to survive) in surface waters (see Figure 5.5, p. 110, for a detailed description of the process). When excess nitrogen and/or phosphorus enters a water body, it fertilizes algae and aquatic plants, boosting their growth. As the crowded algae die off and settle on the sediments, bacteria consume them. Because this decomposition requires oxygen, dissolved oxygen levels in the water decline. These levels can drop too low to support fish and shellfish, leading to dramatic changes in aquatic ecosystems.

Eutrophication (**FIGURE 15.20**) is a natural process, but nutrient input from runoff from farms, golf courses, lawns, and sewage can dramatically increase the rate at which it occurs. We can reduce nutrient pollution by specially treating wastewater to remove nutrients, reducing fertilizer applications, using phosphate-free detergents, and planting vegetation and protecting natural areas around streams to capture nutrients on land before they enter waterways (p. 128).

Biodegradable wastes

Just as adding excess nutrients to waterways can lead to hypoxia, introducing large quantities of biodegradable materials into waters also decreases dissolved oxygen levels. When human wastes, animal manure, paper pulp from paper mills, or yard wastes (grass clippings and leaves) enter waterways, bacterial decomposition escalates as organic material is metabolized, and this lowers dissolved oxygen levels in the water. **Wastewater** is water affected by human activities and is one source of biodegradable wastes. It includes water from toilets, showers, sinks, dishwashers, and washing machines; water used in manufacturing or industrial cleaning processes; and stormwater runoff. The widespread practice of treating wastewater to remove organic matter before it is released into waterways (pp. 415–416) has greatly reduced impacts from biodegradable wastes in rivers in industrialized nations. Oxygen depletion remains a major problem in some industrializing nations, however, where wastewater treatment is less common, and up to 90% of wastewater is released into waterways without treatment.

Sediment

Eroded soils, called **sediments,** can be carried to rivers by runoff and transported long distances by river currents (**FIGURE 15.21**). Clear-cutting, mining, clearing land for development, and cultivating farm fields all expose soil to wind and water erosion (pp. 229–230). Some water bodies, such as the Colorado River and China's Yellow River, are naturally sediment-rich, but many others are not. When a clear-water river receives a heavy influx of eroded sediment, aquatic habitat changes dramatically, and fish adapted to clear water may be killed. We can reduce sediment pollution by better managing farms and forests and avoiding large-scale disturbance of vegetation.

(a) Oligotrophic water body

(b) Eutrophic water body

FIGURE 15.20 Pollution of freshwater bodies by excess nutrients accelerates the process of eutrophication. An oligotrophic water body **(a)** with clear water and low nutrient content may eventually become a eutrophic water body **(b)** with high nutrient content and abundant algae and plant life.

Plastic debris While pollution from discarded plastics in the oceans (pp. 433–435) receives a great deal of attention, waterways and groundwater can also be contaminated with small pieces of plastic, called *microplastics*. Microplastics form when discarded plastics in the environment are broken down by factors such as heat or ultraviolet light, and are washed into waterways. Another major source of microplastics in freshwater systems are microbeads, small pieces of plastic that are added to some cosmetics, soaps, and toothpastes. Microbeads enter rivers when they pass through wastewater treatment plants. They can also enter groundwater and surface waters when sludge from wastewater treatment plants is applied to agricultural fields. Microplastics have been detected in tap water produced from surface waters, water from groundwater wells, and even in bottled water. Plastics harm aquatic life when they are ingested by animals and can release toxic chemicals—such as bisphenol A and phthalates (p. 359)—into aquatic habitats and waterways used by people as sources of drinking water.

The best approach for reducing plastic pollution is to safely dispose of plastic wastes and to ban the use of microbeads in consumer products around the world, as was done in the United States with the Microbead-Free Water Act of 2015. With over 330 million metric tons (360 million tons) of plastic produced annually today—and production expected to triple by 2050—preventing plastic pollution in the world's waters will be a continuing challenge for the foreseeable future.

Thermal pollution Water's ability to hold dissolved oxygen decreases as temperature rises, so some aquatic organisms may not survive when human activities raise water temperatures. When we withdraw water from a river and use it to cool a power plant or other industrial facility, we transfer heat from the facility back into the river where the water is returned. People also raise water temperatures by removing streamside vegetation that shades water.

Too little heat can also cause problems. On the Mississippi and many other dammed rivers, water at the bottoms of reservoirs is

FIGURE 15.21 Sediments wash into the Pacific Ocean from a river in Costa Rica. Farming, construction, and other human activities can cause elevated levels of soil to enter waterways, affecting water quality and aquatic wildlife.

THE SCIENCE behind the story

What's Killing Smallmouth Bass in the Potomac River Watershed?

Vicki Blazer, USGS

There's a mystery to be solved in the waterways that feed the Potomac River—what's killing smallmouth bass (*Micropterus dolomieu*) (**FIGURE 1**) in this historic bass fishery? And like any good investigation, researchers are following the clues and piecing together information to find out "whodunit" or, in this case, "whatdunit."

The story began in 2002 with a large fish kill of smallmouth bass in the Shenandoah and South Branch Potomac rivers. These rivers, in rural areas of western Virginia and West Virginia, respectively, join to form the Potomac River, which then flows eastward past Washington, D.C., and into the Chesapeake Bay (**FIGURE 2**). Surrounded by rolling hills, forests, and agricultural land, the waterways seemed to be one of the last places you'd expect to see massive die-offs of fish.

The usual suspects that cause fish kills—a spill of toxic chemicals or a malfunctioning wastewater treatment plant—were not present. Many of the smallmouth bass captured in the river were covered in blisters, sores, and patches of fungus; their gills were inflamed; and they were listless and underweight. The fishes' immune systems were seemingly being weakened by something in the water, enabling bacteria, parasites, and other normally manageable pathogens to kill huge numbers of fish. Scientists were unable to identify a "smoking gun" responsible for the ailments and mortality they were observing and, at one point, there were as many as 20 plausible explanations in play.

One of the first clues to solving the mystery came from the laboratory of Vicki Blazer, a researcher with the U.S. Geological Survey (USGS) National Fish Health Research Laboratory in West Virginia. While examining under a microscope tissues from specimens of smallmouth bass from the South Branch Potomac River in 2004, she noted that a large percentage of male fish contained oocytes—immature eggs—in their testes. This is an example of an intersex condition, a situation where individuals of one sex exhibit characteristics of both sexes at some point in their lifetime. The phenomenon is not unheard of in smallmouth bass, as anywhere from 10% to 20% of males can display this particular intersex condition in any given year, even in pristine habitats. But rates in the South Branch Potomac River were drastically higher than these "normal" levels, with 50–75% of

FIGURE 1 Smallmouth Bass (*Micropterus dolomieu*).

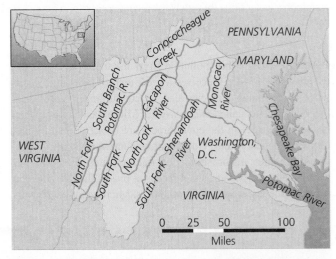

FIGURE 2 Fish kills in the rural waterways of the western Potomac River basin in the early 2000s spurred interest in examining the role of endocrine-disrupting chemicals on freshwater fish. *Data from U.S. Fish & Wildlife Service, Chesapeake Bay Field Office, 2006. Intersex fish: Endocrine disruption in smallmouth bass.*

male fish displaying intersex characteristics (**FIGURE 3**). In light of this, Blazer hypothesized that smallmouth bass in the river were being exposed to chemicals that were disrupting the endocrine systems of the fish, causing males to develop female characteristics. That is, the bodies of male fish would be prompted to develop eggs when they were exposed to a chemical in the water that mimicked the female hormone estrogen.

Endocrine-disrupting chemicals (pp. 360, 371) are compounds that fool the endocrine systems of animals by mimicking the hormones that regulate growth, development, physiology, and sexual maturation. When organisms are exposed to these

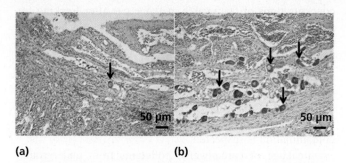

(a)　　　　　　　　　　　**(b)**

FIGURE 3 **Microscope images of testicular tissue from male smallmouth bass showing a low rate of intersex condition (a) and an elevated rate of intersex condition (b).** Immature eggs (oocytes) are the circular structures, some of which are indicated with arrows. *Source: Blazer, V. S. et al., 2013. Reproductive health indicators of fishes from Pennsylvania watersheds: Association with chemicals of emerging concern.* Environmental Monitoring and Assessment *186: 6471–6491.*

chemicals in the environment or in their food or water, it can lead to changes in sexual development, physiological processes, and a host of other biochemical processes that enable organisms to successfully function and reproduce.

Such endocrine-disrupting chemicals enter waterways from a variety of sources. Many pharmaceutical drugs contain endocrine disruptors, and these compounds enter sewer systems when they are excreted in people's urine or unused medication is flushed down a toilet. Agriculture introduces endocrine disruptors through pesticide runoff from farm fields; waterway contamination with wastes from poultry, hog, and livestock facilities where the animals are administered hormones to facilitate rapid growth; and runoff when hormone-laden manure is applied to cropland as a fertilizer. Plastic trash or plastic microbeads from soaps and detergents can also introduce these compounds into waterways as they degrade. Given that there are many chemicals that act as endocrine disruptors and these compounds exert effects at extremely low doses, they are very difficult and costly to detect in water samples from waterways, so scientists cannot easily screen waters to see if these chemicals are present.

Furthermore, while elevated rates of intersex condition may affect reproduction within a population of smallmouth bass, intersex condition in and of itself doesn't lead to the sickness and death observed in these rural waterways. Clearly, further evidence was needed to connect the presence of endocrine-disrupting chemicals with immune suppression in smallmouth bass if that was indeed the cause.

While research into the effects of endocrine disruptors on the immune system of fish continued, so did the fish kills. A 2006 fish die-off killed an estimated 80% of the smallmouth bass in the Shenandoah River, several million fish by some estimates. In 2007, fish kills occurred in the nearby but unconnected upper James and Cowpasture rivers, indicating that the factors killing fish in the Shenandoah and South Branch Potomac rivers were

not unique to those waterways. The USGS continued its work and attempted to identify possible sources of endocrine disruptors by correlating the rate of intersex condition in fish collected from the Potomac Basin with the characteristics of the section of the river in which they were found. This analysis determined that fish showed the highest rates of intersex condition at collection sites where human population density was highest and agriculture was the most intensive, providing leads on the possible origins of the endocrine disruptors affecting smallmouth bass.

As leads from USGS began filtering in, the case connecting immune suppression in freshwater fish with exposure to endocrine-disrupting chemicals strengthened. By 2011, it was clear that exposure to endocrine-disrupting chemicals that mimic estrogen adversely affected the functioning of the immune system in fish, and led to increased susceptibility to pathogenic bacteria, funguses, and viruses. Some of that work came out of the West Virginia USGS laboratory, when a team of USGS scientists found, in 2009, that another bass species in the Potomac watershed (largemouth bass, *Micropterus salmoides*) were showing elevated rates of intersex condition, and that this bass produced fewer immune-related proteins when they were exposed to estrogen than when they were not. The study found that injecting fish with estrogen resulted in reduced production of hepcidin, an iron-regulating hormone that functions in the immune system in fish. Said lead researcher Laura Robertson, "Our research suggests that estrogen-mimicking compounds may make fish more susceptible to disease by blocking production of hepcidin and other immune-related proteins that help protect fish against disease-causing bacteria."

Although the mystery of the cause of the fish kills in the Potomac Basin now had a likely suspect—immune suppression by endocrine-disrupting chemicals—the source of those compounds has been tougher to nail down. While Blazer's research pointed to agriculture chemicals and runoff and wastewater releases from cities and town, scientists have been unable to attribute all these instances of high rates of intersex condition to any one specific chemical and suggest a complex mixture of synthetic chemicals may be involved.

This issue isn't unique to the Potomac River. Population declines in smallmouth bass, coupled with high rates of intersex condition, have also been found in the Susquehanna River in Pennsylvania. And elevated rates of intersex condition—an indication of exposure to chemicals that may weaken the immune system in some fish—have been observed in the Colorado River basin in the southwestern United States, the Columbia River basin in the Northwest, the Mississippi River Basin in the Midwest, and the Mobile River in the Southeast. And since exposure to endocrine-disrupting chemicals has been shown to affect the sexual development and physiological functioning of humans (see Figure 14.2, p. 360), solving the mystery of how these chemicals affect wildlife in our rivers—the same rivers from which many of us draw our drinking water—grows in importance each day.

colder than water at the surface. When dam operators release water from the depths of a reservoir, downstream water temperatures drop suddenly due to the influx of cold water. In some river systems, these pulses of cold water have favored cold-loving invasive fish species over native species adapted to normal river temperatures.

Groundwater pollution is a significant problem

Many pollutants that affect surface waters also affect groundwater. Groundwater pollution is difficult to detect. Moreover, pollutants may reside in groundwater longer than in surface waters. Because groundwater is not exposed to sunlight, contains fewer microbes and minerals, and holds less dissolved oxygen and organic matter than surface waters, pollutants decompose more slowly. For example, concentrations of the herbicide alachlor decline by half after 20 days in soil, but in groundwater this decomposition takes almost four years.

Some chemicals that are toxic at high concentrations, including aluminum, fluoride, nitrates, and sulfates, occur naturally in groundwater. However, groundwater pollution resulting from industrial, agricultural, and urban wastes—such as heavy metals, petroleum products, solvents, and pesticides—can leach through soil and seep into aquifers. For example, nitrate from fertilizers has leached into aquifers across the United States and Canada. Nitrate in drinking water has been linked to cancers, miscarriages, and "blue-baby" syndrome, a condition in which the oxygen-carrying capacity of infants' blood is reduced.

Fracking for natural gas has been implicated in contaminating groundwater aquifers and rendering water unusable for people who live near fracking sites. This issue gained national attention in part from the 2010 documentary film *Gasland,* which profiled how people's well water in Pennsylvania become discolored, clouded with sediment, took on unpleasant odors and flavors, and became infused with high concentrations of flammable methane gas after fracking for natural gas commenced in the area (**FIGURE 15.22**). The issue continues to generate controversy, as the fracking industry contends that the observed changes are natural in origin, while residents point their fingers squarely at fracking and the chemicals used in fracking operations. As the chemical mixes used in fracking are considered proprietary, companies are not required to disclose them, even if groundwater contamination is suspected.

Leakage of carcinogenic pollutants from underground tanks—such as chlorinated solvents, gasoline, and industrial chemicals—also poses a threat to groundwater. The EPA manages a nationwide cleanup program to unearth and repair leaky tanks, and by 2017 the agency had confirmed leakage from over 535,000 tanks from across the United States, and had completed cleanup of more than 465,000 of them.

The leakage of radioactive compounds from underground tanks is also a source of groundwater pollution. Currently, 67 of the 177 underground storage tanks at the Hanford Nuclear Reservation in Washington State have been confirmed to be leaking radioactive waste into the soil. This site is the most radioactively contaminated area in the United States, storing 60% of the United States' high-level radioactive waste. Cleanup of the radioactive material stored at present in the site's aging underground tanks is not scheduled for completion until 2047.

Legislative and regulatory efforts have helped to reduce pollution

As numerous as our freshwater pollution problems may seem, it is important to remember that many were worse a few decades ago, when, for example, Ohio's Cuyahoga River repeatedly caught fire (p. 171). Citizen activism and government response during the 1960s and 1970s in the United States resulted in legislation such as the Federal Water Pollution

FIGURE 15.22 Residents of Dimock, Pennsylvania, contend that fracking operations for natural gas have fouled their drinking water wells. People noticed that the appearance, smell, and taste of their well water changed after fracking commenced nearby.

Control Act of 1972, amended and later renamed the Clean Water Act in 1977. These acts made it illegal to discharge pollution from a point source without a permit, set standards for industrial wastewater and contaminant levels in surface waters, and funded construction of sewage treatment plants. Thanks to such legislation, point-source pollution in the United States was reduced, and rivers and lakes became notably cleaner.

In the past several decades, however, enforcement of water quality laws has grown weaker, as underfunded and understaffed state and federal regulatory agencies have faced challenges when enforcing existing laws. A comprehensive investigation by *The New York Times* in 2009 revealed that violations of the Clean Water Act had been rising and numbered more than 100,000 per year (to say nothing of undocumented instances) in the years preceding the article's publication. The EPA and the states act on only a tiny percentage of these violations, the *Times* found. As a result, 1 in 10 Americans have been exposed to unsafe drinking water—for the most part unknowingly, because many pollutants cannot be detected by smell, taste, or color.

The Great Lakes of Canada and the United States represent an encouraging success story in fighting water pollution. In the 1970s, these lakes, which hold 18% of the world's surface fresh water, were badly polluted with wastewater, fertilizers, and toxic chemicals. Algal blooms fouled beaches, and lake Erie was so polluted it was pronounced "dead." Today, efforts of the Canadian and U.S. governments to regulate pollutant inputs into the lakes have paid off. According to Environment Canada, releases of seven toxic chemicals are presently down by 71% compared to releases several decades ago, municipal phosphorus has decreased by 80%, and chlorinated pollutants from paper mills are down by 82%. Levels of PCBs and DDE are down by 78% and 91%, respectively. Bird populations are rebounding, and Lake Erie is now home to the world's largest walleye fishery. The Great Lakes' troubles are by no means over—sediment pollution is still heavy, algal blooms still plague Lake Erie, and fish are not always safe to eat. However, the progress so far shows how conditions can improve when citizens push their governments to take action.

We treat our drinking water

Technological advances as well as government regulation have greatly improved drinking water quality. The treatment of drinking water is a widespread and successful practice in industrialized nations today. Before being sent to your tap, water from a reservoir or aquifer is treated with chemicals to remove particulate matter; passed through filters of sand, gravel, and charcoal; and/or disinfected with small amounts of an agent such as chlorine to combat pathogenic bacteria. The U.S. EPA sets standards for more than 90 drinking water contaminants, which local governments and private water suppliers are obligated to meet.

We treat our wastewater

Wastewater treatment is also now a mainstream practice in industrialized nations. Recall that wastewater includes water that carries sewage, water from residences, water used in manufacturing or industrial cleaning processes, and stormwater runoff. Natural systems can process the pollutants in moderate amounts of wastewater, but the large and concentrated amounts that our densely populated areas generate can harm ecosystems and pose health threats if released into waterways. Thus, attempts are now widely made to treat wastewater before it is released into the environment.

In rural areas, **septic systems** are the most popular method of wastewater disposal. In a septic system, wastewater runs from the house to an underground septic tank, inside which solids and oils separate from water. The clarified water proceeds by gravity or pressure to a drain field of perforated pipes laid horizontally in gravel-filled trenches underground. Microbes decompose pollutants in the wastewater that these pipes emit. Periodically, solid waste from the septic tank is pumped out and taken to a landfill.

In more densely populated areas, municipal sewer systems carry wastewater from homes and businesses to centralized treatment locations. There, pollutants are removed by physical, chemical, and biological means (**FIGURE 15.23**, p. 416). At a treatment facility, large items are removed from the incoming water with screens and the wastewater is then subjected to **primary treatment**, the physical removal of contaminants in settling tanks or clarifiers. This process removes about 60% of solids suspended in the wastewater, which then proceeds to **secondary treatment**, in which water is stirred and aerated so that aerobic bacteria consume and degrade the small pieces of organic pollutants that were not captured in primary treatment. Roughly 90% of suspended solids may be removed after secondary treatment. Finally, the clarified water is treated with chlorine, and occasionally with ultraviolet light, to kill any bacteria that remain in the water. Most often, the treated water, called *effluent*, is piped into rivers or the ocean following treatment, but in some communities it is filtered through wetlands to enhance its purification (see **SUCCESS STORY**, p. 417). However, many municipalities are recycling "reclaimed" water for lawns and golf courses, for irrigation, or for industrial purposes such as cooling water in power plants.

As water is purified throughout the treatment process, the solid material removed is termed *sludge*. Sludge is sent to digesting vats, where microorganisms decompose much of the matter. The result, a wet solution of "biosolids," is then dried and disposed of in a landfill, incinerated, or used as fertilizer on cropland. Methane-rich gas created by the decomposition process can be captured and used to generate electricity. Each year, about 6 million dry tons of sludge are generated in the United States.

WEIGHING the issues

Sludge on the Farm

About half the biosolids produced from sewage sludge each year find new life as fertilizer on farmland. This practice makes productive use of the sludge, increases crop output, and conserves landfill space, but many people have voiced concern over the potential accumulation of toxic metals and proliferation of dangerous pathogens in agricultural soil. Do you think this practice represents an efficient use of resources or an unnecessary risk? What further information would you want to know to inform your decision?

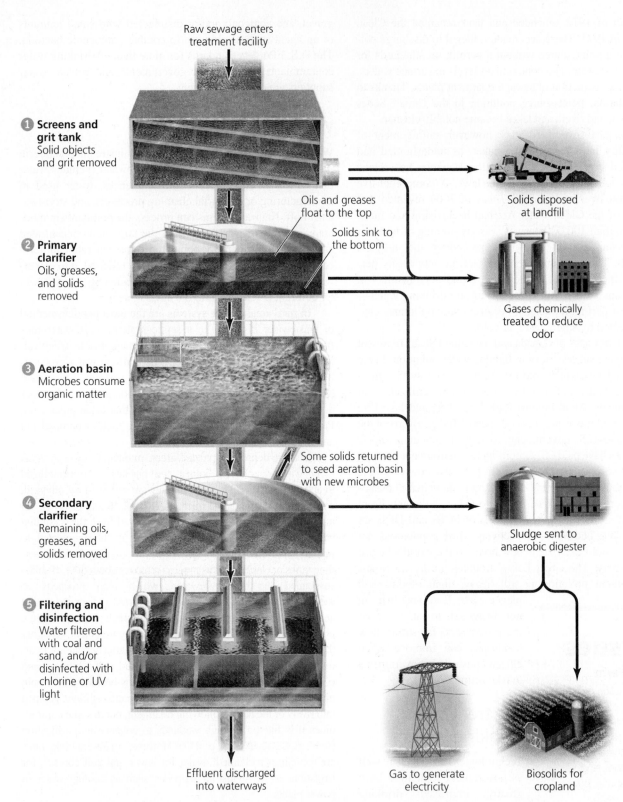

Raw sewage enters
treatment facility

1 Screens and grit tank
Solid objects and grit removed

Solids disposed at landfill

Oils and greases float to the top

Solids sink to the bottom

2 Primary clarifier
Oils, greases, and solids removed

Gases chemically treated to reduce odor

3 Aeration basin
Microbes consume organic matter

Some solids returned to seed aeration basin with new microbes

4 Secondary clarifier
Remaining oils, greases, and solids removed

Sludge sent to anaerobic digester

5 Filtering and disinfection
Water filtered with coal and sand, and/or disinfected with chlorine or UV light

Effluent discharged into waterways

Gas to generate electricity

Biosolids for cropland

FIGURE 15.23 Shown here is a generalized process from a modern, environmentally sensitive wastewater treatment facility. Wastewater initially passes through screens to remove large debris and into grit tanks to let grit settle ❶. It then enters tanks called primary clarifiers ❷, in which solids settle to the bottom and oils and greases float to the top for removal. Clarified water then proceeds to aeration basins ❸ that oxygenate the water to encourage decomposition by aerobic bacteria. Water then passes into secondary clarifier tanks ❹ for removal of further solids and oils. Next, the water may be purified ❺ by chemical treatment with chlorine, passage through carbon filters, and/or exposure to ultraviolet light. The treated water (called *effluent*) may then be piped into natural water bodies, used for urban irrigation, flowed through a constructed wetland, or used to recharge groundwater. In addition, most treatment facilities use anaerobic bacteria to digest sludge removed from the wastewater. Biosolids from digesters may be sent to farm fields as fertilizer, and gas from digestion may be used to generate electric power.

Using Wetlands to Aid Wastewater Treatment

Long before people built the first wastewater treatment plants, natural wetlands filtered and purified water. Recognizing this, engineers have begun manipulating wetlands—and have even constructed new wetlands—to employ as tools to cleanse wastewater. In this approach, wastewater that has gone through primary or secondary treatment at a conventional facility is pumped into the wetland, where microbes living amid the algae and aquatic plants decompose the remaining pollutants. Water cleansed in the wetland can then be released into waterways or allowed to percolate underground.

One of the first constructed wetlands was established in Arcata, a town on northern California's scenic Redwood Coast. This 35-hectare (86-acre) engineered wetland system was built in the 1980s after residents objected to a $50 million plan to build a large treatment plant that would have pumped treated wastewater into the ocean. The wetland, which cost just $7 million to build, not only treats wastewater but also

serves as a haven for wildlife and human recreation. In fact, the Arcata Marsh and Wildlife Sanctuary has brought the town's waterfront back to life, with more than 200,000 people visiting each year. Additionally, wildlife has flourished in the area, with 100 species of plants, 300 species of birds, 6 species of fish, and a diversity of mammals and invertebrates populating the wetlands. The practice of treating wastewater with artificial wetlands is growing fast; today, more than 500 artificially constructed or restored wetlands in the United States perform this valuable service.

The Arcata Marsh and Wildlife Sanctuary is the site of an artificially engineered, constructed wetland system.

→ **Explore the Data** at **Mastering Environmental Science**

CENTRAL CASE STUDY
connect & continue

TODAY, communities in west Texas, and many other parts of the United States, struggle to meet the competing demands for fresh water from agriculture, burgeoning cities, and industry. The emerging technology of hydraulic fracturing, while comprising a tiny amount of the total water used in the United States, can be locally significant in the many arid parts of the United States that overlie shale plays of oil and natural gas. This added water demand, particularly in places such as Texas that have minimal regulations on water withdrawals from aquifers, adds to existing stresses on water supplies and accelerates these areas toward dangerous levels of water overextraction.

In some parts of Texas, residents have aimed to avoid this outcome and voted to create groundwater conservation districts, groups empowered by the state to enact restrictions on groundwater extraction to make water use sustainable and conserve aquifers for the future. Nearly 100 such groups exist in Texas, but many are underfunded and often lack the ability to adequately monitor the amount of groundwater being extracted in their jurisdiction. The lack of resources also makes them unable to withstand lengthy legal battles, should their restrictions on water extractions be challenged in court by water or fracking companies—companies with ample resources to support long-term litigation. Given these limitations and the drastic increase in water use for fracking that is predicted in the coming years, it is feared that groundwater conservation districts may not be able to adequately manage local water supplies, and

some residents are calling on state officials to intervene.

Water shortages aren't the only downside to fracking. It remains unclear what effect hydraulic fracturing operations may have on nearby groundwater aquifers that are used by people for drinking water and agricultural irrigation. Residents near fracking wells in Pennsylvania contend that fracking has introduced contaminants into their drinking water that may be endangering their health, but investigations by state and federal agencies remain inconclusive as fracking operations continue to expand across the state.

Although fracking is not the largest use of water or the biggest source of water pollution in the United States, it is part of a larger challenge faced by all nations—how best to manage increasingly scarce water supplies among the myriad of human demands for water, while conserving some for the vital aquatic systems that supply invaluable ecosystem services. As human populations continue to grow and water supplies become less predictable due to climate change's effects on precipitation levels and temperatures, humanity no longer has the luxury of time to put these affairs in order if we are to leave a sustainable world to our children and grandchildren.

- **CASE STUDY SOLUTIONS** Texas' governor has put you in charge of water policy for the western part of the state.

The state has recently experienced severe drought, its aquifers have been overpumped, and many wells have run dry. Agricultural production in the state is down, and farmers are clamoring for more water for their crops. Meanwhile, more water is needed for burgeoning urban populations in cities such as Midland and Odessa. Under these conditions, hydraulic fracking operations in west Texas are looking to expand production—and with it their water use. Would you recommend approving this expansion? The expansion would increase economic activity in the state but put further strain on already scarce water supplies. Would you recommend approving the expansion with stipulations, and if so, what would those be?

- **LOCAL CONNECTIONS** Fracking is undertaken only in areas where the local geology contains a great deal of shale rock that holds significant reserves of natural gas or petroleum. Where are the closest fossil-fuel-bearing shale deposits in your region? Are those areas water-stressed? Do special economic considerations or ecological considerations exist that would affect any decisions to allow hydraulic fracturing there?

- **EXPLORE THE DATA** What proportion of fracking wells in the United States are located in areas already experiencing water stress? → **Explore Data** relating to the case study on **Mastering Environmental Science.**

REVIEWING **Objectives**

You should now be able to:

+ **Describe the distribution of fresh water on Earth and the major types of freshwater systems**

Of all the water on Earth, only about 1% is readily available for our use. Fresh water occurs as groundwater and surface waters, and common freshwater ecosystems include rivers and streams, lakes and ponds, and wetlands. (pp. 392–399)

+ **Discuss how humans use water and alter freshwater systems**

Water from groundwater and surface waters is used in agriculture, industry, and homes. Overextraction of water from these sources can imperil their long-term use and endanger ecosystems. Humans also modify freshwater systems by controlling floods with levees, diverting waters with canals, and creating reservoirs with dams. (pp. 399–405)

+ **Assess problems of water supply and propose solutions to address depletion of fresh water**

Solutions to shortages of fresh water include increasing supplies of fresh water, through processes such as desalination, or reducing demand for water by farms, businesses, and homes. Making water reflect its true costs, by eliminating government subsidies and/or privatizing water utilities, has also been offered as a water-conserving approach. (pp. 406–408)

+ **Describe the major classes of water pollution and propose solutions to address water pollution**

Water pollution stems from point sources and non-point sources, and is caused by toxic chemicals, microbial pathogens, excessive nutrients, biodegradable wastes, microplastics, sediment, and thermal pollution. Legislation and regulation have succeeded in improving water quality in industrialized nations in recent decades. (pp. 408–415)

+ **Explain how we treat drinking water and wastewater**

Municipalities treat drinking water by filtering and disinfection in a multistep process. Septic systems help treat wastewater in rural areas, while cities and towns use municipal wastewater treatment facilities, which treat wastewater physically, biologically, and chemically in a series of steps. (pp. 415–417)

SEEKING Solutions

1. How can we lessen agricultural demand for water? Describe some ways we can reduce household water use.

2. Describe three ways in which your own actions contribute to water pollution. Now describe three ways in which you could diminish these impacts.

3. Have the provisions of the Clean Water Act been effective? Discuss some of the methods we can adopt, in addition to "end-of-pipe" solutions, to prevent water pollution.

4. **THINK IT THROUGH** Having solved the water depletion problem in your state, your next task is to deal with pollution of the groundwater that provides your state's drinking water supply. Recent studies have shown that one-third of the state's groundwater has levels of pollutants that violate EPA standards for human health, and citizens are fearful for their safety. What steps would you consider taking to safeguard the quality of your state's groundwater supply, and why?

5. **THINK IT THROUGH** The Army Corps of Engineers is proposing building a dam on the river near your town, and some local residents are voicing opposition to the plan. You are city manager, and the City Council has asked you to present them with a brief overview of the costs and benefits of such a dam. Which three benefits would you highlight? Which three costs? Why did you choose these particular benefits and costs as most important?

CALCULATING Ecological Footprints

One of the single greatest personal uses of water is for showering. Showerheads installed in homes and apartments built before 1992 dispense at least 5 gallons of water per minute, but low-flow showerheads produced after that year dispense just 2.5 gallons per minute. Given one shower a day and an average daily shower time of 8 minutes, calculate the amounts of water used and saved over the course of a year with old standard versus low-flow showerheads, and record your results in the table.

	ANNUAL WATER USE WITH STANDARD SHOWERHEADS (GALLONS)	ANNUAL WATER USE WITH LOW-FLOW SHOWERHEADS (GALLONS)	ANNUAL WATER SAVINGS WITH LOW-FLOW SHOWERHEADS (GALLONS)
You			
Your class			
Your state			
United States			

1. The EPA is currently promoting showerheads that produce still lower flows of 2 gallons per minute (gpm). How much water would you save per year by using a 2-gpm showerhead instead of a 2.5-gpm showerhead?

2. How much water would you be able to save annually by shortening your average shower time from 8 minutes to 6 minutes? Assume you use a 2.5-gpm showerhead.

3. Compare your answers to those of Questions 1 and 2. Do you save more water by showering 8 minutes with a 2-gpm showerhead or 6 minutes with a 2.5-gpm showerhead?

4. Can you think of any factors *not* being considered in this scenario of water savings? Explain.

Mastering Environmental Science

Marine and Coastal Systems and Resources

Pacific Ocean
Eastern Garbage
Patch

CENTRAL
case study

A Sea of Plastic in the Middle of the Ocean

> *As I gazed from the deck at the surface of what ought to have been a pristine ocean, I was confronted, as far as the eye could see, with the sight of plastic.*
> Captain Charles Moore, surveying the "Great Pacific Garbage Patch"

> **We are producing far too much plastic believing it is disposable. It's not, it's indestructible.**
> Jo Ruxton, researcher and producer of the documentary *A Plastic Ocean*

Captain Charles Moore has been tied to the Pacific Ocean his entire life. As a young man, he sailed its waters with his father, and after completing college decided to make the preservation and restoration of the Pacific Ocean off the southern California coast his life's work. He became interested in developing methods for sampling ocean waters for chemicals, harmful bacteria, and plastics, and he used his oceanographic research vessel, the *Alguita,* to conduct research along the coastal waters of Mexico and southern California.

In 1997, Moore entered the *Alguita* in the Trans-Pac yacht race from Los Angeles, California, to Honolulu, Hawai'i, so that he could test out the vessel after some recent repairs. On the return voyage to California, he sailed across a little-traveled portion of the Pacific Ocean. There, he encountered a huge floating mass of debris he described as a "soupy" collection that included tires, floating plastics, chemical drums, coat hangers, fishing nets, and other items. It wasn't the presence of plastic items in the ocean that was alarming about this discovery. Plastic bottles, bags, straws, and small bits of degraded plastics—much of it carried to offshore ocean waters by rivers—had been washing up on beaches and shorelines around the world from the actions of tides and currents for decades. Rather, it was the presence of *so much plastic* floating in the remote Pacific Ocean *so far from land*.

Such was the first recorded visual confirmation of what is now called the **Great Pacific Garbage Patch** (GPGP), a 1.6-million-km^2 (617,000-mi^2) aggregation of floating debris and plastics in a portion of the Pacific Ocean called the North Pacific Gyre, one of five gyres in the world's oceans (**FIGURE 16.1**, p. 422). **Gyres** are circular rotations of ocean water that are driven by wind patterns—the Coriolis effect that comes from Earth's rotation (pp. 455–456)—and the land masses of the continents. Debris in the world's oceans—particularly floating plastic items that resist breakdown—collects in the circulating waters of the gyres and can be pushed into the calm waters at their centers. Sitting in placid waters surrounded by swirling currents, debris accumulates over time in these parts of the ocean.

Upon completing this chapter, you will be able to:

+ Identify physical, geographic, and chemical aspects of the marine environment

+ Explain how the oceans influence and are influenced by climate

+ Describe major types of marine ecosystems

+ Assess impacts from marine pollution

+ Describe the state of ocean fisheries and the reasons for their decline

+ Evaluate marine protected areas and reserves as solutions for conserving biodiversity

◄ A seagull in Rome, Italy, holds a discarded cigarette butt—whose filter contains plastics

▲ Captain Charles Moore holding a sample of debris from the Great Pacific Garbage Patch

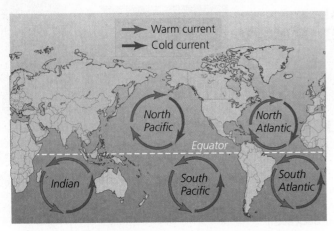

FIGURE 16.1 The five gyres where plastics and other debris collect in the oceans. Floating materials collect in the calm waters inside the circular currents and can remain trapped there for long periods of time.

The North Pacific Gyre has two "garbage patches" within it, the one discovered by Moore at the eastern edge of the gyre and another closer to Asia at the western edge of the gyre. Fed by solid waste from Asia, both the North Pacific and Indian gyres have been found to contain high densities of plastics and other debris. The North Atlantic Gyre, supplied by debris from Europe and the eastern part of North America, also contains substantial quantities of plastic material. Moore visited the South Pacific Gyre in 2017, and this area, once thought to contain low levels of plastic—based on an expedition six years earlier—was found to contain levels of plastics that rivaled those in the GPGP.

Much of the material floating within these gyres and floating in the waters along coastlines is composed of plastic. Plastics are a particularly dangerous type of trash for the oceans. Plastic bags and bottles, discarded fishing nets and line, and small particles of plastic and other trash can harm marine creatures. Organisms can become entangled and drown in discarded fishing nets or large pieces of plastic debris, particularly sea creatures that need to surface for air, such as dolphins and sea turtles (**FIGURE 16.2a**).

Because plastic is designed to not break down, it can drift for centuries before degrading, settling to the ocean bottom, or washing up on beaches. Plastic does degrade slowly in seawater and sunlight—or by microbial degradation—but, in doing so, breaks down into smaller and smaller bits (called **microplastics**) that become more and more numerous. This can actually make things worse: The oceans are now filled with uncountable trillions of tiny particles of plastic floating just under the surface like confetti (**FIGURE 16.2b**), and in the GPGP, the mass of microplastics has been found to be 180 times the mass of plankton and tiny marine creatures in surface waters.

Organisms mistake the small pieces of plastic for plankton or fish eggs, ingest them, and suffer injury or death. A 2010 study found that plankton-eating fish in the GPGP had an average of 2.1 pieces of plastic in their digestive tracts. Fish are not the only organisms affected. Seabird chicks are killed when parents unknowingly feed them pieces of plastic (**FIGURE 16.2c**). One study attributed 40% of premature deaths of albatross

chicks in the Pacific islands of Midway to this cause. Some 267 species are affected by marine plastic debris, leading to the death of an estimated 100,000 marine mammals, such as seals and dolphins, and 1 million seabirds each year.

Some larger pieces of plastic sink, littering the bottom of the ocean—where the low-temperature, low-light conditions are not favorable for degradation—and affecting ocean floor ecosystems. In the North Sea off Norway, for example, underwater surveys have found more than 100 pieces of plastic per square kilometer on the ocean floor.

Floating pieces of plastic can serve as "rafts" and transport species over long distances they could not travel on their own. Seaweeds, sessile invertebrates, and other creatures have become invasive species when carried to new habitats on floating plastic. For example, two species of bryozoans (a predatory, sessile invertebrate that forms polyps on seaweed and stones) are reaching the Florida coast from the Caribbean on floating plastics, where their polyps are damaging ecologically important native seaweed species.

Plastics, of course, can have toxic effects on organisms. Plastics contain harmful substances, such as bisphenol A and phthalates (Chapter 14), which leach into ocean water or the digestive tracts of animals (if the plastic is ingested). Pieces of plastic in the ocean can also chemically "grab" and concentrate persistent organic pollutants (POPs; p. 384) such as DDT and PCBs, thereby contaminating the plastic and increasing its toxicity to any organism that ingests it. The risk from toxicity can be meaningful; one 2017 study that sampled the GPGP reported that 84% of the plastics collected had levels of toxic pollutants above established levels for harm. To make matters worse, many of the harmful chemicals in plastics can bioaccumulate (p. 373) in organisms and be passed through the marine food web.

Because plastics take 50 to 600 years to degrade at sea and there is no viable way to collect the small bits of plastics that litter the oceans, preventing their entry into the oceans is the only key to remedying oceanic plastic pollution (**FIGURE 16.2d**). In 2006, the U.S. Congress responded to ocean pollution by passing the Marine Debris Research, Prevention, and Reduction Act, aiding efforts to keep plastics out of marine waters. Congress strengthened these efforts in 2015 by passing a ban on the sale and distribution of products containing tiny plastic *microbeads*. Added to some toothpastes and shower gels to act as tiny scrubbing agents, trillions of these plastic particles were entering U.S. waters on a daily basis.

Our oceans are accumulating plastic wastes at an alarming rate. The oceans are estimated to currently contain around 200 million metric tons of plastic. Although it is difficult to quantify annual plastic inputs, approximately 8 million metric tons plastic every year are estimated to join the 200 million tons already in the sea. And the vast majority of the impacts of these accumulations are not occurring in the open ocean; rather, they are occurring in the productive coastal marine ecosystems with which humans regularly interact.

Moore's 1997 discovery spawned a movement to rid our oceans of plastics and the dangers they pose. This challenge will not be easy. Plastics are ubiquitous in modern life, and many of our uses of plastic are as single-use products, such

(a) Plastics entangling wildlife

(b) Microplastics from ocean waters

(c) Organisms ingesting plastics

(d) Preventing oceanic plastic pollution

FIGURE 16.2 Plastics affect oceanic life. (a) Like thousands of marine animals every year, this loggerhead sea turtle became entangled in a discarded fishing net. **(b)** Pieces of degraded plastic like these scooped from shallow water in Hawai'i are also dangerous to organisms. **(c)** Plastics are often mistaken for food items and ingested by wildlife, such as by this dead albatross chick. **(d)** Prevention is the best approach for addressing oceanic plastic pollution. Here, floating barriers capture trash in the Los Cerritos Channel in Long Beach, California, to prevent it from entering the Pacific Ocean.

as plastic bags, cups, utensils, and coffee stirrers that are disposed of in massive numbers and readily find their way into the world's waterways and oceans. Reducing our use of plastics, preventing their entry into the ocean, and seeking ways to cleanse the oceans of plastic already there are the only current viable mechanisms for battling the problem. In Moore's words, cleaning the oceans of plastics first requires we drastically lower its inputs: "It's like trying to bail out the bathtub with the spigot on. You're not going to do it until you shut off the spigot" (see **DataGraphic**, p. 434).

The stakes are high. One study estimated that if current trends in plastic pollution continue, by 2050, there will be more plastics, by weight, than fish in our oceans—an outcome we must do everything in our power to prevent.

The Oceans

To understand the multitude of stressors facing our oceans, including plastic pollution, we must understand the system we're dealing with. For example, it's been said that planet "Earth" should more properly be named planet "Ocean." After all, ocean water covers most of the planet's surface. The oceans are an important component of Earth's interconnected aquatic systems (p. 392). Although a small number of rivers empty into inland seas, the vast majority of rivers empty into oceans; thus, the oceans receive most of the inputs of water, sediments, pollutants, and organisms carried by freshwater systems.

The oceans touch and are touched by virtually every environmental system and every human endeavor. They shape our planet's climate, teem with biodiversity, provide us resources, and facilitate our transportation and commerce. The oceans provide the fish you eat and the crude oil you need for your car—or that's needed to power your public transportation—and they influence your weather, wherever you live in the world.

The world's five oceans—Pacific, Atlantic, Indian, Arctic, and Southern—are all connected, making up a single vast body of water. This "world ocean" covers 71% of Earth's surface and contains 97.5% of its water (see Figure 15.3, p. 392). Let's first briefly survey the physical and chemical makeup of the oceans—for although they may look homogeneous from a beach, boat, or airplane, marine systems are actually very complex and dynamic.

Seafloor topography can be rugged

Even though most maps depict oceans as smooth swaths of blue, when we examine what is beneath the waves, we see that the geology of the ocean floor is anything but uniform. Underwater volcanoes shoot forth enough magma to build islands above sea level, such as the Hawaiian Islands (p. 41). Steep canyons as large as Arizona's Grand Canyon lie just offshore of some continents. The lowest spot in the oceans—the Mariana Trench in the South Pacific—is deeper by more than a mile than Mount Everest is high.

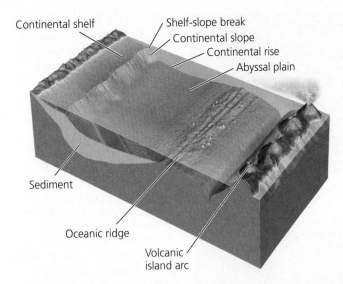

Continental shelf — Shelf-slope break — Continental slope — Continental rise — Abyssal plain

Sediment

Oceanic ridge

Volcanic island arc

FIGURE 16.3 A stylized bathymetric profile shows key geologic features of the submarine environment. Shallow water exists around the edges of continents over the continental shelf, which drops off at the shelf-slope break. The steep continental slope gives way to the more gradual continental rise, and all are underlain by sediments from the continents. Vast areas of seafloor are flat abyssal plain. Seafloor spreading occurs at oceanic ridges, and oceanic crust is subducted in trenches (p. 36). Volcanic activity along trenches may give rise to island chains such as the Aleutian Islands. Features on the left side of this diagram are more characteristic of the Atlantic Ocean, and features on the right side of the diagram are more characteristic of the Pacific Ocean.

Three-dimensional diagrams (**FIGURE 16.3**) that reflect bathymetry (the measurement of ocean depths) and topography (physical geography, or the shape and arrangement of landforms) illustrate underwater geographic features. In bathymetric profile, gently sloping **continental shelves** sit beneath the shallow waters bordering the continents. The continental shelf then drops off at the *shelf-slope break,* where the *continental slope* angles more steeply downward to the deep ocean basin below.

Some island chains, such as the Florida Keys, are formed by coral reefs (pp. 431–432) and lie atop the continental shelf. Others, such as the Aleutian Islands, which curve across the North Pacific from Alaska toward Russia, are volcanic in origin. The Aleutians are also the site of a deep trench that, like the Mariana Trench in the South Pacific, formed at a convergent tectonic plate boundary, where one slab of crust dives beneath another in the process of subduction (pp. 36–37).

Wherever reefs, volcanism, or other processes create physical structure underwater, life thrives. Marine animals make use of physical structure as habitat, and as a result, topographically complex areas often make for productive fishing grounds. Coral reefs are one such example, providing fish, invertebrates, and a diversity of other marine creatures places to forage and hide.

Ocean water contains high concentrations of dissolved salts

Ocean water contains approximately 96.5% H_2O by mass. Most of the remainder consists of ions from dissolved salts (**FIGURE 16.4**). Ocean water is salty primarily because ocean basins are the final repositories for runoff that collects salts from weathered rocks and carries them, along with sediments, to the ocean. Whereas the water in the ocean evaporates, the salts do not, and they accumulate in ocean basins (the salinity of ocean water generally ranges from 33,000 to 37,000 parts per million, whereas the salinity of freshwater runoff typically is less than 500 parts per million). Besides dissolved salts, nutrients such as nitrogen and phosphorus occur in seawater in trace amounts (well under 1 part per million) and play essential roles in nutrient cycling (p. 121) in marine ecosystems. Another aspect of ocean chemistry is dissolved gas content. Roughly 36% of the gas dissolved in seawater is oxygen, which is produced by underwater photosynthetic plants, bacteria, and phytoplankton (p. 82) and also enters waters by diffusion from the atmosphere. Oxygen concentrations are highest in the upper layer of the ocean, reaching 13 ml/L of water. Marine organisms depend on oxygen dissolved in water, just as terrestrial organisms rely on oxygen in air, so most marine life, including fish such as cod, live in these well-oxygenated waters. If dissolved oxygen is depleted from ocean waters, a hypoxic "dead zone" may ensue (pp. 106, 115–117), killing organisms or forcing them to flee to better-oxygenated waters (pp. 109–110). Another gas that is soluble in ocean water is carbon dioxide (CO_2). As we pump excess CO_2 into the atmosphere by burning fossil fuels, more CO_2 diffuses into the oceans. We shall soon see (pp. 437–438)

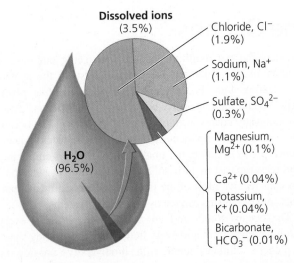

Dissolved ions (3.5%)

Chloride, Cl⁻ (1.9%)

Sodium, Na⁺ (1.1%)

Sulfate, SO_4^{2-} (0.3%)

H_2O (96.5%)

Magnesium, Mg^{2+} (0.1%)

Ca^{2+} (0.04%)

Potassium, K^+ (0.04%)

Bicarbonate, HCO_3^- (0.01%)

FIGURE 16.4 Ocean water consists of 3.5% salt, by mass. Most of this salt is NaCl in solution, so sodium and chloride ions are abundant. A number of other ions are also present.

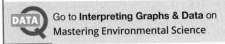

DATA Go to **Interpreting Graphs & Data** on Mastering Environmental Science

how this affects ocean pH (p. 29), turning the water more acidic and posing problems for marine life.

Solar energy stratifies ocean water

Sunlight warms the ocean's surface but does not penetrate deeply, so ocean water is warmest at the surface and becomes colder with depth. Surface waters in tropical regions receive more solar radiation and therefore are warmer than surface waters in temperate or polar regions. Warmer water is less dense than cooler water, but water also becomes denser as it gets saltier. These relationships give rise to different layers of water: Heavier (colder and saltier) water sinks, whereas lighter (warmer and less salty) water remains nearer the surface. Waters of the surface zone are heated by sunlight and stirred by wind such that they are of similar density down to a depth of about 150 m (490 ft). Below this zone lies the **pycnocline,** a region in which density increases rapidly with depth. The pycnocline contains about 18% of the ocean's water by volume, compared to the surface zone's 2%. The remaining 80% of the ocean's water lies in the deep zone beneath the pycnocline. The dense water in this deep zone is sluggish and unaffected by winds, storms, sunlight, and temperature fluctuations.

Despite the daily heating and cooling of surface waters, ocean temperatures are much more stable than air temperatures on land. Midlatitude oceans experience yearly temperature variation of only around 10°C (18°F), and tropical and polar ocean temperatures are more stable than air temperatures measured on land at the same latitudes. The reason for this relative stability is that water has a high **heat capacity,** a measure of the heat energy required to increase temperature by a given amount. It takes more energy to increase the temperature of a given amount of water than it does to increase the temperature of a comparable amount of air. High heat capacity enables ocean water to absorb and hold a tremendous amount of heat. For example, just the top 2.6 m (8.5 ft) of the oceans hold as much heat as the entire atmosphere! By absorbing heat and releasing it to the atmosphere, the oceans help regulate Earth's climate (Chapter 18). They also influence climate by moving heat from place to place via the ocean's surface circulation, a system of currents that move in the pycnocline and the surface zone.

Surface water flows horizontally in currents

Earth's ocean is home to riverlike flows of water driven by density differences, heating and cooling, gravity, and wind. Surface **currents** flow horizontally within the upper 400 m (1300 ft) of water for great distances and in long-lasting patterns across the globe (**FIGURE 16.5**). Warm-water currents carry water heated by the sun from equatorial regions, while cold-water currents carry water cooled in high-latitude regions

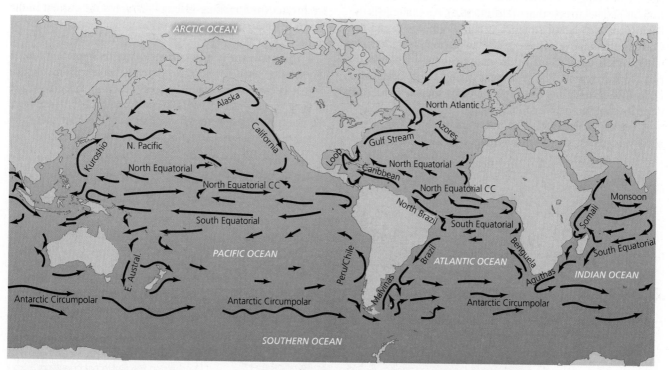

FIGURE 16.5 The upper waters of the oceans flow in surface currents, long-lasting and predictable global patterns of water movement. Warm- and cold-water currents interact with the planet's climate system, and people have used them for centuries to navigate the oceans. *Adapted from Rick Lumpkin (NOAA/AOML).*

Go to **Interpreting Graphs & Data** on **Mastering Environmental Science**

or from the deep ocean. Some surface currents are very slow. Others, like the Gulf Stream, are fast and powerful. From the Gulf of Mexico, the Gulf Stream flows up the Atlantic coast at nearly 2 m/sec, or more than 4 mph. Averaging 70 km (43 mi) across, the Gulf Stream continues across the North Atlantic, bringing warm water to Europe and moderating that continent's climate, which otherwise would be much colder.

Besides influencing climate, ocean currents have aided navigation and shaped human history. Currents helped carry Polynesians to Easter Island (p. 8), Darwin to the Galápagos (p. 51), and Europeans to the New World. Currents transport heat, nutrients, pollution, and marine species from place to place. Recent research examining smaller-scale, lesser understood currents, those driven by winds, is now being used to predict locations in the world's oceans where plastics and other debris might accumulate (see **THE SCIENCE BEHIND THE STORY**, p. 428).

Water also flows vertically in the oceans

Surface winds and heating also create vertical currents in seawater, which circulate water between the deep ocean and the surface. Where horizontal surface currents diverge from one another, cold, deep waters are pulled to the surface in a process called **upwelling.** Upwelled water is rich in nutrients brought up from the ocean bottom, so upwellings are often sites of high primary productivity (p. 114) and lucrative fisheries. Upwellings will occur where strong winds blow away from or parallel to coastlines (**FIGURE 16.6**). An example is the Pacific coast of North America, where north winds and the Coriolis effect move surface waters away from shore, raising nutrient-rich water from below and creating a biologically rich offshore region. The cold water from the deep ocean also chills the air along the coast, giving cities such as San Francisco its cool summers and famous fog.

In areas where surface currents converge, surface water sinks in a process called **downwelling.** Downwelling transports warm surface water rich in dissolved gases to deeper waters, providing an influx of dissolved oxygen for deep-water life and "burying" CO_2 from the atmosphere in the deep ocean. Vertical currents also occur within the deep zone of the ocean, where differences in density can lead to rising and falling convection currents, similar to those in molten rock (p. 35) and in air (p. 455).

Ocean currents affect Earth's climate

The horizontal and vertical movements of ocean water can have far-reaching effects on climate, both regionally and globally. **Thermohaline circulation,** also known as *meridional overturning circulation* (MOC), is a worldwide current system in which warmer, lower-salinity water moves horizontally along the surface and colder, saltier water (which is denser) moves horizontally in the deep ocean (**FIGURE 16.7**). One segment of this worldwide conveyor-belt system includes the warm surface water in the Gulf Stream that flows across the Atlantic Ocean to Europe. Upon reaching Europe, this water releases heat to the air, keeping Europe warmer than it otherwise would be given its latitude. The now-cooler water becomes saltier through evaporation and thus becomes denser and sinks, creating a region of downwelling known as **North Atlantic Deep Water** (NADW). If this system were ever to completely break down, Europe and parts of North America would likely experience a drastic drop in temperature. That seems to have occurred about 13,000 years ago. As the planet was warming after a lengthy ice age, inputs of freshwater from melting glaciers disrupted the current in the North Atlantic, plunging the world into an ice age that lasted some 1300 years.

This conveyor belt of currents plays an important role in moderating changes in surface temperatures from climate change because it essentially "stores" heat from the atmosphere in deep ocean waters. That is, surface waters absorb heat from the warming atmosphere and carry some of this heat with them as they are driven deeper in the ocean by these currents.

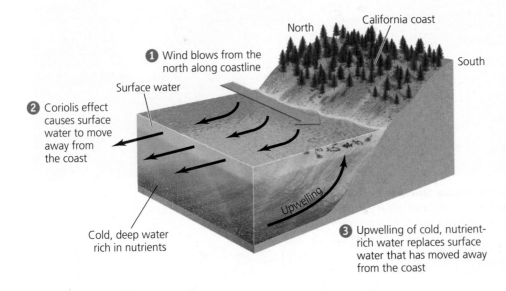

1 Wind blows from the north along coastline

North

California coast

South

Surface water

2 Coriolis effect causes surface water to move away from the coast

Cold, deep water rich in nutrients

Upwelling

3 Upwelling of cold, nutrient-rich water replaces surface water that has moved away from the coast

FIGURE 16.6 Upwelling is the movement of bottom waters upward. It often brings nutrients up to the surface, creating productive areas for marine life. For example, north winds blow along the California coastline **1**, while the Coriolis effect (pp. 455–456) draws wind and water away from the coast **2**. Water is then pulled up from the bottom **3** to replace the water that moves away from shore.

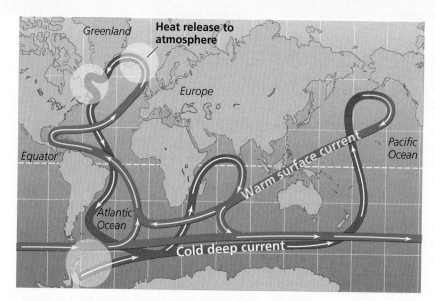

FIGURE 16.7 As part of the oceans' thermohaline circulation, warm surface currents carry heat from equatorial waters toward colder latitudes, such as northern Europe. The water then cools and sinks, forming a global conveyor belt of currents.

Recent research has shown that the portion of the MOC in the Atlantic (the *Atlantic meridional overturning circulation,* or AMOC) alternates naturally between slow phases (when its currents slow down) and rapid phases (when its currents speed up). When the AMOC is in a slow phase—at is was from 1975 to 1998—atmospheric temperatures increase more rapidly because a slower current permits less atmospheric heat to be stored in the ocean bottom than when the AMOC is in a rapid phase—as it has been since 2000. During this more recent period, atmospheric temperatures have increased more slowly due to more atmospheric heat being stored in the ocean. When the AMOC shifts to a slow phase, atmospheric temperatures are expected to once again rise rapidly.

Another interaction between ocean currents and the atmosphere that influences climate is the **El Niño–Southern Oscillation** (ENSO), a systematic shift in atmospheric pressure, sea surface temperature, and ocean circulation in the tropical Pacific Ocean. Under normal conditions, prevailing winds blow from east to west along the equator, from a region of high pressure in the eastern Pacific to one of low pressure in the western Pacific, forming a large-scale convective loop in the atmosphere (**FIGURE 16.8a**). The winds push surface waters westward, causing water to "pile up" in the western Pacific. As a result, water near Indonesia can be 50 cm (20 in.) higher and 8°C warmer than water near South America, elevating the risk of coastal flooding in the Pacific. The westward-moving

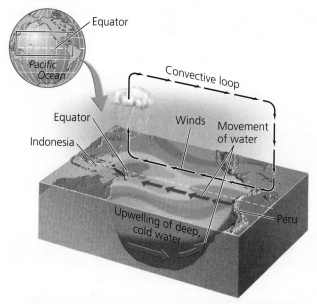

(a) Normal conditions

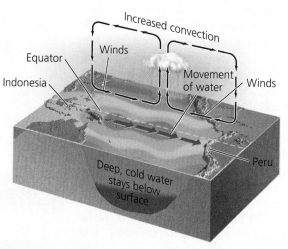

(b) El Niño conditions

FIGURE 16.8 El Niño conditions occur every 2 to 8 years, causing marked changes in weather patterns. In these diagrams, red and orange denote warmer water, and blue and green denote colder water. Under normal conditions **(a),** prevailing winds push warm surface waters toward the western Pacific. Under El Niño conditions **(b),** winds weaken, and the warm water flows back across the Pacific toward South America, like water sloshing in a bathtub. *Adapted from National Oceanic and Atmospheric Administration, Tropical Atmospheric Ocean Project.*

THE SCIENCE behind the story

Can We Predict the Oceans' "Garbage Patches"?

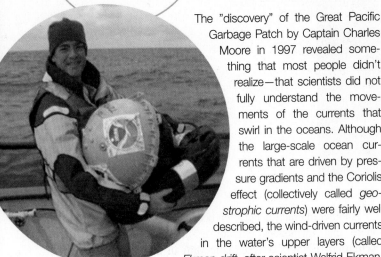

A researcher prepares to deploy a buoy for the Global Drifter Program

The "discovery" of the Great Pacific Garbage Patch by Captain Charles Moore in 1997 revealed something that most people didn't realize—that scientists did not fully understand the movements of the currents that swirl in the oceans. Although the large-scale ocean currents that are driven by pressure gradients and the Coriolis effect (collectively called *geostrophic currents*) were fairly well described, the wind-driven currents in the water's upper layers (called *Ekman drift*, after scientist Walfrid Ekman, who first modeled them in the early 20th century) were poorly understood. Thanks to the work of Nikolai Maximenko and his collaborators, that is no longer the case. Researchers are now more capable of predicting the movements of floating material in the oceans, such as plastics and other debris, and identifying the places where this material will accumulate—allowing us to identify other garbage patches in the world's oceans.

Maximenko, a senior researcher with the International Pacific Research Center at the University of Hawai'i at Mānoa, studies Ekman drift using data gathered by the Global Drifter Program of the U.S. National Oceanic and Atmospheric Administration (NOAA). Since 1979, the program has deployed more than 12,000 buoys, or "drifters," in the world's oceans and tracked their movements to determine patterns in ocean currents. Each drifter is composed of a 30- to 40-cm (12- to 16-in.) surface float, which contains sensors for detecting ocean temperature, salinity, and other properties, as well as a transmitter to send its information to satellites overhead. The float is tethered to a type of anchor, called a subsurface drogue, which hangs at a depth of 15 m (49 ft). The drogue keeps the float upright on the surface and provides an underwater "sail" for currents to push. Drifters typically operate for about 400 days before they stop transmitting, and the program aims to maintain an array of 1250 drifters at any given time in the world's oceans.

In 2008, Maximenko partnered with Peter Niiler of the Scripps Institution of Oceanography to use data from the Global Drifter Program to produce a more detailed map of surface ocean currents. The team combined data on drifter movements with satellite altimetry and wind currents, producing a highly detailed map of both geostrophic currents and Ekman drift. As the issue of oceanic plastic pollution gained attention, Maximenko saw an opportunity to use his map to predict the areas of the ocean where floating debris is likely to accumulate. To do so, he partnered with his colleague Jan Hafner to create a computer model that predicts the movements of drifters over long time periods in the world's oceans. He then ran a simulation in which he uniformly distributed drifters in the world's oceans and saw where they would become concentrated over time.

The simulation's results, published in 2012 in *Marine Pollution Bulletin,* revealed that oceanic debris was likely to accumulate in portions of five subtropical gyres, including the area of the Great Pacific Garbage Patch. The simulation also revealed that 70% of the drifters would remain at sea after 10 years, showing the lengthy life span of floating debris in the oceans. One of the model's predictions, which were first released in 2008 (**FIGURE 1**), was verified when a 2010 study found high concentrations of plastic in the North Atlantic Gyre where Maximenko's model predicted it. The remaining garbage patches have all since been verified at their predicted locations. Maximenko's model was even used to predict the fate of the debris washed into the Pacific Ocean by the tsunami that struck Japan in 2011 (p. 23).

Thanks to the work of Maximenko and his colleagues, we now have a greater understanding of the movements of plastics and other debris in the oceans. Armed with this knowledge, we will be better able to properly assess and, we hope, reduce the impacts of plastics on marine life.

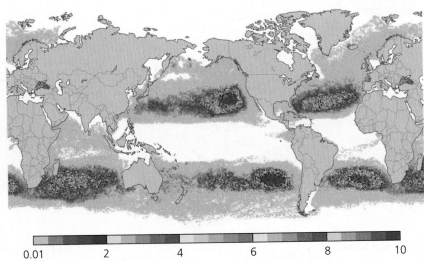

0.01   2   4   6   8   10

Relative change in drifter concentration after 10 years from uniform distribution

FIGURE 1 A computer model predicted where plastics and trash would accumulate after 10 simulated years of ocean current activity. *Source: International Pacific Research Center, IPRC Climate, Vol. 8, No. 2, 2008.*

surface waters allow cold water to rise up from the deep in a nutrient-rich upwelling along the coast of Peru and Ecuador.

El Niño conditions are triggered when air pressure decreases in the eastern Pacific and increases in the western Pacific, weakening the equatorial winds and allowing the warm water to flow eastward toward South America (**FIGURE 16.8b**). This suppresses upwelling along the Pacific coast of the Americas, shutting down the delivery of nutrients that support marine life and fisheries. This phenomenon was called El Niño (Spanish for "little boy" or "Christ Child") by Peruvian fishermen because the arrival of warmer waters usually occurred shortly after Christmas. El Niño events alter weather patterns around the world, creating rainstorms and floods in areas that are generally dry, such as southern California, and causing drought and fire in regions that are typically moist, such as Indonesia. Coastal industries such as Peru's anchovy fisheries are devastated by El Niño events, and the 1982–1983 El Niño alone caused more than $8 billion in economic losses worldwide. Food shortages in Asia caused by the 1997–1998 El Niño plunged an estimated 15% of the people in some nations into poverty. The El Niño event of 2015–2016 was especially strong, with massive heat waves afflicting India and drought ravaging Southeast Asia.

La Niña events are the opposite of El Niño events; in a La Niña event, unusually cold waters rise to the surface and extend westward in the equatorial Pacific when winds blowing to the west strengthen, and weather patterns are affected in opposite ways. ENSO cycles are periodic but irregular, occurring every 2 to 8 years. Scientists are exploring whether warming air and sea temperatures due to climate change (Chapter 18) are increasing the frequency and strength of these cycles.

Marine and Coastal Ecosystems

With their variation in topography, temperature, salinity, nutrients, and sunlight, marine and coastal environments feature a variety of ecosystems. These systems may not give us the water we need for drinking and growing crops, but they teem with biodiversity and provide many other necessary resources. As we survey marine and coastal ecosystems, keep in mind that they are part of a web of interconnected freshwater and marine aquatic systems (see Figure 15.4, p. 393) that exchange water, organisms, sediments, pollutants, and other dissolved substances with one another. Hence, the topics we discuss in this chapter are greatly influenced by those in freshwater ecosystems, and vice versa.

Intertidal zones undergo constant change

Where the ocean meets the land, **intertidal,** or *littoral,* ecosystems (**FIGURE 16.9**) spread between the uppermost reach of the high tide and the lowest limit of the low tide. **Tides** are the periodic rising and falling of the ocean's height at a given

Supratidal zone (splash zone)

Level of high tide

Intertidal zone

Level of low tide

Subtidal zone

FIGURE 16.9 The rocky intertidal zone stretches along rocky shorelines between the lowest and highest reaches of the tides. The intertidal zone provides niches for a diversity of organisms, including sea stars (starfish), crabs, sea anemones, corals, chitons, mussels, nudibranchs (sea slugs), and sea urchins. Areas higher on the shoreline are exposed to the air more frequently and for longer periods, so organisms that tolerate exposure best specialize in the upper intertidal zone. The lower intertidal zone is exposed less frequently and for shorter periods, so organisms less tolerant of exposure thrive in this zone.

location, caused by the gravitational pull of the moon and sun. In most parts of the world, high and low tides occur roughly 6 hours apart, so intertidal organisms spend part of each day submerged in water, part of the day exposed to air and sun, and part of the day being lashed by waves.

The intertidal environment is a tough place to make a living, but it is home to a remarkable diversity of organisms. The rocky intertidal zone is so diverse because environmental conditions such as temperature, salinity, and moisture change dramatically from its high to its low reaches. Life abounds in the crevices of rocky shorelines, which provide shelter and pools of water (tide pools) during low tides. Sessile (stationary, non-swimming) animals such as anemones, mussels, and barnacles live attached to rocks, filter-feeding on plankton in the water that washes over them. Urchins, sea slugs, chitons, and limpets eat intertidal algae or scrape food from the rocks. Sea stars (starfish) creep slowly along, preying on the filter-feeders and herbivores. Crabs clamber around the rocks, scavenging detritus.

Salt marshes line temperate shorelines

Along many of the world's coasts at temperate latitudes, **salt marshes** occur where the tides wash over gently sloping sandy or silty substrates. Rising and falling tides flow into and out of channels called *tidal creeks* and at highest tide spill over onto elevated marsh flats (**FIGURE 16.10**). Marsh flats grow thick with salt-tolerant grasses, as well as rushes, shrubs, and other herbaceous plants.

Salt marshes boast very high primary productivity and provide critical habitat for shorebirds, waterfowl, and many commercially important fish and shellfish species.

FIGURE 16.10 Salt marshes occur in temperate intertidal zones where the substrate is muddy. Tidal waters flow in channels called tidal creeks amid flat areas called benches, sometimes partially submerging the salt-adapted grasses.

Salt marshes also filter pollution and stabilize shorelines against storm surges, an unusual rise in ocean level caused by offshore storms. However, because people desire to live and do business along coasts, we have altered or destroyed vast expanses of salt marshes to make way for coastal development. When salt marshes are destroyed, we lose the ecosystem services they provide, such as the protection they offer inland areas by damping waves and winds from offshore storms. When Hurricane Katrina struck the Gulf Coast in 2005, for instance, the flooding was made worse because vast areas of salt marshes had vanished during the preceding decades as a result of engineering of the Mississippi River, development, and subsidence from oil and gas drilling.

Mangrove forests line coasts in the tropics and subtropics

In tropical and subtropical latitudes, mangrove forests replace salt marshes along gently sloping sandy and silty coasts. **Mangroves** are among the few types of trees that are salt tolerant, and they have unique roots that curve upward like snorkels to obtain oxygen or downward like stilts to support the tree in changing water levels (**FIGURE 16.11**). Fish, shellfish, crabs, snakes, and other organisms thrive among the root networks, and birds feed and nest in the dense foliage of these coastal forests. Besides serving as nurseries for fish and shellfish that people harvest, mangroves also provide wood for fuel and construction, natural dyes, and chemical compounds that have medicinal properties.

From Florida to Mexico to the Philippines, half the world's mangrove forests have been destroyed as people have developed coastal areas. Shrimp farming in particular has greatly impacted mangroves, because mangrove forests are often destroyed to build large pens along the coastline in which shrimp are raised in aquaculture (p. 250). When mangroves are removed, coastal areas lose the ability to slow runoff, filter pollutants, and resist soil erosion. When mangroves are destroyed, offshore systems such as coral reefs and seagrass beds are more readily degraded. Mangrove forests, like salt marshes, also protect coastal communities against storm surges and tsunamis. The 2004 Indian Ocean tsunami (p. 44) devastated areas where mangroves had been removed because of shrimp farming and coastal development, but it caused less damage where mangroves were intact.

Fresh water meets saltwater in estuaries

Many salt marshes and mangrove forests occur in or near **estuaries,** water bodies

where rivers flow into the ocean, mixing fresh water with saltwater. Estuaries are biologically productive ecosystems that experience fluctuations in salinity with the daily and seasonal variations in tides and freshwater runoff. The shallow water of estuaries nurtures sea grass beds and other plant life and provides critical habitat for shorebirds and many commercially important shellfish species.

Estuaries everywhere have been affected by coastal development, water pollution, habitat alteration, and over-fishing. The Chesapeake Bay demonstrates many of the challenges faced by estuaries. Bay waters suffer from seasonal oxygen depletion from cultural eutrophication (p. 110), and its once-thriving oyster populations, which naturally filter the water of pollutants, have been decimated by overharvesting (Chapter 5). Coastal ecosystems such as estuaries have borne the brunt of human impact because two-thirds of Earth's people choose to live within 160 km (100 mi) of the ocean.

Coral reefs are treasure troves of biodiversity

Shallow subtropical and tropical waters are home to coral reefs. A reef is an underwater outcrop of rock, sand, or other material. A **coral reef** is a mass of calcium carbonate composed of the shells of tiny marine animals known as **corals.** A coral reef may occur as an extension of a shoreline; along a *barrier island* paralleling a shoreline; or as an *atoll,* a ring around a submerged island.

Corals are tiny invertebrate animals related to sea anemones and jellyfish. They remain attached to rock or existing reef and capture passing food with stinging tentacles. Corals also derive nourishment from symbiotic algae known as **zooxanthellae,** which inhabit their bodies and produce food through photosynthesis—and provide the diversity of vibrant colors in reefs. Most corals are colonial, and the surface of a

Kelp forests harbor many organisms

Along many temperate coasts, large brown algae, or **kelp,** grow from the floor of continental shelves, reaching up toward the sunlit surface. Some kelp reaches 60 m (200 ft) in height and can grow 45 cm (18 in.) per day. Dense stands of kelp form underwater "forests" (**FIGURE 16.12**). Kelp forests supply shelter and food for invertebrates and fish, which in turn provide food for predators such as seals and sharks. Kelp forests also absorb wave energy and protect shorelines from erosion. Some types of kelp are eaten by people, and kelp provides alginates, chemical compounds that serve as thickeners in consumer products such as cosmetics, paints, paper, soaps, and ice cream.

FIGURE 16.12 "Forests" of tall brown algae known as kelp grow from the floor of the continental shelf. Numerous fish and other creatures eat kelp or find refuge among its fronds.

coral reef consists of millions of densely packed individuals. As corals die, their shells remain part of the reef, and new corals grow atop them. This accumulation of coral shells enables the reef to persist and grow larger over time.

Like kelp forests, coral reefs protect shorelines by absorbing wave energy. They also host tremendous biodiversity (**FIGURE 16.13a**) because they provide complex physical structure (and thus many habitats) in shallow nearshore waters, which are regions of high primary productivity. If you have ever gone diving or snorkeling over a coral reef, you probably noticed the staggering diversity of anemones, sponges, hydroids, tubeworms, and other sessile invertebrates; the innumerable mollusks, flatworms, sea stars, and urchins; and the many fish species that find food and shelter in reef nooks and crannies.

The promotion of biodiversity by coral reefs makes their dramatic decline worldwide particularly alarming. Many reefs have fallen victim to *coral bleaching*, which occurs when zooxanthellae die or abandon the coral, thereby depriving the coral of nutrition. Corals lacking zooxanthellae lose color and frequently die, leaving behind ghostly white patches in the reef (**FIGURE 16.13b**). Once large areas of coral die, species that hide within the reef are exposed to higher levels of predation, and their numbers decline. Without living coral—the foundation of the ecosystem on which so many species rely—biological diversity on reefs declines as organisms flee or perish. Coral bleaching is thought to occur when corals are strongly stressed, with common stressors including exposure to elevated levels of pollutants and increased sea surface temperatures associated with global climate change. The Great Barrier Reef, which stretches for 2300 km (1400 mi) along the coast of Australia, has recently been severely damaged by bleaching. Major bleaching events in 2016 and 2017, attributed to rising sea temperatures, affected some 1500 km (900 mi) of the reef and have decimated large stretches of its northern and middle sections.

Another threat to coral comes from nutrient pollution in coastal waters, which promotes the growth of algae that are smothering reefs in the Florida Keys and many other regions. In addition, coral reefs sustain damage when divers use cyanide to stun fish in capturing them for food or for the pet trade, a common practice in the waters of Indonesia and the Philippines. And as global climate change accelerates, the oceans are becoming increasingly more acidic as excess carbon dioxide from the atmosphere reacts with seawater to form carbonic acid. The resulting acidification threatens to deprive corals of the carbonate ions they need to produce their structural parts (pp. 437–438).

Corals stressed by warming, acidic oceans, and nutrient pollution are also more susceptible to diseases. *Stony coral*

tissue loss disease, which affects many species of coral and is believed to be caused by pathogenic bacteria, is decimating reefs along the southern coast of Florida. In 2014, the disease was only seen in isolated patches in reefs near Miami, but from 2014 to 2019, the disease spread across all the reefs in southern Florida and is now seen in the Florida Keys.

A few coral species thrive in waters outside the tropics and build reefs on the ocean floor at depths of 200 to 500 m (650 to 1650 ft). These little-known reefs, which occur in cold-water areas off the coasts of Norway, Spain, the British Isles, and elsewhere, are only now being studied by scientists. Already, however, many have been badly damaged by bottom-trawling fishing vessels (pp. 440–441). Norway and other countries are now beginning to protect some of these deep-water reefs and the biological diversity they support.

Open-ocean ecosystems vary in their biodiversity

The uppermost 10 m (33 ft) of ocean water absorbs 80% of the solar energy that reaches its surface. For this reason,

(a) Coral reef community

Bleaching is evident in the whitened regions of this coral

(b) Bleached coral

FIGURE 16.13 Coral reefs provide food and shelter for a tremendous diversity (a) of fish and other creatures. Today these reefs face multiple stresses from human impacts. Many corals have died as a result of coral bleaching **(b),** in which corals lose their zooxanthellae.

nearly all the oceans' primary productivity occurs in the well-lit top layer, or **photic zone.** Generally, the warm, shallow waters of continental shelves are the most biologically productive and support the greatest species diversity. Habitats and ecosystems occurring between the ocean's surface and floor are classified as **pelagic,** whereas those that occur on the ocean floor are classified as **benthic.**

Biological diversity in pelagic regions of the open ocean is highly variable in its distribution. Primary production (p. 114) and animal life near the surface are concentrated in regions of nutrient-rich upwelling. Microscopic phytoplankton constitute the base of the marine food chain in the pelagic zone and produce up to half of the atmosphere's oxygen (**FIGURE 16.14**). These photosynthetic algae, protists, and cyanobacteria feed zooplankton (p. 82), which in turn become food for fish, jellyfish, whales, and other free-swimming animals. Predators at higher trophic levels include larger fish, sea turtles, and sharks. Fish-eating birds such as puffins, petrels, and shearwaters feed at the surface of the open ocean, returning periodically to nesting sites on islands and coastlines. As the majority of plastics that enter the oceans float, pelagic organisms are particularly prone to ingesting small pieces of plastic (Figure 16.2b) or becoming entangled in larger items (Figure 16.2a). Benthic organisms are also exposed to plastics when heavier plastics sink or small pieces are driven downward by currents, and this pollution can be particularly damaging in the productive near-shore benthic ecosystems where life thrives.

In the little-known deep-water benthic ecosystems, animals have adapted to tolerate extreme water pressures and to live in the dark without food from autotrophs. Many of these often bizarre-looking creatures scavenge carcasses or organic detritus that sinks from surface waters. Others are predators, and still others attain food from symbiotic mutualistic (p. 80) bacteria. Ecosystems form around hydrothermal vents (pp. 34–35) on the bottom of the deep ocean, where heated water spurts from the seafloor, carrying minerals that precipitate to form rocky structures. Tubeworms,

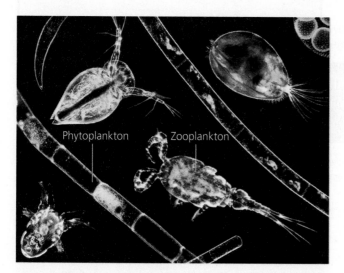

FIGURE 16.14 The uppermost reaches of ocean water contain billions upon billions of phytoplankton—tiny photosynthetic algae, protists, and bacteria that form the base of the marine food chain. This part of the ocean is also home to zooplankton, small animals and protists that dine on phytoplankton.

shrimp, and other creatures in these systems use symbiotic bacteria to derive their energy from chemicals in the heated water rather than from sunlight, a process called *chemosynthesis* (p. 35). They manage to thrive within amazingly narrow zones between scalding-hot and icy-cold water.

Marine Pollution

People have long made the oceans a sink for waste and pollutants. Even into the mid-20th century, coastal U.S. cities would dump trash and untreated sewage into the ocean. The Clean Water Act (p. 173) helped reduce discharges of point-source pollution (p. 409) such as these into the oceans, but a great deal of today's marine pollution comes from countless non-point sources and is much more difficult to regulate.

As long as rivers flow to the sea, much of our pollution on land sooner or later ends up in the oceans. Accordingly, oil, plastic, toxic chemicals, and excess nutrients that run off of the land into waterways eventually make their way into the oceans. Ocean-based sources, such as abandoned nets and gear from fishing boats and sewage and trash from cruise ships, add to the ocean's pollution inputs. The volumes of trash added to Earth's oceans is stunning, and the scope of the problem can be gauged by the amount picked up by volunteers who trek beaches in the Ocean Conservancy's annual International Coastal Cleanup. Over a span of 25 years, more than 65 million kg (144 million lb) of waste has been collected from the world's beaches as part of this program.

FAQ

Do any rivers in the United States dump large quantities of plastics into the ocean?

Yes—the mighty Mississippi River. A 2015 survey of the waters of the Gulf of Mexico off the Louisiana coast near the mouth of the river found plastic particle concentrations that rivaled those measured in the Great Pacific Garbage Patch. The Mississippi River watershed encompasses 31 states and is home to 50 cities and roughly 100 million people, and the river dumps plastics collected over this massive area into the Gulf. This plastic pollution poses problems for the Gulf's productive fisheries and the many beaches that line the Gulf Coast, so in 2018 the Mississippi River Cities and Towns Initiative (a coalition of cities and towns along the river) announced a plan that embraced plastics recycling and other initiatives aimed at reducing plastic inputs into the river by 20% over the next several years.

Plastics are long-lived in marine environments

We have seen that plastics pose dangers to marine creatures when animals become entangled in plastic debris, ingest small pieces of plastic that they mistake for food, and are poisoned by toxic chemicals released into waters or their digestive systems by plastics. One of the biggest issues that make plastic pollution especially damaging is the longevity of plastics in marine environments. Whereas many other pollutants exist for short durations, plastics can pose dangers for decades or centuries. A plastic beverage bottle can take 450 years to

DATAGRAPHIC How Can We Avoid Choking the Oceans with Plastic?

Due to predicted higher human populations and growing affluence, **more plastic is expected to enter oceans in 2050 than today.**

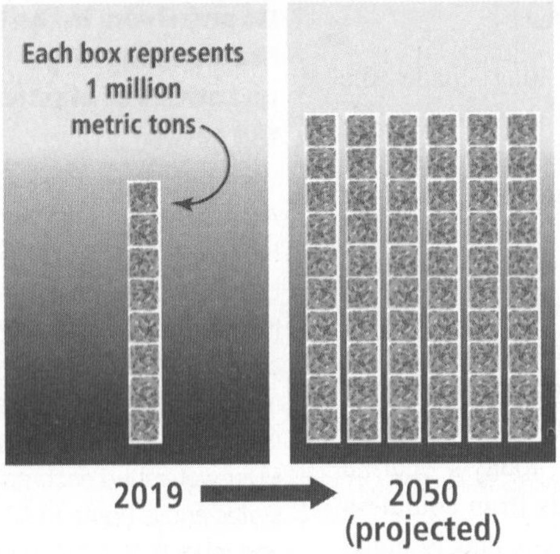

Plastics entering oceans
Each box represents 1 million metric tons
2019
2050 (projected)

Most plastics produced float in seawater, where they can entangle wildlife or where they can fragment and enter the food web.

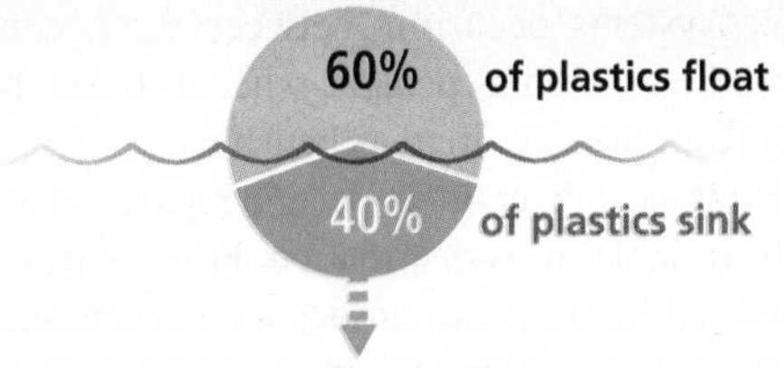

60% of plastics float
40% of plastics sink

Plastics degrade slowly in seawater. It can take decades, even centuries, before the combined actions of sunlight, waves, and biological activity breaks them down completely.

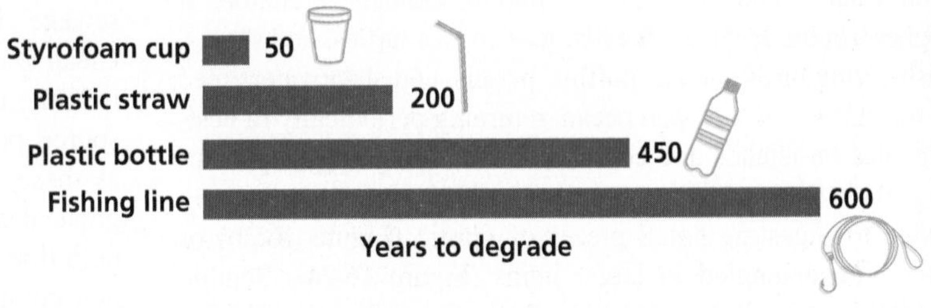

Styrofoam cup 50
Plastic straw 200
Plastic bottle 450
Fishing line 600
Years to degrade

Keeping plastics out of the ocean is the first step towards recovery:

Some U.S. states **charge a refundable deposit** on plastic drink bottles. This drastically increases recycling rates and keeps these plastics out of the oceans.

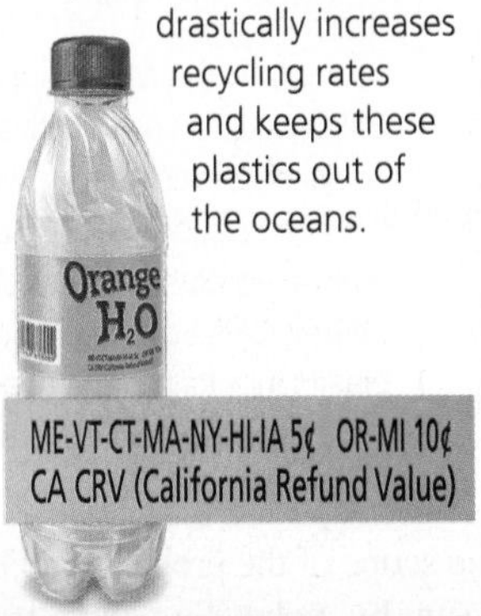

Orange H₂O
ME-VT-CT-MA-NY-HI-IA 5¢ OR-MI 10¢
CA CRV (California Refund Value)

We can **eliminate our use of single-use plastic products,** such as straws and cups.

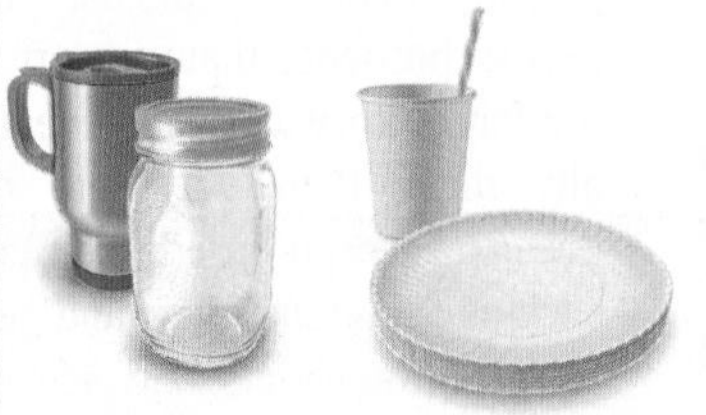

Switch to reusable containers

Fabricate single-use products out of paper, which readily degrades in seawater

We can **enact bans or taxes on plastic grocery bags,** which easily wash or blow into waterways.

Locations with bans or taxes on plastic bags, as of 2020

*

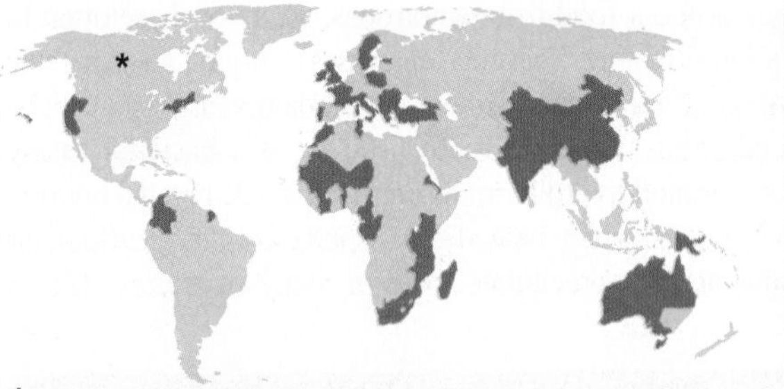

*Canada has announced plans to ban single-use plastics by the year 2021.

Trends suggest that marine plastic pollution will get worse before it gets better, but by embracing approaches to reuse, recycle, and reduce disposal of plastic items, we can help address plastic pollution in the oceans.

Data from Jambeck, J. R., et al., 2015. Plastic Waste Inputs from Land into the Ocean. Science 34: 768 -771. World Economic Forum and Ellen MacArthur Foundation, 2016. The New Plastics Economy: Rethinking the Future of Plastics. Andrady, A. L., 2011. Microplastics in the marine environment. Mar. Pollut. Bull. 62:1596 -1605. Woods Hole Sea Grant, 2018. United Nations Environment Programme, 2018. Single-Use Plastics: A Roadmap for Sustainability.

break down, and a million such bottles are purchased every minute in the world today. An estimated 500 million plastic straws are used daily in the United States, but a straw that finds its way into the ocean can float for 200 years before fully degrading. The plastic bag you use for mere minutes to carry home groceries from the store? It will remain in the ocean for two decades (see **DATAGRAPHIC**).

Up to 30% of the plastics entering the ocean are carried there by rivers—and it turns out that not all the world's rivers are equal in their plastic inputs to the ocean. A 2017 study concluded that 90% of the plastics entering the oceans came from only 10 river systems in East Asia, South Asia, and Africa (**FIGURE 16.15**). All these rivers have two things in common. First, each has large numbers of people living in their

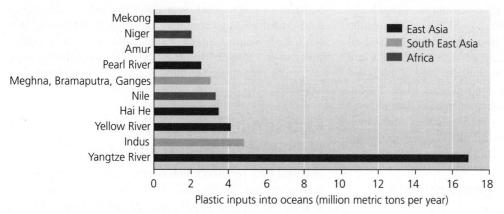

FIGURE 16.15 Ten river systems are responsible for 90% of the plastic inputs from waterways to oceans.

Data from Schmidt, C., et al., 2017. Export of Plastic Debris by Rivers into the Sea. Environmental Science and Technology 51:12246-12253.

watersheds, in some cases hundreds of millions of people. Second, cities and towns in the watersheds of these rivers have inadequate sanitation systems that promote plastic wastes entering waterways. Without coordinated garbage pickup and safe disposal in landfills, as is common in North American and European cities near large rivers that have large populations, wastes are often disposed of in open-air dumps, where they can more easily be washed or blown into nearby waterways. Efforts are already underway in some nations, such as China and India, to reduce plastic inputs in these waterways; bans on the use of plastic bags is one way. Given the tremendous inputs of plastics to the oceans from this handful of rivers, such initiatives could have far-reaching benefits.

Oil pollution comes from spills of all sizes

Oil pollution is another important source of marine pollution. Although large oil spills occur infrequently, their impacts can be staggering near the spill site. The world's attention was drawn to the danger that oil spills pose to fisheries, economies, and ecosystems in 1989 when the oil tanker *Exxon Valdez* ran aground in Prince William Sound, spilling hundreds of thousands of barrels of oil and causing an ecological disaster along the Alaskan coast.

Catastrophic oil pollution made headlines again in 2010 when British Petroleum's *Deepwater Horizon* offshore drilling platform exploded and sank into the Gulf of Mexico off the Louisiana coast (pp. 544–545). Oil gushed from the platform's underwater well, was spread widely by ocean currents, and washed up on coastal areas across the northern Gulf of Mexico. Hundreds of miles of water, sediments, and shoreline along the coasts of Louisiana, Mississippi, Alabama, and Florida were affected. Five years after the accident, populations of oysters, crabs, and sea turtles remained at low levels, and large numbers of marine mammals were beaching themselves on Gulf shores, suggesting that the impacts of this accident may continue for some time.

Given the severity of the *Deepwater Horizon* and other oil spills, it may be surprising to learn that about half of the petroleum entering the world's oceans in a given year originates from natural seeps in the ocean bottom. The impacts on organisms from these natural petroleum releases are less severe than the impacts from spills, however, because the see page from

ocean floor is more diffuse. The spillage that results from an accident involving an oil tanker or oil rig, in contrast, is concentrated in a relatively small area.

Much of the oil entering the ocean from sources other than seeps accumulates in waters from innumerable small and widely spread non-point sources. Shipping vessels and recreational boats can leak oil as they ply ocean waters. Motor oil from vehicles on roads and parking lots is washed into streams by rains and carried to the sea. Spills from oil tankers account for 12% of oil pollution in an average year, and 3% comes from leakage that occurs during the extraction of oil by offshore oil rigs. Although the *Exxon Valdez* spill was catastrophic, the good news is that the amount of oil spilled from tankers worldwide has decreased since 1995, in part because of an increased emphasis on spill prevention and response (**FIGURE 16.16**).

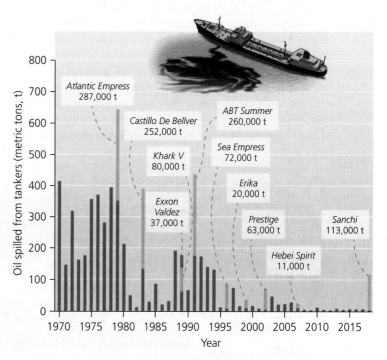

FIGURE 16.16 Less oil is being spilled into ocean waters today in large tanker spills than in the years prior to 1995.
The bar chart shows cumulative quantities of oil spilled worldwide from nonmilitary spills of more than 7 metric tons. Larger spills are identified by vessel name, and spill amounts from these events are indicated with orange or green bars. *Data from International Tanker Owners Pollution Federation Ltd.*

To help deal with the problem of catastrophic oil spills, the U.S. Oil Pollution Act of 1990 created a $1 billion oil pollution prevention and cleanup fund. It also required that by 2015 all oil tankers in U.S. waters be equipped with double hulls as a precaution against puncture. Further, in the wake of the *Deepwater Horizon* spill, the U.S. government enacted more stringent regulations on offshore drilling operations, but some of these regulations were relaxed by the Department of the Interior, the federal agency overseeing offshore oil operations, under President Donald Trump's administration.

Toxic pollutants can contaminate seafood

Aside from the harm that pollutants such as petroleum and plastic can do to marine life, toxic pollutants can make some fish and shellfish unsafe to eat. One prime concern today is mercury contamination. Mercury is a toxic heavy metal (p. 370) released into the environment from coal combustion (p. 535), mine tailings, and other sources. After settling onto land and water, mercury bioaccumulates in animal tissues and biomagnifies as it makes its way up the food chain (p. 373). As a result, fish and shellfish at high trophic levels (pp. 81–82) can contain dangerously elevated levels of mercury. Eating seafood high in mercury is particularly dangerous for young children and for pregnant or nursing mothers because of the neurological damage mercury poisoning causes.

Because seafood is an important part of a healthy diet, nutritionists do not advocate avoiding seafood entirely. However, people in at-risk groups should avoid fish high in mercury (such as swordfish, shark, and albacore tuna) while continuing to eat seafood low in mercury (such as catfish, salmon, and canned light tuna).

Humans are also not immune to the harmful effects of plastics pollution. A 2019 study used the number of microplastics detected in foods such as fish, shellfish, salt and sugar, and beverages such as water and beer, to calculate that the average human ingests approximately 50,000 tiny particles of plastic each year—and this analysis examined foods and drinks that comprise only 15% of the average person's daily caloric intake. People that drink only bottled water were estimated to ingest 130,000 microplastics from water alone, compared to 4000 from people who drank only tap water.

Excess nutrients cause eutrophication and algal blooms

Pollution from fertilizer runoff or other nutrient inputs can create low-oxygen "dead zones" in coastal ecosystems, as we saw with the Chesapeake Bay (Chapter 5) and the Gulf of Mexico (pp. 116–117). The release of excess nutrients into surface waters can spur unusually rapid growth of phytoplankton, causing eutrophication in freshwater and saltwater systems.

Excessive nutrient concentrations sometimes give rise to population explosions among species of microscopic marine algae called *dinoflagellates*. Some dinoflagellates produce powerful toxins that attack the nervous systems of vertebrates. Blooms of these organisms are known as **harmful algal blooms.** Some dinoflagellates produce reddish pigments that discolor surface waters, and blooms of these species are nicknamed **red tides** (**FIGURE 16.17**). Harmful algal blooms can cause illness and death among zooplankton, birds, fish, marine mammals, shellfish, and people as their toxins are passed up the food chain. They also cause economic loss for communities that rely on beach tourism and fishing. The Gulf Coasts of Florida and Alabama recently suffered an extended period of red tides, starting in late 2017 and persisting through 2018. Although red tides typically occur every year, they often don't last many months at a time as this outbreak did. Beaches along the Florida coast were littered with dead fish, manatees and sea turtles that were poisoned by toxins, people experienced respiratory problems from airborne toxins, and many seaside communities suffered economically during the profitable summer months as tourists stayed away.

(a) A red tide-inclucing dinoflagellate

(b) Red tide at a beach in Australia

(c) Fish kill in Florida from red tide in 2018

FIGURE 16.17 Red tides are a type of harmful algal bloom in which tiny dinoflagellate algae (a) produce a pigment that turns the water red (b) and release toxins that can cause fish kills (c) and respiratory problems in humans. Unusually persistent red tides have plagued the Gulf Coast of Florida, Alabama, and Mississippi in recent years.

Light and sound pollution can disorient marine organisms

Although we don't often think of light as a pollutant, the behavior of marine organisms can change when they are exposed to artificial light in places, at times, or in intensities to which they are not used. Many organisms use light as a cue to guide their behaviors, and artificial lights from coastal homes, businesses, and oil rigs have been found to alter the behavior of marine invertebrates, reptiles, fish, and birds.

Sea turtle hatchlings, for example, use cues from lights to orient toward the ocean after hatching from eggs on the beach, and lights near beaches can upset this orientation, leading to elevated mortality in hatchlings as they wander about on land instead of escaping to the sea. Because the darkness reduces their susceptibility to predators, zooplankton migrate upward in the water column to feed at night. If the upper water layers are illuminated by artificial light at night, this foraging behavior is disrupted and the zooplankton feed less, leading to reduced growth and increased mortality.

Reducing the degree of artificial light in coastal areas can lessen the effects of artificial light on marine creatures, but doing so is difficult given the extensive development of coastlines. But since recent research has shown the disrupting role that light pollution can play in marine ecosystems, the issue is now being considered in coastal planning decisions.

Water, like air, can transmit vibrations, and sound travels through oceans as it does on land. Some organisms, such as whales and dolphins, navigate in waters by *echolocation*—the process by which an organism emits sound waves and uses the echoes that reflect off objects to orient themselves. Loud underwater sounds, such as the strong sonar pulses used by the military to detect submarines, can damage the sensitive echolocation organs in these organisms and cause disorientation. Beaked whales, for example, are startled by sonar pulses and actively flee the source of the sound by diving into deeper waters without adequately engaging the physiological mechanisms they possess to prevent decompression sickness—an accumulation of nitrogen bubbles in the blood that occurs with exposure to the strong pressures of the deep ocean. The result is damage to their brains, organs, and central nervous systems, which can result is mass strandings of whales on beaches. Spain banned the use of naval sonar near its Canary Islands in 2004 after many mass strandings of whales occurred after naval activity in the area.

Climate change is altering ocean chemistry

Elevated levels of carbon dioxide in the atmosphere from fossil fuel combustion can pollute ocean water and change its chemical properties—much in the way excess plant nutrients or toxic substances change the chemical properties of seawater and affect marine organisms. The oceans absorb CO_2 from the atmosphere, as we first saw in our study of the carbon cycle (see Figure 5.17, p. 123). As our civilization pumps excess CO_2 into the atmosphere by burning fossil fuels for energy and removing carbon-sequestering vegetation from the land, the buildup of atmospheric CO_2 is causing the planet to grow warmer, setting in motion many changes and consequences (Chapter 18).

The oceans have soaked up roughly one-third of the excess CO_2 that we've added to the atmosphere, and this has slowed global climate change. However, there are two concerns. The first concern is that the ocean's surface water may soon become saturated with as much CO_2 as it can hold. Once it reaches this limit, climate change will accelerate because the oceans will no longer remove large amounts of carbon dioxide from the atmosphere.

The second concern is that as ocean water soaks up CO_2, it becomes more acidic. As **ocean acidification** proceeds, sea creatures such as corals, snails, and mussels have difficulty forming calcium carbonate shells (or "skeletons," in coral) because the carbonate ions (CO_3^{2-}) they need to create them become less available with increasing acidity. Elevated acidity levels can even cause the shells of sea creatures to begin dissolving.

Chemistry tests in the lab show that coral skeletons, for example, begin to erode faster than they are built once the carbonate ion concentration falls below 200 micromoles/kg of seawater. Researchers studying coral reefs in the field are finding the same thing: Reefs are growing only in waters with greater than 200 micromoles/kg of carbonate ion availability. A 2007 study used historical data and computer simulations to model how the viability of coral reefs around the world would vary at differing levels of atmospheric CO_2 and, hence, differing levels of ocean acidity. The results predicted that as atmospheric CO_2 levels rise, coral reefs will shrink in distribution, diversity, and density (**FIGURE 16.18**, p. 438). By the time atmospheric CO_2 levels reach 500 ppm, little area of ocean will be left with conditions to support coral reefs (the shrinking areas shown in blue in the figure).

Because of this threat to coral reefs, scientists are intensifying the study of coral responses to warmer and acidified ocean waters to better inform efforts to conserve ecologically important reef habitats. Research published in 2015, for example, revealed that individual coral colonies vary greatly in their responses to heat and acidity, suggesting that some reefs may be more resilient than others to changes in water temperature and pH. In another study, scientists found that an endangered Caribbean coral was able to sustain growth under stressful conditions by increasing its feeding rate, indicating that some reefs may be better able to persist under increasingly stressful conditions as long as food is plentiful. Various initiatives are currently underway to selectively breed strains of coral that tolerate warmer, more acidic waters and to seed existing reefs with these "hearty" individuals to maintain coral populations on threatened reefs.

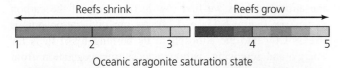

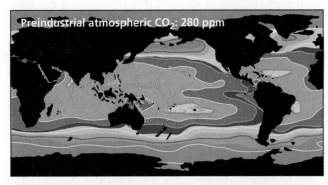

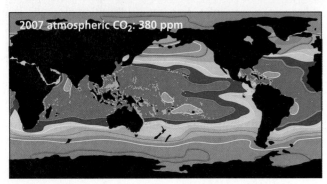

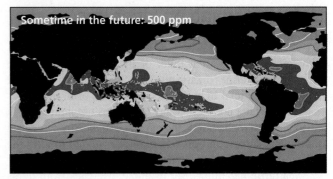

FIGURE 16.18 Increased atmospheric carbon dioxide levels have decreased the number of ocean areas that support coral reefs. Today's shallow-water reef locations (pink dots) are shown superimposed on maps of oceanic *aragonite saturation*—a measure of the concentration in seawater of the carbonate and calcium ions corals need to build their skeletons—at three levels of atmospheric CO_2. Most of Earth's tropical and subtropical oceans were suitable for the growth of coral reefs (blue; **top left panel**) before people began emitting sizable amounts of CO_2 into the atmosphere during the industrial revolution, but in 2007, when atmospheric CO_2 concentrations were 380 ppm **(bottom left panel),** fewer regions were found to be suitable. When the planet reaches 500 ppm **(top right panel),** very few areas of the ocean will have conditions suitable for sustaining coral reefs. *Adapted from Hoegh-Guldberg, O., et al., 2007. Coral reefs under rapid climate change and ocean acidification. Science 318: 1737–1742. Fig 4. Reprinted with permission from AAAS.*

Emptying the Oceans

As severe as the impacts of oceanic pollution on marine organisms can be, most scientists concur that the more worrisome dilemma is the overharvesting of fish to meet human demands for food. Sadly, the old cliché that "there are always more fish in the sea" may not prove to be true. The oceans of today are overfished, and many stocks of wild fish have been severely depleted due to centuries of extensive harvesting.

Worldwide demand for seafood varies greatly by region and in some cases within regions (**FIGURE 16.19**). The average person in North America, Asia, Oceania, and Europe eats moderate to extensive amounts of seafood, but consumption levels in Latin America and Africa are far lower.

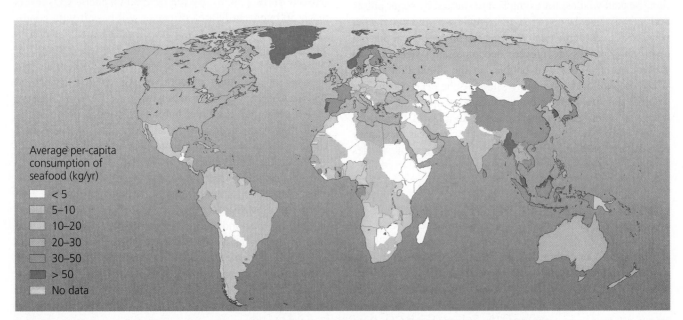

FIGURE 16.19 Societies around the world consume differing amounts of seafood. Cultural factors, affluence, and the availability of seafood are some of the factors that affect consumption rates across nations. *Source: Food and Agriculture Organization of the United Nations, 2018. The state of world fisheries and aquaculture 2018.*

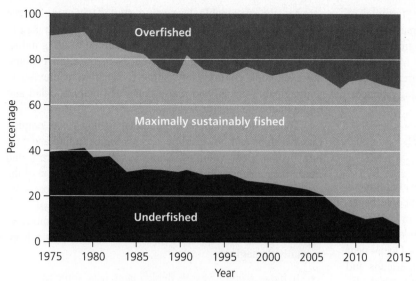

FIGURE 16.20 Global fish stocks that are overfished or maximally harvested increased from 1975 to 2015. The percentage of underfished stocks also fell substantially over the same period. *Source: Food and Agriculture Organization of the United Nations, 2018.* The state of world fisheries and aquaculture 2018.

Within regions, note that the two population powerhouses in Asia—China and India—consume very different amounts of seafood despite both nations having extensive lengths of ocean shoreline.

The oceans and their biological resources have met human needs for thousands of years, but today we are placing unprecedented pressure on marine resources. Around 60% of the world's wild marine fisheries are fully exploited (**FIGURE 16.20**), meaning that we cannot harvest them more intensively without depleting them, according to the United Nations Food and Agriculture Organization (FAO). More than 30% of wild marine fish populations are already overexploited and being driven toward extinction. Only less than one-tenth of the world's wild marine fish populations can yield more than they are already yielding without being driven into decline.

Total global wild fisheries catch, after decades of increases, leveled off after about 1988 (**FIGURE 16.21**) despite increased fishing effort. In 2008, the FAO concluded that "the maximum wild capture fisheries potential from the world's oceans has probably been reached." A comprehensive 2006 study in the journal *Science* predicted that if current trends continue, wild populations of *all* ocean species that we fish for today will collapse by 2048. If these fisheries collapse as predicted, we will lose not only an important food source for people, but also the ecosystem services that healthy fisheries provide. Productivity will be reduced, ecosystems will become more sensitive to disturbance, and the filtering of water by vegetation and organisms (such as oysters) will decline, making harmful algal blooms, dead zones, fish kills, and beach closures more common.

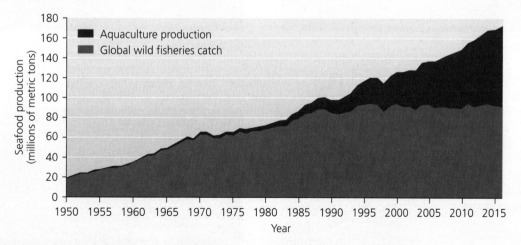

FIGURE 16.21 After rising for decades, world fisheries capture production has stalled. Many scientists fear that a global decline is imminent if conservation measures are not taken. Note how the growth in aquaculture has enabled seafood production to increase, despite flat hauls from capture fisheries in recent decades. *Data from the Food and Agriculture Organization of the United Nations, 2018.* The state of world fisheries and aquaculture 2018.

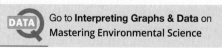

Go to **Interpreting Graphs & Data** on **Mastering Environmental Science**

Note from Figure 16.21 that aquaculture, the "farming" of seafood for human consumption, has grown tremendously in recent decades. We will examine this topic in greater detail shortly, but first we need to examine captures of wild fish because to prevent a collapse of the world's fisheries, it is vital that we turn immediately to more sustainable fishing practices.

Industrialized fishing facilitates overharvesting

It is tempting to blame Earth's problems exclusively on modern civilization, but scientists are learning that people began depleting some marine populations centuries or millennia ago. In 1768, for example, Steller's sea cow, a relative of the manatee, which was once abundant in the Pacific Ocean near Alaska, was hunted to extinction by Native Americans and European whalers for its meat, oil, and blubber. Such depletions of marine populations tended to be localized, however, as people long ago lacked the technology to intensively harvest organisms and exploit marine creatures and fisheries, especially those in the open ocean far from shore.

Mostly sustainable harvesting started changing in the 19th century, when many species of whales were nearly hunted to extinction as part of the commercial whaling industry. Continued technological advances in the 20th century, however, took commercial fishing to what were previously unimaginable levels, and many fisheries finally buckled and broke under the intense pressure.

Today's industrialized commercial fishing fleets employ massive ships and powerful new technologies to find and capture fish in great volumes. Some vessels even process and freeze their catches while at sea. And because no portion of the ocean is unreachable by modern fishing fleets, our impacts are much more rapid and intensive than they were in the past.

The modern fishing industry uses several methods to capture fish at sea that are highly efficient but also environmentally damaging:

- In *purse seining*, vessels deploy large nets, some as long as 1 km (0.6 mi), around schools of fish near the surface (**FIGURE 16.22a**). The nets are suspended in the upper water column by floating buoys on the net top and weights on the net bottom. Once the school of fish is

(a) Purse seining

(b) Driftnetting

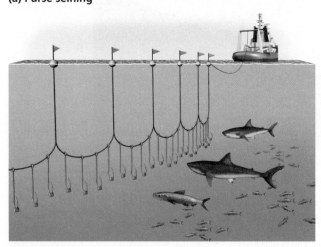

(c) Longlining

(d) Bottom-trawling

FIGURE 16.22 Commercial fishing fleets use several methods of capture. These illustrations are schematic for clarity and do not portray the immense scale that these technologies can attain; for instance, industrial trawling nets can be large enough to engulf multiple jetliners.

encircled, the net is drawn shut like a laundry bag (or the hood on a sweatshirt) by the purse line, a rope running through the top of the net.

- Some vessels set out long driftnets that span large expanses of water (**FIGURE 16.22b**). Chains of these transparent nylon mesh nets are arrayed to drift with currents to capture passing fish, and they are held vertical by floats at the top and weights at the bottom. *Driftnetting* usually targets species that traverse open water in immense schools, such as herring, sardines, and mackerel. Specialized forms of driftnetting are used for sharks, shrimp, and other animals.

- *Longlining* (**FIGURE 16.22c**) involves setting out extremely long fishing lines, up to 80 km (50 mi) long, with up to several thousand baited hooks spaced along their lengths. Tuna and swordfish are among the species targeted by longline fishing.

- *Trawling* entails dragging immense cone-shaped nets through the water, with weights at the bottom and floats at the top. Trawling in open water captures pelagic fish, whereas *bottom-trawling* (**FIGURE 16.22d**) involves dragging weighted nets across the floor of the continental shelf to catch groundfish and other benthic organisms, such as scallops.

Discarded nylon nets, ropes, and lines from fishing vessels have been found to contribute greatly to the mass of plastics choking our oceans. A 2018 study found, for example, that fishing nets comprised *nearly half* of the mass of plastics collected during trawls in the Great Pacific Garbage Patch (see **THE SCIENCE BEHIND THE STORY**, p. 442).

Wild fisheries collapse quickly under intensive harvest

Throughout the world's oceans, today's industrialized fishing fleets are depleting wild marine populations with astonishing speed. In a 2003 study, Canadian fisheries biologists Ransom Myers and Boris Worm analyzed data from FAO archives and found the same pattern for region after region: In just a decade after the arrival of industrialized fishing, catch rates dropped precipitously, with 90% of large-bodied fish and sharks eliminated. Populations then stabilized at 10% of their former levels. Myers and Worm concluded that the oceans today contain only one-tenth of the large-bodied animals they once did.

An example of the consequences of overfishing can be seen in the Atlantic cod fishery. Cod fishing has been the economic engine for hundreds of communities in coastal New England and eastern Canada since European colonization. The Atlantic cod (*Gadus morhua*) is a type of groundfish, a name given to fish that live or feed on the bottom. Atlantic cod inhabit cool ocean waters on both sides of the North Atlantic, and two discrete populations, or *stocks*, are found in North America. One inhabits the Grand Banks off Newfoundland, and another lives on Georges Bank off Massachusetts (**FIGURE 16.23**).

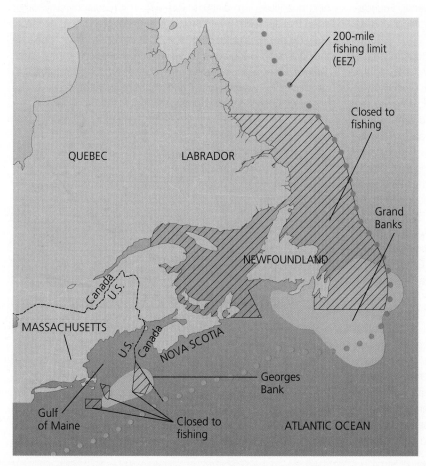

FIGURE 16.23 Populations of Atlantic cod inhabit areas of the northwestern Atlantic Ocean, including the Grand Banks and Georges Bank, regions of shallow water that are especially productive for groundfish. Portions of these and other areas have been closed to fishing because cod populations have collapsed after being overfished.

THE SCIENCE behind the story

Inventorying the Great Pacific Garbage Patch

Julia Reisser, The Ocean Cleanup Foundation

Captain Charles Moore was the first person to visually verify the existence of the Great Pacific Garbage Patch and provide some general information about the nature of the debris and microplastics floating in the calm waters inside the North Pacific Gyre. But although these reports were informative, they did not provide a comprehensive, detailed inventory of the area—Moore's research vessel was simply not outfitted to conduct such rigorous research.

Our knowledge of how plastics behave in the open ocean—where they are subjected to ultraviolet light, heat fluctuations, wave action, and biological activity by microorganisms—is poorly understood, so it wasn't known if plastics from land-based sources, such as beverage containers or plastic bags, were abundant in the GPGP or if the debris was largely from marine sources, such as discarded nets and lines from fishing vessels. In 2015, researchers from the Ocean Cleanup Foundation, led by lead oceanographer Julia Reisser, stepped in to answer these questions.

Reisser's team surveyed the GPGP by conducting more than 650 trawls with 18 research vessels and by surveying the GPGP from the air with high-resolution cameras (taking 7000 images). They inventoried the many items collected from the trawls and detected by aerial surveys and worked with Laurent Lebreton—a fellow researcher at the Ocean Cleanup Foundation—to process this data to generate a model (p. 120) explaining the nature of the debris in the GPGP, changes in debris concentrations over time, and the relative size of the GPGP over time. This study was unique because it was funded, in part, by a crowdsourcing initiative by the Ocean Cleanup Foundation.

Their findings, published in 2018 in the journal *Nature Scientific Reports,* contained a great many surprises that helped scientists better understand the nature of plastics in the GPGP and the fate of plastics in the open ocean. One surprise was the model's estimated mass of floating plastics in the patch—79,000 metric tons, a value four to sixteen times higher than other estimates. The researchers also found that the GPGP was accumulating plastics at an exponential rate far higher than that of waters outside the GPGP, although this conclusion was based on a relatively small number of surveys within the GPGP prior to 2005 (**FIGURE 1**).

Another unexpected finding was that more than 75% of the total mass of debris in the GPGP was from large pieces of debris, objects larger than 5 cm (2 in.) in size. Nearly half (46%) of the mass of collected objects was discarded fishing nets,

called *ghost nets,* far exceeding the team's expectation that these would constitute only 20% of the debris mass.

Microplastics made up only 8% of the mass of the debris collected, but represented 94% of the 1.8 trillion pieces of debris estimated to be floating in the GPGP. So, by mass, fishing gear was the most notable *type* of debris in the patch, but microplastics were the most commonly *found* type of debris. The patch therefore contained sizable quantities of large items that could entangle swimming organisms and huge numbers of tiny plastics that can be swallowed by sea creatures and can release toxins into the surrounding seawater.

The study found plastic products from 12 different nations, with roughly 30% of the plastics coming from Japan and another 30% from China. Most of the plastics were produced since 2000, but the team found 17 items produced in the 1990s, 7 from the 1980s, and one object from 1977—an item that had been floating at sea for nearly 40 years and still had a recognizable production date. An estimated 10% to 20% of the material in the GPGP was attributed to the tsunami that struck northern Japan in 2011 and washed some 4.5 million metric tons of debris into the northern Pacific Ocean.

The study thus shed some light on the general characteristics of the debris in the GPGP, aiding our knowledge of how plastics degrade in the open ocean. Although rigid plastics were found in high quantities, film plastics were not. The researchers hypothesized that such films, with their high surface area to volume ratios, likely do not persist for long periods in the open ocean because they are rapidly degraded by waves and biological activity. Such material, they concluded, is far more likely to be returned to coastlines by waves, tides, and winds right after it enters the ocean from land-based sources.

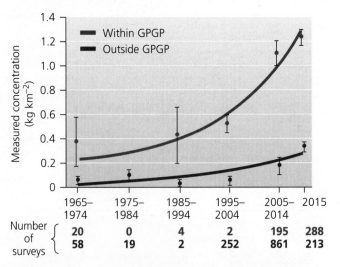

Number of surveys	1965–1974	1975–1984	1985–1994	1995–2004	2005–2014	2015
	20	0	4	2	195	288
	58	19	2	252	861	213

FIGURE 1 **The density of plastic debris is increasing faster inside the Great Pacific Garbage Patch than in surrounding waters.** *Source: Lebreton, L., et al., 2018. Evidence that the Great Pacific Garbage Patch is rapidly accumulating plastic. Nature Scientific Reports 8: 4666.*

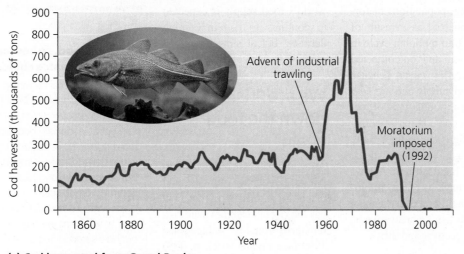

(a) Cod harvested from Grand Banks

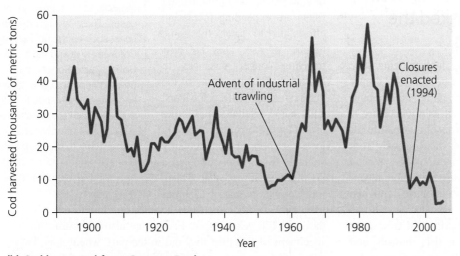

(b) Cod harvested from Georges Bank

FIGURE 16.24 In the North Atlantic off the coast of Newfoundland, commercial catches of Atlantic cod have fluctuated over time. Catches from the Grand Banks increased with intensified fishing by industrial trawlers in the 1960s and 1970s **(a)**, but the fishery subsequently crashed, and moratoria imposed in 1992 and 2003 have not brought it back. A similar pattern is seen in the cod catches on Georges Bank **(b)**; industrial fishing produced 30 years of high catches, followed by a collapse and the closure of some areas to fishing. *Data from (a) Millennium Ecosystem Assessment, 2005. Ecosystems and human well-being: Biodiversity synthesis. Washington, D.C.: World Resources Institute. Used with permission; and (b) O'Brien, L., et al., 2008. An assessment of 19 Northeast groundfish stocks through 2007. Woods Hole, MA: Northeast Fisheries Science Center.*

For centuries, the Grand Banks provided ample fish. With advancing technology, however, ships became larger and more effective at finding fish. By the 1960s, massive industrial trawlers (the majority from Europe) were vacuuming up unprecedented numbers of groundfish off the Canadian coast. In 1977, Canada exercised its legal right to the waters extending 200 nautical miles from shore, kicked out foreign fishing fleets, and claimed most of the Grand Banks for itself. The Canadian fleet then developed the same industrial technologies and revved up its fishing industry like never before.

Then came the crash. Catches dwindled in the 1970s because too many fish had been harvested and because bottom-trawling had destroyed huge expanses of the cod's underwater habitat (**FIGURE 16.24a**). By 1992, the situation was dire: Scientists reported that mature cod were at just 10% of their long-term abundance, and the Canadian government instituted a two-year ban on fishing in the area. People affected by the moratorium on fishing were offered compensation and job training for new skills and were given incentives for early retirement.

Cod stocks on the Grand Banks did not rebound by 1994, so the Canadian government extended the moratorium, enacted bans on all other major cod fisheries, and

scrambled to offer more compensation to displaced fishers. In 2009, after data showed the stock was recovering slightly, a portion of the Grand Banks off the southeastern coast of Newfoundland was reopened to cod fishing. Fishing continues today in this limited area, and researchers and resource managers are monitoring populations closely to see how they fare over time.

Across the border in U.S. waters, cod stocks had similarly collapsed in the Gulf of Maine and on Georges Bank (**FIGURE 16.24b**). In 1994, the National Marine Fisheries Service (NMFS) closed three prime fishing areas on Georges Bank. Over the next several years, the NMFS designed a number of regulations meant to protect and restore the fishery, but these steps were too little, too late. A 2005 report revealed that the cod were not recovering, and, with cod stocks dangerously low, further restrictions were enacted.

There has been some good news in the cod fisheries, however, since the bans initiated in 1992 (Canada) and 1994 (United States). Populations of cod in Canadian waters are rebounding north of Newfoundland, and populations on the Grand Banks are slowly recovering. Research on cod populations reveals only part of the story, however. When cod were driven to low numbers by overharvesting, populations of forage fish, such as capelin, increased ninefold as a result of reduced cod predation. This change put into motion a cycle in which these forage fish then preyed on and outcompeted young cod, further slowing cod recovery. However, because they have outstripped their plankton food supply and are now themselves being harvested by fishermen, populations of forage fish are now in decline, which gives cod a better opportunity to rebound. And there are hopeful signs on Georges Bank as well: Seafloor invertebrates have begun to recover in the absence of trawling, spawning stock of haddock and yellowtail flounder has risen, and sea scallops have increased in biomass 14-fold. These modest but encouraging recoveries show us that protecting oceans and creating no-fishing areas give us hope that depleted fisheries can one day recover and thrive.

The growth of aquaculture—as well as other factors—has masked the collapse of large fisheries

Most people have not noticed the massive damage being done to the world's wild commercial fisheries because the supply of fish and seafood at the grocery store has remained abundant. That is due in large part to the growth in **aquaculture**—the raising of fish, seafood, and other forms of marine life for human consumption (pp. 250–251). The rapid growth from aquaculture (see Figure 16.21, p. 439) since 1990 has enabled the global production of seafood to increase even though captures of wild fish and other marine species have remained stable for decades.

Aquaculture is performed in both marine and freshwater environments by raising animals (such as fish, mussels, and shrimp) and marine plants (such as edible seaweed) in enclosures within bodies of water or by growing them in specialized facilities in giant tanks. The largest production of food by aquaculture, by weight, is for freshwater fish (inland aquaculture), followed by aquatic plants and mollusks (oysters, mussels, and scallops) (**FIGURE 16.25**). Aquaculture's contribution to the

world's food supply is expected to continue to rise in coming decades, but at slower rates of increase than recently seen.

We are "fishing down the food chain"

Aquaculture has boosted supplies of aquatic animals and plants for human food, but how has the overall catch of wild fish remained roughly stable since 1990 (see Figure 16.21, p. 439) despite the wide-scale depletion of wild fish stocks? First, fishing fleets are expending increasing effort just to catch the same number of fish. They are now traveling longer distances to reach less-fished portions of the ocean, fishing in deeper waters, and using satellites and other technologies to more effectively locate schools of commercially valuable fish. For example, a 2010 study showed that British trawlers were working 17 times harder just to catch the same number of cod and other fish as they had 120 years before. Fishers are also harvesting smaller specimens of fish than they did in the past, when older, larger fish were more abundant.

Second, as desirable species become too rare to fish profitably, fleets begin targeting other, less desirable species that are more abundant. Time and again, fleets have depleted popular food fish (such as cod) and shifted to species of lower value (such as capelin, a small forage fish in the smelt family that

FAQ

I love seafood, so how can I make sustainable choices?

To most of us, marine fishing practices may seem a distant phenomenon over which we have no control, especially because more than 80% of seafood sold in the United States is imported. But although we don't have control over the seafood offered in markets, we have full control over which items we buy. Finding out how seafood items were caught is difficult because this information is not made readily available to consumers in most cases. Thus, several organizations, such as the Environmental Defense Fund and the Marine Conservation Society, have devised concise online guides and smartphone apps to help consumers differentiate fish and shellfish that are overfished or whose capture is ecologically damaging from those that are harvested more sustainably.

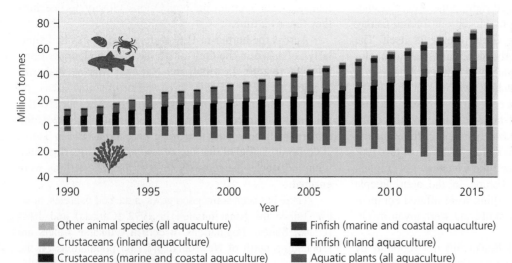

FIGURE 16.25 The production of seafood from aquaculture has risen dramatically since 1990. Both freshwater and marine creatures are raised for human consumption, as are aquatic plants. *Data from the Food and Agriculture Organization of the United Nations, 2018.* The state of world fisheries and aquaculture 2018.

Other animal species (all aquaculture)

Crustaceans (inland aquaculture)

Crustaceans (marine and coastal aquaculture)

Mollusks (all aquaculture)

Finfish (marine and coastal aquaculture)

Finfish (inland aquaculture)

Aquatic plants (all aquaculture)

Saving Dolphins with "Dolphin-Safe" Labels on Tuna

The scale of bycatch and solutions to address it are illustrated by the story of dolphins and tuna. For reasons that are not fully understood, pods of dolphins in the Pacific often swim above large schools of tuna, and fishermen began using dolphins as indicators of large schools of tuna. To catch the tuna, fishing vessels would encircled the tuna—and the dolphins in the water above—with a purse seine net and draw the net in, trapping both tuna and dolphins. Hundreds of thousands of dolphins were needlessly killed each year throughout the 1960s using this method. Starting in the late 1960s, international pressure on tuna fisheries began to mount, however, due in part to the popularity of the 1960s television show *Flipper*—starring a lovable bottlenose dolphin. Congress responded to this public pressure with the passage of the U.S. Marine Mammal Protection Act of 1972, which forced U.S. fishing fleets to modify their gear and fishing practices to allow dolphins to escape from fishing nets. Dolphin bycatch dropped greatly as a result.

However, as other nations' ships began catching more tuna in the Pacific, dolphin bycatch rose again. Because U.S. fleets were suffering a competitive disadvantage because they were operating under more restrictions than fleets from other nations, the U.S. government required that tuna imported from foreign fleets also minimize dolphin bycatch, and it supported ecolabeling efforts (p. 154) to label tuna as "dolphin-safe" if its capture used methods designed to avoid bycatch. These measures helped reduce dolphin deaths from 133,000 in 1986 to fewer than 1500 per year since 1998.

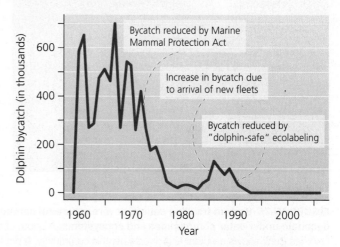

Bycatch has severely decreased dolphin populations.
Bycatch of dolphins by tuna fleets declined first as a result of regulations following the 1972 U.S. Marine Mammal Protection Act and later when "dolphin-safe" ecolabeling encouraged fleets to adopt methods that reduced bycatch. *Data from National Oceanic and Atmospheric Administration, www.noaa.gov.*

Tuna harvests still result in bycatch of dolphins and other non-target species, but this greater than 99% reduction in dolphin mortality from tuna bycatch can rightfully be celebrated as a success in preserving marine biodiversity.

→ **Explore the Data** at **Mastering Environmental Science**

---

feeds largely on plankton). Because this approach often entails catching smaller species that feed at lower trophic levels, this phenomenon has been termed "fishing down the food chain."

Some of these species were undesirable ones that fishermen formerly threw back when fishing for more marketable species but that underwent "image makeovers" to aid their sale to consumers. For example, the species of fish now called "orange roughy" was called "slimehead" by fishermen because of its unique mucous canals. Similarly, the toad-colored "toothfish" that fishermen once threw overboard has found new life as "Chilean sea bass," even though the species is not biologically classified as a sea bass. With intensive harvest, populations of orange roughy and Chilean sea bass are now overfished in many areas, forcing harvesters to "fish *even further* down the food chain."

Industrialized fishing kills nontarget animals and damages ecosystems

Modern fishing practices haul in more than just the species they target. *Bycatch,* the accidental capture of animals while fishing, accounts for the deaths of millions of fish, sharks, marine mammals, and birds each year. The impact of bycatch can be substantial. A 2016 report from NOAA reported that around 11% of all fish harvested in U.S. fisheries were captured unintentionally.

Purse seining and driftnetting capture dolphins, seals, and sea turtles, as well as countless nontargeted fish. Most of these creatures end up drowning (mammals and turtles need to surface to breathe) or dying from air exposure on deck (fish suffocate when kept out of the water). The vaquita porpoise, which is found only in the northern Gulf of California in Mexican waters, is being driven to extinction by driftnetting. Vaquita are captured as bycatch in driftnets and as of 2019, only about 30 individuals remain, down from a population of roughly 600 in 1997.

Driftnetting is now banned in international waters (by a United Nations resolution passed in 1992) because of excessive bycatch, but the practice continues in many national waters. Purse seining has been outlawed in the territorial waters of some nations, but continues to be used in international waters, particularly to catch tuna (see **SUCCESS STORY**).

Similar bycatch problems exist with longline fishing, which kills turtles, sharks, and seabirds like albatrosses—magnificent seabirds with wingspans up to 3.6 m (12 ft). Several methods are being developed to limit bycatch from longline fishing (such as using flagging to scare birds away from the lines), but an

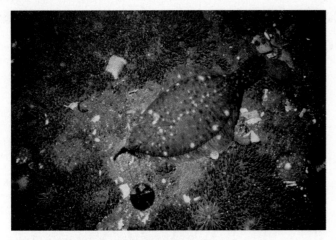

(a) Before trawling at Georges Bank

(b) After trawling at Georges Bank

FIGURE 16.26 Bottom trawling causes severe structural damage to reefs and benthic habitats, and it can decimate underwater communities and ecosystems. A photo of an untrawled location **(a)** on the seafloor of Georges Bank shows a vibrant and diverse benthic community. A photo of the same site after trawling **(b)** shows a flattened expanse of sea bottom with only scarce biological diversity and productivity.

estimated 300,000 seabirds of various species die each year when they become caught on hooks while trying to ingest bait.

Bottom-trawling not only results in bycatch but also can destroy entire ecosystems. The weighted nets crush organisms in their path and leave long swaths of damaged sea bottom. Bottom-trawling is especially destructive to structurally complex areas, such as reefs, that provide shelter and habitat for animals. In recent years, underwater photography has begun to reveal the extent of structural and ecological disturbance done by bottom-trawling (**FIGURE 16.26**). Bottom-trawling is often likened to clear-cutting (pp. 320–321) and strip mining (p. 649). In heavily fished areas, the bottom may be damaged multiple times. On Georges Bank, it is estimated that the average expanse of ocean floor has been trawled three times. Bottom-trawling here is known to destroy young cod as bycatch and is thought to be a main reason the Georges Bank cod stock is not recovering.

Marine Conservation

Because we bear responsibility and stand to lose a great deal if valuable marine ecological systems collapse, scientists have been working to develop solutions to the problems that threaten the oceans. Many have begun by reconsidering the strategies used traditionally in fisheries management.

Loss of marine biodiversity erodes ecosystem services

Overfishing, pollution, habitat change, and other factors that deplete biodiversity can threaten the ecosystem services we derive from the oceans. In the 2006 study that predicted a global fisheries collapse by 2048 (p. 439), the study's authors analyzed all existing scientific literature to summarize the effects of biodiversity loss on ecosystem function and ecosystem services. They found that across 32 different controlled experiments conducted by various researchers, systems with reduced species diversity or

genetic diversity showed less primary and secondary production and were less able to withstand disturbance than areas without such reduced diversity.

The team also found that when biodiversity was reduced, habitats that serve as nurseries for fish and shellfish were also affected, threatening future populations of these species. Moreover, when biodiversity was lower, so was the amount of water purification by filter-feeding organisms (such as oysters), which clean waters of excess nutrients, sediments, and phytoplankton that can lead to eutrophication, harmful algal blooms, dead zones, fish kills, and beach closures.

We can protect unique and biodiverse areas in the ocean

Despite efforts to restrict fish harvests to sustainable levels, many fish and shellfish stocks around the world have crashed. Thus, many scientists and resource managers believe that it is time to shift the focus away from individual species and toward viewing marine resources as elements of larger ecological systems. This perspective means considering the impacts of fishing practices on habitat quality, species interactions, and other factors that may have indirect or long-term effects on populations. One key aspect of such ecosystem-based management (p. 318) is to set aside areas of ocean where systems can function without human interference. Although these areas cannot be protected from impacts such as pollution from floating microplastics, they can successfully limit damage from other human activities.

Hundreds of **marine protected areas** (MPAs) have been established, most of them along the coastlines of developed

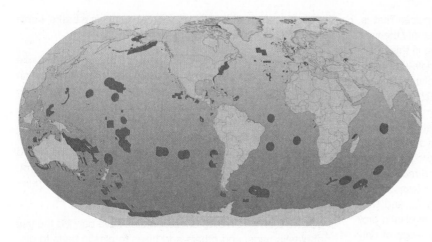

FIGURE 16.27 Marine protected areas are found across the globe. More than 7% of the world's oceans are afforded some sort of protection (highlighted in purple), but critics contend that for many, these protections do little to prevent biodiversity loss, habitat destruction, and ecosystem degradation. *Source: IUCN and UNEP-WCMC (April 2019). The World Database on Protected Areas (WDPA); available at: www.protectedplanet.net.*

countries. Much like our national parks, these areas restrict some human activities within their boundaries (such as oil drilling), but nearly all MPAs allow fishing or other extractive activities. As one report from the National Resources Defense Council put it, MPAs "are dredged, trawled, mowed for kelp, crisscrossed with oil pipelines and fiber-optic cables, and swept through with fishing nets." By 2019, 7.6% of the world's oceans had been designated as MPAs (**FIGURE 16.27**).

Because of the lack of true refuges from fishing pressure, many scientists want to establish areas where fishing is prohibited. Such no-take areas have come to be called **marine reserves.** Designed to preserve ecosystems intact, marine reserves are also intended to improve fisheries. Scientists argue that marine reserves can act as production factories for fish for surrounding areas, because fish larvae produced inside reserves will disperse outside and stock other parts of the ocean. Data from marine reserves around the world have indicated that reserves can work as win-win solutions that benefit ecosystems, fish populations, and fishing economies. A comprehensive review of data from marine reserves in 2001 revealed that just one to two years after their establishment, marine reserves, on average, increased species diversity by 23%, the density of organisms within the reserve by 91%, the total biomass of organisms by 192%, and the average size of each organism by 31%.

That same year, 161 prominent marine scientists signed a "consensus statement" summarizing the effects of marine reserves. Besides boosting fish biomass, total catch, and record-sized fish, the report stated, marine reserves yield several benefits. Within reserve boundaries, for example, mortality and habitat destruction are reduced; the chance of biodiversity loss is lessened; and the reserve produces rapid and long-term increases in the abundance, diversity, and productivity of marine organisms. Outside the reserve, benefits are seen when individuals of protected species spread outside reserves into nearby waters and when larvae of species protected within reserves "seed the seas" outside reserves. If marine reserves can be made to work and to be accepted, they may well seed the seas and help lead us toward solutions to one of our most pressing environmental problems.

CENTRAL CASE STUDY
connect & continue

TODAY, the plastics that are ubiquitous in our modern lifestyle continued to be produced in ever-growing quantities each year, and global demand for plastics is only predicted to increase as human populations grow and more people move to cities and adopt high-consumption urban lifestyles.

Efforts to mechanically remove plastics that have concentrated in ocean waters are showing promise. One of the first initiatives—the Ocean Cleanup, a nonprofit foundation headed by Dutch inventor Boyan Slat (the same organization that surveyed the GPGP)—raised $40 million and deployed a floating plastics collection system in the Great Pacific Garbage Patch in 2018. This initial prototype was unsuccessful though, as it failed to contain the smaller plastics it captured and was heavily damaged by the constant pounding of wind and waves on the open ocean. But an updated prototype—redesigned to resist wave and wind damage and to properly contain the small plastics it captured—was successfully operating in the GPGP in 2019. Although this project and others like it are encouraging, we are likely many years away from deploying large-scale mechanical solutions that will clean the plastics from our oceans—highlighting the continued need to prevent plastics from entering marine environments in the first place. That is, we've got to "turn off the spigot" before we can "empty the tub."

Some governments are aggressively moving in this direction by banning the use of plastics in many single-use products, such as plastic bags, straws, cups, and plates. Canada and the European Union are phasing out by 2021 the use of plastics in single-use products for which alternative materials—usually paper—are available. For other plastic-containing products for which alternatives are not readily available—such as beverage containers and food packaging—programs have been created to attain high rates of recycling and reduce over time the percentage of those products that is plastic. Many nations, as well as the states of Hawai'i, California, and New York, have banned or greatly restricted the use of plastic grocery bags. Such bags, which have an average usage of only 12 minutes, can easily be carried into oceans by wind or water, but can take hundreds of years to degrade in the ocean.

Plastics are not the only danger faced by marine environments, however. Pollution, ocean acidification, and destructive fishing practices degrade coastal and open-ocean ecosystems, and climate change is altering the temperature and chemistry of the world's oceans. The good news is that we have solutions that can save these critical ecosystems; we just need the collective will to see them implemented and maintained. The challenge we face in saving our oceans, and the species that occupy them, is great—but it's a battle worth fighting to protect 71% of our planet.

- **CASE STUDY SOLUTIONS** You are the mayor of a seaside city in the United States that is considering enacting a ban on plastic bags to reduce the level of plastics entering the ocean. You know that other U.S. cities—such as Seattle, Chicago, and Boston—have enacted such bans, but you are fearful of the reaction of some of your constituents. Who in your city do you think would object to such a ban? What groups would likely support it? Why? Provide three strong points that you would use to try to convince skeptical constituents and businesses that banning plastic bags is a worthwhile idea.

- **LOCAL CONNECTIONS** Six U.S. states have banned the use of plastic bags, and others add taxes to plastic bags to discourage their use or actively support programs that promote the recycling and reuse of plastic bags. On the other end of the spectrum, 10 states have passed legislation that prevents cities within those states from enacting local restrictions on the use of plastic bags. Where does your state stand on the issue? Where do you think your state should stand?

- **EXPLORE THE DATA** Which of the world's oceans contains the most plastic? → **Explore Data** relating to the case study on **Mastering Environmental Science.**

REVIEWING Objectives

You should now be able to:

+ **Identify physical, geographic, and chemical aspects of the marine environment**

The five major oceans, with their complex underwater topography, contain 97% of Earth's water and cover nearly three-fourths of its surface. Ocean waters move horizontally in currents, and vertical currents transport water, heat, and dissolved substances from surface waters to the deep ocean, and vice versa. (pp. 423–426)

+ **Explain how the oceans influence and are influenced by climate**

The thermohaline circulation redistributes heat around the world and shapes regional climate. El Niño and La Niña events occur periodically and alter climate. It is hypothesized that global warming could alter existing circulation patterns, which in turn would affect Earth's climate. (pp. 426–429)

+ **Describe major types of marine ecosystems**

Marine and coastal ecosystems include intertidal zones, salt marshes, mangrove forests, estuaries, kelp forests, coral reefs, and pelagic and deep-water open-ocean systems. Many of these systems are highly productive, are rich in biodiversity, and can suffer heavy impacts from human activities. (pp. 429–433)

+ **Assess impacts from marine pollution**

Plastic debris harms marine life and can accumulate in ocean regions where currents converge. Large oil spills from offshore oil rigs or oil tankers can profoundly affect marine ecosystems. Toxic levels of mercury can bioaccumulate in some large fish species. Nutrient pollution can lead to dead zones and harmful algal blooms in the ocean. Absorption of excess carbon dioxide from the atmosphere by the oceans leads to the acidification of waters, which hinders corals, oysters, and other creatures that create carbonate shells or skeletons. (pp. 433–438)

+ **Review the state of ocean fisheries and identify reasons for their decline**

More than 90% of the world's marine fish populations are exploited or overexploited, and the global fish catch has stagnated since the late 1980s despite increased fishing effort and improved technologies. Industrial fishing approaches have overexploited fisheries and resulted in high rates of bycatch and degradation of marine habitats. Aquaculture, the farming of seafood, has growth tremendously in recent decades and bolstered food supplies. (pp. 438–446)

+ **Evaluate marine protected areas and reserves as solutions for conserving biodiversity**

We have established fewer protected areas in the oceans than we have on land. Most marine protected areas allow some level of fishing to occur, but no-take marine reserves that prohibit fish harvests can protect ecosystems, boost fish populations, and make fisheries sustainable. (pp. 446–447)

SEEKING Solutions

1. What benefits do you derive from the oceans? How do your choices affect the oceans? Give specific examples.

2. We have been able to reduce the amount of oil we spill into the oceans, but other petroleum-based products such as plastic continue to litter our oceans and shorelines. Discuss some ways that we can reduce this impact on the marine environment.

3. Describe the trends in global wild fisheries catch from 1950 to 1990 and from 1990 to 2016, and explain several factors that account for these trends.

4. **THINK IT THROUGH** You are mayor of a coastal town where some residents are employed as commercial fishers and others make a living serving ecotourists who come to snorkel and scuba dive at the nearby coral reef. In recent years, several fish stocks have crashed, and ecotourism is dropping off as fish disappear from the increasingly degraded reef. Scientists are urging you to help establish a marine reserve around portions of the reef, but most commercial and recreational fishers are opposed to this idea. What steps would you take to restore your community's economy and environment?

5. **THINK IT THROUGH** You operate an aquaculture operation (p. 250) that raises oysters for sale to seafood restaurants. Ocean acidification poses a major threat to your business by robbing oysters of the ions they need to grow shells, potentially leading to slower growth rates and higher mortality in your oysters. When your U.S. senator makes a campaign stop in your area, you approach her about the issue to see if the federal government can help. How would you lobby your senator to aid your business? Would you request financial help from the government or advocate aggressive action to combat climate change? Or both? What arguments would you make to convince the government to support aquaculture?

CALCULATING Ecological Footprints

The relationship between the ecological goods and services that individuals use and the amount of land area needed to provide those goods and services is relatively well developed. People also use goods and services from Earth's oceans, where the concept of area is less useful. It is clear, however, that our removal of fish from the oceans has an impact, or an ecological footprint.

The table shows data on the mean annual per-person consumption from ocean fisheries for North America, China, and the world as a whole. Using this consumption data, calculate the amount of fish each consumer group would consume per year, given the annual per capita consumption rates, for each of these three regions. Record your results in the table.

Annual Consumption

CONSUMER GROUP	NORTH AMERICA (21.6 KG PER PERSON)	CHINA (41.0 KG PER PERSON)	WORLD (20.2 KG PER PERSON)
You	21.6	41.0	20.2
Your class			
Your state			
United States			
World			

Data: Food and Agriculture Organization of the United Nations, 2018. *The state of world fisheries and aquaculture 2018.*

1. Calculate the ratio of North America's per capita fish consumption rate to that of the world. Compare this ratio to the ratio of the per capita ecological footprints for the United States, Canada, and Mexico (see Figure 1.11, p. 16) versus the world average footprint of 2.7 hectares (ha)/person/year. Can you account for similarities and differences between these ratios?

2. The population of China has grown at an annual rate of 1.1% since 1987, while over the same period fish consumption in China has grown at an annual rate of 9%. Speculate on the reasons behind China's rapidly increasing consumption of fish.

3. What ecological concerns do the combined trends of human population growth and increasing per capita fish consumption raise for you? What role might you play in contributing to these concerns or to their solutions?

Mastering Environmental Science

The Atmosphere, Air Quality, and Air Pollution Control

Clearing the Air in L.A. and Mexico City

> I left L.A. in 1970, and one of the reasons I left was the horrible smog. And then they cleaned it up. That was one of the greatest things the government has ever done for me. You have beautiful days now. It's a much, much nicer place to live.
>
> Actor and comedian Steve Martin, speaking to *Los Angeles Magazine*

> **This city can be a model for others.**
>
> Miguel Ángel Mancera, former mayor of Mexico City

Los Angeles has long symbolized air pollution. For decades, the city was blanketed by smog, that unhealthy mix of air pollutants resulting from fossil fuel combustion. Exhaust from millions of automobiles clogging L.A.'s freeways regularly became trapped by the mountains that surround the city, and the region's warm sunshine would turn the pollutants to smog. In response, Los Angeles worked hard to improve its air quality, succeeding with policy efforts and new technologies. Today, L.A. still suffers the nation's worst smog, but its skies are clearer than in decades.

One city that looked to L.A.'s success as it planned its own responses to smog is Mexico City, the capital of Mexico and one of the world's largest metropolises. Not long ago, Mexico City suffered the most polluted air in the world. On days of poor air quality in the 1990s, residents wore face masks on the streets, teachers kept students inside at recess, and outdoor sports events were canceled. Children drawing pictures would use brown crayons to color the sky. Each year, thousands of deaths and tens of thousands of hospital visits were blamed on pollution. Mexican novelist Carlos Fuentes called his capital "Makesicko City."

As in Los Angeles, traffic has long generated most of the pollution in Mexico City, where motorists in nearly 10 million cars traverse miles of urban sprawl. And like L.A., Mexico City lies in a valley surrounded by mountains, vulnerable to temperature inversions that trap pollutants over the city. Mexico City environmental chemist Armando Retama likens his hometown to "a casserole dish with a lid on top."

Facing these challenges, Mexico City's 21 million people fought back and made notable improvements in air quality. City leaders took bold action to clean up the air, and in recent years, Mexico City has enjoyed a renaissance. As the smog began to clear, revealing beautiful views of the snow-capped peaks that ring the valley, Mexico City became an international model for other cities seeking to fight pollution.

Efforts began in the 1990s, when Mexico City officials shut down an oil refinery, moved heavy industry outside the city, and pushed factories and power plants to shift to cleaner-burning natural gas. City leaders forced a reformulation of the liquefied petroleum gas most city residents used for cooking and heating so that it would burn more cleanly. Policymakers also

Upon completing this chapter, you will be able to:

+ Describe the composition, structure, and function of Earth's atmosphere

+ Relate weather and climate to atmospheric conditions

+ Identify major outdoor air pollutants, and outline the scope of air pollution

+ Assess strategies and solutions for control of outdoor air pollution

+ Explain stratospheric ozone depletion, and identify steps taken to address it

+ Describe acid deposition, discuss its consequences, and explain how we are addressing it

+ Characterize the scope of indoor air pollution, and assess solutions

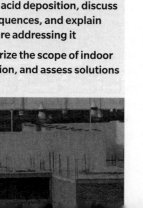

◄ **Mexico City on a smoggy day**

▲ **Testing emissions at a vehicle inspection station in Los Angeles**

mandated that lead be removed from gasoline, that the sulfur content of diesel fuel be reduced, and that catalytic converters (pp. 460–461) be phased in for new vehicles. Each new mayor of the city built on the accomplishments of the previous mayor. Emissions testing was introduced for private cars, and the city upgraded its taxis, buses, and public vehicles to cleaner and more fuel-efficient models. Bus service was expanded and new lines were added to the subway system. To monitor air quality, sampling stations were set up across the city, sending real-time data to city engineers.

In 2010, Mexico City rolled out *ECOBICI,* a bicycle-sharing program, to help free short-distance commuters from dependence on cars. Today, this program is the largest bike-sharing system in North America, with thousands of red-and-white bikes at stations throughout the city; people can rent a bike cheaply at one location and drop it off at another. In another popular initiative, every Sunday morning the city's main boulevard, the *Paseo de la Reforma,* is closed to automobile traffic, creating a safe and pleasant community space for pedestrians, bikers, joggers, and skateboarders.

All these changes paid off with cleaner air. Back in 1991, Mexico City's air was deemed hazardous to breathe on all but 8 days of the year. By 2010–2015, most pollutants had been slashed by more than 75%, and the air was meeting health standards on one of every two days.

Still, as the city's population continued to grow and as car ownership increased, congestion worsened and smog from automobile traffic persisted. City officials had established a program in 1989 called *Hoy No Circula* (No-Driving Day), which prohibited people from driving their car on one weekday per week (based on the last digit of the car's license-plate number). In 2008, leaders expanded the program to include Saturdays. In each case, drivers found ways around the restrictions, sometimes by carpooling but sometimes by buying a second car—often an older, cheaper, more-polluting model. Scientific analysis showed that the Saturday program was ineffective at cutting pollution as a result. Moreover, many people resent the restrictions on driving and complain that there is not enough mass transit to make it practical for most people to commute to work across the vast metropolis by means other than by car.

Today, a new crop of political leaders is pursuing fresh solutions to traffic congestion and seeking to show that Mexico City can continue to serve as a global model for progress against air pollution. Many cities worldwide can use such a model. Nations in Asia and Africa that are industrializing as they build wealth for their citizens are confronting air quality challenges like those that plagued the United States and other wealthy nations a generation and more ago. Cities like Beijing, China, and Delhi, India, are battling pollution even worse than what Mexico City and Los Angeles have faced. We will examine the efforts made in Los Angeles, Mexico City, and elsewhere as we learn about Earth's atmosphere and how to reduce the pollutants we release into it.

The Atmosphere

Every breath we take reaffirms our connection to the **atmosphere,** the layer of gases that envelops our planet. The atmosphere moderates our climate, provides oxygen, helps shield us from meteors and hazardous solar radiation, and transports and recycles water and nutrients.

Earth's atmosphere consists of 78% nitrogen (N_2) and 21% oxygen (O_2) by volume of dry air. The remaining 1% is composed of argon (Ar) and minute concentrations of other gases (**FIGURE 17.1**). The atmosphere also contains water vapor (H_2O) in concentrations that vary with time and place from 0% to 4%.

Over our planet's long history, the atmosphere's composition has changed. When Earth was young, its atmosphere was dominated by carbon dioxide (CO_2), nitrogen, carbon monoxide (CO), and hydrogen (H_2). Then, about 2.7 billion years ago, oxygen began to build up with the emergence of microbial life that emitted oxygen by photosynthesis (p. 33). Today, human activity is altering the quantities of some atmospheric gases, such as carbon dioxide, methane (CH_4), and ozone (O_3). Before exploring how our pollutants affect the air we breathe and how we strive to control pollution, we will begin with an overview of Earth's atmosphere.

The atmosphere is layered

The atmosphere above us that seems so vast is actually just 1/100 of Earth's diameter—a thin coating like the fuzzy skin

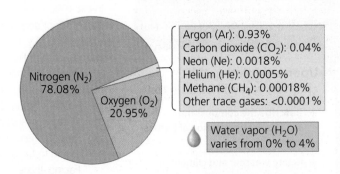

FIGURE 17.1 **Earth's atmosphere consists of nitrogen, oxygen, argon, and a mix of gases at dilute concentrations.**

of a peach. It consists of four layers that differ in temperature, density, and composition (**FIGURE 17.2**).

Within the bottommost layer, the **troposphere,** air movement drives our planet's weather. Although it is thin (averaging 11 km (7 mi) high) relative to the atmosphere's other layers, the troposphere contains three-fourths of the atmosphere's mass because gravity pulls mass downward, making air denser near Earth's surface. Tropospheric air gets colder with altitude, dropping below −50°C (−60°F) at the top of the troposphere. At this point, temperatures stabilize, marking a boundary called the *tropopause.* The tropopause acts like a cap, limiting mixing between the troposphere and the atmospheric layer above it, the stratosphere.

The **stratosphere** extends 11–50 km (7–31 mi) above sea level. It is similar in composition to the troposphere, but is much drier and less dense. Its gases experience little vertical mixing, so

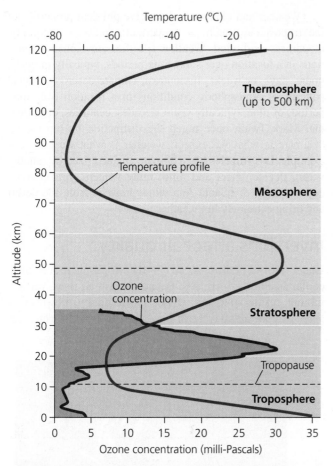

FIGURE 17.2 **The atmosphere is layered.** Temperature **(red line)** drops with altitude in the troposphere, rises with altitude in the stratosphere, drops in the mesosphere, and rises in the thermosphere. The tropopause separates the troposphere from the stratosphere. Ozone **(blue area)** is densest in the lower stratosphere. *Adapted from Jacobson, M.Z., 2002. Atmospheric pollution: History, science, and regulation. Cambridge, U.K.: Cambridge University Press; and Parson, E.A., 2003. Protecting the ozone layer: Science and strategy. Oxford, U.K.: Oxford University Press.*

once substances enter it, they tend to remain for a long time. The stratosphere warms with altitude, because its ozone and oxygen absorb the sun's ultraviolet (UV) radiation (p. 33). Most of the atmosphere's ozone concentrates in a portion of the stratosphere roughly 15–35 km (9–22 mi) above sea level, a region we call the **ozone layer.** By absorbing and scattering incoming UV radiation, the ozone layer greatly reduces the amount of this radiation that reaches Earth's surface. Because UV light can damage living tissue and cause genetic mutations in DNA, living things rely on the protective presence of the ozone layer.

Above the stratosphere stretches the mesosphere, where temperatures decrease with altitude and where incoming meteors burn up. Above the mesosphere is the thermosphere, which extends upward to an altitude of about 500 km (300 mi). Still higher, our atmosphere merges into space in a region called the exosphere.

Pressure, humidity, and temperature vary within the atmosphere

Air moves dynamically within the lower atmosphere as a result of differences in the physical properties of air masses.

Among these properties are pressure and density, relative humidity, and temperature.

Gravity pulls gas molecules toward Earth's surface, causing air to be most dense near the surface and less dense as altitude increases. **Atmospheric pressure,** which measures the force per unit area produced by a column of air, also decreases with altitude, because at higher altitudes there are fewer molecules being pulled down by gravity (**FIGURE 17.3**). At sea level, atmospheric pressure averages 14.7 lb/in.[2] or 1013 millibars (mb). Mountain climbers trekking to Mount Everest, the world's highest mountain, can view their destination from Kala Patthar, a nearby peak, at roughly 5.5 km (18,000 ft) in elevation. At this altitude, pressure is 500 mb, and half the atmosphere's air molecules are above the climber, whereas half are below. A climber who reaches Everest's peak at 8.85 km (29,035 ft) in elevation, where the "thin air" is just over 300 mb, stands above two-thirds of the molecules in the atmosphere! When we fly on a commercial jet airliner at a typical cruising altitude of 11 km (36,000 ft), we are above 80% of the atmosphere's molecules.

Another property of air is **relative humidity,** the ratio of water vapor that air contains at a given temperature to the maximum amount it *could* contain at that temperature. Daytime relative humidity in June in the desert at Phoenix, Arizona, averages only 31% (meaning that the air contains less than a third of the water vapor possible at its temperature), whereas on the tropical island of Guam, relative humidity rarely drops below 88%.

People are sensitive to changes in relative humidity because we perspire to cool our bodies. When humidity is high, the air is already holding nearly as much water vapor as it can, so sweat evaporates slowly, and the body cannot cool itself efficiently. That is why high humidity makes it feel

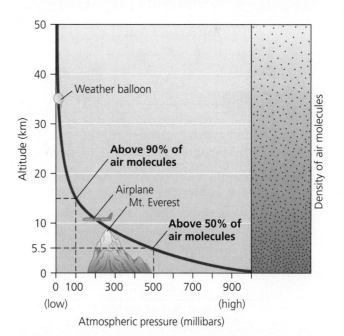

FIGURE 17.3 **As one climbs higher through the atmosphere, gas molecules become less densely packed, and atmospheric pressure decreases.** One needs to be only 5.5 km (3.4 mi) high to be above half the planet's air molecules. *Adapted from Ahrens, C.D., 2007. Meteorology today, 8th ed., FIGURE 1.9. © 2007. Belmont, CA: Brooks/Cole. By permission of Cengage Learning.*

hotter than it actually is. Conversely, low humidity speeds evaporation and makes it feel cooler than it actually is.

The temperature of air also varies with location and time. At the global scale, temperature varies over Earth's surface because the sun's rays strike some areas more directly than others. At more local scales, temperature varies with topography, plant cover, proximity of water to land, and many other factors.

The sun's energy influences weather and climate

An enormous amount of energy from the sun constantly bombards the upper atmosphere—more than 1000 watts/m^2, thousands of times more than the total output of electricity generated by human society. Of this solar energy, about 70% is absorbed by the atmosphere and planetary surface, while the rest is reflected back into space (see Figure 18.2, p. 487).

On the surface of our planet, land and water absorb solar energy and then emit thermal infrared radiation, which warms the air and causes some water to evaporate. As a result, air near Earth's surface tends to be warmer and moister than air at higher altitudes. These differences set into motion a process of **convective circulation** (**FIGURE 17.4**). Warm air, being less dense, rises and creates vertical currents. As air rises into regions of lesser atmospheric pressure, it expands and cools, causing moisture to condense and fall as rain. Once the air cools, it descends and becomes denser, replacing warm air that is rising. The descending air picks up heat and moisture near ground level and begins to rise again, continuing the process. Convective circulation patterns occur in ocean waters (p. 427), in magma beneath Earth's surface (p. 36), and even in a simmering pot of soup. Convective circulation influences both weather and climate.

Weather and climate each involve physical properties of the troposphere, such as temperature, pressure, humidity, cloudiness, and wind. **Weather** specifies atmospheric conditions in a location over short time periods, typically minutes, hours, days, or weeks. In contrast, **climate** describes typical patterns of atmospheric conditions in a location over long periods of time, typically years, decades, centuries, or millennia. Mark Twain once noted the distinction by remarking, "Climate is what we expect; weather is what we get." For example, Los Angeles has a climate characterized by reliably warm, dry summers and mild, rainy winters, yet on some autumn days, dry Santa Ana winds blow in from the desert and bring extremely hot weather.

Inversions affect air quality

Under most conditions, air in the troposphere becomes cooler as altitude increases. Because warm air rises, vertical mixing results (**FIGURE 17.5a**). Occasionally, however, a layer of cool air may form beneath a layer of warmer air.

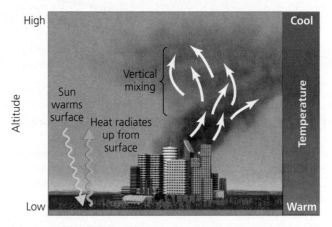

(a) Normal conditions

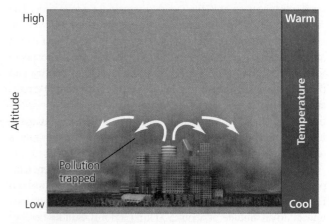

(b) Temperature inversion

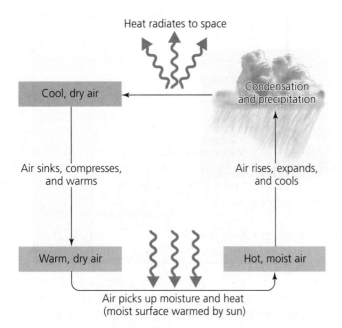

FIGURE 17.4 Convective circulation helps drive weather. Air heated near Earth's surface picks up moisture and rises. Once aloft, this air cools, and moisture condenses, forming clouds and precipitation. Cool, drying air begins to descend, compressing and warming in the process. Warm, dry air near the surface then picks up moisture, beginning the cycle anew.

FIGURE 17.5 Temperature inversions trap air and pollutants. Under normal conditions **(a)**, air becomes cooler with altitude and air of different altitudes mixes, dispersing pollutants upward. In a temperature inversion **(b)**, dense cool air remains near the ground, and air warms with altitude within the inversion layer. Little mixing occurs, and pollutants are trapped.

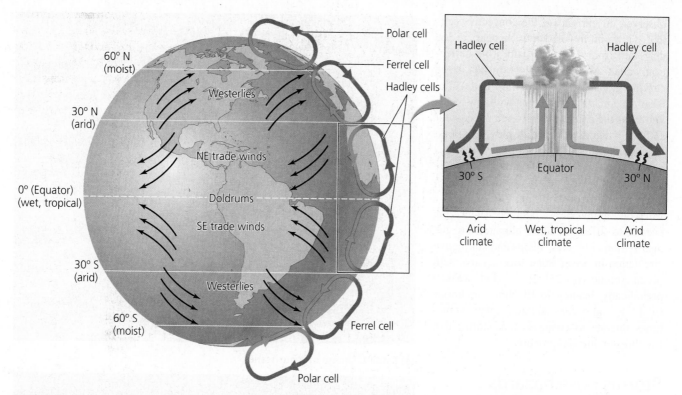

FIGURE 17.6 Large-scale convective cells create global patterns of moisture and wind. These cells give rise to a wet climate in tropical regions, arid climates around 30° latitude, moist climates around 60° latitude, and dry climates near the poles. Surface air movement of these cells interacts with the Coriolis effect to create global wind currents, shown with thin black arrows.

This departure from the normal temperature profile is known as a **temperature inversion,** or thermal inversion (**FIGURE 17.5b**). The band of cooler air in which temperature rises with altitude is called an **inversion layer** (because the normal direction of temperature change is inverted). The cooler air of the inversion layer is denser than the warmer air above, so it resists vertical mixing and remains stable. Temperature inversions can occur in different ways, sometimes involving cool air at ground level and sometimes involving an inversion layer higher above the ground. One common type of inversion occurs in mountain valleys where slopes block morning sunlight, keeping ground-level air within the valley shaded and cool.

Vertical mixing allows pollutants in the air to be carried upward and diluted, but temperature inversions trap pollutants near the ground. As a result, cities such as Los Angeles and Mexico City suffer their worst pollution when inversions prevent pollutants from being dispersed. Both metropolitan areas are encircled by mountains that promote inversions, interrupt air flow, and trap pollutants. Los Angeles experiences inversions most often when a "marine layer" of air cooled by the Pacific Ocean moves inland. In Mexico City in 1996, a persistent temperature inversion sparked a five-day crisis in which pollution killed at least 300 people and sent 400,000 to hospitals. Across the world, inversions frequently concentrate pollution over metropolitan areas in valleys ringed by mountains, from Tehran to Seoul to Río de Janeiro to São Paulo.

Large-scale circulation systems produce global climate patterns

At the global scale, convective air currents contribute to broad climate patterns (**FIGURE 17.6**). Near the equator, solar radiation sets in motion a pair of convective cells known as *Hadley cells.* Here, where sunlight is most intense, surface air warms, rises, and expands. As it does so, it releases moisture, producing the heavy rainfall that gives rise to tropical rainforests. After releasing much of its moisture, this air diverges and moves in currents heading north and south. The air in these currents cools and descends at about 30° latitude north and south. Because the descending air is now dry, the regions around 30° latitude are quite arid, giving rise to deserts. Two additional pairs of convective cells, *Ferrel cells* and *polar cells,* lift moist air and create precipitation around 60° latitude north and south (enabling forests to grow) and cause dry air to descend at 30° latitude and in the polar regions (giving rise to arid tundra). By creating wet climates near the equator, arid climates near 30° latitude, moist regions near 60° latitude, and dry conditions near the poles, these three pairs of convective cells—combined with temperature variation—help explain why biomes tend to be arrayed in latitudinal bands (see Figure 4.19, p. 94).

The Hadley, Ferrel, and polar cells interact with Earth's rotation to produce global wind patterns (see Figure 17.6). As Earth rotates on its axis, regions of the planet's surface near the equator move west to east more quickly than regions near the poles. As a result, from the perspective of an Earth-bound observer, air currents of the convective cells that flow north or south appear to be deflected from a straight path. This deflection, called the **Coriolis effect,** results in the curving global wind patterns in Figure 17.6. Near the equator lies a region with few winds known as the *doldrums*. Between the equator and 30° latitude, *trade winds* blow from east to west. From 30° to 60° latitude, *westerlies* blow from west to east. For centuries, people made use of these patterns to facilitate ocean travel by wind-powered sailing ships.

The atmosphere interacts with the oceans to affect weather, climate, and the distribution of biomes. Winds and convective circulation in ocean water together maintain ocean currents (p. 425). Trade winds weaken periodically, leading to El Niño conditions (p. 427). And oceans and atmosphere sometimes interact to create violent storms that can threaten life and property.

(a) Satellite image of a hurricane

Storms pose hazards

Hurricanes (**FIGURE 17.7a**) form when warm, moisture-laden air over tropical oceans rises and winds rush into these areas of low pressure. In the Northern Hemisphere, these winds turn counterclockwise because of the Coriolis effect. In other regions, such cyclonic storms are called *cyclones* or *typhoons*. The powerful convective currents of these storms draw up immense amounts of water vapor. As the warm, moist air rises and cools, water condenses (because cool air cannot hold as much water vapor as warm air) and falls heavily as rain. In North America, the Gulf Coast and Atlantic Coast are most susceptible to hurricanes.

Tornadoes (**FIGURE 17.7b**) form when a mass of warm air meets a mass of cold air and the warm air rises quickly, setting a powerful convective current in motion. If high-altitude winds are blowing faster and in a different direction from low-altitude winds, the rising column of air may begin to rotate. Eventually, the spinning funnel of rising air may lift up soil and objects in its path with winds up to 500 km per hour (310 mph). In North America, tornadoes are most apt to form in the Great Plains and the Southeast, where cold air from Canada and warm air from the Gulf of Mexico frequently meet.

Understanding how the atmosphere functions helps us predict violent storms and warn people of their approach. Such knowledge also helps us comprehend how our pollution of the atmosphere affects climate, ecological systems, economies, and human health.

(b) Photograph of a tornado

FIGURE 17.7 Hurricanes and tornadoes are cyclonic storms that pose hazards to life and property.

Outdoor Air Quality

Throughout history, we have made the atmosphere a dumping ground for our airborne wastes. Whether from simple wood fires or modern coal-burning power plants, we have generated **air pollutants,** gases and particulate material added to the atmosphere that can affect climate or harm people or other living things. Fortunately, our efforts to control **air pollution,**

(a) Natural fire in California

(c) Mount Saint Helens eruption, 1980

Atlantic Ocean

Africa

500 km

(b) Satellite image of winds blowing dust from Africa to the Americas

FIGURE 17.8 Wildfire, dust storms, and volcanoes are three natural sources of air pollution.

the release of air pollutants, have brought some of our best successes in confronting environmental problems.

In recent decades, public policy and improved technologies have helped us reduce most types of **outdoor air pollution** (often called *ambient air pollution*) in industrialized nations. However, outdoor air pollution remains a problem, particularly in industrializing nations and in urban areas. Scientists with the World Health Organization estimate that each year 4.2 million people die prematurely as a result of health problems caused by outdoor air pollution.

We also face an enormous air pollution issue in our emission of greenhouse gases (p. 488), which drive global climate change. Addressing our release of carbon dioxide, methane, and other gases that warm the atmosphere stands as one of our civilization's most urgent challenges. (We discuss this issue separately and in depth in Chapter 18.)

Some pollution is from natural sources

When we think of outdoor air pollution, we tend to envision smokestacks belching smoke from industrial plants. However, natural processes also pollute the air. Some of these natural impacts are made worse by human activity and land use policies.

Fires from burning vegetation emit soot and gases (**FIGURE 17.8a**). Worldwide, more than 60 million hectares (150 million acres, an area the size of Texas) of forest and grassland burn in a typical year. Fires occur naturally, but human influence can make them more severe. Across North America, the suppression of fire for decades has allowed fuel to build up in forests and eventually feed highly destructive fires (pp. 322–323). In the tropics, many farmers clear forest for farming and grazing using a "slash-and-burn" approach (p. 219). Today, climate change (Chapter 18) is leading to drought in many regions, including the Los Angeles basin, and worsening fires as a result.

Wind-blown dust is another natural pollution source. Dust storms may occur naturally when wind sweeps over arid terrain, but they are made worse by unsustainable farming and grazing practices that strip vegetation from the soil, leading to erosion and desertification (p. 230). Continental-scale dust storms took place in the United States in the 1930s, when soil from the drought-plagued Dust Bowl states blew eastward to the Atlantic (p. 231). Today, trade winds blow soil across the Atlantic Ocean from Africa to the Americas (**FIGURE 17.8b**). Strong westerlies sometimes lift soil from deserts in Mongolia and China and blow it all the way across the Pacific Ocean to North America.

Volcanic eruptions (p. 40) release large quantities of particulate matter, as well as sulfur dioxide and other gases, into the troposphere (**FIGURE 17.8c**). In 2012, Mexico City residents went on alert as the nearby volcano of Popocatepetl erupted, adding to the region's pollution challenges. Ash from volcanic eruptions can ground airplanes, destroy car engines, and pose respiratory health risks. Major eruptions may blow matter into the stratosphere, where it can circle the globe for months or years. Sulfur dioxide reacts with water and oxygen and then condenses into fine droplets, called aerosols (p. 489), which reflect sunlight back into space and thereby cool the atmosphere and surface. The 1991 eruption of Mount Pinatubo in the Philippines ejected nearly 20 million tons of ash and aerosols and cooled global temperatures by 0.5°C (0.9°F).

We create outdoor air pollution

Human activity produces many air pollutants. As with water pollution, the air pollution we create can emanate from point sources or non-point sources (p. 409). A point source describes a specific location from which large quantities of pollutants are discharged (such as a coal-fired power plant). Non-point sources are more diffuse, consisting of many small, widely spread sources (such as millions of automobiles).

Pollutants released directly from a source are termed **primary pollutants.** Ash from a volcano, sulfur dioxide from a power plant, and carbon monoxide from an engine are all primary pollutants. Often primary pollutants react with one another, or with constituents of the atmosphere, and form other pollutants, called **secondary pollutants.** Examples include ozone, formed near ground-level (p. 461) from pollutants in urban smog, or the acids in acid rain (p. 474), formed when certain primary pollutants react with water and oxygen.

Because substances differ in how readily they react in air and in how quickly they settle to the ground, they differ in their **residence time,** the amount of time a substance spends in the atmosphere. Pollutants with brief residence times exert localized impacts over short time periods. Most particulate matter and most pollutants from automobile exhaust stay aloft only hours or days, which is why air quality in a city like Mexico City or Los Angeles rises and falls with daily traffic patterns and changes with weather from day to day. In contrast, pollutants with long residence times can exert impacts regionally or globally for long periods, even centuries (**FIGURE 17.9**). The pollutants that drive climate change and those that deplete Earth's ozone layer (two separate phenomena!— see **FAQ**, p. 474) are each able to cause global and long-lasting impacts because they persist in the atmosphere for so long.

The Clean Air Act addresses pollution

To address air pollution in the United States, Congress has passed a series of laws, most notably the **Clean Air Act.** First enacted in 1963, the Clean Air Act has been amended multiple times, chiefly in 1970 and 1990. This body of legislation funds research into pollution control, sets standards for air quality, and encourages emissions standards for automobiles and for stationary point sources such as industrial plants. It also imposes limits on emissions from new sources, funds a

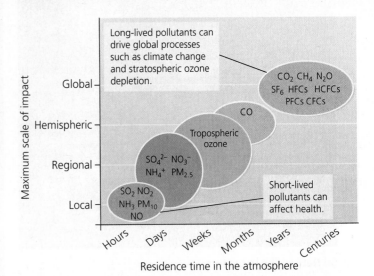

FIGURE 17.9 Substances with short residence times affect air quality locally, whereas those with long residence times affect air quality globally. *Adapted from United Nations Environment Programme, 2007.* Global environment outlook (GEO-4), *Nairobi, Kenya.*

nationwide air quality monitoring system, and enables citizens to sue parties violating the standards. The 1990 amendments introduced an emissions trading program (p. 181) for sulfur dioxide.

Under the Clean Air Act, the U.S. Environmental Protection Agency (EPA) sets nationwide standards for (1) emissions of several key pollutants and (2) concentrations of major pollutants in ambient air. It is largely up to the states to monitor emissions and air quality and to develop, implement, and enforce regulations within their borders. States submit implementation plans to the EPA for approval, and if a state's plans are not adequate, the EPA can take control of enforcement. If a region fails to clean up its air, the EPA can prevent it from receiving federal money for transportation projects.

Agencies monitor emissions

State and local agencies monitor and report to the EPA emissions of a variety of air pollutants known to affect human health. The most significant pollutants are profiled below and in **FIGURE 17.10.** Across the United States in 2017, human activity polluted the air with 86 million tons of these pollutants (117 million tons, if dust and wildfires are included).

Carbon monoxide Carbon monoxide (CO) is a colorless, odorless gas produced primarily by the incomplete combustion of fuel. Vehicles and engines account for most CO emissions in the United States. Other sources include industrial processes, waste combustion, and residential wood burning. In addition, wildfires release a great deal of this pollutant. Carbon monoxide is hazardous because it binds to hemoglobin in red blood cells, preventing the hemoglobin from binding with oxygen. Carbon monoxide poisoning brings about nausea, headaches, fatigue, heart and nervous system damage, and potentially death.

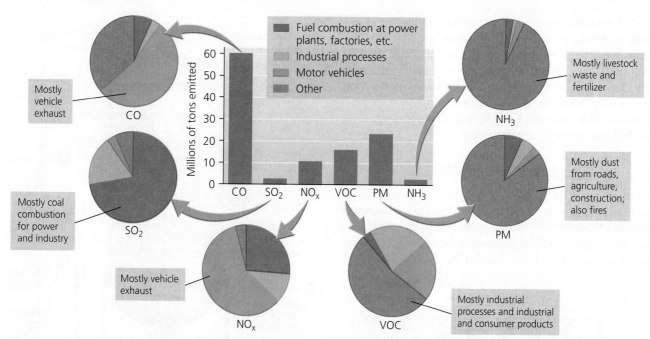

FIGURE 17.10 In 2017 across the United States, ambient air was polluted by 117 million tons of several major pollutants whose emissions are monitored by the EPA and state agencies. The central bar chart shows absolute amounts (by dry weight) of each pollutant. The surrounding pie charts summarize the major sources of each pollutant. *Data from U.S. EPA.*

Sulfur dioxide Sulfur dioxide (SO_2) is a colorless gas with a pungent odor. Most emissions result from the combustion of coal for electricity generation. During combustion, elemental sulfur (S), a contaminant in coal, reacts with oxygen (O_2) to form SO_2. Once in the atmosphere, SO_2 may react to form sulfur trioxide (SO_3) and sulfuric acid (H_2SO_4), which may then settle back to Earth in acid deposition (p. 474).

Nitrogen oxides Nitrogen oxides (NO_X) are a family of compounds that include nitric oxide (NO) and nitrogen dioxide (NO_2). Most U.S. emissions of nitrogen oxides result when nitrogen and oxygen from the atmosphere react at high temperatures during combustion in vehicle engines. Fossil fuel combustion in industry and at power plants accounts for most of the rest. NO_X emissions contribute to smog, acid deposition, and stratospheric ozone depletion.

Volatile organic compounds Volatile organic compounds (VOCs) are carbon-containing chemicals emitted by vehicle engines and a wide variety of solvents, industrial processes, household chemicals, paints, plastics, and consumer items. Examples range from benzene to acetone to formaldehyde. One common group of VOCs consists of hydrocarbons (p. 28) such as methane (CH_4, the primary component of natural gas), propane (C_3H_8, used as a portable fuel), butane (C_4H_{10}, found in cigarette lighters), and octane (C_8H_{18}, a component of gasoline). Most of the VOCs released in the United States come from natural sources; for example, plants produce isoprene and terpenes, compounds that generate a bluish haze that gave the Blue Ridge Mountains their name. VOCs can react to produce secondary pollutants, as occurs in urban smog.

Particulate matter Particulate matter (PM) is composed of solid or liquid particles small enough to be suspended in the atmosphere. Particulate matter includes primary pollutants such as dust and soot, as well as secondary pollutants such as sulfates and nitrates. Scientists classify particulate matter by the size of the particles. Smaller particles are more likely to get deep into the lungs and to pass through tissues, causing damage to the lungs, heart, and brain. PM_{10} pollutants consist of particles less than 10 microns in diameter (one-seventh the width of a human hair). $PM_{2.5}$ pollutants consist of still-finer particles less than 2.5 microns in diameter. Most PM_{10} pollution consists of dust from roads, construction, and agriculture, whereas most $PM_{2.5}$ pollution results from fuel combustion and fires as well as dust.

Ammonia Ammonia (NH_3) is a colorless gas with a sharp and pungent odor. The household cleaning fluid we are familiar with is a diluted solution of this gas dissolved in water, but pure NH_3 in gas form is a waste product of animals and a component of chemical fertilizers. As a result, the main sources of ammonia emissions are feedlots for cattle, hogs, and chickens; and agricultural fertilizers. As a pollutant in the air, ammonia is harmful on its own but is also a precursor to smog, particulate matter, and deposition of nitrogen and acids.

Lead Lead (Pb) is a heavy metal that enters the atmosphere as a particulate pollutant. For decades, the lead-containing compounds tetraethyl lead and tetramethyl lead were added to gasoline to improve engine performance. Unfortunately, exhaust from the combustion of leaded gasoline emits airborne lead, which can be inhaled or can settle on land and water. When lead enters the food chain, it accumulates in body tissues and can cause central nervous system malfunction and many other ailments (p. 366). Since the 1980s, most nations have phased out leaded gasoline (p. 17), and the main remaining U.S. sources of atmospheric lead today are aviation fuel and industrial metal smelting.

We have reduced emissions

Since passage of the Clean Air Act of 1970, the United States has reduced emissions of its major air pollutants substantially (**FIGURE 17.11a**). These dramatic reductions in emissions have occurred despite significant increases in the nation's population, energy consumption, miles traveled by vehicle, and gross

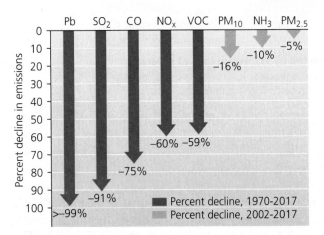

(a) Declines in eight major pollutants

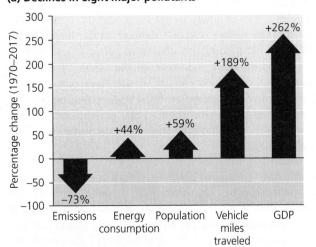

(b) Trends in major indicators

FIGURE 17.11 U.S. emissions have declined sharply since 1970. Reductions **(a)** have been achieved in all the major pollutants tracked by the EPA, despite increases **(b)** in U.S. energy consumption, population, vehicle miles traveled, and gross domestic product. *Data from U.S. EPA.*

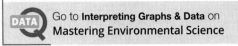

Go to **Interpreting Graphs & Data** on **Mastering Environmental Science**

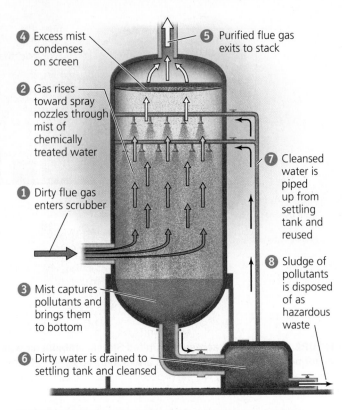

4 Excess mist condenses on screen

5 Purified flue gas exits to stack

2 Gas rises toward spray nozzles through mist of chemically treated water

1 Dirty flue gas enters scrubber

3 Mist captures pollutants and brings them to bottom

7 Cleansed water is piped up from settling tank and reused

8 Sludge of pollutants is disposed of as hazardous waste

6 Dirty water is drained to settling tank and cleansed

FIGURE 17.12 Scrubbers typically remove at least 90% of particulate matter and gases such as sulfur dioxide. Scrubbers and other pollution control devices come in many designs. In this spray-tower wet scrubber, polluted air rises through a chamber while nozzles spray a mist of water mixed with lime or other active chemicals to capture pollutants and wash them out of the air.

domestic product (**FIGURE 17.11b**). Most other industrialized nations have taken their own steps to reduce emissions and have attained similar results.

This success in controlling pollution has resulted from policy steps and from technological advances, each motivated by grassroots social demand for cleaner air. In factories, power plants, and refineries, technologies such as baghouse filters, electrostatic precipitators, and **scrubbers** (**FIGURE 17.12**) have been installed to chemically convert or physically remove airborne pollutants before they are emitted from smokestacks. With motor vehicles, cleaner-burning engines and automotive technologies such as catalytic converters have cut down on pollution from automobile exhaust. In a **catalytic converter,** engine exhaust reacts with metals that convert hydrocarbons, CO, and NO_X into carbon dioxide, water vapor, and nitrogen gas (**FIGURE 17.13**).

Pollution control technologies such as scrubbers and catalytic converters were initially resisted by industry because of added expense but came to be widely adopted once policy measures were in place encouraging or mandating them. Other policies have been influential, as well. Phaseouts of leaded gasoline caused lead emissions to plummet, and the EPA's Acid Rain Program and its

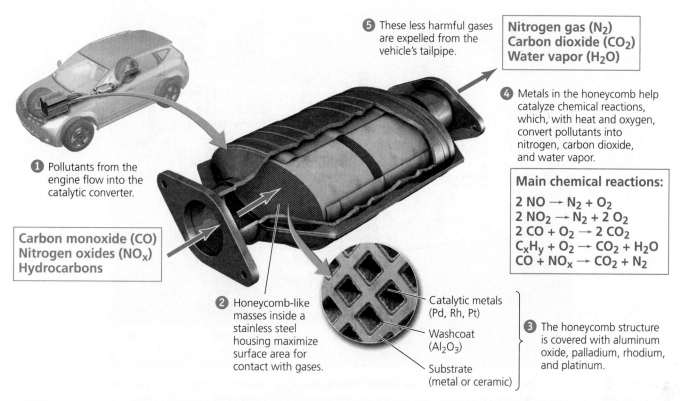

⑤ These less harmful gases are expelled from the vehicle's tailpipe.

Nitrogen gas (N₂)
Carbon dioxide (CO₂)
Water vapor (H₂O)

④ Metals in the honeycomb help catalyze chemical reactions, which, with heat and oxygen, convert pollutants into nitrogen, carbon dioxide, and water vapor.

① Pollutants from the engine flow into the catalytic converter.

Carbon monoxide (CO)
Nitrogen oxides (NOₓ)
Hydrocarbons

Main chemical reactions:

$$2\,NO \rightarrow N_2 + O_2$$
$$2\,NO_2 \rightarrow N_2 + 2\,O_2$$
$$2\,CO + O_2 \rightarrow 2\,CO_2$$
$$C_xH_y + O_2 \rightarrow CO_2 + H_2O$$
$$CO + NO_x \rightarrow CO_2 + N_2$$

② Honeycomb-like masses inside a stainless steel housing maximize surface area for contact with gases.

Catalytic metals (Pd, Rh, Pt)

Washcoat (Al₂O₃)

Substrate (metal or ceramic)

③ The honeycomb structure is covered with aluminum oxide, palladium, rhodium, and platinum.

FIGURE 17.13 Catalytic converters improve air quality by filtering pollutants from vehicle exhaust.

emissions trading system, along with clean coal technologies (p. 548), have reduced releases of SO₂ and NOₓ. As a result of all these emissions reductions, air quality has improved markedly in the United States and in many other nations.

Air quality has improved

In the United States, the EPA and the states monitor outdoor air quality by measuring the concentrations of six **criteria pollutants,** air pollutants judged to pose substantial risk to human health. For each of these, the EPA has established *national ambient air quality standards,* which are maximum concentrations legally allowable in ambient outdoor air. The six criteria pollutants include four of the air pollutants discussed above whose emissions are formally monitored—carbon monoxide, sulfur dioxide, particulate matter, and lead—as well as nitrogen dioxide and tropospheric ozone.

Nitrogen dioxide Nitrogen dioxide (NO₂) is a foul-smelling, highly reactive, reddish brown gas that contributes to smog and acid deposition. Along with nitric oxide (NO), NO₂ belongs to the family of compounds called nitrogen oxides (NOₓ). Nitric oxide reacts readily in the atmosphere to form NO₂, which is both a primary and secondary pollutant.

Tropospheric ozone Ozone in the stratosphere shields us from the dangers of UV radiation, but ozone produced by human activity accumulates low in the troposphere.

Tropospheric ozone (O₃), also called *ground-level ozone,* is a colorless gas and a secondary pollutant, created by the reaction of nitrogen oxides and volatile carbon-containing chemicals in the presence of sunlight.

A major component of photochemical smog (p. 468), tropospheric ozone poses health risks because the O₃ molecule will readily split into a molecule of oxygen gas (O₂) and a free oxygen atom. The oxygen atom may then participate in reactions that can injure living tissues and cause respiratory problems. Tropospheric ozone is the pollutant that most frequently exceeds its national ambient air quality standard.

Across the United States, more than 4000 monitoring stations take hourly or daily air samples to measure pollutant concentrations. The EPA compiles these data and calculates values on its *Air Quality Index (AQI)* for each site. Each of six pollutants—CO, SO₂, NO₂, O₃, PM₁₀, and PM₂.₅—receives an AQI value from 0 to 500 that reflects its current concentration. AQI values below 100 indicate satisfactory air conditions, and values above 100 indicate unhealthy conditions. The highest AQI value from the pollutants on a particular day is reported as the overall AQI value for that day, and these values are made available online and in weather forecasts.

Monitoring has quantified how ambient air quality has improved through time as a result of emissions reductions due to new technologies and clean air regulations. Data show how Los Angeles and other metropolises are making

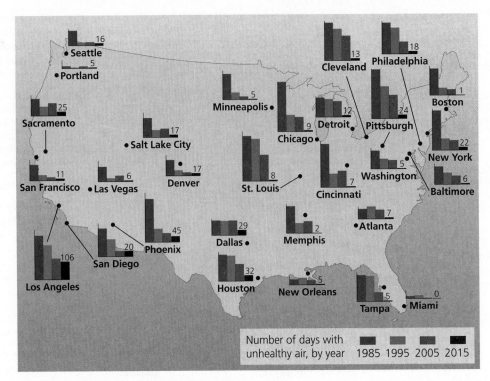

FIGURE 17.14 In most U.S. cities, air has become cleaner than it once was. This map shows numbers of days with unhealthy air from years spanning four decades for 29 metropolitan areas, according to the Air Quality Index (AQI). The AQI combines data on CO, NO$_2$, SO$_2$, O$_3$, and particulate matter. Days with AQI values over 100, shown here, indicate unhealthy air and violations of national ambient air quality standards. *Data from U.S. EPA.*

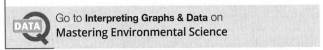

Go to **Interpreting Graphs & Data** on
Mastering Environmental Science

perceptible headway toward cleaner air for their residents (**FIGURE 17.14**). Thanks to the actions of scientists, policymakers, industrial leaders, and everyday people, outdoor air quality today is far better than it was a generation or two ago (see **SUCCESS STORY**).

Still, there remains much room for improvement. Concerns over new pollutants are emerging, greenhouse gas emissions are altering the climate, and people in low-income communities often find themselves living near hotspots of pollution. In fact, many Americans live in areas where pollution continues to reach unhealthy levels. For instance, despite enormous progress, residents of Los Angeles County and four adjacent counties still breathe air that frequently violates national ambient air quality standards for most of the six criteria pollutants. As of 2017, 111 million Americans lived in counties that regularly violated a national ambient air quality standard, most often the one for tropospheric ozone.

Moreover, recent policy trends at the federal level threaten to reverse progress on clean air. For example, the Trump administration disbanded scientific panels advising on air quality, sought to increase coal combustion, rolled back automotive fuel economy requirements, delayed bans on toxic chemicals, and weakened regulations on mercury and other air pollutants. Although many states are resisting these moves and some efforts are tied up in court, these developments make clear that we can never take for granted our progress in gaining cleaner air.

Toxic pollutants pose health risks

Agencies also monitor and regulate emissions of **toxic air pollutants,** substances known to cause cancer; reproductive defects; or neurological, developmental, immune system, or respiratory problems in people and other organisms. Under the 1990 Clean Air Act, the EPA regulates 187 toxic air pollutants produced by a variety of activities, including metal smelting, sewage treatment, and industrial processes. These pollutants range from the heavy metal mercury (from coal-burning power plant emissions and other sources) to VOCs such as benzene (a component of gasoline) and methylene chloride (found in paint stripper). Based on monitoring at 300 sites across the United States, scientists estimate that toxic air pollutants cause cancer in 1 out of every 25,000 Americans (40 cancer cases per 1 million people). Health risks are highest in urban and industrialized areas such as the Los Angeles region. Nationwide, the EPA estimates that Clean Air Act regulations on facilities such as chemical plants, waste incinerators, dry cleaners, and coke ovens have helped reduce emissions of toxic air pollutants since 1990 by 68%.

Clearing the Air for Better Health across the United States

With every breath we take, each of us alive today benefits from America's success in fighting outdoor air pollution. A generation or two ago, poor air quality hampered people's health and well-being across the United States. The Clean Air Act changed that by encouraging pollution reduction and spurring advances in technology. The resulting drop in outdoor air pollution since 1970 represents one of the country's greatest accomplishments in safeguarding human health and environmental quality. As the figure shows, ambient concentrations of all criteria pollutants have declined below their health standards set by the EPA. Lead (not shown) has plummeted from 900% of its standard since 1980. The EPA estimates that between 1970 and 1990 alone, clean air regulations and the resulting technological advances in pollution control saved the lives of 200,000 Americans. This success demonstrates how seemingly intractable problems can be addressed when government and industry apply information from science and are responsive to the public's demands. However, progress gained can also be lost. To protect our hard-won gains and build on our success, we must insist that our policymakers uphold beneficial regulations and resist influence from polluting interests that threaten our health and safety. We also would do well to remember that outdoor air is cleaner than indoor air—so it pays to get off the couch, turn off the screen, and get outside!

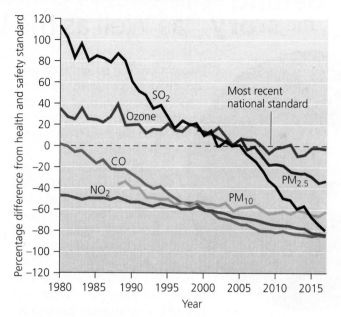

Concentrations of criteria pollutants in ambient air across the United States have fallen in the wake of Clean Air Act regulations. *Data from U.S. EPA.*

→ **Explore the Data** at **Mastering Environmental Science**

Rural areas also confront pollution challenges

Although we tend to focus on cities when discussing air pollution, air quality is also a concern in rural areas. Residents of rural regions suffer health impacts from the drift of dust and airborne pesticides from farms, ammonia from fertilizers, and particulate matter and carbon monoxide from wildfires. Air pollutants also emanate from feedlots (p. 248) where cattle, hogs, or chickens are raised. The huge numbers of animals densely concentrated at feedlots and the voluminous amounts of waste they produce generate methane, hydrogen sulfide, and ammonia. Studies show that people working at and living near feedlots have high rates of respiratory illness. In addition to emissions from such nearby sources, many rural regions downwind from urban areas receive industrial pollutants that drift far from cities, factories, and power plants.

Indeed, some of the worst air quality in the United States occurs in rural regions, including California's Central Valley (the nation's agricultural fruit basket, where dust and pesticide drift are major factors) and areas near natural gas extraction sites, where methane-laced fumes from extraction pollute the air. Recent data show that even the national parks experience nearly as much ozone pollution as U.S. cities, because of long-distance drift of urban smog.

Air pollution remains severe in industrializing nations

Although the United States and other industrialized nations have improved their air quality, outdoor air pollution is growing worse in many countries across the developing world. In these societies, more and more people are driving automobiles, while proliferating factories and power plants emit more pollutants as governments encourage economic growth. At the same time, many people continue to burn traditional sources of fuel, such as wood, charcoal, and coal, for cooking and home heating. As a result, nine people out of ten in the world today live in areas where air quality fails to meet the health guidelines of the World Health Organization (WHO).

Mexico embodies these trends, and despite the progress in its capital, residents of Mexico City and many other Mexican cities and towns continue to endure a variety of health impacts from polluted air (see **THE SCIENCE BEHIND THE STORY**, pp. 464–465).

THE SCIENCE behind the story

Does Air Pollution Affect the Brain, as Well as the Lungs and the Heart?

Dr. Isabelle Romieu

"I know I'm inhaling poison," a 38-year-old candy vendor named Guadalupe told a reporter amid the fumes of a traffic-choked intersection in Mexico City. "But there is nothing I can do."

For as long as we have polluted our air, people have felt effects on their health. But identifying and quantifying those impacts poses a challenge for scientists. For researchers wanting to understand pollution's health impacts—and to design solutions for people like Guadalupe—what better place to go than Mexico City, long home to some of the world's worst air pollution?

A key first step is to determine what's actually in the air. One researcher who led the way is Mario Molina, the Nobel Prize–winning chemist who helped discover the cause of stratospheric ozone depletion and who appears in this chapter's other **The Science behind the Story** feature (pp. 472–473). In 2003 and 2006, Molina organized hundreds of scientists to sample the air in Mexico City, his hometown. The nearly 200 research publications spawned by these efforts clarified many aspects of the city's air quality. One study used machines to identify and record individual particles in real time. It found that metal-rich particulates from trash incinerators were peaking in the morning, whereas smoke from fires outside the city blew in during the afternoon. Other researchers discovered that volatile organic carbons control the amount of tropospheric ozone formed in smog. City officials responded by targeting VOC emissions for reduction, while also discouraging automobile traffic (**FIGURE 1**).

Few people understand Mexico City's air pollution better than Armando Retama, the city's director of atmospheric monitoring. But he may grasp its impacts best when he leaves town. "I can breathe better. I'm not all dry. My eyes aren't irritated. My skin doesn't crack," he says. "We have chronic symptoms that we aren't aware of."

Most known health impacts of urban pollution affect the respiratory system. At high altitudes like Mexico City's, the "thin air" forces people to breathe deeply to obtain enough oxygen, which means that they pull more air pollutants into their lungs than people at lower elevations. Many studies confirm that Mexico City residents show poorer lung function than people from less-polluted areas and that respiratory problems and emergency room visits become more numerous when pollution is severe.

In 2007, a research team led by Isabelle Romieu of Mexico's National Institute of Public Health examined the effects of growing up amid polluted air. Her team measured lung function in 3170 eight-year-old children from 39 Mexico City schools across three years and correlated it with their exposure to tropospheric ozone, nitrogen dioxide, and particulate matter. The team found that children from more-polluted neighborhoods lagged behind those from cleaner ones in the ability to inhale and exhale deeply—indicating smaller, weaker lungs. Romieu and her colleagues also showed that the city's pollution worsens asthma in children. Analyzing data from 200 asthmatic and healthy children, her team found that children in areas with more traffic and pollutants coughed, wheezed, and used medication more often than children in cleaner areas.

Another Mexican researcher, Lilian Calderón-Garcidueñas—now at the University of Montana—compared chest X-ray films and medical records of Mexico City children with those of similar children from less-polluted locations. Her team found hyperinflation and other problems with the lungs of the Mexico City youth. Mexico City children reported many respiratory problems, whereas rural children did not (**FIGURE 2**).

FIGURE 1 During a resurgence of smog in Mexico City in 2016, commuters wore face masks and took advantage of free mass transit once authorities restricted car traffic.

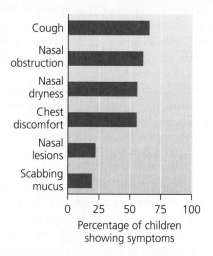

FIGURE 2 Mexico City children show respiratory symptoms from air pollution. These data are from 174 Mexico City children. Of 27 similar children from less-polluted areas outside Mexico City, none showed any of these conditions. *Data from Calderón-Garcidueñas, L., et al., 2003. Respiratory damage in children exposed to urban pollution.* Pediatric Pulmonology *36: 148–161.*

Air pollution also harms the heart and the cardiovascular system, affecting heart rate, blood pressure, blood clotting, blood vessels, and atherosclerosis. Epidemiological studies (p. 375) show that pollution correlates with emergency room admissions for heart attacks, chest pain, and heart failure, as well as death from heart-related causes. This is because tiny particulates can work their way into the bloodstream, causing the heart to reduce blood flow or go out of rhythm. The heart mounts an inflammatory response against pollutant particles laden with dead bacteria in the blood, but if pollution is persistent, the inflammation becomes chronic and stresses the heart. Even young people are at risk. One Mexican research team analyzed the hearts of 21 Mexico City residents who had died at an early age and found that pollution exacts a toll before age 18.

All these impacts of air pollution on the lungs and heart can lead to higher rates of death. Studies by one research team in Mexico City compared death certificate records against air pollution measurements. The team found that death rates rose immediately after severe pollution episodes, especially in response to particulate matter (**FIGURE 3**).

Today, researchers are learning that air pollution also affects our brains. Calderón-Garcidueñas was one of the first to recognize this connection. She noticed that older dogs in polluted Mexico City neighborhoods often seemed lethargic, disoriented, and senile. To test her observations scientifically, she examined the brains of such dogs after they died and compared them to brains of dogs from less-polluted areas. The brains of dogs from polluted parts of the city showed deposits of the protein amyloid ß, the "plaques" that signal Alzheimer's disease.

Subsequent research of hers and others has indicated that pollution appears to damage children's brain tissue in ways similar to Alzheimer's disease. In one study, Calderón-Garcidueñas used brain scans and found that 56% of Mexico City youth had lesions on the prefrontal cortex, whereas fewer than 8% did in a region with clean air. In another study, her team compared 20 children from Mexico City with 10 children from a Mexican city with clean air, measuring their cognitive skills and scanning their brains with magnetic resonance imagery (MRI). The Mexico City children performed more poorly on most cognitive tests of reasoning, knowledge, and memory. The differences in cognition were consistent with differences in volume of white matter in key portions of the brain as revealed by the MRIs.

Such findings led researchers at the University of Southern California to run rigorously controlled experimental tests with lab mice. These scientists are collecting polluted air alongside Los Angeles freeways and pumping the pollutants into the air that lab mice breathe. They then compare the brains of these mice after death to those of mice that breathed clean air. Results thus far show that tiny $PM_{2.5}$ pollutants do indeed pass through tissue layers and harm the brains of the mice.

Today, a number of scientists are conducting long-term epidemiological research to better assess the impacts of air pollution on the human brain. A 2016 review of 18 such studies from six nations found that all but one showed some correlation between air pollution and dementia. Continued research could reveal promising new avenues to fight dementia and Alzheimer's disease.

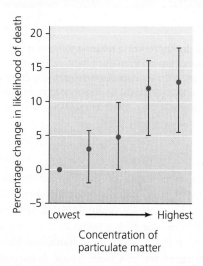

FIGURE 3 Rates of death increase in Mexico City with exposure to air pollution. *Data from Borja-Aburto, V., et al., 1997. Ozone, suspended particulates, and daily mortality in Mexico City.* American Journal of Epidemiology *145: 258–268.*

(a) Beijing, China

(b) Delhi, India

FIGURE 17.15 Industrializing nations today are struggling with severe air pollution in major cities. At Tiananmen Square in Beijing, China **(a)**, children wear face masks during an "airpocalypse" gripping the city. Delhi, India **(b)** has been rated by the World Health Organization as having the poorest air quality in the world.

Currently, inhabitants of the vast sprawling cities of India, China, Pakistan, Bangladesh, and other Asian nations suffer the world's worst air quality (**FIGURE 17.15**). According to 2018 WHO data for $PM_{2.5}$ pollutants, 11 of the 12 most polluted cities on Earth today are in India, and 54 of the most polluted 100 cities are in China. Many Asian nations have fueled their rapid industrial development with coal, the most polluting fossil fuel. Power plants and factories often use outdated, inefficient, heavily polluting technology because it is cheaper and quicker to build. At the same time, car ownership is skyrocketing as incomes rise. Because many cities of the developing world struggle with challenges of urban poverty, they have made relatively few efforts to tackle air pollution, despite its vast toll on health.

Many people in the Middle East and Africa also suffer dangerous air quality, in cities from Cairo to Peshawar to Kampala to Baghdad to Kabul. Urban air quality is nearly as bad in eastern European nations such as Poland and Bulgaria, where old Soviet-era factories pollute the air and where many people burn coal for home heating.

Across India, China, and other nations of southern Asia, pollution from cars, industry, agriculture, and wood-burning stoves has resulted in a 2-mile-thick layer of pollution that hangs over the region throughout the dry season each December through April. This massive layer of brownish haze is dubbed the Asian Brown Cloud or Atmospheric Brown Cloud. Experts estimate that it reduces the sunlight reaching Earth's surface in southern Asia by 10–20%, promotes flooding in some areas and drought in others by altering the monsoon, decreases rice yields by 5–10%, speeds the melting of Himalayan glaciers by depositing dark soot that absorbs sunlight, and contributes to many thousands of deaths each year.

In China's capital of Beijing in the winter of 2013, pollution became so severe that airplane flights were canceled and people wore face masks to breathe. Levels of particulate matter were literally off the charts; a monitor atop the U.S. Embassy designed to measure the Air Quality Index detected record-breaking readings beyond the maximum value of 500, up to 800. During this "airpocalypse," a fire at a factory went unnoticed for three hours because the pollution was so thick that no one could see the smoke. Thousands of people suffered ill health as the pollution soared 30 times past WHO safe limits. Each winter since that time, Beijing's "airmageddon" returned. Conditions became so bad that the national government and the official state media were finally forced to admit the problem and begin a public discussion about solutions.

In many Chinese cities, air pollution is worse than in Beijing, and across China, the health impacts of air pollution are enormous. Breathing the air in some Chinese cities is like smoking two packs of cigarettes a day. Recent research reports blame outdoor air pollution for more than 1 million premature deaths in China each year. One study found that residents of polluted northern China die, on average, five years earlier than residents of southern China, where the air is cleaner. Moreover, prevailing westerly winds carry some of China's pollution across the Pacific Ocean to North America. A 2014 study calculated that pollution from China adds at least one smoggy day per year to Los Angeles's total.

In 2015, China's citizens were transfixed by an online video, *Under the Dome,* that went viral. This powerful documentary, produced and narrated by a well-known Chinese investigative reporter, Chai Jing, confirmed for China's people what they already know: They spend their daily lives trapped under a dome of dangerously polluted air. The Chinese government banned the video yet also proclaimed a commitment to fighting pollution. Many commentators drew a parallel with *Silent Spring,* Rachel Carson's 1962 book (p. 171) that galvanized the environmental movement in the United States.

China's government is now striving to reduce pollution. It has closed down some heavily polluting factories and mines, phased out some subsidies for polluting industries, and installed pollution controls in power plants. It subsidizes efficient electric heaters and gas furnaces for homes to replace dirty, inefficient coal stoves. It has mandated cleaner formulations for gasoline and diesel and is raising standards for fuel efficiency and emissions for cars beyond what the United States requires. It has set up a nationwide air quality monitoring system at least as informative and transparent as that of the United States. In Beijing, mass transit is being expanded, many buses run on natural gas, and heavily polluting vehicles are kept out of the central city. China is also developing wind, solar, and nuclear power more aggressively than any other nation in the world, to substitute for coal-fired power. Global health and development experts are watching to see whether China can clean its air as effectively as Western democracies have and whether India and other nations can follow with success of their own.

Smog poses health risks

Now let's take a closer look at one of the most prevalent types of air pollution: smog. **Smog** is a general term for a mixture of air pollutants that can accumulate as a result of fossil fuel combustion, generally over industrial regions or urban areas with heavy automobile traffic.

Since the onset of the industrial revolution, cities have suffered a type of smog known as **industrial smog** (FIGURE 17.16a). When coal or oil is burned, some portion is completely combusted, forming CO_2; some is partially combusted, producing CO; and some remains unburned and is released as soot (particles of carbon). Moreover, coal contains contaminants such as mercury and sulfur. Sulfur reacts with oxygen to form sulfur dioxide, which can undergo a series of reactions to form sulfuric acid and other compounds. These substances, along with soot, are the main components of industrial smog.

The most severe industrial smog event in the United States occurred in the small town of Donora, Pennsylvania, in 1948 (FIGURE 17.16b). Donora is located in a mountain valley, and one cloudy day after air had cooled during the night, the morning sun did not reach the valley floor to warm and disperse the cold air. The resulting temperature inversion trapped smog containing particulate matter emissions from a steel and wire factory. Twenty-one people died, and more than 6000 people—nearly half the town—became ill.

The world's worst industrial smog crisis occurred in London, England, in 1952, when a high-pressure system settled over the city for several days, trapping sulfur dioxide and particulate matter emitted from factories and coal-burning stoves. The foul conditions killed 4000 people—and by some estimates up to 12,000.

In the wake of "killer smog" episodes such as those in London and Donora, governments began regulating industrial emissions, which greatly reduced industrial smog.

(a) Formation of industrial smog

(b) Donora, Pennsylvania, at midday in its 1948 smog event

FIGURE 17.16 Industrial smog results from fossil fuel combustion. When coal or oil is burned in a power plant or factory, soot (particulate matter of carbon) is released, and sulfur contaminants give rise to sulfur dioxide, which may react with atmospheric gases to produce further compounds **(a).** Carbon monoxide and carbon dioxide are also emitted. Under certain weather conditions, industrial smog can blanket whole regions, as it did in Donora, Pennsylvania **(b),** shown in the daytime during its deadly 1948 smog episode.

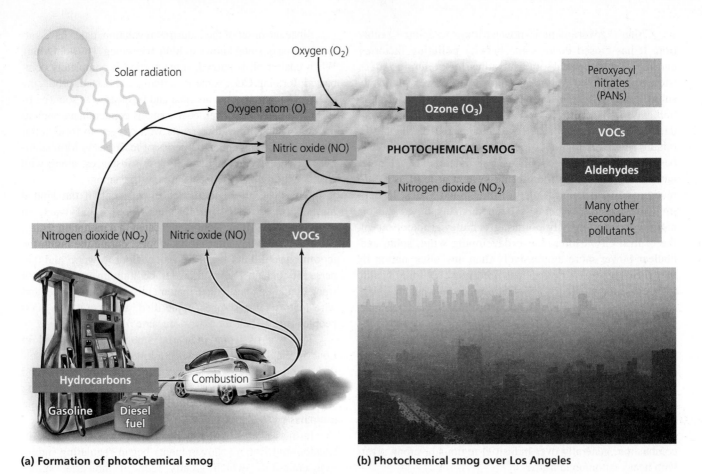

(a) Formation of photochemical smog

(b) Photochemical smog over Los Angeles

FIGURE 17.17 **Photochemical smog results when pollutants from automobile exhaust react amid exposure to sunlight.** Nitrogen dioxide, nitric oxide, and VOCs initiate a series of chemical reactions **(a)** that produce a toxic brew of secondary pollutants, including ozone, peroxyacyl nitrates (PANs), and aldehydes. Photochemical smog is common over Los Angeles **(b)** and many other urban areas, especially those with hilly topography or frequent inversions.

However, in industrializing regions such as China, India, and eastern Europe, coal burning and lax pollution control continue to result in industrial smog that poses significant health risks.

Most smog in urban areas today results largely from automobile exhaust. In Mexico City, vehicles contribute 31% of VOCs, 50% of sulfur dioxide, and 82% of nitrogen oxides. Some of these emissions react with sunlight, so pollution tends to be worst in cities with sunny climates, such as Mexico City and Los Angeles. Such cities suffer from **photochemical smog,** which forms when sunlight drives chemical reactions between primary pollutants and atmospheric compounds, producing a cocktail of more than 100 different chemicals, tropospheric ozone often being the most abundant (**FIGURE 17.17a**). Because it also includes NO_2, photochemical smog generally appears as a brownish haze (**FIGURE 17.17b**).

Hot, sunny, windless days in urban areas create ideal conditions for the formation of photochemical smog. On a typical weekday, exhaust from morning traffic releases NO and VOCs into a city's air. Sunlight then promotes the production of ozone and other secondary pollutants, leading pollution typically to peak in midafternoon. Photochemical

smog irritates people's eyes, noses, and throats, and over time can lead to asthma, lung damage, heart problems, decreased resistance to infection, and even cancer.

We can take steps to reduce smog

Photochemical smog afflicts countless American cities and suburbs, from Atlanta to Newark to Houston to Salt Lake City. Mayors, city councilors, and state and federal regulators everywhere are trying to devise ways to clear their air.

Los Angeles's struggle with air pollution began in 1943, when the city's first major smog episode cut visibility to three blocks. With the city's image as a clean and beautiful coastal haven at risk, civic leaders confronted the problem head-on. Los Angeles's quest to understand and solve its smog problem spurred much of the original research into how photochemical smog forms and how automobiles might burn fuel more cleanly.

Los Angeles's city and county officials began their air pollution control efforts by passing ordinances restricting emissions from power plants, oil refineries, and petrochemical facilities and then targeted emissions from motor vehicles. Because air pollution spreads from place to place, responsibility for pollution control soon moved from the city and county levels to the state and federal levels.

California took the lead among U.S. states in adopting pollution control technology and setting emissions standards for vehicles. In 1967, state leaders established the California Air Resources Board, the first state agency focused on regulating air quality. Today in California and 33 other states, drivers are required to have their vehicle exhaust inspected periodically. These inspection programs require owners to repair cars that emit excessive pollution, and they have cut emissions by 30%.

California's demands also helped lead the auto industry to develop less-polluting cars. A study by the non-profit group Environment California concluded that a new car today generates just 1% of the smog-forming emissions of a 1960s-era car. For this reason, the air is cleaner, even with many more vehicles on the road. In Los Angeles, VOC pollution has declined by 98% since 1960, even though the city's drivers now burn 2.7 times more gasoline. L.A.'s peak smog levels have also decreased substantially since 1980 (**FIGURE 17.18**).

Despite its progress, Los Angeles still suffers the worst tropospheric ozone pollution of any U.S. metropolitan area, according to the American Lung Association. L.A. residents breathe air exceeding California's health standard for ozone on more than 90 days per year. One recent study calculated that air pollution in the L.A. basin and the nearby San Joaquin Valley each year causes nearly 3900 premature deaths and costs society $28 billion (due to hospital admissions, lost workdays, etc.).

Across the world, many cities struggle with photochemical smog and are working to develop solutions. One example is Tehran, the capital of Iran. Here, city leaders now require vehicle inspections, regulate traffic into the city center, and pay drivers to trade in old polluting cars for newer, cleaner ones. The government reduced sulfur in diesel fuel, removed lead from gasoline, converted buses to run on (cleaner-burning) natural gas, and began expanding the subway system.

Of all the world's cities, Mexico City is perhaps most celebrated for its success in reducing smog—once the worst in the world—even as its population, cars, and economic activity continue to grow. As a result of many steps taken (**TABLE 17.1**), smog has been reduced since 1990 by more than half. The city has enjoyed even more success with other pollutants: particulate matter is down by 70%, carbon monoxide

WEIGHING
the **issues**

Smog-Busting Solutions

Does the city you live in, or a major city near you, suffer from photochemical smog or other air pollution? How is this city responding? What policies do you think it should pursue? What benefits might your city enjoy from such policies? Might they cause any unintended consequences?

TABLE 17.1 Steps Mexico City Has Taken to Fight Smog

- Established no-driving days (*Hoy No Circula* program)
- Required vehicle emissions tests twice yearly
- Mandated catalytic converters
- Developed upcoming regulations on truck emissions

- Removed lead from gasoline
- Reduced sulfur in diesel fuel
- Reformulated LPG for cleaner cooking and heating

- Shut down a major oil refinery
- Shifted heavy industries outside city limits
- Tightened pollution regulations; pushed shift to natural gas

- Launched continent's biggest bike-sharing program
- Launched car-sharing program
- Expanded number of buses
- Expanded number of subway lines
- Introduced low-emission buses and taxis

FIGURE 17.18 In the Los Angeles region, tropospheric ozone in photochemical smog has been reduced since the 1970s, thanks to public policy and improved automotive technology. Ozone pollution still violates the federal health standard, however. *Data from South Coast Air Quality Management District.*

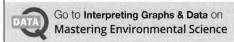

Go to **Interpreting Graphs & Data** on **Mastering Environmental Science**

by 74%, sulfur dioxide by 86%, and lead by 95%. A new federal law coming into force in 2021 should help further by tightening emissions regulations on trucks and buses, making Mexican standards similar to those in the United States, Europe, and Japan.

Should we regulate greenhouse gases as air pollutants?

Even as our society seeks to reduce various air pollutants, we continue to release vast quantities of the greenhouse gases (p. 488) that are driving global climate change (Chapter 18). As a result, our emission of carbon dioxide and other gases that warm the lower atmosphere is arguably today's biggest air pollution problem. Industry and utilities generate many of these emissions, but we all contribute by living carbon-intensive lifestyles. Each year, the average American vehicle driver releases close to 6 metric tons of carbon dioxide, 275 kg (605 lb) of methane, and 19 kg (41 lb) of nitrous oxide, all of them greenhouse gases that drive climate change.

In 2007, the U.S. Supreme Court ruled that the EPA has legal authority under the Clean Air Act to regulate carbon dioxide and other greenhouse gases as air pollutants. President Barack Obama urged Congress to address greenhouse gas emissions through bipartisan legislation. When Congress failed to do so, Obama instructed the EPA to develop regulations for these emissions. In 2011, the EPA introduced moderate carbon emission standards for cars and light trucks, and in 2012, it announced that it would begin phasing in limits on carbon emissions for new coal-fired power plants and cement factories (but not existing ones).

The coal-mining and petrochemical industries objected, and several states joined them in suing the EPA, but a court of appeals unanimously upheld the EPA regulations. The automotive industry supported these regulations. U.S. automakers had begun investing in fuel-efficient vehicles and preferred to have one set of federal emissions standards rather than having to worry about meeting many differing state standards. The public also voiced strong support; 2.1 million Americans sent comments to the EPA in favor of its actions—a record number of public comments for any federal regulation.

In 2015, the EPA launched a regulatory plan for existing power plants, the Clean Power Plan. Under the plan, each state was allowed to choose how to reduce power plant emissions—by upgrading technology, switching from coal to natural gas, enhancing efficiency, promoting renewable energy, or through carbon taxes or cap-and-trade (p. 181) emissions trading programs. The plan aimed to cut carbon dioxide emissions from power plants by 32% below 2005 levels by the year 2030. The EPA estimated that SO_2 and NO_X would be reduced by 20%, that cleaner air would save 3600 lives, and that by 2030 the plan would bring public health and climate benefits worth $54 billion each year.

A number of industries and states challenged the plan in court, and in 2016, the Supreme Court issued a stay of the plan in a controversial 5-to-4 ruling, preventing the EPA from enforcing the plan until the lawsuits were resolved. Once Donald Trump became president, his EPA administrators acted to dismantle the plan amid his administration's aggressive efforts to roll back Obama-era environmental policies.

Advocates of action against climate change will no doubt continue to face formidable opposition from emitting industries and from policymakers who fear that regulations will hamper economic growth. Yet if we were able to reduce emissions of other major pollutants sharply since 1970 while advancing our economy, we can hope to achieve similar results in reducing greenhouse gas emissions. Indeed, although U.S. carbon dioxide emissions have risen significantly since 1970, they fell by 11% from 2007 to 2018 even as the economy grew. This decrease in emissions resulted from a shift from coal to cleaner-burning natural gas and from improved fuel efficiency in automobiles and other technologies.

Ozone Depletion and Recovery

Although ozone at ground level in the troposphere is a pollutant in photochemical smog, ozone is a highly beneficial gas in the stratosphere, where it forms the ozone layer (p. 453; see Figure 17.2). Within the ozone layer, concentrations of ozone reach only about 12 parts per million, but ozone molecules are so effective at absorbing the sun's ultraviolet radiation that even this diffuse concentration helps protect life on Earth's surface from UV radiation's damaging impacts on tissues and DNA.

A generation ago, scientists discovered that our planet's stratospheric ozone was being depleted, posing a major threat to human health and the environment. Years of dynamic research by hundreds of scientists (see **THE SCIENCE BEHIND THE STORY**, pp. 472–473) revealed that certain airborne chemicals destroy ozone and that most of these **ozone-depleting substances** are human-made. Our subsequent campaign to halt degradation of the stratospheric ozone layer stands as one of society's most successful efforts to address a major environmental problem.

Synthetic chemicals deplete stratospheric ozone

Researchers identifying ozone-depleting substances pinpointed primarily **halocarbons**—human-made compounds derived from simple hydrocarbons (p. 28) in which hydrogen atoms are replaced by halogen atoms such as chlorine, bromine, or fluorine. In the 1970s, industry was producing more than 1 million tons per year of one type of halocarbon, **chlorofluorocarbons (CFCs).** CFCs were useful as refrigerants, as fire extinguishers, as propellants for aerosol spray cans, as cleaners for electronics, and for making

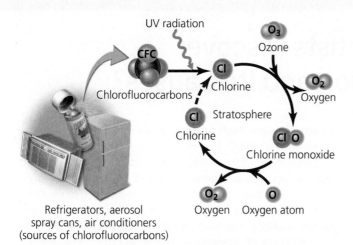

FIGURE 17.19 CFCs destroy ozone in a multistep process, repeated many times. A chlorine atom released from a CFC molecule in the presence of UV radiation reacts with an ozone molecule, forming one molecule of oxygen gas and one chlorine monoxide (ClO) molecule. The oxygen atom of the ClO molecule then binds with a stray oxygen atom to form oxygen gas, leaving the chlorine atom to begin the destructive cycle anew.

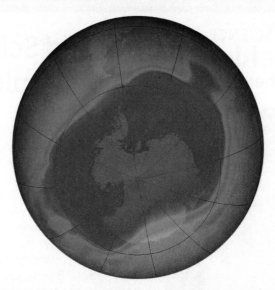

FIGURE 17.20 The "ozone hole" is a vast area of thinned ozone density in the stratosphere over the Antarctic region. It has reappeared seasonally each September in recent decades. Colorized satellite imagery of Earth's Southern Hemisphere from September 24, 2006, shows the ozone hole (purple/blue) at its maximal recorded extent to date.

polystyrene foam. Because CFCs rarely reacted with other chemicals, scientists surmised that they would be harmless to people and the environment.

Unfortunately, the nonreactive qualities that made CFCs ideal for industrial purposes were having disastrous consequences for the ozone layer. Whereas reactive chemicals get broken down quickly, CFCs reach the stratosphere unchanged and can linger there for a century or more. In the stratosphere, intense UV radiation from the sun eventually breaks CFCs into their constituent chlorine and carbon atoms. In a two-step chemical reaction (FIGURE 17.19), each newly freed chlorine atom can split an ozone molecule and then ready itself to split more. During its long residence time in the stratosphere, each free chlorine atom can catalyze the destruction of as many as 100,000 ozone molecules!

The ozone hole appears each year

In 1985, researchers shocked the world when they announced that stratospheric ozone levels over Antarctica in the southern springtime had declined by nearly half during just the previous decade, leaving a thinned ozone concentration that was soon named the **ozone hole** (FIGURE 17.20). During each Southern Hemisphere spring since then, ozone concentrations over this immense Antarctic region have dipped to roughly half their historic levels.

Extensive scientific detective work has revealed why seasonal ozone depletion is focused over Antarctica (and, to a lesser extent, the Arctic). During the dark and frigid Antarctic winter (June to August), temperatures in the stratosphere dip below −80°C (−112°F), enabling unusual high-altitude *polar stratospheric clouds* to form. Many of these icy clouds contain condensed nitric acid, which splits chlorine atoms off from compounds such as CFCs. The freed chlorine atoms accumulate in the clouds, trapped over Antarctica by wind currents that swirl in a circular *polar vortex* (p. 499) that prevents air from mixing with the rest of Earth's atmosphere.

In the Antarctic spring (starting in September), sunshine returns and dissipates the clouds. This releases the chlorine atoms, which begin destroying ozone. The solar radiation also catalyzes chemical reactions, speeding up ozone depletion as temperatures warm. The ozone hole lingers over Antarctica until December, when warmed summer air shuts down the polar vortex, allowing ozone-depleted air to diffuse away and ozone-rich air from elsewhere to stream in. The ozone hole vanishes until the following spring.

By the time researchers had learned all this, plummeting ozone levels had become a serious international concern. Ozone depletion was occurring globally, year-round—not just in Antarctica. Scientists worried that intensified UV exposure at Earth's surface would promote skin cancer, damage crops, and kill off ocean phytoplankton, the base of the marine food chain.

We addressed ozone depletion with the Montreal Protocol

Policymakers responded to the scientific concerns, and international efforts to restrict production of CFCs bore fruit in 1987 with the **Montreal Protocol.** In this treaty, the world's nations agreed to cut CFC production in half by 1998.

THE SCIENCE behind the story

How Did Scientists Discover Ozone Depletion and Its Causes?

Releasing a high-altitude balloon equipped to measure stratospheric ozone

In discovering the depletion of stratospheric ozone and coming to understand the roles of halocarbons and other substances, scientists have relied on historical records, field observations, laboratory experiments, computer models, and satellite technology.

The story starts back in 1924, when British scientist G.M.B. Dobson built an instrument that measured atmospheric ozone concentrations by sampling sunlight at ground level and comparing the intensities of wavelengths that ozone does and does not absorb. By the 1970s, the Dobson ozone spectrophotometer was being used by a global network of observation stations.

Meanwhile, atmospheric chemists were learning how stratospheric ozone is created and destroyed. Ozone and oxygen exist in a natural balance, with one occasionally reacting to form the other and oxygen being far more abundant. Researchers found that certain chemicals naturally present in the atmosphere, such as hydroxyl (OH) and nitric oxide (NO), destroy ozone, making the ozone layer thinner than it would otherwise be. And nitrous oxide (N_2O) produced by soil bacteria can make its way to the stratosphere and produce NO, Dutch meteorologist Paul Crutzen reported in 1970. This last observation was important, because some human activities, such as fertilizer application, were increasing emissions of N_2O.

Following Crutzen's report, American scientists Richard Stolarski and Ralph Cicerone showed in 1973 that chlorine atoms can catalyze the destruction of ozone even more effectively than N_2O can. And two years earlier, British scientist James Lovelock had developed an instrument to measure trace amounts of atmospheric gases and found that virtually all the chlorofluorocarbons (CFCs) humanity had produced in the past four decades were still aloft, accumulating in the stratosphere.

The stage was now set for the key insight. In 1974, American chemist F. Sherwood Rowland and his Mexican postdoctoral associate Mario Molina took note of all the preceding research and realized that CFCs were rising into the stratosphere, being broken down by UV radiation, and releasing chlorine atoms that ravaged the ozone layer (see Figure 17.19, p. 471). Molina and Rowland's analysis, published in the journal *Nature,* earned them the 1995 Nobel prize in chemistry jointly with Crutzen.

The paper also sparked discussion about setting limits on CFC emissions. Industry leaders attacked the research; DuPont's chairman of the board reportedly called it "a science fiction tale . . . a load of rubbish . . . utter nonsense." But measurements in the lab and in the stratosphere by numerous researchers soon confirmed that CFCs and other halocarbons were indeed depleting ozone. In response, the United States and several other nations banned the use of CFCs in aerosol spray cans in 1978. Other uses continued, however, and by the early 1980s, global production of CFCs was again on the rise.

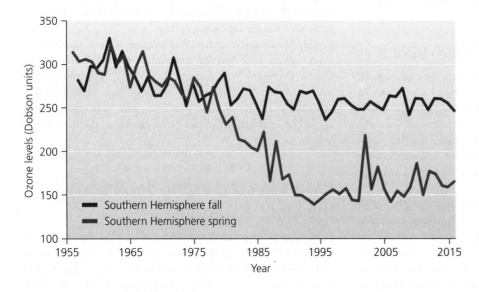

FIGURE 1 Data from Halley, Antarctica, show a decrease in springtime stratospheric ozone concentrations from the 1950s to 1990. Once ozone-depleting substances began to be phased out, ozone concentrations stopped declining. *Data from British Antarctic Survey.*

Then, a shocking new finding spurred the international community to take action. In 1985, Joseph Farman and colleagues analyzed data from a British research station in Antarctica that had been recording ozone concentrations since the 1950s. Farman's team reported in *Nature* that springtime Antarctic ozone concentrations had plummeted by 40–60% just since the 1970s (**FIGURE 1**). Farman's team had beaten a group of National Aeronautics and Space Administration (NASA) scientists to the punch. The NASA scientists were sitting on reams of data from satellites showing a global drop in ozone levels, but they had not yet submitted their analysis for publication.

But why should an "ozone hole" be localized over Antarctica of all places? And why only in the southern spring? To determine what was causing this odd phenomenon, atmospheric chemists Susan Solomon, James Anderson, Crutzen, and others mounted expeditions in 1986 and 1987 to analyze atmospheric gases using ground stations and high-altitude balloons and aircraft. From their data they figured out how the region's polar stratospheric clouds and circulating winds provide ideal conditions for chlorine from CFCs and other chemicals to set in motion the destruction of massive amounts of ozone.

By 1987, the mass of scientific evidence helped convince the world's nations to agree on the Montreal Protocol, which aimed to cut CFC production in half by 1998. Within two years, further scientific evidence and computer modeling showed that more drastic measures were needed. In 1990, the Montreal Protocol was strengthened to include a complete phaseout of CFCs by 2000, in the first of several follow-up agreements. Today, amounts of ozone-depleting substances in the stratosphere are beginning to level off.

As the ozone layer begins a long-term recovery, scientists continue their research. In 2009, a team led by A.R. Ravishankara of the National Oceanic and Atmospheric Administration determined that nitrous oxide (N_2O) had become the leading cause of ozone depletion (**FIGURE 2**). Its emissions are not regulated, so its impacts had come to surpass those that the remaining halocarbons were exerting. Ravishankara's team pointed out that regulating nitrous oxide, which is also a potent greenhouse gas, would help address climate change as well as speed ozone recovery. However, the picture has turned out to be more complicated, because two other greenhouse gases—carbon dioxide and methane—both help increase amounts of stratospheric ozone, so our emission of these gases is actually helping the ozone layer to recover.

Today, scientists are keeping close track of numerous measurements from satellites and from ground-based monitoring stations to gain a detailed chronicle of the ozone layer's slow recovery across the globe (**FIGURE 3**). As atmospheric concentrations of halocarbons continue to decline, the changing balance of gases such as nitrous oxide, methane, and carbon dioxide will play a larger role in determining the speed and extent of the recovery. If all goes well, scientists expect full recovery at midlatitudes by midcentury, followed by full recovery in the polar regions one or two decades later.

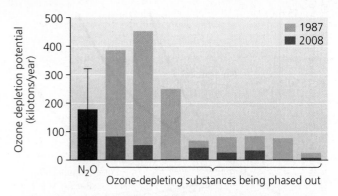

FIGURE 2 **Because most ozone-depleting substances were phased out beginning in 1987, nitrous oxide (N_2O; left bar) has become the primary ozone-depleting substance we emit.** It has less impact than CFCs and other halocarbons did in 1987 (**full-bar values**) but more impact than any other substance today (**lower portion of bars**). *Adapted from Ravishankara, A.R., et al., 2009. Nitrous oxide (N_2O): The dominant ozone-depleting substance emitted in the 21st century. Science 326: 123–125, FIGURE 1. Reprinted by permission of AAAS and the author.*

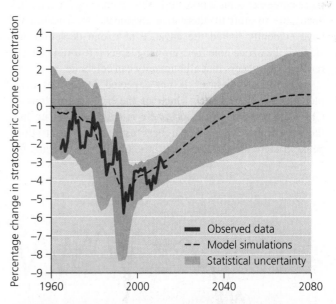

FIGURE 3 **Globally, stratospheric ozone is starting to recover.** Data averaged from multiple satellite and ground-based sources (**red line**) show a global decrease in observed ozone concentrations, followed by a gradual increase. Simulations from many models (**blue dashed line** shows mean; **gray shading** shows statistical uncertainty) indicate that recovery should be complete later this century. *Adapted from World Meteorological Organization, 2014. Scientific assessment of ozone depletion: 2014. Geneva, Switzerland: WMO, Global Ozone Research and Monitoring Project—Report #55.*

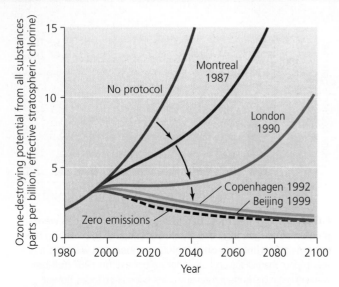

FIGURE 17.21 The Montreal Protocol reduced atmospheric concentrations of pollutants that destroy stratospheric ozone. Follow-up agreements in London, Copenhagen, and Beijing reduced them still more. In this graph, *y*-axis values give a collective measure of ozone-destructive potential from all substances, and most data indicate future projected values. *Data from Emmanuelle Bournay, UNEP/GRID-Arendal, http://maps.grida.no.*

Follow-up agreements (**FIGURE 17.21**) deepened the cuts, advanced timetables for compliance, and added nearly 100 ozone-depleting substances. Most substances covered by these agreements have now been phased out, and industry has been able to shift to alternative chemicals. As a result, we have evidently halted the advance of ozone depletion and stopped the ozone hole from growing larger (**FIGURE 17.22**)—a remarkable success that all humanity can celebrate.

Earth's ozone layer is not expected to recover completely until after 2060. Much of the 5.5 million tons of CFCs emitted into the troposphere is still diffusing up into the stratosphere, and researchers estimate concentrations peaking there around 2020. Because of this time lag and the

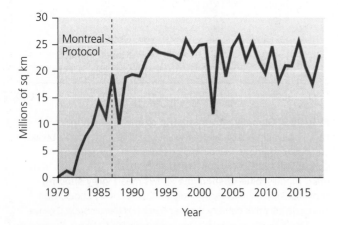

FIGURE 17.22 Phaseouts of ozone-depleting substances since 1987 have halted the growth of the Antarctic ozone hole. *Data from NASA, reflecting averages from 7 Sept. to 13 Oct. each year.*

long residence times of many halocarbons, we can expect many years to pass before our policies have the desired result.

One challenge in restoring the ozone layer is that nations can plead for some ozone-depleting substances to be exempt from the ban. For instance, the United States was allowed to continue using methyl bromide, a fumigant used to control pests on strawberries. Yet despite the remaining challenges, the Montreal Protocol and its follow-up amendments are widely considered our biggest success story in addressing a global environmental problem. The success has been attributed to several factors:

1. **Informative scientific research developed rapidly, facilitated by new and evolving technologies.**

2. **Policymakers engaged industry in helping solve the problem.** Industry became willing to develop replacement chemicals in part because patents on CFCs were running out and firms wanted to profit from next-generation chemicals.

3. **Implementation of the Montreal Protocol after 1987 followed an adaptive management approach** (p. 318), adjusting strategies in response to new scientific information, technological advances, and economic data.

Because of its success in addressing ozone depletion, the Montreal Protocol is widely viewed as an inspiring model for international cooperation needed to resolve other pressing global problems, including biodiversity loss (p. 297), persistent organic pollutants (p. 384), and climate change (p. 516).

Addressing Acid Deposition

Just as stratospheric ozone depletion crosses political boundaries, so does another atmospheric pollution concern—**acid deposition,** the deposition of acidic (p. 29) or acid-forming pollutants from the atmosphere onto Earth's surface. As with ozone depletion, we are enjoying some success in addressing this challenge.

Fossil fuel combustion spreads acidic pollutants far and wide

Acid deposition often takes place by precipitation (commonly referred to as **acid rain**) but also may occur by fog, gases, or the deposition of dry particles. Acid deposition is one type of **atmospheric deposition,** a term that refers more broadly to the wet or dry deposition of a wide variety of pollutants, including mercury, nitrates, organochlorines, and others, from the atmosphere onto Earth's surface.

Acid deposition originates primarily with the emission of sulfur dioxide and nitrogen oxides, largely through fossil fuel combustion by automobiles, power plants, and industrial facilities. Once airborne, these pollutants react with water, oxygen, and oxidants to produce compounds of low pH (p. 29), primarily sulfuric acid and nitric acid. Suspended in the troposphere, droplets of these acids may travel days or weeks for hundreds of kilometers (**FIGURE 17.23**). Depending on climate, 20–80% of all acidic compounds emitted into the atmosphere may fall in precipitation, with the remainder falling as dry deposition.

Acid deposition has many impacts

Acid deposition has wide-ranging detrimental effects on ecosystems (**TABLE 17.2**). Acids leach nutrients such as calcium, magnesium, and potassium ions out of the topsoil, altering soil chemistry and harming plants and soil organisms. This process occurs as hydrogen ions from acid precipitation take the place of calcium, magnesium, and potassium ions in soil compounds, and these valuable nutrients leach into the subsoil, where they become inaccessible to plant roots.

Acid precipitation also "mobilizes" toxic metal ions such as aluminum, zinc, mercury, and copper by chemically converting them from insoluble forms to soluble forms. Elevated soil concentrations of metal ions such as aluminum damage the root tissues of plants, hindering their uptake of water and nutrients. In some areas, acid fog with a pH of 2.3 (equivalent to vinegar and more than 1000 times more acidic than normal

rainwater) has enveloped forests for extended periods, killing trees. Animals are affected by acid deposition, too; populations of snails and other invertebrates typically decline, which reduces the food supply for birds.

When acidic water runs off from land, it affects streams, rivers, and lakes. Thousands of lakes in Canada, Europe, the United States, and elsewhere have lost their fish because acid precipitation leaches aluminum ions out of soil and rock and into waterways. These ions damage the gills of fish and disrupt their salt balance, water balance, breathing, and circulation.

TABLE 17.2 Ecological Impacts of Acid Deposition

ACID DEPOSITION IN NORTHEASTERN FORESTS HAS . . .

- Accelerated leaching of base cations (ions such as Ca^{2+}, Mg^{2+}, NA^+, and K^+, which counteract acid deposition) from soil

- Allowed sulfur and nitrogen to accumulate in soil, where excess N can overfertilize native plants and encourage weeds

- Increased dissolved inorganic aluminum in soil, hindering plant uptake of water and nutrients

- Leached calcium from needles of red spruce, causing trees to die from wintertime freezing

- Increased mortality of sugar maples due to leaching of base cations from soil and leaves

- Acidified 41% of Adirondack, New York, lakes and 15% of New England lakes

- Diminished lakes' capacity to neutralize further acids

- Elevated aluminum levels in surface waters

- Reduced species diversity and abundance of aquatic life, affecting entire food webs

Adapted from Driscoll, C.T., et al., 2001. *Acid rain revisited.* Hubbard Brook Research Foundation. © 2001 C.T. Driscoll. Used with permission.

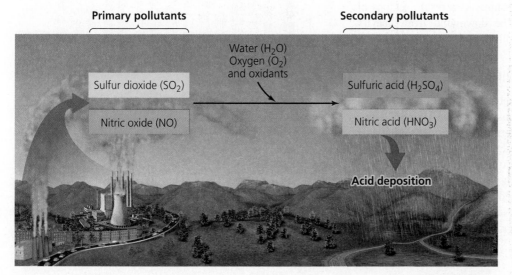

FIGURE 17.23 Acid deposition has impacts far downwind from where pollutants are released. Sulfur dioxide and nitric oxide emitted by industries, utilities, and vehicles react in the atmosphere to form sulfuric acid and nitric acid. These acidic compounds descend to Earth's surface in rain, snow, fog, and dry deposition.

FIGURE 17.24 Acid deposition corrodes statues and buildings. Shown is an Egyptian obelisk known as Cleopatra's Needle, in Central Park, New York City, **(a)** before and **(b)** after significant acid deposition.

(a) Before acid rain damage

(b) After acid rain damage

The severity of all these effects depends not only on the pH of the deposition but also on the acid-neutralizing capacity of the soil, rock, or water that receives the acidic input. Substrates differ naturally in their chemistry and pH, and regions with more alkaline soil, rock, or water have a greater capacity to neutralize acids and thus buffer themselves against acid deposition. This also means that once calcium or similar ions are leached from a soil, the soil becomes more sensitive to acidification.

Besides altering ecosystems, acid precipitation damages crops, erodes stone buildings, and corrodes vehicles, causing billions of dollars in damage. As acid precipitation erases the writing from tombstones and dissolves away features from ancient cathedrals in Europe, sacred temples in Asia, and revered monuments in Washington, D.C., it hastens the loss of cultural amenities that are beyond monetary value (**FIGURE 17.24**).

Because the pollutants leading to acid deposition can travel long distances, their impacts may be felt far from their sources. For instance, much of the pollution from power plants and factories in Pennsylvania, Ohio, and Illinois travels east with prevailing winds and falls out in states such as New York, Vermont, and New Hampshire. As a result, regions of greatest acidification tend to be downwind from heavily industrialized source areas of pollution.

We are addressing acid deposition

The Acid Rain Program established under the Clean Air Act of 1990 helped fight acid deposition in the United States. This program set up a cap-and-trade emissions trading system (p. 181) for sulfur dioxide. Coal-fired power plants were allocated permits for emitting SO_2 and could buy, sell, or trade these allowances. Each year, the overall amount of allowed pollution was lowered. (See p. 514 for further explanation of how such a system works.) The economic incentives created by this program encouraged polluters to

switch to low-sulfur coal, invest in technologies such as scrubbers, and devise other ways to become cleaner and more efficient. During the course of the cap-and-trade program, SO_2 emissions across the United States fell by 67% (**FIGURE 17.25**). As a result, average sulfate loads in precipitation across the eastern United States were 51% lower in 2008–2010 than in 1989–1991 and are still lower today.

The Acid Rain Program also required power plants to reduce nitrogen oxide emissions, with the EPA allowing plants flexibility in how they implemented the reductions. Emissions of NO_X fell significantly as a result, and wet

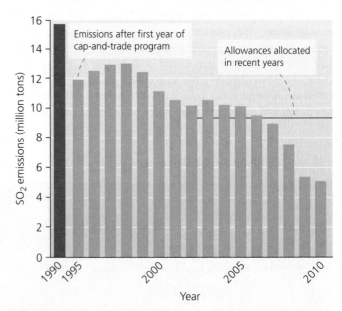

FIGURE 17.25 Sulfur dioxide emissions fell 67% in the wake of an emissions trading system. By 2010, emissions from U.S. power plants participating in this EPA program mandated by the 1990 Clean Air Act had dropped well below the amount allocated in permits **(black line)**. *Data from U.S. EPA.*

nitrogen deposition also declined. Thanks to the declines in SO_2 and NO_X, air and water quality improved throughout the eastern United States (**FIGURE 17.26**). This market-based program spawned similar cap-and-trade programs for other pollutants, including greenhouse gases (p. 488). The Los Angeles region adopted its own cap-and-trade program in 1994. The RECLAIM (Regional Clean Air Incentives Market) program has helped the L.A. basin reduce emissions of sulfur oxides and nitrogen oxides by more than 70%.

Many have attributed the success in reducing acid deposition nationwide to the Acid Rain Program, and the EPA has calculated that the program's economic benefits (in health care expenses avoided, for instance) outweighed its costs by

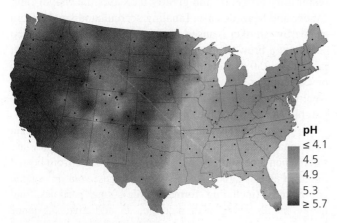

(a) pH of precipitation in 1990

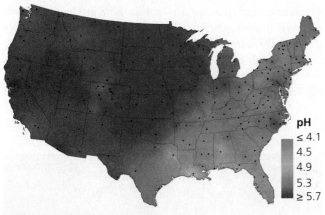

(b) pH of precipitation in 2017

FIGURE 17.26 Precipitation has become less acidic as air quality has improved under the Clean Air Act. Average pH values for precipitation recovered (i.e., increased back toward normal natural values) between **(a)** 1990 and **(b)** 2017. Precipitation remains most acidic in regions near and downwind from areas of heavy industry. Black dots represent monitoring locations. *Data from the National Atmospheric Deposition Program.*

 Go to **Interpreting Graphs & Data** on **Mastering Environmental Science**

40 to 1. However, some experts maintain that pollution declined because cleaner fuels became less expensive and because simultaneous conventional regulation mandated emissions cuts. Indeed, during this time period, European nations using command-and-control regulation (p. 178) instead of emissions trading reduced their SO_2 emissions by even more than the United States did. In 2011, emissions trading ended once the EPA issued its Cross-State Air Pollution Rule, which aimed to limit pollution drifting from upwind states into downwind states.

As with recovery of the ozone layer, there is a time lag before the positive consequences of emissions cuts kick in, so it is taking years for acidified ecosystems to recover. Research indicates that soils across the northeastern United States are showing signs of recovery but remain degraded and vulnerable. Scientists also point out that further pollution reductions are needed if we are to fully restore ecosystems in the Northeast and prevent further damage to property and infrastructure.

Although the United States, Canada, and western Europe are recovering from acid deposition after cutting sulfur emissions, acid deposition is becoming worse in many industrializing nations. Today, China emits the most sulfur dioxide of any nation as a result of coal combustion in power plants and factories that lack effective pollution control equipment. Not surprisingly, China has the world's worst acid rain problem. The Chinese government is tackling the issue, but it faces a challenge as the nation's industry and power demands continue to grow.

Overall, data on acid deposition show progress in controlling outdoor air pollution but also indicate that more can be done. The same can be said for indoor air pollution—a source of human health risks that is less familiar to most of us but equally hazardous.

Indoor Air Quality

Indoor air generally contains higher concentrations of pollutants than does outdoor air. As a result, the health impacts from **indoor air pollution** in workplaces, schools, and homes rival those from outdoor air pollution. WHO attributes nearly 3.8 million premature deaths each year to indoor air pollution (compared with 4.2 million for outdoor air pollution). Indoor air pollution takes more than 10,000 lives each day.

If this seems surprising, consider that the average American spends at least 90% of his or her time indoors. Then consider the dizzying array of consumer products in our homes and offices that play major roles in our daily lives. Many of these products are made of synthetic materials, and novel synthetic substances are not comprehensively tested for health effects before being brought to market (Chapter 14). Furniture, carpeting, cleaning fluids, insecticides, and plastics all exude volatile chemicals into the air.

FIGURE 17.27 In the developing world, many people build fires indoors for cooking and heating, as in this home in Burma. Indoor fires expose people to severe pollution from particulate matter and carbon monoxide.

Ironically, well-meaning attempts to enhance the energy efficiency of buildings have ended up worsening indoor air quality. To reduce heat loss, building managers often seal off ventilation, while designers often construct new buildings with limited ventilation and with windows that do not open. These steps save energy, but they also trap stable, unmixed air—and pollutants—inside. Fortunately, we have many feasible ways to address the primary causes of indoor air pollution.

Burning fuel causes indoor pollution in the developing world

Indoor air pollution has by far the greatest impact in the developing world, where poverty forces millions of families to burn wood, charcoal, animal dung, or crop waste inside their homes for cooking and heating, with little or no ventilation (FIGURE 17.27). In the air of such homes, concentrations of particulate matter commonly surpass U.S. EPA standards by 20 times, researchers have found. As a result, people inhale dangerous amounts of soot, carbon monoxide, and other pollutants on a daily basis. This promotes allergies, sinus infections, cataracts, asthma, bronchitis, and emphysema, and leads to millions of premature deaths by pneumonia, stroke, heart disease, and lung cancer. International health researchers estimate that indoor air pollution from burning fuelwood, dung, and coal is responsible for nearly 7% of all deaths worldwide each year. In addition, the widespread use of kerosene lamps for indoor lighting in villages without electricity exposes people to pollution from fine particles and other substances. Many people are not aware of the health risks from these pollution sources, and most are too poor to have viable alternatives.

Cigarette smoke and radon are the primary indoor pollutants in industrialized nations

In wealthier industrialized nations, the primary indoor air health risks are cigarette smoke and radon (a naturally occurring radioactive gas). Smoking is estimated in the United States alone to cause 160,000 lung cancer deaths per year and secondhand smoke to cause 3000. Radon is the second-leading cause of lung cancer in the developed world, responsible for an estimated 21,000 deaths per year in the United States and for 15% of lung cancer cases worldwide.

Smoking cigarettes and other tobacco products irritates the eyes, nose, and throat; worsens asthma and other respiratory ailments; and greatly increases the risk of lung cancer and heart disease. Inhaling secondhand smoke from a nearby smoker causes the same problems. Tobacco smoke is a brew of more than 4000 chemical compounds, more than 250 of which are known or suspected to be toxic or carcinogenic. Smoking has become less prevalent in wealthy societies as a result of public education campaigns, and many venues now ban smoking.

E-cigarettes (FIGURE 17.28) are thought to be safer than traditional cigarettes, but that does not mean that they are safe. When a person "vapes," or inhales aerosols from heated liquid in an e-cigarette, he or she may ingest volatile organic compounds (including formaldehyde), fine particles, and heavy metals along with nicotine, an addictive substance from tobacco that can harm brain development. In addition, many vaping liquids are flavored with chemicals such as diacetyl, which is linked to a rare but severe lung disease. Because e-cigarettes are a relatively new product and have received limited scrutiny from scientists and regulators, we do not fully know how the long-term consequences of vaping compare with those of traditional cigarette smoking. While research into lung illnesses associated with vaping proceeds, the Centers for Disease Control states that "e-cigarettes are

FIGURE 17.28 "Vaping" e-cigarettes poses substantial health risks. Although probably safer than traditional cigarettes, e-cigarettes deliver many harmful substances into the lungs.

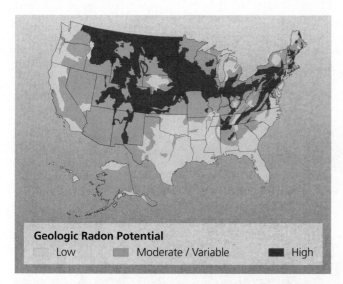

Geologic Radon Potential

| Low | Moderate / Variable | High |

FIGURE 17.29 One's risk from radon depends largely on underground geology. This map shows levels of risk across the United States. *Data from U.S. Geological Survey, 1993. Generalized geologic radon potential of the United States.*

not safe. . . . If you've never smoked or used other tobacco products or e-cigarettes, don't start."

Radon (p. 366) is a radioactive gas resulting from the natural decay of uranium in soil, rock, or water. It seeps up from the ground and can infiltrate buildings. Because radon is colorless and odorless, its presence can be impossible to predict without knowing an area's underlying geology (**FIGURE 17.29**). The only way to determine whether radon is entering a building is to sample air with a test kit. The EPA estimates that 6% of U.S. homes exceed its safety standard for radon. Since the 1980s, close to a million U.S. homes have undergone radon mitigation, and homes today are being built with radon-resistant features.

Many VOCs pollute indoor air

In our daily lives at home, we are exposed to many indoor air pollutants (**FIGURE 17.30**, p. 480). The most diverse are volatile organic compounds (p. 459). These airborne carbon-containing compounds are released by plastics, oils, perfumes, paints, cleaning fluids, adhesives, and pesticides. VOCs evaporate from furnishings, building materials, color film, carpets, laser printers, and sheets of paper. Some products, such as chemically treated furniture, release large quantities of VOCs when new and progressively less as they age. Other items, such as photocopying machines, emit VOCs each time they are used.

Although we are surrounded by products that emit VOCs, they are generally released in very small quantities. Studies find concentrations of VOCs in buildings to be nearly always less than 1 part per 10 million, but that is a much greater concentration than is found outdoors. Moreover, we experience instances of high exposure. The "new car smell" that fills the interiors of new automobiles comes from a complex mix of dozens of VOCs as they outgas from the newly manufactured plastic, metal, and leather components of the car. The smell diminishes with time, but some scientific studies warn of health risks from this mix and recommend that you keep a new car well ventilated.

The implications for human health of chronic exposure to VOCs are far from clear. Because they generally exist in low concentrations and because individuals regularly are exposed to mixtures of many different types, it is extremely difficult to study the effects of any one pollutant. An exception is formaldehyde, a VOC that has clear and known health impacts. Formaldehyde off-gasses from the glues and resins in pressed wood (such as plywood), insulation, and other products. It can irritate mucous membranes, can induce skin allergies, and is considered a likely carcinogen.

Living organisms can pollute

The most widespread source of indoor air pollution in the developed world may be living organisms. Tiny dust mites can worsen asthma and trigger allergies, as can dander (skin flakes) from pets. The airborne spores of some fungi, molds, and mildews can cause allergies, asthma, and other respiratory ailments. Some airborne bacteria can cause infectious disease, including legionnaires' disease. Of the estimated 10,000–15,000 annual U.S. cases of legionnaires' disease, 5–15% are fatal. Heating and cooling systems in buildings make ideal breeding grounds for microbes, providing moisture, dust, and foam insulation as substrates, along with air currents to carry the organisms aloft.

Microbes that induce allergic responses are thought to be a major cause of *building-related illness,* a sickness triggered by indoor pollution. When the cause of such an illness is a mystery and when symptoms are general and nonspecific, the illness is often called **sick building syndrome.** The U.S. Occupational Safety and Health Administration (OSHA) estimates that 30–70 million Americans have suffered ailments related to the building in which they live. We can reduce the prevalence of sick building syndrome by using low-toxicity construction materials and ensuring that buildings are well ventilated.

We can enhance indoor air quality

Using low-toxicity materials, monitoring air quality, keeping rooms clean, and providing adequate ventilation are the keys to alleviating indoor air pollution in most situations. Remedies for fuelwood pollution in the developing world include drying wood before burning (which reduces the amount of smoke produced), cooking outside, shifting to less-polluting fuels (such as natural gas), and replacing inefficient fires with cleaner stoves that burn fuel more efficiently. Installing hoods, chimneys, or cooking windows can increase ventilation for little cost, alleviating most indoor

WEIGHING
the issues

How Safe Is Your Indoor Environment?

Name some potential indoor air quality hazards in your home, work, or school environment. Are these spaces well ventilated? What could you do to improve the safety of the indoor spaces you use?

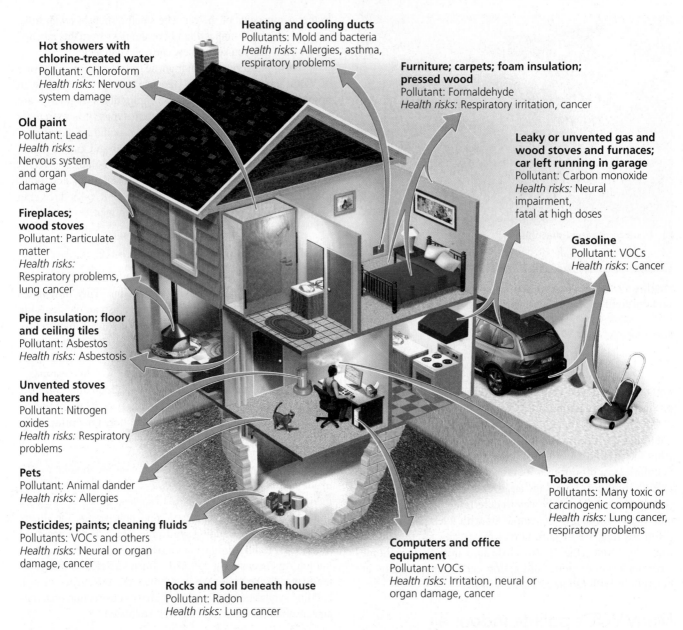

Hot showers with chlorine-treated water
Pollutant: Chloroform
Health risks: Nervous system damage

Old paint
Pollutant: Lead
Health risks: Nervous system and organ damage

Fireplaces; wood stoves
Pollutant: Particulate matter
Health risks: Respiratory problems, lung cancer

Pipe insulation; floor and ceiling tiles
Pollutant: Asbestos
Health risks: Asbestosis

Unvented stoves and heaters
Pollutant: Nitrogen oxides
Health risks: Respiratory problems

Pets
Pollutant: Animal dander
Health risks: Allergies

Pesticides; paints; cleaning fluids
Pollutants: VOCs and others
Health risks: Neural or organ damage, cancer

Heating and cooling ducts
Pollutants: Mold and bacteria
Health risks: Allergies, asthma, respiratory problems

Furniture; carpets; foam insulation; pressed wood
Pollutant: Formaldehyde
Health risks: Respiratory irritation, cancer

Leaky or unvented gas and wood stoves and furnaces; car left running in garage
Pollutant: Carbon monoxide
Health risks: Neural impairment, fatal at high doses

Gasoline
Pollutant: VOCs
Health risks: Cancer

Tobacco smoke
Pollutants: Many toxic or carcinogenic compounds
Health risks: Lung cancer, respiratory problems

Computers and office equipment
Pollutant: VOCs
Health risks: Irritation, neural or organ damage, cancer

Rocks and soil beneath house
Pollutant: Radon
Health risks: Lung cancer

FIGURE 17.30 The typical home contains many sources of indoor air pollution. Shown are common sources, the major pollutants they emit, and some of the health risks they pose.

smoke pollution. In regions without electricity, off-grid solar or wind installations can provide clean power for cooking, heating, and lighting. Because newer fuels and technologies are often more expensive than traditional biomass, government subsidies or microfinance loans are often helpful. Such investments pay off by providing people better health, more time and capacity to increase their incomes, and a better quality of life.

In the industrialized world, we can avoid cigarette smoke, limit our exposure to new plastics and treated wood, and restrict our contact with pesticides, cleaning fluids, and other toxic substances by keeping them in garages or outdoor sheds. The EPA recommends that we test our homes and offices for radon, mold, and carbon monoxide. Because

carbon monoxide is so deadly and so hard to detect, most U.S. states now require new homes to be equipped with alarms that sound if they detect dangerous levels of this gas. In addition, keeping rooms and air ducts clean and free of mildew and other biological pollutants will reduce potential irritants and allergens. Most of all, keeping our indoor spaces well ventilated will minimize concentrations of the pollutants among which we live.

Progress is being made worldwide in alleviating the health toll of indoor air pollution. Researchers calculate that rates of premature death from indoor air pollution dropped nearly 40% from 1990 to 2010 and are still dropping. With continued efforts, we should see further progress in safeguarding people's health.

TODAY, in Mexico City, the city's success in fighting air pollution is being tested as rising population, car ownership, and traffic congestion threaten to overwhelm the substantial progress that past leaders have made in cleaning the air. Road construction is outpacing mass transit improvement as more and more people move into outlying districts yet commute to jobs in the central city. By one recent analysis, Mexico City now suffers the worst traffic congestion in the world; the typical driver spends three hours a day stuck in traffic that moves an average of just 13 mph, wasting 9.5 days per year of each commuter's life. Since 2016, the city has experienced several air quality emergencies on a scale that residents had not witnessed in more than a decade—and smog still contributes to an estimated 4000 deaths each year.

Newly elected political leaders have vowed to rise to the challenge. Claudia Scheinbaum, Mexico City's first female mayor, has a scientific background in emissions research and made pollution control a theme of her 2018 campaign. She aims to make her city a global "center of innovation for public and private transport" by encouraging electric vehicles, improving the bus system, and adding a cable-car system for low-income commuters. Scheinbaum plans to meet with auto manufacturers to try to toughen emissions standards. Her efforts are being embraced at the national level by Mexico's new president, Andrés Manuel López Obrador. The mayor of Mexico City from 2000 to 2005, when great strides were made in improving air quality, López Obrador is viewed as a proponent of environmental sustainability. Residents of Mexico City hoping for cleaner air are welcoming this promising window of political opportunity. Their city's air quality will also benefit from tighter new federal emissions regulations on trucks and buses due to come into effect in 2021.

Just as Mexicans have found that past progress toward cleaner air cannot be taken for granted, people in Los Angeles and across the United States are also now realizing that past success is no guarantee of future success. Since 2017, decades of progress toward cleaner air have been put at risk in the wake of policy rollbacks and reduced enforcement of air quality regulations under the Trump administration. If the United States wishes to continue its march toward cleaner, healthier air, its leaders will need to prioritize this goal and facilitate the technological and policy approaches to achieve it.

Valuing scientific research is also crucial—science has proven key to addressing major air quality issues, including ozone depletion and acid deposition. Science also leads to technology, and a new array of creative technological innovations is emerging that may help us improve air quality. Some involve vacuuming pollutants out of the air and converting them into useful substances. In one such effort, a tower-like structure in the Dutch city of Rotterdam produces gemstones from fine particulate matter. In another effort, hydrogen fuel is produced using solar energy. A third such effort turns the carbon from car exhaust into ink that can be used for printing. New technologies are also letting people monitor pollution levels place to place with great precision, enabling them to decide whether and how to avoid areas of highest risk. For example, one phone app takes current data from nearby air quality monitoring stations and converts the data to an equivalent number of cigarettes being smoked daily. The app has shown that the average Los Angeles County resident inhales pollutants equivalent to smoking half a cigarette a day, that Mexico City residents inhale the equivalent of up to 6.5 cigarettes daily, and that people in Delhi sometimes breathe air equal to 20 cigarettes per day!

As the world's less-wealthy nations industrialize and seek to control air pollution, people in places like Delhi will be relying on continued integration of science, policy, economics, and technology to help reduce indoor and outdoor air pollution of all kinds. And in all societies, people will need to remain vigilant and keep pressure on their leaders to protect public health by working to reduce pollution and improve air quality.

- **CASE STUDY SOLUTIONS** Describe five ways in which Los Angeles or Mexico City has responded to its air pollution challenges. What results have each of these responses produced? Now consider your city or a major city near where you live. Describe at least one approach used by L.A. or Mexico City that you feel would help address air pollution in your city, and explain why.

- **LOCAL CONNECTIONS** Go online and examine the maps and resources available at the EPA's "AirNow" website at www.airnow.gov. What is the current status of the Air Quality Index in the place you live? How does it compare with the AQI in other regions in North America? Why do you think your region may be better or worse than others in its air quality? Now go to the EPA's "AirCompare" website at https://www3.epa.gov/aircompare/, and compare your county with some other county that you think you might like to live in or visit. Describe one way in which pollution or health data for the two counties differs. Taking into account all you've learned in this chapter about sources of air pollutants and the conditions under which pollutants can drift or concentrate, what would you say are some reasons your region experiences certain kinds of pollution? Why might it avoid some types of pollution that other regions suffer?

- **EXPLORE THE DATA** How do researchers link information on air pollution with impacts on health? → **Explore Data** relating to the case study on **Mastering Environmental Science**.

REVIEWING Objectives

You should now be able to:

+ Describe the composition, structure, and function of Earth's atmosphere

The atmosphere consists of 78% nitrogen gas, 21% oxygen gas, and various other gases in minute concentrations. Temperature and other attributes vary across its four layers. Ozone is concentrated in the stratosphere. The atmosphere moderates climate, provides oxygen, conducts and absorbs solar radiation, and transports and recycles nutrients and waste. (pp. 452–454)

+ Relate weather and climate to atmospheric conditions

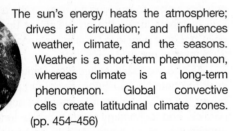

The sun's energy heats the atmosphere; drives air circulation; and influences weather, climate, and the seasons. Weather is a short-term phenomenon, whereas climate is a long-term phenomenon. Global convective cells create latitudinal climate zones. (pp. 454–456)

+ Identify major outdoor air pollutants, and outline the scope of air pollution

Natural sources such as fires, volcanoes, and windblown dust pollute the atmosphere, and human activity can worsen some of their impacts. Most pollution is generated by people, however, and we emit pollutants from point and non-point sources. Major pollutants include carbon monoxide, sulfur dioxide, nitrogen oxides, VOCs, particulate matter, lead, ammonia, nitrogen dioxide, and tropospheric ozone. Photochemical smog, created by chemical reactions of pollutants in the presence of sunlight, impairs visibility and human health widely in urban areas. Industrial smog from fossil fuel combustion persists in urban and industrial areas of many developing nations. Today, industrializing nations such as China and India are experiencing the world's worst air pollution. (pp. 456–468)

+ Assess strategies and solutions for control of outdoor air pollution

Thanks to public policy (such as the U.S. Clean Air Act) and to pollution control technologies (such as scrubbers and catalytic converters), pollutant emissions in the United States and many other wealthy nations have decreased substantially since 1970, and ambient air quality has improved. Cities such as Los Angeles and Mexico City have taken bold steps to address photochemical smog. Governments of nations such as China are beginning to combat air pollution, as well. (pp. 458–470)

+ Explain stratospheric ozone depletion, and identify steps taken to address it

CFCs and other persistent human-made compounds destroy ozone in the stratosphere; ozone depletion is most severe over Antarctica, where an "ozone hole" appears each spring. Thinned ozone concentrations pose dangers to life because they allow more ultraviolet radiation to reach Earth's surface. Fortunately, the Montreal Protocol and its follow-up agreements have proven remarkably successful in reducing emissions of ozone-depleting substances. However, the long residence time of CFCs in the atmosphere is delaying full restoration of stratospheric ozone. (pp. 470–474)

+ Describe acid deposition, discuss its consequences, and explain how we are addressing it

Acid deposition results when pollutants such as SO_2 and NO react in the atmosphere to produce acids that are deposited on Earth's surface. Acid deposition may be wet or dry and may occur a long distance from the source of pollution. Acid deposition damages soils, water bodies, plants, animals, ecosystems, and human property. Regulation, cap-and-trade programs, and technology have all helped reduce acid deposition in North America, and industrializing nations are beginning to tackle the problem, as well. (pp. 474–477)

+ Characterize the scope of indoor air pollution, and assess solutions

Indoor air pollution causes nearly as many deaths and health problems worldwide as outdoor air pollution. Indoor burning of fuelwood is the developing world's primary indoor air pollution risk, whereas cigarette smoke and radon are the worst indoor pollutants in the developed world. VOCs and living organisms commonly pollute indoor air. Using low-toxicity materials, keeping spaces clean, monitoring air quality, and maximizing ventilation all help enhance indoor air quality. (pp. 477–480)

SEEKING Solutions

1. Name one type of natural air pollution, and discuss how human activity can sometimes worsen it. What potential solutions can you think of to minimize this human impact?

2. Explain how and why emissions of major pollutants have been reduced by more than 70% in the United States since 1970, despite increases in population, energy use, and economic activity. Describe at least two ways you think air quality could be further improved.

3. International action through a treaty helped halt stratospheric ozone depletion, but other transboundary pollution issues—such as acid deposition and greenhouse gas pollution—have not yet been addressed as effectively. What types of actions do you think are appropriate for pollutants that cross political boundaries?

4. **THINK IT THROUGH** Suppose you are the head of your county health department and the EPA informs you that your county has failed to meet the national ambient air quality standards for ozone, sulfur dioxide, and nitrogen dioxide. Your county is partly rural but is home to a city of 200,000 people and 10 sprawling suburbs. There are several large and aging coal-fired power plants, a number of factories with advanced pollution control technology, and no public transportation system. Describe at least three steps you would urge the county government to take to meet the air quality standards. Explain specifically what effects you would expect each of these steps to have, and why.

5. **THINK IT THROUGH** You have just taken a job at a medical clinic in your hometown. The nursing staff has asked you to develop a brochure for patients featuring tips on how to minimize health impacts from air pollution (both indoor and outdoor) in their daily lives. List the top five tips you will feature, and explain for each why you will include it in your brochure.

CALCULATING Ecological Footprints

"While only some motorists contribute to traffic fatalities, all motorists contribute to air pollution fatalities." So stated a writer for the Earth Policy Institute, pointing out that air pollution kills far more people than vehicle accidents. According to EPA data, emissions of nitrogen oxides in the United States in 2017 totaled 10.7 million tons (21.4 billion lb). Nitrogen oxides come from fuel combustion in motor vehicles, power plants, and other industrial, commercial, and residential sources, but nearly 6.4 million tons (12.8 billion lb) of the 2017 total came from motor vehicles. The U.S. Census Bureau estimates the nation's population to have been 325.5 million in 2017 and projects that it will reach 373.5 million in 2040. Considering these data, calculate the missing values in the table (1 ton = 2000 lb).

	TOTAL NO$_X$ EMISSIONS (lb)	NO$_X$ EMISSIONS FROM VEHICLES (lb)
You		
Your class		
Your state		
United States	21.4 billion	12.8 billion

Data from U.S. EPA.

1. By what percentage is the U.S. population projected to increase between 2017 and 2040? Do you think NO$_X$ emissions will increase, decrease, or remain the same over that period of time? Why? (You may want to refer to Figure 17.11.)

2. Assume you are an average American driver. Using the 2017 emissions totals, how many pounds of NO$_X$ emissions are you responsible for creating? How many pounds would you prevent if you were to reduce by half the vehicle miles you travel? What percentage of your total NO$_X$ emissions would that be?

3. Describe three ways in which an American driver could reduce his or her NO$_X$ emissions from vehicular travel. What steps could you take to reduce the overall NO$_X$ emissions for which you are responsible?

Mastering Environmental Science

CHAPTER
18

Global Climate Change

case study

Rising Seas Threaten South Florida

FLORIDA

Atlantic Ocean

Gulf of Mexico   Miami •

> Miami, as we know it today, is doomed. It's not a question of if. It's a question of when.
>
> University of Miami geologist Dr. Harold Wanless

> Miami Beach is not going to sit back and go underwater.
>
> Philip Levine, mayor of Miami Beach

t happens now in Miami at least six times a year. Salty water bubbles up from drains, seeps up from the ground, fills the streets, and spills across lawns and sidewalks. Under the dazzling sun of a South Florida sky, floodwaters stall car traffic, creep into doorways, force businesses to close, and keep people from crossing the street. Employees struggle to get to work while tourists stand around, baffled.

The flooding is most severe in Miami Beach, the celebrated strip of glamorous hotels, clubs, shops, and restaurants that rises from a seven-mile barrier island just offshore from Miami. The carefree, affluent image of Miami Beach, with its sun and fun, is increasingly jeopardized by the grimy reality of these unwelcome saltwater intrusions. By 2030, flooding is predicted to strike Miami and Miami Beach about 45 times per year—becoming no longer a curious inconvenience but an existential threat.

These mysterious floods that seem to come out of nowhere are a recent phenomenon, so Miami-area residents have only recently come to realize that the ocean is slowly swallowing their coastal metropolis. The cause? Rising sea levels driven by global climate change.

The world's oceans rose 20 cm (8 in.) in the 20th century as warming temperatures expanded the volume of seawater and caused glaciers and ice sheets to melt, discharging water into the oceans. These processes are accelerating today, and scientists predict that sea level will rise another 26–98 cm (10–39 in.) or more in this century as climate change intensifies.

As sea levels rise, coastal cities across the globe—from Venice to Amsterdam to New York to San Francisco—are facing challenges. In the United States, scientists find that the Atlantic Seaboard and the Gulf Coast are especially vulnerable. The hurricane-prone shores of Florida, Louisiana, Texas, and the Carolinas are at risk, as are coastal cities such as Houston and New Orleans. From Cape Cod to Corpus Christi, millions of Americans who live in shoreline communities are beginning to suffer significant expense, disruption to daily life, and property damage as beaches erode, neighborhoods flood, aquifers are fouled, and storms strike with more force.

Nowhere in America is more vulnerable to sea level rise than Miami and its surrounding communities in South Florida. Six million people live in this region, and three-fourths of them inhabit low-lying coastal areas that hold most of the region's wealth and property. Experts calculate that Miami alone has more than $400 billion in assets at risk from sea level rise—more than any other city in the world (**FIGURE 18.1**, p. 486).

Upon completing this chapter, you will be able to:

+ **Describe Earth's climate system, and explain the factors that influence global climate**

+ **Identify greenhouse gases, and characterize human influences on the atmosphere and on climate**

+ **Summarize how researchers study climate**

+ **Outline current and expected future trends and impacts of climate change in the United States and around the world**

+ **Suggest and assess ways we may respond to climate change**

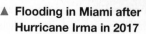

◀ **Bulldozing beach sand off a Fort Lauderdale boulevard after a storm surge**

▲ **Flooding in Miami after Hurricane Irma in 2017**

FIGURE 18.1 In Miami, billions of dollars of property and infrastructure are located within just meters of the ocean.

Hurricanes strike the area frequently, and every centimeter of sea level rise makes the surge of seawater from a storm more expansive, costly, and dangerous. In 2017, Hurricane Irma sent floodwaters coursing through the streets of Miami, Miami Beach, Jacksonville, and other Florida cities. Miami's mayor Tomás Regalado, a Republican, pleaded with federal leaders to recognize that climate change was fueling stronger storms and hurting cities like his. "If this isn't climate change, I don't know what is," Regalado declared. "This is truly [a] poster child for what is to come."

South Florida is highly sensitive to sea level change because its landscape is exceptionally flat; just 1 m (3.3 ft) of sea level rise would inundate more than a third of the region. A 4-m (13-ft) rise in sea level would submerge Miami and reduce the region to a handful of small islands.

The porous limestone bedrock that underlies South Florida also poses a challenge. Pockmarked with holes like Swiss cheese, this permeable rock lets water percolate through. That is why Miami's floods seem to come from out of nowhere; during the highest tides of the year, ocean water is forced inland, where it mixes with fresh water underground and is pushed up as a briny mixture through the limestone directly into yards and streets. As a result, Miami and its neighboring cities cannot simply wall themselves off from a rising ocean, because seawalls won't stop water from seeping up from below.

Moreover, as salt water moves inland, it contaminates the fresh drinking water of South Florida's Biscayne Aquifer. Fort Lauderdale and several other communities are already struggling with saltwater incursion. Florida is building desalination plants to convert seawater to drinking water, but desalination (p. 406) is expensive and consumes large amounts of energy.

Miami-area leaders and citizens are taking action to safeguard their region's future. In 2010, commissioners of Broward, Miami-Dade, Monroe, and Palm Beach counties adopted an agreement to work together on strategies to respond to climate change and its effects in the region. This agreement, the Southeast Florida Regional Climate Change Compact, has garnered wide recognition as a model for regional cooperation on climate issues. Despite a lack of money from the state and federal governments, these policymakers are helping protect assets in the short term by building up dunes and raising building foundations. They are also helping communities adjust in the long term by shifting development inland and ending subsidies for insurance for development in low-lying coastal areas.

In Miami Beach, the city is elevating its main roadways by 3 feet, and businesses are being urged to remodel their first floors. The city raised residential stormwater charges and is spending $400 million installing a system of massive pumps to extract floodwater.

And in Miami in 2017, voters approved a "Miami Forever" bond measure pushed by outgoing Mayor Regalado. Fully $192 million of funds from this bond measure will go toward measures to address flooding, including sea walls, pumps, and storm drain upgrades. In Miami, Miami Beach, and other communities, engineers and policymakers expect these measures to get their cities through the next two or three decades. They recognize, however, that more interventions will be needed later—or people and whole communities will have to retreat from the coast and move inland.

Only time will tell whether South Florida's communities will overcome their challenges and provide a shining example for other regions. In Miami and many other coastal cities, vast sums will be spent on pumps, drains, pipes, seawalls, and other engineering solutions, but ultimately these fixes are only temporary; they may buy time, but they cannot stop the water forever. In the long term, only reducing our emissions of greenhouse gases will halt sea level rise and the many other imminent consequences of global climate change.

Our Dynamic Climate

Climate influences virtually everything around us, from the day's weather to major storms, from crop success to human health, and from national security to the ecosystems that support our economies. If you are a student in your teens or twenties, the accelerating change in our climate may well be *the* major event of your lifetime and the phenomenon that most shapes your future.

Climate change is also the fastest-developing area of environmental science. New scientific studies that refine our understanding of climate are published every week, and policymakers and businesspeople make decisions and take actions just as quickly in response. Because new developments are occurring so rapidly, we urge you to explore beyond this book and, with your instructor, examine the most recent information on climate change and the consequences it may have for your future.

What is climate change?

Climate describes an area's long-term atmospheric conditions, including temperature, precipitation, wind, humidity, barometric pressure, solar radiation, and other characteristics. *Climate* differs from *weather* (p. 454) in that weather specifies conditions over hours or days, whereas climate summarizes conditions over years, decades, or centuries.

Global climate change—generally referred to simply as **climate change**—describes an array of changes in aspects of Earth's climate, such as temperature, precipitation, and the frequency and intensity of storms. People often use the term *global warming* synonymously in casual conversation, but **global warming** refers specifically to an increase in Earth's average surface temperature. Global warming is only one aspect of global climate change, but warming does in turn drive other components of climate change. Some researchers point out that the terms *warming* and *change* are so mild-sounding as to be misleading and that a more accurate term for what Earth is experiencing today would be *climate disruption*.

Over the long term, our planet's climate varies naturally. However, today's disruptive changes are unfolding at an exceedingly rapid rate, and they are creating conditions humanity has never experienced. Earth's climate is a complex and finely tuned system on which life depends, and rapid and disruptive changes in the geologic past have resulted in mass extinctions (pp. 60, 284). Scientists agree that the climate disruption we are witnessing today is being driven by human activities, notably fossil fuel combustion and deforestation. Understanding how and why today's climate is changing—and judging how we might respond—requires understanding how our planet's climate normally functions.

Three factors influence climate

Three natural factors exert the most influence on climate. The first is the sun. Without the sun, Earth would be dark and frozen. The second is the atmosphere. Without this protective layer of gases, Earth would be as much as 33°C (59°F) colder on average, and temperature differences between night and day would be far greater than they are. The third is the oceans, which store and transport heat and moisture.

The sun supplies most of our planet's energy. Earth's atmosphere, clouds, land, ice, and water together absorb about 70% of incoming solar radiation and reflect the remaining 30% back into space (**FIGURE 18.2**). The 70% that is absorbed powers everything from wind to waves to evaporation to photosynthesis.

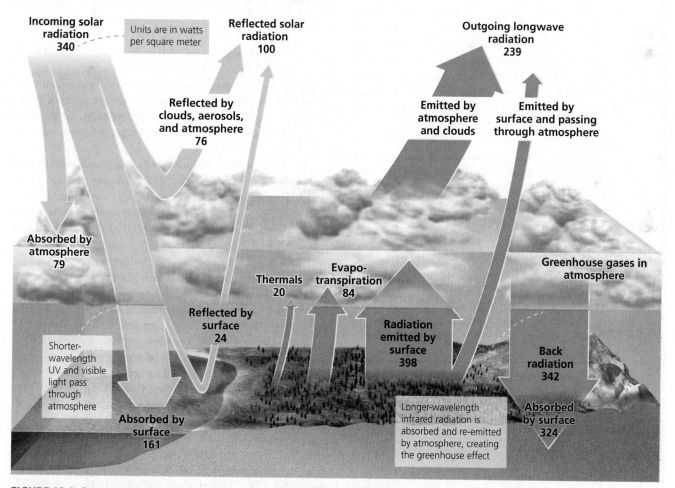

FIGURE 18.2 Our planet receives about 340 watts of energy per square meter from the sun, and Earth naturally reflects or emits this same amount. Earth absorbs nearly 70% of the solar radiation it receives and reflects the rest back into space **(yellow arrows)**. The radiation absorbed is then emitted **(orange arrows)** as infrared radiation, which has longer wavelengths. Greenhouse gases in the atmosphere absorb a portion of this long-wavelength radiation and then emit it, sending some back downward to warm the atmosphere and the surface by the greenhouse effect. *Data from Intergovernmental Panel on Climate Change (IPCC); Stocker, T.F., et al. (Eds.), 2013. Climate change 2013: The physical science basis. Contribution of Working Group I to the fifth assessment report of the IPCC. Cambridge, UK, and New York: Cambridge University Press.*

Greenhouse gases warm the lower atmosphere

As Earth's surface absorbs solar radiation, the surface increases in temperature and emits infrared radiation (p. 33), radiation with wavelengths longer than those of visible light. Atmospheric gases having three or more atoms in their molecules tend to absorb infrared radiation. These include water vapor (H_2O), carbon dioxide (CO_2), methane (CH_4), nitrous oxide (N_2O), and ozone (O_3), as well as halocarbons, a diverse group of mostly human-made gases (p. 470). All these gases are known as **greenhouse gases.** After absorbing radiation emitted from the surface of Earth, greenhouse gases emit infrared radiation. Some of this emitted energy is lost to space, but much of it travels back downward, warming the lower atmosphere (specifically the troposphere; p. 452) and the surface in a phenomenon known as the **greenhouse effect.**

Greenhouse gases differ in their capacity to warm the troposphere. Scientists quantify this difference by measuring *global warming potential,* the relative ability of one molecule of a given greenhouse gas to contribute to warming. **TABLE 18.1** shows global warming potentials for several greenhouse gases. Values are expressed in relation to carbon dioxide, which is assigned a value of 1. For example, at a 20-year time horizon, a molecule of methane is 84 times more potent than a molecule of CO_2. Yet because a methane molecule typically resides in the atmosphere for less time than a CO_2 molecule, methane's global warming potential is reduced at longer time horizons (it is 28 at a 100-year horizon).

Although CO_2 is less potent on a per-molecule basis than most other greenhouse gases, it is far more abundant in the atmosphere. Moreover, greenhouse gas emissions from human activity consist mostly of CO_2; for this reason, CO_2 has caused nearly twice as much warming since the industrial revolution as have methane, nitrous oxide, and halocarbons combined.

Greenhouse gas concentrations are rising fast

The greenhouse effect is a natural phenomenon, and greenhouse gases have been present in our atmosphere throughout Earth's history. That's a good thing: Without the natural greenhouse

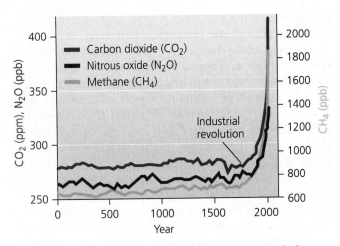

FIGURE 18.3 Since the start of the industrial revolution around 1750, global concentrations of carbon dioxide, methane, and nitrous oxide in the atmosphere have increased markedly. *Data from IPCC, 2013.* Fifth assessment report; *and National Oceanic and Atmospheric Administration (NOAA).*

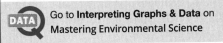

Go to **Interpreting Graphs & Data** on Mastering Environmental Science

effect, our planet might be too cold to support life. Thus, it is not the natural greenhouse effect that concerns scientists today but rather the anthropogenic (human-generated) intensification of the greenhouse effect. By increasing the concentrations of greenhouse gases over the past 250 years (**FIGURE 18.3**), we are intensifying the greenhouse effect beyond what humanity has ever experienced.

We have boosted Earth's atmospheric concentration of carbon dioxide from roughly 278 parts per million (ppm) in the 1700s to more than 400 ppm today (see Figure 18.3). The concentration of CO_2 in our atmosphere now is far higher than it has been in over 800,000 years (four times longer than the human species has existed)—and likely higher than it has been in several *million* years.

Why have atmospheric carbon dioxide levels risen so much? Most carbon is stored for long periods in the upper layers of the lithosphere (rock, sediment, and soil; p. 35). The deposition, partial decay, and compression of carbon-based organic matter (mostly plants and phytoplankton) in wetland or marine areas hundreds of millions of years ago led to the formation of carbon-rich coal, oil, and natural gas (p. 529). Over the past two centuries, we have extracted these fossil fuels from the ground and burned them in our homes, power plants, and automobiles, transferring large amounts of carbon from one reservoir (underground deposits that stored the carbon for millions of years) to another (the atmosphere). This sudden flux of carbon from the lithosphere into the atmosphere is the main reason atmospheric CO_2 concentrations have risen so dramatically.

At the same time, people have cleared and burned forests to make room for crops, pastures, villages, and cities. Forests serve as a reservoir for carbon as plants conduct photosynthesis (p. 33) and store carbon in their tissues. When we clear forests, we reduce the biosphere's ability to remove carbon

TABLE 18.1 Global Warming Potentials of Four Greenhouse Gases

GREENHOUSE GAS	RELATIVE HEAT-TRAPPING ABILITY (IN CO₂ EQUIVALENTS)	
	OVER 20 YEARS	OVER 100 YEARS
Carbon dioxide	1	1
Methane	84	28
Nitrous oxide	264	265
CFC-12 (Dichlorodifluoromethane)	10,800	10,200

Data from IPCC, 2013. *Fifth Assessment Report.*

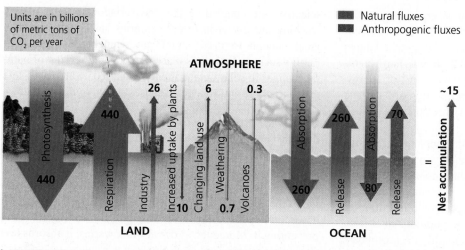

Units are in billions of metric tons of CO_2 per year

ATMOSPHERE

Natural fluxes
Anthropogenic fluxes

LAND

Photosynthesis
440
440

26
Respiration

Industry

Increased uptake by plants
10

Changing land use
6

Weathering
0.7

Volcanoes
0.3

OCEAN

Absorption
260
260
Release

Absorption
70
80
Release

Net accumulation
~15
=

FIGURE 18.4 Human activities are sending more carbon dioxide from Earth's surface to its atmosphere than is moving from the atmosphere to the surface. Shown are all current fluxes of CO_2, with arrows sized according to mass. Green arrows show natural fluxes, and gray arrows show anthropogenic fluxes. *Adapted from IPCC, 2007.* Fourth assessment report.

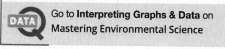

Go to **Interpreting Graphs & Data** on **Mastering Environmental Science**

dioxide from the atmosphere. In this way, deforestation (p. 312) contributes to rising atmospheric CO_2 concentrations. **FIGURE 18.4** summarizes scientists' understanding of the fluxes (natural and anthropogenic) of carbon dioxide among the atmosphere, land, and oceans.

Methane concentrations are also rising; the atmospheric concentration of methane today is 2.5 times higher than in 1750 (see Figure 18.3) and is the highest by far in more than 800,000 years. We release methane by tapping into fossil fuel deposits, raising livestock that emit methane as a metabolic waste product, growing rice crops, and disposing of organic matter in landfills.

We have also driven up atmospheric concentrations of nitrous oxide. This greenhouse gas, a by-product of feedlots, chemical manufacturing plants, auto emissions, and synthetic nitrogen fertilizers, has risen by 22% since 1750 (see Figure 18.3).

Among other greenhouse gases, ozone concentrations in the troposphere have risen more than 40% since 1750, a result of photochemical smog (p. 468). The contribution of some halocarbon gases to global warming is falling because of the Montreal Protocol and controls on their production and use (p. 474), but the impacts of others are rising. Emissions of all greenhouse gases combined are 42% higher than in 1990.

Water vapor is the most abundant greenhouse gas in our atmosphere and contributes most to the natural greenhouse effect. Its concentrations vary locally, but because its global concentration has not changed, it is not thought to have driven industrial-age climate change.

Various factors warm or cool the surface

Whereas greenhouse gases exert a warming effect on the atmosphere, **aerosols**—microscopic droplets and particles aloft in

the atmosphere—can have either a warming or a cooling effect. Soot particles, or "black carbon aerosols," generally cause warming by absorbing solar energy, but most other aerosols cool the atmosphere by reflecting the sun's rays. When sulfur dioxide from fossil fuel combustion enters the atmosphere, it undergoes various reactions, some of which lead to acid deposition (p. 474) and form a sulfur-rich aerosol haze that blocks sunlight. Sulfate aerosols released by major volcanic eruptions can cool Earth's climate for up to several years, as occurred in 1991 with the eruption of Mount Pinatubo in the Philippines (p. 458).

To measure the degree of impact that a given factor exerts on Earth's temperature, scientists calculate its **radiative forcing,** the amount of change in thermal energy that the factor causes. Positive forcing warms the surface, whereas negative forcing cools it. When scientists sum up the effects of all factors, they find that Earth is now experiencing overall radiative forcing of about 2.3 watts/m^2 (**FIGURE 18.5**). This means that our planet today is receiving and retaining roughly 2.3 watts/m^2 more thermal energy than it is emitting into space. (By contrast, the pre-industrial Earth of 1750 was in balance, emitting as much radiation as

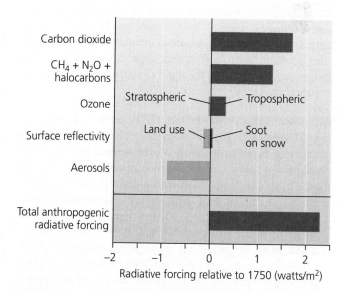

Carbon dioxide

$CH_4 + N_2O$ + halocarbons

Ozone   Stratospheric — — Tropospheric

Surface reflectivity   Land use — — Soot on snow

Aerosols

Total anthropogenic radiative forcing

−2   −1   0   1   2
Radiative forcing relative to 1750 (watts/m^2)

FIGURE 18.5 This graph shows the radiative forcing of major factors that warm or cool our planet. Radiative forcing is expressed as the effect each factor has on temperature today relative to the year 1750, in watts/m^2. Red bars indicate positive forcing (warming), and blue bars indicate negative forcing (cooling). *Data from IPCC, 2013.* Fifth assessment report.

it was receiving.) This extra amount is equivalent to the power converted into heat and light by 200 incandescent lightbulbs (or more than 900 compact fluorescent lamps) across a football field. As Figure 18.2 shows, Earth naturally receives and gives off about 340 watts/m^2 of energy. Although 2.3 may seem like a small proportion of 340, heat from this imbalance accumulates, and over time it is enough to alter climate significantly.

One question researchers are still grappling with is the role of clouds in our changing climate. As tropospheric temperatures increase, physics tells us that more water should evaporate from Earth's surface into the atmosphere. On one hand, more atmospheric water vapor could lead to more warming, which could cause more evaporation, in a positive feedback loop (p. 108) that would amplify the greenhouse effect. On the other hand, more water vapor could enhance cloudiness, which might, in a negative feedback loop (p. 108), slow global warming by reflecting more solar radiation back into space. In this second scenario, depending on whether low- or high-elevation clouds result, the clouds might either shade and cool Earth (negative feedback) or contribute to warming and accelerate evaporation and cloud formation (positive feedback). Scientists are not yet sure which effect might predominate or how the balance might vary from place to place. Because of feedback loops, minor changes in the atmosphere can potentially lead to major effects on climate. This poses challenges for making accurate predictions of future climate change.

Climate varies naturally for several reasons

Aside from atmospheric composition, our climate is influenced by cyclical changes in Earth's rotation and orbit, variation in energy released by the sun, absorption of carbon dioxide by the oceans, and ocean circulation patterns. However, scientific data indicate that none of these four natural factors can fully explain the rapid climate change that we are experiencing today.

Milankovitch cycles In the 1920s, Serbian mathematician Milutin Milankovitch described three types of periodic changes in Earth's rotation and orbit around the sun. Over thousands of years, our planet wobbles on its axis, varies in the tilt of its axis, and experiences change in the shape of its orbit, all in regular long-term cycles of different lengths. These variations, known as **Milankovitch cycles,** alter the way solar radiation is distributed over Earth's surface (**FIGURE 18.6**). By modifying patterns of atmospheric heating, these cycles trigger long-term climate variation, including periodic episodes of *glaciation* during which global surface temperatures drop and ice sheets expand outward from the poles.

Solar output The sun varies in the amount of radiation it emits, over both short and long timescales. However,

scientists are concluding that the variation in solar energy reaching our planet in recent centuries has simply not been great enough to drive significant temperature change on Earth's surface. Estimates place the radiative forcing of natural changes in solar output at only about 0.05 watts/m^2—less than any of the anthropogenic causes shown in Figure 18.5.

Ocean absorption The oceans hold 50 times more carbon than the atmosphere holds. Oceans absorb carbon dioxide from the atmosphere when CO_2 dissolves directly in water and when marine phytoplankton use it for photosynthesis. However, the oceans are absorbing less CO_2 than we are adding to the atmosphere (see Figure 5.17, p. 123). Thus, carbon absorption by the oceans is slowing global warming but is not preventing it. Moreover, as ocean water warms, it absorbs less CO_2 because gases are less soluble in warmer water—a positive feedback effect (p. 108) that accelerates warming of the atmosphere.

Ocean circulation Ocean water exchanges heat with the atmosphere, and ocean currents move energy from place to place. In equatorial regions, the oceans receive more heat from the sun and atmosphere than they emit. Near the poles, the oceans emit more heat than they receive. Because cooler water is denser than warmer water, the cooler water at the poles tends to sink, and the warmer surface water from the equator moves to take its place. This principle underlies global ocean circulation patterns (p. 425).

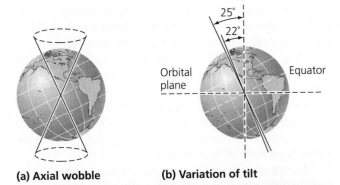

(a) Axial wobble

(b) Variation of tilt

(c) Variation of orbit

FIGURE 18.6 There are three types of Milankovitch cycles: (a) an axial wobble that occurs on a 19,000- to 23,000-year cycle; **(b)** a 3-degree shift in the tilt of Earth's axis that occurs on a 41,000-year cycle; and **(c)** a variation in Earth's orbit from almost circular to more elliptical, which repeats every 100,000 years.

Earth's climate does indeed change naturally over very long periods of time, but there is nothing "natural" about today's sudden climate disruption. We know that human activity is directly causing the unnaturally rapid changes that we are now witnessing. Moreover, humanity has never before experienced the sheer amount of change predicted for this century. Greenhouse gas concentrations are already higher than they've been in more than 800,000 years, and they are rising. The human species, *Homo sapiens,* has existed for just 200,000 years, and our civilization arose only in the past few thousand years during an exceptionally stable period in Earth's climate history. Unless we reduce our emissions, we soon will be challenged by climate conditions our species has never lived through.

The oceans' thermohaline circulation system (p. 426) moves warm tropical water northward, providing Europe a far milder climate than it would otherwise have. Scientists are studying whether freshwater input from Greenland's melting ice sheet might shut down this warm-water flow. Such an occurrence would plunge Europe into much colder conditions.

Multiyear climate variability results from the El Niño–Southern Oscillation (p. 427), which involves systematic shifts in atmospheric pressure, sea surface temperature, and ocean circulation in the tropical Pacific Ocean. These shifts overlie longer-term variability from a phenomenon known as the Pacific Decadal Oscillation. El Niño and La Niña events alter weather patterns in diverse ways, often leading to rainstorms and floods in dry regions and drought and fire in moist regions, all of which have consequences for wildlife, agriculture, and fisheries.

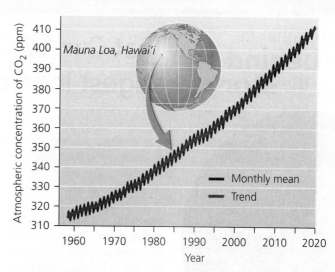

FIGURE 18.7 Atmospheric concentrations of carbon dioxide are rising steeply. Direct long-term measurements began in 1958, when Charles Keeling started collecting these data at Hawaii's Mauna Loa Observatory. *Data from National Oceanic and Atmospheric Administration, Earth System Research Laboratory, Global Monitoring Division, 2019.*

well-mixed air from over vast stretches of ocean blows across the top of Earth's most massive mountain. These data show that atmospheric CO_2 concentrations increased from 315 ppm in 1958 to more than 410 ppm today (**FIGURE 18.7**).

Direct measurements of climate variables such as temperature and precipitation extend back in time somewhat further. Precise and reliable thermometer measurements cover more than a century. Fishermen have recorded the timing of sea ice formation each year, and winemakers have kept meticulous records of precipitation and the length of the growing season. Accurate records of these types extend back, at most, a few hundred years.

Studying Climate Change

To comprehend any phenomenon that is changing, we must study its past as well as its present. Scientists monitor present-day climate, but they also have devised clever means of inferring past change and sophisticated methods to predict future conditions.

Direct measurements tell us about the present

Today, we measure temperature with thermometers, rainfall with rain gauges, wind speed with anemometers, and air pressure with barometers, using computer programs to integrate and analyze this information in real time. With these technologies and more, we document fluctuations in weather day-by-day and hour-by-hour across the globe.

We also measure the chemistry of the atmosphere and the oceans. Direct measurements of carbon dioxide concentrations in the atmosphere reach back to 1958, when scientist Charles Keeling began analyzing hourly air samples from a monitoring station at Hawaii's Mauna Loa Observatory. Here, unpolluted,

Proxy indicators tell us about the past

To understand past climate, scientists decipher clues from thousands or millions of years ago by taking advantage of the record-keeping capacity of the natural world. **Proxy indicators** are types of indirect evidence that serve as proxies, or substitutes, for direct measurement.

For example, Earth's ice caps, ice sheets, and glaciers hold clues to climate history. In frigid regions near the poles and atop high mountains, snow falling year after year compresses into ice. Over millennia, this ice accumulates to great depths, preserving within its layers tiny bubbles of the ancient atmosphere. Scientists can examine the trapped air bubbles by drilling into the ice and extracting long columns, or cores. The layered ice, accumulating season after season for thousands of years, provides a timescale. By studying the chemistry of the bubbles in each layer in these ice cores, scientists can determine atmospheric composition, greenhouse gas concentrations,

THE SCIENCE behind the story

What Can We Learn from the World's Longest Ice Core?

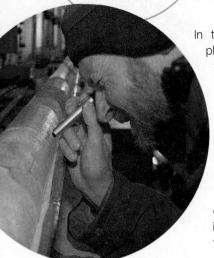

Inspecting an Antarctic ice core

In the most frigid reaches of our planet, snow falling year after year for millennia compresses into ice and stacks up into immense sheets that scientists can mine for clues to Earth's climate history. The ice sheets of Antarctica and Greenland trap tiny air bubbles, dust particles, and other proxy indicators (p. 491) of past conditions. By drilling boreholes and extracting ice cores, researchers can tap into these priceless archives.

Recently, researchers drilled and analyzed the deepest core ever. At a remote and pristine site in Antarctica named Dome C, they drilled down 3270 m (10,728 ft) to bedrock and pulled out more than 800,000 years' worth of ice. The longest previous ice core (from Antarctica's Vostok station) had gone back "only" 420,000 years.

Ice near the top of these cores was laid down most recently, and ice at the bottom is oldest, so by analyzing ice at intervals along the core's length, researchers can generate a timeline of environmental change. To date layers of the ice core, researchers first analyze deuterium isotopes (p. 27) to determine the rate of ice accumulation, referencing studies and models of how ice compacts over time. They then calibrate the timeline by matching recent events in the chronology (e.g., major volcanic eruptions) with independent data sets from previous cores, tree rings, and other sources.

Dome C, a high summit of the Antarctic ice sheet, is one of the coldest spots on the planet, with an annual mean

temperature of −54.5°C (−66.1°F). The Dome C ice core was drilled by the European Project for Ice Coring in Antarctica (EPICA), a consortium of researchers from 10 European nations.

In 2004, this team of 56 researchers published a paper in the journal *Nature,* reporting data on surface air temperature across 740,000 years. The researchers had obtained this temperature data by measuring the ratio of deuterium isotopes to normal hydrogen in the ice, a ratio that is temperature dependent.

From 2005 to 2008, five follow-up papers in the journals *Science* and *Nature* reported analyses of greenhouse gas concentrations from the EPICA ice core and extended the gas and temperature data back to cover all 800,000 years. By analyzing air bubbles trapped in the ice, the researchers quantified atmospheric concentrations of carbon dioxide and methane (red line and blue line, respectively, in **FIGURE 1**).

These data show that, by emitting these greenhouse gases since the industrial revolution, we have brought their atmospheric concentrations well above the highest levels they reached naturally at any time in the past 800,000 years. Today's carbon dioxide spike is too recent and sudden to show up in the ice core, but its concentration (which surpassed 400 ppm in 2015) is far above previous maximum values (of ~300 ppm). The data reveal that we as a society have brought ourselves deep into uncharted territory.

The EPICA results also confirm that temperature swings in the past were tightly correlated with concentrations of greenhouse gases (compare the top two data sets in Figure 1 with the temperature data set at bottom). This finding bolsters the scientific consensus that greenhouse gas emissions are causing our planet to warm today.

Also clear from the data is that temperature has varied with swings in solar radiation due to Milankovitch cycles (p. 490).

temperature, snowfall, solar activity, and even (from trapped soot particles) the frequency of forest fires and volcanic eruptions during each time period. By extracting ice cores from Antarctica, scientists have been able to go back in time 800,000 years, reading Earth's history across eight glacial cycles (see **THE SCIENCE BEHIND THE STORY**).

Researchers also drill cores into beds of sediment beneath bodies of water. Sediments often preserve pollen grains and other remnants from plants that grew in the past (as described in the study of Easter Island; pp. 8–9). Because

climate influences the types of plants that grow in an area, knowing what plants were present can tell us a great deal about the climate at that place and time.

Tree rings provide another proxy indicator. The width of each ring of a tree trunk cut in cross-section reveals how much the tree grew in a particular growing season. A wide ring means more growth, generally indicating a wetter year. Long-lived trees such as bristlecone pines can provide records of precipitation and drought going back several thousand years. Tree rings are also used to study fire history,

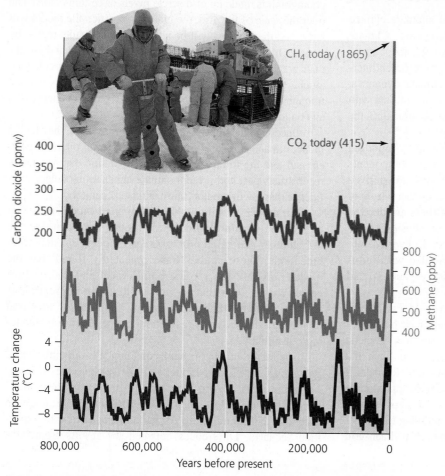

FIGURE 1 Data from the EPICA ice core reveal changes across 800,000 years. Shown are atmospheric carbon dioxide concentration **(red line)**, atmospheric methane concentration **(blue line)**, and deviations from today's average surface temperature **(black line)**. Concentrations of CO_2 and methane rise and fall in tight correlation with temperature. Today's values for CO_2 and methane are included at the top right of the graph, for comparison. *Adapted by permission of Macmillan Publishers Ltd.: Brook, E., 2008. Paleoclimate: Windows on the greenhouse. Nature 453: 291–292, Fig 1a. www.nature.com.*

The complex interplay of Milankovitch cycles produces periodic temperature fluctuations on Earth resulting in periods of glaciation (when temperate regions of the planet are covered in ice)

and in warm interglacial periods (periods between glaciations). The Dome C ice core spans eight glacial cycles.

One finding from the ice core is not easily explained. Intriguingly, early glacial cycles differ from recent cycles (see the black line in Figure 1). In the recent cycles, glacial periods are long and interglacial periods are brief, with a rapid rise and fall of temperature. Interglacial periods thus appear on the graph as tall, thin spikes. In older glacial cycles, the glacial and interglacial periods are of more equal duration, and the interglacial periods are not as warm. This shift in the nature of glacial cycles had been noted before by researchers working with oxygen isotope data from marine fossils (foraminifera). But why cycles should differ before and after the 450,000-year mark, no one knows.

Today, polar scientists are searching for a site that might provide an ice core stretching back even further than 800,000 years. Such a core might provide information to confirm and refine data from marine isotopes, which indicate that it has been several million years since carbon dioxide levels and temperatures have been as high as they are today.

The intriguing patterns revealed by the Dome C ice core show that we still have plenty to learn about Earth's complex climate history. However, the clear relationship between greenhouse gases and temperature evident in the EPICA data suggests that if we want to prevent sudden global warming, we will need to reduce our society's greenhouse emissions.

because a charred ring indicates that a fire took place at the site in that year.

In arid regions such as the U.S. Southwest, packrat middens are a valuable source of climate data. Packrats are rodents that carry seeds and plant parts back to their middens, or dens, in caves and rock crevices sheltered from rain. In arid locations, plant parts may be preserved for centuries inside these middens, allowing researchers to study the past flora of the region.

To assess past ocean conditions, researchers gather data from ancient coral reefs (p. 431). Living corals take in trace elements and isotope ratios (p. 26) from ocean water as they grow, and they incorporate these chemical clues, layer by layer, into growth bands in the reefs they build. Researchers also gather data from the fossilized shells of single-celled marine creatures called foraminifera. These shells preserve isotope ratios of oxygen present in the water when they were formed, which reveal water temperature at the time. Data from foraminifera indicate that it has been several million years since temperatures on Earth have been as warm as they are becoming today.

Models help us predict the future

To predict future climate change, scientists simulate climate processes with sophisticated computer programs. **Climate models** are programs that combine what is known about atmospheric circulation, ocean circulation, atmosphere-ocean interactions, and feedback cycles to simulate climate dynamics. These models require manipulating vast amounts of data with complex mathematical equations—a task not possible until the advent of modern supercomputers.

To create a climate model, scientists begin by inputting mathematical equations describing how various components of Earth's systems function. Some equations are derived from physical laws such as those on the conservation of matter, energy, and momentum (p. 25). Others are derived from observational and experimental data on physics, chemistry, and biology gathered from the field. Converted into computing code, these equations are integrated with information about Earth's landforms, hydrology, vegetation, and atmosphere.

To handle the mind-boggling complexity of Earth's climate system, most models consist of submodels, each addressing a different component—ocean water, sea ice, glaciers, forests, deserts, troposphere, stratosphere, and so on. Each of a model's building blocks must be given equations to make it behave realistically in space and time. Modelers approximate reality by dividing time into periods (called time steps) and by dividing Earth's surface into cells or boxes according to a three-dimensional grid (called grid boxes) (**FIGURE 18.8**). Each grid box contains land, ocean, or atmosphere, much like a digital photograph is made up of discrete pixels of certain colors. The finer the scale of the grid, the greater resolution the model will have, and the better it will be able to predict results region by region. Once the grid is established, the processes that drive climate are assigned to each grid box, with their rates parceled out among the time steps. The model lets the grid boxes interact through time by means of the flux of materials and energy into and out of each grid box.

Once modelers have input all this information, they let the model run and simulate climate from the past through the present and into the future. A computer simulation that accurately reconstructs past and present climate enhances modelers' confidence that it will predict future climate accurately.

A number of studies have compared model runs that include only natural processes, model runs that include only human-generated processes, and model runs that combine both. These studies have repeatedly found that the model runs that incorporate both human and natural processes are the ones that fit real-world climate observations the best (**FIGURE 18.9**). This finding supports the idea that human activities, as well as natural processes, are influencing our climate.

The major human influence on climate is our emission of greenhouse gases, and modelers must select values to enter for future emissions if they want to predict future climate. Generally, they will run their simulations multiple times, each time with a different future emission rate according to a specified scenario. Differences between the results from such

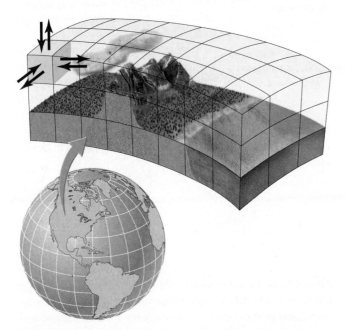

FIGURE 18.8 Climate models divide Earth's surface into a layered grid. Each grid box represents land, air, or water, and interacts with adjacent grid boxes via the flux of materials and energy. *Adapted from Bloom, A.J., 2010. Global climate change: Convergence of disciplines. Sunderland, MA: Sinauer Associates.*

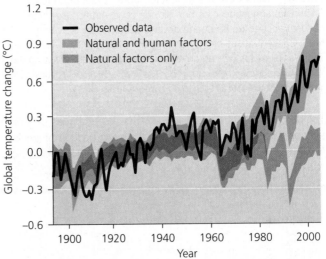

FIGURE 18.9 Models that incorporate both natural and anthropogenic factors predict observed climate trends best. *Adapted from Melillo, J.M., et al. (Eds.), 2014. Climate change impacts in the United States: The third national climate assessment. U.S. Global Change Research Program.*

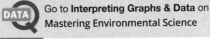

Go to **Interpreting Graphs & Data** on **Mastering Environmental Science**

scenarios tell us what influence these different emission rates would have. You can see results from such a comparison in Figure 18.29 (p. 509).

Impacts of Climate Change

Virtually everyone is noticing changes in the climate these days. Miami-area residents suffer flooding. Texas ranchers and California farmers endure multiyear droughts. Coastal homeowners struggle to obtain insurance against hurricanes and storm surges. People from New York to Atlanta to Chicago to Los Angeles face unprecedented heat waves and cold snaps. Climate change has already had numerous impacts on our planet and on our society. If we continue to emit greenhouse gases, the consequences of climate change will grow more severe.

Scientific evidence for climate change is extensive

For decades, scientists have studied climate change in enormous breadth, depth, and detail. As a result, the scientific literature today is replete with many thousands of independent published studies, and we have gained a rigorous understanding of most aspects of climate change.

To make this vast and growing research knowledge accessible to policymakers and the public, the **Intergovernmental Panel on Climate Change** (IPCC) has taken up the task of periodically reviewing and summarizing it. Established in 1988 by the United Nations Environment Programme and the World Meteorological Organization, the IPCC consists of many hundreds of scientists and national representatives from around the world. Every seven years, this international body has released an updated report. Its *Fifth Assessment Report* was released in 2013–2014. Summarizing thousands of scientific studies, this report documented observed trends in surface temperature, precipitation patterns, snow and ice cover, sea levels, storm intensity, and other factors. It also predicted future trends after considering a range of potential scenarios for future greenhouse gas emissions. The report addressed impacts of climate disruption on wildlife, ecosystems, and society. Finally, it discussed strategies we might pursue in response.

Since then, the IPCC has issued a number of other reports on particular aspects of climate change, each of them authored and peer-reviewed by top scientific experts. The IPCC's *Sixth Assessment Report* is scheduled for release in 2021–2022. You can download and explore any and all IPCC reports yourself; they are publicly accessible online at the IPCC's website.

For the United States, impacts are assessed by the U.S. Global Change Research Program, which Congress created in 1990 to coordinate federal climate research. Its *Fourth National Climate Assessment* was published in two volumes in 2017 and 2018. Produced by more than 300 experts, these mammoth volumes summarized what current research tells us about climate trends and the predicted future impacts of climate change on the United States. Meanwhile, scientists around the world continue to monitor, model, and analyze our changing climate, and new research is being completed faster than ever.

Temperatures continue to rise

Average surface temperatures on Earth—both air temperature over land and sea surface temperature—have risen by about 1.1°C (2.0°F) in the past 100 years (**FIGURE 18.10**). Most of this increase has occurred since 1975, and just since 2000, we have experienced 17 of the 18 warmest years since global

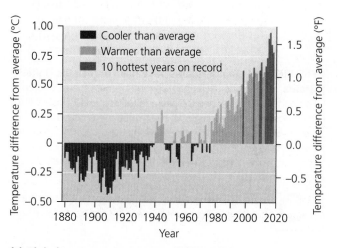

(a) Global temperature measured since 1880

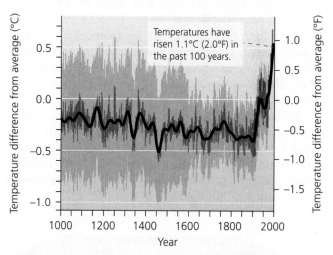

(b) Northern Hemisphere temperature, past 1000 years

FIGURE 18.10 Global temperatures have risen sharply in the past century. Data from thermometers **(a)** show changes in Earth's average surface temperature since 1880. Since 1976, every year has been warmer than average. In **(b)**, proxy indicators **(blue)** and thermometer data **(red)** together show average temperatures in the Northern Hemisphere **(black line)** over the past 1000 years. The gray-shaded zone represents the 95% confidence range. *Data from* **(a)** *NOAA National Centers for Environmental Information; and* **(b)** *IPCC, 2001. Third assessment report.*

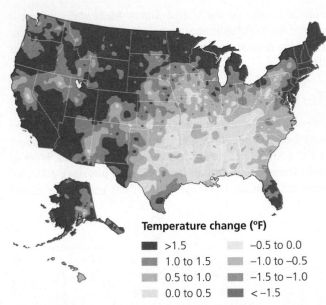

Temperature change (°F)

■ >1.5	▨ −0.5 to 0.0
▨ 1.0 to 1.5	▨ −1.0 to −0.5
▨ 0.5 to 1.0	▨ −1.5 to −1.0
▨ 0.0 to 0.5	■ < −1.5

FIGURE 18.11 Temperatures have risen across the United States. Most of the nation has warmed by more than 1 degree Fahrenheit (0.6°C) when the average for the period 1986–2016 is compared to the average for 1901–1960. *Data from NOAA National Centers for Environmental Information as presented in U.S. Global Change Research Program, 2017. Climate science special report: Fourth national climate assessment, Volume I (Wuebbles, D.J., et al., eds.). USGCRP, Washington, D.C.*

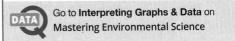 Go to **Interpreting Graphs & Data** on **Mastering Environmental Science**

measurements began 140 years ago! Since the 1960s, each decade has been warmer than the last. If you were born after 1976, you have never in your life lived through a year with average global temperatures lower than the 20th-century average. If you were born after 1985, you have never even lived through a *month* with cooler-than-average global temperatures.

In the United States, temperatures in most areas have risen by more than 1 full degree Fahrenheit in just a few decades (**FIGURE 18.11**). We can expect global surface temperatures to continue rising because we are still emitting greenhouse gases and because the greenhouse gases already

in the atmosphere will continue warming the globe for decades to come. At the end of the 21st century, the IPCC predicts that global temperatures will be 1.0−3.7°C (1.8−6.7°F) higher than today's, depending on how well we control our emissions. Unusually hot days and heat waves will become more frequent than they are today. Future changes in temperature (**FIGURE 18.12**) are predicted to vary from region to region in ways that they already have. For example, polar regions will continue to experience the most intense warming.

Precipitation is changing

A warmer atmosphere speeds evaporation and holds more water vapor, and precipitation has increased worldwide by 2% over the past century. Some regions of the world are receiving less rain and snow than usual, however, while others are receiving more. In the western United States, droughts have become more frequent and severe, harming agriculture, worsening soil erosion, reducing water supplies, and triggering wildfire. In parts of the eastern United States, heavy rain events have increased, leading to floods that have killed dozens of people, left thousands homeless, and inflicted billions of dollars in damage.

Future changes in precipitation (**FIGURE 18.13**) are predicted to intensify regional changes that have already occurred. Many wet regions will receive more rainfall, increasing the risk of flooding, while many dry regions will become drier, worsening water shortages.

Extreme weather is becoming "the new normal"

The sheer number of extreme weather events in recent years—droughts, floods, hurricanes, snowstorms, cold snaps, heat waves—has caught everyone's attention, and weather records are falling left and right. In the United States alone just since 2012, the nation has experienced several major droughts across large regions, several devastating wildfire seasons, an abnormal number of major floods, an unusual number of severe hurricanes, record-breaking heat waves (e.g., in March 2012), and record-setting periods of frigid winter storms (in 2014, 2018, and 2019). Scientific

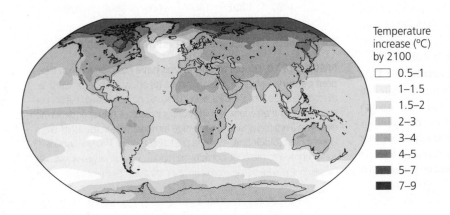

Temperature increase (°C) by 2100

□ 0.5–1	
▨ 1–1.5	
▨ 1.5–2	
▨ 2–3	
▨ 3–4	
▨ 4–5	
■ 5–7	
■ 7–9	

FIGURE 18.12 Surface temperatures are projected to rise for the years 2081–2100, relative to 1986–2005. Land masses are expected to warm more than oceans, and the Arctic will warm the most. This map was generated using an intermediate emissions scenario involving an average global temperature rise of 2.2°C (4.0°F). *Data from IPCC, 2013. Fifth assessment report.*

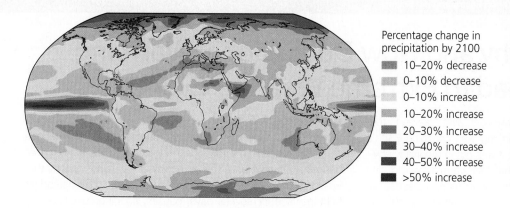

FIGURE 18.13 Precipitation (June–Aug.) is projected to change for the years 2081–2100, relative to 1986–2005. Browner shades indicate less precipitation; bluer shades indicate more. This map was generated using an intermediate emissions scenario involving an average global temperature rise of 2.2°C (4.0°F) by 2100. *Data from IPCC, 2013.* Fifth assessment report.

Percentage change in precipitation by 2100

- 10–20% decrease
- 0–10% decrease
- 0–10% increase
- 10–20% increase
- 20–30% increase
- 30–40% increase
- 40–50% increase
- \>50% increase

data summarized by the U.S. Climate Extremes Index confirm that the frequency of extreme weather events in the United States has roughly tripled since its lowest point in 1970 (**FIGURE 18.14a**).

Scientists are not the only ones to notice these trends. The insurance industry is finely attuned to such patterns, because insurers pay out money each time a major storm, drought, or flood hits. Data from one major insurer, the German firm Munich Re, indicate that since 1980, extreme weather events causing losses have more than tripled worldwide and that monetary losses have increased as well (**FIGURE 18.14b**).

Researchers are still debating precisely why and how global warming is leading to more extreme weather events. But in the past several years, scientists have begun to coalesce around a general answer involving polar warming and weakened jet streams (see **THE SCIENCE BEHIND THE STORY**, pp. 498–499).

With hurricanes, the explanation is now well-understood: The oceans have absorbed more than 90% of our planet's excess energy from global warming, and warmer ocean temperatures impart more energy to hurricanes. This energy drives higher wind speeds and increases the amount of airborne water vapor, resulting in heavier rainfall. As a result, the number of intense hurricanes has increased over the past 40 years. In the United States in 2017, flooding from Hurricane Harvey devastated Houston, the nation's fourth-largest city, along with large regions of Texas and Louisiana. Harvey was followed by Hurricane Irma, which swept through the Caribbean before inundating coastal cities from Miami to Jacksonville to Charleston. Then, Hurricane Maria brought ruin to Puerto Rico. The next year, Hurricane Florence drenched the Carolinas and Hurricane Michael flattened Florida communities, killing more than 60 people.

As global warming intensifies hurricanes, higher sea levels magnify the damage these storms can inflict on coasts. An example is Superstorm Sandy—a hurricane until just before it made landfall—which battered the Atlantic coast in 2012, leaving 160 people dead and thousands homeless amid $65 billion in damage. In New Jersey, coastal communities were deluged with salt water and sand, washing away homes and boardwalks. In New York City, economic activity ground to a halt as tunnels, subway stations, vehicles, and buildings

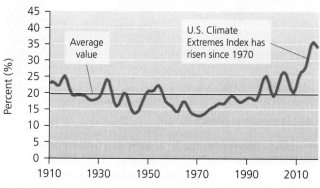

(a) U.S. Climate Extremes Index through time

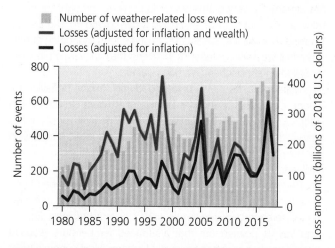

- Number of weather-related loss events
- Losses (adjusted for inflation and wealth)
- Losses (adjusted for inflation)

(b) Weather-related insurance losses worldwide since 1980

FIGURE 18.14 Extreme weather events and their impacts have increased. Values for the U.S. Climate Extremes Index **(a)** have been rising since 1970. Globally, weather-related property losses **(b)** have increased, both in number of events and in financial value. When data are adjusted for society's increased wealth, however, the increase is not as striking. *Data from* **(a)** *U.S. Climate Extremes Index and* **(b)** *NatCatService of Munich Re Group, Munich, Germany.*

THE SCIENCE behind the story

Why and How Does Global Warming Lead to Extreme Weather Events?

Dr. Jennifer Francis

When scientists first realized that global warming was changing Earth's climate, it was not immediately obvious that this warming would also bring about more episodes of extreme weather, such as storms, droughts, floods, wildfires, heat waves, and cold snaps. However, that is what has occurred: As Earth has gotten warmer, extreme weather events have become more frequent and severe. In the past several years, researchers have begun to zero in on an explanation.

You've probably heard of the jet stream and the polar vortex. These two phenomena play starring roles in the drama. But let's start with a quick look at how weather systems traverse the globe, beginning by looking at jet streams in North America.

High above Canada, a strong current of air that encircles the globe—a *jet stream*—blows through the upper troposphere and lower stratosphere (p. 452), from west to east. Another jet stream blows across Mexico and the southern tier of U.S. states. These jet streams are driven by the Coriolis effect (p. 456) and the huge convective cells of Earth's atmosphere (see Figure 17.6, p. 455) that help determine patterns of air movement, heat, and moisture across vast regions. In North America, most weather systems move from west to east, influenced by the jet streams. A relatively straight and fast-blowing jet stream allows weather systems to move through quickly. A slower jet stream, however, bends widely to and fro, like a slow-moving river. As it meanders north and south, its looping curves can stall weather systems by blocking their movement in what is called an *atmospheric blocking pattern.*

Meteorologists have understood jet stream behavior for years, but in 2012, Jennifer Francis of Rutgers University and Stephen Vavrus of the University of Wisconsin applied this knowledge in a research paper to propose a specific mechanism for how and why global warming tends to lead to more extreme weather.

Temperature records show that global warming has been greatest in the Arctic, which has reduced the average temperature difference between Arctic and temperate regions. Francis and Vavrus pointed out that a reduced temperature gradient should weaken the intensity of the Northern Hemisphere's polar jet stream, the one that usually is centered over Canada. As the jet stream slows down and its meandering loops become

longer, these long, lazy loops move west to east more slowly and may get stuck in a north–south orientation for long periods of time, creating an atmospheric blocking pattern. When that happens, Francis and Vavrus explained, a rainy system that would normally move past a city in a day or two may instead be held in place for several days, causing flooding. Likewise, dry conditions over a farming region might last two weeks instead of two days, resulting in drought. Hot spells last longer, and cold spells last longer, too.

FIGURE 1 shows how this scenario played out in March 2012. Once the jet stream entered a blocking pattern, a warm-weather system became stuck in place, roasting the eastern United States with a record-breaking heat wave. A similar blocking pattern over the Atlantic Ocean later that year ended up

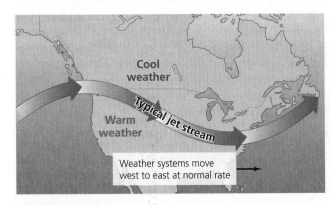

(a) Normal jet stream

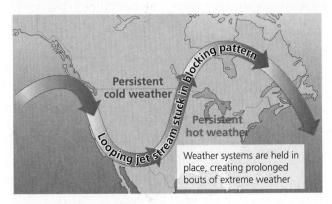

(b) Jet stream in March 2012

FIGURE 1 Changes in the jet stream can cause extreme weather events. Arctic warming can slow the jet stream, causing it to depart from its normal configuration **(a)** and create a blocking pattern **(b)** that stalls weather systems in place, leading to extreme weather events. A blocking pattern like the one illustrated here brought record-breaking heat to the eastern United States in March 2012.

steering Superstorm Sandy into the U.S. East Coast. Researchers soon noted that atmospheric blocking patterns were associated with a 2011 drought in Texas, wildfires in Colorado in 2012, floods and heat waves in Europe, and many other extreme weather events.

As researchers debated how well Francis and Vavrus' general idea explained various specific weather episodes, some of them also noted that—ironically—a weakened jet stream driven by global warming frequently causes episodes of severely cold weather. To the north of the polar jet stream each winter, cold air swirls around in a massive low-pressure system called a *polar vortex*. As long as the jet stream remains strong, it acts as a kind of lasso, corralling the frigid air, and the polar vortex stays contained above the Arctic. But when the jet stream weakens, slows down, and begins meandering, it can allow huge lobes of cold air to reach far south into temperate regions of the United States, Europe, or Asia. The effect is like opening a refrigerator door and letting the cold air spill out. **FIGURE 2** shows how that occurred in January 2019, when vast areas of the United States were plunged into record cold temperatures. Wind-chill readings dipped below −60°F in Minnesota, and Chicago became colder than Antarctica. Similar intrusions of the polar vortex had caused severe conditions in other extreme winters, including 2009–2010, 2013–2014, and 2017–2018, when Europe for a time became colder than the North Pole.

Francis, now at Woods Hole Research Center, is among the scientists studying this phenomenon. She joined climate researcher Judah Cohen, who monitors the polar vortex each day and predicts its behavior for the firm Atmospheric and Environmental Research. In a 2018 study, they demonstrated that cold spells resulting from disruption of the polar vortex had become more frequent in the eastern United States in recent decades. These researchers predict that as global warming proceeds, weakening the polar jet stream more often, the eastern and central United States will experience more and more severe blasts of Arctic air in mid-winter to late winter.

As researchers build knowledge, they are developing the ability to attribute particular weather events to climate change. Rarely is a specific event tied solely to climate change, but in a statistical sense, researchers can often now state that a given storm, flood, or drought was made, say, two or three times more likely by warming air or ocean temperatures. Each year now, the *Bulletin of the American Meteorological Society* publishes a report with analyses of extreme weather events from the previous year. In the 2018 report, 70% of the 146 weather events analyzed were found to be tied, in part, to climate change.

As climate change makes our weather wilder and wilder, will we act to curb global warming? Perhaps not. A study in 2019 examined 2 billion social media postings to analyze discussion of the weather by everyday people. Frances Moore of the University of California–Davis and her team found that people quickly became used to changing weather conditions and that events of a magnitude judged remarkable at one time were no longer judged remarkable just several years later. On average, people's ideas of "normal" weather conditions were based on weather they experienced just two to eight years earlier. As a result of this rapid adjustment in perception, extreme weather becomes "normalized" very quickly—and society may as a result come to accept severe weather without even recognizing how recent and unusual it is.

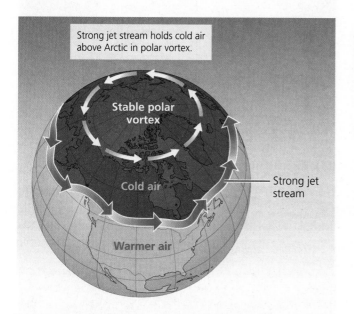

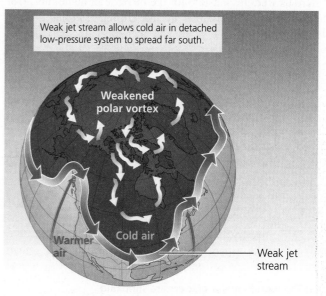

FIGURE 2 **A weakened jet stream in winter can let frigid Arctic air from the polar vortex spill south into the temperate zone.** An event like the one illustrated here brought record-breaking cold to the eastern United States in January 2019.

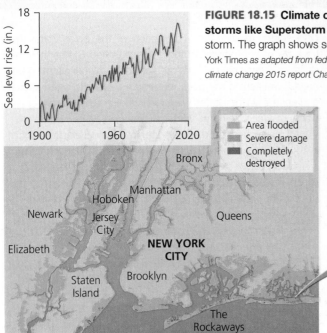

FIGURE 18.15 Climate change contributes to the power and reach of devastating storms like Superstorm Sandy. The map shows areas in New York City flooded by this 2012 storm. The graph shows sea level rise in New York City in the past century. *Map data from* The New York Times *as adapted from federal agencies; graph data from Horton, R., et al., 2015. New York City panel on climate change 2015 report Chapter 2: Sea level rise and coastal storms.* Ann. N.Y. Acad. Sci., *1336: 36–44.*

Shoreline homes toppled by storm surge

flooded. A fire broke out amid flooded homes in Queens and destroyed an entire neighborhood (**FIGURE 18.15**).

For years, researchers had conservatively stated that although climate trends influence the probability of what the weather may be like on any given day, no single particular weather event can be directly attributed to climate change. In the aftermath of Superstorm Sandy, a metaphor spread across the Internet: When a baseball player takes artificial steroids and starts hitting more home runs, you can't attribute any one particular home run to the steroids, but you *can* conclude that steroids were responsible for the increase in home runs. Our greenhouse gas emissions are like artificial steroids that are supercharging our climate and increasing the instance of extreme weather events.

Melting ice has far-reaching effects

As the world warms, mountaintop glaciers are disappearing (**FIGURE 18.16**). Between 1950 and 2018, the World Glacier Monitoring Service estimates that the world's major glaciers, on average, each lost mass equivalent to more than 26 m (85 ft)

(b) Jackson Glacier in 1911

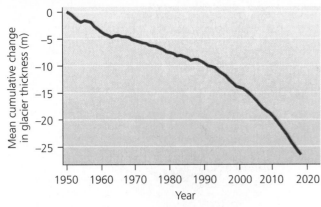

(a) The world's major glaciers are shrinking

FIGURE 18.16 Glaciers are melting rapidly as global warming proceeds. The graph **(a)** shows average declines in thickness of ice (water equivalent) in 40 of the world's major glaciers monitored since 1950. The Jackson Glacier in Glacier National Park, Montana, retreated substantially between **(b)** 1911 and **(c)** 2009. *Data from World Glacier Monitoring Service.*

(c) Jackson Glacier in 2009

FIGURE 18.17 Ice sheets are melting in both Greenland and Antarctica. Huge amounts of ice are being lost **(a)** in each location as glacial ice melts and calves off **(b)** into the ocean. *Data from NASA.*

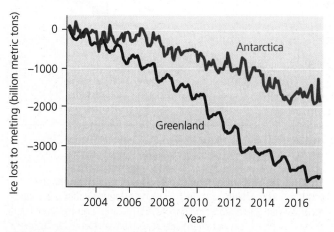

(a) Ice lost to melting in Greenland and Antarctica

(b) Calving glacier

in vertical height of water. Many glaciers on tropical mountaintops have disappeared already. In Glacier National Park in Montana, only 26 of nearly 150 glaciers present at the park's inception in 1910 remain, and scientists estimate that these glaciers will disappear sometime between 2030 and 2080.

Mountains accumulate snow in winter and release meltwater gradually during summer. One out of six people across the world live in regions that depend on mountain meltwater. As a warming climate diminishes mountain glaciers, summertime water supplies are declining for millions of people, which will likely force whole communities to look elsewhere for water or to move.

Warming temperatures are also melting vast amounts of polar ice (**FIGURE 18.17**). In the Arctic, where temperatures have warmed faster than anywhere else, the immense ice sheet that covers Greenland is melting faster and faster. Meanwhile, in Antarctica, coastal ice shelves the size of Rhode Island have disintegrated as a result of contact with warmer ocean water, and research now suggests that the entire West Antarctic ice sheet may be on its way to unstoppable collapse, which would generate a 3-m (10-ft) rise in sea level.

Sea ice is also thinning (**FIGURE 18.18**). As Arctic sea ice melts earlier in the season, freezes later, and recedes from

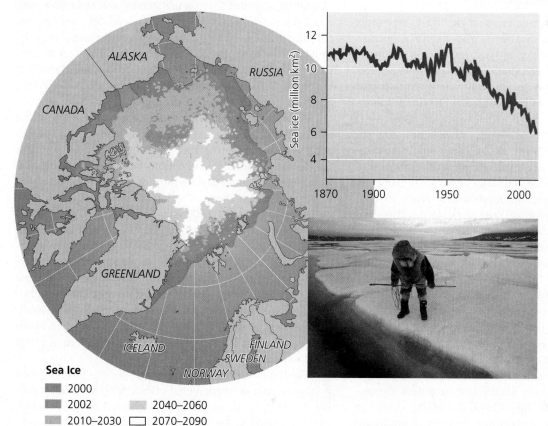

Sea Ice
- 2000
- 2002
- 2010–2030
- 2040–2060
- 2070–2090

FIGURE 18.18 As Arctic sea ice melts, it recedes from large areas. The map shows mean minimum summer extent of sea ice for the recent past, present, and future. The graph shows declines in sea ice averaged from six data sets. *Data from National Center for Atmospheric Research and National Snow and Ice Data Center.*

shore, it becomes harder for Inuit people and for polar bears alike to hunt the seals they each rely on for food. As Arctic sea ice disappears, new shipping lanes are opening up for commerce, and governments and companies are rushing to exploit newly accessible underwater oil and mineral reserves. Already, Russia, Canada, the United States, and other nations are jockeying for position, trying to lay claim to regions of the Arctic as the ice melts.

One reason warming has been greatest in the polar regions—and is accelerating—is that as snow and ice melt, their white surfaces are replaced by darker, less reflective surfaces (such as bare ground, pools of meltwater, or ocean water). These darker colors decrease the surface's *albedo,* or capacity to reflect light. As a result, more of the sun's rays are absorbed at the surface, and the surface warms. In a process of positive feedback, this warming causes more ice and snow to melt, which in turn causes more absorption of radiation and more warming (see Figure 5.2b, p. 108).

Warming Arctic temperatures are also causing *permafrost* (permanently frozen ground) to thaw. As ice crystals within permafrost melt, the thawing soil settles, destabilizing buildings, pipelines, roads, and bridges. A recent study estimates that Alaska will suffer $5.6–$7.6 billion in damage to public infrastructure by 2080 due to climate change. Moreover, when permafrost thaws, it can release methane that has been stored for thousands of years. Because methane is a potent greenhouse gas, its release acts as a positive feedback mechanism that intensifies climate change.

Rising sea levels may affect hundreds of millions of people

As glaciers and ice sheets melt, increased runoff into the oceans causes sea levels to rise. Sea levels also are rising because ocean water is getting warmer, and water expands in volume as it warms. In addition, as we extract groundwater from aquifers (for drinking and to apply to farmland), the wastewater that enters rivers and the excess irrigation water that runs off farmland eventually reach the ocean, adding to sea level rise.

Worldwide, average sea levels have risen 26.0 cm (10.2 in.) since 1880 (**FIGURE 18.19**). These numbers represent vertical rises in water level, and on most coastlines, a vertical rise of a few inches translates into a great many feet of incursion inland. Higher sea levels lead to beach erosion, coastal flooding, intrusion of saltwater into aquifers, and greater impacts from storm surges. A *storm surge* is a temporary and localized rise in sea level generated by a storm. The higher the sea level is to begin with, the farther inland a storm surge can reach.

Regions experience differing amounts of sea level change due to patterns of ocean currents or because land may be rising or subsiding naturally, depending on local geologic conditions. The United States is experiencing varying degrees of sea level rise (**FIGURE 18.20**), with the East Coast and the Gulf Coast most at risk. Miami and South Florida face rising seas in part because global warming has affected the oceans' thermohaline circulation system (pp. 426–427), slowing down the Gulf Stream, the massive current that brings water northward

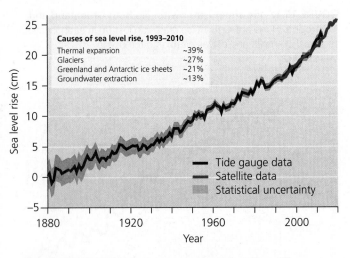

FIGURE 18.19 Global average sea level has risen 26 cm (10.2 in.) since 1880. Sea level is rising because water expands as it warms, glaciers and ice sheets are melting, and much of the groundwater we extract eventually reaches the ocean. *Data from IPCC, 2013.* Fifth assessment report; *CSIRO; and NASA.*

past Florida's Atlantic coast. As the Gulf Stream slows, its water spreads out to the east and west, adding to the amount reaching shore.

So far, New Orleans and southern Louisiana have seen the greatest impacts from sea level rise. There, marshes of the Mississippi River delta continue to disappear as rising seas eat away at coastal vegetation, as dams upriver hold back silt that once maintained the delta, and as the land subsides due to petroleum extraction. All told, more than 2.5 million hectares (1 million acres) of Louisiana's coastal wetlands have vanished since 1940. This wetland loss is depriving New Orleans (much of which is below sea level, safeguarded only by levees) of protection against storm surges. This fact became tragically apparent when Hurricane Katrina flooded much of the city in 2005, killing more than 1800 people and inflicting $80 billion in damage across the region.

Around the world, rising seas are eating away at the salt marshes, dunes, mangrove forests, and coral reefs that serve as barriers protecting our coasts. In the southern Pacific Ocean, small island nations have seen sea levels rise as quickly as 9 mm (about a third of an inch) per year. Rising seas threaten the very existence of countries like the Maldives, a nation of 1200 islands in the Indian Ocean. In the Maldives, four-fifths of the land lies less than 1 m (39 in.) above sea level (**FIGURE 18.21a**). Salt water is contaminating drinking water supplies, and storms intensified by warmer water are eroding beaches and damaging the coral reefs that support the nation's tourism and fishing industries. Residents have evacuated several low-lying islands, and political leaders have looked to buy property in mainland nations in case their people one day need to abandon their homeland.

For these reasons, leaders of the Maldives have played a prominent role in international efforts to fight global warming. In 2009, Maldives President Mohamed Nasheed and his cabinet donned scuba gear and dove into a coastal lagoon,

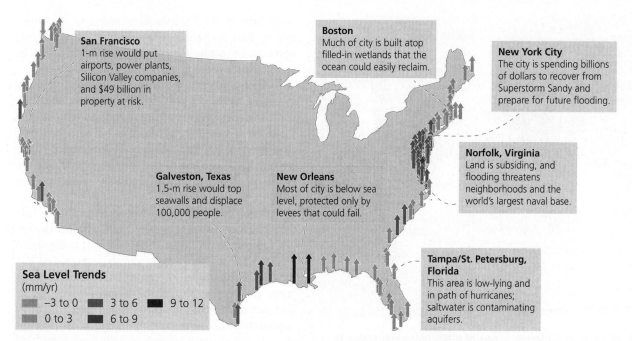

FIGURE 18.20 Rising sea levels are putting many U.S. cities at risk of costly damage. Rates are highest (taller darker blue arrows) where land is subsiding along the Gulf and Atlantic coasts. *Data from National Oceanic and Atmospheric Administration; city profiles adapted from Rising seas: A city-by-city forecast.* Rolling Stone, 20 June 2013.

Within the figure:

San Francisco
1-m rise would put airports, power plants, Silicon Valley companies, and $49 billion in property at risk.

Boston
Much of city is built atop filled-in wetlands that the ocean could easily reclaim.

New York City
The city is spending billions of dollars to recover from Superstorm Sandy and prepare for future flooding.

Galveston, Texas
1.5-m rise would top seawalls and displace 100,000 people.

New Orleans
Most of city is below sea level, protected only by levees that could fail.

Norfolk, Virginia
Land is subsiding, and flooding threatens neighborhoods and the world's largest naval base.

Tampa/St. Petersburg, Florida
This area is low-lying and in path of hurricanes; saltwater is contaminating aquifers.

Sea Level Trends (mm/yr)
-3 to 0 | 3 to 6 | 9 to 12
0 to 3 | 6 to 9

where they held the world's first underwater cabinet meeting—part of a campaign to draw global attention to the impacts of climate change (**FIGURE 18.21b**). Today, people from island nations like the Maldives are no longer so alone in voicing concern over climate change. As major cities such as Miami come face-to-face with the challenges of sea level rise, communities and nations around the world are hastening to come up with solutions.

In its *Fifth Assessment Report,* the IPCC predicted that mean sea level worldwide will rise 26–82 cm (10–32 in.) higher by 2100, depending on our level of emissions. However, research since then is finding that ice in Antarctica and especially Greenland is melting faster and faster, which would lead seas to rise more quickly—perhaps 1–2 m (3.3–6.5 ft) by 2100. More than half of the U.S. population lives in coastal counties, and 3.7 million Americans live within 1 vertical

(a) Malé, capital of the Maldives

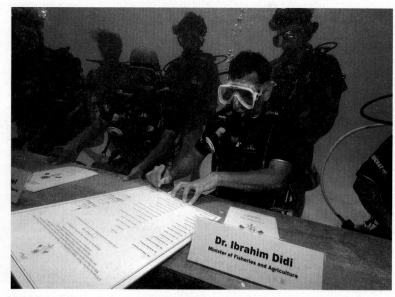

(b) The "underwater cabinet meeting"

FIGURE 18.21 Rising sea levels threaten island nations. The capital of the Maldives **(a)** is crowded onto an island averaging just 1.5 m (5 ft) above sea level. In 2009, the Maldives' president led his cabinet in an underwater meeting **(b)** to focus international attention on the plight of island nations vulnerable to sea level rise.

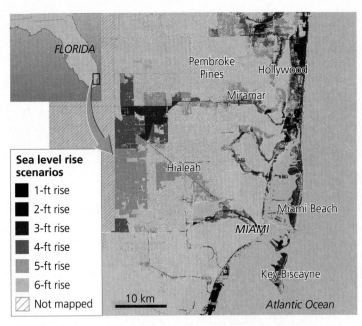

FLORIDA

Pembroke
Pines

Hollywood

Miramar

Sea level rise scenarios

■ 1-ft rise
■ 2-ft rise
■ 3-ft rise
■ 4-ft rise
■ 5-ft rise
■ 6-ft rise
▨ Not mapped

Hialeah

Miami Beach

MIAMI

Key Biscayne

10 km

Atlantic Ocean

FIGURE 18.22 Miami, Florida, is one of many cities vulnerable to sea level rise. Shown are areas of the Miami region that would be flooded by rises in sea level of 1–6 feet. *Data from NOAA Coastal Flood Exposure Mapper, www.coast.noaa.gov/floodexposure/#/map.*

meter of the high tide line. It is estimated that a 1-m rise threatens 180 U.S. cities with losing an average of 9% of their land area. South Florida is judged most at risk (**FIGURE 18.22**). Here, 2.4 million people, 1.3 million homes, and 1.8 million acres are vulnerable, according to experts with the Surging Seas project of Climate Central. In 107 South Florida towns, fully half the population is at risk.

Whether sea levels this century rise 26 cm, 1 meter, or more, hundreds of millions of people will be displaced or will need to invest in costly efforts to protect against high tides and storm surges. And in the long term, if we fail to cut greenhouse gas emissions, sea levels will keep rising. A 2015 study calculated that if we burn all the fossil fuels remaining in the world, all the ice on Antarctica would melt (half of it in just 1000 years), raising sea levels by about 1 foot per decade. If all the world's ice melted, the oceans would be 65 m (216 ft) higher, which would put the entire state of Florida—along with the rest of the Eastern Seaboard and Gulf Coast—under water. The lower Mississippi River all the way up to Memphis would become a gigantic bay, and California's Central Valley, where much of our food is grown today, would be submerged under an inland sea.

Acidifying oceans imperil marine life

As carbon dioxide concentrations in the atmosphere rise, the oceans absorb more CO_2. So far, the oceans have absorbed roughly one-fourth of the CO_2 we have added to the atmosphere. This is altering ocean chemistry, making seawater more acidic—a phenomenon referred to as **ocean acidification** (p. 437).

Ocean acidification threatens marine animals such as corals, clams, oysters, mussels, and crabs, which pull carbonate ions out of seawater to build their exoskeletons of calcium carbonate. As seawater becomes more acidic, carbonate ions become less available, and calcium carbonate begins to dissolve, jeopardizing the existence of these animals (see fuller discussion in Chapter 16, pp. 437–438).

Global ocean chemistry has decreased by 0.1 pH unit, which corresponds to a 26% rise in acidity (hydrogen ion concentration). Scientists are witnessing initial impacts on a variety of marine animals, and acidified seawater has already killed billions of larval oysters in Washington and Oregon, jeopardizing the region's industry. By 2100, scientists predict that seawater will decline in pH by another 0.06–0.32 units, possibly enough to destroy most of our planet's living coral reefs (p. 432). Such destruction could be catastrophic for marine biodiversity and fisheries because so many organisms depend on coral reefs for food and shelter. Indeed, ocean acidification, with the potential loss of marine life, threatens to become one of the most far-reaching impacts of global climate change.

Coral reefs face two additional risks from climate change: Warmer waters contribute to deadly coral bleaching (**FIGURE 18.23**; and see p. 432), and stronger storms physically damage reefs. All these factors concern residents of places like South Florida. South of Miami in the Florida Keys, coral reefs protect coastlines from erosion, offer snorkeling and scuba diving sites for tourism, and provide habitat for fish consumed locally and exported for profit.

Healthy coral before bleaching

Dead coral after bleaching

FIGURE 18.23 Coral bleaching can occur when warmed ocean waters cause corals to expel the symbiotic algae they rely on for food. Without the algae (which give the corals color), the corals become white and may die. This photo shows coral before **(left)** and after **(right)** a bleaching event in American Samoa. The world's coral reefs are also vulnerable to ocean acidification.

Organisms and ecosystems are affected

As global warming proceeds, it is modifying biological phenomena that rely on temperature. In the spring, plants are now leafing out earlier, insects are hatching earlier, and animals are breeding earlier than previously. These shifts can create mismatches in seasonal timing with phenomena mostly driven by length of day, such as bird migration. For example, in eastern North America, forests are greening up earlier in spring, but many migratory songbirds are failing to arrive early enough to keep up with the change—and this can have consequences. For instance, in Europe, birds known as great tits are raising fewer young because the insects they eat are peaking in abundance earlier and the birds have been unable to adjust.

Biologists are also recording spatial shifts in the ranges of species as plants and animals move toward the poles or upward in elevation (toward cooler areas) as temperatures warm (**FIGURE 18.24**). Some organisms will not be able to cope; indeed, the IPCC has estimated that climate change could threaten as many as 20–30% of all plant and animal species with extinction. Many trees may not be able to shift their distributions fast enough. Rare species may be forced out of preserves and into developed areas where they cannot survive. Animals and plants adapted to mountainous environments may be forced uphill until there is nowhere left to go.

Changes in precipitation also have consequences. In regions where heavy rainstorms are increasing, erosion and flooding can pollute and alter aquatic systems. Where rain and snow are decreasing, lakes, ponds, wetlands, and streams are drying up. Overall, the many impacts of climate change on ecological systems will tend to diminish the ecosystem goods and services on which people depend.

Farms and forests experience a mix of consequences

Understanding the effects of climate change on plants is vital, especially in the farms and forests that we manage for food and resources. Plants draw in carbon dioxide for photosynthesis, so it stands to reason that an atmosphere richer in CO_2 might enhance plant growth, resulting in more CO_2 being removed from the air. On the other hand, climate change could inhibit plant growth by promoting drought, fire, or disease, in which case more CO_2 may be released to the air. Scientists studying these questions are finding complex answers, and large-scale outdoor experiments are showing that extra CO_2 can both augment and diminish plant growth.

In the forests that provide our timber and paper products, enriched atmospheric CO_2 can spur growth, but drought, fire, and disease often eliminate these gains. Foresters increasingly find themselves battling catastrophic fires, invasive species, and insect and disease outbreaks, all of which can worsen with longer periods of warm and dry weather. For instance, milder winters and hotter, drier summers are promoting outbreaks of bark beetles that are destroying millions of acres of trees in western North America (p. 324).

For some agricultural crops in the temperate zones, moderate warming may slightly increase production because growing seasons become longer. Research shows that enriched atmospheric carbon dioxide for photosynthesis may boost yields, although the effects appear to be mixed. Moreover, some research shows that crops can become less nutritious when supplied with more CO_2. Perhaps of most importance is that if rainfall continues to shift in space and time, intensified droughts and floods will likely cut into agricultural productivity (**FIGURE 18.25**). Considering all factors together, the IPCC has predicted that global crop yields will increase somewhat initially but that beyond a rise of 3°C (5.4°F) they will decline, worsening hunger in developing nations.

FIGURE 18.24 Animal populations are shifting toward the poles and upward in elevation. Mountain-dwelling animals such as the pika, a unique mammal of western North America, are being forced upslope (into more limited habitat) as temperatures warm. Many pika populations in the Great Basin have disappeared from mountains already.

Pikas are disappearing from mountains after being forced upslope.

FIGURE 18.25 Drought induced by climate change decreases crop yields. Withered corn fields like this one in Illinois were a common sight in 2012, when the U.S. government declared 1000 counties across 26 states to be disaster areas due to drought.

Climate change affects our health, wealth, and security

Drought, flooding, storm surges, and sea level rise are already taking a toll on the lives and livelihoods of millions of people. Ultimately, these impacts have consequences for our health, wealth, and national security.

Health As climate change proceeds, we are facing more heat waves—and heat stress can cause death, especially among older adults. A 1995 heat wave in Chicago killed more than 500 people, and a 2003 heat wave in Europe killed 35,000 people. A warmer climate also exposes us to other health risks:

- Respiratory ailments from air pollution as hotter temperatures promote photochemical smog (p. 468)

- Expansion of tropical diseases, such as malaria and dengue fever, into temperate regions as vectors of infectious disease (such as mosquitoes) spread

- Disease and sanitation problems when floods overcome sewage treatment systems

- Injuries and drowning from worsened storms

Health hazards from cold weather will decrease, but many researchers believe that the increase in warm-weather hazards will more than offset these gains.

Wealth Researchers predict that the economic costs of climate change will substantially outweigh the benefits, especially as climate change grows more severe. They also expect climate change to widen the gap between rich and poor, both within and among nations. Poorer people have less wealth and technology with which to adapt to climate change, and they rely more on resources (such as local food and water) that are sensitive to climate disruption.

Economists have tried to quantify damages from climate change by totaling up its various external costs (pp. 143, 165). Their estimates for the **social cost of carbon,** the economic cost of damages resulting from each ton of carbon dioxide we emit, run the gamut from $10 to $350 per ton, depending on what costs are included and what discount rate (pp. 143, 151) is used. The U.S. government during the Obama administration applied a formal estimate of roughly $40 per ton ($37 in 2007 dollars, rising with inflation) and used this official estimate to decide when and how to impose regulations that affect emissions (p. 151). Other nations and many large corporations use their own estimates.

In terms of overall cost to society, the IPCC has estimated that climate change may impose annual costs of 1–5% of gross domestic product (GDP) globally, with losses being moderate at first but rising as warming intensifies. Most researchers calculate that poor nations will lose proportionally more than rich nations and that tropical nations like Brazil and India will lose far more than high-latitude nations like Canada and Russia. For the United States, the U.S. Environmental Protection Agency (EPA) calculated in 2015 that reducing greenhouse gas emissions would help the United States avoid $235–$334 billion in annual costs by the year 2050 and $1.3–$1.5 trillion by 2100.

Subsequently, the *Fourth National Climate Assessment* (p. 495) predicted the U.S. economy would lose about 1% of GDP annually with 2.2°C (4.0°F) of temperature change, rising to 11% of GDP with 8.1°C (14.5°F) of warming. Regardless of the precise numbers, most economists have concluded that investing money now to fight climate change will save us a great deal of money in the future.

National security The many costs and impacts of climate disruption are beginning to endanger the ability of nations to ensure social stability and protect their citizens from harm. The Pentagon, the White House, the U.S. Navy, the Council on Foreign Relations, and the Central Intelligence Agency have all concluded and publicly reported that climate change is contributing to political violence, war and revolution, humanitarian disasters, and refugee crises (**FIGURE 18.26**). "Climate change will affect the Department of Defense's ability to defend the Nation and poses immediate risks to U.S. national security," the U.S Defense Department stated bluntly in a 2014 report. The report described how storms, rising seas, and other impacts are "threat-multipliers," making small problems larger. Already, extreme weather events have damaged military installations (**FIGURE 18.27**), weakened the economies and infrastructure of allies and trading partners, disrupted flows of oil and gas, and strained emergency response abilities, while Arctic melting has set off a competitive race among nations to claim polar resources.

When environmental conditions worsen, crops may fail and people may suffer acute or chronic economic losses and health impacts. As a result, some individuals in desperate circumstances may leave their homes and become refugees, while others may turn to radical political ideologies or even terrorism. That is why global security experts have linked climate change and drought to the origins of the war in Syria, to conflicts elsewhere in the Middle East and Africa, and to the resulting refugee crisis in Europe. In the years ahead, the world's militaries and emergency responders will be devoting more and more of their efforts toward problems created or made worse by climate disruption.

FIGURE 18.26 Climate change contributes to humanitarian, geopolitical, and national security problems. Prolonged drought associated with climate change undermined agriculture and helped spark the civil war in Syria and the rise of the Islamic State (ISIS), many experts believe. The resulting flow of refugees put social and political strains on European nations that received them.

(a) Naval Station Norfolk, situated along a low-lying coast

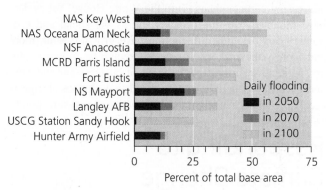

(b) U.S. military bases most vulnerable to sea level rise

FIGURE 18.27 American military leaders are concerned about climate change. Military bases such as Naval Station Norfolk, on the Virginia coast **(a)**, are vulnerable to sea level rise. The graph **(b)** shows percentages of land expected to be lost from selected bases this century due to daily tidal flooding. *Data from Union of Concerned Scientists.*

Impacts will be significant in the United States

Each of us will experience climate change differently, depending on where we live. Across the United States, impacts will be significant. The *Fourth National Climate Assessment,* produced by the U.S. Global Change Research Program in 2017–2018 (p. 495), is an excellent guide to the trends and impacts Americans can expect (**TABLE 18.2**). This report showed that average U.S. temperatures have increased by 1.0°C (1.8°F) since 1901, with the vast majority of this rise occurring in recent decades. Temperatures are predicted to rise by another 1.4°C (2.5°F) by 2050—and an uncertain amount more by the end of this century, depending on our level of emissions (**FIGURE 18.28**, p. 508). Extreme weather events have become more frequent, and the costs they impose on farmers, city-dwellers, coastal communities, and taxpayers across the country are escalating.

Impacts vary by region, and each region of the United States will face its own challenges. For instance, winter and spring precipitation is projected to decrease in the South but increase in the North. Drought may strike in some regions and flooding in others. Sea level rise will affect the Atlantic and Gulf Coasts more than the West Coast. You can learn more about the scientific predictions for your own region by consulting the *Fourth National Climate Assessment,* which is publicly accessible online.

TABLE 18.2 Some Predicted Impacts of Climate Change in the United States

- Average temperatures will rise 1.4°C (2.5°F) further by 2050 and likely more after that.
- Droughts, flooding, and wildfire will worsen; dry areas will get drier and wet areas wetter.
- Extreme weather events will become more frequent. The costs they impose on society will grow.
- Sea level will rise an additional 0.3–1.2 m (1–4 ft) by 2100. Sea level rise will be greater than this global average on the U.S. Atlantic and Gulf coasts.
- Daily tidal flooding is accelerating in 25 coastal cities. Storm surges will erode beaches and coastal wetlands, destroy real estate, and damage infrastructure.
- Health problems due to heat stress, disease, and pollution will rise. Some tropical diseases will spread north.
- Drought, fire, and pest outbreaks will continue to alter forests.
- Ocean acidification will affect marine ecosystems and fisheries.
- Although enhanced CO_2 and longer growing seasons favor crops, increased drought, heat stress, pests, and diseases will decrease most yields.
- Snowpack will decrease in the West; water shortages will worsen in many areas.
- Alpine ecosystems and barrier islands will begin to vanish.
- Melting permafrost will undermine Alaskan buildings and roads.
- Impacts to infrastructure, property, labor productivity, and natural resources may result in annual economic losses of hundreds of billions of dollars by 2100.

Adapted from U.S. Global Change Research Program, 2017. *Climate science special report: Fourth national climate assessment, Volume I* (Wuebbles, D.J., et al., eds.). USGCRP, Washington, D.C.; and U.S. Global Change Research Program, 2018. *Impacts, risks, and adaptation in the United States: Fourth national climate assessment, Volume II* (Reidmiller, D.R., et al., eds.). USGCRP, Washington, D.C.

What is behind the political debate over climate change?

Scientists agree that today's global warming is a result of the well-documented recent increase in greenhouse gas concentrations in our atmosphere. They agree that this rise results primarily from our combustion of fossil fuels for energy and secondarily from the loss of carbon-absorbing vegetation due to deforestation and other changes in land use. They have documented a wide diversity of impacts on the physical properties of our planet, on organisms and ecosystems, and on human well-being.

Yet despite the overwhelming evidence for climate disruption and its impacts, many people, especially in the United States, have long tried to deny that it is happening. Most of these "climate skeptics" or "climate change deniers" now accept that the climate is changing but still express doubt that we are the cause. While most of the world's nations moved

Reduced Emissions Scenario
Projected Temperature Change (°F)
End-of-Century (2071–2099 average)

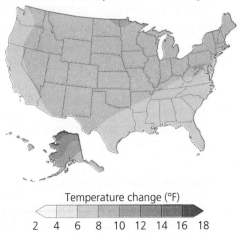

Continued Emissions Scenario
Projected Temperature Change (°F)
End-of-Century (2071–2099 average)

Temperature change (°F)

2 4 6 8 10 12 14 16 18

FIGURE 18.28 Average temperatures across the United States are predicted to rise by the end of this century. Even under a scenario of sharply reduced emissions **(top)**, temperatures are predicted to rise by 4–6°F in the 48 contiguous states and up to 10°F in the Arctic. Under a scenario of business-as-usual emissions **(bottom)**, temperatures are predicted to rise by 8–10°F in many of the 48 contiguous states and up to 16°F in the Arctic. *Data from U.S. Global Change Research Program, 2017.* Climate science special report: Fourth national climate assessment, Volume I *(Wuebbles, D.J., et al., eds.). USGCRP, Washington, D.C.*

forward to confront climate change through international dialogue, public discussion of climate change in the United States remained mired in misleading debates fanned by corporate interests, political think tanks, and a handful of scientists funded by fossil fuel industries, all of whom have aimed to cast doubt on the scientific consensus.

For instance, the oil corporation Exxon-Mobil funded attacks on climate science for years, methodically sowing doubt in the public discourse—after its own in-house scientists had done cutting-edge research back in the 1980s documenting climate change. In the 2010 book *Merchants of Doubt,* science historians Naomi Oreskes and Erik Conway reveal how some of the ideologically motivated individuals who cast doubt on climate science had previously done the

same against scientific conclusions on the risks of tobacco smoke, DDT, ozone depletion, and acid rain.

The views of climate change deniers have long been amplified by the mainstream American news media, which traditionally have sought to present two sides to every issue, even when the arguments of the two sides were not equally supported by evidence. In today's more fragmented media environment, climate change has become a victim of political partisanship, and many of us receive information—and misinformation—from social media and partisan sources online.

Donald Trump's presidency brought opposition to addressing climate change into the White House. Yet as data have mounted over the decades and as the social and economic costs of climate disruption have grown clearer, more and more policymakers, corporate executives, military leaders, national security experts, heads of business and industry, and everyday people have concluded that climate change is escalating and is causing impacts to which we must respond.

Responding to Climate Change

From this point onward, our society will be focusing on how best to respond to the challenges posed by the disruption of our climate. In the big picture, we know exactly what we need to do to bring down atmospheric greenhouse gas concentrations and reverse climate change: *We need to begin withdrawing more carbon from the atmosphere than we put into it.* Doing so will most likely require shifting completely from fossil fuels to clean energy sources *and* finding ways to remove some carbon from the atmosphere reliably and permanently. In the meantime, it will help to reduce fossil fuel consumption by using energy more efficiently. The good news is that everyone can play a part in these vital efforts—not just leaders in government and business, but everyday people from all walks of life, and especially today's youth.

Society can respond in two ways

We can react to climate change with two fundamentally different approaches. One is to address the root cause of the problem, and the other is to protect ourselves from its impacts. Addressing the root cause of climate change means pursuing actions that reduce greenhouse gas emissions or withdraw these gases from the atmosphere. This approach is called **mitigation** because the aim is to mitigate the problem—that is, to alleviate it or lessen its severity. Major strategies for mitigation include improving energy efficiency, switching to clean and renewable energy sources, and preserving and restoring forests. Others include recovering landfill gas and encouraging farm practices that protect soil quality. The sooner society begins reducing emissions, the lower the level at which they will peak, and the less we will alter climate **(FIGURE 18.29)**. Ultimately, we will want to reverse climate change by bringing greenhouse gas concentrations in the atmosphere back to their natural levels.

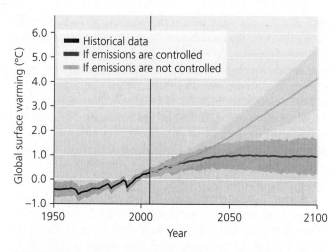

FIGURE 18.29 **The sooner we stabilize our emissions, the less climate change we will cause.** The red line shows temperature change expected if we strongly limit carbon emissions. The orange line shows change expected if we fail to control emissions effectively. These data projections represent averaged output of a large number of climate models under two emissions scenarios studied by the IPCC. *Data from IPCC, 2013.* Fifth assessment report.

FIGURE 18.30 **Miami Beach is trying to adapt to sea level rise by constructing an elaborate system of pumps and drainage pipes.**

The second type of response—cushioning ourselves from the impacts of climate change—is called **adaptation** because the goal is to adapt to change. Installing elaborate pump systems, as Miami and Miami Beach are each doing to pump out their floodwaters, is one example of adaptation (**FIGURE 18.30**). Another is erecting seawalls, as residents of the Maldives and many coastal cities have done. The Netherlands—a very low, flat, coastal nation—has so much experience with seawalls, levees, and other infrastructure to manage seawater that Dutch engineers and consultants are in high demand in Miami and other cities worldwide that find themselves confronting sea level rise. Other examples of adaptation include restricting coastal development, adjusting farming practices to cope with drought, and modifying water management practices to deal with reduced river flows, glacial outburst floods, or salt contamination of groundwater.

We need to pursue adaptation measures because even if we could halt all our emissions right now, the greenhouse gas pollution already in the atmosphere would continue driving global warming for years, with global temperatures rising an estimated 0.6°C (1.0°F) more by the end of the century. Because this amount of change is already locked in, we would be wise to find ways to adapt to its impacts. More importantly, however, we need to pursue mitigation; otherwise, climate change will eventually overwhelm any efforts we might make to adapt. We will spend the remainder of our chapter examining strategies for the mitigation of climate change.

What can each of us do?

Although halting climate change requires action from leaders in policy, business, and engineering, every one of us as individuals can play a role. Just as we each have an ecological footprint (p. 5), we each have a **carbon footprint** that expresses the amount of carbon we are responsible for emitting. We can reduce our carbon footprints by taking steps in our everyday lives (**TABLE 18.3**, p. 510). Taken together, individual decisions and actions scale up to make a difference at the societal level.

How you get around Transportation accounts for more than 35% of U.S. CO_2 emissions, largely because we rely on motor vehicles that run on gasoline. If you can drive less—or live without a car—you will cut down greatly on your carbon footprint. To reduce reliance on cars, some people live nearer to their workplaces. Others use mass transit such as buses, subway trains, and light rail. Still others bike or walk to work.

If you drive, selecting a fuel-efficient vehicle that gets good gas mileage makes a huge difference. Alternatives to the traditional combustion-engine automobile get the best fuel efficiency; they include electric vehicles, gasoline-electric hybrids (p. 555), and vehicles that use hydrogen fuel cells (p. 613) or fuels such as compressed natural gas or biodiesel (p. 578). In any vehicle, simple steps like driving the speed limit and keeping your tires at optimal inflation pressures can boost your fuel efficiency by up to a third.

How you power up From cooking to heating to lighting to surfing the Internet, much of what we do each day depends on electricity, which accounts for 35% of U.S. emissions. One easy way to reduce electricity consumption—and save money at the same time—is to purchase energy-saving products. You can consult EnergyGuide labels and tags from the EPA's Energy Star program when shopping to judge which brands and models will save energy and money (pp. 554–555). In recent years, American consumers, businesses, and industries have saved hundreds of billions of dollars while reducing emissions by adopting energy-efficient appliances, lighting, windows, doors, ducts, insulation, electronics, and heating and cooling systems.

How you get around.
Walk, bike, or take mass transit instead of driving a car. If you drive, select a fuel-efficient vehicle, drive the speed limit, and keep your tires inflated.

How you power up.
Promote efficiency by using energy-saving products. Encourage your institution to obtain energy from solar panels and water heaters, wind turbines, and ground-source heat pumps.

What you eat.
Enjoy a plant-rich diet with fewer animal products, particularly carbon-intensive foods like beef. Waste less food and eat locally grown food.

What you do.
Get involved in campus sustainability efforts, get engaged politically, or get trained for a green-collar job. Be a model for others by living a less wasteful, more sustainable lifestyle.

As individuals and as a society we also can switch to cleaner energy sources. In recent years, the United States has reduced its emissions from electricity generation largely by switching from coal to cleaner-burning natural gas at power plants (see Figure 19.14, p. 539). Cleaner still are nuclear power (Chapter 20) and renewable energy sources such as solar power, wind power, geothermal power, bioenergy, and hydropower (Chapters 20 and 21). As a student on campus, you can help persuade your college or university to purchase renewable power or to generate solar or wind power on campus. As a homeowner, you can install solar panels or ground-source heat pumps.

What you eat One of the most effective ways to lower your carbon footprint is through your diet. Eating animal products such as meat, eggs, and milk is far more energy-intensive than eating lower on the food chain—so a vegetarian or vegan diet will greatly reduce the emissions you are responsible for. The type of meat also makes a difference: Pound for pound, beef leads to eight times more emissions than chicken (see Figure 10.9c, p. 249). Scientists estimate that animal agriculture accounts for 14.5% of the world's greenhouse gas emissions. For many of us, reducing the amount of red meat we eat while shifting to more fruits and vegetables can reduce our carbon footprint more than any other single action we could take, while also enhancing our health.

Eating locally grown food (or buying any locally made product)—as opposed to items shipped from far away—tends to cut down on fuel use from long-distance transportation. Being careful not to waste food also makes a difference. About 40% of U.S. food goes to waste, adding greatly to emissions. Taking smaller portions (encouraged by trayless dining at campus dining halls) and ordering wisely at restaurants help to reduce food waste.

What you do We all can take many other steps to minimize our carbon footprints. Here are just a few:

- Cut back on waste by buying only what you need, reusing items whenever possible, and recycling and composting what you no longer need (Chapter 17).

- Get involved in sustainability efforts on campus (p. 18), such as running recycling programs, enhancing energy and water efficiency, managing gardens and sustainable dining halls, or pressing administrators to build green buildings or invest in renewable energy.

- Get engaged politically by communicating with your representatives in government, supporting candidates for office who will tackle climate issues, and advocating for policies to support clean energy and energy efficiency.

- Be a model for others in your daily life. Encourage resource conservation at your workplace. Consider pursuing a career in a profession that promotes energy efficiency or renewable energy. If you are a parent, pass on to your children what you've learned about addressing climate change. Our actions reverberate with the people around us—so through what you say and do, you can multiply your own efforts.

A new generation is pressing for action

Many young people are finding their voices through a growing global protest movement sparked by Swedish teenager Greta Thunberg (**FIGURE 18.31**). In August 2018, at age 15, Thunberg skipped school for three weeks to stage a solitary protest outside Sweden's parliament, demanding action from policymakers to address climate change. As she continued her protests once a week each Friday, other students joined her, news spread by social media, and a movement called Fridays for the Future blossomed internationally. Students worldwide began staging school climate strikes to demand that political leaders act to reduce emissions. On Friday, September 20, 2019, an estimated 7 million or more youth and adults took to

FIGURE 18.31 Swedish climate activist Greta Thunberg inspired a global youth movement to urge policymakers to address climate change. Her solo school strikes in Sweden led eventually to leadership in efforts worldwide—showing how much difference one committed young person can make. Here she addresses a climate protest outside the White House in Washington, D.C., in 2019.

the streets of thousands of cities across the world in a Global Climate Strike (p. 174).

Thunberg was in New York City for the Global Climate Strike. She had sailed across the Atlantic Ocean in a carbon-neutral voyage to the United States, where she triggered a burst of media attention that made clear for the first time to many older Americans how angry many young people were about the potential for climate change to ravage their futures. Invited to speak to a U.S. Congressional committee, Thunberg told lawmakers, "Don't invite us here to just tell us how inspiring we are.... If you want advice for what you should do, invite scientists. Ask scientists for their expertise. We don't want to be heard. We want the science to be heard."

Thunberg was even more blunt at the United Nations Climate Action Summit in New York City, telling the world leaders gathered there: "For more than 30 years the science has been crystal clear. How dare you continue to look away, and come here saying that you are doing enough? ... You are failing us. But the young people are starting to understand your betrayal. The eyes of all future generations are upon you. And if you choose to fail us, I say we will never forgive you."

Multiple strategies can help us reduce emissions

There is no single "magic bullet" for stopping climate change. Reducing emissions will require steps by many people and institutions across many sectors of our economy. Fortunately, most reductions can be achieved using proven and available technologies and approaches. Waste managers are cutting emissions by generating energy from waste in incinerators, capturing methane seeping from landfills, and encouraging recycling, composting, and the reuse of materials and products (pp. 623, 633). Power producers are capturing excess heat from electricity generation and putting it to use by cogeneration (p. 553). Many farmers and ranchers are sustainably managing their soil to store more carbon or are using new techniques to reduce methane emissions from rice cultivation and cattle and nitrous oxide emissions from fertilizer. Preserving forests, reforesting cleared areas, and pursuing sustainable forestry practices (p. 324) all help to absorb carbon from the air.

As our society transitions away from fossil fuels and toward a "decarbonized" economy run on clean energy sources (Chapters 20 and 21), we are also trying to capture emissions before they leak to the atmosphere. Billions of dollars are being spent to develop *carbon capture and storage,* by which carbon dioxide is removed from emissions and stored below ground under pressure in deep salt mines, depleted oil and gas deposits, or other underground reservoirs (see Figure 19.24, p. 549).

One promising roadmap for how our society can shift to a low-carbon future has been presented by an effort called the Drawdown Project. Launched by entrepreneur and sustainability guru Paul Hawken in 2016, the Drawdown Project has engaged hundreds of experts to come up with what it calls "the most comprehensive plan ever proposed to reverse global warming." The plan proposes 100 strategies (**DATAGRAPHIC**, p. 512) that, if deployed at a large scale over the next 30 years, could together bring us to "drawdown," the point at which atmospheric greenhouse gases peak and begin to decline. These 100 strategies are bold yet realistic—80 of them are already being carried out somewhere in the world and are economically viable. Twenty are considered innovative ideas not yet proven but which "could be veritable game-changers." For each strategy, the Drawdown team has estimated financial costs and benefits and the amount of carbon dioxide the strategy could pull out of the atmosphere. Altogether, the 80 established strategies over 30 years have the potential to pull at least 1.05 trillion tons of CO_2 from the air and to save society $47 trillion.

WEIGHING the **issues**

Taking Climate Change to Court

People desiring stronger action against climate change are now taking their grievances to court. In a lawsuit launched in 2016, the nonprofit Our Children's Trust and 21 young people demanded action on climate change, saying that the U.S. government had "willfully ignored this impending harm" (pp. 161–163). In the Netherlands in 2015, a group named Urgenda and 900 citizens sued the Dutch government for "knowingly contributing" to global warming—and Urgenda and the citizens won. The court agreed that existing Dutch policy would not hold warming to the internationally set 2°C goal and ordered the government to deepen emission cuts. In Peru, a farmer sued a fossil fuel power company, asking monetary compensation for impacts of glacier melt on his community. Likewise, the cities of Oakland and San Francisco have sued oil companies to recover money the cities will need to spend to adapt to sea level rise.

Do you think lawsuits are an appropriate way to strengthen policy responses to climate change? To what degree are leaders in government obligated to protect their constituents from climate change? Do you favor suing fossil fuel companies for compensation for climate change impacts? What ethical or human rights issues, if any, do you think climate change presents? How could these best be resolved?

Experts at Project Drawdown have done the math. Their exhaustive research indicates that **we can reduce greenhouse gases in the atmosphere by the year 2050**, by pursuing 80 proven and existing strategies scaled up to a realistic degree.

Below are the **top ten strategies** that will make the most difference:

Billions of tons of CO_2 reduced using proven strategies

1,051

1,442

1,613

Greenhouse gases in atmosphere

Status quo

Plausible

Drawdown

Optimum

2020   2050

Year

89.7   **Manage disposal of HFC refrigerants,** powerful greenhouse gases being phased out.

84.6   **Expand wind power** to 22% of electricity generation.

70.5   **Reduce food waste** by 50%, which reduces deforestation for agriculture.

66.1   **Shift to plant-rich diets** with less meat, reducing deforestation for agriculture.

61.2   **Restore tropical forests** on 435 million acres to enhance carbon sequestration.

51.5   **Expand education for girls,** slowing population growth by giving women more social and economic opportunity.

51.5   **Expand access to contraception,** slowing population growth by preventing unintended pregnancies.

36.9   **Expand utility-scale solar farms** to 10% of electricity generation.

31.2   **Combine trees and grazing in "silvopasture,"** which stores more carbon than grassland ranching.

24.6   **Expand PV rooftop solar power** to 7% of electricity generation.

These strategies accompany broad efforts to:

• **Modernize the electric grid**

• **Improve batteries for energy storage**

• **Retrofit buildings for energy efficiency**

...And CARBON PRICING can help encourage many of these efforts. $$$

➤ We CAN stop and reverse global warming using a number of proven and existing strategies, if we can muster the social, economic, and political means to scale them up.

Data from Hawken, Paul, ed., 2016. *Drawdown: The most comprehensive plan ever proposed to reverse global warming.* Penguin, New York.

What role should government play?

Amid all the promising strategies and technologies for reducing greenhouse gas emissions, people often disagree on what role government should play to encourage such efforts: Should it mandate change through laws and regulations? Should it design policies that give private entities financial incentives to reduce emissions? Should it impose no policies at all and hope that private enterprise will develop solutions on its own?

In 2007, the U.S. Supreme Court ruled that carbon dioxide was a pollutant that the Environmental Protection Agency (EPA) could regulate under the Clean Air Act (pp. 173, 458). When Barack Obama became president, he instead urged that Congress craft laws to address emissions. In 2009, the House of Representatives passed legislation to create a nationwide cap-and-trade system (pp. 181, 514) in which industries and utilities would compete to reduce emissions for financial gain and under which emissions would be mandated to decrease 17% by 2020. However, legislation did not pass in the Senate. As a result, responsibility for addressing emissions passed to the EPA, which began phasing in emissions regulations on industry and utilities to spur energy efficiency retrofits and renewable energy use.

In 2013, Obama announced that, because of legislative gridlock, he would take steps to address climate change using the president's executive authority. His Climate Action Plan aimed to jump-start renewable energy development, modernize the electrical grid, finance clean coal and carbon storage efforts, improve automotive fuel economy, protect and restore forests, and encourage energy efficiency. It also led to the Clean Power Plan (p. 470), under which the EPA proposed to regulate existing power plants.

Trump's presidency, together with Republican control of Congress, resulted in the sudden reversal of most Obama-era federal policies designed to reduce greenhouse gas emissions. By executive order, Trump halted the Clean Power Plan, and his administration withdrew from engagement in international efforts to curb climate change. However, political leaders in many U.S. states continued to advance policies supporting energy efficiency and the use of clean and renewable energy sources.

We can put a price on carbon

To encourage efforts to reduce emissions across society, there are growing calls for "putting a price on carbon"—using financial incentives as tools to motivate voluntary reductions in emissions. **Carbon pricing** is intended to compensate the public for the external costs (pp. 143, 165) we all suffer from fossil fuel emissions and climate change—those economic losses resulting from the various impacts of climate change on health, property, infrastructure, and more.

Carbon pricing lifts the burden of paying for these impacts off the shoulders of the public and shifts it to the parties responsible for emissions. Supporters of carbon pricing view it as the fairest, least expensive, and most effective approach to reducing emissions. There are two primary approaches to carbon pricing: (1) carbon taxation and (2) carbon trading.

Carbon taxation A **carbon tax** is a type of green tax (p. 179) on the carbon content of fossil fuels. In carbon taxation, governments charge suppliers of fossil fuels a fee for each unit of CO_2 that results from their product. For instance, a government might tax firms that mine, process, sell, or import fossil fuels—such as coal-mining companies, natural gas suppliers, or oil refineries. These parties will pass the cost of the tax along to distributors and retailers, such as gas stations or power utilities, who will, in turn, pass these costs along to consumers in the form of higher prices for products like gasoline or electricity.

Higher prices give consumers motivation to reduce fossil fuel use by switching to cheaper alternatives. For instance, a driver might opt to buy a more fuel-efficient vehicle, a homeowner might choose to live in a city near a bus route rather than in a remote suburb, or a factory might relocate to a site along a railroad line. As consumer demand evolves, businesses, industries, and utilities gain motivation to switch to less carbon-intensive products—a coal-fired power plant might switch to cleaner-burning natural gas, or an oil company might boost its investments in solar and wind energy. As the cost of the tax is passed along through transactions, every party in the economy gains incentives to buy, sell, and use products that cause less carbon pollution.

Although costs are passed along, consumers end up paying nothing extra if carbon taxation takes a **fee-and-dividend** approach. In this approach, funds the government receives through the carbon tax (the "fee") are transferred back to taxpayers in the form of a tax cut or a tax refund (the "dividend"). This way, any costs of a carbon tax that get passed along to consumers will be reimbursed to them (**FIGURE 18.32**) through a reduction in their income taxes or sales taxes. In theory, such a system gives everyone a financial incentive to reduce emissions

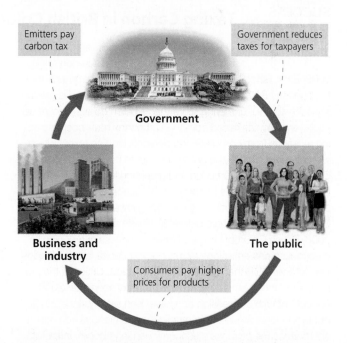

FIGURE 18.32 In revenue-neutral carbon taxation, emitters pay a carbon tax to the government and then charge consumers higher prices for products, but consumers are reimbursed by tax cuts from the government. No parties gain or lose money, yet all parties gain motivation to reduce emissions.

while imposing no financial burden on taxpayers and no drag on the economy. The fee-and-dividend approach is a type of **revenue-neutral carbon tax,** so named because there is no net transfer of revenue from taxpayers to the government. For this reason, the approach is gaining broad appeal across the political spectrum.

Carbon taxes of various types have been introduced in roughly 40 nations and 20 cities and states. Sweden has one of the world's steepest and longest-running carbon taxes and has reduced fossil fuel use greatly as a result (see Figure 20.1, p. 562) while expanding its economy and providing its residents one of the world's highest standards of living. The Canadian province of British Columbia has embarked on revenue-neutral carbon taxation and also is thriving (see **SUCCESS STORY**). Taxes on energy can be politically risky, however; when France's government tried to adopt an energy tax whose costs were perceived to fall on working-class people—after it had reduced income taxes for wealthy people—working-class people staged protests that came to be known as the Yellow Vest Movement, and the government was forced to back down. In the United States, voters in Washington State have twice rejected a carbon tax referendum. But Boulder, Colorado, taxes electricity consumption, Maryland's Montgomery County taxes power plants, and San Francisco Bay Area counties tax businesses for emissions.

Carbon trading The other means of pricing carbon is carbon trading. A **carbon trading** system is an emissions trading system (p. 181) for carbon dioxide or other greenhouse gases. In an emissions trading system, a government sets up a market in permits for the emission of pollutants, and companies, utilities, or industries buy and sell the permits among themselves (**FIGURE 18.33**). In such a market, the price of permits fluctuates freely according to supply and demand. In the approach known as **cap-and-trade** (p. 181), the government sets a cap on the amount of pollution it will allow and then gives, sells, or auctions permits to emitters that allow them to emit a certain fraction of the total amount. Over time, the government lowers the cap to ensure that total emissions decrease. Emitters with too few permits to cover their pollution must reduce their emissions, buy permits from other emitters, or pay for carbon offset credits (p. 518).

As with carbon taxation, carbon trading harnesses the financial incentives of free-market capitalism while granting emitters the freedom to decide how they can best reduce emissions. In theory, once polluters are charged a price for polluting, market forces do the work of reducing pollution in an economically efficient way by allowing business, industries, or utilities flexibility in how to respond.

The world's largest cap-and-trade program is the European Union Emission Trading Scheme. This market began in 2005, but investors soon realized that national governments had allocated too many permits to their industries. The overallocation gave companies little incentive to reduce emissions, so permits lost their value, and market prices collapsed. Europeans addressed these problems by making emitters pay for permits, and the market runs more effectively today. Similar difficulties befell the world's first emissions trading program for greenhouse gas reduction, the Chicago Climate Exchange, which operated from 2003 to 2010 and involved several hundred corporations, institutions, and municipalities.

SUCCESS story — Taxing Carbon in British Columbia

The Canadian province of British Columbia introduced a revenue-neutral carbon tax in 2008. Established by a politically conservative government, the tax began at $10 per metric ton of CO_2 equivalent and is gradually rising to $50, garnering more than a billion dollars in revenue each year. Monies raised from the carbon tax have been replacing revenues from corporate and personal income taxes, which were lowered. So far, the policy appears to be working: British Columbia's fuel consumption and greenhouse gas emissions have declined, even as the province's economy has grown strongly—and its taxpayers are enjoying lower income taxes. Comparing the nine years before 2008 with the nine years after 2008, British Columbia has had greater success in reducing its greenhouse gas emissions, emissions per capita, and emissions per dollar of GDP than has the rest of Canada. Critics of the policy point out that emissions dropped steeply soon after 2008 concurrent with a recession and have risen slightly since 2010, but proponents argue that the policy kept emissions from rising still more as the province's economy began to boom. Interpretations of the data vary, but in the meantime, perceived success in British Columbia led Canada's national government to enact nationwide carbon pricing. The program calls for each of

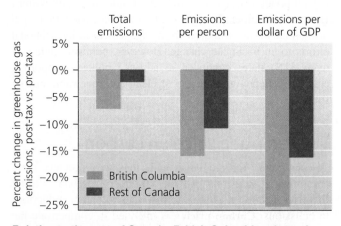

Relative to the rest of Canada, British Columbia released fewer emissions after introducing its carbon tax (2008–2016) than before the tax (1999–2007). *Data from Statistics Canada.*

Canada's provinces either to adopt a national revenue-neutral carbon tax (starting at $20/ton in 2019 and rising to $50/ton in 2022) or to join a cap-and-trade market.

→ **Explore the Data** at Mastering Environmental Science

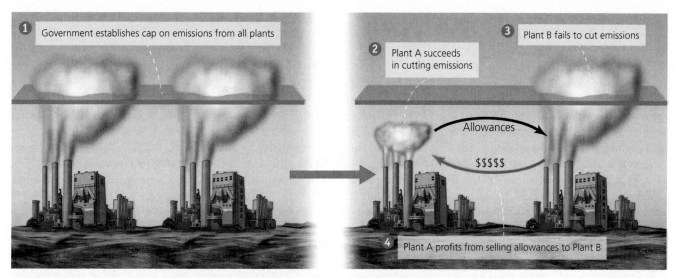

FIGURE 18.33 A cap-and-trade emissions trading system harnesses the efficiency of market capitalism to reduce emissions. In the diagram, Plant A succeeds in reducing its emissions below the cap, and Plant B does not. As a result, Plant B pays money to Plant A to purchase allowances that Plant A no longer needs to use. Plant A profits from this sale, and the government cap is met, reducing pollution overall. Over time, the cap can be lowered to achieve further emissions cuts.

Today, California is running a cap-and-trade program, and the state is on track to reduce emissions to 1990 levels by 2020 while its economy continues to thrive. The Canadian province of Quebec has now joined California's market. Meanwhile, nine northeastern states are participating in the Regional Greenhouse Gas Initiative. In this effort, Connecticut, Delaware, Maine, Maryland, Massachusetts, New Hampshire, New York, Rhode Island, and Vermont run a cap-and-trade program for power plant emissions. From 2005 to 2017, these states cut their CO_2 emissions from power plants by more than 50% even as their economies grew. It is estimated that investment of the auction proceeds will save consumers $8.6 billion in energy costs and eliminate 28 million tons of CO_2 emissions.

China is setting up a national cap-and-trade program, and Mexico may do so as well. All these early experiments are providing lessons for how to set up effective and sustainable trading systems.

WEIGHING the issues

Carbon Tax or Cap-and-Trade?

What advantages and disadvantages do you see in using a carbon tax to reduce greenhouse gas emissions? What pros and cons do you see in using a cap-and-trade system to achieve this goal? What do you think of the idea of a revenue-neutral carbon tax? If you were a U.S. senator, what type of policy would you advance to address emissions in the United States, and why?

International climate negotiations have sought to limit emissions

Disruption of the climate is a global problem, so global cooperation is needed to forge effective solutions. In 1992, most of the world's nations signed the **United Nations Framework Convention on Climate Change.** This treaty outlined a plan for reducing greenhouse gas emissions to 1990 levels by the year 2000 through a voluntary approach. Emissions kept rising, however,

so nations forged a binding agreement to *require* emissions reductions. An outgrowth of the Framework Convention drafted in 1997 in Kyoto, Japan, the **Kyoto Protocol,** mandated signatory nations, by the period 2008–2012, to reduce emissions of six greenhouse gases to levels below those of 1990.

As of 2016, nations that signed the Kyoto Protocol had decreased their emissions by 13.0% from 1990 levels (**FIGURE 18.34**). However, much of this reduction was due to economic contraction in Russia and nations of the former Soviet Bloc following the breakup of the Soviet Union. When these nations are factored out, the remaining signatories

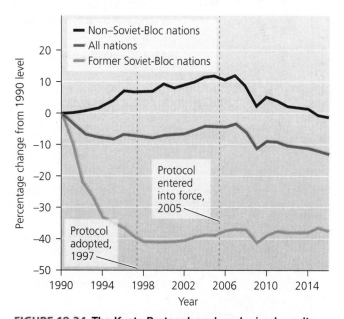

FIGURE 18.34 The Kyoto Protocol produced mixed results. Nations ratifying it reduced their emissions of six greenhouse gases by 13.0% by 2016, but this was largely due to unrelated economic contraction in the former Soviet-Bloc countries. *Data from U.N. Framework Convention on Climate Change. Values do not include influences of land use and forest cover.*

showed only a 1.3% decrease in emissions. Nations not parties to the accord, including China, India, and the United States, increased their emissions.

The United States was the only developed nation not to ratify the Kyoto Protocol. U.S. leaders objected to how it required industrialized nations to reduce emissions but did not require the same of rapidly industrializing nations such as China and India. Proponents of the Kyoto Protocol countered that this approach was justified because industrialized nations had created the climate problem and therefore should take the lead in resolving it.

All along, representatives of the world's nations were meeting at a series of annual conferences, trying to design a treaty to succeed the Kyoto Protocol (**FIGURE 18.35**). Delegates from European nations and small island nations generally took the lead, while China, India, and the United States were reluctant to commit to emissions cuts. At a contentious 2009 conference in Copenhagen, Denmark, nations endorsed a goal of limiting global temperature increase to 2°C but failed to agree on commitments. In Cancun, Mexico, in 2010, nations made progress on a plan, nicknamed *REDD* (p. 317), to help tropical nations reduce forest loss, and developed nations promised to help fund mitigation and adaptation efforts for developing nations. In Doha, Qatar, in 2012, negotiators extended the Kyoto Protocol until 2020, but a number of nations backed out, and the protocol now applies to only about 15% of the world's emissions.

The Paris Agreement produced global consensus

At the 2015 climate conference in Paris, France, the world's nations made stronger commitments than ever before. European nations announced plans to further reduce their emissions. China pledged to cut back on coal-fired power and to establish a cap-and-trade program. Brazil promised to halt deforestation. India agreed to slow its emissions growth, reforest its land, and aggressively develop renewable energy. And the United States committed to emissions cuts it hoped would result from new regulations on coal-fired power plants and from switching from coal to natural gas. All these national commitments were voluntary and unenforceable, but a strong global consensus was forged, creating powerful diplomatic peer pressure on all parties to live up to their pledges. In the resulting **Paris Agreement,** nations declared a joint goal to limit global temperature rise to 2.0°C (3.6°F) above pre-industrial levels and to attempt to keep the rise below 1.5°C (2.7°F).

The impressive progress at the Paris conference resulted because all 197 nations were encouraged to bring their own particular solutions to the table. Success at Paris was also facilitated by an agreement a year earlier between U.S. President Obama and Chinese President Xi Jinping, which broke the impasse between these two largest polluting nations (**FIGURE 18.36**). In their joint announcement, Obama had pledged that the United States would reduce carbon emissions by 28% by 2025, and Xi Jinping had promised that China would halt its emissions growth by 2030.

In 2017, Trump announced that his administration would withdraw the United States from the Paris Agreement. His decision drew scorn and

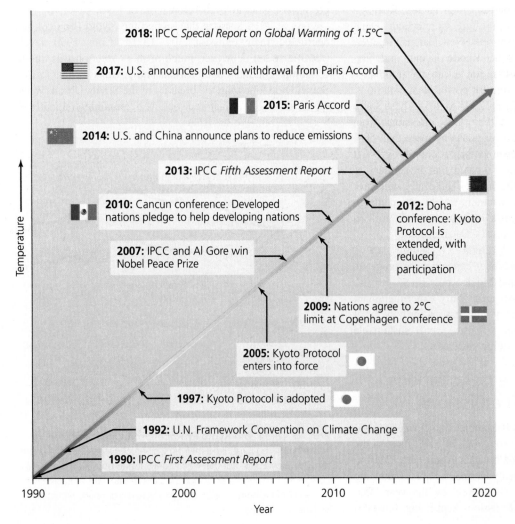

FIGURE 18.35 Negotiators have worked tirelessly for years to forge international agreements to reduce emissions. Shown is a timeline of highlights of climate change science and policy since 1990.

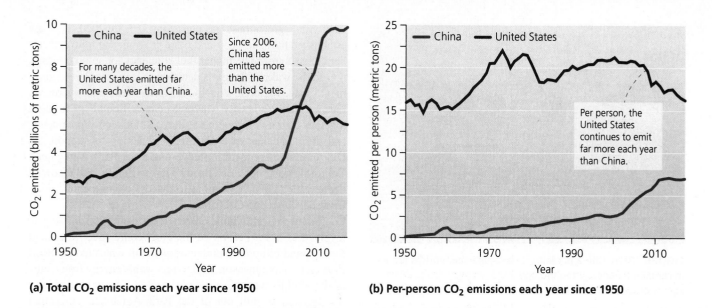

(a) Total CO$_2$ emissions each year since 1950

(b) Per-person CO$_2$ emissions each year since 1950

FIGURE 18.36 The United States and China are the world's biggest emitters of greenhouse gases. In recent years, China **(a)** surpassed the United States in emitting the most. However, China has many more people. On a per-person basis **(b)**, the United States still emits far more than China. *Data from Global Carbon Project and Carbon Dioxide Information Analysis Centre, as presented by Our World in Data (ourworldindata.org).*

outrage from around the world and was opposed in polls by the majority of the American people. What impact a U.S. withdrawal—which by the terms of the agreement cannot formally take effect until 2020—might have is unclear. Many people feared that the decision would cause other nations to abandon the commitments they made in Paris. Yet the resolute unity that nations displayed in the face of Trump's announcement suggested that the rest of the world might instead redouble its commitment to fighting climate change.

Ultimately, the Paris Agreement will be judged by how well nations live up to their pledges. However, those pledges alone will not be adequate to halt climate change, for even if all nations fully meet their commitments, calculations indicate that the average global temperature by 2100 would still rise about 3.0°C (5.4°F) (**FIGURE 18.37**). Many experts now predict that success in mitigating climate change will come largely from business investment, technological advances, and economic incentives, as well as on government initiatives at the national, regional, state, and local levels.

Will emissions cuts hurt the economy?

Many policymakers have opposed mandates to reduce emissions because they fear that reducing emissions will hamper economic growth. This idea is understandable given that our economies have relied so heavily on fossil fuels. However, nations such as Germany, France, and the United Kingdom have reduced their emissions since 1990 while enhancing their economies and providing their citizens very high standards of living. Per person, the emission of greenhouse gases in wealthy nations such as Denmark, New Zealand, Hong Kong, Switzerland, and Sweden is less than half that in the United States.

Indeed, the United States was able to reduce its carbon dioxide emissions by 11% from 2007 to 2018 as a result of

efficiency measures and a shift from coal to natural gas in power plants. The U.S. economy grew during this period, suggesting that cutting emissions need not hinder economic growth. Moreover, in 2015, for the first time in history, global

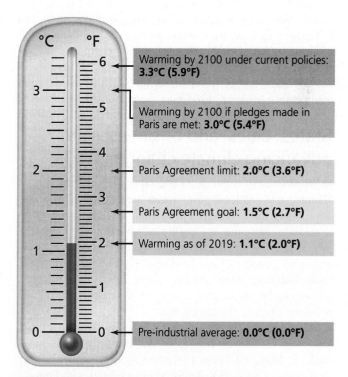

FIGURE 18.37 Commitments made in the 2015 Paris Agreement would limit the global temperature increase to 3.0°C (5.4°F) by 2100. This increase is less than the 3.3°C (5.9°F) rise predicted in the absence of those commitments but greatly exceeds the target of limiting temperature rise to 1.5°C (2.7°F). *Data from Climate Action Tracker.*

FIGURE 18.38 China is racing to become the world's leader in renewable energy technology. Here, workers at a Chinese factory produce photovoltaic solar panels.

carbon emissions fell even while the world economy continued to grow. Encouraging data like these suggest that we may soon reach a historic turning point at which economic growth becomes decoupled from greenhouse gas emissions.

Because resource use and per capita emissions are high in the United States, policymakers and industries often assume that the United States has more to lose economically from restrictions on emissions than developing nations do. However, industrialized nations are also the ones most likely to *gain* economically from major energy transitions because they are best positioned to invent, develop, and market new technologies to power the world in a post–fossil fuel era. Germany, Japan, and China have realized this fact and are now leading the world in production, deployment, and sales of solar energy technology (**FIGURE 18.38**). If the United States does not act quickly to develop such energy technologies for the future, the future could belong to nations like China, Germany, and Japan.

States, cities, and businesses are advancing climate change efforts

In the absence of action at the federal level to address climate change, state and local governments across the United States are moving forward. In South Florida, mayors, county commissioners, and other leaders are collaborating to protect the people and property of their region against flooding, erosion, and other impacts of sea level rise. By regulating development, installing pump systems, raising streets and buildings, and strengthening coastal dunes, these leaders are helping residents adapt to climate change.

Elsewhere, political leaders are trying to limit greenhouse gas emissions. Mayors from more than 1000 cities have signed the U.S. Mayors Climate Protection Agreement, committing their cities to pursue policies to "meet or beat" Kyoto Protocol guidelines. A number of U.S. states have enacted targets or mandates for renewable energy production. The boldest state-level action so far has come in California. In 2006, that state's legislature worked with then-governor Arnold Schwarzenegger to pass the Global Warming Solutions Act, which sought to cut

California's greenhouse gas emissions 25% by 2020. This law established the state's cap-and-trade program and followed earlier efforts to mandate higher fuel efficiency for automobiles.

In the wake of Trump's decision to leave the Paris Agreement, state, regional, and local efforts went into overdrive. Former New York City Mayor Michael Bloomberg brought together dozens of governors, mayors, university presidents, and corporate leaders who promised action to support the U.S. emissions reduction pledge under the Paris Agreement. California governor Jerry Brown flew to Beijing and met with Chinese President Xi Jinping to discuss climate cooperation between China and California.

Moreover, individuals from the private sector became involved as never before. At the Paris conference, Microsoft founder and billionaire philanthropist Bill Gates led an effort to attract private investment into renewable energy from some of the world's wealthiest people. And as Trump mulled over his decision to pull out of the Paris Agreement, the chief executives of many major corporations urged him to stay in it. Business thrives on certainty and stability of policy direction, and most corporations today have already invested a great deal in energy efficiency and emissions reductions—and have found that these efforts help their bottom line.

Ultimately, many businesses, utilities, universities, governments, and individuals are aiming to achieve **carbon-neutrality,** a condition in which no net greenhouse gases are emitted. In practice, achieving carbon-neutrality generally requires buying carbon offsets. In a **carbon offset,** one entity makes a voluntary payment that funds emissions reductions made by a second entity. The emissions reductions thus offset emissions that the first entity releases. For example, a coal-burning power plant could fund a reforestation project to plant trees that will soak up as much carbon dioxide as the coal plant emits. Similarly, a university could fund clean renewable energy projects to make up for fossil fuel energy that the university uses. In principle, carbon offsets are a powerful idea, but rigorous oversight is needed to make sure that offset funds achieve what they are intended for—and that offsets fund only emissions cuts that would not occur otherwise.

Should we engineer the climate?

What if all our efforts to reduce emissions are not adequate to rein in climate change? As climate disruption becomes more severe, increasing numbers of scientists and engineers are reluctantly considering drastic, assertive steps to alter Earth's climate in a last-ditch attempt to reverse global warming—an approach called **geoengineering** (**FIGURE 18.39**).

One geoengineering approach would be to suck carbon dioxide out of the air. To do so, we might enhance photosynthesis in natural systems by planting trees over large areas or by fertilizing ocean phytoplankton with nutrients like iron. A more high-tech method might be to design "artificial trees," structures that chemically filter CO_2 from the air.

A second geoengineering approach would be to cool the planet by blocking sunlight before it reaches Earth. We might deflect sunlight by injecting sulfates or other fine dust particles into the stratosphere, by seeding clouds with seawater, or by deploying fleets of reflecting mirrors on land, at sea, or in space.

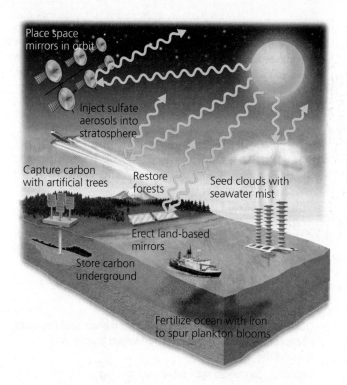

FIGURE 18.39 Geoengineering proposals seek to use technology to remove carbon dioxide from the air or reflect sunlight away from Earth. However, most geoengineering ideas would take years to develop, may not work well, or might cause undesirable side effects. Thus, they are not a substitute for reducing emissions.

Labels in figure:
Place space mirrors in orbit
Inject sulfate aerosols into stratosphere
Capture carbon with artificial trees
Restore forests
Seed clouds with seawater mist
Erect land-based mirrors
Store carbon underground
Fertilize ocean with iron to spur plankton blooms

Scientists were long hesitant even to discuss the notion of geoengineering. The potential methods are technically daunting, would take years or decades to develop, and might pose unforeseen risks. Blocking sunlight does not reduce greenhouse gas concentrations, so ocean acidification would continue. And any method will work only as long as society has the money and the ability to maintain it. If we were to block sunlight for decades and then suddenly stop doing so while greenhouse gas concentrations were higher than ever, the atmosphere would warm exceedingly quickly, likely causing catastrophic climate change. Moreover, many experts are wary of promulgating hope for easy technological fixes, lest politicians lose incentive to reduce emissions through policy.

However, as climate change intensifies, more scientists are becoming willing to contemplate geoengineering as a backup plan. Respected researchers and scientific institutions are beginning to assess the risks and benefits of geoengineering options so that we can be ready to take well-informed action if climate change becomes severe enough to justify it.

We all can address climate change

The disruption of our climate is having intensifying consequences from Miami to the Maldives, from Alaska to Bangladesh, and from New York to the Netherlands. Government policies, corporate actions, international treaties, carbon pricing, technological innovations—and perhaps even geoengineering—all have roles to play in addressing climate change. But in the end, the most influential factor may be the collective decisions of millions of regular people. College students in particular are crucial for driving the personal and societal changes needed to reduce carbon footprints. Today, a groundswell of interest is sweeping across campuses, and many students are pressing their school administrations to seek carbon-neutrality, divest from fossil fuels, and promote renewable energy.

Global climate change may be the biggest challenge we face, but reversing it would be our greatest victory. With concerted action, there is still time to avert the most severe impacts. Through education, innovation, and lifestyle choices, we have the power to turn the tables on climate change and help bring about a bright future for humanity and our planet.

CENTRAL CASE STUDY
connect & continue

TODAY, in South Florida, local leaders and everyday people are investing time, thought, money, and creativity into finding solutions to rising sea levels. Miami's mayor Francis Suarez and other policymakers are engaging the city's neighborhood representatives, businesspeople, and community groups in decisions about how to spend the $192 million in bond money for climate change adaptation recently approved by voters. A citizen oversight board will review and recommend projects to the City Commission over the course of 10 years. A network of community groups, organized as the Miami Climate Alliance, is pressuring policymakers to be inclusive of low-income people and people of color in their decision-making. Alliance leaders urged Suarez in a letter to carry out the bond efforts so as to "elevate the voices and well-being of those who are most vulnerable to sea-level rise, hurricanes, poverty, homelessness, and other factors that threaten our ability to be resilient and thrive as a city."

In Miami Beach, work is proceeding on building pumping systems and raising streets and foundations. Home prices show some evidence of starting to slip in areas experiencing frequent flooding, but by and large, demand is still strong, and real estate values are holding up. Leaders of Miami Beach and other communities in the Southeast Florida Regional Climate Change Compact are working to tailor financial and insurance incentives to guide future development toward upland areas less susceptible to flooding.

At the state level, Florida's government is showing signs of becoming more involved in addressing climate change, after eight years of inaction under former governor Rick Scott. Governor Ron DeSantis, elected in 2018, has established an office to help Florida's communities prepare for sea level rise by providing funding, coordination, and technical assistance. Coastal residents throughout the state will be watching to see if the DeSantis administration follows through on these commitments.

Like people anywhere who love their homes, residents of South Florida are girding themselves for a long battle to protect their land, communities, and quality of life while our global society inches its way toward emissions reductions. For all of us across the globe, taking steps to confront climate change represents the foremost challenge for our future.

• **CASE STUDY SOLUTIONS** You are the city manager for a coastal U.S. city that scientists predict will be hit hard by sea level rise. You have just returned from a professional conference in Florida, where you toured Miami and Miami Beach and learned of the efforts being made there to adapt to climate change. What steps would you take to help your own city prepare for rising sea level? How would you explain the risks and impacts of climate change to your fellow city leaders to gain their support? Of the measures being taken in Florida communities, which would you choose to study closely, which would you want to implement right away, and which would be highest priority in the long run? Explain your choices.

• **LOCAL CONNECTIONS** How might your campus reduce its greenhouse gas emissions? Come up with three specific proposals for ways to reduce emissions on your campus that you feel would be effective and feasible. How would you present these proposals to campus administrators to gain their support?

• **EXPLORE THE DATA** How are researchers making predictions about sea level rise when our understanding of Arctic and Antarctic melting is changing so fast? → **Explore Data** relating to the case study on **Mastering Environmental Science**.

REVIEWING Objectives

You should now be able to:

+ Describe Earth's climate system, and explain the factors that influence global climate

Solar radiation, Milankovitch cycles, ocean absorption, and ocean circulation all influence climate. Greenhouse gases warm the atmosphere by absorbing and emitting infrared radiation. Earth's climate changes naturally over time, but today, human influence is altering it rapidly. Our planet is now experiencing radiative forcing of 2.3 watts/m^2 of thermal energy above what it was experiencing 250 years ago. (pp. 486–491)

+ Identify greenhouse gases, and characterize human influences on the atmosphere and on climate

Greenhouse gases such as carbon dioxide, methane, water vapor, nitrous oxide, and ozone keep our atmosphere warm by exerting a natural greenhouse effect. As we burn fossil fuels and clear forests, we increase atmospheric concentrations of greenhouse gases, and this intensifies the greenhouse effect. Artificial greenhouse gases such as halocarbons are also being added. Aerosol pollution exerts a variable but slight cooling effect, yet overall, human emissions are warming the atmosphere, land, and oceans. (pp. 488–491)

+ Summarize how researchers study climate

Data from proxy indicators—such as ice cores, sediment cores, tree rings, packrat middens, and coral reefs—reveal information about past climate. Direct measurements of temperature, precipitation, and other conditions tell us about current climate. Climate models serve to predict future changes in climate. (pp. 491–495)

+ Outline current and expected future trends and impacts of global climate change in the United States and around the world

Globally, temperatures have warmed by about 1.1°C (2.0°F) in the past century and are predicted to rise 1.0–3.7°C (1.8–6.7°F) more by 2100. Changes in precipitation vary by region. Extreme weather events are becoming more frequent, likely due in part to modification of the jet stream. Melting glaciers will diminish water supplies, and melting ice sheets are adding to sea level rise. Sea level has risen 26.0 cm (10.2 in.) since 1880, and its continued rise poses growing risks of damage to islands and coasts. Of all impacts of our greenhouse gas emissions, ocean acidification may become the most far-reaching. All these changes exert impacts on organisms and ecosystems, as well as on agriculture, forestry, health, economics, and global security. (pp. 494–508)

+ Suggest and assess ways we may respond to climate change

Both mitigation and adaptation are necessary. We can reduce greenhouse gas emissions with a variety of strategies, including energy efficiency, clean and renewable energy sources, new automotive technologies, mass transit, sustainable agricultural methods, and forest protection and restoration. International efforts to devise effective treaties to restrain climate change have fallen short of what is needed, but renewable energy technologies present economic opportunities, and many states, cities, and businesses are acting to reduce emissions. Carbon taxation and carbon trading offer promising avenues to harness market forces and financial incentives in the fight against climate change. (pp. 508–519)

SEEKING Solutions

1. Some people argue that we need "more proof" or "better science" before we commit to substantial changes in our energy economy. How much certainty do you think we need before we should take action regarding climate change? How much certainty do you need in your own life before you make a major decision? Should nations and elected officials follow a different standard? Do you believe that the precautionary principle (pp. 260, 383) is an appropriate standard in the case of global climate change? Why or why not?

2. Suppose you would like to make your own lifestyle carbon-neutral. You plan to begin by reducing the emissions you are responsible for by 25%. What three actions would you take first to achieve your goal?

3. Name one thing you learned in this chapter about a way to address climate change. Describe one thing that surprised you. Thinking about your region, your state, or your country, what means of reducing emissions do you think holds the most promise, and why?

4. **THINK IT THROUGH** You have just been elected governor of a medium-sized U.S. state. Polls show that the public wants you to take action to reduce greenhouse gas emissions, but does not want prices of gasoline or electricity to rise. Your state legislature will support you in your efforts as long as you remain popular with voters. One neighboring state has enacted legislation mandating that one-third of its energy come from renewable sources within 10 years. Another neighboring state has joined a regional emissions trading consortium. A third neighboring state has established a revenue-neutral carbon tax. What actions will you take in your first year as governor, and why? What effects would you expect each action to have?

5. **THINK IT THROUGH** You have been appointed as the U.S. representative to an international conference to negotiate terms of an emissions reduction treaty that would build upon or replace the Paris Agreement. The U.S. government has instructed you to take a leading role in designing the new treaty and to engage constructively with other nations' representatives while protecting America's economic and political interests. What type of agreement will you try to shape? Describe at least three components that you would propose or agree to and at least one that you would oppose.

CALCULATING Ecological Footprints

We all contribute to climate change, because fossil fuel combustion plays such a large role in supporting the lifestyles we lead. Likewise, as individuals, each one of us can help address climate change through personal decisions and actions in how we live our lives. Several online calculators enable you to calculate your own personal carbon footprint, the amount of carbon emissions for which you are responsible. Go to www.nature.org/greenliving/carboncalculator/ or to www.carbonfootprint.com/calculator.aspx, take the quiz, and enter the relevant data in the table.

	CARBON FOOTPRINT (TONS PER PERSON PER YEAR)
World average	
U.S. average	
Your footprint	
Your footprint with three changes (see Question 3)	

1. How does your personal carbon footprint compare to that of the average U.S. resident? How does it compare to that of the average person in the world? Why do you think your footprint differs in the ways it does?

2. As you took the quiz and noted the impacts of various choices and activities, which one surprised you the most?

3. Think of three changes you could make in your lifestyle that would lower your carbon footprint. Now take the quiz again, incorporating these three changes. Enter your resulting footprint in the table. By how much did you reduce your carbon footprint?

4. What do you think would be an admirable yet realistic goal for you to set as a target value for your own carbon footprint? What would you need to do to reach that target?

Mastering Environmental Science

Students Go to **Mastering Environmental Science** for assignments, an interactive e-text, and the Study Area with practice tests, videos, and activities.

Instructors Go to **Mastering Environmental Science** for automatically graded activities, videos, and reading questions that you can assign to your students, plus Instructor Resources.

Fossil Fuels and Energy Efficiency

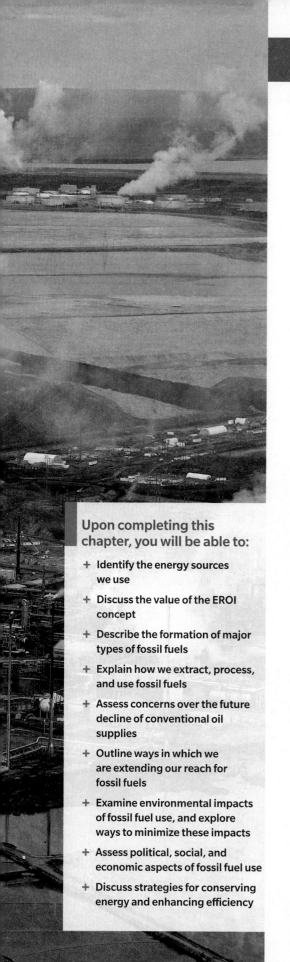

An oil sands processing facility in Alberta, Canada

CENTRAL case study

Alberta's Oil Sands and the Keystone XL Pipeline:
Canadian Camaraderie or Climate Catastrophe?

> " If we can get [oil] up there in Canada and bring it down, it's great. It's going to put a lot of wealth in our tax base.
> Montana rancher Mike Hammond
>
> I see it serving . . . the oil corporations . . . so they can make their billions despite who's in the way, and who's at risk and what's at risk.
> Native American nurse and single mom
> Marina Starr "

Everything about Canada's oil sands is enormous. These fossil fuel deposits cover a region the size of Illinois, underlying boreal forests that span the width of the continent. Open pits miles wide are dug to extract the fuel. Inside these gargantuan pits, million-pound haul trucks with tires 14 feet wide move earth with shovels five stories high, looking like ants when viewed from the rim. The economic value of the extracted oil is huge, as well. And the implications of mining these deposits for oil are momentous: Burning all this fuel would alter the very climate of our planet.

Oil sands, also called **tar sands,** are layers of sand or clay saturated with a viscous, tarry type of petroleum called *bitumen*. Vast areas of these wet blackish deposits underlie a thinly populated region of the Canadian province of Alberta. To some people the oil sands represent wealth and security, a key to maintaining our fossil-fuel-based lifestyle far into the future. To others, they threaten appalling pollution and a severe disruption of Earth's climate.

To extract oil from oil sands, companies clear the forest and strip-mine the land, creating open pits up to 75 m (250 ft) deep. The gooey deposits dug out are mixed with hot water and chemicals to separate the bitumen from the sand, and the bitumen is removed and processed. Toxic tailings (leftover residues of sand and rock particles, chemicals, and water) are dumped into engineered structures that form "lakes" that may cover areas even larger than the mines. Where oil sands occur more deeply underground, hot water is injected down shafts to liquefy, separate, and extract the bitumen in place.

Mining for oil sands began in Alberta in 1967, but for many years, it was hard to make money extracting these low-quality deposits. Rising oil prices after 2003, however, turned it into a profitable venture, and dozens of companies rushed in. Canadian oil sands became the source of up to 2.3 million barrels of oil per day, and each truckload leaving a mine carried bitumen worth close to $20,000. Thanks to the oil sands, Canada today boasts the world's third-largest proven reserves of oil, after Venezuela and Saudi Arabia.

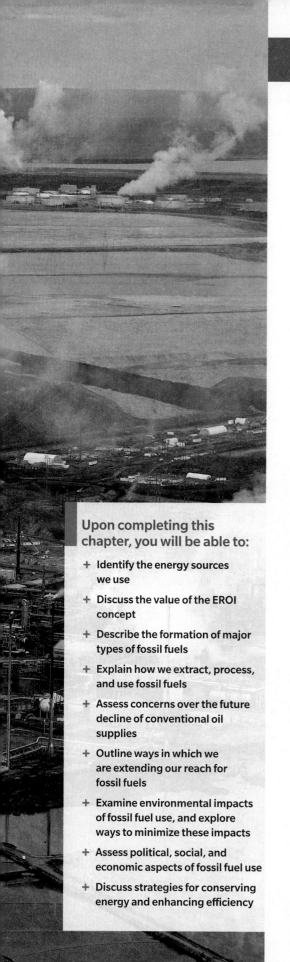

◀ An oil sands processing facility in Alberta, Canada

▲ Protesting the Keystone XL pipeline at the U.S. White House

Upon completing this chapter, you will be able to:

+ Identify the energy sources we use
+ Discuss the value of the EROI concept
+ Describe the formation of major types of fossil fuels
+ Explain how we extract, process, and use fossil fuels
+ Assess concerns over the future decline of conventional oil supplies
+ Outline ways in which we are extending our reach for fossil fuels
+ Examine environmental impacts of fossil fuel use, and explore ways to minimize these impacts
+ Assess political, social, and economic aspects of fossil fuel use
+ Discuss strategies for conserving energy and enhancing efficiency

523

Canada looked for buyers south of its border first, and the TransCanada Corporation (now renamed TC Energy) built the Keystone Pipeline to ship diluted bitumen into the United States. This pipeline began operating in 2010, bringing oil from Alberta nearly 3500 km (2200 mi) to Illinois and Oklahoma (**FIGURE 19.1**). An extension later conveyed the oil to refineries on the Texas coast. As production increased, TransCanada proposed the Keystone XL extension, a 1900-km (1180-mi) stretch of larger-diameter pipeline that would cut across the Great Plains to shave off distance and add capacity to the existing system. This proposed shortcut leg would also transport oil from the newly productive oil fields of the Bakken Formation in North Dakota and Montana.

The Keystone XL pipeline proposal soon met opposition from people living along the proposed route who were concerned about health, water quality, and property rights. It also faced continent-wide opposition from advocates of action to address global climate change.

Pipeline proponents in the United States argued that the Keystone XL project would create jobs for workers in the American heartland and would guarantee a dependable oil supply for decades. They stressed that buying oil from Canada—a stable, friendly, democratic neighbor—would ease U.S. reliance on oil-producing nations with authoritarian governments and poor human rights records, such as Saudi Arabia and Venezuela.

Opponents of the pipeline extension expressed dismay at the destruction of boreal forest and anxiety about transporting oil above the continent's largest aquifer, where spills might contaminate drinking water for millions of people and irrigation water for America's breadbasket. They pointed out that most of the oil would be destined for export, not for use in the United States. Most of all, they sought to prevent extraction of a vast new source of fossil fuels whose combustion would release enormous amounts of greenhouse gases. By encouraging a source of oil that is energy-intensive to extract and that burns less cleanly than conventional oil, they held, the United States would prolong fossil fuel dependence and intensify climate change, when it should instead be transitioning to clean renewable energy.

Because the proposed Keystone XL pipeline would cross an international border, it required a presidential permit from the U.S. Department of State. Starting in 2008, an escalating political drama came to include lawsuits, conflict-of-interest charges, high-stakes quarrels between congressional Republicans and President Barack Obama, and street protests at the White House. In 2015, Obama decided against approving the Keystone XL pipeline. Obama told the nation that his administration had judged that the pipeline "would not serve the national interest of the United States" because its construction would not contribute meaningfully to the U.S. economy, it would not lower gasoline prices for consumers, and it would not enhance the country's energy security. Moreover, he noted, to approve the pipeline on the eve of global climate talks in Paris (p. 516) would undercut U.S. leadership and leverage at this historic forum to address climate change.

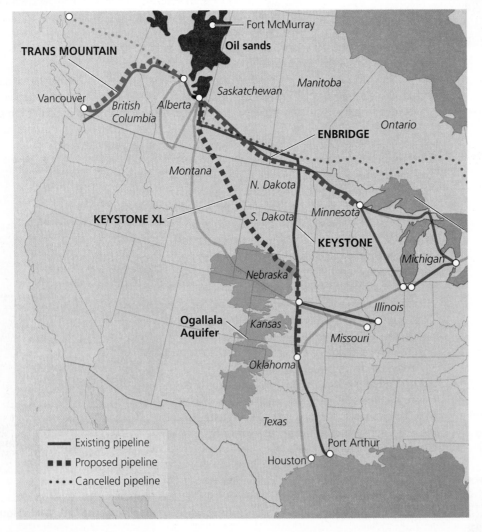

FIGURE 19.1 The Keystone pipeline is one of several pipelines that bring bitumen from Alberta's oil sands into the United States. The proposed construction of the Keystone XL pipeline set off a complex debate. Projects to build or expand other pipelines have been canceled or mired in delay.

Once Donald Trump became president in 2017, he reversed Obama's decision and signed an executive order approving the Keystone XL project. TC Energy prepared to break ground on construction but soon ran into legal obstacles. In 2018, a district judge in Montana blocked construction until further environmental assessment could be conducted. The judge ruled that whereas Obama's decision had been based on extensive research by the State Department, Trump's order lacked a "reasoned explanation" for the reversal and failed to address the impacts of greenhouse gas emissions, impacts on Native American resources, feasibility of spill responses, or the economic viability of the pipeline. Trump responded in 2019 by scrapping his original order and issuing a new permit to enable speedy development of the pipeline project. TC Energy appealed the judge's decision, arguing that the new permit made it moot, and the Ninth Circuit Court of Appeals agreed and threw out the judge's ruling, removing a major obstacle for the project.

Meanwhile, other pipeline projects designed to export Alberta's bitumen were facing trouble. One plan was to expand the Trans Mountain pipeline, which delivers oil from Alberta west through the Rocky Mountains to Vancouver, British Columbia, where it can be loaded on ships and exported to China and around the world. Widespread opposition to this project led the private company planning it to abandon the idea. Canada's government, under Prime Minister Justin Trudeau, purchased the Trans Mountain project in an effort to save it, but a court ruling halted construction, ordering the government to renegotiate with Native American tribes and take steps to protect marine life against impacts. Trudeau's administration did so, yet the pipeline's many opponents in British Columbia are finding new ways to fight the project. At the same time, to the east, a project to pipe oil across Canada to the Atlantic coast was canceled amid opposition from eastern provinces, while an expansion of the Enbridge Line 3 pipeline through Minnesota was running into resistance and delays.

All these difficulties meant that existing pipeline capacity was maxed out. Increasingly, bitumen was being shipped by rail instead (which is more expensive and arguably less safe), but ultimately, companies found themselves extracting more bitumen than they could ship out and sell. Prices fell, and so did investment in oil sands extraction companies. Alberta's Premier Rachel Notley announced mandated cuts in oil production in an effort to prop up prices.

Amid all these happenings, advocates for addressing climate change grew hopeful that slowing the extraction of oil sands would buy more time for society to switch to clean renewable sources of energy and that most of Canada's oil sands would remain in the ground. Meanwhile, the broader debate over the role of fossil fuels in our economy continued. These energy sources power our civilization and have enabled our modern standard of living—yet as climate change worsens, we face the need to wean ourselves from them and shift to clean, renewable energy sources. The way in which we handle this complex transition will shape the quality of our lives and the future of our society and our planet.

Sources of Energy

Humanity has devised many ways to harness the renewable and nonrenewable forms of energy available on our planet (**TABLE 19.1**). We use these energy sources to heat and light our homes; power our machinery; fuel our vehicles; produce plastics, pharmaceuticals, and synthetic fibers; and provide the many comforts and conveniences to which we've grown accustomed.

Nature offers a variety of energy sources

Most of Earth's energy comes from the sun. We can harness energy from the sun's radiation directly by using solar power technologies. Solar radiation also helps drive wind and the water cycle, enabling us to harness wind power and hydroelectric power. And, of course, sunlight drives photosynthesis (p. 33) and the growth of plants, from which we take wood and other biomass as a fuel source. When plants and other organisms die and are buried in sediments under particular conditions for many millions of years, their matter and their stored chemical energy may be slowly transformed into **fossil fuels,** highly combustible substances formed from the remains of organisms from past geologic ages. We rely on three main fossil fuels, in the form of a solid (coal), liquid (oil), and gas (natural gas).

TABLE 19.1 Energy Sources We Use

ENERGY SOURCE	DESCRIPTION	TYPE OF ENERGY
Coal	Fossil fuel extracted from ground (solid)	Nonrenewable
Oil	Fossil fuel extracted from ground (liquid)	Nonrenewable
Natural gas	Fossil fuel extracted from ground (gas)	Nonrenewable
Nuclear energy	Energy from atomic nuclei of uranium	Nonrenewable
Biomass energy	Energy stored in plant matter from photosynthesis	Renewable
Hydropower	Energy from running water	Renewable
Solar energy	Energy from sunlight directly	Renewable
Wind energy	Energy from wind	Renewable
Geothermal energy	Earth's internal heat rising from core	Renewable
Tidal and wave energy	Energy from tides and ocean waves	Renewable

In addition, a great deal of energy emanates from Earth's core, making geothermal power available for our use. Energy also results from the gravitational pull of the moon and sun, which powers ocean tides. Finally, an immense amount of energy resides within the bonds among protons and neutrons in atoms, and this energy provides us with nuclear power. (We explore all these energy sources as alternatives to fossil fuels in Chapters 20 and 21.)

Energy sources such as sunlight, geothermal energy, and tidal energy are considered perpetually renewable because they are readily replenished, so we can keep using them without depleting them (p. 4). In contrast, energy sources such as coal, oil, and natural gas are considered nonrenewable. These fossil fuels take so long to form that, once depleted, they cannot be replaced within any time span useful to our civilization. It takes a thousand years for the biosphere to generate the amount of organic matter that must be buried to produce a single day's worth of fossil fuels for our society. To replenish the fossil fuels we have burned so far would take many millions of years. At our rising rate of consumption, we will use up Earth's easily accessible store of conventional fossil fuels in just decades. For this reason, and because fossil fuels exert severe environmental impacts, renewable energy sources increasingly are being developed as alternatives to fossil fuels.

We rely mostly on fossil fuels

Since the industrial revolution, fossil fuels have replaced biomass as our society's dominant source of energy. Global consumption of coal, oil, and natural gas has risen for years and is now at its highest level ever (**FIGURE 19.2**). We have favored fossil fuels as an energy source because their high energy content makes them efficient to burn, ship, and store. A single

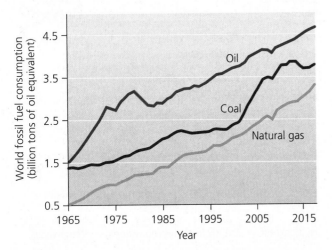

FIGURE 19.2 Annual global consumption of fossil fuels has risen greatly over the past half-century. Oil remains our leading energy source. *Data from BP p.l.c., 2019.* Statistical review of world energy 2019.

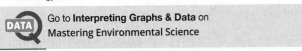

Go to **Interpreting Graphs & Data** on **Mastering Environmental Science**

TABLE 19.2 Nations with the Largest Proven Reserves of Fossil Fuels

COAL (% world reserves)		OIL (% world reserves)		NATURAL GAS (% world reserves)	
United States	23.7	Venezuela*	17.5	Russia	19.8
Russia	15.2	Saudi Arabia	17.2	Iran	16.2
Australia	14.0	Canada*	9.7	Qatar	12.5
China	13.2	Iran	9.0	Turkmenistan	9.9
India	9.6	Iraq	8.5	United States	6.0

*Most reserves in Venezuela and Canada consist of oil sands, which are included in these figures.
Data from BP p.l.c., 2019. *Statistical review of world energy 2019.*

gallon of oil contains as much energy as a person would expend in nearly 600 hours of human labor.

Some regions of the globe have substantial reserves of coal, oil, or natural gas, whereas others have very few. Half the world's proven reserves of crude oil lie in the Middle East. The Middle East is also rich in natural gas, as is Russia. The United States possesses the most coal of any nation (**TABLE 19.2**).

We use fossil fuels for transportation, manufacturing, heating, and cooking, as well as to generate **electricity,** a secondary form of energy that is convenient to transfer over long distances and apply to a variety of uses. Each type of fuel has its own mix of uses, and each contributes in different ways to our economies and our daily needs. For instance, oil is used mostly for transportation, whereas coal is used mostly to generate electricity.

Societies differ in how they use energy, however. Industrialized nations apportion roughly one-third to transportation, one-third to industry, and one-third to all other uses. In contrast, industrializing nations devote more energy to subsistence activities such as growing and preparing food and heating homes. Moreover, people in industrializing countries often rely on manual or animal energy sources instead of automated ones. For instance, most rice farmers in Southeast Asia plant rice by hand, but industrial rice growers in California use airplanes. Because industrialized nations rely more on mechanized equipment and technology, they use more fossil fuels. In the United States, oil, coal, and natural gas together supply 80% of energy demand (**FIGURE 19.3**).

Rates at which we consume energy also vary from place to place. A world map of per-person consumption shows that people in more industrialized regions generally consume far more energy than do people of less industrialized regions (**FIGURE 19.4**, p. 528). Per person, the most industrialized nations use more than 50 times more energy than do the least industrialized nations. The United States claims only 4.4% of the world's population, but it consumes fully one-sixth of the world's energy.

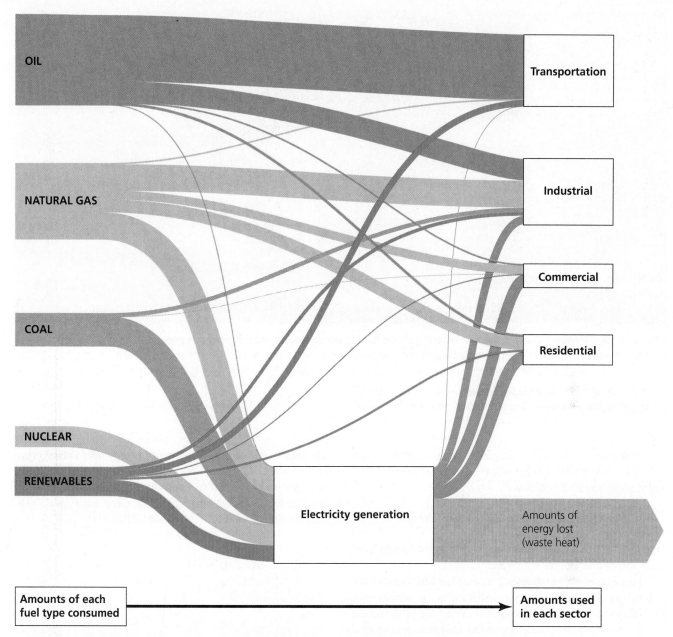

FIGURE 19.3 Total energy flow of the United States. Amounts are represented by the thickness of each bar. Energy consumed (domestic production plus imports minus exports) is shown on the left, and destinations of this energy are shown on the right. Portions of each energy source are used directly in the residential, commercial, industrial, and transportation sectors. Other portions are used to generate electricity, which in turn powers these sectors. The large amounts of energy lost as waste heat are shown on the right. *Adapted from Lawrence Livermore National Laboratory. Data are for 2018, from U.S. Energy Information Administration.*

It takes energy to make energy

We do not simply get energy for free. To harness, extract, process, and deliver energy, we need to invest substantial inputs of energy. For instance, mining oil sands in Alberta requires using powerful vehicles and heavy machinery, as well as constructing an immense infrastructure of roads, pipelines, waste ponds, storage tanks, water intakes, processing facilities, housing for workers, and more—all requiring the use of energy. Natural gas must be burned to

heat the water that is used to separate the bitumen from the sand. Processing and piping the oil away from the extraction site, and then refining it into products we can use, require further energy inputs. Thus, when evaluating an energy source, it is important to subtract the costs in energy that we invest from the benefits in energy that we receive. **Net energy** expresses the difference between energy returned and energy invested:

$$\text{Net energy} = \text{Energy returned} - \text{Energy invested}$$

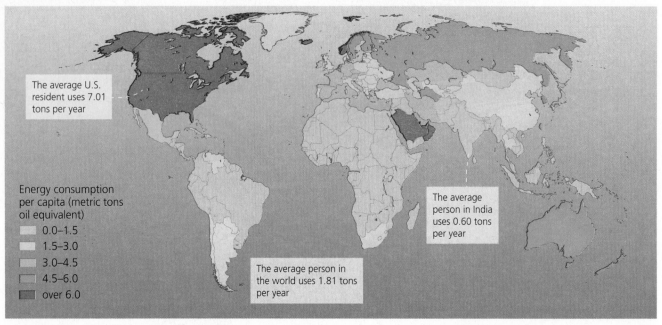

The average U.S. resident uses 7.01 tons per year

The average person in India uses 0.60 tons per year

The average person in the world uses 1.81 tons per year

Energy consumption per capita (metric tons oil equivalent)
- 0.0–1.5
- 1.5–3.0
- 3.0–4.5
- 4.5–6.0
- over 6.0

FIGURE 19.4 People in wealthy industrialized nations tend to consume the most energy per person. This map combines all types of energy, standardized to metric tons of "oil-equivalent" (the amount of fuel needed to result in the energy gained from burning one metric ton of crude oil). *Data from BP p.l.c., 2019. Statistical review of world energy 2019.*

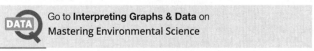

Go to **Interpreting Graphs & Data** on Mastering Environmental Science

When we assess energy sources, it is also useful to use a ratio known as **EROI (energy returned on investment).** EROI ratios are calculated as follows:

$$EROI = Energy\ returned/Energy\ invested$$

Higher EROI ratios mean that we receive more energy from each unit of energy that we invest.

Fossil fuels are widely used because their EROI ratios have historically been high. However, EROI ratios can change over time. Ratios rise as the technologies to extract and process fuels become more efficient, and they fall as resources are depleted and become harder to extract. For example, EROI ratios for "production" (extraction and processing) of conventional oil and natural gas in the United States declined from roughly 24:1 in the 1950s to about 11:1 by 2010 (**FIGURE 19.5**). So, we used to gain 24 units of energy for every unit of energy expended, but now we gain only 11. EROI ratios for both the discovery and the production of oil and gas have declined because we found and extracted the easiest deposits first and now must work harder and harder to find and extract the remaining amounts. For the Alberta oil sands, EROI ratios are even lower, averaging around 4:1, because oil sands are a low-quality fuel that requires a great deal of energy to extract and process.

Where will we turn for energy?

Since the onset of the industrial revolution, coal, oil, and natural gas have powered the astonishing advances of our civilization.

These extraordinarily rich sources of energy have helped bring us a standard of living our ancestors could scarcely have imagined. Yet because fossil fuel deposits are finite and nonrenewable, easily accessible supplies of the three main fossil fuels have dwindled, and EROI ratios have fallen.

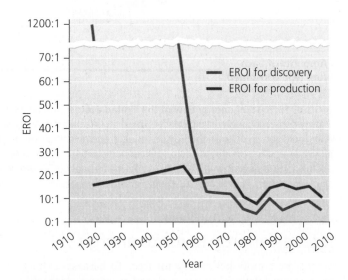

FIGURE 19.5 EROI values for the discovery and production of oil and gas in the United States have declined over the past century. *Data from Guilford, M., et al., 2011. A new long term assessment of energy return on investment (EROI) for U.S. oil and gas discovery and production. pp. 115–136 in Hall, C., and D. Hansen (Eds.), Sustainability, Special Issue, 2011, New studies in EROI (energy return on investment).*

In response, today we are devoting enormous amounts of money, energy, and technology to extend our reach for fossil fuels. We are using potent extraction methods to free gas and oil tightly bound in rock layers. We are returning to sites abandoned once further extraction became too difficult and deploying powerful new machinery and techniques to squeeze more fuel from them. We are drilling deeper underground, farther offshore, and into the Arctic seabed. And we are pursuing new types of fossil fuels, such as Alberta's oil sands.

There is, however, a different way we can respond to the ongoing depletion of conventional fossil fuel resources: We can hasten the development of clean and renewable energy sources to replace them (Chapters 20 and 21). By transitioning to clean and renewable alternatives, we can gain energy that is sustainable in the long term while reducing pollution and its health impacts and the emission of greenhouse gases that drive climate change.

Fossil Fuels: Their Formation, Extraction, and Use

To grapple effectively with the energy issues we face, it is important to understand how fossil fuels are formed, how we locate deposits, how we extract these resources, and how our society puts them to use.

Fossil fuels are formed from ancient organic matter

The fossil fuels we burn today in our vehicles, homes, industries, and power plants were formed from the tissues of organisms that lived 100–500 million years ago. The energy these fuels contain originated with the sun and was converted to chemical-bond energy by photosynthesis. The chemical energy in these organisms' tissues then became concentrated as the tissues decomposed and their compounds were altered and compressed (**FIGURE 19.6**).

Fossil fuels are formed only when organic material is broken down in an *anaerobic* environment, one with little or no oxygen. Such environments include the bottoms of lakes, swamps, and shallow seas. Over millions of years, organic matter that accumulates at the bottoms of such water bodies may be converted into crude oil, natural gas, or coal, depending on (1) the chemical composition of the material, (2) the temperatures and pressures to which it is subjected, (3) the presence or absence of anaerobic decomposers, and (4) the passage of time.

Because fossil fuels form only under certain conditions, they occur in isolated deposits. For example, Alberta's oil sands hold rich reserves of oil that surrounding regions do not. Geologists searching for fossil fuels drill cores and conduct ground, air, and seismic surveys to map underground rock formations and predict where fossil fuel deposits might occur (see **THE SCIENCE BEHIND THE STORY**, pp. 532–533).

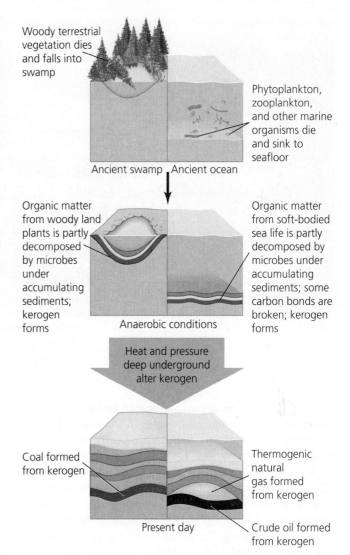

FIGURE 19.6 Fossil fuels form over many millions of years after organisms die and end up in oxygen-poor conditions. This process can occur when trees fall into lakes and are buried by sediment or when phytoplankton and zooplankton drift to the seafloor and are buried **(top).** Organic matter that undergoes slow anaerobic decomposition deep under sediments forms kerogen **(middle).** Coal results when plant matter is compacted so tightly that little decomposition occurs **(bottom left).** Geothermal heating may turn kerogen into crude oil and natural gas **(bottom right),** which come to reside in porous rock layers beneath dense, impervious layers.

Coal The world's most abundant fossil fuel is **coal,** a hard, blackish substance formed from organic matter (generally woody plant material) compressed under very high pressure, creating dense, solid carbon structures. Coal typically results when water is squeezed out of such material as pressure and heat increase over time and when little decomposition takes place. The proliferation 300–400 million years ago of swamps where organic material was buried created coal deposits in many regions of the world.

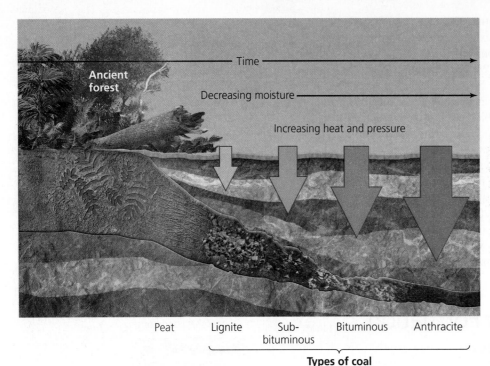

Coal varies from deposit to deposit in the amount of water, carbon, and potential energy it contains. Organic material that is broken down anaerobically but remains moist and near the surface is called *peat*. As peat decomposes further, as it is buried under sediments, as heat and pressure increase, and as time passes, moisture is squeezed out and carbon compounds compress more tightly together, forming coal. The more coal is compressed, the denser is its carbon content and the greater is its energy content per unit volume (**FIGURE 19.7**).

To extract coal from deposits near the surface, we use strip mining, in which heavy machinery scrapes away huge amounts of earth. For deposits deep underground, we use subsurface mining, digging vertical shafts and blasting out networks of horizontal tunnels to follow seams, or layers, of coal. (Strip mining and subsurface mining are illustrated in Figure 23.7, p. 649.) We are also mining coal on immense scales in the Appalachian Mountains, blasting away entire mountaintops in a process called mountaintop removal mining (pp. 538, 652–655).

Oil and natural gas The thick blackish liquid we know as **oil** consists of a mix of many types of hydrocarbon molecules (p. 28)—simple organic compounds consisting solely of carbon and hydrogen atoms. The term **crude oil** refers specifically to oil extracted from the ground before it is refined. **Natural gas** is a gas consisting primarily of methane (CH_4) and lesser, variable amounts of other volatile hydrocarbons. Oil is also known as **petroleum,** although this term is commonly used to refer to oil and natural gas collectively.

Both oil and natural gas are formed from organic material (especially dead plankton) that drifted down through coastal marine waters millions of years ago and became buried in sediments on the ocean floor. This organic material was transformed by time, heat, and pressure into today's natural gas and crude oil. Natural gas may form directly, but it may also form from coal or oil altered by heating. As a result, natural gas is often found above deposits of oil or seams of coal and is often extracted along with those fuels.

Underground pressure tends to drive oil and natural gas upward through cracks and fissures in porous rock until they become trapped under a dense, impermeable rock layer. Oil and gas companies employ geologists to study rock formations to identify promising locations. Once a location is identified, a company conducts exploratory drilling, by boring small holes to great depths. If enough oil or gas is encountered, extraction may begin. Because oil and gas are under pressure while in the ground, they rise to the surface when a deposit is tapped. Once pressure is relieved and some portion has risen to the surface, the remainder will need to be pumped out.

Unconventional fossil fuels Besides the three conventional fossil fuels—coal, oil, and natural gas—other types of fossil fuels exist, often called "unconventional" because we are not (yet) using them as widely. Three examples of unconventional fossil fuels are oil sands, oil shale, and methane hydrate.

As we've seen, oil sands consist of moist sand and clay containing 1–20% bitumen, a thick and heavy form of petroleum that is rich in carbon and poor in hydrogen. Oil sands result from crude oil deposits that have been degraded and chemically altered by water erosion and bacterial decomposition. The leading scientific hypothesis to explain Alberta's oil sands is that geologic changes tens of millions of years ago as the Rocky Mountains were uplifted caused crude oil to migrate northeastward and upward until it saturated rock and soil in what is now northeastern Alberta. Microorganisms

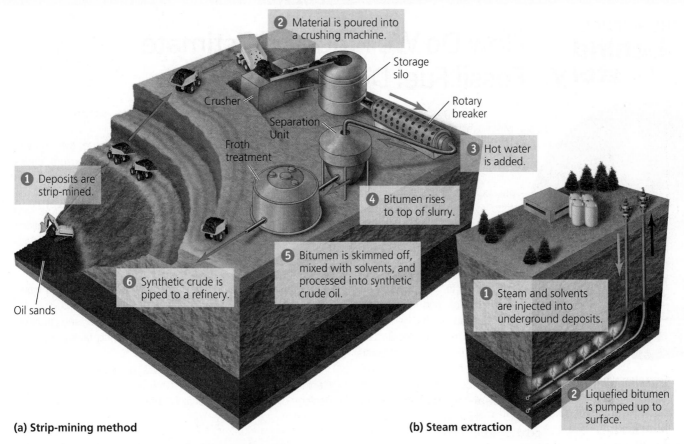

(a) Strip-mining method

① Deposits are strip-mined.

Oil sands

② Material is poured into a crushing machine.

Crusher

Storage silo

Separation Unit

Froth treatment

Rotary breaker

③ Hot water is added.

④ Bitumen rises to top of slurry.

⑤ Bitumen is skimmed off, mixed with solvents, and processed into synthetic crude oil.

⑥ Synthetic crude is piped to a refinery.

(b) Steam extraction

① Steam and solvents are injected into underground deposits.

② Liquefied bitumen is pumped up to surface.

FIGURE 19.8 Oil sands are extracted by two processes. Near-surface deposits of oil sands **(a)** are strip-mined. Deeper deposits of oil sands **(b)** are liquefied and extracted through well shafts.

(which are abundant near the surface but absent at depth) began to consume the oil, particularly the lighter components, leaving degraded heavy bitumen. Oil from oil sands is extracted by two main methods (**FIGURE 19.8**). Bitumen from either process must then be chemically refined and processed to create synthetic crude oil (called *syncrude*).

The second type of unconventional fossil fuel, **oil shale**, is sedimentary rock (p. 38) filled with organic matter that can be processed into a liquid form of petroleum called **shale oil.** Oil shale is formed by the same processes that form crude oil but occurs when the organic matter was not buried deeply enough or subjected to enough heat and pressure to form oil. Oil shale is extracted using strip mines or subsurface mines. It can be burned directly like coal, or it can be processed in several ways. One way to process oil shale is to bake it in the presence of hydrogen and in the absence of air to extract liquid petroleum (a process called *pyrolysis*).

The third unconventional fossil fuel, **methane hydrate,** is an ice-like solid consisting of molecules of methane embedded in a crystal lattice of water molecules. Methane hydrate occurs in sediments in the Arctic and on the ocean floor because it is stable at temperature and pressure conditions found there. Most methane in these gas hydrates formed from bacterial decomposition in anaerobic environments, but some resulted from heat and pressure deeper below the surface.

Economics determines how much will be extracted

As we develop more powerful technologies for locating and extracting fossil fuels, the proportions of these fuels that are physically accessible to us—the *technically recoverable* portions—tend to increase. However, whereas technology determines how much fuel *can* be extracted, economics determines how much *will* be extracted. Extraction becomes increasingly expensive as a resource is removed, so companies rarely find it profitable to extract the entire amount. Instead, a company will consider the costs of extraction (and other expenses) and balance these costs against the income it expects from sale of the fuel. Because market prices of fuel fluctuate, the portion of fuel from a given deposit that is *economically recoverable* fluctuates as well. As market prices rise, economically recoverable amounts approach technically recoverable amounts.

The amount of a fossil fuel that is technologically *and* economically feasible to remove under current conditions is called its **proven recoverable reserve.** Proven recoverable reserves increase as extraction technology improves or as market prices of the fuel rise. Proven recoverable reserves decrease as fuel deposits are depleted or as market prices fall (making extraction unprofitable). Some examples of proven recoverable reserves are shown in Table 19.2 (p. 526).

THE SCIENCE behind the story

How Do We Find and Estimate Fossil Fuel Deposits?

Petroleum geologists study seismic data to determine where oil or gas might be found.

Drilling for oil or gas is risky business: Most wells are unproductive, and a company that doesn't pick its spots effectively could soon go bankrupt. That's why oil and gas companies turn to scientists to help them figure out where to drill.

The oil and gas industry employs petroleum geologists who study underground rock formations to predict where deposits of oil and natural gas might lie. Because the organic matter that gave rise to fossil fuels was buried in sediments, geologists know to look for sedimentary rock that may act as a source. They also know that oil and gas tend to seep upward through porous rock until being trapped by impermeable layers.

To map subsurface rock layers, petroleum geologists first survey the landscape on the ground and from airplanes and satellite imagery, studying rocks on the surface. Because rock layers often become tilted over geologic time, these strata may protrude at the surface, giving geologists an informative "side-on" view.

But to really understand what's deep beneath the surface, scientists conduct seismic surveys. In seismic surveying, a base station creates powerful vibrations at the surface by exploding dynamite, thumping the ground with a large weight, or using an electric vibrating machine (**FIGURE 1**). These vibrations send seismic waves down and outward in all directions through the ground, just as ripples spread when a pebble is dropped into a pond.

As they travel, the waves encounter layers of different types of rock. Each time a seismic wave encounters a new type of rock with a different density, some of the wave's energy is reflected off the boundary. Other

wave energy may be refracted, or bent, sending refraction waves upward.

As reflected and refracted waves return to the surface, devices called seismometers (also used to measure earthquakes) record their strength and timing. Scientists collect data from seismometers at multiple surface locations and run the data through computer programs. By analyzing how long it takes all the waves to reach the various receiving stations and how strong they are at each site, researchers can triangulate and infer the densities, thicknesses, and locations of underlying geologic layers.

Seismic surveying is similar to how we use sonar in water or how bats use echolocation as they fly. It is also used for finding coal deposits, salt and mineral deposits, and geothermal energy hotspots, as well as for studying faults, aquifers, and engineering sites.

Using data from such techniques, geologists with the U.S. Geological Survey (USGS) in 1998 assessed the subsurface geology of the Arctic National Wildlife Refuge on Alaska's North

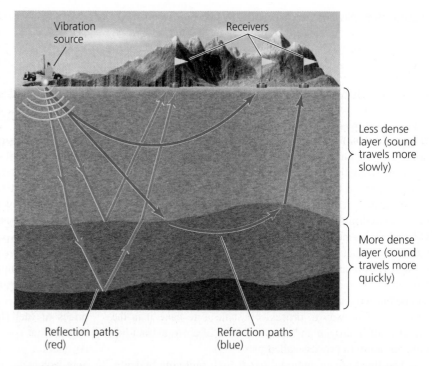

FIGURE 1 Seismic surveying provides clues to the location and size of fossil fuel deposits. Powerful vibrations are created, and receivers measure how long it takes seismic waves to reach other locations. Scientists interpret the patterns of wave reception to infer the densities, thicknesses, and locations of underlying rock layers.

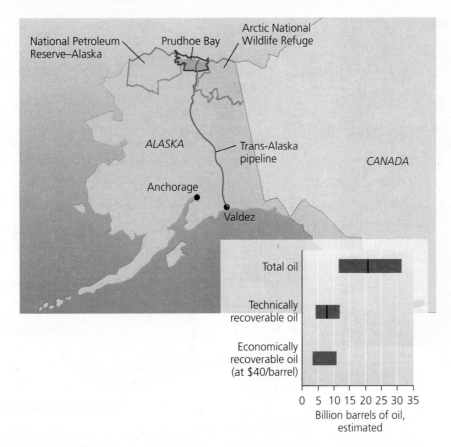

FIGURE 2 **Surveys have given us data on how much oil may underlie Alaska's North Slope.** Of the total oil scientists estimate to exist in the portion of the Arctic National Wildlife Refuge that could someday be opened to drilling, only some is technically recoverable, and less is economically recoverable. *Data from U.S. Geological Survey.*

Slope to predict how much oil it may hold (**FIGURE 2**). Over three years, dozens of scientists conducted fieldwork and combined their results with a reanalysis of 2300 km (1400 mi) of seismic survey data that industry had collected in the 1980s.

After studying their resulting subsurface maps, USGS scientists concluded, with 95% certainty, that between 11.6 and 31.5 billion barrels of oil lay underneath the region of the refuge that Congress has debated opening for drilling. The mean estimate of 20.7 billion barrels is enough to supply the United States for three years at its current rate of consumption.

However, some portion of oil from any deposit is impossible to extract using current technology, so geologists estimate *technically recoverable* amounts of fuels. In its estimate for the Arctic Refuge, the USGS calculated technically recoverable oil to total 4.3–11.8 billion barrels, with a mean estimate of 7.7 billion barrels (just more than one year of U.S. consumption).

The portion of this oil that is *economically recoverable* depends on the costs of extracting it and the price of oil on the world market. USGS scientists calculated that at a price of

$40 per barrel, 3.4–10.8 billion barrels would be economically worthwhile to recover. At higher prices, the economically recoverable amount would be closer to the technically recoverable amount.

In 2002, the USGS conducted similar analyses for the National Petroleum Reserve–Alaska, a vast region of tundra to the west of the Arctic Refuge and Prudhoe Bay that the U.S. government set aside in 1923 as an emergency reserve of petroleum. USGS scientists estimated that this region contained 9.3 billion barrels of technically recoverable oil. Further surveying in recent years has increased the estimated amount.

After Congress in 2017 approved opening an area of the Arctic Refuge to oil and gas drilling, federal agencies under the Trump administration have pushed to allow companies to begin intensive seismic surveying, using fleets of "thumper trucks" (see Figure 19.18, p. 542) and the newest technology to get a better idea of how much oil lies beneath the tundra. Only once that work is done will it be known whether the oil will be economical enough to extract.

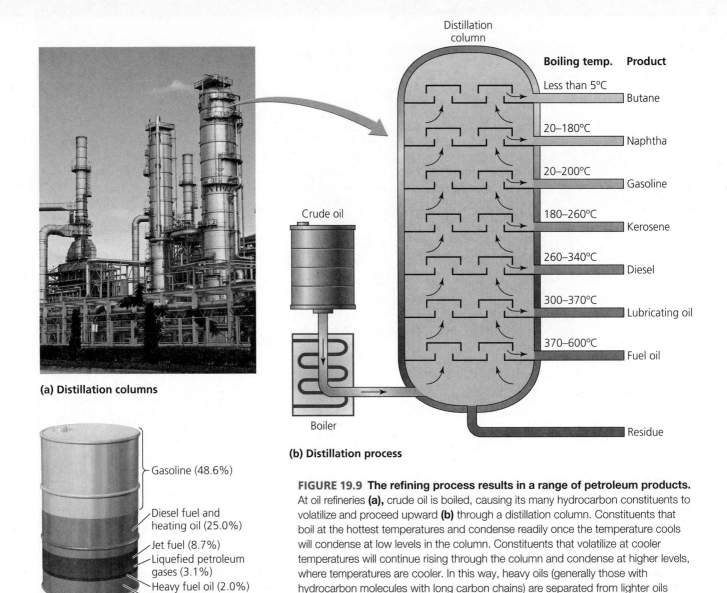

(a) Distillation columns

Crude oil

Distillation column

Boiling temp.	Product
Less than 5°C	Butane
20–180°C	Naphtha
20–200°C	Gasoline
180–260°C	Kerosene
260–340°C	Diesel
300–370°C	Lubricating oil
370–600°C	Fuel oil
	Residue

Boiler

(b) Distillation process

Gasoline (48.6%)

Diesel fuel and heating oil (25.0%)

Jet fuel (8.7%)

Liquefied petroleum gases (3.1%)

Heavy fuel oil (2.0%)

Other (12.5%)

(c) Typical output of refined oil

FIGURE 19.9 The refining process results in a range of petroleum products. At oil refineries **(a),** crude oil is boiled, causing its many hydrocarbon constituents to volatilize and proceed upward **(b)** through a distillation column. Constituents that boil at the hottest temperatures and condense readily once the temperature cools will condense at low levels in the column. Constituents that volatilize at cooler temperatures will continue rising through the column and condense at higher levels, where temperatures are cooler. In this way, heavy oils (generally those with hydrocarbon molecules with long carbon chains) are separated from lighter oils (generally those with short-chain hydrocarbon molecules). Shown in **(c)** are percentages of each major category of product typically generated from a barrel of crude oil. *Data (c) from U.S. Energy Information Administration.*

Refining gives us a diversity of fuels

Once we extract oil or gas, it must be processed and refined (**FIGURE 19.9**). Because crude oil is a complex mix of hundreds of types of hydrocarbons, we can create many types of petroleum products by separating its various components. The many hydrocarbon molecules in crude oil have carbon chains of different lengths (p. 28). Chain length affects a substance's chemical properties, which has consequences for human use, such as whether a given fuel burns cleanly in a car engine. Through the process of **refining** at a refinery, hydrocarbon molecules are separated by size and are chemically transformed to create specialized fuels for heating, cooking, and transportation and to create lubricating oils, asphalts, and the precursors of plastics and other petrochemical products.

The amount of a fossil fuel "produced" (extracted and processed) from a nation's reserves and the amount that a nation consumes each depend on many factors. **TABLE 19.3** shows amounts produced and amounts consumed (as a percentage of global production and consumption) by leading nations.

Fossil fuels have many uses

Each major type of fossil fuel has its own mix of uses.

Coal People have burned coal to cook food, heat homes, and fire pottery for thousands of years. Coal-fired steam engines helped drive the industrial revolution by powering factories, trains, and ships, and coal fueled the furnaces of the steel industry. Today, we burn coal largely to generate electricity. In coal-fired power plants, coal combustion

TABLE 19.3 Top Producers and Consumers of Fossil Fuels

PRODUCTION (% world production)		CONSUMPTION (% world consumption)	
COAL			
China	46.7	China	50.5
United States	9.3	India	12.0
Indonesia	8.3	United States	8.4
OIL			
United States	16.2	United States	20.5
Saudi Arabia	13.0	China	13.5
Russia	12.1	India	5.2
NATURAL GAS			
United States	21.5	United States	21.2
Russia	17.3	Russia	11.8
Iran	6.2	China	7.4

Data from BP p.l.c., 2019. *Statistical review of world energy 2019.*

converts water to steam, which turns turbines to create electricity (**FIGURE 19.10**).

Natural gas We use natural gas to generate electricity in power plants, to heat and cook in our homes, and for many other purposes. Converted to a liquid at low temperatures (liquefied natural gas, or LNG), it can be shipped long distances in refrigerated tankers. Versatile and clean-burning, natural gas emits just half as much carbon dioxide per unit of energy released as coal and one-third less than oil. For this reason, many energy experts view natural gas as a climate-friendly "bridge fuel" that can help us transition from today's polluting fossil fuel economy toward a clean renewable energy economy. However, many other experts worry that investing in natural gas will simply delay our transition to renewables and instead deepen our reliance on fossil fuels.

Oil The modern use of oil for energy began after the 1850s, when crude oil gushed from the world's first commercial oil wells drilled in Ontario and Pennsylvania. Over the next 40 years, drillers in Pennsylvania supplied half the world's oil and helped establish a fossil-fuel-based economy that would hold sway for decades to come. Today, our global society consumes nearly 750 L (200 gal) of oil each year for every man, woman, and child. Most is used as fuel for vehicles, including gasoline for cars, diesel for trucks, and jet fuel for

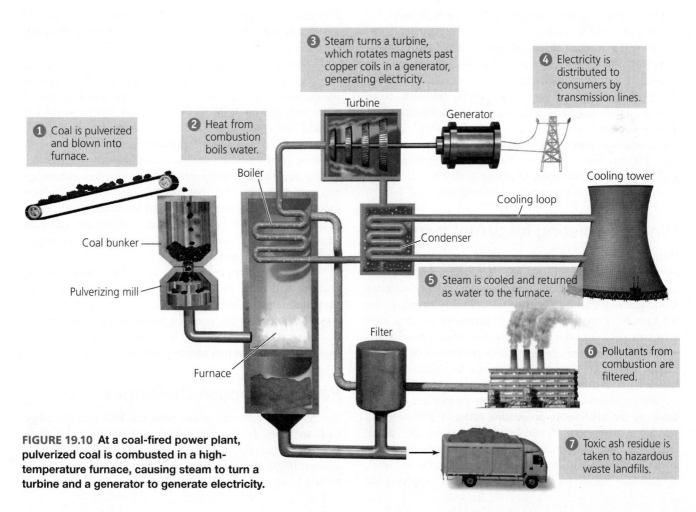

FIGURE 19.10 At a coal-fired power plant, pulverized coal is combusted in a high-temperature furnace, causing steam to turn a turbine and a generator to generate electricity.

1. Coal is pulverized and blown into furnace.
2. Heat from combustion boils water.
3. Steam turns a turbine, which rotates magnets past copper coils in a generator, generating electricity.
4. Electricity is distributed to consumers by transmission lines.
5. Steam is cooled and returned as water to the furnace.
6. Pollutants from combustion are filtered.
7. Toxic ash residue is taken to hazardous waste landfills.

Turbine

Generator

Cooling tower

Cooling loop

Boiler

Coal bunker

Condenser

Pulverizing mill

Filter

Furnace

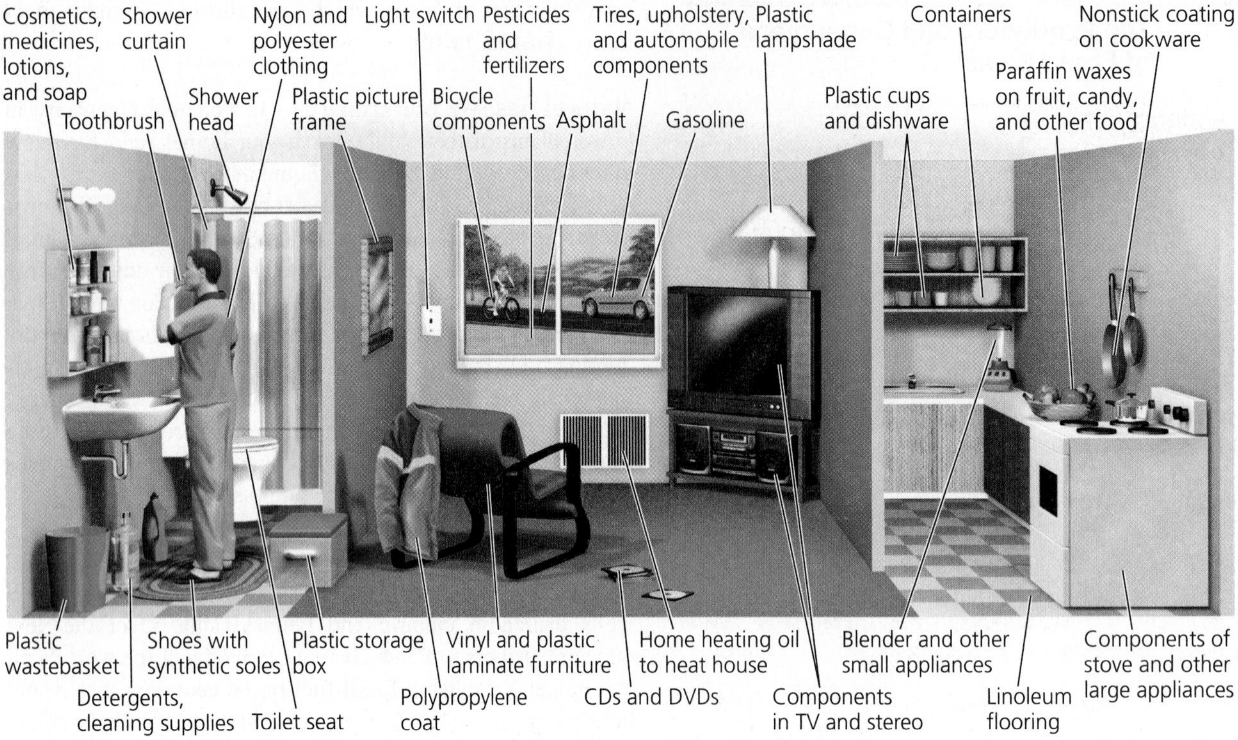
Cosmetics, medicines, lotions, and soap
Shower curtain
Nylon and polyester clothing
Light switch
Pesticides and fertilizers
Tires, upholstery, and automobile components
Plastic lampshade
Containers
Nonstick coating on cookware
Toothbrush
Shower head
Plastic picture frame
Bicycle components
Asphalt
Gasoline
Plastic cups and dishware
Paraffin waxes on fruit, candy, and other food
Plastic wastebasket
Shoes with synthetic soles
Plastic storage box
Vinyl and plastic laminate furniture
Home heating oil to heat house
Blender and other small appliances
Components of stove and other large appliances
Detergents, cleaning supplies
Toilet seat
Polypropylene coat
CDs and DVDs
Components in TV and stereo
Linoleum flooring

FIGURE 19.11 Petroleum products are everywhere in our daily lives. Petroleum is used to make many of the fabrics we wear, the materials we consume, and the plastics in countless items we use every day.

airplanes. Fewer homes burn oil for heating these days, but industry and manufacturing continue to use a great deal.

Refining techniques and chemical manufacturing have greatly expanded our uses of petroleum to include a wide array of products and applications, from plastics to lubricants to fabrics to pharmaceuticals. In today's world, petroleum-based products are all around us in our everyday lives (**FIGURE 19.11**). Take a moment to explore Figure 19.11, and reflect on all the conveniences in your own life that result from petroleum products. The fact that we use petroleum to create so many items and materials we have come to rely on makes it vital that we take care to conserve our remaining oil reserves.

We are depleting fossil fuel reserves

Because fossil fuels are nonrenewable, the total amount available on Earth declines as we use them. Many scientists and oil industry analysts calculate that we have already extracted more than half the world's conventional oil reserves. So far, we have used up about 1.4 trillion barrels of oil, and most estimates hold that about 1.3 trillion barrels of proven recoverable reserves remain. Adding proven reserves of oil from oil sands in Canada and Venezuela brings the estimated total remaining to about 1.73 trillion barrels.

To assess how long this remaining oil will last, analysts calculate the **reserves-to-production ratio,** or R/P ratio, by dividing the amount of remaining reserves by the annual rate of "production" (extraction and processing). At current levels of production (34.6 billion barrels globally per year), 1.73 trillion barrels would last 50 more years. Applying the R/P ratio to natural gas, we find that the world's proven reserves of this resource would last 51 more years. For coal, the latest R/P ratio estimate is 132 years.

The actual number of years remaining for these fuels could turn out to be less than these figures suggest if our demand and production continue to rise. Alternatively, the actual number of years may end up being *more* than these figures suggest if demand and consumption fall due to increased efficiency or higher market prices. The actual number of years may also turn out to be greater because proven recoverable reserves increase as new deposits are discovered and as extraction technology becomes more powerful. For example, enhanced technology for oil and gas extraction in the United States has expanded the nation's proven reserves of these fuels considerably in recent years, enabling the United States to become the world's largest producer of both oil and natural gas.

Eventually, however, extraction of any nonrenewable resource will come to a peak and then decline. In general, extraction tends to decline once reserves are depleted halfway. If demand for the resource holds steady or rises while extraction declines, a shortage will result. With oil, this scenario has come to be nicknamed **peak oil.**

Peak oil will pose challenges

To understand concerns about peak oil, let's turn the clock back to 1956. In that year, Shell Oil geologist M. King Hubbert calculated that U.S. oil production would peak around 1970. His prediction was ridiculed at first, but it proved to be accurate; U.S. production peaked in that very year (**FIGURE 19.12a**). This peak in production came to be known as *Hubbert's peak*. Today, however, U.S. oil production has risen dramatically to surpass this amount, as hydraulic fracturing (pp. 390, 539) has enabled us to extract formerly inaccessible deposits.

For the world as a whole, many scientists and industry experts are calculating that global production of oil will soon begin to decline (**FIGURE 19.12b**). Since about 2005, production of conventional oil has, in fact, been declining, and we have had to seek out a variety of novel petroleum sources to compensate for this decline.

Discovery or development of wholly new sources of oil can delay the peak by boosting our proven reserves. That is what is happening as we exploit Canada's oil sands. People who count on a series of new sources becoming available in the future tend to believe that we can continue relying on oil for a long time. Indeed, R/P ratios for oil have risen over recent decades. However, at some point we will reach peak oil—as well as "peak gas" and "peak coal." The question is when—and whether we will be prepared to deal with the resulting challenges.

Predicting an exact date for peak oil is difficult because many companies and governments do not reveal their data on oil reserves. Moreover, estimates differ as to how much oil we can continue extracting from existing deposits. For these reasons, estimates vary for the timing of an oil production peak, although most studies predict dates before 2035.

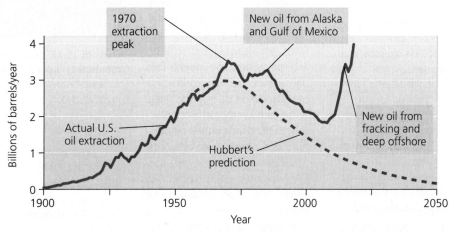

(a) Hubbert's prediction of peak in U.S. crude oil extraction, along with actual data

FIGURE 19.12 Peak oil describes a peak in production. U.S. production of crude oil peaked in 1970 **(a),** just as geologist M. King Hubbert had predicted. However, drilling in Alaska and the Gulf of Mexico enhanced production during the subsequent decline, and production today has risen sharply after deep offshore drilling and hydraulic fracturing made new deposits accessible. As for global oil production, many analysts calculate that it may soon peak. Shown in **(b)** is one projection, from a recent analysis by scientists at the Association for the Study of Peak Oil. These data show that conventional sources of oil peaked around 2005 and that since then unconventional sources have driven the continued overall rise. Paler shades of color indicate future data predicted by the researchers. *Data from **(a)** Hubbert, M.K., 1956. Nuclear energy and the fossil fuels. Shell Development Co. Publ. No. 95, Houston, TX; and U.S. Energy Information Administration; and **(b)** Campbell, C.J., and Association for the Study of Peak Oil. By permission of Dr. Colin Campbell.*

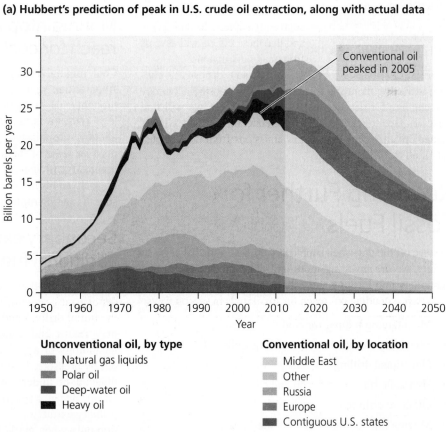

(b) Modern prediction of peak in global oil extraction

FAQ

Hasn't "peak oil" been debunked? We're extracting more oil all the time, right?

An eventual peak is inevitable with any nonrenewable resource, so the notion of peak oil is sound. Its timing is a major question, though—and much depends on how one defines "oil." If one counts only conventional crude oil, data indicate that we passed the global production peak back around 2005. If one also includes "unconventional" oil from difficult-to-access sources such as oil sands, deep-water offshore oil, polar oil, and "tight oil" freed by fracking (p. 390), production worldwide has been roughly flat since 2005. If one also lumps in various additional petroleum sources (such as liquids condensed from natural gas), production is still rising. In the big picture, conventional oil has already been declining for more than a decade, and today, we increasingly rely on a host of petroleum sources that are more difficult and expensive to access.

Whenever an oil peak occurs, if demand for oil continues to rise as supply falls, a spike in oil prices could trigger economic ripple effects that could profoundly affect our lives. Writer James Howard Kunstler has sketched a frightening scenario of a post-peak world during what he has called "the long emergency": Lacking cheap oil with which to transport goods long distances, today's globalized economy would collapse into isolated local economies. Large cities would need to run urban farms to feed their residents, and with less mechanized farming and fewer petroleum-based fertilizers and pesticides, we might feed only a fraction of the world's people. The American suburbs would be hit particularly hard because of their dependence on the automobile. In contrast, more optimistic observers argue that as oil supplies dwindle, rising prices will create powerful incentives for businesses, governments, and individuals to conserve energy and to develop alternative energy sources (Chapters 20 and 21)—and that these efforts will save us from major disruptions.

If we discover and exploit enough new deposits to continue extracting more and more oil, we might postpone our day of reckoning for decades. If we do so, however, we will find ourselves face-to-face with a bigger concern: runaway climate change driven by greenhouse gas emissions from the combustion of all that additional oil!

Reaching Further for Fossil Fuels

To stave off the day when supplies of oil, gas, and coal begin to decline, we are investing more and more money, effort, and technology into locating and extracting new fossil fuel deposits. We are extending our reach for fossil fuels in several ways:

- Mountaintop mining for coal
- Secondary extraction from existing wells
- Directional drilling
- Hydraulic fracturing for oil and gas
- Offshore drilling in deep waters
- Moving into ice-free waters of the Arctic
- Exploiting new "unconventional" fossil fuel sources

FIGURE 19.13 In mountaintop removal mining for coal, entire mountain peaks are leveled, and fill is dumped into adjacent valleys. Shown is an aerial view spanning many square miles in West Virginia.

All these pursuits are expanding the amount of fossil fuel energy available to us. However, as we extend our mining and drilling efforts into less-accessible places to obtain fuel that is harder to extract, we reduce the EROI ratios of our fuels, intensify pollution, and worsen climate change.

Mountaintop mining extends our reach for coal

Coal mining has long been an economic mainstay of the Appalachian region, but **mountaintop removal mining** has brought coal extraction—and its impacts—to a whole new level (**FIGURE 19.13** and p. 652). In this method, entire mountaintops are blasted away to access seams of coal. The massive scale of mountaintop removal mining makes it economically efficient. However, it can cause staggering volumes of rock and soil to slide downslope, polluting or burying streams and disrupting life for people living nearby.

Secondary extraction yields additional fuel

At a typical oil or gas well, as much as two-thirds of a deposit may remain in the ground after primary extraction, the initial drilling and pumping of oil or gas. So, companies may return and conduct secondary extraction using new technology or approaches to force the remaining oil or gas out by pressure. In secondary extraction for oil, solvents are injected or underground rocks are flushed with water or steam. Because secondary extraction is more expensive than primary extraction, most deposits undergo secondary extraction only when market prices of oil and gas are high enough to make the process profitable.

Directional drilling reaches more fuel with less impact

Engineers have brought us a long way from the days when we simply sank a vertical shaft into the ground to suck up oil and gas. Today's **directional drilling** technology allows drillers to bore down vertically and then curve to drill horizontally, thus enabling them to follow horizontal layered deposits. Directional drilling allows access to a large underground area (up to several thousand meters in radius) around each drill pad. As a result, fewer drill pads are needed. Directional drilling enhances efficiency and profits for drillers, and it also reduces the surface footprint of drilling, thereby lessening some of its environmental impacts.

Hydraulic fracturing expands our access to oil and gas

For oil and natural gas trapped tightly in shale or other rock, petroleum companies now use hydraulic fracturing. **Hydraulic fracturing** (also called **fracking**) involves pumping chemically treated water under high pressure into deep layers of rock to crack them. Sand or small glass beads are injected to hold the cracks open as the water is withdrawn. Gas or oil then travels upward, with pressure and pumping, through the newly created system of fractures (see Figure 15.1, p. 390).

Hydraulic fracturing has been used to extract natural gas from formations of shale rock such as the Marcellus Shale—a vast gas-bearing formation beneath Pennsylvania, New York, and neighboring states. Likewise, fracking of the Bakken Formation—layers of shale and dolomite that underlie parts of North Dakota, Montana, and Canada—has allowed us to extract *tight oil,* conventional oil held tightly in or near shale (which differs from shale oil, a petroleum liquid from specially processed oil shale).

By unlocking formerly inaccessible deposits of shale gas and tight oil, fracking ignited a boom in oil and gas production in the United States. The new flow of oil reduced U.S. dependence on foreign imports and led, by 2016, to a glut on the world market that brought oil prices down to less than one-fourth of their 2008 high. The new flow of natural gas lowered the price of that fuel and enabled many power plants to switch from coal to gas. Because natural gas is much cleaner-burning than coal, carbon dioxide emissions from electricity generation have been reduced substantially (**FIGURE 19.14**). Considering that during this period the U.S. economy and population each grew, this drop in emissions is a notable achievement in the fight against global warming.

Yet despite the benefits it has delivered, fracking has sparked debate everywhere it has occurred (**FIGURE 19.15**). Above the Marcellus Shale deposit, fracking has affected the landscapes, economies, politics, and everyday lives of people in Pennsylvania, Ohio, New York, and neighboring states. The choices people face between financial gain and protecting their health, drinking water, and environment have been dramatized in popular films such as *Promised Land* and *Gasland.* In North

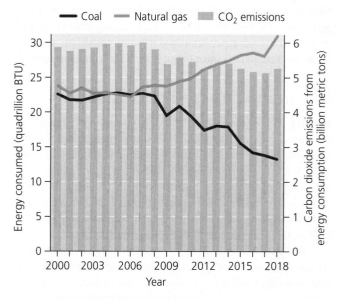

FIGURE 19.14 Fracking has led natural gas to replace coal at many U.S. power plants. Americans' consumption of coal and natural gas used to be roughly equal, but from 2007 to 2018, use of natural gas rose by 31% while use of coal fell by 42%. This sudden divergence reduced greenhouse gas emissions. U.S. emissions of carbon dioxide from energy consumption (**bars** in the graph) peaked in 2007 and have since dropped by 12%. *Data from U.S. Energy Information Administration.*

FIGURE 19.15 Hydraulic fracturing is expanding U.S. production of oil and natural gas but is also sparking debates in communities where it takes place. This drill rig is fracking for shale gas on private land near homes in rural Pennsylvania.

Dakota, fracking for oil supercharged the economy and drew people from around the nation for high-paying jobs, but it also polluted the landscape, drew down water resources, and left many workers jobless once oil prices fell and rigs shut down.

We are drilling farther offshore

Roughly 35% of the oil and 10% of the natural gas extracted in the United States today come from offshore sites, primarily in the Gulf of Mexico and secondarily off southern California. The Gulf today is home to 90 drilling rigs and 3500 production platforms—each of them an engineering marvel built to withstand wind, waves, and ocean currents. Geologists estimate that most U.S. gas and oil remaining occurs offshore and that deep-water sites in the Gulf of Mexico alone may hold 59 billion barrels of oil.

We have been drilling in shallow water for several decades, but as oil and gas are depleted at shallow-water sites and as drilling technology improves, the industry is moving into deeper and deeper water. This kind of drilling poses risks, however, and our ability to drill in deep water has outpaced our capacity to deal with accidents.

In 2008, responding to rising gasoline prices and seeking to lessen dependence on foreign oil, the U.S. Congress lifted a long-standing moratorium on offshore drilling along much of the nation's coastline. The Obama administration then opened vast areas for drilling, including most waters from Delaware to Florida, more of the Gulf of Mexico, and most waters off Alaska's North Slope. However, just weeks after this announcement, the *Deepwater Horizon* oil spill occurred (p. 544). Public reaction forced the Obama administration to backtrack, canceling offshore drilling projects it had approved and putting a hold on further approvals until new safety measures could be devised. Later, the administration sought a middle path, opening access to areas holding 75% of technically recoverable offshore oil and gas reserves while banning drilling offshore from states that did not want it. Drilling leases were expanded off Alaska and in the Gulf of Mexico, but not along the East and West Coasts.

Melting ice is opening the Arctic

Today, all eyes are on the Arctic. As global climate change melts the sea ice that covers the Arctic Ocean (p. 501), new shipping lanes are opening, and nations and companies are jockeying for position, scrambling to lay claim to areas of ocean where fossil fuels might lie beneath the seafloor. However, offshore drilling in Arctic waters poses severe pollution and safety risks. Frigid temperatures, ice floes, winds, waves, and brutal storms make conditions challenging and accidents likely. If a spill were to occur, icebergs, pack ice, storms, cold, and wintertime darkness would hamper response efforts, while frigid water temperatures would slow the natural breakdown of oil.

So far, Royal Dutch Shell has been the only company to pursue offshore drilling in Alaska's stormy waters—and it met with one mishap after another. One drilling rig ran aground while being towed during a storm. Another drilling ship nearly met the same fate in the Aleutian Islands. A containment dome

FIGURE 19.16 "Kayaktivists" in Seattle protested Shell's plans to drill for oil in the Arctic Ocean by blockading a drilling rig in 2015.

intended to control leaks was crushed during testing. An ice-breaker ran aground and had to be towed for repairs all the way to Portland, Oregon, where it faced media-savvy demonstrators protesting its arrival by rowing in kayaks and dangling from a bridge. This followed protests in Seattle over Shell's plans to house drilling rigs in Seattle's port (**FIGURE 19.16**). Meanwhile, Arctic storms halted Shell's drilling. In 2015, Shell gave up and withdrew from the Arctic after spending $7 billion in efforts to drill there.

We are exploiting new fossil fuel sources

Three sources of "unconventional" fossil fuels—oil sands, oil shale, and methane hydrate—are abundant and together could theoretically supply our civilization for centuries. The oil sands of Alberta—and those in Venezuela, which hold even more oil—have already increased the amount of oil available to us.

The world's known deposits of oil shale may contain 3 trillion barrels of oil (more than all the world's conventional crude oil), but most of it will not be easily extracted. About 40% of known reserves are in the United States, mostly on public land in Colorado, Wyoming, and Utah. Shale oil is costly to extract, and its EROI is very low, with estimates ranging from 4:1 down to just 1.1:1. Historically, oil prices have been too low to make processing shale oil profitable, but each time oil prices rise, shale oil attracts attention.

As for methane hydrate, scientists estimate there are enormous amounts of this substance on Earth, holding perhaps twice as much carbon as all known deposits of oil, coal, and natural gas combined. However, we do not yet know whether extraction is safe and reliable. Japan recently extracted methane hydrate from the seafloor by sending down a pipe and lowering pressure within it so that the methane turned to gas and rose to the surface. Yet if extraction were to destabilize a methane hydrate deposit on the seafloor, a landslide and tsunami could result and lead to a sudden release of methane, a potent greenhouse gas, into the atmosphere.

Oil sands, oil shale, and methane hydrate are abundant, but they are no panacea for our energy challenges. Their net energy values and EROI ratios are very low, because they are expensive to extract and process. These fuels also exert severe environmental impacts on land and water. Moreover, burning them would likely emit more greenhouse gases than our use of coal, oil, and natural gas currently does, worsening air pollution and climate change.

Addressing Impacts of Fossil Fuel Use

Our society's love affair with fossil fuels and the many petrochemical products we develop from them has helped ease constraints on travel, lengthen our life spans, and boost our material standard of living beyond what our ancestors could have dreamed. Yet continued reliance on fossil fuels poses growing risks to human health, environmental quality, and social, political, and economic stability. The predicted impacts from climate change alone are so great that the International Energy Agency's chief economist recently joined a growing chorus of scientists in concluding that we will need to leave most fossil fuels in the ground if we are to avoid dangerous climate disruption. Let's now briefly survey the main impacts of fossil fuel use—beginning with extraction, continuing through emissions, and ending with social issues—and assess ways to minimize these impacts.

Extraction of fossil fuels brings many environmental impacts

The means by which we remove fossil fuel resources from the earth have direct impacts on landscapes and natural habitats (FIGURE 19.17), as well as a diversity of consequences for ecological processes and environmental quality.

Coal The mining of coal exerts substantial impacts on natural systems and human well-being (p. 649). Subsurface coal mining has long posed health and safety risks to miners. Miners are vulnerable to accidents, and they breathe coal dust and hazardous gases in confined spaces, which often leads to respiratory diseases. Strip mining, for its part, destroys large swaths of habitat and causes extensive soil erosion. It also results in chemical runoff into waterways through acid mine drainage (p. 650), whereby sulfide minerals in freshly exposed rock surfaces react with oxygen and rainwater to produce sulfuric acid, which can leach toxic metals from rocks. Many nations require mining companies to restore affected areas after mining. This reclamation is beneficial but rarely is able to recreate the ecological communities that existed prior to mining (p. 656).

FIGURE 19.17 Oil sands extraction dramatically alters Alberta's forested landscape. Shown is a portion of the Suncor mine near Fort McMurray.

Mountaintop removal mining (pp. 538, 652) exerts impacts that exceed even those of strip mining. When countless tons of rock and soil are removed from the top of a mountain, material slides downhill, where immense areas of habitat can be degraded or destroyed and creek beds can be clogged and polluted, affecting wildlife, ecosystems, and local residents. (We explore all these issues of coal mining in Chapter 23.)

Oil and natural gas To drill for conventional oil or gas on land, road networks must be constructed, and many sites may be explored in the course of prospecting. The extensive infrastructure needed to support a full-scale drilling operation typically includes access roads, transport pipelines, housing for workers, waste piles for excavated soil, and ponds to collect toxic sludge—all of which tend to disturb wildlife, fragment habitats, and pollute the soil, air, and water. These impacts can be readily seen at most drilling sites and have been well documented on the Alaskan tundra, where oil and gas have been extracted from the Prudhoe Bay region for decades.

To protect natural resources in remote wilderness areas of Alaska's North Slope, policymakers in 1980 put most of the Arctic National Wildlife Refuge off-limits to drilling but reserved for future decision-making an area of coastal plain named the 1002 Area (after Section 1002 of the congressional bill that established it). After decades of debate, Congress in 2017 narrowly approved opening the 1002 Area to petroleum extraction, via a provision attached to unrelated tax legislation. Federal agencies under the Trump administration aggressively pushed through environmental reviews and sought to begin exploration and intensive seismic surveying (p. 532) in 2019.

The potential development of Alaska's North Slope for oil and gas raises many issues. Wilderness advocates point to scientific evidence for a wide range of impacts, based on studies in the nearby Prudhoe Bay region. They point out

FIGURE 19.18 Since Congress in 2017 approved oil and gas extraction in the 1002 Area of the Arctic National Wildlife Refuge, debate has flared over the many likely impacts. Scars are still visible on the tundra from seismic surveying done in the 1980s, and new surveys with fleets of "thumper trucks" could threaten polar bears, which raise their young in dens under the snow. Development may also affect caribou and the native Gwich'in people who hunt them.

that seismic surveying decades ago left hundreds of miles of scars on the tundra that are still visible today—and they fear that new surveys (which need to be done in winter when snow and ice cover the ground) will collapse polar bear dens. Infrastructure and noise from oil development is expected to affect the thousands of caribou that migrate to the coastal plain to give birth to their calves—as well as the native Gwich'in people who rely on the caribou herd for sustenance (**FIGURE 19.18**).

Fortunately, directional drilling (p. 539) has become widely used today. This method provides much greater horizontal reach than traditional drilling, so fewer drill pads are needed and less infrastructure is required, reducing the footprint on the landscape.

Hydraulic fracturing Extracting oil or natural gas by hydraulic fracturing exerts all the traditional impacts of drilling and poses additional concerns. In fracking, chemicals are mixed with pressurized water and sand and injected deep underground, giving rise to risks of water pollution that are not yet completely understood. One risk is that fracking fluids may leak out of drilling shafts and into aquifers that people rely on for drinking water. Another concern is that methane may contaminate groundwater used for drinking if it travels up fractures or leaks through a shaft.

Hydraulic fracturing also gives rise to air pollution as methane and volatile toxic components of fracking fluids seep up from drilling locations. Some of the unhealthiest air in America has been documented in a remote region of Wyoming near fracking operations. Many residents of areas near fracking sites have experienced polluted air and fouled drinking water; more research is needed to assess the extent of such pollution and to quantify health risks.

Fracking also consumes immense volumes of fresh water (pp. 389–391). Injected water often returns to the surface laced with salts, radioactive elements such as radium, and toxic chemicals such as benzene from deep underground. This wastewater may be sent to sewage treatment plants that are not designed to handle the contaminants and that do not test for radioactivity. In Pennsylvania, millions of gallons of drilling waste from Marcellus Shale fracking sites have been sent to treatment plants, which then release their water into rivers that supply drinking water for people in Pittsburgh, Harrisburg, and other cities. In response to public outcry and government pressure, the oil and gas industry is beginning to reduce its consumption of clean water by reusing its wastewater in multiple injections.

Finally, fracking is now known to cause earthquakes (pp. 42–43). Most have been minor, but they do raise questions about whether all the social and economic costs of fracking have been considered.

Oil sands Impacts from the extraction of oil sands surpass those of conventional fossil fuel extraction. Because everything is cleared across huge areas when oil sands are mined (see Figure 19.17), this results in the large-scale loss of boreal forest (in Alberta) or tropical forest (in Venezuela). Deforestation has severe consequences for species, habitats, ecosystems, soil, water, climate, and people (Chapter 12). Industry representatives counter that they are mandated to attempt restoration after mining is finished. However, effective reclamation has not yet been demonstrated, and regions denuded by the very first oil sand mine in Alberta more than 30 years ago have still not recovered.

Three barrels of water are required to produce each barrel of oil from oil sands, and the toxic wastewater that results is discharged into huge reservoirs. The Syncrude mine's tailings reservoir near Fort McMurray, Alberta, for example, is so massive that it is held back by the world's second-largest dam. Migratory waterfowl land on water bodies like Syncrude's reservoir and are killed as the oily water gums up their feathers and impairs their ability to insulate themselves.

Finally, oil from oil sands burns less cleanly than conventional oil. Estimates of the average difference range from 14–21%. So, using oil sands as a source of oil will create 14–21% more pollution and climate warming—without producing additional energy.

Fuel can leak during transport

Once fossil fuels are extracted, they must be transported for processing and refining and then transported again to consumers. Between transport episodes, they may be stored for long periods. At each stage, there is the risk of leakage.

Pipeline spills Oil is commonly conveyed by pipeline, and pipelines are subject to corrosion, vandalism, and equipment malfunction, any of which can result in spills that contaminate soil and water. Oil from Canada's oil sands (which is more corrosive than conventional crude oil) has spilled from a number of pipelines, fouling the Kalamazoo River in Michigan; a residential neighborhood of Mayflower, Arkansas; and other areas. People living along the proposed route of the Keystone XL pipeline worry that if oil were to spill from that pipeline, it would sink into the region's porous ground and quickly reach its shallow water table, contaminating the Ogallala Aquifer. This aquifer (p. 396) provides 2 million Americans with drinking water and irrigates a large portion of U.S. agriculture.

Another high-profile pipeline project is the Dakota Access Pipeline, a 1900-km (1200-mi) underground pipeline proposed to bring oil from Bakken Formation drilling sites in North Dakota to a tank farm in Illinois. In 2016–2017, thousands of people joined Native American protests against the pipeline on the Standing Rock Indian Reservation (**FIGURE 19.19**). The Standing Rock Sioux objected to disturbance to sacred burial grounds they believed would result where the pipeline would cross their land. They also maintained that the pipeline would put their water supply at risk of pollution, where it would cross under the Missouri River. After months of protests, legal wrangling, and media attention, the U.S. Army Corps of Engineers in the final weeks of the Obama administration denied an easement for pipeline construction. The next month, however, shortly after his inauguration, Trump ordered expedited construction of the pipeline.

FIGURE 19.19 Thousands of demonstrators protested the Dakota Access Pipeline on the Standing Rock Sioux Reservation in 2016–2017. They contended that oil spilled from the pipeline could contaminate water supplies and degrade sacred land.

FIGURE 19.20 Transporting oil by rail poses risks. Explosive crashes, such as this one in Lac-Mégantic, Quebec, have occasionally resulted as North American oil extraction has surpassed pipeline capacity.

FIGURE 19.21 The explosion at the *Deepwater Horizon* drilling platform in 2010 unleashed the world's largest accidental oil spill. Vessels tried unsuccessfully to put out the blaze.

Rail transport As oil extraction in North America has soared in recent years, the amount needing to be transported has surpassed pipeline capacity. As a result, more and more oil is being transported by rail instead, in pressurized tank cars. Tragically, a series of explosive derailments of trains carrying North Dakota crude oil and Albertan oil sands oil has illustrated the dangers of transporting oil by train (**FIGURE 19.20**). Worst has been the 2013 explosion in Lac-Mégantic, Quebec, which killed 47 people and destroyed the town's center. In response to spills and explosions, the Obama administration developed regulations to upgrade the safety of tanker cars. Industry complained of additional cost, whereas safety advocates said the steps were not strong enough.

Coal is generally transported by rail in open cars, which releases coal dust into the air. In the Pacific Northwest, clean energy advocates are opposing the transport of coal by train from the interior West where it is mined to coastal terminals to be shipped to Asia. Pollution along the route is one concern; another is that it facilitates China's reliance on coal, a driver of global climate change.

Methane leaks Any time natural gas leaks into the air, its main component, methane, is added to the atmosphere. Methane is a powerful greenhouse gas (p. 488), so its leakage worsens global warming and climate change. Methane can leak at various points during the extraction, transport, and processing of natural gas—and it can leak from aging pipes that bring natural gas to consumers. A 2015 study of the Boston region found that 2.7% of natural gas piped to homes and businesses was escaping into the air—higher than the 1% previously assumed. By comparison, researchers estimate that 3.6–7.9% of natural gas extracted by shale gas fracking operations leaks into the air.

A great deal of methane also escapes to the atmosphere from oil drilling sites where gas accompanies oil deposits yet pipelines for the gas do not exist. At many such sites, natural gas is flared off—the stream of gas is set on fire. This process produces carbon dioxide, however, which is also a greenhouse gas.

Marine oil spills Oil spilled from tanker ships or offshore drilling platforms can foul coastal waters and beaches. In 2010, British Petroleum's *Deepwater Horizon* offshore drilling platform exploded and sank off the coast of Louisiana, killing 11 workers (**FIGURE 19.21**). Emergency shut-off systems failed, and oil gushed out of a broken pipe on the ocean floor a mile beneath the surface at a rate of 30 gallons per second. The flow of oil and gas continued out of control for three months, spilling roughly 4.9 million barrels (206 million gallons) of oil. The crisis proved difficult to control because people had never had to deal with a spill so deep underwater. It revealed that offshore drilling presents serious risks of environmental impact that may be difficult to address.

As the oil spread through the Gulf of Mexico and washed ashore, the region suffered a wide array of impacts (**FIGURE 19.22**). Of the countless animals killed, most conspicuous were birds, which cannot regulate their body temperature once their feathers become coated with oil. However, the underwater nature of the BP spill meant that unknown numbers of fish, shrimp, corals, and other marine animals also were killed, affecting ecosystems in complex ways. Plants in coastal marshes died, and the resulting erosion put New Orleans and other coastal cities at greater risk from storm surges and flooding. Gulf Coast fisheries, which supply much of the nation's seafood, were hit hard, with thousands of fishers and shrimpers put out of work. Beach tourism suffered, and indirect economic and social impacts have lasted for years. Throughout this process, scientists studied aspects of the spill and its impacts on the region's people and natural systems (see **THE SCIENCE BEHIND THE STORY**, pp. 546–547).

(a) Brown pelican coated in oil

(b) Beach cleanup

FIGURE 19.22 Impacts of the *Deepwater Horizon* spill were many. This brown pelican, coated in oil **(a),** was one of many thousands of animals killed. For months, workers and volunteers labored **(b)** to clean oil from the Gulf's beaches.

The *Deepwater Horizon* spill was the largest accidental oil spill in world history, but there have been many others. Notable is one that has been leaking oil for more than 15 years. People responding to the *Deepwater Horizon* spill kept noticing an unrelated slick nearby. Eventually, they determined that it was a little-known ongoing spill that had begun in 2004 when Hurricane Ivan set off submarine landslides that destroyed a platform owned by Taylor Energy. Since then, the damaged wells have been releasing an estimated 300–700 barrels of oil each day, and scientists estimate that the flow could continue at this rate for a century if the wells are not capped.

In the amount of oil released, the Taylor and *Deepwater Horizon* spills eclipsed the *Exxon Valdez* tanker spill of 1989, but the ecological impacts of the *Exxon Valdez* event were greater due to its occurrence in a biologically rich nearshore location. In the *Exxon Valdez* spill, oil from Alaska's North Slope, piped to the port of Valdez through the trans-Alaska pipeline, caused severe damage to ecosystems and economies in Alaska's Prince William Sound after the tanker ran aground. Three decades later, a layer of oil remains in the region's beaches, and some animal populations have not recovered.

The *Exxon Valdez* spill led U.S. policymakers to tighten regulation and improve spill-response capacity. However, if drilling moves into Arctic waters, our capacity to control accidents could be sorely tested by the harsh conditions. In U.S. Arctic waters open to oil and gas leasing, some sites are 1000 miles from the nearest Coast Guard station.

Fortunately, pollution from large spills has declined in recent decades (see Figure 16.16, p. 435), thanks to government regulations (such as requiring double-hulled ships) and improved spill-response efforts. Most water pollution from oil today results from innumerable, small non-point sources to which we all contribute: oil from automobiles, homes, industries, gas stations, and businesses that runs off roadways and enters streams and rivers (see Figure 15.19, p. 409).

Fossil fuel emissions pose health risks

A variety of airborne pollutants emitted from fossil fuels affect human health. Combusting coal can emit mercury that bioaccumulates in organisms' tissues, poisoning animals as it moves up food chains (p. 373) and posing health risks to people. Gasoline combustion in automobiles releases pollutants that irritate the nose, throat, and lungs, as well as cancer-causing hydrocarbons such as benzene and toluene. Gases such as hydrogen sulfide can evaporate from crude oil, irritate the eyes and throat, and cause asphyxiation. Crude oil also may contain trace amounts of poisons such as lead and arsenic. As a result, workers at drilling operations, at refineries, and in other jobs that entail frequent exposure to oil can develop serious health problems, including cancer.

The combustion of oil in vehicles and coal in power plants releases sulfur dioxide and nitrogen oxides, which contribute to smog (pp. 467–468) and acid deposition (p. 474). Air pollution from fossil fuel combustion is intensifying in developing nations as they industrialize, but it has been reduced in developed nations as a result of laws such as the U.S. Clean Air Act and government regulations to protect public health (Chapter 17). In these nations, public policy has encouraged industry to develop and install technologies that reduce pollution, such as catalytic converters that cleanse vehicle exhaust (see Figure 17.13, p. 461).

Clean coal technologies aim to reduce air pollution from coal

At coal-fired power plants, scientists and engineers are seeking ways to cleanse coal exhaust of sulfur, mercury, arsenic, and other impurities. **Clean coal technologies** refer to an array of techniques, equipment, and approaches that

THE SCIENCE behind the story

What Were the Impacts of the Gulf Oil Spill?

A scientist rescues an oiled Kemp's ridley sea turtle.

President Barack Obama echoed the perceptions of many Americans when he called the *Deepwater Horizon* oil spill in the Gulf of Mexico "the worst environmental disaster America has ever faced." But what has scientific research told us about the actual impacts of the Gulf oil spill?

We will never have all the answers, because the deep waters affected by the spill are especially difficult to study. Yet the intense and focused scientific response to the spill demonstrates the dynamic way in which science can assist society.

As the spill was taking place, government agencies called on scientists to help determine how much oil was leaking. Researchers eventually determined the rate reached 62,000 barrels per day. Using underwater imaging, aerial surveys, and shipboard water samples, researchers tracked the movement of oil up through the water column and across the Gulf. These data helped predict when and where oil might reach shore, thereby serving to direct prevention and cleanup efforts. Meanwhile, as engineers struggled to seal off the well using remotely operated submersibles, researchers assisted government agencies in assessing the fate of the oil (**FIGURE 1**).

University of Georgia biochemist Mandy Joye, who had studied natural seeps in the Gulf for years, documented that the leaking wellhead was creating a plume of oil the size of Manhattan. She also found evidence of low oxygen concentrations, or hypoxia (p. 106), because some bacteria consume oil and gas, depleting oxygen from the water and making it uninhabitable for fish and other creatures.

Joye and other researchers feared that the thinly dispersed oil might devastate plankton (the base of the marine food chain) and the tiny larvae of shrimp, fish, and oysters (the pillars of the fishing industry). Scientists taking water samples documented sharp drops in plankton during the spill, but it will take years to learn whether the impact on larvae will diminish populations of adult fish and shellfish. Studies on the condition of living fish in the region have shown gill damage, tail rot, lesions, and reproductive problems at much higher levels than is typical.

What was happening to life on the seafloor was a mystery, because only a handful of submersible vehicles in the world are able to travel to the crushing pressures of the deep sea. Luckily, a team of researchers led by Charles Fisher of Penn State University was scheduled to embark on a regular survey of deep-water coral across the Gulf of Mexico in late 2010. Using the three-person submersible *Alvin* and the robotic vehicles *Jason* and *Sentry*, the team found healthy coral communities at sites far away from the Macondo well but found dying corals and brittlestars covered in a brown material at a site 11 km (7 miles) from the Macondo well.

Eager to determine whether this community was contaminated by the oil spill, the researchers added chemist Helen White of Haverford College to their team, and returned to the site a month later, thanks to a National Science Foundation program that funds rapid response research. On this trip, chemical

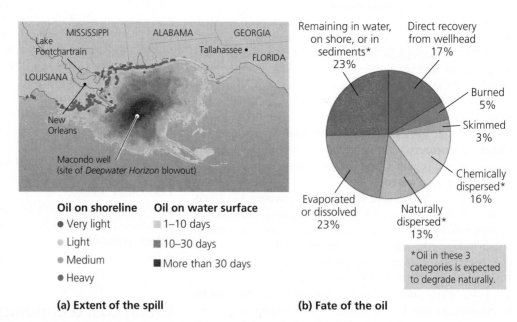

(a) **Extent of the spill**

Oil on shoreline
- Very light
- Light
- Medium
- Heavy

Oil on water surface
- 1–10 days
- 10–30 days
- More than 30 days

(b) **Fate of the oil**

Remaining in water, on shore, or in sediments* 23%
Direct recovery from wellhead 17%
Burned 5%
Skimmed 3%
Chemically dispersed* 16%
Naturally dispersed* 13%
Evaporated or dissolved 23%

*Oil in these 3 categories is expected to degrade naturally.

FIGURE 1 Scientists helped track oil from the *Deepwater Horizon* spill. The map **(a)** shows areas polluted by oil. The pie chart **(b)** gives a breakdown of the oil's fate. These data helped inform studies of the oil's impacts on marine life and human communities. *Source: (a) NOAA Office of Response and Restoration; (b) NOAA.*

analysis of the brown material showed it to match oil from the British Petroleum (BP) spill.

Other questions revolve around impacts of the chemical dispersant that BP used to break up the oil, a compound called Corexit 9500. Work by biologist Philippe Bodin following the *Amoco Cadiz* oil spill in France in 1978 had found that Corexit 9500 appeared more toxic to marine life than the oil itself. BP threw an unprecedented amount of this chemical at the *Deepwater Horizon* spill, injecting a great deal directly into the path of the oil at the wellhead. This chemical caused the oil to dissociate into trillions of tiny droplets that dispersed across large regions. Many scientists worried that it caused the oil to affect more plankton, larvae, and fish than it would have otherwise.

Impacts of the oil on birds, sea turtles, and marine mammals were more easily seen but still difficult to assess. Hundreds of these animals were cleaned and saved by wildlife rescue teams. Officially confirmed deaths numbered 6104 birds, 605 turtles, and 97 mammals, but a much larger number surely succumbed to the oil and were never found; some researchers estimated that up to 50 times more fatalities occurred. What impacts this mortality may have on populations in coming years is unclear. (After the *Exxon Valdez* spill in Alaska in 1989, populations of some species rebounded, but populations of others have never come back.) Researchers have been following the movements of marine animals in the Gulf with radio transmitters to try to learn what effects the oil may have had.

As images of oil-coated marshes saturated the media, researchers worried that widespread death of marsh grass would leave the shoreline vulnerable to severe erosion by waves. Louisiana has already lost vast areas of coastal wetlands to dredging, sea level rise, sinking of land due to the removal of oil and gas, and silt capture by dams on the Mississippi River. Fortunately, researchers found that oil did not penetrate to the roots of most plants and that oiled grasses were sending up new growth. Indeed, Louisiana State University researcher Eugene Turner said that loss of marshland from the oil "pales in comparison" with marshland lost each year in Louisiana due to other factors.

The ecological damage caused by the spill had measurable consequences for people. The region's mighty fisheries were shut down, forcing thousands of people out of work. The government tested fish and shellfish for contamination and reopened fishing once they were found to be safe, but consumers balked at buying Gulf seafood. Beach tourism remained low all summer as visitors avoided the region. Together, losses in fishing and tourism totaled billions of dollars.

Scientists expect some consequences of the Gulf spill to be long-lasting. Oil from the similar *Ixtoc* blowout off Mexico's coast in 1979 still lies in sediments near dead coral reefs, and fishers there say it took 15–20 years for catches to return to normal. After the *Amoco Cadiz* spill, it took seven years for oysters and other marine species to recover. In Alaska, oil from the 1989 *Exxon Valdez* spill remains embedded in beach sediments today.

However, many researchers are hopeful about the Gulf of Mexico's recovery from the *Deepwater Horizon* spill. The Gulf's warm waters and sunny climate speed the natural breakdown of oil. In hot sunlight, volatile components of oil evaporate from the surface and degrade in the water, so fewer toxic compounds reach marine life. In addition, bacteria that consume hydrocarbons thrive in the Gulf because some oil has always seeped naturally from the seafloor and because leakage from platforms, tankers, and pipelines is common. These microbes give the region a natural self-cleaning capacity.

Researchers continue to conduct a wide range of scientific studies (**FIGURE 2**). A consortium of federal and state agencies has coordinated research and restoration efforts in the largest ever Natural Resource Damage Assessment, a process mandated under the Oil Pollution Act of 1990. Answers to questions will come to light as long-term impacts become clear.

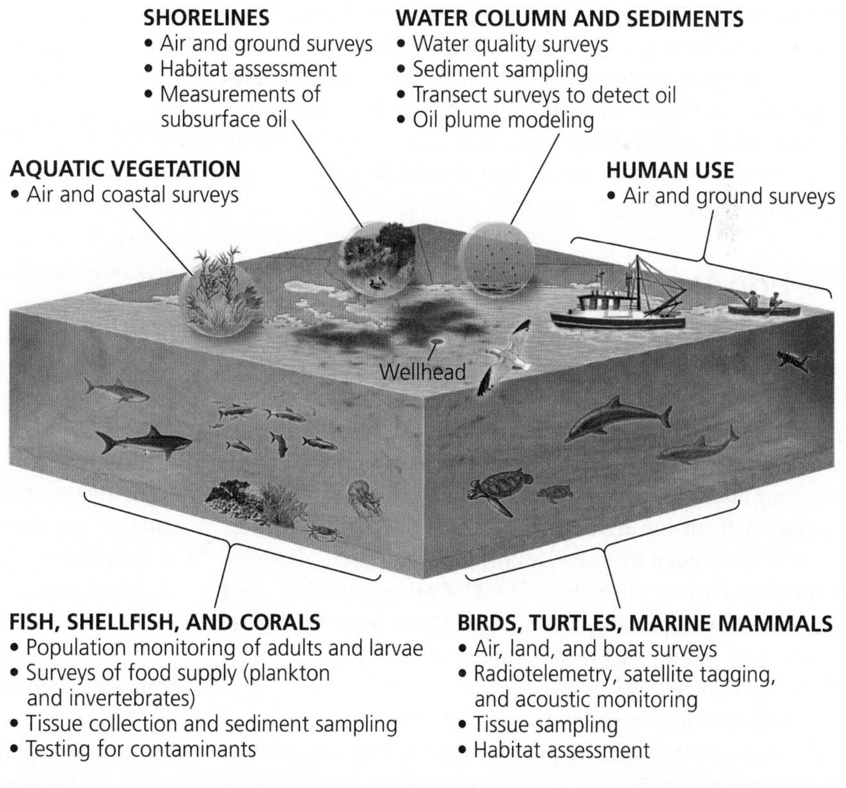

SHORELINES
• Air and ground surveys
• Habitat assessment
• Measurements of subsurface oil

WATER COLUMN AND SEDIMENTS
• Water quality surveys
• Sediment sampling
• Transect surveys to detect oil
• Oil plume modeling

AQUATIC VEGETATION
• Air and coastal surveys

HUMAN USE
• Air and ground surveys

Wellhead

FISH, SHELLFISH, AND CORALS
• Population monitoring of adults and larvae
• Surveys of food supply (plankton and invertebrates)
• Tissue collection and sediment sampling
• Testing for contaminants

BIRDS, TURTLES, MARINE MAMMALS
• Air, land, and boat surveys
• Radiotelemetry, satellite tagging, and acoustic monitoring
• Tissue sampling
• Habitat assessment

FIGURE 2 Thousands of researchers continue to help assess damage to natural resources from the *Deepwater Horizon* oil spill. They have been surveying habitats, collecting samples and testing them in the lab, tracking wildlife, monitoring populations, and more.

aim to remove chemical contaminants during the generation of electricity from coal. Among these technologies are scrubbers, devices that chemically convert or physically remove pollutants (see Figure 17.12, p. 460). Some scrubbers use minerals such as magnesium, sodium, or calcium in reactions to remove sulfur dioxide (SO_2) from smokestack emissions. Others use chemical reactions to strip away nitrogen oxides (NO_X), breaking them down into elemental nitrogen and water. Multilayered filtering devices are used to capture tiny ash particles.

Another "clean coal" approach is to dry coal that has high water content to make it cleaner-burning. We can also gain more power from coal with less pollution through *gasification*, in which coal is converted into a cleaner synthesis gas (called *syngas*) by reacting it with oxygen and steam at a high temperature. Syngas from coal can be used to turn a gas turbine or to heat water to turn a steam turbine.

The U.S. government and the coal industry have each invested billions of dollars in clean coal technologies for new power plants. These efforts have helped reduce air pollution from sulfates, nitrogen oxides, mercury, and particulate matter (pp. 458–463). If the many older plants that still pollute our air were retrofitted with these technologies, pollution could be reduced even more. That was a goal of the Obama administration's Clean Power Plan (pp. 174, 513), launched in 2015 but scuttled by the Supreme Court and the Trump administration.

The coal industry spends a great deal of money fighting regulations on its practices. As a result, many power plants have little in the way of pollution control technologies. Many energy analysts emphasize that "clean coal" technologies may make for *cleaner* coal but will never result in energy that is completely clean. They argue that coal is an inherently dirty way of generating power and should be replaced outright with cleaner energy sources.

Carbon emissions drive climate change

In addition to the many health impacts from fossil fuel pollution, emissions from the combustion of coal, oil, and natural gas are driving global climate change. When we burn fossil fuels, we alter fluxes in Earth's carbon cycle (p. 122). We essentially remove carbon from a long-term reservoir underground and release it into the air. This occurs as carbon from the hydrocarbon molecules of fossil fuels unites with oxygen from the atmosphere during combustion, producing carbon dioxide (CO_2). Carbon dioxide is a greenhouse gas (p. 488), and CO_2 released from fossil fuel combustion warms our planet and drives changes in global climate (Chapter 18).

Because climate change is having diverse, severe, and widespread ecological and socioeconomic impacts, carbon dioxide pollution (**FIGURE 19.23**) is becoming recognized as the single biggest negative consequence of fossil fuel use. Methane is also a potent greenhouse gas that drives climate warming. Switching to new fossil fuel sources may only make emissions worse; oil sands are estimated to generate

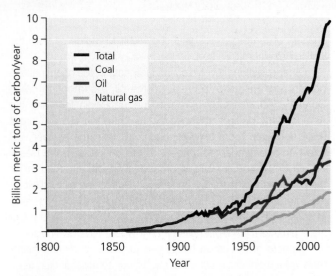

FIGURE 19.23 Emissions from fossil fuel combustion have risen dramatically as nations have industrialized and as population and consumption have grown. In this graph, global emissions of carbon from carbon dioxide are subdivided by source (coal, oil, or natural gas). *Data from Carbon Dioxide Information Analysis Center, Oak Ridge National Laboratory, U.S. Department of Energy, Oak Ridge, TN.*

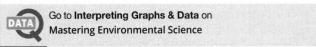

Go to **Interpreting Graphs & Data** on **Mastering Environmental Science**

14–20% more greenhouse gas emissions than conventional oil, and shale oil is still more polluting.

Can we capture and store carbon?

Because fossil fuel combustion is intensifying the greenhouse effect and worsening climate change, many current efforts focus on **carbon capture and storage** (p. 511). This approach involves capturing carbon dioxide emissions, converting the gas to a liquid, and then sequestering (storing) it in the ocean or underground in a geologically stable rock formation (**FIGURE 19.24**).

Carbon capture and storage is being attempted at a variety of facilities. The world's first coal-fired power plant to approach zero emissions opened in 2008 in Germany. This plant removes its sulfate pollutants and captures its carbon dioxide and then compresses the CO_2 into liquid form, trucks it away, and injects it 900 m (3000 ft) underground into a depleted natural gas field.

In North Dakota, the Great Plains Synfuels Plant gasifies its coal and then sends half the CO_2 through a pipeline into Canada, where a Canadian oil company injects it into an oilfield to help it pump out the remaining oil. The North Dakota plant also captures, isolates, and sells seven other types of gases for various purposes.

The highest-profile effort at carbon capture and storage has suffered a rocky history. Beginning in 2003, the U.S. Department of Energy teamed up with seven energy companies to build a prototype of a near-zero-emissions coal-fired

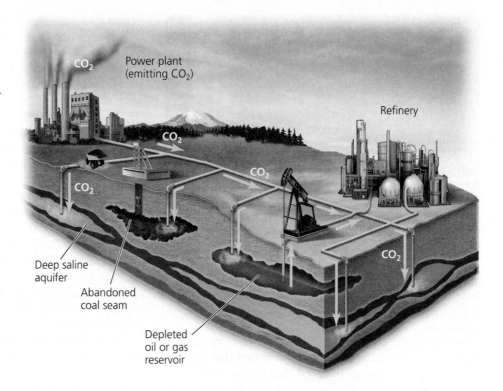

FIGURE 19.24 Carbon capture and storage schemes propose to inject liquefied carbon dioxide emissions underground. The CO_2 may be injected into depleted fossil fuel deposits, deep saline aquifers, or oil or gas deposits undergoing secondary extraction.

CO₂

Power plant (emitting CO_2)

Refinery

CO_2

CO_2

CO_2

CO_2

Deep saline aquifer

Abandoned coal seam

Depleted oil or gas reservoir

power plant. The *FutureGen* project (and its successor, the $1.65-billion *FutureGen 2.0*) aimed to design, construct, and operate a coal-burning power plant that captures 90% of its carbon dioxide emissions and sequesters the CO_2 deep underground beneath layers of impermeable rock. It was hoped that this showcase project, located in downstate Illinois, could be a model for a new generation of power plants across the world, but financial challenges plagued it, and in 2015, the project was suspended.

At this point, carbon capture and storage remains too unproven to be the central focus of a clean energy strategy. We do not know how to ensure that carbon dioxide will stay underground once injected there. Injection might in some cases contaminate groundwater supplies or trigger earthquakes. Injecting carbon dioxide into the ocean would worsen the tremendous problem of ocean acidification that threatens the future of coral reefs and marine life (pp. 437, 504). Moreover, carbon capture and storage is energy-intensive and decreases the EROI of coal, adding to its cost and the amount we consume. Finally, many renewable energy advocates fear that pursuing this approach takes the burden off emitters and prolongs our dependence on fossil fuels rather than facilitating a shift to renewables.

WEIGHING
the **issues**

Clean Coal and Carbon Capture

Do you think we should spend billions of dollars to try to find ways to burn coal more cleanly and to sequester carbon emissions from fossil fuels? Or is our money better spent on developing new clean and renewable energy sources, even if they don't yet have enough infrastructure to generate power at the scale that coal can? What pros and cons do you see in each approach?

We all pay external costs

The costs of addressing the many health and environmental impacts of fossil fuel production and use are generally not internalized in the market prices of fossil fuels. Instead, we all pay these external costs (pp. 143, 165) through medical expenses, costs of environmental cleanup, and impacts on our quality of life. The prices we pay at the gas pump or on our monthly utility bill do not even cover the financial costs of fossil fuel production. Rather, fossil fuel prices have been kept inexpensive as a result of government subsidies and tax breaks to petroleum producers. The profitable and well-established fossil fuel industries still receive far more financial support from taxpayers than do the young and struggling renewable energy sources (Figure 7.13, p. 180; Figure 21.7, p. 596). In this way, we all pay extra for fossil fuel energy through our taxes, generally without even realizing it. Ending subsidies could help level the economic playing field for energy sources, and adopting carbon taxes (p. 513) could help address external costs.

Fossil fuel extraction has mixed consequences for local people

Wherever fossil fuels are extracted, people living nearby must weigh the environmental, health, and social drawbacks of extractive development against the potential financial benefits. In North America, communities where fossil fuel extraction takes place often experience a flush of high-paying jobs and economic activity, and for many people, these economic benefits far outweigh other concerns. Perceptions can change with time, however. Economic booms

FIGURE 19.25 Pollution often affects residents of regions where fossil fuel extraction has taken place. This Pennsylvania homeowner can set fire to her tap water because it is contaminated with methane from nearby fracking for shale gas.

generally prove temporary, lasting only as long as the resource holds out and market prices for fuel stay high. At some point, all booms go bust, and once the jobs and money have left a community, its residents may be left with a polluted environment for generations to come (**FIGURE 19.25**).

Fort McMurray has been the hub of Alberta's oil sands boom. This remote northern frontier town's population skyrocketed from 2000 in the 1960s to 80,000 as people flocked there looking for jobs. An additional 40,000 oilfield workers were being flown in and housed to work shifts, 3-weeks on, 10 days off, at the height of the recent boom. Most incoming residents have been men, averaging 32 years of age, and the city boasts the highest birth rate in Canada. Salaries are high, but so are rents and home prices. Like all boomtowns, Fort McMurray's population outgrew its infrastructure, and services became stretched thin. And like all boomtowns, it experienced a bust when the price of its principal resource fell. For Fort McMurray, the bust came in 2015 as world oil prices fell amid oversupply. Unable to make a profit on oil sands at the low prices, companies began cutting off contracts and orders and laying off workers. Across Alberta in 2015, more than 35,000 people in the energy sector lost their jobs. Still, after $200 billion of investment, many companies are in the game for the long haul and are continuing to extract oil sands, hoping that prices will soon rise and pipeline capacity will be increased. As oil prices fluctuate in future years, Fort McMurray may see more booms and more busts.

Along the route out of Alberta and through the United States via the proposed Keystone XL pipeline, TC Energy estimates that the project would create 6500 construction jobs for two years plus 7000 one-year jobs for manufacturers of supplies. For landowners, the pipeline project would have mixed consequences. The corporation negotiated with thousands of landowners along the Keystone XL route, offering them money for the right to install the pipeline

across their land. Many were happy to accept payments, but landowners who declined the offers found their land rights taken away by **eminent domain**—the policy by which courts set aside private property rights to make way for projects judged to be for the public good. The landowner is paid an amount determined by a court to be fair, and the court's decision cannot be appealed.

In Alaska, the oil industry maintains public support for drilling by paying the Alaskan government a portion of its revenues. Since the 1970s, the state of Alaska has received more than $70 billion in oil revenues. One-fourth of these revenues are placed in the Permanent Fund, an investment fund that pays yearly dividends to all citizens. Since 1982, each Alaska resident has received annual payouts ranging from $331 to $2072. However, in recent years, a problem has come up: Alaskan officials had become so reliant on oil-industry money that they levied no personal income taxes or sales taxes and ran the state budget into a hole. Today, Alaskans are facing severe cutbacks in public services as a result.

Distribution of resource-extraction revenue among citizens, as occurs in Alaska, is unusual. In most parts of the world where fossil fuels are extracted, local residents suffer pollution without compensation. When multinational corporations pay developing nations for access to extract oil or gas, the money generally does not trickle down from government officials to the people who live where the extraction takes place. Moreover, oil-rich developing nations such as Ecuador, Venezuela, and Nigeria tend to have few environmental regulations, and governments may not enforce regulations if doing so would jeopardize the large sums of money associated with oil development.

In Ecuador, local people brought suit against Chevron for environmental and health impacts from years of oil extraction in the nation's rainforests. An Ecuadorian court in 2011 found the oil company guilty and ordered it to pay $9.5 billion for cleanup. Chevron refused, and the court battle proceeded to the United States, where a judge threw out the ruling. In 2018, an international arbitration court in The Hague ruled in favor of Chevron.

In Nigeria, the Shell Oil Company extracted $30 billion of oil from land of the native Ogoni people. Oil spills, noise, and gas flares caused chronic illness among the Ogoni, but oil profits went to Shell and to Nigeria's military dictatorships, while the Ogoni remained in poverty with no running water or electricity. Ogoni activist Ken Saro-Wiwa worked for fair compensation to the Ogoni. After 30 years of persecution by Nigeria's government, he was arrested in 1994, given a trial universally regarded as a sham, and put to death by military tribunal.

Wherever in the world fossil fuel extraction comes to communities, people find themselves divided over whether the short-term economic benefits are worth the long-term health and environmental impacts. Today, this debate is occurring from Pennsylvania and New York to North Dakota and Colorado, everywhere the petroleum industry is fracking for oil and gas. The debate has gone on for years in Appalachia over mountaintop removal mining. And along the Gulf of Mexico, the oil and gas industry employs 100,000 people and helps fund local economies—yet still more people are employed in tourism, fishing, and service industries, all of which were hurt by the *Deepwater Horizon* spill. There are no easy answers, but impacts could be reduced if fossil fuel industries were to put more health and environmental safeguards in place for workers and residents.

Dependence on foreign energy affects economic security

Putting all your eggs in one basket is always a risky strategy. Because virtually all our modern technologies and services depend in some way on fossil fuels, we are susceptible to supplies becoming costly or unavailable. Nations with few fossil fuel reserves of their own are especially vulnerable. For instance, nations such as Germany, France, South Korea, Japan, and China consume far more energy than they produce and thus rely heavily on imports. **FIGURE 19.26** contrasts a selection of high-producing and high-consuming nations for oil.

When a nation relies on foreign supplies of oil, nations selling oil can control energy prices, forcing nations buying oil to pay more as supplies dwindle. This dynamic became

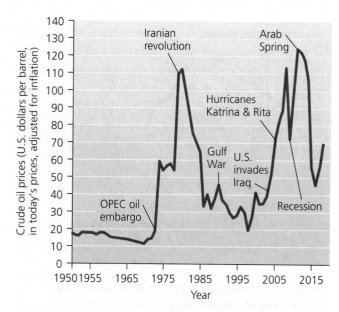

FIGURE 19.27 World oil prices have gyrated over the decades. Often, this fluctuation has resulted from political and economic events in oil-producing countries, particularly in the Middle East. Prices shown are adjusted for inflation and expressed in 2018 dollars. *Data from BP p.l.c., 2019.* Statistical review of world energy 2019.

clear in 1973, when the Organization of Petroleum Exporting Countries (OPEC) resolved to stop selling oil to the United States, which had begun relying on foreign imports in the wake of its 1970 oil production peak. The predominantly Arab nations of OPEC opposed U.S. support for Israel in the Arab–Israeli Yom Kippur War and sought to raise prices by restricting supply. OPEC's embargo created panic in the West and caused oil prices to skyrocket (**FIGURE 19.27**), spurring inflation. Fear of oil shortages drove Americans to wait in long lines at gas pumps. A similar supply shock occurred in 1979 in response to the Iranian revolution.

With the majority of world oil reserves located in the politically volatile Middle East, crises in this region have dramatically affected oil supplies and prices time and again. For U.S. leaders, this quandary has enhanced the allure of using fracking to boost domestic oil and gas production. By supplying more of its own energy, the United States gains leverage internationally and becomes less susceptible to foreign entanglements. U.S. proponents of the Keystone XL pipeline used this reasoning to argue for importing petroleum from the oil sands of Canada, a stable, friendly, democratic neighboring country. However, as President Obama and others pointed out, this oil would lessen U.S. reliance on Middle Eastern nations only if it were consumed at home, whereas the oil industry's plan had always been to export it.

In recent years, the United States has diversified its sources of imported petroleum considerably and now receives most of it from non–Middle Eastern nations, including Canada, Mexico, and Colombia. Within a decade, the United

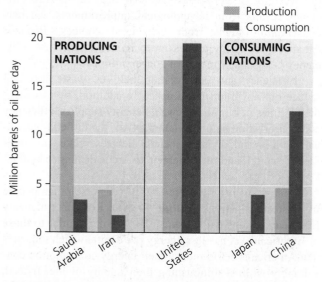

FIGURE 19.26 Some nations produce and export oil, whereas others rely on imports. Saudi Arabia and Iran produce far more oil than they consume and are able to export oil to high-consumption countries such as Japan and China. The United States relied heavily on foreign imports for many years but recently has increased its production to cover most of its domestic demand. *Data from U.S. Energy Information Administration.*

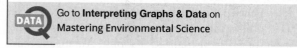

Go to **Interpreting Graphs & Data** on **Mastering Environmental Science**

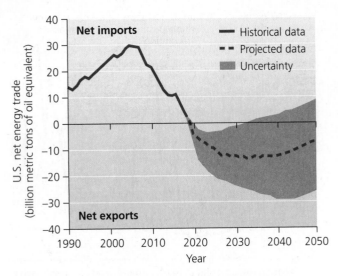

FIGURE 19.28 The United States has imported most of its energy (coal, oil, and natural gas combined) from other nations for decades, but it is expected very soon to become a net exporter of energy. *Data from U.S. Energy Information Administration, Annual Energy Outlook 2019.*

States is expected to become a net exporter of petroleum (meaning that the country will export more than it imports). It is expected to become a net exporter of energy overall even sooner (**FIGURE 19.28**).

Diversifying sources of foreign energy was one way in which the United States responded to the supply shocks of the 1970s. U.S. leaders also enacted conservation measures, funded research into renewable energy, and established an emergency stockpile (which today stores one month of oil), called the Strategic Petroleum Reserve, deep underground in salt caverns in Louisiana. They also called for secondary extraction at old wells and encouraged the development of additional domestic sources.

Since then, the desire to reduce reliance on foreign oil by boosting domestic supply has driven the expansion of offshore drilling into deeper and deeper water. Today, it is driving the push to drill for oil in Arctic waters, despite the risks. It is also driving the plan to open the Arctic National Wildlife Refuge in Alaska to oil extraction, despite arguments that drilling there would spoil America's last true wilderness while adding little to the nation's oil supply. As domestic oil and gas production rises with enhanced drilling, the United States becomes freer to make geopolitical decisions without being hamstrung by dependence on foreign energy. At the same time, however, climate change is

WEIGHING the **issues**

Drill, Baby, Drill?

Do you think the United States should encourage hydraulic fracturing by subsidizing it and loosening regulations? Do you think the United States should open more of its offshore waters to oil and gas extraction? In each case, what benefits and costs do you foresee? Would the benefits likely exceed the costs? How strictly should government regulate oil and gas extraction once drilling begins? Give reasons for your answers.

posing new national security concerns (p. 506). As we extend our reach for fossil fuels, our society will continue to debate the complex mix of social, political, economic, and environmental costs and benefits.

Energy Efficiency and Conservation

Fossil fuels are in limited supply, and their use has health, environmental, political, and socioeconomic consequences. For these reasons, many people have concluded that fossil fuels are not a sustainable solution to our energy needs. They see a need to shift to clean and renewable sources of energy that exert less impact on climate and human health. As our society transitions to renewable energy, it will benefit us to extend the availability and affordability of fossil fuels. We can do so by conserving energy and by improving energy efficiency.

Efficiency and conservation bring benefits

Energy efficiency describes the ability to obtain a given amount of output while using a lesser amount of energy input. **Energy conservation** describes the practice of reducing wasteful or unnecessary energy use. In general, efficiency results from technological improvements, whereas conservation stems from behavioral choices. Because enhanced efficiency allows us to reduce energy use, efficiency is a primary means of conservation.

Efficiency and conservation help us waste less and reduce our environmental impact. In addition, by extending the lifetimes of our nonrenewable energy supplies, efficiency and conservation help alleviate many of the difficult individual choices and divisive societal debates related to fossil fuels, from oil sands development to Arctic drilling to hydraulic fracturing.

Americans use far more energy per person than people in most other nations (**FIGURE 19.29a**). Residents of many European nations enjoy standards of living similar to those of Americans yet use less energy per capita, which indicates that Americans could reduce their energy consumption considerably without diminishing their quality of life. Indeed, per-person energy consumption *has* declined slightly in the United States over the last three decades (see Figure 19.29a)—during a period of sustained economic growth.

The United States also cut in half its **energy intensity,** or energy use per dollar of Gross Domestic Product (GDP) (**FIGURE 19.29b**), over the last three decades. Lower energy intensity indicates greater efficiency, and these data show that the United States now gets twice as much economic bang for its energy buck. Thus, although the United States continues to burn through more energy per dollar of GDP than most other industrialized nations, Americans have achieved tremendous gains in efficiency already and should be able to make further progress.

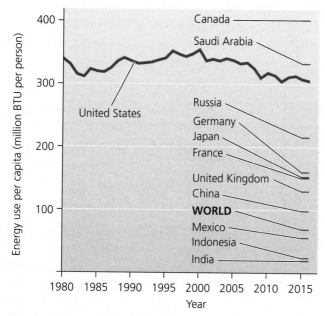

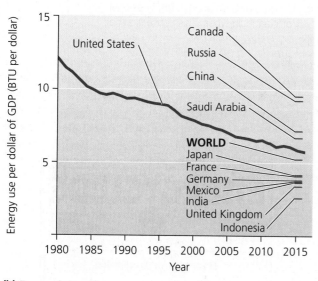

(a) Per capita energy consumption

(b) Energy intensity

FIGURE 19.29 The United States trails other developed nations in energy efficiency but has made much progress. U.S. per-person energy use **(a)** has fallen slightly since 1980 but remains higher than that of most other nations. U.S. energy intensity **(b)** has fallen steeply and now approaches that of other developed nations. Energy intensity is calculated here as energy use per inflation-adjusted 2010 dollars of GDP, using purchasing power parities, which control for differences among nations in purchasing power. *Data from U.S. Energy Information Administration.*

Personal actions and efficient technologies are two routes to conservation

As individuals, we can make conscious choices to reduce our energy consumption by driving less, dialing down thermostats, turning off lights when rooms are not in use, and investing in energy-efficient vehicles, devices, and appliances. For any given individual or business, reducing energy consumption saves money while helping conserve resources.

As a society, we can conserve energy by developing technologies and strategies to make energy-consuming devices and processes more efficient. Currently, more than two-thirds of the fossil fuel energy we use is simply lost, as waste heat, in automobiles and power plants (see Figure 19.4).

One way we can improve the efficiency of power plants is through **cogeneration,** in which excess heat produced during electricity generation is captured and used to heat nearby workplaces and homes and to produce other kinds of power. Cogeneration can almost double the efficiency of a power plant. The same is true of coal gasification (p. 548) and of combined cycle generation, in which coal is treated to create hot gases that turn a gas turbine, while exhaust from this turbine heats water to drive a steam turbine.

In homes, offices, and public buildings, a significant amount of heat is needlessly lost in winter and gained in summer because of poor design and inadequate insulation (**FIGURE 19.30**). Improvements in design can reduce the energy required to heat and cool buildings. Such improvements may involve passive solar design (p. 597), better insulation, a building's location, the vegetation around it, and even the color of its roof (lighter colors keep buildings cooler by reflecting the sun's rays).

In most homes and businesses, a small but significant portion of electricity is squandered to "vampire" power loss, energy that electronic devices in standby mode consume while switched off. However, recent government regulation has limited how much standby power manufacturers can design devices to consume, and this is reducing vampire power loss.

FIGURE 19.30 A thermogram reveals heat loss from buildings. It records energy in the infrared portion of the electromagnetic spectrum (p. 33). In this image, one house is uninsulated; its red color signifies warm temperatures where heat is escaping. Green shades signify cool temperatures, where heat is being conserved. Also note that in all houses, more heat is escaping from windows than from walls.

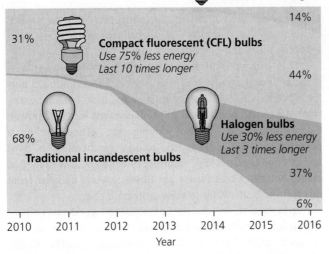

LED bulbs
Use 85% less energy
Last 25 times longer

14%

31%

Compact fluorescent (CFL) bulbs
Use 75% less energy
Last 10 times longer

44%

68%

Traditional incandescent bulbs

Halogen bulbs
Use 30% less energy
Last 3 times longer

37%

6%

2010 2011 2012 2013 2014 2015 2016
Year

(a) Conversion to more efficient bulbs in the United States

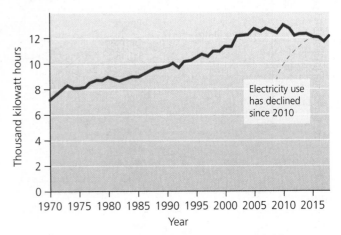

Electricity use
has declined
since 2010

Thousand kilowatt hours

1970 1975 1980 1985 1990 1995 2000 2005 2010 2015
Year

(b) U.S. residential electricity demand per household

FIGURE 19.31 Lighting has become far more energy-efficient. In the United States since 2010 **(a),** traditional incandescent lightbulbs have been replaced by a succession of increasingly efficient bulb types. This conversion has contributed to a drop in U.S. residential energy use **(b)** since 2010. *Data from U.S. Energy Information Administration, and adapted from Popovich, N., 2019. America's light bulb revolution.* New York Times, 8 March 2019.

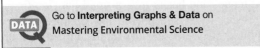 Go to **Interpreting Graphs & Data** on **Mastering Environmental Science**

Many appliances and consumer products have been reengineered through the years to enhance efficiency. A striking example is the rapid transition to energy-efficient lighting that is currently underway. As new, more efficient, technologies were developed that improved upon the traditional pear-shaped incandescent lightbulb, government mandates for greater efficiency drove a phaseout of incandescent bulbs and their replacement by successive waves of more-efficient lights (**FIGURE 19.31a**). Compact fluorescent bulbs initially were the sole substitute, but many consumers came to prefer halogen bulbs for their warmer light quality. Most recently, light-emitting diodes (LEDs) have become affordable and are being adopted quickly. Partly as a result of these trends, residential electricity use in U.S. households has been declining in recent years, following decades of steep increases (**FIGURE 19.31b**).

Federal standards for energy-efficient appliances have already reduced per-person home electricity use below what it was in the 1970s. "EnergyGuide" labels (**FIGURE 19.32**) have helped. These yellow-and-black product labels show data on test results for energy efficiency and are mandated under a federal program requiring manufacturers of certain types of appliances to post these data for consumers. The U.S. EPA's Energy Star program has also helped. This program certifies appliances, electronics, doors and windows, and other products that surpass efficiency standards (see **SUCCESS STORY**). These two labeling programs enable consumers to take energy use into account when shopping,

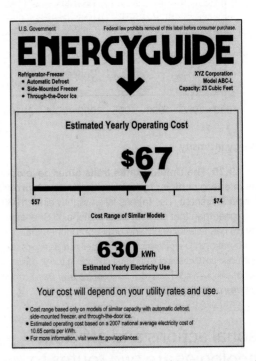

FIGURE 19.32 EnergyGuide labels provide consumers with information on energy efficiency. These yellow-and-black labels on ovens, water heaters, and other appliances allow us to compare the performance of one brand or model versus others. In the example shown here, a refrigerator uses an estimated 630 kilowatt-hours of electricity per year. It has an estimated yearly operating cost of $67, which is intermediate compared with similar brands and models.

SUCCESS story

Improving Energy Efficiency

When shopping for electronics, appliances, or home and office equipment, most of us would like to choose energy-efficient models. But how can we know how efficient a given TV, microwave oven, computer printer, or refrigerator actually is? Luckily, U.S. consumers can look for the Energy Star label. This familiar blue-and-white label tells us that a product has been independently certified as performing above federal energy-efficiency standards. By helping consumers identify and choose energy-efficient brands and models, the Energy Star program is reducing electricity demand and saving consumers money. For each extra dollar we spend on an Energy Star product, we save an average of $4.50 on energy costs while preventing more than 35 pounds of greenhouse gas emissions, according to the Environmental Protection Agency, which runs the program. From 1992 through 2017, the Energy Star program helped Americans save $480 billion in utility bills. It achieved this while saving 4 trillion kilowatt-hours of electricity and reducing greenhouse gas emissions by more than 3 billion tons (an amount equivalent to the annual emissions from more than 600 million cars). Today Energy Star has expanded to include certification programs for entire homes, commercial buildings, and industrial plants. Armed with information from this simple label, each and every one of us is empowered to make our own decisions about conserving energy through our purchasing behavior.

The familiar label signifying that an item has surpassed federal energy-efficiency standards

→ **Explore the Data** at Mastering Environmental Science

and each program has been credited with reducing energy consumption and carbon emissions by enormous amounts. These programs also save U.S. consumers many billions of dollars; studies show in case after case that savings on utility bills more than offset the slightly higher prices of energy-efficient products.

Automobile fuel efficiency is a key to conservation

Automotive technology represents perhaps our best opportunity to conserve large amounts of fossil fuels. Automakers can enhance fuel efficiency for conventional gasoline-powered vehicles by using lightweight materials, continuously variable transmissions, and more efficient engines. Additionally, an increasing array of fuel-efficient alternative vehicles (p. 509) is available: electric cars, hybrid cars (cars that run on gasoline plus electricity generated while driving), plug-in hybrids (hybrid cars that can be charged by a power cord while not in use), and vehicles that use hydrogen fuel cells (p. 613). Among electric/gasoline hybrids, current models obtain fuel-economy ratings of up to about 50 miles per gallon (mpg)—twice that of the average American car. Many fully electric vehicles now obtain fuel-economy ratings of more than 100 mpg.

One of the ways in which U.S. leaders responded to the OPEC embargo of 1973 was to mandate an increase in the fuel efficiency of automobiles. Automakers responded by boosting fuel efficiency more than 60% between 1975 and 1982 (**FIGURE 19.33**). Over the following three decades, however, as market prices for oil fell, many of the conservation and efficiency initiatives of this time were abandoned. Without high market prices and a threat of shortages, people lost the economic motivation to conserve, and U.S. policymakers repeatedly failed to raise the corporate average fuel efficiency (CAFE) standards, which set benchmarks for auto manufacturers to meet. The average fuel efficiency of new vehicles fell from 22.0 mpg in 1987 to 19.3 mpg in 2004 as sales of sport-utility vehicles increased relative to sales of passenger cars.

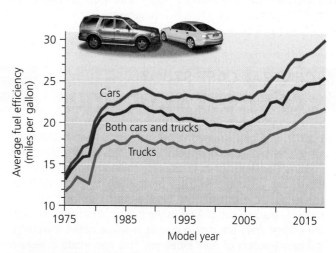

FIGURE 19.33 Automotive fuel efficiencies are influenced by public policy. Fuel efficiency for automobiles in the United States rose dramatically in the late 1970s as a result of legislative mandates but then stagnated without further laws to improve fuel economy. Legislation in 2007 once again improved fuel efficiency. *Data from U.S. Environmental Protection Agency.*

In 2007, Congress mandated that automakers raise average fuel efficiency to 35 mpg by the year 2020. Then, when automakers requested a government bailout during the 2008–2009 recession, President Obama persuaded them to agree to boost average fuel economies to 54.5 mpg by 2025. The required technologies would add more than $2000 to the average price of a car, but drivers would save perhaps $6000 in fuel costs over the car's lifetime. These policies resulted in a substantial increase in average fuel efficiencies (see Figure 19.33). However, the Trump administration has since moved to revoke these policies, so the 54.5-mpg goal will likely not come to pass.

WEIGHING the **issues**

More Miles, Less Gas

If you drive an automobile, what gas mileage does it get? How does it compare to the vehicle averages in Figure 19.33? If your vehicle's fuel efficiency were 10 mpg greater and if you drove the same amount, how many gallons of gasoline would you no longer need to purchase each year? How much money would you save?

Do you think U.S. leaders should mandate further increases in the CAFE standards? Should the government raise taxes on gasoline sales as an incentive for consumers to conserve energy? What effects on economics, on health, and on environmental quality might each of these steps have?

U.S. policymakers could do more to encourage oil conservation. The United States has kept taxes on gasoline extremely low, relative to most other nations. Americans pay two to three times *less* per gallon of gas than drivers in many European countries. (In fact, gasoline in the United States is sold more cheaply than bottled water!) As a result, U.S. gasoline prices do not account for the substantial external costs (pp. 143, 165) that oil production and consumption impose on society. Some experts estimate that if all costs to society were taken into account, the price of gasoline would exceed $13 per gallon. Instead, our artificially low gas prices diminish our economic incentive to conserve.

The rebound effect cuts into efficiency gains

Energy efficiency is a vital pursuit, but it may not always save as much energy as we expect because gains in efficiency from better technology may be partly offset if people engage in more energy-consuming behavior as a result. For instance, a person who buys a fuel-efficient car may choose to drive more because he or she feels it's okay to do so (or because he or she can afford to) now that less gas is being used per mile. This phenomenon is called the **rebound effect,** and studies indicate that it is widespread and significant. In some instances, the rebound effect may completely erase efficiency gains. As our society pursues energy efficiency, the rebound effect will be an important factor to consider.

We need both conservation and renewable energy

Despite concerns over the rebound effect, energy efficiency and conservation efforts are vital to creating a sustainable future for our society. It is often said that reducing our energy use is equivalent to finding a new oil reserve. Some estimates hold that energy conservation and efficiency in the United States could save 6 million barrels of oil a day—nearly the amount gained from all offshore drilling and considerably more than would be gained from Canada's oil sands. In fact, conserving energy is *better* than finding a new reserve because it alleviates health and environmental impacts while at the same time extending our future access to fossil fuels.

Regardless of how effectively we conserve, we will still need energy to power our civilization, and the energy will need to come from somewhere. Most energy experts maintain that the only sustainable way of guaranteeing ourselves a reliable, long-term supply of energy is to ensure sufficiently rapid development of renewable energy sources (Chapters 20 and 21).

CENTRAL CASE STUDY
connect & continue

TODAY, the long-simmering battles over oil sands development and pipeline construction have moved into the court system. Coalitions of local property owners, Native American tribes, and climate advocates are continuing to mount legal challenges to each project that seeks to expand pipeline capacity leading out of Alberta. As these cases move through the courts, they are delaying, and in some cases thwarting, the construction of new pipelines. This has made investors reluctant to invest in the expansion of oil sands ventures—and the resulting drop in investment has taken a toll on the industry's production and profits.

As of 2019, two of the five pipeline projects intended to increase the transport of bitumen out of Alberta have been canceled, while the other three are in limbo. The project

widely expected to go most smoothly, the Enbridge Line 3 pipeline, faces trouble in Minnesota, where a court blocked approval of the pipeline expansion because environmental impact studies failed to address the risk of oil spills in Lake Superior's watershed. In Michigan, the state's attorney general threatened to shut down Enbridge's existing pipeline if the company does not quickly make upgrades for safety and spill prevention to the 66-year-old structure, which crosses the Straits of Mackinac between Lake Huron and Lake Michigan. And in British Columbia, the Trans Mountain pipeline expansion supported by the Canadian government faces widespread opposition from residents concerned about oil spills and impacts on marine life.

Amid these setbacks to the oil sands industry, Alberta's voters in 2019 elected as their province's premier a staunch supporter of the industry, Jason Kenney, who aims to do all he can to boost oil sands production. Kenney vowed to eliminate the carbon tax and the cap on greenhouse gas emissions that Alberta's previous premier had agreed to accept as part of a new national policy pushed by Prime Minister Justin Trudeau. Trudeau's emissions reduction policies are also being fought in court by the governments of Ontario and Saskatchewan, and if these provinces win, Alberta will have free rein to promote fossil fuel development without constraints from the national government.

As for the Keystone XL project, a decade-long legal battle in Nebraska was resolved in 2019 when the Nebraska Supreme Court ruled that the pipeline could proceed through the state. However, pipeline opponents brought suit in Montana, arguing that legally required studies of environmental impacts are being skipped or ignored. Private land will need to be taken from landholders by eminent domain, and TC Energy faces the prospects of protests and civil disobedience if and when construction commences.

The divergent views that people hold on Canada's oil sands and the pipelines that deliver their oil reflect our society's confounding relationship with fossil fuels. Coal, oil, and natural gas helped us build the complex industrialized societies we enjoy, yet now amid Earth's changing climate they threaten ever-higher costs to our health, environment, and economy. In all likelihood, the long-running debate over oil sands extraction and transport will have taken a few more twists and turns by the time you read this book. We encourage you to research the latest developments online, discuss them with your instructor, and wrestle thoughtfully with the issues.

- **CASE STUDY SOLUTIONS** Summarize the main arguments for and against the Keystone XL pipeline extension. What problems might the pipeline help solve? What problems might it create? Do you personally think Canada should continue to develop Alberta's oil sands? Should the United States have approved construction of the Keystone XL pipeline extension? Give reasons for your answers.

- **LOCAL CONNECTIONS** What are students, faculty, or administrators on your campus doing to encourage energy efficiency? What is being done on your campus to conserve energy or to reduce fossil fuel use? Is there a call for your college or university to divest its investment holdings from fossil fuel corporations? What is being done to encourage use of renewable energy? What additional steps, if any, regarding fossil fuel use and policies do you think your institution should take, and why?

- **EXPLORE THE DATA** How do researchers link data on fossil fuel use with impacts of global warming and climate change? → **Explore Data** relating to the case study on **Mastering Environmental Science**.

REVIEWING Objectives

You should now be able to:

+ **Identify the energy sources we use**

Many energy sources are available, but since the industrial revolution, nonrenewable fossil fuels—coal, oil, and natural gas—have become our primary sources of energy. (pp. 525–527)

+ **Discuss the value of the EROI concept**

The concepts of net energy and EROI allow us to compare the amount of energy obtained from a source with the amount invested in its extraction and processing. This information can help us decide which energy sources are most efficient and productive. (pp. 527–528)

+ **Describe the formation of major types of fossil fuels**

Fossil fuels are formed over millions of years as buried organic matter is chemically transformed by heat, pressure, or anaerobic decomposition. Coal results from organic matter that undergoes compression but little decomposition. Crude oil is a thick, liquid mixture of hydrocarbons formed under high temperature and pressure. Natural gas consists mostly of methane and can form directly or from coal or oil altered by heating. Oil sands contain bitumen, a tarry substance formed from oil degraded by bacteria, whereas shale oil and methane hydrate are other unconventional sources with future potential. (pp. 529–531)

+ **Explain how we extract, process, and use fossil fuels**

Scientists locate fossil fuel deposits by analyzing subterranean geology. Coal is mined underground and strip-mined from the surface, whereas we drill wells to pump out oil and gas. Oil sands may be strip-mined or dissolved underground and extracted through well shafts. Refineries separate components of crude oil to produce a wide variety of fuels. Coal and natural gas are used principally to generate electricity. Oil powers transportation and is used to create a diversity of petroleum-based products. (pp. 531–536)

+ **Assess concerns over the future decline of conventional oil supplies**

Any nonrenewable resource can be depleted, and we have depleted nearly half the world's conventional oil. Once we pass the peak of oil production, a gap between rising demand and falling supply may pose immense challenges for society. (pp. 536–538)

+ **Outline ways in which we are extending our reach for fossil fuels**

The conventional fossil fuels that remain unexploited are harder to access and more expensive to extract, so we are applying technology and new approaches to extract them, including mountaintop removal mining, secondary extraction, directional drilling, hydraulic fracturing, deep offshore drilling, and Arctic exploration. We are also mining unconventional fossil fuels such as oil sands. (pp. 538–541)

+ **Examine environmental impacts of fossil fuel use, and explore ways to minimize these impacts**

In various ways, coal mining, oil and gas drilling, and oil sands extraction can alter ecosystems, pollute air and waterways, and pose health risks. Hydraulic fracturing magnifies some of these impacts. Fuels can also spill during transport, affecting soil, water, wildlife, and health. Emissions from fossil fuel combustion pollute air, pose health risks, and drive climate change. Public policy and advances in pollution control technology have reduced emissions, but much remains to be done. If we could safely capture carbon dioxide and sequester it underground at a large-enough scale, a primary drawback of fossil fuels would be resolved, but carbon capture and storage remains unproven so far. (pp. 541–550)

+ **Assess political, social, and economic aspects of fossil fuel use**

Fossil fuels impose a variety of external costs. Fossil fuel extraction creates jobs but leaves pollution. People living in areas of extraction experience a range of consequences. Today's societies are heavily reliant on fossil fuel energy, and nations that consume more fossil fuels than they produce are vulnerable to supply restrictions. (pp. 549–552)

+ **Discuss strategies for conserving energy and enhancing efficiency**

Energy conservation involves both personal choices and efficient technologies. Efficiency in power plant combustion, lighting, consumer appliances, and automotive fuel efficiency all play roles in conserving energy. The rebound effect, however, can partly negate our conservation efforts. Conservation lengthens our access to fossil fuels and reduces environmental impact, but to build a sustainable society, we will also need to shift to renewable energy sources. (pp. 552–556)

SEEKING Solutions

1. Describe three specific health or environmental impacts resulting from fossil fuel production or consumption. For each impact, what steps could governments, industries, or individuals take to alleviate the impact? What might prevent them from taking such steps? What could encourage them to take such steps?

2. Contrast the experiences of the Ogoni people of Nigeria with those of the citizens of Alaska. How have they been similar, and how have they been different? Do you think businesses or governments should take steps to ensure that local people benefit from oil drilling operations? How could they do so?

3. Based on data in this chapter, give two reasons why it is reasonable to expect that Americans should be able to use energy more efficiently in the future as the U.S. economy expands.

4. **THINK IT THROUGH** You are the mayor of a rural Nebraska town along the route of the proposed Keystone XL pipeline. Some of your town's residents are eager to have jobs they believe the pipeline will bring. Others are fearful that oil leaks from the pipeline could contaminate the water supply. Some of your town's landowners are looking forward to receiving payments from TC Energy for use of their land, whereas others dread the prospect of noise, pollution, and trees being cut on their property. If the company receives too much local opposition, it says it may move the pipeline route away from your town. What information would you seek from TC Energy, from your state regulators, and from scientists and engineers before deciding whether support for the pipeline is in the best interest of your town? How would you make your decision? How might you try to address the diverse preferences of your town's residents?

5. **THINK IT THROUGH** You are elected governor of the state of Florida as the federal government is debating opening new waters to offshore drilling for oil and natural gas. Drilling in Florida waters would create jobs for Florida citizens and revenue for the state in the form of royalty payments from oil and gas companies. However, there is always the risk of a catastrophic oil spill, with its ecological, social, and economic impacts. Would you support or oppose offshore drilling off the Florida coast? Why? What, if any, regulations would you insist be imposed on such development? What questions would you ask of scientists before making your decision? What factors would you consider in making your decision?

CALCULATING Ecological Footprints

Scientists at the Global Footprint Network calculate the energy component of our ecological footprint by estimating the amount of ecologically productive land and sea required to absorb the carbon released from fossil fuel combustion. This amount translates into 5.7 ha (14.1 acres) of the average American's 8.1-ha (20.0-acre) ecological footprint. Another way to think about our footprint, however, is to estimate how much land would be needed to grow biomass with an energy content equal to that of the fossil fuel we burn.

Assume that you are an average American who burns about 6.0 metric tons of oil-equivalent in fossil fuels each year and that average terrestrial net primary productivity (p. 111) can be expressed as 0.0037 metric ton/ha/year. Calculate how many hectares of land it would take to supply our fuel use by present-day photosynthetic production.

	HECTARES OF LAND FOR FUEL PRODUCTION
You	1649
Your class	
Your state	
United States	

1. Compare the energy component of your ecological footprint calculated in this way with the 5.7 ha calculated using the method of the Global Footprint Network. Explain why results from the two methods may differ.

2. Earth's total land area is approximately 15 billion hectares. Compare this area to the hectares of land for fuel production from the information in the table you generated.

3. In the absence of stored energy from fossil fuels, how large a human population could Earth support at the level of consumption of the average American, if all of Earth's area were devoted to fuel production? Do you consider this scenario to be realistic? Provide two reasons why or why not.

Mastering Environmental Science

Students Go to **Mastering Environmental Science** for assignments, an interactive e-text, and the Study Area with practice tests, videos, and activities.

Instructors Go to **Mastering Environmental Science** for automatically graded activities, videos, and reading questions that you can assign to your students, plus Instructor Resources.

CHAPTER

20

Conventional Energy Alternatives

SWEDEN

Norway   Finland

EUROPE

CENTRAL case study

Will Sweden Free Itself of Fossil Fuels?

"
Sweden will become one of the first fossil [fuel]-free . . . states in the world.
Swedish Prime Minister Stefan Löfven, 2015

If [Sweden] phases out nuclear power, then it will be virtually impossible for the country to keep its climate-change commitments.
Yale University economist William Nordhaus, 1997
"

Flashback to 1986: On the morning of April 28, alarms went off at a nuclear power plant in Sweden, as sensors detected unusually high levels of radiation. However, the radioactivity was not coming from within the plant. Instead, it had been blown through the atmosphere from the Soviet Union. The workers at this Swedish power plant had discovered the outside world's first evidence of the disaster at Chernobyl, more than 1200 km (750 mi) away in what is now the nation of Ukraine. Chernobyl's nuclear reactor had exploded two days earlier, the result of human error and unsafe engineering, but the Soviet government had not yet admitted it.

Fast-forward to 2011: On March 11, seismometers in Sweden and throughout the world recorded vibrations from the shaking of the massive Tohoku Earthquake off the coast of Japan. Soon thereafter, a tsunami inundated the Japanese coast, destroying entire towns, killing more than 18,000 people, and disabling nuclear reactors at the Fukushima Daiichi power plant. For the next several weeks, the world stood on edge as Japanese authorities frantically tried to control the leakage of radiation from the plant and avert further catastrophe.

The events at Chernobyl and Fukushima had broad and long-lasting repercussions. Each event altered the course of the world's use of nuclear energy, one of our main alternatives to fossil fuels. While people in Ukraine and Japan still struggle with the legacies of these events, the global debate over nuclear power affects us all.

In Sweden in the days after Chernobyl, low levels of radioactive fallout rained down on the countryside, contaminating crops and cows' milk. For many Swedes, the Chernobyl catastrophe confirmed the decision they had made collectively six years earlier, in a 1980 referendum, to phase out their country's nuclear power program, shutting down all nuclear plants by 2010.

But trying to phase out nuclear power proved difficult. Sweden had turned to nuclear power as it sought to reduce its use of coal, oil, and natural gas because of the health and environmental impacts of fossil fuels. As a result, Sweden became one of the few nations in the world to decrease its reliance on fossil fuels. The nation has cut its fossil fuel use by more than half since the 1970s while boasting a thriving economy that continues to provide its people with one of the highest living

Upon completing this chapter, you will be able to:

+ Discuss reasons for seeking alternatives to fossil fuels

+ Summarize the contributions to world energy supplies of conventional alternatives to fossil fuels

+ Describe nuclear energy, and explain how we harness it for electrical power

+ Assess the benefits and drawbacks of nuclear power, and discuss the societal debate over this energy source

+ Describe established and emerging sources and techniques involved in harnessing bioenergy, and assess bioenergy's benefits and shortcomings

+ Outline the scale, methods, and impacts of hydroelectric power

◀ **A technician inside a nuclear power plant near Stockholm, Sweden**

▲ **Using biogas to fuel vehicles in Kristianstad, Sweden**

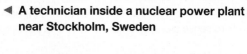

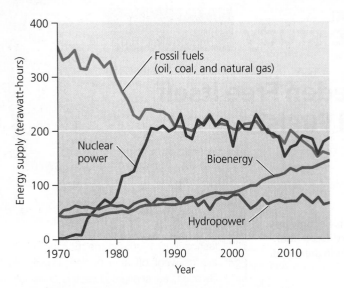

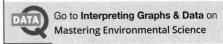

FIGURE 20.1 Sweden has reduced its use of fossil fuels (mostly oil) by more than half since 1970, largely by replacing fossil fuels with nuclear power. *Data from Swedish Energy Agency.*

DATA Go to **Interpreting Graphs & Data** on **Mastering Environmental Science**

standards on Earth (**FIGURE 20.1**). Sweden did so largely by replacing fossil-fuel-fired electricity with nuclear power—and as 2010 drew nearer, Sweden was still relying on 10 nuclear reactors for more than 30% of its energy supply and more than 40% of its electricity. If nuclear plants were to be shut down as planned, without reverting to fossil fuels, the nation would need to boost its investment in alternative energy sources dramatically.

As a result, Sweden chose to throw its weight behind research and development of renewable energy sources. Hydroelectric power was already supplying a great deal of the nation's electricity and could not be expanded much more. Instead, the government hoped that energy from biomass sources and wind power could fill the gap. So, Sweden manipulated market forces to make fossil fuels more expensive and renewable energy more affordable. It boosted bioenergy and wind by applying a carbon tax (p. 513) to fossil fuels and by subsidizing renewable energy through a certificate program, in which producers and users of fossil fuel electricity were required to buy certificates from producers and users of renewable electricity.

Sweden soon built itself into an international leader in renewable energy and today gets more than half its energy from renewable sources (**FIGURE 20.2a**). Still, renewables were taking longer to develop than hoped, so policymakers repeatedly postponed the nuclear phaseout. In 2009, just before the original self-imposed deadline, the Swedish government announced that it was reversing its policy and would not phase out nuclear power after all. Instead, new power plants would be built as existing ones needed replacing.

In announcing this decision, Sweden's leaders said it would be fiscally and socially irresponsible to dismantle the nation's nuclear program without a ready replacement. Without an abundance of clean renewable power, a nuclear phaseout would mean a return to fossil fuels or would require cutting down immense areas of forest

simply to combust biomass. Policymakers cited Sweden's international obligations to hold down its carbon emissions under the Kyoto Protocol (p. 515). They reminded citizens that nuclear power is free of atmospheric pollution and is an effective way to minimize the greenhouse gas emissions that drive climate change.

Then Fukushima occurred. As the drama played out in Japan, anti-nuclear protestors all over the world staged demonstrations. Many national governments reassessed their commitments to nuclear power, ran safety checks of existing plants, and halted plans for new plants.

The Swedish government's shifting policies have reflected changing public sentiments. The majority of Swedes have always seen the clear benefits of fighting pollution and climate change by developing energy sources such as nuclear power, bioenergy, and hydropower, and they are proud of having moved so decisively away from fossil fuels. Yet each nuclear accident has eroded otherwise strong public support for nuclear power.

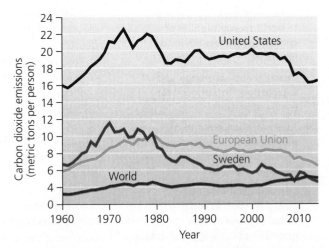

(a) Renewable energy

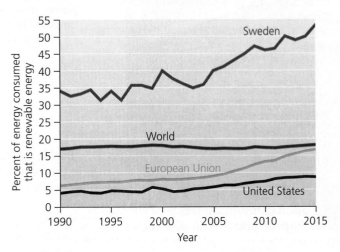

(b) Carbon dioxide emissions per person

FIGURE 20.2 Sweden is a world leader in the energy transition away from fossil fuels. Over the past few decades, Sweden's adoption of renewable energy **(a)** has greatly exceeded that of the European Union as a whole, the United States, and the world. Likewise, it has reduced its per-person carbon dioxide emissions **(b)** below the EU, U.S., and world averages. *Data from World Bank and Carbon Dioxide Information Analysis Center.*

In 2015, Sweden's new prime minister, Stefan Löfven, announced that his government would aim one day to make Sweden completely free of fossil fuels. This proclamation was accompanied by an announcement of $550 million in new public funding toward renewable energy sources such as wind power, solar power, hydroelectric power, and bioenergy.

Today, Sweden continues to excel on the world stage in renewable energy production and in reducing carbon emissions (**FIGURE 20.2b**). Time will tell the fate of nuclear power in Sweden, but Swedish citizens and leaders continue to support the development of renewable energy sources to keep their nation a world leader in the shift away from fossil fuels.

Alternatives to Fossil Fuels

Fossil fuels drove the industrial revolution and helped create the unprecedented material prosperity we enjoy today. Our global economy is still powered largely by fossil fuels; more than 80% of our energy comes from oil, coal, and natural gas (**FIGURE 20.3a**). These three fuels also generate nearly two-thirds of the world's electricity (**FIGURE 20.3b**). However, easily extractable supplies of these nonrenewable energy sources are in decline, and we are expending more and more energy and money to extract them (pp. 538–541). Moreover, the use of coal, oil, and natural gas drives global climate change and entails many other impacts to human health, society, and the environment (Chapters 17, 18, and 19).

For these reasons, most energy experts accept that we need to shift from fossil fuels to energy sources that are less easily depleted and that bring fewer health and environmental impacts. Developing alternatives to fossil fuels also brings economic, social, and national security benefits by diversifying our mix of energy options, thus lessening price volatility and dependence on foreign fuel imports.

We have developed a range of alternatives to fossil fuels (see Table 19.1, p. 525). Most of these energy sources are renewable and have less impact on health and the environment than oil, coal, or natural gas. Most alternative energy sources remain more expensive than fossil fuels, at least in the short term and when external costs of fossil fuels (pp. 143, 165, 549) are not included in market prices. Yet that is changing rapidly, and some alternative energy sources are now cheaper than fossil fuels in many areas. As technologies develop and as we invest in infrastructure to better transmit power from renewable sources, prices will continue to fall, helping us transition to these alternative energy sources.

Nuclear power, bioenergy, and hydropower are conventional alternatives

Three alternative energy sources are currently the most developed and most widely used: nuclear power, hydroelectric power, and energy from biomass. These energy sources are well established, and each of them already plays a substantial role in our energy and electricity budgets. As a result, we can therefore describe nuclear power, hydropower, and biomass energy (bioenergy) as "conventional alternatives" to fossil fuels.

Each of these three conventional energy alternatives is generally considered to exert less environmental impact than fossil fuels but more impact than "new renewable" alternatives

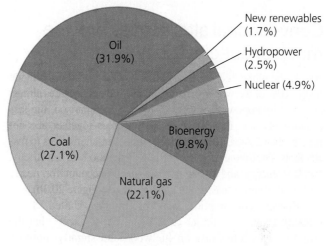

(a) World energy supply, by source

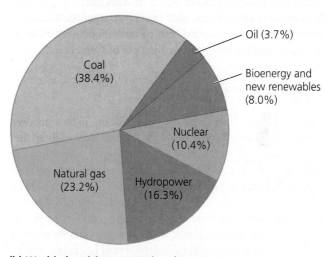

(b) World electricity generation, by source

FIGURE 20.3 Fossil fuels dominate the global energy supply. Together, oil, coal, and natural gas account for 81% **(a)** of the world's primary energy supply. Nuclear power and hydroelectric power contribute substantially to global electricity generation **(b)**, but fossil fuels still power almost two-thirds of our electricity. *Data from International Energy Agency, 2018.* Key world energy statistics 2018. *Paris: IEA.*

such as solar, wind, geothermal, and ocean power (Chapter 21). Yet as we will see, each of our conventional energy alternatives involves a unique and complex mix of benefits and drawbacks. Nuclear power is typically considered a nonrenewable energy source, and hydropower and bioenergy are generally described as renewable, but the reality is more complicated.

Sweden has shown that it is possible for a wealthy and advanced economy to replace fossil fuels gradually with these conventional alternative sources while continuing to raise living standards for its citizens. Since 1970, Sweden has slashed its fossil fuel use from 81% to 26% of its national energy budget (see Figure 20.1). Today, the three conventional alternatives—nuclear power, bioenergy, and hydropower—together provide Sweden with two-thirds of its energy and 90% of its electricity.

Conventional alternatives provide much of our electricity

Across the globe, fuelwood and other bioenergy sources provide nearly 10% of the world's energy, nuclear power provides almost 5%, and hydropower provides 2–3%. By comparison, the less established "new renewable" energy sources together account for less than 2% (see Figure 20.3a). However, alternatives to fossil fuels contribute greatly to our generation of electricity. Nuclear energy and hydropower together account for nearly 27% of the world's electricity generation (see Figure 20.3b).

Energy consumption patterns in the United States (**FIGURE 20.4a**) are similar to those globally, except that the United States relies less on fuelwood and slightly more on fossil fuels and nuclear power than most other countries. A graph showing trends in energy consumption in the United States since 1950 (**FIGURE 20.4b**) reveals two things. First, conventional alternatives play small yet significant roles in overall energy use. Second, use of conventional alternatives has been growing more slowly than use of fossil fuels.

Nuclear Power

Nuclear power occupies a unique position in our modern debate over energy. It does not pollute the air as does fossil fuel combustion, and therefore it offers us a powerful means of reducing greenhouse gas emissions and fighting climate change. Yet nuclear power's great promise has been clouded by nuclear weaponry, the thorny dilemma of radioactive waste disposal, and the long shadow of accidents at Chernobyl and Fukushima. As a result, public safety concerns and the costs of addressing them have constrained the spread of nuclear power.

Fission releases nuclear energy in reactors to generate electricity

Nuclear energy is the energy that holds together protons and neutrons (pp. 25–26) in the nucleus of an atom. We can harness this energy by converting it to thermal energy inside **nuclear reactors,** facilities contained within nuclear power plants. This thermal energy is then used to generate electricity by heating water to produce steam that turns turbines. The generation of electricity using nuclear energy in this way is what we call **nuclear power.**

The reaction that drives the release of nuclear energy inside nuclear reactors is **nuclear fission,** the splitting apart of atomic nuclei (**FIGURE 20.5**). In fission, the nuclei of large,

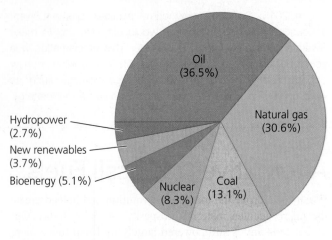

(a) U.S. energy consumption, by source

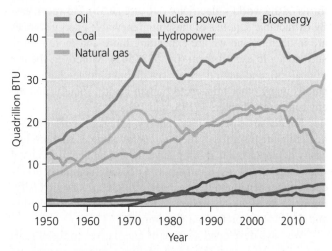

(b) U.S. energy consumption, 1950–2018

FIGURE 20.4 Fossil fuels predominate in the United States. Together, oil, natural gas, and coal account for 80% **(a)** of U.S. energy consumption. The line graph **(b)** shows amounts of each energy source used in the United States since 1950. *Data from U.S. Energy Information Administration, 2019.*

Go to **Interpreting Graphs & Data** on **Mastering Environmental Science**

heavy atoms, such as uranium or plutonium, are bombarded with neutrons. Ordinarily, neutrons move too quickly to split nuclei when they collide with them, but if neutrons are slowed down, they can break apart nuclei. In a nuclear reactor, the neutrons bombarding uranium are slowed down with a substance called a *moderator,* most often water or graphite. Each split nucleus emits energy in the form of heat, light, and radiation and also releases neutrons. These resulting neutrons (two to three per nucleus in the case of uranium-235) can in turn bombard other uranium-235 (^{235}U) atoms, resulting in a self-sustaining chain reaction.

If not controlled, this chain reaction becomes a runaway process of positive feedback (p. 108)—the process that creates the explosive power of a nuclear bomb. Inside a nuclear power plant, however, fission is controlled so that, on average, only

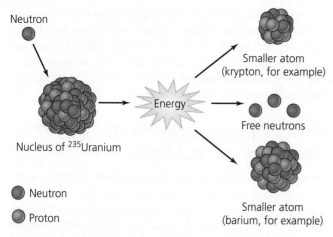

Neutron

Nucleus of ^{235}Uranium

Energy

Smaller atom
(krypton, for example)

Free neutrons

Smaller atom
(barium, for example)

● Neutron
● Proton

FIGURE 20.5 Nuclear fission drives nuclear power. In nuclear fission, the nucleus of an atom of uranium-235 is bombarded with a neutron. The collision splits the uranium atom into smaller atoms and releases two or three neutrons, along with energy in the form of heat, light, and radiation.

one of the two or three neutrons emitted with each fission event goes on to induce another fission event. To soak up the excess neutrons produced when uranium nuclei divide, *control rods* made of a metallic alloy that absorbs neutrons are placed among the water-bathed fuel rods of uranium. Engineers move these control rods into and out of the water to maintain the fission reaction at the desired rate. In this way, the chain reaction maintains a constant output of energy at a controlled rate.

All this takes place within the reactor core and is the first step in the electricity-generating process of a nuclear power plant (**FIGURE 20.6**). The reactor core is housed within a

reactor vessel; and the vessel, steam generator, and associated plumbing loops are often protected within a containment building. Containment buildings, with their meter-thick concrete and steel walls, are constructed to prevent leaks of radioactivity due to accidents or natural catastrophes such as earthquakes. Not all nations require containment buildings, which points out the key role that government regulation plays in protecting public safety.

First developed commercially in the 1950s, nuclear power experienced most of its growth during the 1970s and 1980s. The United States generates the most electricity from nuclear power of any nation—more than one-fourth of the world's production—yet less than 20% of U.S. electricity comes from nuclear power. A number of other nations rely more heavily on nuclear power (**TABLE 20.1**, p. 566). France leads the list, receiving 72% of its electricity from nuclear power, and nuclear power is growing in several Asian nations. As of 2019, 451 nuclear power plants were operating in 30 nations.

Nuclear energy comes from processed and enriched uranium

We use the element uranium for nuclear power because its atoms are radioactive. Radioactive isotopes, or radioisotopes (p. 26), emit subatomic particles and high-energy radiation as they decay into lighter radioisotopes until they ultimately become stable isotopes. The isotope uranium-235 decays into a series of daughter isotopes, eventually forming lead-207. Each radioisotope decays at a rate determined by that isotope's *half-life* (p. 27), the time it takes for half of the atoms to give off radiation and decay. The half-life of ^{235}U is about 700 million years.

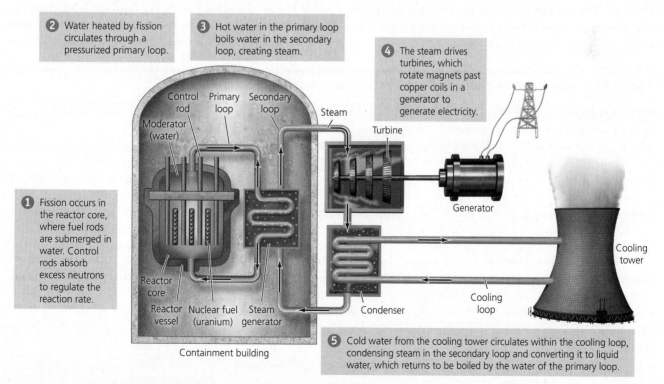

❷ Water heated by fission circulates through a pressurized primary loop.

❸ Hot water in the primary loop boils water in the secondary loop, creating steam.

❹ The steam drives turbines, which rotate magnets past copper coils in a generator to generate electricity.

❶ Fission occurs in the reactor core, where fuel rods are submerged in water. Control rods absorb excess neutrons to regulate the reaction rate.

Control rod Primary loop Secondary loop Steam

Moderator (water)

Turbine

Generator

Reactor core

Reactor vessel Nuclear fuel (uranium) Steam generator

Condenser

Cooling loop

Cooling tower

Containment building

❺ Cold water from the cooling tower circulates within the cooling loop, condensing steam in the secondary loop and converting it to liquid water, which returns to be boiled by the water of the primary loop.

FIGURE 20.6 In a pressurized light water reactor (the most common type of nuclear reactor), radioactive uranium fuel rods heat water, and steam turns turbines to generate electricity.

TABLE 20.1 Top Producers of Nuclear Power

NATION	NUCLEAR POWER CAPACITY (gigawatts)	NUMBER OF REACTORS	ELECTRICITY FROM NUCLEAR POWER (%)
United States	100.0	99	20.0
France	63.1	58	71.6
Japan	39.8	42	3.6
China	34.5	39	3.9
Russia	26.1	35	17.8
South Korea	23.1	25	27.1
Canada	13.6	19	14.6
Ukraine	13.1	15	55.1
Germany	10.8	8	11.6
Sweden	9.1	9	39.6

Data from International Atomic Energy Agency.

We obtain uranium from various minerals in naturally occurring ore (rock that contains minerals of economic interest (p. 646)) that we extract by large-scale mining. Uranium-containing minerals are uncommon, and uranium ore is in finite supply, so nuclear power is generally considered a nonrenewable energy source.

More than 99% of the uranium in nature occurs as the isotope uranium-238. Uranium-235 (with three fewer neutrons) makes up less than 1% of the total. Because ^{238}U does not emit enough neutrons to maintain a chain reaction when fissioned, we use ^{235}U for commercial nuclear power. Therefore, we must process the ore we mine to enrich the concentration of ^{235}U to at least 3%. The enriched uranium is formed into pellets of uranium dioxide (UO_2), which are incorporated into metallic tubes called *fuel rods* that are used in nuclear reactors.

After several years in a reactor, the decayed uranium fuel no longer generates adequate energy, so it must be replaced with new fuel. In some countries, the spent fuel is reprocessed to recover the remaining energy. However, reprocessing is costly relative to the market price of uranium, so most spent fuel is disposed of as radioactive waste (p. 572–573).

Nuclear power delivers energy more cleanly than fossil fuels

Using fission, nuclear power plants generate electricity without creating air pollution from stack emissions. In contrast, burning fossil fuels generally emits sulfur dioxide, which contributes to acid deposition; particulate matter, which impairs human health; and carbon dioxide (CO_2) and other greenhouse gases, which drive climate change. Of course, the construction of nuclear plants has a large carbon footprint, but the actual power-generating process is essentially emission-free.

Considering all the steps involved, from construction through mining through power generation, researchers from the International Atomic Energy Agency have calculated that nuclear power releases from 4 to 150 times fewer emissions than fossil fuel combustion. Scientists estimate that nuclear power helps the United States avoid emitting 600 million metric tons of carbon dioxide each year, equivalent to the CO_2 emissions of almost all passenger cars in the nation and 11% of total U.S. CO_2 emissions. Globally, using nuclear power in place of fossil fuels helps the world avoid emissions of 2.5 billion metric tons of carbon dioxide per year, about 7% of global CO_2 emissions. This is why expanding nuclear power would be one of our society's most effective means of fighting climate change.

Nuclear power has additional advantages over fossil fuels—coal in particular (TABLE 20.2). For residents living downwind from power plants, scientists calculate that nuclear power poses far fewer chronic health risks from pollution than does coal.

TABLE 20.2 Risks and Impacts of Coal-Fired versus Nuclear Power Plants

TYPE OF IMPACT	COAL	NUCLEAR
Land and ecosystem disturbance from mining	Extensive, on surface or underground	Less extensive
Greenhouse gas emissions	Considerable emissions	None from plant operation; much less than coal over the entire life-cycle
Other air pollutants	Sulfur dioxide, nitrogen oxides, particulate matter, and other pollutants	No pollutant emissions
Radioactive emissions	No appreciable emissions	No appreciable emissions during normal operation; possibility of emissions during severe accident
Occupational health among workers	More known health problems and fatalities	Fewer known health problems and fatalities
Health impacts on nearby residents	Air pollution impairs health	No appreciable known health impacts under normal operation
Effects of accident or sabotage	No widespread effects	Potentially catastrophic widespread effects
Solid waste	More generated	Less generated
Radioactive waste	None	Radioactive waste generated
Fuel supplies remaining	Should last several hundred more years	Uncertain; supplies could last longer or shorter than coal supplies

For each type of impact, the more severe impact is highlighted in red.

For instance, nuclear power prevents the emission of half a million tons of nitrogen oxide and 1.4 million tons of sulfur dioxide each year that might otherwise be generated by coal-fired power plants. And because uranium generates far more power than coal by weight or volume, less uranium than coal needs to be mined. As a result, uranium mining causes less damage to landscapes and generates less solid waste than coal mining. Moreover, in the course of normal operation, nuclear power plants have proved safer for workers than coal-fired plants.

Supporting the assessment that nuclear power is safer in many ways than fossil fuel power, a 2015 research paper published in the journal *Energy Policy* examined Sweden's plan to phase out nuclear power and calculated the costs that this phase-out would have if the nation had to replace the lost power with coal-fired power. These scientists concluded that replacing Sweden's nuclear plants with coal would produce 1.8 billion tons of extra carbon dioxide emissions and cause more than 51,000 extra deaths due to the health impacts of coal pollution. Already Sweden's nuclear plants, by replacing coal, had prevented 2.1 billion tons of CO_2 emissions and saved 61,000 lives, the researchers calculated.

Nuclear power also has drawbacks, however. One main drawback is that the waste it produces is radioactive, and arranging for safe disposal of this waste has proved challenging. Another major drawback is that if an accident occurs at a power plant, the consequences can potentially be catastrophic.

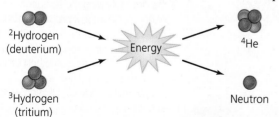
Fusion remains a dream

For as long as scientists and engineers have generated power from nuclear fission, they have tried to figure out how to harness nuclear fusion as well. **Nuclear fusion**—the process that drives our sun's vast output of energy and the force behind hydrogen bombs (thermonuclear bombs)—involves forcing together the small nuclei of lightweight elements under extremely high temperature and pressure. The hydrogen isotopes deuterium and tritium can be fused together to create helium, releasing a neutron and a tremendous amount of energy (**FIGURE 20.7**).

However, overcoming the mutually repulsive forces of protons in a controlled manner is difficult, and fusion requires temperatures of many millions of degrees Celsius. As a result, researchers have not yet developed this process for commercial power generation. Despite billions of dollars of funding and decades of research, fusion experiments in the lab still require scientists to input more energy than they produce from the process. That is, they experience a loss in net energy (p. 527) and a ratio of energy returned on investment (EROI; p. 528) lower than 1.

If we could find a way to control fusion in a reactor, the potential payoffs would be immense. We could produce vast amounts of energy using water as a fuel, and the process would create only low-level radioactive wastes, without pollutant emissions or the risk of dangerous accidents, sabotage, or weapons proliferation. A consortium of nations is collaborating to build a prototype fusion reactor, the International Thermonuclear Experimental Reactor (ITER), in France. It aims to achieve an EROI of 10:1. Construction is scheduled for completion in 2025, after which experiments would be conducted.

Nuclear power poses small risks of large accidents

Although nuclear power delivers energy more cleanly than fossil fuels, the possibility of catastrophic accidents has spawned a great deal of public anxiety. Three events have been influential in shaping public opinion about nuclear power: Three Mile Island, Chernobyl, and Fukushima.

Three Mile Island In Pennsylvania in 1979, a combination of mechanical failure and human error at the **Three Mile Island** plant caused coolant water to drain from the reactor vessel, temperatures to rise inside the reactor core, and metal surrounding the uranium fuel rods to start melting, releasing radiation. This course of events is termed a **meltdown,** and at Three Mile Island (**FIGURE 20.8**), it proceeded through half of one reactor core. Area residents stood ready to be evacuated as the nation held its breath, but fortunately, most radiation remained inside the containment building.

FIGURE 20.8 Pennsylvania's Three Mile Island nuclear power plant suffered a partial meltdown in 1979. This "near miss" alerted the world that a major accident could potentially occur.

^{2}Hydrogen (deuterium)

^{3}Hydrogen (tritium)

Energy

^{4}He

Neutron

FIGURE 20.7 Can we harness nuclear fusion? In nuclear fusion, two small atoms, such as the hydrogen isotopes deuterium and tritium, are fused together, releasing energy along with a helium nucleus and a free neutron. So far, scientists have not been able to fuse atoms without supplying far more energy than the reaction produces, so this process is not used commercially.

The accident at Three Mile Island was brought under control within days, and the damaged reactor was shut down, but multi-billion-dollar cleanup efforts stretched on for years. Three Mile Island is best regarded as a near miss; the meltdown could have been far worse had it proceeded through the entire stock of uranium fuel or had the containment building not contained the radiation. Residents have shown no significant health impacts in the years since, but the event raised safety concerns worldwide. In fact, it was Three Mile Island that inspired Sweden's referendum in 1980 that resulted in its national vote for a phaseout of nuclear power.

Chernobyl In 1986, the **Chernobyl** power plant in Ukraine (part of the Soviet Union at the time) suffered the most severe nuclear accident yet (**FIGURE 20.9a**). Engineers had turned off safety systems to conduct tests, and human error, along with unsafe reactor design, led to explosions that destroyed the reactor and sent clouds of radioactive debris billowing into the atmosphere. For 10 days, radiation escaped from the plant while emergency crews risked their lives putting out fires (and some responders later died from radiation exposure). Most residents of the area remained at home for these 10 days, exposed to high levels of radiation, before the Soviet government belatedly began evacuating more than 100,000 people.

The accident killed 31 people directly and sickened thousands more. Exact numbers are uncertain because of inadequate data and the difficulty of determining long-term radiation effects (see **THE SCIENCE BEHIND THE STORY,** pp. 570–571).

(a) The destroyed reactor at Chernobyl, 1986

(c) The containment dome under construction, 2015

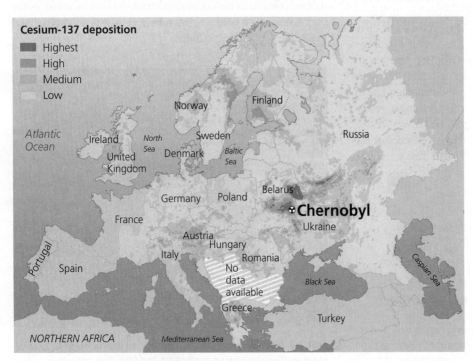

Cesium-137 deposition
- Highest
- High
- Medium
- Low

(b) The spread of radioactive fallout from Chernobyl, 1986

FIGURE 20.9 The world's worst nuclear accident unfolded in 1986 at Chernobyl. The destroyed reactor **(a)** was later encased in a massive concrete sarcophagus to contain further radiation leakage. Radioactive fallout **(b)** was deposited across Europe; complex patterns of cesium-137 deposition resulted from atmospheric currents and rainstorms in the days following the accident. Although Chernobyl produced 100 times more fallout than the U.S. bombs dropped on Hiroshima and Nagasaki in World War II, it was distributed over a much wider area. Thus, levels of contamination outside Ukraine, Belarus, and western Russia were relatively low. Today, an international team has built a huge new confinement structure **(c)** that has been slid into place to encase the deteriorating sarcophagus around the plant. *Data (b) from chernobyl.info, Swiss Agency for Development and Cooperation, Bern, Switzerland.*

Health authorities estimate that most of the 6000-plus cases of thyroid cancer diagnosed in people who were children at the time resulted from radioactive iodine spread by the accident.

Atmospheric currents carried radioactive fallout from Chernobyl across much of the Northern Hemisphere, particularly Ukraine, Belarus, and parts of Russia and Europe (**FIGURE 20.9b**). Fallout was greatest where rainstorms brought radioisotopes down from the radioactive cloud. Parts of Sweden received high amounts of fallout, and the accident reinforced the Swedish public's anxieties about nuclear power. A poll taken after the event showed that nearly half of Swedish citizens had come to regret their own nation's investment in nuclear power.

Following the catastrophe at Chernobyl, workers erected a gigantic concrete sarcophagus around the demolished reactor, scrubbed buildings and roads, and removed irradiated materials. However, the landscape for at least 30 km (19 mi) around the plant remains contaminated today, the demolished reactor is still full of dangerous fuel and debris, and radiation is leaking from the hastily built, quickly deteriorating sarcophagus. In 2016, an international team finished building an enormous confinement structure (**FIGURE 20.9c**) and slid it on rails into place around the old sarcophagus to prevent a re-release of radiation.

Fukushima Daiichi On March 11, 2011, a magnitude-9.0 earthquake struck eastern Japan and sent an immense tsunami roaring onshore (pp. 23–25). More than 18,000 people were killed, and many thousands of buildings were destroyed. This natural disaster affected the operation of several of Japan's nuclear power plants, most notably **Fukushima Daiichi**. At this plant, located on the seacoast, the earthquake shut down power and the 14-m (46-ft) tsunami topped a 5.7-m (19-ft) seawall and flooded the plant's emergency power generators (**FIGURE 20.10a**). Without electricity, workers could not use moderators and control rods to cool the uranium fuel, and the fuel began to overheat as fission proceeded, uncontrolled.

Amid the chaos across the region, help was slow to arrive, so workers flooded the reactors with seawater in a desperate effort to prevent meltdowns. After several explosions and fires, three reactors experienced full meltdowns, and three others were seriously damaged. Parts of the plant remained inaccessible for months because of radioactive water. It will likely require decades to fully clean up the site.

Radioactivity was released during and after the events at Fukushima at levels about one-tenth of those from Chernobyl. Thousands of Fukushima-area residents were evacuated and screened for radiation effects, and restrictions were placed on food and water from the region. Much of the radiation spread by air or water into the Pacific Ocean (**FIGURE 20.10b**), and trace amounts were detected around the world. In the years following the event, a slow flow of radioactive groundwater from beneath the plant has continued to leak into the ocean. Long-term health effects on the region's people are being studied (see **The Science behind the Story,** pp. 570–571).

In the aftermath of the disaster, Japan's government idled all 50 of the nation's nuclear reactors and embarked on safety inspections. Efforts to restart the reactors over the following months and years were met with public debate and street protests. Across the world, many nations reassessed their own

(a) The tsunami barrels toward the Fukushima reactors

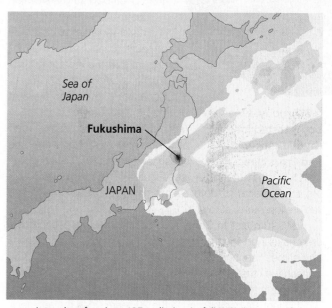

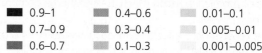

Intensity of cesium-137 radiation in fallout relative to intensity at the plant site (1 = 100%, 0.9 = 90%, etc.)

■ 0.9–1	■ 0.4–0.6	0.01–0.1
■ 0.7–0.9	0.3–0.4	0.005–0.01
0.6–0.7	0.1–0.3	0.001–0.005

(b) Most radiation drifted eastward over the ocean

FIGURE 20.10 The Fukushima Daiichi crisis was unleashed after an earthquake generated a massive tsunami. The tsunami **(a)** tore through a seawall and inundated the plant's nuclear reactors. Most of the radiation that escaped from the plant drifted over the ocean, as shown in this map **(b)** of cesium-137 isotopes in the days following the accident. *Data (b) from Yasunari, T.J., et al. 2011. Cesium-137 deposition and contamination of Japanese soils due to the Fukushima nuclear accident.* Proc. Natl. Acad. Sci. USA *108: 19530–19534.*

nuclear programs. In Sweden, where substantial majorities of citizens had long supported their nation's production of nuclear power, public opinion polls showed Swedes to be divided over whether to keep nuclear power or phase it out. Of all nations, Germany reacted most strongly, shutting down half of its nuclear power plants and deciding to phase out the

THE SCIENCE behind the story

What Health Impacts Have Resulted from Chernobyl and Fukushima?

A Chernobyl-area boy is tested for thyroid cancer in a Red Cross clinic.

In the wake of the meltdowns at Japan's Fukushima Daiichi nuclear power plant in 2011, Japanese authorities evacuated residents and tried to keep people safe, while medical scientists from around the world rushed to study how the release of radiation might affect human health. Both looked back to lessons learned from Chernobyl 25 years earlier.

Determining long-term health impacts of radiation exposure is enormously difficult. Most scientists expect that the reactor failures at Fukushima will have fewer health consequences than those at Chernobyl, because less radiation was emitted at Fukushima and because most of it drifted over the ocean away from populated areas (see Figure 20.10b). However, judging the impacts of both events have been challenging and require long-term study.

Let's turn first to Chernobyl. The hundreds of researchers who have tried to pin down Chernobyl's health impacts have sometimes reached differing conclusions. In an effort to find consensus, the World Health Organization (WHO) engaged 100 experts to review all studies through 2006 and summarize what scientists had learned in the 20 years since the accident. In 2008, the United Nations Scientific Committee on the Effects of Atomic Radiation (UNSCEAR) issued a comprehensive report as well.

The most severe effects were documented in emergency workers who battled to contain the incident in its initial days. Medical staff treated and recorded the progress of 134 workers hospitalized with acute radiation sickness (ARS). Radiation destroys cells in the body, and if the destruction outpaces the body's abilities to repair the damage, the person will soon die. Symptoms of ARS include vomiting, fever, diarrhea, thermal burns, mucous membrane damage, and weakening of the immune system. In total, 28 people died from ARS soon after the accident. Those who died had the greatest estimated exposure to radiation.

The major longer-term health consequence of Chernobyl's radiation has been thyroid cancer in children. The thyroid gland is where our bodies concentrate iodine, and one of the most common radioactive isotopes released early in the disaster was iodine-131 (^{131}I). Healthy children have large and active thyroid glands, so they are especially vulnerable to thyroid cancer induced by radioisotopes.

Expecting that thyroid cancer might be a problem, medical workers measured iodine activity in the thyroid glands of several hundred thousand people in Russia, Ukraine, and Belarus following the accident. They also measured food contamination and surveyed people about what they ate. These data showed that drinking milk from cows that had grazed on contaminated grass was the main route of exposure to ^{131}I, although fresh vegetables also contributed.

As doctors had feared, rates of thyroid cancer rose among children in regions of highest exposure (**FIGURE 1**). Multiple studies found linear dose-response relationships (p. 376) in data from Ukraine and Belarus. By 2006, medical professionals estimated the number of cases of thyroid cancer in children at 6000 and rising. Fortunately, treatment of thyroid cancer has a high success rate, so as of that time, only 15 children had died from it.

Studies addressing other health aspects of the accident have found limited impact. Some research has shown an increase in cataracts due to radiation, especially among emergency workers. But neither the WHO nor the United Nations assessments found evidence that rates of leukemia or any other cancer (aside from thyroid cancer) had risen among people exposed to Chernobyl's radiation. Still, some cancers may appear decades after exposure, so it is possible that some illnesses have yet to arise.

When the Fukushima crisis struck, Japanese authorities applied lessons from Chernobyl. They screened people evacuated from the region for radiation (**FIGURE 2**), and they provided children with iodine tablets that supply the thyroid gland with stable iodine, preventing it from taking up radioactive iodine.

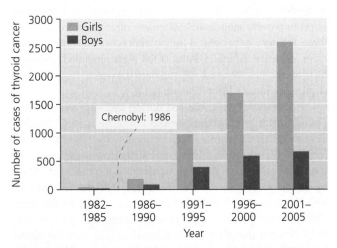

FIGURE 1 The incidence of thyroid cancer in girls and boys younger than age 18 rose in regions of Ukraine, Belarus, and Russia that experienced the heaviest fallout of radioactive iodine from Chernobyl. *Data from U.N. Scientific Committee on the Effects of Atomic Radiation, 2008. Sources and effects of ionizing radiation, Vol. 2. New York, 2011.*

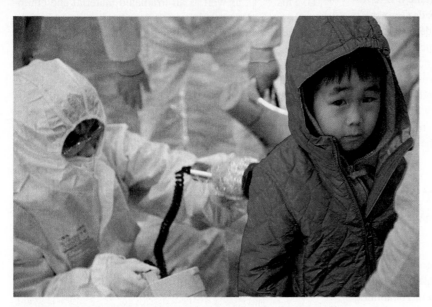

FIGURE 2 Children from the Fukushima region were screened for radiation and later monitored for health impacts.

about health impacts from Fukushima in the years following the event. Each report concluded that there was no clear evidence yet for cancers or other radiation-related health impacts from the Fukushima event and that radiation levels and exposure were low enough that any increases in cancer risk were very small. They cautioned that it will prove difficult to measure statistically rare instances of cancer in a large population exposed to very low doses.

Researchers will try, however. They will be assisted by Japan's government, which budgeted $1.2 billion for coordinated research over 30 years into long-term health effects of low-dose radiation. Questionnaires were sent to all 2 million residents of Fukushima Prefecture, asking where they were in the days after the accident and what they ate and drank. More than 500,000 of these surveys were returned. Thyroid exams are being given to all 360,000 children and teens from the region, and 20,000 pregnant women and their babies are being closely monitored. All 200,000 people evacuated from the area are getting regular exams, and mental health support is being offered.

The most notable finding of the Fukushima Health Management Survey thus far is that the disruption caused to people's lives from stress, fear, and all the difficulties of being evacuated from their homes has had noticeable health effects. Researchers have documented that people from Fukushima experienced greater psychological distress than people elsewhere in Japan and that they have suffered from increased incidences of a variety of common lifestyle disorders, including obesity, hypertension, diabetes, liver dysfunction, and lipid imbalance. Ironically, radiation seems to have been the least that people had to worry about in the aftermath of the Fukushima event.

Results from thyroid examinations of the 360,000 young people have stirred debate. Ultrasound imaging turned up nodules or cysts in nearly 48% of children but "suspicious or malignant" nodules in only 0.06%. Most medical researchers believe that the 48% figure is due to the fact that intensive screening with ultrasound imaging is bound to turn up large numbers of minor and harmless tumors that people would otherwise never have even known they had. Yet the data have alarmed many parents and children and added to their psychological trauma.

As more results of the Fukushima Health Management Survey come in, they should provide valuable information on the health effects of low-dose radiation and the psychological and physical impacts that result from the disruption of people's everyday lives after traumatic events. This knowledge, in turn, may help us respond more effectively if some future event occurs.

They also restricted agriculture near the plant, stopping contaminated food and milk from entering the market. These measures to prevent people's uptake of ^{131}I apparently were effective. In late March 2011, 1000 evacuated children were tested for ^{131}I exposure, and none showed evidence of a high dose affecting the thyroid.

As the crisis was brought under control, some researchers began predicting what long-term health impacts might arise. Using data on cancer rates from studies on survivors of the atomic bombs dropped on Hiroshima and Nagasaki, researchers extrapolated to predict cancer rates at Fukushima's much lower radiation doses. Frank Von Hippel calculated that the Fukushima accident exposed about 1 million people to more than 1 curie per square kilometer of radiation from cesium-137, which should result in a 0.1% increase in cancer risk among them, or 1000 cancer cases.

In other work, John Ten Hoeve and Mark Jacobson of Stanford University used an atmospheric model together with data on radiation, population dispersion, and weather to estimate radiation levels people received. In the journal *Energy and Environmental Science,* they predicted that Fukushima's radiation would eventually produce 125 cancer-related deaths and 178 nonfatal cancer cases. Large uncertainties attended both of Hoeve and Jacobson's estimates; the projected number of total cases ranged from 39 to 2900. Meanwhile, their Stanford colleague Burton Richter wrote to the journal that—far from suggesting that nuclear power is dangerous—the number of predicted fatalities from Fukushima is far less than Japan would have experienced from air pollution had its electricity been produced by fossil fuels instead!

As with Chernobyl, both the WHO and UNSCEAR issued status reports reviewing what was scientifically understood rest by 2022. Sweden's environment minister at the time criticized Germany's decision, saying it would impede Germany's ability to move away from fossil fuels—and indeed, Germany has been burning more coal for power since that time. In 2015, Japan began restarting some of its nuclear plants, following four years of high electricity prices due to the import of replacement power fired by coal and natural gas. Yet despite the prospect of economic recovery, the majority of Japanese citizens remained opposed to restarting the nuclear plants, and most plants will not come back online until after 2025.

We are managing risks from nuclear power and nuclear weapons with some success

It is fortunate that we have not experienced more accidents on the scale of Fukushima or Chernobyl—yet smaller-scale incidents have occurred. A 1999 accident at a plant in Tokaimura, Japan, killed two workers and exposed more than 400 to leaked radiation. In Sweden in 2006, the Forsmark plant north of Stockholm narrowly avoided a meltdown after two generators failed to start up following a power outage. Each year, several U.S. reactors undergo emergency shutdowns in response to tornadoes, hurricanes, earthquakes, and flooding.

We can be thankful that the designs of most modern reactors are safer than Chernobyl's. Designs for future plants promise even more safety features. And in most emergencies at reactors around the world, safety systems have functioned well. For instance, Japan's Onagawa power plant was closer to the epicenter of the 2011 quake than was the Fukushima plant, yet its safety systems protected it from serious damage.

However, as plants around the world age, they require more maintenance. Moreover, radioactive material could be stolen from plants and used in terrorist attacks. This possibility has been especially worrisome in the cash-strapped nations of the former Soviet Union, where hundreds of former nuclear sites have gone without adequate security for years. Finally, there is the concern that some nations may try to use peaceful power generation as a cover to develop nuclear weapons.

To address concerns about stolen fuel and to reduce the world's nuclear weapons stockpiles, the United States and Russia embarked on a remarkably successful program nicknamed "Megatons to Megawatts." Between 1993 and 2013, the United States purchased weapons-grade uranium and plutonium from Russia, let Russia process the materials into less-enriched fuel, and then shipped this fuel to the United States for peaceful use in power generation. As a result, up to 10% of the electricity in the United States in recent years has been generated from fuel recycled from Russian warheads that used to be atop missiles pointed at U.S. cities!

Waste disposal remains a challenge

Even if nuclear power could be made completely safe, we still would be left with the conundrum of what to do with spent fuel rods and other radioactive waste. Recall that fission uses ^{235}U as fuel, leaving as waste the 97% of uranium that is ^{238}U. This ^{238}U, as well as all irradiated material and equipment that is no longer being used, must be disposed of safely. The half-lives of uranium, plutonium, and many other radioisotopes are inconceivably long—700 million years for ^{235}U and 4.5 *billion* years for ^{238}U. Because this waste will continue emitting radiation for as long as our civilization exists, it must be placed in unusually stable and secure locations where radioactivity will not harm future generations.

Currently, radioactive waste from power generation is being held in temporary storage at nuclear power plants. To minimize radiation leakage, spent fuel rods are sunken in pools of cooling water (**FIGURE 20.11a**) or encased in thick casks of steel, lead, and concrete (**FIGURE 20.11b**). In total, U.S. power plants are storing more than 70,000 metric tons of high-level radioactive waste (such as spent fuel)—enough to fill a football field to the depth of 7 m (23 ft)—as well as much more low-level radioactive waste (such as contaminated clothing and equipment). This waste is held at more than 120 sites spread

(a) Wet storage

(b) Dry storage

FIGURE 20.11 U.S. nuclear waste is stored at nuclear power plants, because no central repository yet exists. Spent fuel rods are kept in wet storage **(a)** in pools of water, which keep them cool and reduce radiation release, or in dry storage **(b)** in thick-walled casks layered with lead, concrete, and steel.

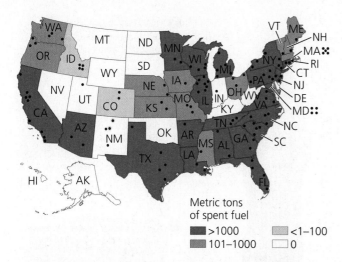

FIGURE 20.12 **High-level radioactive waste from civilian reactors is currently stored at more than 120 sites in 39 states across the United States.** In this map, dots indicate storage sites, and colors indicate the amount of waste stored in each state. *Data from Nuclear Energy Institute.*

Metric tons of spent fuel

- >1000
- 101–1000
- <1–100
- 0

across 39 states (**FIGURE 20.12**). The majority of Americans live within 125 km (75 mi) of temporarily stored waste.

Because storing waste at many dispersed sites creates a large number of potential hazards, nuclear waste managers would prefer to send all waste to a single, central repository that can be properly maintained and guarded. Sweden is far ahead of most nations in this regard and selected the Forsmark power plant site as its single location for final disposal (**FIGURE 20.13**). If expansion of the facility is approved and constructed as planned, all the nation's spent fuel rods and other high-level waste will be buried in copper canisters 500 m (1640 ft) underground within stable bedrock.

In the United States, a lengthy search homed in on Yucca Mountain, a remote site in the Nevada desert. Here, waste was to be stored in a network of tunnels 300 m (980 ft) underground, and $13 billion was spent on development of the site. Scientists and policymakers chose this location because they determined that it was unpopulated, has minimal risk of earthquakes, receives little rain that could contaminate groundwater with radioactivity, has a deep water table atop an isolated aquifer, and is on federal land that can be protected from sabotage. However, some scientists and antinuclear activists challenged these conclusions, and most Nevadans opposed the choice. In 2010, the Obama administration ended support for the project, although lawsuits have challenged this decision and the Trump administration may attempt to revive it.

Until the United States establishes a central repository for radioactive waste from commercial nuclear power plants, this waste will remain spread among many locations. However, one concern with a centralized repository is that waste would need to be transported there from the many current storage sites and from each nuclear plant in the future. Because this process would involve many thousands of shipments by rail and truck across hundreds of public highways through almost every state of the union, some people worry that the risk of an accident is unacceptably high.

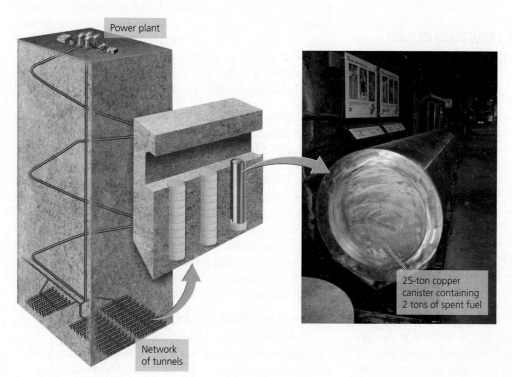

Power plant

25-ton copper canister containing 2 tons of spent fuel

Network of tunnels

FIGURE 20.13 **Sweden leads the world in attempts to establish a central repository for nuclear waste.** Plans call for placing waste in copper canisters and burying them in a network of tunnels deep beneath the Forsmark nuclear power plant. Each canister will be surrounded by clay, and tunnels will be backfilled and sealed.

Nuclear power's growth has slowed

Dogged by concerns over waste disposal, safety, and cost overruns, nuclear power's growth has slowed. Since 1990, nuclear power has grown by only 25% worldwide, much slower than electricity generation overall. Public anxiety in the wake of Chernobyl made utilities less willing to invest in new nuclear power plants, and reaction to Fukushima stalled a resurgence in the industry. Building, maintaining, operating, and ensuring the safety of nuclear facilities is enormously expensive, and almost every nuclear plant has overrun its budget. In addition, plants have aged more quickly than expected because of problems that were underestimated, such as corrosion in coolant pipes. The plants that have been shut down—well over 100 to date—have served on average less than half their expected lifetimes. Moreover, decommissioning (shutting down) a plant is sometimes more expensive than the original construction.

As a result of these financial issues, electricity from nuclear power remains more expensive than electricity from fossil fuels. Governments are still subsidizing nuclear power to keep electricity prices down for ratepayers, but many private investors lost interest long ago. In the United States, the nuclear industry stopped building plants following Three Mile Island, and public opposition scuttled many plants that were under construction. Of the 259 U.S. nuclear reactors ordered since 1957, nearly half have been canceled. At its peak in 1990, the United States had 112 operating reactors; in 2019, it had 99.

Nonetheless, nuclear power remains one of the few currently viable alternatives to fossil fuels with which we can generate large amounts of electricity in short order. This is why a growing number of environmental advocates propose expanding nuclear capacity using a new generation of reactors designed to be safer and less expensive. For a nation wishing to cut its pollution and greenhouse gas emissions quickly and substantially, nuclear power is in many respects the leading option.

A new generation of reactors could transform and revitalize nuclear power

Today, a variety of new reactor designs are being developed by government agencies and private companies, with an eye toward revitalizing nuclear power by making it less expensive, more efficient, safer, and more reliable. One main approach is scaling down reactors to a smaller size. Another is using novel types of fuels and cooling systems.

Small size, modular design *Small modular reactors* (SMRs) are built at factories and could be used individually or in combination at power plants (**FIGURE 20.14**). By prefabricating these reactors in a centralized location and then installing the number needed on-site, the costs are reduced and the financial risk for investors is lowered. The United States may see its first operational SMR in the mid-2020s. An SMR design from an Oregon-based company, NuScale, was approved by the Nuclear Regulatory Commission in 2018, and if all goes smoothly, 12 reactors will be installed in Idaho with the support of the U.S. Department of Energy and a consortium of 45 power utilities. Each of the 12 reactors will be 65 ft long,

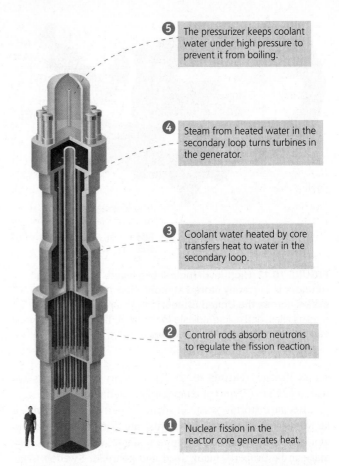

⑤ The pressurizer keeps coolant water under high pressure to prevent it from boiling.

④ Steam from heated water in the secondary loop turns turbines in the generator.

③ Coolant water heated by core transfers heat to water in the secondary loop.

② Control rods absorb neutrons to regulate the fission reaction.

① Nuclear fission in the reactor core generates heat.

FIGURE 20.14 Small modular reactors are compact in size, self-contained, and can be transported and installed where needed. They are among a new generation of innovative nuclear reactors that engineers are developing to be more productive, less wasteful, safer, and more reliable.

produce 50 MW of power (about 1/20th the power of a conventional nuclear reactor), and be able to run for a year without refueling. These SMRs will use no pumps or motors, relying instead on gravity and conduction for cooling. The simplicity of the design sped the approval process, and if this Idaho project is successful, it could spur a flood of similar efforts.

Novel fuels and coolants A variety of designs for nuclear reactors already exist; the pressurized light water reactor shown in Figure 20.6 is just one of several types currently used in operating commercial plants. They are Generation II reactors. (Generation I reactors were early prototypes.) In recent years, the nuclear industry has improved on existing technology to expand generating capacity and enhance safety, creating Generation III reactors that produce power more efficiently and safely. However, the future of nuclear power may rest with the radical departures in design of Generation IV reactors, which use a diversity of novel fuels, coolants, and approaches. With these new reactor designs, engineers and investors are seeking to:

- **Boost efficiency** by gaining more energy (up to 300 times more) from the same amount of fuel.

- **Reduce radioactive waste** by reusing spent fuel in a closed cycle or by using fuels with shorter half-lives.

- **Enhance safety** by enabling reactors to shut down automatically in the event of a problem and by eliminating the need for high pressurization, uninterrupted electric power, and a continuous source of water for cooling.

- **Reduce costs,** in part because fewer safety concerns may make containment domes unnecessary, minimizing expenses and regulatory hurdles.

A diversity of Generation IV reactor designs are in various stages of development across the globe. Seven of these designs have been selected for targeted development by a consortium of nations called the Generation IV International Forum with the aim of constructing operational reactors by 2030. Some of these designs have obtained substantial funding from Bill Gates and other private investors hoping to address climate change with new technologies.

One of the seven main Generation IV designs is a sodium-cooled fast reactor that uses liquid sodium as a coolant (eliminating the need for pressurization) and can run on a variety of fuels, including recycled spent fuel. China is building such a reactor and hopes to begin operation in 2023.

Another design, the molten salt reactor, features uranium fuel dissolved in fluoride salt coolants that circulate in graphite channels. The salts automatically prevent overheating (even if electrical power is lost) because as salts expand with heat they separate the uranium atoms, reducing collisions and slowing the fission reaction. A Canadian company plans to build this kind of plant in Ontario by 2030.

Yet another design, the very-high-temperature gas reactor, cooled by helium, features uranium fuel encased in graphite blocks or in carbon and silicon carbide spheres the size of billiard balls. The high temperatures achieved by this design enable heat and hydrogen fuel to be produced as useful by-products. Variants in which helium passes through a bed of the spheres are known as "pebble-bed" reactors. China had plans to connect a prototype pebble-bed reactor to its grid in 2019. Other Generation IV reactor designs use liquified lead as a coolant and thorium as a fuel.

Development and deployment of Generation IV reactors will take years, and there are plenty of economic and technical uncertainties. In the long run, if the process succeeds, one or more of these new approaches may help us address climate change in coming decades.

With many open questions for nuclear power's future and with climate change worsening, our society must determine where we will turn for clean and sustainable energy. Increasingly, people are looking to renewable sources of energy: energy sources that are replenished naturally and cannot be depleted by our use. Many promising renewable sources are still in rapid phases of development (Chapter 21), but two of them—bioenergy and hydroelectric power—are already well developed and widely used.

Bioenergy

Bioenergy—also known as *biomass energy*—is energy obtained from biomass. **Biomass** consists of organic material derived from living organisms or organisms that have recently died, and it contains chemical energy (pp. 31–32) that originated with sunlight and photosynthesis. We can harness bioenergy from many types of plant matter, including wood from trees, charcoal from wood charred in the absence of oxygen, and matter from agricultural crops, as well as from combustible animal waste products such as cattle manure.

The great attraction of bioenergy is that—in principle—it is renewable and releases no net carbon dioxide into the atmosphere. Although burning biomass emits plenty of carbon, photosynthesis had pulled this same amount of carbon from the atmosphere to create the biomass just years, months, or weeks before. Therefore, in theory, when we replace fossil fuels with bioenergy, we reduce net carbon flux to the atmosphere, helping alleviate global warming (Chapter 18). However, in practice it is not that simple, and judging the sustainability of any given bioenergy strategy requires careful consideration of what biomass source we are using and how we gain energy from it.

We gain bioenergy from many sources

To a poor farmer in Africa, bioenergy comes from cutting wood from trees or collecting livestock manure by hand and burning it to heat and cook for her family. To an industrialized farmer in Iowa, bioenergy means shipping his grain to a hi-tech refinery that converts it to liquid fuel to run automobiles. The diversity of sources and approaches involved in bioenergy (**TABLE 20.3**) gives us many ways to address our energy demands.

More than 1 billion people use wood from trees as their principal energy source. In rural regions of developing nations, people (generally women) gather fuelwood to burn in

TABLE 20.3 **Sources and Uses of Bioenergy**

DIRECT COMBUSTION FOR HEATING
- Wood cut from trees (fuelwood)
- Charcoal
- Manure from farm animals

BIOPOWER FOR GENERATING ELECTRICITY
- Crop residues (such as cornstalks) burned at power plants
- Forestry residues (wood waste from logging) burned at power plants
- Processing wastes (solid or liquid waste from sawmills, pulp mills, and paper mills) burned at power plants
- "Landfill gas" burned at power plants
- Livestock waste from feedlots for gas from anaerobic digesters

BIOFUELS FOR POWERING VEHICLES
- Corn grown for ethanol
- Bagasse (sugarcane residue) grown for ethanol
- Soybeans, rapeseed, and other crops grown for biodiesel
- Used cooking oil for biodiesel
- Plant matter treated with enzymes to produce cellulosic ethanol
- Algae grown for biofuels

FIGURE 20.16 Forestry residues are a major source of material for biopower in some regions. Woody debris commonly accumulates as a byproduct of logging operations, and this debris can be collected and used as source material for biopower.

FIGURE 20.15 Well over a billion people in developing countries rely on wood from trees for heating and cooking. In principle, biomass is renewable, but in practice, it may not be if forests are overharvested.

their homes for heating, cooking, and lighting (**FIGURE 20.15**). Although fossil fuels and electricity are replacing traditional energy sources as developing nations industrialize, fuelwood, charcoal, and manure still account for one-third of energy use in these nations—and up to 90% in the poorest nations.

These traditional biomass sources are renewable only if they are not overharvested. Harvesting fuelwood at unsustainably rapid rates can lead to deforestation (p. 312), soil erosion (p. 229), and desertification (p. 230), thereby damaging landscapes, diminishing biodiversity, and impoverishing societies. Heavily populated arid regions that support meager woodlands are most vulnerable to overharvesting, and these include many regions of Africa and Asia. Burning fuelwood and other biomass for cooking and heating also poses health hazards from indoor air pollution (see Figure 17.27, p. 478).

Although much of the world still relies on fuelwood, charcoal, and manure, new and innovative bioenergy approaches are being developed using a variety of materials (see Table 20.3). Some of these biomass sources can be burned in power plants to produce **biopower**, generating heat and electricity. Other sources can be converted into **biofuels**, liquid fuels used primarily to power automobiles. Because many of these novel biofuels and biopower strategies depend on technologies resulting from extensive research and development, they are being developed primarily in wealthier industrialized nations, such as Sweden and the United States.

We use biomass to generate electricity

We harness biopower by combusting biomass to produce heat or generate electricity in the same way that we burn coal for power (see Figure 19.10, p. 535). We can do so using a variety of sources and techniques.

Waste products Some waste products can be used as sources for biopower. The forest products industry generates large amounts of woody debris in logging operations (**FIGURE 20.16**) and at sawmills, pulp mills, and paper mills. Sweden's efforts to promote bioenergy have focused largely on using forestry residues. Because so much of that nation is forested and the timber industry is a major part of the Swedish economy, plenty of forestry waste is available.

Other waste sources used widely in many countries for biopower include animal waste from feedlots and residue from agricultural crops, such as cornstalks and corn husks. In addition, the anaerobic bacterial breakdown of waste in landfills produces methane, and this "landfill gas" is being captured and sold as fuel (p. 632). Methane and other gases can also be produced in a controlled way in anaerobic digestion facilities. This *biogas* can then be burned in a power plant's boiler to generate electricity (see **SUCCESS STORY**).

Bioenergy crops We are beginning to grow certain types of plants as crops to generate biopower. These plants include fast-growing grasses such as bamboo, fescue, and switchgrass, as well as trees such as specially bred willows and poplars. Many of these plants are also being grown to produce liquid biofuels.

Combustion strategies Power plants built to combust biomass operate like those fired by fossil fuels: Combustion heats water, creating steam to turn turbines and generators and thereby generating electricity. Much of the biopower produced

Turning Waste into Energy

The Swedish city of Kristianstad is best known as the home of Absolut Vodka. But Kristianstad is now gaining attention for its capacity to produce energy from waste. Back in 1999, this city of 81,000 people—the hub of an agricultural and food-processing region—aimed to free itself of a dependence on fossil fuels. It built a power plant that burns forestry waste, then constructed a facility to turn waste into biogas (a mix of methane and other gases that results from breaking down organic matter amid a lack of oxygen). Kristianstad's biogas plant receives household garbage, crop waste, food industry waste, and animal manure and uses anaerobic digestion to turn this waste into biogas. In addition, the city's landfill and wastewater treatment plant each collect methane, adding to the city's biogas supply. The biogas is burned to generate heat and electricity for area homes and businesses. It is also refined and used to fuel hundreds of cars, trucks, and buses engineered to run on biogas. In total, this waste-to-energy system replaces about 7% of Kristianstad's gasoline and diesel fuel each year, all its district heating, and much of its electricity, while excess biogas is sold to neighboring communities. A by-product of biogas production is liquid organic fertilizer, and nearly 100,000 tons of this fertilizer are sold to area farms each year.

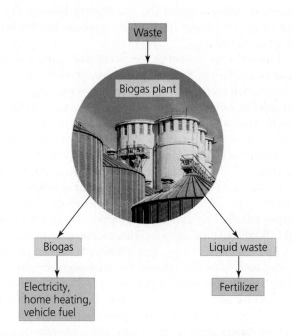

Kristianstad's biogas plant uses organic waste to produce electricity, home heating, vehicle fuel, and fertilizer.

➔ **Explore the Data** at Mastering Environmental Science

---

so far comes from power plants that use cogeneration (p. 553) to generate both electricity and heating. These plants are often located where they can take advantage of forestry waste.

In some coal-fired power plants, wood chips or pellets are combined with coal in a high-efficiency boiler in a process called *co-firing*. We can substitute biomass for up to 15% of the coal with only minor equipment modification and no appreciable loss of efficiency. Co-firing is a relatively easy and inexpensive way for utilities to expand their use of renewable energy.

We also harness biopower by gasification (p. 548), in which biomass is vaporized at extremely high temperatures in the absence of oxygen, creating gases that turn a turbine to propel a generator. Another method of heating biomass in the absence of oxygen results in pyrolysis (p. 531), producing a liquid fuel called pyrolysis oil that can be burned to generate electricity.

Scales of production At small scales, farmers, ranchers, or villages can operate modular biopower systems that use livestock manure to generate electricity. Small household biodigesters provide portable, decentralized energy production for remote rural areas. At large scales, the forest products industry is using its waste to generate power, and industrialized farmers are growing bioenergy crops. In Sweden, nearly one-fourth of the energy supply now comes from biomass, and biomass provides more than four times more fuel for electricity generation than coal, oil, and natural gas combined. Waste liquids from pulp

mills, solid wood waste, municipal solid waste, biogas, and other sources are all used. In the United States, several dozen biomass-fueled power plants are now operating, and several dozen coal-fired plants are pursuing co-firing.

Benefits and drawbacks By enhancing energy efficiency and putting waste products to use, biopower conserves resources and helps move utilities and industries in a sustainable direction. Biopower helps alleviate climate change as well by reducing carbon dioxide emissions, while capturing landfill gas reduces emissions of methane, a potent greenhouse gas. Biopower also benefits human health by cutting down on pollution; as it replaces coal in power plants, emissions of sulfur dioxide are reduced because plant matter, unlike coal, contains no appreciable sulfur content. In addition, biomass resources tend to be geographically dispersed, so using them can help support rural economies and reduce dependence on imported fuels.

A disadvantage is that when we burn plant matter for power, we deprive the soil of nutrients and organic matter it would have gained from the plant matter's decomposition. If we draw fertility from soil and never return it, the soil becomes progressively depleted. That is also the case when we burn plant matter as a liquid fuel, as discussed next. The depletion of soil fertility is a major long-term problem for bioenergy and is one reason why relying solely on bioenergy for our energy demands is not a sustainable option.

Ethanol can power automobiles

Liquid fuels from biomass sources are powering millions of vehicles on today's roads. The two primary biofuels developed so far are ethanol (for gasoline engines) and biodiesel (for diesel engines).

Ethanol is the alcohol in beer, wine, and liquor. It is produced as a biofuel by fermenting biomass, generally from carbohydrate-rich crops, in a process similar to brewing beer. In fermentation, carbohydrates are converted to sugars and then to ethanol. Spurred by the 1990 Clean Air Act amendments, a 2007 congressional mandate, and generous government subsidies, ethanol is widely added to gasoline in the United States to conserve oil and reduce automotive emissions. In 2018 in the United States, 61 billion L (16.1 billion gal) of ethanol were produced—49 gal for every American—mostly from corn (**FIGURE 20.17a**). This amount has grown rapidly, and more than 200 U.S. ethanol production facilities are operating today.

Any vehicle with a gasoline engine runs well on gasoline blended with up to 10% ethanol, but automakers are also producing *flexible-fuel vehicles* that run on E-85, a mix of up to 85% ethanol and 15% gasoline. More than 20 million such cars are on U.S. roads today. Most gas stations do not yet offer E-85, so drivers often fill these cars with conventional gasoline, but this situation is changing as infrastructure for ethanol increases. In Brazil, almost all new cars are flexible-fuel vehicles, and ethanol from crushed sugarcane residue (called *bagasse*) accounts for half of all fuel that Brazil's drivers use.

Together, Brazil and the United States currently produce nearly 85% of the world's ethanol supply (**FIGURE 20.17b**).

(a) Corn grown for ethanol

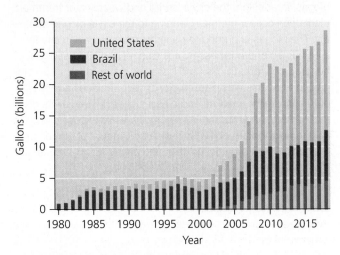

(b) Global ethanol production, 1980–2018

FIGURE 20.17 Ethanol is booming. Nearly 40% of the U.S. corn crop **(a)** is used to produce ethanol to add to gasoline. Next to the United States, Brazil produces most of the rest of the world's ethanol, from bagasse (sugarcane residue). Ethanol production **(b)** has grown rapidly in recent years. *Data (b) from Renewable Fuels Association.*

 Go to **Interpreting Graphs & Data** on **Mastering Environmental Science**

FAQ

Why not keep adding more ethanol to gasoline?

Increasing the proportion of ethanol in formulations of gasoline helps conserve oil and reduce reliance on foreign imports. However, producing more corn ethanol requires that more land be converted to corn production, with all the environmental impacts and fossil fuel use that industrial agriculture entails. Already, 40% of the U.S. corn crop is used to make ethanol. (Some by-products of ethanol production are used in livestock feed; with this use accounted for, 28% of the U.S. corn crop goes solely to ethanol.) The more corn we divert from food to ethanol, the more food prices rise. For these reasons and more, researchers are instead studying other plants as more efficient sources of ethanol and are trying to produce cellulosic ethanol from crop and forestry wastes.

Ethanol is not our most sustainable energy choice

The enthusiasm for corn-based ethanol shown by U.S policy-makers is not widely shared by scientists. Growing corn to produce ethanol takes up millions of acres of land and intensifies the use of pesticides, fertilizers, and fresh water. It also requires substantial inputs of fossil fuel energy (for operating farm equipment, making petroleum-based pesticides and fertilizers, transporting corn to processing plants, and heating water in refineries to distill ethanol). In the end, corn ethanol yields only a modest amount of energy relative to the energy that needs to be input. The EROI ratio (p. 528) for corn-based ethanol is variable and controversial, but recent estimates place it around 1.3:1 (see **THE SCIENCE BEHIND THE STORY**, pp. 580–581). In other words, we need to expend 1 unit of energy just to gain 1.3 units of energy from ethanol. The EROI of Brazilian bagasse ethanol is considerably higher, but the extremely low ratio for corn ethanol makes this fuel inefficient. For all these reasons, very few experts view corn ethanol as an effective path to sustainable energy use.

Biodiesel powers diesel engines

Drivers of diesel-fueled vehicles can use **biodiesel**, a fuel produced from vegetable oil, used cooking grease, or animal fat. The oil or fat is mixed with small amounts of ethanol or

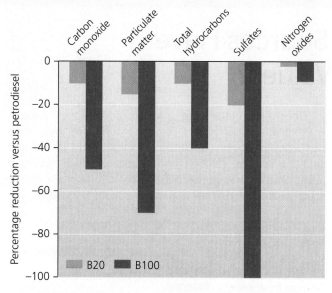

FIGURE 20.18 **Burning biodiesel in a diesel engine emits less pollution than burning conventional petroleum-based diesel.** Shown are percentage reductions in several major pollutants that one can attain by using B20 (a mix of 20% biodiesel and 80% petroleum-based diesel) and B100 (pure biodiesel). *Data from U.S. Environmental Protection Agency.*

 Go to **Interpreting Graphs & Data** on **Mastering Environmental Science**

FIGURE 20.19 **Appalachian State University students produced enough biodiesel to drive this minibus to Washington, D.C., for the National Sustainable Design Expo.**

methanol in the presence of a chemical catalyst. In Europe, where most biodiesel is used, rapeseed oil is the oil of choice, whereas American biodiesel producers use mostly soybean oil. Vehicles with diesel engines can run on 100% biodiesel (called B100), or biodiesel can be mixed with conventional diesel derived from petroleum; a 20% biodiesel mix (called B20) is common.

Replacing diesel with biodiesel cuts down on emissions (**FIGURE 20.18**). Biodiesel's fuel economy is nearly as good, it costs just slightly more than diesel, and it is nontoxic and biodegradable. Growing numbers of people are fueling their cars with biodiesel from waste oils—and college students are creating biodiesel from food waste from dining halls and fast-food restaurants (**FIGURE 20.19**). Some buses, recycling trucks, and state and federal fleet vehicles now run on biodiesel or biodiesel blends. Indeed, some enthusiasts have taken biofuel use further: Eliminating the processing step that biodiesel requires, they use straight vegetable oil in their diesel engines (which requires modifying the engine).

Using waste oil as a biofuel is sustainable, but most biodiesel today, like most ethanol, comes from crops grown specifically for the purpose—which has environmental impacts. Growing soybeans in Brazil and oil palms in Southeast Asia hastens the loss of tropical rainforest (pp. 314–316). Growing soybeans in the United States and rapeseed in Europe takes up large areas of land as well. Because the major crops grown for biodiesel and for ethanol exert heavy impacts on the land, farmers and agricultural scientists are experimenting with a variety of other crops, from wheat, sorghum, cassava, and sugar beets to hemp, jatropha, and the grass miscanthus.

Novel biofuels are being developed

One promising next-generation biofuel crop is algae (better known to most of us as green pond scum!). Several species of these photosynthetic organisms produce large amounts of lipids (fats, oils, or waxes) that can be converted to biodiesel. Alternatively, carbohydrates in algae can be fermented to create ethanol. In fact, we can use algae to produce a variety of fuels, even jet fuel (**FIGURE 20.20**).

FIGURE 20.20 **Commercial aircraft are beginning to use biofuels.** KLM and several other airlines have been powering flights with various biofuels for several years. United Airlines has used algae-based biofuel in a 30% blend on flights between Houston and Chicago since 2011. In 2015, it began fueling flights between Los Angeles and San Francisco with fuel derived from waste oils, farm waste, and landfill trash.

THE SCIENCE behind the story

Which Energy Sources Have the Best EROI Values?

Dr. Charles Hall, a pioneer of EROI research

As our society develops energy sources as alternatives to fossil fuels, being able to compare these sources to one another so we can make informed decisions about which sources to prioritize will be helpful. One important aspect to compare is the measure called EROI, or energy returned on investment (p. 528), a ratio that indicates how much energy is produced for each unit of energy that is invested in the energy-producing activity.

It's fairly straightforward to measure how much energy is produced when we burn a fuel or when water or wind turns a turbine. The difficult part is calculating how much energy is invested across all the many aspects of finding, extracting, transporting, and using a resource, and in manufacturing the equipment needed to do so. To fully and accurately measure the denominator in an EROI ratio, the researcher must conduct a thorough life-cycle analysis (pp. 267, 633), summing up all the energy invested at each stage in the life-cycle of the process. As more researchers begin to do so, we are gaining better data that help us compare energy sources.

Charles Hall of the State University of New York College of Environmental Science and Forestry in Syracuse, New York, has been a pioneer in calculating EROI values. In 2014, he and students Jessica Lambert and Stephen Balogh surveyed recent research and compiled EROI estimates for major energy sources (**FIGURE 1**). Their summary, published in the journal

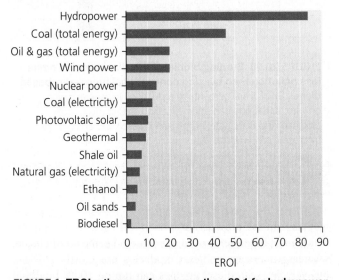

FIGURE 1 EROI ratios vary from more than 80:1 for hydropower down to 2:1 for biodiesel. *Adapted from Hall, C., et al. 2014. EROI of different fuels and the implications for society. Energy Policy 64: 141–152.*

Algae can be grown at a large scale in outdoor circulating ponds or, in a more controlled and intensive way, in closed tanks or in closed transparent tubes called photobioreactors (**FIGURE 20.21**). Algae grow faster and produce more oil than terrestrial biofuel crops, and they can grow in seawater, saline water, or nutrient-rich wastewater from sewage treatment plants. Because carbon dioxide speeds their growth, algae can even make use of smokestack emissions. Indeed, placing algae farms next to power plants as a source of carbon capture (pp. 548–549) is a promising combination. However, producing biofuels from algae is expensive and private investors have lost money on their bets thus far, so further development is needed to bring costs down.

Relying on any monocultural crop for energy may not be sustainable. With that in mind, researchers are refining techniques to produce **cellulosic ethanol** by using enzymes to produce ethanol from cellulose, the substance that gives structure to all plant material. This approach would be a major advance because ethanol made from corn or sugarcane uses starch, which is nutritionally valuable to us, and

FIGURE 20.21 Algae are a leading candidate for next-generation biofuels. Here algae grow at a demonstration facility.

Energy Policy, revealed large differences in efficiency among our major energy sources.

Hydroelectric power had the best EROI value, averaging 84:1. So, for each unit of energy we invest in hydropower, we produce 84 units of energy. Once dams, reservoirs, and transmission lines are constructed, the process of passing running water through turbines year after year is highly efficient.

Nuclear power places above the middle of the pack in the Hall team's study, with an EROI of 14:1. Nuclear power produces a large amount of electricity, but it also requires substantial investment in uranium mining and in the facilities, equipment, and maintenance of power plants.

Solar power technologies tend to have moderately low EROI values, but as these young renewable technologies are further developed, we can expect EROI values to rise. Indeed, wind power is already among the most efficient energy sources, with an EROI around 20:1.

At the low end of the spectrum in the Hall team's study were biodiesel, oil sands, and ethanol, with EROIs of only 2:1, 4:1, and 5:1, respectively. Mining oil sands and processing their bitumen requires huge amounts of energy expenditure (Chapter 19). The biofuels ethanol and biodiesel have low values because most come from crops like corn and soybeans grown in industrial monocultures. Large amounts of equipment, fertilizer, pesticides, and irrigation—all powered by or made from fossil fuels—are used to grow these crops and transport the yields to refineries.

Ethanol has received more research attention than most other fuels in regard to EROI, and results have varied. Early attempts at making cellulosic ethanol resulted in EROI values below 1:1 (energy was lost), whereas, in contrast, one study showed an EROI of 48:1 for ethanol from molasses in India.

Earlier work by Hall found EROI values of ethanol from Brazilian sugarcane to average 5:1 but those from U.S. corn ethanol to average only 1.3:1. Invariably, researchers have found the EROI of corn ethanol to be remarkably low, which is a large part of the reason many environmental scientists oppose U.S. government subsidies for corn ethanol.

In one 2011 research paper, Hall and two colleagues sought to clarify why different studies had come up with different EROI estimates for corn ethanol. By carefully comparing the methods of each study, they determined that the differences were due to how the researchers measured energy investment in the many steps involved in producing ethanol. They recommended that future researchers be more transparent and explicit about their methods as they try to work out the EROI values of newer biofuels, such as those from switchgrass and cellulosic ethanol.

The 2014 review by Hall and his students made clear that EROI values for fossil fuels were historically very high but have been dropping through the years (see Figure 19.5, p. 528). This drop has occurred because we have already extracted the easily reached deposits and now need to expend more and more energy to reach the less-accessible deposits that remain. Newer, unconventional, fossil fuel resources such as oil sands and shale oil have low EROI values. So, if we switch to these sources as conventional oil begins to run out, we will deplete them rapidly (while producing a great deal of pollution), because a great deal of energy will need to be devoted to extracting and processing them.

As studies of EROI become more numerous and higher in quality, they will help scientists, engineers, and policymakers compare energy sources. Such comparisons will allow us to make better decisions as we chart our energy future and shift from fossil fuels to alternative and renewable energy sources.

---

proponents of food security voice ethical concerns about using crops for fuel when millions of people continue to die from hunger. Cellulose, in contrast, is of no food value to people yet is abundant in all plants. Converting plant material to fuel would still deprive the soil of organic matter from decomposition, but if we can produce cellulosic ethanol in a commercially feasible way, ethanol could be made from low-value crop waste (such as corn stalks and husks) rather than from high-value crops.

Certain plants hold promise as crops specifically for cellulosic ethanol. Switchgrass (**FIGURE 20.22**) could potentially be a sustainable choice for the United States because it

WEIGHING the **issues**

Biofuels

Do you think producing and using ethanol from corn is a good idea? Do the benefits outweigh the drawbacks? Explain why or why not. Do you think we should invest billions of dollars into developing next-generation biofuels such as algae and cellulosic ethanol? Can you suggest ways of using biofuels that would minimize environmental impacts?

FIGURE 20.22 Switchgrass is being studied as a crop to produce cellulosic ethanol and also provides fuel for biopower.

is a native grass of the North American prairies. It could provide wildlife habitat while serving as a crop, especially if planted in a polyculture mixed with other species. Cellulosic ethanol from switchgrass has an EROI ratio of 5:1 (much better than corn ethanol's), and growing this perennial grass could prevent the soil erosion and depletion of soil carbon that would result from removing too much crop waste for ethanol.

Is bioenergy carbon-neutral?

In principle, energy from biomass is carbon-neutral, releasing no net carbon dioxide into the atmosphere, because burning biomass releases carbon dioxide that plants recently pulled from the air by photosynthesis. However, burning biomass for energy is not carbon-neutral if forests are destroyed to plant bioenergy crops. Forests sequester more carbon (in vegetation and soil) than croplands do, so cutting forests to plant crops increases carbon flux to the atmosphere. Bioenergy also is not carbon-neutral if we consume fossil fuel energy to produce the biomass (e.g., by using tractors, fertilizers, and pesticides to grow biofuel crops).

International climate change policy so far has largely failed to encourage sustainable bioenergy approaches. The Kyoto Protocol (p. 515) required nations to submit data on emissions from energy use and from changes in land use, but only the emissions from energy use were "counted" toward judging how well nations were controlling their emissions under the treaty. Negotiators during the Paris Agreement (p. 516) improved on this situation, but much work remains. In the meantime, researchers are trying to develop means of using bioenergy that are truly renewable and carbon-neutral. With continued research and careful decision making, our many bioenergy options may yet provide promising avenues for sustainably replacing fossil fuels.

Hydroelectric Power

Next to biomass, we draw more renewable energy from the motion of water than from any other resource. In **hydroelectric power,** or **hydropower,** we use the kinetic energy of flowing river water to turn turbines and generate electricity. We examined hydroelectric power and its environmental impacts in our discussion of freshwater resources (pp. 404–405). Now we will take a closer look at hydropower as an energy source.

Hydropower uses three approaches

Most hydroelectric power today comes from impounding water in reservoirs behind concrete dams that block the natural flow of river water, allowing for the controlled release of that water through the dam. Because water is stored behind dams, this approach is called the **storage** technique. As reservoir water passes through a dam, it turns the blades of turbines, which cause a generator to generate electricity

(**FIGURE 20.23**). Electricity generated in the powerhouse of a dam is transmitted by transmission lines to the electrical grid that serves consumers, while the water flows to the riverbed below the dam and continues downriver. By storing water in reservoirs, dam operators can ensure a steady and predictable supply of electricity, even during periods of naturally low river flow.

An alternative approach is the **run-of-river** technique, which generates electricity without greatly disrupting a river's flow. One method is to divert a portion of a river's flow through a pipe or channel, passing it through a powerhouse and then returning it to the river (**FIGURE 20.24**, p. 584). The diversion can be carried out with or without a small reservoir that pools water temporarily, and the pipe or channel can be run along the surface or underground. Another method is to siphon river water directly through a dam to turn turbines and then return the water to the river. Run-of-river systems are useful in remote areas far from electrical grids and in regions without the economic resources to build and maintain large dams, but some such systems can be large indeed. The run-of-river approach cannot guarantee reliable water flow in all seasons, but it reduces many impacts of the storage technique by leaving most water in the river channel.

To better control the timing of flow, pumped-storage hydropower can be used. In **pumped storage,** water is pumped from a lower reservoir to a higher reservoir during times when demand for power is weak and prices are low. When demand is strong and prices are high, water is sent downhill through a turbine, generating electricity. Although energy must be input to pump the water, pumped storage can be profitable. When paired with intermittent energy sources such as solar and wind power, it can help balance a region's power supply by compensating for dips in power availability (see Figure 21.26, p. 612).

Hydropower is clean and renewable, yet has impacts

Hydroelectric power has three clear advantages over fossil fuels. First, it is renewable; as long as precipitation falls from the sky and fills rivers and reservoirs, we can use water to turn turbines. Second, hydropower is efficient. It has been shown to have an EROI ratio of more than 80:1, higher than any other modern-day energy source. Third, hydropower emits no carbon dioxide or other pollutants into the atmosphere, which helps safeguard air quality, climate, and human health. Fossil fuels *are* used in constructing and maintaining dams—and large reservoirs may release the greenhouse gas methane as a result of anaerobic decay in deep water. But overall, hydropower accounts for only a small fraction of the greenhouse gas emissions typical of fossil fuel combustion.

Although it is renewable, efficient, and produces little air pollution, hydropower does exert negative impacts. Damming rivers (p. 404) destroys habitat for wildlife because

(a) Ice Harbor Dam, Snake River, Washington

(b) Generators inside McNary Dam, Columbia River

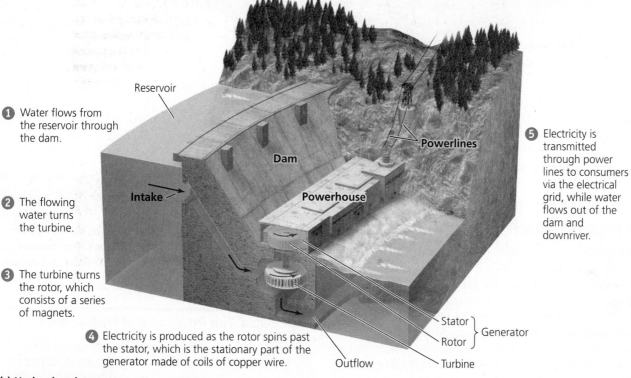

1 Water flows from the reservoir through the dam.

2 The flowing water turns the turbine.

3 The turbine turns the rotor, which consists of a series of magnets.

4 Electricity is produced as the rotor spins past the stator, which is the stationary part of the generator made of coils of copper wire.

5 Electricity is transmitted through power lines to consumers via the electrical grid, while water flows out of the dam and downriver.

Reservoir

Dam

Intake

Powerhouse

Powerlines

Stator ⎫
Rotor ⎭ Generator

Outflow

Turbine

(c) Hydroelectric power

FIGURE 20.23 We generate hydroelectric power with large dams. Inside these dams **(a)**, flowing water turns massive turbines, which rotate generators **(b)** to generate electricity. Water is funneled from the reservoir through the dam and its powerhouse **(c)** and into the riverbed below.

ecologically rich riparian areas above dam sites are submerged and those below often are starved of water. Because water discharge is regulated to optimize electricity generation, the natural flooding cycles of rivers are disrupted. Suppression of flooding prevents river floodplains from receiving fresh nutrient-laden sediments. Instead, sediments become trapped behind dams, where they begin filling the reservoir. Dams also cause thermal pollution (p. 411): Water downstream may become unusually warm if water levels are made unnaturally shallow, yet periodic flushes of cold water

occur from the release of reservoir water. Such thermal shocks, together with habitat alteration, have diminished or eliminated native fish populations in many dammed waterways. In addition, dams generally block the passage of fish and other aquatic creatures, effectively fragmenting the river and reducing biodiversity in each stretch. All these ecological impacts generally bring negative social and economic consequences for local communities. (We discussed the benefits, drawbacks, and impacts of dams more fully in Chapter 15; pp. 404–405.)

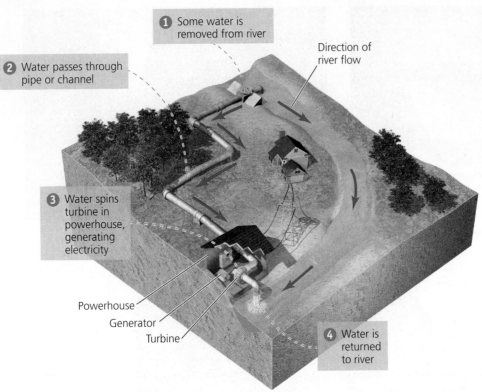

① Some water is removed from river

② Water passes through pipe or channel

③ Water spins turbine in powerhouse, generating electricity

Direction of river flow

Powerhouse

Generator

Turbine

④ Water is returned to river

FIGURE 20.24 Run-of-river systems divert a portion of a river's water. Water may be piped downhill through a powerhouse and released downriver, as shown.

Hydropower is widely used but may not expand much more

Hydroelectric power accounts for one-sixth of the world's electricity production (see Figure 20.3b). For nations with large amounts of river water and the economic resources to build dams, hydropower has been a keystone of their development and wealth. Sweden receives 11% of its total energy and 40% of its electricity from hydropower. Norway, Brazil, Venezuela, Canada, and several other nations obtain large amounts of their energy from hydropower (**TABLE 20.4**).

The great age of dam building for hydroelectric power (as well as for flood control and irrigation) began in the 1930s in the United States, when the federal government constructed dams as public projects to bring electricity to rural communities and to employ people during the Great Depression. U.S. dam construction peaked in 1960, when 3123 dams were completed in a single year. American engineers subsequently exported their dam-building expertise and technology to nations of the developing world. In India in 1963, Prime Minister Jawaharlal Nehru called dams "the temples of modern India," referring to their central importance in that nation's industrialization.

Today, the world is witnessing some gargantuan hydroelectric projects. China's recently completed Three Gorges Dam (p. 404) is the world's largest. Its reservoir displaced 1.3 million people, and the dam is now generating as much electricity as dozens of coal-fired or nuclear plants.

However, hydropower has less potential for expansion than most other renewable energy sources. The main reason is that most of the world's large rivers that offer excellent opportunities for hydropower are already dammed. For instance, in the wake of Sweden's 1980 referendum to phase out nuclear

TABLE 20.4 Top Producers of Hydropower

NATION	HYDROPOWER PRODUCED (PERCENTAGE OF WORLD TOTAL)	PERCENTAGE OF NATION'S ELECTRICITY GENERATION FROM HYDROPOWER
China	28.6	19.2
Canada	9.3	58.0
Brazil	9.1	65.8
United States	7.0	6.8
Russia	4.5	17.1
Norway	3.5	96.2
India	3.3	9.3
Japan	2.0	8.0
Venezuela	1.6	60.1
Turkey	1.6	24.5

Data from the International Energy Agency.

power, many Swedes had hoped that hydropower could compensate for the electrical capacity that would be lost. However, Sweden had already dammed so many of its rivers that it could not gain much additional hydropower by erecting more dams.

In addition, as people have grown more aware of the ecological impacts of dams, residents in some regions are resisting dam construction. Such is the case in Sweden, where many citizens have made it clear that they want some rivers to remain undammed, preserved in their natural state. As a result, in Sweden, hydropower's contribution to the national energy budget has remained little changed for more than 40 years. In the United States, 98% of rivers appropriate for dam construction already are dammed, and

many of the remaining 2% are protected under the Wild and Scenic Rivers Act. Some dams that are no longer functional or profitable are being dismantled so that river habitats can be restored (p. 405).

Overall, hydroelectric power will likely continue to increase in industrializing nations that have yet to dam their rivers, but in industrialized nations hydropower growth will likely slow. The International Energy Agency forecasts that hydropower's share of electricity generation will decline very slightly by 2035, whereas the share of other renewable energy sources will nearly triple. As we shall see, newer renewable energy sources have impressive potential for rapid growth (Chapter 21).

CENTRAL CASE STUDY
connect & continue

TODAY, people everywhere are feeling the impacts of fossil fuel extraction and combustion on health, environmental quality, and the climate, and many nations are seeking to diversify their energy portfolios with alternative energy sources. Sweden has been leading the way for years and is now accelerating its efforts. The nation met its goal to get 50% of its energy from renewable sources by 2020. Indeed, it achieved that goal eight years early, and today its energy mix is 55% renewable.

In 2017, Sweden hastened its goal to become carbon-neutral by 2050 by five years, committing to zero net carbon emissions by 2045. The law committing the nation to the new carbon-neutral goal passed overwhelmingly in the Swedish Parliament. It established a climate council and mandated that an action plan by updated every four years.

To achieve the goal of carbon-neutrality by 2050, Sweden plans first and foremost to tackle its transportation sector, promoting the use of electric vehicles and greatly expanding the use of biofuels to replace petroleum. Overall, the nation plans to cut its emissions by 85% and then aims to offset the remainder by planting trees domestically and by funding reforestation projects and forest conservation measures in other countries.

What might seem like a green pipe dream in many other societies seems eminently feasible in Sweden, which has invested in hydroelectric power, bioenergy, and nuclear power so effectively in recent decades that it has managed to cut its fossil fuel use by more than half. Whether or not Sweden eventually becomes the first nation in history to free itself of fossil fuels, its experience thus far demonstrates that having a vibrant, healthy, wealthy society does not require a reliance on fossil fuels.

Because Sweden has shown time and again that it can walk its talk, it has made itself an effective model for other nations to follow. Indeed, just a year after Sweden's carbon-neutrality announcement, the European Union as a whole also set as a formal goal reaching carbon-neutrality by 2050.

As Sweden navigates its ongoing societal debate over the safety of nuclear power, it can boast that it is already saving thousands of its citizens' lives each year by using far less coal, oil, or gas to generate its electricity. Regardless of the outcome of the debate over nuclear power, today Swedes—and growing numbers of people everywhere—are looking forward to a world powered by clean and renewable energy.

- **CASE STUDY SOLUTIONS** Sweden's prime minister has appointed you as the nation's energy minister and has asked you to recommend and direct energy policies for Sweden. Based on everything you have learned from this chapter, describe how you would approach the question of whether or not to phase out Sweden's use of nuclear power. What specific questions would you ask of scientists and energy experts before making a final policy recommendation? What would you like the nation's energy mix to look like 10 years in the future? And 25 years in the future? What actions would need to be taken to reach your goals?

- **LOCAL CONNECTIONS** Do some research to find out where your campus gets its energy. What percentage of total energy on your campus comes from fossil fuels? From nuclear power? From bioenergy? From hydropower? From other renewable sources such as wind or solar power? What about electricity: What percentages of electricity come from each of these sources? Are students on your campus making any attempts to urge a shift away from fossil fuels and toward alternative energy sources? If so, what is being done? What more do you think could be done or should be done?

- **EXPLORE THE DATA** Each region of the United States has a different mix of energy sources. Do you know how your region compares to others? → **Explore Data** relating to the case study on **Mastering Environmental Science**.

REVIEWING Objectives

You should now be able to:

+ **Discuss reasons for seeking alternatives to fossil fuels**

Fossil fuels are nonrenewable resources, and we are gradually depleting their easily accessible reserves. Burning fossil fuels causes air pollution that results in many environmental and health impacts and drives global climate change. (p. 563)

+ **Summarize the contributions to world energy supplies of conventional alternatives to fossil fuels**

Alternatives to fossil fuels that are most widely used include nuclear power, bioenergy, and hydroelectric power. Biomass provides 10% of global primary energy use, nuclear power provides 5%, and hydropower provides 2–3%. Nuclear power generates more than 10% of the world's electricity, and hydropower generates 16%. (pp. 563–564)

+ **Describe nuclear energy, and explain how we harness it for electrical power**

We gain nuclear power by converting the energy of subatomic bonds into thermal energy, using uranium isotopes. Uranium is mined, enriched, processed into pellets and fuel rods, and used in nuclear reactors. By controlling the reaction rate of nuclear fission, nuclear power plant engineers produce heat that powers electricity generation. (pp. 564–566)

+ **Assess the benefits and drawbacks of nuclear power, and discuss the societal debate over this energy source**

Nuclear power is a clean energy source because it does not emit the pollutants that fossil fuels do. It thereby reduces health impacts and is a key component of addressing climate change. However, for many people the risk of a catastrophic power plant accident, like those at Chernobyl and Fukushima, outweighs these benefits. The disposal of nuclear waste also remains a dilemma: Both temporary storage and single-repository approaches involve risks. Economic factors and cost overruns have slowed the nuclear industry's growth, but a new generation of reactors currently in development may potentially revitalize the industry. (pp. 566–575)

+ **Describe established and emerging sources and techniques involved in harnessing bioenergy, and assess bioenergy's benefits and shortcomings**

Fuelwood remains the major source of bioenergy, especially in developing nations. We use biomass to generate electrical power in several ways with a variety of source materials. Biofuels such as ethanol and biodiesel are used to power vehicles; some crops are grown for this purpose, and waste oils are also used. Bioenergy is renewable and in principle adds no net carbon to the atmosphere. However, overharvesting of wood can lead to deforestation, and growing crops solely for fuel can be inefficient and ecologically damaging. New biofuels from algae and cellulosic ethanol may hold promise for more sustainable biofuel production. (pp. 575–582)

+ **Outline the scale, methods, and impacts of hydroelectric power**

Hydroelectric power is generated when river water runs through a powerhouse and turns turbines. Three major approaches (storage, run-of-river, and pumped storage) each offer advantages. Hydropower produces little air pollution, but dams and reservoirs alter riverine ecology and local economies. Hydropower is the world's largest source of electricity outside of fossil fuels, but its long-term growth potential appears more limited than that of other renewable sources. (pp. 582–585)

SEEKING Solutions

1. Nuclear power has by now been widely used for a half-century, and the world has experienced only two major accidents (Chernobyl and Fukushima) responsible for any significant number of health impacts. Would you call that a good safety record? Should we maintain, decrease, or increase our reliance on nuclear power? Why might safety at nuclear power plants be better in the future? Why might it be worse?

2. How serious a problem do you think the disposal of radioactive waste represents? Describe in detail how you would like to see this issue addressed.

3. There are many different sources of biomass and many ways of harnessing energy from biomass. Describe a biomass source that you think seems particularly beneficial to society, and explain why. Now describe a biomass source that you view as particularly problematic, and explain why. What bioenergy sources and strategies do you think our society should focus on investing in?

4. **THINK IT THROUGH** You are the head of the national department of energy in a nation that has just experienced a minor accident at one of its nuclear plants. A partial meltdown released radiation, but the radiation was fully contained inside the containment building, and there were no health impacts on area residents.

However, residents are terrified, and the media are highlighting the risks of nuclear power. Your nation relies on its five nuclear plants for 25% of its energy and 50% of its electricity needs. It has no fossil fuel deposits and recently began a promising but still-young program to develop renewable energy options. What will you tell the public at your next press conference, and what policy steps will you recommend taking to ensure a safe and reliable national energy supply?

5. **THINK IT THROUGH** You are an investor seeking to invest in alternative energy. You are considering buying stock in companies that (1) construct nuclear reactors, (2) build turbines for hydroelectric dams, (3) build corn ethanol refineries, (4) are ready to supply farm and forestry waste for cellulosic ethanol, and (5) are developing algae farms. For each of these companies, what questions would you research before deciding how to invest your money? How do you expect you might apportion your investments, and why?

CALCULATING **Ecological Footprints**

Our three conventional energy alternatives vary tremendously in their EROI (energy returned on investment; p. 528). Examine the data on EROI for each of the energy sources as provided in the figure from **The Science behind the Story** on p. 580, and enter the data in the table.

ENERGY SOURCE	EROI
Coal (electricity)	12
Natural gas (electricity)	
Nuclear power	
Hydropower	
Photovoltaic solar	
Wind power	
Ethanol	
Biodiesel	

1. How many units of energy would you generate by investing 1 unit of energy into producing hydropower? To generate that same amount of energy, about how many units of energy would you need to invest into producing nuclear power? Roughly how many units of energy would you need to invest into producing electricity from coal if you wanted to generate that same amount of energy?
2. Based on EROI values, is it more efficient to obtain energy from electricity from natural gas or from wind power? Which source would you guess has a larger ecological footprint, based on EROI values?
3. Let's say you wanted to generate 100 units of energy from biodiesel. About how many units of energy would you need to invest? Explain your calculations.
4. Based on EROI values alone, which energy sources would you advocate that we further develop and use? Which would you urge society to avoid? What other issues, besides EROI, are worth considering when comparing energy sources?

Mastering Environmental Science

Students Go to **Mastering Environmental Science** for assignments, an interactive e-text, and the Study Area with practice tests, videos, and activities.

Instructors Go to **Mastering Environmental Science** for automatically graded activities, videos, and reading questions that you can assign to your students, plus Instructor Resources.

New Renewable Energy Sources

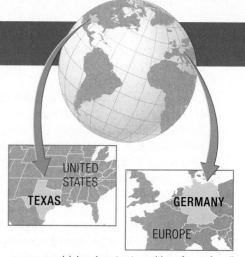

Harnessing Wind and Sun in Texas and Germany

> [Renewable energy] will provide millions of new jobs. It will halt global warming. It will create a more fair and just world. It will clean our environment and make our lives healthier.
>
> Hermann Scheer, energy expert and German parliament member

> It's almost like you're on top of the world on a turbine. When you get out near the blades, it's just you and the sky.
>
> Texas wind power technician Kaitlin Sullivan

As our world begins to transition from fossil fuels to renewable energy, Texas and Germany are showing us two different paths forward. This U.S. state and this European nation together demonstrate that there are multiple ways to move toward a renewable energy future.

Texas became a global leader in wind power by encouraging companies to chase profits while harnessing its rich wind resources. Germany became a global leader in solar power—despite its relatively modest solar resources—by enacting policies to address energy security, pollution, and climate change. Each location ignited a boom in electricity from renewable energy with its own mix of policy steps and free-market economic incentives.

Let's first visit Texas, which might seem an unlikely place for a renewable energy revolution. For decades, Texas has been the epicenter of the U.S. petroleum industry, leading the nation in the production of oil and natural gas. But Texas has had another energy resource in abundance all along—wind.

In the western parts of the state, winds blow strongly and consistently across vast regions of open land. Entrepreneurs in the wind industry recognized the huge potential here and also were enticed by the state's pro-business, anti-regulatory outlook. The lack of regulatory barriers for private firms to extract energy resources soon made Texas known as the "Wild West of wind energy."

Texas policymakers gave wind power a boost when the legislature passed a law in 1999, signed by Governor George W. Bush (who later became U.S. president), mandating utilities to produce a quota of renewable energy. In 2005, the state government, under Governor Rick Perry, increased the quota. Assisted by minimal regulation at the state level and a tax credit at the federal level, the wind industry began erecting thousands of turbines on land leased from ranchers and farmers, who received one-time payments or periodic royalties from the energy revenue. Soon Texas became the leading producer of wind-powered electricity in the United States, meeting its renewable energy quota 15 years early.

Wind power grew so fast in Texas that during periods of strongly blowing wind and low demand, it began to outstrip

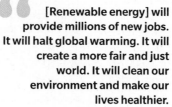

Upon completing this chapter, you will be able to:

+ Identify the major sources of renewable energy, and assess their recent growth and future potential

+ Describe solar energy and how we harness it, and evaluate its advantages and disadvantages

+ Describe wind power and how we harness it, and evaluate its advantages and disadvantages

+ Describe geothermal energy and how we harness it, and evaluate its advantages and disadvantages

+ List ocean energy sources, and describe their potential

+ Survey approaches for energy storage, and discuss the potential role of hydrogen and fuel cells

◀ **Turbines at the Lone Star Wind Farm in West Texas**

▲ **Installing PV solar panels**

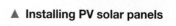

589

the capacity of existing transmission lines to carry the power. Each time that occurred, the statewide grid manager, the Electric Reliability Council of Texas (ERCOT), would need to curtail wind power generation.

Perry and the legislature responded by investing $7 billion of taxpayers' money into expanding the transmission grid. Most Texans live in cities like Dallas, Fort Worth, San Antonio, Austin, and Houston in the eastern part of the state, whereas most wind power is harnessed in the western part of the state. Under Texas Senate Bill 20 passed in 2005, ERCOT and the Public Utility Commission of Texas were charged with linking power-producing regions in the west with power-consuming areas in the east. They designated "Competitive Renewable Energy Zones" in five western regions with the most power potential, encouraged companies to develop wind farms in these regions, and then let them compete to sell power to ERCOT at the lowest prices. Once the grid was enhanced, with 5800 km (3600 mi) of new and expanded transmission lines, the curtailments ended and wind power soared (**FIGURE 21.1**).

By 2018, wind power had come to provide nearly 19% of Texas's electricity, and the percentage is still growing. Thanks to wind power, urban and suburban Texans enjoy some of the country's cheapest electricity rates, while rural communities in West Texas have been infused with jobs and income.

Germany took a somewhat different path to its leadership in solar power. Far from being resource-rich like Texas is with wind, this northern European nation receives less sun each year than Alaska! Yet Germany made itself a global leader in photovoltaic (PV) solar technology, which produces electricity from sunshine. Today, Germany obtains more than 7% of its electricity from solar power and generates the most solar power per person of any nation.

To boost solar power in such a cool and cloudy country, Germany's political leaders enacted bold policies that used economic incentives to promote solar power and other forms of renewable energy. The nation's **feed-in tariff** system required utilities to buy power (at guaranteed premium prices under long-term contract) from anyone who could generate power from renewable energy sources and feed it into the electrical grid. In response, German homeowners and businesses rushed to install PV panels and began selling their excess solar power to utilities at a profit. Germany now gets fully one-third of its electricity from renewable sources.

Germany's push for renewable energy dates back to 1990. In the wake of the disaster at the Chernobyl nuclear power plant in the Soviet Union (p. 568), Germany decided to phase out its own nuclear power plants. However, if they were shut down, the nation would lose virtually all its clean energy and would become utterly dependent on coal, oil, and gas imported from Russia and the Middle East.

Enter Hermann Scheer, a German parliament member and an expert on renewable energy. While everyone else assumed that solar, wind, and geothermal energy were costly and risky, Scheer saw them as a great economic opportunity. In 1990, Scheer helped push through a landmark law establishing feed-in tariffs. Ten years later, the law was revised and strengthened: The Renewable Energy Sources Act of 2000 aimed to promote renewable energy production and use, enhance the security of

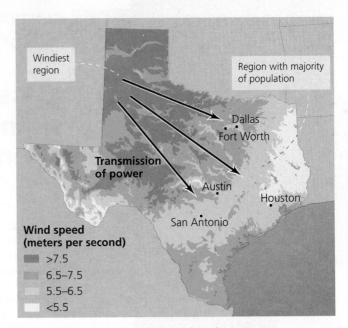

(a) Average wind speed at 80 m altitude, in Texas

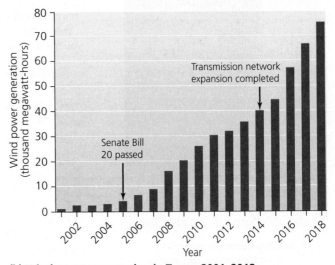

(b) Wind power generation in Texas, 2001–2018

FIGURE 21.1 Texas has ignited a boom in wind power. Texas invested $7 billion to build transmission lines **(a)** to bring power from windy regions of West Texas to the cities and suburbs of East Texas. Generation of wind power soared **(b)** as the transmission network was being enhanced. *Data from (a) National Renewable Energy Laboratory and (b) U.S. Energy Information Administration.*

the energy supply, reduce carbon emissions, and lessen the many external costs (pp. 143, 165) of fossil fuel use.

Under this law, each renewable energy source was assigned its own payment rate according to market considerations. However, because utilities pass along to consumers the costs of feed-in tariffs, these costs soon grew to 15% of the average German resident's electric bill. To ease this burden on ratepayers, the German government in 2010 decided to slash PV solar tariff rates.

When German consumers heard that the solar tariff rate would be reduced, sales of PV modules skyrocketed as people rushed to lock in contracts at the existing rate. Then from 2013 on, solar installations in Germany slowed as feed-in tariff rates became low and consumers lost the economic incentive to invest in solar (**FIGURE 21.2**).

By reducing the subsidies, Germany's leaders aimed to encourage technological innovation for efficiency within the solar industry, thereby creating a stronger industry that can sustain

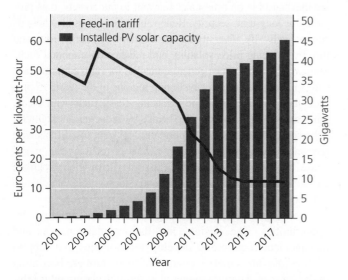

FIGURE 21.2 PV solar power rose steeply in Germany, thanks to feed-in tariffs. The rapid rise in installed capacity of PV solar then slowed as Germany's feed-in tariffs were reduced. *Data from International Energy Agency, 2016.* Photovoltaic power systems programme annual report 2018; *and AGEE-Stat, Federal Ministry for Economic Affairs and Energy, Germany.*

growth over the long term and outcompete foreign companies for international business. Indeed, boosted by domestic demand, German renewable energy industries have become global leaders, designing and selling technologies around the world while employing 350,000 people.

By 2050, Germany aims to obtain 80% of its electricity from renewable sources. To achieve this historic energy transition, which Germans call the *Energiewende,* the government has been allotting more public money to renewable energy than any other nation—more than $25 billion annually in recent years. However, the transition has been bumpy. In response to public protests following the Fukushima nuclear disaster in Japan in 2011 (pp. 23, 569), the German government shut down 7 of its 15 nuclear power plants and promised to phase out the rest by 2022. As a result, the electricity supply from nuclear power fell sharply, causing rates to rise and leading the country to burn more coal for power.

Although fossil fuels still dominate their overall energy budgets, Germany and Texas are each improving their air quality and helping fight climate change by replacing some fossil fuels with renewable energy. And because solar and wind resources vary over time, both Germany and Texas are working to confront the issue of sometimes having too much renewable power and at other times not having enough. For the future, Germany and Texas are each seeking to keep renewables growing at a steady and predictable rate and to ensure stability in energy supplies.

The similarities and differences in the experiences of Texans and Germans have much to teach us about ways we can pursue renewable energy and harness its many benefits. Other states and nations are paying attention as they seek to transition away from fossil fuels at this crucial time in our society's energy history.

Shifting to Renewable Energy

Texas and Germany are playing starring roles in a larger global shift toward renewable energy. Across the world, nations are seeking to move away from fossil fuels while ensuring a reliable, affordable, sustainable supply of energy. They are doing so because the economic and social costs, national security risks, and health and environmental impacts of fossil fuel dependence (Chapter 19) are all intensifying.

The two renewable energy sources that are most widely used are bioenergy, the energy from combustion of biomass (wood and other plant matter), and hydropower, the energy from running water. (We cover these two energy sources in depth in Chapter 20.) These well-established, conventional alternatives to fossil fuels are renewable, but they can be depleted with overuse, and they exert some undesirable environmental impacts.

In this chapter, we explore a group of alternative energy sources that are often called "new renewables" or "modern renewables." These diverse sources include renewable energy from the sun, from wind, from Earth's geothermal heat, and from ocean water. These energy sources are not actually new. In fact, they are as old as our planet, and people have used

them for millennia. We commonly refer to them as "new" or "modern" because (1) they are just beginning to be used on a wide scale in today's industrial society, (2) they are harnessed using technologies still in a rapid phase of development, and (3) they will likely play much larger roles in the future.

"New" renewable sources are growing fast

As with other energy sources, the new renewables provide energy for three types of applications: (1) to generate electricity, (2) to heat air or water, and (3) to fuel vehicles. So far, their contribution to our society's overall energy budget remains small, but their potential is enormous. Today, we obtain just under 2% of our global energy from the new renewable energy sources, whereas fossil fuels still provide 81% of the world's energy (see Figure 20.3a, p. 563).

The new renewable sources are making a similarly small contribution to our global generation of electricity thus far (see Figure 20.3b, p. 563). At this point, 24% of our electricity comes from sources of renewable energy overall, but hydropower accounts for two-thirds of this amount.

Nations and regions vary in the renewable sources they use. In the United States, most renewable energy comes from biomass and hydropower. As of 2018, wind power accounted for 22.0% of renewable energy, solar energy for 8.3%, and geothermal energy for 1.9% (**FIGURE 21.3a**). Of electricity generated in the United States from renewables overall, hydropower accounts for nearly 40%, whereas contributions as of 2018 from new renewable sources included 37.0% from wind power, 13.0% from solar power, and 2.3% from geothermal power (**FIGURE 21.3b**).

Although they currently make up a small proportion of our energy budget, the so-called new renewable energy sources are growing quickly. Over the past four decades, solar, wind, and geothermal energy sources have grown far faster than has the overall energy supply. The long-term leader in growth among all energy sources is wind power, which has expanded by nearly 50% *each year* since the 1970s. In recent years, solar power has grown faster than wind. Because these sources started from such low levels of use, however, it will take them some time to catch up to conventional sources.

Renewable energy offers advantages

Renewable energy sources offer substantial benefits for individuals and for society. Unlike fossil fuels, the new renewable sources are inexhaustible on timescales relevant to our society, thus providing us long-term security. Renewable alternatives also help us economically by diversifying an economy's energy mix and thereby reducing price volatility and reliance on imported fuels. Some renewable sources generate income for rural communities, and some help people in developing regions of the world to produce their own energy. And replacing fossil fuels with clean, renewable energy benefits our health by reducing air pollution (Chapter 17). Perhaps most important, replacing fossil fuels with renewable energy is our primary means of reducing the greenhouse gas emissions (**FIGURE 21.4**) that drive global climate change (Chapter 18). In all likelihood, stopping climate change will require a full transition to clean, renewable energy.

Shifting to renewable energy also creates employment opportunities. The design, installation, maintenance, and management required to develop technologies and rebuild and operate our society's energy infrastructure are becoming major sources of employment today, through **green-collar jobs.**

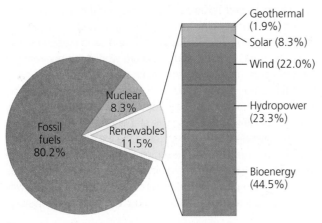

(a) U.S. consumption of renewable energy, by source

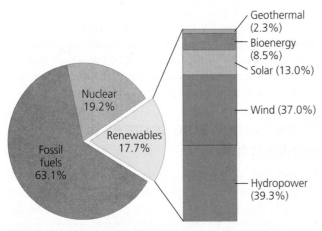

(b) U.S. electricity generation from renewable sources

FIGURE 21.3 Renewable energy sources contribute a small but growing portion of the energy we consume. In the United States, 11.5% of total energy consumption (transport, heating, electricity, etc.) **(a)** is from renewable energy—mostly bioenergy, hydropower, and wind. Of electricity generated in the United States **(b),** 17.7% comes from renewable sources, predominantly hydropower and wind. *Data are for 2018, from Energy Information Administration.*

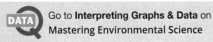

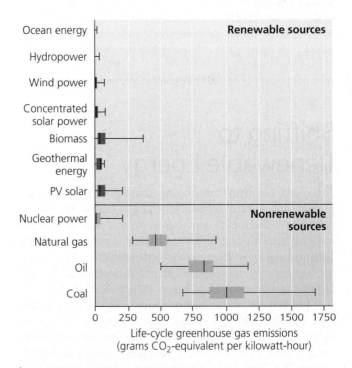

FIGURE 21.4 Renewable energy sources release far fewer greenhouse gas emissions than do fossil fuels. Shown are medians and ranges of estimates from scientific studies of each source when used to generate electricity. *Data from Intergovernmental Panel on Climate Change, 2012.* Renewable energy sources and climate change mitigation. Special report. *New York: Cambridge University Press.*

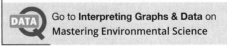

JOBS CREATED
PV solar: 3,605,000
Biofuels: 2,063,000
Hydropower: 2,054,000
Wind power: 1,160,000
Solar heating: 801,000
Biomass: 787,000
Other: 462,000

FIGURE 21.5 Renewable energy creates new green-collar jobs. As of 2019, nearly 11 million people worldwide were employed in jobs connected to renewable energy. *Data from REN21, 2019. Renewables 2019: Global status report. Paris, REN21, UNEP.*

Already nearly 11 million people work in renewable energy jobs around the world (**FIGURE 21.5**).

Whether we can—or even should—give up fossil fuels entirely and switch to 100% renewable energy remains a matter of some debate among energy experts. Many are doubtful we can make a complete transition soon, in part because we lack the infrastructure needed to transport huge volumes of power from renewable sources to consumers inexpensively at nationwide scales. Yet some recent research makes a strong case that we can transition quickly and would gain substantial benefits by doing so (see **THE SCIENCE BEHIND THE STORY**, pp. 594–595).

FAQ

Isn't renewable energy too expensive and untested to rely on?

Not at all. For decades now, renewable energy sources and technologies have been supplying power to millions of people. That's true for the established renewable sources of hydropower and bioenergy (Chapter 20), but also for most "new renewable" sources—from geothermal power and ground-source heat pumps to solar water heating and PV solar cells to onshore and offshore wind power. All renewable energy sources have become far more affordable than they used to be, and in many locations, electricity from wind is already cheaper than electricity from fossil fuels.

Policy and investment can accelerate our transition from fossil fuels

Most renewable energy remains more expensive than fossil fuel energy. However, prices are falling fast, and some sources have become cost-competitive with fossil fuels for electricity (**FIGURE 21.6a**). Today, electricity from wind power and from utility-scale solar power is cheaper than electricity from fossil fuels and nuclear power across large portions of the United States (**FIGURE 21.6b**). As more governments, utilities, corporations, and consumers promote and use new renewable energy, market prices of new renewables should continue to fall, further hastening their adoption.

Feed-in tariffs like Germany's hasten the spread of renewable energy by creating financial incentives for businesses and individuals. Governments are also setting goals or mandating that certain percentages of power come from renewable sources. As of 2019, nearly all the world's nations and 31 U.S. states had set official targets for renewable energy use. Governments also invest in research and development of technologies, lend money to renewable energy startups, and offer tax credits and tax rebates to companies and individuals who produce or buy renewable energy. In Texas, wind power

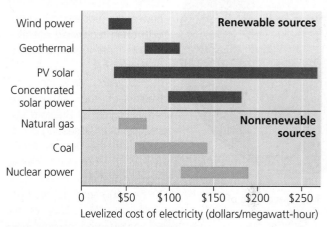

(a) Electricity costs for renewable and nonrenewable sources

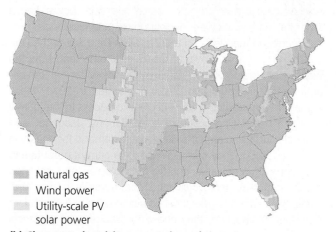

(b) Cheapest electricity sources in each county

FIGURE 21.6 Renewable energy is becoming cost-competitive with nonrenewable energy. Shown **(a)** are ranges for the price of electricity from major sources as of 2018. Prices for renewable sources are lower than shown here if subsidies are included or if external costs (pp. 143, 165) are considered. The map **(b)** shows the lowest-cost electricity source in each county of the United States, according to the most recent data (without including external costs). Wind power and utility-scale solar power are cheapest across large areas of the nation. *Data from (a) Lazard, 2018. Lazard's levelized cost of energy analysis—version 12.0. New York: Lazard; and (b) University of Texas at Austin Energy Institute, Levelized cost of electricity in the United States, Version 1.4.0, http://calculators. energy.utexas.edu/lcoe_map/#/county/tech.*

 Go to Interpreting Graphs & Data on Mastering Environmental Science

THE SCIENCE behind the story

Can We Power the World with Renewable Energy?

Mark Jacobson of Stanford University

Despite the rapid growth of renewable energy, many experts remain skeptical that we will ever be able to replace fossil fuels entirely. Yet some recent scientific research has aimed to outline in detail how we might power our society completely with clean renewable energy—without fossil fuels, biofuels, or nuclear power.

For years, Mark Jacobson, director of the Atmosphere/Energy Program at Stanford University, has been assessing energy sources with the goal of finding solutions to pollution, climate change, and energy insecurity. In 2009, he published a full life-cycle analysis (pp. 267, 633) of the social, health, and environmental impacts of all major energy sources and found that new renewable sources based on wind, water, and solar power had a lower impact than biofuels, nuclear power, and fossil fuels.

Jacobson then teamed up with Mark Delucchi of the University of California, Davis to examine whether and how the world could meet 100% of its energy needs with clean renewable energy from the sun, wind, and water. In 2011, these researchers published a pair of scientific papers in the journal *Energy Policy*. Using government projections, Jacobson and Delucchi first calculated the likely global demand for energy in 2030 and 2050. They then examined the current outputs and limitations of renewable energy technologies and selected those that were technically and commercially proven and established. Because electrical power is more energy-efficient than fuel combustion, they chose to propose only electrical technologies (e.g., battery-electric vehicles rather than gasoline-powered vehicles).

The researchers then calculated what it would take to manufacture these technologies at the needed scale and to build infrastructure for renewable energy storage and transmission throughout the world. Because sources such as solar power and wind power are intermittent (varying from hour to hour and day to day), Jacobson and Delucchi judged what balance of sources was needed to compensate for intermittency and ensure a consistent, reliable energy supply.

Once all the math was done, Jacobson and Delucchi concluded that the world *can*, in fact, fully replace fossil fuels, nuclear power, and biofuels and meet all its energy demands with clean renewable sources alone. They proposed a quantitative breakdown of the various wind, water, and solar technologies needed to power the world in 2030 (**TABLE 1**).

To achieve this transition, society would need to greatly expand its transmission infrastructure, construct fleets of fuel-cell-powered vehicles and ships, and more. To deal with intermittency of sun and wind and prevent gaps in energy supplies, we would need to link complementary combinations of sources across large regions and, when there is oversupply, use the extra energy to produce hydrogen fuel. Achieving all that in a few decades would clearly be ambitious, but Jacobson and Delucchi argue that most barriers to achieving a transition to renewables are social and political, not technological or economic.

With an eye toward impacts, Jacobson and Delucchi also added up the total area of land that all this new renewable energy infrastructure would require. The researchers calculated that 0.74% of Earth's land surface would be occupied directly by energy infrastructure, with 0.41% being newly required beyond what is already taken up. An additional 1.18% of area would be needed for spacing between structures (mostly wind turbines), but half of this area could be over water if half our wind power is offshore, and most land between turbines could be used for farming or grazing.

Jacobson and Delucchi estimated that the overall cost of energy in their proposed scenario would be roughly the same as the cost of energy today. One main challenge, however, is the

TABLE 1 Renewable Energy Infrastructure Needed to Power the World in 2030

TECHNOLOGY	NUMBER OF PLANTS OR DEVICES NEEDED	PERCENTAGE OF GLOBAL DEMAND SATISFIED
Wind turbines	3,800,000	50
Wave devices	720,000	1
Geothermal plants	5,350	4
Hydroelectric plants	900	4
Tidal turbines	490,000	1
Rooftop PV systems	1,700,000,000	6
Solar PV plants	40,000	14
CSP plants	49,000	20

Data from Jacobson, M.Z., and M.A. Delucchi, 2011. Providing all global energy with wind, water, and solar power, Part I: Technologies, energy resources, quantities and areas of infrastructure, and materials. *Energy Policy* 39: 1154–1169.

limited availability of a handful of rare-earth metals (such as platinum, lithium, indium, tellurium, and neodymium) used in certain materials and equipment for wind, water, and solar technologies. Although additional reserves of these metals may be discovered and mined, we will likely need to enhance efforts to recycle them.

Jacobson and Delucchi's work drew both praise and criticism. Critics, such as Australian energy expert Ted Trainer, believed that Jacobson and Delucchi underestimated the costs of their proposal, overestimated efficiency gains from electrification, and failed to offer persuasive quantitative evidence that intermittency can be overcome.

More recently, Jacobson and Delucchi worked with eight colleagues to design plans for how each of the 50 U.S. states might power itself entirely with wind, water, and solar sources. This research team evaluated each state's resources and energy demands and published a study in the journal *Energy and Environmental Science* in 2015 that summarized its state-specific "roadmaps." The researchers determined a mix of renewable sources and technologies that could feasibly meet the energy demand forecast for the United States in 2050. **FIGURE 1** shows their overall plan for transitioning to 100% renewable energy across the United States. Footprints of the infrastructure required would be 0.47% of land area (0.42% new), with spacing area of 2.4% (1.6% new).

Across all 50 states, the team estimated that a full conversion to renewable energy would raise the number of energy jobs from 3.9 million to 5.9 million and would eliminate up to 62,000 premature deaths caused by pollution each year. The proposal would also save an estimated $3.3 trillion in costs from predicted climate change impacts worldwide due to U.S emissions in 2050. As a result, they calculated, the average U.S. resident in 2050 would save $260 in energy costs, $1500 in health costs, and $8300 in climate change costs each year.

In a related 2015 paper in *Proceedings of the National Academy of Sciences,* Jacobson and his colleagues ran simulation models aiming to demonstrate how hydropower and energy

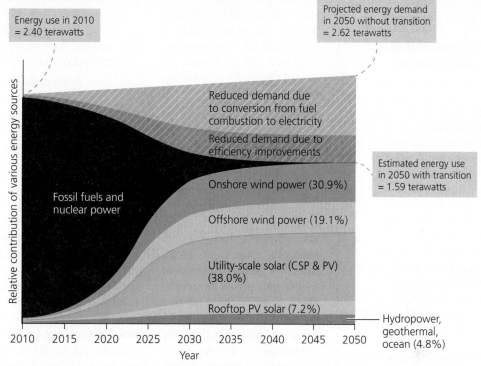

FIGURE 1 **Fossil fuels and nuclear power can be phased out and replaced entirely by renewable energy by 2050 in the United States, research indicates.** This graph summarizes the transition to renewables proposed by Jacobson's team. Heights of each colored band at a given year show percentages of total U.S. energy demand to be met by each energy source in that year. *Data from Jacobson, M., et al., 2015. 100% clean and renewable wind, water, and sunlight (WWS) all-sector energy roadmaps for the 50 United States.* Energy Environ. Sci. *8: 2093–2117.*

DATA Go to **Interpreting Graphs & Data** on **Mastering Environmental Science**

storage systems could solve the challenge of peaks and troughs from intermittent wind and solar energy production. Their analysis purported to show that this challenge could be overcome, enabling a full transition to an energy economy run purely on wind, water, and solar.

Many researchers disagreed with this conclusion. In 2017, a team of 21 scientists headed by energy expert Christopher Clack authored a paper in the same journal vehemently rebutting many aspects of the Jacobson team's analysis. They found fault with its methods and maintained that many assumptions made were unrealistic. Clack's team felt that energy research overall supports the idea that we can transition to an economy dominated by renewable energy but that we will need to continue using nuclear power and fossil fuels to some degree.

Jacobson's team responded to its critics, and a lengthy debate between the two camps has ensued. Such open debate—and the further research it spurs—is precisely how science as a whole moves forward. As a result, we can look forward to better and better scientific guidance as our society accelerates its transition to renewable energy.

has flourished not only because wind is plentiful and land was available, but also because the state created a welcoming business environment with few regulations, set official targets for renewables, and invested $7 billion to enhance grid connections between producing regions and population centers.

When a government boosts an industry with such policies, the private sector often responds with investment of its own, because private investors recognize an enhanced chance of profit. Indeed, global investment (public plus private) in renewable energy has hovered around $300 billion annually in recent years—six times the amount just a decade earlier. In a feedback process, investment may breed success, and success may breed further investment.

Yet the economics of renewable energy have been erratic. Technologies evolve quickly, and policies vary from place to place and can change unpredictably. Wind and solar companies in the United States have had difficulty planning their investment decisions because Congress has never extended tax credits over a long enough time period to provide businesses certainty. As a result, renewable markets have been volatile, prices have fluctuated, growth has often come in rapid bursts, and many promising companies have met with bankruptcy.

Critics of public subsidies for renewable energy complain that funneling taxpayer money to particular energy sources is inefficient and skews the market. Instead, they propose, we should let energy sources compete freely. However, proponents of renewable energy point out that governments have long subsidized fossil fuels and nuclear power far more than they now subsidize renewables. As a result, there has never been a level playing field or a truly free market.

In one study, researchers dug into data on the U.S. government's energy subsidies and tax breaks (p. 180) over the past century. The report revealed three main findings:

1. Together, oil and gas have received 75 times more subsidies—and nuclear power has received 31 times more—than new renewable energy sources (**FIGURE 21.7a**). These figures alone are not surprising, because oil, gas, and nuclear power have been major contributors to the U.S. energy supply for a longer period than solar, wind, and geothermal sources.

2. Per year, oil and gas have received 13 times more subsidies than new renewable energy sources, and nuclear power has received more than 9 times more than new renewable energy (**FIGURE 21.7b**). These figures are notable, because a per-year comparison controls for the amount of time each source has been used.

3. Even in the earliest years of each energy source (when subsidies are most useful), oil, gas, and nuclear power received far more in subsidy support than the new renewables have received. In fact, in only one single year have solar, wind, and geothermal combined *ever* received as much support as the *lowest* amount ever offered to oil, gas, or nuclear power.

More recent research concludes that across the world today, for every $1 in taxpayer money that goes toward renewable energy, somewhere between $2 and $4 goes to fossil fuels. On average, each person on Earth today pays $40 per year in subsidies to fossil fuel industries. Meanwhile, according to one recent estimate, the average person suffers roughly $700 in external costs each year from fossil fuels.

Many people believe that the subsidies showered on nonrenewable energy sources in the United States have helped enhance the economy, national security, and international influence of the country by establishing thriving global

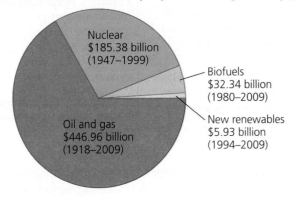

(a) Total subsidies

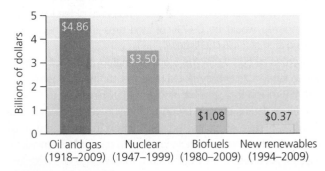

(b) Per-year subsidies

FIGURE 21.7 Fossil fuels and nuclear power have received far more in U.S. government subsidies (mostly tax breaks) than have renewable energy sources. This is true both for **(a)** total amounts over the past century and for **(b)** average amounts per year. *Data are in 2010 dollars, from Pfund, N., and B. Healey, 2011. What would Jefferson do? The historical role of federal subsidies in shaping America's energy future. DBL Investors.*

energy industries dominated by U.S. firms. Yet by this logic, if the United States wants to be a global leader to rival nations such as China and Germany in the transition to clean and renewable energy, it will likely need to direct greater political and financial support toward these new energy sources.

Solar Energy

The sun releases astounding amounts of energy by converting hydrogen to helium through nuclear fusion (p. 567). The tiny proportion of this energy that reaches Earth is enough to drive most processes in the biosphere, helping make life possible on our planet. Each day, Earth receives enough **solar energy,** or energy from the sun, to power human consumption for a quarter of a century. On average, each square meter of Earth's surface receives about 1 kilowatt of solar energy—17 times the energy of a lightbulb. As a result, a typical home has enough roof area to meet all its power needs with rooftop panels that harness solar energy.

We can collect solar energy using passive or active methods

The simplest way to harness solar energy is by **passive solar energy collection.** In this approach, buildings are designed to maximize absorption of sunlight in winter yet to minimize it in summer (**FIGURE 21.8**). South-facing windows maximize the capture of winter sunlight. Overhangs shade windows in summer, when the sun is high in the sky and

Plants buffer house from temperature swings.

Overhang shades summer sun from above.

Thermal mass in floors and walls absorbs heat and then slowly releases it.

South-facing porch lets in low-angle sunlight in winter.

FIGURE 21.8 Passive solar design elements can be seen in this house built by college students. Stanford University students and staff designed and constructed this solar-powered house as part of the sixth biannual Solar Decathlon. In this event run by the U.S. Department of Energy, college and university teams compete to design and build the best houses fully powered by solar energy.

cooling is desired. Planting vegetation around a building buffers it from temperature swings. Passive solar techniques also use materials that absorb heat, store it, and release it later. Such *thermal mass* (e.g., of straw, brick, or concrete) often makes up floors, roofs, and walls or can be used in portable blocks. Passive solar approaches are an important component of green building design (p. 348). By heating buildings in cold weather and cooling them in warm weather, passive solar methods conserve energy and reduce costs.

Active solar energy collection makes use of technology to focus, move, or store solar energy. For instance, *flat plate solar collectors* heat water and air for homes and businesses (**FIGURE 21.9**). These panels generally consist of dark-colored, heat-absorbing metal plates mounted in flat glass-covered boxes, most often on rooftops. Water, air, or antifreeze runs through tubes that pass through the collectors, transferring heat from the collectors to the building or its water tank.

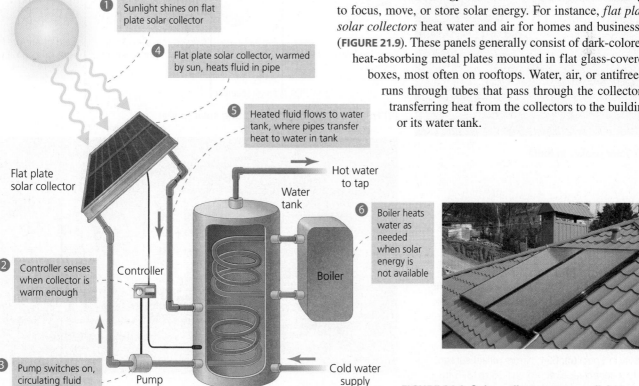

① Sunlight shines on flat plate solar collector

④ Flat plate solar collector, warmed by sun, heats fluid in pipe

⑤ Heated fluid flows to water tank, where pipes transfer heat to water in tank

Flat plate solar collector

② Controller senses when collector is warm enough

Controller

③ Pump switches on, circulating fluid through system

Pump

Water tank

Hot water to tap

⑥ Boiler heats water as needed when solar energy is not available

Boiler

Cold water supply

FIGURE 21.9 Solar collectors use sunlight to heat water for homes and businesses.

More than 300 million households and businesses worldwide heat water with solar collectors. China is the world's leader in this technology, having installed 70% of the world's solar collectors. The United States is a very distant second, and most American solar heating units are used for swimming pools. Germany is third; feed-in tariffs motivated Germans to install 200,000 new systems in one year (2008) alone. Solar heating systems are especially useful in isolated locations, including remote rural communities in the developing world. In Gaviotas, a remote town on the high plains of Colombia in South America, residents use active solar technology for heating, cooling, and water purification. Gaviotas is but one example illustrating that solar energy need not be confined to wealthy communities or to regions that are always sunny.

Concentrating sunlight focuses energy

We can intensify solar energy by gathering sunlight from a wide area and focusing it on a single point. That is the principle behind solar cookers, simple portable ovens that use reflectors to focus sunlight onto food and cook it (**FIGURE 21.10a**). Such cookers are proving extremely useful in the developing world.

At much larger scales, utilities are using the principle behind solar cookers to generate electricity at large centralized facilities and transmit power to homes and businesses via the electrical grid. Such **concentrated solar power** (CSP) is being harnessed by several methods (**FIGURE 21.10b**) in sunny regions in Spain, the U.S. Southwest, and elsewhere. The dominant technology used so far is the parabolic trough approach (leftmost diagram in Figure 21.10b), in which curved mirrors focus sunlight onto synthetic oil in pipes. The superheated oil is piped to an adjacent facility where it heats water, creating steam that drives turbines to generate electricity. In another approach, hundreds of mirrors concentrate sunlight onto a receiver atop a tall "power tower" (**FIGURE 21.10c**). From this central receiver, heat is transported by air or fluids (often molten salts) through pipes to a steam-driven generator to create electricity. CSP facilities can harness light from lenses or mirrors spread across large areas of land, and the lenses or mirrors may move to track the sun's movement.

Concentrated solar power holds promise for producing tremendous amounts of energy. The International Energy Agency has estimated that just 260 km^2 (100 mi^2) of Nevada desert dedicated to CSP facilities could generate enough electricity to power the entire U.S. economy. Indeed, the world's deserts soak up enough sun in just six hours to power humanity's energy demands for an entire year! To harness this potential, German scientists, industrialists, and investors a decade ago spearheaded an ambitious effort to create an immense CSP

(a) Solar cooker in India

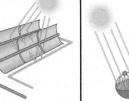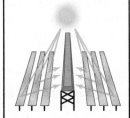

Curved reflectors heat liquid in horizontal tubes

Each curved reflector focuses light on its own small receiver

Curved mirrors reflect light onto absorber tube

Field of mirrors focuses light on central power tower

(b) Four methods of concentrating solar power

FIGURE 21.10 By concentrating solar energy, we can provide heat and electricity. Solar cookers **(a)** focus solar radiation to cook food. Utilities concentrate solar power with several approaches **(b)** to generate electricity at large scales. At the Planta Solar 10 (PS10) facility in Spain **(c),** 624 mirrors reflect sunlight onto a central receiver atop a 115-m (375-ft) power tower.

(c) The PS10 power tower facility in Spain

facility in the Sahara Desert. The $775-billion DESERTEC project called for thousands of mirrors to harness the Sahara's sunlight and transmit electricity to Europe, the Middle East, and North Africa. Critics of the project argued that it would be vulnerable to sandstorms and political disputes and would be less reliable and more expensive than decentralized production from rooftop panels. Such concerns led investors to back out of the project, and today its future is uncertain.

Although CSP has great potential, CSP developments (which typically spread across huge stretches of land) pose considerable environmental impacts. All land beneath the mirrors is cleared and graded, destroying sensitive habitat for desert species, and maintenance requires enormous amounts of water, which is scarce and precious in desert regions. In recent years, global electricity generation by CSP plants has grown almost as fast as electricity generation from photovoltaic rooftop panels.

PV cells generate electricity

The most direct way to generate electricity from sunlight involves photovoltaic (PV) systems. **Photovoltaic (PV) cells** convert sunlight to electrical energy when light strikes one of a pair of plates made primarily of silicon, a semiconductor that conducts electricity. The light causes one plate to release electrons, which are attracted by electrostatic forces to the opposing plate. Connecting the two plates with wires enables the electrons to flow back to the original plate, creating an electrical current (direct current, DC), which can be converted into alternating current (AC) and used for residential and commercial electrical power (**FIGURE 21.11a**). Small PV cells may power your watch or your calculator. Atop the roofs of homes and other buildings, PV cells are arranged in modules, which make up panels that can be gathered together in arrays (**FIGURE 21.11b**).

Researchers are experimenting with variations on PV technology, including **thin-film solar cells,** photovoltaic materials compressed into ultra-thin sheets. Thin-film solar cells are lightweight and far less bulky than the standard crystalline silicon cells shown in Figure 21.11. Although less efficient at converting sunlight to electricity, they are cheaper to produce. Thin-film technologies can be incorporated into roofing shingles and potentially many other types of surfaces, even highways. For these reasons, some people view thin-film solar technologies as a promising direction for the future.

Photovoltaic cells of all types can be connected to batteries that store the accumulated charge until needed. Alternatively, producers of PV electricity can sell their power to their local utility if they are connected to the regional electrical grid. In parts of 46 U.S. states, homeowners can sell power to

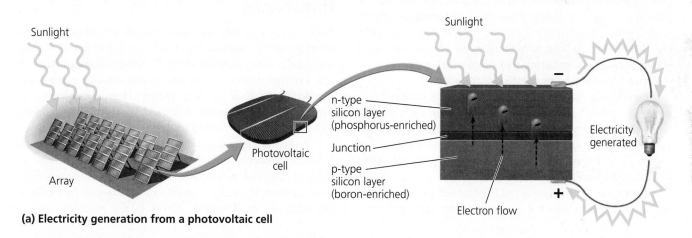

(a) Electricity generation from a photovoltaic cell

(b) Rooftop PV panels in Freiburg, Germany

FIGURE 21.11 A photovoltaic (PV) cell converts sunlight to electrical energy. When sunlight hits the silicon layers of the cell **(a),** electrons are knocked loose from some of the silicon atoms and tend to move from the boron-enriched "p-type" layer toward the phosphorus-enriched "n-type" layer. Connecting the two layers with wiring remedies this imbalance as electrical current flows from the n-type layer back to the p-type layer. This direct current (DC) is converted to alternating current (AC) to produce usable electricity. The photo **(b)** shows PV panels on rooftops of homes in Germany.

their utility in a process called **net metering,** in which the value of the power the consumer provides is subtracted from the consumer's utility bill. Feed-in tariff systems like Germany's go a step further by paying producers more than the market price of the power, thereby offering power producers the hope of turning a profit.

Solar energy offers many benefits

The sun will continue burning for another 4–5 billion years, which makes it inexhaustible as an energy source for human civilization. Moreover, the amount of solar energy reaching Earth should be enough to power our civilization once we deploy technology adequate to harness it. These advantages of solar energy are clear, but the technologies themselves also provide benefits. PV cells and other solar technologies use no fuel, are quiet and safe, contain no moving parts, and require little maintenance. An average unit can produce energy for 20–30 years.

Solar systems allow for local, decentralized control over power. In places where PV systems are connected to a regional electrical grid, homeowners can sell excess solar power to their utility through feed-in tariffs or net metering. More remote homes, businesses, and rural communities can use solar power to produce electricity without being near a power plant or connected to a grid. Likewise, one can use flat plate solar collectors to heat water or air just about anywhere. And in many industrializing nations, inexpensive solar cookers (see Figure 21.10a) help families of any income level cook food without gathering fuelwood, which eases people's daily workload and helps reduce deforestation.

Developing, manufacturing, and deploying solar technology also create new green-collar jobs. Currently, among renewable energy sources, PV technology employs the most people, resulting in nearly 3.6 million jobs worldwide (see Figure 21.5). PV installation is the fastest-growing job in the United States today.

Finally, a major advantage of solar power over fossil fuels is that solar power does not emit greenhouse gases and other air pollutants. The manufacture and transport of equipment *does* currently require fossil fuels, but once up and running, a PV system produces no emissions (see Figure 21.4). In Germany, carbon dioxide emissions from energy sources have fallen by 25% since 1990, and emissions of seven other major pollutants (CH_4, N_2O, SO_2, NO_X, CO, volatile organic compounds, and dust) have been reduced by 12–95%. At least half of these reductions are attributed to solar power and other new renewable energy paid for under the feed-in tariff system.

To estimate the economic and environmental results of installing a PV system, consumers can access online calculators offered by the U.S. National Renewable Energy Laboratory and other sources. As of 2019, these calculators estimated that installing a standard 5-kilowatt PV system atop a home in Fort Worth, Texas, to provide most of the home's annual power needs would save the homeowners $833 each year on energy bills and would prevent more than 5 tons of carbon dioxide emissions per year—as much CO_2 as results from burning 525 gallons of gasoline. Even in overcast Seattle, Washington, a 5-kilowatt system producing half a home's energy needs will save $426 per year and prevent more than 3.6 tons of CO_2 emissions (equal to the emissions from burning 375 gallons of gas).

Location, timing, and cost can be drawbacks

Solar energy currently has three main disadvantages—yet all three are being resolved. One limitation is that not all regions are equally sunny (**FIGURE 21.12**). People in Seattle or Anchorage will find it more challenging to rely on solar energy than people in Phoenix or San Diego. However, observe in Figure 21.12 that Germany receives less sunlight than Alaska, and yet Germany is a world leader in solar power.

A second limitation is that solar energy is an intermittent resource. Power is only produced when the sun shines, so daily or seasonal variation in sunlight can limit stand-alone solar systems if storage capacity in batteries or fuel cells is not adequate or if backup power is not available from a

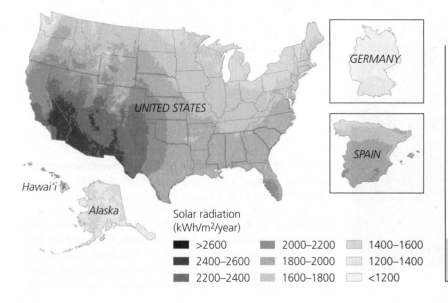

Solar radiation (kWh/m²/year)

- >2600
- 2400–2600
- 2200–2400
- 2000–2200
- 1800–2000
- 1600–1800
- 1400–1600
- 1200–1400
- <1200

FIGURE 21.12 Solar radiation varies from place to place. Harnessing solar energy is more profitable in sunny regions such as the southwestern United States than in cloudier regions such as Alaska and the Pacific Northwest. However, compare solar power leaders Germany and Spain with the United States. Spain is similar to Kansas in the amount of sunlight it receives, and Germany is cloudier than Alaska, suggesting that solar power can be used with success just about anywhere. *Data from National Renewable Energy Laboratory, U.S. Department of Energy.*

 Go to **Interpreting Graphs & Data** on **Mastering Environmental Science**

municipal electrical grid. However, at the utility scale, nonrenewable energy sources or renewable pumped-storage hydropower (p. 582) can help compensate for periods of low solar power production. And for individuals, battery storage is fast becoming efficient and affordable; indeed, the majority of German homeowners installing solar panels are now also storing excess energy in batteries to use later.

The third drawback of current solar technology is the upfront cost of the equipment. Because of the investment cost, solar power remains the most expensive way to produce electricity (see Figure 21.6a). The high costs result because the technologies are still developing. Moreover, solar power is competing against energy sources (fossil fuels and nuclear power) that have remained relatively cheap as a result of decades of taxpayer support through subsidies and whose external costs (pp. 143, 165) are not included in market prices.

However, recent declines in price and ongoing improvements in efficiency of solar technologies have been impressive, even in the absence of significant funding from government and industry. At their advent in the 1950s, solar technologies had efficiencies of around 6% while costing $600 per watt. Today, PV cells are showing up to 20% efficiency commercially and 46% efficiency in lab research, suggesting that future solar cells could be more efficient than any solar energy technologies we have today. Solar systems continue to become more affordable and now can sometimes pay for

themselves in less than 10 years. After that time, they provide energy virtually for free as long as the equipment lasts.

Solar energy is expanding

Although active solar technology dates from the 18th century, it was pushed to the sidelines as fossil fuels came to dominate our energy economy. Funding for research and development of solar technology has been erratic. After the 1973 oil embargo (p. 551), the U.S. Department of Energy funded the installation and testing of more than 3000 PV systems, providing a boost to fledgling companies in the solar industry. But later, as oil prices declined, so did government funding for solar power.

Largely because of the lack of investment, solar energy contributes just 0.93%—93 parts in 10,000—of the U.S. energy supply and just 1.6% of U.S. electricity generation. Even in Germany, which gets more of its energy from solar than any other nation, the percentage is only 7%. However, solar energy use has grown by nearly a third each year worldwide over the past four decades. Solar energy is proving especially attractive in developing countries rich in sun but poor in infrastructure, where hundreds of millions of people still live without electricity. Solar energy use is also growing rapidly on college and university campuses, where it plays a major role in campus sustainability efforts (see **SUCCESS STORY**).

SUCCESS story
Solar Energy on Campus

Among the many sustainability initiatives on college and university campuses today, the installation of solar energy technology has been one of the most significant. On hundreds of campuses, solar power and heating have been cutting emissions, saving on energy costs, and providing teaching and learning opportunities. Students have played a leading role in solarizing some campuses—such as at Northwestern University, where students planned a project, raised $117,000 for it, and then oversaw its installation. At Arizona State University, PV panels and solar heating systems at 89 locations on campus have replaced enough fossil fuel energy to reduce annual carbon dioxide emissions by the equivalent of nearly 5000 cars. At the University of Arizona, solar panels covering a parking structure do double duty by shading cars from the hot desert sun. At the University of Tennessee in Knoxville, PV panels generate electricity for charging stations for electric vehicles. Colorado State University integrates its solar development with the local community's economy and is a leader in research into photovoltaics. And in a major milestone, Butte College in northern California became the first U.S. campus to become "grid-positive," generating more electricity than it consumes. The college used special government loans to install 25,000 PV panels on rooftops, in fields, and atop parking lots. These schools offer just a few examples of how institutions of higher learning are turning to solar energy to make their campuses operate more

Stanford University students designed and built this solar-powered race car, which competed internationally in a race across the Australian desert.

sustainably. All these steps serve educational functions, too, as students learn and observe firsthand the many ways in which renewable energy can power our society now and in the future.

→ **Explore the Data** at Mastering Environmental Science

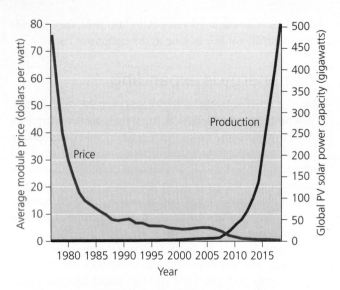

FIGURE 21.13 **Global production of PV solar power has grown rapidly, and prices have fallen sharply.** *Data from REN21, 2019. Renewables 2019: Global status report. Paris: REN21, UNEP; Bloomberg New Energy Finance; and pvXchange.*

PV technology is the fastest-growing power source today, having recently doubled every two years (**FIGURE 21.13**). Germany led the world in installation of PV technology until China (with its much larger economy) passed it in 2015—and German rooftops host more than one-sixth of all PV cells in the world. Germany's investment began in 1998 when Hermann Scheer spearheaded a "100,000 Rooftops" program to install PV panels atop 100,000 German roofs. The popular program ended up easily surpassing this goal, and today more than *1.5 million* German rooftops have PV systems.

China leads the world in yearly production of PV cells, followed by Germany and Japan, while U.S. firms now account for only 2% of the industry. In fact, the Chinese government's support of its solar industry led to so much production that supply has outstripped global demand in recent years. Highly subsidized Chinese firms have been selling solar products abroad at low prices (often at a loss), driving U.S. and European solar manufacturers out of business. In response, the United States and European nations slapped tariffs on Chinese

imports, while both sides filed complaints with the World Trade Organization in an escalating global trade dispute.

Despite volatility in the industry as firms try to deal with swings in national policies, global production of PV cells continues to rise sharply while prices fall (see Figure 21.13). At the same time, efficiencies are increasing, making each unit more powerful. Use of solar technology should continue to expand as prices fall, technologies improve, and governments enact economic incentives to spur investment. Similar trends are apparent for wind power, another major and growing source of renewable energy.

Wind Power

As the sun heats the atmosphere, it causes air to move, producing wind. We harness **wind power** (electricity generated from wind) by using **wind turbines,** mechanical assemblies that convert wind's kinetic energy (p. 31), or energy of motion, into electrical energy.

Wind turbines convert kinetic energy to electrical energy

Today's wind turbines have their roots in Europe, where wooden windmills were used for 800 years to grind grain and pump water. Thousands of American farms and ranches still use windmills to pump water; it is thought that there are 80,000 in Texas alone. The first wind turbine built to generate electricity was constructed in the late 1800s in Cleveland, Ohio. However, it was not until after the 1973 oil embargo that governments and industry in North America and Europe began funding research and development for wind power.

In a modern wind turbine, wind turns the blades of the rotor, which rotate machinery inside a compartment called a *nacelle* that sits atop a tower (**FIGURE 21.14**). Inside the nacelle are a gearbox, a generator, and equipment to monitor and control the turbine's activity. Today's towers average 80 m (260 ft) in height, and the largest are taller than the Statue of Liberty. Higher is generally better, to minimize turbulence (and potential damage) while maximizing wind speed.

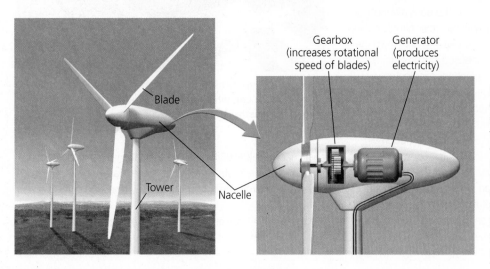

FIGURE 21.14 **A wind turbine converts wind's energy of motion into electrical energy.** Wind spins a turbine's blades, turning a shaft that extends into the nacelle. Inside the nacelle, a gearbox converts the rotational speed of the blades, which can be up to 20 revolutions per minute (rpm) or more, into much higher speeds (more than 1500 rpm), providing adequate motion for the generator to produce electricity.

Engineers design turbines to rotate back and forth in response to changes in wind direction so that the motor faces into the wind at all times. Some turbines are designed to generate low levels of electricity by turning in light breezes. Others are programmed to rotate only in strong winds, generating large amounts of electricity in short time periods. Slight differences in wind speed yield substantial differences in power output for two reasons. First, the energy content of wind increases as the square of its velocity; thus, if wind velocity doubles, energy quadruples. Second, an increase in wind speed causes more air molecules to pass through the wind turbine per unit time, making power output equal to wind velocity cubed. Thus, a doubled wind velocity results in an eightfold increase in power output.

Turbines are often erected in groups; such a development is called a **wind farm.** Today's largest wind farms contain hundreds of turbines. Texas boasts five of the world's ten largest wind farms; the largest, Roscoe, contains 627 turbines spread across 650 km^2 (250 mi^2) of land.

Wind power is growing fast

Wind currently provides just a small proportion of the world's power needs, but wind power is growing fast (**FIGURE 21.15**). Five nations account for almost three-fourths of the world's wind power output (**FIGURE 21.16a**), but dozens of nations now produce wind power. Germany had long produced the most, but the United States overtook it in 2008, and China surpassed the United States two years later.

Denmark leads the world in obtaining the highest percentage of its energy from wind power. In this small European nation, wind farms supply nearly half of Danish electricity needs (**FIGURE 21.16b**). Germany is sixth in this respect, and the United States is 15th. Texas generates the most wind power of all U.S. states and generates more wind power than all but five nations. Iowa, South Dakota, and Kansas each obtain more than 30% of their electricity from wind.

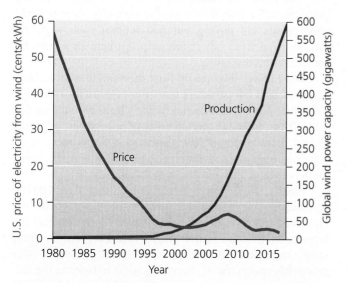

FIGURE 21.15 Global production of wind power has been doubling every three years in recent years, and prices have fallen. *Data from Global Wind Energy Council; and U.S. Department of Energy, EERE, 2018.* 2017 Wind technologies market report.

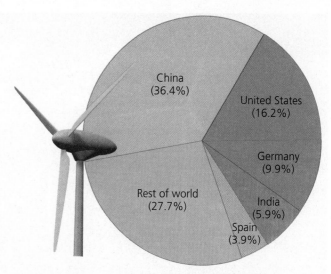

(a) Percentage of global wind power in each nation

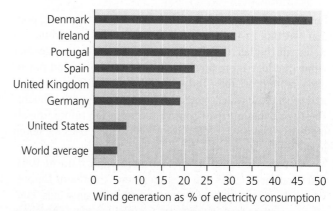

(b) Leading nations in proportion of electricity from wind power

FIGURE 21.16 Several nations are leaders in wind power. Most of the world's wind power capacity **(a)** is concentrated in a handful of nations led by China, the United States, and Germany. Yet tiny Denmark **(b)** obtains the highest percentage of its electricity needs from wind. *Data from (a) World Wind Energy Association; and (b) U.S. Department of Energy, EERE, 2018.* 2017 Wind technologies market report.

Wind power's growth in the United States has been haphazard because Congress has never committed to a long-term federal tax credit for wind development but instead has passed a series of short-term renewals, leaving the industry uncertain about how to invest. However, experts agree that wind power's growth will continue, because only a small portion of this resource is currently being tapped and because wind power at favorable locations already generates electricity at prices lower than fossil fuels (see Figure 21.6b). Recent reports estimate that with adequate development, the United States could meet 20% of its electricity demands with wind power by 2030 and 35% by 2050.

Offshore sites are productive

Wind speeds on average are 20% greater over water than over land, and air is less turbulent (more steady) over water.

FIGURE 21.17 More and more wind farms are being developed offshore. Offshore winds tend to be stronger and steadier than winds over land.

For these reasons, offshore wind turbines are becoming popular (**FIGURE 21.17**). Costs to erect and maintain turbines in water are higher than on land, but the stronger, steadier winds produce more power and make offshore wind potentially more profitable. Today's offshore wind farms are limited to shallow water, where towers are sunk into sediments singly or using a tripod configuration. In the future, towers may also be placed in deep water on floating pads anchored to the seafloor.

Denmark erected the first offshore wind farm in 1991, and soon more came into operation across northern Europe, where the North Sea and Baltic Sea offer strong winds. Germany raised its feed-in tariff rate for offshore wind sharply in 2009, and within just five years, 14 new wind farms were operating. By 2019, more than 4500 wind turbines were powering 105 wind farms in the waters of 11 European nations.

In the United States, a small project in Rhode Island waters is the nation's first commercial offshore wind farm. As of 2019, roughly 20 more offshore wind developments were in the planning stages, mostly off the North Atlantic coast. Several wind farms are being proposed off the Texas coast to take advantage of the favorable wind conditions there.

Wind power has many benefits

Like solar power, wind power produces no emissions once the equipment is manufactured and installed. As a replacement for fossil fuel combustion in the average U.S. power plant, running a 1-megawatt wind turbine for 1 year prevents the release of more than 1500 tons of carbon dioxide, 6.5 tons of sulfur dioxide, 3.2 tons of nitrogen oxides, and 60 lb of mercury, according to the U.S. Environmental Protection Agency. The amount of carbon pollution that all U.S. wind turbines together prevent from entering the atmosphere is equal to the emissions from 34 million cars or from combusting the cargo of 23 hundred-car freight trains of coal each and every day.

Under optimal conditions, wind power appears efficient in its energy returned on investment (EROI; pp. 528, 580). Studies find that wind turbines produce roughly 20 times more energy than they consume—an EROI value superior to that of most energy sources. Wind farms also use less water than do conventional power plants.

Wind turbine technology can be used on many scales, from a single tower for local use to large farms that supply entire regions. Small-scale turbine development can help make local areas more self-sufficient, just as solar energy can. Even college and university campuses are installing wind turbines. Carleton College in Minnesota was the first campus to do so, in 2004, and today Carleton gets up to 70% of its electricity from its two turbines.

Another benefit of wind power is that farmers and ranchers can make money leasing their land for wind development. A single turbine can bring in anywhere from $2000 to $25,000 in annual royalties while occupying just a quarter-acre of land. Most of the land can still be used for agriculture. Royalties from the wind power company provide the farmer or rancher revenue while also increasing property tax income for the rural community. Across Texas, wind leases are paying out $60 million each year, and income from wind power has helped a number of ranches survive recent droughts.

Wind power involves up-front expenses to erect turbines and to expand infrastructure to transmit electricity, but unlike fossil fuel power plants, wind turbines incur no ongoing fuel costs. Thus, although startup costs of wind farms generally exceed those of fossil fuel plants, wind farms incur fewer expenses once up and running, which is why wind power often is cheaper than electricity from fossil fuels.

Finally, wind power creates jobs. More than 100,000 Americans and nearly 1.2 million people globally are employed in jobs related to wind power. More than 100 colleges and universities offer programs and degrees that train people in the skills needed for jobs in wind power and other renewable energy fields. Servicing wind turbines is the second-fastest-growing job in the United States—and the fastest-growing one in Texas, where 24,000 wind-related jobs are providing lucrative employment for young people in rural areas.

Wind power has limitations

Like solar energy, wind is an intermittent resource; we have no control over when it will occur. This limitation is alleviated, though, if wind is one of several sources contributing to a utility's power generation (see **THE SCIENCE BEHIND THE STORY**, pp. 606–607). Pumped-storage hydropower (p. 582) can help compensate during windless times, and batteries or hydrogen fuel (p. 613) can store energy generated by wind and release it later when needed.

Just as wind varies from time to time, it varies from place to place. Global wind patterns combine with local topography—mountains, hills, water bodies, forests, cities—to make some places windier than others. Resource planners and wind power companies study wind patterns revealed by meteorological research before planning a wind farm. A map of average wind speeds across the United States (**FIGURE 21.18a**) reveals that mountainous regions, offshore sites, and areas of the Great Plains are best. Based on such data, the wind power industry has located much of its generating capacity in states with high wind speeds (**FIGURE 21.18b**) and is expanding in the Great Plains and mountain states. Provided that wind farms are strategically erected in optimal locations, an estimated 15% of U.S. energy demand could be met using only 43,000 km^2 (16,600 mi^2) of land (with less than 5% of this land area actually occupied by turbines, equipment, and access roads).

However, most of North America's people live near the coasts, far from the Great Plains and mountain regions that have the best wind resources. Thus, continent-wide transmission networks would need to be strategically and thoughtfully enhanced—as they have been in Texas—to send wind-generated electricity to these coastal population centers, or numerous offshore wind farms would need to be developed.

When wind farms *are* proposed near population centers, local residents often oppose them. Turbines are generally located in exposed, conspicuous sites, and although many people find the structures beautiful, others dislike them and feel that they clutter the landscape. Although polls show wide public approval of existing wind projects and of the concept of wind power, newly proposed wind projects often elicit the **not-in-my-backyard** (NIMBY) syndrome among people living nearby. For instance, America's first offshore wind farm went to Rhode Island because an earlier proposal for one in nearby Nantucket Sound off Massachusetts failed after facing years of opposition from many wealthy residents of Cape Cod, Nantucket, and Martha's Vineyard.

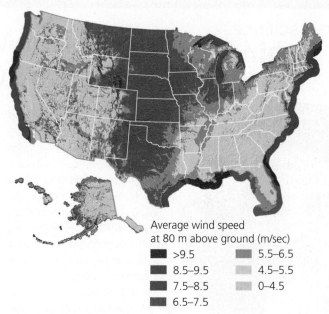

Average wind speed
at 80 m above ground (m/sec)

- ⬛ >9.5
- ⬛ 8.5–9.5
- ⬛ 7.5–8.5
- ⬛ 6.5–7.5
- ⬛ 5.5–6.5
- ⬛ 4.5–5.5
- ⬛ 0–4.5

(a) Annual average wind power

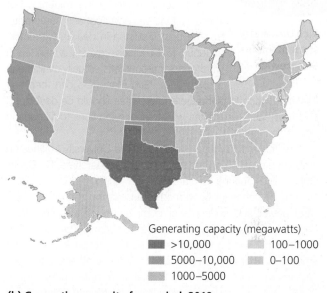

Generating capacity (megawatts)

- ⬛ >10,000
- ⬛ 5000–10,000
- ⬛ 1000–5000
- ⬛ 100–1000
- ⬛ 0–100

(b) Generating capacity from wind, 2019

FIGURE 21.18 Wind speed varies from place to place. Maps of average wind speeds **(a)** help guide the placement of wind farms. Another map **(b)** shows wind power generating capacity installed in each U.S. state through 2019. *Sources: (a) U.S. National Renewable Energy Laboratory; (b) American Wind Energy Association, 2019. 1st Quarter 2019 Market Report.*

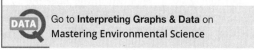

Go to **Interpreting Graphs & Data** on **Mastering Environmental Science**

WEIGHING the **issues**

Wind and NIMBY

If you could choose to get your electricity from a wind farm or a coal-fired power plant, which would you choose? How would you react if the electric utility proposed to build the wind farm that would generate your electricity atop a ridge running in back of your neighborhood, such that the turbines would be clearly visible from your living room window? Would you support or oppose the development? Why? If you would oppose it, where would you suggest the farm be located? Do you think anyone might oppose it in that location?

Wind turbines also pose a hazard to birds and bats, which may be killed if they fly into the blades or are hit by the intense pressure waves these blades create as they rotate. Large open-country raptors such as golden eagles are known to be at risk, and turbines located on ridges along migratory flyways are likely most damaging. One strategy for protecting birds and

THE SCIENCE behind the story

How Well Do Wind and Solar Complement One Another in Texas?

Rice University student Joanna Slusarewicz

A major critique of wind power and solar power has been that they are intermittent resources. Wind does not always blow, and sunshine reaches the ground only during daytime when it is not cloudy. Because wind and sunshine are only intermittently available, we might risk shortages of power if we were to rely on them for all of our electricity needs.

Storing energy—with batteries, hydrogen fuel, or pumped storage hydropower, for example—can help compensate for intermittency, but these options can add cost and complexity. So, the best first step in dealing with intermittency is to assess exactly when each wind or solar resource is actually available in a given area. Grid managers can then use this information to plan where future sites should be developed and how best to link sites with transmission lines to fill gaps in power generation.

For Texas, two researchers from Rice University in Houston embarked on such an assessment. The researchers, Dr. Daniel Cohan and Joanna Slusarewicz, an undergraduate student working in Cohan's lab, wanted to assess the degree to which wind power and solar power complement one another from hour to hour, day to day, and season to season. They set out to test whether wind and solar sources tend to peak at different times and thereby compensate for one another's shortfalls. Slusarewicz and Cohan hoped that their data—published in 2018 in the journal *Renewables: Wind, Water and Solar*—could help the Electric Reliability Council of Texas (ERCOT) maximize "complementarity" and ensure a reliable supply of renewable power at all times.

Slusarewicz and Cohan began by selecting 12 sites across Texas—five sites with existing wind farms and seven sites with existing solar facilities (**FIGURE 1**). Three of the wind sites were in West Texas, and two were in southern Texas near the Gulf Coast. The scientists then compiled data collected by federal agencies from 2007 to 2013 on solar irradiance, wind strength, and other weather variables for the 12 sites.

Using these data, they assessed total power generation, variability, and reliability for each site. A key measure was the *capacity factor,* or percentage of a site's possible power output that is actually achieved, averaged over time.

When the researchers looked at how the generation of wind power and solar power varied throughout a typical 24-hour period, their findings were unsurprising yet striking: Solar power averaged strong in the daytime and, of course, was absent at night. In contrast, wind power was generated throughout the 24 hours but was greater at night than during the day. In this way, wind and solar were complementary to each other: wind was strong when solar was weak, and vice versa.

Comparing two sample days—the summer solstice (June 21) and the winter solstice (December 21), the data show solar power to be greater in June, as expected (**FIGURE 2**). The comparison also shows that in summer, the coastal wind sites provide a great deal of power in the late afternoon and early evening—a time of high electricity demand when millions of air conditioners are running in the Texas heat.

Slusarewicz and Cohan also analyzed seasonal patterns across a typical year. They found that wind sites averaged higher capacity factors than solar sites but that this difference was greatest in winter and least in late summer. However, when it comes to the reliability of an electric grid, what matters most is

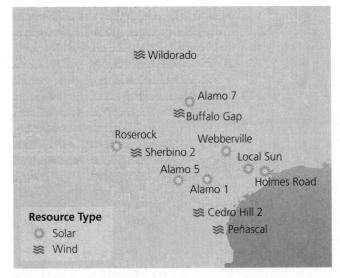

FIGURE 1 Twelve sites across Texas were studied. These included 7 solar sites, three West Texas wind sites, and two southern coastal wind sites (Cedro Hill 2 and Peñascal). *Data from Slusarewicz, J.H. and D.S. Cohan, Assessing solar and wind complementarity in Texas.* Renewables: Wind, Water and Solar *5: 7.*

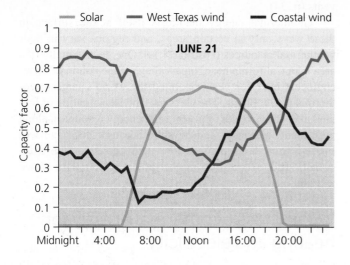

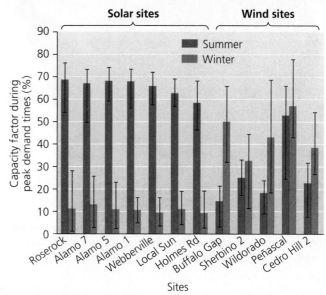

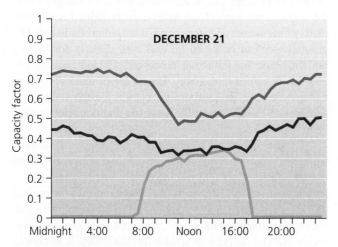

FIGURE 2 Solar power peaks in midday, whereas wind power tends to be higher at night. In summer (shown by graph for June 21, at top), coastal sites have stronger wind in late afternoon and evening than they do in winter (shown by graph for December 21, at bottom). *Data from Slusarewicz, J.H., and D.S. Cohan, Assessing solar and wind complementarity in Texas.* Renewables: Wind, Water and Solar 5: 7.

FIGURE 3 Solar sites provided power to meet peak demand in summer, whereas wind power served this purpose in winter. *Data from Slusarewicz, J.H., and D.S. Cohan, Assessing solar and wind complementarity in Texas.* Renewables: Wind, Water and Solar 5: 7.

how well energy sources can supply power at times of peak demand. In this respect, the data were clear. Solar sites had high capacity factors at peak demand times in the summer but low production in the winter. Wind sites showed the opposite pattern, with higher capacity factors at peak demand times in the winter (**FIGURE 3**).

The researchers delved deeper by conducting detailed tests of different paired combinations of sites, measuring how much sites varied, and assessing patterns when power production is enhanced with technology to allow solar panels to track the sun and turbines to swing to better follow the wind. All their analyses led to essentially the same conclusion: In Texas, pairing wind power with solar power gives quite good complementarity— when one resource goes down, the other tends to go up. The data also suggested that having a diversity of wind sites was important, because their power production varied more than the solar sites.

Slusarewicz and Cohan hope that their data can be used by ERCOT to help with decisions on where and how to build and interlink future renewable energy projects. The data do not show that Texas is ready to shift to 100% renewables quite yet, but they do indicate that wind and solar can go a long way toward replacing fossil fuels while ensuring reliability of electrical power at all times. The research suggests that together, wind and solar power from a diversity of sites across Texas can produce a steady and reliable production of electricity and help stabilize the grid.

In principle, these findings should apply to many parts of the world, because patterns of wind strength and solar irradiance seem complementary in many regions. However, few locations have as much sun, wind, *and* diversity of areas with differing conditions as Texas does. Thus, other states and nations would be wise to assess their own conditions in detail to determine how confidently they can rely on the complementarity of wind and solar resources.

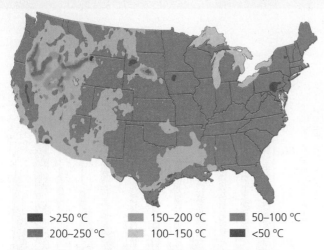

>250 °C **150–200 °C** **50–100 °C**
200–250 °C **100–150 °C** **<50 °C**

FIGURE 21.19 Geothermal resources in the United States are greatest in the western states. This map shows water temperatures 3 km (1.9 mi) belowground. *Data from Idaho National Laboratory.*

bats is to select sites that are not on migratory flyways or amid prime habitat for species likely to fly into the blades, but more research on wildlife impacts and how to prevent them is urgently needed.

Geothermal Energy

Geothermal energy is thermal energy that arises from beneath Earth's surface. The radioactive decay of elements (p. 27) amid high pressures deep in the interior of our planet generates heat that rises to the surface through magma (molten rock, p. 34) and through cracks and fissures. Where this energy heats groundwater, natural spurts of heated water and steam rise up from below and may erupt through the surface as terrestrial geysers or as submarine hydrothermal vents (p. 34).

Geothermal energy manifests itself at the surface in these ways only in certain areas, and regions vary in their geothermal resources (**FIGURE 21.19**). One geothermally rich area is in California near Napa Valley's wine country. There, engineers have for years operated the world's largest geothermal power plants, The Geysers. The island nation of Iceland also has a wealth of geothermal energy resources, along with many volcanoes and geysers, because the island was formed from lava that extruded and cooled at the Mid-Atlantic Ridge (p. 36), along the spreading boundary of two tectonic plates.

We harness geothermal energy for heating and electricity

Geothermal energy can be harnessed directly from geysers at the surface, but most often, wells must be drilled down hundreds or thousands of meters toward heated groundwater. Hot groundwater can be piped up and used directly for heating buildings and for industrial processes. Iceland heats nearly 90% of its homes in this way. Direct use of naturally heated water is efficient and inexpensive, but it is feasible only where geothermal energy is readily available and does not need to be transported far.

Geothermal power plants harness the energy of naturally heated underground water and steam to generate electricity (**FIGURE 21.20**). Generally, a power plant brings water at temperatures of 150–370°C (300–700°F) or more to the surface and converts it to steam by lowering the pressure in specialized compartments. The steam turns turbines to generate electricity. At The Geysers in California, electricity is generated for 725,000 homes.

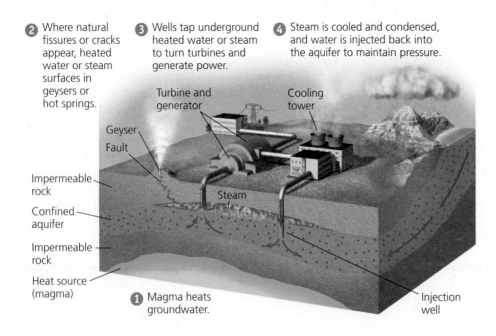

2 Where natural fissures or cracks appear, heated water or steam surfaces in geysers or hot springs.

3 Wells tap underground heated water or steam to turn turbines and generate power.

4 Steam is cooled and condensed, and water is injected back into the aquifer to maintain pressure.

Turbine and generator

Cooling tower

Geyser
Fault

Impermeable rock

Confined aquifer

Impermeable rock

Steam

Heat source (magma)

1 Magma heats groundwater.

Injection well

FIGURE 21.20 Geothermal power plants generate electricity using naturally heated water from underground.

FIGURE 21.21 The Nesjavellir power station in Iceland uses geothermal energy to provide heated water and to generate electricity.

At Iceland's Nesjavellir power station (**FIGURE 21.21**), steam piped in from wells heats water from a lake. The heated water is sent through an insulated pipeline to the capital city of Reykjavik, where residents use it for washing and space heating. Globally, one-third of the geothermal energy we use is for electricity, and two-thirds is used for direct heating.

Heat pumps make use of temperature differences

Heated groundwater is available only in certain areas, but we can take advantage of the mild temperature differences that exist naturally between the soil and the air just about anywhere. Soil varies in temperature from season to season less than air does, because it absorbs and releases heat more slowly and because warmth and cold do not penetrate deeply belowground. Just several inches below the surface, temperatures are nearly constant year-round. Geothermal heat pumps, or **ground-source heat pumps,** make use of this fact.

Ground-source heat pumps provide heating in the winter by transferring heat from the ground into buildings and provide cooling in the summer by transferring heat from buildings into the ground. This heat transfer is accomplished with a network of underground plastic pipes that circulate water and antifreeze (**FIGURE 21.22**). Because heat is simply moved from place to place rather than being produced using outside energy inputs, heat pumps can be highly energy-efficient.

More than 600,000 U.S. homes use ground-source heat pumps. Compared with conventional electric heating and

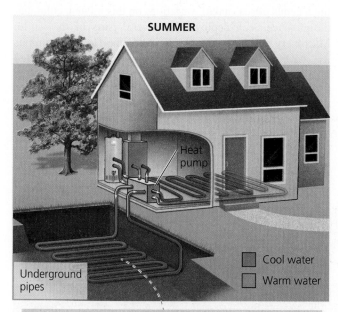

SUMMER

Heat pump

Underground pipes

☐ Cool water
☐ Warm water

In summer, soil underground is **cooler** than surface air. Water flowing through the pipes transfers heat from the house to the ground, cooling air in ducts or a radiant cooling system under the floor.

WINTER

In winter, soil underground is **warmer** than surface air. Water flowing through the pipes transfers heat from the ground to the house, warming air in ducts, water in a tank, or a radiant heating system under the floor.

FIGURE 21.22 Ground-source heat pumps provide an efficient way to heat and cool air and water in a home. A network of pipes filled with water and antifreeze extends underground. Soil is cooler than air in the summer **(left),** and warmer than air in the winter **(right),** so by running fluid between the house and the ground, these systems adjust temperatures inside.

cooling systems, ground-source heat pumps heat spaces 50–70% more efficiently, cool them 20–40% more efficiently, can reduce electricity use by 25–60%, and can reduce emissions by up to 70%.

Geothermal power has pros and cons

All forms of geothermal energy—direct heating, electrical power, and ground-source heat pumps—greatly reduce emissions relative to fossil fuel combustion. By one estimate, each megawatt of electricity produced at a geothermal power plant prevents the emission of 7.0 million kg (15.5 million lb) of carbon dioxide each year. Geothermally heated water can release dissolved gases, including carbon dioxide, methane, ammonia, and hydrogen sulfide, but these dissolved gases generally occur in small quantities, and many facilities use filters to cut down on their emission.

In principle, geothermal energy is renewable, because using it does not affect the amount of thermal energy produced underground. However, not every geothermal power plant will be able to operate indefinitely. If a plant uses heated water more quickly than groundwater is recharged, it will eventually run out of water. That problem was occurring at The Geysers in California, which began operating in 1960. In response, operators began injecting municipal wastewater into the ground to replenish the supply. Many geothermal plants worldwide are now injecting water back into aquifers to help maintain pressure and sustain the resource.

A second reason geothermal energy is not always renewable is that patterns of geothermal activity in Earth's crust shift naturally over time. As a result, an area that produces hot groundwater now may not always do so. In addition, some hot groundwater is laced with salts and minerals that corrode equipment and pollute the air. These factors may shorten the lifetime of plants, increase maintenance costs, and add to pollution.

The greatest limitation of geothermal power is that it is restricted to regions where we can tap energy from naturally heated groundwater. Places such as Iceland, northern California, and Yellowstone National Park (with its famed geysers) are rich in naturally heated groundwater, but most areas of the world are not.

Enhanced geothermal systems might widen our reach

To broaden where we can harness geothermal energy, engineers have been developing **enhanced geothermal systems** (EGS), in which we drill deeply into dry rock, fracture the rock, and pump in cold water. The water becomes heated deep underground and is then drawn back up and used to generate power. In theory, we could use EGS in many locations. Germany, for instance, has little heated groundwater, but feed-in tariffs have enabled an EGS facility to operate profitably there.

EGS technology shows significant promise, and a 2006 report estimated that heat resources below the United States could power the world's energy demands for several millennia. However, EGS also appears to trigger minor earthquakes. Unless we can develop ways to use EGS safely and reliably, our use of geothermal power will remain localized.

Ocean Energy Sources

The oceans are home to several underexploited sources of energy resulting from continuous natural processes. Of the four approaches being developed, three involve motion, and one involves temperature.

We can harness energy from tides, waves, and currents

Just as dams on rivers use flowing fresh water to generate hydroelectric power (p. 582), we can use kinetic energy from the natural motion of ocean water to generate electrical power.

Scientists and engineers are working to harness the motion of ocean waves and convert their energy into electricity. Many designs for machinery to harness **wave energy** have been invented, although few have begun commercial use. Some designs for offshore facilities involve floating devices that move up and down with the waves. An example is the snake-like wave energy converter shown in **FIGURE 21.23**. Built by the Scottish company Pelamis (a Latin word denoting a genus of sea snake), variations on this jointed, columnar design have been deployed in several parts of the world. Machinery and hydraulic fluids inside the floating columns use wave motion to generate electricity, which is transmitted to shore via undersea cables.

FIGURE 21.23 Various technologies harness energy from ocean waves. This wave energy converter stretches across the water off the coast of Scotland. As waves flex segments of the machinery, hydraulic equipment in the joints generates electricity and sends it by undersea cables to shore, where it powers 500 homes.

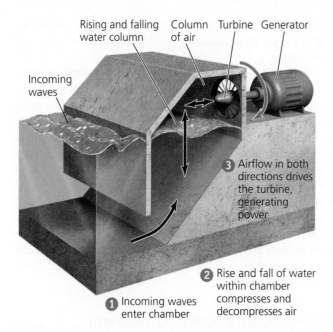

FIGURE 21.24 **Coastal facilities harness energy from ocean waves.** In one design, as waves enter and exit a chamber ❶, the air inside is alternately compressed and decompressed ❷, creating airflow that rotates turbines ❸ to generate electricity.

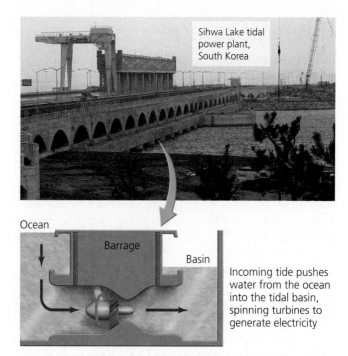

FIGURE 21.25 **We can harness tidal energy by allowing ocean water to spin turbines.** At the world's largest tidal power facility, water flows from the ocean into a huge enclosed basin as the tide rises, spinning turbines to generate electricity.

Wave energy is greatest on the open ocean, but transmitting electricity to shore is expensive; thus, many efforts to harness wave energy are situated along coasts. Some designs for coastal onshore facilities funnel waves from large areas into narrow channels and elevated reservoirs, from which water then flows out, generating electricity as hydroelectric dams do. Other coastal designs use rising and falling waves to push air into and out of chambers, turning turbines (**FIGURE 21.24**). The first commercially operating wave energy facility began operating in 2011 in Spain, using technology tested in Scotland. Demonstration projects exist in Europe, Japan, and Oregon.

We are also developing ways of harnessing energy from tides. The rise and fall of ocean tides (p. 429) twice each day move large amounts of water past any given point on the world's coastlines. Differences in height between low and high tides are greatest in long, narrow bays such as Alaska's Cook Inlet or the Bay of Fundy between New Brunswick and Nova Scotia. Such locations are best for harnessing **tidal energy** by erecting dams across the outlets of tidal basins. In the most common design, the incoming tide flows through sluice gates and is trapped behind them. Then, as the outgoing tide passes through the gates, it turns turbines to generate electricity. The world's largest tidal generating station, South Korea's Sihwa Lake facility, generates electricity from the incoming tide (**FIGURE 21.25**). Some designs generate electricity from water moving in both directions.

The Sihwa Lake power station opened in 2011 and is just larger than the La Rance tidal facility in France, which has operated for half a century. Smaller facilities operate in Canada, China, Russia, and the United Kingdom. The first U.S. tidal station began operating in 2012 in Maine, and one is now being installed in New York City's East River. Five more tidal stations are planned in South Korea, but some have been delayed amid concerns over environmental impacts. Tidal stations release few or no pollutant emissions, but they can affect the ecology of estuaries and tidal basins.

A third way to harness marine kinetic energy is to use the motion of ocean currents (p. 425), such as the Gulf Stream. Underwater turbines have been erected in European waters to test this approach.

The ocean stores thermal energy

Each day, the tropical oceans absorb solar radiation with the heat content of 250 billion barrels of oil—enough to provide 20,000 times the electricity used daily in the United States. The ocean's sun-warmed surface is warmer than its deep water, and **ocean thermal energy conversion** (OTEC) relies on this gradient in temperature.

In one approach, warm surface water is piped into a facility to evaporate chemicals, such as ammonia, that boil at low temperatures. These evaporated gases spin turbines to generate electricity. Cold water piped up from ocean depths then condenses the gases, so they can be reused. In another approach, warm surface water is evaporated in a vacuum, and its steam turns turbines and then is condensed by cold water. Because ocean water loses salts as it evaporates, the water can be recovered, condensed, and sold as desalinized fresh water for drinking or agriculture. So far, no OTEC facility operates commercially, but research is being conducted in Hawai'i and elsewhere.

Energy Storage

Each renewable energy source we have discussed can be used to generate electricity more cleanly than is possible with fossil fuels. However, it is challenging to store electricity easily and cheaply in large quantities for use when and where it is needed. That is why most vehicles rely on gasoline for power. As renewable energy sources that are intermittent—such as solar power and wind power—come to comprise larger proportions of our energy budgets, it is becoming increasingly important that we develop ways to store large quantities of electricity conveniently and efficiently at low cost. "Decarbonizing" our energy economy and shifting fully to renewable energy—as we will need to do if we are to halt climate change—will require that we devise new means of storing energy and that we scale up existing means. Fortunately, we have many known and proven avenues for **energy storage,** the capture and accumulation of energy for use at a later time.

Batteries use chemical reactions to store energy

The means of storing energy that is most familiar to us is to use a **battery,** a device that employs chemical reactions to generate electricity. There are many types of batteries, including disposable batteries like those that power a flashlight, rechargeable lead-acid batteries like those that power conventional automobiles, and rechargeable lithium-ion batteries like those that power your cell phone and most electric vehicles. Research and development into new and better types of batteries is active, and costs of batteries are falling rapidly. These improvements are beginning to make possible the kind of large-scale storage that will be necessary to convert our electric grids to accommodate rising amounts of solar and wind power.

In Texas, the Notrees Wind Storage Demonstration Project uses a 36-megawatt facility of batteries to store energy from wind and transmit it as needed, helping ensure an even, stable, reliable power supply. In Germany, more than 120,000 people have now invested in battery storage systems along with their solar PV panels. A well-outfitted German may obtain all the electricity for his or her home or small business from rooftop panels and be able to charge an electric car as well. Excess electricity then feeds into the battery for use at night or on a cloudy day. Once the battery is topped off, any further excess is sent automatically to the grid and collects a feed-in tariff payment. All along, smart software allows the homeowner or business owner to monitor energy production, use, storage, and sale.

Pumped storage hydropower is our primary means of energy storage

Currently, our principal way of storing energy is pumped storage hydropower (p. 582; **FIGURE 21.26**). In pumped storage, water is pumped uphill from one reservoir to another when demand for electricity is low and prices are cheap and then is allowed to flow downhill through turbines to generate electricity when demand is strong and prices are high. By using excess electricity generated by wind and solar sources during windy and sunny time periods to pump the water, pumped storage puts this "extra" renewable energy to use and helps utilities maximize electricity generation when needed to satisfy consumer demand.

Water is pumped uphill using excess power when demand is low.

Upper reservoir

Water flows downhill to generate electricity when demand is high.

Lower reservoir

FIGURE 21.26 Pumped storage hydropower is a major means of storing energy. As seen here at Kinzua Dam on the Allegheny River in Pennsylvania, water is pumped to the upper reservoir when demand is low and then flowed through turbines to generate electricity when demand (and prices) are high.

We have a variety of other ways to store energy

There are other ways of storing energy, many of them well established and economical. For instance, *flywheels* are devices that rotate at high speeds in a vacuum. They store energy in the form of rotational movement and then generate electricity when needed by slowing down to release the energy. Flywheels last a long time with little maintenance and come in all sizes for use at points of generation or points of distribution. Energy can also be stored in compressed air in tanks. If excess electricity from wind or solar power is used to compress the air, the air can be decompressed later to generate more electricity. We also have many ways of storing thermal energy and using heat or cold to drive electricity generation.

We will likely need to expand our use of energy storage in multiple ways. One recent analysis of U.S. electricity needs during a time of unusually high demand—the January 2019 polar vortex event (p. 499)—calculated that we will need to increase our energy storage capacity by 25 times to be able to handle such an event with a grid run fully on solar and wind power.

Hydrogen and fuel cells offer a promising avenue

Some energy experts envision that hydrogen—the simplest and most abundant element in the universe—together with electricity could serve as the basis for a clean, safe, and efficient energy system. In such a system, electricity generated from intermittent wind or solar power would be used to produce hydrogen. **Fuel cells**—essentially, hydrogen batteries (**FIGURE 21.27**)—would then use hydrogen to produce electricity to power vehicles, computers, cell phones, home heating, and more.

Hydrogen gas (H_2) tends not to exist freely on Earth. Instead, hydrogen atoms bind to other molecules, becoming incorporated in everything from water to organic molecules. To obtain hydrogen gas for fuel, we must force these substances to release their hydrogen atoms, which requires an input of energy. Scientists are studying several ways of producing hydrogen. In the process of **electrolysis,** electricity is input to split hydrogen atoms from the oxygen atoms of water molecules:

$$2H_2O \rightarrow 2H_2 + O_2$$

Electrolysis produces pure hydrogen, and it does so without emitting the carbon- or nitrogen-based pollutants of fossil fuel combustion. Once isolated, hydrogen gas can be used as a fuel to produce electricity with a fuel cell. The chemical reaction involved in a fuel cell is simply the reverse of that for electrolysis. Two hydrogen molecules and one oxygen molecule are each split, and their atoms bind to form two water molecules:

$$2H_2 + O_2 \rightarrow 2H_2O$$

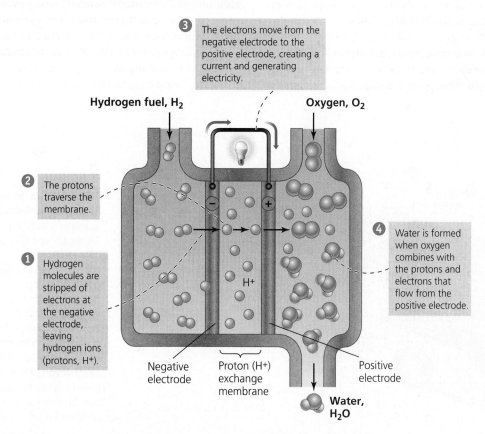

❸ The electrons move from the negative electrode to the positive electrode, creating a current and generating electricity.

Hydrogen fuel, H_2

Oxygen, O_2

❷ The protons traverse the membrane.

❶ Hydrogen molecules are stripped of electrons at the negative electrode, leaving hydrogen ions (protons, H+).

❹ Water is formed when oxygen combines with the protons and electrons that flow from the positive electrode.

H+

Negative electrode

Proton (H+) exchange membrane

Positive electrode

Water, H_2O

FIGURE 21.27 Hydrogen fuel drives electricity generation in a fuel cell. Water and heat are the only waste products that result.

Figure 21.27 shows how this process occurs within one common type of fuel cell. As shown in the diagram, hydrogen gas (usually compressed and stored in an attached fuel tank) is introduced into one side of the cell, and the movement of the hydrogen's electrons from one electrode to the other electrode creates the output of electricity.

Hydrogen gives us many things to consider

Basing an energy system on hydrogen could help shift us away from fossil fuels and reduce the greenhouse gas emissions that drive climate change. However, whether producing hydrogen will cause pollution over its life-cycle depends on the source of the electricity used for electrolysis. If coal is burned to generate the electricity, the process will not reduce emissions. However, if the electricity is produced by a clean-energy source, hydrogen production by electrolysis would create much less pollution and greenhouse warming than relying on fossil fuels.

The environmental impact of hydrogen production also depends on the source material for the hydrogen. Besides water, hydrogen can be obtained from biomass and from fossil fuels, which generally requires less energy input but results in pollution. For instance, extracting hydrogen from the methane (CH_4) in natural gas produces one molecule of the greenhouse gas carbon dioxide for every four molecules of hydrogen gas:

$$CH_4 + 2H_2O \rightarrow 4H_2 + CO_2$$

NASA's space programs have used fuel-cell technology since the 1960s, and auto companies have developed vehicles that run on hydrogen. More than a decade ago, the nation of Iceland set out to move toward a "hydrogen economy." Iceland achieved several early steps in its 30- to 50-year plan to phase out fossil fuels, such as converting buses in the capital city of Reykjavik to run on hydrogen fuel. But the global economic downturn in 2008–2009 and delays in manufacturing hydrogen cars stymied its progress, and future efforts are uncertain. Meanwhile, Germany launched a number of hydrogen-fueled buses (**FIGURE 21.28**). Today, certain cities in Germany, Japan, China, Brazil, Canada, the United Kingdom, California, and elsewhere feature modest fleets of hydrogen buses and cars dependent on a small number of hydrogen filling stations.

One major drawback of hydrogen at this point is a lack of infrastructure to make use of it. To convert a nation such as Germany or the United States to hydrogen would require massive and costly development of facilities to transport, store, and provide the fuel.

A benefit of hydrogen is that we will never run out of it because it is so abundant. Hydrogen can be clean and nontoxic to use, and—depending on its source and the source of electricity for its extraction—it may produce few greenhouse gases and other pollutants. Hydrogen fuel cells are also energy-efficient. Depending on the type of fuel cell, 35–70% of the energy released in the reaction can be used—or up to 90% if the system is designed to capture heat as well as electricity. These rates are comparable or superior to those of most nonrenewable alternatives. In addition, fuel cells are silent and nonpolluting. Unlike batteries (which also produce electricity through chemical reactions), fuel cells generate electricity whenever hydrogen fuel is supplied, without ever needing recharging. For all these reasons, hydrogen fuel cells remain promising for fueling vehicles—something that, with more investment and infrastructure, might happen one day on a large scale.

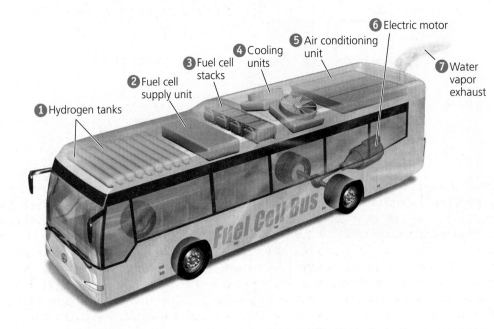

FIGURE 21.28 In one type of hydrogen-fueled bus, hydrogen is stored in nine fuel tanks ❶. The fuel cell supply unit ❷ controls the flow of hydrogen, air, and cooling water into the fuel cell stacks ❸. Cooling units ❹ and the air conditioning unit ❺ dissipate waste heat produced by the fuel cells. Electricity generated by the fuel cells is changed from direct current (DC) to alternating current (AC) by an inverter, and the electricity is transmitted to the electric motor ❻, which powers the operation of the bus. The vehicle's exhaust ❼ consists simply of water vapor.

❻ Electric motor

❺ Air conditioning unit

❹ Cooling units

❸ Fuel cell stacks

❷ Fuel cell supply unit

❼ Water vapor exhaust

❶ Hydrogen tanks

Fuel Cell Bus

connect & continue

TODAY, as renewable energy technologies are being rapidly adopted across the world, states and nations are looking to Texas and Germany for two prime examples of how public policy and economic incentives can accelerate a transition from fossil fuels to renewable energy. In Germany, feed-in tariffs have led to the generation of so much electricity from PV solar, wind, and other renewable sources that debates have shifted from concerns about adequate supplies to questions about grid capacity, timing, and pricing. Yet Germans also wonder how much renewable energy will continue to expand once all the feed-in tariffs expire.

Germany's leaders have now revised the national energy policy, replacing feed-in tariffs for large-scale producers with an auction system. Under the new system, the government decides how much new capacity of each type of renewable energy it will allow each year and then auctions off permits for this development to the lowest bidders. Proponents of this policy change predict that it will lower costs for ratepayers, strengthen industries through competition, and allow time for the grid to expand along with the new mix of energy sources. Critics of the auction system say that it favors powerful firms over small startup companies, creates risk and uncertainty that will discourage private investment, and threatens to stifle Germany's progress toward its renewable energy targets. Small-scale PV solar installations are still eligible for feed-in tariffs under the revised law, so ordinary homeowners are unaffected, but the impacts of this policy change on Germany's energy situation remain to be seen.

In Texas, all eyes are on the federal government to see whether Congress renews the Production Tax Credit for wind power once it expires in 2020. Nonetheless, most energy experts predict that wind power will continue to expand in Texas and that it will remain the cheapest source of electrical power, even in the absence of subsidies.

Wind generation in Texas continues to increase—and prices continue to fall—for other reasons as well. ERCOT is introducing increasingly sophisticated strategies for managing demand and for forecasting weather at wind farm sites in 5-minute increments. Turbines are becoming more powerful and efficient—which also means that wind technicians are needed to upgrade turbine equipment as well as maintain it. Jobs servicing wind turbines are expected to double by 2026, helping breathe more life into rural economies. Texans do not expect another multibillion-dollar infusion into the statewide grid anytime soon, but future grid enhancements could involve decentralizing and allowing power to flow both ways between utilities and customers, as in Germany and as in U.S. states with net metering.

As Texas and Germany blaze their own paths forward, other states and nations are showing leadership in renewable energy as well. As of 2019, more than 100 nations, states, and provinces had implemented some sort of feed-in tariff for renewable energy. In North America, Vermont, Ontario, and the city of Gainesville, Florida, have established feed-in tariff systems similar to Germany's, while California, Hawai'i, Maine, New York, Oregon, Rhode Island, Washington, and utilities in several other states conduct more limited programs.

A major step is being taken by California, which beginning in 2020 is mandating that every new home that is built include PV solar panels on its roof. Although many details remain to be worked out, the impact of this groundbreaking policy could be significant and help boost the U.S. solar industry, improve energy security, and decrease carbon emissions. With steps like these around the world, our society is moving steadily toward a future fueled by renewable energy.

- **CASE STUDY SOLUTIONS** Explain how Texas ignited a boom in wind power within its borders. Now explain how Germany accelerated its development of PV solar power. What specific steps did Texas and Germany each take, and what have been the results so far in each place? How have their approaches been similar, and how have they been different? What future questions and challenges is each place confronting? Do you think the United States as a whole would be better off adopting an approach like Texas's or an approach like Germany's to promote renewable energy across the nation? Explain in detail which you would choose and why.

- **LOCAL CONNECTIONS** Find out where your college or university gets its energy. What percentage of total energy comes from fossil fuels, and what percentages come from each type of renewable source? What percentage of electricity comes from fossil fuels, and what percentages come from each type of renewable source? Does your campus produce any solar power or wind power from panels or turbines on-site? Does it purchase renewable energy from the regional utility? Has your institution made any efforts to increase the proportion of renewable energy it buys or produces? If not, what steps do you think your school could be taking to do so?

- **EXPLORE THE DATA** How fast is renewable energy expanding in different parts of the world? → **Explore Data** relating to the case study on **Mastering Environmental Science**.

REVIEWING Objectives

You should now be able to:

+ **Identify the major sources of renewable energy, and assess their recent growth and future potential**

The "new renewable" energy sources include solar, wind, geothermal, and ocean energy sources. These energy sources are not truly "new" but are in a stage of rapid development. They currently provide far less energy and electricity than we obtain from fossil fuels or other conventional sources, but they are growing quickly. Relative to fossil fuels, the new renewables alleviate air pollution, reduce greenhouse gas emissions, and help diversify a society's energy mix. Government subsidies have long favored nonrenewable energy, but investment and public policies such as feed-in tariffs can accelerate our transition to renewable sources. (pp. 591–597)

+ **Describe solar energy and how we harness it, and evaluate its advantages and disadvantages**

We can harness energy from the sun's radiation using passive methods or by active methods involving powered technology. Solar technologies include flat plate collectors for heating water and air, mirrors to concentrate solar rays, and photovoltaic (PV) cells to generate electricity. PV solar is today's fastest-growing energy source. Solar energy is perpetually renewable, creates no emissions, and enables decentralized power. However, solar radiation varies from place to place and time to time, and harnessing solar energy remains expensive. (pp. 597–602)

+ **Describe wind power and how we harness it, and evaluate its advantages and disadvantages**

Energy from wind is harnessed using wind turbines mounted on towers. Turbines are often erected in arrays at wind farms located on land or offshore, in locations with optimal wind conditions. Wind power is one of today's fastest-growing energy sources. Wind energy is renewable, turbine operation creates no emissions, wind farms can generate economic benefits, and the cost of wind power is competitive with that of electricity from fossil fuels. However, wind is an intermittent resource and is adequate only in some locations. Turbines kill some birds and bats, and wind farms often face opposition from local residents. (pp. 602–608)

+ **Describe geothermal energy and how we harness it, and evaluate its advantages and disadvantages**

Thermal energy from the naturally occurring radioactive decay in Earth's core rises toward the surface and heats groundwater. This energy may be harnessed by geothermal power plants and used to directly heat water and air or to generate electricity. Ground-source heat pumps heat and cool homes and businesses by circulating water whose temperature is moderated at shallow depths underground. Geothermal energy can be efficient, clean, and renewable, but naturally heated water occurs near the surface only in certain areas, and this water may be exhausted if overpumped. Enhanced geothermal systems allow us to gain geothermal energy from more regions, but the approach can trigger minor earthquakes. (pp. 608–610)

+ **List ocean energy sources, and describe their potential**

Ocean energy sources include the motion of tides, waves, and currents, as well as the thermal heat of ocean water. Ocean energy is perpetually renewable and holds much promise, but technologies have seen limited commercial development thus far. (pp. 610–611)

+ **Survey approaches for energy storage, and discuss the potential role of hydrogen and fuel cells**

Expanding energy storage capacity is vital for the transition to renewable energy. Batteries and pumped storage hydropower are among a number of means of storing energy for later generation of electricity. Hydrogen can serve as a fuel to store and transport energy and to power vehicles. Hydrogen fuel cells generate electricity by controlling an interaction between hydrogen and oxygen, and they produce only water as a waste product. Hydrogen infrastructure requires much more development, but hydrogen can be clean, safe, and efficient. (pp. 611–614)

SEEKING Solutions

1. For each source of renewable energy discussed in this chapter, what factors stand in the way of an expedient transition to it from fossil fuel use? In each case, what could be done to ease a shift toward these renewable sources? Would market forces alone suffice to bring about this transition, or would we also need government policies? Do you think such a transition would be good for our economy? Why or why not?

2. Do some research online to find out what energy sources produce the most (a) energy and (b) electricity in your own state. Create diagrams like those in Figure 21.3, showing a quantitative breakdown of the energy and electricity your state's residents use. Which renewable energy sources does your state use more than the United States as a whole, and which does it use less? What might be the reasons for these patterns?

3. Explain the circumstances under which using hydrogen fuel could be helpful in moving toward a low-emission future. If you could advise policymakers of Iceland, Germany, or the United States on hydrogen, what would you tell them?

4. **THINK IT THROUGH** You are the president of a nation the size of Germany, and your legislature is asking you to propose a national energy policy. Your country is located along a tropical coastline. Your geologists do not yet know whether there are fossil fuel deposits or geothermal resources under your land, but your country gets a lot of sunlight and a fair amount of wind, and broad, shallow shelf regions line its coasts. Your nation's population is moderately wealthy but is growing fast, and importing fossil fuels from other nations is becoming expensive. What approaches would you propose in your energy policy? Name some specific steps you would urge your legislature to fund. Are there trade relationships you would seek to establish with other countries? What questions would you fund scientists to research?

5. **THINK IT THROUGH** You are the CEO of a company that develops wind farms. Your staff has presented you with three options, listed as follows, for sites for your next development. Describe at least one likely advantage and at least one likely disadvantage you would expect to encounter with each option. What further information would you like to know before deciding which to pursue?

   - Option A: A remote rural site in West Texas
   - Option B: A ridge-top site in the suburbs of Philadelphia
   - Option C: An offshore site off the Florida coast

CALCULATING Ecological Footprints

Assume that average per-person residential consumption of electricity is 12 kilowatt-hours per day (4380 KWh/year) and that photovoltaic panels have an electrical output of 15% incident solar radiation (i.e., that for every unit of the sun's energy they receive, they can generate 15% of that amount in electricity). Also assume that PV panels cost $250 per square meter. Now refer to Figure 21.12 on p. 600 and estimate the area and cost of the PV panels needed to provide all the residential electricity used by each person in the table.

	AREA OF PHOTOVOLTAIC CELLS	COST OF PHOTOVOLTAIC CELLS
You		
A resident of southwest Arizona		
A resident of Alaska		
A resident of northern Germany		

1. What additional information would you need to increase the accuracy of your estimates for the areas in the table?

2. Considering the distribution of solar radiation in the United States, where do you think it will be most feasible for society to greatly increase the amount of electricity generated from photovoltaic solar power?

3. The purchase price of a photovoltaic system is considerable. What other costs and benefits should you consider, in addition to the purchase price, when contemplating "going solar"?

Mastering Environmental Science

Students Go to **Mastering Environmental Science** for assignments, an interactive e-text, and the Study Area with practice tests, videos, and activities.

Instructors Go to **Mastering Environmental Science** for automatically graded activities, videos, and reading questions that you can assign to your students, plus Instructor Resources.

CHAPTER

22

Managing Our Waste

CENTRAL
case study

OHIO

Miami • • Ohio
University   University

A Mania for Recycling on Campus

> An extraterrestrial observer might conclude that conversion of raw materials to wastes is the real purpose of human economic activity.
> Gary Gardner and Payal Sampat, Worldwatch Institute

> Recycling is one of the best environmental success stories of the late 20th century.
> U.S. Environmental Protection Agency

At the time of year when NCAA basketball fever sweeps America's campuses, there's another kind of March Madness now taking hold: a mania for recycling.

It began in 2001, when waste managers at two Ohio campuses got the idea to use their schools' long-standing athletics rivalry to jump-start their recycling programs. Ed Newman of Ohio University, in Athens, and Stacy Edmonds Wheeler of Miami University, across the state in Oxford, challenged each other to see whose campus could recycle more in a 10-week competition. Come April, Miami University had taken the prize, recycling 41.2 pounds per student. Recyclemania was born.

Students at other colleges and universities heard about the event and wanted to get in on the action, and year by year more schools joined. Today, Recyclemania pits several hundred institutions against one another, involving several million students and staff across North America. The event has grown to have a board of directors and major corporate sponsors.

Each year student leaders rouse their campuses to compete in 2 divisions and 11 different categories over 8 weeks in February and March. Every week during the competition, recycling bins are weighed and campuses report their data, which are compiled online at the Recyclemania website as the competition proceeds. The all-around winner gets a funky trophy made of recycled materials (a figure nicknamed "Recycle Dude," whose body is a rusty propane tank)—and, more important, global bragging rights for a year.

In the spring of 2019, some 300 colleges and universities slugged it out. In the end, the battlefield was littered with stories of the victors and the vanquished (**FIGURE 22.1**, p. 620). Loyola Marymount University in Los Angeles, California, took top honors, recycling an impressive 89% of its waste, topping runners-up Berkshire Community College in Massachusetts and Kendall College of Art and Design of Ferris State University in Michigan. Loyola Marymount also took the prize for most recyclables per person, with a hefty 78.7 pounds per student. Rutgers University in New Jersey claimed top honors in the Total Recycling category, which measures total weight of items recycled, topping out at a staggering 2,575,073 pounds. And Knox College in Illinois won in the Food Organics category, after its donation of 5000 pounds of food to a local food bank gained it high scores in preventing organic waste.

Campuses also compete to see which can collect the most of certain types of items per person. In 2019, Union College of New York collected the most

Upon completing this chapter, you will be able to:

+ Summarize major approaches to managing waste, and compare and contrast the types of waste we generate

+ Discuss the nature and scale of the waste dilemma

+ Evaluate source reduction, reuse, composting, and recycling as approaches for reducing waste

+ Describe landfills and incineration as conventional waste disposal methods

+ Discuss industrial solid waste and principles of industrial ecology

+ Assess issues in managing hazardous waste

◄ Students at Pacific Lutheran University competing in the Recyclemania tournament

▲ The world's biggest collegiate waste management event

619

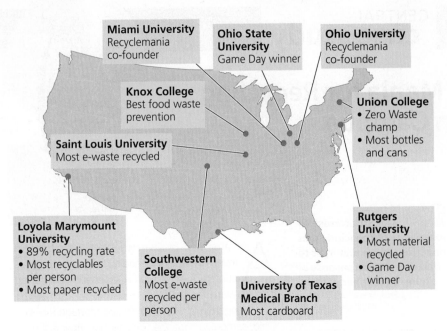

FIGURE 22.1 Eight schools were top winners among 300 participating in Recyclemania in 2019. The event began at Ohio University and Miami University and takes place each spring.

Miami University
Recyclemania
co-founder

Ohio State University
Game Day winner

Ohio University
Recyclemania
co-founder

Knox College
Best food waste prevention

Union College
• Zero Waste champ
• Most bottles and cans

Saint Louis University
Most e-waste recycled

Loyola Marymount University
• 89% recycling rate
• Most recyclables per person
• Most paper recycled

Southwestern College
Most e-waste recycled per person

University of Texas Medical Branch
Most cardboard

Rutgers University
• Most material recycled
• Game Day winner

resources and the manufacture of new goods. Each year, students in the event help prevent the release of at least 100,000 tons of carbon dioxide—equal to the emissions output of 21,000 cars or the electricity use of 13,000 homes. By focusing the attention of administrators on waste issues, Recyclemania facilitates the expansion of campus waste reduction programs. Most important, it gets a new generation of young people revved up about the benefits of recycling.

Recyclemania is the largest of a number of campus-based competitions in the name of sustainability. In an event called the Campus Conservation Nationals, schools have competed against one another for savings in water use and energy use. In its biggest year in 2015, more than 345,000 students in 1374 buildings at 125 colleges and universities took part, saving 394,000 gallons of water (equal to 2500 hours in the shower) and 1.9 million kilowatt-hours of electricity (equivalent to taking 182 homes off the power grid for a year).

Thanks in part to Recyclemania, recycling is the most widespread activity among campus sustainability efforts (pp. 18–19). These diverse efforts include water conservation, energy efficiency, green buildings, transportation options, sustainable food in dining halls, and campus gardens. Students restore native plants and habitats, promote renewable energy, and advocate for carbon neutrality on campus. A growing movement, campus sustainability is thriving because for students, faculty, staff, and administrators, it's satisfying to do the right thing and pitch in to help make campuses more sustainable. And it's even more fun when you can compete and show that you can do it better than your rival school across the state!

bottles and cans, Loyola Marymount gathered the most paper, and the University of Texas Medical Branch amassed the most cardboard. Southwestern College in Kansas recycled the most electronic waste per capita and Saint Louis University in Missouri recycled the most total e-waste. Union College topped Harvard University of Massachusetts in the "Race to Zero Waste" for minimizing its waste in one building during one month. Finally, Ohio State University and Rutgers University were champions in the competition to see who could best reduce waste at a home basketball game.

By encouraging all this recycling and waste reduction, Recyclemania cuts down on pollution from the mining of new

Approaches to Waste Management

As the world's population rises and as we produce and consume more material goods, we generate more waste. **Waste** refers to any unwanted material or substance that results from a human activity or process. Waste can degrade water quality, soil quality, air quality, and human health. Waste also indicates inefficiency, so reducing waste can save money and resources. For these reasons, waste management has become a vital pursuit.

For management purposes, we divide waste into several categories. **Municipal solid waste** is nonliquid waste that comes from homes, institutions, and small businesses. **Industrial solid waste** includes waste from production of consumer goods, mining, agriculture, and petroleum extraction and refining. **Hazardous waste** refers to solid or liquid waste that is toxic, chemically reactive, flammable, or corrosive. Another type of waste is wastewater (p. 410), water we use in our households, businesses, industries, or public facilities and drain or flush down

our pipes, as well as the polluted runoff from streets and storm drains. (We discuss wastewater in Chapter 15, pp. 415–417.)

There are three main components of **waste management:**

1. Minimizing the amount of waste we generate
2. Recovering discarded materials and recycling them
3. Disposing of waste safely and effectively

We have several ways to reduce the amount of material in the **waste stream,** the flow of waste as it moves from its sources toward disposal destinations (**FIGURE 22.2**). Minimizing waste at its source—called **source reduction**—is the preferred approach. We can achieve source reduction when manufacturers use materials more efficiently or when consumers buy fewer goods, buy goods with less packaging, or use those goods longer. Reusing goods you already own, purchasing used items, and donating your used items for others all help reduce the amount of material entering the waste stream.

The next-best strategy in waste management is **recovery,** which consists of recovering, or removing, waste from the waste

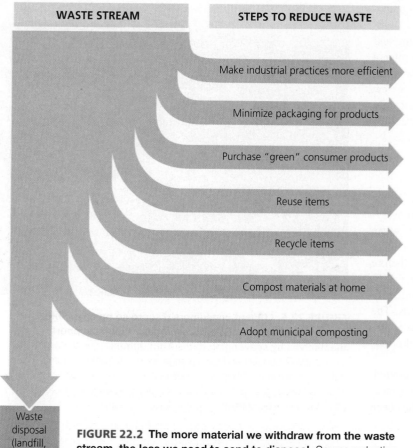

WASTE STREAM	STEPS TO REDUCE WASTE

Make industrial practices more efficient

Minimize packaging for products

Purchase "green" consumer products

Reuse items

Recycle items

Compost materials at home

Adopt municipal composting

Waste disposal (landfill, incinerator)

FIGURE 22.2 The more material we withdraw from the waste stream, the less we need to send to disposal. Source reduction (top three steps) is the most effective way to minimize waste.

converting it to mulch or humus (p. 217) through natural biological processes of decomposition. Recycling and composting are fundamental features of the way natural systems function; all materials in nature are broken down at some point, and matter cycles through ecosystems (Chapter 5). People have taken these concepts from nature and applied them in our society to help cut down on waste and conserve resources.

Regardless of how well we decrease the waste stream through source reduction and recovery, there will likely still be some waste left to dispose of. Disposal methods include burying waste in landfills and burning waste in incinerators. The linear movement of products from their manufacture to their disposal is often described as "cradle-to-grave." As much as possible, however, the modern waste manager attempts to follow a **cradle-to-cradle** approach instead—one in which the materials from products are recovered and reused to create new products. We will first examine how waste managers use source reduction, recovery, and disposal to manage municipal solid waste, and then we will turn to industrial solid waste and hazardous waste.

Municipal Solid Waste

Municipal solid waste is what we commonly refer to as "trash" or "garbage." In the United States, paper, food scraps, yard trimmings, and plastics are the principal components of municipal solid waste, together accounting for two-thirds of what enters the waste stream (**FIGURE 22.3a**). Paper is recycled and yard trimmings are composted at high rates, so after recycling and composting reduce the waste stream, food scraps and plastics are left as the largest components of U.S. municipal solid waste (**FIGURE 22.3b**).

stream. Recovery includes recycling and composting. **Recycling** is the process of collecting used goods and sending them to facilities that extract and reprocess raw materials that can then be used to manufacture new goods. **Composting** is the practice of recovering organic waste (such as food and yard waste) by

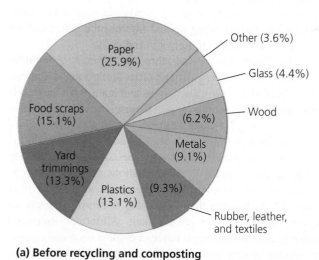

(a) Before recycling and composting

Paper (25.9%)
Other (3.6%)
Glass (4.4%)
Food scraps (15.1%)
Wood
(6.2%)
Metals (9.1%)
Yard trimmings (13.3%)
Plastics (13.1%)
(9.3%)
Rubber, leather, and textiles

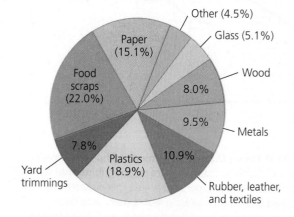

(b) After recycling and composting

Paper (15.1%)
Other (4.5%)
Glass (5.1%)
Food scraps (22.0%)
Wood
8.0%
9.5%
Metals
7.8%
Plastics (18.9%)
10.9%
Yard trimmings
Rubber, leather, and textiles

FIGURE 22.3 Components of the municipal solid waste stream in the United States. Paper products make up the greatest portion by weight **(a),** but after recycling and composting removes many items **(b),** the waste stream becomes one-third smaller. Within this reduced waste stream, food scraps are the largest contributor because a great deal of paper is recycled and yard waste is composted. *Data from U.S. Environmental Protection Agency, 2018.* Advancing sustainable materials management: 2015 fact sheet. *Washington, D.C.: EPA.*

In developing nations, food scraps are often the primary component, and paper makes up a smaller proportion.

Most municipal solid waste comes from packaging and nondurable goods (products meant to be discarded after a short period of use). In addition, consumers throw away old durable goods and outdated equipment as they purchase new products. Plastics, which came into wide consumer use only after 1970, have accounted for the greatest relative increase in the waste stream during the past several decades.

Consumption leads to waste

As we acquire more goods, we generate more waste. In the United States since 1960, waste generation (before recovery) has nearly tripled (**FIGURE 22.4**), and per-person waste generation has risen by 67%. Today, Americans produce more than 260 million tons of municipal solid waste (before recovery)—close to 1 ton per person. The average U.S. resident generates 2.0 kg (4.5 lb) of trash per day—considerably more than people in most other industrialized nations. The relative wastefulness of the American lifestyle, with its excess packaging and reliance on nondurable goods, has led critics to label the United States the "throwaway society."

However, Americans are beginning to turn this around. Thanks to source reduction and reuse (especially by businesses looking to cut costs), total waste generation has been

FIGURE 22.5 Affluent consumers discard so much usable material that some people in developing nations support themselves by scavenging items from dumps. Tens of thousands of people used to scavenge each day from this dump outside Manila in the Philippines, selling material to junk dealers for 100–200 pesos (US$2–$4) per day. The dump was closed in 2000 after an avalanche of trash killed hundreds of people.

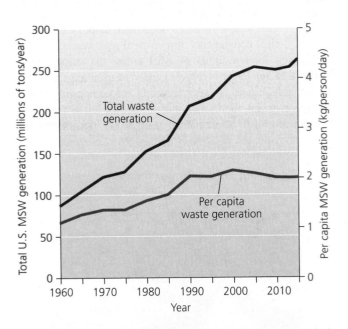

FIGURE 22.4 Rising U.S. waste generation has now leveled off. Total U.S. waste generation before recycling **(blue line)** nearly tripled after 1960, and U.S. per capita waste generation before recycling **(red line)** rose by 67%. In recent years, total and per-person waste generation have each leveled off, largely thanks to source-reduction efforts. *Data from U.S. Environmental Protection Agency, 2018. Advancing sustainable materials management: 2015 fact sheet. Washington, D.C.: EPA.*

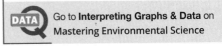

Go to **Interpreting Graphs & Data** on **Mastering Environmental Science**

roughly flat since about 2005. Americans now generate less waste per person than they have since the late 1980s.

In developing nations, people consume fewer resources and goods and, as a result, generate less waste. However, consumption is intensifying in developing nations as they become more affluent, and these nations are generating more and more waste. This growth in waste reflects rising material standards of living, but it also results from an increase in packaging, manufacturing of nondurable goods, and production of inexpensive, poor-quality goods that wear out quickly. As a result, trash is piling up and littering the landscapes of countries from Mexico to Kenya to Indonesia.

Like Americans in the "throwaway society," wealthy consumers in developing nations often discard items that can still be used. In fact, at many dumps and landfills in the developing world, poor people support themselves by selling items that they scavenge (**FIGURE 22.5**).

In many industrialized nations in addition to the United States, per capita rates of waste generation have begun to decline in recent years. Wealthier nations also can afford to invest more in waste collection and disposal, so they are often better able to manage their waste and minimize impacts on human health and the environment. Moreover, enhanced recycling and composting—fed by a conservation ethic growing among a new generation on today's campuses—have been removing more material from the waste stream (**FIGURE 22.6**). As of 2015, U.S. waste managers were recovering 34.7% of the waste stream for composting and recycling, incinerating 12.8%, and sending the remaining 52.5% to landfills.

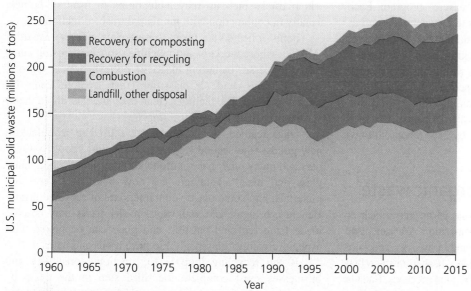

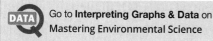

FIGURE 22.6 As recycling and composting have grown in the United States, the proportion of waste going to landfills has declined. As of 2015, 52.5% of U.S. municipal solid waste went to landfills and 12.8% to incinerators, whereas 34.7% was recovered for composting and recycling. *Data from U.S. Environmental Protection Agency, 2018.* Advancing sustainable materials management: 2015 fact sheet. *Washington, D.C.: EPA.*

DATA Go to **Interpreting Graphs & Data** on **Mastering Environmental Science**

Reducing waste is our best option

Reducing the amount of material entering the waste stream is the preferred strategy for managing waste. Recall that preventing waste generation in this way is known as *source reduction*. This preventative approach avoids costs of disposal and recycling, helps conserve resources, minimizes pollution, and can save consumers and businesses money.

One means of source reduction is to reduce the materials used to package goods. Packaging helps preserve freshness, prevent breakage, protect against tampering, and provide information—yet much packaging is extraneous. Consumers can give manufacturers incentive to reduce packaging by choosing minimally packaged goods, buying unwrapped fruit and vegetables, and buying food from the bulk sections of stores. Manufacturers can switch to packaging that is more recyclable. They can also reduce the size or weight of goods and materials, as they already have with aluminum cans, plastic soft drink bottles, personal computers, and much else.

Recently, many policymakers have taken aim at a major source of waste—plastic grocery bags. These lightweight polyethylene bags can persist for centuries in the environment, choking and entangling wildlife (especially in the oceans; Chapter 16) and littering the landscape—yet Americans discard 100 billion of them each year. A number of cities, the states of Maine and Vermont, and more than 20 nations have now enacted bans or limits on their use. Financial incentives are also effective. When Ireland began taxing these bags, their use dropped 90%. IKEA stores began charging for them and saw similar drops in usage. Many businesses now give discounts if you bring your own reusable canvas bags.

Increasing the longevity of goods also helps reduce waste. Because companies seek to maximize sales, they often have a financial incentive to produce short-lived goods that need to be replaced frequently. As a result, increasing the longevity of goods is largely up to the consumer. If consumer demand for goods that last longer is great enough, manufacturers will respond.

Reusing items helps reduce waste

To reduce waste, you can save items to use again or substitute disposable goods with durable ones. **TABLE 22.1** presents a sampling of actions we all can take to reduce waste. Habits as simple as bringing your own coffee cup to coffee shops or bringing sturdy reusable fabric bags to the grocery store can, over time, have an impact. You can also donate unwanted items and shop for used items yourself at yard sales and resale centers. More than 6000 reuse centers exist in the United States, including stores run by organizations such as Goodwill Industries and the Salvation Army. Besides reducing waste, reusing items saves money. Used items are often every bit as functional as new ones, and they are often much less expensive.

On some campuses, students collect unwanted items and resell them or donate them to charity. Students at the University

---

TABLE 22.1 Some Everyday Things You Can Do to Reduce and Reuse

- Donate used items to charity
- Reuse boxes, paper, plastic wrap, plastic containers, aluminum foil, bags, wrapping paper, fabric, packing material, etc.
- Rent, borrow, or lend items instead of buying them
- Bring reusable cloth bags shopping
- Make double-sided photocopies
- Keep electronic documents rather than printing items out
- Bring your own coffee cup to coffee shops
- Pay a bit extra for durable, long-lasting, reusable goods rather than disposable ones
- Buy rechargeable batteries
- Select goods with less packaging
- Compost kitchen and yard wastes
- Buy clothing and other items at resale stores and garage sales
- Use cloth napkins and rags, not paper napkins and towels

Adapted from U.S. Environmental Protection Agency.

of Texas at Austin run a "Trash to Treasure" program. Each May, they collect 40–50 tons of items that students discard as they move off campus and then resell these items at low prices in August to arriving students. This program keeps waste out of the landfill, provides arriving students with items they need at low cost, and raises $10,000–$20,000 per year that gets plowed back into campus sustainability efforts. Hamilton College in New York runs a similar program, called "Cram & Scram," that reduces Hamilton's landfill waste by 28% (about 90 tons) each May.

Composting recovers organic waste

Composting is the conversion of organic waste into mulch or humus (p. 217) through natural decomposition. We can place waste in compost piles, underground pits, or specially constructed containers. As waste is added, heat from microbial action builds in the interior and decomposition proceeds. Banana peels, coffee grounds, grass clippings, autumn leaves, and other organic items can be converted into rich, high-quality compost through the actions of earthworms, bacteria, soil mites, sow bugs, and other detritivores and decomposers (pp. 81–82, 216). The compost is then used to enrich soil. Home composting is a prime example of how we can live more sustainably by mimicking natural cycles and incorporating them into our daily lives.

On campus, composting is becoming popular. Ball State University in Indiana shreds surplus furniture and wood pallets and makes them into mulch to nourish campus plantings. Ithaca College in New York composts much of its food waste, saving thousands of dollars each year in landfill disposal fees. The compost is used on campus plantings, and student-run experiments showed that the plantings grew better with the compost mix than with chemical soil amendments.

Municipal composting programs—more than 3500 across the United States at last count—divert yard debris out of the waste stream and into central composting facilities, where it decomposes into mulch that community residents can use for gardens and landscaping. Increasingly, these programs are also accepting food scraps for composting. About one-fifth of the U.S. waste stream is made up of materials that can easily be composted. Composting reduces landfill waste, enriches soil, enhances soil biodiversity, helps soil resist erosion, makes for healthier plants and more pleasing gardens, and reduces the need for chemical fertilizers.

Recycling consists of three steps

Recycling, too, offers many benefits. It involves collecting used items and breaking them down so that their materials can be reprocessed to manufacture new items. Recycling today in the United States diverts about 68 million tons of materials away from incinerators and landfills.

The recycling loop consists of three basic steps. The first step is to collect and process used goods and materials, as is being done on so many campuses. Some towns and cities designate locations where residents can drop off recyclables or receive money for them. Others offer curbside recycling, in which trucks pick up recyclable items in front of homes, usually in conjunction with municipal trash collection.

Items collected are taken to **materials recovery facilities** (MRFs), where workers and machines sort items using automated processes including magnetic pulleys, optical sensors, water currents, and air classifiers that separate items by weight and size. The facilities clean the materials, shred them, and prepare them for reprocessing.

Once readied, these materials are used to manufacture new goods—the second step in the recycling loop. Newspapers and many other paper products use recycled paper, many glass and metal containers are now made from recycled materials, and some plastic containers are of recycled origin. Benches, bridges, and walkways in city parks may now be made from recycled plastics, and glass can be mixed with asphalt (creating "glassphalt") to pave roads and paths.

If the recycling loop is to function, consumers and businesses must complete the third step in the cycle by purchasing ecolabeled products (p. 154) made from recycled materials. By buying recycled goods, consumers provide economic incentive for industries to recycle materials and for recycling facilities to open or expand.

Recycling has grown rapidly

Today, nearly 10,000 curbside recycling programs across all 50 U.S. states serve 70% of all Americans. These programs, and the 800 MRFs in operation today, have sprung up only in the past few decades. Recycling in the United States rose from 6.4% of the waste stream in 1960 to 25.8% in 2015 (and 34.7% if composting is included; **FIGURE 22.7**).

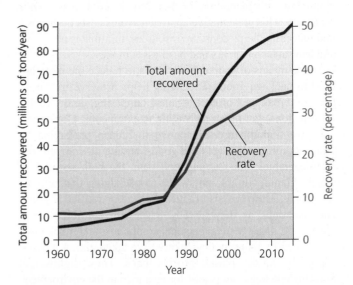

FIGURE 22.7 Recovery has risen sharply in the United States. Today, more than 91 million tons of material are recovered (68 million tons by recycling and 23 million tons by municipal composting), making up one-third of the waste stream. *Data from U.S. Environmental Protection Agency.*

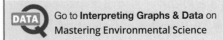

Go to **Interpreting Graphs & Data** on **Mastering Environmental Science**

TABLE 22.2
Recovery Rates for Various Materials in the United States

MATERIAL	PERCENTAGE RECYCLED OR COMPOSTED
Lead-acid batteries	99
Steel cans	71
Newspapers	71
Paper and paperboard	67
Yard trimmings	61
Aluminum cans	55
Tires	40
Glass containers	33
Total plastics	9

Data from U.S. Environmental Protection Agency.

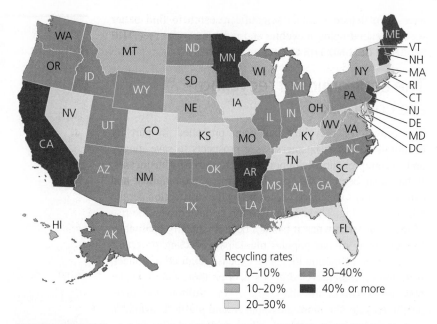

Recycling rates
- 0–10%
- 10–20%
- 20–30%
- 30–40%
- 40% or more

FIGURE 22.8 U.S. states vary greatly in the rates at which their citizens recycle. *Data from Shin, D., 2014.* Generation and disposition of municipal solid waste (MSW) in the United States—A national survey. *New York: Columbia University, Earth Engineering Center.*

Recycling rates vary greatly from one product or material type to another, ranging from nearly zero to almost 100% (**TABLE 22.2**). Recycling rates among states also vary greatly (**FIGURE 22.8**). This variation makes clear that opportunities remain for further growth in recycling.

Many college and university campuses run active recycling programs, although attaining high recovery rates can be challenging in the campus environment. The most recent survey of campus sustainability efforts suggested that the average recycling rate was only 29%. Thus, there appears to be much room for growth. Fortunately, waste management initiatives are relatively easy to conduct on campus because these efforts offer students and faculty many opportunities for small-scale improvements and because people generally enjoy recycling and reducing waste.

Besides participation in Recyclemania, there are many ways to promote recycling on campus. Louisiana State University students initiated recycling efforts at home football games, and over three seasons they recycled 68 tons of refuse that otherwise would have gone to the landfill. "Trash audits" or "landfill on the lawn" events involve emptying trashcans or dumpsters and sorting out recyclable items (**FIGURE 22.9**). When students at Ashland University in Ohio audited their waste, they found that 70% was recyclable, and they used this information to press their administration to support recycling programs. On some campuses, students

WEIGHING the issues

Managing Waste on Your Campus

Does your campus have a recycling program? Does it have composting initiatives? Does it run programs to reduce or reuse materials? Think about the types and amounts of waste generated on your campus. Describe several examples of this waste that you think could be prevented or recycled, and describe how that might be done in each case. If you could do one thing on campus to improve your school's waste management practices, what would it be?

FIGURE 22.9 In a trash audit, students sort through rubbish and separate out recyclables. Events like this "Mt. Trashmore" exercise at Central New Mexico Community College in 2015 show passersby just how many recyclable items are needlessly thrown away.

have even helped conduct scientific research to find better ways of encouraging recycling and reducing waste (see **THE SCIENCE BEHIND THE STORY**, pp. 628–629).

The economics of recycling are complex

The growth of recycling has been propelled in part by economic forces as businesses see prospects to save money and as entrepreneurs see opportunities to start new businesses. It has also been driven by the desire of community and campus leaders to reduce waste and by the satisfaction people take in recycling. These latter two forces have driven the rise of recycling even when it has not been financially profitable. In fact, many of our popular municipal recycling programs are run at an economic loss. The expense required to collect, sort, and process recycled goods is often more than recyclables are worth in the marketplace. In addition, the more people recycle, the more glass, paper, and plastic is available to manufacturers for purchase, which drives down prices.

The low commodity prices of recent years have also posed a challenge to recycling programs. When world oil prices are low, buying new plastic (made from petroleum) can be cheaper than buying recycled plastic; and when market prices of metals are low, buying newly mined metals can be cheaper than buying recycled metals. When recycling is no longer profitable for those in the recycling industry, MRFs may shut down, municipalities may cancel contracts, and recycling companies may go out of business.

Recycling advocates, however, point out that market prices do not take into account external costs (pp. 143, 165)—in particular, the environmental and health impacts of *not* recycling. For instance, it has been estimated that globally, recycling saves enough energy to power more than 6 million households per year. Each year in the United States, recycling and composting together save energy equal to that of 230 million barrels of oil and prevent carbon dioxide emissions equal to those of 39 million cars (**TABLE 22.3**). Recycling aluminum cans saves 95% of the energy required to make the same amount of aluminum from mined virgin bauxite, its source material.

WEIGHING the **issues**

Costs of Recycling and Not Recycling

Should governments subsidize recycling programs if they are run at an economic loss? What types of external costs—costs not reflected in market prices—do you think would be involved in not recycling, say, aluminum cans? Do you think these costs justify sponsoring recycling programs even when they are not financially self-supporting? Why or why not?

China's new policy has upended recycling efforts

The latest economic complication for recycling efforts has come as a result of steps taken by China. In 2018, recycling programs across the world were thrown into disarray when China cut back sharply on its imports of recyclable materials—particularly plastics—from other nations.

TABLE 22.3 Annual Greenhouse Gas Reductions due to Recovery of Various Materials in the United States

MATERIAL	WEIGHT RECOVERED (MILLIONS OF TONS)	EQUAL TO NUMBER OF CARS TAKEN OFF THE ROAD
Paper and paperboard	43.0	31,000,000
Metals	7.9	4,500,000
Textiles	2.3	1,200,000
Wood	2.5	798,000
Plastics	3.0	760,000
Food	1.8	308,000
Yard trimmings	20.6	220,000
Glass	3.2	210,000
Rubber and leather	1.2	127,000

Data from U.S. Environmental Protection Agency.

For years, most of the world's recyclable materials were processed in China. Because China was exporting ships full of trade goods and consumer products across the oceans, it was convenient and inexpensive for nations like the United States to load the emptied ships with discarded paper, plastic, and scrap metal for the return journey back to China. Labor costs were low in China, so it was economical for Chinese workers to recycle the material. And China's booming market of a billion people, many of them gaining wealth and moving into the middle class, was creating a huge consumer base for products made from recycled materials. Because it was so easy to rely on this mutually beneficial trade with China, the United States and most other nations failed to invest in building enough domestic infrastructure to process their own recyclables.

Then in 2018, China largely stopped importing recyclables—particularly plastics—from the United States and other nations. The main stated reason was that the materials contained so much contamination (e.g., food residues on containers and nonrecyclable plastics mixed in with recyclable ones) that China no longer found it economical to process the materials. Additionally, labor costs (e.g., workers' wages) were rising, and as China built its own wealth, its consumers were generating more recyclable waste of their own. Finally, as China sought to become an economic superpower and global leader, it wanted to show environmental leadership and did not want to be viewed as a dumping ground for dirty scrap. Together, all these trends led the Chinese government to announce its policy, named National Sword, banning many foreign recyclables.

The new policy threw the U.S. recycling industry into turmoil, and American waste managers have been struggling

(a) Handling plastic waste at a recycling facility in China

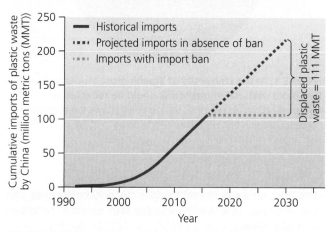

(b) China's import policy could strand huge amounts of plastic waste

FIGURE 22.10 China's recent policy restricting imports of recycled material from other countries delivered a shock to the recycling market. While China recycles increasing amounts of its own plastic waste **(a)**, recyclers across the United States and other nations are stockpiling materials in hopes of later sales, sending materials to landfills, or shutting down recycling programs. If the world cannot find effective ways to respond, **(b)** by 2030, the Chinese ban may result in 111 million metric tons of plastic not being recycled. *Data from Brooks, A., et al., 2018.* The Chinese import ban and its impact on global plastic waste trade. *Science Advances 4: eaat0131.*

being scrapped altogether. As a result of this situation, many millions of tons of plastic, and other materials, will probably never be recycled (**FIGURE 22.10**).

Efforts to rebuild the recycling market are underway

In response to the crisis that has followed China's policy steps, recycling advocates are trying to cut down on contamination in the waste stream by doing a better job of educating consumers. One challenge is the sheer variety of products and packaging, which often makes it confusing to know what is recyclable and what is not. The other main challenge is convincing well-meaning people *not* to try to recycle items that cannot be recycled. People often find it difficult to throw away an item that they wish could be recycled, but these good intentions can lead to unfortunate consequences: The more a recycling stream is contaminated with nonrecyclable materials, the harder it is for workers and machinery at MRFs to process the materials. If loads of recyclables are found to have too high a percentage of contamination when they arrive at a MRF, the entire load may be rejected and sent to the landfill instead. As a result, recycling advocates have urged consumers: "When in doubt, throw it out."

In recent years, contamination in the U.S. recycling stream has surpassed 25%—that is, one out of every four items in the stream was nonrecyclable and had to be removed—putting undue burden on MRFs and their workers. One reason for this high contamination rate was a widespread shift from multiple-stream recycling (asking consumers to sort their items into different bins before recycling them) to single-stream recycling (letting consumers mix all recyclables into a single bin). Many municipalities made this shift in an effort to make it easier for consumers to recycle. The shift helped boost recycling rates, but studies show that it also resulted in more trash being thrown in the recycling stream.

With China now demanding that imported recyclables not exceed a contamination rate of 0.5% (1 part in 200), it seems unlikely that China will begin accepting foreign materials any time soon. We can hope that the present crisis will serve to encourage the United States and other nations to expand their own capacities for processing recyclables domestically. More than a dozen U.S. paper mills have already announced plans to expand processing of recycled paper, but it will take several years for this expansion to occur. A bigger challenge will be expanding processing for plastics (or reducing our reliance on plastics), which continue to be produced in a diversity of forms for a wide variety of products.

If we can successfully develop more technologies and methods to use recycled materials in new ways and if we can encourage more manufacturers to use recycled products, markets for recyclables should continue to expand, and new recycling business opportunities may arise. We are just beginning to shift from an economy that moves linearly—from raw materials to products to waste—to a more sustainable economy that moves circularly, taking a cradle-to-cradle approach and using waste products as raw materials for new manufacturing.

to deal with the consequences. Because the United States has relied on China's MRFs for so long, the United States does not have enough of its own MRFs, and the costs of building a MRF are considerable. Efforts to export recyclables to alternative countries like India, Malaysia, Vietnam, and Thailand mostly failed, as these nations had limited capacity to receive more than they already do. As a result, millions of tons of recyclables across the United States have been stockpiled in warehouses while managers hope for a place to send them in the near future. In many areas, a portion of recyclable materials are simply being sent to the landfill. Municipalities that used to sell scrap plastic for up to $300 per ton now have to pay money to dispose of them instead. Hundreds of recycling programs across the nation and the world are being scaled back or are

Can Campus Research Help Reduce Waste?

Thousands of students on college and university campuses are engaged in efforts to reduce waste. The campus environment also provides opportunities to conduct scientific research on how to better manage waste.

The descriptive research involved in a trash audit (p. 625) is straightforward to conduct and can yield valuable data with practical relevance. At the University of Washington, students and faculty in the UW Garbology Project studied waste on their campus for several years. Working with instructor Dr. Jack Johnson and university waste managers, students sorted through the contents of trashcans, recycling bins, and compost receptacles and discovered that more than 80% of the material thrown in the trash was, in fact, recyclable or compostable (**FIGURE 1**). This information was helpful, because if the university could devise better ways to divert recyclable items and compostable food matter from the waste stream, it could save $225,000 per year on landfill fees.

At Western Michigan University, students in Dr. Harold Glasser's course in 2012 audited food waste in three campus dining halls to test what strategies best minimized waste. They found that a dining hall providing made-to-order servings ended up with 0.23 lb/meal of wasted food, whereas a dining hall with a traditional buffet-style serving produced 0.27 lb/meal of food waste. A third dining hall, which featured trayless dining, performed best, showing only 0.18 lb/meal of waste. The students' data thus seemed to support the idea that people waste less food when trays are not provided.

Students and faculty on multiple campuses have also run manipulative experiments to determine how best to encourage recycling and reduce waste. Such research involves comparing

A student does her part to recycle

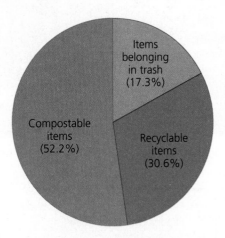

FIGURE 1 At the University of Washington, students found that most material in trashcans could be recycled or composted. As an example, the pie chart shows the average contents of trash from one floor of a building in 2014; only 17% was actually trash. *Data from UW Garbology Project.*

data from an experimental treatment condition (such as after new recycling bins are introduced) with baseline conditions used as a control.

An early such study was that of behavioral scientist Timothy Ludwig and others, who in 1998 examined student behavior with recyclable aluminum drink cans at Appalachian State University. The researchers compared recycling rates when recycling bins were in the hallways (the baseline condition) with recycling rates when the bins were brought into classrooms. Because many students consumed drinks in classrooms, the classroom location proved more convenient, and so recycling increased (**FIGURE 2**). This research showed that making recycling containers easier to find and more convenient to use can boost recycling rates.

Similar results were found by graduate student Ryan O'Connor and colleagues at University of Houston–Clear Lake in 2010. Sampling plastic drink bottles from trashcans and recycling bins in three academic buildings, they found that

Financial incentives help address waste

To encourage recycling, composting, and source reduction, waste managers frequently offer consumers economic incentives to reduce the waste stream. In "pay-as-you-throw" garbage collection programs, municipalities charge

residents for home trash pickup according to the amount of trash they put out. The less waste one generates, the less one has to pay.

Bottle bills are another approach hinging on financial incentives. In the 10 U.S. states and 25 nations that have adopted these laws, consumers pay a deposit on bottles or cans upon purchase—often 5 or 10 cents per container—and

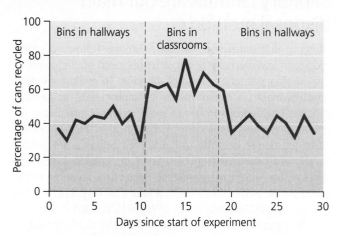

FIGURE 2 At Appalachian State University, recycling rates rose when recycling bins were moved from hallways to classrooms. *Data from Ludwig, T., et al., 1998. Increasing recycling in academic buildings: A systematic replication.* J. Appl. Behavior Analysis *31: 683–686.*

DATA Q Go to **Interpreting Graphs & Data** on **Mastering Environmental Science**

making recycling bins more colorful or adding more bins had no effect on recycling rates but that moving the recycling bins from hallways and common areas into classrooms increased recycling rates greatly.

In 2017, researchers at Western Michigan University challenged the idea that moving bins to places where people use recyclable items was necessarily the best solution. Instead, graduate student Katherine Binder and others tested the hypothesis that removing all trash cans and recycling bins from classrooms and placing them together side-by-side only in common areas (hallways) would lead to better recycling rates. Within a single campus building, these researchers compared recycling rates on floors where this was done with recycling rates on floors where trash cans and recycling bins were left in the classrooms. Their data supported their hypothesis: Recycling rates rose when trash cans and recycling bins were removed from classrooms, forcing people who needed to dispose of items to walk to centrally located areas, where they encountered clearly marked recycling and trash receptacles to choose between. The research team concluded that this

strategy also reduced contamination from incorrectly sorted items and saved money because fewer receptacles were needed.

Researchers have also experimentally tested the influence of educational efforts on recycling behavior. At University of Wisconsin–Stout, student Jessica Van Der Werff ran an experiment in 2008 comparing recycling rates of freshmen who took a recycling workshop during freshman orientation with those who did not. Van Der Werff monitored the trash and recycling from two residence halls throughout the fall semester. She found that students from the residence hall who had attended the workshop showed nearly 40% higher recycling rates (**FIGURE 3**).

Taken together, campus research into waste management has revealed that we can increase recycling rates through education and the strategic location of bins. Students engaged in this work have generated many practical suggestions for reducing waste; they urge that campuses provide enough receptacles, clarify how to sort items correctly, and make it convenient and easy to recycle and compost. By taking such lessons to heart, every campus should be able to significantly reduce its waste stream.

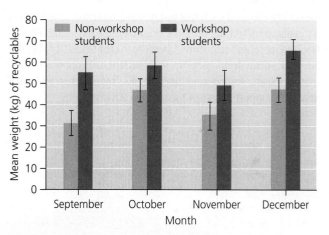

FIGURE 3 At University of Wisconsin–Stout, students who took a workshop on recycling recycled items at a higher rate than those who did not. *Data from Van Der Werff, J., 2008. Teaching recycling: The relationship between education and behavior among college freshmen and its effect on campus recycling rates.* Undergraduate project report, University of Wisconsin–Stout.

then receive a refund when they return them to stores or recycling centers after use. The first bottle bills were passed in the 1970s to cut down on litter, but they have also served to decrease the waste stream. Beverage container recycling rates for states with bottle bills are 2.5 times higher than for states without them (**FIGURE 22.11**, p. 630). U.S. states with bottle bills report that their beverage container litter has decreased

by 69–84%, their total litter has decreased by 30–65%, and their per capita container recycling rates have risen 2.6-fold.

States with bottle bills now face two challenges. One is to amend these laws to include new kinds of containers, particularly the proliferating diversity of plastic containers. The second challenge is to adjust refunds for inflation. In the half-century since Oregon passed the nation's first bottle bill, the value of a

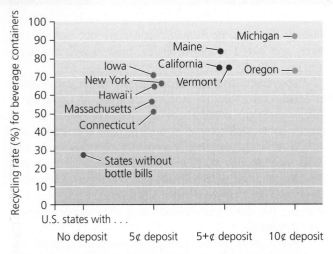

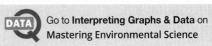

FIGURE 22.11 Bottle bills increase recycling rates, and higher redemption amounts boost these rates further, data suggest. States with bottle bills have much higher recycling rates for beverage containers than states without bottle bills. *Data from Container Recycling Institute, Arlington, VA, 2019.*

DATA Go to **Interpreting Graphs & Data** on **Mastering Environmental Science**

nickel has dropped such that today, the refund would need to be 31 cents to reflect the refund's original intended value. Oregon boosted its refund rate from 5 cents to 10 cents in 2017, but that was not adequate to keep up with inflation. Proponents argue that increasing refund amounts will raise return rates, and available data support this view (see Figure 22.11).

Sanitary landfills are our main disposal method

Material that remains in the waste stream following source reduction and recovery needs to be disposed of, and landfills provide our primary method of disposal. In modern **sanitary landfills,** waste is buried in the ground or piled up in huge mounds engineered to prevent waste from contaminating the environment and threatening public health (**FIGURE 22.12**). Most municipal landfills in the United States are regulated locally or by the states, but they must meet national standards set by the U.S. Environmental Protection Agency (EPA) under the **Resource Conservation and Recovery Act** (p. 173), a major federal law enacted in 1976 and amended in 1984.

In a sanitary landfill, waste is partially decomposed by bacteria and compresses under its own weight to take up less space. Soil is layered along with the waste to speed decomposition, reduce odor, and lessen infestation by pests. Some infiltration of rainwater into the landfill is good, because it encourages biodegradation by bacteria—yet too much is not good, because contaminants can escape if water carries them out.

To protect against environmental contamination, U.S. regulations require that landfills be located away from wetlands and earthquake-prone faults and be at least 6 m (20 ft) above the water table. The bottoms and sides of sanitary landfills must be lined with heavy-duty plastic and 60–120 cm (2–4 ft) of impermeable clay to help prevent contaminants from seeping into aquifers. Sanitary landfills also have systems of pipes, collection ponds, and treatment facilities to collect and treat **leachate,** liquid that results when

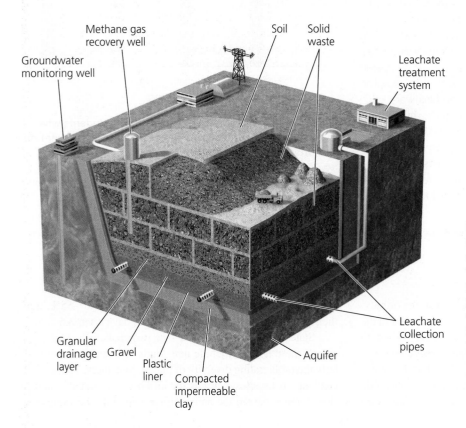

FIGURE 22.12 Sanitary landfills are engineered to prevent waste from contaminating soil and groundwater. Waste is laid in a large depression lined with plastic and impervious clay designed to prevent liquids from leaching out. Pipes draw out these liquids from the bottom of the landfill. Waste is layered with soil, filling the depression, and then is built into a mound until the landfill is capped. Landfill gas produced by anaerobic bacteria may be recovered, and waste managers monitor groundwater for contamination.

FAQ

How much does garbage decompose in a landfill?

You might assume that a banana peel you throw in the trash will soon decay away to nothing in a landfill. However, it just might survive longer than you do! That's because surprisingly little decomposition occurs in landfills. Researcher William Rathje, a retired archaeologist known as the "Indiana Jones of Solid Waste," made a career out of burrowing into landfills and examining their contents to learn about what we consume and what we throw away. His research teams would routinely come across whole hot dogs, intact pastries that were decades old, and grass clippings that were still green. Newspapers 40 years old were often still legible, and the researchers used them to date layers of trash.

substances from the trash dissolve in water as rainwater percolates downward.

Once a landfill is closed, it is capped with an engineered cover consisting of layers of plastic, gravel, and soil. Managers are required to maintain leachate collection systems for 30 years after a landfill has closed, and regulations require that groundwater be monitored regularly for contamination.

Despite improvements in liner technology and landfill siting, however, liners can be punctured, and leachate collection systems eventually cease to be maintained. Moreover, landfills are kept dry to reduce leachate, but the bacteria that break down material thrive in wet conditions. Dryness, therefore, slows waste decomposition. In fact, the low-oxygen conditions of most landfills turn trash into a sort of time capsule. Researchers examining landfills often find some of their contents perfectly preserved, even after years or decades.

In 1988, the United States had nearly 8000 landfills, but today it has fewer than 1800. Waste managers have consolidated the waste stream into fewer landfills of larger size. In many cities, landfills that were closed are now being converted into public parks or other uses (**FIGURE 22.13**). The world's largest landfill conversion project is at New York City's former Fresh Kills Landfill. This site, on Staten Island, was the primary repository of New York City's garbage for half a century, and its mounds rose higher than the nearby Statue of Liberty. Today, New York is transforming the site into a world-class public park—a verdant landscape of ball fields, playgrounds, jogging trails, rolling hills, and wetlands teeming with wildlife that, once completed, will be almost three times bigger than Central Park.

Incinerating trash reduces pressure on landfills

Just as sanitary landfills are an improvement over open dumping, incineration in specially constructed facilities is better than open-air burning of trash. **Incineration,** or combustion, is a controlled process in which garbage is burned at very high temperatures (**FIGURE 22.14**, p. 632). At incineration facilities, waste is generally sorted and metals are removed. Metal-free waste is chopped into small pieces to aid combustion and then is burned in a furnace. Incinerating waste reduces its weight by up to 75% and its volume by up to 90%.

The ash remaining after trash is incinerated contains toxic components and therefore must be disposed of in hazardous-waste landfills (p. 637). Moreover, when trash is burned, hazardous chemicals—including dioxins, heavy metals, and polychlorinated biphenyls (PCBs) (Chapter 14)—may be created and released into the atmosphere. Such emissions inspired a backlash against incineration from citizens concerned about health hazards.

Most industrialized nations now regulate incinerator emissions, and some have banned incineration outright. Engineers have also developed technologies to reduce emissions. Scrubbers (see Figure 17.12, p. 460) chemically treat the gases produced in combustion to remove hazardous components and neutralize acidic gases, such as sulfur dioxide and hydrochloric acid, turning them into water and salt.

THEN: Fresh Kills Landfill in operation

NOW: Fresh Kills Landfill site today

FIGURE 22.13 Old landfills, once capped, can serve other purposes. Visitors to Freshkills Park in New York City can enjoy a panoramic view of the Manhattan skyline from atop what used to be an immense mound of trash.

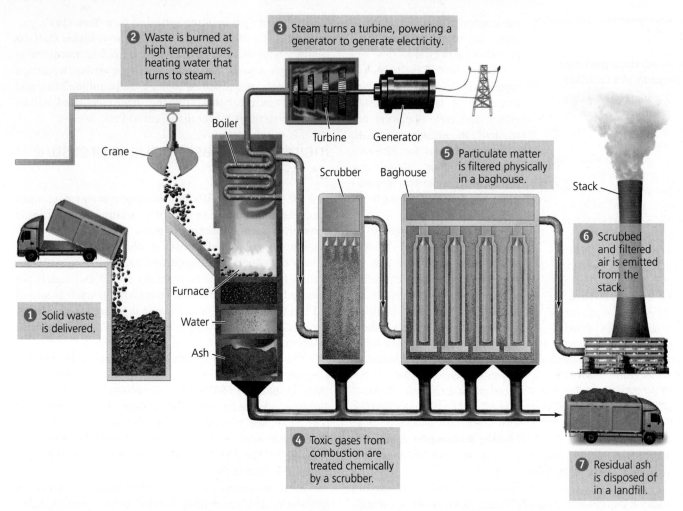

2 Waste is burned at high temperatures, heating water that turns to steam.

3 Steam turns a turbine, powering a generator to generate electricity.

Boiler

Crane

Turbine   Generator

Scrubber   Baghouse

5 Particulate matter is filtered physically in a baghouse.

Stack

6 Scrubbed and filtered air is emitted from the stack.

Furnace

Water

Ash

1 Solid waste is delivered.

4 Toxic gases from combustion are treated chemically by a scrubber.

7 Residual ash is disposed of in a landfill.

FIGURE 22.14 In a waste-to-energy (WTE) incinerator, solid waste is combusted, greatly reducing its volume and generating electricity at the same time.

Scrubbers generally accomplish this either by spraying liquids formulated to neutralize the gases or by forcing the gases through dry lime.

Particulate matter, called *fly ash,* contains some of the worst dioxin and heavy metal pollutants in incinerator emissions. To physically remove these tiny particles, facilities may use a huge system of filters known as a *baghouse.* In addition, burning garbage at especially high temperatures can destroy certain pollutants, such as PCBs. Even all these measures, however, do not fully eliminate toxic emissions.

We can gain energy from trash

Incineration reduces the volume of waste, but it can serve to generate electricity as well. Most incinerators now are in fact **waste-to-energy** (WTE) **facilities,** which use the heat produced by waste combustion to boil water, creating steam that drives electricity generation or that fuels heating systems. When burned, waste generates about 35% of the energy generated by burning coal. Roughly 80 WTE facilities are operating across the United States (mostly in the Northeast and South), with a total capacity to process 95,000 tons of waste per day.

Revenues from power generation, however, are often not enough to offset the considerable financial cost of building and running incinerators. Because it can take many years for a WTE facility to become profitable, companies that build and operate these facilities sometimes require communities contracting with them to guarantee the facility a minimum amount of garbage. On occasion, such long-term commitments have interfered with communities' subsequent efforts to reduce their waste through recycling and source reduction.

Combustion in WTE plants is not the only way to gain energy from waste. Deep inside landfills, bacteria decompose waste in an oxygen-deficient environment. This anaerobic decomposition produces **landfill gas,** a mix of gases roughly half of which is methane (pp. 28, 530). Landfill gas can be collected, processed, and used in the same way as natural gas (p. 530). Today, hundreds of landfills are collecting landfill gas and selling it for energy.

We can recycle material from landfills

Landfills can offer us useful by-products beyond landfill gas. With improved technology for sorting rubbish and recyclables, businesses and entrepreneurs are weighing the economic benefits and costs of rummaging through landfills to salvage materials of value that can be recycled. Steel, aluminum, copper, and other metals are abundant enough in some landfills to make salvage operations profitable when market prices for the metals are high enough. For instance, Americans throw out so many aluminum cans that at 2019 prices for aluminum, the nation buries $5.5 billion of this metal in landfills each year. If we could retrieve all the aluminum from U.S. landfills, it would exceed the amount the world produces from a year's worth of mining ore.

Landfills also offer soil mixed with organic waste that can be mined and sold as premium compost. In addition, old landfill waste can be incinerated in newer, cleaner-burning WTE facilities to produce energy. Some companies are even looking into gaining carbon offset credits (p. 518) by harvesting methane (a greenhouse gas that contributes to climate change) leaking from open dumps in developing nations.

Such approaches have been tried in places from New York to Israel to Sweden to Singapore. The costs of mining landfills and meeting regulatory requirements while commodity prices change unpredictably have meant that investing in landfill mining has been risky, but that could change in the future if prices of commodities rise and landfill mining technologies improve.

Industrial Solid Waste

Industrial solid waste includes waste from factories, agriculture, ore mining, petroleum extraction, and more. Each year, industrial facilities in the United States generate more than 7 billion tons of waste, according to the EPA, about 97% of which is wastewater. Thus, very roughly, 230 million or so tons of solid waste are generated by 60,000 facilities each year—an amount approaching that of municipal solid waste.

Regulation and economics each influence industrial waste generation

Most methods and strategies of waste disposal, reduction, and recycling by industry are similar to those for municipal solid waste. Businesses that dispose of their own waste on site must design and manage their landfills in ways that meet state, local, or tribal guidelines. Other businesses pay to have their waste disposed of at municipal disposal sites. Whereas the federal government regulates municipal solid waste, it is state or local governments that regulate industrial solid waste (with federal guidance). Regulation varies greatly from place to place, but in most cases, state and local regulation of industrial solid waste is less strict than federal regulation of municipal solid waste. In many areas, industries are not required to have permits, install landfill liners or leachate collection systems, or monitor groundwater for contamination.

The amount of waste generated by a manufacturing process is a good measure of its efficiency; the less waste produced per unit or volume of product, the more efficient that process is, from a physical standpoint. However, physical efficiency is not always reflected in economic efficiency. Often it is cheaper for industry to manufacture its products or perform its services quickly but messily. That is, it can be cheaper to generate waste than to avoid generating waste. In such cases, economic efficiency is maximized, but physical efficiency is not. Because our market system awards only economic efficiency, often industry has no financial incentive to achieve physical efficiency. The frequent mismatch between these two types of efficiency is a major reason the output of industrial waste is so great.

Rising costs of waste disposal enhance the financial incentive to decrease waste. Once either the government or the market makes the physically efficient use of raw materials economically efficient as well, businesses will gain financial incentives to reduce their waste.

Industrial ecology seeks to make industry more sustainable

To reduce waste, growing numbers of industries today are experimenting with industrial ecology. A holistic approach that integrates principles from engineering, chemistry, ecology, and economics, **industrial ecology** seeks to redesign industrial systems to reduce resource inputs and to maximize both physical and economic efficiency. Industrial ecologists would reshape industry so that nearly everything produced in a manufacturing process is used, either within that process or in a different one (see **SUCCESS STORY**, p. 635).

The larger idea behind industrial ecology is that industrial systems should function more like ecological systems, in which organisms use almost everything that is produced. This principle brings industry closer to the ideal of ecological economists, in which economies function in a circular fashion rather than a linear one (p. 147). It means taking a cradle-to-cradle approach, in which products and manufacturing systems are designed to maximize reuse and recycling of materials into new products.

Industrial ecologists pursue their goals in several ways:

- They examine the entire life-cycle of a given product—from its origins in raw materials, through its manufacturing, to its use, and finally to its disposal—and look for ways to make the process more efficient. This strategy is called **life-cycle analysis** (p. 267).

- They try to identify how waste products from one manufacturing process might be used as raw materials for another. For instance, used plastic beverage containers can be shredded and reprocessed to make items such as benches, tables, and decks.

- They seek to eliminate environmentally harmful products and materials from industrial processes.

- They study the flow of materials through industrial systems to look for ways to create products that are more durable, recyclable, or reusable. For instance, they seek to design computers, automobiles, and appliances to be easily disassembled so more of their components can be reused or recycled.

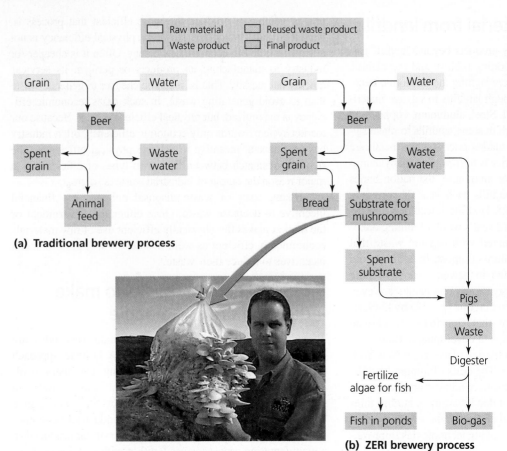

Raw material   Reused waste product
Waste product   Final product

(a) Traditional brewery process

(b) ZERI brewery process

Businesses are adopting industrial ecology

Attentive businesses are taking advantage of the insights of industrial ecology to save money while reducing waste. For example, American Airlines switched from hazardous to non-hazardous materials in its Chicago facility, decreasing its need to secure permits from the EPA. The company used more than 50,000 reusable plastic containers to ship goods, reducing packaging waste by 90%. Its Dallas–Fort Worth headquarters recycled enough aluminum cans and white paper in five years to save $205,000 and recycled 3000 broken baggage containers into lawn furniture. A program to gather suggestions from employees brought more than 700 ideas to reduce waste—and 15 of these ideas saved the company more than $8 million.

The Swiss Zero Emissions Research and Initiatives (ZERI) Foundation sponsors dozens of innovative projects worldwide that attempt to create goods and services without generating waste. One example involves breweries in Canada, Sweden, Japan, and Namibia. Brewers in these projects take waste from the beer-brewing process and use it to fuel other processes (**FIGURE 22.15**). As a result, the brewer can make money from bread, mushrooms, pigs, gas, and fish, as well as beer, all while producing little waste. By attempting to create closed-loop systems, ZERI projects cut down on waste while increasing output and income, often generating new jobs as well.

Few businesses have taken industrial ecology to heart as much as the Atlanta-based international modular carpet tile company Interface, which founder Ray Anderson set on the road to sustainability years ago. Interface asks customers to return used carpet tiles for recycling and for reuse as backing for new carpet. It modified its design and production methods to reduce waste. It adapted its boilers to use landfill gas for energy. Through such steps, Anderson's company cut its waste generation by 80%, its fossil fuel use by 45%, and its water use by 70%—all while saving $30 million per year, holding prices steady for its customers, and raising profits by 49%.

Hazardous Waste

Hazardous wastes are diverse in their chemical composition and may be liquid, solid, or gaseous. By EPA definition, hazardous waste is waste that is one of the following:

- *Ignitable.* Likely to catch fire (e.g., gasoline or alcohol).
- *Corrosive.* Apt to corrode metals in storage tanks or equipment (e.g., strong acids or bases).
- *Reactive.* Chemically unstable and readily able to react with other compounds, often explosively or by producing noxious fumes (e.g., ammonia reacting with chlorine bleach).
- *Toxic.* Harmful to human health when inhaled, ingested, or touched (e.g., pesticides or heavy metals).

Hazardous wastes are diverse

Industry, mining, households, small businesses, agriculture, utilities, and building demolition all create hazardous waste. Industry produces the most, but in developed nations, industrial

Creating an Industrial Ecosystem

One place the ideals of industrial ecology have come to life is the city of Kalundborg, Denmark. Here, starting in 1972, dozens of private and public enterprises gradually formed a network of business relationships that are conserving resources while saving money. Anchoring the Kalundborg Eco-Industrial Park is a coal-fired power plant, the Asnaes Power Station. It sends its excess steam to a nearby Statoil petroleum refinery and a Novo-Nordisk pharmaceutical factory, which use the steam to run their operations. The Statoil refinery sends Asnaes its wastewater, cooling water, and waste gas, which the power plant uses to generate electricity, and sells sulfur to a local acid manufacturer. The power plant sends its waste fly ash to a cement company and sells gypsum removed from its waste gas by a scrubber to a Gyproc factory that makes drywall. Power plants also routinely create large amounts of waste heat, and in Kalundborg, this heat is piped to more than 3000 homes as district heating and to a regional fish farm. Treated sludge from both the fish farm and the pharmaceutical plant is sent to area farms as fertilizer. By efficiently using one another's waste products, the Kalundborg Eco-Industrial Park has reduced pollution, cut greenhouse gas emissions, and conserved resources such as water, coal, and oil—all while saving hundreds of millions of dollars for the enterprises involved.

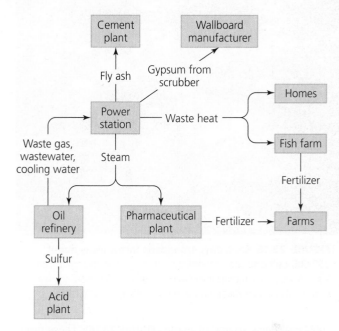

In a model for industrial ecology, networked enterprises in Kalundborg, Denmark, use one another's waste materials as resources.

➜ **Explore the Data** at Mastering Environmental Science

---

waste disposal is often highly regulated. This regulation has reduced the amount of hazardous waste entering the environment from industrial activities. As a result, households are now the largest source of unregulated hazardous waste.

Household hazardous waste includes a wide range of items, including paints, batteries, oils, solvents, cleaning agents, lubricants, and pesticides. Americans generate 1.6 million tons of household hazardous waste annually, and the average home contains close to 45 kg (100 lb) of it in sheds, basements, closets, and garages. Although many hazardous substances become less hazardous over time as they degrade chemically, two types are particularly hazardous because their toxicity persists over time: organic compounds and heavy metals.

Organic compounds and heavy metals pose hazards

In our daily lives, we rely on synthetic organic compounds (human-made carbon-based chemicals) and compounds derived from petroleum to resist bacterial, fungal, and insect activity. Plastic containers, rubber tires, pesticides, solvents, and wood preservatives are useful to us precisely because they resist decomposition. We use these products to protect buildings from decay, kill pests that attack crops, and keep stored goods intact. However, the capacity of the compounds in these products to resist decay is a double-edged sword, for it also makes them

persistent pollutants. Many synthetic organic compounds are toxic because they are readily absorbed through the skin and can act as mutagens, carcinogens, teratogens, and endocrine disruptors (p. 370).

Heavy metals such as lead, chromium, mercury, arsenic, cadmium, tin, and copper are used widely in industry for wiring, electronics, metal plating, metal fabrication, pigments, and dyes. Heavy metals enter the environment when paints, electronic devices, batteries, and other materials are disposed of improperly. Lead from fishing weights and from hunting ammunition accumulates in rivers, lakes, and forests. In older homes, lead from pipes contaminates drinking water, and lead paint remains a toxic problem, especially for infants and children who encounter lead paint in their home environments. Heavy metals that are fat-soluble and break down slowly are prone to bioaccumulate and biomagnify (p. 373). In California's Coast Range, for instance, mercury that has washed downstream from abandoned mercury mines enters lakes and rivers, is consumed by bacteria and invertebrates, and accumulates in increasingly large quantities up the food chain, poisoning organisms at higher trophic levels and making fish unsafe to eat.

E-waste has grown

Today's proliferation of computers, printers, smartphones, tablets, TVs, and other electronic technology has created a

FIGURE 22.16 Each day, Americans throw away about 350,000 cell phones. Phones that enter the waste stream can leach toxic heavy metals into the environment. Alternatively, we can recycle phones for reuse and to recover their valuable metals.

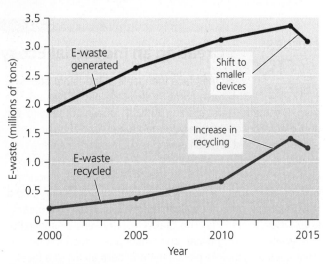

FIGURE 22.17 Increasing amounts of electronic waste are being recycled. The total amount of electronic waste generated each year in the United States has risen, but the shift to mobile devices and tablets has helped decrease this amount since 2011. *Data from U.S. Environmental Protection Agency.*

substantial new source of waste (**FIGURE 22.16**). These products have short life spans before people judge them obsolete, and most are discarded after just a few years. The amount of this **electronic waste**—often called **e-waste**—has grown rapidly and now makes up more than 1% of the U.S. solid waste stream by weight. More than 7 billion electronic devices have been sold in the United States since 1980, and U.S. households discard more than 300 million per year—two-thirds of them still in working order.

Most electronic items we discard have ended up in conventional sanitary landfills and incinerators. However, electronic products contain heavy metals and toxic flame-retardants, and research indicates that e-waste is hazardous and should be treated as such. The EPA and a number of states are now taking steps to keep e-waste out of conventional landfills and incinerators and instead redirect it to hazardous-waste sites.

Fortunately, the downsizing of many electronic items and the shift toward mobile devices and tablets mean that fewer raw materials by weight are now going into electronics being manufactured—and as a result, U.S. e-waste generation appears to have recently leveled off. In addition, more and more electronic waste today is being recycled (**FIGURE 22.17**; and see **Success Story** in Chapter 23, p. 659). Americans now recycle 40% of their e-waste, by weight. Increasingly, used electronics are collected by businesses, nonprofit organizations, or municipal services and are processed for reuse or recycling. Campus e-waste recycling drives are proving especially effective (see Table 1.3, p. 18).

Devices collected for recycling are shipped to facilities and taken apart, and the parts and materials are refurbished and reused in new products. There are serious concerns, however, about health risks that recycling may pose to workers doing the disassembly. Wealthy nations ship much of their e-waste to developing countries, where low-income workers disassemble the devices and handle toxic materials with minimal safety regulations. Environmental justice and workplace safety concerns will need to be resolved if electronic recycling is to be conducted safely and responsibly.

Another challenge is that the recent conversion of television and computer monitor technology from cathode-ray tubes to LCD and plasma screens has meant that there is no longer much demand for recycled cathode-ray tubes. As a result, old cathode-ray tubes (rich in toxic lead) are piling up in recyclers' warehouses and are at risk of never being recycled.

Besides keeping toxic substances out of our waste stream, e-waste recycling helps us recover rare and lucrative trace metals used in electronics. A typical cell phone contains up to a dollar's worth of precious metals (p. 659). By one estimate, 1 ton of computer scrap contains more gold than 16 tons of mined ore from a gold mine, and 1 ton of iPhones contains more than 300 times more. Every ounce of metal we can recycle from a manufactured item is an ounce of metal we don't need to mine from the ground. Thus, "mining" e-waste for metals helps reduce the environmental impacts of mining the earth. For example, several recent Olympic Games have produced their gold, silver, and bronze medals (**FIGURE 22.18**) partly from metals recovered from recycled and processed e-waste!

We regulate the disposal of hazardous waste

For many years, we discarded hazardous waste without special treatment. In some cases, people did not know that certain substances were harmful to human health. In other cases, it was assumed that the substances would disappear or be sufficiently diluted in the environment. A number of well-publicized pollution incidents in the 1970s—such as the resurfacing of toxic chemicals in a residential area at Love Canal in upstate New York years after their burial—convinced the public that hazardous waste deserves special attention and treatment.

Many communities now designate sites or special collection days to gather household hazardous waste or designate

FIGURE 22.18 Medals for winning athletes at the 2020 Olympic Games in Tokyo, Japan, were made partly from precious metals recycled from discarded e-waste.

FIGURE 22.19 Many communities designate collection sites or collection days for household hazardous waste. Here, workers handle waste from a collection event in Brooklyn, New York.

facilities for the exchange and reuse of hazardous substances (**FIGURE 22.19**). Once consolidated, the hazardous waste is transported for treatment and disposal.

Under the Resource Conservation and Recovery Act, the EPA sets standards by which states manage hazardous waste. The act also requires large generators of hazardous waste to obtain permits. Finally, it mandates that hazardous materials be tracked "from cradle to grave." As hazardous waste is generated, transported, and disposed of, the producer, carrier, and disposal facility must each report to the EPA the type and amount of material generated; its location, origin, and destination; and the way it is handled. This process is intended to prevent illegal dumping and to encourage the use of reputable waste carriers and disposal facilities.

Because current U.S. law makes disposing of hazardous waste quite costly, irresponsible companies sometimes illegally dump waste, creating health risks for residents and financial headaches for local governments forced to deal with the mess (**FIGURE 22.20**). At the international scale, companies from industrialized nations often find it cheaper to pay cash-strapped developing nations to take hazardous waste—or cheaper still, to dump it illegally. In nations with lax environmental and health regulations, workers and residents are often uninformed of or unprotected from the health dangers of this waste. This global environmental justice issue (p. 139) continues despite the Basel Convention, an international treaty crafted to limit such practices.

High costs of disposal, however, have also encouraged conscientious businesses to invest in reducing their hazardous waste. Many hazardous materials can be broken down by incineration at high temperatures in cement kilns. Other hazardous materials can be treated with bacteria that break down harmful components and synthesize them into new compounds. In a process called **bioremediation** (p. 30), various plants, fungi, and microbes have been used (and sometimes specially bred or engineered) to take up specific contaminants from soil and then break down organic contaminants into safer compounds or concentrate heavy metals in their tissues. The organisms are eventually harvested and disposed of.

We use three disposal methods for hazardous waste

We have developed three primary means of hazardous-waste disposal: landfills, surface impoundments, and injection wells. These methods do nothing to diminish the hazards of the substances, but they help keep the waste isolated from people, wildlife, and ecosystems. Design and construction standards for landfills that receive hazardous waste are stricter than those for ordinary sanitary landfills. Hazardous waste landfills must have several impervious liners and leachate removal systems and must be located far from aquifers.

FIGURE 22.20 Unscrupulous individuals and businesses sometimes dump hazardous waste illegally to avoid disposal costs.

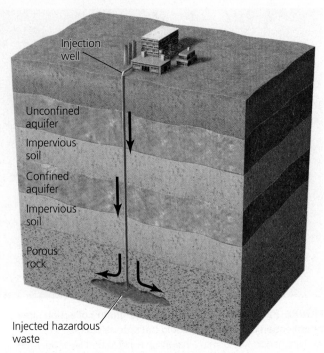

FIGURE 22.21 Liquid hazardous waste is pumped deep underground by deep-well injection. The well must be drilled below any aquifers, into porous rock isolated by impervious clay.

Liquid hazardous waste, or waste in dissolved form, may be stored in **surface impoundments,** shallow depressions lined with plastic and an impervious material, such as clay. The liquid or slurry is placed in the pond and water is allowed to evaporate, leaving a residue of solid hazardous waste on the bottom. This process is repeated, and eventually the dry residue is removed and transported elsewhere for permanent disposal. Impoundments are not ideal. The underlying layer can crack and leak waste. Some material may evaporate or blow into surrounding areas. Rainstorms may cause waste to overflow and contaminate nearby areas. For these reasons, surface impoundments are used only for temporary storage.

In **deep-well injection,** a well is drilled deep beneath the water table into porous rock, and wastes are injected into it (**FIGURE 22.21**). The process aims to keep waste deep underground, isolated from groundwater and human contact. However, wells can corrode and can leak wastes into soil, contaminating aquifers, and deep-well injection may very occasionally induce earthquakes. Roughly 34 billion L (9 billion gal) of hazardous waste are placed in U.S. injection wells each year.

Radioactive waste is especially hazardous

Radioactive waste is particularly dangerous to human health and is persistent in the environment. The dilemma of disposal has dogged the nuclear energy industry for decades (p. 572). The United States has no designated site to dispose of its commercial nuclear waste if Yucca Mountain in Nevada is removed from consideration (p. 573). Instead, waste will continue to accumulate at the many nuclear power plants spread throughout the nation (see Figure 20.12, p. 573).

Currently, a site in the Chihuahuan Desert in New Mexico serves as a permanent disposal location for radioactive waste from military sources. The Waste Isolation Pilot Plant was the world's first underground repository for transuranic waste from nuclear weapons development. The mined caverns holding the military waste are located 655 m (2150 ft) below ground in a huge salt formation thought to be geologically stable. This site became operational in 1999 and receives shipments of waste from 23 other locations.

Contaminated sites are being cleaned up, slowly

Many thousands of former military and industrial sites remain contaminated with hazardous waste in the United States and virtually every other nation on Earth. For most nations, dealing with these messes is simply too difficult, time-consuming, and expensive. In 1980, however, the U.S. Congress passed the Comprehensive Environmental Response Compensation and Liability Act (CERCLA; p. 173). This law established a federal program to clean up U.S. sites polluted with hazardous waste. The EPA administers this cleanup program, called the **Superfund.** Under EPA auspices, experts identify sites polluted with hazardous chemicals, take action to protect groundwater, and clean up the pollution. Later laws also charged the EPA with cleaning up **brownfields,** lands whose reuse or development is complicated by the presence of hazardous materials.

High-profile events in two locations received extensive media coverage and were instrumental in leading to the Superfund legislation. Outside of Louisville, Kentucky, at a site called *Valley of the Drums,* hazardous waste began leaking from 100,000 metal drums, contaminating waterways with 140 types of chemicals. And in *Love Canal,* a residential neighborhood in Niagara Falls, New York, more than 800 families were evacuated in 1978–1980 after toxic chemicals buried by a company and the city in past decades rose to the surface (**FIGURE 22.22**). The chemicals contaminated homes and an elementary school and apparently led to birth defects, miscarriages, and other health impacts.

FIGURE 22.22 In Love Canal, an angry homeowner erected this sign of protest before being evacuated. Outrage over the contamination of this neighborhood in Niagara Falls helped lead to the Superfund program.

Once a Superfund site is identified, EPA scientists evaluate how near the site is to homes, whether wastes are confined or likely to spread, and whether the pollution threatens drinking water supplies. Sites judged to be harmful are placed on the National Priorities List and ranked according to the risk to human health that they pose. Cleanup proceeds as funds are available. Throughout the process, the EPA is required to hold public hearings to inform area residents of its findings and to receive feedback.

The objective of CERCLA was to charge the polluting parties for the cleanup of their sites, according to the *polluter-pays principle* (p. 165). For many sites, however, the responsible parties cannot be found or held liable, and in such cases—roughly 30% so far—cleanups have been covered by taxpayers and from a trust fund established by a federal tax on industries producing petroleum and chemical raw materials.

Congress let the tax expire, however, and the trust fund went bankrupt in 2004, so taxpayers are now shouldering the entire burden. As the remaining cleanup jobs become more expensive, fewer are being completed.

As of 2019, 1337 Superfund sites remained on the National Priorities List, and only 413 had been cleaned up or otherwise removed from the list. The average cleanup has cost more than $25 million and has taken nearly 15 years. Many sites are contaminated with hazardous chemicals we have no effective way to deal with. In such cases, cleanups simply aim to isolate waste from human contact, either by building trenches and barriers around a site or by excavating contaminated material and shipping it to a hazardous-waste disposal facility. For all these reasons, the current emphasis in the United States and elsewhere is on preventing hazardous-waste contamination in the first place.

CENTRAL CASE STUDY
connect & continue

TODAY, our society can celebrate great progress in addressing its waste management challenges: Recycling and composting efforts now allow Americans to divert one-third of all solid waste away from disposal. Students on college and university campuses are making great contributions in accelerating these trends. The enthusiasm of students for recycling is apparent each year in the success of Recyclemania—and this competition is just the tip of the iceberg. Most campuses have their own recycling and waste reduction programs, which continue to grow and evolve as students and staff find new and innovative ways of inspiring people to reduce waste.

If you would like to encourage waste reduction efforts on your own campus, there are many sources of information and advice you can consult, most of them easily accessible online. Longtime campus sustainability leaders such as the American Association for Sustainability in Higher Education and the National Wildlife Federation's Campus Ecology program offer a plethora of online resources for waste reduction and many other pursuits. A number of private waste management companies, including the largest one in the United States, Waste Management, Inc., are now also teaming up with campuses to help with the logistics of creating and maintaining workable waste strategies and programs.

One new and active student-focused resource is the Post-Landfill Action Network (PLAN), a group connecting colleges and universities that want to work together to educate, inspire, and empower students to make positive change on their campuses. PLAN organizes an annual Students for Zero Waste Conference, as well as various workshops and events around the country. Established by students from the University of New Hampshire, the group now links several dozen U.S. schools.

Engagement and successes on campus have helped fuel advances in waste management throughout society at large. Still, our prodigious consumption habits continue to create voluminous amounts of products and packaging, some of which is difficult or impossible to recycle. Our waste management efforts are marked by a number of challenges, including the cleanup of Superfund sites, the safe disposal of hazardous and radioactive

waste, the proliferation of plastics and electronic waste, and the need to respond to China's recent ban on imported recyclable materials. These dilemmas make clear that the best solution is to reduce our generation of waste and to pursue a cradle-to-cradle approach. Finding ways to reduce, reuse, and efficiently recycle the materials and goods that we use stands as a key ongoing challenge for our society.

- **CASE STUDY SOLUTIONS** Does your college or university participate in Recyclemania? If so, describe how it has done so, what events it has staged, how successful these events were, and how this success might be improved. If not, peruse the Recyclemania website (http://recyclemania.org), and describe how you think your school might compete effectively in Recyclemania. What events, programs, or strategies do you think would be most effective on your campus to give it a shot at winning one of the categories in Recyclemania?

- **LOCAL CONNECTIONS** On your campus, what efforts are being made to encourage recycling? Is there a composting program? How do the dining halls deal with food waste? Does the campus host any waste reduction events? Is there a student group helping address waste issues? Consult several online sources, such as the ones discussed in this Connect & Continue section, to learn more about the wide range of activities and strategies that students, faculty, and staff are pursuing on many campuses to help reduce waste. What is being done on other campuses that your school is not doing? Describe at least three waste reduction strategies or activities you would like to see your campus pursue, and explain why. What would you need to do to get such an effort off the ground?

- **EXPLORE THE DATA** Recycling and composting have made great strides, but much more progress could be made. → **Explore Data** relating to the case study on **Mastering Environmental Science.**

REVIEWING Objectives

You should now be able to:

+ **Summarize major approaches to managing waste, and compare and contrast the types of waste we generate**

Source reduction, recovery, and disposal are the three main components of waste management. Source reduction is preferred, and recovery is the next best. Municipal solid waste comes from homes, institutions, and small businesses. Industrial solid waste comes from manufacturing, mining, agriculture, and petroleum extraction and refining. Hazardous waste is toxic, chemically reactive, flammable, or corrosive. (pp. 620–621)

+ **Discuss the nature and scale of the waste dilemma**

Developed nations generate far more waste than developing nations, but developed nations are beginning to decrease their waste, whereas waste in developing nations is increasing as population and consumption grow. (p. 622)

+ **Evaluate source reduction, reuse, composting, and recycling as approaches for reducing waste**

Reducing waste before it is generated is the best management approach. Reusing items helps reduce waste. Composting creates organic matter for gardening and farming, whereas recycling removes 26% of the U.S. waste stream. Bottle bills are one way in which financial incentives can motivate people to reduce waste. The economics of recycling are complex, however, and

recent policy changes by China have increased challenges for recycling efforts by restricting international trade in recyclables. (pp. 623–630)

+ **Describe landfills and incineration as conventional waste disposal methods**

Sanitary landfills are our main disposal method. Properly built and maintained, they guard against contamination of groundwater, air, and soil. Incinerators reduce waste volume by burning it. Pollution control technology removes most pollutants from incinerator emissions, but some pollutants escape, and toxic ash needs to be disposed of in landfills. To make the most of our waste, we are harnessing energy from landfill gas and generating electricity from incineration, and we are also starting to recycle materials from landfills. (pp. 630–633)

+ **Discuss industrial solid waste and principles of industrial ecology**

Regulations differ, but industrial solid waste management is similar to that for municipal solid waste. Industrial ecology provides ways for industry to enhance efficiency and studies how industrial systems can mimic ecological systems with a cradle-to-cradle approach. (pp. 633–635)

+ **Assess issues in managing hazardous waste**

Organic compounds, heavy metals, electronic waste, and radioactive waste are common types of hazardous waste. Hazardous waste is regulated and monitored, yet illegal dumping occurs, and no fully satisfactory method of disposing of hazardous waste has yet been devised. The Superfund program cleans up hazardous-waste sites, but cleanup is a long and expensive process. (pp. 634–639)

SEEKING Solutions

1. How much waste do you generate? Look into your waste bin at the end of the day and categorize and measure the waste there. List all other waste you may have generated in other places throughout the day. How much of this waste could you have avoided generating? How much could have been reused or recycled?

2. Of the various waste management approaches covered in this chapter, which ones are your community or campus pursuing? Would you suggest pursuing any new approaches? If so, which ones, and why?

3. Can manufacturers and businesses benefit from source reduction if consumers were to buy fewer products as a result? How? Given what you know about industrial ecology, what do you think the future of sustainable manufacturing may look like?

4. **THINK IT THROUGH** You are the president of your college or university. Your students participate in Recyclemania each year and want you to make the school a leader in waste reduction and industrial ecology. Consider the industries and businesses in your community and the ways they interact with facilities on your campus. Bearing in mind the principles of industrial ecology, can you think of any novel ways in which your school and local businesses might mutually benefit from one another's services, products, or waste materials? Are there waste products from one business, industry, or campus facility that another might put to good use? What steps would you propose to take as president?

5. **THINK IT THROUGH** You are the CEO of a major corporation that produces containers for soft drinks and a wide variety of other consumer products. Your company's shareholders are asking that you improve the company's image—while not cutting into profits—by taking steps to reduce waste. What steps would you consider taking?

CALCULATING Ecological Footprints

For years, the "State of Garbage in America" survey has documented the capacity of Americans to generate prodigious amounts of municipal solid waste (MSW). According to the most recent survey, on a per capita basis, Missouri residents generate the least MSW (4.5 lb/day), and Hawai'i residents (and that state's many tourists) generate the most (15.5 lb/day). The average for the entire country is 6.8 lb MSW per person per day. Calculate the amount of MSW generated in 1 day and in 1 year by each of the groups listed, if they were to generate MSW at each of the rates shown in the table.

GROUPS GENERATING MUNICIPAL SOLID WASTE	AMOUNT OF MSW GENERATED, AT THREE PER CAPITA GENERATION RATES					
	U.S. AVERAGE (6.8 LB/DAY)		MISSOURI (4.5 LB/DAY)		HAWAI'I (15.5 LB/DAY)	
	DAY	YEAR	DAY	YEAR	DAY	YEAR
You	6.8	2482				
Your class						
Your state						
United States						
World						

Data from Shin, D., 2014. *Generation and disposition of municipal solid waste (MSW) in the United States—A national survey*. New York: Columbia University, Earth Engineering Center.

1. Suppose your town of 50,000 people has just approved construction of a landfill nearby. Estimates are that it will accommodate 1 million tons of MSW. Assuming that the landfill is serving only your town and that your town's residents generate waste at the U.S. average rate, for how many years will it accept waste before filling up? How much longer would a landfill of the same capacity serve a town of the same size in Missouri?

2. One study has estimated that the average world citizen generates 1.47 pounds of trash per day. How many times more does the average U.S. citizen generate?

3. The same study showed that the average resident of a low-income nation generates 1.17 pounds of waste per day and that the average resident of a high-income nation generates 2.64 pounds per day. Why do you think U.S. residents generate so much more MSW than people in other "high-income" countries, when standards of living in those countries are comparable?

Mastering Environmental Science

Minerals and Mining

D.R.
CONGO

Region of
coltan mining

Mining for . . .
Cell Phones?

" The conflict in the
Democratic Republic of the
Congo has become mainly
about access, control, and trade
of five key mineral resources:
coltan, diamonds, copper,
cobalt, and gold.
Report to the United Nations Security
Council, April 2001

Coltan . . . is not helping the
local people. In fact, it is the
curse of the Congo.
Ghanaian-Canadian journalist Kofi
Akosah-Sarpong "

You walk across your college campus to your next class. In the process, you pull out your cell phone and text a friend—and likely give very little thought to the technology that makes this text possible. What you probably don't know is that inside your phone is a little-known metal called tantalum—just a tiny amount—and without it, no cell phone can operate. Half a world away, a miner in the heart of Africa toils all day in a jungle streambed, sifting sediment for nuggets of coltan ore, which contain tantalum.

In bedeviling ways, tantalum links our glossy, global high-tech economy with one of the most abused regions on Earth. Democratic Republic of Congo (D.R. Congo) has long been embroiled in a sprawling regional conflict fueled by ethnic tensions and access to valuable natural resources. This conflict has involved six nations and various rebel militias and has claimed more than 5 million lives since 1998. The current conflict is the latest chapter in the sad history of a nation rich in natural resources—copper, cobalt, gold, diamonds, uranium, and timber—whose impoverished people have been repeatedly robbed of the control of those resources.

Tantalum (Ta), element number 73 on the periodic table (**APPENDIX C**), is at the heart of the current battle over resources in D.R. Congo. As noted, we rely on this metal for our cell phones, but it is also vital for production of computer chips, DVD players, video game consoles, and digital cameras. Tantalum is ideal for capacitors (the components that store energy and regulate current in miniature circuit boards) because it is highly heat resistant and readily conducts electricity.

Tantalum comes from a dull blackish mineral called tantalite, which often occurs with a mineral called columbite—so the ore is referred to as columbite–tantalite, or coltan for short. In eastern D.R. Congo, miners dig craters in rainforest streambeds, panning for coltan much as early California miners panned for gold.

As information technology boomed in the late 1990s, global demand for tantalum rose, and market prices for the metal shot up. High prices led some Congolese miners to mine coltan by choice, but many more were forced to work as miners.

Upon completing this chapter, you will be able to:

+ Describe the types of mineral resources and their uses

+ Explain how minerals are processed prior to use

+ Describe the major methods of mining

+ Discuss the environmental and social impacts of mining

+ Explain mine reclamation efforts and mining policy

+ Evaluate ways to encourage sustainable use of mineral resources

◄ Children mining for coltan ore

▲ The endangered okapi, one of the casualties of mining in eastern D.R. Congo

643

Open war broke out in D.R. Congo in 1998 when rebel groups, supported by forces from neighboring Rwanda and Uganda, attempted to overthrow the government of President Laurent-Désiré Kabila. The country was fragmented by conflict between government forces and various rebel groups, with 11 other African nations becoming involved in the hostilities. In mineral-rich eastern D.R. Congo, fighting was particularly intense. Local farmers were chased off their land; villages were burned; and civilians were raped, tortured, and killed. Soldiers and rebels seized control of mining operations. They forced farmers, refugees, prisoners, and children to work and enriched themselves by selling coltan to traders, who in turn made profits by selling it to processing companies in the United States, Europe, and Asia. These companies refine and sell tantalum powder to capacitor manufacturers, which in turn sell capacitors to Nokia, Motorola, Sony, Intel, Apple, Dell, and other high-tech corporations that use the capacitors in their products.

The conflict also caused ecological havoc as people streamed into national parks to avoid the fighting and locate new sources of minerals. The result was the clearing of rainforests for fuelwood and the killing of wildlife for food, including forest elephants, endangered gorillas, and okapi—a rare zebra-like relative of the giraffe categorized as endangered in 2013. Miners disturbed streambeds in the search for coltan, increasing erosion rates and choking streams with sediments.

A grass-roots activist movement urged international action to stop the violence and exploitation in D.R. Congo and advanced the slogan, "No blood on my cell phone!" In 2001, an expert panel commissioned by the United Nations concluded that coltan riches were fueling, financing, and prolonging the war. The panel urged a UN embargo on coltan and other minerals from regions of D.R. Congo where conflict flourished.

Corporations rushed to assure consumers that they were not using tantalum from eastern D.R. Congo—noting that the region was producing less than 10% of the world's supply. Meanwhile, some observers believed that an embargo could hurt the long-suffering Congolese people. The mining life may be miserable, they said, but it pays better than most jobs in a land where the average income is only 20 cents a day.

In 2001, Kabila was assassinated and his son, Joseph, assumed power. A 2002 peace treaty led to the withdrawal of foreign troops from D.R. Congo, but conflict with rebel groups within the country continues, bankrolled by the mineral riches of the region. Although success by Congolese troops and an African-led UN intervention brigade against a major rebel group in eastern D.R. Congo has helped reduce conflict, the region nevertheless remains largely lawless. Today, the national government has little control over eastern D.R. Congo, enabling continued exploitation of the wealth of mineral riches, forests, and endangered animals in the region.

Steps are now being taken to help support legitimate Congolese mines while preventing the exploitation that has defined mining in D.R. Congo in the recent past. Industry groups, working with national governments and nongovernmental aid organizations, are creating a certification system for conflict-free coltan to aid consumers in avoiding products that contain conflict minerals.

These efforts and others offer an opportunity to significantly reduce trade in conflict minerals while promoting trade of minerals sourced from legitimate mines in poor nations such as D.R. Congo. It is hoped that ongoing regulatory efforts to certify minerals will establish a framework that not only satisfies the world's demand for mineral resources but also protects the people and ecosystems that provide them.

Earth's Mineral Resources

Coltan provides just one example of how we extract raw materials from beneath our planet's surface and turn them into products we use every day. We mine and process a wide array of mineral resources in the modern world. Indeed, without these resources—which we use to make everything from building materials to fertilizers—civilization as we know it could not exist. Just consider a typical scene from a student lounge at a college or university (FIGURE 23.1) and note how many items are made with elements from the minerals we extract.

Both gradual geologic processes and catastrophic geologic hazards influence the distribution of rocks and minerals in the lithosphere and their availability to us. In the lithosphere (p. 35), the region that includes the uppermost layers of rock near Earth's surface, the rock cycle (p. 37) creates new rock and alters existing rock. Plate tectonics (pp. 35–37) builds mountains; shapes the geography of oceans, islands, and continents; and gives rise to earthquakes and volcanoes. Throughout the world, geologic processes are fundamental to shaping the world around us and affecting the distribution of the mineral resources our modern society requires.

Rocks provide the minerals we use

A **rock** is any solid aggregation of minerals. A **mineral,** in turn, is any naturally occurring solid inorganic substance (a pure element or compound) with a regular crystalline structure, a defined chemical composition, and distinct physical properties (p. 37). For instance, the mineral tantalite consists of the chemical elements tantalum (Ta), oxygen (O), iron (Fe), and manganese (Mn) (see APPENDIX C for a periodic table of chemical elements). Tantalite occurs most commonly in pegmatite, a type of igneous rock (p. 38) similar to granite. In addition to tantalite, pegmatite generally contains the minerals feldspar, quartz, and mica, and on occasion it even includes gemstones and other rare minerals.

In some cases, the rock itself is a resource, such as the gravel that is used to make concrete. In other cases, valuable

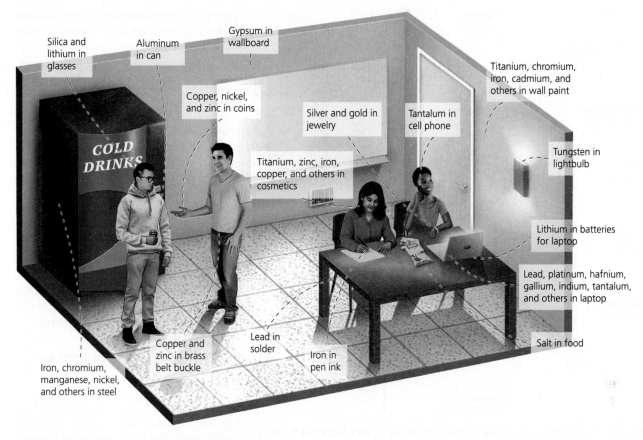

FIGURE 23.1 Elements from minerals that we mine are everywhere in the products we use in our everyday lives. This scene from a typical college student lounge points out just a few of the many elements from minerals that surround us.

mineral resources, such as iron or phosphate, are embedded within rocks. To secure these minerals, miners remove the mineral-containing rock and then separate the desired mineral from the surrounding rock. In either case, the process begins by removing rocks, fossil fuels (such as coal), or mineral-bearing rock from the lithosphere.

We obtain minerals by mining

We obtain the minerals we use through the process of mining. The term *mining* in the broad sense describes the extraction of any resource that is nonrenewable on a timescale that can keep pace with human consumption. We indeed mine for minerals, and because we extract fossil fuels and groundwater from the earth faster than they can replenish, we are said to be *mining* those resources as well.

When used specifically in relation to minerals, **mining** refers to the systematic removal of rock, soil, or other material for the purpose of extracting minerals of economic interest. Because most minerals of interest occur widely spread out in low concentrations, miners and mining geologists first try to locate concentrated sources of minerals before mining begins.

Concentrated deposits of minerals form in several different ways, causing them to be unevenly distributed on Earth. In some cases, minerals are concentrated in magma and so accumulate in rock layers beneath magma chambers and in areas with large quantities of igneous rocks (p. 38). That is true of many metals, including tantalum, iron, nickel, and platinum. In other cases, hot groundwater dissolves minerals from large areas of rock and then concentrates them into one area when the water cools and the minerals precipitate out of solution. This type of concentration occurs for minerals containing sulfur, zinc, and copper. Geologic processes, such as plate tectonics and rock weathering (p. 217), also play a role in creating mineral reserves and making them accessible for mining.

We use mined materials extensively

The average American consumes 17,440 kg (38,449 lb) of new minerals and fuels every year, according to 2018 estimates by the U.S. Minerals Education Coalition. More than half of the annual mineral and fuel use is from the coal, oil, and natural gas used to supply our intensive demands for energy. Much of the remaining mineral use is attributable to the sand, gravel, and stone used in constructing our buildings, roads, bridges, and parking lots. Metal use is dwarfed by our utilization of fuels and stone, but the average American will still use more than a ton of aluminum over his or her lifetime.

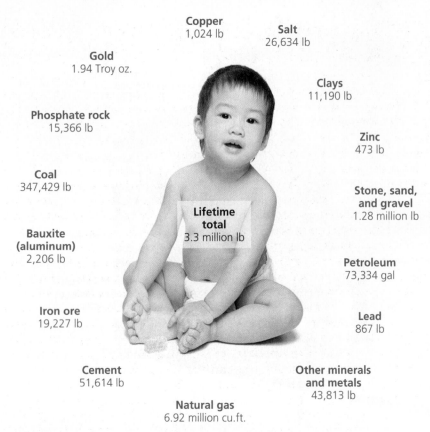

Gold
1.94 Troy oz.

Copper
1,024 lb

Salt
26,634 lb

Phosphate rock
15,366 lb

Clays
11,190 lb

Coal
347,429 lb

Zinc
473 lb

Stone, sand,
and gravel
1.28 million lb

Lifetime
total
3.3 million lb

Bauxite
(aluminum)
2,206 lb

Petroleum
73,334 gal

Iron ore
19,227 lb

Lead
867 lb

Cement
51,614 lb

Other minerals
and metals
43,813 lb

Natural gas
6.92 million cu.ft.

FIGURE 23.2 At current rates of use, a baby born in the United States in 2018 is predicted to use 1.37 million kg (3.03 million lb) of minerals over his or her lifetime. *Data from Minerals Education Coalition, 2018.*

At current rates of use, a child born in 2018 in the United States will use more than 1.37 million kg (3.03 million lb) of minerals and fuel during his or her lifetime (**FIGURE 23.2**). This level of consumption clearly shows the potential of recycling and reuse (such as recycling stone and gravel from old highways into new construction) to make our modern, mineral-intensive lifestyle more sustainable.

Metals are extracted from ores

Some minerals can be mined for metals. As we have seen, the tantalum used in electronic components comes from the mineral tantalite (**FIGURE 23.3**). A **metal** is a substance that typically is lustrous, opaque, and malleable and can conduct heat and electricity. Most elemental metals are not found in a pure state in Earth's crust but instead are present within **ore,** a mineral or grouping of minerals from which we extract metals. Copper, iron, lead, gold, and aluminum are among the many economically valuable elemental metals we extract from mined ore.

We process metals after mining ore

Extracting minerals from the ground is the first step in putting them to use. However, most metals need to be processed in some way to become useful for our products. For example, after ores are mined, the metal-bearing rock is pulverized and washed, and the desired minerals are then isolated using chemical or physical means. For iron, this process, known as **smelting,** involves heating ore-bearing rocks to extremely high temperatures in a blast furnace and collecting the molten iron when it separates

(a) Coltan ore

(b) Capacitors containing tantalum

FIGURE 23.3 Tantalum is used to manufacture electronics. Coltan ore **(a)** is mined from the ground and then processed to extract the pure metal tantalum. This metal is used in capacitors **(b)** and other electronic components.

FIGURE 23.4 A worker guides molten iron out of a blast furnace. The metal is separated from the surrounding rock by melting ore at high temperatures and collecting the heavier metal.

FAQ

How do geologists find mineral deposits that are deep in the earth?

Searching for reserves of underground minerals, also called prospecting, can be pursued in a number of ways. The earliest prospectors explored promising areas on foot, looking for exposed seams of mineral-containing rocks or for minerals carried into streams by runoff. Today, geologists direct vibrations into underground rock strata and capture the vibrations with sensors as the vibrations bounce off underground rock layers and reflect back to the surface. This process enables scientists to visualize the underlying rock layers and identify likely locations for reserves, just as they do for fossil fuel deposits (pp. 529–533). Geologists also measure the magnetic fields in rock layers to look for metal ores, and they conduct chemical analyses of stream water to detect minerals of interest. If promising sites are located, cores can be drilled deep into the ground and inspected for the desired mineral before actual mining begins.

from the surrounding minerals (FIGURE 23.4). For aluminum, bauxite ore is first treated with chemicals to extract alumina (an aluminum oxide), and then an electrical current is used to generate pure aluminum from alumina. With coltan, processing facilities use acid solvents to separate tantalite from columbite. Other chemicals are then used to produce metallic tantalum powder. This powder can be consolidated by various melting techniques and shaped into wire, sheets, or other forms. Sometimes we mix, melt, and blend one metal with another metal or a nonmetal substance to form an **alloy.** For example, steel is an alloy of the metal iron that has been mixed with a small quantity of carbon.

Processing minerals exerts environmental impacts. First, most processing methods are water-intensive and energy-intensive, straining precious water resources and generating greenhouse gases that contribute to global climate change. Second, many chemical reactions and heating processes used for extracting metals from ore emit air pollution—smelting plants in particular can be hotspots of toxic air pollution (Chapter 17).

Pollution from mineral processing is also a concern. Soil and water commonly become polluted by **tailings,** portions of

ore left over after metals have been extracted. Heavy metals may leach from ore tailings as well as chemicals applied in the extraction process. For instance, we use the chemical cyanide to extract gold from ore, and sulfuric acid to extract copper, and both cyanide and sulfuric acid are highly toxic to organisms (Chapter 14).

Mining operations often store toxic slurries of tailings in large reservoirs called **surface impoundments** (FIGURE 23.5). Their walls are designed to prevent leaks or collapse, but accidents can occur if the structural integrity of the impoundment is compromised.

FIGURE 23.5 This surface impoundment in West Virginia holds coal tailings from surface mining operations. The impoundment is contained by an earthen dam 117 m (385 ft) tall that can hold more than 10 billion L (2.8 billion gal) of waste. An elementary school, visible in the lower left portion of the image, lies downslope from the impoundment.

There are approximately 18,000 tailing impoundments worldwide, and a 2017 report by the United Nations found that the number of failures had increased from 1960 to 2010, with 40 significant impoundment failures occurring in the decade leading up to the report. In 2000, a breach of an impoundment near Inez, Kentucky, released more than 1 billion liters (250–300 million gal) of coal slurry (water used to wash mined coal that contained numerous toxic compounds), blackening 120 km (75 mi) of streams, killing aquatic wildlife, and affecting drinking water supplies for many communities for weeks. The failure of two impoundments at an iron ore mine in Brazil in 2015 buried nearby villages in a toxic slurry of water and mining waste, claiming 19 lives and fouling the nearby Doce River for 800 km (500 mi). In January 2019, an impoundment at an iron ore mine in the same region of Brazil suddenly collapsed, releasing a slurry that swept away people, buried a nearby village, and claimed an estimated 330 lives.

We also mine nonmetallic minerals and fuels

We also mine and use many minerals that do not contain metals. **FIGURE 23.6** illustrates the nation of origin and uses for some economically important mineral resources, both metallic and nonmetallic.

Sand and gravel, the most commonly mined mineral resources, are used as fill and as construction materials for the manufacturing of products such as concrete. Each year, more than $7 billion of sand and gravel are mined in the United States. Phosphate rock provides us phosphorus for fertilizers and industrial chemicals. We mine limestone, salt, potash, and other minerals for a number of diverse purposes.

The processing of nonmetals can also have significant environmental impacts. Producing cement, for example, consumes huge quantities of fossil fuels because its production requires the heating of its constituent parts to nearly 1500°C (2700°F). If the cement industry were its own country, its emissions of greenhouse gases would be third globally—only surpassed by the emissions of China and the United States. Gemstones are treasured for their rarity and beauty. For instance, diamonds have long been prized—and like coltan, they have fueled resource wars. Besides the conflict in eastern Congo, the diamond trade has acted to fund, prolong, and intensify wars in Angola, Sierra Leone, Liberia, and elsewhere as armies exploit local people for mine labor and then sell the diamonds for profit. That is why you may hear the phrase "blood diamonds," just as coltan has been called a "conflict mineral."

We also mine substances we use for fuel. Uranium ore is a mineral from which we extract the metal uranium, which we use in the production of nuclear power (pp. 564–567). One of the most common fuels we mine is coal.

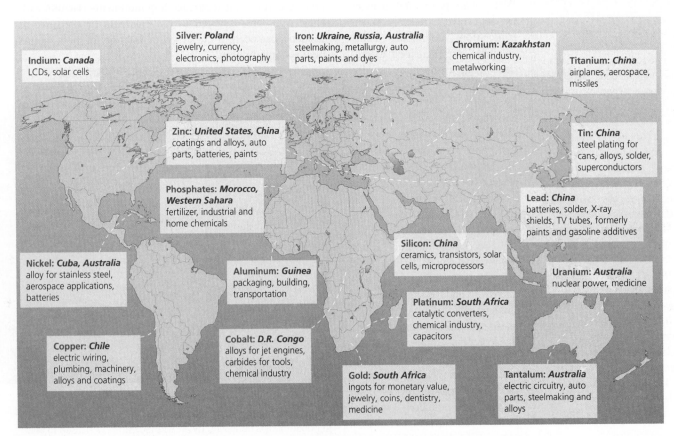

FIGURE 23.6 The minerals we use come from all over the world. Shown is a selection of economically important minerals, mostly metals, together with their major uses and their main nation of origin. Note that only a minority of mined minerals, and their uses and origins, is shown.

FAQ

Concrete is everywhere, so why isn't it in the list of mined materials we all use?

Actually, concrete is on the list of mined materials—but only in its constituent parts. To create concrete, you combine crushed rock, called *aggregate,* with cement and water. Cement is made by heating—to extremely high temperatures—mined materials that contain calcium, silicon, aluminum, and iron. Typically, limestone rock, clays, and iron ore or iron waste from steel mills are used in this process to create a powdery material. When this material is combined with water and aggregate, a chemical reaction is initiated that binds all the components together to form a slurry that can be poured into molds of nearly any shape. Within the molds, steel or iron rods provide a reinforcing framework, and the concrete dries into a rigid product capable of supporting massive weight loads—hence our use of concrete in buildings, roadways, bridges, and dams. So, although concrete doesn't meet the definition of a mineral, it is represented on the list by the categories "Stone, sand, and gravel," "Clays," "Cement," and "Iron ore."

Coal (pp. 529–530) is the modified remains of ancient swamp plants and is composed primarily of the mineral carbon. Other fossil fuels—petroleum, natural gas, and alternative fossil fuels such as oil sands, oil shale, and methane hydrates—are also organic and are extracted from the earth (Chapter 19).

Mining Methods and Their Impacts

Mining for minerals is an important industry that provides jobs for people and revenue for communities in many regions. Mining supplies us raw materials for countless products we use daily, so it is necessary for the lives we lead. In 2018, raw materials from mining contributed $82 billion to the U.S. economy, and after processing, mineral materials contributed $766 billion. About 145,000 Americans were employed directly in mining for coal and metals in 2018, and the mining industry, together with processors and manufacturers of products from mined materials, employed more than 1.3 million people.

But as we have seen, mining also exacts a price in environmental and social impacts. The extraction of minerals—and the cleaning and processing of mined materials—often involves the degradation of large areas of land, thereby exerting severe impacts on the environment and on the people living near the mining locations.

Depending on the nature of the mineral deposit, any of several mining methods may be used to extract the resource from the ground. If multiple methods are appropriate for a given resource, companies typically select a method based on its economic efficiency. We will examine major mining approaches commonly used throughout the world and will also take note of the impacts of each approach as we proceed.

Strip mining removes surface layers of soil and rock

When a resource occurs in shallow horizontal deposits near the surface, the most effective mining method is often **strip mining,** whereby layers of surface soil and rock are removed from large areas to expose the resource. Heavy machinery removes the overlying soil and rock (termed *overburden*) from a strip of land, and the resource is extracted. This strip is then refilled with the overburden that had been removed, and miners proceed to an adjacent strip of land and repeat the process. Strip mining is commonly used for coal (**FIGURE 23.7a**) and oil sands (pp. 523–525), and sometimes for sand and gravel. Strip mining can be economically efficient, but it obliterates natural communities over large areas, and the soil in refilled areas can easily erode away.

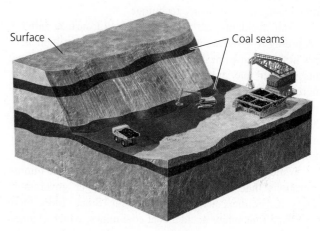

(a) Strip mining

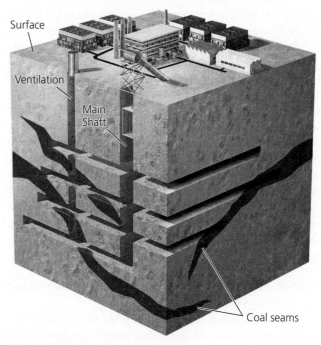

(b) Subsurface mining

FIGURE 23.7 Coal mining illustrates two types of mining approaches. In strip mining **(a),** soil is removed from the surface in strips, exposing seams from which coal is mined. In subsurface mining **(b),** miners work belowground in shafts and tunnels blasted through the rock. These passageways provide access to underground seams of coal or minerals.

In subsurface mining, miners work underground

When a resource occurs in concentrated pockets or seams deep underground, and the earth allows for safe tunneling, mining companies pursue **subsurface mining.** In this approach, shafts are excavated deep into the ground, and networks of tunnels are dug or blasted out to follow deposits of the mineral (**FIGURE 23.7b**). Miners remove the resource systematically and transport it to the surface. We use subsurface mining for metals such as zinc, lead, nickel, tin, gold, copper, and uranium, as well as for diamonds, phosphate, salt, and potash. In addition, a great deal of coal is mined using the subsurface technique.

Fatal accidents are not unusual in subsurface mining. According to government figures, 931 miners lost their lives in Chinese coal mines in 2014 alone, and critics argue that this number may be underestimated. As recently as 2000, more than 7600 miners were perishing annually in Chinese mines, but fatalities have dropped significantly in the past 20 years as government regulators strengthened safety rules and shut down many smaller mines where accidents were more common.

Although the United States has fewer subsurface mining operations than China and relatively stronger government safety inspections, accidents still occur. Since 2006, there have been four major accidents at coal mines, which claimed 52 lives, including an explosion at an underground coal mine in 2010 in West Virginia that killed 29 miners. This number seems small, but there are relatively few people employed in mining, and this accident rate makes mining one of the United States' statistically most dangerous ways to make a living.

Occasionally, mines can have impacts even long after the mines have been closed. Subsurface and strip mining can pollute waterways through the process of **acid mine drainage,** which occurs when sulfide minerals in newly exposed rock surfaces react with oxygen and rainwater to produce sulfuric acid. As the sulfuric acid runs off, it leaches metals from underlying rocks, and many of these metals are toxic to organisms. Acid drainage can affect fish and other aquatic organisms when it runs into streams (**FIGURE 23.8**) and can pollute groundwater supplies people use for drinking water or irrigating crops. Although acid drainage is a natural phenomenon, mining greatly accelerates this process by exposing many new rock surfaces at once. Releases of acid mine drainage from mines can last for many years, sometimes for centuries, after the mines close. Mines in Spain from the era of the Roman Empire, some 2000 years ago, still leach sulfuric acid into waterways today.

Closed subsurface mines can have other, long-lasting effects on the environment. For example, coal veins in abandoned mines underneath Centralia, Pennsylvania, caught fire in 1961. The once-thriving city became a ghost town as nearly all its residents accepted buyouts in the 1980s and relocated when it became clear that the smoldering fires beneath them could not be contained; indeed, the fires continue to burn to this day (**FIGURE 23.9**). Natural disasters or accidents can lead to catastrophic releases of toxin-laden groundwater that accumulates in abandoned

FIGURE 23.8 Acidic drainage contaminates a stream in western Pennsylvania. The orange color is due to iron from the drainage settling out on the soil surface and forming rust.

subsurface mines, as occurred in 2015 when millions of gallons of heavy metal–laden mine drainage from an abandoned gold mine in Colorado were accidentally released into the nearby Animas River, killing aquatic wildlife and contaminating water supplies for communities in Colorado, Utah, New Mexico, and

FIGURE 23.9 Smoldering mine fires beneath Centralia, Pennsylvania, have led to the creation of a "ghost town." A nearly one-mile-long section of State Route 61 was closed because subsidence due to mine fires caused the road to buckle and crack. The roadway has since become a tourist attraction for people visiting the area.

FAQ

Why would anyone work in a mine when it's such dangerous work?

Subsurface mining is the most dangerous form of mining and indeed one of society's most dangerous occupations. Besides risking injury or death from dynamite blasts, natural gas explosions, and collapsing shafts and tunnels, miners inhale toxic fumes and coal dust, which can lead to respiratory diseases, including fatal black lung disease.

Many of the people who work in mines do so because they have few other options. Underground mining often occurs in economically depressed areas, such as Appalachia in the United States, where mining is one of the few jobs that pay well. And for most mining jobs, people can begin working right out of high school. So, although the work is dangerous, many miners are willing to accept those risks to provide for themselves and their families because relatively fewer reliable, well-paying career opportunities are available.

the Navajo nation. The U.S. government estimates that as many as 500,000 abandoned mines exist in the United States and that full cleanup of these sites could cost taxpayers up to $70 billion.

Open pit mining can create immense holes in the ground

When a mineral is spread widely and evenly throughout a rock formation or when the earth is unsuitable for tunneling, the method of choice is **open pit mining.** This technique essentially involves digging a gigantic hole and removing the desired ore, along with waste rock that surrounds the ore. The ore-bearing rock is sent off for processing, and the waste rock is dumped in massive heaps outside the pit. Open pit mines are terraced so that workers and massive machinery can move about, and the pit is expanded until the resource runs out or becomes unprofitable to mine. Open pit mining is used to

extract copper, iron, gold, diamonds, and coal, among other resources. We also use this technique to extract clay, gravel, sand, and stone such as limestone, granite, marble, and slate, but in these cases we generally call the pits *quarries.* Some open pit mines are inconceivably enormous. The world's largest, the Bingham Canyon copper mine near Salt Lake City, Utah, is 4 km (2.5 mi) across and 1.2 km (0.75 mi) deep (**FIGURE 23.10**).

Open pit mines are so large because huge volumes of waste rock need to be removed to extract relatively small amounts of ore, which in turn contain still smaller traces of valuable minerals. The sheer size of these mines means that the degree of habitat loss and aesthetic degradation is considerable. Another impact is chemical contamination from acid drainage as water runs off the waste heaps or collects in the pit. Once mining is complete, abandoned pits generally fill up with groundwater, which soon becomes toxic as water and oxygen react with sulfides from the ore. Acidic water from the pit can harm wildlife and can percolate into aquifers and spread throughout the region.

The Berkeley Pit, a former copper mine near Butte, Montana, is today one of the largest Superfund toxic waste cleanup sites (pp. 638–639) in the United States. After its closure in 1982, the pit filled with groundwater and became so acidic (pH of 2.2) and concentrated with toxic metals that microbiologists discovered new species of microbes in the water—the harsh conditions were so rare in nature that scientists had never before encountered microbes adapted to them. In 2016, a flock of roughly 10,000 snow geese (*Anser caerulescens*), diverted from their typical migration routes by a winter storm, landed in the toxic waters of the mine pit. Dead geese soon began turning up in the local area, and autopsies of dead birds found lesions throughout their

FIGURE 23.10 The Bingham Canyon open pit mine outside Salt Lake City, Utah, is the world's largest human-made hole in the ground. This immense mine produces mostly copper.

FIGURE 23.11 Mining in D.R. Congo destroys vast swathes of natural areas. Here, workers toil in the pit of a gold mine in the northeastern part of the country under the watchful eye of a rebel soldier. Illegal mines even operate within a protected, forested reserve for the endangered okapi.

digestive systems, a finding consistent with the birds having ingested mine pit water with elevated levels of sulfuric acid and toxic metals. An estimated 3000 to 4000 birds perished in this event.

Open pit mining on a smaller scale is used in D.R. Congo (**FIGURE 23.11**). Miners extract coltan reserves from large pits and then separate the mineral from the surrounding rock and soil by placing it in the flowing waters of streams or rivers. Doing so washes large amounts of soil and debris into waterways, making them uninhabitable for fish and other life for many miles downstream. It also disturbs stream banks, causing erosion and harming ecologically important riparian plant communities. Gold miners using this method in northern California's rivers in the decades following the Gold Rush washed so much river-borne debris all the way to San Francisco Bay that a U.S. district court ruling in 1884 finally halted this mining practice.

Mountaintop mining reshapes ridges and can fill valleys

When a resource occurs in underground seams near the tops of ridges or mountains, mining companies may practice **mountaintop removal mining,** in which several hundred vertical feet of mountaintop may be blasted off to enable strip-mining of the now-exposed resource (**FIGURE 23.12**). In mountaintop removal mining, a mountain's forests are clear-cut, the timber is sold, topsoil is removed, and then rock is repeatedly blasted away to expose the coal for extraction. The destructive nature of this mining method has led to it being

FIGURE 23.12
Mountaintop mining removes entire mountaintops to obtain the coal underneath.
The rocks removed during the process are dumped into adjacent valleys, burying streams, promoting flooding, and contaminating drinking water supplies.

described as "strip mining on steroids." This method of mining is used in the Appalachian Mountains of the eastern United States to secure coal (see **THE SCIENCE BEHIND THE STORY**, pp. 654–655).

Afterward, overburden is placed back onto the mountaintop. This waste rock is unstable and typically takes up more volume than the original rock, however, so a great deal of waste rock is generally dumped into adjacent valleys (a practice called *valley filling*). Mountaintop mining produces less than 1% of the United States' coal, but removal operations in the Appalachian Mountains have blasted away an area larger than the state of Delaware and buried nearly 3200 km (2000 mi) of streams.

Mountaintop mining has greatly increased the land area disturbed by surface mining for coal in the Appalachian coal mining region. From 1985 until 2005, mining companies disturbed about 10 m^2 (108 ft^2) of land to produce a metric ton of coal. By 2015, however, that area had risen to nearly 30 m^2 (322 ft^2) per ton **(FIGURE 23.13)**, indicating that the easiest to obtain coal seams—the thickest seams covered by the least amount of overbearing rock—have largely been mined out. If mountaintop mining is used in Appalachia to extract the lower-quality coal reserves that remain, these trends suggest the waste generation from those mining activities will be staggering.

Scientists are finding that dumping tons of debris into valleys degrades or destroys immense areas of habitat, clogs streams and rivers, and pollutes waterways with acid drainage. With slopes deforested and valleys filled with debris, erosion intensifies, mudslides become frequent, and flash floods ravage the lower valleys. Studies have found that mountaintop mining, and the valley fills that are a part of the process, lower the biological diversity of macroinvertebrates (such as larval insects) in Appalachian streams, decrease the abundance of salamanders, negatively affect fish, and alter the communities of microbes inhabiting stream sediments and waters. Worsening the environmental impact of mountaintop removal mining is that the Appalachian forests that are cleared in mountaintop mining are some of the richest forests for biodiversity in the nation—these forests fail to reestablish on formerly mined sites and convert to ecosystems dominated by grasses and shrubs.

People living in communities near the mining sites experience social and health impacts. Explosions that are a part of clearing the mountaintops crack house foundations and drinking water wells, loose rock tumbles down into yards and homes, and floods tear through properties when mining operations block or divert streams. Coal and rock dust causes respiratory ailments, and contaminated water unleashes a variety of health problems. Studies have shown, for example, that people in mountaintop mining areas exhibit elevated levels of birth defects, lung cancer, heart disease, kidney disease, pulmonary disorders, hypertension, and early mortality.

Critics of mountaintop removal mining argue that valley filling violates the Clean Water Act (p. 173) because runoff flowing through waste rocks in valleys often contains high levels of salts and toxic metals that degrade water quality and affect aquatic organisms. Although hundreds of permits for mountaintop mining were issued during the Bill Clinton and George W. Bush administrations, in 2010 the U.S. Environmental Protection Agency (EPA) announced new guidelines that prohibit valley fills unless strict measures of water quality can be attained.

Restrictions on damage to aquatic systems from mining were updated and formalized when in 2016, after seven years of study and input, the Obama administration implemented new regulations for mining that addressed water quality issues. They said, among other things, that (1) a mining company wishing to open a surface or subsurface mine—including a mountaintop removal mine—could not significantly degrade the waterways near the mine site and (2) the company had to do an assessment of local waterways prior to mining, monitor those waterways while mining, and then restore waterways to an approximation of their original state once mining is complete.

Environmentalists were not thrilled with the new regulations, as they believed they did not go far enough to protect aquatic ecosystems, but an independent analysis concluded that if these regulations were implemented, more than 400 km (250 mi) of waterways each year would see improved water quality from 2017 to 2040. With the seating of President Donald Trump—a vocal supporter of the coal industry—in 2017, the coal industry lobbied aggressively to remove these regulations. Their efforts were successful, and the new regulations were repealed in 2017 using the Congressional Review Act, a provision that enables simple majorities in both chambers of Congress to overturn recently passed regulations, as long as the president agrees.

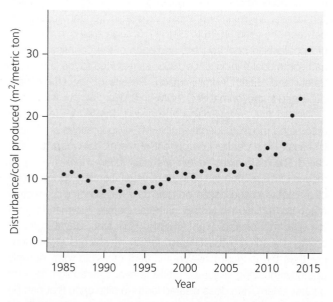

FIGURE 23.13 Since 2010, the amount of land that was disturbed to secure a metric ton of coal in Appalachia has risen dramatically. *Source: Pericak, A.A., et al., 2018. Mapping the yearly extent of surface coal mining in Central Appalachia using Landsat and Google Earth Engine. PLOS ONE 13(7): e0197758, Figure 6. https://doi.org/10.1371/journal.pone.0197758.*

THE SCIENCE
behind
the **story**

Mapping Mountaintop Mining's "Footprint" in Appalachia

The study area in the Appalachian coal-mining region

Studies have shown that mountaintop removal mining (MTRM) and the destructive practices this method employs—blasting off mountaintops and filling adjacent valleys with debris—has the potential to change the biological, geological, and hydrological characteristics of Appalachian ecosystems and appears to negatively affect the health of people living in the proximity of MTRM operations. In 2016, under President Barack Obama, the U.S. Department of the Interior awarded a $1 million grant to the National Academies of Sciences, Engineering, and Medicine to further investigate the links between human health and MTRM activities. A panel of experts was assembled to review relevant research to better inform regulators, resource managers, mining companies, and residents of mining regions about the potential threats to human health from living near active or former mining sites. If connections could be made between MTRM operations and human health impacts, it was hoped that preventative steps could be employed by MTRM companies to minimize adverse effects on people living near their mines.

To conduct analyses linking factors to one another, it is imperative that researchers are able to connect—in both space and time—MTRM operations and measures of human health in nearby communities. Although data from health professionals can be used to catalog the health of people living near mining sites, mapping the extent of MTRM operations, particularly on short timescales, has historically been problematic.

The most efficient way to identify MTRM operations in the rugged landscape of the Appalachian Mountains has been by examining satellite images of the same areas over time and detecting when new mines are formed, how long they are active, and when they cease operations. Such images have been readily available for nearly half a century from the Landsat satellite program. Co-managed by the U.S. Geological Survey and the National Aeronautics and Space Administration, it has been mapping the globe since 1972. Landsat produces images of the same location roughly every 16 days, so researchers can identify mining sites by examining these images and looking for changes in the landscape that accompany MTRM.

Analyzing time sequences of images over any large land area is extremely time-consuming and expensive, even with the aid of supercomputers, which greatly limits the number of images that can be examined and the wealth of data that can be obtained. But thanks to collaborations between Duke University, SkyTruth (a non-profit organization that uses satellite imagery to study environmental impacts), Google Earth, Appalachian Voices (an organization dedicated to sustainability in Appalachia), and West Virginia University, scientists now have a new tool that enables satellite imagery to be rapidly and inexpensively analyzed.

Previous work by SkyTruth used traditional methods of analysis with satellite images from the Landsat satellite program to document the extent of MTRM in Appalachia from 1976 to 2005. SkyTruth's report, released in 2009, provided a map of MTRM operations in Appalachia—but the map could only show when a given mine was active at a coarse ten-year timescale, making refined connections between active mining operations and indicators of human and ecosystem health problematic. Today, there are no such constraints. A new approach enables researchers to generate, in a cost-efficient manner, the first year-by-year map of the spread of MTRM in Appalachia.

To make the mapping process cheaper and faster, the new team used a tool that was unavailable to researchers in 2009—the Google Earth Engine platform and its library of Landsat satellite images (https://earthengine.google.com). This technology enabled the researchers to map the extent of MTRM disturbance to the landscape over an 83,000 km^2 (32,000 mi^2) mountainous region encompassing 74 counties in Kentucky, Tennessee, Virginia, and West Virginia in central Appalachia between the years 1985 to 2015. This large area is in the heart of Appalachian coal country, and each of the counties studied had some coal mining since 1983, as reported to the U.S. Mine Safety and Health Administration. Results of the study were published in the journal *PLOS ONE* in 2018.

Within the Earth Engine application, the team created an automated model to identify active MTRM operations at a yearly timescale. The central component of the model was a statistic called the *normalized difference vegetation index* (NDVI), a means of measuring the amount of vegetation in a specific pixel of a satellite image based on detected wavelengths of electromagnetic radiation. Landsat satellites gather information from the spectrum of visible light, but also from other wavelengths of energy, including "near infrared" radiation, which is slightly longer than the longest wavelengths (red) of visible light. Actively photosynthesizing vegetation reflects near infrared light to a far greater extent than does cleared land—a difference that can be quantified, pixel by pixel, in each Landsat image. The NVDI of pixels in mined and non-mined areas in the region differ greatly, and the researchers used this measure to assign each pixel the status of *likely active mine* (an area in which mining is occurring, or recently ceased) or *likely non-mine* (undisturbed land or land revegetated after mining operations ceased). To avoid overestimating the degree of mining, the model "masked out" image pixels of urban areas, roads, and small areas of cleared land.

After further refinements and testing, the Earth Engine model calculated that some 2900 km^2 (1100 mi^2) of land in the study area was disturbed by surface mining between 1985 and 2015. When disturbance data from 1976 to 1985 were added to this figure, a sum total of 5900 km^2 (2275 mi^2) of land in the region was found to have been affected by MTRM operations since the mid-1970s. In fact, MTRM operations were the single-largest cause of land use change in the region and affected a stunning 7% of the land in the study area.

What else does analysis of results from the new model reveal about MTRM in Appalachia? The data show that land area under active mining increased yearly from 1985 to 2010, but has been declining ever since (**FIGURE 1a**). In the same periods, cumulative mined area rose and then began to level off (**FIGURE 1b**). The area of land newly converted to a mine in a given year increased slightly from 1985 to around 2000 and then declined precipitously (**FIGURE 1c**). Data on production of coal as reported to the U.S. Mine Safety and Health Administration by mines in the region also show a peak around 2005, followed by a substantial decline to the lowest level of production in the study period in 2015 (**FIGURE 1d**).

Governmental action during the study period, among other factors, likely underlies some of the observed trends. Whereas previous presidential administrations granted permits for MTRM operations, the Obama administration began implementing more stringent review of MTRM permits in 2009, and the EPA announced in 2010 new regulations on MTRM as they pertained to the Clean Water Act (p. 173). These changes may help explain the steep decline in the land area actively being mined or converting to mining from 2009 onward.

It is unclear if downward trends will continue. Just as the Obama administration departed from the policies of previous administrations and strengthened restrictions on MTRM operations and coal-burning power plants, starting in 2017 the Trump administration loosened many of the Obama administration's policies. If MTRM operations indeed expand in the near future due to relaxations in governmental regulations, the degree of land disturbance is likely to drastically increase.

Also in 2017, the Department of the Interior abruptly ceased funding the National Academies' study. So although the Google Earth Engine tool was able to provide a means to determine the impacts of MTRM operations at small spatial scales and small timescales, thus enabling these data to be matched with health information from local health professionals and see how MTRM may be affecting people living near mining sites, a more comprehensive examination of any relationship between these factors may have to wait for policy changes under a future presidential administration.

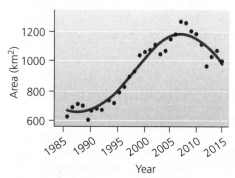

(a) Land area under active mining

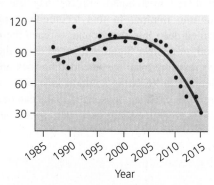

(b) Cumulative land area affected by mining

(c) Land area newly converted to mining

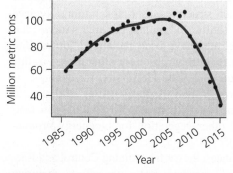

(d) Coal production from surface mining

FIGURE 1 Year-by-year data from 1985 to 2015 show distinct temporal patterns that relate to a number of factors affecting the coal industry and the regulation of mountaintop removal mining operations. Part (d) coal production data are as reported by mines to the U.S. Mine Safety and Health Administration. *Source: Pericak, A.A., et al., 2018. Mapping the yearly extent of surface coal mining in Central Appalachia using Landsat and Google Earth Engine. PLOS ONE 13(7): e0197758, Figure 3. https://doi.org/10.1371/journal.pone.0197758.*

Placer mining uses running water to isolate minerals

After having been displaced from weathered rock and carried along by runoff to waterways, some metals and gems accumulate in riverbed deposits. To search for these metals and gems, miners sift through material in modern or ancient riverbed deposits, generally using running water to separate lightweight mud and gravel from heavier minerals of value. This technique is called **placer mining** (pronounced "plasser").

Placer mining is the method used by many of Congo's miners, who wade through streambeds, sifting through sediments by hand with a pan or simple tools, searching for high-density minerals that settle to the bottom while low-density material washes away. Today's African miners practice small-scale placer mining similar to the method used by American miners who ventured to California in the Gold Rush of 1849 and later to Alaska in the Klondike Gold Rush of 1896–1899 (**FIGURE 23.14**). Indeed, placer mining for gold is still practiced in areas of Alaska and Canada, although miners in the regions use large dredges and heavy machinery today.

Solution mining dissolves and extracts underground resources

When a deposit is especially deep and the resource can be dissolved in a liquid, miners may use a technique called **solution mining.** In this technique, a narrow borehole is drilled deep into the ground to reach the deposit, and water, acid, or another liquid is injected down the borehole to leach the resource from the surrounding rock and dissolve it in the liquid. The resulting solution is sucked out, and the desired resource can then be isolated. Salts can be mined in this way; water is pumped into deep salt caverns, the salt dissolves in

the water, and the salty solution is extracted. Sodium chloride (table salt), lithium, boron, bromine, magnesium, potash, copper, and uranium can be mined in this way.

Solution mining generally exerts less environmental impact than other mining techniques, because less area at the surface is disturbed. The primary potential impacts involve accidental leakage of acids into groundwater surrounding the borehole, and the contamination of aquifers with acids, heavy metals, or uranium leached from the rock.

Some mining occurs in the ocean

The oceans hold many minerals useful to our society. We extract some minerals from seawater, such as magnesium from salts held in solution. We extract other minerals from the ocean floor, often using large vacuum-cleaner–like hydraulic dredges. Valuable minerals found on or beneath the seafloor include calcium carbonate (used in making cement), copper, zinc, silver, and gold ore. Many minerals are concentrated in manganese nodules, small ball-shaped accretions that are scattered across parts of the ocean floor. More than 1.5 trillion tons of manganese nodules may exist in the Pacific Ocean alone, and their reserves of metal may exceed all terrestrial reserves. The logistical difficulty of mining them, however, has kept their extraction too expensive thus far.

As land resources become scarcer and as undersea mining technology develops, mining companies may turn increasingly to the seas. Impacts of undersea mining are largely unknown, but increases in such mining would undoubtedly destroy marine habitats and organisms that have not yet been studied. It would also likely cause some metals to diffuse into the water column at toxic concentrations and enter the food chain.

Restoration helps reclaim mine sites

Because of the environmental impacts of mining, governments of the United States and other nations now require that mining companies restore, or reclaim, surface-mined sites following mining. The aim of such restoration, or **reclamation,** is to restore the site to an approximation of its pre-mining condition.

To restore a site, companies are required to remove buildings and other structures used for mining, replace overburden, fill in shafts, and replant the area with vegetation. Former mine sites have been reclaimed for wildlife habitat (**FIGURE 23.15**) or repurposed for other uses. Virginia Polytechnic Institute and State University, for example, is coordinating a project that is investigating the use of reclaimed coal mines in southwestern Virginia to grow fast-growing trees and other vegetation for bioenergy production (pp. 575–582).

In the United States, the Surface Mining Control and Reclamation Act of 1977 mandates restoration efforts, requiring companies to post bonds to cover reclamation costs before mining can be approved, thus ensuring that if the company fails to restore the land for any reason, the government will have the money to step in and do so. Most other nations exercise less

FIGURE 23.14 Prospectors in the American West in the nineteenth century secured gold by placer mining. Sediment from stream bottoms was collected in metal pans, and a mixing motion allowed the sediment to be poured off while the heavier gold settled to the bottom of the pan.

FIGURE 23.15 More mine sites are now being restored. Here, bison graze on land reclaimed from a tar sands mining operation in Alberta, Canada (foreground of image). The active portion of the mine is visible in the background.

WEIGHING
the **issues**

Restoring Mined Areas

Mining has severe environmental impacts, but restoring mined sites to their pre-mining condition can be costly and difficult. How extensively should mining companies be required to restore a site after a mine is shut down, and what criteria should we use to guide restoration? Should we require nearly complete restoration? What should our priorities be—to minimize water pollution, impacts on human health, biodiversity loss, soil damage, or other factors? What measures should we use to evaluate the results of restoration? Should the amount of restoration we require depend on how much money the company made from the mine? Explain your recommendations.

oversight regarding reclamation, and in nations such as D.R. Congo, there is essentially no regulation.

The mining industry has made great strides in reclaiming mined land. But even on sites that are restored, impacts from mining can be severe and long-lasting because the water and soil is often acidic from acid drainage and can contain elevated levels of metals that are toxic to native plant life. Such changes often make it difficult to regain the same biotic communities that were naturally present on the sites before mining. Today, researchers are hybridizing varieties of wild grasses to create new strains of plants that can tolerate the soil conditions on reclaimed sites, with the hope that such plants will pave the way for the reintroduction of native vegetation. Establishing vibrant plant communities on reclaimed sites is vital, because the plants stabilize the soil, prevent erosion, and can help establish conditions that favor the reestablishment of native vegetation and functioning ecosystems.

An 1872 law still guides U.S. mining policy

The ways in which mining companies stake claims and use land in the United States is guided by a law that is nearly 150 years old. The **General Mining Act of 1872** encourages people and companies to prospect for minerals on federally owned land by allowing any U.S. citizen or any company with permission to do business in the United States to stake a claim on any plot of public land open to mining. The person or company owning the claim gains the sole right to take minerals from the area. The claim-holder can also patent the claim (buy the land) for $5 or less per acre—still at 1872 prices! Regardless of the profits claim-holders might make on minerals they extract, the law requires no payments of any kind to the public and, until recently, no restoration of the land after mining.

The General Mining Act of 1872 was enacted partly in response to the chaos of the multiple gold rushes of the nineteenth century, and was designed to bring some order to mining activities. It also aimed to promote mining at a time when the government was trying to hasten settlement of the West in an orderly way. The law may have made good sense in 1872, but the United States has changed a great deal since then, and many question the law's suitability for today's nation.

Supporters of the policy say that it is appropriate and desirable to continue encouraging the domestic mining industry, which must undertake substantial financial risk and investment to locate resources that are vital to our economy. Critics counter that the policy gives away valuable public resources to private interests nearly for free and have tried unsuccessfully over the past two decades to repeal or modify this act.

The General Mining Act of 1872 covers a wide variety of metals, gemstones, uranium, and minerals used for building materials. In contrast, fossil fuels, phosphates, sodium, and sulfur are governed by the Mineral Leasing Act of 1920. This law sets terms for leasing public lands that vary according to the resource being mined, but in all cases, unlike the General Mining Act, the terms also include the ongoing payment of rents for the use of the land and the payment of royalties on profits.

Toward Sustainable Mineral Use

Mining exerts many environmental impacts, but we also have another concern to keep in mind: Minerals are not inexhaustible resources (p. 4). Like fossil fuels, they form far more slowly than we use them, and if we continue to mine them, they will eventually be depleted. As a result, it will benefit us to find ways to conserve the supplies we have left and to make them last.

Minerals are nonrenewable resources in limited supply

Some minerals we use are abundant and will likely never run out, but others are rare enough that they could soon become unavailable. For instance, geologists in 2019 calculated that the world's known reserves of tantalum would last about 110 more years at today's rate of consumption. If demand for tantalum increases as expected, it could run out faster.

Most pressing may be dwindling supplies of the chemical element indium (In). This obscure metal (atomic number 49 in the periodic table, **APPENDIX C**), which is used in LCD screens and other electrical components, might last only another 25 years. A lack of indium and the metal gallium (Ga) would threaten the production of high-efficiency cells for solar power. Platinum (Pt) is dwindling, too, and if it became unavailable, it would be harder to develop fuel cells and catalytic converters for vehicles. However, because of supply concerns and price volatility, industries now are working intensely to develop ways of substituting other materials for indium, and platinum's high market price encourages recycling, which may keep it available, albeit as an expensive metal.

FIGURE 23.16 shows estimated years remaining for selected minerals at today's consumption rates. Calculating how long a given

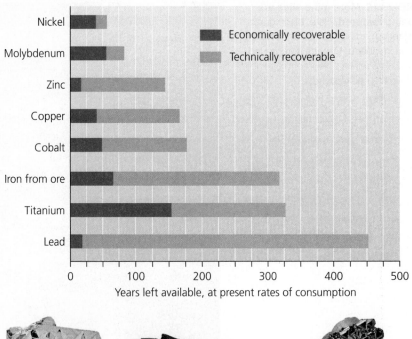

FIGURE 23.16 Minerals are nonrenewable resources, so supplies of metals are limited. The numbers of remaining years that certain metals are estimated to be economically recoverable at current prices are shown in red in the graph. The entire lengths of the bars (red plus orange) show how long certain metals are estimated to be available at present rates of consumption, using current technology on all known deposits, whether economically recoverable or not. *Data are for 2018, from U.S. Geological Survey, 2019. Mineral commodity summaries 2019. Reston, VA: USGS.*

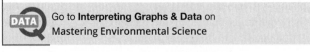

Go to **Interpreting Graphs & Data** on **Mastering Environmental Science**

Recycling Metals from E-Waste

By some estimates, about 500 million old cell phones are currently lying inactive in people's homes and offices, and upgrades and improvements render more than 130 million additional cell phones obsolete each year in the United States alone. So, what can you do with your old cell phone? Aware of the problems posed by e-waste, more and more people are donating their retired cell phones to recycling programs and, by doing so, are giving the valuable metals and minerals within them a second life.

For example, when you turn in your old phone for recycling rather than discarding it, the phone may be dismantled and the various parts recycled for their metals. Alternatively, the phone may be refurbished and resold—people in African nations in particular readily buy used cell phones because they are inexpensive and land-line phone service does not always exist in poor and rural areas. Either way, by recycling your cell phone, you're helping to extend the availability of our finite resources.

Today, only about 10% of old cell phones are recycled, meaning that we have a long way to go. But as more of us recycle our phones, computers, and other electronic items, we'll be closer to sustainably reusing tantalum and other valuable metals while decreasing the amount of e-waste that enters the waste stream.

→ **Explore the Data** at **Mastering Environmental Science**

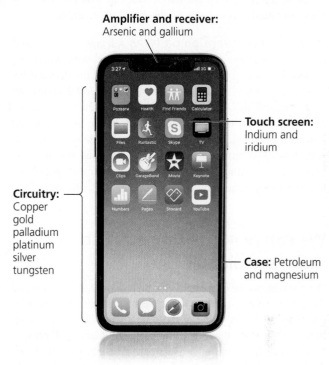

Amplifier and receiver: Arsenic and gallium

Touch screen: Indium and iridium

Circuitry: Copper gold palladium platinum silver tungsten

Case: Petroleum and magnesium

Your cell phone contains a diversity of mined materials from around the world.

---

mineral resource will be available to us is beset by a great deal of uncertainty. For example, as minerals become scarcer, demand for them increases and price rises. Higher market prices make it more profitable for companies to mine the resource, so they become willing to spend more to reach further deposits that were not economically worthwhile originally.

Estimates of mineral resources may increase or decrease over time, for several reasons. We'll discuss each of them in turn.

Discovery of new reserves As we discover new deposits of a mineral, the known reserves—and thus the number of years this mineral is available to us—increase. For this reason, some previously predicted shortages have not come to pass, and we may have access to these minerals for longer than currently estimated. For example, in 2010 geologists associated with the U.S. military discovered that Afghanistan holds immense mineral riches that were previously unknown. The newly discovered reserves of iron, copper, niobium, lithium, and many other metals are estimated to be worth more than $1 trillion—enough to realign the entire Afghan economy around mining. It is important to note, however, that such riches are not guaranteed to make Afghanistan a wealthy nation; history teaches us that regions rich in nonrenewable resources, such as D.R. Congo and Appalachian regions of the United States, have often been unable to prosper from them.

New extraction technologies Just as rising prices of scarce minerals encourage companies to expend more effort to access difficult-to-reach deposits, rising prices also may favor the development of enhanced mining technologies that can reach more minerals at less expense. If more powerful technologies are developed, they may increase the amount of minerals that are technically feasible to mine.

Recycling Advances in recycling technologies and the extent to which minerals are currently being recycled have helped extend the lifetimes of many mineral resources and will likely continue to do so as the technology and the demand for recycled materials grows (see **SUCCESS STORY**). Considering Earth's finite supplies of mineral resources, conserving them through reuse and recycling can benefit us today—by preventing price hikes that result from reduced supply—while safeguarding these valuable resources for future generations.

Changing social and technological dynamics New societal developments and new technologies in the marketplace can modify demand for minerals in unpredictable ways. Just as cell phones and computer chips boosted demand for tantalum, fiber-optic cables decreased demand for copper as they replaced copper wiring in communications applications. Today, lithium-ion batteries are replacing nickel-cadmium batteries in many devices. Synthetically produced diamonds are driving down prices of natural diamonds and extending their availability. Additionally, health concerns sometimes motivate change: We have replaced toxic substances such as lead and mercury with safer materials in many applications, for example.

Changing consumption patterns Changes in rates and patterns of consumption also alter the speed with which we exploit mineral resources. For instance, while the demand for most minerals had been steadily rising for several decades, the global economic recession from 2007 to 2009 depressed demand for most minerals, causing a decrease in mining and mineral consumption during those years. The demand for minerals has since returned to previous levels, and this trend is expected to continue as China, India, and other populous, industrializing nations rapidly increase their consumption.

We can make our mineral use more sustainable

We can address the challenges of a finite supply of minerals and the environmental damage associated with their extraction by encouraging the recycling of these resources. Electronic waste, or e-waste—from discarded computers, printers, cell phones, handheld devices, and other electronic products—is rising fast. Recycling old electronic devices helps keep the hazardous substances associated with them out of landfills while also assisting us to conserve valuable minerals such as tantalum, for which the mining industry estimates that recycling accounts for 20–25% of this metal's availability for use in manufacturing. Municipal recycling programs handle used items that consumers place in recycling bins, thus diverting metals from the waste stream (pp. 620–621). Currently, around 35% of metals in the U.S. municipal solid waste stream are diverted for recycling. For example, 80% of the lead we consume today comes from recycled materials, in particular, recycled car batteries. Similarly, 33% of our copper comes from recycled copper sources such as pipes and wires. We recycle steel, iron, platinum, and other metals from auto parts. Altogether, we have found ways to recycle much of our gold, lead, iron and steel scrap, chromium, zinc, aluminum, and nickel. **TABLE 23.1** shows minerals that currently boast high recycling rates in the United States.

In many cases, recycling can also decrease energy use substantially. For instance, making steel by re-melting recycled iron and steel scrap requires much less energy than producing steel from virgin iron ore. Because this practice saves money, the steel industry today is designed to make efficient use of iron and steel scrap. More than half its scrap comes from discarded consumer items such as cars, cans, and appliances. Similarly, more than 40% of the aluminum in the United States today is recycled. Recycling is beneficial because it takes more than 20 times more energy to extract virgin aluminum from ore (bauxite) than it does to obtain it from recycled sources—and avoids the releases of the greenhouse gases (pp. 488–489) that would accompany this added energy investment.

TABLE 23.1 Recycled Minerals in the United States

MINERAL	U.S. RECYCLING RATE
Gold	Slightly less is recycled than is consumed
Iron and steel scrap	Nearly 100% for vehicles, 88% for appliances, 71–98% for construction materials, 70% for cans
Lead	71% consumed comes from recycled post-consumer items
Nickel	52% consumed is from recycled nickel
Zinc	25% produced is recovered, mostly from recycled materials used in processing
Chromium	29% is recycled in stainless steel production
Copper	35% of U.S. supply comes from various recycled sources
Aluminum	28% produced comes from recycled post-consumer items
Tin	22% consumed is from recycled tin
Germanium	30% consumed worldwide is recycled. Optical device manufacturing recycles more than 60%
Molybdenum	About 30% gets recycled as part of steel scrap that is recycled
Cobalt	29% consumed comes from recycled scrap
Niobium (columbium)	Perhaps 20% gets recycled as part of steel scrap that is recycled
Silver	18% consumed is from recycled silver. U.S. recovers as much as it produces
Bismuth	All scrap metal containing bismuth is recycled, providing less than 5% of consumption
Diamond (industrial)	2% of production is from recycled diamond dust, grit, and stone

Data are for 2018, from U.S. Geological Survey, 2019. *Mineral commodity summaries 2019.* Reston, VA: USGS.

TODAY, D.R. Congo remains embroiled in civil unrest and its mineral and biological resources continue to be exploited. In late 2018, the Congolese people went to the polls to vote for a new president in what was hoped to be the first peaceful democratic transition of power in the nation's history. But even though election observers determined candidate Martin Fayulu to be victorious, the nation's electoral commission declared and certified one of his opponents, Félix Tshisekedi, the winner, despite widespread allegations of fraud and voting irregularities. Joseph Kabila, former long-time president barred from running for another term due to constitutional restrictions, is expected to reemerge in 2023 when the nation once again holds elections. It has been reported that from 1999 to 2002, the Kabila government transferred some $5 billion of assets from state-run mining interests to private companies it controlled, showing the financial power to be gained by controlling access to these valuable resources.

In the opening case study, we saw that procuring metals and minerals from D.R. Congo for products in our own country can have profound impacts on people and ecosystems far away, a realization that prompted the current movement toward the certification of minerals and gemstones as "conflict free." These initiatives are vital, because conflict minerals are also an emerging problem in regions other than central Africa. A thriving black market in coltan is emerging in remote portions of the northern Amazon jungle, aided by the refugee crisis and chaos in the region associated with ongoing political unrest in Venezuela. Afghanistan's newly discovered mineral riches, coupled with the nation's long history of civil war, suggest that it, too, could become a significant source of conflict minerals.

Manufacturers, governments, and nongovernmental organizations are collaborating to make certification efforts practical and meaningful in the complex, multilayered global trade in metals and minerals. Through these efforts, and efforts to make our mineral use more sustainable by maximizing recovery and recycling, we are working toward a future in which we can simultaneously meet our demand for these vital geologic resources while minimizing environmental impacts and adverse effects on the people who extract and produce them—like the many toiling in mines in eastern D.R. Congo.

- **CASE STUDY SOLUTIONS** The story of coltan in D.R. Congo is just one example of how an abundance of exploitable resources can often worsen or prolong military conflicts in nations that are too poor or ineffectively governed to protect these resources. In such resource wars, civilians often suffer the most as civil society breaks down. Suppose you are the head of an international aid agency that has earmarked $10 million to help address conflicts related to mining in D.R. Congo. You have access to government and rebel leaders in Congo and neighboring countries, to ambassadors of the world's nations in the United Nations, and to representatives of international mining corporations. Based on what you know from this chapter, what steps would you consider taking to help improve the situation in D.R. Congo?

- **LOCAL CONNECTIONS** D.R. Congo may produce a great deal of the tantalum used in cellular phones around the world, but how do residents of D.R. Congo compare to other nations with respect to the proportion of their population that owns a cell phone? How do their usage rates compare to those in the United States and in other nations with similar or very different rates of industrialization?

- **EXPLORE THE DATA** What nations are the biggest producers of tantalum, and how has production changed over time? → **Explore Data** relating to the case study on **Mastering Environmental Science.**

REVIEWING Objectives

You should now be able to:

+ **Describe the types of mineral resources and their uses**

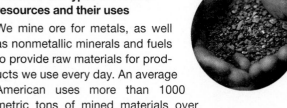

We mine ore for metals, as well as nonmetallic minerals and fuels to provide raw materials for products we use every day. An average American uses more than 1000 metric tons of mined materials over his or her lifetime. (pp. 644–646)

+ **Explain how minerals are processed prior to use**

Metal ore and valuable minerals must typically be separated from surrounding minerals through physical or chemical means. Alloys are produced when we combine a metal with another metal or a nonmetallic substance, such as carbon. (pp. 646–649)

+ **Describe the major methods of mining**

Many methods are available to miners to secure valuable minerals and fuels. Strip mining removes surface layers of soil and rock to mine resources from above, and subsurface mining tunnels underground to reach deeper resources. Open pit mining involves digging gigantic holes. Placer mining uses running water to isolate high-density minerals. Mountaintop removal coal mining removes immense amounts of rock from mountaintops and dumps them into valleys below. Solution mining uses liquids to dissolve minerals in place. (pp. 649–653)

+ **Discuss the environmental and social impacts of mining**

Many methods of mining completely remove vegetation, soil, and natural habitat. Acid drainage occurs when water leaches compounds from rocks exposed by mining, forming a low-pH solution that can be high in toxic metals. Mountaintop removal for coal destroys forests and adjacent valleys and streams. People living near mines experience adverse health and environmental impacts, but mining can offer employment in economically depressed regions. (pp. 649–653)

+ **Explain mine reclamation efforts and mining policy**

Mine reclamation efforts are challenging and sometimes fall short of effective ecological restoration due to the toxic soil and water conditions left behind by mining operations. Some mining effects, such as acid drainage, are difficult to contain and long-lasting. Mining policy in the United States is still being dictated by a law passed in 1872. (pp. 654–658)

+ **Evaluate ways to encourage sustainable use of mineral resources**

As minerals are nonrenewable resources, conserving supplies is vital to future generations. Reuse and recycling by industry and consumers are the keys to more sustainable practices of mineral use. (pp. 658–660)

SEEKING Solutions

1. List three impacts of mining on the natural environment, and describe how particular mining practices can lead to each of these impacts. How are these impacts being addressed? Can you think of additional solutions to prevent, reduce, or mitigate these impacts?

2. List three impacts of mining on people's health, lifestyles, or well-being, and describe how particular mining practices can lead to each of these impacts. How are these impacts being addressed? Can you think of additional solutions to prevent, reduce, or mitigate these impacts?

3. You have won a grant from the EPA to work with a mining company to develop a more effective way of restoring a mine site that is about to be closed. Describe a few preliminary ideas for carrying out restoration better than it is typically being done. Now describe a field experiment you would like to run to test one of your ideas.

4. **THINK IT THROUGH** In your public speaking class, you have been assigned to present (1) one argument in favor of retaining the General Mining Act of 1872, (2) one argument in favor of repealing or reforming the law, and (3) a conclusion regarding your personal view of what to do regarding this legislation. Would you retain, repeal, or reform the 1872 law? Why? What arguments would you make in (1) or (2) that inform your ultimate decision?

5. **THINK IT THROUGH** As you finish your college degree, you learn that the mountains behind your childhood home in the hills of Kentucky are slated to be mined for coal using the mountaintop removal method. Your parents, who still live there, are worried for their health and safety and do not want to lose the beautiful forested creek and ravine behind their property. However, your brother is out of work and could use a steady, well-paying mining job. What would you attempt to do in this situation?

CALCULATING Ecological Footprints

As we saw in Figure 23.15, the supplies of some metals are limited enough that, at today's prices, these metals could be available to us for only a few more decades. After that, prices will rise as they become scarcer. The number of years of total availability (at all prices) depends on a number of factors: On one hand, metals will be available for longer if new deposits are discovered, new mining technologies are developed, or recycling efforts are improved. On the other hand, if our consumption of metals increases, the number of years we have left to use them will decrease.

Currently, the United States consumes metals at a much higher per-person rate than the world does as a whole. If one goal of humanity is to lift the rest of the world up to U.S. living standards, pressures on mineral supplies will sharply increase.

The following table shows currently known economically recoverable global reserves for several metals, together with the amount used per year worldwide. For each metal, calculate and enter in the fourth column the years of supply left at current prices by dividing the reserves by the amount used annually.

The fifth column shows the amount that the world would use if everyone in the world consumed the metal at the rate that Americans do. Now calculate the years of supply left at current prices for each metal if the world were to consume the metals at the U.S. rate, and enter these values in the sixth column.

METAL	KNOWN ECONOMIC RESERVES (THOUSAND METRIC TONS)[1]	AMOUNT CONSUMED PER YEAR (THOUSAND METRIC TONS)[2]	YEARS OF ECONOMIC SUPPLY LEFT	AMOUNT USED PER YEAR IF EVERYONE CONSUMED AT U.S. RATE (THOUSAND METRIC TONS)	YEARS OF ECONOMIC SUPPLY LEFT IF EVERYONE CONSUMED AT U.S. RATE
Titanium	940,000	6100	154.1	25,558	36.8
Copper	830,000	21,000			
Zinc	230,000	13,000			
Nickel	89,000	2300			
Tin	4700	310			
Antimony	1500	140			
Silver	560	27			
Gold	54	3.3			

Data are for 2018, from U.S. Geological Survey, 2019. *Mineral commodity summaries 2019*. Reston, VA: USGS.

[1] "Known economic reserves" include extractable amounts under current economic conditions. These amounts exclude additional ("technically recoverable") reserves that could be mined at greater cost (compare Figure 23.15).

[2] World consumption data are assumed equal to world production data.

1. Which of these eight metals will last the longest under current economic conditions and at current rates of global consumption? For which of these metals will economic reserves be depleted fastest?

2. If the average citizen of the world consumed metals at the rate that the average U.S. citizen does, the economic reserves of which of these eight metals would last the longest? Which would be depleted fastest if everyone consumed at the U.S. rate?

3. In this chart, our calculations of years of supply left do not factor in population growth. How do you think population growth will affect these numbers?

4. Describe two general ways that we could increase the years of supply left for these metals. What do you think it will take to accomplish them?

Mastering Environmental Science

How to Interpret Graphs

Presenting data in ways that help make trends and patterns visually apparent is a vital element of science. For scientists, businesspeople, journalists, policymakers, and others, the primary tool for expressing patterns in data is the graph. Thus, the ability to interpret graphs is a skill that you will want to cultivate. This appendix guides you in how to read graphs, introduces a few key conceptual points, and surveys the most common types of graphs, giving rationales for their use.

Navigating a Graph

A graph is a diagram showing relationships among *variables*, which are factors that can change in value. The most common types of graphs relate values of a *dependent variable* to those of an *independent variable*. As explained in Chapter 1 (p. 10), a dependent variable is so named because its values "depend on" the values of an independent variable. In other words, as the values of an independent variable change, the values of the dependent variable change in response. In a manipulative experiment (p. 11), changes that a researcher specifies in the value of the independent variable *cause* changes in the value of the dependent variable. In observational studies, there may be no causal relationship, and scientists may plot a correlation (p. 11). In a positive correlation, values of one variable go up or down along with values of another. In a negative correlation, values of one variable go up when values of the other go down. Whether we are graphing a correlation or a causal relationship, the values of the independent variable are known or specified by the researcher, whereas the values of the dependent variable are unknown until the research has taken place. The values of the dependent variable are what we are interested in observing or measuring.

By convention, independent variables are generally represented on the horizontal axis, or *x axis*, of a graph, while dependent variables are represented on the vertical axis, or *y axis*. Numerical values of variables generally become larger as one proceeds rightward on the *x axis* or upward on the *y axis*. Note that the tick marks along the axes must be uniformly spaced so that when the data are plotted, the graph gives an accurate visual representation of the scale of quantitative change in the data.

As a simple example, **FIGURE A.1** shows data from the Breeding Bird Survey that reflect population growth of the Eurasian collared dove following its introduction to North America. The *x* axis shows values of the independent variable, which in this case is time, expressed in units of years. The dependent variable, presented on the *y* axis, is the average number of doves detected on each route. For each year, a data point is plotted on the graph to show the average number of doves detected. In this particular graph, a line (dark red curve) was then drawn through the actual data points (orange dots), showing how closely the empirical data match an exponential growth curve (p. 66), a theoretical phenomenon of importance in ecology.

Now that you're familiar with the basic building blocks of a graph, let's survey the most common types of graphs you'll see, and examine a few vital concepts in graphing.

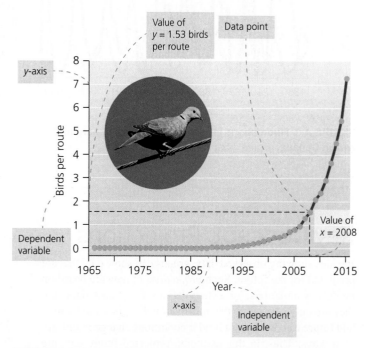

FIGURE A.1 Exponential population growth, demonstrated by the Eurasian collared dove in North America in recent years. (Figure 3.16, p. 66)

Mastering Environmental Science

Once you've explored this appendix, take advantage of the graphing resources on **Mastering Environmental Science.** The GraphIt Tutorials allow you to practice building graphs. The **Interpreting Graphs and Data** exercises guide you through critical-thinking questions on graphed data from recent research in environmental science. The **Data Analysis Questions** help you hone your skills in reading graphed data. All these features will help you expand your comprehension and use of graphs.

Graph Type: Line Graph

A line graph is used when a data set involves a sequence of some kind, such as a series of values that occur one by one and change through time or across distance. In a line graph, a line runs from one data point to the next. Line graphs are most appropriate when the *y* axis expresses a continuous numerical variable, and the *x* axis expresses either continuous numerical data or discrete sequential categories (such as years). **FIGURE A.2** shows values for the size of the ozone hole over Antarctica in recent years. Note how the data show that the size of the hole increases until 1987, when the Montreal Protocol (p. 471) came into force, and then begins to stabilize afterward.

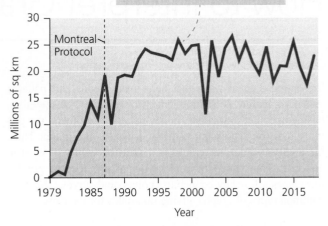

Yearly data show an increase in ozone hole size before the Montreal Protocol and stabilization afterwards

FIGURE A.2 Size of the Antarctic ozone hole before and after a treaty that was designed to address it. (Figure 17.22, p. 474)

One useful technique is to plot two or more data sets together on the same graph. This allows us to compare trends in the data sets to see whether they may be related and, if so, the nature of that relationship. In **FIGURE A.3**, recorded numbers of a predator species rise and fall immediately following those of its prey, suggesting a possible connection.

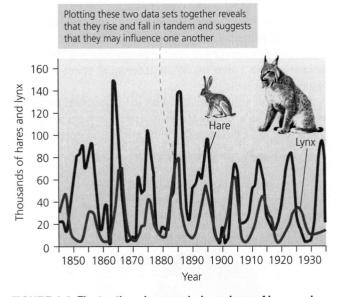

Plotting these two data sets together reveals that they rise and fall in tandem and suggests that they may influence one another

FIGURE A.3 Fluctuations in recorded numbers of hare and lynx in Canada. (Figure 4.5, p. 79)

Key Concept: Projections

Besides showing observed data, graphs can show data that are predicted for the future. Such *projections* of data are based on models, simulations, or extrapolations from past data, but they are only as good as the information that goes into them—and future trends may not hold if conditions change in unforeseen ways. Thus, in this textbook, projected future data are shown with dashed lines, as in **FIGURE A.4**, to indicate that they are less certain than data that have already been observed. Be careful when interpreting graphs in the popular media and on the Internet, however; often newspapers, magazines, websites, and advertisements will show projected future data in the same way as known past data!

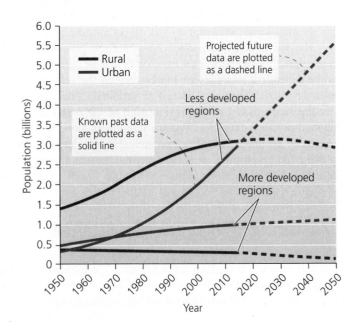

Projected future data are plotted as a dashed line

Less developed regions

Known past data are plotted as a solid line

More developed regions

FIGURE A.4 Past population change and projected future population change for rural and urban areas in more developed and less developed regions. (Figure 13.1, p. 338)

Graph Type: Bar Chart

A bar chart is most often used when one variable is a category and the other is a number. In such a chart, the height (or length) of each bar represents the numerical value of a given category. Higher or longer bars signify larger values. In **FIGURE A.5**, the bar for the category "Respiratory infections" is higher than that for "Malaria," indicating that respiratory infections cause more deaths each year (the numerical variable on the *y* axis) than does malaria.

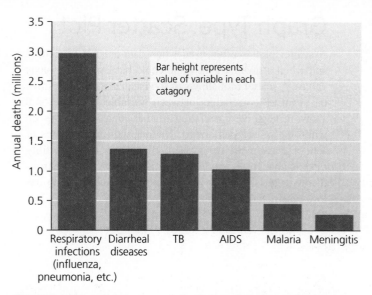

FIGURE A.5 Leading causes of death from infectious disease. (Figure 14.4b, p. 364)

As we saw with line graphs, it is often instructive to graph two or more data sets together to reveal patterns and relationships. A bar chart such as **FIGURE A.6** lets us compare two data sets (oil production and oil consumption) both within and among nations. A graph that does double duty in this way allows for higher-level analysis (in this case, suggesting which nations depend on others for petroleum imports). Most bar charts in this book illustrate multiple types of information at once in this manner.

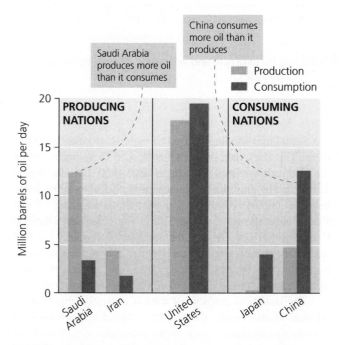

FIGURE A.6 Oil extraction and consumption by selected nations. (Figure 19.26, p. 551)

Graph Type: Pie Chart

A pie chart is used when we wish to compare the numerical proportions of some whole that are taken up by each of several categories. Each category is represented visually like a slice from a pie, with the size of the slice reflecting the percentage of the whole that is taken up by that category. For example, **FIGURE A.7** compares the percentages of genetically modified crops worldwide that are soybeans, corn, cotton, and canola.

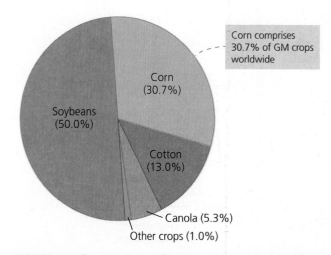

FIGURE A.7 Genetically modified crops grown worldwide, by type. (Figure 10.19a, p. 258)

Graph Type: Scatter Plot

A scatter plot is often used when data are not sequential and when a given *x*-axis value could have multiple *y*-axis values. A scatter plot allows us to visualize a broad positive or negative correlation between variables. **FIGURE A.8** shows a negative correlation (that is, one value goes up while the other goes down): Nations with higher rates of school enrollment for girls tend to have lower fertility rates. Cuba, for example, has a high rate of school enrollment for girls and low fertility, whereas South Sudan has low enrollment and high fertility.

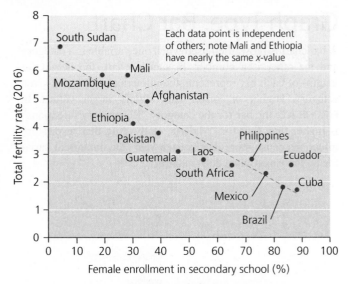

FIGURE A.8 Fertility rate and education of girls, by nation.
(Figure 8.15, p. 206)

Key Concept: Statistical Uncertainty

Most data sets involve some degree of uncertainty. When a graphed value represents the *mean* (average) of many measurements, the researcher may want to use mathematical techniques to show the degree to which the raw data vary around this mean. Results from such statistical analyses may be expressed in a number of ways, and the two graphs in this section show methods used in this book.

In a bar chart or scatter plot (**FIGURE A.9**), thin black lines called *error bars* may be shown extending above and/or below each mean data value. In this example of likelihood of death from air pollution, error bars show the most variation at the highest measured concentration of pollutants and no variation at the lowest measured concentration.

Sometimes shading is used to express variation around a mean. The black data line in **FIGURE A.10** shows mean global sea level readings from tide gauges since 1880. This data line is surrounded by gray shading indicating statistical variation. Note how the amount of statistical uncertainty is exceeded by the sheer scale of the sea level rise. This gives us confidence that sea level is truly rising, despite the statistical uncertainty we find around mean values each year. Note also how the amount of uncertainty has decreased through time. This reflects improvements in technology, enabling more accurate measurements.

The statistical analysis of data is critically important in science. In this book, we provide a broad and streamlined introduction to many topics, so we often omit error bars from our graphs and leave out details of statistical significance from our discussions. Bear in mind that this is for clarity of presentation only; the research we discuss analyzes its data in far more depth than any textbook could possibly cover.

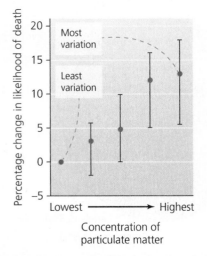

FIGURE A.9 Likelihood of death due to air pollution.
(17-SBS1 Figure 3, p. 465)

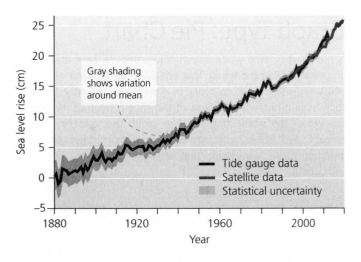

FIGURE A.10 Change in global sea level, measured since 1880.
(Figure 18.19, p. 502)

Metric System

MEASUREMENT	UNIT AND ABBREVIATION	METRIC EQUIVALENT	METRIC TO ENGLISH CONVERSION FACTOR	ENGLISH TO METRIC CONVERSION FACTOR
Length	1 kilometer (km)	= 1000 (10^3) meters	1 km = 0.62 mile	1 mile = 1.61 km
	1 meter (m)	= 100 (10^2) centimeters	1 m = 1.09 yards	1 yard = 0.914 m
			= 3.28 feet	1 foot = 0.305 m
			= 39.37 inches	= 30.5 cm
	1 centimeter (cm)	= 10 millimeters	1 cm = 0.394 inch	1 inch = 2.54 cm
		= 0.01 (10^{-2}) meter		
	1 millimeter (mm)	= 0.01 (10^{-2}) centimeter	1 mm = 0.039 inch	
Area	1 square meter (m^2)	= 10,000 square centimeters	1 m^2 = 1.196 square yards	1 square yard = 0.8361 m^2
			= 10.764 square feet	1 square foot = 0.0929 m^2
	1 hectare (ha)	= 10,000 square meters	1 ha = 2.47 acres	1 acre = 0.405 hectare
	1 square kilometer (km^2)	= 1,000,000 square meters	1 km^2 = 0.386 square mile	1 square mile = 2.59 km^2
Mass	1 metric ton (t)	= 1000 kilograms	1 t = 1.103 tons	1 ton = 0.907 t
	1 kilogram (kg)	= 1000 grams	1 kg = 2.205 pounds	1 pound (lb) = 0.4536 kg
	1 gram (g)	= 1000 milligrams	1 g = 0.0353 ounce	1 ounce = 28.35 g
	1 milligram (mg)	= 0.001 gram		
Volume (solids)	1 cubic meter (m^3)	= 1,000,000 cubic centimeters	1 m^3 = 1.3080 cubic yards	1 cubic yard = 0.7646 m^3
	1 cubic centimeter (cm^3 or cc)	= 0.000001 cubic meter	1 cm^3 = 0.0610 cubic inch	1 cubic inch = 16.387 cm^3
	1 cubic millimeter (mm^3)	= 0.001 cubic centimeter		
Volume (liquids and gases)	1 kiloliter (kl or kL)	= 1000 liters	1 kL = 264.17 gallons	1 gallon = 3.785 L
	1 liter (l or L)	= 1000 milliliters	1 L = 0.264 gallon	1 quart = 0.946 L
			= 1.057 quarts	= 946 ml
	1 milliliter (ml or mL)	= 0.001 liter	1 ml = 0.034 fluid ounce	1 fluid ounce = 29.57 ml
		= 1 cubic centimeter	= approx. 1/5 teaspoon	1 teaspoon = approx. 5 ml
Temperature	Degrees Celsius (°C)		$°C = \dfrac{5}{9}(°F - 32)$	$°F = \dfrac{9}{5}(°C) + 32$
Energy and Power	1 gigawatt (GW)	= 1,000,000,000 (10^9) watts		
	1 megawatt (MW)	= 1,000,000 (10^6) watts		
	1 kilowatt (kW)	= 1000 (10^3) watts		
	1 watt (W)	= 0.001 kilowatt		
		= 1 joule/second		
	1 kilowatt-hour (kWh)	= 3,600,000 joules		
		= 3412 BTU		
		= 860,400 calories		
	1 calorie (cal)	= The amount of energy needed to raise the temperature of 1 gram (1 cm^3) of water by 1 degree Celsius		
	1 joule	= 0.239 cal		
		= 2.778×10^{-7} kilowatt−hours		
Pressure	1 atmosphere (atm)	= 1013.25 millibars (mbar)		
		= 14.696 pounds per square inch (psi)		
		= 760 millimeters of mercury (mmHg)		

Periodic Table of the Elements

Key (example):

6	Atomic number
C	Chemical symbol
12.011	Atomic weight
Carbon	Name

Representative (main group) elements — Transition metals — **Representative (main group) elements**

Period	IA	IIA	IIIB	IVB	VB	VIB	VIIB	VIIIB	VIIIB	VIIIB	IB	IIB	IIIA	IVA	VA	VIA	VIIA	VIIIA
1	1 **H** 1.0079 Hydrogen																	2 **He** 4.003 Helium
2	3 **Li** 6.941 Lithium	4 **Be** 9.012 Beryllium											5 **B** 10.811 Boron	6 **C** 12.011 Carbon	7 **N** 14.007 Nitrogen	8 **O** 15.999 Oxygen	9 **F** 18.998 Fluorine	10 **Ne** 20.180 Neon
3	11 **Na** 22.990 Sodium	12 **Mg** 24.305 Magnesium											13 **Al** 26.982 Aluminum	14 **Si** 28.086 Silicon	15 **P** 30.974 Phosphorus	16 **S** 32.066 Sulfur	17 **Cl** 35.453 Chlorine	18 **Ar** 39.948 Argon
4	19 **K** 39.098 Potassium	20 **Ca** 40.078 Calcium	21 **Sc** 44.956 Scandium	22 **Ti** 47.88 Titanium	23 **V** 50.942 Vanadium	24 **Cr** 51.996 Chromium	25 **Mn** 54.938 Manganese	26 **Fe** 55.845 Iron	27 **Co** 58.933 Cobalt	28 **Ni** 58.69 Nickel	29 **Cu** 63.546 Copper	30 **Zn** 65.39 Zinc	31 **Ga** 69.723 Gallium	32 **Ge** 72.61 Germanium	33 **As** 74.922 Arsenic	34 **Se** 78.96 Selenium	35 **Br** 79.904 Bromine	36 **Kr** 83.8 Krypton
5	37 **Rb** 85.468 Rubidium	38 **Sr** 87.62 Strontium	39 **Y** 88.906 Yttrium	40 **Zr** 91.224 Zirconium	41 **Nb** 92.906 Niobium	42 **Mo** 95.94 Molybdenum	43 **Tc** 98 Technetium	44 **Ru** 101.07 Ruthenium	45 **Rh** 102.906 Rhodium	46 **Pd** 106.42 Palladium	47 **Ag** 107.868 Silver	48 **Cd** 112.411 Cadmium	49 **In** 114.82 Indium	50 **Sn** 118.71 Tin	51 **Sb** 121.76 Antimony	52 **Te** 127.60 Tellurium	53 **I** 126.905 Iodine	54 **Xe** 131.29 Xenon
6	55 **Cs** 132.905 Cesium	56 **Ba** 137.327 Barium	57* **La** 138.906 Lanthanum	72 **Hf** 178.49 Hafnium	73 **Ta** 180.948 Tantalum	74 **W** 183.84 Tungsten	75 **Re** 186.207 Rhenium	76 **Os** 190.23 Osmium	77 **Ir** 192.22 Iridium	78 **Pt** 195.08 Platinum	79 **Au** 196.967 Gold	80 **Hg** 200.59 Mercury	81 **Tl** 204.383 Thallium	82 **Pb** 207.2 Lead	83 **Bi** 208.980 Bismuth	84 **Po** 209 Polonium	85 **At** 210 Astatine	86 **Rn** 222 Radon
7	87 **Fr** 223 Francium	88 **Ra** 226.025 Radium	89** **Ac** 227.028 Actinium	104 **Rf** 267 Rutherfordium	105 **Db** 268 Dubnium	106 **Sg** 269 Seaborgium	107 **Bh** 270 Bohrium	108 **Hs** 269 Hassium	109 **Mt** 278 Meitnerium	110 **Ds** 281 Darmstadtium	111 **Rg** 281 Roentgenium	112 **Cn** 285 Copernicium	113 **Nh** 286 Nihonium	114 **Fl** 289 Flerovium	115 **Mc** 289 Moscovium	116 **Lv** 293 Livermorium	117 **Ts** 293 Tennessine	118 **Og** 294 Oganesson

Rare earth elements

***Lanthanides**

58 **Ce** 140.115 Cerium	59 **Pr** 140.908 Praseodymium	60 **Nd** 144.24 Neodymium	61 **Pm** 145 Promethium	62 **Sm** 150.36 Samarium	63 **Eu** 151.964 Europium	64 **Gd** 157.25 Gadolinium	65 **Tb** 158.925 Terbium	66 **Dy** 162.5 Dysprosium	67 **Ho** 164.93 Holmium	68 **Er** 167.26 Erbium	69 **Tm** 168.934 Thulium	70 **Yb** 173.04 Ytterbium	71 **Lu** 174.967 Lutetium

****Actinides**

90 **Th** 232.038 Thorium	91 **Pa** 231.036 Protactinium	92 **U** 238.029 Uranium	93 **Np** 237.048 Neptunium	94 **Pu** 244 Plutonium	95 **Am** 243 Americium	96 **Cm** 247 Curium	97 **Bk** 247 Berkelium	98 **Cf** 251 Californium	99 **Es** 252 Einsteinium	100 **Fm** 257 Fermium	101 **Md** 258 Mendelevium	102 **No** 259 Nobelium	103 **Lr** 262 Lawrencium

The periodic table arranges elements by atomic number and atomic weight into horizontal rows called *periods* and vertical columns called *groups*.

Elements of each group in Class A have similar chemical and physical properties. This reflects the fact that members of a particular group have the same number of valence shell electrons, which is indicated by the group's number. For example, group IA elements have one valence shell electron, group IIA elements have two, and group VA elements have five. In contrast, as you progress across a period from left to right, properties of the elements change, varying from the very metallic properties of groups IA and IIA to the nonmetallic properties of group VIIA to the inert elements (noble gases) in group VIIIA. This reflects changes in the number of valence shell electrons.

Class B elements, or transition elements, are metals and generally have one or two valence shell electrons. In these elements, some electrons occupy more distant electron shells before the deeper shells are filled.

In this periodic table, elements with symbols printed in black exist as solids under standard conditions (25°C and 1 atmosphere of pressure); elements in red exist as gases; and those in dark blue as liquids. Elements with symbols in green do not exist in nature and must be created by some type of nuclear reaction.

Geologic Time Scale

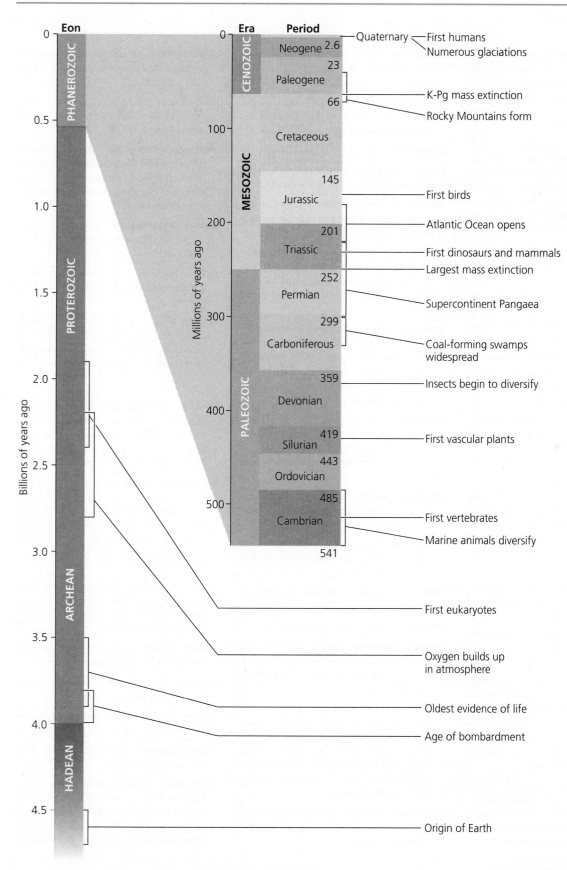

- First humans
- Numerous glaciations
- K-Pg mass extinction
- Rocky Mountains form
- First birds
- Atlantic Ocean opens
- First dinosaurs and mammals
- Largest mass extinction
- Supercontinent Pangaea
- Coal-forming swamps widespread
- Insects begin to diversify
- First vascular plants
- First vertebrates
- Marine animals diversify
- First eukaryotes
- Oxygen builds up in atmosphere
- Oldest evidence of life
- Age of bombardment
- Origin of Earth

Glossary

acid deposition The settling of acidic or acid-forming pollutants from the *atmosphere* onto Earth's surface. This process may take place by precipitation, fog, gases, or the deposition of dry particles. One type of *atmospheric deposition*. Compare *acid rain*.

acid mine drainage A process in which sulfide minerals in newly exposed rock surfaces react with *oxygen* and rainwater to produce sulfuric acid, which causes chemical *runoff* as it *leaches* metals from the rocks. Acid drainage is a natural phenomenon, but mining greatly accelerates it by exposing many new surfaces.

acid rain *Acid deposition* that takes place through rain.

acidic The property of a *solution* in which the concentration of *hydrogen* (H^+) *ions* is greater than the concentration of hydroxide (OH^-) ions. Compare *basic*.

active solar energy collection An approach in which technological devices are used to focus, move, or store *solar energy*. Compare *passive solar energy collection*.

acute exposure Exposure to a *toxicant* occurring in high amounts for short periods of time. Compare *chronic exposure*.

adaptation (1) The process by which traits that lead to increased reproductive success in a given environment evolve in a *population* through *natural selection*. (2) A trait that confers greater likelihood that an individual will reproduce. (3) With regard to *climate change*, the pursuit of strategies to protect ourselves from the impacts of *climate change*. Compare *mitigation*.

adaptive management The systematic testing of different management approaches to improve methods over time.

aerosols Very fine liquid droplets or solid particles aloft in the atmosphere.

age structure The relative numbers of individuals of different ages within a *population*. Age structure can have a strong effect on rates of population growth or decline and is often expressed as a ratio of age classes, consisting of organisms (1) not yet mature enough to reproduce, (2) capable of reproduction, and (3) beyond their reproductive years.

agricultural revolution The shift around 10,000 years ago from a hunter-gatherer lifestyle to an agricultural way of life in which people began to grow crops and raise domestic animals. Compare *industrial revolution*.

agriculture The practice of cultivating *soil*, producing crops, and raising livestock for human use and consumption.

air pollutant A gas or particulate material added to the atmosphere that can affect climate or harm people or other living things.

air pollution The release of *air pollutants*.

airshed The geographic area that produces air pollutants likely to end up in a waterway.

allergen A *toxicant* that overactivates the immune system, causing an immune response when one is not necessary.

alloy A substance created by fusing a *metal* with other metals or nonmetals. Bronze is an alloy of the metals copper and tin, and steel is an alloy of iron and the nonmetal *carbon*.

ammonia (NH_4) A colorless gas with a sharp and pungent odor, released from *feedlots*, fertilizers, and other sources. One of several major *air pollutants* whose emissions are monitored by the *Environmental Protection Agency* and state agencies.

ammonification The process in which decomposers, particularly bacteria, break down organic matter and release compounds containing ammonia or ammonium ions (NH_4^+).

anthropocentrism A human-centered view of our relationship with the *environment*. Compare *biocentrism; ecocentrism*.

aquaculture The cultivation of fish, shrimp, algae, or other aquatic organisms for food in controlled environments.

aqueduct An artificial channel for transporting water. Also called a canal.

aquifer An underground water reservoir.

artificial selection *Natural selection* conducted under human direction. Examples include the selective breeding of crop plants, pets, and livestock.

asbestos Any of several types of *mineral* that form long, thin microscopic fibers—a structure that allows asbestos to insulate buildings for heat, muffle sound, and resist fire. When inhaled and lodged in lung tissue, asbestos scars the tissue and may eventually lead to lung cancer or *asbestosis*.

asbestosis A disorder resulting from lung tissue scarred by acid following prolonged inhalation of *asbestos*.

asthenosphere A layer of the upper *mantle*, just below the *lithosphere*, consisting of especially soft rock.

atmosphere The layer of gases surrounding planet Earth. Compare *biosphere; hydrosphere; lithosphere*.

atmospheric deposition The wet or dry deposition onto land of a wide variety of pollutants, including mercury, nitrates, and organochlorines. *Acid deposition* is one type of atmospheric deposition.

atmospheric pressure The weight (or gravitational force) per unit area produced by a column of air.

atom The smallest component of an *element* that maintains the chemical properties of that *element*.

autotroph An organism that can use the *energy* from sunlight to produce its own food. Includes green plants, algae, and cyanobacteria. Also known as a primary producer.

Bacillus thuringiensis (Bt) A naturally occurring soil bacterium that produces a protein that kills certain insects. A common means of controlling crop pests. Its genes have been engineered into some crop plants.

background extinction rate The average rate of *extinction* that occurred before the appearance of humans. For example, the *fossil record* indicates that for both birds and mammals, one *species* in the world typically became extinct every 500–1000 years. Compare *mass extinction event*.

basic The property of a *solution* in which the concentration of *hydroxide* (OH^-) *ions* is greater than the concentration of *hydrogen* (H^+) ions. Compare *acidic*.

battery A device for *energy storage* that employs chemical reactions to generate *electricity*.

bedrock The continuous mass of solid rock that makes up Earth's *crust*.

benthic Of, relating to, or living on the bottom of a water body. Compare *pelagic*.

bioaccumulation The buildup of *toxicants* in the tissues of an animal.

biocentrism A philosophy that ascribes relative values to actions, entities, or properties on the basis of their effects on all living things or on the integrity of the biotic realm in general. The biocentrist evaluates an action in terms of its overall impact on living things, including—but not exclusively focusing on—human beings. Compare *anthropocentrism; ecocentrism*.

biodiesel Diesel fuel produced by mixing vegetable oil, used cooking grease, or animal fat with small amounts of *ethanol* or methanol (wood alcohol) in the presence of a chemical catalyst. A type of *biofuel*.

biodiversity The variety of life across all levels of biological organization, including the diversity of *species, genes, populations, communities,* and *ecosystems*. Also called *biological diversity*.

biodiversity hotspot An area that supports an especially great diversity of *species,* particularly species that are *endemic* to the area.

bioenergy *Energy* harnessed from plant and animal matter, including wood from trees, charcoal from burned wood, and combustible animal waste products, such as cattle manure. (*Fossil fuels* are not considered bioenergy sources because their organic matter has not been part of living organisms for millions of years and has undergone considerable chemical alteration since that time.) Also called *biomass energy*.

biofuel Fuel produced from *biomass* sources and used primarily to power automobiles. Examples include *ethanol* and *biodiesel*.

biogeochemical cycle See *nutrient cycle*.

biological control Control of pests and weeds with organisms that prey on or parasitize them rather than with chemical *pesticides*. Commonly called biocontrol.

biological diversity See *biodiversity*.

biological hazard Human health hazards that result from ecological interactions among organisms. These include *parasitism* by viruses, bacteria, or other *pathogens*. Compare *infectious disease; chemical hazard; cultural hazard; physical hazard*.

biomagnification The magnification of the concentration of *toxicants* in an organism caused by its consumption of other organisms in which toxicants have *bioaccumulated*.

biomass (1) In *ecology,* organic material that makes up living organisms; the collective mass of living matter in a given place and time. (2) In energy, organic material derived from living or recently living organisms, containing chemical *energy* that originated with *photosynthesis*.

biome A major regional complex of similar plant *communities;* a large *ecological* unit defined by its dominant plant type and vegetation structure.

biophilia An inherent love for and fascination with nature and an instinctive desire people have to affiliate with other living things. Defined by biologist E.O. Wilson as "the connections that human beings subconsciously seek with the rest of life."

biopower Power attained by combusting *biomass* sources to generate *electricity*.

bioprospecting Searching for organisms that might provide new drugs, medicines, foods, or other products of value or interest.

bioremediation A process by which living organisms are used to degrade toxic substances or render them inert. Generally plants, fungi, or microbes are used to take up specific contaminants from soil or water and then break organic contaminants down into safer compounds or concentrate heavy metals in their tissues.

biosphere The cumulative total of all the planet's living organisms and the nonliving portions of the *environment* they inhabit.

biosphere reserve A tract of land with exceptional *biodiversity* that couples preservation with *sustainable development* to benefit local people. Biosphere reserves are designated by UNESCO (the *United Nations Educational, Scientific and Cultural Organization*) following application by local stakeholders.

biotechnology The material application of biological *science* to create products derived from organisms. The creation of *transgenic* organisms is one type of biotechnology.

birth control The effort to control the number of children one bears, particularly by reducing the frequency of pregnancy. Compare *contraception; family planning*.

bisphenol A (BPA) An *endocrine-disrupting chemical* found in plastics.

boreal forest A *biome* of northern coniferous forest that stretches in a broad band across much of Canada, Alaska, Russia, and Scandinavia. Also known as *taiga,* boreal forest consists of a limited number of *species* of evergreen trees, such as black spruce, that dominate large regions of forests interspersed with occasional bogs and lakes.

bottle bill A law establishing a program whereby consumers pay a deposit on bottles or cans upon purchase—often 5 or 10 cents per container—and then receive a refund when they return them to stores after use. Bottle bills reduce litter, raise *recycling* rates, and decrease the *waste stream*.

bottleneck In environmental science, a step in a process that limits the progress of the overall process.

breakdown product A compound that results from the degradation of a toxicant.

brownfield An area of land whose redevelopment or reuse is complicated by the presence or potential presence of hazardous material.

campus sustainability A term describing a wide array of efforts taking place on college and university campuses through which students, faculty, staff, and administrators are working to reduce the environmental impacts and *ecological footprints* of their institutions.

canopy The upper level of tree leaves and branches in a *forest*.

cap-and-trade An *emissions trading* system in which government sets a cap by determining an allowable level of *pollution* and then issues polluting parties permits to pollute up to this level. A party receives credit for amounts it does not emit and can then sell this credit to other parties.

captive breeding The practice of keeping members of *threatened* or *endangered species* in captivity so that their young can be bred and raised in controlled *environments* and subsequently reintroduced into the wild.

carbohydrate An *organic compound* consisting of *atoms* of carbon, hydrogen, and oxygen.

carbon The chemical *element* with six protons and six neutrons. A key element in *organic compounds*.

carbon capture and storage Technologies or approaches to remove *carbon dioxide* from emissions of power plants or other facilities and sequester, or store, it (generally in liquid form) underground under pressure in locations where it will not seep out, in an effort to mitigate *global climate change*.

carbon cycle A major *nutrient cycle* consisting of the routes that *carbon atoms* take through the nested networks of environmental *systems*.

carbon dioxide (CO_2) A colorless gas used by plants for *photosynthesis,* given off by *respiration,* and released by burning *fossil fuels*. A primary *greenhouse gas* whose buildup contributes to *global climate change*.

carbon footprint The cumulative amount of *carbon,* or *carbon dioxide,* that a person or institution emits, and is indirectly responsible for emitting, into the *atmosphere,* contributing to *global climate change*. Compare *ecological footprint*.

carbon monoxide (CO) A colorless, odorless gas produced primarily by the incomplete combustion of fuel. An EPA *criteria pollutant*.

carbon-neutrality The state in which an individual, business, or institution emits no net carbon to the atmosphere. It may be achieved by reducing carbon emissions or employing *carbon offsets* to offset emissions.

carbon offset A voluntary payment to another entity intended to enable that entity to reduce the *greenhouse gas* emissions that one is unable or unwilling to reduce oneself. The payment thus offsets one's own emissions.

carbon pricing The practice of putting a price on the emission of *carbon dioxide,* either through *carbon trading* or a *carbon tax,* as a means to address *global climate change*. Carbon pricing compensates the public for the external costs of fossil fuel use by shifting costs to emitters and creates financial incentives to reduce emissions.

carbon tax A type of *green tax* charged to entities that pollute by emitting *carbon dioxide*. Carbon taxation is one approach to *carbon pricing* and gives polluters a financial incentive to reduce emissions in an effort to address *global climate change*. Compare *carbon trading; fee-and-dividend; revenue-neutral carbon tax*.

carbon trading A form of *emissions trading* that focuses on the emission of *carbon dioxide*. In a carbon trading market, emitters buy and sell permits to emit CO₂. Carbon trading is one approach to *carbon pricing* and gives polluters a financial incentive to reduce emissions in an effort to address *global climate change*. Compare *carbon tax*.

carcinogen A chemical or type of radiation that causes cancer.

carrying capacity The maximum *population size* of a given organism that a given *environment* can sustain.

case history Medical approach involving the observation and analysis of individual patients.

Cassandra A worldview (or a person holding the worldview) that predicts doom and disaster as a result of our environmental impacts. Compare *Cornucopian*.

catalytic converter Automotive technology that chemically treats engine exhaust to reduce *air pollution*. Reacts exhaust with metals that convert *hydrocarbons*, CO, and NO_X into *carbon dioxide, water* vapor, and *nitrogen* gas.

cellular respiration The process by which a *cell* uses the chemical reactivity of *oxygen* to split glucose into its constituent parts, water and *carbon dioxide*, and thereby release chemical energy that can be used to form chemical bonds or to perform other tasks within the cell. Compare *photosynthesis*.

cellulosic ethanol *Ethanol* produced from the cellulose in plant tissues by treating it with enzymes. Techniques for producing cellulosic ethanol are under development because of interest in making ethanol from low-value crop waste (residues such as corn stalks and husks), rather than from the sugars of high-value crops.

chaparral A *biome* consisting mostly of densely thicketed evergreen shrubs occurring in limited small patches. Its "Mediterranean" *climate* of mild, wet winters and warm, dry summers is induced by oceanic influences. In addition to ringing the Mediterranean Sea, chaparral occurs along the coasts of California, Chile, and southern Australia.

character displacement A phenomenon resulting from *competition* among *species* in which competing species evolve characteristics

that better adapt them to specialize on the portion of the resource they use. The species essentially become more different from one another, reducing their competition.

chemical hazard Chemicals that pose human health hazards. These include *toxins* produced naturally, as well as many of the disinfectants, *pesticides,* and other synthetic chemicals that our society produces. Compare *biological hazard; cultural hazard; physical hazard*.

chemistry The study of the different types of *matter* and how they interact.

chemosynthesis The process by which bacteria in *hydrothermal vents* use the chemical energy of hydrogen sulfide (H₂S) to transform inorganic *carbon* into *organic compounds*. Compare *photosynthesis*.

Chernobyl Site of a nuclear power plant in Ukraine (then part of the Soviet Union), where in 1986 an explosion caused the most severe *nuclear reactor* accident the world has yet seen. The term is also often used to denote the accident itself. Compare *Fukushima Daiichi; Three Mile Island*.

chlorofluorocarbon (CFC) A type of *halocarbon* consisting of only chlorine, fluorine, carbon, and hydrogen. CFCs were used as refrigerants, as fire extinguishers, as propellants for aerosol spray cans, as cleaners for electronics, and for making polystyrene foam. They were phased out under the *Montreal Protocol* because they are *ozone-depleting substances* that destroy stratospheric *ozone*.

chronic exposure Exposure for long periods of time to a *toxicant* occurring in low amounts. Compare *acute exposure*.

citizen science The observation, study, and collection of data for scientific purposes by nonprofessional volunteers. Also called community science.

city planning The professional pursuit that attempts to design cities in such a way as to maximize their efficiency, functionality, and beauty. Also known as *urban planning*. Compare *regional planning*.

classical economics Founded by Adam Smith, the study of the behavior of buyers and sellers in a capitalist market economy. Holds that individuals acting in their own self-interest may benefit society provided that their behavior is constrained by the rule of law and by private property rights and operates within competitive markets. Compare *neoclassical economics*.

clay *Sediment* consisting of particles less than 0.002 mm in diameter. Compare *sand; silt*.

Clean Air Act U.S. *legislation* to control *air pollution*, first enacted in 1963 and amended multiple times since, most significantly in 1970 and 1990. Funds research into pollution control, sets standards for air quality, encourages emissions standards for automobiles

and for stationary point sources such as industrial plants, imposes limits on emissions from new sources, funds a nationwide air quality monitoring system, enables citizens to sue parties violating the standards, and introduced an *emissions trading* program for *sulfur dioxide*.

clean coal technologies An array of techniques, equipment, and approaches to remove chemical contaminants (such as sulfur) during the process of generating *electricity* from *coal*.

clear-cutting The harnessing of timber by cutting all the trees in an area. Although it is the most cost-efficient method, clear-cutting is also the most ecologically damaging.

climate The pattern of atmospheric conditions that typifies a geographic region over long periods of time (typically years, decades, centuries, or millennia). Compare *weather*.

climate change See *global climate change*.

climate diagram A visual representation of a region's average monthly temperature and *precipitation*. Also known as a climatograph.

climate model A computer program that combines what is known about weather patterns, atmospheric circulation, atmosphere-ocean interactions, and feedback mechanisms, to simulate *climate* processes.

coal A solid blackish *fossil fuel* formed from organic matter (generally woody plant material) that was compressed under very high pressure and with little decomposition, creating dense, solid carbon structures.

coevolution The process by which two or more species evolve in response to one another. Parasites and hosts may coevolve, as may flowering plants and their pollinators.

cogeneration A practice in which the extra heat generated in the production of *electricity* is captured and put to use heating workplaces and homes, as well as producing other kinds of power.

colony collapse disorder A mysterious malady afflicting honeybees, which has destroyed roughly one-third of all honeybees in the United States annually in recent years. Likely caused by chemical *insecticides, pathogens* and parasites, *habitat* and resource loss, or combinations of these factors.

command-and-control A top-down approach to policy in which a legislative body or a regulating agency sets rules, standards, or limits and threatens punishment for violations of those limits.

community In *ecology,* an assemblage of *populations* of interacting organisms that live in the same area at the same time.

community-based conservation The practice of engaging local people to conserve land and wildlife in their own region.

community ecology The scientific study of patterns of species diversity and interactions among *species,* ranging from one-to-one interactions to complex interrelationships involving entire *communities.*

community-supported agriculture (CSA) A system in which consumers pay farmers in advance for a share of their yield, usually in the form of weekly deliveries of produce.

competition A species interaction in which multiple organisms seek the same limited resource(s). Competition can take place among members of the same species or among members of different species.

competitive exclusion An outcome of interspecific competition in which one *species* excludes another species from resource use entirely.

compost A mixture produced when *decomposers* break down organic matter, such as food and crop waste, in a controlled environment.

composting The conversion of organic *waste* into mulch or humus by encouraging, in a controlled manner, the natural biological processes of decomposition.

compound A *molecule* whose *atoms* are composed of two or more *elements.*

concentrated solar power (CSP) A means of generating *electricity* at a large scale by focusing sunlight from a large area onto a smaller area. Several approaches are used.

concession The right to extract a resource, granted by a government to a corporation. Compare *conservation concession.*

confined aquifer A water-bearing, porous layer of rock, sand, or gravel that is trapped between an upper and lower layer of less permeable substrate, such as clay. The water in a confined aquifer is under pressure because it is trapped between two impermeable layers. Also called an artesian aquifer. Compare *unconfined aquifer.*

conservation biology A scientific discipline devoted to understanding the factors, forces, and processes that influence the loss, protection, and restoration of *biodiversity.*

conservation concession A type of *concession* in which a conservation organization purchases the right to prevent resource extraction in an area of land, generally to preserve habitat in developing nations.

conservation ethic An ethic holding that people should put natural resources to use but that we have a responsibility to manage them wisely. It promotes the prudent, efficient, and sustainable extraction and use of natural resources for the good of present and future generations. Compare *preservation ethic.*

Conservation Reserve Program U.S. policy in farm bills since 1985 that pays farmers to stop cultivating highly erodible cropland and instead place it in conservation reserves planted with grasses and trees.

conservation tillage *Agriculture* that limits the amount of tilling (plowing, disking, harrowing, or chiseling) of *soil.* Compare *no-till.*

consumer An organism that consumes other living organisms.

consumptive use Use of *fresh water* in which water is removed from a particular *aquifer* or surface water body and is not returned to it. *Irrigation* for *agriculture* is an example of consumptive use. Compare *non-consumptive use.*

continental collision The meeting of two tectonic plates of continental *lithosphere* at a *convergent plate boundary,* wherein the continental *crust* on both sides resists *subduction* and instead crushes together, bending, buckling, and deforming layers of rock and forcing portions of the buckled crust upward, often creating mountain ranges.

continental shelf The gently sloping underwater edge of a continent, varying in width from 100 m (330 ft) to 1300 km (800 mi), with an average slope of 1.9 m/km (10 ft/mi).

contingent valuation A technique that uses surveys to determine how much people would be willing to pay to protect a resource or to restore it after damage has been done.

contour farming The practice of plowing furrows sideways across a hillside, perpendicular to its slope, to help prevent the formation of rills and gullies. The technique is so named because the furrows follow the natural contours of the land.

contraception The deliberate attempt to prevent pregnancy despite sexual intercourse. Compare *birth control.*

control The portion of an *experiment* in which a *variable* has been left unmanipulated, to serve as a point of comparison with the *treatment.*

controlled experiment An *experiment* in which a *treatment* is compared against a *control* in order to test the effect of a *variable.*

convective circulation A circular *current* (of air, water, magma, etc.) driven by temperature differences. In the atmosphere, warm air rises into regions of lower *atmospheric pressure,* where it expands and cools and then descends and becomes denser, replacing warm air that is rising. The air picks up heat and moisture near ground level and prepares to rise again, continuing the process.

Convention on Biological Diversity A 1992 international treaty that aims to conserve *biodiversity,* use biodiversity in a *sustainable* manner, and ensure the fair distribution of biodiversity's benefits.

Convention on International Trade in Endangered Species of Wild Fauna and Flora (CITES) A 1973 treaty facilitated by the *United Nations* that protects *endangered* species by banning the international transport of their body parts.

conventional law International law that arises from *conventions,* or treaties, that nations agree to enter into. Compare *customary law.*

convergent evolution The evolutionary process by which very unrelated species acquire similar traits as they adapt to similar selective pressures from similar environments.

convergent plate boundary The area where tectonic plates converge or come together. Can result in *subduction* or *continental collision.* Compare *divergent plate boundary; transform plate boundary.*

coral Tiny marine animals that build *coral reefs.* Corals attach to rock or existing reef and capture passing food with stinging tentacles. They also derive nourishment from photosynthetic symbiotic algae known as *zooxanthellae.*

coral reef A mass of calcium carbonate composed of the skeletons of tiny colonial marine organisms called *corals.*

core The innermost part of Earth, made up mostly of iron, that lies beneath the *crust* and *mantle.*

Coriolis effect The apparent deflection of north–south air *currents* to a partly east–west direction, caused by the faster spin of regions near the equator than of regions near the poles as a result of Earth's rotation.

Cornucopian A worldview (or a person holding the worldview) that we will find ways to make Earth's natural resources meet all our needs indefinitely and that human ingenuity will see us through any difficulty. Compare *Cassandra.*

correlation Statistical association (positive or negative) among variables. The association may be causal or may occur by chance.

corridor A passageway of protected land established to allow animals to travel between islands of protected *habitat.*

cost-benefit analysis A method commonly used in *neoclassical economics* in which estimated costs for a proposed action are totaled and then compared to the sum of benefits estimated to result from the action.

covalent bond A type of chemical bonding where atoms share electrons in chemical bonds. An example is a water molecule, which forms when an oxygen atom shares electrons with two hydrogen atoms.

cradle-to-cradle An approach to *waste management* and industrial design in which the materials from products are recovered and reused to create new products.

criteria pollutant One of six *air pollutants*—carbon monoxide, sulfur dioxide, nitrogen dioxide, tropospheric ozone, particulate matter, and *lead*—for which the *Environmental Protection Agency* has established maximum allowable concentrations in ambient outdoor air because of the threats they pose to human health.

crop rotation   The practice of alternating the kind of crop grown in a particular field from one season or year to the next.

cropland   Land that people use to raise plants for food and fiber.

crude oil   *Oil* in its natural state, as it occurs once extracted from the ground but before processing and refining.

crust   The lightweight outer layer of the Earth, consisting of *rock* that floats atop the malleable *mantle,* which, in turn, surrounds a mostly iron *core.*

cultural hazard   Human health hazards that result from the place we live, our socioeconomic status, our occupation, or our behavioral choices. These include choosing to smoke cigarettes, or living or working with people who do. Also known as *lifestyle hazard.* Compare *biological hazard; chemical hazard; physical hazard.*

culture   The overall ensemble of knowledge, beliefs, values, and learned ways of life shared by a group of people.

current   The flow of a liquid or gas in a certain direction.

customary law   International law that arises from long-standing practices, or customs, held in common by most *cultures.* Compare *conventional law.*

dam   Any obstruction placed in a river or stream to block the flow of water so that water can be stored in a *reservoir.* Dams are built to prevent floods, provide drinking water, facilitate *irrigation,* and generate *electricity.*

Darwin, Charles   (1809–1882) English naturalist who proposed the concept of *natural selection* as a mechanism for *evolution* and as a way to explain the great variety of living things. Compare *Wallace; Alfred Russel.*

data   Information, generally quantitative information.

debt-for-nature swap   A transaction in which a conservation organization pays off a portion of a developing nation's international debt in exchange for a promise by the nation to set aside reserves, fund environmental education, and better manage protected areas.

decomposer   An organism, such as a fungus or bacterium, that breaks down leaf litter and other nonliving matter into simple constituents that can be taken up and used by plants. Compare *detritivore.*

deep-well injection   A *hazardous waste* disposal method in which a well is drilled deep beneath an area's *water table* into porous rock below an impervious *soil* layer. Wastes are then injected into the well so that they will be absorbed into the porous rock and remain deep underground, isolated from *groundwater* and human contact. Compare *surface impoundment.*

deforestation   The clearing and loss of *forests.*

demographer   A social scientist who studies the size; density; distribution; age structure; sex ratio; and rates of birth, death, immigration, and emigration of human populations. See *demography.*

demographic fatigue   An inability on the part of governments to address overwhelming challenges related to population growth.

demographic transition   A theoretical *model* of economic and cultural change that explains the declining death rates and birth rates that occurred in Western nations as they became industrialized. The model holds that industrialization caused these rates to fall naturally by decreasing mortality and by lessening the need for large families. Parents would thereafter choose to invest in quality of life rather than quantity of children.

demography   A *social science* that applies the principles of *population ecology* to the study of statistical change in human *populations.*

denitrifying bacteria   Bacteria that convert the nitrates in *soil* or water to gaseous *nitrogen* and release it back into the *atmosphere.*

density-dependent   The condition of a *limiting factor* whose effects on a *population* become stronger or weaker depending on the *population density.* Compare *density-independent.*

density-independent   The condition of a limiting factor whose effects on a population are independent of (not affected by) population density. Compare *density-dependent.*

deoxyribonucleic acid   (DNA) A double-stranded *nucleic acid* composed of four nucleotides, each of which contains a sugar (deoxyribose), a phosphate group, and a nitrogenous base. DNA carries the hereditary information for living organisms and is responsible for passing traits from parents to offspring. Compare *ribonucleic acid (RNA).*

dependent variable   The *variable* that is affected by manipulation of the *independent variable* in an *experiment.*

deposition   The arrival of eroded *soil* at a new location. Compare *erosion.*

desalination   The removal of salt from seawater to generate fresh water for human use. Also known as *desalinization.*

descriptive science   Research in which scientists gather basic information about organisms, materials, systems, or processes that are not yet well known. Compare *hypothesis-driven science.*

desert   The driest *biome* on Earth, with annual *precipitation* of less than 25 cm. Because deserts have relatively little vegetation to insulate them from temperature extremes, sunlight readily heats them in the daytime, but daytime heat is quickly lost at night, so temperatures vary widely from day to night and from season to season.

desertification   A form of *land degradation* in which more than 10% of a land's productivity is lost due to *erosion,* soil compaction, forest removal, *overgrazing,* drought, *salinization, climate* change, water depletion, or other factors. Severe desertification can result in the expansion of desert areas or creation of new ones. Compare *land degradation; soil degradation.*

detritivore   An organism, such as a millipede or soil insect, that scavenges the waste products or dead bodies of other community members. Compare *decomposer.*

development   The use of natural resources for economic advancement (as opposed to simple subsistence, or survival).

directional drilling   A drilling technique (e.g., for oil or natural gas) in which a drill bores down vertically and then bends horizontally to follow layered deposits for long distances from the drilling site. This technique enables extracting more *fossil fuels* with less environmental impact on the surface.

discounting   A practice in *neoclassical economics* by which short-term costs and benefits are granted more importance than long-term costs and benefits. Future effects are thereby "discounted," because the idea is that an impact far in the future should count much less than one in the present.

disturbance   An event that affects environmental conditions rapidly and drastically, resulting in changes to the *community* and *ecosystem.* A disturbance can be natural or may be caused by people.

divergent plate boundary   The area where tectonic plates push apart from one another as *magma* rises upward to the surface, creating new *lithosphere* as it cools and spreads. A prime example is the Mid-Atlantic Ridge. Compare *convergent plate boundary; transform plate boundary.*

dose   The amount of *toxicant* a test animal receives in a dose-response test. Compare *response.*

dose-response analysis   A set of experiments that measure the *response* of test animals to different *doses* of a *toxicant.* The response is generally quantified by measuring the proportion of animals exhibiting negative effects.

dose-response curve   A curve that plots the *response* of test animals to different *doses* of a *toxicant,* as a result of *dose-responseanalysis.*

downwelling In the ocean, the flow of warm surface water toward the ocean floor. Downwelling occurs where surface *currents* converge. Compare *upwelling*.

drainage basin See *watershed*.

Dust Bowl An area that loses huge amounts of *topsoil* to wind *erosion* as a result of drought and/or human impact. First used to name the region in the North American Great Plains severely affected by drought and topsoil loss in the 1930s. The term is now also used to describe that historical event and others like it.

dynamic equilibrium The state reached when processes within a *system* are moving in opposing directions at equivalent rates so that their effects balance out.

e-waste See *electronic waste*.

earthquake A release of energy that occurs as Earth relieves accumulated pressure between masses of *lithosphere* and that results in shaking at the surface.

ecocentrism A philosophy that considers actions in terms of their damage or benefit to the integrity of whole ecological systems, including both living and nonliving elements. For an ecocentrist, the well-being of an individual is less important than the long-term well-being of a larger integrated ecological system. Compare *anthropocentrism* and *biocentrism*.

ecolabeling The practice of designating on a product's label how the product was grown, harvested, or manufactured so that consumers can judge which brands use more sustainable processes.

ecological economics A school of *economics* that applies the principles of *ecology* and *systems* thinking to the description and analysis of *economies*. Compare *environmental economics; neoclassical economics*.

ecological footprint A concept that measures the cumulative area of biologically productive land and water required to provide the resources a person or population consumes and to dispose of or recycle the waste the person or population produces. The total area of Earth's biologically productive surface that a given person or population "uses" once all direct and indirect impacts are summed together.

ecological modeling The practice of constructing and testing *models* that aim to explain and predict how ecological *systems* function.

ecological restoration The practice of attempting to reverse the effects of human disruption of ecological systems and to restore *communities* to their condition before the disruption. This practice applies principles of *restoration ecology*.

ecologist A scientist who studies *ecology*.

ecology The *science* that deals with the distribution and abundance of organisms, the interactions among them, and the interactions between organisms and their nonliving *environments*.

economic growth An increase in an economy's activity—that is, an increase in the production and consumption of goods and services.

economics The study of how we decide to use scarce resources to satisfy demand for *goods* and *services*.

economy A social *system* that converts resources into *goods* and *services*.

ecosystem In *ecology*, an assemblage of all organisms and nonliving entities that occur and interact in a particular area at the same time.

ecosystem diversity The number and variety of ecosystems in a particular area. One way to express *biodiversity*. Related concepts consider the geographic arrangement of *habitats, communities*, or *ecosystems* at the landscape level, including the sizes, shapes, and interconnectedness of patches of these entities.

ecosystem ecology The scientific study of how the living and nonliving components of *ecosystems* interact.

ecosystem services Processes that naturally result from the normal functioning of ecological systems and from which human beings draw benefits. Examples include nutrient cycling, air and water purification, climate regulation, *pollination*, waste recycling, and more.

ecosystem-based management The attempt to manage the harvesting of resources in ways that minimize impact on the *ecosystems* and ecological processes that provide the resources.

ecotone A transitional zone where *ecosystems* meet.

ecotourism Tourism involving natural areas and outdoor recreation. Often viewed as providing financial incentives for conservation efforts.

ED₅₀ ED_{50} (effective dose–50%) The amount of a *toxicant* it takes to affect 50% of a *population* of test animals. Compare LD_{50}; *threshold dose*.

edge effect An impact on organisms, populations, or communities that results because conditions along the edge of a habitat fragment differ from conditions in the interior.

El Niño An exceptionally strong warming of the eastern Pacific Ocean that occurs every 2–8 years and depresses local fish and bird *populations* by altering the marine *food web* in the area. Originally, the name that Spanish-speaking fishermen gave to an unusually warm surface *current* that sometimes arrived near the Pacific coast of South America around Christmastime. Compare *La Niña*.

El Niño–Southern Oscillation (ENSO) A systematic shift in atmospheric pressure, sea surface temperature, and ocean circulation in the tropical Pacific Ocean. ENSO cycles give rise to *El Niño* and *La Niña* conditions.

electricity A secondary form of *energy* that can be transferred over long distances and applied for a variety of uses.

electrolysis A process in which electrical current is passed through a *compound* to release *ions*. Electrolysis offers one way to produce *hydrogen* for use as fuel: Electrical current is passed through water, splitting the water *molecules* into hydrogen and *oxygen atoms*.

electron A negatively charged particle that moves about the nucleus of an *atom*.

electronic waste Discarded electronic products such as computers, monitors, printers, televisions, DVD players, cell phones, and other devices. *Heavy metals* in these products mean that this *waste* may be judged hazardous. Also known as *e-waste*.

element A fundamental type of *matter*; a chemical substance with a given set of properties, which cannot be broken down into substances with other properties. Chemists currently recognize 92 elements that occur in nature, as well as more than 20 others that have been artificially created.

emergent property A characteristic that is not evident in a *system*'s components.

eminent domain A policy in which a government pays landowners for their land at market rates and landowners have no recourse to refuse. In eminent domain, courts set aside private property rights to make way for projects judged to be for the public good.

emissions trading The practice of buying and selling government-issued marketable permits to emit pollutants. Under a *cap-and-trade* system, the government determines an acceptable level of *pollution* and then issues permits to pollute. A company receives credit for amounts it does not emit and can then sell this credit to other companies. Compare *cap-and-trade*.

endangered In danger of becoming extinct in the near future.

Endangered Species Act The primary *legislation*, enacted in 1973, for protecting *biodiversity* in the United States. It forbids the government and private citizens from taking actions (such as developing land) that would destroy *threatened* and *endangered* species or their *habitats*, and it prohibits trade in products made from *threatened* and *endangered* species.

endemic Native and restricted to a particular geographic region. An endemic *species* occurs in one area and nowhere else on Earth.

endocrine-disrupting chemical A *toxicant* that interferes with the *endocrine (hormone) system*.

energy The capacity to change the position, physical composition, or temperature of *matter;* a force that can accomplish work.

energy conservation The practice of reducing *energy* use as a way of extending the lifetime of our *fossil fuel* supplies, of being less wasteful, and of reducing our impact on the *environment*. Conservation can result from behavioral decisions or from technologies that demonstrate *energy efficiency*.

energy efficiency The ability to obtain a given result or amount of output while using less energy input. Technologies permitting greater energy efficiency are one main route to *energy conservation*.

energy intensity A measure of energy use per dollar of *Gross Domestic Product (GDP)*. Lower energy intensity indicates greater efficiency.

energy storage The capture and accumulation of *energy* produced at one place or time for use (generally by generation of electricity) at another place or time where or when demand for it is greater.

enhanced geothermal systems (EGS) A recently developed approach whereby engineers drill deeply into rock, fracture it, pump in water, and then pump it out once it is heated belowground. This approach would enable us to obtain *geothermal energy* in many locations.

environment The sum total of our surroundings, including all of the living things and nonliving things with which we interact.

environmental economics A school of *economics* that modifies the principles of *neoclassical economics* to address environmental challenges. Most environmental economists believe that we can attain *sustainability* within our current economic systems. Compare *ecological economics; neoclassical economics*.

environmental ethics The application of *ethical standards* to environmental questions and to relationships between people and nonhuman entities.

environmental health The study of environmental factors that influence human health and quality of life and the health of *ecological* systems essential to environmental quality and long-term human well-being.

environmental history The scholarly study of how people have interacted with the natural world through time, including both how humans have affected the *environment* and how the *environment* has affected society.

environmental impact statement (EIS) A report of results from detailed scientific studies that assess the potential effects on the *environment* that would likely result from a development project or other action undertaken by the government.

environmental justice The fair and equitable treatment of all people with respect to environmental policy and practice, regardless of their income, race, or ethnicity. This principle is a response to the perception that minorities and the poor suffer more pollution than the majority and the more affluent.

environmental literacy A basic understanding of Earth's physical and living systems and how we interact with them. Some people take the term further and use it to refer to a deeper understanding of society and the environment or a commitment to advocate for *sustainability*.

environmental policy *Public policy* that pertains to human interactions with the *environment*. Environmental policy generally aims to regulate resource use or reduce *pollution* to promote human well-being or protect natural systems.

Environmental Protection Agency (EPA) An administrative agency of the U.S. government charged with conducting and evaluating research, monitoring environmental quality, setting standards, enforcing those standards, assisting the states in meeting standards and goals for environmental protection, and educating the public.

environmental science The scientific study of how the natural world functions, how our *environment* affects us, and how we affect our environment.

environmental studies An academic *environmental science* program that emphasizes the social sciences as well as the natural sciences.

environmental toxicology The study of *toxicants* that come from or are discharged into the *environment,* including the study of health effects on humans, other animals, and *ecosystems*.

environmentalism A social movement dedicated to protecting the natural world and, by extension, people from undesirable changes brought about by human action.

epidemiological study A study that involves large-scale comparisons among groups of people, usually contrasting a group known to have been exposed to some *toxicant* and a group that has not.

EROI An abbreviation for "energy returned on investment." The ratio determined by dividing the quantity of *energy* returned from a process by the quantity of energy invested in the process. Higher EROI ratios mean that more energy is produced from each unit of energy invested. Compare *net energy*.

erosion The removal of material from one place and its transport to another by the action of wind or water. Compare *deposition*.

estuary An area where a river flows into the ocean, mixing *fresh water* with saltwater.

ethanol The alcohol in beer, wine, and liquor, produced as a *biofuel* by fermenting biomass, generally from *carbohydrate*-rich crops such as corn or sugarcane.

ethical standard A criterion that helps differentiate right from wrong.

ethics The academic study of good and bad, right and wrong. The term can also refer to a person's or group's set of moral principles or values.

European Union (EU) Political and economic organization formed after World War II to promote Europe's economic and social progress. As of 2019, the EU consisted of 28 member nations.

eutrophic Term describing a water body that has high-nutrient and low-oxygen conditions. Compare *oligotrophic*.

eutrophication The process of *nutrient* enrichment, increased production of organic matter, and subsequent *ecosystem* degradation in a water body.

evaporation The conversion of a substance from a liquid to a gaseous form.

even-aged Condition of timber plantations—generally *monocultures* of a single *species*—in which all trees are of the same age. Most *ecologists* view plantations of even-aged stands more as crop *agriculture* than as ecologically functional *forests*. Compare *uneven-aged*.

evolution Genetically based change in *populations* of organisms across generations. Changes in *genes* may lead to changes in the appearance, physiology, and/or behavior of organisms across generations, often by the process of *natural selection*.

executive order Specific legal instructions for a U.S. government agency issued directly by the president.

experiment An activity designed to test the validity of a *hypothesis* by manipulating *variables*. See *controlled experiment*.

exponential growth The increase of a *population* (or of anything) by a fixed percentage each year. This results in a J-shaped curve on a graph. Compare *logistic growth*.

external cost A cost borne by someone not involved in an economic transaction. Examples include harm to citizens from *water pollution* or *air pollution* discharged by nearby factories.

extinction The disappearance of an entire *species* from Earth. Compare *extirpation*.

extirpation The disappearance of a particular *population* from a given area, but not the entire *species* globally. Compare *extinction*.

exurb A region surrounding a city and beyond the suburbs, generally inhabited by affluent individuals seeking even more space than the suburbs provide.

family planning The effort to plan the number and spacing of one's children to offer children and parents the best quality of life possible.

farmers' market A market at which local farmers and food producers sell fresh, locally grown items.

fee-and-dividend A *carbon tax* program in which proceeds from the tax are paid to consumers as a tax refund or "dividend." This strategy seeks to prevent consumers from losing money if polluters pass their costs along to them.

feedback loop A circular process in which a *system*'s output serves as input to that same system. See *negative feedback loop; positive feedback loop.*

feed-in tariff A payment made by a utility to a business or homeowner who produces *electricity* (generally from renewable sources) and feeds it into the electrical grid. Feed-in tariff systems are set up—and rates are set—by governments, generally as a means of encouraging *renewable energy.* Compare *net metering.*

feedlot A huge indoor or outdoor pen designed to deliver *energy*-rich food to animals living at extremely high densities. Also called a factory farm or concentrated animal feeding operation.

fertilizer A substance that promotes plant growth by supplying essential *nutrients* such as *nitrogen* or *phosphorus.* See also *inorganic fertilizer; organic fertilizer.*

first law of thermodynamics The physical law stating that *energy* can change from one form to another, but cannot be created or lost. The total energy in the universe remains constant and is said to be conserved.

flooding The spillage of water over a river's banks due to heavy rain or snowmelt.

floodplain The region of land over which a river has historically wandered and periodically floods.

flux The movement of nutrients among *reservoirs* in a *nutrient cycle.*

food chain A linear series of feeding relationships. As organisms feed on one another, energy is transferred from lower to higher *trophic levels.* Compare *food web.*

food security The guaranteed availability of an adequate, safe, nutritious, and reliable food supply to all people at all times.

food web A visual representation of feeding interactions within an *ecological community* that shows an array of relationships between organisms at different *trophic levels.* Compare *food chain.*

forensic science The scientific analysis of evidence to make an identification or answer a question relating to a crime or an accident. Often called forensics.

forest Any ecosystem characterized by a high density of trees.

forest type A category of *forest* defined by its predominant tree *species.*

forestry The professional management of *forests.*

fossil The remains, impression, or trace of an animal or plant of past geologic ages that has been preserved in rock or *sediments.*

fossil fuel A *nonrenewable natural resource,* such as *crude oil, natural gas,* or *coal,* produced by the decomposition and compression of organic matter from ancient life. Fossil fuels have provided most of society's *energy* since the *industrial revolution.*

fossil record The cumulative body of *fossils* worldwide, which paleontologists study to infer the history of past life on Earth.

fracking See *hydraulic fracturing.*

Frank R. Lautenberg Chemical Safety for the 21st Century Act U.S. legislation, enacted in 2016, that updates the *Toxic Substances Control Act* and directs the EPA to monitor and regulate industrial chemicals.

free rider A party that fails to invest in conserving resources, controlling *pollution,* or carrying out other responsible activities and instead relies on the efforts of other parties to do so. For example, a factory that fails to control its emissions gets a "free ride" on the efforts of other factories that do.

fresh water Water that is relatively pure, holding very few dissolved salts.

fuel cell An *energy storage* device that can store and transport energy to produce *electricity,* much as a *battery* can. A *hydrogen* fuel cell generates electricity by the input of hydrogen fuel and *oxygen,* producing only water as a waste product.

Fukushima Daiichi Japanese nuclear power plant severely damaged by the tsunami associated with the March 2011 Tohoku earthquake that rocked Japan. Most radiation drifted over the ocean away from population centers, but the event was history's second most serious nuclear accident. Compare *Chernobyl; Three Mile Island.*

full cost accounting An accounting approach that attempts to summarize all costs and benefits by assigning monetary values to entities without market prices and then generally subtracting costs from benefits. Examples include the *Genuine Progress Indicator* and the Happy Planet Index. Also called true cost accounting.

fundamental niche The full *niche* of a species. Compare *realized niche.*

gene A stretch of *DNA* that represents a unit of hereditary information.

gene editing An array of new methods allowing scientists in the lab to change particular stretches of *DNA* in the genome of a target organism in order to express desirable traits in the organism. More targeted and direct than recombinant DNA technology. CRISPR is one common method. Also called genome editing.

General Mining Act of 1872 U.S. law that legalized and promoted *mining* by private individuals on public lands for just $5 per acre, subject to local customs, with no government oversight.

generalist A *species* that can survive across a wide array of *habitats* or that can use a wide array of resources. Compare *specialist.*

genetic diversity A measure of the differences in *DNA* composition among individuals within a given *species.*

genetic engineering Any process scientists use to manipulate an organism's genetic material in the lab by adding, deleting, or changing segments of its *DNA.*

genetically modified food Food derived from a *genetically modified organism.* Often abbreviated as GM food.

genetically modified organism (GMO) An organism that has been *genetically engineered,* often using recombinant DNA technology.

gentrification The transformation of a neighborhood to conditions (such as expensive housing and high-end shops and restaurants) that cater to wealthier people. Often results in longtime lower-income residents being "priced out" of their homes or apartments.

Genuine Progress Indicator (GPI) An *economic* indicator that attempts to differentiate between desirable and undesirable economic activity. The GPI accounts for benefits such as volunteerism and for costs such as environmental degradation and social upheaval. Compare *Gross Domestic Product (GDP).*

geoengineering Any of a suite of proposed strategies to cool Earth's climate by removing carbon dioxide from the atmosphere or reflecting sunlight away from Earth's surface. Such ideas are controversial and are not ready to implement.

geographic information system (GIS) Computer software that takes multiple types of data (for instance, on geology, hydrology, vegetation, animal species, and human development) and overlays them on a common set of geographic coordinates. GIS is used to create a complete picture of a landscape and to analyze how elements of the different datasets are arrayed spatially and how they may be correlated. A common tool of geographers, landscape ecologists, resource managers, and conservation biologists.

geology The scientific study of Earth's physical features, processes, and history.

geothermal energy Thermal *energy* that arises from beneath Earth's surface, ultimately from the radioactive decay of elements amid high pressures deep underground. Can be used to generate electrical power in power plants, for direct heating via piped water, or in *ground-source heat pumps.*

global climate change Systematic change in aspects of Earth's *climate,* such as temperature, *precipitation,* and storm intensity. Generally refers today to the current warming trend in global temperatures and the many climatic changes associated with it. Compare *global warming.*

global warming An increase in Earth's average surface temperature. The term is most frequently used in reference to the pronounced warming trend of recent decades. Global warming is one aspect of *global climate change* and in turn drives other components of climate change.

globalization The ongoing process by which the world's societies have become more interconnected, linked in many ways by diplomacy, commercial trade, and communication technologies.

Great Pacific Garbage Patch A portion of the North Pacific *gyre* where currents concentrate plastics and other floating debris that pose danger to marine organisms.

green building (1) A structure that minimizes the ecological footprint of its construction and operation by using sustainable materials, using minimal energy and water, reducing health impacts, limiting pollution, and recycling waste. (2) The pursuit of constructing or renovating such buildings.

green-collar job A job resulting from an employment opportunity in a more sustainably oriented *economy,* such as a job in *renewable energy.*

Green New Deal A nonbinding resolution introduced to the U.S. House of Representatives in 2019 that put forth ambitious goals for addressing climate change and meeting social justice goals by converting the U.S. economy to renewable energy and creating green energy jobs.

Green Revolution An intensification of the industrialization of *agriculture* in the developing world in the latter half of the 20th century that dramatically increased crop yields produced per unit area of farmland. Practices include devoting large areas to *monocultures* of crops specially bred for high yields and rapid growth; heavy use of *fertilizers, pesticides,* and *irrigation* water; and sowing and harvesting on the same parcel of land more than once per year or per season.

green tax A levy on environmentally harmful activities and products aimed at providing a market-based incentive to correct for *market failure.* Compare *subsidy.*

greenbelt A long and wide corridor of park land, often encircling an entire urban area.

greenhouse effect The warming of Earth's surface and *atmosphere* (especially the *troposphere*) caused by the *energy* emitted by *greenhouse gases.*

greenhouse gas A gas that absorbs infrared radiation released by Earth's surface and then warms the surface and *troposphere* by emitting *energy,* thus giving rise to the *greenhouse effect.* Greenhouse gases include *carbon dioxide* (CO_2), water vapor, *ozone* (O_3), nitrous oxide (N_2O), halocarbon gases, and *methane* (CH_4).

greenwashing A public relations effort by a corporation or institution to mislead customers or the public into thinking it is acting more sustainably than it actually is.

greenway A strip of park land that connects parks or neighborhoods; often located along a river, stream, or canal.

Gross Domestic Product (GDP) The total monetary value of final goods and services produced in a country each year. GDP sums all economic activity, whether good or bad, and does not account for benefits such as volunteerism or for *external costs* such as environmental degradation and social upheaval. Compare *Genuine Progress Indicator (GPI).*

gross primary production The *energy* that results when *autotrophs* convert solar energy (sunlight) to energy of chemical bonds in sugars through *photosynthesis.* Autotrophs use a portion of this production to power their own metabolism, which entails oxidizing *organic compounds* by *cellular respiration.* Compare *net primary production.*

ground-source heat pump A pump that harnesses *geothermal energy* from near-surface sources of earth and water to heat and cool buildings. Operates on the principle that temperatures belowground are less variable than temperatures aboveground.

groundwater Water held in *aquifers* underground. Compare *surface water.*

gyre An area of the ocean where currents converge and floating debris accumulates.

Haber-Bosch process A process to synthesize ammonia on an industrial scale. Developed by German chemists Fritz Haber and Carl Bosch, the process has enabled humans to double the natural rate of *nitrogen fixation* on Earth and thereby increase *agricultural* productivity, but it has also dramatically altered the *nitrogen cycle.*

habitat The specific *environment* in which an organism lives, including both biotic (living) and abiotic (nonliving) elements.

habitat fragmentation The process by which an expanse of natural *habitat* becomes broken up into discontinuous fragments, often as a result of farming, logging, road building, or other types of human development and land use.

habitat selection The process by which organisms select *habitats* from among the range of options they encounter.

habitat use The process by which organisms use *habitats* from among the range of options they encounter.

half-life The amount of time it takes for one-half the atoms of a *radioisotope* to emit radiation and decay. Different radioisotopes have different half-lives, ranging from fractions of a second to billions of years.

halocarbon A class of human-made chemical *compounds* derived from simple *hydrocarbons* in which *hydrogen* atoms are replaced by halogen atoms such as bromine, fluorine, or chlorine. Many halocarbons are *ozone-depleting substances* or *greenhouse gases.*

harmful algal bloom A *population* explosion of toxic algae caused by excessive *nutrient* concentrations.

hazardous waste Liquid or solid *waste* that is toxic, chemically reactive, flammable, or corrosive. Compare *industrial solid waste; municipal solid waste.*

heat capacity A measure of the heat energy required to increase the temperature of a given amount of a substance by a given amount.

herbicide A synthetic chemical used to kill plants. Compare *insecticide, pesticide.*

herbivory A species interaction in which animals consume plants.

heterotroph An organism that consumes other organisms. Includes most animals, as well as fungi and microbes that decompose organic matter. Also known as a *consumer.*

homeostasis The tendency of a *system* to maintain constant or stable internal conditions.

humus A dark, spongy, crumbly mass of material made up of complex organic compounds, resulting from the partial decomposition of organic matter.

hurricane A type of cyclonic storm that forms over the ocean but can do damage upon its arrival on land.

hydraulic fracturing A process to extract shale gas or tight oil, in which a drill is sent deep underground and angled horizontally into a shale formation; water, sand, and chemicals are pumped in under great pressure, fracturing the rock; and gas migrates up through the drilling pipe as sand holds the fractures open. Also called hydrofracking or simply *fracking.*

hydrocarbon An *organic compound* consisting solely of *hydrogen* and *carbon atoms.*

hydroelectric power The generation of *electricity* using the *kinetic energy* of moving water, generally by using turbines within dams. Also called *hydropower.*

hydrogen The chemical *element* with one proton. The most abundant element in the universe. Also, a possible fuel for our future economy.

hydrogen bond A weakly attractive interaction between *molecules* due to the attraction of partial positive and partial negative charges.

hydrologic cycle The flow of water—in liquid, gaseous, and solid forms—through our biotic and abiotic *environment*.

hydroponics The practice of growing plants without soil. In hydroponic systems, nutrients are supplied to plant roots directly via water-based solutions, generally inside sealed greenhouses. Used most often with leafy greens and tomatoes.

hydropower See *hydroelectric power*.

hydrosphere All water—salt or fresh, liquid, ice, or vapor—in surface bodies, underground, and in the *atmosphere*. Compare *biosphere; lithosphere*.

hypothesis A statement that attempts to explain a phenomenon or answer a *scientific* question. Compare *theory*.

hypothesis-driven science Research in which scientists pose questions that seek to explain how and why things are the way they are. Generally proceeds in a somewhat structured manner, using *experiments* to test *hypotheses*. Compare *descriptive science*.

hypoxia The condition of extremely low dissolved *oxygen* concentrations in a body of water.

igneous rock One of the three main categories of rock. Formed from cooling *magma*. Granite and basalt are examples of igneous rock. Compare *metamorphic rock; sedimentary rock*.

incineration A controlled process of burning solid *waste* for disposal in which mixed garbage is combusted at very high temperatures. Compare *sanitary landfill*.

independent variable The *variable* that a scientist manipulates in an *experiment*.

indoor air pollution *Air pollution* that occurs indoors.

industrial agriculture *Agriculture* that uses large-scale mechanization and *fossil fuel* combustion, enabling farmers to replace horses and oxen with faster and more powerful means of cultivating, harvesting, transporting, and processing crops. Other aspects include large-scale *irrigation* and the use of *inorganic fertilizers*. Use of chemical *herbicides* and *insecticides* reduces *competition* from weeds and *herbivory* by insects. Compare *traditional agriculture*.

industrial ecology A holistic approach to industry that integrates principles from engineering, chemistry, *ecology, economics,* and other disciplines and seeks to redesign industrial *systems* so as to reduce resource inputs and minimize inefficiency.

industrial revolution The shift beginning in the mid-1700s from rural life, animal-powered agriculture, and manufacturing by artisans to an urban-centered society powered by *fossil fuels*. Compare *agricultural revolution*.

industrial smog "Gray-air" *smog* caused by the incomplete combustion of *coal* or oil when burned. Compare *photochemical smog*.

industrial solid waste Nonliquid *waste* that is not especially hazardous and that comes from production of consumer goods, mining, *petroleum* extraction and *refining,* and *agriculture*. Compare *hazardous waste; municipal solid waste*.

industrial stage The third stage of the *demographic transition* model, characterized by falling birth rates that close the gap with falling death rates and reduce the rate of *population* growth. Compare *pre-industrial stage; post-industrial stage; transitional stage*.

infant mortality rate The number of deaths of infants under 1 year of age per 1000 live births in a population.

infectious disease A disease in which a pathogen attacks a host. Compare *noninfectious disease*.

inorganic fertilizer A *fertilizer* that consists of mined or synthetically manufactured mineral supplements. Inorganic fertilizers are generally more susceptible than *organic fertilizers* to *leaching* and *runoff* and may be more likely to cause unintended off-site impacts.

insecticide A synthetic chemical used to kill insects. Compare *herbicide; pesticide*.

instrumental value Value ascribed to something for the pragmatic benefits it brings us if we put it to use. Also called utilitarian value.

integrated pest management (IPM) The use of multiple techniques in combination to achieve long-term suppression of pests, including *biological control,* use of *pesticides,* close monitoring of *populations, habitat* alteration, *crop rotation, transgenic* crops, alternative tillage methods, and mechanical pest removal.

intercropping Planting different types of crops in alternating bands or other spatially mixed arrangements.

interdisciplinary Involving or borrowing techniques from multiple traditional fields of study and bringing together research results from these fields into a broad synthesis.

Intergovernmental Panel on Climate Change (IPCC) An international panel of *climate* scientists and government officials established in 1988 by the *United Nations Environment Programme* and the World Meteorological Organization. The IPCC's mission is to assess and synthesize scientific research on *global climate change* and to offer guidance to the world's policymakers, primarily through periodic published reports.

intertidal Of, relating to, or living along shorelines between the highest reach of the highest *tide* and the lowest reach of the lowest tide.

intrinsic value Value ascribed to something for its intrinsic worth; the notion that the thing has a right to exist and is valuable for its own sake. Also called inherent value.

introduced species A species introduced by human beings from one place to another (whether intentionally or by accident). Some introduced species may become *invasive species*.

invasive species A *species* that spreads widely and rapidly becomes dominant in a *community,* interfering with the community's normal functioning.

inversion layer A band of air in which temperature rises with altitude (i.e., in which the normal direction of temperature change is inverted). Cool air at the bottom of the inversion layer is denser than the warm air above, so it resists vertical mixing and remains stable. A key feature of a *temperature inversion*.

ion An electrically charged *atom* or combination of atoms.

ionic bond A type of chemical bonding where electrons are transferred between atoms, creating oppositely charged ions that bond due to their differing electrical charges. Table salt, sodium chloride, is formed by the bonding of positively charged sodium ions with negatively charged chloride ions.

ionizing radiation A high-energy form of radiation that can damage the cells of living things. Sources of this radiation include the sun and radioactive particles from *nuclear energy* and natural sources.

IPAT model A formula that represents how humans' total impact (I) on the *environment* results from the interaction among three factors: *population* (P), affluence (A), and technology (T).

irrigation The artificial provision of water to support *agriculture*.

island biogeography theory *Theory* initially applied to oceanic islands to explain how *species* come to be distributed among them. Researchers have increasingly applied the theory to islands of *habitat* (patches of one type of habitat isolated within "seas" of others). Aspects of the theory include *immigration* and *extinction* rates, the effect of island size, and the effect of distance from the mainland. Full name is the equilibrium theory of island biogeography.

isotope One of several forms of an *element* having differing numbers of *neutrons* in the nucleus of its *atoms*. Chemically, isotopes of an element behave almost identically, but they have different physical properties because they differ in mass.

kelp Large brown algae, or seaweed, that can form underwater "forests," providing habitat for marine organisms.

keystone species A *species* that has an especially far-reaching effect on a *community*.

kinetic energy *Energy* of motion. Compare *potential energy*.

Kyoto Protocol An international agreement drafted in 1997 that called for reducing, by 2012, emissions of six *greenhouse gases* to levels lower than their levels in 1990. It was extended to 2020 until a replacement treaty can be reached. An outgrowth of the *United Nations Framework Convention on Climate Change*.

La Niña An exceptionally strong cooling of surface water in the equatorial Pacific Ocean that occurs every 2–8 years and has widespread climatic consequences. Compare *El Niño*.

lab-grown meat Animal muscle tissue cultivated in a lab through *in vitro* tissue culture, to provide meat for people to eat, without the impacts of livestock agriculture. Still new, expensive, and developing, but holds promise. Also called cultured meat.

land degradation A general deterioration of land that diminishes its productivity and biodiversity, impairs the functioning of its ecosystems, and reduces the ecosystem services the land can offer us. Compare *soil degradation; desertification*.

land trust A local or regional organization that preserves lands valued by its members. In most cases, land trusts purchase land outright with the aim of preserving it in its natural condition.

landfill gas A mix of gases, consisting of roughly half *methane*, that is produced by anaerobic decomposition deep inside *landfills*.

landscape ecology The study of how landscape structure affects the abundance, distribution, and interaction of organisms. This approach to the study of organisms and their *environments* at the landscape scale focuses on broad geographic areas that include multiple *ecosystems*.

landslide The collapse and downhill flow of large amounts of rock or soil. A severe and sudden form of *mass wasting*.

lava *Magma* that is released from the *lithosphere* and flows or spatters across Earth's surface.

law of conservation of matter The physical law stating that *matter* may be transformed from one type of substance into others, but that it cannot be created or destroyed.

LD$_{50}$ (lethal dose–50%) The amount of a *toxicant* it takes to kill 50% of a *population* of test animals. Compare *ED$_{50}$; threshold dose*.

leachate Liquid that results when substances from waste dissolve in water as rainwater percolates downward. Leachate may sometimes seep through liners of a *sanitary landfill* and leach into the *soil* underneath.

leaching The process by which *minerals* dissolved in a liquid (usually water) are transported to another location (generally downward through *soil horizons*).

lead (Pb) A heavy metal that may be ingested through water or paint or that may enter the *atmosphere* as a particulate pollutant through combustion of leaded gasoline or other processes. Atmospheric lead deposited on land and water can enter the *food chain,* accumulate within body tissues, and cause *lead poisoning* in animals and people. An EPA *criteria pollutant*.

lead poisoning Poisoning by ingestion or inhalation of the heavy metal *lead,* causing an array of maladies including damage to the brain, liver, kidney, and stomach; learning problems and behavioral abnormalities; anemia; hearing loss; and even death. Lead poisoning can result from drinking water that passes through old lead pipes or ingesting dust or chips of old lead-based paint.

Leadership in Energy and Environmental Design (LEED) The leading set of standards for certification of a *green building*.

legislation A law or laws passed by Congress (or state legislatures).

Leopold, Aldo (1887–1949) American scientist, scholar, philosopher, and author. His book *The Land Ethic* argued that humans should view themselves and the land itself as members of the same *community* and that humans are obligated to treat the land ethically.

levee A long raised mound of earth erected along a river bank to protect against floods by holding rising water in the main channel. Synonymous with dike.

life-cycle analysis A quantitative analysis of inputs and outputs across the entire life cycle of a product—from its origins, through its production, transport, sale, and use, and finally its disposal—in an attempt to judge the sustainability of the process and make it more ecologically efficient.

life expectancy The average number of years that individuals in particular age groups are likely to continue to live.

light pollution Pollution from urban or suburban lights that obscures the night sky, impairing people's visibility of stars.

limiting factor A physical, chemical, or biological characteristic of the *environment* that restrains *population* growth.

lipids A class of chemical compounds that do not dissolve in water and are used in organisms for energy storage, for structural support, and as key components of cellular membranes.

lithosphere The outer layer of Earth, consisting of *crust* and uppermost *mantle* and located just above the *asthenosphere*. More generally, the solid part of Earth, including the rocks, *sediment,* and *soil* at the surface and extending down many miles underground. Compare *atmosphere; biosphere; hydrosphere*.

loam *Soil* with a relatively even mixture of *clay-, silt-,* and *sand*-sized particles.

logistic growth The pattern of population growth that results as a *population* at first grows exponentially and then is slowed and finally brought to a standstill at *carrying capacity* by *limiting factors*. This results in an S-shaped curve on a graph. Compare *exponential growth*.

macromolecule A very large molecule, such as a *protein, nucleic acid, carbohydrate,* or *lipid*.

macronutrient Elements and compounds required in relatively large amounts by organisms. Examples include *nitrogen, carbon,* and *phosphorus*.

magma Molten, liquid rock.

malnutrition The condition of impaired health due to an imbalance of *nutrients* in the diet. Includes both *undernutrition* and *overnutrition*.

mangrove A tree with a unique type of roots that curve upward to obtain *oxygen,* which is lacking in the mud in which they grow, or that curve downward to serve as stilts to support the tree in changing water levels. Mangrove forests grow on the coastlines of the tropics and subtropics.

mantle The malleable layer of rock that lies beneath Earth's *crust* and surrounds a mostly iron *core*.

marine protected area (MPA) An area of the ocean set aside to protect marine life from fishing pressures. An MPA may be protected from some human activities but be open to others. Compare *marine reserve*.

marine reserve An area of the ocean designated as a "no-fishing" zone, allowing no extractive activities. Compare *marine protected area*.

market failure The failure of markets to take into account the *environment*'s positive effects on *economies* (e.g., *ecosystem services*) or to reflect the negative effects of economic activity on the environment and thereby on people (*external costs*).

mass extinction event The *extinction* of a large proportion of the world's *species* in a very short time period due to some extreme and rapid change or catastrophic event. Earth has seen five mass extinction events during the past half-billion years.

mass transit A public transportation system for a metropolitan area that moves large numbers of people at once. Buses, trains, subways, streetcars, trolleys, and light rail are types of mass transit.

mass wasting The downslope movement of soil and rock due to gravity. Compare *landslide.*

materials recovery facility (MRF) A *recycling* facility where items are sorted, cleaned, shredded, and prepared for reprocessing into new items.

matter All material in the universe that has mass and occupies space. See *law of conservation of matter.*

maximum sustainable yield The maximal harvest of a particular *renewable natural resource* that can be accomplished while still keeping the resource available for the future.

meltdown The accidental melting of the uranium fuel rods inside the core of a *nuclear reactor,* causing the release of radiation.

meta-analysis A scientific analysis that gathers together results from all scientific studies on a particular research question and statistically analyzes their data for significant patterns or trends that hold across all of them together. An analysis of analyses.

metal A type of chemical *element,* or a mass of such an element, that typically is lustrous, opaque, and malleable and that can conduct heat and electricity. Some *alloys,* such as brass or steel, are also referred to as metals in everyday language.

metamorphic rock One of the three main categories of rock. Formed by great heat and/or pressure that reshapes crystals within the rock and changes its appearance and physical properties. Common metamorphic rocks include marble and slate. Compare *igneous rock; sedimentary rock.*

metapopulation A network of subpopulations, most of whose members stay within their respective landscape *patches,* but some of whom move among patches or mate with members of other patches.

methane hydrate An ice-like solid consisting of molecules of *methane* embedded in a crystal lattice of water molecules. Most is found in sediments on the continental shelves and in the Arctic. Methane hydrate is an unconventional *fossil fuel.*

micronutrient Elements and compounds required in relatively small amounts by organisms. Examples include zinc, copper, and iron.

microplastics Small pieces of plastic, typically less than 5 mm in size.

migration The movements of individuals between populations of organisms.

Milankovitch cycle One of three types of variations in Earth's rotation and orbit around the sun that result in slight changes in the relative amount of solar radiation reaching Earth's surface at different latitudes. As the cycles proceed, they change the way solar radiation is distributed over Earth's surface and contribute to changes in *atmospheric* heating and circulation that have triggered *glaciations* and other changes in *climate.*

mineral A naturally occurring solid *element* or inorganic *compound* with a crystal structure, a defined chemical composition, and distinct physical properties. Compare *ore; rock.*

mining (1) In the broad sense, the extraction of any resource that is nonrenewable on the timescale of our society (such as *fossil fuels* or *groundwater*). (2) In relation to *mineral* resources, the systematic removal of *rock, soil,* or other material for the purpose of extracting minerals of economic interest.

mitigation The pursuit of strategies to lessen the severity of climate change, notably by reducing emissions of greenhouse gases. Compare *adaptation.*

model A simplified representation of a complex natural process, designed by scientists to help understand how the process occurs and to make predictions.

molecule A combination of two or more *atoms.*

monoculture The uniform planting of a single crop over a large area. Characterizes *industrial agriculture.* Compare *polyculture.*

Montreal Protocol International treaty ratified in 1987 in which 180 (now 197) nations agreed to restrict production of *chlorofluorocarbons (CFCs)* in an effort to halt stratospheric *ozone* depletion. It is a protocol of the Vienna Convention for the Protection of the Ozone Layer. The Montreal Protocol is widely considered the most successful effort to date in addressing a global *environmental* problem.

mosaic In *landscape ecology,* a spatial configuration of *patches* arrayed across a landscape.

mountaintop removal mining A large-scale form of *coal* mining in which entire mountaintops are blasted away so as to extract the resource. Although economically efficient, large volumes of rock and soil generally slide downhill, causing extensive impacts on surrounding *ecosystems* and human residents.

Muir, John (1838–1914) Scottish immigrant to the United States who eventually settled in California and made the Yosemite Valley his wilderness home. Today, he is most strongly associated with the *preservation ethic.* Muir argued that nature deserved protection for its own *intrinsic value* (an *ecocentrist* argument) but also claimed that nature facilitated human happiness and fulfillment (an *anthropocentrist* argument).

multiple use A principle guiding management policy for *national forests* specifying that forests be managed for recreation, wildlife habitat, mineral extraction, water quality, and other uses, as well as for timber extraction.

municipal solid waste Nonliquid *waste* that is not especially hazardous and that comes from homes, institutions, and small businesses. Compare *hazardous waste; industrial solid waste.*

mutagen A *toxicant* that causes *mutations* in the *DNA* of organisms.

mutation An accidental change in *DNA* that may range in magnitude from the deletion, substitution, or addition of a single nucleotide to a change affecting entire sets of chromosomes. Mutations provide the raw material for evolutionary change.

mutualism A species interaction in which all participating organisms benefit from their interaction. Compare *parasitism.*

National Environmental Policy Act (NEPA) A U.S. law enacted on January 1, 1970, that created an agency called the Council on Environmental Quality and requires that an *environmental impact statement* be prepared for any major federal action.

national forest An area of forested public land managed by the U.S. Forest Service. The system consists of 191 million acres (more than 8% of the nation's land area) in many tracts spread across all but a few states.

National Forest Management Act *Legislation* passed by the U.S. Congress in 1976 mandating that plans for renewable resource management be drawn up for every *national forest.* These plans were to be explicitly based on the concepts of *multiple use* and *maximum sustainable yield* and be open to broad public participation.

national park A scenic area set aside for recreation and enjoyment by the public and managed by the National Park Service. The U.S. national park system today numbers more than 400 sites totaling 84 million acres and includes national historic sites, national recreation areas, national wild and scenic rivers, and other areas.

national wildlife refuge An area of public land set aside to serve as a haven for wildlife and also sometimes to encourage hunting, fishing, wildlife observation, photography, environmental education, and other uses. The system of more than 560 sites is managed by the U.S. Fish and Wildlife Service.

natural capital Earth's accumulated wealth of *natural resources* and *ecosystem services.*

natural gas A *fossil fuel* consisting primarily of *methane* (CH_4) and including varying amounts of other volatile hydrocarbons.

natural resource Any of the various substances and *energy* sources that we take from our environment and that we need in order to survive.

natural sciences Academic disciplines that study the natural world. Compare *social sciences*.

natural selection The process by which traits that enhance survival and reproduction are passed on more frequently to future generations of organisms than traits that do not, thereby modifying the genetic makeup of populations through time. Natural selection acts on genetic variation and is a primary driver of *evolution*.

negative feedback loop A *feedback loop* in which output of one type acts as input that moves the *system* in the opposite direction. The input and output essentially neutralize each other's effects, stabilizing the system. Compare *positive feedback loop*.

neoclassical economics A mainstream economic school of thought that explains market prices in terms of consumer preferences for units of particular commodities and that uses *cost-benefit analysis*. Compare *classical economics; ecological economics; environmental economics*.

neonicotinoid A new class of chemical *insecticide*. Neonicotinoids are sprayed on plants or used to coat seeds. When seeds are treated, the poison becomes systemic throughout the plant, dispersing through its tissues as it grows and making the tissues toxic to insects.

net energy The quantitative difference between *energy* returned from a process and *energy* invested in the process. Positive net *energy* values mean that a process produces more energy than is invested. Compare *EROI*.

net metering Process by which homeowners or businesses with *photovoltaic* systems or *wind turbines* can sell their excess *solar energy* or *wind power* to their local utility. Whereas *feed-in tariffs* award producers with prices above market rates, net metering offers market-rate prices.

net primary production The *energy* or biomass that remains in an ecosystem after *autotrophs* have metabolized enough for their own maintenance through *cellular respiration*. Net primary production is the energy or biomass available for consumption by *heterotrophs*. Compare *gross primary production; secondary production*.

net primary productivity The rate at which *net primary production* is produced. See *productivity; gross primary production; net primary production; secondary production*.

neurotoxin A *toxicant* that assaults the nervous system. Neurotoxins include heavy metals, *pesticides*, and some chemical weapons developed for use in war.

neutron An electrically neutral (uncharged) particle in the nucleus of an *atom*.

new forestry A set of *ecosystem-based management* approaches for harvesting timber that explicitly mimic natural disturbances. For instance, "sloppy clear-cuts" that leave a variety of trees standing mimic the changes a *forest* might experience if hit by a severe windstorm.

new urbanism An approach among architects, planners, and developers that seeks to design neighborhoods in which homes, businesses, schools, and other amenities are within walking distance of one another so that families can meet most of their needs close to home without the use of a car.

niche The functional role of a *species* in a *community*.

nitrification The conversion by bacteria of ammonium ions (NH_4^+) first into nitrite ions (NO_2^-) and then into nitrate ions (NO_3^-).

nitrogen The chemical *element* with seven *protons* and seven *neutrons*. The most abundant element in the *atmosphere*, a key element in *macromolecules*, and a crucial plant *nutrient*.

nitrogen cycle A major *nutrient cycle* consisting of the routes that *nitrogen atoms* take through the nested networks of environmental *systems*.

nitrogen dioxide (NO_2) A foul-smelling reddish brown gas that contributes to *smog* and *acid deposition*. It results when atmospheric *nitrogen* and *oxygen* react at the high temperatures created by combustion engines. An EPA *criteria pollutant*.

nitrogen fixation The process by which inert *nitrogen* gas combines with *hydrogen* to form ammonium ions (NH_4^+), which are chemically and biologically active and can be taken up by plants.

nitrogen-fixing bacteria Bacteria that live independently in the soil or water, or those that form *mutualistic* relationships with many types of plants and provide *nutrients* to the plants by converting gaseous *nitrogen* to a usable form.

nitrogen oxide (NO_X) One of a family of compounds that includes nitric oxide (NO) and *nitrogen dioxide (NO_2)*.

no-till *Agriculture* that does not involve tilling (plowing, disking, harrowing, or chiseling) the *soil*. The most intensive form of *conservation tillage*.

noise pollution Undesired ambient sound.

nonconsumptive use Use of *fresh water* in which the water from a particular *aquifer* or surface water body either is not removed or is removed only temporarily and then returned. The use of water to generate electricity in hydroelectric *dams* is an example. Compare *consumptive use*.

nongovernmental organization (NGO) An organization not affiliated with any national government, and frequently international in scope, that pursues a particular mission or advocates for a particular cause.

noninfectious disease A disease that develops as a result of the interaction of an individual organism's genes, lifestyle, and environmental exposures, rather than by pathogenic infection. Compare *infectious disease*.

nonmarket value A value that is not usually included in the price of a *good* or *service*.

non-point source A diffuse source of *pollutants*, often consisting of many small sources. Compare *point source*.

nonrenewable natural resources *Natural resources* that are in limited supply and are formed much more slowly than we use them. Compare *renewable natural resources*.

North Atlantic Deep Water (NADW) The deep portion of the *thermohaline circulation* in the northern Atlantic Ocean.

not-in-my-backyard (NIMBY) Syndrome in which people do not want something (e.g., a polluting facility) near where they live, even if they may want or need the thing to exist somewhere.

novel community An ecological *community* composed of a novel mixture of organisms, with no current analog or historical precedent. Also known as a no-analog community.

nuclear energy The *energy* that holds together *protons* and *neutrons* within the nucleus of an *atom*. Several processes, each of which involves transforming *isotopes* of one *element* into isotopes of other elements, can convert nuclear energy into thermal energy, which is then used to generate *electricity*. See also *nuclear fission; nuclear power; nuclear reactor*.

nuclear fission The conversion of the *energy* within an *atom*'s nucleus to usable thermal energy by splitting apart atomic nuclei. Compare *nuclear fusion*.

nuclear fusion The conversion of the *energy* within an *atom*'s nucleus to usable thermal energy by forcing together the small nuclei of lightweight *elements* under extremely high temperature and pressure. Developing a commercially viable method of nuclear fusion remains an elusive goal.

nuclear power The use of *nuclear energy* to generate *electricity*, accomplished using *nuclear fission* within *nuclear reactors* in power plants.

nuclear reactor A facility within a nuclear power plant that initiates and controls the process of *nuclear fission* to generate *electricity*.

nucleic acid A *macromolecule* that directs the production of *proteins*. Includes *DNA* and *RNA*.

nutrient An *element* or *compound* that organisms consume and require for survival.

nutrient cycle The comprehensive set of cyclical pathways by which a given *nutrient* moves through the *environment*.

ocean acidification The process by which today's oceans are becoming more *acidic* (attaining lower *pH*) as a result of increased *carbon dioxide* concentrations in the atmosphere. Ocean acidification occurs as ocean water absorbs carbon dioxide from the air and forms carbonic acid. This process impairs the ability of corals and other organisms to build exoskeletons of calcium carbonate, imperiling coral reefs and the many organisms that depend on them.

ocean thermal energy conversion (OTEC) An *energy* source (not yet commercially used) that involves harnessing the solar radiation absorbed by tropical ocean water by strategically manipulating the movement of warm surface water and cold deep water.

oil A *fossil fuel* produced by the slow underground conversion of *organic compounds* by heat and pressure. Oil is a mixture of hundreds of different types of *hydrocarbon* molecules characterized by *carbon* chains of different lengths. Compare *crude oil; petroleum.*

oil sands *Fossil fuel* deposits that can be mined from the ground, consisting of moist sand and clay containing 1–20% bitumen. Oil sands represent *crude oil* deposits that have been degraded and chemically altered by water erosion and bacterial decomposition. Also called *tar sands.*

oil shale *Sedimentary rock* filled with kerogen that can be processed to produce liquid *petroleum.* Oil shale is formed by the same processes that form *crude oil* but occurs when kerogen was not buried deeply enough or subjected to enough heat and pressure to form oil.

oligotrophic Term describing a water body that has low-nutrient and high-oxygen conditions. Compare *eutrophic.*

open pit mining A *mining* technique that involves digging a gigantic hole and removing the desired *ore* along with waste *rock* that surrounds the ore.

ore A *mineral* or grouping of minerals from which we extract *metals.*

organic agriculture *Agriculture* that uses no synthetic *fertilizers* or *pesticides* but instead relies on biological approaches such as *composting* and *biological control.*

organic compound A *compound* made up of *carbon atoms* (and, generally, *hydrogen atoms*) joined by covalent bonds and sometimes including other *elements,* such as *nitrogen, oxygen,* sulfur, or *phosphorus.* The unusual ability of carbon to build elaborate *molecules* has resulted in millions of different organic compounds showing various degrees of complexity.

organic fertilizer A *fertilizer* made up of natural materials (largely the remains or wastes of organisms), such as animal manure, crop residues, charcoal, fresh vegetation, and *compost.* Compare *inorganic fertilizer.*

outdoor air pollution *Air pollution* that occurs outdoors. Also called ambient air pollution.

overgrazing The consumption by too many animals of plant cover, impeding plant regrowth and the replacement of *biomass.* Overgrazing can worsen damage to *soils,* natural *communities,* and the land's productivity for further grazing.

overnutrition A condition of excessive food intake in which people receive more than their daily needs of calories, proteins, vitamins, minerals, or carbohydrates. Compare *undernutrition; malnutrition.*

overshoot The amount by which humanity's resource use, as measured by its *ecological footprint,* has surpassed Earth's long-term capacity to support human life.

oxygen The chemical *element* with eight *protons* and eight *neutrons.* A key *element* in the *atmosphere* that is produced by *photosynthesis.*

ozone-depleting substance One of a number of airborne chemicals, such as *halocarbons,* that destroy *ozone* molecules and thin the *ozone layer* in the *stratosphere.*

ozone hole Term popularly used to describe the thinning of the stratospheric *ozone layer* that has been occurring over Antarctica each year as a result of *chlorofluorocarbons (CFCs)* and other *ozone-depleting substances.*

ozone layer A portion of the *stratosphere,* roughly 15–35 km (9–22 mi) above sea level, that contains most of the *ozone* in the *atmosphere.*

paradigm A dominant philosophical and theoretical framework within a scientific discipline.

parasitism A species interaction in which one organism, the parasite, depends on another, the host, for nourishment or some other benefit while simultaneously doing the host harm. Compare *mutualism; predation.*

parent material The base geologic material in a particular location.

Paris Agreement An international agreement reached in 2015 that called for limiting global temperature rise to 2.0°C (3.6°F) above pre-industrial levels and attempting to keep the rise below 1.5°C (2.7°F). Each nation proposed its own strategies for reducing its *greenhouse gas* emissions. An outgrowth of the *United Nations Framework Convention on Climate Change.*

particulate matter Solid or liquid particles small enough to be suspended in the *atmosphere* and able to damage respiratory tissues when inhaled. Includes *primary pollutants,*

such as dust and soot, as well as *secondary pollutants,* such as sulfates and nitrates. An EPA *criteria pollutant.*

passive solar energy collection An approach in which buildings are designed and building materials are chosen to maximize direct absorption of sunlight in winter and to keep the interior cool in the summer. Compare *active solar energy collection.*

patch In *landscape ecology,* spatial areas within a landscape. Depending on a researcher's perspective, patches may consist of habitat for a particular organism, or communities, or ecosystems. An array of patches forms a *mosaic.*

pathogen A parasite that causes disease in its host.

pathway inhibitor A *toxicant* that interrupts vital biochemical processes in organisms by blocking one or more steps in important biochemical pathways. Compounds in the *herbicide* atrazine kill plants by blocking key steps in the process of *photosynthesis.*

peak oil Term used to describe the point of maximum production of *petroleum* in the world (or for a given nation), after which oil production declines. It is expected to be roughly the midway point of extraction of the world's oil supplies.

peer review The process by which a scientific manuscript submitted for publication in an academic journal is examined by specialists in the field, who provide comments and criticism (generally anonymously) and judge whether the work merits publication in the journal.

pelagic Of, relating to, or living between the surface and floor of the ocean. Compare *benthic.*

pest A pejorative term for any organism that damages crops that we value. The term is subjective and defined by our own economic interests and is not biologically meaningful. Compare *weed.*

pesticide A substance (generally a synthetic chemical) used to kill insects (called an *insecticide*), plants (called an *herbicide*), or fungi (called a fungicide).

petroleum See *oil.* However, the term is also used to refer to both *oil* and *natural gas* together.

pH A measure of the concentration of *hydrogen ions* in a *solution.* The pH scale ranges from 0 to 14: A solution with a pH of 7 is neutral; solutions with a pH below 7 are *acidic,* and those with a pH higher than 7 are *basic.* Because the pH scale is logarithmic, each step on the scale represents a 10-fold difference in hydrogen ion concentration.

phosphorus cycle A major *nutrient cycle* consisting of the routes that *phosphorus atoms* take through the nested networks of environmental *systems.*

photic zone In the ocean or a freshwater body, the well-lit top layer of water where *photosynthesis* occurs.

photochemical smog "Brown-air" *smog* formed by light-driven reactions of *primary pollutants* with normal atmospheric *compounds* that produce a mix of more than 100 different chemicals, *tropospheric ozone* often being the most abundant among them. Compare *industrial smog*.

photosynthesis The process by which *autotrophs* produce their own food. Sunlight powers a series of chemical reactions that convert *carbon dioxide* and water into sugar (glucose), thus transforming low-quality *energy* from the sun into high-quality energy the organism can use. Compare *cellular respiration*.

photovoltaic cell A device designed to collect sunlight and directly convert it to electrical *energy*. Light striking one of a pair of metal plates in the cell causes the release of *electrons*, which are attracted by electrostatic forces to the opposing plate. The flow of electrons from one plate to the other creates an electrical current. The basis of PV solar technology.

phthalates A class of endocrine-disrupting chemicals found in fire retardants and plasticizers.

phylogenetic tree A treelike diagram that represents the history of divergence of *species* or other taxonomic groups of organisms.

physical hazard Physical processes that occur naturally in our environment and pose human health hazards. These include discrete events such as *earthquakes, volcanic eruptions,* fires, floods, blizzards, *landslides,* hurricanes, and droughts, as well as ongoing natural phenomena such as ultraviolet radiation from sunlight. Compare *biological hazard; chemical hazard; cultural hazard*.

Pinchot, Gifford (1865–1946) The first professionally trained American forester, Pinchot helped establish the U.S. Forest Service. Today, he is the person most closely associated with the *conservation ethic*.

pioneer species A *species* that arrives earliest following an ecological *disturbance,* beginning the ecological process of *succession* in a terrestrial or aquatic *community*.

placer mining A *mining* technique that involves sifting through material in modern or ancient riverbed deposits, generally using running water to separate lightweight mud and gravel from heavier *minerals* of value.

plate tectonics The process by which Earth's surface is shaped by the extremely slow movement of tectonic plates, or sections of *crust*. Earth's surface includes about 15 major tectonic plates. Their interaction gives rise to processes that build mountains, cause *earthquakes,* and otherwise influence the landscape.

poaching The illegal killing of wildlife, usually for meat or body parts.

point source A specific spot—such as a factory—where large quantities of *air pollutants* or *water pollutants* are discharged. Compare *non-point source*.

policy A formal set of general plans, principles, rules, or guidelines that are intended to guide decision making and that may direct individual, organizational, or societal behavior.

pollination The process by which male sex cells of a flowering plant (pollen) fertilize female sex cells of a flowering plant (ova, or egg cells). This may occur by wind, or it may result from a plant-animal interaction in which one organism (e.g., a bee or a hummingbird) transfers pollen from flower to flower.

polluter-pays principle Principle specifying that the party responsible for producing *pollution* should pay the costs of cleaning up the pollution or mitigating its impacts.

polybrominated diphenyl ethers (PBDEs) Synthetic compounds that provide fire-retardant properties and are used in a diverse array of consumer products, including computers, televisions, plastics, and furniture. Released during production, disposal, and use of products, these chemicals persist and accumulate in living tissue and appear to be *endocrine-disrupting chemicals*.

polyculture The planting of multiple crops in a mixed arrangement or in close proximity. An example is some traditional Native American farming that mixed maize, beans, squash, and peppers. Compare *monoculture*.

polymer A chemical *compound* or mixture of compounds consisting of long chains of repeated *molecules*. Important biological molecules, such as *DNA* and *proteins,* are examples of polymers.

population A group of organisms of the same *species* that live in the same area at the same time. Species are often composed of multiple populations.

population density The number of individuals within a *population* per unit area. Compare *population size*.

population distribution The spatial distribution of organisms in an area. Three common patterns are random, uniform, and clumped.

population ecology The scientific study of the quantitative dynamics of population change and the factors that affect the distribution and abundance of members of a *population*.

population growth rate The rate of change in a *population*'s size per unit time (generally expressed in percent per year), taking into accounts births, deaths, immigration, and emigration. Compare *rate of natural increase*.

population size The number of individual organisms present at a given time in a *population*.

positive feedback loop A *feedback loop* in which output of one type acts as input that moves the *system* in the same direction. The input and output drive the system further toward one extreme or another. Compare *negative feedback loop*.

post-industrial stage The fourth and final stage of the *demographic transition* model, in which both birth and death rates have fallen to a low level and remain stable there, and *populations* may even decline slightly. Compare *industrial stage; pre-industrial stage; transition stage*.

potential energy *Energy* of position. Compare *kinetic energy*.

precautionary principle The idea that one should not undertake a new action until the ramifications of that action are well understood.

precipitation Water that condenses out of the *atmosphere* and falls to Earth in droplets or crystals.

precision agriculture The use of technology to precisely monitor crop conditions, crop needs, and resource use to maximize production while minimizing waste of resources.

predation A species interaction in which one *species* (the predator) searches for, captures, and ultimately kills its prey. Compare *parasitism*.

prediction A specific statement, generally arising from a *hypothesis,* that can be tested directly and unequivocally.

pre-industrial stage The first stage of the *demographic transition* model, characterized by conditions that defined most of human history. In pre-industrial societies, both death rates and birth rates are high. Compare *industrial stage; post-industrial stage; transitional stage*.

prescribed fire The practice of burning areas of *forest* or grassland under carefully controlled conditions to improve the health of *ecosystems,* return them to a more natural state, reduce fuel loads, and help prevent uncontrolled catastrophic fires.

preservation ethic An ethic holding that we should protect the natural *environment* in a pristine, unaltered state. Compare *conservation ethic*.

primary forest Natural *forest* uncut by people. Compare *secondary forest*.

primary pollutant A hazardous substance, such as soot or *carbon monoxide,* that is emitted into the *troposphere* directly from a source. Compare *secondary pollutant*.

primary production The conversion of solar energy to the energy of chemical bonds in sugars during *photosynthesis,* performed by *autotrophs*. Compare *secondary production*.

primary source A source that presents novel information and stands on its own. Compare *secondary source*.

primary succession A stereotypical series of changes as an ecological *community* develops over time, beginning with a lifeless substrate. In terrestrial *systems,* primary succession begins when a bare expanse of rock, *sand,* or *sediment* becomes newly exposed to the atmosphere and *pioneer species* arrive. Compare *secondary succession.*

primary treatment A stage of *wastewater* treatment in which contaminants are physically removed. Wastewater flows into tanks in which sewage solids, grit, and particulate matter settle to the bottom. Greases and oils float to the surface and can be skimmed off. Compare *secondary treatment.*

probability A quantitative description of the likelihood of a certain outcome.

producer An organism that uses energy from sunlight to produce its own food. Includes green plants, algae, and cyanobacteria.

productivity The rate at which plants convert solar *energy* (sunlight) to *biomass. Ecosystems* whose plants convert solar energy to biomass rapidly are said to have high productivity. See *net primary productivity; gross primary production; net primary production.*

protein A *macromolecule* made up of long chains of amino acids.

proton A positively charged particle in the nucleus of an *atom.*

proven recoverable reserve The amount of a given *fossil fuel* in a deposit that is technologically and economically feasible to remove under current conditions.

proxy indicator A source of indirect evidence that serves as a proxy, or substitute, for direct measurement and that sheds light on past climate. Examples include data from ice cores, sediment cores, tree rings, packrat middens, fossilized foraminifera, and coral reefs.

public policy *Policy* made by governments, including those at the local, state, federal, and international levels; it may consist of *legislation, regulations,* orders, incentives, and practices intended to advance societal well-being. See also *environmental policy.*

public trust doctrine A legal philosophy that holds that natural resources such as air, water, soil, and wildlife should be held in trust for the public and that government should protect them from exploitation by private interests. This doctrine has its roots in ancient Roman law and in the Magna Carta.

pumped storage A technique used to generate *hydroelectric power,* in which water is pumped from a lower reservoir to a higher reservoir when power demand is weak and prices are low. When demand is strong and prices are high, water is allowed to flow downhill through a turbine, generating *electricity.* Compare *run-of-river; storage.*

pycnocline A zone of the ocean beneath the surface in which density increases rapidly with depth.

radiative forcing The amount of change in thermal energy that a factor (such as a *greenhouse gas* or an *aerosol*) causes in influencing Earth's temperature. Positive forcing warms Earth's surface, whereas negative forcing cools it.

radioactive The quality by which some *isotopes* "decay," changing their chemical identity as they shed atomic particles and emit high-energy radiation.

radioisotope A radioactive *isotope* that emits subatomic particles and high-*energy* radiation as it "decays" into progressively lighter isotopes until becoming a stable isotope.

radon A highly *toxic,* radioactive, colorless gas that seeps up from the ground in areas with certain types of bedrock and that can build up inside basements and homes with poor air circulation.

rainshadow effect The process by which arid regions are formed in the "rainshadow" of a mountain range. As moist air is blown over a mountain range, the air releases precipitation as it rises and cools; once the air has passed over the mountains, it is depleted of moisture, creating the arid rainshadow region.

rangeland Land used for grazing livestock. Also called pastureland.

rate of natural increase The rate of change in a *population*'s size resulting from birth and death rates alone, excluding migration. Compare *population growth rate.*

REACH Program of the European Union that shifts the burden of proof for testing chemical safety from national governments to industry and requires that chemical substances produced or imported in amounts of over 1 metric ton per year be registered with a new European Chemicals Agency. *REACH,* which stands for Registration, Evaluation, Authorisation, and Restriction of Chemicals, went into effect in 2007.

realized niche The portion of the *fundamental niche* that is fully realized (used) by a *species.*

rebound effect The phenomenon by which gains in efficiency from better technology are partly offset when people engage in more energy-consuming behavior as a result. This common psychological effect can sometimes reduce conservation and efficiency efforts substantially.

recharge zone An area where water infiltrates Earth's surface and reaches an *aquifer* below.

reclamation The act of restoring a *mining* site to an approximation of its pre-mining condition. To reclaim a site, companies are required to remove all mining structures, replace *overburden,* fill in mine shafts, and replant the area with vegetation.

recovery Waste management strategy composed of *recycling* and *composting.*

recycling The process by which materials are collected and then broken down and reprocessed to manufacture new items.

Red List An updated list of *species* facing high risks of *extinction.* The list is maintained by the International Union for the Conservation of Nature (IUCN).

red tide A *harmful algal bloom* consisting of algae that produce reddish pigments that discolor surface waters.

Reducing Emissions from Deforestation and Forest Degradation A proposed international program, still being developed, to encourage the conservation of *forests* globally for the purpose of reducing *greenhouse gas* emissions to control *climate change.* A key mechanism is the transfer of funds from wealthy nations to poorer forest-rich nations. Initially abbreviated as REDD, the program became known as REDD+ as it expanded in scope.

refining The process of separating *molecules* of the various *hydrocarbons* in *crude oil* into different-sized classes and transforming them into various fuels and other petrochemical products.

regime shift A fundamental shift in the overall character of an ecological community, generally occurring after some extreme *disturbance,* and after which the community may not return to its original state. Also known as a phase shift.

regional planning Planning across regions to maximize efficiency, functionality, and beauty in ways similar to *city planning,* but across broader geographic areas and usually in concert with multiple municipal governments. Compare *city planning; urban planning.*

regulation A specific rule issued by an administrative agency based on the more broadly written statutory law passed by Congress (or state legislatures) and enacted by the president (or governor). Regulations are intended to achieve the objectives of the *legislation* from which they derive.

regulatory taking The deprivation of a property's owner, by means of a law or *regulation,* of most or all economic uses of that property.

relative humidity The ratio of water vapor that air contains at a given temperature to the maximum amount it could contain at that temperature.

relativist An ethicist who maintains that *ethics* do and should vary with social context. Compare *universalist.*

renewable natural resources *Natural resources* that are virtually unlimited or that are replenished by the *environment* over relatively short periods (hours to weeks to years). Compare *nonrenewable natural resources*.

replacement fertility The *total fertility rate (TFR)* that maintains a stable *population* size.

reproductive window The portion of a woman's life between sexual maturity and menopause during which she may become pregnant.

reserves-to-production ratio The total remaining reserves of a *fossil fuel* divided by the annual rate of production (extraction and processing). Abbreviated as R/P ratio.

reservoir (1) An artificial water body behind a dam that stores water for human use. (2) A location in which nutrients in a *biogeochemical cycle* remain for a period of time before moving to another reservoir. Can be living or nonliving entities. Compare *flux; residence time*.

residence time (1) In a *biogeochemical cycle*, the amount of time a nutrient typically remains in a given *reservoir* before moving to another. Compare *flux*. (2) In the *atmosphere*, the amount of time a gas molecule or a *pollutant* typically remains aloft.

resilience The ability of an ecological *community* to change in response to disturbance but later return to its original state. Compare *resistance*.

resistance The ability of an ecological *community* to remain stable in the presence of a disturbance. Compare *resilience*.

Resource Conservation and Recovery Act U.S. law (enacted in 1976 and amended in 1984) that specifies, among other things, how to manage *sanitary landfills* to protect against environmental contamination. Often abbreviated as RCRA.

resource management Strategic decision making about how to extract resources so that resources are used wisely and conserved for the future.

resource partitioning The process by which *species* adapt to *competition* by evolving to use slightly different resources, or to use shared resources in different ways, thus minimizing interference with one another.

response The type or magnitude of negative effects an animal exhibits in response to a *dose* of *toxicant* in a *dose-response analysis*. Compare *dose*.

restoration ecology The study of both the historical conditions of ecological *communities* as they existed before humans altered them and the restoration of those same ecological communities to such earlier conditions. Principles of restoration ecology are applied in the practice of *ecological restoration*.

revenue-neutral carbon tax A type of *fee-and-dividend* program in which funds from the *carbon tax* that a government collects are disbursed to citizens in the form of payments or tax refunds. It is "revenue-neutral" because the government neither gains nor loses revenue in the end.

revolving door The movement of individuals working in the private sector to government agencies that regulate that particular portion of the private sector, or vice versa. May create conflicts of interest.

ribonucleic acid (RNA) A usually single-stranded *nucleic acid* composed of four nucleotides, each of which contains a sugar (ribose), a phosphate group, and a nitrogenous base. RNA carries the hereditary information for living organisms and is responsible for passing traits from parents to offspring. Compare *deoxyribonucleic acid (DNA)*.

risk The mathematical probability that some harmful outcome (for instance, injury, death, *environmental* damage, or *economic* loss) will result from a given action, event, or substance.

risk assessment The quantitative measurement of *risk*, together with the comparison of risks involved in different activities or substances.

risk management The process of considering information from scientific *risk assessment* in light of economic, social, and political needs and values, to make decisions and design strategies to minimize *risk*.

rock A solid aggregation of *minerals*.

rock cycle The very slow process in which *rocks* and the *minerals* that make them up are heated, melted, cooled, broken, and reassembled, forming *igneous, sedimentary,* and *metamorphic* rocks.

run-of-river Any of several methods used to generate *hydroelectric power* without greatly disrupting the flow of river water. Run-of-river approaches eliminate much of the *environmental* impact of large *dams*. Compare *pumped storage; storage*.

runoff The water from *precipitation* that flows into streams, rivers, lakes, and ponds, and (in many cases) eventually to the ocean.

salinization The buildup of salts in surface *soil* layers.

salt marsh Flat land that is intermittently flooded by the ocean where the *tide* reaches inland. Salt marshes occur along temperate coastlines and are thickly vegetated with grasses, rushes, shrubs, and other herbaceous plants.

salvage logging The removal of dead trees following a natural disturbance. Although it may be economically beneficial, salvage logging can be ecologically destructive because *snags* provide food and shelter for wildlife and because removing timber from recently burned land can cause *erosion* and damage to *soil*.

sand *Sediment* consisting of particles 0.005–2.0 mm in diameter. Compare *clay; silt*.

sanitary landfill A site at which solid waste is buried in the ground or piled up in large mounds for disposal, designed to prevent the waste from contaminating the *environment*. Compare *incineration*.

savanna A *biome* characterized by grassland interspersed with clusters of acacias and other trees. Savanna is found across parts of Africa (where it was the ancestral home of our *species*), South America, Australia, India, and other dry tropical regions.

science (1) A systematic process for learning about the world and testing our understanding of it. (2) The accumulated body of knowledge that arises from this dynamic process.

scientific method A formalized method for testing ideas with observations that involves a more-or-less consistent series of interrelated steps.

scrubber Technology to chemically treat gases produced in combustion so as to reduce smokestack emissions. These devices typically remove hazardous components and neutralize acidic gases, such as *sulfur dioxide* and hydrochloric acid, turning them into water and salt.

second law of thermodynamics The physical law stating that the nature of *energy* tends to change from a more-ordered state to a less-ordered state; that is, *entropy* increases.

secondary forest *Forest* that has grown back after *primary forest* has been cut. Consists of second-growth trees.

secondary pollutant A hazardous substance produced through the reaction of primary pollutants with one another or with other constituents of the *atmosphere*. Compare *primary pollutant*.

secondary production The total *biomass* that *heterotrophs* generate by consuming *autotrophs*. Compare *primary production*.

secondary source A source that presents information about primary sources. Compare *primary source*.

secondary succession A stereotypical series of changes as an *ecological community* develops over time, beginning when some *disturbance* disrupts or dramatically alters an existing community. Compare *primary succession*.

secondary treatment A stage of *wastewater* treatment in which biological means are used to remove contaminants remaining after *primary treatment*. Wastewater is stirred up in the presence of *aerobic* bacteria, which degrade organic pollutants in the water. The wastewater then passes to another settling tank, where remaining solids drift to the bottom. Compare *primary treatment*.

sediment The eroded remains of *rocks*.

sedimentary rock One of the three main categories of rock. Formed when dissolved *minerals* seep through *sediment* layers and act as a kind of glue, crystallizing and binding together sediment particles. Sandstone and shale are examples of sedimentary rock. Compare *igneous rock; metamorphic rock.*

seed bank A storehouse for samples of seeds representing the world's crop diversity.

seed-tree Timber harvesting approach that leaves small numbers of mature and vigorous seed-producing trees standing so that they can reseed a logged area.

selection system Method of timber harvesting whereby single trees or groups of trees are selectively cut while others are left, creating an *uneven-aged* stand.

septic system A *wastewater* disposal method, common in rural areas, consisting of an underground tank and series of drainpipes. Wastewater runs from the house to the tank, where solids precipitate out. The water proceeds downhill to a drain field of perforated pipes laid horizontally in gravel-filled trenches, where microbes decompose the remaining waste.

sex ratio The proportion of males to females in a *population.*

shale gas Natural gas trapped deep underground in tiny bubbles dispersed throughout formations of shale, a type of sedimentary rock. Shale gas is often extracted by *hydraulic fracturing.*

shale oil A liquid form of petroleum extracted from deposits of *oil shale.*

shelterbelt A row of trees or other tall perennial plants that are planted along the edges of farm fields to break the wind and thereby minimize wind *erosion.*

shelterwood Timber harvesting approach that leaves small numbers of mature trees in place to provide shelter for seedlings as they grow.

sick building syndrome A building-related illness produced by indoor *pollution* in which the specific cause is not identified.

silicon The chemical *element* with 14 *protons* and 14 *neutrons.* An abundant element in rocks in Earth's crust.

silt *Sediment* consisting of particles 0.002–0.005 mm in diameter. Compare *clay; sand.*

sink In a *nutrient cycle,* a *reservoir* that accepts more *nutrients* than it releases.

sinkhole An area where the ground has given way with little warning as a result of subsidence caused by depletion of water from an *aquifer.*

slash-and-burn A mode of agriculture frequently used in the tropics in which natural vegetation is cut and then burned, adding nutrition to the soil, before farming begins. Generally farmers move on to another plot once the soil fertility is depleted.

SLOSS Abbreviation for "single large or several small." The debate over whether it is better to make reserves large in size and few in number or many in number but small in size.

smart growth A *city planning* concept in which a community's growth is managed in ways intended to limit *sprawl* and maintain or improve residents' quality of life.

smelting A process in which *ore* is heated beyond its melting point and combined with other *metals* or chemicals to form metal with desired characteristics. Steel is created by smelting iron ore with carbon, for example.

smog Term popularly used to describe unhealthy mixtures of air *pollutants* that often form over urban and industrial areas as a result of fossil fuel combustion. See *industrial* smog; *photochemical smog.*

snag A dead tree that is still standing. Snags are valuable for wildlife.

social cost of carbon An estimate of the total economic cost of damages resulting from the emission of *carbon dioxide* (from *fossil fuel* burning, *deforestation,* etc.) and resulting *global climate change,* on a per-ton basis. Estimates vary widely, but the U.S. government has recently used an estimate of roughly $40/ton CO_2.

social sciences Academic disciplines that study human interactions and institutions. Compare *natural sciences.*

socially responsible investing Investing in companies that have met criteria for environmental or social sustainability.

soil A complex plant-supporting *system* consisting of disintegrated rock, organic matter, air, water, *nutrients,* and microorganisms.

soil degradation A deterioration of soil quality and decline in soil productivity, resulting primarily from forest removal, cropland agriculture, and *overgrazing* of livestock. Compare *land degradation; desertification.*

soil horizon A distinct layer of *soil.*

soil profile The cross-section of a *soil* as a whole, including all *soil horizons* from the surface to the *bedrock.*

solar energy Energy from the sun. Solar energy is perpetually renewable and may be harnessed in several ways.

solution mining A *mining* technique in which a narrow borehole is drilled deep into the ground to reach a *mineral* deposit, and water, acid, or another liquid is injected down the borehole to *leach* the resource from the surrounding rock and dissolve it in the liquid. The resulting solution is then sucked out, and the desired resource is isolated.

source In a *nutrient cycle,* a *reservoir* that releases more *nutrients* than it accepts.

source reduction The reduction of the amount of material that enters the *waste stream* to avoid the costs of disposal and *recycling,* help conserve resources, minimize *pollution,* and save consumers and businesses money.

specialist A *species* that can survive only in a narrow range of *habitats* or that depends on very specific resources. Compare *generalist.*

speciation The process by which new *species* are generated. In one common mechanism, allopatric speciation, species form in the aftermath of the physical separation of populations over some geographic distance.

species A distinct type of organism. A *population* or group of populations of a particular type of organism whose members uniquely share certain characteristics and can breed freely with one another and produce fertile offspring. Different biologists may have different approaches to diagnosing species boundaries.

species coexistence An outcome of interspecific competition in which no competing *species* fully excludes others and the species continue to live side by side.

species diversity The number and variety of *species* in a given region.

species-area curve A graph showing how number of *species* varies with the geographic area of a landmass or water body. Number of species commonly doubles as area increases 10-fold.

sprawl The unrestrained spread of urban or suburban development outward from a city center and across the landscape. Often specified as growth in which the area of development outpaces *population* growth.

steady-state economy An *economy* that does not grow or shrink but remains stable.

storage Technique used to generate *hydroelectric power,* in which large amounts of water are impounded in a reservoir behind a concrete *dam* and then passed through the dam to turn *turbines* that generate *electricity.* Compare *pumped storage; run-of-river.*

stratosphere The layer of the *atmosphere* above the *troposphere;* it contains the *ozone layer* and extends 11–50 km (7–31 mi) above sea level.

strip mining The use of heavy machinery to remove huge amounts of earth to expose *coal* or *minerals,* which are mined out directly. Compare *subsurface mining.*

subcanopy The middle level of trees in a *forest,* beneath the *canopy.*

subduction The *plate tectonic* process by which denser *crust* slides beneath lighter crust at a *convergent plate boundary.* Often results in *volcanism.*

subsidy A government grant of money or resources to a private entity, intended to support and promote an industry or activity. A tax break is one type of subsidy. Compare *green tax*.

subsurface mining Method of *mining* underground deposits of *coal, minerals,* or fuels, in which shafts are dug deeply into the ground and networks of tunnels are dug or blasted out to follow coal seams. Compare *strip mining*.

suburb A smaller community located at the outskirts of a city.

succession A stereotypical series of changes in the composition and structure of an *ecological community* through time. See *primary succession; secondary succession*.

sulfur dioxide (SO_2) A colorless gas that can result from the combustion of *coal*. In the *atmosphere,* it may react to form sulfur trioxide and sulfuric acid, which may return to Earth in *acid deposition*. An EPA *criteria pollutant*.

Superfund A program administered by the *Environmental Protection Agency* in which experts identify sites polluted with hazardous chemicals, protect *groundwater* near these sites, and clean up the *pollution*. Established by the Comprehensive Environmental Response Compensation and Liability Act (CERCLA) in 1980.

surface impoundment (1) A disposal method for *hazardous waste* or mining waste in which waste in liquid or slurry form is placed into a shallow depression lined with impervious material such as clay and allowed to evaporate, leaving a solid residue on the bottom. (2) The site of such disposal. Compare *deep-well injection*.

surface water Water located atop Earth's surface. Compare *groundwater*.

sustainability A guiding principle of *environmental science* entailing conserving resources, maintaining functional ecological systems, and developing long-term solutions such that Earth can sustain our civilization and all life for the future, allowing our descendants to live at least as well as we have lived.

sustainable agriculture *Agriculture* that can be practiced in the same way and in the same place far into the future. Sustainable agriculture does not deplete *soils* faster than they form, nor does it reduce the clean water, genetic diversity, pollinators, and other resources essential to long-term crop and livestock production.

sustainable development Development that satisfies our current needs without compromising the future availability of *natural capital* or our future quality of life.

Sustainable Development Goals A set of 17 goals for *sustainable development* for humanity globally, agreed to by representatives of the world's nations at the United Nations in 2015.

sustainable forest certification A form of *ecolabeling* that identifies timber products that have been produced using *sustainable* methods. The Forest Stewardship Council (FSC) and several other organizations issue such certification.

symbiosis A relationship between different *species* of organisms that live in close physical proximity. People most often use the term "symbiosis" when referring to a *mutualism,* but symbiotic relationships can be either parasitic or mutualistic.

synergistic effect An interactive effect (as of *toxicants*) that is more than or different from the simple sum of their constituent effects.

system A network of relationships among a group of parts, elements, or components that interact with and influence one another through the exchange of *energy,* matter, and/or information.

tailings Portions of *ore* left over after *metals* have been extracted in *mining*.

tar sands See *oil sands*.

temperate deciduous forest A *biome* consisting of midlatitude *forests* characterized by broad-leafed trees that lose their leaves each fall and remain dormant during winter. These forests occur in areas where *precipitation* is spread relatively evenly throughout the year: much of Europe, eastern China, and eastern North America.

temperate grassland A *biome* whose vegetation is dominated by grasses and that features more extreme temperature differences between winter and summer and less *precipitation* than *temperate deciduous forests*. Also known as steppe or prairie.

temperate rainforest A *biome* consisting of tall coniferous trees; cooler and less *species*-rich than *tropical rainforest* and milder and wetter than *temperate deciduous forest*.

temperature inversion A departure from the normal temperature distribution in the *atmosphere,* in which a pocket of relatively cold air occurs near the ground, with warmer air above it. The cold air, denser than the air above it, traps *pollutants* near the ground and can thereby cause a buildup of *smog*. Also called a thermal inversion.

teratogen A *toxicant* that causes harm to the unborn, resulting in birth defects.

terracing The cutting of level platforms, sometimes with raised edges, into steep hillsides to contain water from *irrigation* and *precipitation*. Terracing transforms slopes into series of steps like a staircase, enabling farmers to cultivate hilly land while minimizing their loss of *soil* to water *erosion*.

theory A widely accepted, well-tested explanation of one or more cause-and-effect relationships that has been extensively validated by a great amount of research. Compare *hypothesis*.

thermodynamics Study of the relationships between different types of energy.

thermohaline circulation A worldwide system of ocean currents in which warmer, fresher water moves along the surface and colder, saltier water (which is denser) moves deep beneath the surface.

thin-film solar cell A photovoltaic material compressed into an ultra-thin lightweight sheet that may be incorporated into various surfaces to produce photovoltaic solar power.

threatened Likely to become *endangered* soon.

Three Mile Island Nuclear power plant in Pennsylvania that in 1979 experienced a partial *meltdown*. The term is often used to denote the accident itself, the most serious *nuclear reactor* malfunction that the United States has thus far experienced. Compare *Chernobyl; Fukushima Daiichi*.

threshold dose The amount of a *toxicant* at which it begins to affect a *population* of test animals. Compare ED_{50}; LD_{50}.

tidal energy *Energy* harnessed by erecting a *dam* across the outlet of a tidal basin. Water flowing with the incoming or outgoing *tide* through sluices in the dam turns turbines to generate *electricity*.

tide The periodic rise and fall of the ocean's height at a given location, caused by the gravitational pull of the moon and sun.

topsoil That portion of the *soil* that is most nutritive for plants and is thus of the most direct importance to *ecosystems* and to *agriculture*. A *soil horizon* also known as the A horizon.

tornado A type of cyclonic storm in which warm air rises quickly in a funnel, potentially lifting up soil and objects and threatening life and great damage to property.

total fertility rate (TFR) The average number of children born per female member of a *population* during her lifetime.

toxic air pollutant An *air pollutant* known to cause cancer, reproductive defects, or neurological, developmental, immune system, or respiratory problems in humans or to cause substantial ecological harm by affecting the health of nonhuman animals and plants. The *Clean Air Act* identifies 187 toxic air pollutants, ranging from the heavy metal mercury to *volatile organic compounds (VOCs)* such as benzene and methylene chloride.

Toxic Substances Control Act (TSCA) A 1976 U.S. law that directs the *Environmental*

Protection Agency to monitor thousands of industrial chemicals and gives the EPA authority to regulate and ban substances found to pose excessive risk.

toxicant A substance that acts as a poison to humans or wildlife.

toxicity The degree of harm a chemical substance can inflict.

toxicology The scientific field that examines the effects of poisonous chemicals and other agents on humans and other organisms.

toxin A *toxic* chemical stored or manufactured in the tissues of living organisms. For example, a chemical that plants use to ward off *herbivores* or that insects use to deter predators.

traditional agriculture *Agriculture* in which human and animal muscle power, along with hand tools and simple machines, performs the work of cultivating, harvesting, storing, and distributing crops. Compare *industrial agriculture*.

tragedy of the commons The process by which publicly accessible resources open to unregulated use tend to become damaged and depleted through overuse. The term was coined by Garrett Hardin and is widely applicable to resource issues.

transform plate boundary The area where two tectonic plates meet and slip and grind alongside one another, creating *earthquakes*. For example, the Pacific Plate and the North American Plate rub against each other along California's San Andreas Fault. Compare *convergent plate boundary; divergent plate boundary*.

transgene A *gene* that has been extracted from the *DNA* of one organism and transferred into the DNA of an organism of another *species*.

transgenic Term describing an organism that contains a *transgene*, or *DNA* from another *species*.

transit-oriented development A development approach in which compact communities in the *new urbanism* style are arrayed around stops on a major rail transit line.

transitional stage The second stage of the *demographic transition* model, which occurs during the transition from the *pre-industrial stage* to the *industrial stage*. It is characterized by declining death rates but continued high birth rates. See also *post-industrial stage*. Compare *industrial stage; post-industrial stage; pre-industrial stage*.

transpiration The release of water vapor by plants through their leaves.

treatment The portion of an *experiment* in which a *variable* has been manipulated in order to test its effect. Compare *control*.

triple bottom line An approach to sustainability that attempts to meet environmental, economic, and social goals simultaneously.

trophic cascade A series of changes in the *population* sizes of organisms at different *trophic levels* in a *food chain*, occurring, for example, when predators at high trophic levels indirectly promote populations of organisms at low trophic levels by keeping species at intermediate trophic levels in check. Trophic cascades may become apparent when a top predator is eliminated from a system.

trophic level Rank in the feeding hierarchy of a *food chain*. Organisms at higher trophic levels consume those at lower trophic levels.

tropical dry forest A *biome* that consists of deciduous trees and occurs at tropical and subtropical latitudes where wet and dry seasons each span about half the year. Widespread in India, Africa, South America, and northern Australia. Also known as *tropical deciduous forest*.

tropical rainforest A *biome* characterized by year-round rain and uniformly warm temperatures. Tropical rainforests have dark, damp interiors; lush vegetation; and highly diverse biotic communities. Found in Central America, South America, Southeast Asia, west Africa, and other tropical regions.

troposphere The bottommost layer of the *atmosphere;* it extends to 11 km (7 mi) above sea level.

tropospheric ozone *Ozone* that occurs in the *troposphere*, where it is a *secondary pollutant* created by the interaction of sunlight, heat, *nitrogen oxides,* and volatile *carbon*-containing chemicals. A major component of *photochemical smog*, it can injure living tissues and cause respiratory problems. An EPA *criteria pollutant*. Also called ground-level ozone.

tsunami An immense swell, or wave, of ocean water triggered by an *earthquake, volcano,* or *landslide* that can travel long distances across oceans and inundate coasts.

tundra A *biome* that is nearly as dry as *desert* but is located at very high latitudes along the northern edges of Russia, Canada, and Scandinavia. Extremely cold winters with little daylight and moderately cool summers with lengthy days characterize this landscape of lichens and low, scrubby vegetation.

unconfined aquifer A water-bearing, porous layer of rock, *sand,* or gravel that lies atop a less permeable substrate. The water in an unconfined aquifer is not under pressure because there is no impermeable upper layer to confine it. Compare *confined aquifer*.

undernutrition A condition of insufficient nutrition in which people receive fewer calories, proteins, vitamins, minerals, or carbohydrates than are needed on a daily basis for a healthy diet. Compare *overnutrition; malnutrition*.

understory The layer of a *forest* consisting of small shrubs and trees above the forest floor and below the *subcanopy,* usually shaded by foliage above it.

uneven-aged Term describing stands consisting of trees of different ages. Uneven-aged stands more closely approximate a natural *forest* than do *even-aged* stands.

United Nations (UN) Organization founded in 1945 to promote international peace and to cooperate in solving international economic, social, cultural, and humanitarian problems.

United Nations Framework Convention on Climate Change An international treaty signed in 1992 outlining a plan to reduce emissions of *greenhouse gases*. Gave rise to the *Kyoto Protocol*.

universalist An ethicist who maintains that there exist objective notions of right and wrong that hold across cultures and situations. Compare *relativist*.

upwelling In the ocean, the flow of cold, deep water toward the surface. Upwelling occurs in areas where surface *currents* diverge. Compare *downwelling*.

uranium The chemical *element* with 92 *protons* and 92 *neutrons*. Uranium is used as a fuel source to produce electricity with *nuclear energy*.

urban ecology A scientific field of study that views cities explicitly as *ecosystems*. Researchers in this field apply the fundamentals of *ecosystem ecology* and *systems* science to urban areas.

urban growth boundary A line on a map established to separate areas zoned to be high density and urban from areas intended to remain low density and rural. The aim is to control *sprawl*, revitalize cities, and preserve the rural character of outlying areas.

urban heat island effect The phenomenon whereby a city becomes warmer than outlying areas because of the concentration of heat-generating buildings, vehicles, and people and because buildings and dark paved surfaces absorb heat and release it at night.

urban planning See *city planning*. Compare *regional planning*.

urbanization A population's shift from rural living to city and suburban living.

variable In an *experiment*, a condition that can change. See *dependent variable* and *independent variable*.

vector An organism that transfers a *pathogen* to its host. An example is a mosquito that transfers the malaria pathogen to humans.

volatile organic compound (VOC) One of a large group of potentially harmful organic chemicals used in industrial processes. One of several major *air pollutants* whose emissions are monitored by the *Environmental Protection Agency* and state agencies.

volcano A site where molten rock, hot gas, or ash erupts through Earth's surface, often creating a mountain over time as cooled *lava* accumulates.

Wallace, Alfred Russel (1823–1913) English naturalist who proposed, independently of *Charles Darwin,* the concept of *natural selection* as a mechanism for *evolution* and as a way to explain the great variety of living things.

waste Any unwanted material or substance that results from a human activity or process.

waste management Strategic decision making to minimize the amount of *waste* generated, to maximize the *recovery* of discarded materials, and to dispose of waste safely and effectively.

waste stream The flow of *waste* as it moves from its sources toward disposal destinations.

waste-to-energy facility An incinerator that uses heat from its furnace to boil water to create steam that drives *electricity* generation or that fuels heating systems.

wastewater Any water that is used in households, businesses, industries, or public facilities and is drained or flushed down pipes, as well as the polluted *runoff* from streets and storm drains.

water A *compound* composed of two hydrogen atoms bonded to one oxygen atom, denoted by the chemical formula H_2O.

water pollution The release of matter or *energy* into waters that causes undesirable impacts on the health and well-being of humans or other organisms. Water pollution can be physical, chemical, or biological.

water table The upper limit of *groundwater* held in an *aquifer*.

waterlogging The saturation of *soil* by water, in which the *water table* is raised to the point that water bathes plant roots. Waterlogging deprives roots of access to gases, essentially suffocating them and damaging or killing the plants.

watershed The entire area of land from which water drains into a given river. Also called *drainage basin.*

wave energy *Energy* harnessed from the motion of ocean waves. Many designs for machinery to harness wave energy have been invented, but so far, few are operating commercially.

weather The local physical properties of the *troposphere,* such as temperature, pressure, humidity, cloudiness, and wind, over relatively short time periods (typically minutes, hours, days, or weeks). Compare *climate.*

weathering The process by which rocks and *minerals* are broken down, turning large particles into smaller particles. Weathering may proceed by physical, chemical, or biological means.

weed A pejorative term for any plant that competes with our crops. The term is subjective and defined by our own economic interests and is not biologically meaningful. Compare *pest.*

wetland A system in which the soil is saturated with water and that generally features shallow standing water with ample vegetation. These biologically productive systems include freshwater marshes, swamps, bogs, and seasonal wetlands such as vernal pools.

wilderness area Federal land that is designated off-limits to development of any kind but is open to public recreation, such as hiking, nature study, and other activities that have minimal impact on the land.

wildland-urban interface A region where urban or suburban development meets forested or undeveloped lands.

wind farm A development involving a group of *wind turbines*.

wind power A source of *renewable energy,* in which *kinetic energy* from the passage of wind through *wind turbines* is used to generate *electricity*.

wind turbine A mechanical assembly that converts the wind's *kinetic energy,* or energy of motion, into electrical energy.

work When a force acts on an object, causing it to be displaced (to move in space).

World Bank Institution founded in 1944 that serves as one of the globe's largest sources of funding for economic development, including such major projects as *dams, irrigation* infrastructure, and other undertakings.

world heritage site A location internationally designated by the *United Nations* for its cultural or natural value. There are more than 1000 such sites worldwide.

World Trade Organization (WTO) Organization based in Geneva, Switzerland, that represents multinational corporations and promotes free trade by reducing obstacles to international commerce and enforcing fairness among nations in trading practices.

worldview A way of looking at the world that reflects a person's (or a group's) beliefs about the meaning, purpose, operation, and essence of the world.

xeriscaping Landscaping using plants that are adapted to arid conditions.

zoning The practice of classifying areas for different types of development and land use.

zooxanthellae Symbiotic algae that inhabit the bodies of *corals* and produce food through *photosynthesis.*

Selected Sources and References for Further Reading

Chapter 1

Bahn, Paul, and John Flenley. 1992. *Easter Island, Earth island*. Thames and Hudson, London.

Blakeslee, Sandra, 2010. *Evaluating information—Applying the CRAAP test*. Meriam Library, California State Library, Chico

Blomqvist, L., et al. 2013. Does the shoe fit? Real versus imagined ecological footprints. *PLOS Biology* 11:1–6.

Brown, Lester R. 2009. *Plan B 4.0: Mobilizing to save civilization*. Earth Policy Institute and W.W. Norton, New York.

Brown, Lester R. 2011. *World on the edge: How to prevent environmental and economic collapse*. Earth Policy Institute and W.W. Norton, New York.

Campus Ecology Resource Center. National Wildlife Federation. www.nwf.org/EcoLeaders.

Diamond, Jared. 2005. *Collapse: How societies choose to fail or succeed*. Viking, New York.

Flenley, John, and Paul Bahn. 2003. *The enigmas of Easter Island*. Oxford University Press.

Global Footprint Network. www.footprintnetwork.org. Oakland, Calif.

Harrison, Paul, and Fred Pearce, eds. 2000. *AAAS atlas of population & environment*. University of California Press, Berkeley.

Hunt, Terry L., and Carl P. Lipo. 2006. Late colonization of Easter Island. *Science* 311: 1603–1606.

Hunt, Terry L., and Carl P. Lipo. 2011. *The statues that walked: Unraveling the mystery of Easter Island*. Simon & Schuster.

Kuhn, Thomas S. 1962. *The structure of scientific revolutions*, 2nd ed., 1970. University of Chicago Press, Chicago.

Lomborg, Bjorn. 2001. *The skeptical environmentalist: Measuring the real state of the world*. Cambridge University Press.

Millennium Ecosystem Assessment. 2005. *Ecosystems and human well-being: General synthesis*. Millennium Ecosystem Assessment and World Resources Institute.

Ponting, Clive. 1991. *A green history of the world: The environment and the collapse of great civilizations*. Penguin Books, New York.

Popper, Karl R. 1959. *The logic of scientific discovery*. Hutchinson, London.

Redman, Charles R. 1999. *Human impact on ancient environments*. University of Arizona Press, Tucson.

Robertson, Margaret. 2014. *Sustainability principles and practice*. Routledge, Abingdon, U.K.

Sagan, Carl. 1997. *The demon-haunted world: Science as a candle in the dark*. Ballantine Books, New York.

TEEB. 2010. *The economics of ecosystems and biodiversity: Mainstreaming the economics of nature: A synthesis of the approach, conclusions, and recommendations of TEEB*. The Economics of Ecosystems and Biodiversity.

U.N. Environment Programme. 2019. *Global environment outlook—GEO-6 : Healthy planet, healthy people*. Cambridge Univ. Press, U.K., and UNEP, Nairobi.

Valiela, Ivan. 2001. *Doing science: Design, analysis, and communication of scientific research*. Oxford University Press.

Van Tilburg, Jo Anne. 1994. *Easter Island: Archaeology, ecology, and culture*. Smithsonian Institution Press, Washington, D.C.

Wackernagel, Mathis, and William Rees. 1996. *Our ecological footprint: Reducing human impact on the Earth*. New Society Publishers, Gabriola Island, British Columbia, Canada.

Wackernagel, Mathis, et al. 2002. Tracking the ecological overshoot of the human economy. *PNAS* 99: 9266–9271.

Worldwatch Institute. 2015. *State of the world 2015: Confronting hidden threats to sustainability*. Worldwatch Institute, Washington, D.C.

WWF. 2018. *Living planet report 2018: Aiming higher*. (Grooten, M. and Almond, R.E.A., eds.) WWF, Gland, Switzerland.

Chapter 2

Berry, R. Stephen. 1991. *Understanding energy: Energy, entropy and thermodynamics for every man*. World Scientific Publishing Co.

Christopherson, Robert W. 2011. *Geosystems: An introduction to physical geography*, 8th ed. Prentice Hall, Upper Saddle River, N.J.

Craig, James R., David J. Vaughan, and Brian J. Skinner. 2010. *Earth resources and the environment*, 4th ed. Benjamin Cummings, San Francisco.

Keller, Edward A., and Nicholas Pinter. 2001. *Active tectonics: Earthquakes, uplift, and landscape*, 2nd ed. Prentice Hall, Upper Saddle River, N.J.

Keller, Edward A. 2011. *Introduction to environmental geology*, 5th ed. Prentice Hall, Upper Saddle River, N.J.

Keller, Edward A., and Robert H. Blodgett. 2011. *Natural hazards: Earth's processes as hazards, disasters, and catastrophes*, 3rd ed. Prentice Hall, Upper Saddle River, N.J.

Keranen, Katie M., et al. 2013. Potentially induced earthquakes in Oklahoma, USA: Links between wastewater injection and the 2011 Mw 5.7 earthquake sequence. *Geology* 41: 699-702.

Lancaster, Mike, et al. 2010. *Green chemistry: An introductory text*. Royal Society of Chemistry, London.

Manahan, Stanley E. 2009. *Environmental chemistry*, 9th ed. Lewis Publishers, CRC Press, Boca Raton, Fla.

McMurry, John E. 2011. *Organic chemistry*, 8th ed. Brooks/Cole, San Francisco.

Montgomery, Carla. 2013. *Environmental geology*, 10th ed. McGraw-Hill, New York.

Rubinstein, J.L., and A.B. Mahani, 2015. Myths and facts on wastewater injection, hydraulic fracturing, enhanced oil recovery, and induced seismicity. *Seismological Research Letters* 86: 1–8.

Skinner, Brian J., and Stephen C. Porter. 2003. *The dynamic earth: An introduction to physical geology*, 5th ed. John Wiley & Sons, Hoboken, N.J.

Tarbuck, Edward J., Frederick K. Lutgens, and Dennis Tasa. 2011. *Earth science*, 13th ed. Prentice Hall, Upper Saddle River, N.J.

Timberlake, Karen C. 2011. *Chemistry: An introduction to general, organic, and biological chemistry*, 11th ed. Pearson Benjamin Cummings, San Francisco.

Tkavc, R., et al., 2018. Prospects for fungal bioremediation of acidic radioactive waste sites: Characterization and genome sequence of *Rhodotorula taiwanensis* MD1149. *Frontiers in Microbiology* 8: 2528.

Van Ness, H.C. 1983. *Understanding thermodynamics*. Dover Publications, Mineola, New York.

Zalasiewicz, Jan, et al. 2008. Are we now living in the Anthropocene? *GSA Today* 18(2) (Feb. 2008): 4–8.

Chapter 3

Alvarez, Luis W., et al. 1980. Extraterrestrial cause for the Cretaceous-Tertiary extinction. *Science* 208: 1095–1108.

Atkinson, Carter T., and Dennis A. LaPointe. 2009. Introduced avian diseases, climate change, and the future of Hawaiian honeycreepers. *J. Avian Med. Surg.* 23: 53–63.

Baldwin, Bruce, and Michael Sanderson. 1998. Age and rate of diversification of the Hawaiian silversword alliance (Compositae). *Proc. Natl. Acad. Sci. USA* 95: 9402–9406.

Begon, Michael, Colin R. Townsend, and John L. Harper. 2006. *Ecology: From individuals to ecosystems*, 4th ed. Wiley-Blackwell Publishing, U.K.

Bromham, Lindell, and Marcel Cardillo. 2019. *Origins of biodiversity: An introduction to macroevolution and macroecology.* Oxford Univ. press, Oxford, U.K.

Camp, Richard J., et al. 2009. *Passerine bird trends at Hakalau Forest National Wildlife Refuge, Hawaii.* Hawaii Cooperative Studies Unit Tech. Rep. HCSU-011. University of Hawaii at Hilo.

Camp, Richard J., et al. 2010. Population trends of forest birds at Hakalau Forest National Wildlife Refuge, Hawaii. *Condor* 112: 196–212.

Camp, Richard J., et al. 2016. Evaluating abundance and trends in a Hawaiian avian community using state-space analysis. *Bird Conservation Intl.* 26: 225–242.

Darwin, Charles. 1859. *The origin of species by means of natural selection.* John Murray, London.

Endler, John A. 1986. *Natural selection in the wild.* Monographs in Population Biology 21, Princeton University Press, Princeton, N.J.

Futuyma, Douglas J., and Kirkpatrick, Mark. 2017. *Evolution.* 4th ed. Sinauer Associates (Oxford Univ. Press), Oxford, U.K.

Herron, Jon C., and Scott Freeman. 2014. *Evolutionary analysis.* 5th ed. Pearson, London, U.K.

Krebs, Charles J. 2016. *Why ecology matters.* University of Chicago Press, Chicago.

Lerner, Heather R.L., et al. 2011. Multilocus resolution of phylogeny and timescale in the extant adaptive radiation of Hawaiian honeycreepers. *Current Biology* 21: 1838–1844.

Lovejoy, Thomas, et al., 2019. *Biodiversity and climate change: Transforming the biosphere.* Yale Univ. Press, New Haven, Conn.

MacLeod, Norman. 2013. *The great extinctions: What causes them and how they shape life.* Firefly Books, Richmond Hill, Ontario.

Olson, Storrs L., and Helen F. James. 1982. Fossil birds from the Hawaiian islands: Evidence for wholesale extinction by man before Western contact. *Science* 217: 633–635.

Paxton, Eben H., et al. 2018. Rapid colonization of a Hawaiian restoration forest by a diverse avian community. *Restoration Ecology* 26: 165–173.

Price, Jonathan P., and David A. Clague. 2002. How old is the Hawaiian biota? Geology and phylogeny suggest recent divergence. *Proc. Roy. Soc. B* 269: 2429–2435.

Raup, David M. 1991. *Extinction: Bad genes or bad luck?* W.W. Norton, New York.

Ricklefs, Robert E., and Dolph Schluter, eds. 1993. *Species diversity in ecological communities.* University of Chicago Press, Chicago.

Scott, J. Michael, et al. 1988. Conservation of Hawaii's vanishing avifauna. *BioScience* 38: 238–252.

Smith, Thomas M., and Robert L. Smith. 2014. *Elements of ecology,* 9th ed. Pearson, London, U.K.

Steadman, David W. 1995. Prehistoric extinctions of Pacific island birds: Biodiversity meets zooarchaeology. *Science* 267: 1123–1131.

Urry, Lisa., et al. 2016. *Campbell Biology,* 11th ed. Pearson, London, U.K.

U.S. Fish and Wildlife Service. 2010. *Hakalau Forest National Wildlife Refuge Comprehensive Conservation Plan.* USFWS, Hilo, Hawaii.

Wagner, Warren L., and Vicki A. Funk. 1995. *Hawaiian biogeography: Evolution on a hot spot archipelago.* Smithsonian Institution Press, Washington, D.C.

Wilson, Edward O. 2002. *The future of life.* Alfred A. Knopf, New York.

Chapter 4

Annin, Peter. 2018. *The Great Lakes water wars.* Island Press, Washington, D.C.

Asian Carp Regional Coordinating Committee. 2019. *2019 Asian carp action plan.* Asian Carp Regional Coordinating Committee.

Baskin, Yvonne. 2002. *A plague of rats and rubbervines: The growing threat of species invasions.* Island Press, Washington, D.C.

Breckle, Siegmar-Walter. 2002. *Walter's vegetation of the Earth: The ecological systems of the geo-biosphere,* 4th ed. Springer-Verlag, Berlin.

Bronstein, Judith L. 1994. Our current understanding of mutualism. *Quarterly Journal of Biology* 69: 31–51.

Clewell, Andre F., and James Aronson. 2013. *Ecological restoration: Principles, values, and structure of an emerging profession,* 2nd ed. Island Press, Washington, D.C.

Connell, Joseph H., and Ralph O. Slatyer, 1977. Mechanisms of succession in natural communities. *American Naturalist* 111: 1119–1144.

Crisafulli, Charles M., and Virginia H. Dale, eds. 2018. *Ecological responses at Mount St. Helens: Revisited 35 years after the 1980 eruption.* Springer, New York.

Egan, Dan. 2017. *The death and life of the Great Lakes.* W.W. Norton, New York.

Estes, James A., et al. 2011. Trophic downgrading of planet Earth. *Science* 333: 301–306.

Hobbs, Richard J., Eric S. Higgs, and Carol M. Hall. 2013. *Novel ecosystems: Intervening in the new ecological world order.* Wiley-Blackwell, West Sussex, U.K.

Marris, Emma. 2013. *The rambunctious garden: Saving nature in a post-wild world.* Bloomsbury, USA, New York.

Menge, Bruce A., et al. 1994. The keystone species concept: Variation in interaction strength in a rocky intertidal habitat. *Ecological Monographs* 64: 249–286.

Mittelbach, Gary G., and Brian J. McGill. 2019. *Community ecology.* 2nd ed. Oxford Univ. Press, Oxford, U.K.

Molles, Manuel, and Anna Sher. 2019. *Ecology: Concepts and applications,* 8th ed. McGraw-Hill, Boston.

Orion, Tao, and David Holmgren. 2015. *Beyond the war on invasive species: A permaculture approach to ecosystem restoration.* Chelsea Green Publishing Co., White River Junction, Vt.

Palmer, Margaret A., et al., eds. 2016. *Foundations of restoration ecology,* 2nd ed. Island Press, Washington, D.C.

Pearce, Fred. 2015. *The new wild: Why invasive species will be nature's salvation.* Beacon Press, Boston.

Phelps, Quinton E., et al., 2017. Incorporating basic and applied approaches to evaluate the effects of invasive Asian carp on native fishes: A necessary first step for integrated pest management. *PLOS ONE* 12(9): e0184081.

Power, Mary E., et al. 1996. Challenges in the quest for keystones. *BioScience* 46: 609–620.

Reeves, Andrew, 2019. *Dispatches from the Asian carp crisis.* ECW Press, Toronto.

Ricklefs, Robert, and Rick Relyea. 2018. *Ecology: The economy of nature,* 8th ed. W.H. Freeman, New York.

Rothlisberger, John D., et al. 2012. Ship-borne nonindigenous species diminish Great Lakes ecosystem services. *Ecosystems* 15: 462–476.

Sass, Greg G., et al. 2014. Invasive bighead and silver carp effects on zooplankton communities in the Illinois River, Illinois, USA. *J. Great Lakes Res.* 40: 911–921.

Sax, Dov F., John J. Stachowicz, and Steven D. Gaines, eds. 2012. *Species invasions: Insights into ecology, evolution, and biogeography.* Sinauer, Sunderland, Mass.

Shea, Katriona, and Peter Chesson. 2002. Community ecology theory as a framework for biological invasions. *Trends in Ecology and Evolutionary Biology* 17: 170–176.

Stern, Charles V., et al. 2014. *Asian carp and the Great Lakes region.* Congressional Research Service, Washington, D.C.

Strayer, David L. 2010. Alien species in fresh waters: Ecological effects, interactions with other stressors, and prospects for the future. *Freshwater Biology* 55(Suppl 1): 152–174.

Thompson, John N. 1999. The evolution of species interactions. *Science* 284: 2116–2118.

USDA Pacific Northwest Research Station. 2010. Mount St. Helens 30 years later: A landscape reconfigured. *Science Update,* Issue #19 (12 pp). U.S. Department of Agriculture, Washington, D.C.

U.S. Fish and Wildlife Service. 2012. The cost of invasive species. USFWS, Washington, D.C.

Van Andel, Jelte, and James Aronson. 2012. *Restoration ecology: The new frontier,* 2nd ed. Wiley-Blackwell, West Sussex, U.K.

Weigel, Marlene, ed. 1999. *Encyclopedia of biomes.* UXL, Farmington Hills, Michigan.

Woodward, Susan L. 2003. *Biomes of Earth: Terrestrial, aquatic, and human-dominated.* Greenwood Publishing, Westport, Conn.

Zhang, Hongyan, et al. 2016. Forecasting the impacts of silver and bighead carp on the Lake Erie food web. *Trans. Am. Fish. Soc.* 145: 136–162.

Chapter 5

Chesapeake Bay Foundation. 2018. *2018 State of the Bay report.* Chesapeake Bay Foundation, Annapolis, MD.

Committee on the Mississippi River and the Clean Water Act. 2009. *Nutrient control actions for improving water quality in the Mississippi River basin and northern Gulf of Mexico.* National Academies Press, Washington, D.C.

Diaz, Robert J., and Rutger Rosenberg. 2008. Spreading dead zones and consequences for marine ecosystems. *Science* 321: 926–929.

Field, Christopher B., et al. 1998. Primary production of the biosphere: Integrating terrestrial and oceanic components. *Science* 281: 237–240.

Gruber, Nicolas, and James N. Galloway. 2008. An Earth-system perspective of the global nitrogen cycle. *Nature* 451: 293–296.

Jacobson, Michael, et al. 2000. *Earth system science from biogeochemical cycles to global changes.* Academic Press.

Jones, C., et al. 2010. *How Nutrient Trading Could Help Restore the Chesapeake Bay.* WRI Working Paper. Washington D.C.: World Resources Institute.

Mississippi River/Gulf of Mexico Watershed Nutrient Task Force. 2015. *Report to Congress.* U.S. EPA, Washington, D.C.

Mitsch, William J., et al. 2001. Reducing nitrogen loading to the Gulf of Mexico from the Mississippi River Basin: Strategies to counter a persistent ecological problem. *BioScience* 51: 373–388.

National Science and Technology Council, Committee on Environment and Natural Resources. 2003. *An assessment of coastal hypoxia and eutrophication in U.S. waters.* National Science and Technology Council, Washington, D.C.

Rabalais, Nancy N., et al. 2002. Beyond science into policy: Gulf of Mexico hypoxia and the Mississippi River. *BioScience* 52: 129–142.

Raloff, Janet. 2004. Dead waters: Massive oxygen-starved zones are developing along the world's coasts. *Science News* 165: 360–362.

Raloff, Janet. 2004. Limiting dead zones: How to curb river pollution and save the Gulf of Mexico. *Science News* 165: 378–380.

Ricklefs, Robert E. 2010. *The economy of nature*, 6th ed. W.H. Freeman and Co., New York.

Schlesinger, William H. 2013. *Biogeochemistry: An analysis of global change*, 3rd ed. Academic Press, London.

Schulte, D.M., et al. 2009. Unprecedented restoration of a native oyster metapopulation. *Science* 325: 1124–1128.

Smith, Thomas M., and Robert L. Smith. 2015. *Elements of ecology*, 9th ed. Pearson Education, San Francisco.

Turner, Monica, et al. 2003. *Landscape ecology in theory and practice: Pattern and process.* Springer.

Vitousek, Peter M., et al. 1997. Human alteration of the global nitrogen cycle: Sources and consequences. *Ecological Applications* 7: 737–750.

Chapter 6

Ammons, Elizabeth, and Modhumita Roy. 2015. *Sharing the Earth: An international environmental justice reader.* Univ. of Georgia Press, Athens, Ga.

Arriagada, Rodrigo A., et al. 2015. Do payments pay off? Evidence from participation in Costa Rica's PES program. *PLOS ONE* 10(7): e0131544.

Attfield, Robin. 2019. *Environmental ethics: A very short introduction.* Oxford Univ. Press, Oxford, U.K.

Balmford, Andrew, et al. 2002. Economic reasons for conserving wild nature. *Science* 297: 950–953.

Barbour, Ian G. 1992. *Ethics in an age of technology.* Harper Collins, San Francisco.

Beddoe, Rachael, et al. 2009. Overcoming systemic roadblocks to sustainability: The evolutionary redesign of worldviews, institutions, and technologies. *PNAS* 106: 2483–2489.

Bell, Jasmine. 2016. 5 things to know about communities of color and environmental justice. Center for American Progress. April 25, 2016.

Blewitt, John. 2014. *Understanding sustainable development*, 2nd ed. Routledge, Abingdon, U.K.

Börner, Jan, et al. 2017. The effectiveness of payments for environmental services. *World Development* 96: 359–374.

Brown, Lester. 2001. *Eco-economy: Building an economy for the Earth.* Earth Policy Institute and W.W. Norton, New York.

Carson, Richard T., et al. 1994. Valuing the preservation of Australia's Kakadu Conservation Zone. *Oxford Economic Papers* 46: 727–749.

Cole, Luke W., and Sheila R. Foster. 2001. *From the ground up: Environmental racism and the rise of the environmental justice movement.* New York University Press.

Costanza, Robert, et al. 1997. The value of the world's ecosystem services and natural capital. *Nature* 387: 253–260.

Costanza, Robert, et al. 2014. Changes in the global value of ecosystem services. *Global Env. Change* 26: 152–158.

Costanza, Robert, et al. 2014. *An introduction to ecological economics*, 2nd ed. CRC Press, Boca Raton, Fla.

Daily, Gretchen, ed. 1997. *Nature's services: Societal dependence on natural ecosystems.* Island Press, Washington, D.C.

Daly, Herman E. 1996. *Beyond growth: The economics of sustainable development.* Beacon Press, Boston.

Daniels, Amy E., et al. 2010. Understanding the impacts of Costa Rica's PES: Are we asking the right questions? *Ecological Economics* 69: 2116–2126.

Esty, Daniel C., and Andrew S. Winston. 2009. *Green to gold: How smart companies use environmental strategy to innovate, create value, and build competitive advantage* (revised and updated ed.). John Wiley & Sons, Hoboken, N.J.

Fondo Nacional de Financiamento Forestal (FONAFIFO). Pago de Servicios Ambientales. https://www.fonafifo.go.cr/es/servicios/pago-de-servicios-ambientales/.

Fox, Stephen. 1985. *The American conservation movement: John Muir and his legacy.* University of Wisconsin Press, Madison.

Francis, Pope. 2015. *Encyclical letter Laudato Si' of the Holy Father Francis on care for our common home.* Vatican Press, The Vatican.

Garrigues, Lisa Gale. 2019. Why Costa Rica tops the Happiness Index. *YES! Magazine* Jan. 31, 2019.

Gaworecki, Mike, and Zuzana Burivalova. 2017. Cash for conservation: Do payments for ecosystem services work? *MongaBay*. https://news.mongabay.com/2017/10/cash-for-conservation-do-payments-for-ecosystem-services-work/.

Global Green Growth Institute. 2016. *Bridging the policy and investment gap for payment for ecosystem services: Learning from Costa Rican experience and roads ahead.* GGGI, Seoul, South Korea.

Goodstein, Eban. 1999. *The tradeoff myth: Fact and fiction about jobs and the environment.* Island Press, Washington, D.C.

Goodstein, Eban, and Stephen Polasky. 2014. *Economics and the environment*, 7th ed. John Wiley & Sons, Hoboken, N.J.

Hawken, Paul. 1994. *The ecology of commerce: A declaration of sustainability.* Harper Business, New York.

Hawken, Paul, Amory Lovins, and L. Hunter Lovins. 1999. *Natural capitalism.* Little, Brown, and Co., Boston.

HM Treasury. 2006. *Stern review on the economics of climate change.* HM Treasury and Cambridge University Press.

Kahneman, Daniel, et al. 2006. Would you be happier if you were richer? A focusing illusion. *Science* 312: 1908–1910.

Keller, David R. 2010. *Environmental ethics: The big questions.* Wiley-Blackwell, Hoboken, N.J.

Keohane, Nathaniel O., and Sheila M. Olmstead. 2016. *Markets and the environment.* 2nd ed. Island Press, Washington, D.C.

Kinzig, Ann P., et al. 2011. Paying for ecosystem services—promise and peril. *Science* 334: 603–604.

Klein, Naomi. 2015. *This changes everything: Capitalism versus the climate.* Simon and Schuster, New York.

Leopold, Aldo. 1949. *A Sand County almanac, and sketches here and there.* Oxford University Press.

Martinich, Jeremy, and Allison Crimmins. 2019. Climate damages and adaptation potential across diverse sectors of the United States. *Nature Climate Change* 9: 397–404.

McCauley, Douglas J. 2006. Selling out on nature. *Nature* 443: 27–28.

Meadows, Donella, Jørgen Randers, and Dennis Meadows. 2004. *Limits to growth: The 30-year update.* Chelsea Green Publishing Co., White River Junction, Vt.

Millennium Ecosystem Assessment. 2005. *Ecosystems and human well-being: Opportunities and challenges for business and industry.* Millennium Ecosystem Assessment and World Resources Institute.

Nash, Roderick F. 1989. *The rights of nature.* University of Wisconsin Press, Madison.

Nash, Roderick F. 1990. *American environmentalism: Readings in conservation history*, 3rd ed. McGraw-Hill, New York.

National Research Council, Board on Sustainable Development. 1999. *Our common journey: A transition toward sustainability.* National Academies Press, Washington, D.C.

Nordhaus, William. 2007. Critical assumptions in the Stern Review on Climate Change. *Science* 317: 201–202.

Overseas Development Institute. 2011. *Costa Rica's sustainable resource management: Successfully tackling tropical deforestation.* ODI Publications, London.

Pagiola, Stefano. 2007. Payments for environmental services in Costa Rica. *Ecological Economics* 65: 712–724.

Phan, Thu-Ha Dang, et al. 2018. Do payments for forest ecosystem services generate double dividends? An integrated impact assessment of Vietnam's PES program. *PLOS ONE* 13(8): e0200881.

Pojman, Louis P., et al. 2016. *Environmental ethics: Readings in theory and application,* 7th ed. Wadsworth, Belmont, Calif.

Porras, Ina, et al. 2014. *Ecosystems for sale: Land prices and payments for ecosystem services in Costa Rica.* International Institute for Environment and Development, London.

Purdy, Jedediah. 2016. Environmentalism was once a social justice movement. *The Atlantic.* Dec. 7, 2016.

Renner, Michael. 2008. *Green jobs: Working for people and the environment.* Worldwatch Report 177. Worldwatch Institute, Washington, D.C.

Ricke, Katherine, et al. 2018. Country-level social cost of carbon. *Nature Climate Change* 8: 895–900.

Ricketts, Taylor, et al. 2004. Economic value of tropical forest to coffee production. *PNAS* 101: 12579–12582.

Sachs, Jeffrey D., and Ban Ki-moon. 2015. *The age of sustainable development.* Columbia University Press, New York.

Salzman, James. 2018. The global status and trends of payments for ecosystem services. *Nature Sustainability* 1: 136–144.

Schmidtz, David, and Dan C. Shara. 2018. *Environmental ethics: What really matters, what really works,* 3rd ed. Oxford University Press, Oxford, U.K.

Singer, Peter, ed. 2011. *Practical ethics.* Cambridge University Press, Cambridge, U.K.

Smith, Adam. 1776. *An inquiry into the nature and causes of the wealth of nations.* 1993 ed., Oxford University Press.

Smith, Stephen. 2011. *Environmental economics: A very short introduction.* Oxford University Press, Oxford, U.K.

Stone, Christopher D. 1972. Should trees have standing? Towards legal rights for natural objects. *Southern California Law Review* 1972: 450–501.

Sze, Julie. 2020. *Environmental justice in a moment of danger.* American Studies Now: Critical Histories of the Present (Book 11). Univ. of California Press, Berkeley, Calif.

Talberth, John, et al. 2007. *The genuine progress indicator 2006: A tool for sustainable development.* Redefining Progress, Oakland, Calif.

TEEB. 2010. *The economics of ecosystems and biodiversity: Mainstreaming the economics of nature: A synthesis of the approach, conclusions, and recommendations of TEEB.* The Economics of Ecosystems and Biodiversity.

TEEB. 2012. *Nature and its role in the transition to a green economy.* The Economics of Ecosystems and Biodiversity.

Thompson, Andrea. 2018. Here's how much climate change could cost the U.S. *Scientific American.* Dec. 3, 2018.

Tercek, Mark R., and Jonathan S. Adams. 2015. *Nature's fortune: How business and society thrive by investing in nature.* Island Press, Washington, D.C.

Tietenberg, Tom, and Lynne Lewis. 2018. *Environmental and natural resource economics* 11th ed. Routledge, Abingdon, U.K.

United Nations. 2012. *Report of the United Nations Conference on Sustainable Development.* Rio de Janeiro, Brazil, 20–22 June 2012. http://www.uncsd2012.org/content/documents/814UNCSD%20REPORT%20final%20revs.pdf

United Nations. *Sustainable Development Goals.* www.sustainabledevelopment.un.org.

Vucetich, John A., et al. 2015. Evaluating whether nature's intrinsic value is an axiom of or anathema to conservation. *Conservation Biology* 29: 321–332.

Walker, Gordon. 2012. *Environmental justice: Concepts, evidence, and politics.* Routledge, New York.

White, Lynn. 1967. The historic roots of our ecologic crisis. *Science* 155: 1203–1207.

World Commission on Environment and Development. 1987. *Our common future.* Oxford University Press.

Worldwatch Institute. 2012. *State of the world 2012: Moving toward sustainable prosperity.* Worldwatch Institute, Washington, D.C.

Wunder, Sven, et al. 2008. A comparative analysis of payments for environmental services programs in developed and developing countries. *Ecological Economics* 65: 834–852.

Wunscher, Tobias, et al. 2008. Spatial targeting of payments for environmental services: A tool for boosting conservation benefits. *Ecological Economics* 65: 822–833.

Chapter 7

Blumm, Michael C., and Mary Christina Wood. 2017. "No ordinary lawsuit": Climate change, due process, and the public trust doctrine. *Amer. Univ. Law Rev.* 67(1): 1–87.

Bogojevic, Sanja. 2013. *Emissions trading schemes: Markets, states, and laws.* Hart Publishing, Oxford, U.K.

Carrington, Damian. 2018. Can climate litigation save the world? *The Guardian.* March 20, 2018.

Coady, David, et al. 2017. How large are global fossil fuel subsidies? *World Development* 91: 11–27.

Council on Environmental Quality. 2014. A citizen's guide to the NEPA: Having your voice heard. Executive Office of the President, Washington, D.C.

Dietz, Thomas, et al. 2003. The struggle to govern the global commons. *Science* 302: 1907–1912.

Environmental Law Institute. 2009. *Estimating U.S. government subsidies to energy sources: 2002–2008.* ELI, Washington, D.C.

Fogleman, Valerie M. 1990. *Guide to the National Environmental Policy Act.* Quorum Books, New York.

Hansjurgens, Bernd, ed. 2005. *Emissions trading for climate policy: US and European perspectives.* Cambridge University Press, Cambridge, U.K.

Hardin, Garrett. 1968. The tragedy of the commons. *Science* 162: 1243–1248.

Harvey, Hal. 2018. *Designing climate solutions: A policy guide for low-carbon energy.* Island Press, Washington, D.C.

Hasemyer, David. 2018. Fossil fuels on trial: Where the major climate change lawsuits stand today. *Inside Climate News.* Sept. 10, 2018.

Houck, Oliver. 2003. Tales from a troubled marriage: Science and law in environmental policy. *Science* 302: 1926–1928.

Intergovernmental Panel on Climate Change. 2013. *Climate change 2013: The physical science basis. Working Group I contribution to the Fifth Assessment Report of the IPCC.* (Stocker, Thomas F., et al., eds.). IPCC, Geneva, and Cambridge University Press, Cambridge.

Klein, Naomi. 2019. *On fire: The (burning) case for a Green New Deal.* Simon and Schuster, New York.

Kubasek, Nancy K., and Gary S. Silverman. 2013. *Environmental law,* 8th ed. Pearson Prentice Hall, Upper Saddle River, N.J.

Leslie, Jacques. 2017. In the face of a Trump environmental rollback, California stands in defiance. *Yale Environment 360.* Feb 21, 2017.

Moran, Jack. 2018. Meet Eugene's young climate warriors. *Associated Press.* Sept. 17, 2018. https://www.apnews.com/f6cc9c5368ba424c88a-795916b685e9e.

Myers, Norman, and Jennifer Kent. 2001. *Perverse subsidies: How misused tax dollars harm the environment and the economy.* Island Press, Washington, D.C.

The National Environmental Policy Act. https://ceq.doe.gov/laws-regulations/laws.html.

Next 10. 2018. *2018 California green innovation index.* 10th ed. www.Next10.org.

Nordhaus, Ted, and Michael Shellenberger. 2007. *Break Through: From the death of environmentalism to the politics of possibility.* Houghton Mifflin, Boston.

OECD. 2012. Inventory of estimated budgetary support and tax expenditures for fossil fuels 2013. OECD Publishing.

Our Children's Trust. https://www.ourchildrenstrust.org.

Percival, Robert V., et al. 2018. *Environmental regulation: Law, science, and policy,* 8th ed. Wolters Kluwer, Alphen aan den Rijn, Netherlands.

Popovich, Nadja, et al. 2019. 84 environmental rules being rolled back under Trump. *New York Times* Aug. 29, 2019.

Roberts, David. 2018. Friendly policies keep U.S. oil and coal afloat far more than we thought. *Vox.* July 26, 2018.

Robinson, Mary. 2018. *Climate justice: Hope, resilience, and the fight for a sustainable future.* Bloomsbury Publishing, London, U.K.

Rosen, Julia. 2018. Scientists take opposing sides in youth climate trial. *Science.* Oct. 24, 2018.

Rosenbaum, Walter. 2019. *Environmental politics and policy*, 11th ed. CQ Press, Congressional Quarterly, Inc., Washington, D.C.

Sabin Center for Climate Change Law, and Arnold and Porter. *Climate Change Litigation Database*. http://climatecasechart.com.

Sayler, Zoe. 2019. How 21 meddling kids could force a climate turnaround. *Grist*. July 22, 2019.

Schwartz, John. 2019. Judges give both sides a grilling in youth climate case against the government. *New York Times* June 4, 2019.

Shellenberger, Michael, and Ted Nordhaus. 2004. *The death of environmentalism: Global warming politics in a post-environmental world*. Presented at the Environmental Grantmakers Association meeting, October 2004.

Tietenberg, Tom H. 2006. *Emissions trading: Principles and practice*. 2nd ed. Resources for the Future, Washington, D.C.

Tietenberg, Tom, and Lynne Lewis. 2018. *Environmental and natural resource economics* 11th ed. Routledge, Abingdon, U.K.

Torres, Gerald. 2017. No ordinary lawsuit: The public trust and the duty to confront climate disruption—commentary on Blumm and Wood. *Amer. Univ. Law Rev.* 67(35): 49–73.

U.S. Environmental Protection Agency. Summary of the National Environmental Policy Act. http://www2.epa.gov/laws-regulations/summary-national-environmental-policy-act.

U.S. Office of Management and Budget, 2017. *2017 draft report to Congress on the benefits and costs of federal regulations and agency compliance with the Unfunded Mandates Reform Act*. Executive Office of the President of the United States, Washington, D.C.

Vig, Norman J., and Michael E. Kraft, eds. 2019. *Environmental policy: New directions for the twenty-first century*, 10th ed. CQ Press, Congressional Quarterly, Inc., Washington, D.C.

Wood, Mary Christina, and Charles W. Woodward IV. 2016. Atmospheric trust litigation and the constitutional right to a healthy climate system: Judicial recognition at last. *Wash. J. Env. Law & Policy* 6: 633–683.

World Bank. *World development indicators*. World Bank, Washington, D.C. http://data.worldbank.org/products/wdi.

Worldwatch Institute. 2014. *State of the world 2014: Governing for sustainability*. Worldwatch Institute, Washington, D.C.

Chapter 8

Ball, Philip. 2008. Where have all the flowers gone? [China's one-child legacy.] *Nature* 454: 374–375.

Cohen, Joel E. 1995. *How many people can the Earth support?* W.W. Norton, New York.

De Souza, Roger-Mark, et al. 2003. Critical links: Population, health, and the environment. *Population Bulletin* 58(3). Population Reference Bureau, Washington, D.C.

Ehrlich, Paul. 1968. *The population bomb*. 1997 reprint, Buccaneer Books, Cutchogue, New York.

Ehrlich, Paul R., and Anne H. Ehrlich. 1990. *The population explosion*. Touchstone, New York.

Engelman, Robert. 2008. *More: Population, nature, and what women want*. Island Press, Washington D.C.

Gribble, James N. and J. Bremner. 2012. Achieving a demographic dividend. *Population Bulletin* 67 (2). Population Reference Bureau, Washington, D.C.

Haberl, Helmut, et al. 2007. Quantifying and mapping the human appropriation of net primary production in earth's terrestrial ecosystems. *Proceedings of the National Academy of Sciences of the United States of America* 104: 12942–12947.

Haub, Carl, and O.P. Sharma. 2006. India's population reality: Reconciling change and tradition. *Population Bulletin* 61(3). Population Reference Bureau, Washington, D.C.

Holdren, John P., and Paul R. Ehrlich. 1974. Human population and the global environment. *American Scientist* 62: 282–292.

Imhoff, Marc L., et al. 2004. Global patterns in human consumption of net primary production. *Nature* 429: 870–873.

Kearney, Melissa S. and Phillip B. Levine. 2014. *Media influences on social outcomes: The impact of MTV's 16 and Pregnant on teen childbearing*. The National Bureau of Economic Research, Cambridge, MA.

La Ferrara, E., et al., 2012. Soap operas and fertility: Evidence from Brazil. *American Economic Journal: Applied Economics* 4: 1–31.

Malthus, Thomas R. *An essay on the principle of population*. 1983 ed. Penguin USA, New York.

Notestein, Frank. 1953. Economic problems of population change. Pp. 13–31 in *Proceedings of the Eighth International Conference of Agricultural Economists*. Oxford Univ. Press.

Population Reference Bureau. 2018. *2018 World population data sheet*. Population Reference Bureau, Washington, D.C.

Riley, Nancy E. 2004. *China's population: New trends and challenges*. *Population Bulletin* 59(2). Population Reference Bureau, Washington, D.C.

U.N. Population Division. 2019. *World population prospects: The 2019 revision*. UNPD, New York.

U.N. Population Fund. 2010. *The millennium development goals report 2010*. UNFPA.

U.S. Census Bureau International Database. http://www.census.gov/population/international/data/idb/.

Venter, O., et al., 2016. Sixteen years of change in the global terrestrial human footprint and implications for biodiversity conservation. *Nature Communications* 7:12558

Wackernagel, Mathis, and William Rees. 1996. *Our ecological footprint: Reducing human impact on the earth*. New Society Publishers, Gabriola Island, British Columbia, Canada.

WWF. 2018. *Living planet report 2018: Aiming higher*. (Grooten, M. and Almond, R.E.A., eds.) WWF, Gland, Switzerland.

Chapter 9

Ashman, Mark R., and Geeta Puri. 2002. *Essential soil science: A clear and concise introduction to soil science*. Wiley-Blackwell, Hoboken, N.J.

Bee City USA. https://www.beecityusa.org.

Buchmann, Stephen L., and Gary Paul Nabhan. 1996. *The forgotten pollinators*. Island Press/Shearwater Books, Washington, D.C./Covelo, California.

Crop Trust. Crop diversity: Why it matters. www.croptrust.org/our-mission/crop-diversity-why-it-matters/.

Curtin, Charles G. 2002. Integration of science and community-based conservation in the Mexico/U.S. borderlands. *Conservation Biology* 16: 880–886.

Egan, Timothy. 2006. *The worst hard time: The untold story of those who survived the great American dust bowl*. Mariner Books, Houghton-Mifflin, Boston and New York.

Flack, Sarah 2016. *The art and science of grazing: How grass farmers can create sustainable systems for healthy animals and farm ecosystems*. Chelsea Green Publishing Co., White River Junction, Vt.

Garcia-Tejero, Iva, et al. 2011. *Water and sustainable agriculture*. Springer, New York.

Gemmill-Herren, Barbara. 2016. Pollination services to agriculture: Sustaining and enhancing a key ecosystem service. FAO and Routledge, Abingdon, U.K.

Glanz, James. 1995. *Saving our soil: Solutions for sustaining Earth's vital resource*. Johnson Books, Boulder, Colo.

Goulson, Dave, et al. 2015. Bee declines driven by combined stress from parasites, pesticides, and lack of flowers. *Science* 347: 1435–1444.

Hanson, Thor. 2018. *Buzz: The nature and necessity of bees*. Basic Books, New York.

Havlin, John L., et al. 2013. *Soil fertility and fertilizers*, 8th ed. Pearson Education, Upper Saddle River, N.J.

Hayes, Rhonda Fleming. 2016. *Pollinator friendly gardening: Gardening for bees, butterflies, and other pollinators*. Voyageur Press, New York.

Helmer, Jodi. 2019. *Protecting pollinators: How to save the creatures that feed our world*. Island Press, Washington, D.C.

Huggins, David R., and John P. Reganold. 2008. No-till: How farmers are saving the soil by parking their plows. *Scientific American* 299(1): 70–71.

Imeson, Anton. 2012. *Desertification, land degradation, and sustainability: Paradigms, processes, principles, and policies*. Wiley-Blackwell, Sussex, U.K.

Imhoff, Daniel, and Christina Badaracco. 2019. *The Farm Bill: A citizen's guide*. Island Press, Washington, D.C.

Iowa State University Extension. 2009. *Considerations in selecting no-till*. Iowa State University Extension and NRCS.

IPBES (Potts, Simon G., et al., eds.). 2016. *The assessment report on pollinators, pollination, and food production*. Intergovernmental Science–Policy Platform on Biodiversity and Ecosystem Services, Bonn, Germany.

Jenny, Hans. 1941. *Factors of soil formation: A system of quantitative pedology*. McGraw-Hill, New York.

Kremen, Claire, et al. 2002. Crop pollination from native bees at risk from agricultutral intensification. *Proc. Natl. Acad. Sci. USA* 99: 16812–16816.

Lal, Rattan, ed. 1998. *Soil quality and agricultural sustainability*. Ann Arbor Press, Chelsea, Mich.

Lengnick, Laura. 2015. Resilient agriculture: Cultivating food systems for a changing climate. New Society Publishers, Vancouver, B.C.

Lowe, Jaime. 2018. The super bowl of beekeeping. *New York Times Magazine*. Aug. 15, 2018.

Losey, John E., and Mace Vaughan. 2006. The economic value of ecological services provided by insects. *BioScience* 56: 311–323.

Malpai Borderlands Group. Malpai Borderlands Group. www.malpaiborderlandsgroup.org.

Millennium Ecosystem Assessment. 2005. *Ecosystems and human well-being: Desertification synthesis*. Millennium Ecosystem Assessment and World Resources Institute.

Molden David. 2007. Water for food, water for life; A comprehensive assessment of water management in agriculture. Routledge, Abingdon, U.K.

Montgomery, David R. 2007. *Dirt: The erosion of civilizations*. University of California Press, Berkeley and Los Angeles.

Montgomery, David R. 2007. Soil erosion and agricultural sustainability. *PNAS* 104: 13268–13272.

Natural Resources Conservation Service. Soils. NRCS, USDA. www.soils.usda.gov.

Nordhaus, Hannah. 2011. *The beekeeper's lament: How one man and half a billion honey bees help feed America*. Harper Perennial, New York.

Ohlson, Kristin. 2014. *The soil will save us: How scientists, farmers, and foodies are healing the soil to save the planet*. Rodale Inc., Emmaus, Penn.

Pearson, Gwen. 2015. You're worrying about the wrong bees. *Wired*. April 29, 2015.

Plaster, Edward. 2013. *Soil science and management*, 6th ed. Cengage learning, Boston.

Pollinator Health Task Force. 2015. *National strategy to promote the health of honey bees and other pollinators*. The White House, Washington, D.C.

Pywell, Richard, et al. 2015. Wildlife-friendly farming increases crop yield: Evidence for ecological intensification. *Proc. Roy. Soc. London B* 282: 20151740.

Schulte, Lisa A., et al. 2017. Prairie strips improve biodiversity and the delivery of multiple ecosystem services from corn–soybean croplands. *Proc. Natl. Acad. Sci. USA* 114: 11247–11252.

Soil Science Society of America. Soil basics. www.soils.org/discover-soils/soil-basics.

Trimble, Stanley W., and Pierre Crosson. 2000. U.S. soil erosion rates—myth and reality. *Science* 289: 248–250.

Troeh, Frederick R., and Louis M. Thompson. 2005. *Soil and soil fertility*, 6th ed. Blackwell Publishing, London.

U.S. Department of Agriculture. 2019. *2017 Census of agriculture*. USDA, Washington, D.C.

U.S. Geological Survey. 2014. *Conservation Reserve Program (CRP) contributions to wildlife habitat, management issues, challenges, and policy choices: An annotated bibliography*. U.S. Department of the Interior, Washington, D.C.

Vaughan, Mace, et al. 2015. *Farming for bees: Guidelines for providing native bee habitat on farms*. Xerces Society, Portland, Ore.

Wilkinson, Bruce H. 2005. Humans as geologic agents: A deep-time perspective. *Geology* 33: 161–164.

Xerces Society. http://xerces.org.

Xerces Society. 2011. *Attracting native pollinators*. Storey Publishing, North Adams, Mass.

Chapter 10

Barilla Center for Food and Nutrition (Danielle Nierenberg, ed.). 2018. *Nourished planet: Sustainability in the global food system*. Island Press, Washington, D.C.

Benbrook, Charles M. 2012. Impacts of genetically engineered crops on pesticide use in the U.S.—the first sixteen years. *Environmental Sciences Europe* 24: 24, 13 pp.

Brar, Satinder Kaur. 2012. *Biocontrol: Management, processes, and challenges*. Nova Science Publishers, Hauppauge, N.Y.

Brown, Lester R. 2012. *Full planet, empty plates: The new geopolitics of food scarcity*. Earth Policy Institute, Washington, D.C.

Campus Farmers Network. www.campusfarmers.org.

Center for Food Safety. 2007. *Monsanto vs. U.S. farmers: November 2007 update*. Center for Food Safety, Washington, D.C.

Cerdeira, Antonio L., and Stephen O. Duke. 2006. The current status and environmental impacts of glyphosate-resistant crops: A review. *J. Environ. Qual.* 35: 1633–1658.

Delate, Kathleen, et al. 2015. A review of long-term organic comparison trials in the U.S. *Sustainable Agriculture Research* 4: 5–14.

Diamond, Jared. 1999. *Guns, germs, and steel: The fates of human societies*. W.W. Norton, New York.

Diamond, Jared, and Peter Bellwood. 2003. Farmers and their languages: The first expansions. *Science* 300: 597–603.

Else, Jessica, and Ken Foster, eds. 2018. *How to feed the world*. Island Press, Washington, D.C.

Fedoroff, Nina, and Nancy Marie Brown. 2004. *Mendel in the kitchen: A scientist's view of genetically modified foods*. National Academies Press, Washington, D.C.

Flint, Mary Louise. 2010. *IPM in practice: Principles and methods of integrated pest management*. University of California Division of Agriculture and Natural Resources.

Food and Agriculture Organization of the United Nations. 2006. *Livestock's long shadow: Environmental issues and options*. FAO, Rome.

Food and Agriculture Organization of the United Nations. 2013. *Tackling climate change through livestock: A global assessment of emissions and mitigation opportunities*. FAO, Rome.

Food and Agriculture Organization of the United Nations. 2018. *The state of world fisheries and aquaculture 2018*. FAO Fisheries and Aquaculture Department, Rome.

Food and Agriculture Organization of the United Nations. 2019. *The state of food insecurity and nutirtion in the world 2019*. FAO, Rome.

Gardner, Gary, and Brian Halweil. 2000. *Underfed and overfed: The global epidemic of malnutrition*. Worldwatch Paper 150. Worldwatch Institute, Washington, D.C.

Gurian-Sherman, Doug. 2009. *Failure to yield: Evaluating the performance of genetically engineered crops*. Union of Concerned Scientists, Cambridge, Mass.

Halweil, Brian. 2008. *Farming fish for the future*. Worldwatch Report 176. Worldwatch Institute, Washington, D.C.

Haspel, Tamar. 2018. The public doesn't trust GMOs. Will it trust CRISPR? *Vox*. July 26, 2018.

James, Clive. 2019. *Global status of commercialized biotech/GM crops: 2018*. International Service for the Acquisition of Agri-biotech Applications.

Jones Jr., J. Benton. 2016. *Hydroponics: A practical guide for the soilless grower*. 2nd ed. CRC Press, Boca Raton, Fla.

Klümper, Wilhelm, and Matin Qaim. 2014. A meta-analysis of the impacts of genetically modified crops. *PLOS ONE* 9: e111629.

Kristiansen, P., et al., eds. 2006. *Organic agriculture: A global perspective*. CABI Publishing, Oxfordshire, U.K.

Lappé, Frances Moore, and Joseph Collins. 2015. *World hunger: 10 myths*. Grove Press, New York.

Liebig, Mark A., and John W. Doran. 1999. Impact of organic production practices on soil quality indicators. *J. Environ. Qual.* 28: 1601–1609.

Maeder, Paul, et al. 2002. Soil fertility and biodiversity in organic farming. *Science* 296: 1694–1697.

Mahgoub, Salah E.O. 2015. *Genetically modified foods: Basics, applications, and controversy*. CRC Press, Boca Raton, Fla.

Manning, Richard. 2000. *Food's frontier: The next green revolution*. North Point Press, New York.

Nierenberg, Danielle, and Laura Reynolds. 2012. *Innovations in sustainable agriculture: Supporting climate-friendly food production*. Worldwatch Report 188. Worldwatch Institute, Washington, D.C.

Norris, Robert F., et al. 2003. *Concepts in integrated pest management*. Prentice Hall, Upper Saddle River, N.J.

Paull, John. 2015. The threat of genetically modified organisms (GMOs) to organic agriculture: A case study update. *Agriculture & Food* 3: 56–63.

Pearce, Fred. 2002. The great Mexican maize scandal. *New Scientist* 174: 14.

Pedigo, Larry P., and Marlin Rice. 2014. *Entomology and pest management*, 6th ed. Waveland Press, Long Grove, Ill.

Polak, Paul. 2005. The big potential of small farms. *Scientific American* 293(3): 84–91.

Pollan, Michael. 2007. *The omnivore's dilemma*. Penguin Press, New York.

Pollan, Michael. 2009. *In defense of food: An eater's manifesto*. Penguin Press, New York.

Ponisio, Lauren C., et al. 2014. Diversification practices reduce organic to conventional yield gap. *Proc. Roy. Soc. B* 282: 20141396

de Ponti, Tomek, et al. 2012. The crop yield gap between organic and conventional agriculture. *Agricultural Systems* 108: 1–9.

Pretty, Jules. 2007. *Sustainable agriculture and food*. Earthscan, London.

Pringle, Peter. 2003. *Food, Inc.: Mendel to Monsanto—The promises and perils of the biotech harvest*. Simon and Schuster, New York.

Quist, David, and Ignacio H. Chapela. 2001. Transgenic DNA introgressed into traditional maize landraces in Oaxaca, Mexico. *Nature* 414: 541–543.

Rodale Institute. 2011. *The farming systems trial: Celebrating 30 years*. Rodale Institute, Kutztown, Penn.

Sánchez–Bayo, Francisco. 2014. The trouble with neonicotinoids. *Science* 346: 806–807.

Science. 2013. Smarter pest control. (special section). Pp. 730–759 in *Science* 341 (16 Aug. 2013).

Seufert, V., et al. 2012. Comparing the yields of organic and conventional agriculture. *Nature* 485: 229–232.

Shiva, Vandana. 2000. *Stolen harvest: The hijacking of the global food supply*. South End Press, Cambridge, Mass.

Smil, Vaclav. 2001. *Feeding the world: A challenge for the twenty-first century*. MIT Press, Cambridge, Mass.

Smil, Vaclav. 2013. *Should we eat meat? Evolution and consequences of modern carnivory*. Wiley-Blackwell, Hoboken, N.J.

Stewart, C. Neal. 2004. *Genetically modified planet: Environmental impacts of genetically engineered plants*. Oxford University Press, New York.

Sustainable Agriculture Network. 2010. *The new American farmer: Profiles of agricultural innovation*, 2nd ed. Sustainable Agriculture Network, Beltsville, MD.

Sustainable Agriculture Research and Education (SARE). 2010. *Exploring sustainability in agriculture*. SARE.

Thorburn, Craig. 2015. The rise and demise of integrated pest management in rice in Indonesia. *Insects* 6: 381–408.

University of Michigan Sustainable Food Program. http://sustainability.umich.edu/umsfp.

U.S. Department of Agriculture. 2019. *2017 Census of agriculture*. USDA, Washington, D.C.

U.S. Department of Agriculture Advisory Committee on Biotechnology and 21st Century Agriculture (AC21). 2012. *Enhancing coexistence: A report of the AC21 to the Secretary of Agriculture*. USDA, Washington, D.C.

U.S. National Academy of Sciences. 2010. *The impact of genetically engineered crops on farm sustainability in the United States*. National Academies Press, Washington, D.C.

Wall Street Journal. 2015. Can organic food feed the world? Essays by Catherine Badgley and Steve Savage. *Wall Street Journal*, 12 July 2015.

Weasel, Lisa. 2008. *Food fray: Inside the controversy over genetically modified food*. AMACOM Books, New York.

Weber, Christopher L., and H. Scott Mathews. 2008. Food-miles and the relative climate impacts of food choices in the United States. *Environ. Sci. Technol.* 42: 3508–3513.

Willer, Helga, and Julia Lernoud, eds. 2019. *The world of organic agriculture 2019*. Research Institute of Organic Agriculture (FiBL), Frick, and IFOAM—Organics International, Bonn.

Worldwatch Institute. 2011. *State of the world 2011: Innovations that nourish the planet*. Worldwatch Institute, Washington, D.C.

Xie, Jian, et al. 2011. Ecological mechanisms underlying the sustainability of the agricultural heritage rice-fish coculture system. *Proc. Natl. Acad. Sci. USA* 108: E1381–E1387.

Chapter 11

Baker, C. Scott, Frank Cipriano, and Stephen R. Palumbi. 1996. Molecular genetic identification of whale and dolphin products from commerical markets in Korea and Japan. *Molecular Ecology* 5: 671–685.

Balmford, Andrew, et al. 2002. Economic reasons for conserving wild nature. *Science* 297: 950–953.

Barnosky, Anthony D., et al. 2004. Assessing the causes of late Pleistocene extinctions on the continents. *Science* 306: 70–75.

Baskin, Yvonne. 1997. *The work of nature: How the diversity of life sustains us*. Island Press, Washington, D.C.

Brodie, Jedediah F., et al., eds. 2013. *Wildlife conservation in a changing climate*. University of Chicago Press, Chicago.

Chivian, Eric, and Aaron Berstein, eds. 2008. *Sustaining life: How human health depends on biodiversity*. Oxford University Press, New York.

CITES Secretariat. Convention on International Trade in Endangered Species of Wild Fauna and Flora. www.cites.org.

Collins, James P., and Martha L. Crump. 2009. *Extinction in our times: Global amphibian decline*. Oxford University Press, U.K.

Conniff, Richard, 2012. A bitter pill. *Conservation*, 9 Mar. 2012.

Convention on Biological Diversity. www.cbd.int.

Cooper, John E., and Margaret E. Cooper. 2013. *Wildlife forensic investigation: Principles and practice*. CRC Press, Boca Raton, Fla.

Daily, Gretchen C., ed. 1997. *Nature's services: Societal dependence on natural ecosystems*. Island Press, Washington, D.C.

Galatowitsch, Susan. 2012. *Ecological restoration*. Sinauer Associates, Sunderland, Mass.

Gaston, Kevin J., and John I. Spicer. 2004. *Biodiversity: An introduction*, 2nd ed. Blackwell, London.

Groom, Martha J., et al. 2005. *Principles of conservation biology*, 3rd ed. Sinauer Associates, Sunderland, Mass.

Groombridge, Brian, and Martin D. Jenkins. 2002. *Global biodiversity: Earth's living resources in the 21st century*. UNEP, World Conservation Monitoring Centre, and Aventis Foundation; World Conservation Press, Cambridge, U.K.

Groombridge, Brian, and Martin D. Jenkins. 2002. *World atlas of biodiversity: Earth's living resources in the 21st century*. University of California Press, Berkeley.

Hallman, Caspar A., et al. 2017. More than 75 percent decline over 27 years in total flying insect biomass in protected areas. *PLOS ONE* 12(10): e0185809.

Hambler, Clive, and Susan M. Canney. 2013. *Conservation*, 2nd ed. Cambridge University Press., U.K.

Hoelzel, A. Rus. 2015. Can DNA foil the poachers? *Science* 349: 34–35.

Holdo, Ricardo, M. 2011. Predicted impact of barriers to migration on the Serengeti wildebeest population. *PLoS ONE* 6(1):e16370.

Hopcraft, J. Grant C. 2015. Conservation and economic benefits of a road around the Serengeti. *Conservation Biology* 29: 932–936.

IPBES (Intergovernmental Science–Policy Platform on Biodiversity and Ecosystem Services). www.ipbes.net/assessment-reports.

IPBES (Montanarella, Luca, et al., eds.). 2018. *The assessment report on land degradation and restoration*. Intergovernmental Science–Policy Platform on Biodiversity and Ecosystem Services, Bonn, Germany.

IPBES (Rice, Jake, et al., eds.). 2018. *The regional assessment report on biodiversity and ecosystem services for the Americas*. Intergovernmental Science–Policy Platform on Biodiversity and Ecosystem Services, Bonn, Germany.

Jarvis, Brooke. 2018. The insect apocalypse is here. *New York Times Magazine*. Nov. 27, 2018.

Kaufman, Leslie. 2012. Zoos' bitter choice: To save some species, letting others die. *New York Times*, 27 May 2012.

Kolbert, Elizabeth. 2014. *The sixth extinction: An unnatural history*. Henry Holt and Co., New York.

Laurance, William F., et al. 2015. Estimating the environmental costs of Africa's massive "development corridors." *Current Biology* 25: 3202–3208.

Lister, Bradford C., and Andres Garcia. 2018. Climate-driven declines in arthropod abundance restructure a rainforest food web. *Proc. Natl. Acad. Sci.* 115: E10397–E10406.

Louv, Richard. 2005. *Last child in the woods: Saving our children from nature-deficit disorder*. Algonquin Books, Chapel Hill, N.C.

Lovejoy, Thomas, et al., 2019. *Biodiversity and climate change: Transforming the biosphere*. Yale Univ. Press, New Haven, Conn.

MacArthur, Robert H., and Edward O. Wilson. 1967. *The theory of island biogeography*. Princeton University Press, Princeton, N.J.

Marris, Emma. 2013. *The rambunctious garden: Saving nature in a post-wild world*. Bloomsbury, USA, New York.

Maxwell, Sean, et al. 2016. The ravages of guns, nets, and bulldozers. *Nature* 536: 143–145.

Mburu, John, and Regina Birner. 2007. Emergence, adoption, and implementation of collaborative wildlife management or wildlife partnerships in Kenya: A look at conditions for success. *Society and Natural Resources* 20: 379–395.

McKie, Robin. 2018. Where have all our insects gone? *The Guardian*. June 17, 2018.

Millennium Ecosystem Assessment. 2005. *Ecosystems and human well-being: Biodiversity synthesis*. Millennium Ecosystem Assessment and World Resources Institute.

Mooney, Harold A., and Richard J. Hobbs, eds. 2000. *Invasive species in a changing world*. Island Press, Washington, D.C.

Mora, Camilo, et al. 2011. How many species are there on Earth and in the ocean? *PLoS Biology* 9(8): e1001127.

Ogutu, Joseph O., et al. 2011. Continuing wildlife population declines and range contraction in the Mara region of Kenya during 1977–2009. *Journal of Zoology* 285: 99–109.

Pandya, Rajul, and Kenne Ann Dibner, eds. 2018. *Learning through citizen science: Enhancing opportunities by design*. U.S. National Academy of Sciences Press, Washington D.C.

Pearce, Fred. 2015. *The new wild: Why invasive species will be nature's salvation*. Beacon Press, Boston.

Primack, Richard B., and Anna Sher. 2016. *An introduction to conservation biology*. Oxford Univ. Press.

Quammen, David. 1996. *The song of the dodo: Island biogeography in an age of extinction*. Touchstone, New York.

Rohr, Jason R., et al. 2008. Evaluating the links between climate, disease spread, and amphibian declines. *PNAS* 105:17436–17441.

Rosenzweig, Michael L. 1995. *Species diversity in space and time*. Cambridge University Press.

Roskov, Y., et al. 2019. Species 2000 & ITIS Catalogue of Life. Digital resource at www.catalogueoflife.org/col. Species 2000: Naturalis, Leiden, the Netherlands.

Sánchez–Bayo, Francisco, and Kris A.G. Wyckhuys. 2019. Worldwide decline of the entomofauna: A review of its drivers. *Biological Conservation* 232: 8–27.

Scheele, Ben C. 2019. Amphibian fungal panzootic causes catastrophic and ongoing loss of biodiversity. *Science* 363: 1459–1463.

Sepkoski, John J. 1984. A kinetic model of Phanerozoic taxonomic diversity. *Paleobiology* 10: 246–267.

Simberloff, Daniel. 1998. Flagships, umbrellas, and keystones: Is single-species management passé in the landscape era? *Biological Conservation* 83: 247–257.

Soulé, Michael E. 1986. *Conservation biology: The science of scarcity and diversity*. Sinauer Associates, Sunderland, Mass.

Sutherland, William J. 2018. A 2018 horizon scan of emerging issues for global conservation and biological diversity. *Trends Ecol. Evol.* 33: 47–58.

Takacs, David. 1996. *The idea of biodiversity: Philosophies of paradise*. Johns Hopkins University Press, Baltimore.

TEEB. 2010. *The economics of ecosystems and biodiversity: Mainstreaming the economics of nature: A synthesis of the approach, conclusions, and recommendations of TEEB*. The Economics of Ecosystems and Biodiversity.

Thompson, Ken. 2014. *Where do camels belong? Why invasive species aren't all bad*. Greystone Books, Vancouver, B.C.

U.S. Environmental Protection Agency. Summary of the Endangered Species Act. www.epa.gov/laws-regulations/summary-endangered-species-act.

U.S. Fish and Wildlife Service. Endangered species program. www.fws.gov/endangered.

Wallace-Wells, David. 2019. *The uninhabitable Earth: Life after warming*. Tim Duggan Books.

Ward, Peter D. 2007. *Under a green sky: Global warming, the mass extinctions of the past, and what they can tell us about our future*. Smithsonian Books and Collins, Washington, D.C., and New York.

Wasser, S.K., et al. 2015. Genetic assignment of large seizures of elephant ivory reveals Africa's major poaching hotspots. *Science* 349: 84–87.

Western, David. 2003. Conservation science in Africa and the role of international collaboration. *Conservation Biology* 17: 11–19.

Williams, Florence. 2018. *The nature fix: Why nature makes us happier, healthier, and more creative*. W.W. Norton, New York.

Wilson, Edward O. 1984. *Biophilia*. Harvard University Press, Cambridge, Mass.

Wilson, Edward O. 1992. *The diversity of life*. Harvard University Press, Cambridge, Mass.

Wilson, Edward O. 2002. *The future of life*. Alfred A. Knopf, New York.

Wittemyer, George. 2014. Illegal killing for ivory drives global decline in African elephants. *Proc. Natl. Acad. Sci. USA* 111: 13117–13121.

World Conservation Union. IUCN Red List. www.iucnredlist.org.

WWF. 2018. *Living planet report 2018: Aiming higher*. (Grooten, M. and Almond, R.E.A., eds.) WWF, Gland, Switzerland.

Chapter 12

Abatzoglou, John T., and A. Park Williams. 2016. Impact of anthropogenic climate change on wildfire across western U.S. forests. *Proc. Natl. Acad. Sci. USA* 113: 11770–11775.

Agronne, Dianna M. 2013. *Deforestation and climate change*. Nova Science Publishers, Hauppauge, N.Y.

Aubertin, Catherine, and Estienne Rodary, eds. 2011. *Protected areas, sustainable land?* Ashgate Publishing, Surrey, U.K., and Burlington, Vermont.

Bettinger, Pete, et al. 2008. *Forest management and planning*. Academic Press, New York.

Bratkovich, Steve, et al. 2012. *Forests of the United States: Understanding trends and challenges*. Dovetail Partners Inc., Minneapolis, Minn.

Clary, David. 1986. *Timber and the Forest Service*. University Press of Kansas, Lawrence.

Curtis, P.G., et al. 2018. Classifying drivers of global forest loss. *Science* 361: 1108–1111.

Donato, Daniel C., et al. 2006. Post-wildfire logging hinders regeneration and increases fire risk. *Science* 311: 352.

Duncan, Dayton, and Ken Burns. 2009. *The national parks: America's best idea*. Alfred A. Knopf, New York.

Ferraz, Goncalo N., et al. 2007. A large-scale deforestation experiment: Effects of patch area and isolation on Amazon birds. *Science* 315: 238–241.

Food and Agriculture Organization of the United Nations. 2018. *State of the world's forests 2018*. FAO, Rome.

Food and Agriculture Organization of the United Nations. 2015. *Global forest resources assessment*. FAO, Rome.

Forest Stewardship Council. www.fsc.org.

Fritts, Rachel. 2018. What's causing deforestation? New study reveals global drivers. *Mongabay*. Sept. 14, 2018.

Gorte, Ross W., and Pervaze A. Sheikh. 2013. *Deforestation and climate change*. Congressional Research Service, Washington, D.C.

Haddad, Nick M., et al. 2015. Habitat fragmentation and its lasting impact on Earth's ecosystems. *Science Advances* 1(2): e1500052, 20 Mar. 2015.

Hansen, M.C., et al., 2013. High-resolution global maps of 21st-century forest cover change. *Science* 342: 850–853.

Harris, Larry D. 1984. *The fragmented forest: Island biogeography theory and the preservation of biotic diversity*. University of Chicago Press, Chicago.

Hilty, Jodi A., et al. 2019. *Corridor ecology: Linking landscapes for biodiversity consrvation and climate adaptation*. 2nd ed. Island Press, Washington, D.C.

Jensen, Sara E., and Guy R. McPherson. 2008. *Living with fire: Fire ecology and policy for the twenty-first century*. University of California Press, Berkeley and Los Angeles.

Juniper, Tony. 2019. *Rainforest: Dispatches from Earth's most vital frontlines*. Island Press, Washington, D.C.

Lai, K.K. Rebecca, et al. 2019. What satellite imagery tells us about the Amazon rain forest fires. *New York Times*. Aug. 24, 2019.

Land Trust Alliance. 2016. *2015 National land trust census report: A look at voluntary land conservation in America*. Land Trust Alliance, Washington, D.C.

Laurance, William F., et al. 2011. The fate of Amazonian forest fragments: A 32-year investigation. *Biological Conservation* 144(1): 56–67.

Lindenmayer, David B., et al. 2008. *Salvage logging and its ecological consequences*. Island Press, Washington, D.C.

Lockwood, Michael, et al. 2006. *Managing protected areas: A global guide*. IUCN and Earthscan Publishing, London.

Lovejoy, Thomas E., and Carlos Nobre. 2018. Amazon tipping point. *Science Advances* 4: eaat2340.

Lustgarten, Abraham. 2018. Palm oil was supposed to help save the planet. Instead it unleashed a catastrophe. *New York Times Magazine.* Nov. 20, 2018.

MacArthur, Robert H., and Edward O. Wilson. 1967. *The theory of island biogeography.* Princeton University Press, Princeton, N.J.

MacDicken, Kenneth, ed. 2015. Changes in global forest resources from 1990 to 2015. (Special issue). *Forest Ecology and Management* 352: 1–145.

McKenzie, Donald, Carol Miller, and Donald A. Falk, eds. 2011. The landscape ecology of fire. *Ecological Studies* 213. Springer, New York.

National Forest Management Act of 1976. www.fs.fed.us/emc/nfma/includes/NFMA1976.pdf.

National Interagency Fire Center. www.nifc.gov.

Nepstad, D., et al., 2014. Slowing Amazon deforestation through public policy and interventions in beef and soy supply chains. *Science* 344: 1118–1123.

Newmark, William D. 1987. A land-bridge perspective on mammal extinctions in western North American parks. *Nature* 325: 430.

Noss, Reed F., et al. 2006. Managing fire-prone forests in the western United States. *Frontiers in Ecology and the Environment* 4: 481–487.

Oswalt, Sonja N., et al. 2019. *Forest resources of the United States, 2017: A technical document supporting the Forest Service 2020 RPA assessment.* Gen. Tech. Rep. WO-97. USDA Forest Service, Washington, D.C.

Pimm, Stuart L. 1998. The forest fragment classic. *Nature* 393: 23–24.

Pye, Oliver, and Jayati Bhattacharya, eds. 2012. *The palm oil controversy in Southeast Asia: A transnational perspective.* ISEAS Publishing, Singapore.

Pyne, Stephen. 2001. *Fire: A brief history.* University of Washington Press, Seattle.

Runte, Alfred. 1979. *National parks and the American experience.* University of Nebraska Press, Lincoln.

Schwartz, John. 2015. As fires grow, a new landscape appears in the West. *New York Times,* Sept. 21, 2015.

Science. 2015. Forest health in a changing world. (Special issue). Pp. 800–836 in *Science* 349 (Aug. 21, 2015).

Sedjo, Robert A. 2000. *A vision for the U.S. Forest Service.* Resources for the Future, Washington, D.C.

Smith, David M., et al. 1996. *The practice of silviculture: Applied forest ecology,* 9th ed. Wiley, New York.

Smithsonian Tropical Research Institute. Biological Dynamics of Forest Fragments Project. https://forestgeo.si.edu/research-programs/affiliated-programs/biological-dynamics-forest-fragments-project-bdffp.

Soulé, Michael E., and John Terborgh, eds. 1999. *Continental conservation.* Island Press, Washington, D.C.

Stegner, Wallace. 1954. *Beyond the hundredth meridian: John Wesley Powell and the second opening of the West.* Houghton Mifflin, Boston.

Struzik, Edward. 2017. *Firestorm: How wildfire will shape our future.* Island Press, Washington, D.C.

U.N. Environment Programme and World Conservation Monitoring Centre. 2018. *Protected planet report 2018: Tracking progress towards global targets for protected areas.* Cambridge, U.K.

Williams, Michael. 2006. *Deforesting the Earth: From prehistory to global crisis: An abridgment.* University of Chicago Press, Chicago.

Wilson, Edward O. 2016. *Half Earth: Our planet's fight for life.* Liveright, New York.

Xavantina, Nova. 2019. The Amazon is approaching an irreversible tipping point. *The Economist.* Aug. 1, 2019.

Young, Raymond A., and Ronald L. Giese. 2002. *Introduction to forest ecosystem science and management,* 3rd ed. John Wiley & Sons, New York.

Chapter 13

Abbott, Carl. 2001. *Greater Portland: Urban life and landscape in the Pacific Northwest.* University of Pennsylvania Press.

Abbott, Carl. 2002. Planning a sustainable city. Pp. 207–235 in Squires, Gregory D., ed. *Urban sprawl: Causes, consequences, and policy responses.* Urban Institute Press, Washington, D.C.

Adler, Frederick R., and Colby J. Tanner. 2013. *Urban ecosystems: Ecological principles for the built environment.* Cambridge University Press, Cambridge, U.K.

Bloomberg, Michael. 2015. City century: Why municipalities are the key to fighting climate change. *Foreign Affairs,* Sept./Oct. 2015. Council on Foreign Relations, Washington, D.C.

Brockerhoff, Martin P. 2000. An urbanizing world. *Population Bulletin* 55(3). Population Reference Bureau, Washington, D.C.

Calthorpe, Peter. 2010. *Urbanism in the age of climate change.* Island Press, Washington D.C.

Chakrabarti, Vishaan. 2013. *A country of cities: A manifesto for an urban America.* Metropolis Books, New York.

Chester, Mikhail, et al. 2013. Infrastructure and automobile shifts: Positioning transit to reduce life-cycle environmental impacts for urban sustainability goals. *Environmental Research Letters* 8: 015041 (10 pp).

Christensen, Nick. 2015. *A 50-year map: Years of research, public input, led to agreement on rserves plan.* Metro News, 30 Nov. 2015. Portland, Ore. www.oregonmetro.gov/news/50-year-map-years-research-public-input-led-agreement-reserves-plan.

Cronon, William. 1991. *Nature's metropolis: Chicago and the great West.* W.W. Norton, New York.

Douglas, Ian and Philip James. 2015. *Urban ecology: An introduction.* Routledge, Abingdon, U.K.

Duany, Andres, et al. 2001. *Suburban nation: The rise of sprawl and the decline of the American dream.* North Point Press, New York.

Duany, Andres, et al. 2009. *The smart growth manual.* McGraw-Hill Professional, New York.

Ellin, Nan. 2012. *Good urbanism.* Island Press, Washington, D.C.

Ewing, Reid, et al. 2002. *Measuring sprawl and its impact.* Smart Growth America.

Forman, Richard T. T. 2014. *Urban ecology: Science of cities.* Cambridge University Press, Cambridge, U.K.

Garvin, Alexander. 2016. *What makes a great city.* Island Press, Washington, D.C.

Girardet, Herbert. 2004. *Cities people planet: Livable cities for a sustainable world.* Academy Press.

Harnik, Peter. 2010. *Urban green: Innovative parks for resurgent cities.* The Trust for Public Land and Island Press, Washington, D.C.

Jacobs, Jane. 1992. *The death and life of great American cities.* Vintage.

Langdon, Philip. 2017. *Within walking distance: Creating livable communities for all.* Island Press, Washington, D.C.

Litman, Todd. 2012. *Rail transit in America: A comprehensive evaluation of benefits.* Victoria Transport Policy Institute and American Public Transportation Association.

Litman, Todd. 2015. *Evaluating public transit benefits and costs: Best practices guidebook.* Victoria Transport Policy Institute, Victoria, B.C.

Long Term Ecological Research Network: Baltimore Ecosystem Study. https://lternet.edu/site/baltimore-ecosystem-study.

Long Term Ecological Research Network: Central Arizona—Phoenix LTER. https://lternet.edu/site/central-arizona-phoenix-lter.

Metro.www.metro-region.org.

Metro. 2010. *Urban growth report 2009–2030: Employment and residential.* Metro, Portland, Ore.

Moon-Miklaucic, Christopher, et al. 2019. *The evolution of bike sharing: 10 questions on the emergence of new technologies, opportunities, and risks.* World Resources Institute, Jan. 2019.

New Urbanism. www.newurbanism.org.

New York, City of. 2015. *One NYC 2050: Building a strong and fair city.* City of New York, New York. www.nyc.gov/onenyc.

Newman, Peter, et al. 2017. *Resilient cities: Overcoming fossil fuel dependence.* 2nd ed. Island Press, Washington, D.C.

Northwest Environment Watch. 2004. *The Portland exception: A comparison of sprawl, smart growth, and rural land loss in 15 U.S. cities.* Northwest Environment Watch, Seattle.

Plastrik, Peter, and John Cleveland. 2018. *Life after carbon: The next global transformation of cities.* Island Press, Washington, D.C.

Portland, City of. 2016. Adopted 2035 Comprehensive Plan. www.portland-oregon.gov/bps/70936.

Speck, Jeff. 2018. *Walkable city rules: 101 steps to making better places.* Island Press, Washington, D.C.

TEEB. 2011. *TEEB manual for cities: Ecosystem services in urban management.* The Economics of Ecosystems and Biodiversity.

U.N. Population Division. 2019. *World urbanization prospects: The 2018 revision.* UNPD, New York.

U.S. Green Building Council. www.usgbc.org.

Worldwatch Institute. 2007. *State of the world 2007: Our urban future.* Worldwatch Institute, Washington, D.C.

Worldwatch Institute. 2016. *State of the world 2016: Can a city be sustainable?* Worldwatch Institute, Washington, D.C.

Chapter 14

Ames, Bruce N., et al. 1990. Nature's chemicals and synthetic chemicals: Comparative toxicology. *Proceedings of the National Academy of Sciences of the United States of America* 87: 7782–7786.

Bernhardt, E. S., et al., 2017. Synthetic chemicals as agents of global change. *Frontiers in Ecology and the Environment* 15: 84–90.

Carson, Rachel. 1962. *Silent spring.* Houghton Mifflin, Boston.

Carwhile, Jenny R., et al. 2011. Canned soup consumption and urinary bisphenol A: A randomized crossover trial. *Journal of the American Medical Association* 306: 2218–2220.

Colburn, Theo, Dianne Dumanoski, and John P. Myers. 1996. *Our stolen future.* Penguin USA, New York.

Consumer Reports. 2009. Concern over canned foods: Our tests find wide range of bisphenol A in soups, juice, and more. *Consumer Reports* Dec. 2009.

Curtis, Kathleen, and Bobbi Chase Wilding. 2010. *Is it in us? Chemical contamination in our bodies.* Body Burden Working Group and Commonweal Biomonitoring Resource Center.

European Commission, Environment Directorate General. 2007. *REACH in brief.* European Commission.

Guillette, Louis J. Jr., et al. 2000. Alligators and endocrine disrupting contaminants: A current perspective. *American Zoologist* 40: 438–452.

Harrison, P., and F. Pearce, 2000. *AAAS atlas of population and environment.* University of California Press, Berkeley, CA.

Hayes, Tyrone, et al. 2003. Atrazine-induced hermaphroditism at 0.1 PPB in American leopard frogs (*Rana pipiens*): Laboratory and field evidence. *Environmental Health Perspectives* 111: 568–575.

Landis, Wayne G., et al. 2010. *Introduction to environmental toxicology*, 4th ed. CRC Press, Boca Raton, Fla.

Lang, Iain A., et al. 2008. Association of urinary bisphenol A concentration with medical disorders and laboratory abnormalities in adults. *Journal of the American Medical Association* 300: 1303–1310.

Loewenberg, Samuel. 2003. E.U. starts a chemical reaction. *Science* 300: 405.

McGinn, Anne Platt. 2000. *Why poison ourselves? A precautionary approach to synthetic chemicals.* Worldwatch Paper #153. Worldwatch Institute, Washington, D.C.

Millennium Ecosystem Assessment. 2005. *Ecosystems and human well-being: Health synthesis.* World Health Organization.

Moeller, Dade. 2011. *Environmental health*, 4th ed. Harvard Univ. Press.

Myers, Samuel S. 2009. *Global environmental change: The threat to human health.* Worldwatch Report #181. Worldwatch Institute, Washington, D.C.

National Center for Environmental Health; U.S. Centers for Disease Control and Prevention. 2005. *Third national report on human exposure to environmental chemicals.* NCEH Pub. No. 05-0570, Atlanta.

National Safety Council. 2018. *Injury Facts, 2018.* National Safety Council, Itasca, IL.

Our Stolen Future. www.ourstolenfuture.com.

Pirages, Dennis. 2005. Containing infectious disease. Pp. 42–61 in *State of the world 2005.* Worldwatch Institute, Washington, D.C.

Renner, Rebecca. 2002. Conflict brewing over herbicide's link to frog deformities. *Science* 298: 938–939.

Traynor, K., et al., 2016. In-hive pesticide exposome: Assessing risks to migratory honey bees from in-hive pesticide contamination in the eastern United States. *Nature Scientific Reports* 6:33207.

United Health Foundation, 2017. *America's health rankings, 2017 edition.* United Health Foundation, Minnetonka, MN.

Vogel, Sarah. 2009. The politics of plastic: The making and unmaking of bisphenol A "safety". *Framing Health Matters, American Journal of Public Health Supplement 3,* vol. 99 #S3, S559–S566.

vom Saal, Frederick S., et al. 2012. The estrogenic endocrine disrupting chemical bisphenol A (BPA) and obesity. *Molecular and Cellular Endocrinology* 354: 74–84.

World Health Organization. 2009. *Global health risks: Mortality and burden of disease attributable to selected major risks.* WHO, Geneva, Switzerland.

World Health Organization. 2016. *World health statistics 2016.* WHO, Geneva, Switzerland.

Zogorski, John S., et al. 2006. *Volatile organic compounds in the nation's ground water and drinking-water supply wells.* USGS National Water-Quality Assessment Program Circular 1292.

Zota, A.R., et al., 2016. Recent fast food consumption and bisphenol A and phthalates exposures among the U.S. population in NHANES, 2003–2010. *Environmental Health Perspectives* 124: 1521–1528.

Chapter 15

American Rivers. 2002. *The ecology of dam removal: A summary of benefits and impacts.* American Rivers, Washington, D.C.

Cook, Benjamin I., et al., 2015,*Science Advances* 1(1): e1400082

Gleick, Peter H., and H.S. Cooley. 2009. Energy implications of bottled water. *Environ. Res. Lett.* 4: 014009 (6 pp).

Gleick, Peter H., et al. 2009. *The world's water 2008–2009: The biennial report on freshwater resources.* Island Press, Washington, D.C.

Gleick, Peter H. 2010. *Bottled and sold: The story behind our obsession with bottled water.* Island Press, Washington, D.C.

Harrison, P., and F. Pearce, 2000. *AAAS atlas of population and the environment*, edited by the American Association for the Advancement of Science, © 2000 by the American Association for the Advancement of Science.

Millennium Ecosystem Assessment. 2005. *Ecosystems and human well-being: Wetlands and water synthesis.* Millennium Ecosystem Assessment and World Resources Institute.

Naidenko, Olga, et al. 2008. *Bottled water quality investigation: 10 major brands, 38 pollutants.* Environmental Working Group. www.ewg.org.

Postel, Sandra. 2005. *Liquid assets: The critical need to safeguard freshwater ecosystems.* Worldwatch Paper #170. Worldwatch Institute, Washington, D.C.

Postel, Sandra. 2017. *Replenish: The virtuous cycle of water and prosperity.* Island Press, Washington, D.C.

Reisner, Marc. 1986. *Cadillac desert: The American West and its disappearing water.* Viking Penguin, New York.

TEEB. 2013. *The economics of ecosystems and biodiversity (TEEB) for water and wetlands.* The Economics of Ecosystems and Biodiversity. IEEP, London and Brussels, Ramsar Secretariat, Gland.

Trujillo, Alan P., and Harold V. Thurman. 2013. *Essentials of oceanography*, 11th ed. Prentice Hall, Upper Saddle River, N.J.

U.N. Environment Programme. 2012. "Water." Chapter 4 in *Global environment outlook 5 (GEO-5).* UNEP, Nairobi.

U.N. Environment Programme. 2008. *Water quality for ecosystem and human health*, 2nd ed.

UNEP Global Environment Monitoring System (GEMS)/Water Programme, Burlington, Ontario.

U.N. World Water Assessment Programme. 2012. *U.N. world water development report: Managing water under uncertainty and risk.* Paris, New York, and Oxford, UNESCO and Berghahn Books.

U.S. Environmental Protection Agency. 2009. *Water on tap: What you need to know.* EPA 816-K-09-002. Office of Water, Washington, D.C.

U.S. Government Accountability Office (GAO). 2009. *Bottled water: FDA safety and consumer protections are often less stringent than comparable EPA protections for tap water.* GAO Report to Congressional Requesters, GAO-09-610.

Worldwatch Institute. 2011. *State of the world 2011: Innovations that nourish the planet.*

Worldwatch Institute, Washington, D.C.

Chapter 16

Abbing, Michiel Roscam. 2019. *Plastic soup: An atlas of ocean pollution.* Island Press, Washington, D.C.

Allsopp, Michelle, et al. 2007. *Oceans in peril: Protecting marine biodiversity.* Worldwatch Report 174. Worldwatch Institute, Washington, D.C.

Andrady, A. L. 2011. Microplastics in the marine environment. *Mar. Pollut. Bull.* **62**: 1596–1605.

Food and Agriculture Organization. 2016. *The state of world fisheries and aquaculture 2016.* FAO Fisheries and Aquaculture Department, Rome.

Frank, Kenneth T., et al. 2005. Trophic cascades in a formerly cod-dominated ecosystem. *Science* 308: 1621–1623.

Garrison, Tom. 2012. *Oceanography: An invitation to marine science*, 8th ed. Brooks/Cole, San Francisco.

Gell, Fiona R., and Callum M. Roberts. 2003. Benefits beyond boundaries: The fishery effects of marine reserves. *Trends in Ecology and Evolution* 18: 448–455.

Halweil, Brian. 2006. *Catch of the day: Choosing seafood for healthier oceans.* Worldwatch Paper #172. Worldwatch Institute, Washington, D.C.

Hoegh-Guldberg, O., et al. 2007. Coral reefs under rapid climate change and ocean acidification. *Science* 318: 1737–1742.

Jackson, Jeremy B.C., et al. 2001. Historical overfishing and the recent collapse of coastal ecosystems. *Science* 293: 629–638.

Jambeck, J. R., et al. 2015. Plastic Waste Inputs from Land into the Ocean. *Science* 34: 768–771.

Lebreton, L., et al., 2018. Evidence that the Great Pacific Garbage Patch is rapidly accumulating plastic. *Nature Scientific Reports* 8: 4666.

Lotze, Heike L., et al. 2006. Depletion, degradation, and recovery potential of estuaries and coastal seas. *Science* 312: 1806–1809.

Moore, Charles James. 2008. Synthetic polymers in the marine environment: A rapidly increasing, long-term threat. *Environmental Research* 108: 131–139.

Morrissey, John F. and James L. Sumich. 2010. *Introduction to the biology of marine life,* 10th ed. Jones & Bartlett, Boston.

Myers, Ransom A., and Boris Worm. 2003. Rapid worldwide depletion of predatory fish communities. *Nature* 423: 280–283.

National Center for Ecological Analysis and Synthesis (NCEAS) and Communication Partnership for Science and the Sea (COMPASS), sponsors. 2001. *Scientific consensus statement on marine reserves and marine protected areas.* www.nceas.ucsb.edu/consensus.

National Geographic. 2017. Charles Moore is now a two-time Garbage Patch discoverer (and I can tell you what a Garbage Patch looks like). https://blog.nationalgeographic.org/2017/07/28/charles-moore-is-now-a-two-time-garbage-patch-discoverer-and-i-can-tell-you-what-a-garbage-patch-looks-like.

National Research Council. 2003. *Oil in the sea III: Inputs, fates, and effects.* National Academies Press, Washington, D.C.

Norse, Elliott, and Larry B. Crowder, eds. 2005. *Marine conservation biology: The science of maintaining the sea's biodiversity.* Island Press, Washington, D.C.

Orr, James C. 2005. Anthropogenic ocean acidification over the twenty-first century and its impact on calcifying organisms. *Nature* 437: 681–686.

Pauly, Daniel, et al. 2002. Towards sustainability in world fisheries. *Nature* 418: 689–695.

Pauly, Daniel, et al. 2003. The future for fisheries. *Science* 302: 1359–1361.

Roberts, Callum M., et al. 2001. Effects of marine reserves on adjacent fisheries. *Science* 294: 1920–1923.

Rosenberg, A., et al. 2006. Rebuilding U.S. fisheries: Progress and problems. *Frontiers in Ecology and the Environment* 4(6).

TEEB. 2012. *Why value the oceans—A discussion paper.* The Economics of Ecosystems and Biodiversity.

United Nations Environment Programme. 2018. Single-Use Plastics: A Roadmap for Sustainability. https://wedocs.unep.org/bitstream/handle/20.500.11822/25496/singleUsePlastic_sustainability.pdf.

United Nations Food and Agriculture Organization. 2018. *The state of world fisheries and aquaculture 2018.* New York, NY.

Weber, Michael L. 2001. *From abundance to scarcity: A history of U.S. marine fisheries policy.* Island Press, Washington, D.C.

World Economic Forum and Ellen MacArthur Foundation. 2016. *The New Plastics Economy: Rethinking the Future of Plastics.* www.ellenmacarthurfoundation.org/our-work/activities/new-plastics-economy/2016-report.

World Resources Institute. 2005. *Millennium Ecosystem Assessment, 2005. Ecosystems and human well-being: Biodiversity synthesis.* World Resources Institute, Washington, D.C.

Worm, Boris, et al. 2006. Impacts of biodiversity loss on ocean ecosystem services. *Science* 314: 787–790.

Chapter 17

Ahrens, C. Donald, and Robert Henson. 2018. *Meteorology today,* 12 ed. Cengage Learning, Boston.

Akimoto, Hajime. 2003. Global air quality and pollution. *Science* 302: 1716–1719.

Borja-Aburto, Victor H., et al. 1997. Ozone, suspended particulates, and daily mortality in Mexico City. *Am. J. Epidemiology* 145: 258–268.

Bruce, Nigel, Rogelio Perez-Padilla, and Rachel Albalak. 2000. Indoor air pollution in developing countries: A major environmental and public health challenge. *Bull. World Health Organization* 78: 1078–1092.

Calderón-Garcidueñas, Lilian, et al. 2003. Respiratory damage in children exposed to urban pollution. *Pediatric Pulmonology* 36: 148–161.

Cooper, C. David, and F. C. Alley. 2010. *Air pollution control: A design approach,* 4th ed. Waveland Press, Long Grove, Ill.

Davis, Devra. 2002. *When smoke ran like water: Tales of environmental deception and the battle against pollution.* Basic Books, New York.

Davis, Devra L., et al. 2002. A look back at the London smog of 1952 and the half century since. *Environ. Health Perspect.* 110: A734.

Davis, Lucas W. 2017. Saturday driving restrictions fail to improve air quality in Mexico City. *Nature Scientific Reports* 7: 41652.

Driscoll, Charles T., et al. 2001. *Acid rain revisited: Advances in scientific understanding since the passage of the 1970 and 1990 Clean Air Act Amendments.* Hubbard Brook Research Foundation.

Economist. 2018. Why India is one of the most polluted countries on Earth. *Economist.* Dec. 6, 2018.

Fabian, Peter, and Martin Dameris. 2014. *Ozone in the atmosphere: Basic principles, natural and human impacts.* Springer, Berlin.

Godish, Thad, et al. 2014. *Air quality,* 5th ed. CRC Press, Boca Raton, Fla.

González, Andrés. 2018. Mexico City cleaned up its act but Mexicans don't believe it. *Huffington Post.* Jan. 23, 2018.

Hall, Jane V., et al. 2008. *The benefits of meeting federal clean air standards in the South Coast and San Joaquin Valley air basins.* William and Flora Hewlett Foundation.

Hoffman, Matthew J. 2005. *Ozone depletion and climate change: Constructing a global response.* SUNY Press, New York.

Jacobs, Chip, and William J. Kelly. 2008. *Smogtown: The lung-burning history of pollution in Los Angeles.* Overlook Press, New York.

Jacobson, Mark Z. 2002. *Atmospheric pollution: History, science, and regulation.* Cambridge University Press, Cambridge, U.K.

Jenkins, Jerry C., et al. 2007. *Acid rain in the Adirondacks: An environmental history.* Comstock, Ithaca, N.Y.

Kaiman, Jonathan. 2013. Chinese struggle through 'airpocalypse' smog. *The Observer,* 16 Feb. 2013.

Kerrigan, Saoirse. 2018. 15 projects that could quash air pollution around the world. *Interesting Engineering.* July 3, 2018.

Likens, Gene E. 2004. Some perspectives on long-term biogeochemical research from the Hubbard Brook ecosystem study. *Ecology* 85: 2355–2362.

Loomis, Dana, et al. 1999. Air pollution and infant mortality in Mexico City. *Epidemiology* 10: 118–123.

Lutgens, Frederick K., et al. 2018. *The atmosphere: An introduction to meteorology,* 14th ed. Pearson Education, London, U.K.

McDonnell, Patrick J. 2016. An old nemesis returns: Smoggy days in Mexico City. *Los Angeles Times,* 19 May 2016.

Molina, Mario J., and F. Sherwood Rowland. 1974. Stratospheric sink for chlorofluoromethanes: Chlorine atom catalyzed destruction of ozone. *Nature* 249: 810–812.

Parson, Edward A. 2003. *Protecting the ozone layer: Science and strategy.* Oxford University Press.

Solomon, Susan, et al. 2016. Emergence of healing in the Antarctic ozone layer. *Science* 353: 269–274.

U.S. Environmental Protection Agency. www.epa.gov/environmental-topics/air-topics.

U.S. Environmental Protection Agency. Summary of the Clean Air Act. www.epa.gov/laws-regulations/summary-clean-air-act.

U.S. Environmental Protection Agency. 2016. *2013 program progress: Clean Air Interstate Rule, Acid Rain Program, and former NOx Budget Trading Program.* EPA, Washington, D.C.

U.S. Environmental Protection Agency. *National air quality: Status and trends of key air pollutants.* EPA, Washington, D.C. www3.epa.gov/airtrends/aqtrends.html.

U.S. Environmental Protection Agency. 2018. *2014 National air toxics assessment.* EPA, Washington, D.C. www.epa.gov/national-air-toxics-assessment/nata-overview.

Vallero, Daniel. 2014. *Fundamentals of air pollution,* 5th ed. Academic Press, Cambridge, U.K.

Wong, Edward. 2015. As Beijing shuts down over smog alert, worse-off neighbors carry on. *New York Times,* 9 Dec. 2015.

World Health Organization. 2016. *WHO global urban ambient air pollution database (update 2016).* www.who.int/phe/health_topics/outdoorair/databases/cities.

World Meteorological Organization. *Scientific assessment of ozone depletion: 2014.* WMO, Global Ozone Research and Monitoring Project, Report #55, Geneva, Switzerland, 416 pp.

Chapter 18

Al, Stefan. 2018. *Adapting cities to sea-level rise: Green and gray strategies.* Island Press, Washington, D.C.

Alley, Richard B. 2000. *The two-mile time machine: Ice cores, abrupt climate change, and our future.* Princeton University Press, Princeton, N.J.

Baptiste, Nathalie. 2016. That sinking feeling: The politics of sea-level rise and Miami's building boom. *American Prospect,* 19 Feb. 2016.

Bianco, Nicholas M., and Franz T. Litz. 2010. *Reducing greenhouse gas emissions in the United States using existing federal authorities and state action.* WRI Report, World Resources Institute, Washington, D.C.

Bloom, Arnold J. 2009. *Global climate change: Convergence of disciplines.* Sinauer Associates, Sunderland, Mass.

Boswell, Michael R., et al. 2019. *Climate action planning: A guide to creating low-carbon, resilient communities.* Island Press, Washington, D.C.

Burney, Nelson E., ed. 2010. *Carbon tax and cap-and-trade tools: Market-based approaches for controlling greenhouse gases.* Nova Science Publishers, Hauppauge, N.Y.

Burroughs, William James. 2007. *Climate change: A multidisciplinary approach*, 2nd ed. Cambridge Univ. Press, Cambridge, U.K.

Carbon Tax Center. Where carbon is taxed. www.carbontax.org/where-carbon-is-taxed.

Center for Climate and Energy Solutions. www.c2es.org.

Climate Central. www.climatecentral.org.

Cohen, Judah., et al., 2018. Warm Arctic episodes linked with increased frequency of extreme winter weather in the United States. *Nature Communications* 9: 869.

Davenport, Coral. 2015. A climate deal, 6 fateful years in the making. *New York Times,* 13 Dec. 2015.

Dorger, Samanda. 2019. These U.S. cities are the most at risk from rising seas. *TheStreet.* Jan. 22, 2109.

EPICA community members. 2004. Eight glacial cycles from an Antarctic ice core. *Nature* 429: 623–628.

Field, Christopher, et al., eds. 2012. *Managing the risks of extreme events and disasters to advance climate change adaptation.* Special Report of the Intergovernmental Panel on Climate Change. IPCC and Cambridge University Press.

Fitzpatrick, Matthew C., and Robert R. Dunn. 2019. Contemporary climatic analogs for 540 North American urban areas in the late 21st century. *Nature Communications* 10: 614.

Flannery, Tim. 2005. *The weather makers: The history and future impact of climate change.* Text Publishing, Melbourne, Australia.

Francis, Jennifer A., and Stephen J. Vavrus. 2012. Evidence linking Arctic amplification to extreme weather in mid-latitudes. *Geophysical Research Letters* 39: L06801, 6 pp.

Gertner, Jon. 2017. Is it O.K. to tinker with the environment to fight climate change? *New York Times Magazine.* April 18, 2017.

Goodell, Jeff. 2013. Goodbye, Miami. *Rolling Stone.* 20 June 2013.

Gore, Al. 2006. *An inconvenient truth: The planetary emergency of global warming and what we can do about it.* Rodale Press and Melcher Media, New York.

Gore, Al. 2017. *An inconvenient sequel: Truth to power: Your action handbook to learn the science, find your voice, and help solve the climate crisis.* Rodale Books, New York.

Hansen, James, Makiko Sato, and Reto Ruedy. 2012. Perception of climate change. *Proc. Natl. Acad. Sci. USA* E2415-E2423, 6 Aug. 2012.

Harvey, Hal. 2018. *Designing climate solutions: A policy guide for low-carbon energy.* Island Press, Washington, D.C.

Hawken, Paul, ed. 2017. *Drawdown: The most comprehensive plan ever proposed to reverse global warming.* Penguin Books, New York.

InsideClimate News. www.insideclimatenews.org.

Intergovernmental Panel on Climate Change. www.ipcc.ch.

Intergovernmental Panel on Climate Change. 2013. *Climate change 2013: The physical science basis. Working Group I contribution to the Fifth Assessment Report of the IPCC.* (Stocker, Thomas F., et al., eds.). IPCC, Geneva, and Cambridge University Press, Cambridge.

Intergovernmental Panel on Climate Change. 2014. *Climate change 2014: Impacts, adaptation, and vulnerability. Working Group II contribution to the Fifth Assessment Report of the IPCC.* (Field, Christopher B., and Vicente R. Barros, et al., eds.). IPCC, Geneva, and Cambridge University Press, Cambridge.

Intergovernmental Panel on Climate Change. 2014. *Climate change 2014: Mitigation of climate change. Working Group III contribution to the Fifth Assessment Report of the IPCC.* (Edenhofer, Ottmar, et al., eds.). IPCC, Geneva, and Cambridge University Press, Cambridge.

Intergovernmental Panel on Climate Change. 2014. *Climate change 2014: Synthesis report. Contribution of Working Groups I, II, and III to the Fifth Assessment Report of the IPCC* (Core writing team: Pachauri, R.K., and L.A. Meyer, eds.). IPCC, Geneva, Switzerland.

Intergovernmental Panel on Climate Change. 2018. *Global warming of 1.5°C: An IPCC special report on the impacts of global warming of 1.5°C above pre-industrial levels and related global greenhouse gas emission pathways, in the context of strengthening the global response to the threat of climate change, sustainable development, and efforts to eradicate poverty.* (Masson-Delmotte, V., et al., eds.) IPCC, Geneva, Switzerland.

International Energy Agency. 2015. *Energy and climate change.* World Energy Outlook Special Report. IEA, Paris.

Jackson, Robert B., et al. 2016. Reaching peak emissions. *Nature Climate Change* 6: 7–10.

Jonzén, Niclas, et al. 2006. Rapid advance of spring arrival dates in long-distance migratory birds. *Science* 312: 1959–1961.

Kamp, David. 2015. Can Miami Beach survive global warming? *Vanity Fair,* 11 Nov. 2015.

Klein, Naomi. 2015. *This changes everything: Capitalism versus the climate.* Simon and Schuster, New York.

Klein, Naomi. 2019. *On fire: The (burning) case for a Green New Deal.* Simon and Schuster, New York.

Kolbert, Elizabeth. 2015. The siege of Miami. *New Yorker*, 21 & 28 Dec. 2015.

Kraska, James, ed. 2013. *Arctic security in an age of climate change.* Cambridge University Press, Cambridge, U.K.

Leahy, Stephen. 2018. Here's what you need to known about carbon pricing. *Ensia.* Oct. 31, 2018.

Loria, Kevin. 2018. Miami is racing against time to keep up with sea-level rise. *Business Insider.* April 12, 2018.

Mann, Michael, and Lee R. Kump. 2015. *Dire predictions: Understanding global warming*, 2nd ed. DK Publishing and Pearson Education, New York.

Marshall, George. 2015. *Don't even think about it: Why our brains are wired to ignore climate change.* Bloomsbury USA, New York.

Martinich, Jeremy, and Allison Crimmins. 2019. Climate damages and adaptation potential across diverse sectors of the United States. *Nature Climate Change* 9: 397–404.

Mayewski, Paul A., and Frank White. 2002. *The ice chronicles: The quest to understand global climate change.* University Press of New England, Hanover, N.H.

McKibben, Bill. 2012. Global warming's terrifying new math. *Rolling Stone,* 19 July 2012.

Miami-Dade Sea Level Rise Task Force. 2014. *Miami-Dade Sea Level Rise Task Force Report and Recommendations.* Miami, Fla.

NOAA Climate.gov. www.climate.gov.

NOAA State of the Climate. www.ncdc.noaa.gov/sotc.

Oreskes, Naomi, and Erik Conway. 2010. *Merchants of doubt.* Bloomsbury, London.

Oreskes, Naomi, and Erik Conway. 2014. *The collapse of Western civilization: A view from the future.* Columbia University Press, New York.

Osborne, E., et al., 2018. *Arctic report card 2018.* National Oceanographic and Atmospheric Administration, Washington, D.C.

Pacala, Stephen, and Robert Socolow. 2004. Stabilization wedges: Solving the climate problem for the next 50 years with current technologies. *Science* 305: 968–972.

Parmesan, Camille, and Gary Yohe. 2003. A globally coherent fingerprint of climate change impacts across natural systems. *Nature* 421: 37–42.

Real Climate: Climate science from climate scientists. www.realclimate.org.

Rich, Nathaniel. 2019. *Losing Earth: A recent history.* MCD Books.

Ricke, Katherine, et al. 2018. Country-level social cost of carbon. *Nature Climate Change* 8: 895–900.

Roberts, David. 2016. The political hurdles facing a carbon tax—and how to overcome them. *Vox.* April 26, 2016.

Romm, Joseph. 2015. *Climate change: What everyone needs to know.* Oxford University Press, Oxford, U.K.

The Royal Society. 2005. *Ocean acidification due to increasing atmospheric carbon dioxide.* The Royal Society, London, June 2005.

Schneider, Stephen H., and Terry L. Root, eds. 2002. *Wildlife responses to climate change: North American case studies.* Island Press, Washington, D.C.

Service, Robert. 2012. Rising acidity brings an ocean of trouble. *Science* 337: 146–148.

Solovitch, Sara. 2016. How Miami Beach is keeping the Florida dream alive—and dry. *Politico,* 14 Mar. 2016.

Southeast Florida Regional Climate Change Compact. www.southeastfloridaclimatecompact.org.

Southeast Florida Regional Climate Change Compact Sea Level Rise Working Group. 2015. *Unified sea level rise projection for southeast Florida.* A document prepared for the Southeast Florida Regional Climate Change Compact Steering Committee. 35 pp.

Surging Seas: Sea-level rise analysis by Climate Central. http://sealevel.climatecentral.org.

Thompson, Andrea. 2018. Here's how much climate change could cost the U.S. *Scientific American.* Dec. 3, 2018.

Tompkins, Forbes, and Christina DeConcini. 2014. *Sea-level rise and its impact on Miami-Dade County.* World Resources Institute, Washington, D.C.

Toomey, Diane. 2015. How British Columbia gained by putting a price on carbon. *Yale Environment 360,* 30 Apr. 2015.

Union of Concerned Scientists. 2018. *Underwater: Rising seas, chronic floods, and the implications for U.S. coastal real estate.* Union of Concerned Scientists, Cambridge, Mass.

United Nations. The Paris Agreement. https://unfccc.int/process-and-meetings/the-paris-agreement/the-paris-agreement.

U.N. Environment Programme. 2011. *Bridging the emissions gap.* A UNEP Synthesis Report.

U.S. Department of Defense. 2014. *2014 Climate change adaptation roadmap.* Office of the Deputy Under Secretary of Defense for Installations and Environment, Alexandria, Va.

U.S. Environmental Protection Agency. 2015. *Climate change in the United States: Benefits of global action.* EPA 430-R-15-001, June 2015. EPA, Washington, D.C.

U.S. Global Change Research Program. 2017. *Climate science special report: Fourth national climate assessment, volume I.* (Wuebbles, D.J., et al., eds.) USGCRP, Washington, D.C.

U.S. Global Change Research Program. 2018. *Impacts, risks, and adaptation in the United States: Fourth national climate assessment, volume II.* (Reidmiller, D.R., et al., eds.) USGCRP, Washington, D.C.

Wallace-Wells, David. 2019. *The uninhabitable Earth: Life after warming.* Tim Duggan Books.

Washington Post. 2019. How we can combat climate change: 11 policy ideas to protect the planet. *Washington Post.* Jan. 2, 2019.

Watts, Nick, et al. 2017. The *Lancet* countdown on health and climate change: From 25 years of inaction to a global transformation for public health. *Lancet.* http://dx.doi.org/10.1016/S0140-6736(17)32464-9.

Weiss, Jessica. 2016. Miami Beach's $400 million sea-level rise plan is unprecedented, but not everyone is sold. *Miami New Times,* 19 Apr. 2016.

Wilcox, Jennifer. 2012. *Carbon capture.* Springer, New York.

World Bank. 2012. *Turn down the heat: Why a 4° warmer world must be avoided.* A report for the World Bank by the Potsdam Institute for Climate Impact Research and Climate Analytics. World Bank, Washington D.C.

World Meteorological Organization. 2019. *WMO statement on the status of the global climate in 2018.* WMO, Geneva, Switzerland.

Xu, Yanyang, et al. 2018. Global warming will happen faster than we think. *Nature* 564: 30–32.

Chapter 19

Al-Fattah, Saud M., et al. 2011. *Carbon capture and storage: Technologies, policies, economics, and implementation strategies.* CRC Press, Boca Raton, Fla.

American Petroleum Institute. 2009. *Offshore access to oil and natural gas reserves.* American Petroleum Institute, Washington, D.C.

Avery, Samuel. 2013. *The pipeline and the paradigm: Keystone XL, tar sands, and the battle to defuse the carbon bomb.* Ruka Press, Washington, D.C.

Banerjee, Neela. 2019. Court sides with Trump on Keystone XL permit, but don't expect fast progress. *Inside Climate News.* June 7, 2019.

British Petroleum. 2019. *BP statistical review of world energy 2019.* BP, London.

Energy Information Administration, U.S. Department of Energy. www.eia.doe.gov.

Energy Information Administration, U.S. Department of Energy. Monthly energy review. Data updated monthly online at: www.eia.doe.gov/total-energy/data/monthly.

Energy Information Administration, U.S. Department of Energy. 2019. *Annual energy outlook 2019.* DOE/EIA, Washington, D.C.

Findley, David. 2010. *Do-it-yourself home energy audits: 140 simple solutions to lower energy costs, increase your home's efficiency, and save the environment.* McGraw-Hill, New York.

Hall, Charles A.S., and Doug Hansen. 2011. New Studies in EROI (Energy Return on Investment). *Sustainability,* special issue. MDPI AG, Basel, Switzerland, 2011.

Hall, Charles A.S., et al. 2014. EROI of different fuels and the implications for society. *Energy Policy* 64: 141–152.

Haerens, Margaret. 2010. *Offshore drilling (opposing viewpoints).* Greenhaven Press, Farmington Hills, Mich.

Herring, Horace, and Steve Sorrell, eds. 2009. *Energy efficiency and sustainable consumption: The rebound effect.* Palgrave Macmillan, London.

Hubbert, M. King. 1956. *Nuclear energy and the fossil fuels.* Publication No. 95, Shell Development Company, Houston, Tex.

International Energy Agency. 2018. *Key world energy statistics 2018.* IEA, Paris.

International Energy Agency. 2018. *World energy outlook 2018.* IEA, Paris.

International Energy Agency. 2018. *World energy investment 2018.* IEA, Paris.

Jackson, Robert B., et al. 2014. The environmental costs and benefits of fracking. *Annu. Rev. Environ. Resources* 39: 327–362.

Kitasei, Saya. 2010. *Powering the low-carbon economy: The once and future roles of renewable energy and natural gas.* Worldwatch Report 184. Worldwatch Institute, Washington, D.C.

Kunstler, James H. 2005. *The long emergency.* Atlantic Monthly Press, New York.

Kusnetz, Nicholas. 2019. Canada's tar sands province elects a combative new leader promising oil and pipeline revival. *Inside Climate News.* April 18, 2019.

Leffler, William L., et al., 2011. *Deepwater petroleum exploration and production: A nontechnical guide,* 2nd ed. Pennwell Corp. Publishing, Tulsa, Okla.

Levant, Ezra. 2011. *Ethical oil: The case for Canada's oil sands.* McClelland & Stewart, Toronto, Ontario.

Lovins, Amory B. 2005. More profit with less carbon. *Scientific American* 293(3): 74–83.

Lovins, Amory B., et al. 2004. *Winning the oil endgame: Innovation for profits, jobs, and security.* Rocky Mountain Institute, Snowmass, Colo.

McGlade, Christophe, and Paul Ekins. 2015. The geographical distribution of fossil fuels unused when limiting global warming to 2°C. *Nature* 517: 187–190.

Miller, Bruce G. 2010. *Clean coal engineering technology.* Butterworth-Heinemann, Elsevier.

National Academies of Sciences, Engineering, and Medicine. 2018. *Negative emissions technologies and reliable sequestration: A research agenda.* National Academies Press, Washington, D.C.

National Commission on the BP Deepwater Horizon Oil Spill and Offshore Drilling. 2011. *Deep water: The Gulf oil disaster and the future of offshore drilling.* Report to the President. Oil Spill Commission.

National Oceanic and Atmospheric Administration. 2012. *Natural resource damage assessment for the Deepwater Horizon oil spill: April 2012 status update.* NOAA, Washington, D.C.

OECD. 2012. Inventory of estimated budgetary support and tax expenditures for fossil fuels 2013. OECD Publishing.

Parfomak, Paul, et al., 2012. *Keystone XL pipeline project: Key issues.* Congressional Research Service CRS Report for Congress 7-5700 R41668.

Pirani, Simon. 2018. *Burning up: A global history of fossil fuel consumption.* Pluto Press, London, U.K.

Rao, Vikram. 2012. *Shale gas: The promise and the peril.* RTI International, RTI Press, Research Triangle Institute, N.C.

Renne, John L., and Billy Fields. 2013. *Transport beyond oil: Policy choices for a multimodal future.* Island Press, Washington, D.C.

Ristinen, Robert A., et al., 2016. *Energy and the environment,* 3rd ed. John Wiley & Sons, New York.

Smil, Vaclav. 2015. *Natural gas: Fuel for the 21st century.* Wiley, Hoboken, N.J.

Smil, Vaclav. 2018. *Oil: A beginner's guide.* 2nd ed. Oneworld Publications, London, U.K.

Smil, Vaclav. 2018. *Energy and civilization: A history.* MIT Press, Cambridge, Mass.

Spellman, Frank R. 2013. *Environmental impacts of hydraulic fracturing.* CRC Press, Boca Raton, Fla.

Sumper, Andreas, and Angelo Baggini. 2012. *Electrical energy efficiency: Technologies and applications.* John Wiley & Sons, Chichester, U.K.

Union of Concerned Scientists. 2017. *A dwindling role for coal.* Union of Concerned Scientists, Cambridge, Mass.

U.S. Department of State. 2014. *Final supplemental environmental impact statement for the Keystone XL project.* Jan. 2014.

U.S. Environmental Protection Agency. 2019. *The 2018 EPA automotive trends report: Greenhouse gas emissions, fuel economy, and technology since 1975.* EPA-420-R-19-002. EPA, Washington, D.C.

U.S. Geological Survey. 2001. *Arctic National Wildlife Refuge, 1002 Area, petroleum assessment, 1998, including economic analysis.* USGS Fact Sheet FS-028-01.

U.S. Geological Survey. 2002. *Petroleum resource assessment of the National Petroleum Reserve Alaska (NPRA).* USGS, Washington, D.C.

U.S. Government Accountability Office (GAO). 2007. *Crude oil: Uncertainty about future oil supply makes it important to develop a strategy for addressing a peak and decline in oil production.* Report to Congressional Requesters.

White, Helen K., et al. 2012. Impact of the *Deepwater Horizon* oil spill on a deep-water coral community in the Gulf of Mexico. *Proc. Natl. Acad. Sci. USA* 109: 20303-20308.

White House, The, U.S. 2015. Statement by the President on the Keystone XL Pipeline. www.whitehouse.gov/the-press-office/2015/11/06/statement-president-keystone-xl-pipeline.

Wilcox, Jennifer. 2012. *Carbon capture.* Springer, New York.

Yergin, Daniel. 1991. *The prize: The epic quest for oil, money, and power.* Simon and Schuster, New York.

Yergin, Daniel. 2011. *The quest: Energy, security, and the remaking of the modern world.* Penguin Press, London.

Chapter 20

Ansar, Atif, et al. 2014. Should we build more large dams? The actual costs of hydropower megaproject development. *Energy Policy* 69: 43–56.

Biomass Research and Development Board. 2008. *National biofuels action plan.* Biomass Research and Development Initiative, U.S. DOE and USDA.

British Petroleum. 2019. *BP statistical review of world energy 2019.* BP, London.

The Chernobyl Forum. 2006. *Chernobyl's legacy: Health, environmental and socio-economic impacts* and *recommendations to the governments of Belarus, the Russian Federation and Ukraine. The Chernobyl Forum: 2003–2005.* Second revised version. World Health Organization and International Atomic Energy Agency, Vienna.

Earley, Jane, and Alice McKeown. 2009. *Red, white, and green: Transforming U.S. biofuels.* Worldwatch Report 180. Worldwatch Institute, Washington, D.C.

Energy Efficiency and Renewable Energy, U.S. Department of Energy. www.eere.energy.gov.

Energy Information Administration, U.S. Department of Energy. www.eia.doe.gov.

Energy Information Administration, U.S. Department of Energy. Monthly energy review. Data updated monthly online at: www.eia.doe.gov/total-energy/data/monthly.

European Commission/International Atomic Energy Agency/World Health Organization. 1996. One decade after Chernobyl: Summing up the consequences of the accident. Summary of the conference results. Vienna, Austria, 8–12 April 1996. EC/IAEA/WHO.

Flavin, Christopher. 2008. *Low-carbon energy: A roadmap.* Worldwatch Report 178. Worldwatch Institute, Washington, D.C.

Hall, Charles A.S. 2016. *Energy return on investment: A unifying principle for biology, economics, and sustainability.* Springer, Berlin.

Hall, Charles A.S., and Doug Hansen. 2011. New Studies in EROI (Energy Return on Investment). *Sustainability,* special issue. MDPI AG, Basel, Switzerland, 2011.

Hall, Charles A.S., et al. 2014. EROI of different fuels and the implications for society. *Energy Policy* 64: 141–152.

International Atomic Energy Agency. 2016. *Nuclear power and sustainable development.* IAEA Information Series 02-01574/FS Series 3/01/E/Rev.1. Vienna.

International Atomic Energy Agency. 2006. *Environmental consequences of the Chernobyl accident and their remediation: Twenty years of experience.* Report of the U.N. Chernobyl Forum Expert Group "Environment." IAEA, Vienna.

International Energy Agency. 2007. *Biomass for power generation and CHP.* IEA, Paris.

International Energy Agency. 2018. *Key world energy statistics 2018.* IEA, Paris.

International Energy Agency. 2015. *Technology roadmap: Nuclear energy.* IEA, Paris.

International Energy Agency. 2018. *World energy outlook 2018.* IEA, Paris.

Ishikawa, Tetsuo, et al., 2015. The Fukushima Health Management Survey: Estimation of external doses to residents in Fukushima Prefecture. *Nature Scientific Reports* 5: 12712.

Jackson, R.B., et al. 2018. Global energy growth is outpacing decarbonization. *Environ. Res. Lett.* 13: 120401.

Lynas, Mark. 2014. Nuclear 2.0: *Why a green future needs nuclear power.* UIT Cambridge Ltd.

Murphy, David J., and Charles A.S. Hall. 2010. Year in review—EROI or energy return on (energy) invested. *Annals of the New York Academy of Sciences* 1185: 102–118.

National Renewable Energy Lab, U.S. Department of Energy. www.nrel.gov.

Nature. 2006. Special report: Chernobyl and the future. *Nature* 440: 982–989.

Normile, Dennis. 2011. Fukushima revives the low-dose debate. *Science* 332: 908–910.

Nuclear Energy Agency. 2002. *Chernobyl: Assessment of radiological and health impacts.* 2002 update of *Chernobyl: Ten years on.* OECD, Paris.

Pearce, Fred. 2006. Fuels gold: Are biofuels really the greenhouse-busting answer to our energy woes? *New Scientist,* 23 Sept. 2006: 36–41.

Qvist, Staffan A., and Barry W. Brook. 2015. Environmental and health impacts of a policy to phase out nuclear power in Sweden. *Energy Policy* 84: 1–10.

Randolph, John, and Gilbert M. Masters. 2018. *Energy for sustainability: Foundations for technology, planning, and policy.* 2nd ed. Island Press, Washington, D.C.

REN21 Renewable Energy Policy Network for the 21st Century. 2019. *Renewables 2019 global status report.* REN21 Secretariat, Paris.

Renewable Fuels Association. 2010. *Fueling a high octane future: 2016 ethanol industry outlook.* RFA, Washington, D.C.

Rosenthal, Elisabeth. 2010. Using waste, Swedish city cuts its fossil fuel use. *New York Times,* 10 Dec. 2010.

Sawin, Janet L., and William R. Moomaw. 2009. *Renewable revolution: Low-carbon energy by 2030.* Worldwatch Report 182. Worldwatch Institute, Washington, D.C.

Schneider, Mycle, et al. 2015. *The world nuclear industry status report 2015.* Mycle Schneider Consulting, Paris and London.

Smil, Vaclav. 2016. *Energy transitions: Global and national perspectives.* 2nd ed. Praeger, New York.

Smil, Vaclav. 2018. *Energy and civilization: A history.* MIT Press, Cambridge, Mass.

Spadaro, Joseph V., et al. 2000. Greenhouse gas emissions of electricity generation chains: Assessing the difference. *IAEA Bulletin* 42(2).

Swedish Bioenergy Association (SVEBIO). 2003. *Focus: Bioenergy*. Nos. 1–10. SVEBIO, Stockholm.

Swedish Energy Agency. www.energimyndigheten.se/en.

Swedish Energy Agency. 2012. *Sustainable biofuels 2011*. Swedish Energy Agency, Eskilstuna, Sweden.

Swedish Energy Agency. 2019. *Energy in Sweden 2019*. Swedish Energy Agency, Eskilstuna, Sweden.

Ten Hoeve, John E., and Mark Z. Jacobson. 2012. Worldwide health effects of the Fukushima Daiichi nuclear accident. *Energy & Environmental Science* 5: 8743-8757.

U.N. Environment Programme. 2009. *Assessing biofuels*. UNEP, Nairobi, Kenya.

U.N. Scientific Committee on the Effects of Atomic Radiation. 2014. *Sources, effects, and risks of ionizing radiation.* UNSCEAR 2013 Report to the General Assembly with Scientific Annexes, Vol. I. United Nations, New York.

Wesoff, Eric. 2017. Hard lessons from the great algae biofuel bubble. *Greentech Media.* April 19, 2017.

World Health Organization. 2006. *Health effects of the Chernobyl accident and special health care programmes.* Report of the U.N. Chernobyl Forum Expert Group "Health." WHO, Geneva.

World Health Organization. 2013. *Health risk assessment from the nuclear accident after the Great East Japan Earthquake and Tsunami.* WHO, Geneva.

World Nuclear Association. 2019. *Nuclear power in Sweden.* www.world-nuclear.org/information-library/country-profiles/countries-o-s/sweden.aspx.

World Nuclear Association. 2019. Generation IV nuclear reactors. www.world-nuclear.org/information-library/nuclear-fuel-cycle/nuclear-power-reactors/generation-iv-nuclear-reactors.aspx.

Worldwatch Institute. 2007. *Biofuels for transport: Global potential and implications for sustainable agriculture and energy in the 21st century.* Worldwatch Institute, Washington, D.C.

Yasunari, Teppei J. 2011. Cesium-137 deposition and contamination of Japanese soils due to the Fukushima nuclear accident. *Proc Natl. Acad. Sci. USA* 108: 19530–19534.

Chapter 21

American Wind Energy Association. 2019. *U.S. wind industry annual market report 2019.* AWEA, Washington, D.C.

Barbose, Galen, et al. 2018. *Tracking the sun: Installed price trends for distributed photovoltaic systems in the United States — 2018 edition.* Lawrence Berkeley National Laboratory, Berkeley, Calif.

Brown, Lester R., et al. 2015. *The great transition: Shifting from fossil fuels to solar and wind energy.* Eath Policy Institute and W.W. Norton, New York.

Davidson, Osha Gray. 2012. *Clean break: The story of Germany's energy transformation and what Americans can learn from it.* Inside Climate News.

Dunn, Seth. 2000. The hydrogen experiment. *World Watch* 13: 14–25.

Economist. 2012. Germany's energy transformation: Energiewende. *The Economist*, 28 July 2012.

Edenhofer, Ottmar, et al., eds. 2012. *Renewable energy sources and climate change mitigation.* Special Report of the Intergovernmental Panel on Climate Change. IPCC and Cambridge University Press.

Eddy, Melissa. 2019. Germany lays out a path to quit coal by 2038. *New York Times.* Jan. 26, 2019.

Energy Efficiency and Renewable Energy, U.S. Department of Energy. www.eere.energy.gov.

Energy Efficiency and Renewable Energy, U.S. Department of Energy. 2019. *2018 wind technologies market report.* EERE, Washington, D.C.

Energy Efficiency and Renewable Energy, U.S. Department of Energy. 2019. *2018 Offshore wind technologies market report.* EERE, Washington, D.C.

Energy Information Administration, U.S. Department of Energy. www.eia.doe.gov.

Energy Information Administration, U.S. Department of Energy. Monthly energy review. Data updated monthly online at: www.eia.doe.gov/totalenergy/data/monthly.

Energy Information Administration, U.S. Department of Energy. 2019. *Annual energy outlook 2019.* DOE/EIA, Washington, D.C.

Federal Ministry for the Environment, Nature Conservation, and Nuclear Safety [Germany]. www.erneuerbare-energien.de/en.

Federal Ministry for the Environment, Nature Conservation, and Nuclear Safety [Germany]. 2015. *Offshore wind energy: An overview of activities in Germany.* Berlin.

Federal Ministry of Economics and Technology [Germany]. www.renewables-made-in-germany.com/index.php?id=50&L=1.

Frankfurt School of Finance and Management, U.N. Environment Programme, and Bloomberg New Energy Finance. 2018. *Global trends in renewable energy investment 2018.* Frankfurt School, Frankfurt, Germany.

Global Wind Energy Council. 2018. *Global wind report 2017.* GWEC, Brussels, Belgium.

Hall, Charles A.S., and Doug Hansen. 2011. New Studies in EROI (Energy Return on Investment). *Sustainability*, special issue. MDPI AG, Basel, Switzerland, 2011.

Heinberg, Richard, and David Fridley. 2016. *Our renewable future.* Island Press, Washington, D.C.

Hernández, Kristian. 2018. An uncertain future for America's wind energy capital. *Center for Public Integrity* and *Mother Jones.* Oct. 9, 2018.

Hockenos, Paul. 2019. In Germany, consumers embrace a shift to home batteries. *Yale Environment 360.* March 18, 2019.

International Energy Agency. 2013. *Technology roadmap: Wind energy. 2013 edition.* IEA, Paris.

International Energy Agency. 2018. *Key world energy statistics 2018.* IEA, Paris.

International Energy Agency. Renewables information. www.iea.org/classicstats/relateddatabases/renewablesinformation.

International Energy Agency. 2015. *Technology roadmap: Hydrogen and fuel cells.* IEA, Paris.

International Energy Agency. 2019. *World energy outlook 2019.* IEA, Paris.

International Energy Agency Photovoltaic Power Systems Programme. 2019. *PVPS annual report 2018.* Imprimerie St. Paul, Fribourg, Switzerland.

International Renewable Energy Agency. 2015. *Renewable energy target setting.* IRENA, Masdar City, Abu Dhabi, UAE.

Irfan, Umair. 2019. Texas's wind and sunlight complement each other exceptionally well. That's huge for its grid. *Vox.* Jan. 8, 2019.

Jacobson, Mark Z. 2009. Review of solutions to global warming, air pollution, and energy security. *Energy & Environmental Science* 2: 148–173.

Jacobson, Mark Z., and Mark A. Delucchi. 2009. A path to sustainable energy by 2030. *Scientific American* 301(5): 58–65.

Jacobson, Mark Z., and Mark A. Delucchi. 2011. Providing all global energy with wind, water, and solar power, Part I: Technologies, energy resources, quantities and areas of infrastructure, and materials. *Energy Policy* 39: 1154–1169.

Jacobson, Mark Z., et al., 2015. 100% clean and renewable wind, water, and sunlight (WWS) all-sector energy roadmaps for the 50 United States. *Energy Environ. Sci.* 8: 2093–2117.

Kitasei, Saya. 2010. *Powering the low-carbon economy: The once and future roles of renewable energy and natural gas.* Worldwatch Report 184. Worldwatch Institute, Washington, D.C.

Klein, Naomi. 2019. *On fire: The (burning) case for a Green New Deal.* Simon and Schuster, New York.

Lazard. 2018. *Lazard's levelized cost of energy analysis—Version 12.0.* Nov. 2018. Lazard.

McNamee, Gregory. 2008. *Careers in renewable energy: Get a green energy job.* Pixy Jack Press, Masonville, Colo.

Morris, Craig, and Martin Pehnt. 2015. *Energy transition: The German energiewende.* Heinrich Boll Foundation, Berlin.

National Renewable Energy Lab, U.S. Department of Energy. www.nrel.gov.

National Renewable Energy Lab. 2019. *Q4 2018/Q1 2019 solar industry update.* NREL/PR-6A20-73992. U.S. DOE, Washington D.C.

Ochs, Alexander, and Shakuntala Makhijan. 2012. *Sustainable energy roadmaps: Guiding the global shift to domestic renewables.* Worldwatch Report 187. Worldwatch Institute, Washington, D.C.

OECD. 2012. Inventory of estimated budgetary support and tax expenditures for fossil fuels 2013. OECD Publishing.

Peake, Stephen. 2018. *Renewable energy: Power for a sustainable future.* 4th ed. Oxford University Press USA, New York.

Pfund, Nancy, and Ben Healey. 2011. *What would Jefferson do? The historical role of federal subsidies in shaping America's energy future.* DBL Investors, San Francisco.

Pontin, Jason. 2018. We gotta get a better battery. But how? *Wired.* Sept. 17, 2019.

REN21 Renewable Energy Policy Network for the 21st Century. 2019. *Renewables 2019 global status report.* REN21 Secretariat, Paris.

Quashning, Volker. 2010. *Renewable energy and climate change.* Wiley-IEEE Press.

Ristinen, Robert A., et al., 2016. *Energy and the environment,* 3rd ed. John Wiley & Sons, New York.

St. John, Jeff. 2019. Texas grid operator reports fuel mix is now 30% carbon-free. *GreenTech Media.* Jan. 23, 2019.

Sisson, Patrick. 2018. A mighty wind: How Texas's pro-energy, anti-regulation philosophy conspired to create a regional wind boom. *Curbed.* Oct. 24, 2018.

Slusarewicz, Joanna H., and Daniel S. Cohan. 2018. Assessing solar and wind complementarity in Texas. *Renewables* 5: 7.

Smil, Vaclav. 2016. *Energy transitions: Global and national perspectives.* 2nd ed. Praeger, New York.

Storrow, Benjamin. 2017. The rise of wind power in Texas. *Scientific American.* April 14, 2017.

Tricks, Henry. 2018. What would it take to decarbonise the global economy? *Economist.* Nov. 29, 2018.

Union of Concerned Scientists. 2015. *How energy storage works.* Union of Concerned Scientists, Cambridge, Mass.

University of Texas at Austin Energy Institute. 2019. *Levelized cost of electricity in the United States: Version 1.4.0.* http://calculators.energy.utexas.edu/lcoe_map/#/county/tech.

Weisman, Alan. 1998. *Gaviotas: A village to reinvent the world.* Chelsea Green Publishing Co., White River Junction, Vt.

Wirth, Harry. 2019. Recent facts about photovoltaics in Germany. Fraunhofer ISE, Freiburg, Germany.

World Future Council. 2010. *Feed-in tariffs—Boosting energy for our future.* Earthscan Publications, London.

Chapter 22

Ayres, Robert U., and Leslie W. Ayres. 1996. *Industrial ecology: Towards closing the materials cycle.* Edward Elgar Press, Cheltenham, U.K.

Bevi. 2018. *The road to zero waste universities: It takes more than just a campus recycling program.* Bevi. Sept. 19, 2018. www.bevi.co/blog/zero-waste-universities.

Brooks, Amy L., et al. 2018. The Chinese import ban and its impact on global plastic waste trade. *Science Advances* 2018 4: eaat0131.

Campbell, Stu. 1998. *Let it rot! The gardener's guide to composting,* 3rd ed. Storey Publishing, North Adams, Mass.

Container Recycling Institute. www.container-recycling.org.

Douglas, Ed. 2009. There's gold in them there landfills. *New Scientist,* 1 Oct. 2008.

Graedel, Thomas E., and Braden R. Allenby. 2009. *Industrial ecology and sustainable engineering.* Prentice Hall, Upper Saddle River, N.J.

Higgs, Micaela Marini. 2019. America's new recycling crisis, explained by an expert. *Vox.* April 2, 2019.

Hieronymi, Klaus, et al., eds. 2012. *E-waste management: From waste to resource.* Routledge, Abingdon, U.K.

Javorsky, Nicole. 2019. How American recycling is changing after China's National Sword. *CityLab.* April 1, 2019.

Larmer, Brook. 2018. E-waste offers an economic opportunity as well as toxicity. *New York Times Magazine.* July 5, 2018.

Leonard, Annie. 2010. *The story of stuff.* Free Press, New York.

Lilienfeld, Robert, and William Rathje. 1998. *Use less stuff: Environmental solutions for who we really are.* Ballantine, New York.

Ludwig, T., et al., 1998. Increasing recycling in academic buildings: A systematic replication. *J. Appl. Behavior Analysis* 31: 683–686.

Manahan, Stanley E. 1999. *Industrial ecology: Environmental chemistry and hazardous waste.* Lewis Publishers, CRC Press, Boca Raton, Fla.

McCormick, Erin, et al. 2019. Where does your plastic go? Global investigation reveals America's dirty secret. *The Guardian.* June 17, 2019.

McDonough, William, and Michael Braungart. 2002. *Cradle to cradle: Remaking the way we make things.* North Point Press, New York.

Moscone, Arianna. 2014. *Waste not, want not: A student manual to create zero waste college campuses.* ScholarWorks, Univ. of Massachusetss Amherst.

New York City Department of Parks and Recreation. Fresh Kills Park. www.nycgovparks.org/sub_your_park/fresh_kills_park/html/fresh_kills_park.html.

Pichtel, John. 2014. *Waste management practices: Municipal, hazardous, and industrial,* 2nd ed. CRC Press, Boca Raton, Fla.

Post-Landfill Action Network. 2017. *PLAN plastic-free campus manual.* Post-Landfill Action Network.

Rathje, William, and Colleen Murphy. 2001. *Rubbish! The archeology of garbage.* University of Arizona Press.

Recyclemania. www.recyclemania.org.

Scott, Nicky. 2007. *Reduce, reuse, recycle: An easy household guide.* Chelsea Green Publishing Co., White River Junction, Vt.

Shin, Dolly. 2014. *Generation and disposition of municipal solid waste (MSW) in the United States—a national survey.* M.S. thesis, Columbia University Earth Engineering Center, New York.

Townsend, Timothy. 2011. Environmental issues and management strategies for waste electronic and electrical equipment. *J. Air & Waste Manage. Assoc.* 61: 587–561.

U.S. Environmental Protection Agency. National overview: Facts and figures on materials, wastes, and recycling. www.epa.gov/facts-and-figures-about-materials-waste-and-recycling/national-overview-facts-and-figures-materials.

U.S. Environmental Protection Agency. Summary of the Comprehensive Environmental Response, Compensation, and Liability Act (Superfund). https://www.epa.gov/laws-regulations/summary-comprehensive-environmental-response-compensation-and-liability-act.

U.S. Environmental Protection Agency. Summary of the Resource Conservation and Recovery Act. www.epa.gov/laws-regulations/summary-resource-conservation-and-recovery-act.

U.S. Environmental Protection Agency. 2016. *Electronic products generation and recycling in the United States, 2013 and 2014.* EPA Office of Resource Conservation and Recovery, Washington, D.C.

U.S. Environmental Protection Agency. 2018. *Advancing sustainable materials management: 2015 fact sheet.* EPA530-F-18-004. EPA Office of Resource Conservation and Recovery, Washington, D.C.

van Haaren, R., et al. 2010. The state of garbage in America. *BioCycle* 51: 16–23.

Chapter 23

Essick, Kristi. 2001. Guns, money, and cell phones. *The Industry Standard Magazine,* 11 Jun. 2001.

Christopherson, Robert W. 2011. *Geosystems: An introduction to physical geography,* 8th ed. Prentice Hall, Upper Saddle River, N.J.

Gordon, R.B., et al. 2006. Metal stocks and sustainability. *Proceedings of the National Academy of Sciences of the United States of America* 103(5): 1209–1214.

Hendryx, Michael, and Melissa M. Ahern. 2009. Mortality in Appalachian coal mining regions: The value of statistical life lost. *Public Health Reports* 124: 541–550.

Mooallem, Jon. 2008. The afterlife of cellphones. *New York Times Magazine* 13 Jan. 2008.

Pericak, A.A., et al., 2018. Mapping the yearly extent of surface coal mining in Central Appalachia using Landsat and Google Earth Engine. *PLOS ONE* 13(7): e0197758.

Perkins, Dexter. 2010. *Mineralogy,* 3rd ed. Prentice Hall, Upper Saddle River, N.J.

Rogich, D.G., and G.R. Matos. 2008. *The global flows of metals and minerals.* U.S. Geological Survey Open-File Report 2008–1355, 11 pp.

Sibley, Scott F., ed. 2004. *Flow studies for recycling metal commodities in the United States.* USGS Circular 1196-A-M. U.S. Geological Survey, Reston, Va.

Skinner, Brian J., and Stephen C. Porter. 2003. *The dynamic earth: An introduction to physical geology,* 5th ed. John Wiley & Sons, Hoboken, N.J.

Sullivan, Daniel E. 2006. *Recycled cell phones—A treasure trove of valuable metals.* USGS Fact Sheet 2006-3097. U.S. Geological Survey, Denver, CO.

Tarbuck, Edward J., Frederick K. Lutgens, and Dennis Tasa. 2011. *Earth science,* 13th ed. Prentice Hall, Upper Saddle River, N.J.

U.N. Security Council. 2001. *Report of the panel of experts on the illegal exploitation of natural resources and other forms of wealth of the Democratic Republic of the Congo*. U.N. Security Council, 12 Apr. 2001.

U.N. Security Council. 2007. *Interim report of the group of experts on the Democratic Republic of the Congo, pursuant to Security Council resolution 1698 (2006)*. U.N. Security Council, 25 Jan. 2007.

U.S. Department of the Interior. 2003. *Surface coal mining reclamation: 25 years of progress, 1977–2002*. Office of Surface Mining, Washington D.C.

U.S. Geological Survey, 2019. Mineral commodity summaries 2019.USGS, Reston, VA.

Campus Sustainability Resources

350.org. *Fossil Free: A campus guide to fossil fuel divestment.* https://gofossil-free.org/wp-content/uploads/2014/05/350_FossilFreeBooklet_LO4.pdf.

Association for the Advancement of Sustainability in Higher Education. www.aashe.org.

Association for the Advancement of Sustainability in Higher Education. Campus Sustainability Hub https://hub.aashe.org.

Association for the Advancement of Sustainability in Higher Education. 2010. *Creating a campus sustainability revolving loan fund: A guide for students.* www.aashe.org/wp-content/uploads/2016/11/Campus-Sustainability-Revolving-Loan-Fund.pdf.

Association for the Advancement of Sustainability in Higher Education. 2010. *Sustainability curriculum in higher education: A call to action.* http://wwwp.oakland.edu/Assets/upload/docs/AIS/Conference/2010_Documents_A_Call_to_Action.pdf.

Association for the Advancement of Sustainability in Higher Education. 2018. *Aggregating higher education demand for renewables: A primer.* www.aashe.org/publications/aggregating-higher-education-demand-renewables-primer/.

Association for the Advancement of Sustainability in Higher Education. 2019. *2019 sustainable campus index.* www.aashe.org/sustainable-campus-index/.

Ball State University: Greening of the Campus [conference series.] http://cms.bsu.edu/academics/centersandinstitutes/goc.

Barth, Matthias. 2013. *Implementing sustainability in higher education: Learning in an age of transformation.* Routledge, New York.

Bartlett, Peggy, and Geoffrey W. Chase, eds. 2004. *Sustainability on campus: Stories and strategies for change.* MIT Press, Cambridge, Mass.

Bartlett, Peggy, and Geoffrey W. Chase, eds. 2013. *Sustainability in higher education: Stories and strategies for transformation.* MIT Press, Cambridge, Mass.

Best Colleges, The. 2019. 11 college recycling programs that put all others to shame. www.thebestcolleges.org.

Bevi. 2018. *The road to zero waste universities: It takes more than just a campus recycling program.* Bevi. Sept. 19, 2018. www.bevi.co/blog/zero-waste-universities.

Campus Ecology. 2008. *Campus environment 2008: A national report card on sustainability in higher education.* Campus Ecology program of the National Wildlife Federation.

Campus Farmers Network. www.campusfarmers.org.

National Wildlife Federation Ecoleaders. Campus Greening. www.nwf.org/campus-ecology.

National Wildlife Federation Ecoleaders. Case studies. www.nwf.org/EcoLeaders/Campus-Ecology-Resource-Center/Case-Studies.

Campus Transport Management: Trip reduction programs on college, university, and research campuses. Online TDM Encyclopedia. Victoria Transport Policy Institute. www.vtpi.org/tdm/tdm5.htm.

Carlson, Scott. 2006. In search of the sustainable campus: With eyes on the future, universities try to clean up their acts. *Chronicle of Higher Education* 53: A10.

Creighton, Sarah Hammond. 1998. *Greening the ivory tower: Improving the environmental track record of universities, colleges, and other institutions.* MIT Press, Cambridge, Mass.

Cross, Stan, et al. 2011. *Going underground on campus: Tapping the Earth for clean, efficient heating and cooling.* National Wildlife Federation Ecoleaders. www.nwf.org/EcoLeaders/Campus-Ecology-Resource-Center/Reports/Going-Underground-on-Campus.

Erickson, Christina. 2012. *Student sustainability educators: A guide to creating and maintaining an eco-rep program on your campus.* National Wildlife Federation Ecoleaders. www.nwf.org/EcoLeaders/Campus-Ecology-Resource-Center/Reports/Student-Sustainability-Educators.

Erickson, Christina, and David J. Eagan. 2009. *Generation E: Students leading for a sustainable, clean energy future.* Campus Ecology program of the National Wildlife Federation.

Filho, Walter Leal, et al., eds. 2015. *Implementing campus greening initiatives: Approaches, methods, and perspectives.* World Sustainability Series. Springer, Berlin.

Fossil Free. www.gofossilfree.org. 350.org.

Indvik, Joe, et al. 2013. *Green revolving funds: An introductory guide to implementation and management.* Sustainable Endowments Institute and Association for the Advancement of Sustainability in Higher Education, Cambridge, Mass.

Jones, Kristy, et al. 2015. *The campus wild: How college and university green landscapes provide havens for wildlife and "lands-on" experiences for students.* National Wildlife Federation Ecoleaders. www.nwf.org/EcoLeaders/Campus-Ecology-Resource-Center/Reports/The-Campus-Wild.

Keniry, Julian. 1995. *Ecodemia: Campus environmental stewardship at the turn of the 21st century.* National Wildlife Federation, Washington, D.C.

Koester, Robert J., James Eflin, and John Vann. 2006. Greening of the campus: A whole-systems approach. *Journal of Cleaner Production* 14: 769–779.

Martin, James, and James E. Samels. 2014. *The sustainable university: Green goals and new challenges for higher education.* Johns Hopkins University Press, Baltimore.

Moscone, Arianna. 2014. *Waste not, want not: A student manual to create zero waste college campuses.* ScholarWorks, Univ. of Massachusetss Amherst.

Newman, Julie. 2009. *Reaching beyond compliance: The challenges of achieving campus sustainability.* VDM Verlag.

Post-Landfill Action Network. 2017. *PLAN plastic-free campus manual.* Post-Landfill Action Network.

Princeton Review. 2018. *Guide to 399 green colleges, 2018.* The Princeton Review, Princeton, N.J. www.princetonreview.com/college-rankings/green-guide.

Recyclemania. www.recyclemania.org.

Second Nature. The Presidents' climate leadership commitments. https://secondnature.org/signatory-handbook/the-commitments/.

Sierra Club. 2018. The top 20 coolest schools 2018. www.sierraclub.org/sierra/cool-schools-2018/top-20-coolest-schools-2018. Sierra Club.

Simpson, Walter, ed. 2008. *The green campus: Meeting the challenge of environmental sustainability.* APPA Association of Higher Education.

Simpson, Walter. 2009. *Cool campus! How-to guide for college and university climate action planning.* AASHE, Lexington, Ky.

Sustainable Endowments Institute: College Sustainability Report Card. www.greenreportcard.org.

Sustainable Endowments Institute. 2012. *Greening the bottom line.* Sustainable Endowments Institute, Cambridge, Mass.

Thomashow, Michael. 2014. *The nine elements of a sustainable campus.* MIT Press, Cambridge. Mass.

Toor, Will, and Spenser W. Havlick. 2004. *Transportation and sustainable campus communities: Issues, examples, solutions.* Island Press, Washington, D.C.

University Leaders for a Sustainable Future. www.ulsf.org.

University of New Hampshire Sustainability Institute. Campus calculator [carbon calculator for campuses]. https://sustainableunh.unh.edu/calculator.

U.S. Department of Energy: Solar Decathlon. www.solardecathlon.gov.

U.S. Green Building Council. 2014. LEED campus guidance v2009. https://www.usgbc.org/leedaddenda/100002248. U.S. Green Building Council, Washington, D.C.

Photo Credits

Cover Seng Chye Teo/Moment Unreleased/Getty Images

Part 1 Gareth Mccormack/Getty Images

Chapter 1 Opening Photo NASA **Figure 1.1A** FotoSpeedy/Getty Images **Figure 1.1B** anyaberkut/123RF **Figure 1.1C** Catalin Petolea/Alamy Stock Photo **Figure 1.2** Roman Borodaev/Shutterstock **Table 1.1A** Shutterstock **Table 1.1B** Simon Belcher/Alamy Stock Photo **Table 1.1C** Dougal Waters/Getty Images **Table 1.1D** Zensen/Agencja Fotograficzna Caro/Alamy Stock Photo **The Science behind the Story (mini)** Courtesy of Carl Lip and Terry Hunt **The Science behind the Story (Figure 1)** Creative RF/Getty Images **Figure 1.8** ALIKI SAPOUNTZI/aliki image library/Alamy Stock Photo **Table 1.2** jacoblund/Getty Images **Table 1.3A** Jim West imageBROKER/Newscom **Table 1.3B** Daderot **Table 1.3C** Edmund D. Fountain/ZUMAPRESS/Newscom **Table 1.3D** Jim West/ZUMA Press/Newscom **Table 1.3E** The Washington Post/Getty Images **Table 1.3F** Michael Doolittle/Alamy Stock Photo **Table 1.3G** Washington Imaging/Alamy Stock Photo **Table 1.3H** Prabjit Virdee **Table 1.3I** James Marabello/Alamy Stock Photo **Table 1.3J** James Ennis **Connect & Continue** NASA **Reviewing Objectives 1** Roman Borodaev/Shutterstock **Reviewing Objectives 2** ALIKI SAPOUNTZI/aliki image library/Alamy Stock Photo **Reviewing Objectives 3** Jim West imageBROKER/Newscom

Chapter 2 Opening Photo The Asahi Shimbun/Getty Images **Central Case Study (mini)** Kyodo Kyodo/Reuters Pictures **Figure 2.1** Tomohiko Kano/Mainichi Shimbun/AP Photo **Success Story** STEVE LINDRIDGE/Alamy Stock Photo **The Science behind the Story (mini)** Ted S. Warren/AP Photo **The Science behind the Story (Figure 1)** Science History Images/Alamy Stock Photo **Figure 2.9A** Phil Date/Shutterstock **Figure 2.9B** Jupiterimages/Stockbyte/Getty Images **Figure 2.17A** ES3N/iStock/Getty Images **Figure 2.17B** Jack Dykinga/Nature Picture Library **Figure 2.17C** Tom Till/Alamy Stock Photo **Figure 2.18A** Harry Laub/imageBROKER/Alamy Stock Photo **Figure 2.18B** Perry Correll/Shutterstock **Figure 2.18C** Jochen Tack/Alamy Stock Photo **Figure 2.21B** Jim Sugar/RGB Ventures/SuperStock/Alamy Stock Photo **The Science behind the Story (mini)** Cornell University Photography **Figure 2.22** PHILIPPE LOPEZ/AFP/Getty Images **Central Case Study** PHILIPPE LOPEZ/AFP/Getty Images **Reviewing Objectives** Jim Sugar/RGB Ventures/SuperStock/Alamy Stock Photo

Chapter 3 Opening Photo Jay Withgott **Central Case Study (mini)** Jack Jeffrey/Photo Resource Hawaii/Alamy Stock Photo **Figure 3.2A** Caters News/ZUMA Press/Newscom **Figure 3.2B** Jack Jeffrey **Figure 3.2C** John Elk/Lonely Planet Images/Getty Images **Figure 3.2D** Christian Dworschak/imageBROKER/Alamy Stock Photo **Figure 3.4A** Chris Curtis/123RF **Figure 3.4B** A Jagel/Blickwinkel/Age Fotostock **The Science behind the Story (mini)** Heather Lerner **Figure 3.9A** AlessandroZocc/Shutterstock **Figure 3.9B** Luis Hidalgo/REUTERS/Alamy Stock Photo **Figure 3.10A** Jack Jeffrey/Photo Resource Hawaii/Alamy Stock Photo **Figure 3.10B** Bernard Castelein/Nature Picture Library/Alamy Stock Photo **Figure 3.11** Fotosearch RM/Age Fotostock **Figure 3.12A** NASA **Figure 3.12B** NadyaEugene/Shutterstock **Figure 3.12C** Danita Delimont/Alamy Stock Photo **Figure 3.12D** Jeff Hunter/Photographer's Choice/Getty Images **Figure 3.12E** David Fleetham/Alamy Stock Photo **Figure 3.12F** M Swiet Productions/Moment/Getty Images **Figure 3.13** C. Douglas Peebles/Douglas Peebles Photography/Alamy Stock Photo **Figure 3.14A** Florilegius/Alamy Stock Photo **Figure 3.14B** North Wind Picture Archives/Alamy Stock Photo **Figure 3.15A** Darren Green Photography/Alamy Stock Photo **Figure 3.15B** Art Wolfe/The Image Bank/Getty Images **Figure 3.15C** Patrick Poendl/Shutterstock **The Science behind the Story (mini)** Jack Jeffrey **The Science behind the Story (Figure 1-A)** Jack Jeffrey **The Science behind the Story (Figure 1-B)** Jack Jeffrey **The Science behind the Story (Figure 2-A)** Dubi Shapiro/AGAMI Photo Agency/Alamy Stock Photo **The Science behind the Story (Figure 3-A)** Jack Jeffrey **The Science behind the Story (Figure 3-B)** Jack Jeffrey **Figure 3.16** Mandar Khadilkar/ephotocorp/Alamy Stock Photo **Figure 3.18** Mandar Khadilkar/ephotocorp/Alamy Stock Photo **Figure 3.20** Doug Perrine/Bluegreen Pictures/Alamy Stock Photo **Figure 3.21** Benning, T. L., LaPointe, D., Atkinson, C. T., & Vitousek, P. M. (2002). Interactions of climate change with biological invasions and land use in the Hawaiian Islands: modeling the fate of endemic birds using a geographic information system. Proceedings of the National Academy of Sciences, 99(22), 14246-14249. **Central Case Study** Jay Withgott **Reviewing Objectives 1** Caters News/ZUMA Press/Newscom **Reviewing Objectives 2** M Swiet Productions/Moment/Getty Images **Reviewing Objectives 3** Art Wolfe/The Image Bank/Getty Images **Reviewing Objectives 4** Doug Perrine/Bluegreen Pictures/Alamy Stock Photo

Chapter 4 Opening Photo United States Fish and Wildlife Service **Central Case Study (mini)** Zoonar/Buntbarsch/Zoonar GmbH/Alamy Stock Photo **Figure 4.4** Arto Hakola/Alamy Stock Photo **Figure 4.6A** Arto Hakola/Shutterstock **Figure 4.6B** Minerva Studio/Shutterstock **Figure 4.6C** Darlyne A. Murawski/National Geographic Image Collection/Alamy Stock Photo **Figure 4.7** James L. Amos/National Geographic **Figure 4.8** blickwinkel/Alamy Stock Photo **Figure 4.9** Rolf Nussbaumer/Rolf Nussbaumer Photography/Alamy Stock Photo **The Science behind the Story (mini)** Dr. Quinton Phelps **Figure 4.15** Dan Callister/Alamy Stock Photo **The Science behind the Story (mini)** Dr. Virginia Dale **The Science behind the Story (Figure 1-A)** USGS/Alamy Stock Photo **The Science behind the Story (Figure 1-B)** Images & Volcans/Science Source **The Science behind the Story (Figure 1-C)** Joseph Martinez/Alamy Stock Photo **Figure 4.17** Chris Helzer **Figure 4.18** US Army Photo/Alamy Stock Photo **Success Story (mini)** Fermi National Accelerator Laboratory **Figure 4.21** luri/Shutterstock **Figure 4.22** Gerrit Vyn/Nature Picture Library **Figure 4.23** Zhukova Valentyna/Shutterstock **Figure 4.24** softlight69/123RF **Figure 4.25** Courtesy of Geoffrey Gallice **Figure 4.26** Darrell Gulin/DigitalVision/Getty Images **Figure 4.27** Oleg Znamenskiy/123RF **Figure 4.28** Radius Images/Alamy Stock Photo **Figure 4.29** Bill Brooks/Alamy Stock Photo **Figure 4.30** Earl Scott/Science Source/Getty Images **Central Case Study** United States Fish and Wildlife Service **Reviewing Objectives 1** James L. Amos/National Geographic **Reviewing Objectives 2** Joseph Martinez/Alamy Stock Photo **Reviewing Objectives 3** Dan Callister/Alamy Stock Photo **Reviewing Objectives 4** Darrell Gulin/DigitalVision/Getty Images

Chapter 5 Opening Photo Patrick Semansky/AP Photo **Central Case Study (mini)** MLADEN ANTONOV/AFP/Getty Images **Figure 5.1A** montian noowong/123RF **Figure 5.1B** K.D. Leperi/Alamy Stock Photo **Figure 5.1C** Edwin Remsberg/Alamy Stock Photo **Figure 5.3A** Serg_v/123RF **Figure 5.3B** tomas1111/123RF **Figure 5.3C** Marc Daneau/EyeEm/Getty Images **Figure 5.3D** Ron and Patty Thomas/iStock/Getty Images **The Science behind the Story (mini)** Russell P. Burke/U.S. Army Corps of Engineers **The Science behind the Story (Figure 2)** David Schulte/U.S. Army Corps of Engineers **Figure 5.8** Image by Reto Stöckli, based on data provided by the MODIS Science Team **The Science behind the Story** Dr. Nancy N. Rabalais **Figure 5.11B** Nikolay Dimitrov/YAY Micro/Age Fotostock **Figure 5.11C** Fotosearch/Age Fotostock **Figure 5.11D** Clint Farlinger/Alamy Stock Photo **Figure 5.11E** Ichiro Murata/Shutterstock **Figure 5.13** Ashley Cooper/Global Warming Images/Alamy Stock Photo **Figure 5.14** Jay Fleming Photography **Success Story** David Harp/Chesapeake Photos.com **Figure 5.21A** Bob Nichols/USDA **Figure 5.21B** James Brunker/Alamy Stock Photo **Central Case Study** Patrick Semansky/AP

Photo **Reviewing Objectives 1** Fotosearch/Age Fotostock **Reviewing Objectives 2** James Brunker/Alamy Stock Photo

Chapter 6 Opening Photo John Coletti/AWL Images/Getty Images **Central Case Study (mini)** Ondrej Prosicky/Shutterstock **Figure 6.2A** Martin Shields/Science Source **Figure 6.2B** Marco Simoni/Image Source/Alamy Stock Photo **Figure 6.3A** Yaacov Dagan/Alamy Stock Photo **Figure 6.3B** Juan Carlos Vindas/Moment Open/Getty Images **Figure 6.3C** John Coletti/Photolibrary/Getty Images **Figure 6.4** Library of Congress **Figure 6.5** Library of Congress/Corbis Historical/Getty Images **Figure 6.6** Library of Congress/Corbis Historical/Getty Images **Figure 6.7A** John Moore/Getty Images **Figure 6.7B** U.S. Environmental Protection Agency **Figure 6.7C** Terray Sylvester/The Flint Journal-MLive.com/AP Photo **Table 6.1A** Zhang qingyun/ICHPL Imaginechina/AP Photo **Table 6.1B** Barcroft Media/Getty Images **Table 6.1C** eppicphotography/iStock/Getty Images **Table 6.1D** Rick Wiking/Reuters **The Science behind the Story (mini)** MShieldsPhotos/Alamy Stock Photo **Figure 6.10** LeoSad/Shutterstock **Figure 6.12A** Jakub Cejpek/123RF **Figure 6.12B** belizar/Shutterstock **Figure 6.12C** James Randklev/Photographer's Choice RF/Getty Images **Figure 6.12D** Robert Glusic/Corbis/Getty Images **Figure 6.12E** gary corbett/Alamy Stock Photo **Figure 6.12F** Image Source/Getty Images **Figure 6.12G** Art Wolfe Stock/Robert Harding **The Science behind the Story** Sean Rayford/Getty Images **Figure 6.14** Mel Melcon/Los Angeles Times/Getty Images **Figure 6.16A** Tim UR/Shutterstock **Figure 6.16B** gmlykin/Shutterstock **Figure 6.16C** Anatoly Tiplyashin/Shutterstock **Figure 6.16D** United States Department of Agriculture **Figure 6.16E** Environmental Protection Agency **Figure 6.16F** **Figure 6.17** Gareth Davies/Getty Images **Central Case Study (mini)** John Coletti/AWL Images/Getty Images **Reviewing Objectives 1** Martin Shields/Science Source **Reviewing Objectives 2** John Coletti/Photolibrary/Getty Images **Reviewing Objectives 3** Library of Congress **Reviewing Objectives 4** Zhang qingyun/ICHPL Imaginechina/AP Photo **Reviewing Objectives 5** James Randklev/Photographer's Choice RF/Getty Images

Chapter 7 Opening Photo ZUMA Press, Inc./Alamy Stock Photo **Central Case Study** ZUMA Press, Inc./Alamy Stock Photo **Table 7.1A** Robin Loznak/Our Children's Trust **Table 7.1B** Robin Loznak/Our Children's Trust **Table 7.1C** Robin Loznak/Our Children's Trust **Table 7.1D** Robin Loznak/Our Children's Trust **Table 7.1E** Robin Loznak/Our Children's Trust **Table 7.1F** Robin Loznak/Our Children's Trust **Table 7.1G** Robin Loznak/Our Children's Trust **Table 7.1H** Robin Loznak/Our Children's Trust **Table 7.1I** Robin Loznak/Our Children's Trust **Table 7.1J** Robin Loznak/Our Children's Trust **Table 7.1K** Robin Loznak/Our Children's Trust **Table 7.1L** Robin Loznak/Our Children's Trust **Table 7.1M** Robin Loznak/Our Children's Trust **Table 7.1N** Robin Loznak/Our Children's Trust **Table 7.1O** Robin Loznak/Our Children's Trust **Figure 7.2** ZUMA Press, Inc./Alamy Stock Photo **The Science behind the Story (mini)** Robin Loznak/ZUMA Wire/Alamy Stock Photo **Figure 7.5A** Bettmann/Getty Images **Figure 7.5B** CORBIS/Corbis Historical/Getty Images **Figure 7.6** Erich Hartmann/Magnum Photos **Figure 7.7** Bettmann/Getty Images **Figure 7.8** NASA **Table 7.2A** VDWI Indu/Alamy Stock Photo **Table 7.2B** Picsfive/Shutterstock **Table 7.2C** Stephen Miller/Alamy Stock Photo **Table 7.2D** Toa55/Shutterstock **Table 7.2E** Anatoliy Sadovskiy/123RF **Table 7.2F** Marina Lohrbach/Shutterstock **Table 7.2G** endopack/Getty Images **Table 7.2I** Gregory Smith/Corbis Historical/Getty Images **Figure 7.9A** ZUMA Press, Inc./Alamy Stock Photo **Figure 7.9B** Axel Heimken/dpa/AFP/Getty Images **Figure 7.10** Jennifer Szymaszek/REUTERS **Figure 7.12** David R. Frazier/DanitaDelimont.com "Danita Delimont Photography"/Newscom **Central Case Study Connect & Continue** ZUMA Press, Inc./Alamy Stock Photo **Reviewing Objectives 1** ZUMA Press, Inc./Alamy Stock Photo **Reviewing Objectives 2** Marina Lohrbach/Shutterstock **Reviewing Objectives 3** ZUMA Press, Inc./Alamy Stock Photo **Reviewing Objectives 4** Jennifer Szymaszek/REUTERS

Part 2 Starcevic/Getty Images

Chapter 8 Opening Photo Manfred Gottschalk/Lonely Planet Images/Getty Images **Central Case Study (mini)** Carlos Barria/Reuters **Figure 8.1A** Tim Graham/Alamy Stock Photo **Figure 8.1B** Dmitry Kalinovsky/123RF **Figure 8.1C** iportret/Alamy Stock Photo **The Science behind the Story (mini)** Oscar Venter **The Science behind the Story (mini)** Eliana La Ferrara **The Science behind the Story (Figure 1)** Felipe Dana/AP Photo **Figure 8.18A** kali9/Getty Images **Figure 8.18B** Diana Mewes/Cephas Picture Library/Alamy Stock Photo **Central Case Study** Manfred Gottschalk/Lonely Planet Images/Getty Images **Reviewing Objectives 1** iportret/Alamy Stock Photo **Reviewing Objectives 2** kali9/Getty Images

Chapter 9 Opening Photo Jessa Kay Cruz/The Xerces Society **Central Case Study (mini)** Bryan Reynolds/Alamy Stock Photo **Figure 9.1** Jessa Kay Cruz/The Xerces Society **Figure 9.3** Nigel Cattlin/Science Source **Figure 9.6A** Kevin Foy/Alamy Stock Photo **Figure 9.6B** Fletcher & Baylis/Science Source **Figure 9.6C** sima/Shutterstock **Figure 9.6D** Biological Resources Division **Figure 9.7A** Tony Hertz/Alamy Stock Photo **Figure 9.7B** aerial-photos.com/Alamy Stock Photo **Figure 9.8A** Igor Stevanovic/123RF **Figure 9.8B** v_zaitsev/Getty Images **Figure 9.8A** Igor Stevanovic/123RF **Figure 9.8B** v_zaitsev/Getty Images **Figure 9.9A** sarka/Shutterstock **Figure 9.9B** Rachel Husband/Alamy Stock Photo **Table 9.1A** Edward Phillips/123RF **Table 9.1B** Willy Deganello/123RF **Table 9.1C** Scott Spakowski/500px/Getty Images **Table 9.1D** Sarah Foltz Jordan/The Xerces Society **Figure 9.12** Neil Phillips/Cephas Picture Library/Alamy Stock Photo **Figure 9.13** Eric Mader/The Xerces Society **Figure 9.14** Nancy Lee Adamson/The Xerces Society **The Science behind the Story** Courtesy of Maureen Page **Figure 9.15A** Lena Wurm/Alamy Stock Photo **Figure 9.15B** Florian Kopp/Alamy Stock Photo **Figure 9.16** MORVAN/SIPA/Newscom **Figure 9.17** Jake Lyell/Alamy Stock Photo **Figure 9.18** USDA **Figure 9.20** Franklin D. Roosevelt Library (NLFDR), 4079 Albany Post Road, Hyde Park, NY, 12538-1999/U.S. National Archives and Records Administration **Figure 9.21** USDA **Figure 9.22A** Danita Delimont/Alamy Stock Photo **Figure 9.22B** Kevin Horan/The Image Bank/Getty Images **Figure 9.22C** Keren Su/The Image Bank/Getty Images **Figure 9.22D** Ingna Spence/Alamy Stock Photo **Figure 9.22E** David Wall/Alamy Stock Photo **Figure 9.22F** Bill Barksdale/Alamy Stock Photo **Figure 9.23** Anthony Bannister/Avalon/Photoshot License/Alamy Stock Photo **Success Story** Barney low/Alamy Stock Photo **Figure 9.25** USDA **Central Case Study Connect & Continue** Jessa Kay Cruz/The Xerces Society **Reviewing Objectives 1** Nigel Cattlin/Science Source **Reviewing Objectives 2** sarka/Shutterstock **Reviewing Objectives 3** Courtesy of Maureen Page **Reviewing Objectives 4** USDA

Chapter 10 Opening Photo University of Michigan **Central Case Study (mini)** University of Michigan **Figure 10.2** Jean-Marc Bouju/AP Photo **Figure 10.4** Art Rickerby/The LIFE Picture Collection via Getty Images/Getty Images **Figure 10.6A** Tomsickova Tatyana/Shutterstock **Figure 10.6B** Nigel Cattlin/Alamy Stock Photo **Figure 10.10A** David Parry/pool/REUTERS **Figure 10.10B** Simon Dawson/Bloomberg/Getty Images **Figure 10.11A** Garry Adams/Photolibrary/Getty Images **Figure 10.11B** Jim West/Alamy Stock Photo **Figure 10.12** Wang Huabin/AP Images **Figure 10.14** PiggingFoto/Shutterstock **Figure 10.16A** Australia Lands Department **Figure 10.16B** Australia Lands Department **Table 10.1A** Pigdevil Photo/Shutterstock **Table 10.1B** mexrix/Shutterstok **Table 10.1C** Peter Zijlstra/Shutterstock **Table 10.1D** dabjola/Shutterstock **Table 10.1E** Krasowit/Shutterstock **Table 10.1F** Luis Carlos Jimenez del rio/Shutterstock **Table 10.1G** Oksana Shufrych/Shutterstock **Table 10.1H** JIANG HONGYAN/Shutterstock **Table 10.1I** Maks Narodenko/Shutterstock **Table 10.1J** Sergio33/Shutterstock **The Science behind the Story** Heinz Lohmann Stiftung/Engel & Zimmermann AG **Figure 10.21** JAY DIRECTO/AFP/Getty Images **Figure 10.22A** Richard Ellis/Alamy Stock Photo **Figure 10.22B** USDA **The Science behind the Story** Rodale Institute **The Science behind the Story (Figure 3A)** Rodale Institute **The Science behind the Story (Figure 3B)** Rodale Institute **Figure 10.25** fotoluk1983/iStock/Getty Images **Figure 10.27A** Hiroyuki Uchiyama/Moment Open/Getty Images **Figure 10.27B** CSP_johnnym26/Agefotostock **Figure 10.27C** Arco Images GmbH/Alamy Stock Photo **Figure 10.27D** Rudmer Zwerver/Shutterstock **Figure 10.27E** International Photobank/Alamy Stock Photo **Central Case Study Connect & Continue** University of Michigan **Reviewing Objectives 1** Art Rickerby/The LIFE Picture Collection via Getty Images/Getty Images **Reviewing Objectives 2** Simon Dawson/Bloomberg/Getty Images **Reviewing Objectives 3** PiggingFoto/Shutterstock **Reviewing**

Index